FROM ANIMALS TO ANIMATS 5

Complex Adaptive Systems

John H. Holland, Christopher G. Langton, and Stewart W. Wilson, advisors

Adaptation in Natural and Artificial Systems: An Introductory Analysis with Applications to Biology, Control, and Artificial Intelligence, John H. Holland

Toward a Practice of Autonomous Systems: Proceedings of the First European Conference on Artificial Life, edited by Francisco J. Varela and Paul Bourgine

Genetic Programming: On the Programming of Computers by Means of Natural Selection, John R. Koza

Genetic Programming: The Movie, John R. Koza and James P. Rice

From Animals to Animats 2: Proceedings of the Second International Conference on Simulation of Adaptive Behavior, edited by Jean-Arcady Meyer, Herbert L. Roitblat, and Stewart W. Wilson

Intelligent Behavior in Animals and Robots, David McFarland and Thomas Bösser

Advances in Genetic Programming, edited by Kenneth E. Kinnear, Jr.

Genetic Programming II: Automatic Discovery of Reusable Programs, John R. Koza

Genetic Programming II Video: The Next Generation, John R. Koza

Turtles, Termites, and Traffic Jams: Explorations in Massively Parallel Microworlds, Mitchel Resnick

From Animals to Animats 3: Proceedings of the Third International Conference on Simulation of Adaptive Behavior, edited by Dave Cliff, Philip Husbands, Jean-Arcady Meyer, and Stewart W. Wilson

Artificial Life IV: Proceedings of the Fourth International Workshop on the Synthesis and Simulation of Living Systems, edited by Rodney A. Brooks and Pattie Maes

Comparative Approaches to Cognitive Science, edited by Herbert L. Roitblat and Jean-Arcady Meyer

Artificial Life: An Overview, edited by Christopher G. Langton

Evolutionary Programming IV: Proceedings of the Fourth Annual Conference on Evolutionary Programming, edited by John R. McDonnell, Robert G. Reynolds, and David B. Fogel

An Introduction to Genetic Algorithms, Melanie Mitchell

Catching Ourselves in the Act: Situated Activity, Interactive Emergence, and Human Thought, Horst Hendriks-Jansen

Toward a Science of Consciousness: The First Tucson Discussions and Debates, edited by Stuart R. Hameroff, Alfred W. Kaszniak, and Alwyn C. Scott

Genetic Programming: Proceedings of the First Annual Conference, edited by John R. Koza, David E. Goldberg, David B. Fogel, and Rick L. Riolo

Evolutionary Programming V: Proceedings of the Fifth Annual Conference on Evolutionary Programming, edited by Lawrence J. Fogel, Peter J. Angeline, and Thomas Bäck

Elements of Artificial Neural Networks, Kishan Mehrotra, Chilukuri K. Mohan, and Sanjay Ranka

From Animals to Animats 4: Proceedings of the Fourth International Conference on Simulation of Adaptive Behavior, edited by Pattie Maes, Maja J. Mataric, Jean-Arcady Meyer, and Stewart W. Wilson

Advances in Genetic Programming, Volume 2, edited by Peter J. Angeline and Kenneth E. Kinnear, Jr.

Growing Artificial Societies: Social Science from the Bottom Up, Joshua M. Epstein and Robert Axtell

Artificial Life V: Proceedings of the Fifth International Workshop on the Synthesis and Simulation of Living Systems, edited by Christopher G. Langton and Katsunori Shimohara

An Introduction to Natural Computation, Dana H. Ballard

Fourth European Conference on Artificial Life, edited by Phil Husbands and Inman Harvey

Toward a Science of Consciousness 2: The 1996 Tucson Discussions and Debates, edited by Stuart R. Hameroff, Alfred W. Kaszniak, and Alwyn C. Scott

Artificial Life VI: Proceedings of the Sixth International Workshop on the Synthesis and Simulation of Living Systems, edited by Christoph Adami, Richard K. Belew, Hiroaki Kitano, and Charles E. Taylor

An Introduction to Fuzzy Sets: Analysis and Design, Witold Pedrycz and Fernando Gomide

The Simple Genetic Algorithm: Foundations and Theory, Michael D. Vose

From Animals to Animats 5: Proceedings of the Fifth International Conference on Simulation of Adaptive Behavior, edited by Rolf Pfeifer, Bruce Blumberg, Jean-Arcady Meyer, and Stewart W. Wilson

FROM ANIMALS TO ANIMATS 5

Proceedings of the Fifth International Conference
on Simulation of Adaptive Behavior

edited by
Rolf Pfeifer, Bruce Blumberg, Jean-Arcady Meyer, and Stewart W. Wilson

A Bradford Book

The MIT Press
Cambridge, Massachusetts
London, England

Printed and bound in the United States of America.

This publication may be ordered from The MIT Press, 55 Hayward Street, Cambridge, Massachusetts 02142 USA; by telephone: 1-800-356-0343 (toll-free) or 617-625-8569; by fax: 617-625-6660; by e-mail: mitpress-orders@mit.edu or http://mitpress.mit.edu.

ISSN 1089-4365
ISBN 978-0-262-66144-7 (pb.: alk. paper)

The MIT Press is pleased to keep this title available in print by manufacturing single copies, on demand, via digital printing technology.

These proceedings are dedicated to the memory of Henry (Harry) Bradford Stanton

CONTENTS

LEARNING

MOTOR LEARNING

COLLECTIVE BEHAVIOR

EVOLUTION

PREFACE

The papers in this book were presented at the Fifth International Conference on Adaptive Behavior (SAB98), held at University of Zurich, Zurich, Switzerland, on August 17–21, 1998. The objective of the biennial SAB conference is to bring together researchers from a wide range of backgrounds, including ethology, theoretical biology, psychology, artificial life, machine learning, and robotics, to further our understanding of the behaviors and underlying mechanisms that allow natural and artificial animals ("animats") to adapt and survive in uncertain environments.

Adaptive behavior research is distinguished by its focus on the modeling and creation of complete animal-like systems, which—however simple at the moment—may be one of the best routes to understanding intelligence in natural and artificial systems. The field received initial recognition on the occasion of the first SAB conference, which was held in Paris in September 1990. Subsequent SAB conferences (Hawaii, December 1992, Brighton, England, August 1994, and Cape Cod, USA, September 1996) drew increasing numbers of papers and participants. In 1992, the MIT Press introduced the quarterly journal *Adaptive Behavior*, and The International Society for Adaptive Behavior (ISAB) was established in 1995—both events further marking the emergence of adaptive behavior as a full-fledged scientific discipline. The present proceedings are a comprehensive and up-to-date resource for the latest progress in this exciting field.

The 74 papers presented at the conference and published here were selected from 150 submissions after a two-pass review process designed to ensure high and consistent overall quality. The authors focus on well-defined models, computer simulations, and robotics demonstrations that help characterize and compare various organizational principles or architectures capable of inducing adaptive behavior in real animals or synthetics agents. The papers are ordered according to the scale at which adaptive behavior takes place, ranging from immediate adaptation in sensorimotor control, to learning within an animatís lifetime, to adaptive behavior exhibited by successive generations of animats, and finally to adaptive collective behavior of animats in groups.

In addition to the presentations of accepted papers, a number of distinguished researchers gave keynote lectures. The SAB98 Keynote Lecturers included:

- Rüdiger Wehner, Professor and Director of the department of Zoology at the University of Zurich.
- Thomas Christaller, Professor of Artificial Intelligence at the University of Bielefeld and director of the GMD Institute of System Design Technology.
- Rodney Douglas, Professor and Director of the department of Neuro-Informatics at the University of Zurich and Swiss Federal Institute of Technology.

The inclusion of focused workshops and tutorials was a major innovation of SAB98. Led by some of the leading researchers in the field, these workshops and tutorials gave participants an unparalleled opportunity to explore a particular topic in depth. The tutorials included:

- *Language as a complex adaptive system*
  Lecturer: Luc Steels (VUB AI Lab, Brussels, and Sony CSL, Paris)
- *Modeling mate choice and sexual selection*
  Lecturers: Peter M. Todd (MPI, Berlin) and Geoffrey F. Miller (Univ. College of London)
- *How to build animats*
  Lecturer: Andreas Birk (VUB AI Lab)

- *From sensory-motor map learning to higher order cognition*
  Lecturers: Asada Minoru (Osaka Univ.) and Jun Tani (Sony CSL, Paris)
- *Mechanisms of navigation and spatial memory*
  Lecturers: Hanspeter Mallot (MPI, Tübingen) and Matthias O. Franz (MPI, Tübingen)
- *Embodied cognitive science*
  Lecturers: Rolf Pfeifer (AI Lab, Zurich) and Christian Scheier (AI Lab, Zurich)
- *Walking machines*
  Lecturer: Rüdiger Dillmann (Univ. of Karlsruhe)
- *Evolution of autonomous agents*
  Lecturer: Dave Cliff (MIT AI Lab)
- *Observable operator models of stochastic systems*
  Lecturer: Herbert Jaeger (GMD)

Meanwhile, the workshops included:

- *Adaptive behavior using recurrent neural nets*
  Organizers: Jun Tani (Sony, CSL, Paris), Takashi Ikegami (Univ. of Tokyo), and Jordan Pollack (Brandeis Univ.)
- *Socially situated intelligence*
  Organizers: Bruce Edmonds (Manchester Metropolitan Univ.) and Kerstin Dautenhahn (Univ. of Reading)
- *Coevolution of symbol systems and behavior*
  Organizers: Lawrence Hunter (National Library of Medicine) and Jordan Pollack (Brandeis Univ.)
- *Predicting the consequences of action*
  Organizers: Andreas Birk (VUB AI Lab), Rene te Boekhorst (IFI, Zurich), and Herbert Jaeger (GMD)
- *Grounding emotions in adaptive systems*
  Organizers: Delores Canamero (IIIA-CSIC), Chisato Numaoka (Sony CSL, Paris), and Paolo Petta (OFAI)
- *Biology and robotics*
  Organizer: John Hallam (Univ. of Edinburgh)
- *Biologically-inspired walking*
  Organizers: Rüdiger Dillmann (Univ. of Karlsruhe) and Wilfred Ilg (Univ. of Karlsruhe)
- *Embodied models in neurobiology*
  Organizer: O. Sporns (Neurosciences Institute, San Diego)
- *Adaptive internet agents*
  Organizer: Geoffrey Miller (Univ. College of London)

The conference and its proceedings could not have happened without the substantial help of a wide range of people. First and foremost, we would like to thank the members of the Program Committee and others who thoughtfully reviewed the submissions and worked with many of the authors to ensure a high standard in the accepted papers. Our Program Committee and other reviewers were:

P. Angeline, USA
R. Arkin, USA
M. Asada, Japan
H. Asoh, Japan
J. Atema, USA
R. Beer, USA
L. Berthouze, Japan
L. Berthoz, France
L. Booker, USA
S. Bullock, UK
R. Brooks, USA
T. Christaller, Germany
D. Cliff, USA
T. Collett, UK
H. Cruse, Germany
J. Deneubourg, Belgium
Y. Faihe, Switzerland
J. Ferber, France
D. Floreano, Switzerland
N. Franceschini, France
T. Fukuda, Japan
P. Gaudiano, USA
P. Gaussier, France
S. Giszter, USA
D. Goldberg, USA
J. Hallam, UK
F. Hara, Japan
I. Harvey, UK
C. Hemelrijk, Switzerland
I. Horswill, USA
P. Husbands, UK
K. Ito, Japan
H. Jaeger, Germany
L. Kaelbling, USA
Y. Kakazu, Japan
Y. Kuniyoshi, Japan
D. Lambrinos, Switzerland
H.P. Mallot, Germany
K.F. MacDorman, Japan
M. Mataric, USA
J. Millan, Italy
G. Miller, UK
R. Möller, Switzerland
J. Müller, Switzerland
T. Nakamura, Japan
C. Paul, Switzerland
R. Salomon, Switzerland
G. Sandini, Italy
R. Schaad, Switzerland
C. Scheier, Switzerland
K. Skiyama, Japan
W. Smart, USA
T. Smithers, Spain
E. Spier, UK
O. Sporns, USA
L. Steels, Belgium
L. Stein, USA
Y. Takahashi, Japan
J. Tani, Japan
R. te Boekhorst, Switzerland
F. Toates, UK
P. Todd, Germany
T. Unemi, Japan
W. Uttal, USA
D. Waltz, USA
B. Webb, UK
B. Werger, USA
S. Yamada, Japan
M. Yano, Japan

We are indebted to our sponsors:

- SwissAir, the official carrier for SAB98
- rosys
- Computer Science Department, University of Zurich
- Applied AI systems
- Hewlett-Packard
- K-team
- Neuronics
- Personal Robot Technologies, Inc.
- Starseed Enterprise

Above all, we would like to acknowledge the significant contributions of three individuals who have worked tirelessly to make SAB98 a success: Hiroshi Kobayashi, Ralf Salomon, and Lukas Lichtensteiger. In particular, Hiroshi played an instrumental role in the review process; Ralf was responsible for organizing the tutorials and workshops; Lukas was instrumental in overseeing the logistics.

We would also like to thank Professor Herbert Roitblat for his work as finance chair. Finally, we are once again especially indebted to Jean Solé for the artistic concept of the SAB96 poster and the proceedings cover.

We invite readers to enjoy and profit from the papers in this book, and look forward to the next conference SAB00.

Rolf Pfeifer, Program Chair
Bruce Blumberg, Jean-Arcady Meyer, Stewart W. Wilson

FROM ANIMALS TO ANIMATS 5

THEORETICAL ISSUES—PHILOSOPHICAL AND PSYCHOLOGICAL CONSIDERATIONS

Bringing Representation Back to Life

Andy Clark* and Michael Wheeler**
* Philosophy-Neuroscience-Psychology Research Program
Department of Philosophy
Washington University
Campus Box 1073, One Brookings Drive
St.Louis, MO 63130-4899, USA
andy@twinearth.wustl.edu
** Department of Experimental Psychology
University of Oxford
South Parks Road
Oxford, OX1 3UD, UK
michaelw@psy.ox.ac.uk ***

Abstract

The primary expression of biological intelligence consists in the production of an ongoing sequence of actions that flexibly respond to a constant stream of incoming environmental stimuli. Recent research into artificial autonomous agents (animats) suggests that a great deal of such adaptive behaviour may be grounded not in the systematic activity of internal representations, but in complex interactions involving neural, bodily, and environmental factors. In this paper we investigate the threat to representational explanatory strategies posed by this form of causal complexity in behaviour-generating systems. First we try to identify the true nature and extent of the threat, by considering some telling examples of animat research. Then we pursue an extended analogy between the role of neural states in the causal determination of adaptive behaviour, and the role of genes in the causal determination of phenotypic traits. On the basis of this analogy, we develop a concept of *genic representation* that, we suggest, might offer a way forward. A crucial part of the genic strategy is to appeal to the *normal ecological context* of the inner states in question, in order to balance (a) the fact that adaptive behaviour is often the result of complex multi-factor causation, against (b) the idea that the notion of internal representation plays a special role in explaining such cases.

1. Introduction: Once More into the Breach

The recent wave of research into artificial autonomous agents — henceforth *animats* (Wilson, 1985) — has already made some important contributions to the scientific understanding of intelligence. One of these contributions has been to cast doubt on the disarmingly seductive picture, regularly assumed in mainstream artificial intelligence (AI) and cognitive science, according to which intelligent action is invariably organised into a *sense-model-plan-act cycle* (a term due to Brooks, 1991). On the sense-model-plan-act view, intelligent action begins when a perception-module inside the agent constructs a representation of the external world, from sensory input. This representation is delivered to some sort of reasoning system (classical or connectionist) that computes what the agent (given its internally represented goals and purposes) ought to do next. Finally, a further representation, this time of the desired actions, is constructed and turned into a set of movement-instructions for the agent's motor-mechanisms to follow. Then the cycle starts all over again, with another round of sensing.

One striking thing about the sense-model-plan-act view is that *neither the agent's non-neural body, nor the agent's environment, plays a significant causal role in the production of action.* Of course, the agent's environment provides the information for perception, and that information arrives via the agent's sensors. But after that initial input-stage, and assuming that everything goes 'to plan', neither environment nor sensors play any further part in the successful completion of the behaviour. And although the agent's motor-mechanisms are obviously crucial in enabling it to move, they merely follow the set of movement-instructions issued by the central planner; they do not play any active part in *generating* the behavioural solution. As most animat builders will tell you, there is something wrong-headed about this picture, since both the agent's embodiment and the agent's situatedness (its environmental embeddedness) can be exploited to produce cheap, efficient, robust and flexible solutions to adaptive problems (see next section), and

*** The order of authors is arbitrary

we should expect natural evolution to find such canny solutions.[1] So much seems clear. But these important observations have often been coupled with a further, more radical and more controversial thought. As we shall see, it has been argued that once the embodiment and the situatedness of the agent are taken seriously, the concept of internal representation appears no longer to be the fundamental theoretical primitive for the science of intelligence that it once seemed to be, back in those heady, innocent days of SHRDLU, Shakey and NetTalk.

The current debate surrounding the status of internal representations is a complicated business.[2] In some ways this is a peculiar state of affairs, because the basic idea of internal representation seems straightforward enough: an internal representation is an inner (paradigmatically neural) state whose basic function is to *stand in for* some other (usually external) state of affairs in the agent's inner goings-on. This formulation makes it clear that the 'standing-in-for' relation is at the very heart of representation. Consider, for example, how the primary examples of representational systems — *external* representational schemes such as public language — work: cognitive agents use symbols (e.g., the words of the language) to stand for other things. However, in the case of *internal* representations, there is no conceptual room for inner homunculi that literally understand such-and-such an internal state as standing for such-and-such a state of affairs. Representational systems of the internal variety must operate by means of the direct causal properties of the states and processes involved. What is needed, then is a way of making the notion of representation work, without committing a kind of homuncular fallacy.

At a first pass, what might seem to do the trick here is the notion of standing-in that is provided by the capacity of certain states and processes to figure in episodes of environmentally decoupled problem-solving — cases of what one might call *off-line intelligence*. This notion of standing-in requires that the states and processes concerned constitute a kind of model of the target domain, viz an inner arena in which the unfolding of states marches in step with the way things unfold in the wider world (for further details and development of this idea, see Clark and Grush, forthcoming). This concept of *strong internal representation* is thus the notion of an item or items that figure in an inner model capable of supporting a kind of vicarious problem-solving: one in which moves can be tried out before committing the real flesh and blood to action. Under these conditions, the adaptive role of 'standing in for' can be cashed out in terms that do not invoke an inner understander, but which make clear contact with the primary (external, public) usage nonetheless.

The problem with this strong notion of representation lies, we think, in its scope. We have no doubt that environmentally decoupled off-line reasoning takes place, that it is widespread, and that it is worthy of scientific investigation. But it seems fair to say that the *primary* character of biological intelligence is expressed not (as much traditional AI and cognitive science may have often supposed) in the off-line context, but rather in the production of flexible and appropriate *on-line* responses to an ongoing stream of incoming environmental stimuli (e.g., avoiding predators or catching prey). Such on-line intelligence is, of course, the focus of most animat research, and the home ground for the sorts of embodied and situated solutions that are often thought to undermine representational explanation. And it is precisely here that strong internal representation threatens to break down. Our goal in this paper, then, is to present a different notion of internal representation, one which has just as much potential as its off-line cousin to be explanatorily useful and empirically robust, but one which applies precisely to the supposedly recalcitrant on-line cases.[3] To open up the investigative space, we shall, in the next section, show how two classic examples of animat research pull us in opposite directions, on the issue of internal representation.

2. Crickets Don't, Flies Do

The first of our examples is work by Webb in which mobile robots are used in an attempt to understand how female crickets track species-specific auditory mating calls produced by the male (see, e.g., Webb, 1994).

The female cricket has two ears (one on each of her two front legs). A tracheal tube of fixed length connects these ears to each other, and to other openings (called 'spiracles') on her body. So sound arrives at the ear-drums both externally (i.e., directly from the sound source) and internally (i.e., via the trachea). According to Webb, what happens when the male's auditory signal is picked-up is as follows: On the side nearer the sound source, the externally arriving sound travels less distance than the internally arriving sound, whereas, on the side further away from the sound source, the two sounds travel the same distance. This means that the external and internal arrivals of sound are out of phase at the ear-drum on the side nearer to the sound source, but in phase at the ear-drum on the side further away from the sound source. The practical consequence of this difference is that the amplitude of ear-drum vibration is higher on the side nearer the sound source. In short,

[1] Our expectation here should be reinforced by the evidence from evolutionary robotics that *artificial evolution* often produces these sorts of deeply embodied and situated solutions (see, e.g., Harvey, Husbands, & Cliff, 1994).

[2] For other treatments of the issue by the present authors, see, e.g., Clark (1997), Clark and Toribio (1994), and Wheeler (1994, forthcoming).

[3] Many of the issues discussed here are pursued at greater length in (Wheeler and Clark, submitted).

when the male's signal is picked up, there is a direction-dependent intensity difference at the female's ears, with the side closer to the sound source having the stronger response.

All the female has to do now, in order to find the male, is to keep moving in the direction indicated by the ear-drum with the higher amplitude. This is achieved via the activity of two dedicated interneurons in the female's nervous system, one of which is connected to the female's right ear, the other of which is connected to her left. Each of these neurons fires when its activation level reaches a certain threshold, but because the neuron which is connected to the ear-drum with the higher vibration amplitude will be receiving the stronger input, that neuron will reach threshold more quickly, and hence fire first. So the idea is that, at each burst of sound, the female moves towards the side of the neuron that fires first. In effect, therefore, her tracking mechanism responds only to the *beginning* of a sound burst. It is here that the specific *temporal pattern* of the male's signal becomes important. On the one hand, the gaps between the syllables of the male's song must not be too long, otherwise the information about his location will arrive too infrequently for the female's mechanisms to track him reliably. On the other hand, the gaps between the syllables of the male's song must not be too short. Each of the two dedicated interneurons in the female's nervous system has a decay time during which, after firing, it gradually returns to its non-activated rest-state. During this recovery period, a neuron is nearer to its firing threshold than if it were at rest. In consequence, if a neuron receives input during the decay time, it will fire again more quickly than if it receives that input while at rest. So if the gaps between the syllables of the male's song were shorter than the total decay time of the neurons, or if the song were continuous, then it would often be unclear which of the two neurons had fired first.

In the present context, the key aspect of this proposal is that the observed adaptive success results from a system of organised interactions which involve significant causal contributions from the female cricket's nervous system (the dedicated interneurons), her body (the fixed-length trachea), and her environment (the specific structure of the male's signal). In what follows, we shall use the term *causal spread* to describe any situation in which an outcome is generated by a network of multiple, co-contributing causal factors that extends across readily identifiable systemic boundaries. So, using this terminology, we can say that the distributed mechanism which underlies the female cricket's on-line intelligent behaviour features large-scale causal spread that extends across the boundaries between brain, body, and environment. This sort of mechanism, the style and power of which will be familiar to most animat researchers, is already at odds with the nervous-system-centred approach of most cognitive science. However, things are yet more radical, because Webb is openly critical of the idea that the concept of internal representation will help us to understand that mechanism.

> It could be argued that the robot does contain 'representations', in the sense of variables that correspond to the strength of sensory inputs or signals for motor outputs. But does conceiving of these variables as 'representations of the external world' and thus the mechanism as 'manipulation of symbols' actually provide an explanatory function? It is not necessary to use this symbolic interpretation to explain how the system functions: the variables serve a mechanical function in connecting sensors to motors, a role epistemologically comparable to the function of the gears connecting the motors to the wheels. (Webb, 1994, p.53)

It seems to us that one might argue, in a relatively direct fashion, and with some force, from the presence of causal spread to this sort of anti-representational conclusion. The argument would go something like this:[4] If behaviour is properly seen as coming about through an extended causal process, in which interactions cross the boundaries between brain, body, and environment, then there is little or no warrant to think of any neural states that happen to be involved as fully *specifying* behaviour (that is, as encoding the instructions for what to do), because a neural state that has been identified as causally contributing to some behavioural outcome cannot count as fully specifying that outcome, if various additional causal contributions made by non-neural bodily processes, and/or environmental factors play substantial roles in determining the form of the behaviour.

With this argument in place, let's turn now to our second example of animat research. In work by Franceschini, Pichon, and Blanes (1992), detailed neurobiological knowledge of the visual system of the housefly was used to guide the design of a mobile robot that is capable of navigating its way to a goal (a light source), in real time, whilst avoiding obstacles. This feat is achieved via a series of simple movements, each made according to a specific direction-heading generated in the following way. The robot is endowed with a compound eye featuring a layer of photosensors connected to a parallel array of analogue elementary motion detectors (EMDs). This EMD-array drives an obstacle avoidance layer (a parallel-processing, analogue network) that integrates the motion signals from the EMDs to help control the steering of the robot. Since the only information available to the obstacle avoidance layer concerns motion, as registered by

[4] We do not claim that this is an accurate reconstruction of Webb's own reasoning.

the EMD-array, in an environment of stationary obstacles, the robot is, for the purposes of obstacle avoidance, blind at rest. However, the EMD-array is able to use relative motion information, generated by the robot's own bodily movements during the preceding step, to build a temporary 'snap map' of detected obstacles. This map, which is made available to the obstacle-avoidance layer, is constructed in a polar coordinate system defined in relation to the compound eye. (Since the eye is mounted in a fixed position, it rotates with the wheels, so a frame of reference centred on the eye is also one centred on the direction of movement.) Then, in a local and short-lived motor-map, information concerning the angular bearings of the obstacles (as supplied by the snap map) is fused with information concerning the angular bearing of the goal (as supplied by a supplementary visual system), and the direction-heading of the next movement is generated. This heading is as close as possible to the one that would take the robot straight towards the goal, adjusted so that the robot avoids all detected obstacles.

Franceschini et al.'s robot receives no visual feedback *during* each simple movement. However, notice that, in the period that matters to us — the overall period between the beginning and the end of the task — such feedback is absolutely crucial to the way in which the system achieves the target behaviour. The path taken to the goal is never fully specified by centrally located 'neural' processes, but emerges from the ongoing dynamic feedback between brain, body and environment. So factors located in each of the relevant causal domains — brain, body, and environment — contribute to the distinctive form of the goal-seeking/obstacle-avoidance behaviour, and, moreover, make their contributions during the real-time execution of that behaviour. Like the adaptive behaviour of Webb's robotic cricket, then, the adaptive behaviour of this robotic fly is generated by an underlying mechanism that features causal spread. However, there is an important difference. The inner structures that play their part in explaining the robot's navigational capacities appear to be illuminatingly interpretable as — indeed they were conceived as — *maps*. They are not, of course, maps in the familiar sense of symbolic representations of an objective environment which are stored and recalled for the off-line advance planning of a route. Rather, they are short-lived, task-specific, egocentric (i.e., robot-centred) structures, which are built on-line, during each round of sensing and action. Still, nothing about this particular profile of features seems to undermine, in any obvious way, their status *as maps*. On the face of it, then, those structures appear to be usefully conceptualised as a form of internal representation.[5]

So, on the one hand, we have evidence (from Webb) that if a behaviour-generating system features causal spread, then that system will be resistant to explanatory strategies that appeal to internal representations. On the other hand, we have equally good evidence (from Franceschini et al.) that internal representations can be explanatorily potent and scientifically identifiable theoretical entities, even in systems where causal spread is rampant. The tension here is obvious. So where should we go next, in order to find a resolution? The answer, we think, is developmental biology!

3. Genes to the Rescue

The received orthodoxy in contemporary biology is that genes should be understood as *encoding for* — i.e., as *representing* — phenotypic traits. Thus an organism's genotype is standardly characterised as a *set of instructions for*, or as *a program for*, the building of that organism's phenotype. This basic claim, that there is something which one might reasonably think of as a genetic program, has a distinct tendency to slide, imperceptibly and without due warning, into the following, much stronger claim, which we shall call *strong instructionism about genes*: if one could find out the complete sequence of an organism's DNA, then, in principle, one would be able to use that information *alone* to 'compute' — and so predict in every relevant detail — that organism's phenotypic form.

Now, putting strong instructionism to one side for a moment, consider the following data from the world of plants: Despite the enormous variety of leaf arrangements that we see in the higher plants, there are really only three generic forms in nature. The most common of these is a spiral arrangement, in which successive leaves on the stem appear at a fixed angle of rotation relative to each other. Amazingly, these angles of rotation tend to take one of only a few values, of which the most common is 137.5 degrees. An intriguing explanation of this data has been proposed by Goodwin (1994, pp.105-119), who, in general, seeks an understanding of biological development in terms of the emergence of structure through self-organisation in complex dynamical systems. Goodwin begins his account here by observing that, as leaf-tissue grows in the stem, it places pressure on an elastic surface layer of cells. This pressure, it seems, causes the surface-cells to manufacture organic barricades made of cellulose. Where the next leaf will grow is determined by the fact that, as a result of exactly where the stress has been placed, and exactly how the cellulose barricades have been laid down, the resistance to growth will be stronger in some areas of the surface layer than in others. So, Goodwin argues, the final pattern of leaves (the phenotypic form) is explained by a sequence of mechanical interactions between (i) the growing leaves under the surface layer of cells, and (ii) the cellulose barricades. He proceeds to describe various modelling studies (carried

[5] In fact, they seem to be examples of what Mataric (1991) calls *active-constructive representations*, and what Agre (1988) calls *deictic representations*.

out by other researchers working on this problem) which show (a) that the leaf arrangements observed in nature are stable, generic patterns produced by this sort of system, and (b) that, given certain parameter-values (e.g., the growth rates) and certain naturally observed initial conditions, the dominant generic pattern generated by the self-organising dynamics of this sort of system is, indeed, the most common arrangement found in nature. For our purposes, the crucial take-home-message to be gleaned from this example is that, if Goodwin is right, then the key causal factors that generate the developmental outcome — mechanical stresses in elastic sheets of cells, spatial displacements of growing forms in the face of physical resistance — are *extra-genetic* bodily processes. This is clear evidence that the causal system which underlies development often extends outwards, from the genotype, into the organic body. So developmental systems too can feature causal spread. But how far does this developmental manifestation of the phenomenon go?

Mississippi alligators lay their eggs in a nest of rotting vegetation which produces heat in varying quantities. Eggs that develop at lower temperatures (within some overall range) end up producing females. Those that develop at higher temperatures end up producing males. Eggs in a clutch will pass through the critical developmental window at various different temperatures, meaning that a mixture of females and males will be born. This environmental method of regulating sex ratio (a ratio which, for reasons of population-survival, needs to stay somewhere near 50:50 in the population) might seem a little hit and miss, but it works well enough (for more details, see Goodwin, 1994, p.38). The lesson is clear: environmental features are active causal factors in biological development. Thus the developmental systems that causally determine phenotypic traits, just like the behaviour-generating systems that produce on-line intelligence, can exhibit the most extreme, environment-involving variety of causal spread.

Earlier we observed that genes are typically thought of as being representational entities. So now we might wonder how this representational account of genes fares, once causal spread in the developmental system is taken on board. If strong instructionism were the only position available to the representationalist about genes, then, we think, the representational account of genes would be in trouble, because if phenotypic forms are the products of extra-genetic self-organising processes in the body of the developing organism, in interaction with the environment, then knowing the entire sequence of an organism's DNA will simply be insufficient to predict that organism's phenotypic form. One will need to know, in addition, the principles of biological (self-)organisation which are at work, plus the values of the relevant environmental variables. However, on this occasion at least, the representationalist lives to fight another day, because strong instructionism is *not* her only option.

To bring into view what is, to us at least, a compelling alternative, we can turn, once again, to Goodwin's self-organisation-based account of development. According to Goodwin, the role that genes play in development is to partially select or stabilise the developmental dynamics that, through self-organisation, actualise one of the alternative generic forms that are possible for a particular organism. And genes do this by setting parameters for the self-organising systems that generate phenotypic form. As Goodwin himself puts it, "During reproduction, each species produces gametes with genes defining parameters that specify what morphogenetic trajectory the zygote will follow" (Goodwin, 1994, p.102). Thus although genes do not fully specify the final phenotypic form (strong instructionism is false), they nevertheless *code for* developmental parameters, and, therefore, by extension, one can say that genes code for phenotypic traits. Of course the extension here does not come for free. One has to work for it. But we think that it can be done, by introducing an appeal to the idea of the gene's, or genes', *normal ecological context*. This is the developmental background of non-genetic, causally active (biochemical and environmental) states and processes against which the genes in question operate. Thus, in saying that particular genes represent particular traits, we assume (a) that any principles of self-organisation that we usually expect to operate during development will, indeed, operate, and operate as normal, and (b) that any expected values of the relevant environmental variables will obtain. In other words, it seems legitimate to treat certain genes as coding for a particular phenotypic trait, just so long as the presence of those genes can be seen to yield — in a rich, normal ecological setting — that very phenotypic outcome.

Now let's get back to the part played by neural states in generating adaptive behaviour, and see if we can understand that contribution as being a representational one, by thinking of it as analogous to the part played by genes in generating phenotypic form. So far, a *genic* account of internal representation would look like this: Neural states which are implicated in the production of on-line intelligent behaviour do not fully specify external states of affairs or behaviours. They operate within an assumed, normal ecological context of other causally active (bodily and environmental) states and processes. Neural states are representational in that they encode certain parameters which stabilise the complex neural and bodily dynamics that, in interaction with the environment, yield appropriate behavioural outcomes.[6]

[6] The idea that neural states contribute to the generation of behaviour by acting as parameters for self-organising dynamical systems is also at work in Kelso's notion of 'specific parametric influences' (Kelso, 1995) and in Keijzer's concept of 'internal control parameters' (Keijzer, 1997). However, both Kelso and Keijzer are essentially hostile to rep-

It is important that the crucial appeal to normal ecological context here be prevented from running around untethered. The strategy of treating certain causal factors as mere background against which an instruction set acts has to be applied with selectivity and care. Otherwise there is a real risk of the genic account degenerating into triviality. Imagine, for example, taking a single line from a 15 line program for finding all the prime numbers between 1 and 1000, and claiming that that line of code is a perfectly good set of instructions for solving the prime-number-finding problem, albeit one that works only in the context of the other 14 lines of code. The claim is ridiculous, of course, but how do we prevent it from being licensed by our proposed account? The most important rule to be followed, if one wishes to ensure that one's appeal to ecological context will generate a representational story with legitimate explanatory credentials, is probably this: make sure that by treating the targeted neural contribution as a kind of context-exploiting algorithm for producing behaviour, one explains a good portion of the observed behavioural richness and flexibility. If all the richness and flexibility is located in the operation of other non-neural bodily and environmental factors, then there will be little or no explanatory advantage in treating the neural contribution as a representational one. After all, it is surely the richness and flexibility of behaviour that internal representations are supposed to explain. Thus the single line of code is not usefully thought of as specifying the solution to the prime-number-finding problem in a context provided by the other 14 lines. However, if the full 15 lines of code determine a whole sequence of steps that will collectively solve the problem, just so long as there are certain other causal factors (e.g., a working compiler) in operation, then it is explanatorily acceptable to depict these other causal factors as the 'ecological' backdrop against which the code operates, and to depict the (full) code itself as a representational solution to the prime-number-finding problem.

If our story so far is on the right track, then causal spread does not directly undermine the claim that certain factors in the extended causal web constitute a kind of coded set of instructions (i.e., a body of representations) for generating adaptive behaviour. This is good news for those of us who think that there is still some mileage left in representational forms of explanation. However, we are not yet out of the woods.

4. The Challenge from Explanatory Spread

Consider the possibility that causal spread entails (what we shall call) *explanatory spread.* On such a view, any non-neural features of the agent's body (e.g., tracheal tubes), or any features of the agent's environment (e.g., reliable structures), that figure as full-blooded causal factors in the generation of some adaptive behaviour, will have an explanatory status equal to that of any neural features that so figure. For extra-neural causal factors in the causal story to achieve this explanatory equality, it is not enough that those factors simply be deemed essential to certain forms of behavioural success. Indeed, such an observation verges on the trivial. It is, in fact, another aspect of a point made only a moment ago, that what matters here is that the non-neural factors, along with the neurally realised representations, must reveal themselves as bona fide sources of behavioural flexibility and richness. It is in cases where this condition is met that we have to take seriously the thought that the inference from causal spread to explanatory spread might be justified. And if that inference is justified, then whilst neural states can rightly be said to encode for behaviours, they should not, in any way, be given explanatory privilege in a proper understanding of how those behaviours are generated.

The move from causal spread to explanatory spread is a tempting one; but, given what has been said so far, it results in the following tension: if the neural factors really constitute *instructions*, and the non-neural factors don't, then why isn't the status of those neural factors *as* instructions enough to give them a special explanatory role? The correct thing to say, we think, is that causal co-determination does not imply explanatory parity. The truth behind the suggestion of parity is that the other factors matter and, moreover, that they can matter in ways that we have traditionally thought only the representational factors mattered. Of course, simply saying that causal spread does not imply explanatory spread is one thing; blocking any such implication is another. What is needed, it seems, is a solid understanding of how the representational and the non-representational contributions to adaptive success differ.

Earlier, we introduced the idea that a weak concept of representation might be found in the notion of parameter-setting. Although this idea is (we believe) an important part of the truth, the fact is that the present challenge from explanatory spread exposes a problem, namely that, as a criterion for representation, the notion of parameter-setting is, on its own, far too permissive with respect to the amount of the extended causal system that it might potentially lead us to count as repre-

resentational interpretations of such states. The reason for this hostility becomes clearer, once one realises that these theorists are just as unhappy with a representational interpretation of genes (see, e.g., (Kelso, 1995, p.183) and (Keijzer, 1997, pp.220-6)), and that that unhappiness can be traced to the fact that both tend to cash out the notion of a genetic program in strong-instructionist terms. Assuming that these thinkers carry their strong-instructionist intuitions about what it means for something to be a program over to the behavioural case — which we think they do — we are not surprised that they cannot see any space for a representational understanding of neural states.

sentational. To see how and why this excessive liberality comes about, let's go back to genes for a moment. Here, the inference from causal spread to explanatory spread would suggest that whilst genes can rightly be said to encode for phenotypic traits, they should not be given any explanatory privilege in a proper understanding of biological development. Now recall that, on our present understanding, the representational status of genes is secured by the idea that genes encode developmental parameters. However, although this idea helps us to understand the role that genes, as representations, might play within the overall developmental system, it fails to tell us why genes are the *only* factors in that system that ought to be treated as representations. In addition to genes, certain environmental factors (e.g., developmental temperature) and certain non-genetic factors (e.g., non-genetic properties of the cellular machinery), factors which partially determine the developmental outcome, might legitimately be conceptualised as supplying parameters for the form-generating dynamical system. If we want to preserve the intuition that genes are the only representational entities here, then representation has to be parameter-setting *plus something else.*[7]

On the strength of the genic analogy, it seems that the very same 'something else' that would enable us to preserve the intuition that genes are the only bona fide representational entities in the developmental system might well also help us to distinguish neural states as representational in a similar fashion (i.e., from amongst other parameter-setting factors that might exist in behaviour-generating systems). So what might that 'something else' be?

One tempting, but ultimately insufficient move is to appeal directly to a kind of teleological asymmetry (see Clark, 1998). DNA structures have often been shaped and moulded, by evolution, ***precisely so as*** to 'tip the balance' in favour of certain adaptively desirable outcomes. The same is true of neural states, only here the shaping and moulding might be achieved not only by evolution, but also by learning. Genes or neural states might thus constitute instructions insofar as they (unlike the rest of the ecological backdrop) have been selected *in order that* such-and-such a response or behaviour should occur. To see why this won't work, consider a robot arm that must reach out and grip a part. The multi-factor story hereabouts might include reference to (a) the force of gravity, (b) a spring-like assembly in the arm and (c) a set of instructions for reaching (in that ecological context). The appeal to teleology distinguishes the force of gravity from the rest. But it leaves the spring-like assembly and the putative instruction set on a par. For both may have been selected precisely so as to enable skilled reaching.

The most we can say, perhaps, is that the central nervous system constitutes an especially potent locus of adaptive plasticity, since it is a resource in which minor and cheap-to-implement changes can make a major difference to behaviour. But even if this is right, it still remains unclear why we should depict the neural states as representational. A push in the back may, after all, be a cheap and easy-to-implement means of tipping an ecological balance in favour of a certain outcome (a cliff-face plummeting scenario, for example). But the push is not thereby revealed as a context-bound instruction to plummet. It is just an applied force that tips the scales in a certain (downwards) direction. The real difficulty here is, in fact, exposed by Webb's claim (mentioned earlier) that certain variables in her cricket-robot — variables which correspond to sensory or motor signals — should not be thought of as representations, because their role in connecting sensors to motors is, in effect, equivalent to the role of the gears in connecting the motors to the wheels. The burning question is thus uncovered: when is an element of an extended causal nexus an internally represented instruction, rather than a mere force selected so as to tip a delicate ecological balance in an adaptively valuable direction? That elusive 'something else' is still avoiding us; so let's press on by deploying, once again, our favourite strategy.

Proteins are strings of amino acids which constitute much of the physical stuff of the organic body, and orchestrate the chemical processes in the living cell that underlie development. The mechanism by which DNA indirectly controls the synthesis of proteins has two stages. The first is *transcription*, in which the organism's DNA acts as a template for the manufacture of a different sort of molecule called messenger RNA (mRNA). The second is *translation*, during which these mRNA molecules de-

[7] Similar considerations indicate why causal co-variation is clearly not enough to warrant talk even of weak or genic internal representation. If causal co-variation were treated not only as a necessary condition for a state or process to be representational, but as a sufficient condition too, then the distinction between the (apparently) representational and the (apparently) non-representational features of the underlying causal system would be fatally blurred. For example, genes are certainly not the *only* causal factors in the developmental system that might causally co-vary with phenotypic traits. If one could hold the genotype and the chemical/mechanical processes in the developing body constant, and vary the environment, then one would, it seems, find causal co-variations between environmental variables and phenotypic traits. Similarly, if one could hold the genotype and the environment constant, and vary the chemical/mechanical processes in the developing body, then one would, it seems, find causal co-variations between chemical/mechanical variables and phenotypic traits. So, to avoid collapsing the distinction between the representational (the genic) and the non-representational (the remaining) aspects of the developmental system, a stronger criterion for representation-hood than causal co-variation is required.

termine the manufacture of different proteins. There are three features of the overall process here which, we think, contribute to explaining why the system in question is compellingly understood as being representational in nature: (i) a certain kind of *arbitrariness*, (ii) the presence of *consumer* subsystems that exploit the putative representations for their content, and (iii) the presence of some kind of *combinatoric structure*. Each of these features needs to be cashed out.

Molecules of mRNA are divided into triplets of nucleotide molecules known as *codons*, and it is believed that every instance of a particular mRNA codon results in an instance of the same amino acid being added to a protein. However, there is nothing in current biological knowledge to suggest a convincing physical/chemical reason why the mappings from particular mRNA codons to particular amino acids could not have been set up differently (Crick, 1968). Thus, and this is our first property of interest, it appears that the mappings here are essentially *arbitrary*, in the sense that the equivalence class of different components that could play each distinct role in the overall process is fixed not by non-informational physical or chemical properties of those components, but precisely by the ability of those components to carry specifiable bodies of information.

The next point turns on the machinery that carries out *translation*: In the cell there are protein-manufacturing-sites called *ribosomes*, and molecules of another sort of RNA called *transfer RNA (tRNA)*. Molecules of tRNA are single triplets of nucleotides attached to single amino acids. What happens during translation is that an mRNA molecule becomes attached to a ribosome, and then passes through it, one codon at a time. When a new codon moves into place, the ribosome (through trial and error) locates a molecule of tRNA that, according to certain rules (the base-pairing rules), features a particular nucleotide triplet. The ribosome then strips off the amino acid from the other end of the tRNA molecule, and adds it to the protein which it is building. Stripped of its amino acid, the tRNA molecule floats off into the cytoplasm, only to be 'recharged' with 'the right' amino acid. It seems that the machinery of ribosomes and tRNA exploits the mRNA in such a way that although the putative representations are not literally understood by that machinery as information-bearers, they are nonetheless consumed by it in a way which makes sense only if we depict the machinery as requiring the informational content that those representations carry. Moreover, it seems compelling to think of the distributed ribosome-tRNA subsystem as *decoding* the information that was transcribed from the DNA into the mRNA (which is why, of course, the phenomenon is known as translation). The presence of such consumer-decoder subsystems is a second indication that we are dealing with a representational process.[8]

Our first two properties provide, we suspect, necessary conditions that any system would have to meet, in order to be representational in nature. Arbitrariness (the fact that any suitable information-bearer could play the same functional role) is what forces us to depict certain states or processes in contentful terms; content-based consumption emerges as a kind of 'inner de-coding' in which some specific information-bearer is enabled to guide behaviour. By contrast, the third property we wish to highlight is (probably) only typical of representational systems. The relationship which exists between sequences of mRNA codons and sequences of amino acids suggests the presence of an entire system of states that exhibits a kind of *structured combinatorics*. The presence of such combinatorics contributes, we think, to the appearance of a full inner *code* as against an isolated inner sign. Notice however that the combinatorics need not necessarily be of the *richly* structured variety beloved of classical AI. What is important is that it should be possible for structurally related inner states to guide different-but-related behaviours.

To get a better grip on these proposed features of internal representation, and to glimpse how they might plausibly transfer to the behavioural case, let's consider a robot which was built some years ago by Longuet-Higgins, and which has been discussed by Johnson-Laird (1983) and, more recently, by Keijzer (1997). The robot avoids falling off the edge of tables, but seemingly has no table edge detector. Instead, the main wheels drive two small wheels that hold a piece of sandpaper under the base plate, and things are set up so that the small wheels, relative to the sandpaper, echo the location of the real robot relative to the tabletop. Whenever a small wheel hits a ridge at the edge of the sandpaper, a circuit is closed and the robot turns. As a result, the small

[8] This picture of content-based consumption must be fleshed out with enormous care, so as not to fall foul of the kind of homuncular fallacy mentioned earlier (in section 1.). Thus it must not suggest that the contents of internal representations are actually understood by the consuming subsystems concerned. It seems to us that much recent anti-representationalist sentiment in and around the animat movement trades on the mistaken idea that a representational state must present itself *to* an understanding system *as* having a certain content. Such a view is clearly at work in arguments which insist that such-and-such an internal state cannot be a representation since it doesn't represent anything to the system itself. For example, Brooks and Stein (1993, p.2) caution that the firing of a particular neuron, even if it is non-accidentally associated with, say, the presence of red objects, is not thereby revealed as a representation of redness *for* the system itself. Instead, it represents redness to us, the external observers. For detailed discussions of these and related issues, see, e.g., Wheeler (forthcoming) and Wheeler and Clark (submitted).

wheels and sandpaper model the big robot and table-top. By ceding control to the inner model, the big robot avoids falling off the edge. Commenting on this set-up, Johnson-Laird writes that:

> The small wheels and the piece of sandpaper are not intrinsically a model but they become a model in the robot because of their function as an arbitrarily selected symbolic notation that is used to register the position of the robot in the world. (Johnson-Laird, 1983, p.404)

It seems clear that the inner elements are here being counted as *arbitrary* because the relevant equivalence class of inner systems (the set of different systems that could play the same adaptive role) is fixed not by the first order physical properties of the components (the thickness and weight of the sandpaper etc.) but by their capacity, when appropriately organised, to act as the bearers of a certain (specifiable) body of information. These states bear that information insofar as they support a body of internal dynamics that is fit to track or model a specifiable set of external states of affairs. The inner states thus allow a certain kind of information (one that is not directly transduced by the robot's sensors) to guide the robot's behaviour. The equivalence class of inner processes that could play the same role includes anything that can bear that information and make it effective within the larger system. In addition, the circuit closer is illuminatingly understood as a *consumer* subsystem, one which is targeted on a certain inner state not because of the state's intrinsic physical properties, but because of the information that that state happens to bear. The counterfactuals (which fix the equivalence classes) make it plain: had the designer (which might as well have been evolution) exploited some other inner state as an indicator of the table edge, then the consumer process would have been targeted on that other inner state instead (cf. Millikan, 1989). Finally, notice that the inner processing of the Longuet-Higgins robot (like that of the Franceschini et al. robot) does not rely on classical combinatorics. Instead, it uses a kind of structured inner model to support a variety of content-related encodings. The suggestion, then, is that this triple appeal to arbitrariness, content-based consumption and combinatorial structure might provide a robust and workable means of consistently distinguishing representations from mere forces in the causal systems of interest.

5. Conclusions

Neural states which are implicated in the production of environmentally-embedded adaptive behaviour do not fully specify external states of affairs or behaviours. They operate within an assumed, normal ecological context of other causally active (bodily and environmental) states and processes. However, that does not prevent us from treating the neural states in question as representations. By *encoding* certain parameters, those states stabilise the complex neural and bodily dynamics that, in interaction with the environment, yield appropriate behavioural outcomes. However, if this concept of *genic representation* is to be made secure against the ravages of explanatory spread, then further conditions on what will count as such a representation are needed. We have suggested three candidates: arbitrariness, content-based consumption, and combinatorial structure.

Genic representation is meant to do duty in a variety of cases in which inner resources are delicately keyed so as to make maximal use of bodily and environmental factors and forces in the promotion of ongoing adaptive success (e.g. the Franceschini et al. maps). The concept is still very much 'under development', but we hope and believe that it will play its part in helping us to understand the complex, interactive exchanges between brain, body, and world that conspire together to produce environmentally-embedded adaptive behaviour.

Acknowledgements

Michael Wheeler is a Junior Research Fellow of Christ Church, Oxford, and a Post-Doctoral Research Fellow of the McDonnell-Pew Centre for Cognitive Neuroscience, Oxford.

Many thanks to Susan Hurley, and audiences at the 1997 Annual September Conference of the British Society for the Philosophy of Science (held at Oxford University), the Department of Experimental Psychology at Oxford University, and the School of Cognitive and Computing Sciences at the University of Sussex for invaluable comments on formative presentations of this work.

References

Agre, P. (1988). *The Dynamic Structure of Everyday Life*. Ph.D. thesis, Department of Electrical Engineering and Computer Science, Massachusetts Institute of Technology.

Brooks, R. A. (1991). Intelligence without reason. In *Proceedings of the Twelth International Joint Conference on Artificial Intelligence*, pp. 569–95 San Mateo, California. Morgan Kauffman.

Brooks, R. A., & Stein, L. A. (1993). Building brains for bodies. Artificial Intelligence Lab Report 1439, MIT.

Clark, A. (1997). *Being There: Putting Brain, Body, and World Together Again*. MIT Press / Bradford Books, Cambridge, Mass. and London, England.

Clark, A. (1998). Twisted tales: Causal complexity and cognitive scientific explanation. *Minds and Machines*, *8*, 79–99.

Clark, A., & Grush, R. Towards a cognitive robotics. Forthcoming, in *Adaptive Behavior*.

Clark, A., & Toribio, J. (1994). Doing without representing. *Synthese*, *101*, 401–31.

Cliff, D., Husbands, P., Meyer, J.-A., & Wilson, S. W. (Eds.). (1994). *From Animals to Animats 3: Proceedings of the*

Third International Conference on Simulation of Adaptive Behavior, Cambridge, Mass. MIT Press / Bradford Books.

Crick, F. (1968). The origin of the genetic code. *Journal of Molecular Biology*, *38*, 367–379.

Franceschini, N., Pichon, J. M., & Blanes, C. (1992). From insect vision to robot vision. *Philosophical Transactions of the Royal Society, series B*, *337*, 283–94.

Goodwin, B. (1994). *How the Leopard Changed its Spots: the Evolution of Complexity.* Phoenix, London.

Harvey, I., Husbands, P., & Cliff, D. (1994). Seeing the light: Artificial evolution, real vision. In (Cliff, Husbands, Meyer, & Wilson, 1994), 392-401.

Johnson-Laird, P. (1983). *Mental Models.* Cambridge University Press, Cambridge.

Keijzer, F. (1997). *The Generation of Behavior: On the Function of Representation in Organism-Environment Dynamics.* Ph.D. thesis, University of Leiden.

Kelso, J. A. S. (1995). *Dynamic Patterns.* MIT Press / Bradford Books, Cambridge, Mass. and London, England.

Mataric, M. (1991). Navigating with a rat brain: a neurobiologically inspired model for robot spatial representation. In (Meyer & Wilson, 1991), 169-75.

Meyer, J.-A., & Wilson, S. (Eds.). (1991). *From Animals to Animats: Proceedings of the First International Conference on Simulation of Adaptive Behavior*, Cambridge, Mass. MIT Press / Bradford Books.

Millikan, R. (1989). Biosemantics. *Journal of Philosophy*, *86:6*, 281–97.

Webb, B. (1994). Robotic experiments in cricket phonotaxis. In (Cliff et al., 1994), 45-54.

Wheeler, M. Explaining the evolved: Homunculi, modules, and internal representation. Forthcoming, in the *Proceedings of the First European Conference on Evolutionary Robotics*, Springer-Verlag.

Wheeler, M. (1994). From activation to activity: Representation, computation, and the dynamics of neural network control systems. *Artificial Intelligence and Simulation of Behaviour Quarterly*, *87*, 36–42.

Wheeler, M., & Clark, A. Genic representation: Reconciling content and causal complexity. Submitted for Journal Publication.

Wilson, S. W. (1985). Knowledge growth in an artificial animal. In Grefenstette, J. J. (Ed.), *Proceedings of an International Conference on Genetic Algorithms and their Applications*, pp. 16–23 Pittsburg, PA and Hillsdale, New Jersey. Lawrence Erlbaum Associates.

Some Armchair Worries about Wheeled Behavior

Fred A. Keijzer

Section of Experimental and Theoretical Psychology
Leiden University, P.O. Box 9555, 2300 RB Leiden
The Netherlands
E-mail: keijzer@rulfsw.leidenuniv.nl

Abstract

Autonomous agents research aims to understand adaptive behavior by building models that exhibit such behavior. In this paper, I describe some theoretical worries about the tendency to simplify the sensory-motor control of these behaving models as much as possible. Wheeled robots provide a good example of this tendency. The worries derive from the idea that a complex sensory-motor system is a necessary requirement for reliable functional behavior in a natural environment. The evidence on which I base this conviction is three-fold: (a) when animals, but not wheeled robots, move across a surface this results from variable spatio-temporal patterns across a complex sensory-motor system, (b) a theoretical analysis according to which the stability of distal behavior arises as a result of the variability of proximal behavior, and (c) the possibility to interpret the behavior of animals, but not that of wheeled robots, as a process of self-organization. The use of wheeled robots therefore seems to side step a core problem for understanding adaptive behavior.

1. Introduction

Research on autonomous agents aims to understand adaptive behavior by building models—robots or animats—that exhibit such behavior. This paper presents some theoretical worries about current attempts to understand adaptive behavior in this way.

These worries derive from the fact that organismal and robot behavior differ considerably in the ratio of sensory-motor complexity to the complexity of the functional characteristics of the behavior. Behaving multicellular organisms embody a nervous system and a complex array of sensors and muscles and their opposing rigid structures. In other words, a complex neuro-musculo-skeletal system is at work behind every form of adaptively relevant, functional behavior—even when the latter remains very basic. On the other hand, within autonomous agent research (AAR) most robots—of course not all (e.g. Brooks, 1989; Beer, 1995)—are built with an eye to achieving complex adaptive relations with an environment while relying on a minimally complex sensory-motor system.

As compared to classical cognitive science, AAR pays a great deal of attention to the mediating details of behavior. The recent interest in adaptive behavior is based on the conviction that the concrete, physical sensory-motor interactions between a specific, embodied behaving system and the characteristics of the environment together generate the phenomena of adaptive behavior. Nevertheless, the actual use of detailed and complex sensory-motor interactions in modeling adaptive behavior is kept to a minimum. It is thought to be sufficient that they produce the 'really interesting' functional characteristics of behavior, the characteristics that make behavior intelligent and adaptive.

This state of affairs worries me. I am led to think that studying the wider, functionally adaptive aspects of behavior apart from their naturally complex mediation is insufficient, and even a mistake. In this paper, I will argue that the most important bottle-neck for building robots whose behavior is on a par with that of biological organisms lies in coming to grips with the biological capacity for complex sensory-motor coordinations. AAR should pay much more attention to the organizing principles—the mechanisms—of organismal behavior because a complex sensory-motor system is a prerequisite for the very occurrence, as well as the stability, of the functional characteristics of behavior. I claim that *the core problem for autonomous agent research is to model functionally relevant behavior by means of a complex sensory-motor system*. The major challenge at present is not to model many different kinds of functionally relevant behavior, it is to model the organizing principles on which such functional behavior depends.

To contain this discussion about the importance of sensory-motor complexity, I will focus on one particular example of behavior where this issue comes to the fore very markedly: moving about over a firm surface. In AAR this behavioral task is often accomplished by wheeled robots, whereas in nature very different solutions dominate. What I want to focus on here is, of course, not the use of wheels in particular. Wheeled behavior is merely a good example of the tendency to simplify the sensory-motor processes in order to attend to the 'adaptive' or 'intelligent' aspects of behavior.

2. Movement over hard surfaces

Moving about over a firm surface is a common accomplishment for both animals and robots. Most robots effect this behavior by driving about on wheels. Two commonly used robots, the *Nomad 200* (Sharkey & Heemskerk, 1997; Milán, 1994) and *Khepera* (Floreano & Mondada, 1994), provide good examples of the sensory-motor system that allows a wheeled robot to move. Despite their differences in size (they are respectively 90 cm and 3 cm high) and approximate weight (respectively 60 kilograms and 70 grams), both are statically stable artifacts that rest on wheels. Electric motors set the wheels in motion and so propel the robot forward, backward or let it turn about. Because these are good robots—in the traditional robotic sense of being precise and dependable—there is a reliable one to one relation between a small number of different signals send to the motor and the movement of the robot itself.

I state the obvious here to stress the point that the use of wheels makes movement across flat, hard surfaces a trivially easy task. Wheels make it possible to investigate functional behavior (e.g. goal seeking, learning, imitation, social interactions) for which moving about is a prerequisite, without putting too much work in the necessary sensory-motor system.

How does the wheeled behavior of robots compare to the accomplishments of real animals? When considering movement over a hard surface the end result is the same in both cases, but the means by which it is accomplished differ considerably.

The first thing to notice is the wide variety of mechanisms used to produce forward motion. Legs may be the most familiar means by which animals move about, but there are many other options. In particular, soft-bodied animals exploit many different strategies.

Some animals, such as small, free-living flatworms and nemertines, use *cilia* to crawl (Barnes, Calow & Olive, 1993; Trueman, 1975). Cilia are relatively short, bristle-like cell organelles that project from cells at the underside of the animal. A constant propulsive force is generated by the coordinated movement of a large number of cilia. Each cilium normally has the same pattern and frequency of beating, but individual cilia are at different phases of this cycle—so as not to interfere with one another's powerstrokes—and waves of activity can be seen to pass across the ciliated surface. There are strong size-imposed limitations on ciliary locomotion and its use remains restricted to small animals of usually less than 1 mm long.

Looping is a form of movement used by slightly bigger animals. It derives from muscle power and occurs in elongated, 'tube-like' animals. Some nemertines use looping as a fast escape mechanism, but it has been perfected by leeches (Trueman, 1975). Looping is based on the combined use of circular and longitudinal muscles, the forces of which antagonize one another through a hydrostatic skeleton, a relatively fluid body mass of fixed volume. The longitudinal muscles make the animal short and thick, the circular muscles turn it thin and elongated. Leeches have suckers at their posterior and anterior extremities which attach the leech to the substrate. When the posterior sucker is attached, the leech uses its circular muscles to stretch its body forward, after which it places its anterior sucker on the substrate at almost a full body length in the direction of travel. Then it loosens its posterior sucker, flexes its longitudinal muscles to draw in its body, and places its posterior sucker just behind the anterior one, after which the cycle repeats itself.

A similar musculo-skeletal organization is used to move by *pedal locomotory waves*. These are waves of contraction that pass along the body. The cyclical contraction of circular muscles followed by that of longitudinal muscles, as in looping, is here replaced by the successive contraction of different *regions* of the longitudinal muscles (Trueman, 1975). "A muscular locomotory wave consists essentially of raising a part of the body from the substratum, moving it forwards and reattaching it." (ibid., p.22). The waves that move across the body can be both direct (in the direction of the organism's movement) or retrograde (in the direction the organism is moving from). In the retrograde case, the thinning and elongating parts of the body are lifted free from the substratum and move forward while the thick and shortened parts of the body form the so-called *points d'appui* which anchor the body to the substrate. When the waves are direct the fully elongated body parts form the *points d'appui*, while the thickened parts of the body are the ones that move forward. In this case, the thin portions of the body usually extend stiff, bristle-like appendages to prevent slippage (Barnes, *et al.*, 1993).

There are many more ways in which animals move over hard surfaces: snakes undulate, and many animals walk. The details of these various ways of moving are all different. Still there are commonalities which group all these forms of locomotion together and contrast them with robot movement provided by wheels. Forward movement as it is produced by biological organisms occurs as collective property of many small-scaled components, together making up a neuro-musculo-skeletal system. Forward movement is here not the result of a simple, singular event which sets a motor on or off, as is the case in many robots.

Of course, everybody doing adaptive behavior research will be perfectly aware that driving across a hard surface differs in this respect from ciliating, looping, crawling or walking over it. However, the dominant idea is that the details of such mediating processes can be ignored when one studies the functional aspects of behavior. The use of wheels allows one to study some highly interesting aspects of adaptive behavior (motivation, learning, social relations, imitation etc.)

which would be extremely difficult if one had to implement a complex neuro-musculo-skeletal system first.

Given the difficulty of modeling both the full complexity of the mediating processes involved in animal behavior as well as the functional regularities which result from these, what is wrong with a division of labor and focus on one of these issues at a time? Why worry? My nagging doubts here are that an important aspect of natural behavior is left out of consideration. This doubt is reinforced by some old theoretical ideas concerning the relation between sensory-motor coordinations and functional organism-environment interactions.

3. The many faces of behavior

Behavior is a multi-facetted phenomenon. It is both difficult to define and to analyze. This "does not arise from any lack of ways to do it but from an embarrassment of riches. We can describe an action as a set of muscle twitches, or as a sequence of movements of limbs and other parts, or as a sequence of goal-directed actions, or in even larger units" (Miller, Galanter & Pribram, 1960, p.13).

How is this riches to be dealt with? Ever since Tolman in the 1930s, all psychologists and biologists are familiar with the distinction between molecular and molar behavior—molecular behavior referring to little units and molar behavior to large ones—but this distinction does not play an important role in practical psychological research. As Van der Heijden states:

> it is not the complete behaviour that is observed and registered, but only some specific, quantifiable, aspects, such as the latency and/or correctness of the response. Other aspects, e.g. the duration and the detailed execution of the observable behaviour are simply neglected (Van der Heijden, 1992, p.14).

Confronted with the "embarrassment of riches" of actual behavior, researchers—both psychologists and biologists—tend to concentrate on specific characteristics and leave the rest to other researchers. The choice of these characteristics is constrained by the available experimental and modeling possibilities, but also by preconceptions about what are the most crucial aspects of adaptive behavior. One may agree that the topic of interest is the modeling of adaptive behavior, but this commitment leaves unspecified which aspects of adaptive behavior are chosen to be modeled.

3.1 Brunswik's Lens Model

To provide some handles on the many different aspects of behavior it will be useful to turn to some older ideas, first developed by Fritz Heider (1959) and later elaborated by Egon Brunswik (1952; Campbell, 1966). These provide an instance of the general distinction between molecular and molar behavior, which is particularly congenial for elucidating my worries about wheeled behavior.

Heider and Brunswik were both concerned with the structure of the environment as a relevant factor for the adaptive behavior of organisms. In the 1920s, Heider made a distinction between the 'near' and 'far' environments of an organism and addressed the implications of this distinction for adaptation (Heider, 1959). Brunswik, influenced by Heider, used the notions of proximal and distal to mark the distinction between distal objects or events (the far environment of the organism), and the proximal stimuli (the near environment) which actually reach the organism as a pattern of sensory stimulation (Brunswik, 1952).

The near environment of an organism consists of proximal stimulation which impinges on its body directly. This stimulation in itself tends to be neutral as far as adaptation is concerned. The presence or absence of light- or soundwaves, or of air-born particles does not usually make a dramatic impact on the life-processes of an organism. The near, proximal environment provides a relatively neutral medium in which the organism can move about. The neutral proximal lights, sounds and smells do, however, convey cues about the existence and whereabouts of distal, nonlocal features of the environment.

Heider and Brunswik stated that the adaptively relevant elements in an environment consist primarily of the distal objects or 'things' therein. These are usually at some distance from the organism and provide the possible nourishment, threats or mates on which survival and reproduction depend. Of course, it is only when these distal events or things come close to the organism that they become a matter of life or death. However, such moments—when the tiger is already at your throat—are relatively rare, and when they occur these normally distal—but now proximal—entities also become irrelevant as far as behavior is concerned. Nothing further can be done at such a moment (except to eat or being eaten).

The background assumption is that a certain time-frame is necessary in which an organism still has the opportunity to act. During this time, an organism uses the proximal stimuli as cues to guide it to or away from significant distal objects or events. Behavior then consists of bringing about (or avoiding) specific close encounters with such—most of the time—distal stimuli. As Brunswik says: "Organisms may in this sense be characterized as 'stabilizers' of [distal] events or of relationships" (1952, p.20).

The problem with using proximal stimuli as cues for distal stimuli is their unreliability as signs of distal states. The relation between the proximal stimuli and distal events is not one to one. Similar proximal stimulus patterns can correspond to different distal states of affairs, and, the other way round, the same distal state of affairs can give rise to very divergent proximal patterns. On the behavioral side, the same principle applies. The behavioral means available to the organism are also proximal: it can make movements. A series of 'properly' controlled movements can result in the achievement of an adaptively relevant event. However, the proximal set of movements that leads toward a similar distal event (feeding, escape from predators) will differ every time.

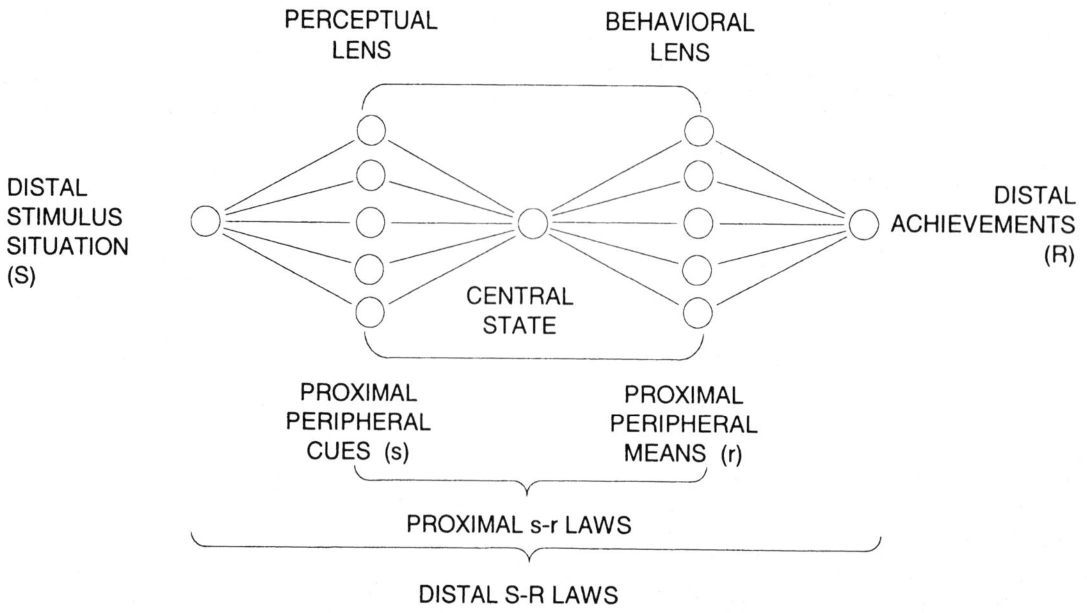

Figure 1. Brunswik's Lens Model (adapted from Brunswik, 1952).

Heider and Brunswik state that the problem to be solved by an organism amounts to the following: given a state in which the organism is immersed in a highly variable, proximal environment—as well at the sensory as at the effector side—how does it achieve a consistent coordination with relatively invariant distal events? This basic question can be illustrated with Brunswik's expository "Lens Model" (Brunswik, 1952; Campbell, 1966).

The Lens Model conveys the idea of proximal variability and distal stability when the organism is observed over several behavioral trials. This particular figure displays another distinction. It consists of two 'lenses'. The first one displays how a particular distal object is, on different occasions, connected to a particular central (mental) state by different proximal sensor states. The second lens portrays how the presence of a central state leads to the achievement of a distal, functional achievement through variable proximal movements on different occasions. The global picture is subsequently one in which a locally, or proximally, operating system—the s-r part can be seen as a single 'lens'—enforces stable distal S-R relations.

Brunswik's talk about proximal s-r laws and distal S-R laws may be less congenial nowadays than it was almost fifty years ago. However, the picture can be easily modified and made to concur with the current insight that responses feed back to the stimulus situation. In that case, the laws become couplings between the organism and its environment at two (actually a continuum of) different scales: sensory-motor couplings for proximally guided behavior and perception-action couplings for distally oriented behavior.

To summarize, a central aspect of behavior consists of an ever varying proximal mediation which stabilizes distal events—and so turns them into distal regularities. To return to the example of forward movement, this distal task is accomplished in many different proximal ways—ciliating, looping, undulating, walking—the execution of which is itself always variable. The net result may in all cases be the same, this result is always accomplished in slightly different ways.

3.2 Behavior as results rather than means

When behavior is divided into relatively low-level proximal characteristics—what might be called the means—and higher-level, distal characteristics—the results, or the ends of behavior—an interesting question presents itself: Which of these two deserves our most particular attention? The question is particularly important for modeling adaptive behavior. Not every aspect of behavior can be replicated in a simulation or a robot, and choices will have to be made.

I think that within current studies of adaptive behavior there is a strong tendency to favor and focus on behavior's distal aspects: Behavior is interpreted in terms of adaptively relevant *results*, rather than in terms of proximally mediating mechanisms—the *means*—that bring these results about.

First, it should be noted that there is a general human tendency to describe behavior in distal terms whenever these apply. Vallacher and Wegner (1987) investigated in what manner people describe what they are doing. They state that "when both a lower and a higher level act identity are available," (e.g. 'lifting a mug to my mouth' and 'drinking coffee') "there is a tendency for the higher level to become prepotent." (p.5) Distal descriptions remain dominant as long as they apply, that is, as long as the distal result of the behavior occurs in a regular way. When the process leading up to the distal result is disturbed, more proximal descriptions come to the fore. For example, one experiment on which they based their conclusions was done with subjects who first had to drink a cup of coffee. One group drank from a normal cup, the other from a very unwieldy one, weighing approximately 500 grams. The first group applied more general descriptions to this activity (such as "getting energized" and "promoting my caffeine habit") while the group with the unwieldy cup identified it in terms like "drinking a liquid" and "lifting a cup to my lips". The message: when distal descriptions apply, then these will be the ones that are preferentially used by human observers.

The general tendency to prefer distal descriptions of behavior seems to have had a large impact on the study of behavior. The Lens Model provides an explanation. The most salient regularities of behavior appear at a distal scale. Distal behavior is easily identifiable because it is perturbation resistant and repeatable. Similar goals are reached time and again. At a proximal scale, there is a wide scattering of behavior. The mediating process is highly variable and there are many different proximal routes leading to the same distal behavior. Unless one is specifically interested in the specific sensory-motor details of specific animals or robots, it does not seem to matter much how exactly the distal behavior comes about. As long as it does.

This practical reason for focussing on behavior's functional regularities is reinforced by the often held conviction that these regularities are also the ones that a behavioral science most particularly needs to explain. Its functionality makes adaptive behavior different from mere physical movements. Brunswik already stated that the more important psychological generalizations were to be found at a distal level, and this attitude can still be witnessed in current work on modeling adaptive behavior. It leads to a serious underestimation of the sensory-motor problems involved in generating such distal regularities.

McFarland and Bösser (1993) offer an example of this tendency within AAR. In their book on intelligent behavior in animals and robots, they described intelligence as "an assessment of performance in terms of some functional criteria" (p.24). Even though they stress the importance of the body which actuates adaptive behavior, their interest centers on ways to optimize choices between multiple—already existing—options for distal behavior. For example, they discuss how an imaginary housekeeping robot chooses between several "simple behavioral tasks" like "gotoN, gotoK, gotoF, gotoW, getmop, returnmop, getcloth, windowclean, floorclean, and sleep" (p.217). Creating a mechanism that could physically perform such tasks is not treated as a problem. According to McFarland and Bösser, 'a' mechanism will do. They even want to leave the kind of mechanism involved in generating behavior unspecified as "it is unsatisfactory to pin the label 'intelligence' on certain mechanisms and not others" (p.16). The problem how any mechanism can be constructed which generates real, biological-style behavior is passed over and treated as solved.

McFarland and Bösser's treatment of intelligent behavior is just one example how behavioral research is skewed toward the distal, functional aspects of behavior, at the cost of more proximal aspects. Using wheeled robots to investigate adaptive behavior is important in this context. The primacy awarded to distal behavior and the use of wheeled robots mutually reinforce one another. The focus on distal behavior makes the exact execution of distal behavior less relevant so that wheeled robots—although biologically implausible—become a plausible way to model adaptive behavior. In return, wheeled robots make it relatively easy to model distally oriented behavior. With a minimally complex sensory-motor system, a wheeled robot is capable of performing intuitively interesting distal tasks, for example, making it possible to envision a housekeeping robot. If such a robot had to walk, crawl or loop, just moving about would become a serious achievement, leaving little place for housekeeping tasks.

At this place my basic worry about wheeled behavior comes to the fore. Wheels allow one to study distal behavior by simplifying the proximal mediating process: What if the proximal mediating mechanism is a much more central aspect of adaptive behavior than many tend to think? More specifically, what if a fairly complex proximal mediating process is an essential ingredient for bringing about natural adaptive behavior?

3.3 The importance of sensory-motor complexity

Brunswik's Lens Model points not only to behavior's distal aspects. The Lens Model also visualizes the proximal mediation of these distal aspects. Rather then putting one of the two in a privileged position, the Lens Model suggests an interpretation of behavior in which the relation between proximal and distal behavior is a key issue for understanding and modeling adaptive behavior.

The Lens Model visualizes distal behavioral regularities as a result of a 'focussing process' that is somehow mediated by proximal behavior. The more easily identifiable distal behavior occurs only because there is a proximal focussing process which is highly variable and context dependent, and therefore difficult to pin down.

The focussing metaphor implies that a proximal behavioral 'lens' is required to acquire stable distal behavior. To interpret this metaphor it is good to remember Ashby's law of requisite variety: "only variety can destroy variety" (Ashby, 1956, p.207). The behaving system needs complex behavioral capacities at a proximal scale—call it room for manoeuvreing—to counteract the many proximal disturbances that will always occur in a natural environment. Only by bringing into play many different options for proximal behavior—that is when the sensory-motor system has many degrees of freedom—can stable distal regularities be ensured under natural circumstances.

When the 'diameter' of the proximal lens is too small—that is when the behaving system has too few sensory-motor degrees of freedom available to counteract disturbances—the proximal countermeasures will be insufficient and the distal regularities will break down or not occur at all.

The use of wheeled robots produces an extreme simplification of the proximal sensory-motor couplings related to moving about. The capacity for maintaining intuitively convincing distal perception-action couplings is therefore severely limited. In terms of the Lens Model, the researcher collapses the proximal lens by selecting a single proximal behavior (driving) which tends to accomplish the distal task of moving about. The resulting distal behavior is brittle. Wheeled robots are restricted to uninterrupted flat surfaces and their capacity to move about breaks down as soon as it encounters steep slopes, gaps, muck, or merely any rough surface with stones and treeroots.

When a wheeled robot encounters a natural terrain, it can no longer be expected to achieve forward movement. In comparison, nature's solutions to the problem of forward motion are not so easily compromised. These solutions are all relatively complex from a proximal point of view. They are also capable of negotiating naturally occurring terrain. To name just one case, looping leeches are not restricted to flat hard surfaces. They can move vertically, negotiate gaps and outperform anything on wheels many times over when it comes to moving about under natural conditions.

A complex mediating mechanism is thus highly important to establish and maintain adaptive behavior that deals effectively with a distal environment. It seems to me therefore that *a key issue of adaptive behavior is not merely to achieve distal goals, but to achieve these goals under always varying proximal circumstances.* Proximal sensory-motor variability is necessary to give rise to distal stability.

3.4 A challenge for modeling adaptive behavior

Researchers working on the simulation and modeling of adaptive behavior will certainly agree that good models of adaptive behavior need to incorporate both proximal and distal behavior. But the claim that it will be necessary to make the proximal behavior sufficiently complex goes beyond this general agreement. This claim has implications that may be less readily shared by all.

One implication for modeling behavior is that wheeled robots do not capture the characteristic signature of natural adaptive behavior: proximal variability leading to distal stability. Wheeled robots achieve the distal effect of moving about with only minimal proximal variability. As a model this does not suffice because a key factor of adaptive behavior is edited out of the model.

When modeling adaptive behavior, the mediating mechanism is not an arbitrary interchangeable producer of movements. It has to fulfil more stringent demands relating to flexibility and adaptivity by being capable of coordinating a large number of degrees of freedom. This implies that adaptive behavior research ought to pay much more attention to *behavioral mechanisms* than is often the case. It is not sufficient that a robot achieves a distal behavioral task. The robot should perform this task in a way which resembles an animal's way of performing this distal task: The robot's performance should be based on the same organizing principles as animal behavior.

This message might be difficult to accept. At least two reasons can be brought against the idea. Firstly, it seems strange to uphold that robots who accomplish the same functional behavior as animals or humans are to be distinguished because of their supposedly 'different organizing principles' or underlying mechanisms. Why would it matter how a task was accomplished, as long as the job is done? Secondly, referring to different organizing principles seems to be rather vague and mystifying. Any reference to such principles should be given a more clear meaning before they can be taken seriously.

The first point can be dealt with rather quickly. Postulating a theoretical difference between the behavior of animals and that of wheeled robots looks only awkward under the assumption that the same tasks *are* accomplished in both cases. However, the argument is that robot and animal behavior have different characteristics. A robot may achieve a distal task under a narrow range of proximal circumstances while an animal can achieve the same distal goal under a much wider range of proximal circumstances. Robots and animals do therefore not accomplish the same distal tasks at all. The difference made between animal and (current) robot behavior on the basis of underlying mechanisms is also reflected in the functional behavioral regularities of animals and robots.[1]

The difference made between the behavior of animals and robots is based on the conviction that animals, but not robots—at least not as they have been built so far—rely on still unspecified organizing principles which form a necessary precondition for achieving the distal behavioral regularities which are so characteristic of animal behavior. Saying something about those principles requires more work.

4. Behavior as a self-organizing process

Stating that the proximal behavior of animals and robots derives from different organizing processes may sound either mysterious or trivial. Interpreted as the obvious difference between organics and electronics the case looks trivial. Of course there is a difference here, but this difference is of a kind which many would think should not matter. There do not seem

---

[1] Elsewhere (Keijzer, 1998), I signal the distinction between animal and robot behavior by making a difference between movements and behavior: Robots (at least those that have been build up to now) merely move, they do not behave. Only animals have so far been capable of true behavior. I admit that it is difficult to draw a crisp borderline between genuine behavior and mere movement, but the distinction is a good way to make the point that not everything that moves automatically behaves.

to be any principled reasons for claiming that artificial creatures, build from electronic and mechanical components, are forever incapable of generating behavior in the same ways as biologicals do. I agree. The idea that animal and (current) robot behavior reflects different organizing principles refers to a more abstracted level of organization: the behavior of animals, but not that of robots, consists of a process of self-organization.

The notion of self-organization is used often and with different meanings. My usage derives mainly from the work of Haken (1987) and Kelso (1995), but see also Goodwin (1994) and Kauffman (1993). Self-organization occurs when a large number of dispersed, small-scaled processes (the microscopy) become coordinated and so generate a pattern at a much larger scale (the macroscopy). A standard example consists of the large-scale convection rolls that arise in a container with fluid when the heat gradient between the cold top and the warm bottom exceeds a certain treshold. Individual molecules, at first each with its own independent direction of movement, start to move together on a large scale. Another example consists of swarms of fireflies where the light flashes of the individual flies become synchronized so that the swarm flashes its light on and of as a whole.

The notion of self-organization is very intriguing, unfinished and one which could have wide repercussions for the study of adaptive behavior (see e.g. Kelso, 1995). As it cannot be adequately described in a short space, I will simply endorse the following points: (1) The world is organized along multiple spatio-temporal scales, each having its own, relatively independent, set of dynamical relations. For example, one can both look at colliding billiard balls or consider the molecules making up these billiard balls. (2) Generally, dynamical relations are studied at a particular scale. In contrast, self-organization is a process where two or more of these—usually independent—dynamics influence one another and become coordinated. (3) When self-organization occurs, the specific characteristics that arise at a macroscopic scale depend on the interactions between *many* small-scaled—microscopic—events. I take these three points to characterize what happens (among other things) when a process of self-organization occurs (see also Keijzer, in press).

These points tie up nicely with animal behavior in its form of forward motion across a firm surface. Forward movement involves at least two different scales of dynamical relations. There are a large number of small-scaled neuro-musculo-skeletal events and there is the large-scale movement of the organism. The biological cases of forward movement described above all consist of a large scale pattern (coordinated cilia, looping, pedal locomotory waves, coordinated legs) that arises as a result of many, smaller scaled interacting parts.

For example, inside worms a neural network is connected to a dispersed system of muscle fibers which are antagonized by a hydrostatic skeleton. Under the influence of the neural network the muscle fibers contract and extend again. At the much larger scale of the whole worm-body, the pedal locomotory waves can be observed that make the worm move over the substrate. This large scale set of dynamical relations can be studied independent from the small-scaled happenings inside the body. However, in forward movement the otherwise separate dynamics are closely related and together the many small-scaled neuro-musculo-skeletal events organize themselves as the large-scale locomotory waves of the worm as a whole. These locomotory waves can be interpreted as a macroscopic order—similar to the convection rolls in heated water—which arises as a result of a mutual influence with a microscopic level of organization.

To summarize, at the heart of behavior lies a capacity to organize spatio-temporal pattern, both across a nervous system and across a musculo-skeletal system. Certain of those spatio-temporal patterns result in forward movement on a hard surface.

The congeniality of self-organization and 'simple' forms of behavior is of course not a proof that self-organization actuality takes place in behavior. At present there is only a suggestive, general correspondence which will need a lot of 'future work' to turn it into a succinct and factual statement. Still, the link looks promising. Self-organization might provide the mechanism that is necessary to maintain distally stable behavior on the basis of microscopic variability. This is exactly what is necessary for adaptive behavior in a natural environment.

In addition, there are intriguing options for 'steering' such processes of self-organization by means of 'internal control parameters' situated within the set of small-scaled dynamical relations. This possibility has interesting implications for the debate on representations (Keijzer, in press), but this goes beyond the aim of the present paper. The message that I want to confer right now is merely that a process of self-organization makes the claim more solid that the behavior of animals derives from specific organizing principles, principles which are not necessarily replicated by any artifact capable of moving about.

Well, do wheeled robots rely on a process of self-organization in order to move about? It seems unlikely. In self-organization there are no direct links between small-scaled and large-scaled events, many small-scaled events make up a similar large-scale event on different occasions, but never in exactly the same way. In wheeled robots, the macroscopic forward motion does derive directly from a few small-scaled signals which control the motors and so what happens on a larger scale. The small-scaled (neural) processing merely selects which signal is sent to the motors, it does not generate a pattern on a bodily scale. The principles involved in self-organization and in wheeled robots are dissimilar in this respect. In other words, if the behavior of animals consists of a process of self-organization, then the behavior of animals and wheeled robots do not reflect the same abstract organizing principles.

5. What are the implications if these worries are justified?

My worries about wheeled behavior hover around the idea that animal behavior requires complex sensory-motor interactions in order to maintain stable functional (distal) behavior in a natural environment. The use of wheels for robots is an example of sidestepping the issue of proximal behavior's complexity in order to study distal behavioral regularities. Averting this issue leads to the neglect of, what I think is a central aspect of natural behavior: how do variable proximal (fast and short-term) sensory-motor couplings maintain stable distal (slower and longer-term) perception-action couplings.

Assuming for now that the present analysis is on the right track, what are the implications? Foremost, it should be stressed that adaptive behavior is not a clear and circumscribed research topic. Rather, what adaptive behavior actually amounts to is an open research question. The tendency to emphasize the functional, distal aspects of behavior ought to be balanced by attention to the mediating mechanisms on which biological systems depend to generate these distal aspects. Behavior should not be treated as a singular (distal) output but rather as the generation of spatio-temporal patterns along a proximal-distal dimension.

Additionally, it seems that more elaborate description and theory is required to explicate behavioral phenomena (Golani, 1992; Jacobs, et al., 1988). The casual observation of an organism or a robot performing some intelligent-looking task does not suffice, and neither does the painstaking and highly systematic observance of outward animal behavior such as is done in ethology. There should be a characterization of the many interlocking aspects of behavior, ranging from the neural dynamics, the sensory-musculo-skeletal system, the actually made movements, to the distal functional regularities of behavior. The big theoretical question is, how does all of this fit together?

The notion of self-organization seems to offer a route to work on this question concerning the organizing principles involved in such a multiply-scaled system. In turn, self-organization is itself a concept that needs working on, and behavior might be a good testing-ground for developing this notion further. The combination of self-organization and behavior thus looks like an interesting cluster for theory development in autonomous agent research.

In this perspective, a theory of behavior explains how the spatio-temporal patterns of behavior come about. To gain insight in these processes, it is relatively unimportant whether the modelled behavior is very 'intelligent' in any traditional sense. What matters is whether a model offers insight into the underlying multiply-scaled organization. The most simple forms of behavior are the most promising in this respect. For example, the most simple nervous systems in existence—the neural nets of *Cnidarians* (such as jelly fish)—do not show any cephalization or ganglions, but form a mesh which somehow coordinates and synchronizes large-scale movement across the bell of the jelly fish, making it contract and expand as a whole in a rhythmic way (Golding, 1993). From the perspective advocated here, this would be a highly interesting system to work on.

None of this involves a break with the major ideas of AAR. It is rather a plea to proceed along a path that has already been chosen. A core idea of AAR was named *emergent functionality* by Steels (1991). The message of emergent functionality is that complex functional behavior can 'emerge' from relatively simple interactions between an organism and an environment. No complex internal knowledgeable system is needed. The present argument presses this approach a step further by making the claim that the 'simple interactions between an organism and an environment'are themselves actually quite complex and 'emerge' from the simultaneous activity of many small-scaled processes dispersed over an extended neuro-musculo-skeletal system.

AAR started out with the claim that it would be beneficial for understanding intelligence to study systems capable of solving simple (non-cognitive) behavioral tasks, examples being wall-following and goal-seeking. AAR always opposed the idea that such behavior is too simple to be of much interest. The argument here is that AAR will have to look at behavior which is much simpler still, if it wants to gain insight in how real animals behave.

Literature

Ashby, W.R. (1956). *An introduction to cybernetics*. London: Methuen.

Barnes, R.S.K., Calow, P. & Olive, P.J.W. (1993). *The invertebrates: A new synthesis* (2nd ed.). Oxford: Blackwell.

Beer, R.D. (1995). A dynamical systems perspective on agent-environment interaction. *Artificial Intelligence, 72*, 173-215.

Brooks, R.A. (1989). A robot that walks: Emergent behaviors from a carefully evolved network. *Neural Computation, 1*, 253-262.

Brunswik, E. (1952). *The conceptual framework of psychology*. Chicago: University of Chicago Press.

Brusca, R.C. & Brusca, G.J. (1990). *Invertebrates*. Sunderland, MA: Sinauer.

Campbell, D.T. (1966). Pattern matching as an essential in distal knowing. In K.R. Hammond (Ed.), *The psychology of Egon Brunswik* (pp. 81-106). New York: Holt, Rinehart and Winston.

Floreano, D. & Mondada, F. (1994). Automatic creation of an autonomous agent: genetic evolution of a neural-network driven robot. In D. Cliff, P. Husbands, J-A Meyer & S.W. Wilson (Eds.), *From animals to animats 3* (pp. 421-430). Cambridge, MA: MIT Press.

Golani, I. (1992). A mobility gradient in the organization of vertebrate movement. *Behavioral and Brain Sciences, 15*, 249-308.

Golding, D.W. (1993). Control systems. In R.S.K. Barnes, P. Calow & P.J.W. Olive *The invertebrates: A new synthesis* (2nd ed.) (pp. 402-452). Oxford: Blackwell.

Goodwin, B.C. (1994). *How the leopard changed its spots: The evolution of complexity*. New York: Scribner's Sons.

Haken, H. (1987). Synergetics: an approach to self-organization. In F.E. Yates (Ed.), *Self-organizing systems: The emergence of order* (pp. 417-437). New York: Plenum Press.

Heider, F. (1959). On Perception and Event Structure and the Psychological Environment. *Psychological Issues, 1(4),* 1-123.

Jacobs, W.J. *et al.* (1988). Observations. *Psychobiology, 16,* 3-19.

Kauffman, S.A. (1993). *The origins of order: Self-organization and selection in evolution.* New York: Oxford University Press.

Keijzer, F.A. (1998). *The generation of behavior: On the function of representation in organism-environment dynamics.* Leiden University. unpublished doctoral thesis.

Keijzer, F.A. (in press). Doing without representations which specify what to do. *Philosophical Psychology.*

Kelso, J.A.S. (1995). *Dynamic patterns: The self-organization of brain and behavior.* Cambridge, MA: MIT Press.

McFarland, D. & Bösser, T. (1993). *Intelligent behavior in animals and robots.* Cambridge, MA: MIT Press.

Millán, José del R. (1994). Learning efficient reactive behavioral sequences from basic reflexes in a goal-directed autonomous robot. In D. Cliff, P. Husbands, J-A Meyer & S.W. Wilson (Eds.), *From animals to animats 3* (pp. 266-274). Cambridge, MA: MIT Press.

Miller, G.A., Galanter, E. & Pribram, K.H. (1960). *Plans and the structure of behavior.* New York: Holt, Rinehart and Winston.

Sharkey, N.E. & Heemskerk, J.N.H. (1997). The neural mind and the robot. In A.J. Brown (Ed.), *Neural network perspective on cognition and adaptive robotics* (pp. 169-194). London: Institute of Physics press.

Steels, L. (1991). Towards a theory of emergent functionality. In J-A. Meyer & S.W. Wilson (Eds.), *From animals to animats* (pp. 451-461). Cambridge, MA: MIT Press.

Trueman, E.R. (1975). *The locomotion of soft-bodied animals.* London: Arnold.

Vallacher, R.R. & Wegner, D.M. (1987). What do people think they're doing? Action identification and human behavior. *Psychological Review, 94,* 3-15.

Van der Heijden, A.H.C. (1992). *Selective attention in vision.* London: Routledge.

Beyond physical grounding and naive time: investigations into short-term memory for autonomous agents

Mark Rylatt and Chris Czarnecki

School of Computing Sciences, De Montfort University, Leicester, LE1 9BH, U.K.
Tel: (0116) 2551551 E-mail: rylatt@dmu.ac.uk, cc@dmu.ac.uk

Abstract

In this paper we present some experiments comparing different connectionist dynamic memory models in an interactive context, including a novel version of a second-order architecture, along with an extended discussion of some foundational issues to which they relate. The scenario is a fine-grained simulation of a sensor-based, non-holonomic mobile robot moving in a two dimensional environment in which are set paradigmatic short-term memory experiments involving delayed response tasks for the robotic subject. We set out our position initially in relation to symbolic AI, connectionism and *nouvelle AI.* In particular, we argue that despite its practical successes, the last named approach does not escape from a foundational error that in different guises undermines the programmatic aspirations of all three. We suggest that this representational problem is inherent in the implementational granularity of systems such as the subsumption architecture and argue for the notion of *embedding in time,* as a first step around the impasse.

1. Introduction

Despite the impressive early successes of reactive and behaviour-based approaches to the autonomous control of mobile robots, such as the subsumption architecture (Brooks, 1986), there is a growing recognition that they represent only a partial solution even to practical problems like navigation (see, for example, (Pal and Kar, 1996). We shall argue that, in the guise of *nouvelle AI* (Brooks, 1990), they also offer an unsatisfactory solution to the foundational difficulties relating to the problem of representational contents. faced by Artificial Intelligence (AI). In (Rylatt et al, 1995) we sketched some early ideas on this theme and in (Rylatt, et al, in press) we survey alternative approaches linked by their use of neural networks in a variety of forms. Here, we revise, develop, and focus those ideas into an argument for a particular general approach to the aforementioned foundational issues - in essence a change of perspective that we call *embedding in time*. First, we present our view that the problems faced by *nouvelle AI* stem from an intractable substrate that prevent it from sharing this perspective. In order to make clear how our view differs from conventional connectionism, we then discuss some recent critiques of connectionist claims, and show how we seek to avoid the representational errors we believe it shares with symbolic AI. We go on to present a preliminary evaluation of candidate connectionist architectures and introduce a hybrid of a second-order network (Pollack, 1995) and the simple dynamic memory architecture (Port et al., 1995) devised for the experiments described later. The experiments we describe belong to a larger research project, working towards an adaptive autonomous agent control system based on neural networks for extended task performance. But these particular experiments investigate a paradigmatic problem of delayed response suggested in a different form by Ulbricht (1996). Through this medium, we intend to highlight the different set of entailments arising from an interactive conception of connectionism. This is in contrast with what we see as an essentially traditional connectionist account when applied to a similar problem at a much lower level of abstraction. The control architecture used in this set of experiments is kept much simpler than the one we have described elsewhere for extended task performance - for example, see (Rylatt et al, 1997) - in order to focus on these separable concerns.

2. Physical grounding is not enough

We start from some reflections on the well-known aphorism - almost a *cri de coeur* for advocates of behaviour-based control - that 'the world is its own best model' (Brooks, 1991a). Tracing the idea back to the seminal work (Brooks, 1986) it can be read simply as an alternative approach intended to circumvent the problems faced by conventional planner-based architectures trying to achieve real-time performance. These architectures of course rely on a central world model that requires continuous updating to maintain consistency with the environment. Instead, Brooks' early creatures worked on the principle of a tight control loop with 'little slack' between input and output. By the time of the later paper however the idea, restated as the *physical grounding hypothesis* (PGH, our acronym), had become the foundation of a putative *nouvelle AI.*

The PGH requires that 'intelligent systems must have their representations grounded in the physical world'. But on analysis (we shall argue), this does not amount to much more than the prescription for building intelligent systems that accompanied it, merely requiring them to be connected to the world via sensors and effectors rather than keyboards.

It seems to us that the PGH, baldly stated, contained a hidden agenda, for Brooks moved rapidly to a position holding that once a commitment to the PGH is made, 'the need for symbolic representations fades away entirely'. At the very least we detect a paradox in the PGH: traditional symbol systems were dismissed but the kind of representations that would replace them and, presumably, be subject to, or emerge from, the grounding process, were not discussed. Indeed, any attempt to interpret the low-bandwidth communications between modules in the subsumption architecture as minimal representations was rejected (Brooks, 1991b). We acknowledge that Brooks defended himself against the charge of denying that intelligence uses any form of representation - indeed his position in the last cited paper (written much earlier, despite the publication date) seemed less equivocal. It claimed no more than that representation is the 'wrong unit of abstraction for the bulkier parts of intelligence' (that is, sensorimotor abilities). So that, by his lights, there is no contradiction in holding that this part of intelligence - the 'hardest part according to an arguably simplistic interpretation of how intelligence evolved (Brooks, 1991a) - can indeed be achieved without representations. But, to reiterate, what kind of representation was he talking about?

For a definition we look to (Brooks, 1991b) were we find a reference to 'explicit, manipulable internal representations' and what we may view as a *nouvelle AI* promissory note for the ultimate achievement of human-level machine intelligence without them. To explain how it could be cashed, Brooks fell back on an argument that can be fairly summarised as follows. Essentially, we only recognise thought in others (other intelligent systems) through the introspection of our own intelligence. By the same yardstick eventually we shall witness the emergence of intelligence in our situated agents as the epiphenomenon of their interactions with the same complex world that gave us our existence proof of intelligence.

We interpret this as a radical position: it appears to hold that representational contents are observer dependent and that they have no other solution; in our view, it amounts to observer idealism[1]. Others have ignored this questionable philosophical basis for *nouvelle AI*. For example, Bickhard and Terveen (1995) suggest - in the context of their account of a radical alternative conception of representation (called *interactivism* and hereinafter referred to as such) - that Brooks has merely failed to exploit the representational potential (as they construe it) in his architectures. We go further, however, and suggest that observer idealism is a necessary consequence for *nouvelle AI,* enforced by the entailments of its developmentally arbitrary substrate, as we shall explain below. While we cannot deny the presence of a functional organisation to support some kind of non-symbolic 'representation', we expect that this will always be of an ad hoc nature. An example of this is Mataric (1992) describing the robot Toto's memory, consisting of a dynamic graph of 'landmark recogniser modules' corresponding to gross sonar configurations (having user semantics such as 'wall right') and compass readings.

We regard this characteristically ad hoc, engineering approach to be a consequence of an implementational granularity that is not amenable to the integration of truly interactive subsystems between input and output. Our perspective requires such subsystems to model intrinsically (rather than ad hoc) dynamical processes, that could give rise to non-conceptual contents for the agent, rather than for the observer. So we agree with (Clark, 1993), that implicit knowledge of this kind is an essential forerunner of manipulable representations. Likewise, we expect it to arise in the dynamics of neural networks, provided that they are not construed merely as stateless input-output systems, and that they conform to one of the requirements of interactivism, namely that the system should interact with an environment so that past outputs affect subsequent inputs. We think that PGH style grounding leads only to isolated and specifically engineered demonstrations of representational efficacy (however technically impressive and practically valuable) and that these would in no way imply the existence of contents-for-the-machine in the above sense. We hold that the loop through the world is not enough - there must also be many recursive loops through a tractable medium internal to the agent that effectively *embed the agent in time* as well as in space.

3. Connectionist hallucinations

We argued above that *nouvelle AI* is constrained by its substrate to offer essentially engineering solutions to problems. In contrast, an immediate, though perhaps superficial, advantage of connectionist architectures is that, unlike systems such as the subsumption architecture (Brooks, 1986) they readily support and, indeed, *integrate* the essential ability to learn (in some sense) and to generalise. Even so, we do not use this to argue that, in the state-of-the-art at least, they measure up to the guessed-at computational implications of human intelligence. From the interactivist perspective, the ability of connectionist systems to implicitly and emergently differentiate input conditions (even if the

---

[1] A philosophical position, that shifts all accounts of representational content to mysterious and unknowable observers. It is arguable that such observers are akin to the homunculi that haunt traditional accounts of representation in the guise of interpreters (of encoded symbol structures) and hence the position is open to the same objections of infinite regress.

nature of this emergence can in some cases be questioned - see below) is its most fundamental distinction from symbol systems. The latter can 'only' (though of course sometimes very successfully at the project level) manipulate pre-existent symbols emanating from the designer's domain ontology. We, however, go on to suggest that this intrinsic capacity of neural networks is also in contrast with *nouvelle AI,* an approach that uses ad hoc means even at the perceptual level - using the subsumption architecture as an example again, we adduce the use of a potential fields abstraction for obstacle avoidance (Brooks, 1986). At this level, we begin to detect some foundational light, though perhaps only a glimmer. It seems that interaction with the environment alone is insufficient as the basis to determine potentially viable architectures (in a programmatic sense). True interaction must intimately involve basic mechanisms, a flow of internal system processes unimpeded by observer-determined abstraction barriers[2]; and connectionism, unlike *nouvelle AI*, appears to provide at least some of the essentials for such a model.

However, interactivists mostly reject connectionism, but (we note) base that rejection mainly on examples of traditional (non-situated) connectionist examples. We, on the other hand, take a somewhat less pessimistic view of the field as a whole. We agree with Dorffner (1997), that it is the most promising basis for a new AI, despite some interpretative tendencies that, in the limit, can amount to foundational programmatic difficulties comparable to those of symbolic AI. We also observe that the descriptive language used by connectionism, to give its experiments a naturalistic flavour, often amounts to a kind of 'ecological correctness'. Beneath it are environments as unrealistic as those of most symbolic AI research, and 'sensory inputs' in the form of symbol-like encodings. And we broadly agree with recent critiques that reject claims made for the emergence of concepts in simple first order networks - see, for example (Verschure, 1997). These claims for conceptual content may amount to a 'hallucination'. But we retain some sympathy with the view held by (Clark, 1993) (if we interpret him correctly), that the domain mastery of such networks may amount at least to a preliminary sketch of a model for non-conceptual content. Fleshing out that sketch is a process at least begun by situating and embodying (in some sense) such systems - that is, making them interactive; but we argue below that this is only part of the way to the full picture.

4. Embedding in time

In this section we develop the notion that systems require more than a spatial embedding if they are to develop even the non-conceptual, for-the-machine, contents that we think a prerequisite for an artificial mentality. To explain this, we observe that much of the emphasis in behaviour-based research was on embedding or situating agents in space so that they would not suffer from the closed system brittleness of traditional, disembodied AI systems. This brittleness is, we believe, reducible to the problem of representational contents, and symbolic AI's difficulties in this respect, we contend, arise from the lack of a temporal essentiality that is denied to its practical manifestations by its theoretical foundation. The absence of any conception of time is a feature of Turing machine theory emphasised by Burkhard and Teeven (1995), in their critique of symbolic AI. Physical Turing machines such as computers, are of course provided with clocks to drive their sequential processing, but these are *engineering* accesories extraneous to the model.

It might be argued that subsumption robots of *nouvelle AI* are more intimately involved with time. But this would be to confuse real-time behaviour - obviously something these creatures are good at - with *real* time (in the sense that we define below); and of this they too have no conception. The clocks that augment the finite state machines (FSM) constituting their substrate are examples of engineered time. In practical terms this means making run-time decisions based on deliberations actually prefigured at design-time using some (usually inflexible) means to arbitrate conflicts between concurrently active modules; temporal effects and dependencies are extraneous to the model. We do not deny the advances made by such systems, we only suggest that claims for them as a complete solution to the programmatic aims of AI need re-examination. We reiterate that, like symbolic AI systems, their temporal imperviousness means that they cannot enjoy for-the-machine representational content (leaving open the question of whether any kind of machine can).

We must concede that most connectionist research, too, has been concerned with spatial patterns or with giving essentially temporal structures, such as natural language, a wholly spatial representation. Considering the case of scientific research into human audition, Port et al. (1995) argue that the standard models suffer from a predication on what they call *naive time.* It is the notion that what we call real or biological time, consisting in the information available to an organism intrinsically bound with timing, uses a representation of absolute time, that is time measured by reference to some external clock. One consequence of this is that such models usually incorporate a temporal buffer by means of which time within a certain window is, in effect made spatial. There appears to be no direct biological evidence for this approach, but buffered models are common in connectionist research into processes that unfold over time. Indeed, in the context of human language processing experiments, the shortcomings of the temporal buffer approach, and of explicit time models, were described by (Elman, 1990). Although his work was not situated (in the sense of having real or realistically simulated sensors or motor outputs), Elman's alternative *implicit* model of time, represented by his simple recurrent network (SRN), con-

[2] Brooks refers to subsumption's behavioural layers as abstraction barriers (Brooks, 1986)

tributes strongly to our ideas. From our perspective, we interpret it thus: the *effects* of time on processing in the system solely constitute the system's *model* of time, and thus time is no longer an engineering accessories extraneous to the model. In the SRN this is achieved through simple feedback connections between layers. According to Elman, 'representations', formed in the hidden units of the SRN (for the moment we shall not argue about how these are to be construed), are involved in mapping both current external stimuli and prior internal states to outputs. They thus intimately bind spatial and temporal patterns in a task-related manner. We hold that this connectionist ability, to model simple, dynamic memory, at the level of granularity of its processing elements, differentiates neural networks from the system of both symbolic AI, and *nouvelle AI*. Both, in some manifestations, and *at some level,* can also be viewed as dynamical systems of a kind.

Before we consider the implications of a more thoroughgoing dynamical systems interpretation of connectionism, it is instructive to descend from programmatic to project level where we can draw some pragmatic corroboration of the account we have set out thus far. We observe that, without dynamic memory, connectionist architectures can become situated in the same (from our perspective, limited) sense as *nouvelle A I* systems. For example, it is now commonplace that a limited degree of obstacle avoidance can be achieved usng a feedforward network as a mobile robot controller, trained with a series of input vectors gathered by driving the real or simulated robot through a cluttered environment. However, (Pal and Kar, 1996) demonstrated the advantage of using recurrent neural networks to provide an integral means of avoiding the cyclical robotic behaviour often observed in similar situations encountered by behaviour-based controllers.

We shall now return to the foundational question of representational content, and whether connectionist systems will ever enjoy it. The negative assessment of (Burkhard and Terveen, 1995) is that connectionist representations always have an underlying symbolic interpretation. By their account, this consists in the compulsion to regard as encodings the activation patterns that differentiate input patterns in typical connectionist architectures. We agree that such encodings amount to temporally impervious correspondences and that that they do not assist our notion of embedding systems in time. However, we believe that situated forms of connectionism are moving away from this insistence on regarding representations as emergent atoms. Some versions, to some extent and at least implicitly, seem to share the anti-representational stance of *nouvelle AI,* and perhaps seek to replace its reactive components with neural networks construed as stateless input/output devices. An example of this is (Saunders et al., 1994)). If internal state is eschewed it may lead to the same foundational errors. However, biological evidence that separate areas of the brain deal with environmentally guided behaviour and behaviour related to internal representations (Guigon and Burnod, 1996), may justify combining the two approaches in modular architectures

The distinction between traditional accounts and the one we are moving towards is most clear at the dynamical systems level of description. We observe initially that the dynamic space of the standard, connectionist feedforward model is organised around local points or regions of attraction. The characteristic behaviour of this model encourages interpretations similar to the folk psychology account of stable referential relations between representational states. However, in recurrent models of the kind we advocate, the dynamics may be characterised by trajectory attractors. In (Peschl, 1993) we find support for the view that traditional representational misconceptions should be abandoned and replaced with a conception of representations in terms of such dynamic state trajectories[3] through the activation space of cognitive systems.

5. Remembering what to do next

We have suggested that a modified but not entirely radical redirection of connectionism may lead to new insights into the vexed question of representational contents. In this section, we provide a rationale for the practical investigations described below. We draw broad inspiration from neuroscience and focus on the association between short-term memory processing areas and behaviour guided by internal representations (Guigon and Burnod, 1996). This is the basis for what we regard as (connectionist) paradigmatic short-term memory experiments, with a simulated autonomous mobile robot as subject. This interpretation should be clear from the nature of the task we have set, at least in its latest form. It requires a delayed response from the subject - the robot is given an instruction stimulus that it must memorise in order to make an appropriate response when it subsequently receives a 'go' signal. In terms of the robot's sensory capacity, the stimuli are differentiable patterns that the robot perceives continuously while moving in a simulated environment. Its responses are temporally extended, motor control decisions, so there are multiple dimensions along which performance can be assessed. Our experiments are more ecological than standard psychological tests and clearly pitched at a much lower level of abstraction than traditional connectionist research. However, before we describe them in more detail, we acknowledge an antecedent of this work. We explain how our re-interpretation of it makes good on our earlier suggestion that there is a sense in which traditional connectionism can still have relevance for a more situated AI, and how it demonstrates that there can

[3] If the cognitive system is modelled using a discrete time dynamical system the term is more correctly *itineraries,* but as is well known, continuous time dynamical systems can be approximated to an arbitrary degree of accuracy by controlling the granularity of the model, so the argument holds for connectionist models too.

be a salient and mutual transfer of ideas between different levels of description. We also use the original problem as the medium to explain some of the task-specific considerations guiding the design of our hybrid model (introduced above) and to highlight some of the problems that emerge when levels of description change.

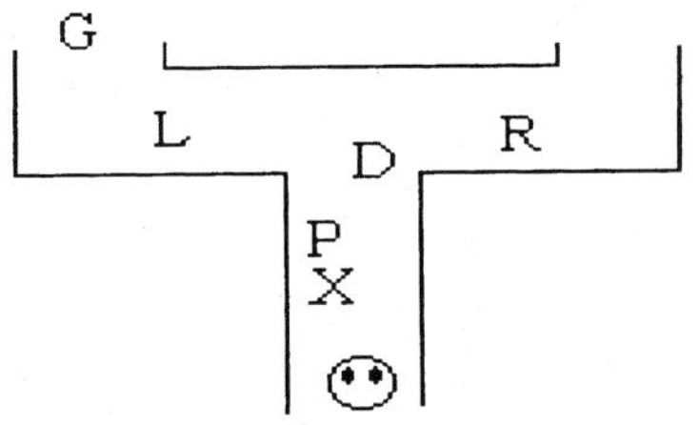

Figure 1: *Ulbricht's problem (see text).*

As originally cast by Ulbricht (1996) (we shall throughout this section refer simply to 'Ulbricht'), this problem involved an agent set in the abstraction of an environment depicted in figure 1. In this form it has a close relationship with work reported by (Port et al, 1995), and a more distant relationship with (Elman, 1990), and Pollack (1995). We make this observation because the agent's percepts are modelled at the level of symbols; and the unfolding of its perceptions in time, by means of strings of those symbols. All these examples are concerned, at some level of description, with either the recognition or prediction of grammars composed of such strings. In Ulbricht's scenario, instead of being interpreted as 'accept' states signifying the recognition of a string as a member of a particular grammar or 'language', the outputs of the network represent motor commands which might be interpreted in the following way, 'go straight ahead while X is perceived, continue straight ahead while P is presented , turn right when D is seen', etc. 'X' can be construed as some environmental feature that the agent is able to differentiate and employ (as a memory trace) in order to make the control decision at 'D' (maybe another environmental feature such as the perception of a fork in the route) and 'P' simply represents a unit interval following the last perception of 'X' during a sequence of which the agent is not receiving input that is relevant to the decision at 'D' (that is the symbol 'P' has no information content relevant to 'D'). Sequential presentations of these symbols can be interpreted as the constant sampling of a continuous signal, with repetitions of symbols representing varying rates of presentation. Two non-trivial problems arising from this are of particular interest for the neural network model and these can be described as the ability to handle long-term dependencies (for example many repetitions of 'P' before 'D') and to generalise across rates of presentation (that is, to ignore variations in rates so that, for example, the model can predict 'D' following 'PPP' having been trained only with 'P').

Ulbricht proposed a novel recurrent network - the *input state network* - for the solution of these two problems in terms of the scenario we have just described. Interestingly she concluded also that a jointly tested SRN was not able to handle either of them. However, the novel network depends rather heavily on the symbolic encoding of the input states. In fact it relies on a transformation of such standard localist connectionist encodings (for example, 10000000 represents 'A') into a randomly distributed representation - from our point of view, a somewhat curious inversion of a naturalistic view of perception. These transformed representations are presented to novel input units that have a so-called 'flexible' decay rate. But this entails a one-to-one interconnection so that each input element is matched with a unit having a different, but fixed, rate of decay. From our point of view, this actually proves to be inflexible, because a naturally distributed representation (resulting from a vector of raw sensor readings) does not fit into this scheme. Each unit would effectively be subject to an arbitrary bias. Although, in this particular problem, inputs carry the information rather than outputs, the intermediate state representation of the hidden units must clearly also carry some useful information.

6. Temporal processing models

We have thus far developed our argument around a single example (the SRN) in a subclass of connectionist architectures with recurrent connections and some reflections on the novel 'input state network'. At the project level, several different architectures belonging to this subclass have been studied. In (Mozer, 1993) can be found a taxonomy in terms of form, content and adaptability - all relevant aspects for application decisions. The processing algorithm is another factor, relevant both to project level and to our programmatic concerns, is. From our developmental perspective on autonomous agents, we manifestly require models with manageable computational overheads. Special purpose training algorithms for recurrent models, such as backpropagation through time (BPTT) and real-time recurrent learning (RTRL), have serious drawbacks. For example, according to (Doya, 1994), RTRL (despite its name) requires $O(n^3)$ memories $O(n^4)$ computations and BPTT requires $O(nT)$ memories. Additionally, according to Pollack, (1995) the BPTT assumption is unstable, an observation that encourages us to focus on the simpler dynamic models. The latter are represented by such architectures and associated learning paradigms as the SRN, the Jordon network (Jordon, 1986), the simple dynamic memory of (Port et al, 1995), second order architectures developed by Pollack (1995) and (Ziemke, 1996a), and hybrids (Ziemke, 1996b). We also avoid approaches that require a temporal buffer, such as the time delay neural network (TDNN), partly for the naturalistic reasons implicit in our discussion of naive time above, and partly due to the issue of memory adaptability raised by Mozer (1993). In this respect TDNNs are classified as static whereas the bufferless models are at least to some degree adaptable, an advantage for autonomous agent applications that face relatively unconstrained problem domains. The reliance on backpropagation (BP), evident in all these versions may be thought a disadvantage for programmatic concerns. However, we justify continuing to rely on this convenient workhorse algorithm by reference to (Clark, 1993). Here we find support, both for the underlying paradigm of gradient-descent error minimisation as a fundamental cognitive principle, and for BP interpreted at the psychological level as a mechanism that supports it. This is of particular advantage for our own project level concerns because, although we use supervised learning in the paradigmatic STM 'laboratory' experiments described below, for demonstrations of extended task performance we rely on complementary reinforcement backpropagation learning (CRBP). We hope that this near-unified approach at the learning paradigm level will facilitate the interplay of insights and findings between controlled and more ecologically valid experiments. Illustrated in figure 2 is the hybrid architecture we propose. It is based on the second-order or quadratic network described by Pollack (1995) and the simple dynamic memory of (Port et al, 1995). It also has similarities with the self-adapting recurrent network (SARN) of (Ziemke 1996) (a hybrid of a second-order network and the SRN). We leave an explanation of its task-oriented features until we describe the task itself - here we will outline its general aspects. Referring again to figure 2, it will be seen that this model has two levels of feedback (dotted arrows). Firstly, feedback occurs from the input layer to itself - we have indicated this by showing an extra 'layer' at a level before the input layer, similar to the context layer of an SRN. We believe that this re-conceptualisation may blur the distinction between 'layer feedback' and 'unit feedback' drawn by Ulbricht (1996). But in fact it shows the fully interconnected (solid arrows) nature of the self-recurrent input layer quite clearly, conforming to the activation equation (Port et al, 1995):

$$y_i(t+1) = squash[\alpha y_i(t) + \sum w_{ij} y_j + input + bias] \quad (1)$$

where α is a decay rate, y is the feedback and *input* represents direct sensory input. The second level of feedback is from the hidden layer and this incorporates the second-order aspects. We show the architecture as a single network but it functions as two subnets in a master-slave relationship. Feedback, from the hidden layer at time $t-1$, is construed as the input layer of the master network, which is then fed forward through a linear function to its output layer. The units of this output layer, in our model, represent the weights of the slave net hidden layer. The terms 'quadratic' and 'simple' reflect the multiplicative order of the connections. It will be seen that the master subnet will have a number of weights equal to the multiple of the simple weights between the layers it connects and the number of units in its input layer. The model is trained using BP in the manner found to be successful by Elman (1990) and Pollack (1995) which effectively backpropagates error one step in time. In this case, as in Ziemke (1996), the hidden layer weight vector, adjusted by the slave net at time t - 1, serves as the target vector for the master net at time t. Thus, the master weights multiplex the activation function of the slave hidden layer, giving (from our perspective) richer representational potential.

In our initial experiments (Rylatt, et al, 1998), we concentrated on the SRN in an attempt to determine whether its advantages in other contexts could be shown to persist in our version of Ulbricht's scenario. Despite her findings that it was unable to solve the two constituent problems, our conclusions were that, at least with some processing enhancements, it was in fact able to demonstrate a limited ability to handle our version of these problems. This perhaps confirms her conclusion that an 'inappropriate' training procedure may have been a factor in its failure, rather than its inadequacy as a general model for temporal processing.

The success of the input state network consisted in being specialised for the problem it addressed, focussing on the temporal unfolding of task-related *input-level* information. We decided to investigate the possibility that the SRN could be enhanced for this task by inducing the hidden layer to provide more input-relevant information to be fed back as temporal context. The use of feedforward networks to compress input into hidden layer representations by training the network to reproduce its own inputs at the output layer (autoassociation) is described in (Kadaba et al, 1990). We set up the Elman network to perform the autoassociation task concurrently with the task of associating percepts with motor outputs by adding extra output units equal to the number of inputs. In training, the inputs would be used as the target vector for the additional outputs so that a compressed input representation would form part of the hidden layer state fed back as context. Additionally, in order to control the amount of state information processed recurrently, we added a constant gain to the context units. In Elman's original SRN the feedback is modelled as connections to the context layer with fixed weights of 1.0. However, short term memory can be modelled generally in RNNs as the convolution of the input sequence with a so-called kernel function (Mozer, 1993):

$$\int_0^t K(t-\tau)X(\tau) \quad (2)$$

where K is the kernel function and X the input function. Applying this idea to the SRN, the equation for the context layer activation can be written :

$$c_j = (1-\gamma)h_j(t) + \gamma c_j(t-1) \quad (3)$$

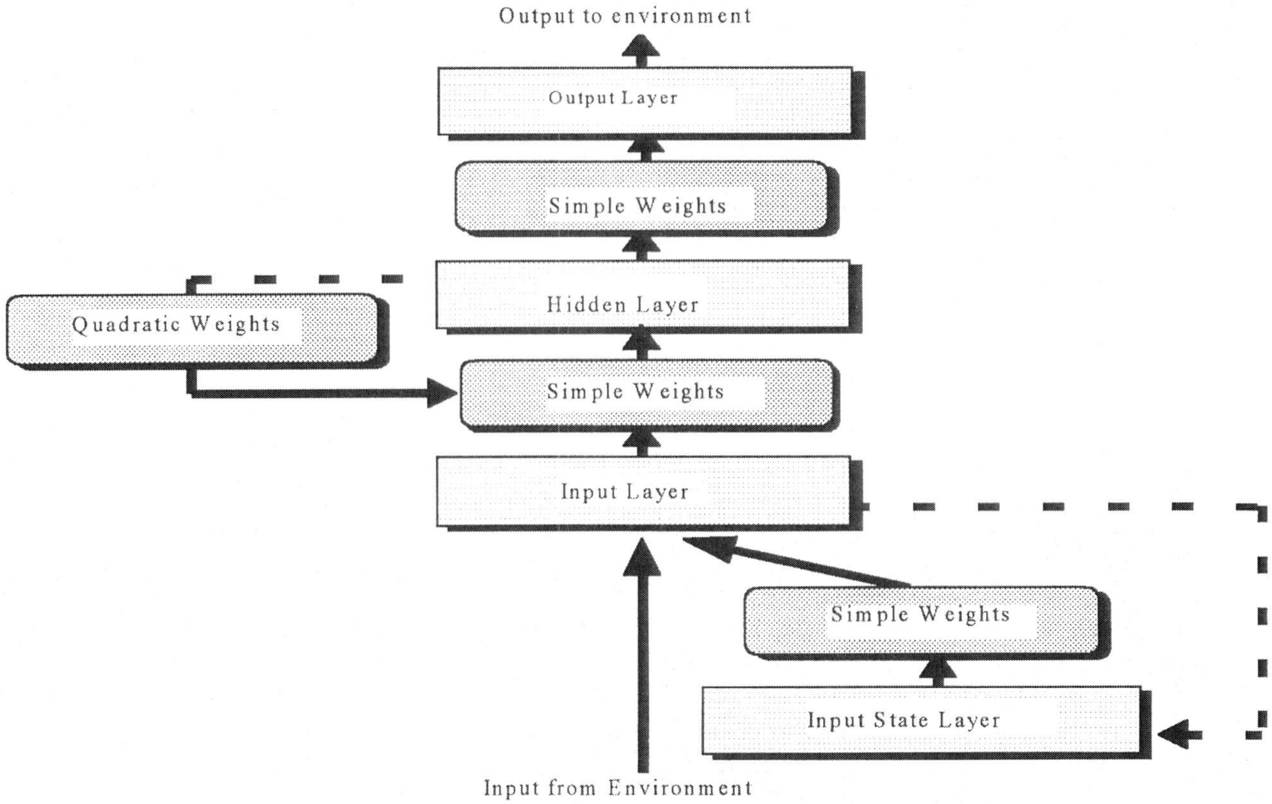

Figure 2:*Hybrid architecture based on second-order network and the simple dynamic memory model (see text)*

where c_j is the jth context unit, h is the jth hidden unit and γ is the gain constant. If γ is set to zero, the effect is the same as that of the standard Elman context unit, that is, of a high definition, low depth memory unit (using the terminology in (Mozer,1993)). An increase in the gain will tune the context units by reducing the definition of the memory, a decrease will increase the depth of the memory.

7. Experiments with time

7.1 Fine grained simulation

For these experiments, we used an integrated mobile robot and neural network simulator (IMRANNS). This was designed and implemented primarily for the development and testing, in fine-grained simulation, of modular neural network mobile robot control architectures - see, for example (Rylatt et al. 1997). A screen shot of the interface, in figure 3, shows one of the trials described later (the robot is shown receiving the 'go' signal in the form of a light beam in the upper left environment window). This illustrates the kind of feedback available to the user in terms of animated activation diagrams and Hinton diagrams.

Although our aim was to cast the problem at a much lower level of abstraction, our simulated environment is still relatively impoverished because the robot has limited perceptual abilities and the environment is two-dimensional. Nevertheless, we believe that similar simulations have transmigrated successfully to real robot demonstrations, according to Meeden (1996) and Pal and Kar (1996). For our experiments, we have used an idealisation of active sonars as range finders, similar to the description in (Stein, 1995). The latter provided adequately realistic, approximate distance information for landmark recognition from a cognitive sketch map, although we employ an original scan line algorithm rather than Bresenham's algorithm. A similar sonar model can be found in (Pal and Kar, 1996), mentioned above. More accurate range profiles, of the general kind we model, would doubtless be obtainable, if necessary, using laser range finders of the kind used by Tani and Fukumura (1994). So there is no reason to suppose our simulation is unrealistic in its sensory demands. Additionally we use the robot's simulated photoreceptors, but as this is merely to provide a 'go' stimulus (see later), the modelling accuracy is not critical.

7.2 Design of experiments

As we suggested at the beginning of this section, recasting the problem at a lower level of abstraction carries entailments that have design implications for the proposed

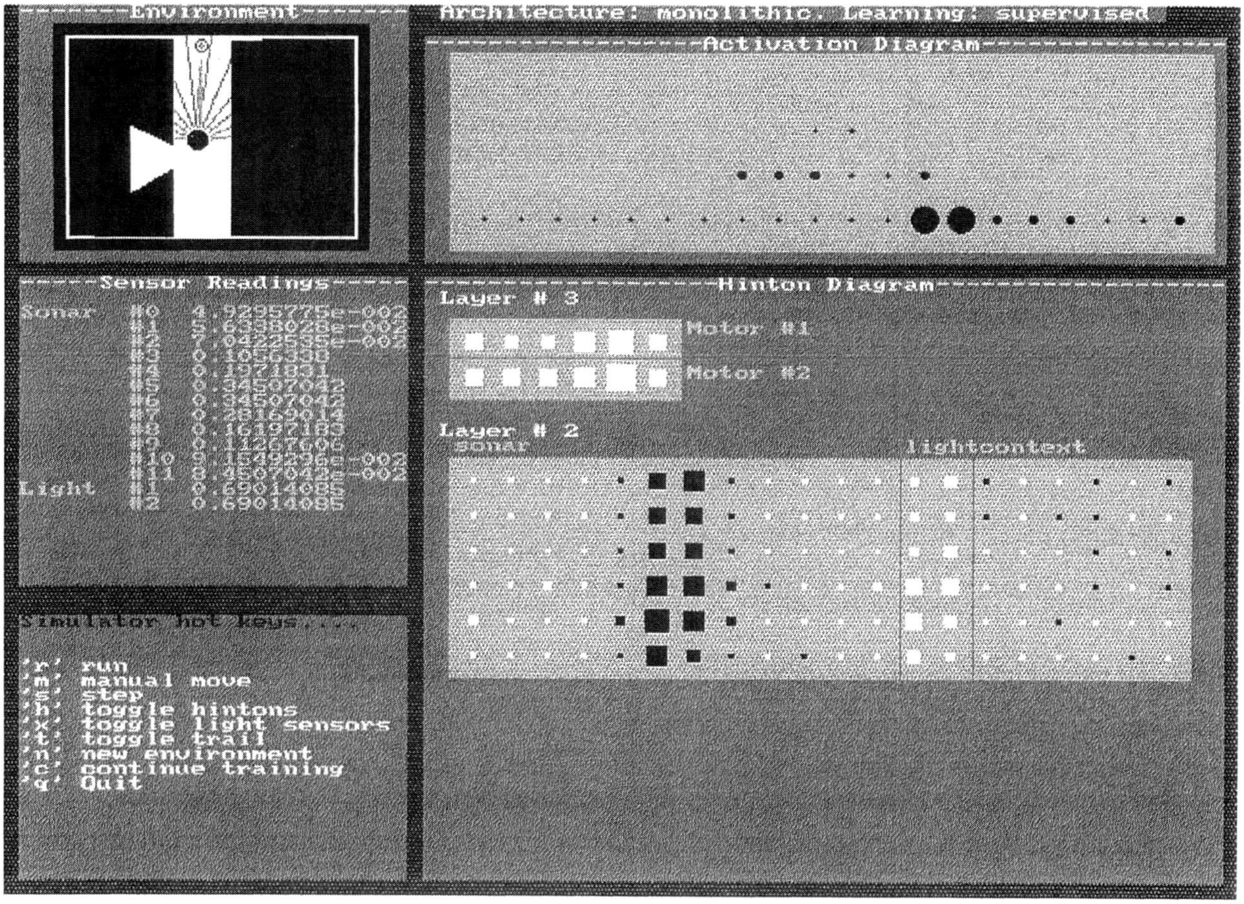

Figure 3: *Screen shot of IMRANNS.*

experiments. For example, referring to figure 1, because the robot has sensors oriented towards the front there would be a considerable time lapse between the robot's last perception of 'D' and its first turning movement to negotiate the intersection. This is so although the simulated robot is holonomic; non-holonomic types would pose even greater problems. The long dependency (in neural network terms) of these temporal events presents a difficult - for example, even the non-situated input state network (see above) seems to be reliable only up to around six time steps into the past.

Figure 4: *Mirror-image environments (see text).*

Figure 4 shows the mirror image environments used for training and testing in the first set of experiments. In this the robot perceives the instructional stimulus (a 'notch' in the 'solid' wall to delimit a zone that should give a distinctive sonar or lasar range finder profile in contrast to the flat expanse on either side). In the first experiments, the 'sign' was located either to the right or to the left of the central aisle. The delayed response task is to execute a 'U' turn, the direction being determined by which side the stimulus was perceived. There is no 'go' signal, the robot simply learns and executes the complete tour so in these experiments only long-term dependencies can be investigated. Apart from the reflection about the vertical axis of symmetry, these two environments are identical in all respects. This is to ensure that the neural network does not distinguish accidentally unbalanced features, and use these as the basis for its decisions, rather than its memory of the distinct range-finder profiles. We took the same precautions for the second set of experiments, designed to investigate both long-term dependencies and generalisation across time combined with a more demanding recognition object as the stimulus. In these, the instruction stimulus is perceived as one of two distinct range-finder profiles (each a distinct shape on the

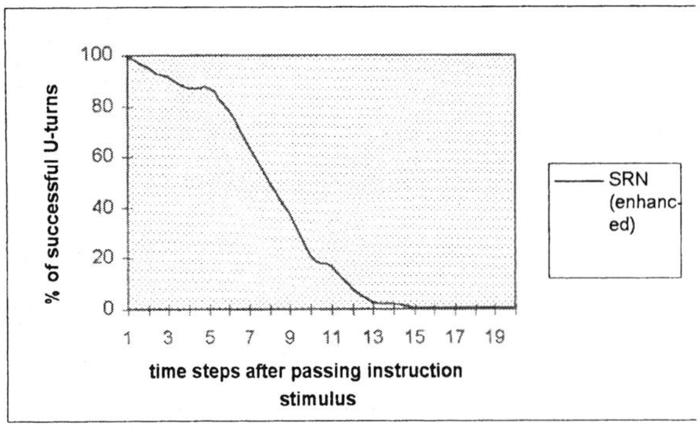

Figure 5: *First experiment (see text).*

same side of the aisle) in another pair of environments. A 'go' signal was given to the robot by means of a light beam as seen in the screen shot. We added more range finders to the robot for this task (figure 3).

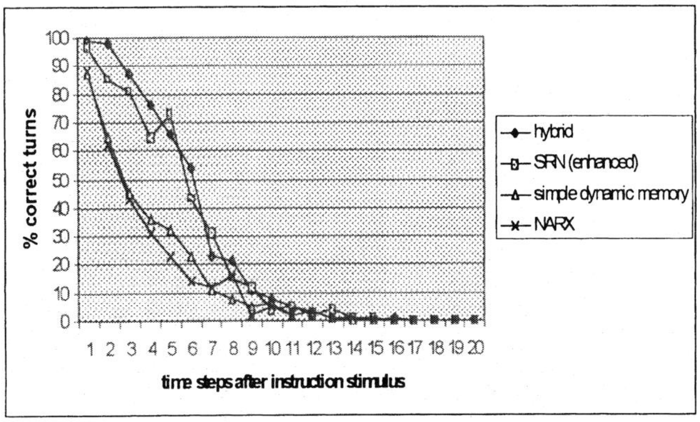

Figure 6: *Second experiment (see text).*

The training procedure used was as follows: continuous samples of the robot's sensory inputs and corresponding motor responses were collected automatically as the robot was moved manually along the centre of the aisle in each of the two environments. If the experiment did not involve a 'go' signal, after a fixed number of steps following the instruction stimulus has disappeared from 'view', the robot was turned about its axis. In this case, the total number of steps before the turning point was varied across multiple training sets. When there was a 'go' signal this was incorporated in the training set - the light was 'flashed' for one time step at a given number of steps after extinction of the stimulus. The network was then trained on each training set, and tested on the robot in each environment respectively to confirm that the appropriate behaviour had been learned in response to each distinct stimulus. As expected, learning this behaviour proved to be a relatively trivial problem for all the nets tested.

The procedure for training on the overall task was initiated by concatenating the two training sets obtained from the paired environments for a given trial. The network was then trained using this expanded training set.

After training the robot was tested in each of paired environments and its ability to perform the required delayed response was assessed.

7.3 Results

Results, averaged across trials for the first experiment, based on just the enhanced SRN (described above) as subject, are summarised in the chart in figure 5. As there is a 50 per cent chance of turning in either direction, the results have been re-scaled to set chance-level performance at zero. Clearly this SRN has some ability to handle long-term dependencies in a continuous domain - cf. the findings of Ulbricht, (1996) showing the complete inability of the SRN network to accommodate in a near-traditional connectionist 'environment'.

The second set of experiments involved several different recurrent models (discussed above). These comparative results are summarised in figure 6, again re-scaled as above. We found that these results largely confirmed our expectations that input level information is most useful to the agent, but the reasonable performance of the NARX network (with three output feedback delay lines) is somewhat surprising and merits further investigation. We hope to report more detailed, statistical analysis of the results in another paper. Here we use the results more as general support for our programmatic contentions.

8. Concluding discussion

In this paper we have developed the notion that situated autonomous agents need to be embedded in time as a prerequisite for enjoying the genuine representational contents we believe necessary for the programmatic aims of AI. We have argued that the temporally impervious substrates of symbolic AI and *nouvelle* AI cannot achieve such temporal embedding. Drawing broad inspiration from neuroscience, we have focussed on the association between short-term memory processing areas and behaviour guided by internal representations. We have suggested that the main qualification connectionism may have as the new paradigm for AI lies in the possibility that a subclass of its models may possess some of the desiderata we identified. We have investigated models with some novel characteristics in a simulation of situated domains, concluding that several have potential for the kind of seamless temporal processing essential to fully embed an autonomous agent in the space-time continuum. This is very much preliminary work and we acknowledge that our models so far lack other essential attributes, for example internal goals (Bickhard and Terveen, 1995). Ways of providing these, perhaps through reinforcement learning, are under investigation.

References

Bickhard, M.H. and Terveen, L. (1995). *Foundational issues in artificial intelligence and cognitive science: impasse and solution*, North Holland, 1995.

Brooks, R.A. (1986). A robust layered control system for a mobile robot, *IEEE Journal of Robotics and Automation*, RA-2, 14-23.

Brooks, R.A. (1990). Elephants don't play chess. *Robotics and Autonomous Systems*, 6, 3-15.

Brooks, R.A. (1991a). Intelligence without representation, *Artificial Intelligence*, 47, 139-159.

Brooks, R.A. (1991b). Intelligence without reason, *Proc. of the Twelfth Int. Conf. on Artificial Intelligence* (IJCAI), vol. 1., 569-595.

Clarke, A. (1993). *Associative engines: connectionism, concepts, and representational change*. Bradford Books, MIT Press, Cambridge, M.A.

Dorffner, G. (1997). Neural networks and a new AI - questions and answers. In: Dorffner, G. (Ed.) *Neural Networks and a New Artificial Intelligence,* International Thompson Computer Press.

Dreyfus, H. and Dreyfus, S. (1986). *Mind over machine: the power of human intuition in the era of the computer.* Free Press.

Elman, J. (1990). Finding structure in time. *Cognitive Science*, 14, 179-211.

Guigon, E. and Burnod, Y. Short-term Memory. In: Arbib (ed.) *The Handbook Brain Theory and Neural Networks*, MIT Press, Cambridge, MA.

Jordon, M.I. (1986). Attractor dynamics and parallelism in a connectionist sequential machine. *Proc. of the 8th. Conf. on Cognitive Science*, 531-546.

Kadaba, N., Nygard, K.E., Juell, P.L. and Kanga, L. (1990). Modular backpropagation network for large domain pattern classification. *Proc. of IJCNN*, Washington D.C., Vol. II., 551 - 554,.

Lin, T., Horne, B.G., Tino, P. and Giles, L. (1996). Learning long-term dependencies is not as difficult with NARX recurrent networks, *IEEE Transactions on Neural Networks.*

Ludik, J., Prins, W., Meert, K. and Catfolis, T. (1997). A comparative study of fully and partially recurrent Networks. *Proc. of the 1997 IEEE Int. Conf. on Neural Networks, Houston* (ICNN97), Texas, Vol. I , 292-297.

Mataric, M.J. (1992). Integration of behaviours into goal-driven, behavior-based robots. *IEEE Journal of Robotics and Automation* 8, 3, 304-312.

Meeden, L., (1996). An incremental approach to developing intelligent neural network controllers for robots. *IEEE Transactions on Systems, Man and Cybernetics*, 26, (Special Issue on Learning Autonomous Robots).

Mozer, M.C., (1993). Neural net architectures for temporal sequence processing, In: A.Weigend and N.Gerschenfeld (eds.), *Predicting the Future and Understanding the Past*, Addison Wesley.

Pal, P.K. and Kar, A., (1996). Sonar-based mobile robot navigation through supervised learning in a neural net. *Autonomous Robots*, 3, 355-374.

Peschl, M.F. (1993). Autonomy vs. environmental dependency in neural knowledge representation." In R.A. Brooks and Pattie Maes (eds), *Artificial Life IV: Proceedings of the Fourth International Workshop on the Synthesis and Simulation of Living Systems*, 417-423.

Pollack, J.B. (1995). The induction of dynamical recognisers. In: Port, R.F. and van Gelder, T., editors, *Mind as Motion: Explorations in the Dynamics of Cognition.* MIT Press, Cambridge, MA.

Port, R.F., Cummins, F. and McAuley, J.D. (1993). Naive time, temporal patterns and human audition. In: Port, R.F. and van Gelder, T., editors, *Mind as Motion: Explorations in the Dynamics of Cognition.* MIT Press, Cambridge, MA.

Rylatt, R. M., Czarnecki, C. A. and Routen, T. W. (1995). A perspective on the future of behaviour-based robots. *Mobile robotics workshop notes*, Tenth Biennial Conference on AI and Cognitive Science, Sheffield, U.K

Rylatt, R.M., Czarnecki, C.A. and Routen, T.W. (1997). A partially recurrent gating network approach to learning action selection by reinforcement. *Proc. of the 1997 IEEE Int Conf. on Neural Networks, Houston* (ICNN97), Texas, Vol. III , 1689-1698.

Rylatt, R. M., Czarnecki, C. A. (1998). On line adaptive controller for mobile robots. Int. ICSC Symposium on the Engineering of Intelligent Systems (EIS 98), Feb. 11- 13 University of La Laguna, Tenerife, Spain.

Rylatt, R.M., Czarnecki, C.A. and Routen, T.W. (in press). Connectionist learning in behaviour-based mobile robots: a survey. *Artificial Intelligence Review*, Kluwer Academic Publishers.

Stein, L.A. (1995). Imagination and situated cognition. In: K.M. Ford, C.Glymour and P.J. Hayes (eds.) *Android Epistemology*, AAAI Press/Press, Menlo Park, C.A.

Saunders, G.M., Kolen, J.F. and Pollack, J.B. (1994). The importance of leaky levels for behaviour-based AI. . *From Animals to Animats 3: Proceedings of the Third Int. Conf. on the Simulation of Adaptive Behaviour*, MIT Press / Bradford Books, Cambridge, M.A..

Tani, J. and Fukumura, N. (1994). Learning goal directed sensory-based navigation of a mobile robot. *Neural Networks.* Vol. 7, No, 3. 553-564.

Ulbricht, C. (1996). Handling time-warped sequences with neural networks. *From Animals to Animats 4: Proc. of the Fourth Int. Conf. on Simulation of Adaptive Behaviour*, 180-192. MIT Press / Bradford Books, Cambridge, MA.

Verschure, P.M. (1997). Connectionist explanation: taking positions in the mind-brain dilemma. In: Dorffner, G. (Ed.) *Neural Networks and a New Artificial Intelligence*, International Thompson Computer Press, London.

Ziemke, T. (1996a). Towards adaptive behaviour system integration using connectionist infinite state automata. *From Animals to Animats 4: Proceedings of the Fourth Int. Conf. on the Simulation of Adaptive Behaviour*, 145-154. MIT Press / Bradford Books, Cambridge, M.A.

Ziemke,T. (1996b) Towards adaptive perception in autonomous robots using second-order recurrent networks, *Proc. of the First Euromicro Workshop on Advanced Mobile Robots*, Kaiseslautern, 89-98.

Exploiting embodiment for category learning

Christian Scheier and Rolf Pfeifer

AILab, University of Zurich

Winterthurerstrasse 190, CH-8057 Zurich, Switzerland

scheier@ifi.unizh.ch

Abstract

An autonomous mobile agent in the real world is exposed to a continuously changing high-dimensional stream of sensory stimulation. If it is to learn category distinctions, it must have a way to reduce the dimensionality of this high-dimensional data stream. Learning is very hard because the invariances are typically not found in the sensory data directly — the classical problem of object constancy: it is a so-called type-2 problem. Rather than trying to improve the learning algorithms — which is the standard approach — the embodied cognitive science view suggests a different approach which focuses on the nature of the data: the agent is not passively exposed to a given data distribution, but, by exploiting its body and through the interaction with the environment, it can actually *generate* the data. More specifically, it can generate correlated data that has the property that it can be easily learned due to redundancies resulting from the appropriate interactions with the environment. Through this interaction, the former type-2 problem is transformed into a type-1 problem, thus reducing the complexity of the learning task by orders of magnitude. These ideas are all illustrated with case studies involving physical and simulated autonomous agents.

1. Introduction

Recently, a new paradigm of cognitive science has emerged. The hallmark of this new approach is the focus on the situated and embodied nature of intelligence. Research in so-called behavior-based AI (Arkin, 1998; Brooks, 1991), embodied neurobiology (Chiehl and Beer, 1997; Almassy et al., in press), and embodied cognitive science (Ballard et al., 1997; Clark, 1997; Pfeifer and Scheier, in press) has challenged the traditional view according to which intelligence is an abstract, symbolic process independent of physical implementation. The embodied approach to intelligence is in accordance with recent empirical findings in a number of disciplines such as developmental psychology (e.g., Thelen and Smith, 1994), cognitive psychology (e.g., Glenberg, 1997), or the neurosciences (e.g., Salinas and Romo, 1998). It has been shown, for example, that people tend to minimize memory storage in natural tasks by exploiting their eye movements (e.g., Bensinger et al., 1995), that individuals are oblivious to significant changes in their visual field and thus are unlikely to store detailed sensory information in a sensory buffer (a phenomenon called "change blindness", Simons and Levin, 1997) or that blindfolded subjects which are asked to mentally rotate previously seen objects by 90 degrees have a much higher accuracy when they are allowed to actually rotate 90 degrees, compared to pointing accuracy when only mental rotation is involved (Montello and Presson, 1993).

One core idea underlying this work is that perception and action are tightly coupled, and, consequently, that perception should be studied in a behavioral context (see also Milner and Goodale, 1995). This idea goes back to Gibson's (1966) ecological approach to perception and has shed new light on fundamental aspects of perceptual processing. For example, in a recent modeling study, Almassy et al. (in press) show that complex neuronal responses only evolve when an agent is allowed to actively explore its environment.

While this work is an important step towards a new understanding of cognition, the mere coupling of perception and action, and the patterns resulting from behavior in an environment, are not sufficient, we argue, to solve some of the fundamental problems of cognition. One such problem is category learning in the real world. More concretely, there are two problems that so far have not been resolved satisfactorily. First, an agent in the real world, in particular a mobile agent, is exposed to a continuously changing stream of sensory stimulation which strongly depends on the agent's own movements. What sensory stimulation should the agent react to and what should it learn? This is the focus of attention problem. Second, the sensory stimulation from one and the same object varies greatly depending on distance, viewing angle, lighting conditions, etc. This is the problem of object constancy (also called the scaling problem). In the traditional approach, the goal is to find a mapping of the sensory stimulation which—in the case of real-world data—is subject to large variations, onto some internal category representation. The surprising point about these problems is that they are hard for robots, but — subjectively — very easy for humans, almost effortless. An illustration of this phenomenon is a study by Moses et al. (1994) where they have used face images from different individuals, views (5 per individual), and illumination conditions (4 per individual). They compared the different images by using various similarity measures. The main result from these comparisons was that the differences induced by changes in the viewing condition were large compared with the differences between different individuals. Humans, however, that had to recognize the same images had a very high (97%) recognition rate when trained with a single image of each one of the individuals and tested on all other images. Thus, what is easy for humans, turns out to be difficult for machines, in particular mobile robots.

Figure 1 shows the activations of the two front IR sensors of the mobile robot—the Khepera™ robot—that will be used in the experiments presented below, for two objects at 11 adjacent angles and distances. The sensor readings vary significantly as the robot changes direction and distance relative to one and the same object. Another problem involved in real world category learning is that data for two categories (small and large cylinders in this case) can have a significant overlap. This can also be seen in figure 1. The IR sensor readings for the two categories used in the experiments presented below are highly correlated both in time and (sensor) space.

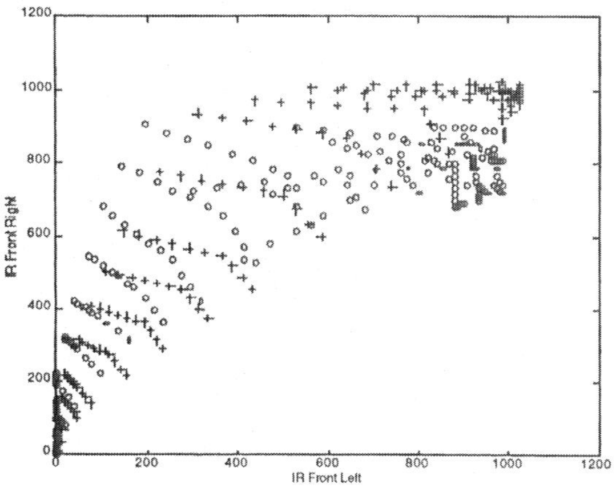

Figure 1 Category learning problem. Data of left front IR (x-axis) and right front IR (y-axis) sensors of a Khepera robot for large (crosses) and small (circles) cylindrical objects. The data were taken from different 11 angles (robot facing the object, and 5 degree rotation to each side) and 31 distances. The data of each object are scattered in sensor space (object constancy problem), and the data belonging to the two different objects overlap.

Clearly, these problems make the task of category learning in the real-world a highly non-trivial one. The main problem is that regions of input space that correspond to one and the same object are not located nearby, but are rather scattered throughout the whole sensor space. Moreover, regions corresponding to different objects may be located next to each other or may even be overlapping. The goal of this paper is to show that despite these problems, real-world categories can be learned by simple agents if they are allowed to interact with the objects in particular ways. In other words, the goal is to show how embodiment can be used to solve the category learning problem.

2. Geometric seperability and two learning problems

Clark and Thornton (1997) have recently introduced the concept of type-2 problems to denote high-dimensional spaces in which regularities are "hidden" or appear only "marginal". They distinguish type-2 (intractable) problems from type-1 problems where the regularities are apparent in the input data. Roughly speaking, a problem is of type 1 if the regularities can be found in the data set itself. Regularity means that the input data can be mapped onto categories (the output). In other words, in type-1 data the output can be predicted with high probability from the input by statistical analysis of the data. If the data are of type-2 they have to be subjected to a transformation before the outputs can be predicted from the inputs with high probability. This idea is illustrated in table 1.

Table 1 Illustration of type-1 and type-2 problems. Left: Type 2, right: type 1. Adapted from Clark and Thornton (1997).

x1	x2		y1	x3		y1
1	2	==>	1	1	==>	1
2	2	==>	0	0	==>	0
3	2	==>	1	1	==>	1
3	1	==>	0	2	==>	0
2	1	==>	1	1	==>	1
1	1	==>	0	0	==>	0

For example, the conditional probability that the output is 1, given that x1 is 1 is 0.5, or the conditional probability that the output is 1 given that the input x2 is 2 is 0.67 (see table 1, left). Most of the conditional probabilities are close to chance. However if the data is transformed, i.e. if we calculate x3 as the difference between the values of x1 and x2 the conditional probabilities are 1 (table 1, right). For example, the conditional probability $p(y1=0|x3=0)=1$. If we interpret x1, and x2 as sensory stimuli and y1 as a category representation, we can say that in the first case, the mapping is not visible in the data, whereas in the transformed case it is easily discernible. Note that the fact that a data set is not of type 1 does not automatically imply that it is of type 2 — there may be no regularities in the data set whatsoever, not even "hidden".

We can now reformulate the core problem of categorization as follows: the main problem in category learning of real-world objects is to transform type-2 problems into type-1 problems. There are two main strategies how this can be achieved. First, the internal processing of the input can be improved. Clark and Thornton (1997), for example, suggest that already learned representations can be used to reduce the complexity of learning subsequent patterns. In other words, the early knowledge is a kind of "lens" through which input data are recoded (Clark and Thornton, 1997, p. 63). The idea is that instead of learning the whole input space at once, the problem can be broken down and made more tractable by "starting small", i.e., by learning some basic properties of the input, and then use the already learned knowledge to "filter" the subsequent input. Another possibility to increase the power of internal processing is to incorporate simple constraints such as symmetry in the network structure (Abu-Mostafa, 1992). This effectively means that the number of free parameters is reduced which leads to better generalization. The constraints reflect a priori assumptions about the data distributions in the environment (e.g. that they are symmetric). For more details, see Meier (1997).

The second approach to transform type-2 into type-1 problems is directly derived from the basic tenets of embodied cognitive science. It consists in exploiting processes of system-environment interaction. In other words, the idea is to exploit the fact that a mobile agent can actively structure its input by manipulating the world. We call this way of interacting with the world sensory-motor coordination (e.g. Pfeifer and Scheier, 1997; Pfeifer and Scheier, in press). We suggest that this manipulation can serve to transform a problem that would be of type 2 (without sensory-motor coordination) into one of type-1. The notion of type-1 and type-2 data sets can be quantified by means of the geometric seperability index (GSI) introduced by Thornton (1997). Geometric seperability is a generalization of linear seperability. The GSI measures how close regions in sensor space belonging to the same object are to each other. It is defined, for a given task, to be the proportion of nearest neighbor inputs with identical output (e.g., category):

$$GSI(f) = \frac{\sum_{i=1}^{N} f(x_i) + f(x_i') + 1 \bmod 2}{N}$$

where f is a binary target function, x is the data set consisting of N vectors, and x_i' is the nearest neighbor of x_i. The GSI has some nice properties which make it an interesting tool for further investigating the differences between type-1 and type-2 learning problems. Thornton (1997) has shown that the GSI can predict with very high accuracy the performance of even the most sophisticated (boundary-based) learning algorithms by simply computing the geometric seperability of the data set. Note also that the GSI is zero for all parity problems (such as the 'XOR' problem). The important point here is that type-2 data sets have low GSI values, while type-1 data sets have large GSI values (Thornton, 1997).

We have computed the GSI for the data set consisting of the IR sensor readings taken at 31 distances (from just visible to right in front of the object) at 200 different angles. Figure 2 shows the GSI for all 200 angles. The mean GSI for these data is 0.47 (+/- 0.11) indicating that these data are of type 2.

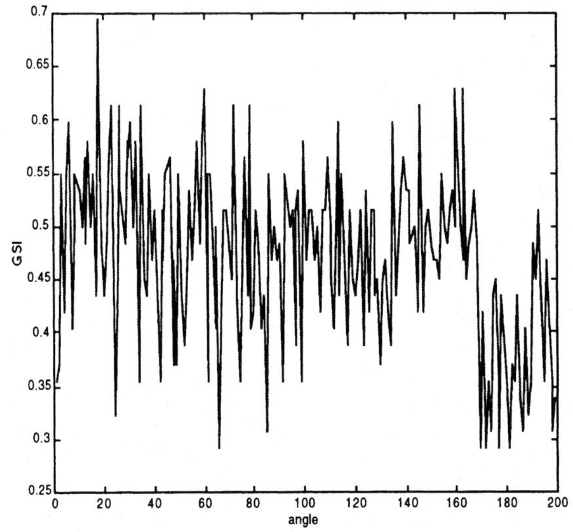

Figure 2 Geometric seperability index (GSI) for real-world categories. Sensor data were from large and small cylinders recorded at 31 distances and 200 angles. At angle 0, the objects are 90 degrees to the right of the robot, at angle 90, the objects are in front of the robot.

This is further supported by a recent study by Nolfi (1996) where data very similar to the ones considered here (two objects—walls and a cylinder—measured by IR sensors of a Khepera robot) where fed into a backpropagation network. The network's performance was very poor, and converged to the correct category in only 22% of the patterns. In other words, the data were of type-2 since not even an universal function approximator such as a backpropagation net could find an underlying mapping between the inputs and categories in the raw data.

These results indicate (a) that category learning for these data is not possible by considering static data only, and (b) that additional mechanisms are required to further simplify the learning problem. In other words, the agents have to be equipped with additional ways to interact with the objects so as to transform the type-2 data into type-1 data (or increase the GSI).

3 Experiments

3.1 Designed sensory-motor coordination

We have shown in previous papers that mechanisms of *sensory-motor coordination* can be used to make agents learn categories in the real world (e.g., Pfeifer and Scheier, 1997; Scheier and Pfeifer, 1995; Scheier and Lambrinos, 1996). Sensory-motor coordination, in essence, involves agent-environment interactions that change the input space so as to facilitate category learning. One such mechanism—circling—is illustrated in figure 3.

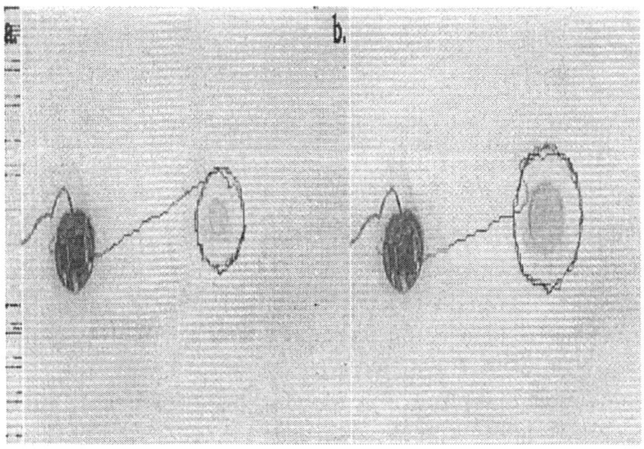

Figure 3 Sensory-motor coordination in a mobile robot. The robot circles around small (a) and large (b) objects

Clearly, this type of behavior reduces the object constancy problem because objects are always viewed from a similar perspective. More importantly, it induces regularities that enable category learning by means of simple Hebbian learning. We have shown in previous studies that agents can learn to distinguish objects such as the ones shown in figure 3. The mean GSI resulting from this circling behavior is 0.95 (+/- 0.04) which is significantly higher than the one resulting, for the same two categories, from static patterns (see above).

3.2 Evolved sensory-motor coordination

In order to further test the idea that mechanisms of sensory-motor coordination can be used to solve the category learning problem we have used methods from evolutionary robotics to evolve the control architecture for agents facing the same category learning task. More specifically, the agents had to learn to distinguish between small and large cylinders by avoiding small cylinders, and staying close to large ones. The agents in this case were simulated, but the sensory readings were taken from a look-up table containing the sensory readings of the real robot for the two cylinders shown in figure 3. We tried to answer two questions: (1) Will the evolved agents use means of sensory-motor coordination to solve this task? (2) What will the resulting GSI look like?

Two types of neural controllers were evolved: feed-forward and recurrent. Neural networks consisted of 12 neurons. Weights were represented by a 6-bit binary string, where the first bit determined whether the connection was to be used or not, and the remaining bits coded the value of the weight in the range -1 to 1. Moreover, for each neuron a threshold value was encoded with 6 bits. Thus, the total length of the chromosome was 936 bits (12 * 12 * 6 bits + 6 * 12 bits). Fitness was incremented whenever the agent was close to a large object, and decremented whenever the agent was close to a small object or a wall. The population size was 30, and in a typical run 100 generations were evolved. In each generation, 50 genes (bits in chromosome) are randomly changed. Figure 4 shows the mean fitness of the fittest individual of each generation. On average, fitness stabilized at a high level after about 40 generations, meaning that the agents were able to make reliable distinctions between the categories.

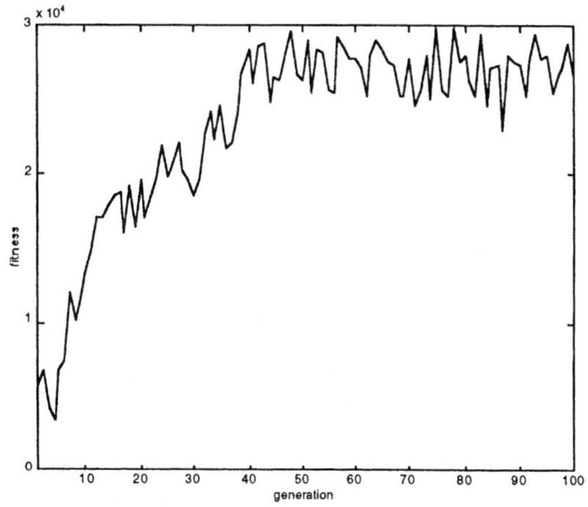

Figure 4. Fitness of fittest individual for 100 generations in the category learning task. Data are means over 10 runs.

The fittest individuals in 86% of the 30 runs used some sort of active interaction with the objects to solve the categorization task (i.e., the motor speeds were non-zero at all times). The remaining 14% approached the object, and then stopped in front of the object. The performance of these latter agents was, however, significantly less stable. In 69% of the runs, circling evolved. In essence, two types of circling behaviors were observed (Figure 5).

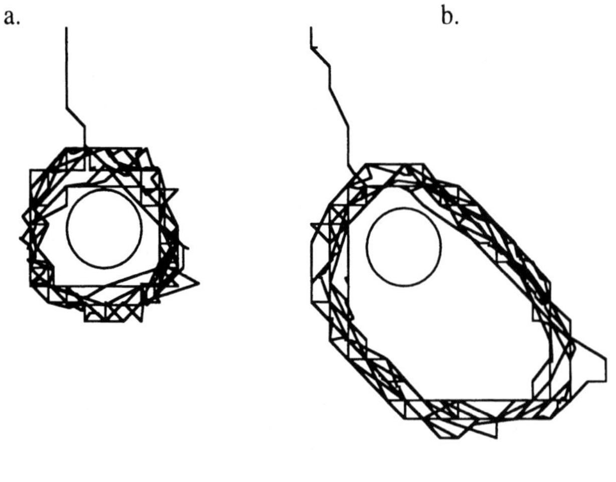

Figure 5 Evolved circling. (a) Circling that evolved in agents with feedforward networks; (b) circling that evolved in agents with recurrent networks.

First, agents that were controlled by feedforward nets stayed always close to the object. Second, agents with recurrent nets made larger circles or ellipses, sometimes loosing direct sensor contact with the object. This answers our first question: the evolutionary process came up with a mechanism of sensory-motor coordination that was essentially the same that was found to enable agents designed by the authors to learn real-world categories. The next important question was, of course, how the agents solved the problem. To this end, the resulting sensory data were analyzed. Figure 6 shows the IR sensor activation for the two types of circling.

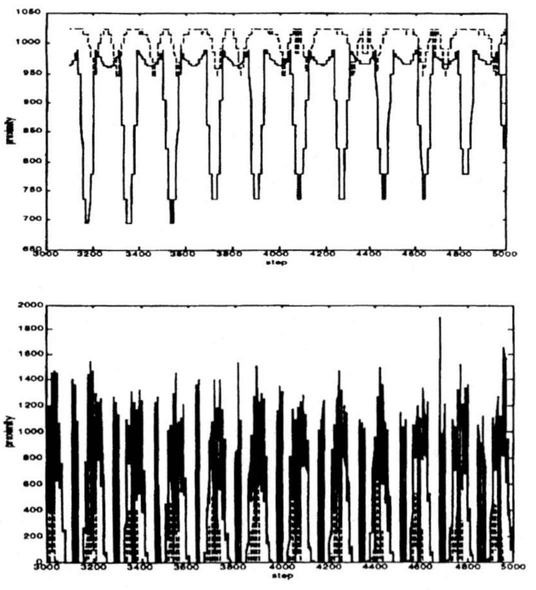

Figure 6: IR sensor readings for two types of circling. Top: Activation of rightmost IR sensor for agent with feedforward net (see Figure 5a). Line: sensor activation for small object; dashed line: sensor activation for large object. Bottom: Activation of right-most IR sensor for agent with recurrent net (moving in ellipses).

The circling induces regular patterns in both cases. This is revealed by the autocorrelations that show that even for the seemingly erratic patterns in figure 6 (bottom), a clear cyclic regularity is contained in the IR readings (data not shown). These patterns enable an easy distinction between

both objects. This is reflected in the resulting GSI values which are, again, very high with a mean at 0.94 (+/- 0.14) for the agents with feedforward, and 0.92 (+/- 0.18) for the agents with recurrent nets, respectively. The underlying neuronal dynamics are quite complex, and far from trivial to understand. . For example, there are neurons that show cyclic behavior in the long-term as well as in the short-term. This type of "fractal" fluctuations on different time-scales were observed in 4 of the 12 neurons. We are currently analyzing these data using techniques of multivariate time series analysis. Ultimately, the goal is to relate the neuronal dynamics with the resulting behavior, and the weight patterns.

4 Discussion

In this paper we showed how embodiment, i.e., specific interactions with objects, can be exploited to induce regularities in input space that enable category learning. The particular mechanism that we studied, and that evolved in the GA experiments, was circling. We have shown that the patterns resulting from this circling behavior have a high GSI which in turn implies that they can be learned by even the simple networks used in the present paper. In other words, the circling behavior transforms the former type-2 data into type-1 data.

We suggest that many of the learning problems faced by mobile robots operating in the real world are of type-2, and should thus be addressed from the perspective of sensory-motor coordination. Instead of increasing the power of the learning algorithms themselves our results suggest that a viable alternative is to exploit the robot-object interaction in order to simplify (or enable) learning. The GSI provides a useful tool in this endeavor as it allows the quantification of one core aspect of the learning problem: separability. Moreover, the GSI can be used to compare different architectures with respect to their success in transforming type-2 into type-1 problems. It will be of significant interest to compare, for example, different mechanisms of sensory-motor coordination with respect to their success in maximizing the GSI of a given learning problem. It will also be of interest to investigate how the morphology of a robot itself can be exploited to maximize the GSI. For example, different positions of the sensors with respect to each other, and with respect to the targets, might dramatically change the resulting GSI and thus learning performance. The present study is rather limited in its sensory-motor setup. Future studies need to increase the sensory-motor complexity of the agents in order to be able to study more interesting means of sensory-motor coordination. Of particular interest are experiments that compute the GSI for various images resulting from a CCD camera. Finally, the GSI itself should be extended to include more than two categories.

Acknowledgments

The authors would like to thank Dimitrios Lambrinos for producing the photographs in figure 3, and Yakuo Ishiguro for providing the simulator for the GA experiments. This research project is funded in part by grant number 11-40564.94 from Swiss National Science Foundation.

References

Abu-Mostafa, Y.S. (1992). Hints and the VC dimension. *Neural Computation,* **5**, 278-288.

Almassy, N., Edelman, G.M., and Sporns, O. (in press). Behavioral constraints in the development of neuronal properties: a cortical model embedded in a real world device. *Cerebral Cortex.*

Arkin, R. (1998). *Behavior-based robotics.* Cambridge, MA: MIT Press.

Ballard, D, Hayhoe, M., Pook, K., and Rao, R. (1997). Deictic Codes for the Embodiment of Cognition. *Behavioral and Brain Sciences*, **23**, 233-265.

Bensinger, D.G., Hayhoe, M., and Ballard, D. (1995). Visual memory in a natural task. *Investigative Ophthalamology & Visual Science.* **36**, S14-S14

Brooks, R. (1991). Intelligence without representation. *Artificial Intelligence*, 47, 139-159.

Chiel, H.J., and Beer, R.D. (1997). The brain has a body: adaptive behavior emerges from interactions of nervous system, body and environment. *Trends in Neuroscience*, 20, 553-557.

Clark, A. (1997). *Being there: Putting brain, body and world together again.* Cambridge, MA: MIT Press.

Clark, A. and Thornton, C. (1997). Trading spaces: computation, representation and the limits of uninformed learning. *Behaviour and Brain Sciences*, 20 (pp. 57-90). Cambridge University Press.

Gibson, J. (1966). *The senses considered as perceptual systems.* Boston: Houghton Mifflin.

Glenberg, A.M. (1997). What is memory for? *Behavioral and Brain Sciences*, 20, 1-55.

Meier, D. (1997). *Generalization and constraints in learning machines.* Unpublished PhD Thesis, Department of Computer Science, University of Zurich

Milner, A.D., and Goodale, M.A. (1995). *The visual brain in action.* Oxford, UK: Oxford University Press.

Montello, D.R., and Presson, C.C. (1993). Movement and orientation in surrounding and imaginal spaces. Manuscript (cited in Glenberg, A.M. (1996). What is memory for? *Behavioral and Brain Sciences*, 20, 1-55.)

Moses, Y., Adini, Y., and Ullman, S. (1994). Face recognition: the problem of compensating for illumination changes. *Proceedings of the European Conference on Computer Vision*, 286-296.

Nolfi, S. (1996). Adaptation as a more powerful tool than decomposition and integration. Technical Report 96-03, Department of Neural Systems and Artificial Life, Institute of Psychology, C.N.R., Rome, Italy.

Pfeifer, R., and Scheier, C. (1997). Sensory-motor coordination: the metaphor and beyond. In R. Pfeifer, and R. Brooks (eds.)*: Robotics and Autonomous Systems*, 20, Special Issue on "Practice and Future of Autonomous Agents", 157-178.

Pfeifer, R. and Scheier, C. (in press). *Unterstanding intelligence.* Cambridge, Mass.: MIT Press.

Salinas, E. and Romo, R. (1998). Conversion of sensory signals into motor commands in primary motor cortex. *J. Neuroscience*, 18(10, 499-511.

Scheier, C. and Lambrinos, D. (1996). Categorization in a real-world agent using haptic exploration and active perception. In: *Proceedings of SAB 96.* Cambridge, Mass.: A Bradford Book, MIT Press, 65-75.

Scheier, C. and Pfeifer, R. (1995). Classification as sensory-motor coordination. In *Proc. European Conference on Artificial Life, ECAL-95*, pp. 656-667.

Simons, D., and Levin, D. (1997). Failure to detect changes to attended objects. *Investigative Ophthalamology & Visual Science.* **38**, 3273-3273

Thelen, E. and Smith, L. (1994). *A dynamic systems approach to the development of cognition and action.* Cambridge, Mass.: MIT Press, Bradford Books.

Thornton, C. (1997). Separability is a learner's best friend. In J.A. Bullinaria, D.W. Glasspool and G. Houghton (Eds.), *Proceedings of the Fourth Neural Computation and Psychology Workshop: Connectionist Re-presentations* (pp. 40-47). London: Springer-Verlag.

Learning To Do Without Cognition

Emmet Spier*, David McFarland**

* Centre for Computational Neuroscience and Robotics
Sussex University, Innovation Centre, Brighton BN1 9SB, UK
emmet@cogs.sussex.ac.uk

** Animal Robotics Laboratory, Department of Zoology
Oxford University, South Parks Road, Oxford OX1 3PS, UK
david.mcfarland@balliol.ox.ac.uk

Abstract

In this paper we show that a phenomenon in animal learning theory (the outcome devaluation effect) for which there is dispute over whether explicit representations and symbolic reasoning is required for its performance, does not require such things. This is done using a reactive motivational model, previously inspired from ethological thought, to which some simple reinforcement learning rules are attached. An instantation of the model is used as the control system of an animat in a spatial computer simulation and it succeeds in learning the necessary parameters to allow the behaviour sequencing system to exhibit the phenomenon.

1 Introduction

How complex can a reactive animat's behaviours get before some begin to appeal for a return to the well established rational techniques in classical artificial intelligence? This paper offers an analysis and performance of a phenomenon in animal learning theory that provokes controversy about the type and complexity of the cognitive machinery necessary for exhibiting the task. It extends a reactive motivational model called the *drk* model (Spier and McFarland, 1996) with some simple reinforcement learning rules and uses an instantation of this model as the control system of an animat in a computer simulation of a Skinner box. Such a model is amenable to implementation on real robots (McFarland and Spier, 1997). In section **2** there is an overview of the outcome devaluation effect, the phenomenon in question, and a discussion of previously offered explanations and their implications. Subsequently, section **3** provides a brief overview of the *drk* model, section **4** offers a theoretical account of how the *drk* model could explain the outcome devaluation effect and section **5** develops a simple set of learning rules that can be used with the *drk* model. The model is then tested, in sections **6** and **7**, within a two-dimensional spatial computer simulation. Section **8** makes an assessment of what has been learnt from this work.

2 The outcome devaluation effect

The outcome devaluation effect was first described by Krieckhaus and Wolf (1968). Since then much work, mainly involving rat experiments, has been expended on understanding this phenomenon (Holland and Straub, 1979; Adams and Dickinson, 1981; Colwill and Rescorla, 1985; being indicative). Typically these experiments involve training a rat in a Skinner box to press a lever for one particular reinforcer and to pull a chain for another, distinguishable, one. Such tools are known as manipulanda. The reinforcers would generally be food items such as glucose solution or food pellets and the actions the rat makes to cause them to be delivered are called instrumental or operant actions. Once the rat reaches a satisfactory performance criterion it is then taken to a separate room, and, after being allowed free access to consume a selected reinforcer, it is then injected with lithium chloride solution (LiCl). This will cause the rat to become sick and, as such, allows the exploitation of a phenomenon known as food aversion conditioning (Garcia, Kimeldorf and Koelling, 1955), where rats avoid consuming a particular food after a close temporal experience with the food and a sickness. After recovery, the rats are then tested (without reinforcement) in the Skinner box to see if they will lever press or chain pull. Half of a four group balanced experimental design (this one following Adams and Dickinson, 1981) can be summarised as :

LP → R 1 CP → R 2	R 1 → LiCl	LP v CP ?
LP → R 1 CP → R 2	R 2 → LiCl	LP v CP ?

with the other two groups being those that receive the opposite instrumental contingencies (LP stands for lever-press, CP stands for chain-pull and R 1 and R 2 are the two reinforcers).

Both Adams and Dickinson (1981) and Colwill and Rescorla (1985) found that the rats response to the manipulandum associated with the delivery of the 'devalued' reinforcer was significantly attenuated. In the above account the 'outcome devaluation' was facilitated by food aversion conditioning; the phenomena have also been described under the same name when the animal is just fed to satiation with the particular substance (resource), a specific satiety (Colwill and Rescorla, 1985; Shipley and Colwill, 1996). The results are broadly similar.

These results are particularly interesting because at the very least they indicate that there must be some hidden state within the rats that can change between the two Skinner box sessions, one where they would activate the manipulandum and another, under exactly the same external physical conditions, where they would not.

Dickinson (1980 and subsequently 1989) gives a detailed consideration of possible internal mechanisms that could exhibit such a behaviour. He outlines the two forms which he calls either procedural or declarative; where a procedural system follows a behaviourist (Watson 1919; Skinner, 1938; Hull, 1943) conception of a behavioural account and a declarative system follows in a Tolman (1932) type cognitive conception. Concisely put, a procedural account assumes that the structure of the representation of the knowledge directly reflects the use to which it will be put, much like associative forms of knowledge.

Declarative knowledge is represented in a form which corresponds to propositions about relationships between events in the world. In the most pared down incarnation these could take the form of logical statements such as, 'lever-press causes food-delivery', where lever-press and food-delivery are symbols explicitly representing the concepts lever press and food delivery. Declarative propositions require some additional machinery to process the symbols and translate the appropriate ones into actions. One important diagnostic criterion for explicit knowledge is that the information can also be misread, whereas the hardwired nature of implicit knowledge permits no such interpretation. The advantage of a declarative system is that through the use of its explicit symbols it can integrate its knowledge to arrive at an action that the simple implicit stimulus response system might not be able to produce.

Utilising this integrative power of a declarative model makes accounting for outcome devaluation very easy. The rat can be considered to have formed three statements concerning the apparatus: 'lever-press causes food-pellet', 'chain-pull causes glucose-solution' and, after food aversion conditioning, 'food-pellet causes sickness'. A simple integration, utilising some kind of machinery that joins the two statements containing the common symbol food-pellet, resolves to produce a further statement, 'lever-press causes sickness'. A comparison of the two resultant operant statements, 'lever-press causes sickness', and, 'chain-pull causes glucose-solution', with some kind of, say, hedonistic selection method[1] will then permit the statement, 'chain-pull causes glucose' to activate some action programs, or at least actions at the same level of description as a procedural account.

Outcome devaluation seems to be much harder to account for from a simple stimulus-response procedural approach. The training period of the experiment might install such new reflexes (conditioned responses, building upon the ideas in associative theory e.g. Shanks, 1995) as, 'see lever, press lever', and 'see chain, pull chain' while relying on previously existing unconditioned responses such as, 'see food, eat food'. After food aversion conditioning a further conditioned response would be established, 'see food-pellet, avoid food-pellet'. With such a configuration of responses it is understandably difficult to conceive how the rat might be able to act according to the observed phenomenon. There seems to be no way of integrating the aversion conditioned response with the operant responses to suppress them; since no reinforcement is supplied in the test session such an agent would see 'lever, press lever' without hindrance. Should the rat use this variant of a stimulus-response model then it would be unable to exhibit the outcome devaluation phenomenon.

In light of such thought, Dickinson (1985) states the strong position,

> "My colleagues and I have argued that teleological control of instrumental behaviour [of which outcome devaluation was his example] cannot be explained, at least at the psychological level, in terms of internal associations which have just excitatory or inhibitory properties. Rather, we argue the knowledge about the action-goal relation must be encoded in a propositional-like form so that it can be operated on by a practical inference process to generate the instrumental performance."

Here he makes two claims, the first being that, 'just excitatory or inhibitory,' properties cannot account for outcome devaluation and secondly that it must be explained by a model similar in kind to the declarative model described above. The subsequent body of this paper demonstrates that by the incorporation of simple learning rules into an already established procedural model (Spier and McFarland, 1996) outcome devaluation-like phenomena can be exhibited.

[1] Presumably one would have to postulate a common currency into which 'sickness' and 'glucose-solution' could be translated (c.f. McFarland and Sibly, 1975).

3 The *drk* model

Previous work discussed at this conference (Spier and McFarland, 1996) developed a motivational model for behavioural sequencing that would not only respond to consumatory cues but could also respond to appetitive or instrumental ones. Figure 1 depicts the architecture of the model which is more concisely expressed in matrix notation. Here we consider that an agent must satisfy two internal state variables[2], say h for hunger and t for thirst; this is expressed by the deficit vector $d = (d_h, d_t)$. In the case of an animat the 'physiological' state space would be those factors that the agent could measure that were correlated with its fundamental tasks, McFarland and Spier (1997) discuss a robot with a state space consisting of energy and work-performed. From previous experience the agent has built a table r of associations that link externally sensed resources to changes in its internal state (their *availabilities*), e.g.

$$r = \begin{pmatrix} r_{Ph} & r_{Sh} \\ r_{Pt} & r_{St} \end{pmatrix}$$

where the columns represent the changes in the agent's (d_h, d_t) state from consuming either a resource of type P (food pellet) or S (sugar solution). Lastly, the agent has also constructed a table k of associations between particular behaviours and how these behaviours influence the ease of obtaining any particular resource (the *accessibility*), e.g.

$$k = \begin{pmatrix} k_{PL} & k_{PC} & k_{Pagent} \\ k_{SL} & k_{SC} & k_{Sagent} \end{pmatrix}$$

where any particular k value can be considered to represent the long term expectation of the quantity of the particular reinforcer (P or S) that can be obtained by performing that particular behaviour (which we call a *tool*). Here L denotes a lever-press, C denotes a chain-pull and 'agent' denotes the direct consummatory behaviour towards the resource. So we find that a single element in k, say, k_{SL} represents how the lever-press tool modifies the nominal accessibility of consuming the sugar solution, likewise, k_{PL} would represent the accessibility for the food-pellet obtained by a lever-press. As developed in Spier and McFarland (1996), the motivational tendency to perform any particular behaviour can be calculated by forming the product $[dr]_{resource} k_{resource\ tool}$ where the $[\cdot]_*$ operator takes the matrix $\cdot$ (dr in our case) and returns the row indexed by $*$. The r and k values are modulated by sensed (Pavlovian) cues to their presence, as elaborated later.

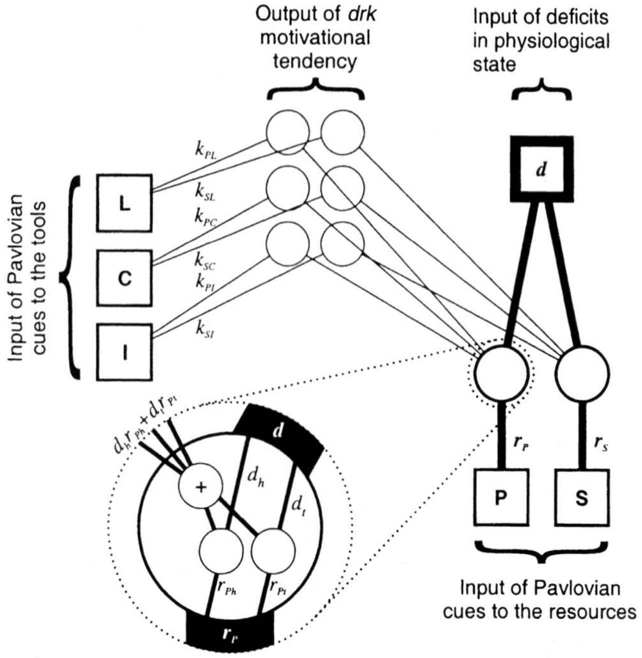

Figure 1 The *drk* model depicted within the connectionist metaphor. The thin edged squares represent a scaleable cue (putatively Pavlovian) that acts as an input to both the r and k pathways. The thick edged square represents the vector output of the physiological state space. The thin edged circles represent a single computational unit which (unless otherwise indicated) multiplies its inputs. The slightly thicker edged circles represent a computational group of which one example is drawn out in full, its output being the *dr* product. Thick lines represent a vector pathway and thin lines a scalar one.

4 A procedural account for outcome devaluation

Using the *drk* model as illustrated above, we will now describe a possible configuration of the d, r and k values, that will be able to exhibit the outcome devaluation effect. With the model so described we can, in the situation where there are no free resources available, form four calculations that can have their magnitudes compared competitively (picking the largest) to assess which behaviour to perform: $[dr]_P k_{PL}$, $[dr]_P k_{PC}$, $[dr]_S k_{SL}$ and $[dr]_S k_{SC}$ as well as the two calculations with the 'agent' k values for freely available resources.

In a concrete example we could suppose that

$$r = \begin{pmatrix} 1 & 0.5 \\ 0 & 1 \end{pmatrix} \text{ and } k = \begin{pmatrix} 0.5 & 0 & 1 \\ 0 & 0.5 & 1 \end{pmatrix}$$

which, from k, would mean that lever-pressing delivers food-pellets and chain-pulling delivers sugar-solution and, from r, that food-pellets only satisfy hunger whereas sugar-solution satisfies both thirst and, to a lesser extent, hunger. If we consider a 'hungry' agent $d = (20, 5)$ then $[dr]_P = 20$ and $[dr]_S = 15$ then since $k_{PL} = k_{SC}$ there is

[2] Although only a two-dimensional physiological state space is used here, the model can be naturally extended to n-dimensions (Spier, 1997).

no particular preference for either instrumental action so the agent would execute lever-press actions. Hence, incorporating the *drk* model into the 'head' of an agent will permit it to switch between lever-pressing, chain-pulling and consummatory behaviour as might be appropriate.

In the case of food-aversion learning we will postulate that the agent can alter the appropriate vector in its r matrix via an association with the induced sickness. In defense of the abuses of a modeller in performing this act, all of the procedural and declarative accounts described above also assume such a slight of hand step (and much worse) so it seems reasonable for the *drk* model to also make such a move. In our example we could imagine that after food-aversion learning with food-pellets

$$r = \begin{pmatrix} 0.01 & 0.5 \\ 0 & 1 \end{pmatrix} \text{ or even } r = \begin{pmatrix} -1 & 0.5 \\ -1 & 1 \end{pmatrix}$$

if the association is highly adversive. Now given an agent with the same $d = (20, 5)$ we find $[dr]_P = 0.2$ and $[dr]_S = 15$ so in this situation the agent will perform chain-pulling actions.

In the case of the animal experiencing a specific satiety devaluation the *drk* account for outcome devaluation is even simpler. In the above case with $d = (20, 5)$ if we feed the agent to satiation with food pellets then the agent's deficit vector would be $d = (0, 5)$ so now $[dr]_P = 0$ and $[dr]_S = 5$ so, again, the agent will perform chain-pulling actions. This result is dependent on the resources possessing different quantities of the agent's state vector and would not account for, say, differently coloured resources.

Such behaviour as described above is reminiscent of what would be expected in an outcome devaluation experiment and the model is certainly described with the same (or even greater) rigour than those in Dickinson (1980). It is now necessary to consider where this model sits in the procedural-declarative classificatory scheme. The significant difference between the *drk* model and the stimulus response model that Dickinson (1980) presents is that both the appetitive/instrumental cue and the consummatory/reinforcing cue are not only associated but combined to form a co-joint cue strength in a common currency of the agent's motivation. From this combined cue the behaviour is determined, consequently, a change in either cue may control an instrumental act. Likewise, since the 'goal' of the behaviour has through some postulated learning shaped the r and k matrices but not been incorporated into them, it is not necessary for the goal to be 'encoded' in the agent for it to act in a goal-achieving manner. Inasmuch as a 'goal-achieving' system (McFarland and Bosser, 1993) like the *drk* model can exhibit outcome devaluation then it would be incorrect to assume that, as Dickinson (1985) has, outcome devaluation is intrinsically a teleological phenomenon and necessarily requires a declarative type reasoning system for it to be made manifest.

5 Incorporating learning rules into the *drk* model

In the previous section it was assumed that the agent governed by the *drk* rules came pre-supplied with ready-made r and k matrices to provide it with the appropriate configurations to allow the agent to lever-press and chain-pull according to its motivational state. It might be reasonable to challenge the validity of such an assumption. Likewise, the supply of ready-made matrices could be seen as inserting some teleological control though the back door. For a fully procedural account, at the current level of description, it is also necessary to describe how the parameters in the r and k matrices can be generated.

In both the psychological and artificial intelligence literature, mechanisms for learning associations have typically been based upon a simple error correcting learning rule called the 'delta rule', first described by Widrow and Hoff (1960). Sutton and Barto (1981a) have shown that under certain assumptions the delta rule is formally equivalent to the Rescorla-Wagner (1972) learning theory developed to account for certain conditioning phenomena. This is attractive and important because it implies that there is a common set of basic assumptions underlying the learning rules in associative learning and, as such, it is not important to consider each minor variant as a distinct algorithm that needs its own special considerations. Indeed, it is well established that it is possible to incorporate learning into an animat with multiple tasks. Sutton and Barto (1981b) furnish an early example of animat type reinforcement learning work while tying to exhibit the psychological phenomenon of latent learning. Blumberg, Todd and Maes (1996) added it to their animated creature toolkit, Touretzky and Saksida (1996) have incorporated into the symbolic reasoning system of their animat and Humphrys (1996) has extended Q-learning to enable an agent to learn to satisfy various tasks when its state space has been reduced by generalisation techniques. The learning rules developed here were designed to be the simplest possible formulations that, while being compatible with some basic psychological observations, were also able to imitate only those learning requirements that section **2** would require for satisfactory performance of outcome devaluation.

5.1 Rules for learning r and k

This section will develop the actual learning rules that were used in the subsequent description of a computer simulation using a spatial environment. The substantive decisions made here are to consider that the exponential learning asymptote of the availability r would be equal to the availability of the resource as experienced in *ad libitum* conditions. Likewise, the accessibility k would capture the reduction of the resource's availability when it has to be obtained by way of a manipulandum, essentially a reduction in the global availability. As such, we will, for r, use a simple delta learning rule whose learning rate will be modified by the reinforcers temporal distance (its 'eligibility' (Barto, Sutton and Anderson, 1983) – captured in a time decaying 'eligibility trace' r'); and for k we will, additionally, take advantage of the decaying eligibility (k') to capture, in k, the time constraint that manipulating the operant controls places upon the agent (since the action, say lever pressing, is temporally separated from obtaining the reinforcer).

Following on from the specific example of the *drk* model in section 4 we define $r' = (r'_P, r'_S)$ to be the associated eligibility traces of the agent when consuming either food-pellets or sugar solution. We also note the agent's change in state vector $\Delta d = (\Delta d_h, \Delta d_t)$ and will use this 'drive reduction' as the reinforcement signal. We can state the learning rule for the change in r as

$$\Delta r_j := \gamma r'_j(\Delta d - r_j) \tag{1}$$

where j indexes the particular resource that has (not) been consumed, the scalar r'_j decays according to $r'_j(t) := \alpha r'_j(t-1)$ and $r_j = (r_{jh}, r_{jt})$ is the vector that contains elements describing the reinforcer in terms of the agent's 'physiological' state space (α and γ are scalar parameters). When applying (1) to r_j we would treat each component of the r_j vector individually and update each according. Hence we are using r' to ensure that modifications of r only occur in an appropriate temporal window and that its learning asymptote is the state change for the execution of the behaviour, which would be the instantaneous availability of the consumed resource.

Unlike r which learns according to the basic delta learning rule the learning rule for the k matrix elements needs to be somewhat more intricate. This is because, firstly the k value for any particular behaviour is a scalar so we no longer have a one-to-one correspondence between the elements of the Δd vector and k. Secondly, we would like, of course, the k elements to reflect the accessibility of the resources. Within the current model the only constraint upon behaviour is the rate at which the behaviour can take place, that being the delay between the operation of a manipulandum and the delivery of a reinforcer. We would want the accessibility to reflect this. Incorporating the k eligibility k', somewhat carefully, both as a learning rate modifier, and additionally, a reinforcement size modifier we obtain the first approximation to a learning rule for k,

$$\Delta k_j := \gamma r'_j k'_j (k'_j . \|\Delta d\| - k_j). \tag{2}$$

The new components of this learning rule are the use of the Euclidean metric $\|\cdot\|$ of taking the square root of the sum of the squares to create a scalar deficit change from the vector Δd, the use of the product $r'_j k'_j$ as the eligibility function and using k'_j within the brackets to modify the asymptotic learning value to reflect the accessibility. Using $r'_j k'_j$ as the eligibility function implies that k_j^{th} matrix component will only be altered when both the appropriate resource and tool 'elements' have been active in the recent past. One possible problem with using $r'_j k'_j$ directly to modify the learning rate is that with both r'_j and k'_j being smaller than 1, then the product will be very much smaller than 1 and hence the learning rates for k will be much smaller than r for no particularly good reason; the purpose of their introduction was to ensure that the appropriate 'elements' were adjusted, not that they controlled learning rate. To match the learning rate of r we can use $\sigma(k'_j)$ instead of k'_j where

$$\sigma(x) = \begin{cases} = 0 \text{ if } x < 0.05 \\ = 1 \text{ if } x \geqslant 0.05 \end{cases}$$

hence simply providing a sharp step function, either $\sigma(k'_j)$ is 0 or $\sigma(k'_j)$ is 1 permitting $r'_j\sigma(k'_j)$ to alter the learning rate of k equivalently to how r'_j does for r_j; the choice of 0.05 is arbitrary. Hence the final form of the learning rule for k used is

$$\Delta k_j := \gamma r'_j \sigma(k'_j)(k'_j . \|d\| - k_j). \tag{3}$$

5.2 Extinction

A final consideration needs to be taken in regards to the learning rules for a reasonably full simulation of the outcome devaluation experiments. A common method used in learning experiments to ascertain the extent of how well an animal has learnt a particular task is the use of an extinction trial. Here the animal would be placed in its Skinner box in the same conditions under which it was trained; however the effects of operating the manipulanda would be disabled. As such the animal would lever press or chain pull to no avail; such behaviours would diminish and finally stop occurring within the trial session. By comparing how much the animal would perform one of the unrewarded manipulative acts compared to the other act the experimenter can determine which manipulative act was, at least, preferred by the animal. The quantity of manipulative acts performed also gives some measure of how well the association was learnt. Such an assessment could not be undertaken if the manipulanda were operative because it would be almost impossible to discern whether the animal's actions were being reinforced through the delivery of the food items or if they were governed by what the animal had learnt.

It is not an aim of this piece of work to provide even a modicum of a plausible account for extinction learning. Neither is extinction learning necessary for the account of outcome devaluation – it is just a diagnostic technique that experimenters use. It is also a complex phenomenon where some time later the animal may recommence the extinguished behaviour (known as spontaneous recovery). For our purposes and in the spirit of the simple account we have previously developed, extinction will be modelled by the subtraction of a small quantity from the active k matrix element when any particular action

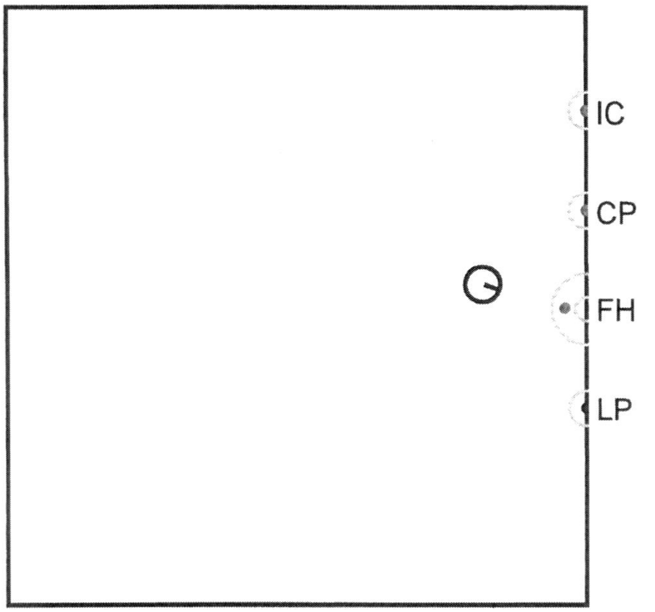

Figure 2 An example of the 2000×2000 unit simulation environment. The agent is depicted by the circle and radius, three manipulanda are shown (LP, CP and IC) as well as the 'food hopper' (FH) where, as can be seen, the reinforcers are deposited after an appropriate action. [For clarity the object sizes are not to scale.]

is completed. For instance, when the agent lever presses for sugar solution then k_{SL} is reduced by a small quantity (say 0.1%). This would have the effect of making the particular behaviour less likely to occur unless it was reinforced – resulting eventually in complete extinction. Of course, with this permanent change in the k value there is no room for phenomena such as spontaneous recovery but this simple approach will permit a comparison of the data presented in the outcome devaluation studies with the simulations performed later.

6 Simulation

The previous sections have provided a mathematical description of the *drk* model that offers an opportunity to account for the outcome devaluation phenomena by taking a particular instantiation of the *drk* model and attaching some simple learning rules which have been developed to capture the kinds of values that such an account requires. This section describes a computer simulation experiment that set out to see if the mathematical description can actually generate its expected behaviour when placed in a less homogeneous environment than a mathematical model (Spier and McFarland, 1997).

6.1 Environment

The simulation environment consisted of a continuous rectangular surface of 2000×2000 units within which resided the agent, the operant controls (manipulanda), a food hopper and two types of reinforcer. The agent possessed a radius of 60 units and the resources one of 15 units. Figure 2 shows a graphical depiction of the actual simulation environment used. The various manipulanda are labelled LP (for lever-press), CP (for chain-pull) and IC (for irrelevant-cue), an extra manipulandum to provide some spurious cue signals in case the behaviour was brittle to the number of salient cues in the environment. In addition FH, the food hopper, which was also a salient cue, demarked an area of the arena into which the reinforcers could be delivered at some random position. The basic simulation design factors were: time is measured in discrete cycles; the agent can move a finite but small distance each cycle; objects are recognised perfectly; the cue strength of any object is a linearly discounted function from a maximum to zero at the maximum sensor range; and any consummatory or manipulative action occupies one cycle.

The agent possessed eight different possible behaviours: do nothing, eat resource A, eat resource B, move five units in a specified direction, lever press, chain pull, operate the irrelevant tool and operate the food hopper. The various manipulandum-directed actions only had an effect upon the simulated world if the agent was within ten units of the particular manipulandum. An implication of this partitioning of behaviours is that the agent had already ascertained how to use the various tools in the environment. Although a significant assumption, for the purposes of demonstrating the outcome devaluation effect in the context of the alternative models reviewed in **2** this assumption seems reasonable since these models also make it. In the experiments reported below, the operation of the irrelevant cue and the food hopper had no effect. The operation of the lever object caused one particular reinforcer to be delivered, say resource A, and the chain object caused the other, say resource B. After an appropriate manipulandum was activated the resource would be delivered at some random location within the outer semi-circle of the food hopper depicted in figure 2. This random positioning would inject a small element of unpredictability into the decision making algorithm of the agent since sometimes it would consume a resource closer to one tool than the other. Each of the two resources, A and B, supplied orthogonal changes in state with A satisfying only hunger and B satisfying only thirst (as such we stop using the corresponding notation of P and S since, at least, sucrose solution satisfies both hunger and thirst). In this case, the change in deficit (h, t) from consuming resource A was $(1, 0)$ and the change from B was $(0, 1)$[3].

[3] Although the simple orthogonal case is described here, the simulation environment and the decision making algorithm could have used non-orthogonal state changes from resources, however, the behaviour of the agent is harder to interpret since it will exhibit progressively more 'trading-off' the greater the non-orthogonality of the resources.

6.2 The decision making algorithm

In line with the initial description of the agent in section **4**, the agent possessed a physiological state space of two dimensions, hunger (h) and thirst (t). As such, the two resources, food pellets (A) and glucose solution (B), also supplied their benefits in terms of hunger and thirst. So we have, as before,

$$d = (h, t) \text{ and } r = \begin{pmatrix} r_{Ah} & r_{Bh} \\ r_{At} & r_{Bt} \end{pmatrix}.$$

For the tool matrix k we need to consider lever pressing (L), chain pulling (C), using the irrelevant cue (I), using the food hopper (F) and direct consumption by the agent ($agent$). As such we find that

$$k = \begin{pmatrix} k_{AL} & k_{AC} & k_{AI} & k_{AF} & k_{Aagent} \\ k_{BL} & k_{BC} & k_{BI} & k_{BF} & k_{Bagent} \end{pmatrix}.$$

There are ten possible appropriate products that the agent can form to evaluate a motivational choice, such as $[dr]_A k_{Aagent}$ for the direct consumption of the resource A; i.e. the agent can attempt to consume each of the two resources using any of the five tools. Within the simulation environment, the r and k values are also adjusted by the cue strength by which the agent can perceive the various items giving a final product calculation for the motivational strength of performing any particular behaviour in the form $q([dr]_B).q(k_{BL})$ where we consider the function $q(\cdot)$ scales the values by their appropriate cue. The cue being calculated by taking the maximum value of either a linearly discounted r or k value with respect to range the objects were from the agent, or a small ambient cue value that was set at the global expectation for encountering the particular object. The cue for the k_{*agent} behaviour was the r_* object.

Every cycle each of these ten products would be calculated and the agent would act according to the product with the highest value. If the tool associated with this product was *agent* then the agent would either move in the direction it perceived the resource was or consume it if it was adjacent to it. If the tool was a manipulandum then the agent would move in the direction it perceived the tool was unless it was adjacent to it at which point it would operate the manipulandum. Initially all the elements in the k matrix were set to a nominal value of 0.1. This small value was enough to initiate unfamiliar tool operations when the tendencies to perform known behaviours were low through small deficits or absent cues to the appropriate tools or resources.

The procedure for testing such an agent was that it was first placed into the simulation environment and initiated. As such, all the learning rules described in **5** were active, adjusting the r and k matrix elements in the agent. After a period when it seemed that there was little change in the matrix values then the outcome devaluation tests would be carried out. In the case of reinforcer devaluation (by LiCl in rats) then as in **4** the r matrix elements of the chosen reinforcer were reduced to zero after the agent had spent some time in an arena with no manipulanda present – this was to prevent associations with the reinforcer devaluation and previous behaviour. As discussed previously, this somewhat magical act is equivalent to the acts made in all other associative or learning accounts. In the case of a specific-satiety devaluation, the agent again in the barren arena had its state variables reduced in multiples of the resource benefit of the chosen reinforcer. Subsequent to either devaluation act the agent was replaced in the arena with the manipulanda present, but deactivated, to assess the agent's performance in extinction.

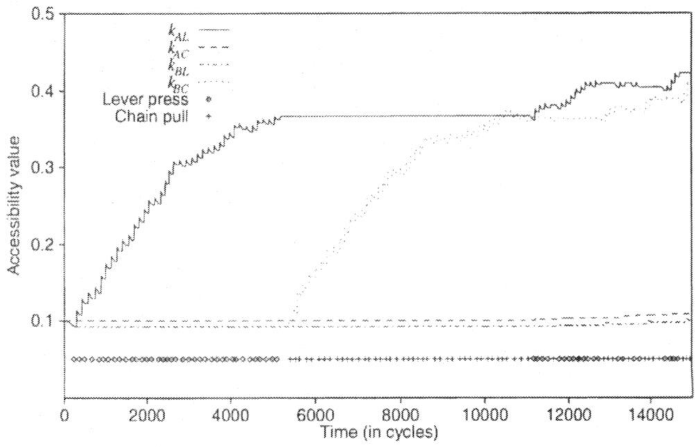

Figure 3 An example of a typical accessibility (k) training session in which a lever-press (LP) delivers resource A and a chain-pull (CP) delivers resource B. Here, as in the text, k_{BL} would stand for the accessibility of obtaining a resource B using a lever-press. The bullets running parallel with the x-axis denote instants when a particular tool was operated (diamonds for A and crosses for B). The traces for the irrelevant cue (IC) are not shown here but are the same as k_{AC} and k_{BL}'s. Further discussion is supplied in the main text.

7 Results

The simulations were carried out in two phases. The first stage was to train the agent in the 'skinner' box so that it could learn the appropriate associations with its k values. The second stage was to test the agent in both the reinforcer devaluation and satiation situations.

Figure 3 shows a typical learning curve for the agent when a lever-press delivers resource A and a chain-pull delivers resource B. During training, initially, only one of the manipulanda was present at any time (the others being 'withdrawn'). The main reason for this was that the naive agent would have nominal accessibility values for all the tools. However, once one tool had a beneficial association it would be used in preference to all other tools, thus preventing the agent from sampling the rest of the environment. Of course, eventually the agent's deficits

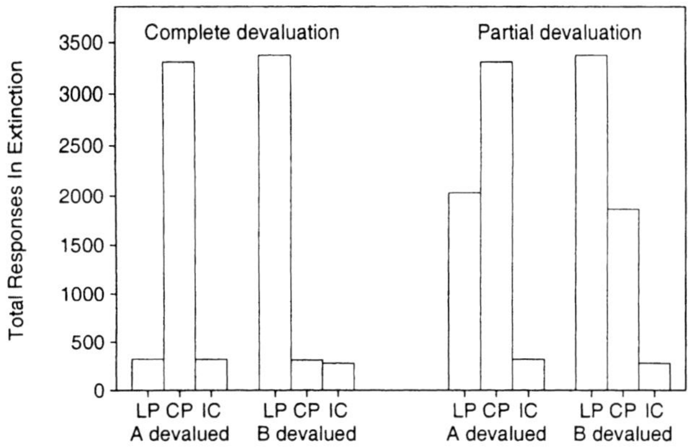

Figure 4 Each of the sets of three bars represents a different experimental condition in a simulated reinforcer devaluation extinction test. The bar heights indicate how many attempts to operate each manipulandum (LP, CP and IC) were executed. The leftmost two sets experienced a 'complete devaluation' where the r value of the chosen reinforcer was devalued to 0. The rightmost set experienced a 'partial devaluation' to 0.2.

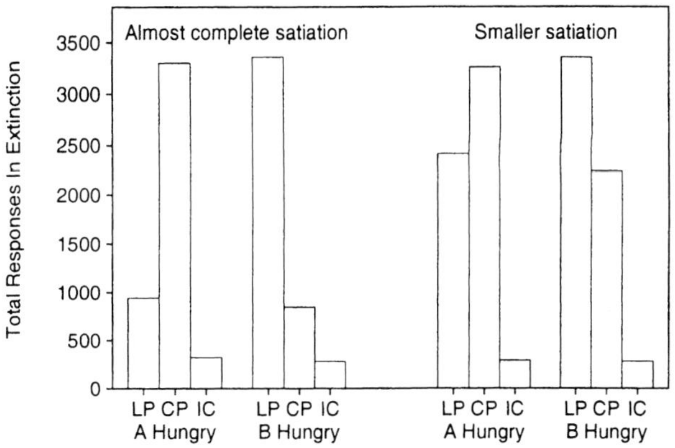

Figure 5 Each of the sets of three bars represents a different experimental condition in a simulated reinforcer satiation extinction test. In the leftmost two sets the agent experienced an 'almost complete satiation' where the chosen deficit state was reduced to a deficit of 20. The rightmost set experienced a 'smaller satiation' to a deficit of 50.

would be fully satiated in a particular dimension and the agent would find the motivational tendency to use that particular tool lower than its as yet unexplored alternative options. However, such a situation would only occur after an extended period of time. This procedure sped up learning all the key associations and the latter third of figure 3 shows the agent finishing off its training session with all the manipulanda operative. As can be seen, the k values follow a negatively accelerated path to their asymptote and the irrelevant k values remain at their nominal values. Also the learning asymptote is around 0.4 where the maximum possible would be one. This reduced value comes from the falling eligibility trace when the agent travels from the manipulandum to the food hopper. The actual k matrix from figure 3 was

$$\begin{pmatrix} k_{AL} & k_{AC} & k_{AI} & k_{AF} & k_{Aagent} \\ k_{BL} & k_{BC} & k_{BI} & k_{BF} & k_{Bagent} \end{pmatrix} =$$

$$\begin{pmatrix} 0.42 & 0.10 & 0.09 & 0.05 & 1.00 \\ 0.09 & 0.40 & 0.1 & 0.05 & 1.00 \end{pmatrix}.$$

Figure 4 shows the results of two different reinforcer devaluation experiments. The height of the bars denote how many attempts were made to operate a particular manipulandum before no further attempts to execute actions against manipulanda were made. The 'complete devaluation' trials shown on the left are when the agent with the training associations established in figure 3 had one of its r values for its resources reduced to 0; the figure shows both situations when resource A and resource B are devalued. The 'partial devaluation' trials shown on the right are of a similar experimental set when the agent had one of its r values reduced to 0.2. As can be seen, in the 'partial devaluation' tests the number of attempts to operate the devalued manipulandum was notably greater than in the 'complete devaluation' case.

Figure 5 shows the results of a similar set of experiments when instead of devaluing the reinforcer the agent had one of its two state variables reduced significantly to simulate a satiation of that particular deficit. The pretrained agent from figure 3 had its initial deficit levels set to $d = (200, 200)$. The leftmost pair of graphs show the results when the chosen deficit variable was reduced to a value 20 and the rightmost pair show results from when the chosen deficit variable was reduced to a value of 50. Again, the smaller the devaluation (satiation), the more the devalued action is executed.

8 Discussion

Figure 3 demonstrated that the extensions made to the *drk* model permitted it to learn appropriate k values (and the r values, although this has not been shown) from the agent's interactions with the environment – it is worthy of note that the resultant simulation k matrix reported in **7** is very similar to the theoretical suggestion in **4**. Figures 4 and 5 clearly demonstrate that the *drk* model can exhibit outcome devaluation with both reinforcer devaluation and specific-satiation. Figure 6 shows data taken from a paper by Colwill and Rescorla that demonstrates both types of outcome devaluation in rats. The data has been transformed to allow easy comparison with the simulation results presented and the shape of all the data sets, simulated and experimental, looks similar.

Of course, we cannot expect the simulation data to match the rat data presented in figure 6. There are too many unknown parameters as well as the rats being notably more complex than the *drk* model. However, the

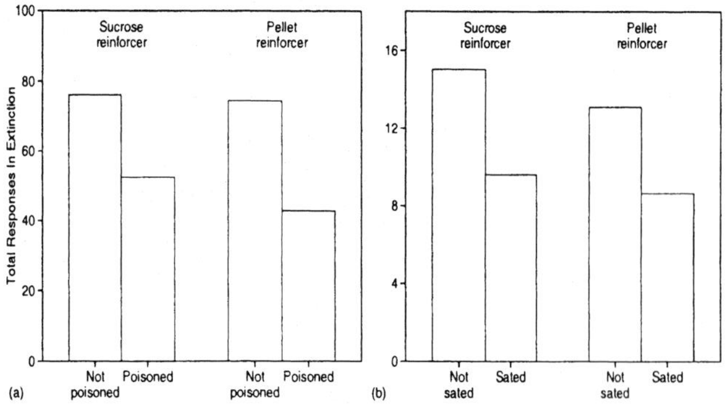

Figure 6 These graphs show data extracted and transformed from Colwill and Rescorla (1985) experiments 2 and 3 that demonstrate outcome devaluation in rats. (a) Shows an outcome devaluation experiment using LiCl as a poison. The left hand side shows the number of responses to the tool that was rewarded by sucrose in training and the right show responses to the tool that was rewarded by Noyes pellets. (b) Shows a reinforcer-specific satiety experiment where the animal was fed to satiety in one of the reinforcers. The left hand side shows the number of responses to the tool rewarded by sucrose in training and the right show responses to the tool rewarded by Noyes pellets. Both graphs (a) and (b) show that the manipulandum associated with the devalued reinforcer is operated less than the alternative.

drk perspective may be able to explain the large difference in the number of responses Colwill and Rescorla's rats made when extinguishing under poisoning or satiation. They attribute the reduced activity in the satiation experiment to a 'very global response suppressive effect' of the satiation process; however, from a motivational perspective we might be inclined to say that the two reinforcers were not particularly orthogonal to each other, so satiation with one reinforcer also has a satiation effect with another. This offers an immediately testable situation where a *drk* perspective would predict that there would be less suppressive effect upon resources that could be independently shown to have a greater orthogonality in the animal's state space.

Less interesting reasons for differences between the two data sets could be attributed to the fact that the simulated agent only has eight possible behaviours and all focused upon consuming the reinforcers; should it have several other behaviour systems in operation then the stark response seen in figures 4 and 5 may have been more muted. Also, the model of extinction, as previously discussed, was very simple and did not interact with the mechanism of behavioural choice (except through the updating of associations), as such, the agent could not utilise its experience of thwarting to modify its actions in the short term, otherwise known as surprise.

The question now arises, whether or not the model presented above would actually satisfy someone who would hold a cognitive position such as in Dickinson (1985). The work of providing the explanation of outcome devaluation falls on two aspects of the *drk* model. The first is that each term (e.g. lever-press, food-pellet) is not just an object with which to form relations but that they also possess a value, innate or learnt, to the animal in terms of the change in the animal's state space. This aspect, although common in ethological ideas, is uncommon in psychological ones – however neural network models can be implicitly using such information encoded within their weights. Even so it seems a perfectly reasonable assumption in this context and additionally has the already established benefit of allowing trading-off and opportunism in behaviour (Spier and McFarland, 1996). The second is that associations with the reinforcer are incorporated twice into the model: firstly as the r component of the drk product; and secondly in the k matrix elements as is always highlighted by the nomenclature $k_{reinforcer\ tool}$. The r association can be seen to be the classically conditioned aspect of the model with the k component being the operant conditioned component within a two-process theory that has a tradition within psychology stretching back to Hull. As such, these additional assumptions over and above those made by the models discussed in **2** seem perfectly reasonable ones to make. Neither would it be reasonable to argue that these additional assumptions provide the model with the 'more complex and powerful forms' of representations that Dickinson (1980) accepts would probably be able to account for outcome devaluation in a procedural manner. Rather, coming from a functional viewpoint has provided a new perspective on the kind of information that these representations could carry and this makes an opportunity for a simpler explanation than postulating some cognitive capacities.

Touretzky and Saksida (1996) invoked a classical reasoning system with representations to model the learning and production of an animal's behaviour in the psychological delayed match to sample (DTMS) task. In light of the work presented here, one might infer that there may not be an obligation to use such cognitive techniques. Plausible extensions to the *drk* model that would admit the production of the DTMS task could take the form of adding an association between the eligibility trace k' and the k value or extending the cue model to incorporate time. Such a model might be more amenable to their label, 'Skinnerbot'.

In summary, the most important thing to note is that it has been shown that a phenomenon for which there has been dispute over the requirement of cognition for its manifestation certainly does not need any cognition. Whether or not cognition is actually used by animals in the production of such behaviours is still an open question.

References

Adams, C.D. and Dickinson, A. (1981). Instrumental Responding Following Reinforcer Devaluation. *Quarterly Journal of Experimental Psychology*, **33B**, pp. 109-121.

Barto, A., Sutton, R. and Anderson, C. (1983). Neuronlike Adaptive Elements that can Solve Difficult Learning Control Problems. *IEEE Transactions on Systems, Man, and Cybernetics*, **SMC-13**, 834–846.

Blumberg, B., Todd, P. and Maes, P. (1996). No bad dogs: Ethological lessions for learning in Hamsterdam. In *From Animals to Animats 4: Proceedings of SAB96.* (ed. Maes, P., Mataric, M., Meyer, J.-A., Pollack, J. and Wilson, S. W.), pp. 295-304. Cambridge, MA: MIT Press.

Colwill, R.M. and Rescorla, R.A. (1985). Postconditioning Devaluation of a Reinforcer Affects Instrumental Responding. *Journal of Experimental Psychology: Animal Behavior Processes*, **11**, pp. 120-132.

Dickinson, A. (1980). *Contemporary Animal Learning Theory.* Cambridge, Cambridge University Press.

Dickinson, A. (1985). Actions and Habits - the Development Of Behavioral Autonomy. *Philosophical Transactions Of the Royal Society Of London Series B- Biological Sciences*, **308**, pp. 67-78.

Dickinson, A. and Dawson, G. (1989). Incentive learning and the motivational control of instrumental performance. *Quarterly journal of experimental psychology*, **41B**, pp. 99-112.

Garcia, J., Kimelforf, D.G. and Koelling, P.A. (1955). Conditioned aversion to saccharine resulting from exposure to gamma radiation. *Science*, **122**, pp. 157-158.

Holland, P. and Straub, K. (1979). Differential Effects of two ways of devaluing the unconditioned stimulus after Pavlovian appetitive conditioning. *Journal of Experimental Psychology: Animal Behavior Processes*, **5**, pp. 65-78.

Hull, C.L. (1943). *Principles of behavior.* New York, Appleton.

Humphrys, M. (1996). Action Selection methods using Reinforcement Learning. In *From Animals to Animats 4: Proceedings SAB96.* (ed. Maes, P., Mataric, M., Meyer, J.-A., Pollack, J. and Wilson, S. W.). Cambridge, MA: MIT Press.

Krieckhaus, E. and Wolf, G. (1968). Acquisition of sodium by rats: Interaction of innate mechanisms and latent learning. *Journal of comparative and Physiological Psychology*, **65**, pp. 197-201.

McFarland, D. and Spier, E. (1997). Basic cycles, utility and opportunism in self-sufficient robots. *Robotics and Autonomous Systems*, **20**, pp. 179-190.

McFarland, D. and Bösser, T. (1993). *Intelligent Behaviour in Animals and Robots.* Cambridge, MA, MIT Press.

McFarland, D. and Sibly, R. (1975). The Behavioural Final Common Path. *Philosophical Transactions of the Royal Society (Series B)*, **270**, pp. 265-293.

Rescorla, R.A. and Wagner, A.R. (1972). A theory of Pavlovian conditioning: Variations in the effectiveness of reinforcement and nonreinforcement. In *Classical conditioning II: current theory and research.* (ed. Black, A. H. and Prokasy, W. F.), pp. 64-99. NY: Appleton-Century-Crofts.

Shanks, D. (1995). *The Psychology of Associative Learning.* Cambridge, CUP.

Shipley, B.E. and Colwill, R.M. (1996). Direct effects on instrumental performance of outcome revaluation by drive shifts. *Animal Learning and Behavior*, **24**, pp. 57-67.

Skinner, B.F. (1938). *The behavior of organisms.* New York, Appleton.

Spier, E. (1997). *From reactive behaviour to adaptive behaviour: Motivational models for behaviour in animals and robots.* D.Phil. thesis, Department of Zoology, Oxford University. Also available at http://www.cogs.sussex.ac.uk/users/emmet .

Spier, E. and McFarland, D. (1996). A Finer-Grained Motivational Model of Behaviour Sequencing. In *From Animals to Animats 4: Proceedings of SAB96.* (ed. Maes, P., Mataric, M., Meyer, J.-A., Pollack, J. and Wilson, S. W.), pp. 255-263. Cambridge, MA: MIT Press.

Spier, E. and McFarland, D. (1997). Possibly Optimal Decision Making under Self-sufficiency and Autonomy. *Journal of theoretical biology*, **189**, pp. 317-331.

Sutton, R. and Barto, A. (1981a). An Adaptive Network that constructs and uses an internal model of its world. *Cognition and Brain Theory*, **4**, pp. 217-246.

Sutton, R. and Barto, A. (1981b). Toward a modern theory of adaptive networks: Expectation and prediction. *Psychological Review*, **88**, pp. 135-170.

Tolman, E.C. (1932). *Purposive behavior in animals and men.* New York, Appleton-Century-Crofts.

Touretzky, D. and Saksida, L. (1996). Skinnerbots. In *From Animals to Animats 4: SAB96.* (ed. Maes, P., Mataric, M., Meyer, J.-A., Pollack, J. and Wilson, S. W.). Cambridge, MA: MIT Press.

Watson, J.B. (1919). *Psychology from the standpoint of a behaviourist.* Philadelphia, Lippincott.

Widrow, B. and Hoff, M.E. (1960). Adaptive switching circuits. In *1960 IRE WESCON Convention Record (pt. 4)*, pp. 96-104.

Learning in psychologically plausible conditions: The Case of an Imaginary Pet Robot

Stephane Zrehen* **,Hiroaki Kitano*,Masahiro Fujita***
*Sony D21 Laboratory, Tokyo
**Center For Neuromorphic Engineering, Caltech 193-36, Pasadena, CA 91125
Email: zrehen@caltech.edu
***Sony Computer Science Laboratory, Tokyo

Abstract

We propose the construction of a pet robot as a test-problem for the psychological plausibility of learning in artificial systems. We analyze the conditions under which animals access the world and adapt their behavior. We finally propose elements of solution for a control architecture for such a machine.

1. Introduction

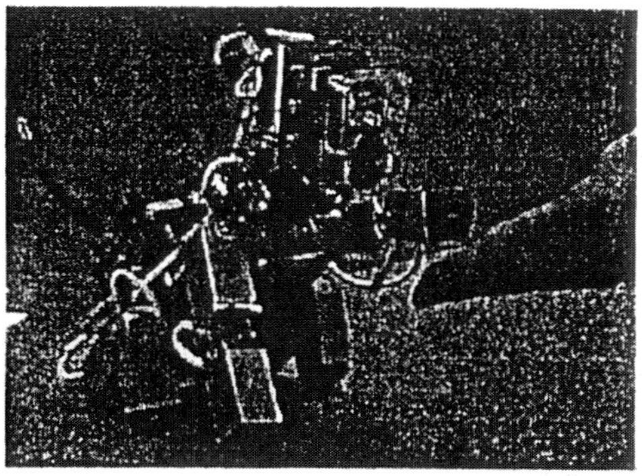

The Sony dog-like robot

In the last year, Sony has demonstrated a small legged robot which definitely reminds one of a little dog This proves the feasibility of building a machine whose movements look natural. In view of these developments, it becomes important to answer the question of what it would take to complete the illusion of an artificial pet. In this paper, we will place ourselves in the position of the person acquiring such a robot expected to behave like a pet. We favor dogs over cats because human interactions with them are richer, and because it is relatively easy to train a dog while it is almost impossible with a cat.

The behavior expected from a pet robot should match as closely as possible that of a real pet. In that sense, psychological plausibility entails a) the possibility for the pet robot owner to attribute a psychology to the robot and b) that pet robots must access and interact with the world -including human beings-and adapt their behaviors in the same conditions as real pets. The first major aspect is that pets take care of themselves most of the time. Unlike current industrial or research robots, they do not give the impression of continuously performing some task. Rather, they seem to mostly participate in activities that they "enjoy", which may change with time.

It is likely that the owner of the pet robot would appreciate seeing the machine get attached to him or her, that is, being recognized as its master. The robot pet's master would also probably expect it to learn how to chase sticks, find its way around the house, behave itself, show interest in fellow robot pets, learn to bark at intruders and show joy when its master comes home. These things take place with real dogs, under minimal control from humans.

All these phenomena should happen as naturally as possible. For example, it would be very unpleasant, or at least very unrealistic, to ask robot owners to pose in front of a white wall for two minutes while the robot learns to recognize them. This is simply not the way it is done with dogs. It would also be unrealistic to endow the robot with a reset button, which would erase all their previous memories, or with a switch that would freeze its movements, or constrain its wandering to a pre-built maze, or give the robot a precise plan of the house, or install a Global Positioning System (GPS) or a camera on the ceiling. Of course, it should be possible to train pet robots, as we train dogs. But training should also be done in the same conditions as with animals, that is, mostly in the form of playing. When playing with an animal, we are faced with all of the animal's behavior, and not just the task we want it to learn. Conversely, the dog faces all of its environment, and not only the part corresponding to the task.

The project of building a pet robot poses problems which are of direct relevance to Animat research. In such a project, most problems addressed by cognitive science, artificial intelligence, machine learning, computer vision and artificial perception in general must be tackled at the same time, inside the same system functioning in real time. Once the physical body of the robot and its sensors and actuators are built, the major problem remaining thus amounts to constructing an artificial brain in charge of most of the elements that a real brain needs to control. Since we still cannot construct machines that breathe, or eat or that could have diseases or heal, all of which require the intervention of the brain in animals, we are left with only cognitive functions, which we may summarize as follows, in large categories:

- the control of the movements
- the interpretation of the sensory-motor events
- the decisions to record certain events and associate them with other events (learning)
- the association of behavior with drives/motivations
- the creation and management of new motivations, such as playing a game or fetching the newspaper
- the expression of emotions and their association with internal or external events
- the emission and interpretation of communication signals connected to the control of behavior.

One key feature of our interactions with animals is that we do not get access to their brains: we cannot add or remove parts of their brains, nor decide what parts of their brains we want to stimulate. We only address their behavior, the change of which can only be suggested, but not forced. Actually, we have no control over what changes actually take place in an animal's brain, unless we perform surgery. We cannot reset an animal, nor force it to focus its attention on all the things we want. So the person acquiring a robot pet should be faced with the same situation as with a newly acquired dog. This person will not have access to its internal circuitry, and will be able to communicate with the robot through its behavior alone.

These constraints share common grounds with many recently developed techniques in the disciplines mentioned above. It is the goal of this paper to devise an informal test for psychological plausibility, corresponding to the illusion of a real animal, and to evaluate current research in the light of this test.

In the next part, we examine in detail the constraints that psychological plausibility imposes, of which a few examples were given above. In the next part, we review current literature in the field of machine learning and adaptive behavior. Then we propose elements of solution for the architecture of the artificial brain for a robot, which allows the display of psychologically plausible behavior.

2. Psychological Plausibility

Psychological plausibility is concerned with the conditions in which animals or humans access the world and learn things about it. It is a different concern than biological plausibility, which addresses the actual neural computations performed in a brain, usually analyzed in response to simple stimuli.

Perceptive access to the world for an animal is limited to the electrical activity of its sensors, which may trigger the activation of other neural networks in its brain. Meaning is created by the brain, which implies that events in the world, the raw data on sensors, must first be translated in neural activity patterns that can be interpreted by the brain.

Learning in a psychologically plausible manner is a direct consequence of the autonomy of animals, and henceforth of all cognitive systems according to the enactivist paradigm (Maturana and Varela, 1987), or to Pfeifer's list of design principles for animats (Pfeifer, 1996). The major aspects that automomy forces on learning are the following:

1. An animal "decides" or "invents" which problems it wants to solve and when. An important aspect is that most learning in animals takes place spontaneously. A dog decides on its own who is its master, or to explore its new home, and what are the things to record to find its way around. When a human trains a dog, there is no guarantee that the animal will lend itself to the game.
2. An animal has the possibility to see the same situation as requiring different behaviors. For instance, a dog faced with a barrier may either decide to learn that it is an obstacle, and should therefore learn to detour or simply go elsewhere. However, it is commonly observed that in certain situations, the same dog can decide to jump over the same barrier. We can deduce from this observation that the functional interpretation of the sensory scene can change according to the animal's motivations. This also implies that the barrier cannot have been learned as a pure obstacle in the first place, but that the different possible interpretations are always available. As a consequence, there cannot be only a direct coupling between perception and action. Rather, sensory scenes must have the ability to be recorded independently from the actions, and the associations with the actions must be triggered, and possibly learned, in accordance with a given motivational state.
3. Learning happens *on-line*, i.e., *while* the animal's behavior is controlled by the brain. A dog learns its way around its new home while exploring it. The consequences are the following:
   - the sensory subspace to which an animal is exposed is limited by its own choice of movements.
   - if learning concerns only one part of the brain –for a visual task for instance–, then the signal for not altering other parts of the brain must originate in the brain itself.
   - if a certain form of learning must appear or disappear at certain stages of development, then the signals for triggering or stopping learning must also originate in the brain.

   In short, we do not have the possibility to take out a part of an animal's brain, expose it to well-chosen stimuli in order to learn a task that we have decided, and plug it back in the brain.

4. There is no way to control which part of an animal's cognitive system is going to be used and possibly altered. In other words, the allocation of resources is done internally, by the brain itself, and the experimenter has no control over this process.
5. We cannot control what exactly will be learned by an animal. We have good reasons to believe that an animal learns to recognize the visual/olfactive/auditory image of its master, but we cannot know what elements this representation contains. Even if neurophysiology can help understand the nature of the representation, we cannot force it to be different.
6. Declarative learning (what) is usually very fast, in one single presentation. When dogs hide their bones in the garden, they do not go back there several thousand times before they can remember the place. In a similar vein, humans recognize other people after meeting them only once. Individual rat learning performances are usually step-wise, suggesting that at some point across trials, learning occurs in one-shot (C.R. Gallistel, personal communication). There is no gradual increase of performance. In a sense, we can say that it takes time to "figure out" the solution. Once this is done, learning is immediate.
7. Learning takes place in an autonomous, internal fashion, possibly under the control of reinforcement signals such as rewards or punishments. When dealing with dogs, we have some knowledge of what they consider as rewards: food, pat on the head..., and as punishments: being dragged outside the room, being talked loudly at, being hit... With babies, it is a little more complex, because we do not really know in advance. We need our careful attention to realize what causes changes in their behavior. With a new type of pet, such as a robot, we may have to find out what are the right messages that they can interpret as reinforcement signals. The only requirement will be some form of invariant recognition of this signal on the part of the robot. It would be unpleasant to have to press a button.

Even when learning is not the issue, autonomy induces certain characteristics on behavior that must be carefully taken into consideration in the project of building a robot. This is most important in communication problems: *we cannot force an animal to understand the meaning of the messages we are trying to convey.*

This problem actually mirrors that of linking observable behavior with brain activity in psychology. Behavior is the externally observable translation of brain activity, and the translation process is not well known. When dealing with a robot, we can only access their behavior, hoping to induce some activity pattern in their brain, which will then be translated back into behavior. For instance, many dogs respond to the words "Sit down!" by sitting down. But this command was not a command for the dog in the first place. The sound signal first needed to be translated into a neural activity pattern that the dog "decided" to interpret as a command, and therefore to actually sit down. The two major consequences of that fact are that a) the dog must be equipped with some segmentation ability that will make it possible to interpret our words as a command and b) that we must adapt *ourselves* to this type of segmentation. For instance, we cannot train a dog to perform any kind of movement upon presentation of a prime number.

The ability to segment the world is a subject of debate in the Artificial Intelligence community, especially among Animat researchers. Following the German biologist Von Uexkull, it is argued that since perception must be directly coupled with actions and that animals cannot be separated from their ecological niche, we have no guarantee that they segment the world along the same lines as we do. This is a reasonable point for biological systems in general whose main cognitive goal is to ensure survival in their own environment with its own characteristics. Nevertheless, we must pay attention to the fact that if our decomposition of the world never coincided with that of dogs, we would not be able to communicate with them at all. This is however an important expectation of a robot pet. We do observe that dogs chase sticks, which they see only briefly before chasing and retrieving them. How can that be interpreted without admitting their ability to recognize the stick after only a brief presentation, or for that matter many objects, as *objects* that can be grabbed in various fashions? There is evidence that pigeons use geon-like visual volumetric primitives (Wasserman et al., 1996), so it is legitimate to suppose a similar mechanisms in other animals as well.

Below, we give a list of constraints on the segmentation properties that seem to be present in animals, with the justification for supposing that they are so:

- the perceptive access animals have of their world is limited to their sensors, and therefore to the physical phenomena these sensors respond to. Dogs and humans to not have sensors for detecting the extremes of electro-magnetic spectrum, and can therefore not extract information from them. We have a poor sense of olfaction, while dogs have an excellent one and a poor sense of vision. This shows that survival while exhibiting intelligent behavior is possible using only a limited number of all the possible physical signals that can be found in nature. Nevertheless, it would be impossible for an animal to communicate in any meaningful way with another whose only sense would be the extremes of electro-magnetic waves. For communication, the signals to which the sensors of two animals respond to must at least partly be of the same nature.
- the segmentation of the sensory scene yields an invariant representation of events in the world. When

someone presents a stick to a dog, before throwing it away, the presentation is uncontrolled. It is very improbable that the dog gets to see all the possible visual configurations of the object during such a short time span. Nevertheless, the dog will be able to retrieve the stick under a new visual angle. Even more striking is the famous case of newborn geese observed by Konrad Lorenz. He noted that these animals "adopt" as their mother the first moving object they see when they come out of their egg. Usually several eggs open at approximately the same time, but the same object, therefore experienced from different points of views, is perceived as having the same meaning for all the animals.

It is reasonable to assume that a very limited number of viewpoints suffices to recognize an object. This strongly suggests that an invariant representation of objects must be learned by animals when they see it. The invariance needs not be total, but enough to account for recognition under familiar viewpoints and limited displacements in the visual field.

– Decomposition of the sensory scene must pre-exist before learning of associations can take place, and be coupled to a saliency property. Let us take the example of a dog in a Pavlovian experiment. How can we explain that the dog eventually learns to salivate to the sound of a bell, and not to its trainer's clothes, or to the day of the week, or the particular intensity on one of its retinal receptors? All these things are present during training. There must therefore be some mechanism for allowing the sorting out of sensory stimuli into "potentially meaningful" and salient elements. It is the recognition of one of these salient elements that can be associated with a behavioral response, or be subject to habituation. Briefly said, associative learning cannot take place at the level of raw sensor data if they are undifferentiated and unstructured.

  Similarly, we observe that babies learn to speak the language of their caretakers. Let us imagine that a baby's mother points to an object and utters the word "bottle". The sensory scene actually faced by the baby is the output of more than one hundred million sensors. How could the baby ever learn to associate a name with the image of the bottle, and not with the particular value of one of these sensors –or of any subset of them–, if the image of a bottle did not have any properties that segment it out of the rest of the scene, and make it more salient? This cannot be explained in terms of trials and errors or by a large number of presentations: We can compute the time it would take to learn to respond to the particular intensity of one given photoreceptor: Admitting just 10 grey levels per receptor, $10^{10000000}$ different images can be made. If we admit an update rate of 50Hz (twice the frequency required for giving the impression of movement), we still end up with an astronomical time, much larger than the age of the universe. The number would be even larger to sort out the activity of an unknown set of photoreceptors.

  It turns out that we do not recognize the difference between the images on a broken TV set, which do not make any sense and have no structure other than some spatio-temporal frequency. Our segmentation properties are limited and constrained in particular ways, which psycho-physics and neurophysiology help to explain.

  In summary, in order to communicate meaningfully with an animal and teach it things which are interesting to us, such as recognizing and chasing a stick, the animal has to be endowed with a segmentation mechanism which functions along the same lines as ours.

– Learning must be performed out of the world decomposition offered by the animal. For instance, dogs recognize their master, which amounts to some form of learning since they were not born knowing who their master would be. We do not know the nature of the representation made by the dog, but we know that it got associated to the master's identity under conditions where there was always a background. Furthermore, no outside observer can tell a dog which part of its sensory scene corresponds to its master. The decision to consider a particular subset of sensory patterns to form the master's representation was taken inside the dog's brain. To use a machine learning vocabulary, this means that the choice of the input patterns to be learned is made by the dog. In a natural situation, we do not carefully choose the stimuli we expose a dog to.

  This remark is particularly relevant when considering the machine learning literature in which the problem is usually to learn one given task. The problem in a real animal or any psychologically plausible creature is not only how to solve a given task, but also *to identify what the task consists of and what data pertain to it* in the first place. Furthermore, the abundance of data surrounding a real animal also poses the question of how to avoid associating too many things: how should learning be regulated?

– In addition to the problem of sorting out which type of sensors and what subset of these sensors is relevant to allow communication with an animal, we must add the problem of understanding the meaning of the situation itself. How does the baby know that his mother is teaching him the name of an object? She could just as well be talking about the weather or what she had for lunch, or saying that the baby should play with that object, or that the baby should give the object to his mother. Obviously, such a scene takes place with the attention of the baby already focused on the object itself, for instance through pointing at the object.

Dogs do not understand pointing, which according to Vygotsky (Vygotsky, 1978) follows the baby's experience in the following terms:

> Initially, this gesture is nothing more than an unsuccessful attempt to grasp something, a movement aimed at a certain object which designates forthcoming activity. The child attempts to grasp an object beyond his reach; his hands, stretched toward that object, remain poised in the air. His fingers make grasping movements. At this initial stage pointing is represented by the child's movement, which seems to be pointing to an object – that and nothing more.
>
> When the mother comes to the child's aid and realizes his movements indicate something, the situation changes fundamentally. Pointing becomes a gesture for others. The child's unsuccessful attempt engenders a reaction not from the object he seeks but *from another person*. Consequently, the primary meaning of that unsuccessful grasping movement is established by others. Only later, when the child can link his unsuccessful grasping movement to the objective situation as a whole, does he begin to understand this movement as pointing. At this juncture there occurs a change in that movement's function: from an object-oriented movement it becomes a movement aimed at another person, a means of establishing relations. *The grasping movement changes to the act of pointing.* As a result of this change, the movement itself is then physically simplified, and what results is the form of pointing that we may call a true gesture. It becomes a true gesture only after it objectively manifests all the functions of pointing for others and is understood by others as such a gesture. Its meaning and functions are created at first by an objective situation and then by people who surround the child.

This explanation points out that the decomposition of the sensory events in the sensory-motor sequence

pointing gesture → mother coming → object in my hand

into meaningful sensory events precedes learning of the temporal/causal association. It also illustrates that the situation is understood in terms of the baby's intentions: to grab the object, which frames the level of description he is then capable of understanding, i.e., associating with the sound of the word. We can analyze similarly the fact that most dogs eventually learn to sit down upon hearing these words. Training can be done by either forcing them to sit while uttering the words, or/and by rewarding them.

A side aspect of psychological plausibility concerns the bodily needs of animals, such as feeding, urinating and mating. It is noteworthy that training of real puppies first aims at regulating these needs. There is no doubt that our sharing some of these needs enhances our sympathy towards animals: we may feel touched by a hungry dog because we know what it feels like to be hungry. The success of the Tamagotchi©, precisely based on this idea of the owner as a caretaker of the virtual creature, even suggests that the physical appearance does not play a major role in attachment. What matters is the fact that the creature expresses needs and desires which the human caretaker can recognize and respond to.

This list of observations can be translated into a set of features and behaviors we expect a pet robot to exhibit. In brief, it appears that for psychological plausibility, a pet robot must address the world in all its complexity, and be endowed with an ability to decompose its sensory-motor space that overlaps with our own: we cannot have specific intervention in its cognitive system.

All these points are directly or indirectly addressed by recent developments in engineering sciences such as computer vision, machine learning and animat research. In the next section, we propose a review of literature in these domains, in view of finding the best tools for designing a robot exhibiting all the behaviors described above.

3. A review of engineering techniques

As was mentioned before, the requirements for a psychologically plausible pet robot overlaps considerably with animat research and machine learning. It also shares grounds with computer vision.

3.1 Animat Research

Pfeifer (1996) proposes a list of design principles for autonomous agents, under the banner of "fungus eaters", artificial creatures that require a complete cognitive system for ensuring their survival in a rich environment. In many respects, this list of principles overlaps with the list of requirements for an artificial pet. In particular, the issue of complete autonomy is crucial in both problems. However, the problem of a pet robot can be different on several aspects and emphasizes issues that could be seen as secondary in fungus eaters.

Pfeifer stresses that animals all have an ecological niche, which, in many cases facilitates meaningful interactions between animals and their environment, in particular when discriminations are to be made between interesting and non-interesting objects. The consequence of that view in animat research has been to design the environment together with the robots' cognitive system. In general, this takes the form of placing in the environment only a few types of objects whose discrimination is easy to make with either specialized sensors or using very simple regularities of the environment. The result is in a sense the invention of toy problems, ironically one of the main criticisms addressed to classical AI. The main difference we see between this view of animat research and the pet robot project is that we expect the robot pet to share much of our own ecological niche, in most of its complexity. Insisting on the coupling of an animal with its niche does not in itself provide the means to tackle the complexity of problems such as object recognition out of a few presentations. On the other hand, classical methods developed in computer vision can help for that purpose. There is nothing fundamentally contrary to the animat approach or to psychological plausibility in computer vision. Object recognition is part of the animat approach, and should not be rejected a priori.

It is usually admitted in the animat community that categories must be built out of sensory-motor interactions with the environment, from the agent's point of view. We tend to agree with this principle, but only to a certain degree. Our position on this matter is the following: categories are made of the association of features with the agent's point of view, which can consist of actions in some cases. The association of an external scene segmented into high-level features with the agent's personal/subjective experience gives the "meaning" of the categories. We thus agree that the meanings accessible to an animat are limited to its own possible interactions with objects. Nevertheless, we believe that object recognition per se must be performed: We expect a pet robot to recognize a shoe as a shoe, which is both something it could grasp–even if this is forbidden, which limits the actual physical interactions it has with shoes–and something that humans wear on their feet and don't like to see torn apart. However, we do not expect it to be able to tell its make, which is totally irrelevant in a pet's world.

We also argue that the features whose grouping constitutes the coding of a category need to be extracted with mechanisms that result from evolution and/or infancy, or for that matter, from design. This principle is used implicitly in Pfeifer's and his co-workers's research. For instance, in (Scheier and Lambrinos, 1996) a neural architecture, based on Edelman's neuronal Darwinism, allows the construction of visuo-haptic categories. Visual and tactile primitives are pre-designed, and it is only their association during lifetime that is learned. Our claim is that it would be impossible to succeed in such experiments if the relevant primitives themselves needed to be discovered during lifetime. We also argue that the set of primitives chosen in that paper (striped patterns on top of Khepera robots) is meant to match only the target objects designed by the authors, and would not allow the recognition of any other objects.

Pfeifer, like most of the animat community, strongly opposes the idea of a central authority in autonomous systems. We argue that a psychologically plausible system must take decisions about such things as what problem to invent and solve next, or when to stop learning a task. This may look like the equivalent of a central authority. But is does not imply the existence of a homunculus sitting inside the agent's brain. Rather that there is a part –possibly a simple neural network and not necessarily an intelligent creature–of the cognitive system where information converges, and *which projects back to these areas.*

3.2 Machine learning

The major developments in the field of statistics and neural networks are concerned with supervised learning. It amounts to training a processing structure to classify correctly a distribution of data after presentation of representative examples for which the correct classification is known by the user.

This approach poses two major problems to the robot pet project:

1. The data have to be chosen carefully by the user for representing the whole input distribution. In an animal, the data are met continuously as it wanders around. There is no guarantee that the set of data it encounters is representative of the whole data set. In addition, this approach supposes that the data corresponding to the task to be learned have already been sorted out from other types of sensor data. In order to implement such a mechanism in a pet robot, networks of this type should be equipped with a filtering mechanism as a pre-processing stage that ensures the proper segmentation.
2. For most learning algorithms such as back-propagation, the learning rate decays as inputs are presented. This ensures stability of the system in the end. If this is not done, then the network continually oscillates, which is not desirable. The problem is that if new inputs must be added to the distribution, then learning must be performed all over again, on the previous input distribution augmented with the new inputs. It is not clear at all how the entire dataset, and not its internal representation as synaptic weights, could be stored by an animal. In addition, learning is too slow and requires two many examples to account for psychologically plausible learning in recognition tasks.

One other major aspect of supervised learning could be perceived as a problem for the robot pet project: the fact that the correct classification must be known in advance. But this is not in itself the essential issue, because it depends on the task and the interpretation of the coding involved. When the task is, for instance, to recognize images of human faces, the problem exists and cannot be solved inside the robot pet: it cannot know on its own if its classification is correct or not. On the other hand, when the problem is to predict the future state of the sensory-motor apparatus, then the correct response can be known by the pet robot (Tani et al., 1997). It can then be used to alter behavior or classify it. The two other problems mentioned nevertheless remain, in particular problem 2.

The main advantage of this technique is that it requires only very simple and scarce intervention by the designer: in the form of a critic who can say if the actions taken by the robot lead to the result or not. The interest in this paradigm is enhanced by the success of Q-learning (Watkins and Dayan, 1992) which ensures the convergence of learning to an optimal policy.

The two major problems identified by researchers in that area are: a) the difficulty of devising a reinforcement function for complex tasks and b) the difficulty of assigning every sensory scene and every action to a predefined state: the so-called "perceptual aliasing problem" (Whitehead and Ballard, 1990).

The methods for coping with these two problems usually amount to decomposing the complex task into simpler ones that are learned independently and following a predefined order (Mahadevan and Connell, 1993) and then to use a segmentation of the sensory data using predefined primitives at a higher level than the raw data. (Asada et al., 1996) This solution usually yields good results for learning how to solve certain tasks that are hard to model, but in the present state it poses certain problems for a pet robot:

- Learning is off-line. The robots must be put back in the same initial conditions several times by the user. This is not a realistic paradigm for a pet robot. These systems should be augmented with additional behavior control procedures that would either force the robot to get back on its own to the desired starting position, or detect the desired starting position when it is encountered, at any moment.
- The decision to stop learning must be taken by the user. Instead, there should be a supervision mechanism internal to the robot for taking this kind of decision.

The Silas project at the MIT Media lab is aiming at designing a virtual dog in a virtual environment. (Blumberg et al., 1996) propose an architecture based on a combination of reinforcement learning and action selection. In that respect, it fulfills most of the demands of psychologically plausible learning. They propose a complex set of appetitive behaviors regulated by internal variables, and learn in which conditions to engage in certain behaviors. The learning paradigm is Temporal Difference (TD) reinforcement learning (Sutton and Barto, 1990) using a virtual trainer, which accounts for learning associations between contiguous events and expected changes of motivational states which play the role of reinforcement signals.

It thus seems that this type of approach is very promising for use in a real artificial pet. However, two problems remain to be solved:

1. The link to the real world must be provided by real sensors. This supposes the implementation of robust primitive extractors, which are not needed in a virtual environment where objects can easily be labeled a priori.
2. The TD learning algorithm, based on slow modifications, does not provide hints about when to stop learning, that is, when to stabilize behavior. This is necessary to learn increasingly complex behaviors.

Unsupervised learning appears to correspond best to the requirements of a pet robot, since it does not rely on the intervention of an intelligent external entity. A fixed global criterion is used, and the data are classified accordingly. Several approaches have been proposed, the main being Kohonen maps (Kohonen, 1982) and ART networks (Carpenter and Grossberg, 1987). Kohonen maps are actually fraught with most of the problems of supervised learning: learning is off-line and does not accommodate new data, and it is slow. On the other hand, ART networks learn incrementally in one step, as the patterns are encountered by the network. The stability/plasticity dilemma is solved in these networks by a single parameter which control learning. The choice of the value of this parameter has to be done outside the network, but in general, the ART framework fits perfectly the requirements of psychological plausibility.

The current research project most relevant to psychologically plausible learning in an artificial agent is the artificial infant Neo project. In (Cohen et al., 1996) an artificial human infant is simulated, and it learns progressively, in a really autonomous fashion, to form concepts grounded in its activities. Learning is based on a kind of statistical Hebbian mechanism that measures cooccurences between percepts and actions. They make a convincing case that useful knowledge can be built in that fashion, starting from a limited set of sensory and motor primitives, and without the need for explicit volition. As we argued earlier, these primitives are necessary for the application they propose. They can be assimilated to simple concepts in that they require invariant representation out of realistic sensors. For instance, they use the recognition of rattle-shaped objects or put-into-mouth movements. If this system was to be implemented in a connectionist architecture, it would require an a priori discrimination of various layers of processing, each coding different sensory properties, in the form of *predefined* primitives: for instance, the visual recognition is divided between color primitives and shape primitives. The problem of building the appropriate filters for such representation needs to be addressed in order to transfer a similar learning scheme to a real robot.

The Neo architecture allows the construction of a two-level hierarchy of concepts and associations, based on general principles of co-occurrence and co-variation. The authors do not provide the means to go beyond this level of abstraction, but they reckon that higher-levels can in principle be obtained using metaphors of the situations observed in the lower levels, as proposed by (Lakoff and Johnson, 1980).

In summary, it appears that the Neo project proposes a very interesting set of ideas for the incremental construction of intelligence. It needs to be augmented with the mechanisms for extracting the desired important primitives out of raw sensor data, and for using the outcome of these early stages of learning for purposive behavior.

3.3 Computer vision

(Biederman, 1987) has proposed a theory of object recognition based on mid-level primitives called "geons" which are a subset of generalized cylinders. In the Recognition By Components (RBC) theory, objects are recognized -and coded in our brains-as spatial arrangements of such mid-level primitives, taking into account qualitative properties as straight or curved or elongated. Objects that can be represented in this fashion are easily recognized, and the flashed image of such an object primes the recognition of the image of the same object seen from a different viewpoint, while pixel correlations are very low. It also appears that three-dimensional properties of such objects, such as their depth profile, are known in the presentation of a single two-dimensional image, while it is not the case with other objects.

This theory is subject to important controversy in the computer vision community, owing mostly to the great difficulty of extracting such primitives out of real images with an artificial system. Nevertheless, we believe that RBC has many merits for the purposes of designing an artificial pet. The first is that most objects can actually represented -at least by humans -in this fashion. Therefore an artificial pet endowed with the capacity to recognize geons and their spatial arrangements would have the capacity to recognize the same objects as ourselves. Second, we have argued earlier that such high-level primitives are necessary for accounting for the speed with which animals learn to recognize objects and situations. We thus believe that such efforts as (Hummel and Biederman, 1992) to design architectures for such purposes, should be pursued and integrated in animats.

4. Elements of a solution for a psychologically plausible artificial pet

The analysis of the previous section points to two major elements of interacting with the environment psychologically plausible conditions:

1. segmentation of the sensory-motor scenes into invariant representation must be done autonomously before associative learning can be applied to its results.
2. definition and choice of the present task and choice of the inputs to be learned belong to the system, without possible external intervention.

We have seen that segmentation can in principle be performed using mid-level primitives. In the vision field, these primitives should give as much structural information about the objects and the environment as possible. This is the subject of investigation of numerous model-based vision researchers, using formal geometrical models of the objects to be recognized. The advent of fast and cheap computers has recently made it possible to use color as well as grey level images.

The ability to segment appropriately the image into objects can be enhanced by using movement information. An animal or a robot evolving in a natural fashion encounters the world through flows of images. Objects being continuous in space belong to the same region, and the relative displacement that occurs when one observes an object belongs to a certain range depending on the distance to the object and its own spatial extent. It seems that very young babies do not really see at a distance of more than one meter, and that in their first months, they saccade on various feature points of an image. (Gaussier, 1992) has devised a vision system based on this principle. Although it gives many of the invariance properties expected from a psychologically plausible system, it can only learn views, and is not equipped with a means to separate the object to be recognized from the background. Optical flow computations, coupled with mid-level representation of the images could in principle overcome this problem.

The bootstrap for cognition can be provided by a sensor responding to a specific simple signal, such as an olfaction sensor. This sense seems to be playing a major role in newborns who can be made quiet by the presence of a cloth impregnated with the mother's breast odor (Cyrulnik, 1980). We can thus imagine a triggering process in which the recording of the main odor takes place when there is something in the animal's mouth, during the first period of life. The presence of this odor above a certain level can then be associated with the visual images that will serve to form the caretaker visual representation. In babies the segmentation problem is facilitated by the fact that visual acuity is tuned to the range of distances between the babies eye when he is being nursed and the mother's face.

As regards learning per se, we have insisted that the task that an animal seems to be solving at any moment is chosen in an autonomous fashion. This implies that there is a kind of selector in the cognitive system that will trigger different feature grouping functions and associations in function of the present task. This trigger can be external as in the case of reflex behaviors, or internal as in the case of game playing: the decision to play combines the opportunity of the situation with the desire to play. The elements for playing can be used for something else: babies often play with their food. In a more general vein, it seems likely that animals build and use a cognitive map of their environment. The same place can be used for different purposes, notably because different things can occupy the same space at different moments.

In most adaptive connectionist systems proposed for controlling animats, two levels are used: one for sensory situations and one for motor actions, whose nodes are associated through the learning procedure. This type of architectures can only account for one task at a time. As a psychologically plausible model of learning requires at least the ability to solve several tasks, using the same set of sensors, an additional control mechanism must be added. Its role is to operate a discrimination between the sensory-motor associations in function of the task to be solved, which must be chosen by another decision mechanism inside the brain. This can easily be done by adding a layer in the system coding the sensory situations. In this additional upper layer, the desired states are coded, and their state of activation can either enable or prevent the activation of the recognition nodes in the inferior layer. The activation of nodes in the upper layer is controlled by the decision mechanism, that is, by the task at hand. When a task is being performed and/or learned, a limited number of recognition nodes in the sensory lower layer can be associated to sensory situations and also to the motor actions that lead to success with regard to the task. When this task is over, the nodes in the upper sensory layer can be inhibited, which thus prevents the activation of the lower layer nodes. This allows a discrimination between nodes, not only in function of the sensory situations that are encountered and which they code, but also in function of the task that is being performed.

In (Zrehen and Gaussier, 1997), an on-line learning connectionist network of this kind was proposed, which allows the association of different places with the satisfaction of different internal drives such as hunger and thirst. This work, however, assumed a fixed set of drives and a priori knowledge of the visual targets that satisfy each of these drives. In order to enhance the psychological plausibility of such a system, the recognition of certain objects as possible targets should be made adaptive, even if it starts from hard-wired recognition mechanisms, that is from reflexes to see as targets a pre-defined set of objects that can be recognized using the available visual primitives. Furthermore, the motivational system should also be able to grow with time.

An important role played by the designer of a robot control architecture is in the coding of reinforcement signal. We have pointed out earlier that a major problem to avoid is to end up making too many associations. A principle that could probably help is John Stuart Mill's law of *concomitant variation*: reinforcement should occur only when there is a *variation of activity* occurring in two or more different areas of the network. That principle can also be used as the basis for the focus of attention, and can ground the notion of causality.

One major remaining problem in the design of a complex psychologically plausible animat concerns the construction of more complex drives, out of a basic set and of experience of interactions with the world. Following Freud's pleasure principle (Freud, 1961), we can admit that the general psychological goal of humans, if not of all animals, is to maximize pleasure and minimize unpleasure. This implies the existence of a processing system measuring pleasure and of another one measuring unpleasure. In the absence of a computational model of qualia we can for the moment equate the activation of one of these two centers with the feeling that humans associate with pleasure or unpleasure. It is also easy to trigger an observable emotional expression of the pleasure state in a robot in response to activation of these centers, for instance by waiving a tail.

The existence of innate sources of pleasure or unpleasure can be modeled by afferent links from sensors measuring such things as the variation of glucose in the blood -a variation in the state of energy for a robot -or the temperature. But in order to account for the possibility of complex forms of pleasure, such as fetching a stick for a dog, we also have to design additional afferent links areas of the cognitive system which code complex sensory-motor situations. The resulting associations can be obtained by a mechanism that learns sensory-motor sequences, and to backtrack these sequences in order to attain a desired situation. Such a mechanism could be applied to a scenario of the following kind for an artificial pet:

The first time it fetches an object thrown by its master, it receives a pat on the head. The sequence

master with object in his hand → I go fetch it →
I receive a pat on the head → feel good

can be learned as it is performed. At a later time, recognizing a similar situation, the desire to feel good through a pat on the head triggers the fetching sequence. This is obviously a very complex problem if it is to be solved in a psychologically plausible manner, since it implies the following list of subproblems to be solved:

- recognize the scene of the master holding an object
- represent the image of the object in an invariant fashion
- set the image of the object as the present target
- deduce from the trajectory where it has fallen approximately. This supposes that the dog already has a notion that objects that are thrown eventually fall somewhere
- get to that place
- set a local search procedure for finding the object
- bring it back to the master

This is but one of the behaviors that one could expect from a realistic pet robot, but it poses interesting challenges in terms of integration.

5. Conclusion

In this paper, we have tried to define the meaning of tackling the world and learning in a psychologically plausible conditions, using the example of an imaginary pet robot. Taking the idea of autonomy to its limits imposes important constraints on what can actually be learned in plausible conditions. We have seen that it looks necessary to dispose of powerful segmentation mechanisms that make potentially meaningful situations/events salient in the robot's sensory field, and also allow immediate invariant recognition. These considerations lead suggest the necessity of using mid-level primitives, akin to Biederman's geons ((Biederman, 1987). However, it might not be possible to rely upon learning during the course of a life-time in order to account for this mid/high-level primitive extraction.

We have also seen that learning must be controlled entirely in an internal fashion, which imposes the choice of adaptive control architectures that globally solve the stability-plasticity dilemma: there should be an internal regulation mechanisms that decides when to learn associations, in order to avoid learning too many of them, which would result in saturating memory. In addition, learning must take place *while* the animal/robot's behavior is controlled by the very architecture undergoing adaptive changes, which limits the type of data accessible to the processing structures. This also imposes the choice of learning algorithms that can cope with such on-line conditions.

The choice of behaviors in which a robot must engage must also be done internally. The different interpretations that can be associated to a given situation must remain available and be associated, by the robot itself, to the appropriate goals.

In summary, it appears that in order to design a robot that will give the illusion of having a psychology, one needs to strike a compromise between the orthodox animat approach with a mid-level primitive extraction and recognition mechanism usually associated with model-based AI. one needs an on-line learning architecture that allows the association of the right *primitives*. When the choice of primitives is appropriate, learning algorithms become very simple.

References

Asada, M., Noda, S., Tawaratsumida, S., and Hosoda, K. (1996). Purposive behavior acquisition for a real robot by vision-based reinforcement learning. *Machine Learning*, 23(2-3):279–.

Biederman, I. (1987). Recognition-by-components: A theory of human image understanding. *Psychological Review*, 94(2):115–147.

Blumberg, B., Todd, P., and Maes, P. (1996). No bad dogs: ethological lessons for learning in hamsterdam. In (Maes et al., 1996).

Carpenter, G. and Grossberg, S. (1987). A massively parallel architecture for a self-organizing neural pattern recognition machine. *CGVIP*, 37:131–143.

Cohen, P. R., Atkin, M. S., and Oates, T. (1996). Neo: Learning conceptual knowledge by sensorimotor interaction with an environment. In Johnson, W. L., editor, *International Conference on Autonomous Agents*, New-York. ACM Press.

Cyrulnik, B. (1980). *Paroles d'homme et memoire de singe.* P.U.F., Paris.

Freud, S. (1961). *Beyond the pleasure principle.* Norton, New-York.

Gaussier, P. (1992). *Simulation d'un systeme visuel comprenant plusieurs aires corticales: Application a l'analyse de scenes.* Electrical engineering, Universite de Paris-Sud.

Harvey, I. and Husbands, P., editors (1997). *European Conference on Artificial Life*, Cambridge. MIT Press.

Hummel, J. E. and Biederman, I. (1992). Dynamic binding in a neural network for shape recognition. *Psychological Review*, 99(3):480–517.

Kohonen, T. (1982). Self-organized formation of topologically correct feature maps. *Biological Cybernetics*, 43:59–69.

Lakoff, G. and Johnson, M. (1980). *Metaphors we live by.* The University of Chicago Press, Chicago.

Maes, P., Mataric, M., Meyer, J.-A., Pollack, J., and Wilson, S., editors (1996). *From Animals to Animats*, Cambridge. MIT Press.

Mahadevan, S. and Connell, J. (1993). *Rapid Task Learning for Real Robot*, volume 43, chapter 5, pages 59–69. Kluwer.

Maturana, H. and Varela, F. J. (1987). *The Tree of Knowledge.* Shambhala, Cambridge.

Pfeifer, R. (1996). Building "fungus eaters": Design principles of autonomous agents. In (Maes et al., 1996).

Scheier, C. and Lambrinos, D. (1996). Learning visuo-haptic categories through exploration. In (Maes et al., 1996).

Sutton, R. and Barto, A. (1990). *Time-Derivative Models of Pavlovian Reinforcement.* MIT Press, Cambridge.

Tani, J., Yamamoto, J., and Nishi, H. (1997). Dynamical interactions between learning, visual attention, and behavior: An experiment with a vision-based mobile robot. In (Harvey and Husbands, 1997), pages 309–318.

Vygotsky, L. S. (1978). *Mind In Society.* Harvard University Press, Cambridge.

Wasserman, E., Gagliardi, J., Karl, Kirpatrick-Steger, Astley, S., and Biederman, I. (1996). The pigeon's recognition of drawings of depth-rotated stimuli. *Journal of Experimental Psychology: Animal Behavior Processes*, 22:205–221.

Watkins, C. and Dayan, P. (1992). Technical note: Q-learning. *Machine Learning*, 8:279–292.

Whitehead, S. and Ballard, D. (1990). Active perception and reinforcement learning. volume 43, pages 129–188.

Zrehen, S. and Gaussier, P. (1997). Grounding symbols for localization using motivations. In (Harvey and Husbands, 1997).

Robots and Representations*

Mark H. Bickhard
Cognitive Science, Psychology, Philosophy
17 Memorial Drive East
Lehigh University
Bethlehem, PA 18015 USA
610-758-3633
610-758-6277 fax
mhb0@lehigh.edu
http://www.lehigh.edu/~mhb0/mhb0.html

Abstract

The design of complex interactive robots inherently yields a form of representation — an interactive form. Interactive representation is, arguably, the foundational form of representation from which all others are derived. It constitutes the emergence of representational truth value *for the system itself*, a criterion not addressed in current literature.

There is a form of representation that arises naturally in the design of complex interactive systems — robots. This form arguably constitutes an emergence of the fundamental form of representation, out of which increasingly complex forms are constructed and derived. Furthermore, this form of representation naturally satisfies an essential meta-epistemological criterion for original representation: system detectable truth value. No alternative approach to representation in the current literature addresses this criterion. Recognizing and exploiting the emergence of this form of representation in robotics and dynamic systems is a rich frontier for exploration.

In standard artificial intelligence and cognitive science models, inputs are received and processed, and, perhaps, outputs emitted. The critical consideration that arises in robot design is the possibility of a closure of this sequence of process such that robot outputs influence subsequent inputs via the environment, and, therefore, influence subsequent internal states and processes in the robot. That is, the critical consideration is the closure of input, processing, and output to include full *inter*action, not just action.

This simple closure introduces several important possibilities. In particular, possible internal states that might be consequent on some action or course of action can be functionally indicated in the robot. Because such possible consequent states in the robot will depend, in part, on the environment, those states, or some one of those states, may or may not actually occur — the environment may or may not yield the appropriate input(s) in response to the output(s) to induce those indicated states in the robot. If none of those indicated states are entered by the robot, then the indications are false, and are *falsified for the robot*. The error in such indications is detectable by and for the system itself.

In effect, to indicate such internal states as consequent on particular interactions on the part of the robot is to implicitly predicate of that environment whatever properties are sufficient to support those indications. It is to anticipate that the environment will in fact respond as indicated, if the interaction is engaged in. Some environments will possess a sufficiency of those properties, and will yield one of the indicated states, while other environments will not possess such properties, and will not yield any of the indicated states. For those environments that do *not* yield an indicated state, to set up such an indication is to set up an implicit predication, an anticipation, that is false, and potentially falsifiable by the system (Bickhard, 1993, in press; Bickhard & Terveen, 1995).

The possibility of error, and especially of system detectable error, is a fundamental meta-epistemological criterion for representation. Whatever representation is, it must be capable of some sort of truth value. Conversely, something *is* representation *for a particular system* only if it is capable of some sort of truth value *for that system*. This is critical because many states and conditions and phenomena are representational — can have truth value — but only for some user or designer or observer outside of the system itself, not for the system itself (Bickhard, 1993; Bickhard & Terveen, 1995).

Moderately complex robots, then, naturally involve a form of representation that is representational for the robot, not just for an observer or analyst or designer or user of the robot. This claim generates five questions: 1) How can notions such as 'indication' in the above discussion be made good in a functional manner in a robot, without committing a logical circularity by presupposing the very representationality that is allegedly being modeled? 2) Why would it be useful for a robot to have such representations of interactive potentialities? 3) How could such a notion of representation possibly be adequate to "normal" representational and cognitive phenomena such as representation of objects; representation of abstractions, such as numbers; language; perception; rationality; and so on? I will only outline an answer to the first of these questions, referring others to other sources. 4) On what basis would a robot set up such indications? And 5) How does this model of representation relate to contemporary research in artificial intelligence, cognitive science, connectionism, and robotics? My responses to this question too will, obviously, be abbreviated.

The Functional Story

First, I need to address the question of how interactive representation could be implemented without presupposing representation. All that is needed are some architectural principles adequate to the model that are themselves strictly functional — not representational. This is, in fact, rather simple. The indicated internal outcome states for an interaction function like final states in an automaton recognizer, but for an automaton that emits outputs to an interactive environment (Bickhard, 1980a). The indication of such states can be implemented with pointers — a pointer, say, to some location that will contain a "1" in the state being indicated. This is certainly not the only architecture that will implement the notions required, but it does suffice.

To indicate the interaction itself, upon which the indications of final states are based, requires only a pointer to the subsystem — perhaps the subroutine or interactive recognizer — that would engage in those interactions. So, a pointer to a subsystem together with a pointer or pointers to final states associated with that subsystem suffices for the implicit predication of interactive representation, but none of these pointers themselves are or require representation. Insofar as there is representation here, it is genuinely emergent in the architectural organization.

The Usefulness of Interactive Representations

Choice. Why would it be useful for a robot to have such indications? For two reasons: First, if there are multiple interactions possible in a particular environment, the indicated internal outcomes of those interactions can be used in selecting which interaction to actually engage in (Bickhard, 1997b). A frog seeing a fly might set up indications of the possibility of tongue-flicking-and-eating, while a frog seeing a shadow of a hawk might set up indications of the possibility of jumping in the water. A frog seeing *both* needs some way to decide, and internal outcome indications provide a basis for such decision — e.g., select the interaction with the indicated outcomes that have the highest priority relative to current set-points. (Note that if the relevant outcomes are presumed to be *represented*, rather than indicated — as must be the case if the outcomes are considered to be external outcomes in the environment — then there is a circularity involved in using such notions to model representation.)

Error. The second reason why such indications might be useful is that they create the possibility of error, and — most importantly — the possibility of the detection of error by the system. Detection of error, in turn, can be useful for guiding heuristics and strategies of interaction, and for evoking and guiding learning processes. Any general form of learning, in fact, *requires* such system detection of error (Bickhard & Terveen, 1995). In slogan form: Only anticipations can be falsified; therefore only anticipations can be learned.

On the Adequacy of Interactive Representation

Interactive, or robotic, representation might seem adequate for the kinds of interactive properties that interactive indications will implicitly predicate of the environment, but there are many other things to be represented that do not prima facie look like interactive properties. To make good on claims of the adequacy of interactive representation as a general form of representation would require a programmatic treatment of many or most of these representational phenomena. There isn't space to even begin that explication here (see, for example, Bickhard, 1980a, 1980b, 1992, 1993, in press, forthcoming; Bickhard & Campbell, 1992; Bickhard & Richie, 1983; Bickhard & Terveen, 1995; Campbell & Bickhard, 1986, 1992), but I will outline an approach to the interactive representation of physical objects in order to indicate that this is at least a plausible programme.

Complexities of Interactive Indications. Before addressing objects per se, I need to outline some forms of complexity that can be involved in interactive indications. The first is that there may be multiple interactive possibilities indicated at a given time. The second is that interactive indications can be conditionalized on each other: interaction **A** with possible outcome **Q** might be indicated, and, if **A** is engaged in and **Q** is in fact obtained, then interaction **B** with possible outcome **R** becomes possible. There are other kinds of complications possible, but branchings and conditionalized iterations of interactive indications will suffice for briefly addressing the problem of object representation.

Webs. Branchings and conditional iterations yield the possibility of interactive indications forming potentially complex webs or nets of indications. In effect, the whole of such a web is indicated as currently possible, but actually reaching some parts of the web will be contingent on perhaps many intermediate interactions and outcomes.

Objects. Some sub-networks in such a complex web may have two critical properties: 1) A subnet may be closed in the sense that, if any part of it is reachable — possible — then all parts of it are. That is, all points (possible states) in such a web indicate the potentiality of all other points in the web, perhaps with necessary intermediate interactions. 2) Such a closed subnet may be invariant under some interesting class of possible interactions, in the sense that many possible interactions will leave the closed subnet invariant in internal structure and still reachable.

A toy block, for example, will offer many possible interactions to a child. The block can be visually scanned, dropped, thrown, chewed, manipulated, and so on. Furthermore, any of such possibilities will indicate the availability of the rest of them: a particular visual scan becomes momentarily unavailable if the block is turned over, but is recoverable if the block is turned back over to its original position. Still further, this entire organization of interactive possibilities remains invariant for many interactions such as manipulations, throwings, hidings, storing in the toy box, locomotions on the part of the child, and so on. It does not remain invariant, however, under burning or crushing.

The point is that a block, for an infant, is no more than such closed and invariant organizations of interactive possibilities. The infant knows nothing of substances or molecules — those all come later in theorizing about such things as blocks. This, I propose, is the general manner in which interactive representation can handle object representations. This is a Piagetian sort of account of object representations (Piaget, 1954), and I suggest similarly Piagetian flavored approaches to other representational issues (though I do not endorse all of Piaget's model: Bickhard, 1988; Bickhard & Campbell, 1989). My basic point in this paper, however, is simply that interactive representation is not immediately blocked by the multitudinous forms of representation that must ultimately be accounted for.

How to Set Up Interactive Indications

Interactive indications might be useful for a robot, but on what basis could they be set up? The general answer is simple: on the basis of previous outcomes of previous interactions. Previous outcomes differentiate environments into those that yield that outcome and those that don't. Conditionalized indications of interactive potentialities are, in effect, conditionalized on those prior differentiations. Any environment that will yield state **Q** from interaction **A** will also — according to the conditional — be an environment in which interaction **B** is possible, with possible consequent state **R**. So, if **Q** is obtained, then direct (not conditionalized) pointers to **B** and **R** should be set up. As with the rest of the model, many complexities can arise here, but this general point suffices to show that the problem of how to set up interactive representations is not itself an aporia.

Relationships to Contemporary Work

Symbol Systems. The symbol system hypothesis involves inputs and subsequent processing (Newell, 1980), but does not involve any necessary notion of closure from outputs back to inputs. It does not involve any recognition of the necessity for *inter*action. Correspondingly, it does not solve the problem of system detectable error. Error can be defined from a user or designer perspective, and it can be useful to do so, but this is still error, therefore representation, for the user or designer, not for the system itself.

In fact, the symbol system hypothesis is one of a large class of approaches to representation that attempt to construe representation in terms of the processing of inputs, with no recognition of the essential importance of full interaction (Bickhard, 1993; Bickhard & Terveen, 1995; Coffa, 1991; Dretske, 1981, 1988; Fodor, 1987, 1990; Hanson, 1990; Loewer & Rey, 1991; Millikan, 1984, 1993; Palmer, 1978; Pylyshyn, 1984; Smith, 1987) Inputs are, in these models, supposed to set up correspondences — usually some special kind of correspondences — between internal states of the system and particular properties or entities in the environment, and *are supposed to represent those properties or entities by virtue of those correspondences.* These approaches have difficulty accounting for error at all, and do not even address the problem of system detectable error. The difficulty in accounting for error arises because, if the correspondence exists, then the representation exists and it is correct, while, if the correspondence does not exist, then the representation does not exist, and so cannot be *in*correct.

There is a great deal of effort being currently devoted to attempting to account for error in some such model (Dretske, 1988; Fodor, 1990; Hanson, 1990; Loewer & Rey, 1991; Millikan, 1993). I will not address the vicissitudes of these attempts (though I do not think that any of them succeed) because none of them even address the stronger criterion of system detectable error. On their own terms, error, if definable at all, is definable from the perspective of some observer or user or designer or analyzer of the system in question, not for the system itself. There are, in addition, many other problematics of this general approach to representation, none of which arise as problematic for the interactive approach (Bickhard, 1980b, 1993; Bickhard & Terveen, 1995).

Connectionism. Interactive representation is a direct alternative to standard symbol manipulation approaches to representation. What about connectionism? The key to the answer to this question is to note that a connectionist net serves to differentiate instances of one input pattern class from instances of some other class of input patterns (Bickhard & Terveen, 1995; Churchland, 1989; McClelland & Rumelhart, 1986; Rumelhart, 1989; Rumelhart & McClelland, 1986; Smolensky, 1988; Waltz & Feldman, 1988). This is precisely what a transducer is supposed to do in standard information processing models (Bickhard & Richie, 1983; Fodor & Pylyshyn, 1981; Loewer & Rey, 1991). A connectionist net can be trained to differentiate new instances of new patterns, while a transducer has to be designed or evolved specifically for its differentiation task, but, beyond the difference between design versus training origins, the tasks that are performed are the same sorts of differentiation tasks.

Connectionism and information processing approaches agree on the next step in the modeling: the differentiations accomplished by transducers or by connectionist nets are taken to be representations of that which has been differentiated. It is this step that immediately encounters a host of problems, not the least of which is how such "representations" could possibly be in error, and how the system itself could detect such error.

In the interactive model, in contrast, such differentiations are useful in order to serve as the basis for setting up interactive indications, and there is no requirement that they be alleged to be representational in order that they serve that function. Sensory input processing, in other words — or connectionist net processing — does not directly generate representations of what is on the environmental end of those inputs, but, instead, sets up indications of what is interactively possible in the environments differentiated by that input processing. Input processing occurs — and serves a necessary function — in both cases, but the interactive model does not require or assume that the system has any representation of what produces the inputs. (Humans do have such representations of the sources of [some of] our inputs, and that must be accounted for, but flatworms don't, and it's not clear that frogs do — primitive forms of representation do not necessarily involve such representations.)

Robotics. Interactive representation cannot exist in standard symbol manipulation or connectionist architectures — there are no closed interactions involved in their purported representations — but it arises naturally in robot design. How does this compare with contemporary robotics literature? The problematics of standard conceptions of representation are clearly recognized (Beer, 1990, 1995; Bickhard, 1996; Brooks, 1991a, 1991b, 1991c; Prem, 1995), and robotic and autonomous agent design problems naturally lead to the sorts of indications that constitute interactive representation — whether or not they are recognized as representational (Kuipers, 1988; Kuipers & Byun, 1991; Maes, 1990; Nehmzow & Smithers, 1991, 1992; Stein, 1994). Recognizing the emergence of representation and representational truth value in those designs, with all the concomitant implications for related domains of cognition and robotic functioning, is a rich frontier for robotics and dynamic systems (Bickhard, 1997a, 1997b, forthcoming; Bickhard & Campbell, 1996a, 1996b; Bickhard & Terveen, 1995; Brooks, 1994; Cherian & Troxell, 1995a, 1995b; Hooker, 1995; Hooker, Penfold, & Evans, 1992; Hooker & Christensen, in preparation).

In particular, the problem of action selection in robots and artificial agents can be solved in the general case only via some sort of selection of actions on the basis of their anticipated consequences. That is, the problem of action selection forces the emergence of interactive representation (Bickhard, 1997b). Once recognized, interactive representation is the framework within which complex representation can be designed and learned, and upon which higher order cognition can be constructed, including rationality and language (Bickhard, 1992, forthcoming, in press; Bickhard & Campbell, 1996a, 1996b; Bickhard & Terveen, 1995). Interactive representation opens the door to genuine higher cognition for artificial agents — just as it did for natural agents (Bickhard, 1992, in press, forthcoming; Bickhard & Campbell, 1992; Bickhard & Terveen, 1995).

Pragmatism. Since the ancient Greeks, mind has been studied primarily as a passive consciousness, processing inputs. The shift to conceiving of mind as an action system is fundamentally due to the philosopher C. S. Peirce (Joas, 1993). Peirce's pragmatism has influenced many people, in spite of its relatively short history, and this literature contains multiple parallels and partial convergences with the interactive model.

Most important are Piaget (e.g., 1954), the later Wittgenstein (1958), Heidegger (1962), and Merleau-Ponty (1962). An action framework, however, does not guarantee viability of the model in detail. The most common subsequent error is a vestigial, sometimes subtle and only implicit, commitment to classical correspondence notions of representation. This can be manifest in arguments *for* such classical notions, even within a dynamic or robotic framework (e.g., Clark, 1997; Clark & Toribio, 1995), or in arguing *against* representation where representation is construed *only* in this classical form (e.g., Brooks, 1991a; Port & van Gelder, 1995), or in arguing for a version of the *idealism* that correspondence conceptions of representation have so often generated (e.g., Maturana & Varela, 1980, 1987; Varela, Thompson, & Rosch, 1991). These and many other positions are discussed in Bickhard & Terveen (1995).

The shift to an action framework is of fundamental importance, and interactivism shares this move with a growing minority of positions in the literature. The interactive model, however, contributes some essential additions and corrections of its own. System detectable error, for example, is not addressed elsewhere, but is, arguably, essential for an adequate naturalization of representation, whether in organisms or machines.

Conclusion

Interactive representation arises naturally in interactive systems. This is evident for robots, in particular, but it is also apparent in simple interactive organisms. Insofar as interactive representation serves as the fundamental form of representation out of which more complex forms are constructed and derived, this serves to connect the most complex human representational abilities with primitive, emergent, evolutionary precursors.

Interactive representation naturally satisfies a critical meta-epistemological criterion — system detectable error — that must be satisfied by any model of original representation, and that is not satisfied, or even addressed, by standard conceptions of representation. As such, interactive representation offers a promising approach to representation in particular and cognitive problems in general.

References

Beer, R. D. (1990). *Intelligence as Adaptive Behavior.* Academic.

Beer, R. D. (1995). Computational and Dynamical Languages for Autonomous Agents. In R. Port, T. J. van Gelder (Eds.) *Mind as Motion: Dynamics, Behavior, and Cognition.* MIT.

Bickhard, M. H. (1980a). A Model of Developmental and Psychological Processes. *Genetic Psychology Monographs, 102*, 61-116.

Bickhard, M. H. (1980b). *Cognition, Convention, and Communication.* New York: Praeger.

Bickhard, M. H. (1988). Piaget on Variation and Selection Models: Structuralism, Logical Necessity, and Interactivism *Human Development, 31*, 274-312.

Bickhard, M. H. (1992). How Does the Environment Affect the Person? In L. T. Winegar, J. Valsiner (Eds.) *Children's Development within Social Contexts: Metatheory and Theory.* (63-92). Erlbaum.

Bickhard, M. H. (1993). Representational Content in Humans and Machines. *Journal of Experimental and Theoretical Artificial Intelligence, 5*, 285-333.

Bickhard, M. H. (1996). Troubles with Computationalism. In W. O'Donohue, R. F. Kitchener (Eds.) *The Philosophy of Psychology.* (173-183). London: Sage.

Bickhard, M. H. (1997a). Is Cognition an Autonomous Subsystem? In S. O'Nuallain, P. McKevitt, E. MacAogain (Eds.). *Two Sciences of Mind.* (115-131). Amsterdam: John Benjamins.

Bickhard, M. H. (1997b). Emergence of Representation in Autonomous Agents. *Cybernetics and Systems: Special Issue on Epistemological Aspects of Embodied Artificial Intelligence, 28*(6), 489-498.

Bickhard, M. H. (forthcoming). Critical Principles: On the Negative Side of Rationality. In Herfel, W., Hooker, C. A. (Eds.) *Beyond Ruling Reason: Non-formal Approaches to Rationality.*

Bickhard, M. H. (in press). Levels of Representationality. *Journal of Experimental and Theoretical Artificial Intelligence.*

Bickhard, M. H., Campbell, R. L. (1989). Interactivism and Genetic Epistemology. *Archives de Psychologie, 57*(221), 99-121.

Bickhard, M. H., Campbell, R. L. (1992). Some Foundational Questions Concerning Language Studies: With a Focus on Categorial Grammars and Model Theoretic Possible Worlds Semantics. *Journal of Pragmatics, 17(5/6),* 401-433.

Bickhard, M. H., Campbell, R. L. (1996a). Topologies of Learning and Development. *New Ideas in Psychology, 14*(2), 111-156.

Bickhard, M. H., Campbell, R. L. (1996b). Developmental Aspects of Expertise: Rationality and Generalization. *Journal of Experimental and Theoretical Artificial Intelligence, 8*(3/4), 399-417.

Bickhard, M. H., Richie, D. M. (1983). *On the Nature of Representation: A Case Study of James J. Gibson's Theory of Perception.* New York: Praeger.

Bickhard, M. H., Terveen, L. (1995). *Foundational Issues in Artificial Intelligence and Cognitive Science — Impasse and Solution.* Amsterdam: Elsevier Scientific.

Brooks, R. A. (1991a). Intelligence without Representation. *Artificial Intelligence, 47*(1-3), 139-159.

Brooks, R. A. (1991b). New Approaches to Robotics. *Science, 253*(5025), 1227-1232.

Brooks, R. A. (1991c). How to Build Complete Creatures Rather than Isolated Cognitive Simulators. In K. VanLehn (Ed.), *Architectures for Intelligence.* (225-239) Erlbaum.

Brooks, R. A. (1994). Session on Building Cognition. Conference on The Role of Dynamics and Representation in Adaptive Behaviour and Cognition. University of the Basque Country, San Sebastian, Spain, December 9, 1994.

Campbell, R. L., Bickhard, M. H. (1986). *Knowing Levels and Developmental Stages.* Basel: Karger.

Campbell, R. L., Bickhard, M. H. (1992). Clearing the Ground: Foundational Questions Once Again. *Journal of Pragmatics, 17(5/6),* 557-602.

Cherian, S., Troxell, W. O. (1995a). Intelligent behavior in machines emerging from a collection of interactive control structures. *Computational Intelligence, 11*(4), 565-592. Blackwell Publishers. Cambridge, Mass. and Oxford, UK.

Cherian, S., Troxell, W. O. (1995b). Interactivism: A Functional Model of Representation for Behavior-Based Systems. In Morán, F., Moreno, A., Merelo, J. J., Chacón, P. *Advances in Artificial Life*: Proceedings of the Third European Conference on Artificial Life, Granada, Spain. (691-703). Berlin: Springer.

Churchland, P. M. (1989). *A Neurocomputational Perspective.* MIT.

Clark, A. (1997). *Being There.* MIT/Bradford.

Clark, A., Toribio, J. (1995). Doing without Representing? *Synthese, 101,* 401-431.

Coffa, J. A. (1991). *The Semantic Tradition from Kant to Carnap.* Cambridge.

Dretske, F. I. (1981). *Knowledge and the Flow of Information.* Cambridge, MA: MIT.

Dretske, F. I. (1988). *Explaining Behavior.* MIT.

Fodor, J. A. (1987). *Psychosemantics.* Cambridge, MA: MIT Press.

Fodor, J. A. (1990). *A Theory of Content and Other Essays.* MIT.

Fodor, J. A., Pylyshyn, Z. (1981). How direct is visual perception?: Some reflections on Gibson's ecological approach. *Cognition, 9,* 139-196.

Hanson, P. P. (1990). *Information, Language, and Cognition.* University of British Columbia Press.

Heidegger, M. (1962). *Being and Time.* New York: Harper & Row.

Hooker, C. A. (1995). *Reason, Regulation, and Realism: Towards a Regulatory Systems Theory of Reason and Evolutionary Epistemology.* SUNY.

Hooker, C. A., Christensen, W. (in preparation). Very Simple Minds.

Hooker, C. A., Penfold, H. B., Evans, R. J. (1992). Towards a Theory of Cognition Under a New Control Paradigm. *Topoi, 11,* 71-88.

Joas, H. (1993). American Pragmatism and German Thought: A History of Misunderstandings. In H. Joas *Pragmatism and Social Theory.* (94-121). University of Chicago Press.

Kuipers, B. J. (1988). The TOUR Model: A Theoretical Definition. From Kuipers, B. J., Levitt, T. Navigation and Mapping in Large-Scale Space. *AI Magazine, 9*(2), 25-43.

Kuipers, B. J., Byun, Y. (1991). A Robot Exploration and Mapping Strategy Based on a Semantic Hierarchy of Spatial Representations. *Robotics and Autonomous Systems, 9,* 47-63.

Loewer, B., Rey, G. (1991). *Meaning in Mind: Fodor and his critics*. Blackwell.

Maes, P. (1990). *Designing Autonomous Agents*. MIT.

Maturana, H. R., Varela, F. J. (1980). *Autopoiesis and Cognition*. Dordrecht, Holland: Reidel.

Maturana, H. R., Varela, F. J. (1987). *The Tree of Knowledge*. Boston: New Science Library.

McClelland, J. L., Rumelhart, D. E. (1986). *Parallel Distributed Processing: Explorations in the Microstructure of Cognition: Vol. 2 Psychological and Biological Models*. Cambridge: MIT.

Merleau-Ponty, M. (1962). *Phenomenology of perception*. New York: Humanities Press.

Millikan, R. (1984). *Language, Thought, and Other Biological Categories*. MIT.

Millikan, R. G. (1993). *White Queen Psychology and Other Essays for Alice*. MIT.

Nehmzow, U., Smithers, T. (1991). Mapbuilding Using Self-Organizing Networks in "Really Useful Robots." In J.-A. Meyer, S. W. Wilson (Eds.) *From Animals to Animats*. (152-159). MIT.

Nehmzow, U., Smithers, T. (1992). Using Motor Actions for Location Recognition. In F. J. Varela, P. Bourgine (Eds.) *Toward A Practice of Autonomous Systems*. (96-104). MIT.

Newell, A. (1980). Physical Symbol Systems. *Cognitive Science, 4*, 135-183.

Palmer, S. E. (1978). Fundamental aspects of cognitive representation. In E. Rosch & B. B. Lloyd (Eds.), *Cognition and categorization*. Hillsdale, NJ: Erlbaum.

Piaget, J. (1954). *The Construction of Reality in the Child*. New York: Basic.

Port, R., van Gelder, T. J. (1995). *Mind as Motion: Dynamics, Behavior, and Cognition*. Cambridge, MA: MIT Press.

Prem, E. (1995). Grounding and the Entailment Structure in Robots and Artificial Life. In Morán, F., Moreno, A., Merelo, J. J., Chacón, P. *Advances in Artificial Life*: Proceedings of the Third European Conference on Artificial Life, Granada, Spain. (39-51). Berlin: Springer.

Pylyshyn, Z. (1984). *Computation and Cognition*. MIT.

Rumelhart, D. E. (1989). The Architecture of Mind: A Connectionist Approach. In M. I. Posner (Ed.) *Foundations of Cognitive Science*. MIT, 133-160.

Rumelhart, D. E., McClelland, J. L. (1986). *Parallel Distributed Processing. Vol. 1: Foundations*. MIT.

Smith, B. C. (1987). *The Correspondence Continuum*. Stanford, CA: Center for the Study of Language and Information, CSLI-87-71.

Smolensky, P. (1988). On the Proper Treatment of Connectionism. *Behavioral and Brain Sciences, 11*, 1-74.

Stein, L. A. (1994). Imagination and Situated Cognition. *Journal of Experimental and Theoretical Artificial Intelligence, 6*, 393-407.

Varela, F. J., Thompson, E., Rosch, E. (1991). *The Embodied Mind*. MIT.

Waltz, D., Feldman, J. A. (1988). *Connectionist Models and Their Implications*. Norwood, NJ: Ablex.

Wittgenstein, L. (1958). *Philosophical Investigations*. New York: Macmillan.

* Thanks are due to the Henry R. Luce Foundation for support during the preparation of this paper.

PERCEPTION AND VISUALLY GUIDED BEHAVIOR

Dynamic Categorization of Explorative Behaviors for Emergence of Stable Sensorimotor Configurations

Luc Berthouze, Yukifumi Shigematsu, Yasuo Kuniyoshi
Electrotechnical Laboratory (ETL), AIST, MITI.
Umezono 1-1-4, Tsukuba, Japan 305.
berthouz@etl.go.jp

Abstract

A prerequisite for autonomous acquisition of novel behaviors would be the ability to evolve internal representation of embodied interaction structures, in other words, sensory-motor categorization. It requires a unified approach to categorical learning of spatio-temporal information. This paper proposes a novel artificial neural architecture for such processing. It is based on a 'spiking neuron' model and a temporal learning rule, and is capable of spatio-temporal association and categorization. The architecture has been applied to an example of visuo-ocular interaction with its surrounding by a binocular active vision system. Its sensory-motor interactions are triggered by a set of a priori reflexes and the neural architecture successfully categorized resulting structure of the interaction. Results from simulation experiments are presented.

1. Introduction

To understand the dynamics of a system engaged in interaction with its surrounding, the observation and categorization of its explorative behaviors is an important pre-requisite. Indeed, like infants' visual behaviors shortly after birth, those behaviors are not necessarily goal-oriented and bodily motion is seldomly constrained (apart from natural constraints such as absence of muscle strength etc...). Reflexes shape/constrain the sensorimotor space to be explored by activating basic body mechanisms (for example, coordinating eye and head motion with VOR reflex (see below)). The use of such information could prove valuable for robots facing an overly complex sensorimotor space (high number of degrees of freedom, highly dynamic environment etc...). Indeed, in a previous work (see (Berthouze and Kuniyoshi, 1998)), we showed that distributed interacting control modules, each embedding a basic innate knowledge about its joint, could lead to the emergence of sensorimotor coordination (emergence of of couplings between each controller). While the emerged coordination proved efficient in realizing the visuomotor tasks such as gaze fixation or saccadic control, its (internal) stability could not be guaranteed. Rapidly changing interaction patterns originate in overall reconfiguration (chaotic-like dynamics) of the control network. This result naturally follows from the complexity (multiplicity of local minima) of the space of possible sensorimotor configuration. While such reset could be perceived as an interesting property since it contributes to the exploration of new sensorimotor space), it is desirable that only stable sensorimotor configuration emerge from the robot's environmental interaction [1]. The hypothesis investigated in this paper is that if it is possible to have the system categorize[2] its sensorimotor couplings during explorative behaviors, then appropriate use of this categorization could guide the adaptive interaction between distributed controllers, eventually bootstrapping the emergence of stable configuration.

Naturally, such categorization must not be designed but should emerge from the system's interaction with the environment. In the following sections, we present the dynamical system approach that we undertook to address this issue. We describe the structure and mechanisms of our novel neural architecture before providing a detailed analysis of simulation results with a 4DOF redundant nonlinear vision head.

2. Categorizing Sensorimotor Patterns In Explorative Phases

The architecture proposed here follows from observations made on the brain which show that grouping or isolation (spatio-temporal associative memory) of multi-modal information was well done by temporal learning processing. In conventional neural networks, cells are static and

[1] It must be understood here that the stability discussed here is not related to stabilizing the dynamics of the system in its environment but rather stabilizing the internal configuration of the system so that it can interact with its environment without loss of its integrity.

[2] In this paper, we understand *categorization* in a way very similar to (Thelen and Smith, 1994), also mentionned by (Pfeifer and Scheier, 1996): Global functions of categorization are memory, learning and performance which arise dynamically from reentrant mapping of motor activity along with sensory information from many modalities... Both signals are completely coupled and act together to form the global maps that are the basis of further development...(p.160).

it is necessary to add recurrent loops or time-delay elements to treat time sequence signals. Instead, (Shigematsu et al., 1996) proposed the SAM model, described in Section 2.1, which, because of their dynamic structure, are capable of temporal information processing. The temporal learning rule, a variant of the well-known Hebbian Learning rule, was shown consistent with physiological results of synaptic plasticity.

2.1 SAM model and Temporal Learning Rule

The SAM model (SAM standing for "Spike Accumulation and delta-Modulation", see (Shigematsu et al., 1996)), consists of three parts:

- Leaky-integration of the input pulses:

$$U_i(t) = \sum_j W_{ij}(t).X_j(t) + \alpha.V_i(t-1) \quad (1)$$

where j is the index of each synapse of the i-th cell, $V_i(t)$ is the internal potential of the i-th cell at instant t (see description below) and $X_j(t)$ is the activation of the j-th input cell.

- Binary-thresholding for pulses generation

$$Y_i(t) = g[U_i(t) - T] \quad (2)$$

where $g[z] = 1$ if $z >= 0$; $g[z] = 0$ otherwise; T is the firing threshold and $Y_i(t)$ is the generated pulse.

- Subtraction of constant from internal potential at spike firing

$$V_i(t) = U_i(t) - p.Y_i(t) \quad (3)$$

where p is a constant that controls the loss of internal potential after spike firing.

Before we present the temporal learning rule, we provide with Figure 1 an illustration of a cell's response under various configurations. In the left chronogram, (from time 0 to time 80), the delay (from 1 to 3 time-units in this case) in response to a stimulus can be adjusted by tuning the firing threshold T. After firing, sharp drops in internal potential can be noted. It results in a cell firing only once even after multiple excitations, i.e. slow dynamics. Such property is obtained with high value of p (subtraction constant at spike firing). In the right part of the chronogram, the opposite setting is used: small loss of potential after spike firing and low firing threshold enables the cell to generate train of spike firings after even a single stimulation.

Temporal Learning Rule : At each time step, the weights $W_{ij}(t)$ are updated according to the temporal learning rule introduced by (Shigematsu et al., 1996).

The temporal learning rule (detailed below) is a revised Hebbian rule with synaptic history. Whereas the classical Hebbian Rule reinforces connections if responses of both input and output cells are simultaneously active,

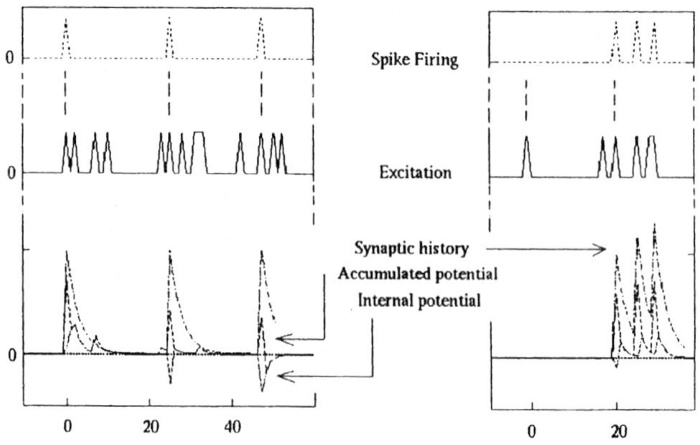

Figure 1 SAM model behavior under various settings. Details are provided in the text.

independantly of the history of the input signals, as formulated by Equation 4,

$$\Delta W_{ij} = C_{ij}.X_j(t).Y_i(t) \quad (4)$$

the temporal learning rule (1) strengthens connections when an input pulse train is an intense burst signal or a one shot signal stimulated simultaneously with another input signal and (2) weakens them when the input signal has a very low frequency or is out-of-phase with other input signals.

Mathematically, the synaptic history is an accumulated value of the input signal received by the i-th cell, from the j-th cell, at a synaptic site of the dendritic portion of the ith-cell [3]. It is computed by:

$$H_{ij}(t) = X_j(t) + q.H_{ij}(t-1) \quad (5)$$

where q is a decay coefficient ($0 < q < 1$). After one input pulse, this input history will decrease exponentially to zero.

When the i-th cell is fired, a new connection efficacy (from each synapse j) is calculated according to the following equations:

$$\begin{aligned} &\text{If} H_{ij}(t) \geq H_1 \text{then} \\ &\qquad \Delta W_{ij} = c_1.(W_{max} - W_{ij}).Y_i(t) \\ &\text{If} H_{ij}(t) < H_1 \text{and} H_{ij}(t) > H_2 \text{then} \\ &\qquad \Delta W_{ij} = c_2.(W_{min} - W_{ij}).Y_i(t) \end{aligned}$$

where c_1, c_2 are parameters fixing the learning speed, W_{max}andW_{min} are respectively maximum and minimum values of the connection efficacy and H_1andH_2 are two thresholds respectively set for enhancement and depression. When the after-effect of an input signal is nearly

[3] (Shigematsu et al., 1996) relates it to an after-effect such as an intra-cellular calcium ion concentration that can be observed in biological systems.

zero and the history is smaller than H_2, the efficacy is not changed, i.e. the memory in the efficacy is not erased by non-correlated signals. This learning calculation is applied to firing cells only.

2.2 Delta-Sigma Modulation

We use the delta-sigma modulation to transform continuous signals (such as a continuous flow of perceptual datas) into pulse trains, which can be integrated by our neural cell model. We follow the procedure reported by (Panter, 1965).

The message $s(t)$ is first integrated and then applied to a delta modulator so that the input to the pulse modulator $\epsilon(t)$ is the difference between the integrated message signal $\int s(t)dt$ and the integrated output pulses $\int p(t)dr$. Because such configuration requires an integrator with very large dynamic range, a simplified version can be built by combining both integrators such that the input to the single integrator is the difference signal $\Delta(t) = s(t) - p(T)$. Then, the integrator stays within a certain limit of proper system operation.

3. Spatio-temporal Associative Memory

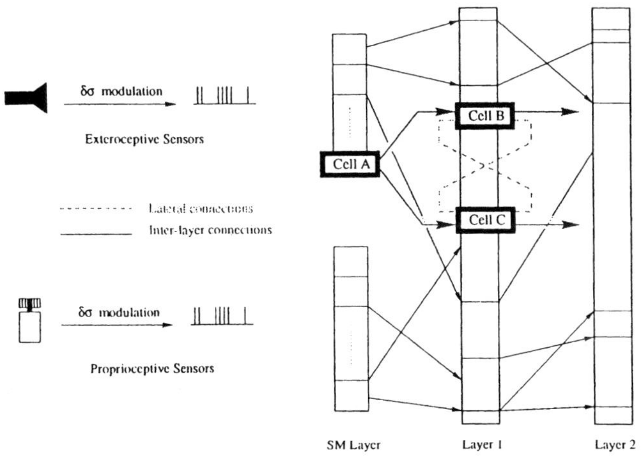

Figure 2 Spatio-temporal Associative Memory: Data from exteroceptive/proprioceptive sensors are modulated into pulse train by $\delta - \sigma$ modulation. Each layer is fully connected to the next one (from left to right) with initially random weights. Lateral connections exist within both layer 1 and layer 2. The highlighted subsystem (cell A, cell B, cell C) can be assimilated to a coupled oscillator and confers to the overall architecture very high dynamic properties.

As shown in Figure 2, our architecture consists of two main components:

- Sensorimotor layer (left, SM layer): The flow of sensorimotor patterns, as generated by both the exteroceptive and proprioceptive sensors (in this paper, respectively two cameras (2 signals each) and motor encoders (6 joints)), is transformed in trains of pulses through the *delta-sigma* modulation described in Section 2.2.
- Associative layers (right, layer 1–2): From one to two layers were used in the experiments reported in this paper. Layers of increasing index model higher-order association. The number of layers reflect a qualitative estimation of the complexity of the sensorimotor couplings present in the system.

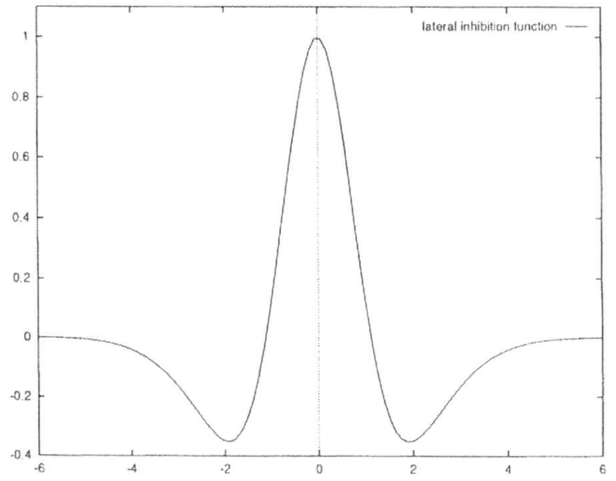

Figure 3 Lateral inhibition function for neurons in intermediate layers (also called "Mexican hat" function.

Layer 1 receives a full complement of inputs from the input layer (SM layer). Each cell of layer 1 is equipped with lateral connections whose weight is proportional to the inhibition function depicted in Figure 3. While lateral cells located in a close neighborhood are connected with a weak connection, more distant lateral cells are inhibited. Grounded in biological systems, this lateral inhibition generates a high dynamics of the layer activity. As an illustration, we highlight the sub-system formed by cell A, B and C (see Figure 2). This subsystem is equivalent to an oscillator. It follows that input-layer (SM layer) and layer 1 (as well as layer 2) form a collection of coupled oscillators whose interesting dynamics properties have been observed (see (Matthews et al., 1991) for example).

Layer 2 receives a full complement of inputs from layer 1. As with layer 1, the connections are initially randomly determined but positive. The range of those values are much lower than for layer 1 so as to generate a slower dynamics. High initial values make layer 2 highly sensitive to variations of layer 1 and tend to create many small unstable clusters. The choice follows from an analysis of experimental results such as depicted by Figure 4.

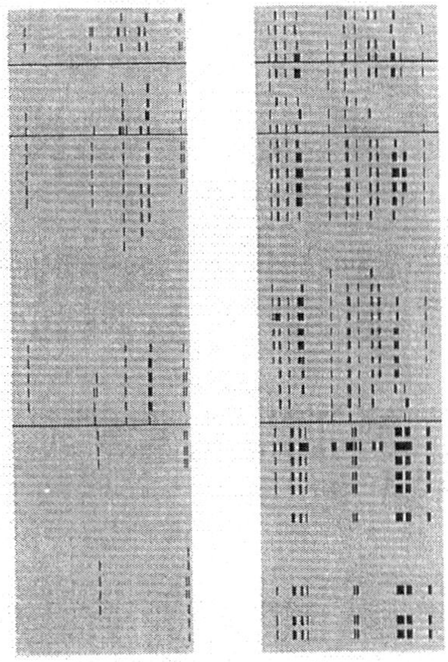

Figure 4 Two different speeds of dynamics: Because of its high dynamic properties, a slight change in the combination (firing threshold, weight) induces drastic changes in the sensitivity of the network (i.e. its synaptic activity) to excitation.

4. Experiments

Explorative behaviors are modeled as a stochastic distribution of several visuomotor behaviors, each of them involving various sensorimotor couplings:

- Random head motion: head and eye joints are randomly actuated, visual stimuli being random as well. This atypical visuomotor behavior is included so that the sensitivity of the architecture to uncorrelated sensorimotor patterns can be evaluated.
- Head motion: head joint (pan and tilt) are actuated without intervention of the VOR reflex described below, i.e. eyes are not coordinated. The interest of this behavior is two-fold: (1) the eyes being mounted on a common tilt, tilt actuation generates similar perceptual patterns in each eye; (2) horizontal neck motion will generate flow patterns similar to horizontal eye motions.
- Eye motion: only eye joints (vergences) are activated. Flow patterns should be similar to flow patterns generated by the above behavior. Correlation of both behaviors would ground the understanding of the redundancy between the joints as well as knowledge about body structure.
- Vestibulo-ocular reflex: as in animal VOR, this reflex stabilizes the retinal images during a neck movement by compensatory eye movements. For a target at infinite distance (approximation made in this experiment), it is achieved by causing the motion of the eyes to be equal and opposite to the motion of the head. It can be seen as a continuous transformation from head velocity to eye velocity.
- Nystagmus reflex: when the eyes reach an extreme position (i.e., an object of attention reaches the periphery of the field of view with an outward direction), they are rapidly flicked back to a new starting position. Unlike VOR, it is a one-shot simultaneous coordinated control of all joints.

Those visuomotor behaviours drive the ocular system, generating sensorimotor patterns which are observed by the neural architecture through exteroceptive/proprioceptive datas. The neural network is initialized as indicated in the Table 1. Because layer 2 must learn higher-order couplings, the emphasis is put on slowing down its dynamics. p is set high so that internal potential drops sharply after firing. The lateral inhibition curve is set so as to activate connections with cells located at most 3 cells apart. Further cells inhibited according to the inhibition function detailed above.

A typical experiment consists of up to 10,000 behavior switchings (those switchings are controlled externally). Figure 5 displays the result of the spike-firing monitor we developed for this experiment. The upper part of the monitor displays the activity of the cells (visual input (cells 0 – 3), motoric command (cells 4 – 9), layer 1 (cell 10 – 24)) over about 500 time-units (an average 100 switchings of behaviors). At each time-unit, the value of the highest connection efficacy over the whole network is calculated and serves as an index of the learning stability. After stabilization, each input is intensively fired (burst firing) in order to measure the state of the network and the pathways associated with that burst firing are recorded. As shown in the lower display of Figure 5, well-defined clusters appear. Each of them correspond to a spatio-temporal correlation between this sensorimotor component and one/many of the other components. A different view on this categorization is provided by Figure 6 where the distribution of peak connection efficacies between input layer and layer 1 is projected on a 2D plane. With this representation, back-projecting the clusters on the input layer makes the reading of existing first-order correlations between inputs straightforward.

In Figure 7, we analyze the operating factors (namely: accumulated potential, internal potential and synaptic history) of a cell that belongs to one of the clusters shown above. This analysis consists of alternative sequences of non-correlated and (temporally or spatially) correlated excitations. In Figure 7, we distinguish two areas (Area 1 and Area 2) of particular interest: in Area1, the absence of cell firing indicates a good robustness to non-correlated inputs. Meanwhile, high sensitivity and stability to correlated inputs is observed in Area 2. After two consecutive correlated excitations, the cell generates a pulse-train. Non-correlated excitations input during the

	Input layer	Layer 1	Layer 2	
n	4 + 6	15	20	Number of cells at each level
a	0.5	0.5	0.5	Decay constant for internal potential
q	0.6	0.6	0.6	Decay constant for synaptic history
p	0.6	0.6	0.9	Subtraction constant (strong loss of int. potential at spike firing
T	0.4	0.4	0.5	Firing threshold
W_ij		0.35	0.07	Connection efficacy initialization (slow dynamics in layer 2)
H_1		0.8	1.6	Threshold for connection efficacy enhancement
H_2		0.3	0.2	Threshold for connection efficacy depression
c_1		0.04	0.04	Learning speed for connection enhancement
c_2		0.01	0.01	Learning speed for connection depression
W_{max}		2.0	2.0	Maximum connection efficacy
W_{min}		-1.0	-1.0	Minimum connection efficacy

Table 1 Initialization of the neural architecture for a typical experiment.

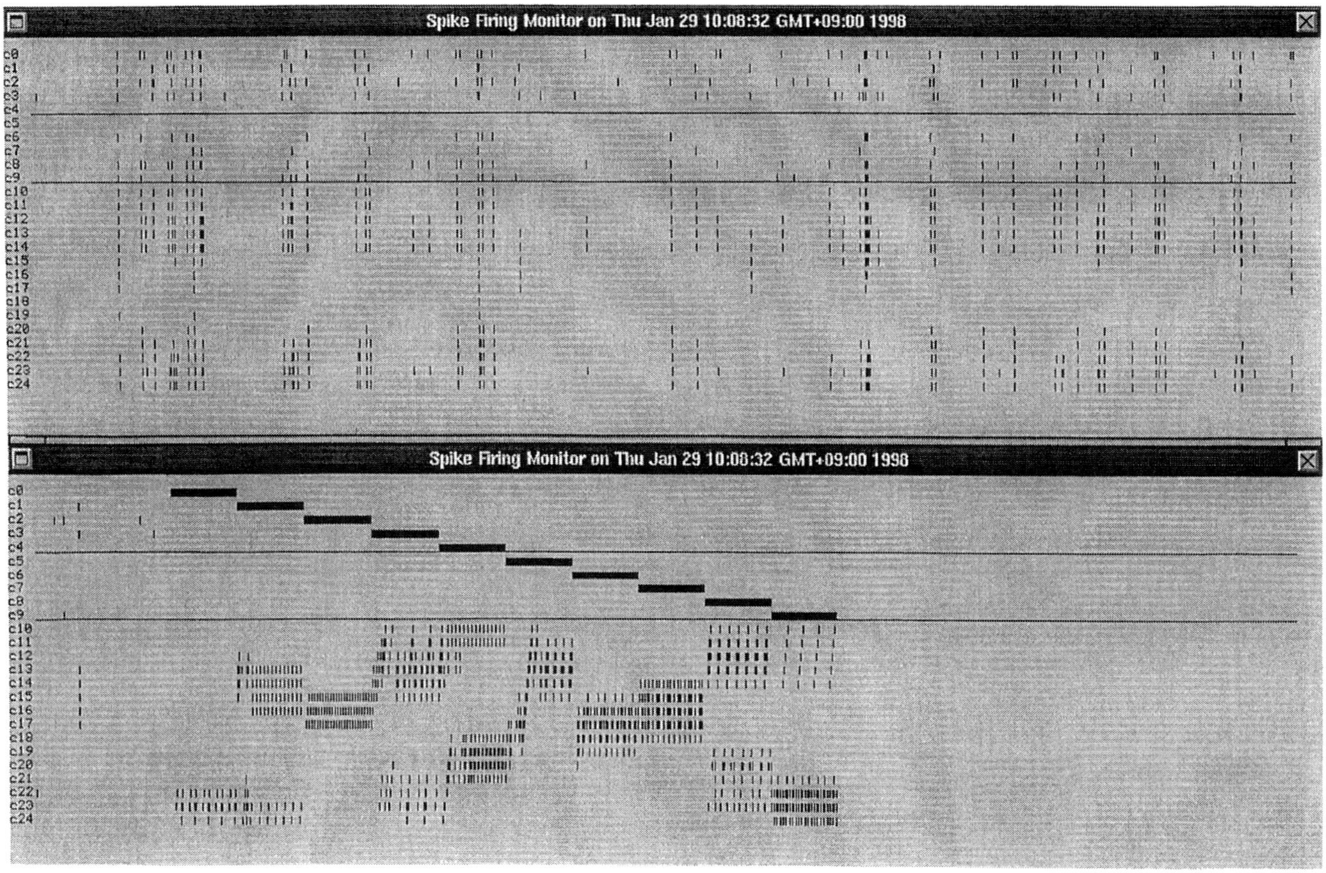

Figure 5 Histogram of the activity of each cell during a typical experiment. See details in the text.

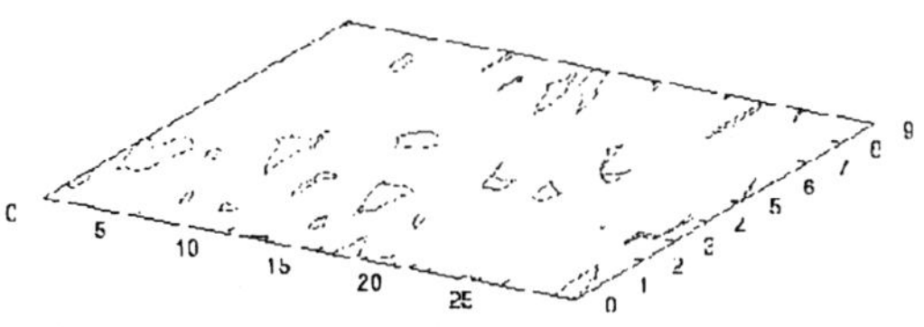

Figure 6 Emergence of clusters in the first intermediate layer. Right-hand axis corresponds to input layer (sensorimotor pattern), while left-hand axis corresponds to cells in the layer 1.

pulse-train, while not directly affecting the cell firing, result in a sharp drop of accumulated potential which eventually stop the train pulse.

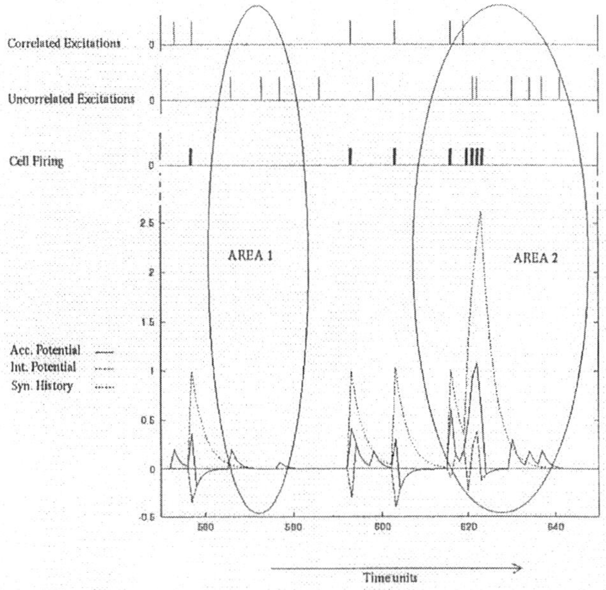

Figure 7 High-sensitivity of cluster member to correlated excitation. For interpretation of areas 1 and 2, see text.

5. Conclusion

Whenever a robot is to live in the real-world, it faces very complex dynamic patterns of interaction: physical relationships with a dynamically changing environment, various sensorimotor couplings (both internal and external). When the number of degrees of freedom increases, the robot (as well as the designer/programmer) faces an overly complex sensorimotor reachable space. Animal systems show periods of explorative behaviors (grooming, visual observation (gazing etc...) and postural exploration) which, because they are unconstrained (except by body limitations), can provide the animal system with knowledge about its own configuration (dynamics, kinematic, etc...). We suggest in this paper that robots could also make use of such knowledge. To address the issue, we propose a novel neural architecture, whose cells are capable of temporal information processing and whose structure support the generation of spatio-temporal associative memories. Through this neural architecture, the categorization of sensorimotor patterns generated by explorative behaviors is achieved. We show that, although the network is capable of highly complex dynamics, the emerging categorization is stable and highly sensitive to the system's sensorimotor couplings. We believe such categorization will prove valuable whenever a robot needs emergence of sensorimotor coordination while actively interacting with the environment.

References

Berthouze, L. and Kuniyoshi, Y. (1998). Emergence and categorization of coordinated visual behavior through embodied interaction. *Autonomous Robots and Machine Learning (joint issue).* to appear.

Matthews, P., Mirollo, R., and Strogatz, S. (1991). Dynamics of a large system of coupled nonlinear oscillators. *Physica D*, 52:293–331.

Panter, P. (1965). *Modulation, Noise, and Spectral Analysis*, chapter Delta Modulation (DM). McGraw-Hill.

Pfeifer, R. and Scheier, C. (1996). *From Perception to Action*, chapter From Perception to Action: The Right Direction?, pages 1–11.

Shigematsu, Y., Ichikawa, M., and Matsumoto, G. (1996). *Perception, Memory and Emotion: Frontiers in Neuroscience*, chapter Reconstitution Studies on Brain Computing with the Neural Network Engineering, pages 581–599. Elsevier Science Ltd.

Thelen, E. and Smith, L. (1994). *A Dynamic Systems Approach to the Development of Cognition and Action.* Cambridge, Mass.: MIT Press, Bradford Books.

Evolution of visually-guided approach behaviour in recurrent artificial neural network robot controllers

Zoltán Biró & Tom Ziemke
Connectionist Research Group, Dept. of Computer Science, University of Skövde
P.O. Box 408, S-54128 Skövde, Sweden
zoltan@ida.his.se, tom@ida.his.se

Abstract

Analysis of internal structures of embodied and situated agents may provide insights into the mechanisms underlying adaptive behaviour. This paper is concerned with the evolution and analysis of visually-guided approach behaviour in a simulated robotic agent controlled by a recurrent artificial neural network, whose connection weights have been evolved using evolutionary algorithms. Analysis of the evolved behaviours and their network-internal mechanisms reveals a behavioural structure and organization resembling a Brooksian subsumption architecture. The task decomposition, as well as the resulting individual behaviours and their integration, however, are realized as network-internal state space dynamics, evolved in the course of agent-environment interaction, i.e. with a minimum of designer intervention.

1 Introduction

Contemporary research in both adaptive behaviour and embodied cognition identifies dynamical coupling and interaction between agents and the environments they are situated in, as the main source of behaviour (e.g. Brooks, 1991; Varela, 1991; Wheeler, 1994; Beer, 1995). Braitenberg (1984) demonstrated that, regardless of the complexity of the observed behaviour, it is a fallacy to presuppose the existence of internal cognitive or psychological structures (such as internal world models and goals) corresponding to all phenomena which can be attributed to an agent from the observer's point of view.

In the context of constructing robotic agent controllers, this fallacy can be avoided by, to as large extent as possible, replacing deliberate design decisions, which are often based on a priori assumptions, by e.g. evolutionary self-organization (e.g. Floreano, 1997). Cliff *et al.* (1992), for example, showed that employing evolutionary methods to evolve controller dynamics, behaviours can emerge without explicit reference to the factors underlying these behaviours.

Much current research, both in adaptive behaviour and embodied cognition, also emphasizes the inseparability of perception and action, and the emergence of agent-internal (cognitive) "structures ... from the recurrent sensorimotor patterns that enable action to be perceptually guided" (Varela et al., 1991, p. 173). The synthesis and analysis of such emergent structures in the mechanisms underlying complex behaviour in artificial embodied and situated agents could provide insight into the nature of adaptive behaviour.

The work presented in this paper is based on Beer's (1996) work on the evolution of dynamical neural networks for visually-guided behaviour. The focus is here put on the analysis of emergent organization of the internal mechanisms of recurrent artificial neural network (ANN) controllers evolved to control visually-guided behaviour in a simulated robotic agent.

Section 2 provides a short description of the methodology, section 3 describes the results obtained from the experiments, and finally, section 4 summarizes these results.

2 Method

The work presented in this paper is inspired by and based on Beer's (1996) work on evolving dynamical neural network controllers for visually-guided behaviour. According to Beer, visually-guided behaviour is "minimally cognitive" behaviour, and therefore an appropriate target of investigation, as it is rich of phenomena complex enough to raise a wide range of cognitively interesting issues while remaining computationally tractable. The experiments underlying the results presented in this paper solely concern a simple robotic agent's visually-guided approach of an object.

The experiments were conducted in an extensively modified version of the Khepera simulator version 2.0 (Michel, 1996). The simulator was modified to allow later validation on a real Nomad 200 mobile robot. The modifications also contributed to make the agent's sensor configuration more similar to that of the agent used by Beer (1996).

The agent's sensors had a range of approximately 660 centimeters and an 18 degree sensor beam (Figure 1). Five sensors, uniformly distributed over $8\pi/5$ ($112.5°$), were utilized in the experiments. To more closely resemble real world conditions during the simulations, a ± 5 % random noise was added to the sensor readings, ± 10 % to the motor speed, and ± 5 % to the direction.

The task of the agent was to approach a 25×25 cm object in a 960×960 cm environment surrounded by walls, the agent having a diameter of 55 cm. The object was positioned within an area of 500×500 cm in the centre of the environment (Figure 1).

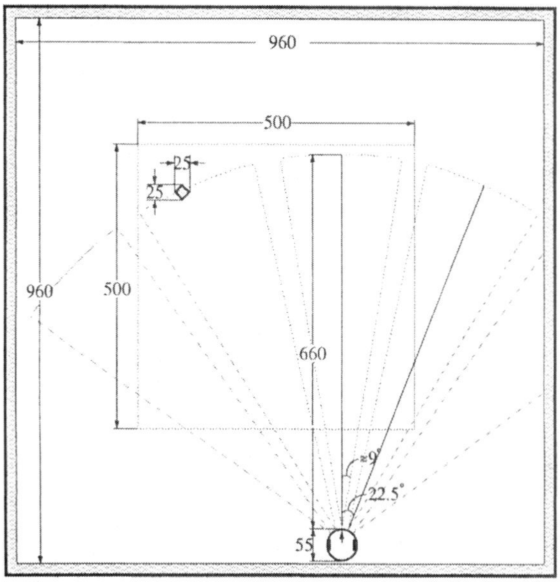

Figure 1: Experimental setup: The agent and its environment.

For each experiment run, the agent's position was determined anew by randomly placing it at an Euclidean distance of 550 cm (ten agent diameters) from the object's position. The heading of the agent was also determined randomly for each run.

The overall design of the controller system was based on evolution of the internal dynamics of controllers by employing evolutionary strategies (ES) for evolving the weights of recurrent control networks which determined the mappings from sensor input to motor output. The control network architecture used here is a simple recurrent network (SRN; Elman, 1990), i.e. a network of three fully connected layers of units, in which the internal state (hidden unit values) at time step t are fed back via recurrent connections and used as extra input in time step $t + 1$ (cf. Figure 2). The logistic activation function was used in a fixed topology (5-3-2) SRN with five input units dedicated for sensor inputs, and two output units for direct control of left and right motors. ANN inputs ranged from 0 (out of range) to 10 (close), and outputs ranged from 0 (zero speed) to 1 (full speed forward).

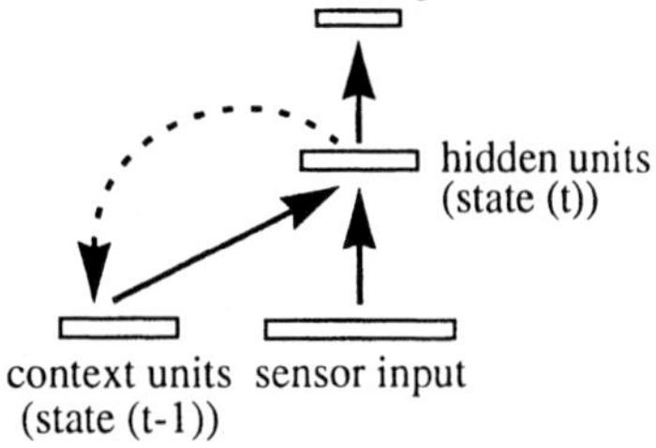

Figure 2: Recurrent control network.

The ES maintained a population of 5 parents (μ) and 35 children (λ), following Bäck & Schwefel's (1993) suggestion of a population ratio $\mu/\lambda \approx 1/7$. A (μ, λ)-ES deterministic selection strategy was employed, where the μ best individuals are selected from λ children to form a new parent population.

For each individual, the ES employed a vector of object variables (encoding all connection weights of a control network) and one mutable strategy parameter representing the magnitude with which the object variables were mutated. The three recombination mechanisms used on object variables were local discrete recombination, local intermediate recombination, and no recombination. Global intermediate recombination was employed on the strategy parameter (see Bäck & Schwefel, 1993, for details).

During a single run, the agent was allowed to move around in the environment until it collided with the object, or for maximally 1500 time steps (corresponding to a maximal Euclidean distance coverage of 6450), after which a fitness measure was calculated. The fitness function (Φ), which was to be minimized, considered the Euclidean distance between the agent and the object. The fitness of five evaluation trials determined the final fitness of each individual.

$$\Phi(a_i(t)) = \frac{\sum_{s=1}^{5} \sqrt{(rx_s - ox_s)^2 + (ry_s - oy_s)^2}}{5}$$

$\Phi(a_i(t))$ is the fitness value for individual i at generation t, s is the evaluation trial, rx, ry are the horizontal and vertical positions of the agent at the end of the evaluation run, ox, and oy the corresponding positions for the object.

3 Results and discussion

Two types of controllers evolved during the experiment runs. These will be referred to as "O1" and "O2" in the following detailed discussion of their properties.

3.1 The O1-type controllers

The O1-type controllers were by far the most common (> 95%) to evolve during the experiments. Although they achieved good performance (i.e. reaching the object or ending up close to it) within the first one hundred generations, their fitness remained unstable. The O1-type controllers may be termed "minimal configuration controllers" as they make use of only one of five sensors for the actual approach task. This "dedicated" sensor was in all observed cases one of the outermost sensors. In addition, there was a general tendency for a turn in the same direction as the dedicated sensor during virtually the entire time. The rare occasions when the opposite motor activation was highest, it resulted in forward motion because of the small activation differences between the two motors.

Hierarchical cluster analysis revealed a clustering of the internal state space (hidden unit activation level) corresponding to different behaviours (cf. Figure 3). The overall behaviour of O1 can be summarized according to the following key behaviour categories:

1. *Passive search.* The agent is passively searching for an object to approach when the object is out of sensor range ("out of sight"). This passive search amounts in a slight turn in the direction of the dedicated sensor which allows a big portion of the environment to be covered by exploiting the limitations of the environment (small, and surrounded by walls).
2. *Object tracking.* The agent moves forward as long as the dedicated sensor reading is high (the object is "in sight") while the reading for the sensor next to the dedicated sensor stays below a certain threshold value.
3. *Obstacle avoidance.* The agent avoids colliding with the walls by making a sharp turn in the same direction as the dedicated sensor. The sharpness depends on the reading on the sensor next to the dedicated sensor. This behaviour thus handles the simplest form of discrimination by exploiting the fact that an object can never be positioned to simultaneously give high sensor readings on two sensors.

Figure 3 illustrates the behavioural clustering of hidden unit activations for an O1 controller for each time step during a sample run.

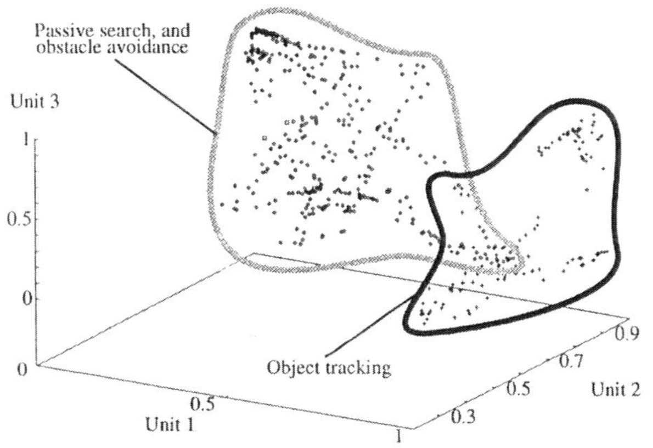

Figure 3: Sample O1 behavioural clustering of hidden unit space.

When the object is lost from sight, it is a result of forward motion (object tracking). This implies that the lost object is positioned, and should subsequently be sought, on the same side as the dedicated sensor. As passive search amounts in turns in the same direction as the dedicated sensor, the agent inherently starts "seeking" for the object. Passive search thus inherently copes with a simple form of the object persistence problem in the limited environment.

Thus, the agent approaches the object by moving in an arc- or spiral-shaped motion trajectory. This is achieved by combining passive search with object tracking. More exactly, object tracking subsumes passive search provided that certain conditions hold. Similarly, obstacle avoidance can subsume object tracking behaviour. The agent's behaviour can be described in terms of a subsumption architecture (Brooks, 1986). The different behaviour modules and their interrelationships are, however, evolved rather than designed (cf. Figure 4).

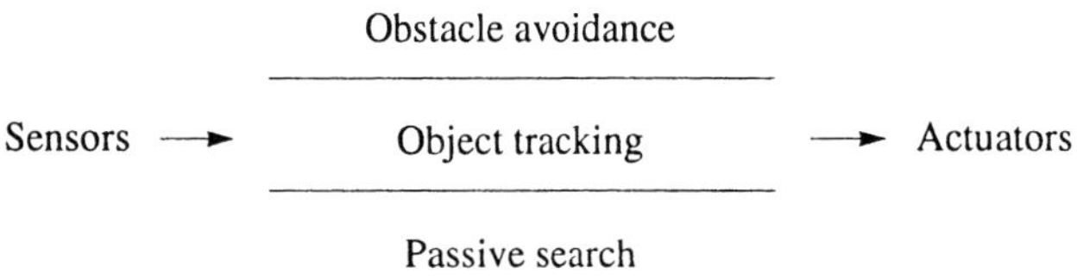

Figure 4: The O1-type controller as a subsumption architecture.

3.2 The O2-type controllers

The O2-type controllers achieved performance, far superior to the O1, by employing (dedicating) four of five sensors for the approach behaviour. Hierarchical cluster analysis of the hidden unit activations unveiled clusters which can be identified with different key behaviours:

1. *Passive search.* The agent is passively searching the environment for an object towards which to orient when the object is out of sensor range ("out of sight"). This passive search amounts in a slight turn. Similar to O1, this behaviour allows a big portion of the environment to be covered by exploiting the limitations of the environment (small, and surrounded by walls).
2. *Active search.* The agent is actively searching the environment if an object is lost from dedicated sensor range, by carrying out a turn in the direction where the object was last sighted. The controller is thus taking advantage of the dynamics of the recurrent ANN in order to produce output according to the dedicated sensor that gave high readings most recently.
3. *Object tracking.* The agent keeps the object in sight through successive forward-left and forward-right turns, i.e. it uses a foveation strategy.
4. *Obstacle avoidance.* The agent exploits the fact that an object cannot be positioned to give high sensor reading on two sensors simultaneously. Contrary to O1, all sensors are utilized in this simple discrimination task.

Figure 5 illustrates the behavioural clustering of hidden unit activations for an O2 controller during a sample run.

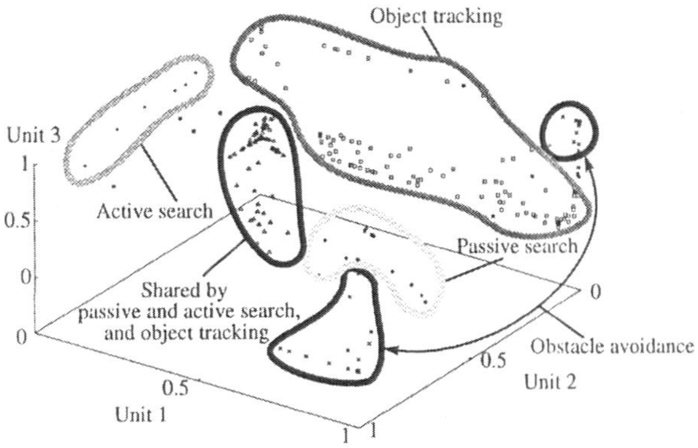

Figure 5: Sample O2 behavioural clustering of hidden unit space.

The agent approaches the object combining object tracking with active search. Similar to O1, the agent's behaviour can be described in terms of a subsumption architecture (cf. Figure 6), where obstacle avoidance behaviour can subsume the object tracking behaviour, which can subsume the active search behaviour, which in turn can subsume the passive search behaviour.

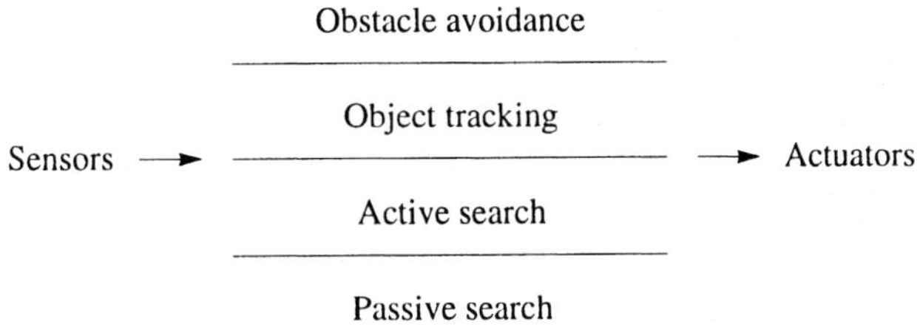

Figure 6: The O2-type controller as a subsumption architecture.

4 Conclusions

As shown in the previous section, both O1 and O2 controllers exhibit an emergent task decomposition into a number of, from the observer's point of view, seemingly separate key behaviours, organized in a subsumption-architecture-like fashion. Unlike in the original subsumption architecture (Brooks, 1986), however, (a) the task decomposition into a set of behaviours, (b) the individual behaviours, and (c) their organization have *evolved* with a minimum of designer intervention, i.e. they are the result of autonomous self-organization rather than design, which has been argued to be a desirable property of robotic agent controllers and a key capacity of autonomous systems by, e.g. Nolfi (1997), and Ziemke (1996a, 1996b).

This behavioural organization was here shown to be realized by means of context- and behaviour-dependently structured internal state spaces (recurrent networks' hidden unit spaces). Here, adaptive behaviour is emergent from dynamic adaptation of one behaviour-producing mechanism to different contexts, rather than arbitration between distinct behaviours actually realized in separate behaviour-producing modules. This once again illustrates (a) that structured and complex (from the observer's point of view) behaviour does not at all have to be the result of corresponding static internal ('cognitive') structures (cf. Braitenberg, 1984), and (b) that behaviours might be best understood as elements of the observer's descriptive vocabulary rather than separate building blocks of an agent's overall behaviour.

References

Bäck, T., Schwefel, H. (1993). An overview of evolutionary algorithms for parameter optimization. *Evolutionary Computation*, **1**: 1-23.

Beer, R.D. (1995). A dynamical systems perspective on agent-environment interaction. *Artificial Intelligence*, **72**: 173-215.

Beer, R.D. (1996). Toward the evolution of dynamical neural networks for minimally cognitive behavior. In Maes, P., Mataric, M., Meyer, J.-A., Pollack J. and Wilson, S., editors, *From Animals to Animats 4: Proceedings of the Fourth International Conference on Simulation of Adaptive Behavior (SAB96)*, pages 421-429. Cambridge, MA: MIT Press

Braitenberg, V. (1984). *Vehicles: Experiments in Synthetic Psychology*. Cambridge, MA: MIT Press.

Brooks, R.A. (1986). A Robust Layered Control System for a Mobile Robot. *IEEE Journal of Robotics and Automation*, **2** : 14-23.

Brooks, R.A. (1991). Intelligence Without Reason. In *Proceedings of the Twelfth International Joint Conference on Artificial Intelligence (IJCAI-91)*, pages 569-595. San Mateo: Morgan Kaufmann.

Elman, J.(1990). Finding structure in time. *Cognitive Science*, **14**: 179-192.

Floreano, D. (1997). Reducing Human Design and Increasing Adaptability in Evolutionary Robotics. In Gomi, T., editor, *Evolutionary Robotics - From Intelligent Robots to Artificial Life (ER'97)*. Kanata, Canada: AAI Books.

Michel, O. (1996). *Khepera simulator version 2.0 user manual*. University of Nice-Sophia Antipolis, Valbonne, France.

Nolfi, S. (1997). Evolving Non-Trivial Behavior on Autonomous Robots: Adaptation is More Powerful Than Decomposition and Integration. In Gomi, T., editor, *Evolutionary Robotics - From Intelligent Robots to Artificial Life (ER'97)*. Kanata, Canada: AAI Books.

Varela, F.J., Thompson, E., Rosch, E. (1991). *The embodied mind: cognitive science and human experience*. Cambridge, MA: MIT Press.

Wheeler, M. (1994). Active Perception in Meaningful Worlds. Cognitive Science Research Paper 327, University of Sussex, UK.

Ziemke, T. (1996a). Towards Adaptive Behaviour System Integration using Connectionist Infinite State Automata. In Maes, P., Mataric, M., Meyer, J.-A., Pollack J. and Wilson, S., editors, *From Animals to Animats 4: Proceedings of the Fourth International Conference on Simulation of Adaptive Behavior (SAB96)*, pages 145-154. Cambridge, MA: MIT Press.

Ziemke, T. (1996b). Towards Autonomous Robot Control via Self-Adapting Recurrent Networks. In von der Malsburg, C., von Seelen, W., Vorbrüggen, J.C. and Sendhoff, B., editors, *Artificial Neural Networks - ICANN 96*. Berlin/Heidelberg, Germany: Springer Verlag.

Simulation and Robot Implementation of Visual Orientation Behaviors of Flies

Susanne A. Huber, Heinrich H. Bülthoff
Max-Planck-Institut für biologische Kybernetik
Spemannstr. 38, 72076 Tübingen, Germany.
susanne.huber@tuebingen.mpg.de
heinrich.buelthoff@tuebingen.mpg.de

Abstract

Flies have developed visuomotor programs for tasks like course stabilization, fixation and approach towards stationary objects, tracking of moving objects, and landing, which are based on the analysis of visual motion information. Theoretical and experimental results suggest that the visuomotor control for course stabilization as well as fixation and approach towards stationary objects may be realized at least partially by one common sensory circuit In order to test this hypothesis we designed simulated agents, with a control architecture that is based on results from fly research. We present agents with a visuomotor controller that regulates the two behaviors of course stabilization and object fixation. To test this controller under real world conditions, we implemented it on a miniature robot. We can show that besides course stabilization and object fixation, the robot additionally approaches stationary objects.

1. Introduction

Researchers have been attracted to the field of insect research for many years, because insects show stimulus-response characteristics that allow insight into the mechanisms of visual information processing as well as into the interaction of sensory input and motor output. Especially the visually controlled orientation behavior of flies is particularly well studied (Reichardt and Poggio, 1976; Heisenberg and Wolf, 1984; Egelhaaf and Borst, 1989 .

A promising approach to help to understand information processing in living systems is the study of artificial systems and their interaction with the environment. The simulation of biological systems and the implementation of behavioral models (Tinbergen, 1953; Brooks, 1986) onto robots seems to be especially useful to the understanding of biological systems. Several examples exist where researchers implemented sensorimotor mechanisms of insects into robots (e.g., Beer, 1990; Cliff, 1992; Cruse, Bartling, Cymbalyk, Dean, and Deifert, 1995; Webb, 1995). Orientation mechanisms inspired by the behavior of flies have been implemented in computer simulation or on mobile robots. For example, visuomotor controllers have been designed that use visual motion information. Mostly these agents use the visual motion information for obstacle avoidance (Franceschini, Pichon, and Blanes, 1992; Duchon and Warren, 1994; Weber, Venkatesh, and Srinivasan, 1997) or tracking of other objects (Cliff, 1992; Missler and Kamangar, 1995).

Results from fly research suggest that the two behaviors of course stabilization and approach towards stationary objects may be realized at least partially by one common behavioral module (Götz and Wenking, 1973; Götz, 1975; Reichardt and Poggio, 1976; Bülthoff, 1982). We use the robotics approach to investigate this hypothesis. Our goal is to build a single visuomotor controller that regulates both behaviors. We modeled the subsequent processing steps in the visual system of flies in detail. Like the robots of Franceschini et al. (1992) and others, the agent uses visual motion information of a 360° horizontal field of view for behavior control. We can demonstrate that our computer-simulated agent is able to control the two behaviors of course stabilization and approach towards stationary objects by evaluating the motion information with a single visuomotor controller. To test the behaviors under real world conditions, we implement the control structure on a miniature robot – the $Khepera^{TM}$ (Mondada, Franzi, and Ienne, 1994).

2. The agent's architecture

2.1 Model of the visual system

We implemented a simplified model (Fig. 1) of visual information processing in flies, i.e. contrast enhancement, reduction of signal redundancy, signal amplification, motion detection and evaluation of the motion signals which takes place in the subsequent neural layers lamina (L), medulla (M) and lobula plate (LP) in the visual section of the fly's brain.

Spatial and temporal lowpass filtering in the retina: The agent has a horizontal array of sensors with a 360° field of view which scans the visual world at the horizon. By ray-tracing, the simulated agent determines (Foley, van Dam, Feiner and Hughes, 1987), the inten-

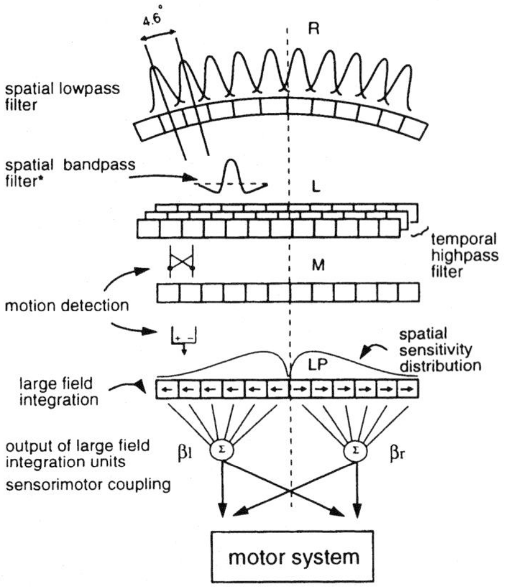

Figure 1 *Model of the visuomotor controller with the functional processing steps that take place in the different layers of the agent's visual system (*the bandpass filter is only realized on the Khepera^{TM} robot, see Section 4.).*

sities of 780 single points which are equally distributed over the 360° horizontal field of view, at the intersections of the lines of sight with the visible surfaces (for the robot implementation see Section 4.). Second, these samples are spatially lowpass filtered by Gaussian filters ($\sigma = 3.8°$) Like in *Drosophila* (Götz 1964), the sensors have an interommatidial angle of $d\varphi = 4.6°$. Hence the array has 78 sensors.

Redundancy reduction and amplification of the signal in the lamina: In *Drosophila* large monopolar cells (LMCs) in the lamina are known to be responsible for signal amplification, local contrast enhancements and reduction of redundant parts of the signal (Laughlin 1987). In this work we modeled the temporal aspects of the LMC cells by applying a temporal highpass filter H ($\tau_H = 20.0$ steps), which eliminates parts of the signal that are steady or slowly changing in time. In a next step the signals are linearly amplified to the full range of 256 grey-level values.

Motion detection in the medulla and lobula plate: For motion perception in insects, Hassenstein and Reichardt (1956) proposed a detector which correlates temporal modulation of image intensities in two neighboring ommatidia. The detector model has two mirror-symmetrical subunits. In each subunit the signals of two input channels are multiplied after the signals have been delayed by two lowpass filters with different time constants ($\tau_1 = 1.5$ steps and $\tau_2 = 5.0$ steps). In a next step the outputs of the two subunits are subtracted to obtain the direction of the motion stimulus.

Integration of the motion signals in the lobula plate: In the fly's lobula plate neurons have been described which are specialized to certain motion patterns. For example the HSE cells receive input from motion detectors that respond stronger to a pattern moving from front to back (progressive motion) than from back to front (regressive motion). The asymmetry results, because the time course of these motion detector subunits is not completely mirror symmetric (Egelhaaf, Borst and Reichardt 1989). This is modeled by a gain of 1.0 for progressive and 0.7 for regressive motion. In addition the horizontal equatorial cells (HSE cells) respond maximal to horizontal progressive motion in the frontolateral field of view (Hausen 1982). Modeling the HSE cells, two large field units integrate the motion information over a 184° field of view in the right and left hemisphere with an overlapping region of 8° in the front. The sensitivity distribution $S(j)$ (j number of sensor[1]) is modeled by the function:

$$S(j) = aj^b e^{-cj} \tag{1}$$

with $a = 0.625$, $b = 0.7$ and $c = 0.15$ and $j \in [-1, 39]$. $S(j)$ is bilaterally symmetrical for the two integration units ($S(j) = S(-j)$) and the maximum of $S(j)$ is at the sensor 5 (at $\varphi_{\max} = 23°$). The outputs (β_l, β_r) of the integration units are coupled via transmission weights to the motor system.

2.2 Model of the motor system:

The agent is modeled as a simple kinematic system with two motors, ignoring its mass and inertia. In order to model the visuomotor control of flies this approximation can be made, because after an initial acceleration, within a short time (< 10 ms) the fly reaches a constant velocity as the force produced by the wings is balanced by the increasing air friction. The motors have a distance of $c =$ 1u, given in units u of the agent's size (1 u = 0.25 cm)).[2]

The velocities v_l and v_r (Fig. 2) of the motors are proportional to the force of the two motors. Each motor produces a constant velocity $v_0 = 0.1$ u/step which is modulated by the outputs of the processed visual information:

$$v_{l,r} = v_0 - T(s_{l,r}) \tag{2}$$

The signals s_l and s_r result from the visual motion information via the control signals m_l and m_r and intrinsic noise n_l and n_r (Fig. 3):

$$s_{l,r} = km_{l,r} + n_{l,r} \tag{3}$$

[1] The sensors ±1 are oriented at the visual angles $\varphi = \pm 2.3°$ off the heading direction and the sensors ±39 at $\varphi = \pm 177.1°$

[2] The force vectors produced by the wings of the fruitfly *Drosophila* have an estimated perpendicular distance from the center of the fly of about 0.2 to 0.3 cm (body-length: 0.3 cm) (Götz, 1964).

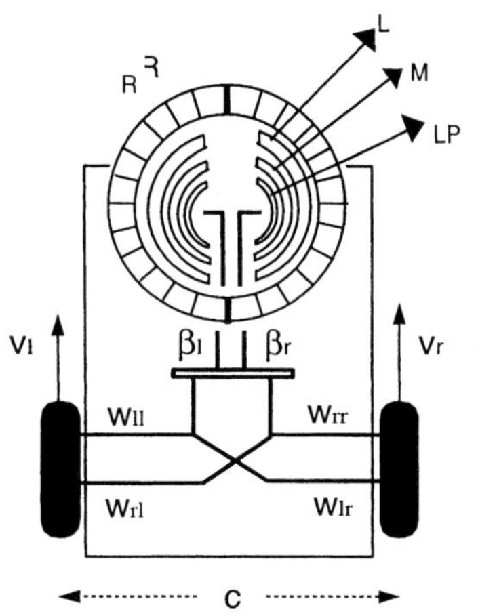

Figure 2 *Sketch of the agent with ring sensor (R), simplified models of the three layers of the visual system: lamina (L), medulla (M), lobula plate (LP), and the transmission weights* ($w_{ll}, w_{lr}, w_{rl}, w_{rr},$) *that couple the outputs of the large field units* (β_l, β_r) *to the motor system.*

where k is a proportionality factor. The control signals m_l and m_r are explained in detail in the next section. As the force produced by the wings of the fly never becomes negative, the velocities v_l and v_r are always above or equal 0 u/step. This is achieved by the sigmoid function $T(s)$ for the left and right motor respectively:

$$T(s) = \begin{cases} -v_0 & \text{if } s < -v_0, \\ s & \text{if } |s| \leq v_0, \\ v_0 & \text{if } s > v_0, \end{cases} \tag{4}$$

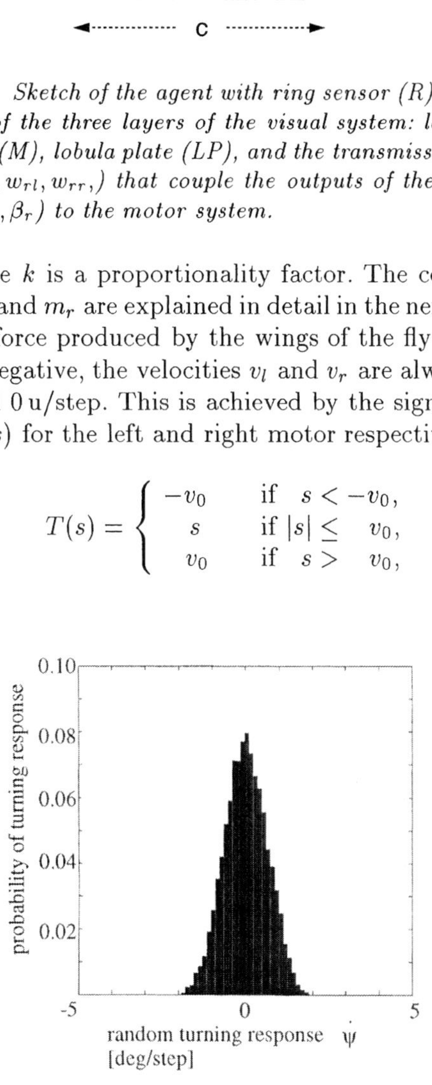

Figure 3 *Noise signal* n_l *of left motor (Gaussian distributed with* $\sigma = 0.64°$/step*).*

The system has two degrees of freedom: translation in the heading direction and rotation around the vertical body-axis. The tranlatory and rotatory velocities are:

$$v_t = \frac{v_r + v_l}{2} \quad \text{and} \quad \dot{\psi} = \frac{v_r - v_l}{c} . \tag{5}$$

2.3 The sensorimotor coupling

Fixation behavior: Reichard and Poggio (1976) showed in behavioral studies with flies that a fly orients itself towards a single black stripe in an otherwise homogeneous arena. This so-called fixation behavior is in contrast to the behavior in a visually homogeneous environment, where a fly turns in all directions with equal probability. Due to intrinsic noise the motor system continuously produces torque, resulting in turning movements in either direction. Reichardt and Poggio (1976) showed that the noise is Gaussian distributed.

One explanation of the fixation behavior is based on the fact that the turning response of flies in open-loop experiments is stronger if a stripe moves from front to back than in the other direction (Heisenberg and Wolf, 1984), due to the asymmetric response of the HS-neurons. As the noisy torque signals of the motor system lead to movements of an object's retinal image, the resulting image flow will cause the fly to orient towards the object, because the positive response to progressive motion is stronger than the negative response to regressive motion.

To model the fixation behavior of flies, the signals that result from the large field integration units are coupled proportionally to the motor system.

$$\begin{aligned} m_l^f(t) &= \omega_{ll}\beta_l + \omega_{lr}\beta_r \\ m_r^f(t) &= \omega_{rl}\beta_l + \omega_{rr}\beta_r \ . \end{aligned} \tag{6}$$

The matrix

$$\mathbf{W} = \begin{pmatrix} \omega_{ll} & \omega_{lr} \\ \omega_{rl} & \omega_{rr} \end{pmatrix} = \begin{pmatrix} 0.9 & -0.4 \\ -0.4 & 0.9 \end{pmatrix} \tag{7}$$

contains the transmission weights for the coupling of the outputs β_l and β_r of the two large field integration units to the left and right motor. We assume bilateral symmetry for the sensorimotor coupling.

Optomotor response: While flying, flies are continuously stabilizing their course. In order to compensate for disturbances, which cause large rotatory image motion, they execute turning movements – the so-called optomotor response – along the direction of the image motion. This behavior is most probably realized by cells that continuously integrate the difference of the output signals from the horizontal cells in the two optic lobes (Heisenberg and Wolf, 1984). A motor signal, purely proportional to the difference of the image motion in the two eyes, has the disadvantage that ambiguous motor signals result with respect to pattern velocity. This is due to the characteristic response of the Reichardt motion detector, which increases with increasing image motion up to an optimum and decreases again for faster motion. It is impossible to decide whether the image motion resulted from a pattern that moves at low or very high

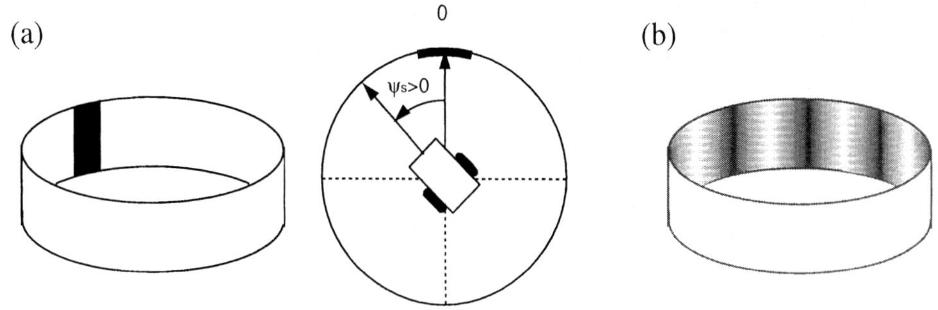

Figure 4 *Sketch of simulated arena. (a) side and top view of the arena with one black stripe mapped onto the wall. The stripe is at 0°, starting orientation of agent is at ψ_s. (b) A sinusoidal pattern is mapped onto the wall.*

speed and full compensation is not possible. In general, a purely proportional controller (without memory) may cause oscillations, because the compensation behavior is active only until the retinal image is stabilized. As the disturbance persists, image flow will be detected again and the compensation behavior is active again until the retinal image is stabilized again etc. This second point, however, does not seem to be a disadvantage for the control system of flies. Warzecha and Egelhaaf (1996) could show that due to the response characteristic of the motion detector large-amplitude fluctuations in the image motion, which are generated when the optomotor system gets unstable, are transmitted with a small gain leading to only relatively small turning responses and thus small image motion.

To model the optomotor response behavior, the signals β_l and β_r that result from the large field units are integrated over time. In simulation the integration is replaced by a summation:

$$B_{l,r}(t) = \sum_{-\infty}^{t} \beta_{l,r}(t') \tag{8}$$

The coupling of these signals to the motor system, leads to the control signals:

$$\begin{aligned} m_l^{or}(t) &= \omega_{ll} B_l(t) + \omega_{lr} B_r(t) \\ m_r^{or}(t) &= \omega_{rl} B_l(t) + \omega_{rr} B_r(t) , \end{aligned} \tag{9}$$

with the transmission weights from Eq. 7.

Optomotor response and object fixation: Experimental results suggest that the behaviors for course stabilization and approach towards stationary objects of flies may be realized at least partially by a common sensory circuit (Götz and Wenking, 1973; Götz, 1975; Reichardt and Poggio, 1976; Bülthoff, 1982). In order to test this we designed agents with a visuomotor controller, that regulates both the optomotor response and the fixation behavior.

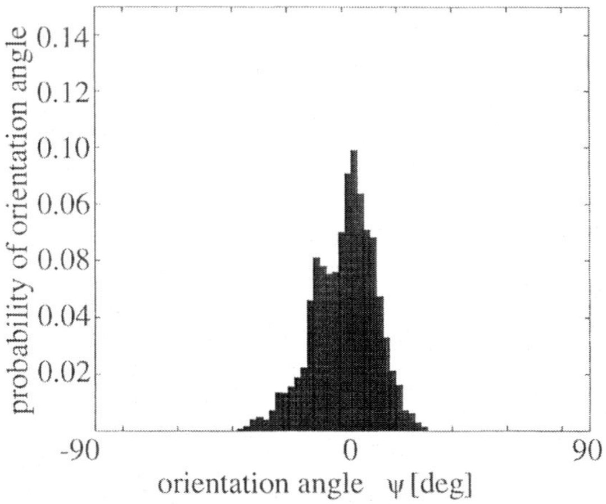

Figure 5 *Histogram of the agent's orientation angle ψ in a drum: with one stripe at $\psi_s = 0.0°$: $\bar{\psi} = 0.95° \pm 11.21°$.*

For the controller the control signals for optomotor response m_{or} and object fixation m_f are combined into a proportional-integral controller:

$$m_{l,r}(t) = k_f m_{l,r}^f(t) + k_{or} m_{l,r}^{or}(t) , \tag{10}$$

with the constant factors $k_f = 0.5$ u/step and $k_{or} = 5.0 \cdot 10^{-4}$ u/step.

3. Experiments with the simulated agent

3.1 Fixation behavior

To compare the performance of the fixation behavior of the agent with that of flies in the experiments of Reichardt and Poggio (1976), we run three experiments. In all three experiments, the agent is fixed in the middle of a drum and therefore has only one degree of freedom, the rotation around its vertical body axis (Fig. 4a). The angular velocity of the turning response $\dot{\psi}$ is given by Eq. 5.

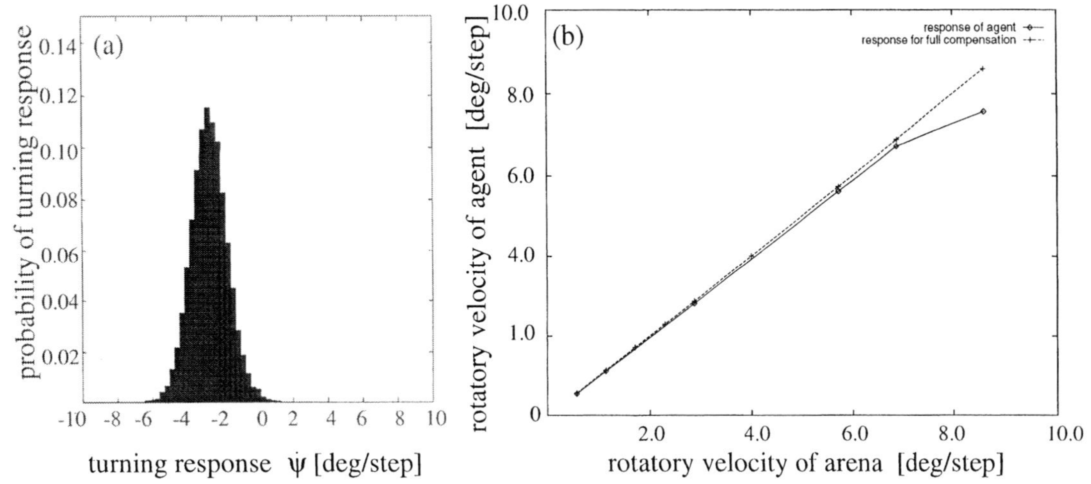

Figure 6 *Optomotor response: (a) Histogram of the agent's optomotor response. The agent is fixed at the center of a rotating drum. Rotation of drum:* $\dot{\psi}_d = 2.9°/step$*; rotation of agent:* $\bar{\dot{\psi}} = 2.8 \pm 1.0°/step$*. (b) The agent's optomotor response in dependency of the angular speed of the arena.*

Experiment: A single black stripe (angular width 17.3°) is presented on the wall of the drum in an otherwise white surrounding. The initial orientation of the agent is $\psi_s = 41.4°$. The simulation runs for 10,000 time steps.

Result: Due to random torques caused by intrinsic noise of the system, the retinal image of the stripe is not stationary. The agent turns towards the object, because the turning response is stronger to progressive than to regressive motion. Within 1100 steps the agent orients it's heading direction towards the stripe. For the rest of the simulation the agent is able to fixate on the stripe (Fig. 5). The slightly overlapping receptive fields of the integration units are necessary for the agent to keep the stripe in the front, because agents which lack this characteristic are not able to fixate on the stripe in front of them.

3.2 Optomotor response behavior

Experiment: Like in the previous experiment, the agent is fixed in the middle of a drum (Fig. 4b) and can only turn around its vertical body axis. In order to test the optomotor response a sinusoidal pattern ($\lambda = 36°$) rotates with a constant angular velocity $\dot{\psi}_d$. The simulations run for 10,000 time steps with different velocities $\dot{\psi}_d$ (Fig. 6b).

Results: In order to stabilize the retinal image and thus its orientation in the drum, the agent produces on average a compensatory turning response of $\bar{\dot{\psi}} \pm \sigma_{\dot{\psi}}$ (Fig. 6). The agent is able to compensate for 97% of the arena's rotation if the angular velocity is lower than 7° /step. Due to the sigmoid function (Eq. 4) the agent is not able to compensate to this extent for higher velocities of the drum.

Experiment: Due to the response characteristic of the Reichardt motion detector, a purely proportional controller, does not lead to an unstable behavior, we also test the performance of the agent with a purely proportional controller (Warzecha and Egelhaaf, 1996):

$$m_{l,r}(t) = k_f m^f_{l,r}(t) \ , \qquad (11)$$

with the constant factors $k_f = 0.5$ u/step.

Result: With a purely proportional controller the agent is able to compensate for only 35% of the external rotation of the drum ($\bar{\dot{\psi}} = 1.0 \pm 1.3°$/step). However, the control signal does not lead to an unstable behavior. The standard deviation is only slightly larger than with the proportional-integral controller.

4. Implementation on the robot

We implemented the controller on a mobile robot – the *Khepera*^{TM} robot. The visual signals result from a small video camera with a conical mirror mounted above it (Chahl and Srinivasan, 1996). The optical axis of the video camera is oriented to the center of the cone. This configuration provides a 360° horizontal field of view extending from 10° below to 10° above the horizon (Franz, Schölkopf, Mallot, and Bülthoff (in press)). The image is sampled on five circles along the horizon within a vertical aperture of 2.1°. The samples are averaged vertically to provide robustness against inaccuracies in the imaging system. Then the samples are horizontally lowpass

Figure 7 *Khepera^{TM} robot with vision module. The optical axis of the camera is oriented vertically upwards, receiving mainly visual input from the image on the conical mirror.*

filtered using a Gaussian filter, resulting in 96 sensors on the horizontal ring, 48 for each eye. The resolution is higher than in the simulation.

Table 1 *Predictive coding: Weights of the lateral inhibition area with center pixel 0 and lateral pixels $\pm3, \pm2, \pm1$ (obtained from 2000 images recorded during typical trajectories of the robot).*

pixel	0	±1	±2	±3
weight	1.000	-0.510	0.003	0.007

For the processing of the video images, we model besides the temporal aspects of the LMCs, their spatial aspects by a spatial bandpass filter obtained by predictive coding – a procedure known for image compression (e.g., Gonzales and Woods, 1992). The weighting function of the filter is formed by three input units (m=3) in either direction. 1 gives an average filter characteristic that is obtained from 2000 images recorded during typical trajectories of the robot. The visual stimuli are then temporally highpass filtered and amplified. Filtering as well as motion detection are the same as in simulation.

The transmission weights that couple the outputs of the large field integration units β_l and β_r to the motor system are the same as in simulation (Eq. 7) and the control signals result from (Eq. 10). The constant factors are set to $k_f = 4.0$ mm/s and $k_{or} = 4.0 \cdot 10^{-3}$ mm/s. Intrinsic noise signals with a Gaussian distribution (Fig. 8) cause random rotations around the vertical axis. The velocity of the motors can be set stepwise to $\pm 8\,n$ mm/s with ($n = 1, ..., 10$) for the left and right motor. The basic velocity is set to $v_0 = 40$ mm/s, which is modulated by the visual motion signals. The robot is updated at a rate of 12 Hz.

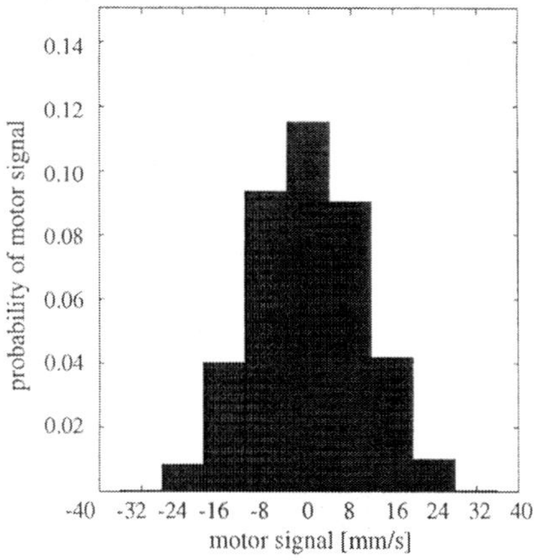

Figure 8 *Gaussian distribution of turning response due to the noise that is added to the motor signals.*

5. Experiments with the robot

5.1 Optomotor response behavior

Experiment: Similar to the fixation experiments with flies, in this experiment the robot is only able to turn around its vertical axis within a circular arena (diameter: 45 cm). For the optomotor response a pattern with black and white stripes is painted onto the wall of the arena ($\lambda = 51.4°$). Instead of a constant rotatory bias of the drum, asymmetric motor signals are added to the left and right motor: ∓ 24 mm/s (this corresponds to $\dot{\psi} = \pm 4.2°$/frame).

Result: In the case of an asymmetric motor signal that leads to a rotation of the robot about the vertical body-axis, the robot produces a compensatory turning response by the internally generated control signals compensating for 97% of the motor asymmetry (Fig. 9a).

5.2 Fixation behavior

Experiment: To test the fixation behavior, a single black stripe (angular width: 25.7°) is painted onto the white wall of the arena. The experiment is started with the stripe being at an orientation of $\psi = 57°$ off the heading direction. Instead of measuring the absolute orientation of the robot in the drum, we used the visual angle of the stripe's retinal image.

Result: In an arena with a single black stripe, the robot is able to orient itself towards the stripe after about 8 s and fixate on it (Fig. 9b). This is in accordance with the result from the simulation.

5.3 Approach behavior

Experiment: In the next experiment the robot was started from various positions and with various orientations in the arena (Fig. 10). The robot has two degrees of freedom: translation in the heading direction and rotation around the vertical body-axis. The rotatory and

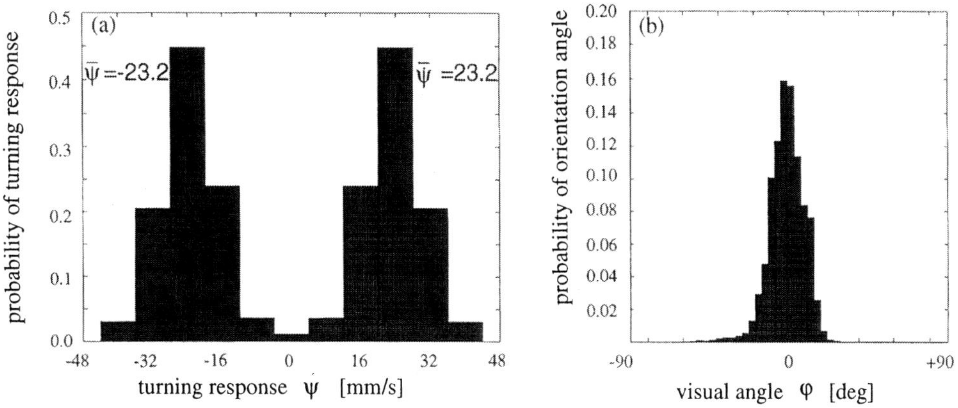

Figure 9 *(a) Histogram of the optomotor response in the presence of a motor asymmetry of the left and right motor* ($\mathbf{v}^a = (v_l^a, v_r^a) = (-24, +24)$ mm/s*) (9700 steps). Compensatory motor signals of the left and right motor result in* $v_l = 23.2 \pm 7.2$ mm/s *and* $v_r = -v_l$. *(b) Histogram (9700 steps) of the visual angle* φ *of the stripe's retinal image:* $\bar{\varphi} = -0.29° \pm 10.36°$.

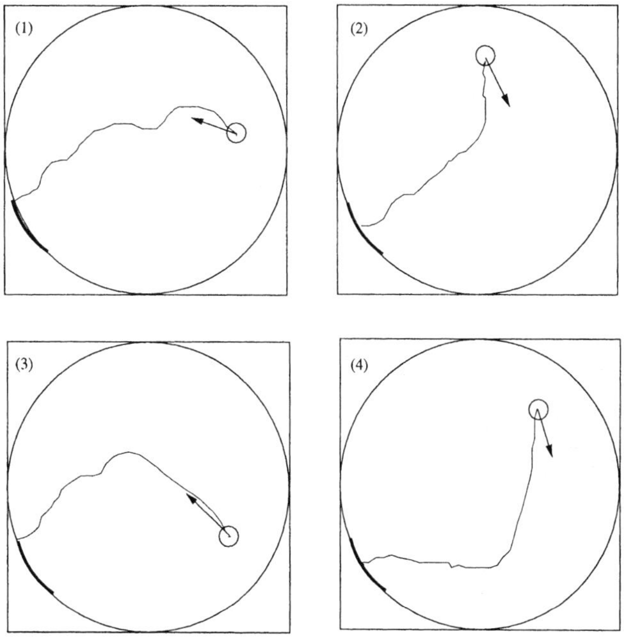

Figure 10 *Approach of single stripe in an arena. The robot starts with the orientations relative to the center point of the stripe* $\Delta\psi = -60°, 60°, 65°, 70°$.

translatory velocities are set according to Eq. 5.

Result: If the robot has two degrees of freedom, i. e. it is enabled to translate and rotate, it approaches the stripe in all cases (Fig. 10). The movement of the robot is guided by the motion that is perceived at the border of the stripe, because from the uniform black stripe no motion information can be derived. Therefore, the agent approaches often one stripe instead of heading directly towards the middle of the stripe. The trajectories have been obtained by a visual tracking system, with a video camera recording the arena from a birds-eye view. The system tracks a red marker on top of the robot.

6. Discussion

We investigated whether one visuomotor controller is sufficient to generate both optomotor response and fixation behavior and we can in fact show that this is possible: The simulated agent as well as the $Khepera^{TM}$ robot show optomotor response and fixation behavior depending on the particular environmental conditions. In addition the robot approaches prominent objects in the arena if it is able to move around.

In order to explain the optomotor response, experimental results on flies suggest a temporal integration stage of the the output signals from the horizontal cells (Heisenberg and Wolf, 1984). We could show that for our agent the integration of the image motion signals leads to an optomotor response behavior comparable to that of flies. Purely proportional control signals are not sufficient to compensate for the drum's rotation. However, like in flies they do not lead to an unstable behavior, because of the motion detectors' response characteristic.

For the fixation of a single stripe, the following perceptual properties are essential: (i) an asymmetric response of the motion detectors to progressive and regressive motion, (ii) the spatial sensitivity function of the large field integration units with overlapping fields in front of the agent. The fixation behavior of the simulated agent is qualitatively comparable to that of flies.

The implementation of the control structure onto the robot was straight forward and the results show that the simulations represent the real world conditions of the robot very well. We slightly changed the proportionality factors of the proportional and integral parts of the controller. The robot is able to fixate on the black stripe as well as to stabilize the retinal image by optomotor response. With two degrees of freedom the robot even approaches the prominent object in the arena.

The implementation of our controller on an analog VLSI chip would be very useful for autonomous robots and for further studies of fly behavior. On such chips, motion detection mechanisms have been developed successfully (Sarpeshkar, Kramer, Indiveri, and Koch, 1996; Harrison and Koch, 1998). These chips integrate both the photosensors and the motion computation on a single chip. Like in biological systems, a parallel processing of visual signals is thus possible. Although the sensors on the chips show a poor signal-to-noise ratio, and the performance decreases significantly at low contrast and low illumination levels. For the simulation of biological processes this does not have to be a disadvantage, as noise is intrinsic in all biological neural systems.

7. Conclusions

The comparison of the results with investigations in flies suggests that the large field cells most probably participate in mediating both the optomotor response and the fixation behavior. Further investigations of the robot's behavior under various experimental conditions have to be done to compare its behavior in more detail with that of flies. It is known from investigations on flies that the large field cells alone are not responsible for the object fixation behavior. Full object fixation is mediated by the contribution of other cells, for example the small field cells (Egelhaaf, 1985) which respond selectively on motion in small parts of their receptive field. In future work our model may be extended by adding small field units.

Our controller has important implications for the design of autonomous agents with a behavior based control architecture. There is one fundamental problem for the design of behavior-based architectures. The design of the agent's architecture involves the definition of a basic set of behaviors that enables the agent to accomplish a given task. The designer has to predict all possible interactions between behavioral modules and the environment. Thus the decomposition of the agent's architecture into behavioral modules has to be done very carefully. As we showed here, in some cases it is not necessary to realize separate modules for different behaviors. Our controller regulates both the optomotor response and the fixation behavior. Which of the two behaviors predominates, depends on the particular environment. Theore, a further detailed understanding of the functionality of sensorimotor processes in biological systems would be very useful for the development of control algorithms for autonomous robots.

Acknowledgments: This work was done in close collaboration with M. O. Franz whom we thank for his interest and valuable criticism. We would also like to thank H. Krapp and A. Borst for stimulating discussions on fly vision and helpful comments on preliminary versions of this article. A longer and more detailed version of this article, in which further tests of the controller are described, is submitted to 'Robots and Autonomous Systems'.

References

R. D. Beer (1990). *Intelligence as adaptive behavior.* Academic Press, San Diego, CA.

R. A. Brooks (1986). A robust layered control system for a mobile robot. *IEEE J. Robot. and Aut.*, 2(1):14 – 23.

H. Bülthoff (1982). *Drosophila* mutants disturbed in visual orientation (ii.). *Biol. Cyb.*, 45:71 – 77.

D. T. Cliff (1992). Neural networks for visual tracking in an artificial fly. In *Proc. 1st Europ. conf. on Art. Life*, pages 78 – 87, Cambridge, MA, MIT Press.

H. Cruse, C. Bartling, G. Cymbalyk, J. Dean, and M. Deifert (1995). A modular artificial neural network for controlling a 6 – legged walking system. *Biol. Cyb.*, 72:421 – 430.

A. P. Duchon and W. H. Warren (1994). Robot navigation from a gibsonian viewpoint. In *Proc. IEEE Int. Conf. Systems, Man and Cyb.*, pages 2272 – 2277, Piscataway, NJ, 1994. IEEE.

M. Egelhaaf (1985). On the neural basis of figure – ground

discrimination by relative motion in the visual system of the fly (ii.). *Biol. Cyb.*, 52:195 - 209.

M. Egelhaaf and A. Borst (1993). Motion computation and visual orientation in flies. *Comp. Biochem. Physiol. A*, 104(4):659 - 673.

J. D. Foley, A. van Dam, S. K. Feiner, and J. F. Hughes (1987). *Computer graphics - principles and practice.* Addison-Wesley, Reading, MA.

N. Franceschini, J. M. Pichon, and C. Blanes (1992). From insect vision to robot vision. *Phil. Trans. R. Soc. Lon. B*, 337:283 - 294.

M. O. Franz, B. Schölkopf, M. A. Mallot, and H. H. Bülthoff (1998). Learning view graphs for robot navigation. *Aut. Robots*, 5:111 - 125.

R. C. Gonzales and R. Woods (1992). *Digital image processing.* Addison - Wesely, Reading, MA.

K. G. Götz (1964). Optomotorische Untersuchung des visuellen Systems einiger Augenmutanten der Fruchtfliege *Drosophila. Kybernetik*, 2(2):77 - 92.

K. G. Götz (1975). The optomotor equilibrium of the *Drosophila* navigation system. *J. Comp. Physiol.*, 99:187 - 210.

K. G. Götz and H. Wenking (1973). Visual control of locomotion in the walking fruitfly *Drosophila. J. Comp. Physiol.*, 85:235 - 266.

R. R. Harrison and C. Koch (1998). An analog VLSI model of the fly elementary motion detector. In *Advances in Neural Information Processing Systems 10*, San Mateo, CA, Morgan Kaufman Publishers.

M. Heisenberg and R. Wolf (1984). *Vision in Drosophila.* Springer Verlag, Berlin.

J. M. Missler and F. A. Kamangar (1995). A neural network for pursuit tracking inspired by the fly visual system. *Neural Networks*, 8(3):463 - 480.

F. Mondada, E. Franzi, and P. Ienne (1994). Mobile robot miniaturization: A tool for investigation in control algorithms. In *Proc. 3rd Int. Symp. Exp. Robotics*, London, Springer Verlag.

W. Reichardt and T. Poggio (1976). Visual control of orientation behavior in the fly. part 1: A quantitative analysis. *Quart. Reviews Biophysics*, 9(3):311 - 375.

R. Sarpeshkar, J. Kramer, G. Indiveri, and C. Koch (1996). Analog vlsi architectures for motion processing: From fundamental limits to system applications. 84(7):969 - 982.

N. Tinbergen (1953). *Instinktlehre - vergleichende Erforschung angeborenen Verhaltens.* Parey Verlag, Berlin, Hamburg.

A.-K. Warzecha and M. Egelhaaf (1996). Intrinsic properties of biological motion detectors prevent the optomotor control system from getting unstable. *Phil. Trans. R. Soc. Lon. B*, 351:1579 - 1591.

B. Webb (1995). Using robots to model animals: a cricket test. *Robotics and Aut. Systems*, 16:117 - 134.

K. Weber, S. Venkatesh, and M. V. Srinivasan (1997). Insect inspired behaviours for the autonomous control of mobile robots. In S. Venkatesh M. V. Srinivasan, editor, *From living eyes to seeing machines*, Oxford, New York, Tokio, Oxford University Press.

How Should a Robot Discriminate Between Objects? A comparison between two methods

Edwin de Jong, Paul Vogt
Vrije Universiteit Brussel
Pleinlaan 2, B-1050 Brussels, Belgium
{edwin, paul}@arti.vub.ac.be

Abstract

In order to discriminate among the different objects in its environment, an agent may develop a primitive notion of concepts based on the sensor data it receives. In this paper, this phenomenon is investigated by having software agents play *discrimination games* with the sensor data of autonomous robots. We have compared the Simple Prototype method and the Adaptive Subspace method. Both methods achieve high discrimination-success rates. The Adaptive Subspace method accomplishes this with a converging and relatively low number of categories. The purpose of these discrimination games is to serve as a basis for lexicon formation experiments. From the experiments in this paper, we conclude that the Adaptive Subspace method is more attractive for discrimination games.

1. Introduction

When a robotic agent investigates its environment it may, depending on its task, need to discriminate between the different objects it encounters. Deciding whether or not two observations concern the same object implies a categorization of the world. Provided a robot has appropriate and sufficiently precise sensors at its disposal, it can classify observations of the objects surrounding it into a number of different categories. We investigate this by letting agents play *discrimination games*. The discrimination games the robots play and the theory behind these one-player games were introduced in (Steels, 1996b); see also (Vogt, 1997). These games are investigated in the context of experiments on lexicon formation, for an overview see (Steels, 1997).

In other experiments at our lab called *naming games* (Steels and Vogt, 1997)(Steels, 1996a), the goal is to let robots communicate about their environment. A problem for these robots is that repeated encounters with the same object generally do not yield identical observations. The differences are caused by changes of the viewing angle, different lighting conditions, reflections, or sensor inaccuracies. In spite of these differences, an agent should categorize its observations such that an object consistently falls into the same category on separate encounters. If this is not the case, the development of meaningful communication about the objects will be severely hindered. Thus, in these experiments, a meaningful conceptualization of the environment will yield a limited number of categories; the smaller the number of categories, the larger the possibility that two agents will each consistently label an object as belonging to a certain category. This in turn facilitates the development of names for such categories which can be used in communication to identify objects. The number of possible categories to be considered grows rapidly with the number of sensors.

We are interested in the question of how a robot should discriminate between objects. In these experiments, what constitutes an object is directly based on the robot's sensory information. This usually, but not always, corresponds to a light source. The agent can autonomously determine whether its categorization allows it to discriminate between the objects in a context. Thus, discrimination success can be calculated by the agent itself, and is not externally given as is the case in reinforcement learning problems. Since no categorization is favored over other categorizations (as is the case with classification), the problem here is unsupervised learning. Many methods for unsupervised learning exist; in this paper, we investigate two possible methods. Both methods adapt the features which determine the agent's categories. Arguments from a cognitive perspective in favor of feature adaptation as opposed to fixed features are discussed in (Schyns et al., 1998).

The first method is based on *Simple Prototypes*. A prototype is a type of feature that can be represented by a sensor-value pair. New prototypes are created whenever discrimination fails. The second method is an adaptive subspace method that resulted from an attempt to combine the Exploration Buckets algorithm (De Jong, 1997) with generalization. This method generalizes over the sensor space by distinguishing only between features that enhance discrimination.

2. The robots and their environment

The experiments are carried out with the two mobile robots shown in figure 1. These robots are controlled by the sensory-motor board SMBII that was developed at the VUB AI-Lab (Vereertbrugghen, 1996). The robots are equipped with several sensors and actuators, as well as a radio-link. The sensors include a white light sensor, a modulated light sensor and three infrared (IR) sensors. One sensor of each type is mounted on the front of the robot and is used for perception. A virtual sensor is introduced to denote the robot itself, since one of the robots may be the subject of a naming game as well. Furthermore, the robots have IR sensors on both sides that are used for IR-taxis and -orientation (Steels and Vogt, 1997) and four bumpers for obstacle avoidance. The actuators include four IR emitters to make each robot visible to the other and two motors to produce the movements of the robots.

Figure 1 The robots used in the experiments.

In the experiments, the robots engage in a series of so-called naming games. As part of the naming game, the robots perform a perception task. This involves rotating 360 degrees and detecting objects using sensors as explained below, thus obtaining a context for the naming game. The objects the robots can detect are made visible by different light sources; the objects include the robots themselves. The light sources correspond to the sensors: there is a white light source, a source emitting light at a modulated frequency and an infrared source. During the perception task, the robots scan their sensors, which are pre-filtered by a threshold to filter out background noise.

The first object in the context is a virtual detection of the robot itself. When a robot encounters an object as it turns, one or more of the sensors will increase until a maximum is reached, after which the sensor value(s) will drop again. Rotating 360 degrees results in a graph like that shown in figure 2. A bounding box is drawn around every hill in the graph in the interval for which the sensor was activated. Since the robot turns (approximately) 360 degrees, the graph is a panoramic view. Therefore, if a sensor is active at both the beginning and the end of the graph, they should be interpreted as a single object.

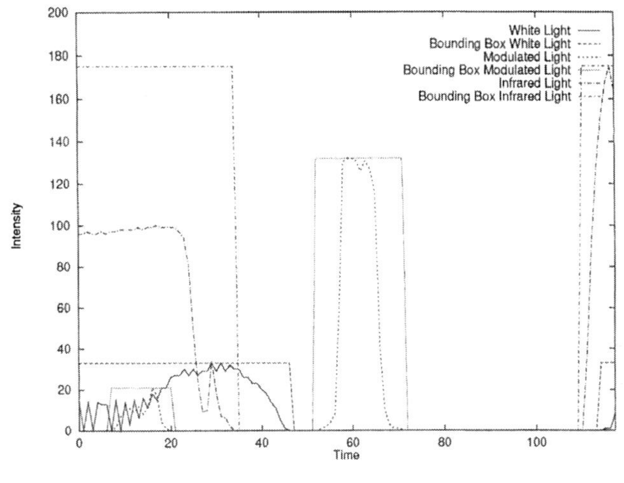

Figure 2 An object as perceived by the robot.

The two outer bounding boxes for infrared in figure 2 are thus part of the same bounding box. A bounding box is interpreted as an object located in the middle of that box. If the positions of two boxes detected are within a distance of 5 time steps, these boxes are considered to represent one object.

For every object, two values for every sensor are obtained as sensory channels. The first one is the filling ratio of the detected hill inside the bounding box and the second value is the height of the hill. The representation for the robot itself is added at the beginning of the context. For reasons of consistency, this virtual sensor also has two sensory channels, thus making the total number of sensory channels 8. This method for object detection differs from previous work. The resulting context contains the detected objects with values for the 8 sensory channels. This data is transmitted to a radio station linked to a PC for further off-line processing.

3. Discrimination Games

In the perception task, a robot has gathered sensor data of the objects it has observed in its environment. Every object can be related to a set of features that the agent has constructed. These objects together constitute the *context*. The agent then chooses one object from the context, called the *topic*. The aim of the discrimination game is to distinguish the topic from every other object in the context. If this discrimination is possible, the game ends in success. If not, it ends in failure, and the agent will try to improve its ability to discriminate by generating or modifying its features. The two methods that have been investigated, take different approaches to relating objects to features and for adapting features. These methods are described in the following two sections.

With the two methods discussed, the agent has a set of features that it adapts based on its experiences in order to increase its chances of success in the discrimination games. The features are quite primitive, in the sense that they directly correspond to a subset of the values of a certain sensory channel. If an object's value v for that sensory channel is a member of this set, the feature is said to *cover* v. Features can be combined into *feature sets*. A feature set covers an object if all features in the set cover the value of the feature's sensory channel for that object. For both methods, two features in a feature set never correspond to the same sensory channel. Thus, the number of features in a feature set is never greater than the number of sensory channels. A feature set is *discriminative* iff it covers the topic but none of the other objects in the context. The discrimination game results in a set of distinctive feature sets. This set may be empty, in which case the discrimination game fails and the system is adapted, as described in the following sections.

4. The Simple Prototype Method

The Simple Prototype method is based on the generation of features which can be viewed in some sense as prototypes. These prototypes should not be confused with the prototypes that are normally used in AI, which are frames for whole objects. The prototypical features discussed here act at the subsymbolic level. Sensor values observed for objects are mapped onto an appropriate prototypical feature. These features are compared in the discrimination games in order to distinguish the topic from other objects in the context. If the discrimination game is unsuccessful, a new feature may be generated for which a sensory channel of the topic is an example. The successful feature sets remain in the memory of the agents. In this way the features are selected for their discriminative powers. So, the method constitutes a selectionist approach; see also (Popescu-Belis, 1997), where Edelman's theory of neuronal group selection (Edelman, 1987) is used. Note that the method is different from the one used in previous work (Vogt, 1997)(Steels, 1996b).

Now, suppose an agent that has a set of prototypical features $\{f_i =< sc_i, v_i, s_i, u_i >\}$, where the sc_i is the attribute, which is the sensory channel on which the feature works, v_i is a value, s_i is the success score and u_i is the use score for f_i. Suppose furthermore that the agent has a context C of objects o_i, where every object has a sensorvalue $o_{i,j}$ pairs for every sensory channel j. The context is obtained from the perception of the robot. The agent has chosen one of the objects from the context to be the topic t. The method can now be described as follows:

- Every object $o_k \in C$ is covered by a set of features

$$\Lambda^{o_k} = \{f_i \mid \forall f_j : (((f_j \neq f_i) \wedge (sc_j = sc_i = l)) \rightarrow (|v_i - o_{k,l}| \leq |v_j - o_{k,l}|)) \wedge (v_i > 0)\}$$

where $o_{k,l}$ is the detected value for sensory channel l. This method for covering is shown schematically in figure 3.

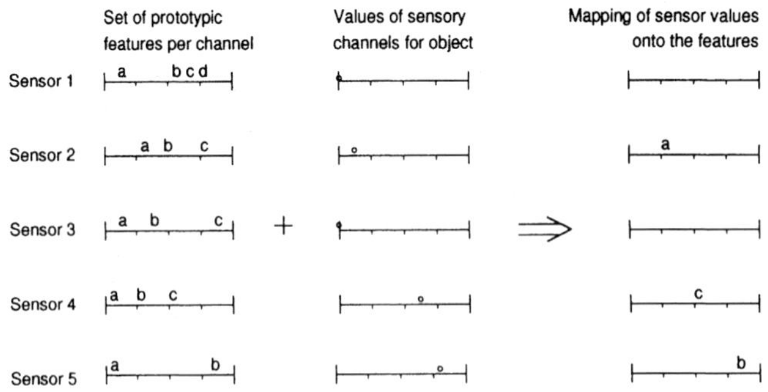

Figure 3 Determining which features cover an object using the Simple Prototype method

- The agent tries to distinguish topic t from the other objects in the context. The result of the discrimination is the set of distinctive feature sets:

$$D^t = \{D^t_n \mid D^t_n \subseteq \Lambda^t \wedge \forall o \in C\backslash\{t\} : (\exists f_j \in D^t_n : \neg(f_j \in \Lambda^o))\}$$

Every D^t_n is associated with a pair of $< s_n, u_n >$, where s_n is a success score and u_n is a usage score. The usage u_n for all $D^t_n \in D^t$ is incremented.
- If $|D^t| \geq 1$, we prefer the set D^t_n for which: (1) $\forall D^t_m \in D^t : |D^t_n| \leq |D^t_m|$, or (2) if this holds for more than one n, we prefer the set for which holds: $s_n/u_n \geq s_m/u_m$ for all $m \neq n$. The success score s_n is incremented for the chosen D^t_n.
- If $D^t = \emptyset$, the features need to be adapted. A new feature $f =< sc, v >$ is generated, so that $sc = sc_i$ and $v = o_{t,i} > 0$ for an arbitrary sensory channel i.

Initially, every agent has exactly one feature for every sensory channel, with a value that is in the middle of the range for that sensory channel.

Let us look at the example given in table 1. Agent r1 has observed 5 objects **o0**,. . .,**o4**. Object **o0**, for example, has value 1 for sensory channels 0 and 1 and value 0 for the remaining channels, and **o1** has values 200 and 1 for sensory channel 4 and 5 respectively.

Feature sets r1:
o0: {sc0-1,sc1-1}
o1: {sc4-127,sc5-127}
o2: {sc2-127,sc3-127}
o3: {sc4-66,sc5-127}
o4: {sc2-127,sc3-127,sc6-127,sc7-127}

In this example, *sc0-1* is a prototypical feature, where *sc0* is the sensory channel and the feature has the value 1; use and success scores are not given. For each object, the features are determined. These together constitute the object's feature set. Object **o0** is covered by features *sc0-1* and *sc1-1*, and object **o1** is covered by *sc4-127* and *sc5-127*. Now, suppose that the agent has chosen object

o1 to be the topic. Since the feature set that covers this object does not cover any other object, this feature set is discriminative. Furthermore, feature *sc4-127* does not cover any other object and is therefore also discriminative. On the other hand, we see feature *sc5-127* in **o3**, and this is not a distinctive feature set. The game is successful and since {*sc4-127*} is the smallest set, this set is preferred.

Now suppose object **o2** was the topic. In this case both features that cover **o2** can also be found in the feature set that covers **o4**. The distinctive feature set would be empty, and a new feature would be made. The feature that would be made is either *sc2-245* or *sc3-57*, since these are the only non-zero sensor values for **o2**.

Since a discrimination game may yield several distinctive feature sets, the number of feature sets may be larger than necessary. This increases the search space considerably, and therefore it is attractive to forget unsuccessful feature sets. Starting from discrimination game 420, every 30 games the agent starts to forget those feature sets that sofar have not been successful ($s = 0$), and that have been acquired more than 300 games before. This way, the unsuccessful feature sets may have the opportunity to become successful.

sc	0	1	2	3	4	5	6	7
o0	1	1	0	0	0	0	0	0
o1	0	0	0	0	200	1	0	0
o2	0	0	245	57	0	0	0	0
o3	0	0	0	0	57	3	0	0
o4	0	0	245	57	0	0	113	130

Table 1 An example of a context with 5 objects. The table lists the values of 8 sensory channels for each object.

5. The Adaptive Subspace Method

The Adaptive Subspace method is based on the principle that different sets of sensory inputs should only be treated as different if this distinction is meaningful. Which distinctions are meaningful is determined by the application. In the case of discrimination games, a distinction should be made if and only if this increases the ability of the agent to discriminate the topic from the other objects, in one or more of the discrimination games it has played. This method results from research on generalization, an important issue in machine learning. An overview is beyond the scope of this paper, but for interesting contributions on generalization, including other subspace methods, see (Landelius, 1994; Murao and Kitamura, 1997; McCallum, 1996). In this paper, orthogonal splits were used; an interesting variation would be to allow non-orthogonal splits as well. Oja discusses subspace methods where, unlike here, subspaces have a lower dimensionality than the original data (Oja, 1983).

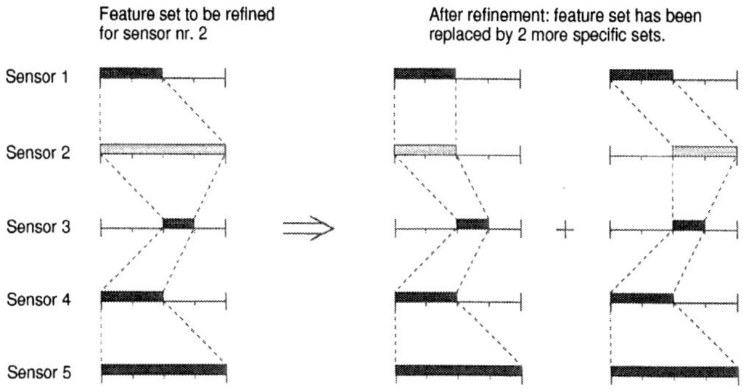

Figure 4 Refining a feature set using the Adaptive Subspace method.

When a discrimination game is played, the agent determines for each object it has observed, by which feature set it is covered. If the feature set of the topic covers no other object, the agent can discriminate the topic from the other objects, and the game is a success. When a game fails, the feature set of the topic is investigated to see whether it should be refined. Refining a feature set amounts to splitting one of the ranges in half and replacing the existing feature set with two new feature sets, one for each half. This is illustrated in figure 4. We will now describe the refinement procedure in greater detail, and then give an example.

- For sensory channel j, feature F_j is a tuple of two integers $< F_{j,l}, F_{j,u} >$, which represent the lower and upper values of an interval. An object o_i is represented by its values for the n sensory channels $o_{i,j}$, where $j \in [0 \ldots n-1]$. F_j covers $o_{i,j}$ if $F_{j,l} \leq o_{i,j} \leq F_{j,u}$. This is written as $covers(F_j, o_{i,j})$. Every feature set contains a feature for each sensory channel. A feature set $F = <F_0, F_1, \ldots F_{n-1}>$ covers an object o_i if $\forall j \in [0 \ldots n-1] : covers(F_j, o_{i,j})$.
- Every point in the statespace is covered by exactly one feature set. This property is not lost during a refinement. The method thus adapts a division of the statespace into regions, hence the name. The contexts C here contains all the objects in a discrimination game except the topic t. The success ratio s is the fraction of the objects in C that are not covered by the feature set F that covers t, and can thus be distinguished from the topic:

$$s = \frac{|\{o | o \in C \wedge \neg covers(F, o)\}|}{|C|}$$

- For each dimension, we compute the success increase that a split in that dimension would yield. After a split, F is replaced by 2 new feature sets FA and FB. The success increase Δs_d for a split in dimension d is calculated as follows:

$$m_j = \frac{F_{j,l} + F_{j,u}}{2}$$

$$FA_j = \begin{cases} <F_{j,l}, F_{j,u}> & if \quad j \neq d \\ <F_{j,l}, m_j> & if \quad j = d \end{cases}$$

$$FB_j = \begin{cases} <F_{j,l}, F_{j,u}> & if \quad j \neq d \\ <m_j, F_{j,u}> & if \quad j = d \end{cases}$$

$$\Delta s_d = \begin{cases} \frac{|\{o|o \in C \wedge covers(FB,o)\}|}{|C|} & if \quad covers(FA,t) \\ \frac{|\{o|o \in C \wedge covers(FA,o)\}|}{|C|} & if \quad covers(FB,t) \end{cases}$$

This is the fraction of the objects in the context (not including the topic) that can be discriminated from the topic after, but not before a split in dimension d.

- The highest success increases of the past h games are stored in a history window named sh. If the inequality

$$\max_d \Delta s_d \geq \max_h sh_h$$

holds, the split in the corresponding dimension is actually performed. Sometimes,

$$\max_d \Delta s_d = 0 \wedge s < 1.0$$

This means no split would have yielded an increase of the success for this discrimination game, but discrimination is nevertheless imperfect. This is the case when all of the objects in the topic's feature set have all of their sensor values in the same halve of the corresponding feature's interval as the topic. In that situation, a *distance split* is performed. During a distance split, the feature set covering the topic is split in the dimension dim with the highest average distance between the sensor values of the objects covered by the feature set and the corresponding sensor value of topic t:

$$dim = arg \max_{j \in [0 \ldots n-1]} \sum_{i \in [1 \ldots |C|]} \frac{||o_{i,j} - t_j||}{i}$$

It is crucial that the refinement in the range of the sensory channel is made only for the current feature set (= subspace). This is what allows the method to handle the higher-dimensional statespaces that result when several sensory channels are used (8 sensory channels or dimensions in these experiments).

We now give an example of discrimination games using the Adaptive Subspace method. Let us assume agent r1 has the following feature sets:

$FS1$ = [0, 255] [0, 255] [0, 255] [0, 255] [0, 255] [0, 255] [0, 127] [0, 255]

$FS2$ = [0, 255] [0, 255] [0, 255] [0, 255] [0, 255] [0, 255] [128, 255] [0, 255]

Furthermore, suppose the agent chooses **o4** (see table 1) to be the topic. Since all objects have a value below 128 for sensory channel 6, all objects are covered by FS1. Thus, none of the 4 objects in the context can be discriminated from the topic, yielding a success score of $\frac{0}{4} = 0$. Since discrimination fails, the agent considers splitting the subspace of the topic, FS1, for each dimension. Dimensions 0 through 5 yield no improvement, but if FS1 is split in dimension 6, it would be replaced by the following two feature sets:

$FS1A$ = [0, 255] [0, 255] [0, 255] [0, 255] [0, 255] [0, 255] [0, 63] [0, 255]

$FS1B$ = [0, 255] [0, 255] [0, 255] [0, 255] [0, 255] [0, 255] [64, 127] [0, 255]

After this split, the topic would be covered by FS1B, but all 4 other objects would be covered by FS1A. Thus, the new success is $\frac{4}{4} = 1$. If the success increase (1 - 0 = 1.0) is greater than or equal to the maximal success increase in the history window, the split is actually performed.

6. Results

In the experiments described here, the Simple Prototype method and the Adaptive Subspace method have been applied to discrimination games with the same robot data. The data set contains 321 contexts. For each context the data contains the sensory channel values of the objects. During an experiment, the sequence of contexts encountered by a robot is repeatedly presented to an agent in the original order. In this section we describe the results of these experiments, and compare the two methods. The robot data has been collected in runs of about 45 minutes (a period limited by the robot's battery). To collect the data, the robots have both been active for 9 hours in total.

Figure 5 shows the success of the Simple Prototype method. The initial success is already quite high. It increases and then continuous to vary, mostly between 0.87 and 1. Figure 6 shows the success of the Adaptive Subspace method. The success steadily increases over time, and a very high performance is attained; after 4,000 games the average success varies mostly between 0.96 and 1. Figure 7 shows the total number of feature sets used as a function of the number of games played for both methods. For the Simple Prototype method, the number of feature sets is rather capricious, and does not seem to converge during the 10,000 games that have been observed. With the Adaptive Subspace method the number of feature sets used does converge, and is substantially smaller.

7. Conclusions

Two methods for discrimination between objects based on sensor data have been compared. Both methods are based on the principle that an agent adapts its categorization of the world to increase its ability to distinguish different objects from each other. The development

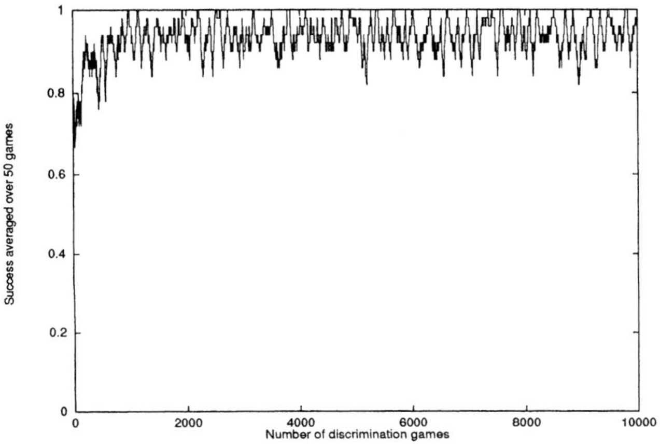

Figure 5 Success using the Simple Prototype method.

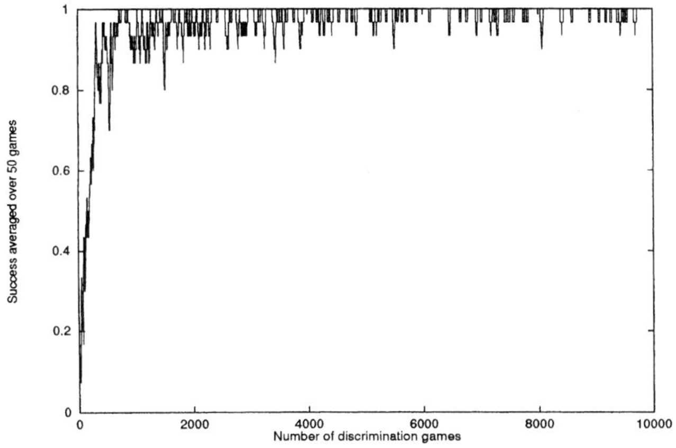

Figure 6 Success using the Adaptive Subspace method.

of discrimination success for both methods makes clear that this objective is attained. Since it is essential that the number of feature sets is as low as possible, the development of this number over time has also been measured. The Simple Prototype method yielded a high success rate, but the number of feature sets does not appear to converge within 10,000 games. The Adaptive Subspace method refines a feature set only after some consideration. This yields an even higher success rate, and has the advantage that the number of feature sets converges to a relatively low number. The question we addressed in this paper is how a robot should discriminate between objects. From the results, we have seen that the Adaptive Subspace method is more attractive with respect to both aims, i.e. a high success rate and a low number of feature sets.

Acknowledgments

The authors want to thank Ruth Aylett, Dave Barnes, Bart de Boer, Michiel de Jong, and Will Lowe for various comments and suggestions.

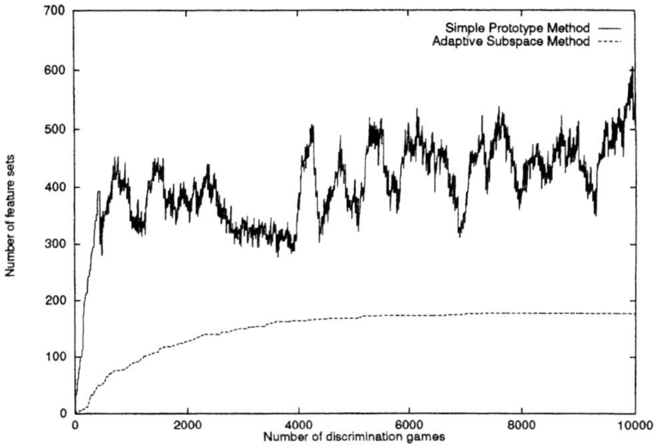

Figure 7 Number of feature sets.

References

De Jong, E. D. (1997). An accumulative exploration method for reinforcement learning. In *Proceedings of the AAAI'97 Workshop on Multiagent Learning, available as AAAI Technical Report WS-97-03.*

Edelman, G. (1987). *Neural Darwinism.* Basic Books Inc., New York.

Landelius, T. (1994). Behaviour representation by growing a learning tree. Master's thesis, Linkoping University.

McCallum, A. K. (1996). Reinforcement learning with selective perception and hidden state. *Ph.D. Thesis.*

Murao, H. and Kitamura, S. (1997). Q-learning with adaptive state space construction. In Birk, A. and Demiris, J., editors, *6th European Workshop on Learning Robots.*

Oja, E. (1983). *Subspace Methods of Pattern Recognition.* Research Studies Press Ltd.

Popescu-Belis, A. (1997). Design of an adaptive multi-agent system based on the "neural darwinism" theory. In *First International Conference on Autonomous Agents*, pages 484–485, Marina del Rey, California.

Schyns, P. G., Goldstone, R. L., and Thibaut, J.-P. (1998). The development of features in object concepts. *Behavioral and Brain Sciences*, In Press.

Steels, L. (1996a). Emergent adaptive lexicons. In Maes, P., editor, *From Animals to Animats 4: Proceedings of the Fourth International Conference On Simulating Adaptive Behavior*, Cambridge MA. The MIT Press.

Steels, L. (1996b). Perceptually grounded meaning creation. In Tokoro, M., editor, *Proceedings of the International Conference on Multi-Agent Systems*, Menlo Park CA. AAAI Press.

Steels, L. (1997). The synthetic modeling of language origins. *Evolution of Communication*, 1(1):1–34.

Steels, L. and Vogt, P. (1997). Grounding adaptive language games in robotic agents. In Husbands, C. and Harvey, I., editors, *Proceedings of the Fourth European Conference on Artificial Life*, Cambridge MA and London. MIT Press.

Vereertbrugghen, D. (1996). Design and implementation of a second generation sensor-motor control unit for mobile robots. Master's thesis, VUB.

Vogt, P. (1997). Perceptual grounding in robots. In Birk, A. and Demiris, J., editors, *6th European Workshop on Learning Robots.*

Evolving Motion-tracking Behaviour For a Panning Camera Head

Nick Jakobi
AnimatLab,
Ecole Normale Supérieure, France
nick.jakobi@ens.fr

Abstract

This paper details experiments in which neural network controllers were evolved in simulation that allowed a simple panning camera head to track patterned objects moving against similarly patterned backgrounds in reality. It begins with a discussion of *minimal simulations*: fast-running, easy-to-build simulations for the evolution of real robot controllers (Jakobi, 1998a; Jakobi, 1998b). The minimal simulation with which the motion-tracking controllers were evolved (along with the rest of the evolutionary machinery) is then described. Experimental results are offered in which an evolved controller was downloaded onto a simple panning camera head and satisfactorily tracked two very differently patterned objects as they moved against similarly patterned backgrounds in a random fashion.

1. Introduction

Various different types of simulation have been used in the past to evolve controllers for robots: in (Jakobi et al., 1995), an empirically verified model of the underlying physics was constructed; in (Miglino et al., 1995), look-up-tables compiled from real-world sensor data were used; and in (Yamanuchi and Beer, 1994), the factory-built simulation came supplied with the robot. However, the fundamental approach underlying the use of all of these simulations is the same: the less differences there are between simulation and reality, the less likely it is that controllers will evolve to depend on them and fail to download into reality. Such an approach is committed to building simulations that are as faithful and accurate a model of the real world as possible.

This paper focuses on a robot task that necessitates the non-trivial use of vision, and accurately modeling how objects appear within a camera image is an extremely computationally expensive and labour intensive activity. Even if the camera image is sub-sampled, as in the experiments reported here, it would still require ray-tracing algorithms, knowledge of the object's light-reflection properties and so on. If constructing a successful simulation for the evolution of visual behaviours necessitated accurately modeling how objects appeared in the camera image, evolving in the real world would probably be preferable.

Fortunately, the theory and methodology of *minimal simulations* put forward in (Jakobi, 1997; Jakobi, 1998a; Jakobi, 1998b) does not advocate that simulations for Evolutionary Robotics should necessarily be as accurate as possible. Instead of trying to eliminate the inevitable differences between simulation and reality they are acknowledged; and instead of leaving it to chance whether controllers evolve to rely on these differences or not, mechanisms are put in place that prevent controllers from evolving to rely on them. Thus controllers are not only evolved in simulation to perform a specific real-world behaviour, according to this approach, they are *also* evolved to be robust to the differences between the simulation and reality. Since such simulations are *not* committed to being as accurate a model of the real world as possible, considerations of computational expense and ease of design may take the place of fidelity as the overriding design criteria. The minimal simulation described below depends fundamentally on this fact for its success. The model of how objects in the camera's field of view affect controller input is so simple as to be almost naive, yet controllers which evolve to be reliably fit within the minimal simulation satisfactorily track randomly patterned objects as they move against similarly patterned backgrounds in reality.

Although Computer Vision researchers may find the results of interest, this is a paper aimed at showing the Evolutionary Robotics community how visual behaviours evolved in minimal simulations can and do transfer onto real robots, and the complexity of evolved controllers should be seen in this context. Controllers have been evolved previously that were capable of 'movement tracking' behaviours (Harvey et al., 1994; Miller and Cliff, 1994; Floreano and Nolfi, 1997), but these have all employed some incidental property of the target object such as its colour or pattern to distinguish it from a background which has a different colour or pattern. Simple strategies can be very effective in these sorts of scenarios. For example, if the target object is black and the background is white then a strategy such as 'always turn towards the darkest part of the image' can be sufficient to track the object as it moves against its background. In the experiments reported here, however, the target object and background were both similarly and randomly patterned. This means two things, firstly that evolving networks could only use the fact that the object was moving to pick it out from the background, and secondly that if the camera started to track the moving object, the camera's motion caused the background to move in the opposite direction within the image. Thus the evolutionary process not only had to track movement caused by the object within the image, it also had to find some way of compensating for the movement within the image caused by the ego-motion of tracking itself. While the minimum mechanisms necessary for performing this sort of behaviour lack the sophistication of the latest hand-designed motion-tracking systems, they are far more complex than those needed for strategies like 'always turn towards the darkest part of the image'.

Section 2. describes the minimal simulation that was used in the experiments reported here to evolve motion-tracking controllers. The rest of the evolutionary machinery, including the neural networks, the encoding scheme, the genetic algorithm and genetic operators is described in section 3.. Experimental results are put forward in section 4. and finally, in section 5., some comments are offered on the paper as a whole.

2. A minimal simulation for evolving motion-tracking behaviour

There is unfortunately not enough room in this paper to explain fully how the minimal simulation used in the experiments reported here derived from the general theory and methodology underlying the minimal simulation approach. This section offers a description of the minimal simulation that is sufficient for re-implementation, but to fully understand why it is the way it is, the reader is referred elsewhere (Jakobi, 1997; Jakobi, 1998a; Jakobi, 1998b). The section begins by describing the simple model of a panning camera-head at the heart of the minimal simulation, and goes on to describe the ways in which this model was varied from trial to trial (according to the methodology) in order to prevent evolving controllers from coming to rely on the differences between the simple model and reality. Finally the exact nature of each fitness test is described.

2.1 The basic model

The panning camera-head used in the experiments reported here consisted of a small camera, connected to a frame-grabber and mounted on a platform whose rotational velocity was under computer-control. In the simulation, both the rotation of the camera and the movement of the target object to be tracked were modeled as horizontal translations. Two variables held the horizontal positions of the centre of the camera's field of view and the centre of the target object, and both the rotation of the camera to the right and left and the movement of the target object to the right and left were modeled as horizontal translations to the right and left. In order to track the target, therefore, the horizontal positions of the left and right edges of the camera's field of view had to contain the horizontal position of at least part of the target object, and the camera had to move horizontally in such a way that the target object never escaped these boundaries.

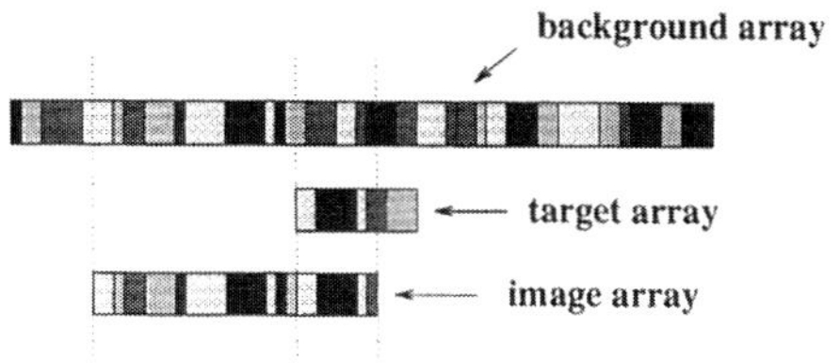

Figure 1 This figure shows how the image array derived in the simulation from the target array superimposed on top of the background array.

Only a sub-sampled region of the real camera image was accessible to controllers: a horizontal strip, three pixels wide and crossing the centre of the camera image with a total angle of acceptance of around 60^0. This was further sub-divided into 32 segments of equal length and the 32 average pixel values of these segments provided input to the controllers. In the simulation, the way in which the background and target object affected the 32 values of the input array was modeled in an extremely simple fashion, and is shown diagrammatically in figure 1. At the start of each run an array of 10000 numbers between 0 and 255 (minimum and maximum grey-level pixel values) was initialized to act as the background. Also, at the start of every fitness trial, a small array of between 6 and 16 random numbers between 0 and 255 was initialized to act as the target object. At each iteration the horizontal position of the left-hand-side of the target object was used as an index into the background array and the target array was superimposed upon the background array at this point. The horizontal position of the left-hand-side of the camera image was then used as an index into the background array and the 32 following values were assigned as the 32 values of the input array. On each iteration, uniform noise was added to each value in the image array in the range ± 5.

2.2 Preventing evolving controllers from relying on simulation/reality differences

In accordance with the minimal simulation methodology (Jakobi, 1998a; Jakobi, 1998b), each fitness test consisted of taking the average score from a set of independent trials and several aspects of the model were varied from trial to trial. This forced evolving controllers to both depend *exclusively* on those aspects of the model that had some counterpart in reality, and to depend on these aspects in such a way that they were robust to inevitable modeling inaccuracies. In particular, the patterns of the target array and the background array encountered by evolving controllers were randomly varied from trial to trial:

- The values that made up the target array were randomly set at the start of each trial in one of two different ways (picked at random): either they were all set to the same random value, or they were all set to different random values.
- The starting positions of the camera image and target object relative to the background array were randomly varied between 0 and 10000.

In this way, evolving controllers could only be reliably fit if they came up with a strategy that depended solely on movement in the image, and not on the shape or colour of the object or the background. In addition, various other parameters of the model were varied from trial to trial:

- The size of the target array was randomly varied between 6 and 16 input elements wide.
- The speed of the target object was randomly set between 0.5 and 1.5 input elements per simulation update.
- The initial direction of the target object, and whether, after the pause stage of each fitness evaluation (see 2.3 below), it would continue in the same or the opposite direction was randomly determined.
- The maximum acceleration of the camera was randomly set between 0.75 and 3 input elements per simulation update squared.

In this way, reliably fit controllers were forced to employ non-brittle strategies that could cope with objects of different sizes traveling at different speeds and in different directions. They were also forced to cope with cameras with different panning motor characteristics.

2.3 The fitness test

Because of the inherently noisy nature of the task, and the degree to which the simple model of 2.1 was varied from trial to trial in order that reliably fit individuals would transfer into reality, several steps were taken to prevent the stochastic effects on returned fitness scores from overpowering the adaptive abilities of the evolutionary process. Each complete fitness evaluation returned the average value scored by an individual in a total of 40 different fitness trials, and each fitness trial was a maximum of 25 simulated seconds in length. This meant that a complete fitness evaluation lasted a maximum of over 16 simulated minutes. To further reduce the effects of trial to trial variation, evolving controllers were only tested on target objects that moved according to random variations of a particular fixed pattern:

1. From a starting position just to the side of the field of view, an object moved towards (and across) the visual field at a randomly determined speed for 10 seconds.

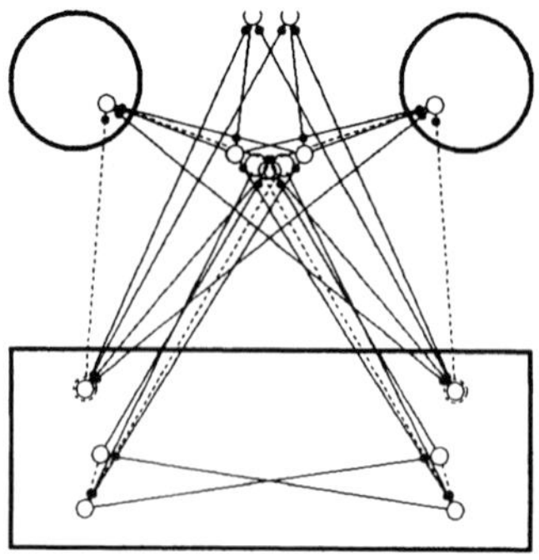

Figure 2 The tracking network.

2. The object stopped moving completely for 5 seconds.
3. The object started moving again at the same speed as before but in a random direction. This phase also lasted 10 simulated seconds.

Although there was nothing special about this movement pattern in particular, it was thought complex enough that controllers which evolved to successfully track objects that moved in this fashion would most probably generalize to track objects that move according to a large range of other more random movement patterns.

At any particular time, controllers were deemed to be successfully tracking an object if any part of the object fell within the field of view of the camera. At each iteration δf was calculated from the difference between the horizontal position of the centre of the target array X_T and the horizontal position of the centre of the image array X_I according to the expression:

$$\delta f = \left\{ \begin{array}{ll} 0 & |X_I - X_T| > 16 \\ 16 - |X_I - X_T| & |X_I - X_T| \leq 16 \end{array} \right\}$$

At the end of each trial, the fitness function returned the final sum of the δfs. If the target object ever completely left the field of view at any point during a fitness trial then the trial was halted at that point, and the sum achieved thus far was returned as the fitness.

3. Evolutionary machinery for evolving motion-tracking behaviour

In this section we describe the evolutionary machinery that, together with the minimal simulation described above, was responsible for evolving neural network control architectures that could perform the behaviour satisfactorily in reality.

3.1 Neural networks

After trying several different types, the neural networks described below were found to be the most successful. However, in order to get this type of network to work satisfactorily, further preprocessing of the 32 input values was necessary. If the value of input i at time t was

$I_i(t)$, then the corresponding value $J_i(t)$ made available to the neural networks was calculated according to the expression

$$J_i(t) = \left\{ \begin{array}{ll} 0 & |I_i(t) - I_i(t-1)| < 15 \\ 255 & |I_i(t) - I_i(t-1)| \geq 15 \end{array} \right\}$$

This meant that high network input could only be caused by sudden changes in the image, such as that resulting from image motion.

Apart from this extra preprocessing stage, the networks used in the experiments reported here were made from binary-valued neurons (either 'on' or 'off') with evolved weights and thresholds. In addition, a time constant was associated with each neuron that specified a period of time that the neuron would remain 'on' after the stimulus that caused it to turn 'on' in the first place had decayed. To update the state of the network at each iteration, the input activity A_j of each of the $j = 1$ *to* N neurons in the network was calculated according to the simple weighted sum:

$$A_j = \sum O_i w_{ij} + J_j - A_{min}$$

where O_i was the output from the i_{th} neuron, w_{ij} was the weight on the connection from the i_{th} neuron to the j_{th} neuron, and J_j was any preprocessed external input to the j_{th} neuron. Each input activity A_j was then normalized to lie between 0 and 1 by reference to its maximum and minimum possible values. Using this normalized value $\bar{A}_j$ to calculate the output from each neuron increased the chance of the threshold lying within the dynamic range of the inputs, and thus increased the chance of random networks doing something rather than nothing. After the normalized input activity of every neuron in the network had been calculated, the output O_j of all of the $j = 1$ *to* N neurons in the network was calculated according to equation

$$O_j = \left\{ \begin{array}{ll} 0 & \bar{A}_j < t_j \\ 1 & \bar{A}_j \geq t_j \quad or \quad T_{\bar{A}_j \geq t_j} < \tau j \end{array} \right.$$

where $T_{\bar{A}_j \geq t_j}$ was the elapsed time since $\bar{A}_j \geq t_j$ was last true, τj was a time constant associated with the j_{th} neuron and t_j was a threshold constant associated with the j_{th} neuron.

Connection weights were in the range ± 2, thresholds were in the range 0 to 1, and time delays were in the range 0 to 16 network updates. The state of the network was updated at the rate of 32 times a second (or the simulated equivalent of a second): twice for each update of the inputs and outputs which were undertaken at the rate of 16 times a second (or the simulated equivalent of a second).

3.2 The encoding scheme

The encoding scheme was a version of the spatially determined encoding scheme described in (Jakobi and Quin, 1998). Along with several previous encoding schemes (Cliff, 1994; Nolfi and Parisi, 1995; Husbands et al., 1994), this scheme uses the idea of a developmental space to allow genotypes to grow under genetic control with a minimum amount of phenotypic disruption.

Figure 2 shows an evolved network and the two dimensional space in which development took place. The location of each neuron within the space was genetically specified and determined its function within the network: a location within either of the two circular regions at the top of the developmental space meant a motor neuron, a location within the large rectangular region at the bottom of the developmental space meant an input neuron, and a location anywhere else meant a hidden neuron. The output signal to the motors was calculated according to the relation $signal = k \times (O_l - O_r)$, where O_l and O_r were the average output values of the neurons in the left and right motor regions respectively and k was a constant equivalent to around 180° per second (3 input elements per simulation update). As explained in section 2. above, there were a total of 32 inputs made available to evolving networks, each corresponding to a certain number of pixels situated on the horizontal line drawn through the middle of the camera image. To determine which of these 32 inputs each input neuron corresponded to, the rectangular input region of the developmental space (see figure 2) was itself divided up into 32 rectangular input sub-regions running horizontally from left to right, each corresponding in order to one of the 32 inputs. Input values were normalised to lie within the range 0 to 1. Inputs and outputs were updated at a speed of 16 times per second (or the simulated equivalent of a second).

Genotypes consisted of a variable number of genes, each encoding the necessary parameters for a neuron. In fact, since bilateral symmetry was imposed on evolving networks by reflecting each neuron across the midline of the developmental space, each gene actually coded the parameters for not one but two neurons. Each gene was 17 integers long, and each integer lay between 0 and 99. The first number of each gene stored the gene's 'age' and played no role in determining the structure of the phenotype (see the explanation below of the genetic operators). The next two numbers of each gene specified the x and y coordinates of the corresponding neuron's position within the developmental space. The fourth number specified the neuron's threshold, and the fifth number specified the neuron's decay constant. The next six numbers specified the target positions and weights for two connections *to* other neurons *from* the neuron in question, and the next two numbers specified the target positions and weights for two connections *from* other neurons *to* the neuron in question.

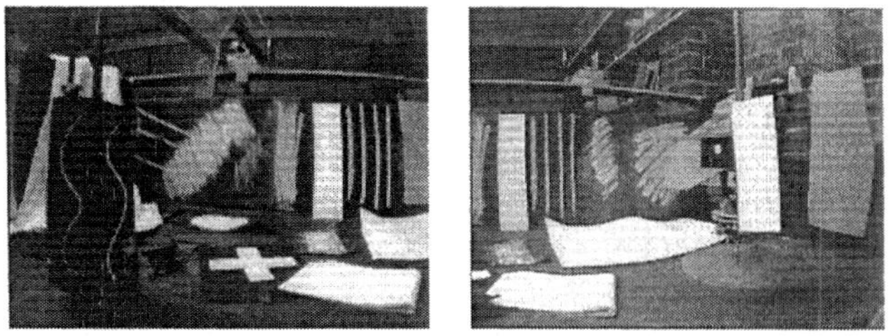

Figure 3 Two views of the gantry robot taken directly from the panning camera head. In the left hand image, the gantry has had a large dark cylinder of paper wrapped around it. In the right hand image, the gantry has had a small white cylinder of paper attached to the side of the gantry facing the camera.

3.3 Genetic algorithm and genetic operators

The genetic algorithm was a simple generational model with truncation selection, elitism and mutation locking (see (Jakobi and Quin, 1998)). It worked as follows. After evaluating every member of the population, the next generation was formed by asexually producing the required number of offspring. At each offspring event, a parent genotype was selected at random from the fittest 50% of the population and subjected to a number of different forms of mutation to produce an offspring. First of all, every unlocked gene on the genotype had a 0.05 chance of undergoing a major 'innovative' mutation which involved being completely randomized and having its age reset to zero. Secondly, every parameter on the genotype, *including* those of genes that were mutation locked had a small chance of undergoing a 'tuning' mutation: the adding of a random offset in the range $\pm 5\%$. The number of tuning mutations per genotype was picked from a Poisson distribution with an expected number of 5. Finally, there was a chance of 0.01 of both adding a completely random gene aged zero or deleting a randomly selected unlocked gene entirely. After enough offspring had been generated, the age of every gene of every individual in the population was increased by 1. Those genes that passed the age of 200 were regarded as mutation-locked and made immune from both innovative mutations and deletion.

4. Experimental results

Out of a total of five runs with the parameter settings given above, only one run evolved networks that could reliably track objects within the simulation, consistently achieving near-optimal fitness on every fitness trial. Of the other four runs, two evolved networks that performed reasonably satisfactory motion-tracking within the simulation, achieving an average of around 80% of maximum fitness on each complete fitness evaluation, and the other two evolved networks that were only capable of reliably achieving around 50% of maximum fitness.

Figure 2 shows one of the networks that evolved to reliably perform motion-tracking behavior within the simulation. The 1300 generations that this network took to evolve required around 24 hours to run on a Sun SPARC Ultra and simulated over 5 years' worth of real world evolution. The network employs a saccading strategy (Carpenter, 1977) that tracks moving objects through a series of jumps and pauses instead of moving smoothly and at the same speed as the object. Furthermore, the strategy is directionally sensitive in that the network will only saccade to track an object if it is moving away *from* the centre of the image *towards* the image boundary, and not if it is moving *towards* the centre of the image *from* the image boundary. This directional sensitivity means that the simulated camera only saccades in the direction of motion of the object.

In order to provide some objective evidence of its ability to transfer across the reality gap, the network was downloaded onto the panning camera head set-up described in section 2. and subjected to a specially devised real-world tracking task involving the gantry robot described in (Harvey et al., 1994). For this task, one of the long walls of the gantry arena was removed and the remaining three walls were decorated with chalk markings and assorted pieces of black, white and grey card (see figure 3). The head of the gantry robot was also decorated, and a short computer program was written that made it parade up and down the arena, parallel to the missing wall, and changing speed and direction at random. The panning camera head was then positioned half way down the length of the missing wall, just outside the arena and facing inwards. The task was to track the patterned object, provided by the gantry robot, as it moved against the similarly patterned background, provided by the remaining three walls of the gantry arena.

Figure 3 shows two images (taken directly from the panning camera head) of the gantry robot decorated as two different patterned objects and viewed against the arbitrarily patterned background provided by the walls of the arena. The network was tested continuously for fifteen minutes on each of these two object patterns and the angular position of both the panning camera head and the gantry robot relative to the panning camera head were automatically and continuously recorded during each test. Figure 4 shows two traces, one for each fifteen minute test, of these angular positions plotted over time. The thick light grey line in each of these traces plots the angular position of the panning camera head's field of view over time. The thin black line plots the relative angular position of the centre of the moving object (i.e. the gantry robot) over time. In total, the network kept the large dark object pattern of figure 3 within the field of view of the tracking camera for 91% of the fifteen minute test period. It kept the small white object pattern of figure 3 within the field of view of the tracking camera for 92% of the fifteen minute test period.

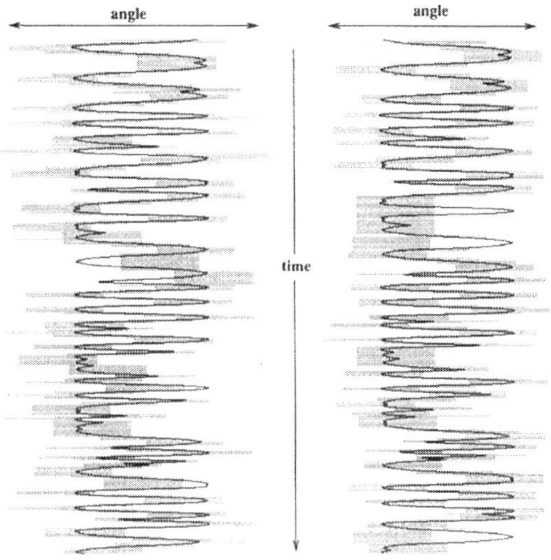

Figure 4 Each of these two traces derive from a continuous fifteen minute test of the ability of the neural network of figure 2 to track moving objects in the real world (see text).

5. Conclusions

The experiments reported in this paper provide further evidence that controllers evolved in an inaccurate and over-simple simulation will cross the reality gap *provided* that the right amount of random variation is included in the simulation in the right way according to the methodology laid out in (Jakobi, 1998a; Jakobi, 1998b). A fast-running, easy to implement model of how objects affected controller input was used to successfully evolve real-world tracking behaviours even though it was inaccurate and over-simple to the point of naivety. The savings this afforded in both construction time and running time are both especially apparent when one considers the computationally intensive ray-tracing algorithms, empirically derived reflection coefficients and so on that are part and parcel of more conventional visual simulations.

The novel neural networks used in these experiments worked well, but only after an extra preprocessing stage had been implemented. This extra stage meant that neurons responded to sudden changes in the values of the 32 pixel fields rather than the values themselves. With all else being equal, therefore, neuronal input was a direct consequence of image motion, and this greatly simplified one crucial aspect of the motion-tracking task. However, even given image motion for free, controllers still evolved that discriminated image motion due to the target object from that due to the movement of the camera, executed well-timed saccades to track the target object, and executed these saccades only in the direction of movement of the target object to minimize the risk of losing it from the field of view. These controllers therefore evolved to perform a non-trivial visual behaviour within the minimal simulation that transferred successfully into reality.

References

Carpenter (1977). *Movements of the Eyes.* Pion Press, London.

Cliff, D. (1994). Ncage. neural control architecture genetic encoding. Cognitive Science Research Paper 325, School of COGS, University of Sussex.

Floreano, D. and Nolfi, S. (1997). Adaptive behaviour in competetive co-evolutionary robotics. In Husbands, P. and Harvey, I., editors, *Proceedings of the 4th European Conference on Artificial Life*, Cambridge, Mass. MIT Press.

Harvey, I., Husbands, P., and Cliff, D. (1994). Seeing the light: Artificial evolution, real vision. In Cliff, D., Husbands, P., Meyer, J.-A., and Wilson, S., editors, *Proc. of 3rd Intl. Conf. on Simulation of Adaptive Behavior, SAB'94*, pages 392–401. MIT Press/Bradford Books.

Husbands, P., Harvey, I., Cliff, D., and Miller, G. (1994). The use of genetic algorithms for the development of sensorimotor control systems. In Gaussier, P. and Nicoud, J.-D., editors, *Proceedings of From Perception to Action Conference*, pages 110–121. IEEE Computer Society Press.

Jakobi, N. (1997). Half-baked, ad-hoc and noisy: Minimal simulations for evolutionary robotics. In Husbands, P. and Harvey, I., editors, *Proc. 4th European Conference on Artificial Life.* M.I.T. Press.

Jakobi, N. (1998a). Evolutionary robotics and the radical envelope of noise hypothesis. *Journal of Adaptive Behaviour*, 6(2):325–368.

Jakobi, N. (1998b). *Minimal Simulations for Evolutionary Robotics.* PhD thesis, School of Cognitive and Computer Sciences, University of Sussex.

Jakobi, N., Husbands, P., and Harvey, I. (1995). Noise and the reality gap: The use of simulation in evolutionary robotics. In Moran, F., Moreno, A., Merelo, J., and Chacon, P., editors, *Proc. 3rd European Conference on Artificial Life*, pages 704–720. Springer-Verlag, Lecture Notes in Artificial Intelligence 929.

Jakobi, N. and Quin, M. (1998). Some problems (and a few solutions) for open-ended evoltuionary robotics. In Husbands, P. and Meyer, J.-A., editors, *Proceedings of Evorobot.* Springer Verlag.

Miglino, O., Lund, H., and Nolfi, S. (1995). Evolving mobile robots in simulated and real environments. *Artifical Life*, 2(4).

Miller, G. F. and Cliff, D. (1994). Protean behavior in dynamic games: Arguments for the co-evolution of pursuit-evasion tactics. In *Proceedings of the third international conference on Simulation of Adaptive Behavior.* MIT Press/Bradford books.

Nolfi, S. and Parisi, D. (1995). Evolving artificial neural networks that develop in time. In Moran, F., Moreno, A., and Merelo, J., editors, *Advances in Artificial Life: Proceedings of the third European conferrence on Artificial Life.*, Berlin. Springer-Verlag.

Yamanuchi, B. and Beer, R. (1994). Integrating reactive, sequential, and learning behaviour using dynamical neural networks. In Cliff, D., Husbands, P., Meyer, J.-A., and Wilson, S., editors, *Proceedings of the Third International Conference on the Simulation of Adaptive Behaviour*, Cambridge, Massachusetts. MIT Press/Bradfork Books.

Look Before You Leap: Peering Behavior for Depth Perception

M. Anthony Lewis **Mark E. Nelson**

Beckman Institute
405 N. Mathews Avenue
University of Illinois
Urbana, Illinois 61801

tlewis@uiuc.edu nelson@vernal.npa.uiuc.edu

Abstract

When presented with a water or an air gap barrier, animals often engage in peering, or side-to-sided head movements, before leaping across the barrier. This strategy is used instead of depth recovery using stereopsis, and likely gives a much better estimate of distance.

In this article we present a neurocomputational model of peering, hosted on a small robot, that explains the essential characteristics of peering reported in the literature. The model builds on recent evidence for non-direction selective movement detectors in insects. Through non-linear transformation of the retinal image, the model produces a 'leap' command without intermediate reconstruction of the external space of the animal.

1. Introduction

About 40 years ago, Wallace (1959) described an investigation into a depth perception mechanism in locusts. The mechanism is called peering. During peering, the animal's body moves in a radial arc centered at the posterior tip of the animal's abdomen. The head translates and, as was shown by Collet (1978), the head remains rotationally stable during translation (see Fig. 1).

These peering movements precede directed leaps by the insect (but apparently do not occur before non-directed escape responses).

It was hypothesized by Wallace that the animal is using motion parallax to judge distance to an object. Wallace reports that Exner (1891) speculates that the articulate eyes of crabs may use the *rate* of movement of an image as a cue to the depth of an object.

Evidence supporting this idea was presented by Wallace and latter in a similar experiment with computer controlled equipment by Sobel (1990). In those experiments, a target was made to move either with or against the direction of the peering animal's head movement.

In the case that the target was moved in opposition to the head movement of the animal, the animal underestimated the distance to the object (as assessed by the leap velocity or leap distance of the animal). If the target was moved with the head movement, the distance to the target was over estimated.

Sobel also tried moving the target in the same direction as the head movement of the animal but *faster.* When he did this, there was a very surprising result. The estimated distance to the object *decreased.* This is a very curious reversal phenomenon.

Sobel's explanation for how the animal computes distance is based on the static geometry of the problem. After a peer, the animal uses two angular measurements made at the extents of the peer to triangulate the object's position.

A neurally plausible mechanism must be able to explain the computation of depth at different speeds of peering. It is known that the speed of peering varies and is not constant for each peer. Secondly, the reversal effect must be explained and the mechanism should rely on known characteristics of image speed transduction in insects.

In this article we present a biologically plausible model for peering. The model is consistent with experiments reported in the literature on peering, as well as new behavioral and anatomical evidence for movement detection pathways in insects.

Furthermore, the model does not involve an intermediate representation of external space. Nor does it require any memory process on the part of the animal. The processed image can be mapped directly to motor behavior without intermediate representation.

In the following section we present the experimental evidence for this hypothesis. In section 3 we describe a model of this behavior. In section 4 we describe a robotic implementation of this behavior. This is followed by experimental results.

2. Experimental Evidence

2.1 Peering Behavior

In Sobel (1990) locust were placed on a platform in the middle of a pie tin filled with water. The visual environment was plain with the exception of a single target, a vertical black stripe against a white background.

The head of the locust was monitored and the stripe could be moved under computer control, tracking the movement of the insect's head.

If the target was moved in the opposite direction of the head movement, the peering effect was exaggerated and the

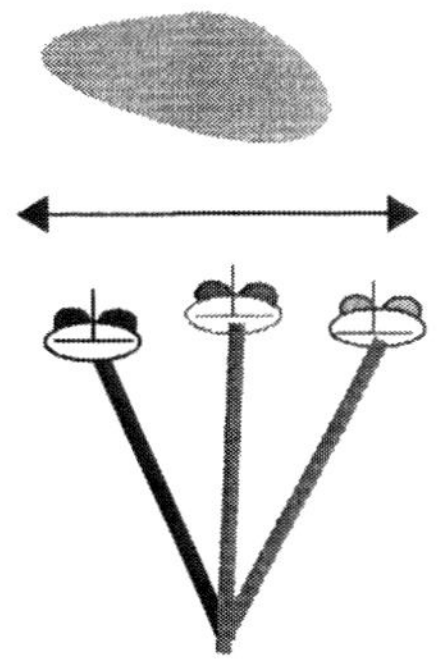

Figure 1. Peering behavior in a locust. The head of the locust is held rotationally stable while it translated perpendicular to the gaze direction.
Retinal speed of the object, normalized by the head translational speed, can predict the distance to the object.

target gave the illusion of being closer. In fact, the locust sometimes reached with their forelimbs to touch the seemingly proximal target.

If the target was moved with the direction of the head movement (but slower) the target appeared farther away. If the target moved faster than the head movement, a curious *reversal* illusion was observed. It was infered by the behavior of the animal that it perceived the target to be closer as speed *increased.*

In monocular locusts, the estimated distance was *doubled.* This results would be expected if the animal simply summed the computations from two sides of the body. This result is not seen in similar experiments in the mantis (Walcher and Kral, 1994).

Sobel suggests that the principle of distance estimation in locusts is based on measurements of angle at the extent of the peers. By triangulation, the distance to the object can be estimated.

However, the suggestion of Exner (1891) has merit, and can be applied to this case (Exner was discussing principle of distance estimation in crustaceans, not insects). In fact, we can place his original suggestion in the context of recent evidence in movement detection systems in insects.

2.2 The Non-Direction Selective Movement Detectors and Motion Parallax

Recently, evidence for non-directional selective movement detector (NDSMD) pathways in insects, first implied by behavioral evidence in bees (Srinivasan and Zhang,1993) have been supported by anatomical evidence in flies (Douglass and Strausfeld, 1996).

The idea is that there are at least *two* movement detection pathways. One pathway is direction selective and supports behaviors such as the optomotor response (turning the head or body to track a rotating environment). A second pathway supports obstacle avoidance and the so-called centering response where an insect flying down a corridor with textured walls will center itself.

The non-direction selective pathway has a higher frequency response than the direction selective pathway. Because of this difference in frequency response, behavioral experiments designed by Srinivasan and Zhang (1993) were able to show that the two pathways are distinct. This was followed by anatomical evidence provided by Douglass and Strausfeld (1996).

3. Neurocomputational Model

A model of obstacle avoidance using non-directional selective movement detectors is presented in Lewis (1998). In that work, the NDSMD is modeled as a filtered temporal derivative of the retinal brightness.

In that work, an argument is made that simple temporal differentiation of brightness will give a reasonable estimate of image motion given sufficiently rich texture of the observed environment.

Another approach is to use a Reichart type detector (Borst and Egelhaaf, 1989) but modified to have a symmetrical response. Figure 2 shows an array of these detectors. The layer 'T' is an array of photodetectors. A delayed and shifted version of the retinal image is correlated with the current image. This results in layer 'S'. Layer 'S' is an estimate of the speed of an object's image on the retina. Because the circuit is symmetric, a given cell in layer 'S' responds equally to leftward *or* rightward image movement.

Once we have this detector, we can estimate the distance to an object. This can be shown in the following way.

3.1 Depth From Parallax

Let us assume a perspective projection:

$$\begin{bmatrix} x \\ y \end{bmatrix} = \frac{\lambda}{Z}\begin{bmatrix} X \\ Y \end{bmatrix} \quad (1)$$

where λ is the focal length of the lens, X, Y, Z is the position of a point in the environment, and x, y is the position of the projection of that point in retinal coordinates. The velocity of the image of a moving point in the world can be found by differentiating (1) with respect to time:

$$\begin{bmatrix} \dot{x} \\ \dot{y} \end{bmatrix} = \frac{\lambda}{Z^2}\begin{bmatrix} Z\dot{X} - X\dot{Z} \\ Z\dot{Y} - Y\dot{Z} \end{bmatrix} \quad (2)$$

If we assume that objects in the environment are fixed in relation to one-and-other and that the observer is moving with relative translational velocity $V_e = \begin{bmatrix} V_x & V_y & V_z \end{bmatrix}^T$ and relative rotational velocity $\Omega_e = \begin{bmatrix} \omega_x & \omega_y & \omega_z \end{bmatrix}^T$ to the environment given in the observer's frame, a point in the environment has relative velocity:

$$\dot{P} = \begin{bmatrix} \dot{X} & \dot{Y} & \dot{Z} \end{bmatrix}^T = -(\Omega_e \times P + V_e) \quad (3)$$

Now substituting into (2):

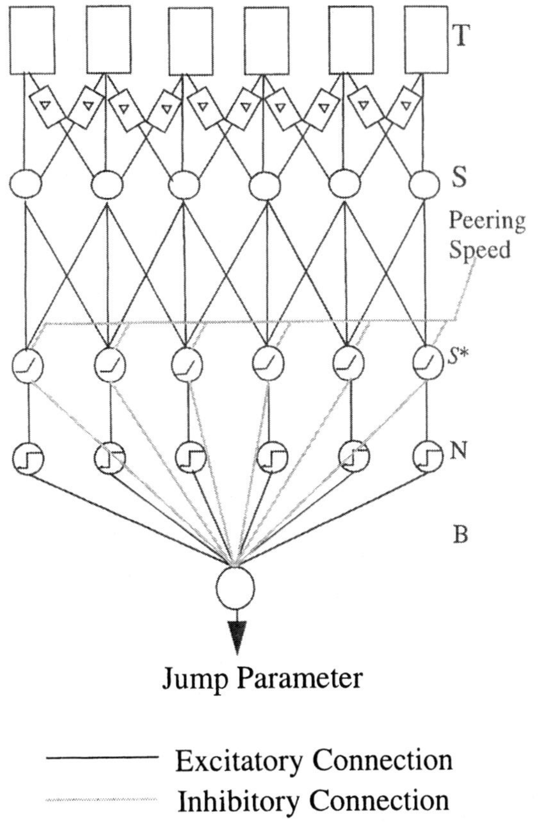

Figure 2. Neural Network for Jump Parameter Computatior from retinal stimulus.

$$\begin{bmatrix}\dot{x}\\ \dot{y}\end{bmatrix} = \frac{1}{Z}\begin{bmatrix}-\lambda & 0 & x\\ 0 & -\lambda & y\end{bmatrix} {}^{c}V_e + \frac{1}{\lambda}\begin{bmatrix} xy & -1-x^2 & \lambda y\\ 1+y^2 & -xy & -\lambda x\end{bmatrix} {}^{c}\Omega_e \qquad (4)$$

This describes the movement of the image point projected onto the retina, the motion field. If a point is on a richly textured surface, the motion field, and the optic flow field will be well approximated by (4).

We assume that our transducer gives the speed but not the direction of the flow field. Taking the absolute value of (4) we have an equation for the computation performed by layer 'S'.

$$\sqrt{s^2} = \sqrt{\frac{\lambda^2 V_z^2}{Z^2}\left[\left(-\alpha + x - \frac{Z}{V_z}\left(xy\omega_x + \omega_y(x^2+1) + y\omega_z\right)\right)^2 + \left(-\beta + y + \frac{Z}{V_z}\left(-\omega_x(y^2+1) + xy\omega_y + x\omega_z\right)\right)^2\right]} \qquad (5)$$

where we have made the substitution: $\left[\dot{X}/\dot{Z}\ \dot{Y}/\dot{Z}\right] \to [\alpha\ \beta]$.

We can see that the terms involving $[\omega_x\ \omega_y\ \omega_z]$ confound our efforts to recover the depth of a point. In the case of peering, the head is stabilized and forward velocity is negligible. Therefore, we can set $[\omega_x\ \omega_y\ \omega_z] = \mathbf{0}$ and $|T_z| = 0$. We now have:

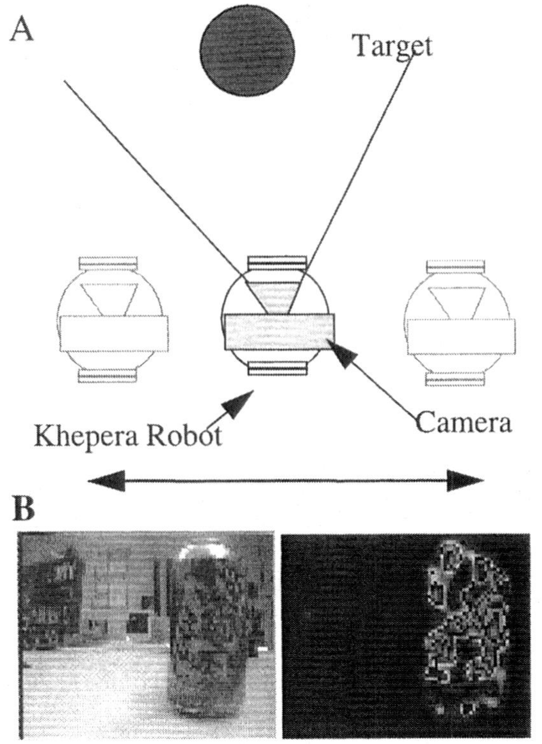

Figure 3. Experimental setup. (A) A Khepera robot with a miniature camera mounted on top. The camera looks sideways, perpendicular to the direction of movement. (B) An example of processed data from layer 'S'. Here the robot peers at a soda can.

$$\frac{1}{|Z|} = \frac{|s|}{\lambda\sqrt{T_x^2 + T_y^2}} \qquad (6)$$

If $|s|$ is the speed estimated by the photo transducer, then if we normalize this quantity by the peering speed of the transducer, we have an estimate of the reciprocal of depth for a point in the image. Let us call that quantity s^*. This is reflected in the layer s^* in Fig 2. Here 'Peering Speed' equals: $\sqrt{T_x^2 + T_y^2}$. If we take an average over the extent of the image then we have the average inverse depth:

$$\frac{1}{Z^*} = \frac{1}{N}\sum_{\forall i,j}\frac{1}{Z(i,j)} = \frac{1}{N}\sum_{\forall i,j} s^* \qquad (7)$$

Inverting, we have

$$Z^* = N/\left(\sum_{\forall i,j} s^*\right) \qquad (8)$$

The computation in eqn (8) is computed in the 'Jump Parameter' neuron. To understand how this is done, we note that if we sum the activity and use that as a shunting inhibition signal, and we apply a hard threshold to this layer (the computation in layer 'N') we essentially count the number of 'on' neurons. The jump parameter neuron sums all of the

Figure 4. Targets used for peering experiments: Foam cup, soda can, utility tool, roll of wire.

'on' neurons to compute 'N' and then divides this by the summation of activity in each s^*. Thus we compute the Jump Parameter which is related to distance (up to a scale factor).

3.2 Reversal Phenomena

The reversal phenomena can be explained by the fact that our model uses NDSMD. Because of this, a target moving faster than the head of an insect will produce the same effect in layer S as a target that is moving slightly slower than the head of the insect.

4. Robot Experiment Setup

A miniature camera was mounted on a Khepera robot base. The camera was positioned so that it looked sideways to the direction of movement (See Fig 3). Thus, as the robot moved in a straight line, the camera translated parallel to the image plane.

Image processing from the camera was done on a SGI Indy workstation. All images were subsampled to 60x80 pixels.

The robot was controlled from a Dual Pentium II workstation via an RS-232 connection. Commands were issued to the workstation based robot controller by a process on the Indy Workstation. For all three processes (Image Processing, Command Issuing, and Robot controller processes), interprocess communication was handled through PVM (Geist et al., 1994).

Performance of the system was measured by varying the target, target distance and peering velocity parameters.

Four different targets (Fig. 4) were placed at various distances from the robot. Targets #1 and #2 where roughly the same size but with different surface texture. Target #3 was about the same height as targets #1 and #2, but was narrower. Target #4 was about the same width as targets #1 and #2 but was much shorter.

The targets were placed at various distances in front of the camera: 2,4,6,8,10, and 12 cm.

Peering velocities of 4 cm/sec and 2 cm/sec were used. The duration of each peer was adjusted so that the total distance traveled was about 8cm.

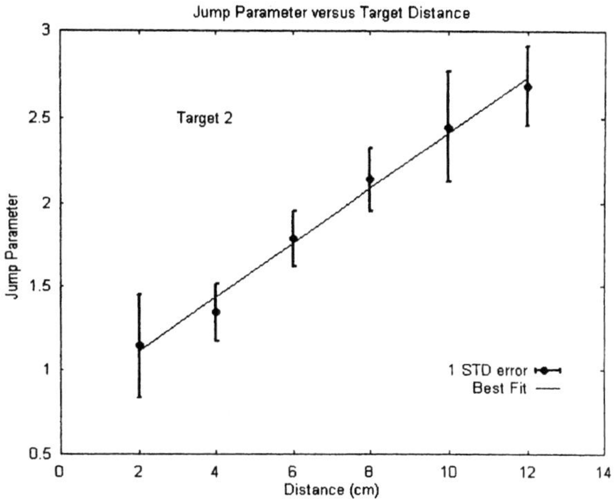

Figure 5. Response of Jump Parameter neuron for target 2 at different distances.

In all cases, the measured variable was the jump parameter as determined by the output of the jump neuron. During a single peer (either to the left or the right), the maximum value of the jump parameter was measured. This value was then recorded as the peer parameter for that particular peer.

Other values such as average jump parameter over a scan could have been used as well. At the ends of the peer, when the robot reversed direction, the velocity fell to zero. It was felt that the maximum value gives the most reliable measurement of distance

It was noticed that often the maximum peer velocity was established close to its final value early on. It seems likely that the peer distance of 8 cm is much more than is needed to determine depth.

In the series of experiments described here, the minimum distance needed for a good estimate of depth was not determined.

5. Experimental Results

5.1 Variation of Jump Parameter with Target Distance

In this experiment, target #2 was placed at various distances in front of the camera. The maximum jump parameter was established for a series of 14 peers at each distance.

In theory, the jump parameter should increase linearly with target distance.The data is plotted in Figure 5 along with a line of best fit (in a least squares sense). As can be seen, the jump parameter did increase monotonically with distance, in a fairly linear fashion.

During each peer, it was noticed that some residual rotation of the camera introduced spurious noise in the measurement. This undoubtedly accounts for some of the spread of the data at each peering distance.

5.2 Variation of Jump Parameter with Different

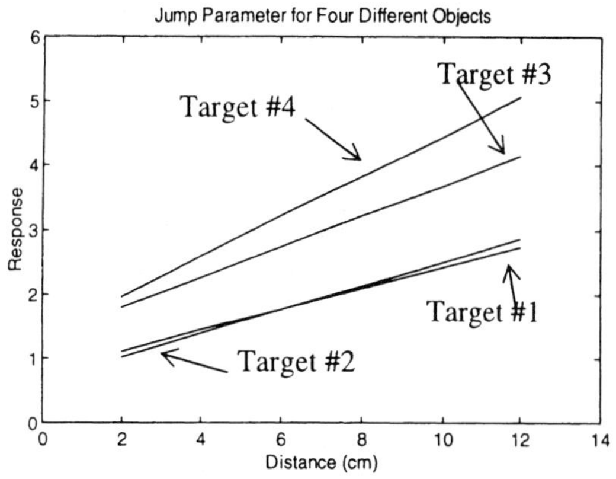

Figure 6. Comparison of jump parameter response verses 4 different targets. The regression lines are plotted and the data points are left out for clarity of presentation.

Targets.

Four different targets were compared and each was placed at the 6 different distances as before. During each peer about 14 data samples were taken.

Because we normalize the activity of the distance measurement layer by the number of 'on' units, the jump parameter should be invariant to target size.

In addition, if sufficient texture is present, the range estimation should be invariant to target texture.

Figure 6 shows the regression lines for each target. Notice target #1 and target #2. These are two objects with different texture but approximately the same size, and they have nearly identical regression lines.

However, targets #3 and #4 have apparently significantly different regression lines. We can speculate that the reason for this is that the background may have been averaged into the reading. Thus we compute a distance that is too large.

5.3 Jump Parameter and Peering Speed Variation

The next experiment was to determine the variation in estimated peering distance with varying peering velocity. In theory, because we normalize the processed image by the peering velocity, the jump parameter should be invariant to peering velocity.

Figure 7 shows the data from this experiment and regression lines. Two targets were compared, target #2, a large target and target #4, a small target.

In both cases of large and small targets, the regression lines are lower at the lower scan speed. This means that the distances at slower scan speed s is over estimated.

An overestimate of distance indicates a reduced response of the movement detectors.

Two possible explanations can account for this. Either there is a non-linearity in the movement detectors or increased peering velocities are associated with increased rotational noise.

These two cases can be distinguished by examination of

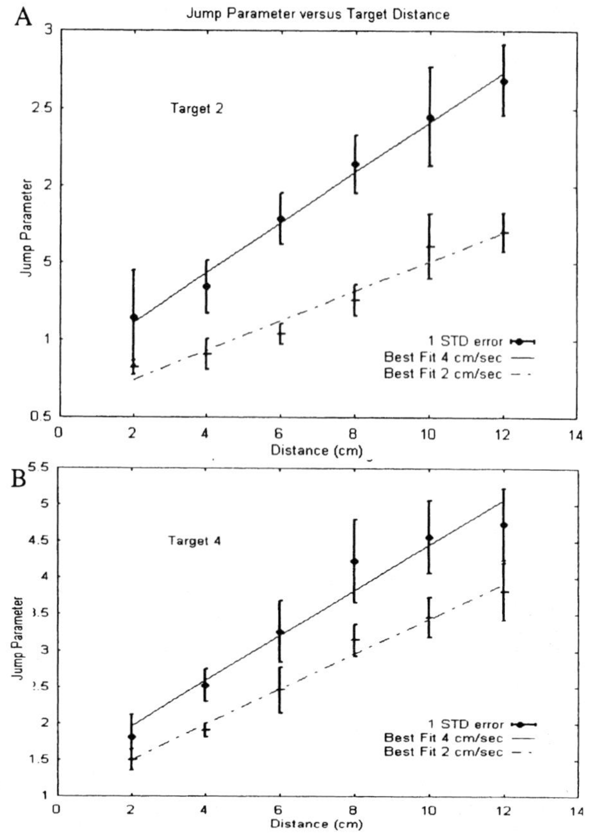

Figure 7. Comparison of jump parameter estimate versus distance. (A) Target #2. (B) Target #4.

the curves. If there is a non-linearity in the movement detectors, then the regression line should show a pronounced curve.

If the background noise is changed, then the line should shift parallel and downward at lower peering speeds.

The second case seems to fit the data. That is, the results indicate that increased peering velocities are associated with increased rotational noise.

6. Discussion

Our results indicate that it is possible to recover a distance estimate to a target using the computational model presented here.

We infer from our results that rotational stability of the head is a problem and will introduced significant errors in distance estimation.

Insects incorporate an optomotor response to rotationally stabilize relative to the environment. In the insect system, the eyes look sideways. During peering, the poles of the optic flow field are perpendicular to the target direction. Any translational movement at the poles must be due to head rotation.

Incorporation of a rotation stabilization mechanism should increase the accuracy target distance estimation and allow the system to accurately judge distances further away.

Our model is able to estimate the distances to objects using biologically plausible mechanisms. It relies on non-directionally selective movement detectors. This is a pathway only recently discovered in insects. The model can explain the reversal phenomena noted by Sobel (1990). In addition, the particular computation performed here is a simpler hypothesis than the computation proposed by Sobel.

Sobel's explanation for how the animal computes distance is based on the static geometry of the problem. After a peer, the animal uses two angular measurements made at the extents of the peer to triangulate the object's position. His computation requires that (A) the animal memorizes an angle at the extremes of movement presumably in some short term storage (B) the animal is able to 'zero' out this short term storage between peers, (C) the animal is able to match different views of the same object (at the end of a peer it must realize that a portion of the image is the same thing that it memorized a short while ago), (D) the animal then uses two angular measurements to compute distance using a non-linear function.

This computation appear difficult to explain using a neural substrate.This appears more complex than the original idea put forth by Exner (1891)— that it is the apparent speed of the object that determines the depth of the object, not the static position of the object at the extents of movement.

Peering is an interesting phenomenon from several points of view. In the behaving animal, peering solves a particularly difficult problem in nature: Depth perception at relatively large distances. At close distances, and especially for moving object (Collet, 1996) stereopsis can recover the depth of a target and is used when an animal strikes it prey. The mechanism of stereopsis has the advantage of stealth and is compatible with a lay and wait strategy where the animal is motionless and waits for unsuspecting prey to approach.

Insects with compound eyes have relatively few photoreceptors. Their omatidia are focused forward into a relatively small binocular region.

Stereopsis for gauging distances to far away objects is not an option for insects. Instead, peering is used.

Peering is potentially of wide use in the robotics community. Peering could be used as a complement to stereopsis in gauging distance to objects. One could image a system were peering is invoked to confirm distance estimation from stereopsis.

7. Conclusion

A neurocomputational model of depth perception using peering was presented that explains how an insect can recover depth using peering behavior. The model is consistent with recent behavioral and anatomical evidence for non-directional selective movement detectors.

The model demonstrates that the memorization of angles to target positions at the extents of a peer is not necessary to recover depth.

The model can also estimate the distance to a variety of objects and at a variety of speeds.

It is speculated that the accuracy of the distance estimate will increase with better rotational stabilization of the camera. In addition, slower scanning speeds may give better results in a real, mechanical system.

The theoretical model can also explain the reversal phenomena described in the literature (Sobel, 1990).

Acknowledgments

The authors acknowledges the useful critique of this work by Narendra Ahuja, John Hart and Lucia Simo. The authors acknowledges the support of ONR grant N000149610657. The authors also acknowledges the loan of the Khepera from UCLA (NSF grant CDA-9303148).

References

A. Borst and M. Egelhaaf (1989), Principles of Visual Motion Detection, Trends in Neurosciences, **12**(8):297-306.

T. S. Collett (1978), Peering— A Locust Behavior Pattern for Obtaining Motion Parallax Information, J. Exp. biol. **76**, pp 237-241.

T.S. Collet (1996), Vision: Simple Stereopis, Current Biology, **6**(11):1392-1395.

J. K. Douglass and N. J. Strausfeld (1996), Visual Motion-Detection Circuits in Flies: Parallel Direction and Non-Direction-Sensitive Pathwasy between the Medulla and the Lobula Plate, J. Neurosci, **16**(15):4551-4562.

Exner, S. (1891), *Die Physiologie der facettirten Augen von Krebsen and Insecten,* pp 206, Leipzig and viena: Franz Deuticke.

A. Geist, A. Beguelin, J. Dongarra, W. Jiang, R. Manchek and V. Sunderam (1994), *PVM Reference here.PVM: Parallel Virtual Machine A Users' Guide and Tutorial for Networked Parallel Computing,* MIT Press, Cambridge.

M. A. Lewis, Visual Navigation In a Robot using ZigZag Behavior, *Advances in Neural Information Processing Systems (NIPS*10),* The MIT Press, Cambridge 1998

E. C. Sobel (1990), The locust's use of motion parallax to measure distance, J. Comp. Physiol A. **167**, pp 579-588.

M. V. Srinivasan and S. W. Zhang (1993), Evidence for Two Distinct Movement-Detecting Mechanisms in Insect Vision, Naturwissenschaften, **80**, pp 38-41.

F. Walcher and K. Kral (1994), Visual deprivation and distance estimation in the preying mantis larva, Physiological Entomology **19**, 230-240.

G. K. Wallace (1959), Visual Scanning in the Desert Locust *Schistocerca Gregaria ForskålJ. Exp. iol. 36, pp 512-525.*

Self-Organizing Internal Representation for Behavior Acquisition of Vision-Based Mobile Robots

Takayuki Nakamura
Nara Inst. of Science and Technology Dept. of Information Systems
8916-5, Takayama-cho, Ikoma, Nara 630-01, Japan
takayuki@is.aist-nara.ac.jp

Abstract

An input generalization problem is one of the most important ones in applying reinforcement learning to real robot tasks. To cope with this problem, we propose a self-partitioning state space algorithm which can make non-uniform quantization of the state space. To show that our algorithm has generalization capability, we apply our method to two tasks in which a soccer robot shoots a ball into a goal and prevent a ball from entering a goal. To show the validity of this method, the experimental results for computer simulation and a real robot are shown.

1. Introduction

Recently, many researchers in robotics (Connel and Mahadevan, 1993) have paid much attention to reinforcement learning methods by which adaptive, reflexive and purposive behavior of robots can be acquired without modeling its environment and its kinematic parameters. A problem in applying reinforcement learning methods to real robot tasks which have continuous state space is that the value function [1] should be able to represent the value in terms of infinitely many state and action pairs. For this reason, function approximators are used to represent the value function when a closed-form solution of the optimal policy is not available.

One approach that have been used to represent the value function is to quantize the state and action spaces into a finite number of cells and collect rewards and punishments in terms of all states and actions. This is one of the simplest forms of generalization in which all the states and actions within a cell have the same value. In this way, the value function is approximated as a table in which each cell has a specific value (e.g., (Connel and Mahadevan, 1993)). However, there is a compromise between the efficiency and accuracy of this table. In order to achieve accuracy, the cell size should be small to provide enough resolution to approximate the value function. But as the cell size gets smaller, the number of cells required to cover the entire state and action spaces grows exponentially, which causes the efficiency of the learning algorithm to become worse because more data is required to estimate the value for all cells. Chapman et. al (Chapman and Kaelbling, 1991) proposed an input generalization method which splits an input vector consisting of a bit sequence of the states based on the already structured actions such as "shoot a ghost" and "avoid an obstacle." However, the original states have been already abstracted, and therefore it seems difficult to be applied to the continuous raw sensor space of real world. Moore et. al (Moore and Atkeson, 1995) proposed a method to resolve the problem of learning to achieve given tasks in deterministic high-dimensional continuous spaces. It divides the continuous state space into cells such that in each cell the actions available may be aiming at the neighboring cells. This aiming is accomplished by a local controller, which must be provided as a prior knowledge of the given task in advance. The graph of cell transitions is solved for shortest paths in an online incremental manner, but a minimax criterion is used to detect when a group of cells is too coarse to prevent movement between obstacles or to avoid limit cycles. The offending cells are split to higher resolution. However, the restriction of this method to deterministic environments might limit its applicability since the real environment is often non-deterministic.

Another approach for representing the value function is to use other types of function approximators, such as neural networks (e.g., (Lin, 1992)), statistical models (Ishiguro et al., 1996; Ueno et al., 1996; Asada et al., 1996; Takahashi et al., 1996) and so on. Boyan and Moore (Boyan and Moore, 1995) used local memory-based methods in conjunction with value iteration; Lin (Lin, 1992) used backpropagation networks for Q-learning; Watkins (Watkins, 1989) used CMAC for Q-learning; Tesauro (Tesauro, 1992) used backpropagation for learning the value function in backgammon. Asada et al. (Asada et al., 1996) used a concentration ellipsoid as a model of cluster (state) of input vectors, inside which a uniform distribution is assumed. They define a state as a cluster of input vectors from which the robot can reach the goal state or the state already obtained by a sequence of one kind action primitive regardless of its length. However, actual distributions are not always uniform.

This paper propose a new method for incrementally dividing a multidimensional continuous state space into some discrete states. This method recursively splits its continuous state space into some coarse spaces called tentative states. It begins by supposing that such tentative states are regarded as the states for Q-learning. In parallel with the learning, it collects statistical evidence regarding immediate rewards r and Q values within this tentative state space. When a tentative state is considered to be *relevant* to the given task by the statistical test for r and Q values based on a minimum description length (hereafter, MDL) criterion (Rissanen, 1989),

[1] *value function* is a prediction of the return available from each state and is important because the robot can use it to decide a next action. See section 2.1 for more details.

this coarse state is divided into finer states. These procedures can make non-uniform quantization of the state space. Our method can be applied to non-deterministic domain because the Q-learning is used to find out the optimal policy for accomplishing the given task.

The remainder of this article is structured as follows: In the next section, we outline the basics of reinforcement learning algorithm. Then, we describe our method to automatically construct the sensor spaces. In section 4., we show the results of the experiments with a simple computer simulation and real robot in which a vision-based mobile robot tries to shoot a ball into a goal and tries to prevent a ball from entering a goal. Finally, we give discussion and concluding remarks.

2. Self-Organizing Internal Representation Based on Reinforcement Learning and MDL Criterion

2.1 Basics of Reinforcement Learning Algorithm

One step Q-learning (Watkins, 1989) has attracted much attention as an implementation of reinforcement learning because it is derived from dynamic programming (Bellman, 1957). Here, we briefly review the basics of the Q-learning (Kaelbling, 1993).

In the Q-learning algorithm, it is assumed that the robot can discriminate the set $\boldsymbol{S}$ of distinct world states, and can take one from the set $\boldsymbol{A}$ of actions on the world. The world is modeled as a Markovian process, making stochastic transitions based on its current state and the action taken by the robot. Let $T(s, a, s')$ be the probability that the world will transit to the next state s' from the current state-action pair (s, a). For each state-action pair (s, a), the *reward* $r(s, a)$ is defined. The general reinforcement learning problem is typically stated as finding a policy that maximizes discounted sum of the reward received over time. A policy f is mapping from $\boldsymbol{S}$ to $\boldsymbol{A}$. The value function $V^f(s_t)$ associated with a given policy f is defined as:

$$V^f(s_t) \equiv E\left[\sum_{n=0}^{\infty} \gamma^n r_{t+n}\right],$$

where s_t is the state of the system at step t and r_t is the reward received at step t given that the agent started in state s_t and executed policy f. γ is the discounting factor, it controls to what degree rewards in the distant future affect the total value of a policy and is just slightly less than 1. The value function measures the expected discounted sum of rewards or expected rewards the robot will receive when it starts from the given state and follows the given policy.

Given definitions of the transition probabilities and the reward distribution, we can solve the optimal policy, using methods from dynamic programming (Bellman, 1957). A more interesting case occurs when we wish to simultaneously learn the dynamics of the world and construct the policy. Watkin's Q-learning algorithm gives us an elegant method for doing this. Let $Q^*(s, a)$ be the expected return or *action-value function* for taking action a in a situation s and continuing thereafter with the optimal policy. It can be recursively defined as:

$$Q^*(s,a) = r(s,a) + \gamma \sum_{s' \in S} T(s,a,s') \max_{a' \in A} Q^*(s', a').$$

Because we do not know T and r initially, we construct incremental estimates of the Q values on line. Starting with $Q(s, a)$ at any value (usually 0), every time an action is taken, update the Q value as follows:

$$Q(s,a) \Leftarrow (1-\alpha) Q(s,a) + \alpha \left[r(s,a) + \gamma \max_{a' \in A} Q(s', a') \right],$$

where α is a leaning rate (between 0 and 1) and γ is the discounting factor which controls to what degree rewards in the distant future affect the total value of a policy (between 0 and 1).

2.2 Function Approximator with Non-uniform Resolution Model

There are some reasons why designing non-uniform function approximators may be more beneficial than designing uniform ones.

- In case that the designers know, up to a certain degree, prior knowledge of the system (for example, what regions of the state-action space will be used more often.), it may be efficient to design the function approximator such that it may use many resources in more heavily transited regions than in regions of the state space that are known to be visited rarely.
- If the amount of resources is limited, a non-uniform function approximator may make better performance and learning efficiency than that achieved with a uniform function approximator just because the former is able to exploit the resources more efficiently than the later.

2.3 Self-Partitioning State Space Algorithm Based on MDL Criterion

In this work, we define the sensor inputs, actions and rewards as follows:

- Sensor input $\boldsymbol{d}$ is described by a N dimensional vector $\boldsymbol{d} = (d_1, d_2, \cdots, d_N)$, each component $d_i (i = 1 \sim N)$ of which represents the measurement provided by the sensor i. The continuous value d_i is provided by the sensor i. Its range $Range(d_i)$ is known in advance. Based on $Range(d_i)$, a measurement d_i is normalized in such a way that d_i can take values in the semi open interval $[0, 1)$.
- The robot continues to take one action primitive at a time until the current state changes. This sequence of the action primitives is called an action. The agent has a set $\boldsymbol{A}$ of possible actions a_j, $j = 1 \sim M$. Such a set is called the action space.
- One of the discrete rewards $r = r_k$, $k = 1 \sim C$ is given to the agent depending on the evaluation of the action taken at a state.

Our method utilizes a hierarchical segment tree in order to represent the non-uniform partitioning of the state space which consists of N dimensional input vector. This representation has an advantage for approximating the non-uniform distribution of sample data. The inner node at i th depth in the j th level keeps the range $b_i^*(j) = [t_i^{low}, t_i^{high})$ of a measurement provided by each sensor i. (Actually, j corresponds to the number of iteration of this algorithm.) $*$ takes a binary value and shows an interval of original range of measurements. Initially, $*$ is set $NULL$. At each inner node in the j the level, $b_i^*(j)$ the range of a measurement of sensor i is partitioned into two equal intervals $b_i^0(j+1) = [t_i^{low}, (t_i^{low}+t_i^{high})/2)$ and $b_i^1(j+1) = [(t_i^{low}+t_i^{high})/2, t_i^{high})$. For example, initially $j = 0$, $b_i^{NULL}(0)$ the range of each dimension i is divided into two equal intervals $b_i^0(0) = [0.0, 0.5)$ and $b_i^1(0) = [0.5, 1.0)$. When sensor input vector $\boldsymbol{d}$ has N dimensions, a segment tree whose depth is N is built (see Fig.1). The leaf node corresponds to the result of classification for observed sensor input vector $\boldsymbol{d}$. As a result, 2^N leaf nodes are generated. These leaf nodes can represent the situations in the agent's environment. The state space represented by the leaf nodes is called "tentative state space" $\boldsymbol{TS}$. Let ts_k, $k = 1 \sim 2^N$ be the component of the tentative state space which is called "tentative state."

Our algorithm works as follows:

1. It starts by assuming that the entire environment is as if it were one state. Initially, the total number of the states $|\boldsymbol{S}|$ and the tentative states $|\boldsymbol{TS}|$ are one and 2^N, respectively.
2. Based on $\boldsymbol{TS}$, our algorithm begins Q-learning. In parallel with this process, it gathers statistics in terms of $r(a_i|ts_k = on)$, $r(a_i|ts_k = off)$, $Q(a_i|ts_k = on)$ and $Q(a_i|ts_k = off)$, which indicate immediate rewards r and discounted future rewards Q in case that individual state is "on" or "off," respectively. In this work, it is supposed that if a N dimensional sensor vector $\boldsymbol{d}$ is classified into a leaf node ts_k, the condition of this node ts_k is defined as "on," otherwise (this means the case that d is classified into the leaf node except ts_k), it is defined as "off."
3. After Q-learning based on $\boldsymbol{TS}$ is converging, our algorithm asks the question whether there are some states in the state description such that the r and Q for states "on"are significantly different from such values for states "off." When the distributions of statistics of ts_k in case of "on" and "off" are different, it is determined that ts_k is *relevant* to the given task. In order to discover the difference between two distributions, our algorithm performs the statistical test based on a MDL criterion. In the section 3.1 and 3.2, these procedures are explained.
4. (a) If there is the state ts'_k adjoining the state ts_k which is shown to be relevant such that the statistical characteristic of Q and actions assigned at the adjoining state are same, merge these two states into one state.
   (b) Otherwise, skip this step.
5. Each leaf nodes ts_k is represented by a combination of intervals each of which corresponds to the range of a measurement provided by each sensor i. These intervals in ts_k which is shown to be relevant are bisected. As a result, in terms of one ts_k, 2^N leaf nodes are generated and correspond to tentative states. These tentative states are regarded as the states in Q-learning at the next iteration.
6. Until our algorithm can't find out the relevant leaf nodes, the procedures 2 $\sim$ are repeated. Finally, a hierarchical segment tree is constructed to represent the partitioning of the state space for achievement of a given task.

After the learning, based on Q values stored at leaf nodes, the agent takes actions for accomplishing the given task.

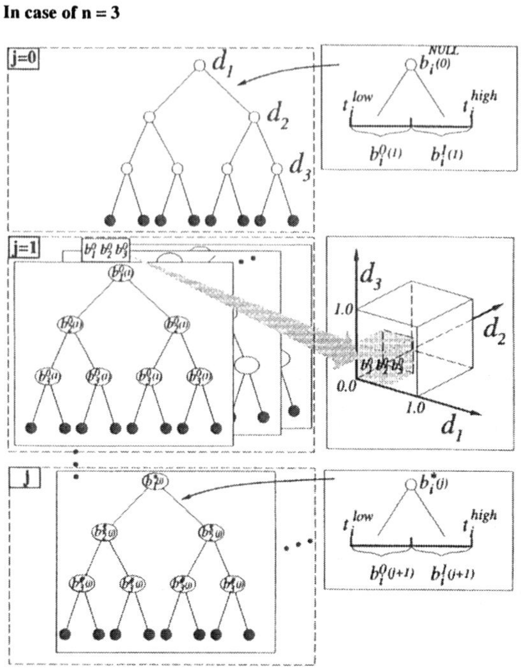

Figure 1 Representation of state space by a hierarchical segment tree

3. The relevance test based on MDL criterion

Here, we explain how to determine whether a state is *relevant* to the task or not. **Fig. 2** shows the difference between the distributions of r or Q values regarding to the state ts_k in case that ts_k is *relevant* or *irrelevant*. As shown in upper part of this figure, when two peaks of the distributions of r or Q values ,which correspond to pair of $r(a_i|ts_k = on)$ and $r(a_i|ts_k = off)$, or pair of $Q(a_i|ts_k = on)$ and $Q(a_i|ts_k = off)$), can be clearly discriminated, it is supposed that ts_k is relevant to the given task because the such state affects the value of state more heavily than the other states does, therefore, it affects how the robot should act at next time step. On the contrary, in case that two peaks are ambiguous as shown in bottom part of this figure, it is considered that ts_k is *irrelevant* to the given task.

Actually, we perform the statistical test with respect r and Q values based on a MDL criterion (Rissanen, 1989) in order to distinguish the distribution of such reinforcement values. Chapman et. al (Chapman and Kaelbling, 1991) used student's t test and McCallum (McCallum, 1996) used Kolmogorov-Smirnov test to find out the difference between two ditributions of discounted future value. Generally, in order to perform conventional statistical test such as their tests, a level of significance should be determined. Since a level of significance reflects the accuracy of its judgement of the test, the acuuracy of discriminating two distributions depends on the level of significance. As a result, the level of significance has a great influence on athe quantization of state space based on the conventional statistical test. There is no criterion for determining a level of significance. On the contrary, our method doesn't need a level of significance something like a threshold. In order to find out the difference between two distributions, our method just compares discription lengths of two distributions based on MDL. The result of quantization of state space based on our method isn't affected by external parameters such a level of significance. In this sense, on the basis of our method, the robot can organize its internal structure by itself.

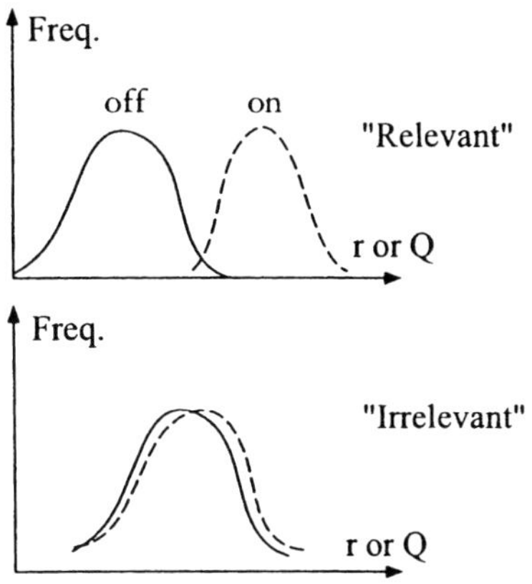

Figure 2 Criterion for determining the relevance of the state

3.1 *The statistical test for* r *value*

Table 1 The distribution of rewards in ts_k

	On	Off
R_1	$n(On, R_1)$	$n(Off, R_1)$
R_2	$n(On, R_2)$	$n(Off, R_2)$
:	:	:
R_C	$n(On, R_C)$	$n(Off, R_C)$
	$n(On)$	$n(Off)$

$n(i)$: the frequency of sample data in the state i $(i = 1, \cdots, S)$

$n(i, r_j)$: the frequency of reward r_j $(j = 1, \cdots, C)$ given in the state i

$p(r_j|i)$: the probability that reward r_j is given in the state i

$$\sum_{j=1}^{C} n(i, r_j) = n(i), \quad \sum_{j=1}^{C} p(r_j|i) = 1, \quad (i = 1, \cdots S).$$

Since the immediate reward r_j value is given at each trial among one of C mutually exclusive rewards r_j $j = 1, 2, \cdots C$, the distribution of r_j in the state ts_k follows a multinominal distribution.

Supposing **Tab. 1** shows the distribution of immediate rewards in case that the state ts_k is "*on*" or "*off*," our algorithm tries to find the difference between the distributions of rewards in two cases of ts_k "on" and "off" based on this table.

The probability $P(\{n(i, r_j)\}|\{p(r_j|i)\})$ that the distribution of the immediate rewards is acquired as shown in **Tab. 1** $\{n(i, r_j)\}$, $(i = 1 \cdots, S, \quad j = 1, \cdots, C)$, can be described by

$$P(\{n(i, r_j)\}|\{p(r_j|i)\}) = \prod_{i=1}^{S} \left\{ \frac{n(i)!}{\prod_{j=1}^{C} n(i, r_j)!} \prod_{j=1}^{C} p(r_j|i)^{n(i, r_j)} \right\}.$$

The likelihood function L of this multinominal distribution can be written as follows:

$$L(\{p(r_j|i)\}) = \sum_{i=1}^{S} \sum_{j=1}^{C} n(i, r_j) \log p(r_j|i) + K,$$

where $K = \log \left\{ \frac{\prod_{i=1}^{S} n(i)!}{\prod_{i=1}^{S} \prod_{j=1}^{C} n(i, r_j)!} \right\}$.

When two multinominal distributions in case that each tentative state ts_k is "on" and "off" can be considered to be same, the probability $p(r_j|i)$ that the immediate reward r_j is given in the state i can be modeled as follows:

$$\text{M1: } p(r_j|i) = \theta(r_j) \quad (i = 1, \cdots, S, \quad j = 1, \cdots, C).$$

Furthermore, its likelihood function L_1 can be written by

$$L_1(\{\theta(r_j)\}) = K + \sum_{j=1}^{C} \left[\left\{ \sum_{i=1}^{S} n(i, r_j) \right\} \log \theta(r_j) \right].$$

Therefore, the maximum likelihood ML_1 can be written by

$$ML_1 = L(\theta(\hat{r_j})) = L\left(\frac{\sum_{i=1}^{S} n(i, r_j)}{\sum_{i=1}^{S} n(i)} \right).$$

where $\theta(\hat{r_j})$ is the maximum likelihood estimation of $\theta(r_j)$. Therefore, the description length $l_{MDL}(M1)$ of the model $M1$ based on MDL principle is

$$l_{MDL}(M1) = -ML_1 + \frac{C-1}{2} \log \sum_{i=1}^{S} n(i).$$

On the contrary, when two multinominal distributions in case of "on" and "off" can be considered to be different, the probability $p(r_j|i)$ can be modeled as follows:

$$\text{M2: } p(r_j|i) = \theta(r_j|i) \quad (i = 1, \cdots, S, \quad j = 1, \cdots, C).$$

Furthermore, its likelihood function L_2 can be written by

$$L_2(\{\theta(r_j|i)\}) = K + \sum_{i=1}^{S} \sum_{j=1}^{C} n(i, r_j) \log \theta(r_j|i).$$

In the same way, the maximum likelihood ML_2 can be written by

$$ML_2 = L(\theta(\hat{r_j})) = L\left(\frac{n(i, r_j)}{n(i)} \right).$$

Therefore, the description length $l_{MDL}(M2)$ of the model $M2$ based on MDL principle is

$$l_{MDL}(M2) = -ML_2 + \frac{S(C-1)}{2} \log \sum_{i=1}^{S} n(i).$$

Based on MDL criterion, we can suppose that discovering the difference between two distributions is equivalent to determining which model is appropriate for representing the distribution of data. In this work, the difference between the distributions is found based on the following conditions:

if $l_{MDL}(M1) \geq l_{MDL}(M2)$
Two distributions are different.
if $l_{MDL}(M2) \geq l_{MDL}(M1)$
Two distributions are same.

3.2 The statistical test for Q value

In order to distinguish the distribution of sampled data of Q values, we perform the statistical test based on a MDL criterion. Let $\boldsymbol{x}^n$ and $\boldsymbol{y}^m$ be the sample data $(x_1, x_2 \cdots, x_n)$ and $(y_1, y_2 \cdots, y_m)$, respectively. $\boldsymbol{x}^n$ and $\boldsymbol{y}^m$ indicate a history of $Q(a_i|ts_k = on)$ and $Q(a_i|ts_k = off)$, respectively. We'd like to know whether these two sample data $\boldsymbol{x}^n$ and $\boldsymbol{y}^m$ come from the two different distributions or the same distribution. Here, we assume the following two model for the distribution of sampled data are $M1$ based on one normal distribution and $M2$ based on two normal distributions.

M1: $N(\boldsymbol{\mu}, \boldsymbol{\sigma}^2)$
M2: $N(\boldsymbol{\mu}_1, \boldsymbol{\sigma}'^2), N(\boldsymbol{\mu}_2, \boldsymbol{\sigma}'^2)$

The normal distribution with mean μ and variance σ is defined by $f(x : \mu, \sigma^2) = \frac{1}{\sqrt{2\pi\sigma^2}} \exp\left\{\frac{-(x-\mu)^2}{2\sigma^2}\right\}$. If both $\boldsymbol{x}^n$ and $\boldsymbol{y}^m$ follow the model $M1$, the probabilistic density function of these sampled data can be written by $\Pi_{i=1}^n f(x_i : \mu, \sigma^2) + \Pi_{i=1}^m f(y_i : \mu, \sigma^2)$. The likelihood function LL_1 of the model $M1$ can be written by

$$LL_1 = \sum_{i=1}^{n} \log f(x_i : \mu, \sigma^2) + \sum_{i=1}^{m} \log f(y_i : \mu, \sigma^2).$$

Therefore, the maximum likelihood MLL_1 can be derived as follows:

$$MLL_1 = -\frac{n+m}{2}(1 + \log 2\pi\hat{\sigma}^2),$$

where $\hat{\sigma}^2$ is the maximum likelihood estimated variance of σ^2 and can be estimated as follows:

$$\hat{\mu} = \frac{1}{n+m}\left\{\sum_{i=1}^{n} x_i + \sum_{i=1}^{m} y_i\right\}$$
$$\hat{\sigma^2} = \frac{1}{n+m}\left\{\sum_{i=1}^{n}(x_i - \hat{\mu})^2 + \sum_{i=1}^{m}(y_i - \hat{\mu})^2\right\}$$

where $\hat{\mu}$ is the maximum likelihood estimated mean of μ. Therefore, the description length $l_{MDL}(M1)$ of the model $M1$ based on MDL principle is

$$l_{MDL}(M1) = -MLL_1 + \log(n+m).$$

On the contrary, if $\boldsymbol{x}^n$ and $\boldsymbol{y}^m$ follow the model $M2$, the probabilistic density function of these sampled data can be written by $\Pi_{i=1}^n f(x_i : \mu_1, \sigma'^2) + \Pi_{i=1}^m f(y_i : \mu_2, \sigma'^2)$. The likelihood function LL_2 of the model $M2$ can be written by

$$LL_2 = \sum_{i=1}^{n} \log f(x_i : \mu_1, \sigma'^2) + \sum_{i=1}^{m} \log f(y_i : \mu_2, \sigma'^2).$$

Therefore, the maximum likelihood MLL_2 can be derived as follows:

$$MLL_2 = -\frac{n+m}{2}(1 + \log 2\pi\hat{\sigma'}^2),$$

where $\hat{\sigma_1}^2$ and $\hat{\sigma_2}^2$ is the maximum likelihood estimated variance of σ_1^2 and σ_2^2, respectively. These statistics can be estimated as follows:

$$\hat{\mu_1} = \frac{1}{n}\sum_{i=1}^{n} x_i, \qquad \hat{\mu_2} = \frac{1}{m}\sum_{i=1}^{m} y_i,$$
$$\hat{\sigma'^2} = \frac{1}{n+m}\left\{\sum_{i=1}^{n}(x_i - \hat{\mu_1})^2 + \sum_{i=1}^{m}(y_i - \hat{\mu_2})^2\right\}.$$

where $\hat{\mu_1}$ and $\hat{\mu_2}$ is the maximum likelihood estimated mean of μ_1 and μ_2, respectively. Therefore, the description length $l_{MDL}(M2)$ of the model $M2$ based on MDL principle is

$$l_{MDL}(M2) = -MLL_2 + \frac{3}{2}\log(n+m).$$

We can recognize the difference between the distributions based on the following condition:

if $l_{MDL}(M1) \geq l_{MDL}(M2)$
$\boldsymbol{x}$ and $\boldsymbol{y}$ arise from the different normal distributions
if $l_{MDL}(M2) \geq l_{MDL}(M1)$
$\boldsymbol{x}$ and $\boldsymbol{y}$ arise from the same normal distribution

4. Experimental Results

To show that our algorithm has a generalization capability, we apply it in order to acquire two different behaviors: one is a shooting behavior and the other is a defending behavior for soccer robots.

4.1 Simulation

We consider an environment shown in Figure 3 (a) where the task for a mobile robot is to shoot a ball into a goal or to defend a shot ball. The environment consists of a ball and a goal, and the mobile robot has a single CCD camera. The robot does not know the location and the size of the goal, the size and the weight of the ball, any camera parameters such as focal length and tilt angle, or kinematics/dynamics of itself.

We performed the computer simulation with the following specifications. The field is 1.52m × 2.74m. The goal posts are located at the center of the left and right line of the rectangle (see Figure 3 (a)) and its height and width are 0.20m and 0.5m, respectively. The robot is 0.10m wide and 0.18m long and kicks a ball of diameter 0.05m. The maximum translation velocity is 5cm/s. The camera is horizontally mounted on the robot (no tilt) and

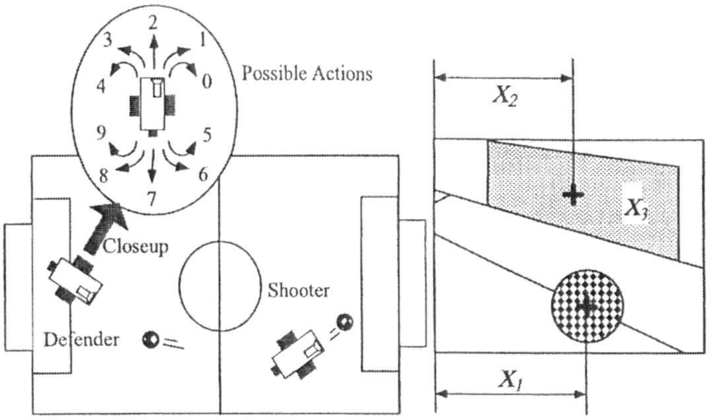

(a) Simulation environment

(b) Input vector as sensor information

Figure 3 Computer Simulation

is in off-centered position. Its visual angle is 36 degrees. The velocities of the ball before and after being kicked by the robot is calculated by assuming that the mass of the ball is negligible compared to that of the robot. The speed of the ball is temporally decreased by a factor 0.8 in order to reflect the so-called "viscous friction." The values of these parameters are determined so that they can roughly simulate the real world.

The robot is driven by two independent motors and steered by front and rear wheels which is driven by one motor. Since we can send the motor control commands such as "move forward or backward in the given direction," all together, we have 10 actions in the action primitive set $\boldsymbol{A}$ as shown in **Fig. 3** (a). Actually, a stop motion does not causes any changes in the environment, we do not take into account this action primitive.

In computer simulation, we take into account the delay due to sensory information processing. The contents of the image processing are color filtering (a ball and a goal are painted in red and blue, respectively), localizing and counting red and blue pixels, and vector calculation. We are using a general graphic workstation for image processing in the real robot experiments. Since it is not specialized on an image processing, it takes about 66 ms to perform these processes. As a result, in the simulation, the robot is assumed to take an action primitive every 66 ms.

The size of the image taken by the camera is 256×240 pixels. An input vector $\boldsymbol{x}$ to our algorithm consists of:

- x_1: the horizontal position of the ball in the image, that ranges from 1 to 256 pixels,
- x_2: the horizontal position of the goal in the image ranging from 1 to 256,
- x_3: the area of the goal region in the image, that ranges from 0 to 256×240 pixels^2

(see Figure 3 (b)). When the ball disappears on right side and left side of the image, its horizontal position is set 0 and 257, respectively. In the same way, the horizontal position of the goal is set 0 and 257 in the lost situation. After the range of these values is normalized in such a way that the range may become the semi open interval $[0, 1)$, they are used as inputs of our method.

A discounting factor γ is used to control to what degree rewards in the distant future affect the total value of a policy. In our case, we set the value a slightly less than 1 ($\gamma = 0.9$). In this work, we set the learning rate $\alpha = 0.25$. In case that the shooting behavior tried to be acquired by our method, as a reward value, 1 is given when the robot succeeded in shooting a ball into a goal, 0.3 is given when the robot just kicked a ball, -0.01 is given when the robot went out of field, 0 is given otherwise. In the same way, in case that the defending behavior tried to be acquired by our method, as a reward value, -0.7 is given when the robot failed in preventing a ball from entering a goal, 1.0 is given when the robot just kicked a ball, -0.01 is given when the robot went out of field, 0 is given otherwise.

In the learning process, Q-learning continues until the sum of estimated Q values seems to be almost convergent. When our algorithm tried to acquire the shooting behavior, our algorithms ended after it iterated the process (Q-learning + statistical test) 8 times. In this case, about $160K$ trials were required to converge our algorithm and the total number of the states is 246. In case that our algorithm tried to acquire the defending behavior, our algorithms ended after it iterated the process (Q-learning + statistical test) 5 times. In this case, about $100K$ trials were required to converge our algorithm and the total number of the states is 141.

Fig. 4 shows the success ratio versus the step of trials in the two learning processes that one is for acquiring a shooting behavior the other is for a defending behavior. We define the success rate as (# of successes)/(# of trials) $\times$ 100(%). As you can see, the bigger the number of iteration is, the higher the success ratio at the final step in each iteration. This means that our algorithm gradually made better segmentation of state space for accomplishing the given task.

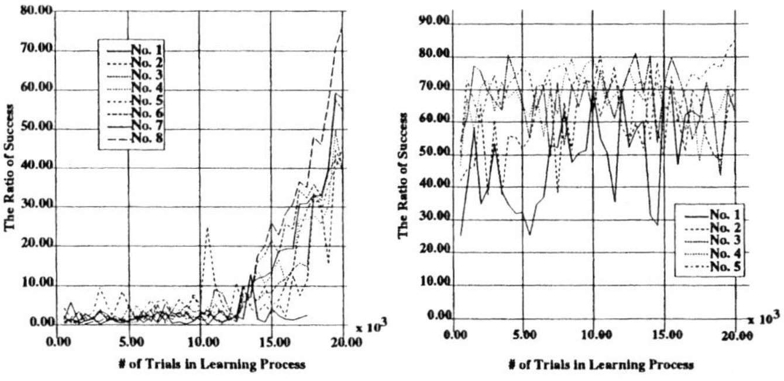

(a) In case of Shooting Behavior

(b) In case of Defending Behavior

Figure 4 The success ratio versus the step of trial

Fig. 5 shows the partitioned state spaces obtained by our method. **Fig. 5** (a) and (b) shows a set of the states where an action is assigned for the shooting behavior and for defending behavior, respectively. In each figure, *Action*2 and *Action*7 corresponds to "moving forward" and "moving backward," respectively. Furthermore, in each figure, *Dim.*1, *Dim.*2 and *Dim.*3 shows the position of ball, the position of goal and the area of goal region, respectively.

One cube in the partitioned state space corresponds to one state. For example, the left in **Fig. 5** (a) shows a group of the cube where the action 2 is assigned as an optimal action. As shown in **Fig. 5** (a), many cubes

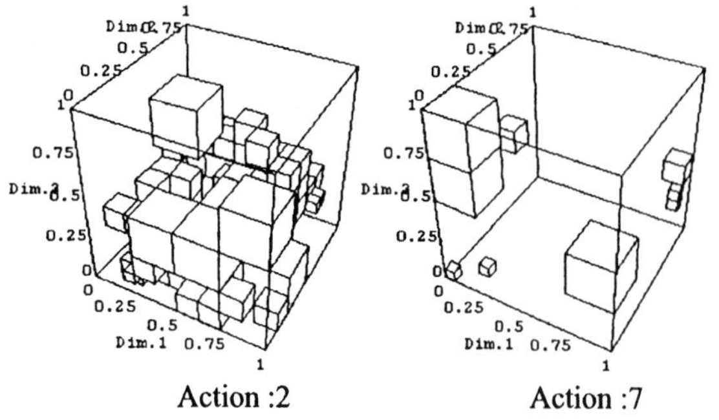

(a) The State Space for Shooting Behavior

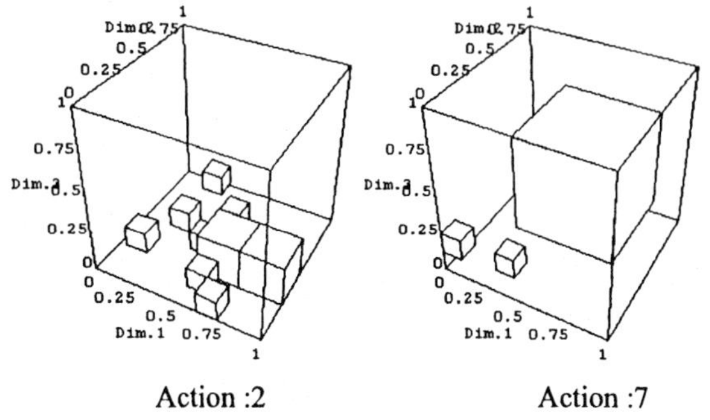

(b) The State Space for Defending Behavior

Figure 5 The Partitioned State Space

where forward actions are assigned concentrate around the center of the entire state space. This means that the robot will take an forward action if the ball and goal are observed around the center of field of its view. This shows very natural behavior for shooting a ball into a goal.

In the right of **Fig. 5** (b), there is one large cube. This means that the the robot will take an backward action if large goal are observed around the center of field of its view. This strategy is plausible behavior for preventing a ball from entering a goal because the robot will have to go back in front of own goal after it moved out there in order to kick a ball.

Fig.6 shows a typical sequence of the shooting behavior in the simulation acquired by our method. The bottom part of each figure in the **Fig. 6** shows the processed image where the white rectangle and dark gray circle indicates the goal and ball region, respectively. Once the robot found the ball, the robot ran to the ball quickly. After that, the robot kicked the ball into the goal.

Fig.7 shows the defending behavior in the simulation acquired by our method. The robot initially looked around because the ball was not observed from its initial position. Then, once the robot found the ball, the robot ran to the ball quickly. After that, the robot kicked the ball to prevent a ball from approaching the goal. The robot followed the ball until the area of opponent's goal grows to a certain extent. Then in order to go back to the front of own goal, the robot took a backward action. It seems that the robot utilizes the area of the opponent's goal to localize its own position in the field. Note that this behavior is just the result of learning and not hand-coded.

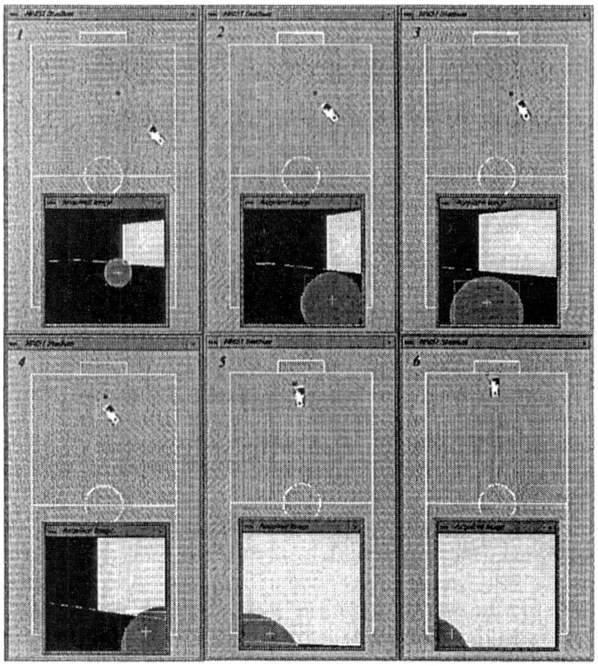

Figure 6 Shooting Behavior

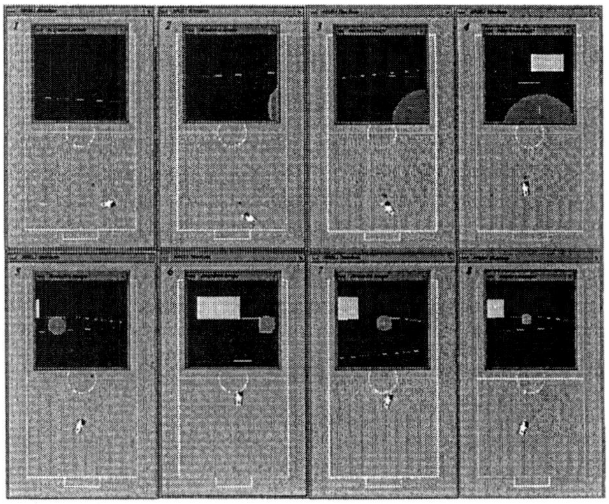

Figure 7 Defending Behavior

4.2 Real Robot Experiments

Fig. 8 shows our real robot system which we have developed to take part in RoboCup-97 competition where several robotic teams are competing on a field. So, the system includes two robots which have the same structure: one for a shooter, the other for a defender. Off-board computer SGI ONYX (R4400/250MHz) perceives the environment through on-board cameras, performs the decision making based on the learned policy and sends motor commands to each robot. A CCD camera

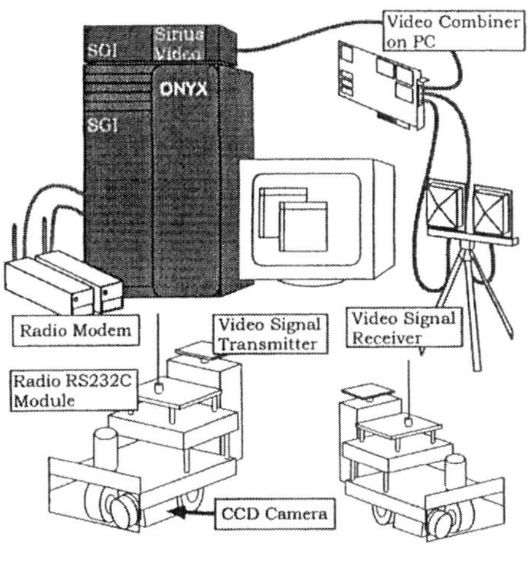

Figure 8 Our robots and system

is set at bottom of each robot in off-centered position . Each robot is controlled by SGI ONYX through radio RS232C. The maximum vehicle speed is about 5cm/s. The images taken by the CCD camera on each robot are transmitted to a video signal receiver. In order to process two images (one is sent from the shooter robot, the other from the defender robot) simultaneously, two video signals are combined into one video signal by a video combiner on PC. Then, the video signal is sent to SGI ONYX for image processing. The color-based visual tracking routine is implemented for tracking and finding a ball and a goal in the image. Our simple tracking method is based on tracking regions with similar color information from frame to frame (see Appendix A for more detail). In our current system, it takes 66 ms to do this image processing for one frame.

Fig. 10 shows how a real robot shoots a ball into a goal based on the state space obtained by our method. 6 images are shown in raster order from the top left to the bottom right in every 2.0 seconds, in which the robot tried to shoot a ball, but failed, then moved backward so as to find a position to shoot a ball, finally succeeded in shooting. Note that the backward motion for retry is just the result of learning and not hand-coded.

Fig. 11 shows how a real robot prevents a ball shot by the opponent's robot from entering a goal based on the state space obtained by our method. 6 images are shown in raster order from the top left to the bottom right in every 2.0 seconds. This sequence was appeared in the RoboCup-97 competition. In the top left image, the white circle shows the position of the defender robot and the white arrow shows the position of the ball. In the time step 1 $\sim$ 2, first, the opponent's robot tried to shoot a ball, then the defender robot prevent a ball from entering a goal. In the time step 3–6, another opponent's robot got a ball cleared by the defender robot and tried to shoot a ball again, then the defender robot defend a goal again. As shown in these figures, the defender robot always moves in front of own goal to find out a shot ball as soon as possible. Note that this behavior is just the result of our learning algorithm and not hand-coded.

5. Concluding Remarks and Discussion

We have proposed a method for self-organizing the state space based on experiences, and shown the validity of the method with computer simulations and real robot experiments. We can regard the problem of state space construction as "segmentation" problem. From a viewpoint of robotics, segmentation of sensory data from the environment depends on the purpose (task), capabilities (sensing, acting, and processing) of the robot, and its environment, and its evaluation should be done based on the robot performance. The state space obtained by our method (**Fig.5** indicates a projection of such a space) corresponds to the subjective representation of the world for the robot to accomplish a given task. Although it seems very limited, such a representation, an inside view of the world for the robot, shows how the robot segments the world. This view is intrinsic to the robot, and based on it the robot might make a subjective decisions when facing with different environments, and further the robot might develop its view through its experiences (interactions with its environment). That is, there might be a possibility that the robot acquires the subjective criterion, and as a result, an emerged behavior can be observed as "autonomous" and/or "intelligent."

There are two kinds of problem in current version of our algorithm:

- If the dimension of the sensor space is higher, the computational time and amount are enormous, and as a result the learning might not converge correctly. In our current implementation, we used input vector consisting of three parameters which were enough to acquired the shooting and defending behaviors. However, there is no guarantee that these parameters are enough in case that the higher-level behaviors such as strategic coordination between two robots are tried to be acquired by our method. Generally, selection of features from raw images is considerably a hard problem. A problem which feature is necessary to accomplish a give task might be much harder when such a feature changes depending on situations. Since use of all possible sensory information seems impossible, selection of features obtained by the given capability for feature detection is more important. An effective method for coping with higher dimension space should be developed, but there is a trade-off between the effectiveness and *a priori* knowledge on the environment and the robot.
- We can regard that as a result of state space construction, the action space also is temporally abstracted by defining an action as a sequence of action primitives. Since our robot has only two degrees of freedom (DOF) spatial abstraction of action space is not necessary. However, spatial abstraction of action space is generally required if the robot has many DOFs. For example, we human beings easily grasp something by controlling a very simple parameter (close or open your hand) although it has many DOFs physically. Both spatial and temporal abstraction of action space is necessary with state space construction.

6. Acknowledgments

The main idea of this paper is thought of while I stayed at AI Lab of Comp. Sci. Dept. of Brown University. I would like to thank Prof. L. P. Kaelbling for their helpful comments during my stay. I also would like to thank Prof. M. Imai (NAIST) for providing research fund and

S. Morita (Japan SGI Cray Corp) for lending SGI ONYX to me.

References

Asada, M., Noda, S., and Hosoda, K. (1996). "Action-Based Sensor Space Categorization for Robot Learning". In *Proc. of IEEE/RSJ International Conference on Intelligent Robots and Systems 1996 (IROS96)*, volume 3.

Bellman, R. (1957). *Dynamic Programming.* Princeton University Press, Princeton, NJ.

Boyan, J. and Moore, A. W. (1995). "Generalization in Reinforcement Learning: Safely Approximating the Value Function". In *Proceedings of Neural Information Processings Systems 7.* Morgan Kaufmann.

Chapman, D. and Kaelbling, L. P. (1991). "Input generalization in delayed reinforcement learning: An alogorithm and performance comparisons". In *Proc. of IJCAI-91*, pages 726–731.

Connel, J. H. and Mahadevan, S., editors (1993). *Robot Learning.* Kluwer Academic Publishers.

Ishiguro, H., Sato, R., and Ishida, T. (1996). "Robot Oriented State Space Construction". In *Proc. of IEEE/RSJ International Conference on Intelligent Robots and Systems 1996 (IROS96)*, volume 3.

Kaelbling, L. P. (1993). "Learning to achieve goals". In *Proc. of IJCAI-93*, pages 1094–1098.

Lin, L.-J. (1992). "Self-improving reactive agents based on reinforcement learning, planning and teaching". *Machine Learning*, 8:293–321.

McCallum, A. K. (1996). "Learning to Use Selective Attention and Short-Term Memory in Sequential Tasks". In *From animals to animats 4*, pages 315–324.

Moore, A. W. and Atkeson, C. G. (1995). "The Parti-game Algorithm for Variable Resolution Reinforcement Learning in Multidimensional State-spaces". *Machine Learning*, 21:199–233.

Rissanen, J. (1989). *Stochastic Complexity in Statistical Inquiry.* World Scientific.

Takahashi, Y., Asada, M., and Hosoda, K. (1996). "Reasonable performance in less learning time by real robot based on incremental state space segmentation". In *Proc. of IEEE/RSJ International Conference on Intelligent Robots and Systems 1996 (IROS96)*, volume 3, pages 1518–1524.

Tesauro, G. (1992). "Practical issues in temporal difference learning". *Machine Learning*, 8:257–277.

Ueno, A., Hori, K., and Nakasuka, S. (1996). "Simultaneous Learning of Situation Classification Based on Rewards and Behavior Selection Based on the Situation". In *Proc. of IEEE/RSJ International Conference on Intelligent Robots and Systems 1996 (IROS96)*, volume 3.

Watkins, C. J. C. H. (1989). *Learning from delayed rewards".* PhD thesis, King's College, University of Cambridge.

A Simple Color-Based Tracking

Our simple tracking method is based on tracking regions with similar color information from frame to frame. We assume the existence of the color models (CM_{ball}, CM_{goal}) for tracking targets, which are estimated at initialization. Actually, we use only hue values (H_{ball} and H_{goal}) of targets as (CM_{ball} and CM_{goal}) to reduce its sensitivity due to changes in illumination. We define a fitness function $\Phi_{target}(x, y)$ at a pixel (x, y) as a criterion for extracting a target region in the image,

$$\Phi_{target}(x,y) = \begin{cases} 1 & H_{target} - \sigma_{target} \le H(x,y) \le H_{target} + \sigma_{target}, \\ 0 & \text{Otherwise} \end{cases}$$

,where $H(x, y)$ and σ_{target} show hue value at (x, y) and a threshold for extraction, respectively. In our current implementation, we set $\sigma_{ball} = 10$ and $\sigma_{goal} = 20$. These values can be calculated based on the standard deviation of H_{ball} and H_{goal} which were estimated at initial estimation process. Based on $\Phi_{target}(x, y)$, the best estimate $(\hat{x}_{target}, \hat{y}_{target})$ for the target's location is calculated as follows:

$$\hat{x}_{target} = \frac{\sum_{(x_i,y_i)\in R} x_i \Phi_{target}(x_i, y_i)}{\sum_{(x_i,y_i)\in R} \Phi_{target}(x_i, y_i)},$$
$$\hat{y}_{target} = \frac{\sum_{(x_i,y_i)\in R} y_i \Phi_{target}(x_i, y_i)}{\sum_{(x_i,y_i)\in R} \Phi_{target}(x_i, y_i)},$$

where R shows the search area. Initially, R implies an entire image plane. After initial estimation for the location of the target, we can know the standard deviations $\sigma(\hat{x}_{target})$ and $\sigma(\hat{y}_{target})$ regarding $(\hat{x}_{target}, \hat{y}_{target})$. Therefore, based on the deviations, R is restricted to a local region during the tracking process as follows:

$$R : \{(x,y)| \\ \hat{x}_{target} - 2.5\sigma(\hat{x}_{target}) \le x \le \hat{x}_{target} + 2.5\sigma(\hat{x}_{target}), \\ \hat{y}_{target} - 2.5\sigma(\hat{y}_{target}) \le y \le \hat{y}_{target} + 2.5\sigma(\hat{y}_{target})\}.$$

$\sum_{(x_i,y_i)\in R} \Phi_{target}(x_i, y_i)$ shows the area of the target in the image. Based on this value, we judge the appearance of the target. If this value is lower than the pre-defined threshold, the target is considered to be lost, then R is set to be the entire image plane for estimation at next time step. We set this threshold for the target area $= 0.05 * S$, where S shows the area of the entire image. **Fig. 9** shows examples of processed images during the task execution. In these figures, a light gray and a dark gray "X" show the estimated location of the ball and the goal, respectively. In the same way, the white rectangles surrounded with dotted line in each image show the estimated area of the ball and the goal.

Figure 9 A example of processed images taken by the robots

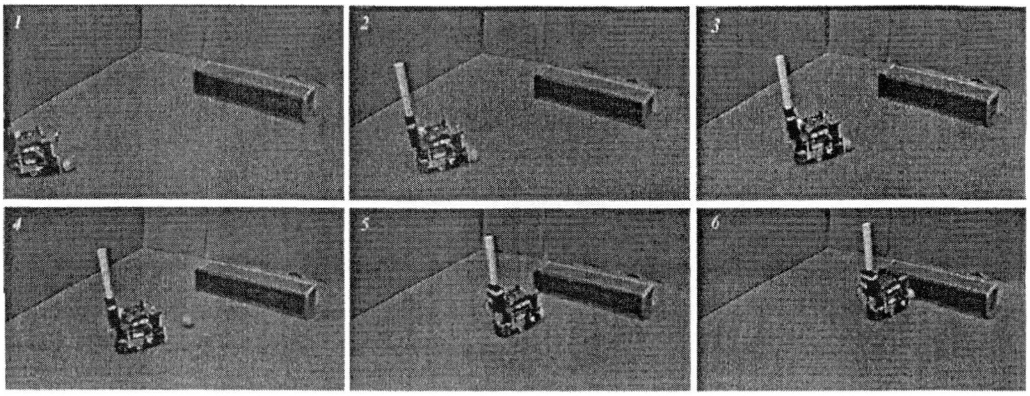

Figure 10: Shooting Behavior on Real Robot

Figure 11: Defending Behavior on Real Robot

Visual Looming as a range sensor for mobile robots

Erol Şahin and Paolo Gaudiano*
Boston University Neurobotics Lab
Dept. of Cognitive and Neural Systems, 677 Beacon St., Boston, MA 02215 USA
E-mail: {erol,gaudiano}@cns.bu.edu

Abstract

This paper describes and evaluates visual looming as a method for monocular range estimation. The looming algorithm is based on the relationship between displacements of the observer relative to an object, and the resulting change in the size of the object's image on the focal plane of the camera. Though the looming algorithm has been described in detail in prior reports, its usefulness for inexpensive, robust ranging has not been realized widely. In this paper we analyze visual looming as a visual range sensor for autonomous mobile robots. Systematic experiments with a *Pioneer 1* mobile robot show that visual looming can be used to extract ranging information much as with sonar. The accuracy of the looming algorithm is found to be significantly more robust than sonar when the object whose distance is being measured is slanted relative to the robot's line of sight. On the other hand, sonar is better suited for objects that cannot be visually segmented from their background, or objects that do not fit entirely within the focal plane. The results suggest that looming can be used as a robust, inexpensive range sensor as a complement to sonar.

1 Introduction

Traditional sensors for mobile robots can be divided along two broad categories: position sensors and visual sensors (*e.g.*, J. Borenstein & Feng, 1996). Position sensors, which include both on-board range sensors (e.g., sonar, infrared, tactile) as well as on-board and off-board localization sensors (e.g., laser, GPS), operate in real-time but provide very low-dimensional information, such as the distance to a target or the location of the robot within a room. In contrast, visual sensors can provide much richer and high-dimensional information about the environment, but few, if any, vision processing schemes exist that can operate in real-time on a mobile platform in an unstructured environment.

*This work is supported by DARPA, the Office of Naval Research and the Navy Research Laboratory through grants ONR-00014-96-1-0772 and ONR-00014-95-1-0409.

The fact that visual sensors contain an enormous amount of information does not necessarily mean that all algorithms based on vision must use all—or even a significant portion—of that information. It is often possible to extract useful low-dimensional information from vision to perform tasks that cannot be done easily with other sensors, or to use vision in a way that mimics the function of other sensors, but with different characteristics. The *visual sonar* algorithm proposed by (Horswill, 1994) is an excellent example: Horswill proposed a method that extracts relative distance information for obstacles located on the floor within the field of view of a robot. Using visual sonar, Horswill was able to achieve fairly robust navigation in an office environment.

In this paper we propose another way of extracting range information on a mobile robot, based on a visual phenomenon known as *looming*. The looming algorithm is useful for autonomous mobile robots as a method of extracting visual range information in real-time at a low computational cost. Furthermore, the looming algorithm is a demonstration of how sensory-motor coordinations can be used effectively for perceptual tasks in autonomous robots (Pfeifer, 1996).

The paper begins with a description of the looming phenomenon. After describing the mathematical formulation of the looming algorithm we present some sample results using a real mobile robot. A systematic comparison of ranging accuracy of looming and sonar is followed by a discussion of the looming algorithm in the context of autonomous agent design.

2 Visual Looming

Visual looming, the expansion of the projection size of an object on the retina, is usually the indication of an approaching object. It is normally perceived as a threat for a possible collision and is sufficient to elicit avoidance and escape behaviors. Caviness *et al.*, (Caviness, Schiff, & Gibson, 1962) reported that behavioral effects of looming show no evidence of habituation or extinction in monkeys. The presence of these behavioral effects in even infant monkeys suggests that detection of visual looming is innate.

Although its behavioral effects have been studied mainly in psychology, looming also has interesting implications for autonomous robotics. Several independent studies have reported the use of the looming effect as a method for obstacle avoid-

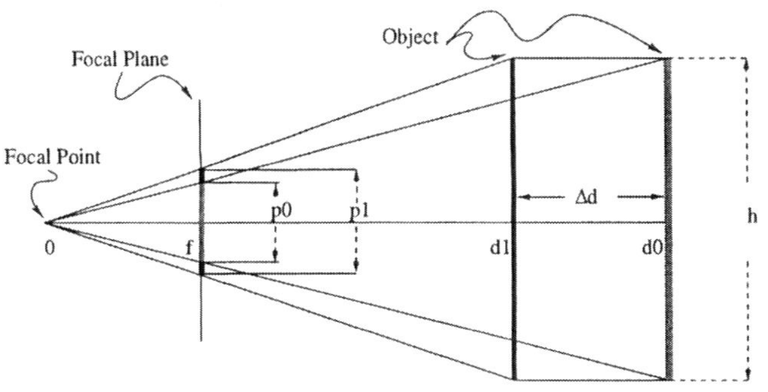

Figure 1: Diagram of the looming relationships in a camera centered coordinate frame.

ance (Joarder & Raviv, 1992) or for extracting the depth of an object (Huttenlocher, Leventon, & Rucklidge, 1995; Williams & Hanson, 1988; Xu, 1992). In particular, (Raviv, 1992) has done an excellent quantitative analysis of visual looming. He defined the looming of a point mathematically and showed how this information can be used for effective obstacle avoidance behavior. In this article we propose that looming can be used for rapid, robust visual ranging with autonomous mobile robots.

3 The Looming algorithm

The looming algorithm is based on the simple fact that objects look larger as they get closer and smaller as they move away. Fig. 1 depicts a scene in which a camera is viewing an object of size h from two different positions, at distances d_0 and d_1 from the object. Note that it is irrelevant whether the displacement is the result of camera movements or object movement. However, because we are interested in mobile robot applications, in the rest of this discussion we presume that the object is stationary while the camera is moving. It is also irrelevant whether the motion is toward or away from the object.

Given a constant focal length f, the size of the projection of the object onto the focal plane depends on the distance between the object and the camera. In the case shown in Fig. 1, p_0 and p_1, respectively, represent the size of the projection of the same object at distances d_0 and d_1. Using similar triangles, it can be easily shown that

$$d_1 = -p_0 \frac{\Delta d}{p_1 - p_0}, \quad d_0 = -p_1 \frac{\Delta d}{p_1 - p_0} \qquad (1)$$

where the net displacement $\Delta d = d_1 - d_0$. We refer to Eq. 1 as the *looming equation*.

The looming equation can be shown to hold under either of the following assumptions: Either (1) the object lies anywhere along a plane parallel to the focal plane, and the displacement is in the direction perpendicular to these planes, or (2) the object is not necessarily parallel to the camera's plane, but the camera is aimed toward the center of the object at both positions (in other words, the camera is moving directly toward or directly away from the center of the object). These assumptions are not problematic in practical use. In typical indoor environments most of the objects are vertically parallel to the camera's focal plane. If the object is not on the camera's line of sight, the robot can be instructed to center the object in its view.

It should be noticed that when the object is slanted relative to the camera's line of sight, small pointing errors will not be significant: the distance estimated by the Eq. 1 will be the average of the distances of the nearest and farthest point of the object projected onto the camera's line of sight, so that as long as the center of the camera's focal plane is somewhere on the object, the distance measured will correspond to a point on the object.

Some important observations can be made from Eq. 1.

1. For two-dimensional objects, the looming equation can be applied to each dimension independently, and the two resulting measurements can then be combined. It is also possible to re-write the equation in terms of the object's area.

2. The looming equation is independent of the implicit parameters of the optical system, such as the focal length, and the pixel size. Hence no camera calibration is required. If the pixel size or the focal length are known, it will be possible to estimate the true object size.

3. The looming equation is independent of the actual object size. The only *caveat* is that the object's projection must fit entirely within the focal plane. In the case of a two-dimensional object, it is sufficient that one dimension fit entirely within the focal plane, so that the algorithm could be used, for instance, to measure the distance to a door even when the image of the door is too tall to fit entirely on the focal plane in the vertical direction.

4. Only two measurements of the object projection separated by a net displacement obeying one of the above conditions are sufficient. No time dependency between projections is required.

An important problem of the algorithm is its noise sensitivity. It can be seen from the looming equation that it is likely to amplify any odometric or visual error when the change in the projection size $p_1 - p_0$ is small. One way to overcome this problem is to use tracking over multiple frames (Chaney, 1994), as we did, allowing for larger displacements. An error analysis of the looming algorithm is beyond the scope of this article but will appear in a later publication. We now turn to experimental results that illustrate the robustness of the looming algorithm.

4 Experimental results with a mobile robot

The biggest problem in the implementation of the looming algorithm is to find a visual preprocessing system that can

(a)

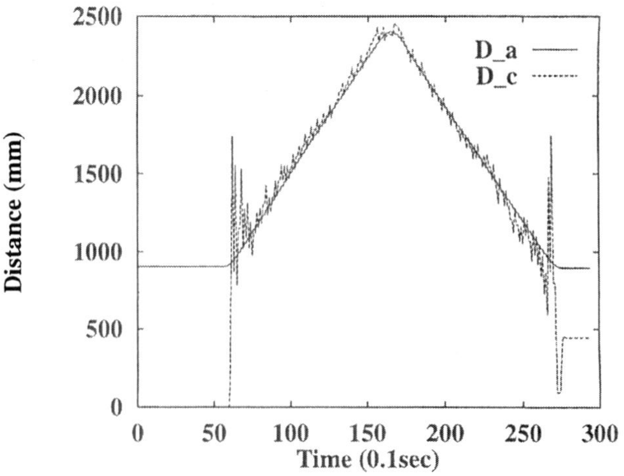

(b)

Figure 2: (**a**): The Pioneer 1 mobile robot, equipped with a color camera and on-board frame grabber (not visible). (**b**): Sample results showing the looming algorithm's ability to estimate distance.

segment an object from its background. In the lack of such a general purpose image segmentation algorithm that can run in real-time, we chose to side-step the problem by using off-the-shelf real-time color tracking hardware, and tested the looming algorithm on colored objects.

Fig. 2(**a**) shows the Pioneer 1 mobile robot (Real World Interface, Inc., Jaffrey, NH) used for our experiments. The robot is equipped with a Chinon CX-062 5mm color camera with 200x250-pixel resolution covering approximately 60 deg of visual field. The camera is connected to a *Cognachrome 2000 Vision System* (Newton Research Labs, Bellevue, WA), a 68332-based frame grabber that can perform a variety of simple operations at frame rate. Among other things, the Cognachrome can be "trained" to locate blobs of arbitrary colors in the image, returning the horizontal and vertical size, as well as the location, of multiple blobs of the desired color.

4.1 Basic results

Figure 2(**b**) illustrates a sample result to demonstrate the performance of the looming algorithm. Here a colored piece of paper was placed on the wall directly in front of the robot. When the robot's displacement is measured through internal odometry, the object's distance can be derived within fractions of a second, and the distance estimate improves over time. The results of Fig. 2(**b**) illustrate typical performance. The solid line in the figure is the real distance, while the dashed line is the distance calculated with the looming algorithm. The figure is obtained as the robot moves back-and-forth at a velocity of 15mm/sec, starting at a distance of about 1,000mm from the colored paper. In this case, we accumulate the distance estimates based on the robot's initial position. Once the robot has moved about 50mm, the distance estimate is already within about 20mm of the actual distance, and remains accurate throughout the back-and-forth movement.

This sample result illustrates basic performance of the looming algorithm using a color tracking system. However, similar results could clearly be obtained with sonar. The usefulness of the looming algorithm becomes evident when we perform a systematic comparison of looming and sonar under a variety of conditions.

4.2 Comparison of looming and sonar as range sensors

One profound limitation of sonar as a ranging sensor is that it usually fails for slanted surfaces. This is due to the nature of sonar, which is fairly directional, and is usually reflected off surfaces at an angle that is equal to the angle of incidence (J. Borenstein & Feng, 1996; Leonard & Durrant-Whyte, 1992). Hence a typical sonar sensor will fail to receive an echo from objects that are slanted at more than approximately 15 deg relative to the line of sight of the sonar. This limitation of sonar is true for most types of surface, with the exception of special materials such as styrofoam that tend to reflect sonar in all directions.

In contrast, the looming algorithm is able to function accurately regardless of an object's slant, as long as the camera is aimed approximately to the center of the object during the robot's displacement. Problems with the looming algorithm may arise if an object is slanted so steeply that its width becomes less than a few pixels wide. But even in this case looming may function well if the object is not steeply slanted along both dimensions, since the looming algorithm can be applied independently for each dimension.

We have carried out a systematic comparison of the performance of looming and sonar when the object whose distance is being measured is slanted by various amounts relative to the sensor's line of sight. Our experimental setup is shown in Fig. 3. A letter sized (216×279 mm) piece of colored paper is attached to a wall at approximately the same height as the

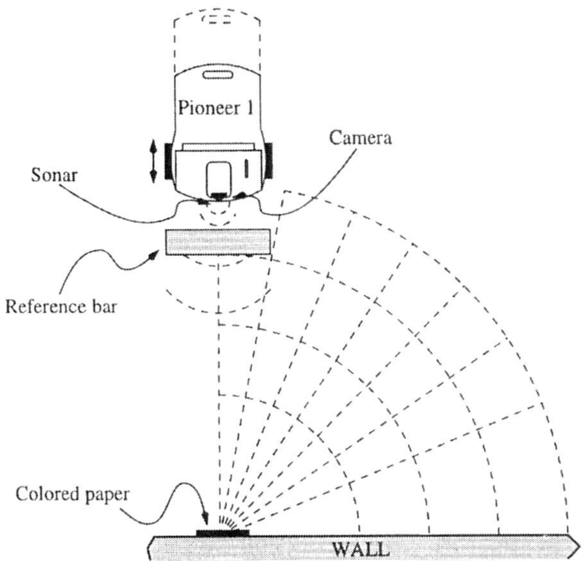

Figure 3: Top view of the experimental setup, showing the Pioneer and the grid of points used for the experiments.

robot's camera. Radial lines are drawn on the floor, emanating from the point aligned vertically with the the center of the paper, at angles of 22.50-90 deg in 11.25 deg increments relative to the wall (meaning that 90 deg corresponds to a line perpendicular to the wall). Each of these lines is marked at distances of 611, 917, 1223 and 1528 mm.

The Pioneer is placed at each one of the resulting 28 points marked on the floor (four different distances at each of the seven angles), aimed in the direction of the piece of paper, as shown in Fig. 3. The distance from the Pioneer to the paper is then measured with the Pioneer's forward-facing sonar and with the looming algorithm.

Distance estimation with the looming algorithm requires the robot to move forward or backward in the direction of the piece of paper. In order to make the data collection automatic and more reliable, a reference bar is attached to the floor at each point on the grid, perpendicular to the underlying radial line, to provide a reference distance and orientation. Initially the robot is aligned with the radial line with both wheels touching the bar. At this point the width and height of the paper (as seen through the camera) are used as a first measurement. Then the robot is instructed to move backwards by 305 mm and a second measurement is recorded. The looming algorithm uses these two measurements to compute the distance of the paper when it was at the bar. Then the robot moves forward until its wheels hit the bar to ensure that the robot is aligned correctly, though the robot's lateral position might be shifted slightly due to odometry errors. This procedure is then repeated twenty times at each point to derive a meaningful average measurement. After every five measurements, the robot is manually shifted back to the reference point to avoid a bias from positional errors. In any event, the shift accumulated during five cycles was never observed to exceed 40 mm. The sonar reading at each point is also repeated 20 times and averaged.

Figure 4 illustrates the results of this experiment. Both parts of the figure include a grid that shows the distance and angle of each of the 28 reference points. Notice that in the figure the vertical line of the grid corresponds to 90 deg, i.e., when the piece of paper is perpendicular to the line of sight of the robot. The figure shows the combined averaged results obtained with sonar (cross symbol), and with the looming algorithm using the paper's width (triangles) or height (squares). In Fig. 4(**a**) we include all points from the three range measurements, while in Fig. 4(**b**) we truncate the sonar results to improve the visibility of the results obtained through looming (using width and height independently) for clarity.

The results illustrate, as expected, that when the paper is slanted by 22.5 deg or more relative to the line of sight of the robot the sonar is unable to detect the wall (the large values shown in the figure correspond to the sonar's maximum range, meaning that an echo was not received). In contrast, the looming algorithm performs quite well up to the maximum slant angle.

Some other things are worth noting here. For one thing, the range measurements obtained with looming using the height of the paper are generally more stable and accurate than those obtained with the width of the paper. This is to be expected because the increasing slant causes foreshortening of the projection in the width but not in the height of the paper. This suggests that, in general, either the height should be used exclusively (at least when the object's vertical projection fits entirely in the focal plane), or else the two measurements should be combined. Also, it appears that the width measurements consistently overestimate distance, whereas the height measurements exhibit some more complex biases. It is possible that some of these effects are due to systematic biases in the color tracking algorithm. A detailed study of these effects and of the algorithm's overall noise sensitivity will be published elsewhere.

5 Discussion

In this article we have shown that looming can be used as an effective way of extracting range information from vision. Its simplicity, robustness and low computational cost, make looming a valuable tool for autonomous mobile robots.

One aspect of looming worth discussing is its use of combined sensory and motor information. As with optic flow, looming could be used to extract time-to-collision of incoming objects simply on the basis of a sequence of frames, i.e., purely sensory data. However, the knowledge that the robot is moving, and of how far it is moving, make it possible to extract the actual distance to the object. The idea that sensory-motor coordination of this type can be useful for autonomous agents has been proposed before, most notably by Pfeifer and colleagues (Pfeifer, 1996).

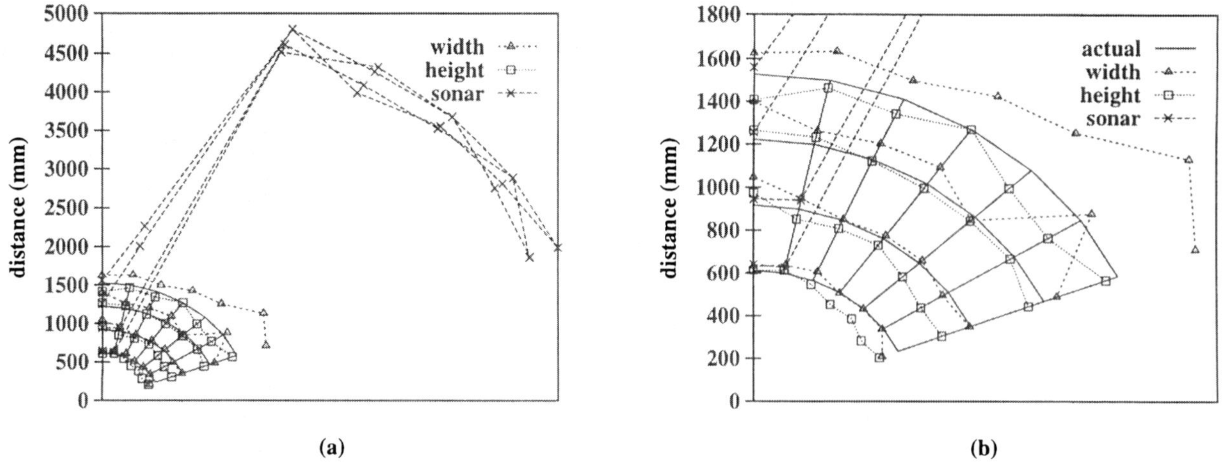

Figure 4: Comparison of sonar and looming as range sensors. Both parts of the figure show the distance estimated using sonar (cross symbols), looming based on object height (squares), or looming based on object width (triangles). The plots are in polar coordinates: each data point corresponds to an estimated distance measured from one of the points that were shown in Fig. 3. The grid of solid lines corresponds to the actual, measured distance and angle of the grid points. The data in part (**b**) are the same as in part (**a**), but the plots are truncated to enhance the visibility of the looming data.

An interesting observation is that the range estimates obtained with looming have the units of the robot's odometric sensors, i.e., if the displacement of the robot is given in wheel revolutions, then the range to an object will have the same units. The exact metric of these units does not matter: for a mobile robot it is equally acceptable, if not better, to use range information in the units of odometric sensors. It is always possible to convert the range information into "our" units by using the radius of the wheels, but while this conversion is convenient for our own representation and interpretation of the data, the data would then have to be converted back into odometric units for navigation.

We propose that using the "proprioceptive" odometry information is a natural way of processing data in mobile robots and may actually lead to greater performance accuracy than the traditional approach of calculating "objective" distance values. Typical odometry errors arise from two distinct sources: systematic errors that can be due for instance to a slightly imprecise knowledge of the robot's kinematics (wheel diameter, inter-wheel distance), and non-systematic errors resulting from wheel slippage and the like (J. Borenstein & Feng, 1996). However, many of the systematic errors are only errors when we try to map them to external, absolute distance values. For instance, if we make a 5% error in measuring wheel diameters, the robot's movements within a fixed frame of reference will seem inaccurate by a proportional amount. However, suppose we want the robot to move exactly half way to a target using looming. If we never translate the odometric information into an external frame of reference, the same 5% error in this case should vanish: the looming algorithm is estimating, essentially, how far the object is in terms of wheel revolutions, and this estimate will be accurate (barring wheel slippage and other non-systematic errors) as long as we don't try to convert it to an external frame of reference.

Another advantage of taking into account the robot's movements is that it is possible to obtain information through looming that could not be obtained with a purely visual sensor. For instance, suppose that an object segmented on the basis of its color is in reality occluded by another object (the occluder) that is placed between the robot and the object being tracked. When the robot moves by a small amount toward the object, the distance estimate obtained through looming will be incorrect: the object will "appear" closer than it is because its projected size has increased more than it should have because of the presence of the occluder[1]. Let's say that the algorithm returns an incorrect estimated distance d_1 after the first step. Now if the robot moves again toward the object by an amount Δd, the looming algorithm will again err, and this time the estimated distance will be *smaller* than $d_1 - \Delta d$. But because the robot *knows* through its odometry that it has only moved by Δd, the algorithm can realize that something is wrong. In addition, with the exception of degenerate cases, the occlusion will not affect both dimensions of the object, so that the algorithm could also notice a discrepancy by comparing the result obtained independently from the two dimensions. Hence the looming algorithm can be extended to include consistency checks across dimensions or even within one dimension across multiple readings, and it can use the results to figure out when there is something wrong with the estimated distances.

In many cases where inconsistencies are found, it should

[1]As the robot moves closer to the object, more of the object will be uncovered from behind the occluder, hence the increase in projection size will be greater than it would have been without the occluder.

be possible to understand the source of the error: for instance, if an occluder is to the left of the midline, the center of the tracked object will shift to the left as the robot approaches the object. Furthermore, in this situation the distance obtained from the height measurement should be consistent across readings, and this would allow the algorithm to discard the horizontal measurement, or even to calculate the location of the occluder. With a similar reasoning it is also possible to detect moving objects: if two subsequent readings are inconsistent, the robot can stop momentarily to check whether the projection size changes in the absence of robot movements.

These considerations again show the power of combining sensory and motor information, and the potential usefulness of looming for autonomous mobile robots.

6 Conclusions

We have proposed that visual looming is a useful complement to sonar for estimating the distance to objects in a mobile robot's environment, and we have presented experimental results to support this proposal.

In their seminal book, Leonard and Durrant-Whyte (1992) provide and excellent analysis of the properties of sonar sensors. The authors conclude that *"sonar is in fact a very good sensor... [but the] first step in using the sonar is to accept the realities of the physics of acoustic sensing... Most of the sonar interpretation algorithms have been built as if sonar is a 'ray-trace scanner"*.

In fact, many navigation schemes rely on sonar as a sort of 'ray-trace scanner'. Examples of such schemes include evidence grids (Elfes, 1987), vector field histograms (Borenstein & Koren, 1989) and frontier-based navigation (Schultz & Adams, 1996; Yamamuchi, Schultz, & Adams, 1997). We propose that although looming cannot replace sonar for basic ranging, it can be useful as a complement to sonar for these and other navigation and localization algorithms.

To summarize, looming is ideally suited for ranging when objects are slanted relative to the robot's (or sonar's) line of sight. Looming is also well suited for objects that may be located on a wall but above or below the sonar's line of sight. Finally, by combining sensory and motor information, looming can be extended to handle complex situations where a purely visual sensor would not be adequate.

References

Borenstein, J., & Koren, Y. (1989). A real-time obstacle avoidance for fast mobile robots. *IEEE Trans. on Systems, Man, and Cybernetics*, *19*(5), 1179–1187.

Caviness, J. A., Schiff, W., & Gibson, J. J. (1962). Persistent fear responses in rhesus monkeys to the optical stimulus of 'looming'. *Science*, *136*, 982–983.

Chaney, R. D. (1994). Multiple-frame motion estimation using simple region feature. In *Proc. of SPIE, Intelligent Robots and Computer Vision XIII: 3D Vision, Product Inspection, and Active Vision*, Vol. SPIE-2354, pp. 10–25.

Elfes, A. (1987). Sonar based real-world mapping and navigation. *IEEE Trans. on Robotics and Automation*, *3*(3), 249–265.

Horswill, I. (1994). Collision avoidance by segmentation. In *Proc. of the 1994 Intl. Conf. on Intelligent Robots and Systems (IROS-94)*. IEEE Press.

Huttenlocher, D. P., Leventon, M. E., & Rucklidge, W. J. (1995). Visually-guided navigation by comparing edge images. In Goldberg, K., Halperin, D., LAtombe, J.-C., & Wilson, R. (Eds.), *Algorithmic Foundations of Robotics*, pp. 85–96. A K Peters.

J. Borenstein, H. E., & Feng, L. (1996). *Navigating mobile robots*. A.K. Peters, Wellesley, MA.

Joarder, K., & Raviv, D. (1992). Autonomous obstacle avoidance using visual fixation and looming. In *Proc. of SPIE*, Vol. 1825, pp. 733–744.

Leonard, J. J., & Durrant-Whyte, H. F. (1992). *Directed sonar sensing for mobile robot navigation*. Kluwer Academic Publishers, Boston.

Pfeifer, R. (1996). Building "fungus eaters": Design principles of autonomous agents. In *From Animals to Animats 4. Proc. of SAB'96*, pp. 3–12.

Raviv, D. (1992). Visual looming. In *Proc. of SPIE*, Vol. 1825, pp. 703–713.

Schultz, A. C., & Adams, W. (1996). Continuous localization using evidence grids. Tech. rep. AIC-96-007, NCARAI. Submitted to Autonomous Robots.

Williams, L. R., & Hanson, A. R. (1988). Translating optical flow into token matches and depth from looming. In *Proc. of the 2nd Intl. Conf. on Computer Vision*.

Xu, S. B. (1992). Qualitative depth from monoscopic cues. In *Proc. of the Intl. Conf. on Image Processing and its Applications*, pp. 437–440.

Yamamuchi, B., Schultz, A. C., & Adams, W. (1997). Integrating exploration and localization for mobile robots. Tech. rep. AIC-97-021, NCARAI. Submitted to Autonomous Robots.

Investigation of a sensorimotor system for saccadic scene analysis: an integrated approach

C. Zetzsche*, K. Schill*, H. Deubel, G. Krieger*, E. Umkehrer*, S. Beinlich***

* Ludwig-Maximilians-Universität München
Institut für Medizinische Psychologie
Goethestr. 31, D-80336 München, Germany
chris@imp.med.uni-muenchen.de

**Ludwig-Maximilians-Universität München
Institut für Psychologie
Leopoldstr. 13, D-80802 München, Germany

Abstract

The sensorimotor system of saccadic scene analysis is investigated in a multi-disciplinary research project. Eye-movements of human observers are recorded for a variety of scenes. Our statistical evaluations indicate that fixations are selected on the basis of higher-order statistical properties. The structure of the polyspectra suggests that nonlinear operators with a selectivity for intrinsically two-dimensional image features serve as "saliency detectors". The bottom-up feature extraction is combined with a knowledge-based system into an integrated architecture, in which a saccadic fixation corresponds to the "best question to ask" given the internal state and the recent experience of the system.

1. Introduction

The analysis of a complex scene by a sequence of rapid saccadic eye movements is a prototypical example for the high degree of efficiency biological agents can achieve with respect to their information processing and sensorimotor control. This efficiency can be regarded to result from a systematically refined evolutionary adaptation to the regularities of the complex terrestrian environment. A common principle underlying these adaptive processes is the optimized allocation of a limited amount of available processing resources to environmental data and behavioral tasks of nearly unlimited complexity.

From the mere amount of data, it would not be affordable for any system to process its complete optical surrounding with maximum spatial resolution. Biological vision has optimized its allocation of limited resources bye use of an inhomogeneous spatial sampling. The majority of the information processing machinery is concentrated into a few degree of the visual field, the *fovea centralis*, whereas the whole remainder, which is of orders larger in area, is only served by a much coarser representation [5]. Such a strategy can only work without the risk of a complete miss of important information because the saccadic system can allocate the foveal processing resources within parts of a second by rapid ballistic movements with velocities of up to 700 deg/sec to the relevant parts of the visual scene. However, while many details of the saccadic system, like the motoric control, the timing parameters, or its reflexive mechanisms are quite well understood, the powerful strategies by which the system manages to select the relevant parts out of complex static scenes are still a challenging research problem.

We approach this problem by a multi-disciplinary approach which involves participants with expertise in experimental psychology, visual psychophysics and modelling, statistical signal processing, and knowledge based system design. By this, we try to cover the research areas which are, in our view, necessary for the analysis, understanding, and modelling of a complete sensorimotor loop like the saccadic eye movement system (Fig. 1). The final goal of this multidisciplinary research project is the provision of a model which can take natural images as input and can derive predictions about the fixations of human observers on these images. Here we provide some intermediate steps towards this goal.

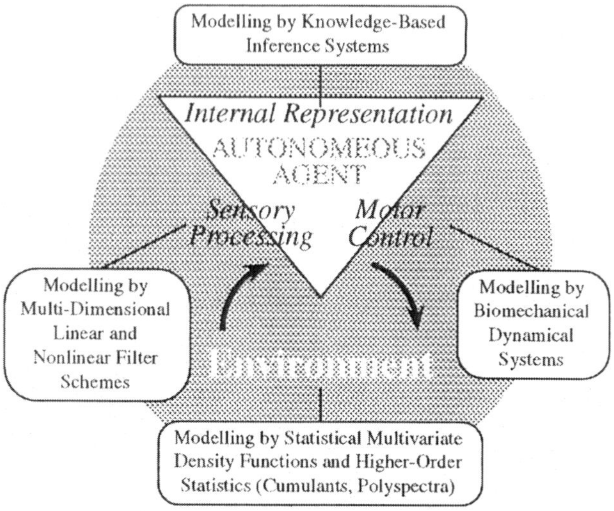

Figure 1 The basic aspects of sensorimotor processing and the respective scientific methods for their analysis and modelling.

For our investigations, we have first obtained an empirical database by recording the eye movements of human observers on a variety of artificial and natural images (sect. 2.). A central concept in our research and modelling being statistical information theory, we start the analysis of these data by an investigation of the differences between the statistical properties of those image regions which were selected by fixations, as opposed to the average statistics of randomly selected regions (sect. 3.). The results of this statistical anlysis are then used to derive a suggestion for a nonlinear "saliency-detector" in early vision (sect. 4.). Since purely bottom-up data-driven approaches are known to be insufficient for the complete description of saccadic fixation strategies we finally consider an architecture which integrates the low-level feature extraction processes with a high-level knowledge-based reasoning system (sect. 5.).

Figure 2 Test images used in our experiments. The small circles indicate the locations of eye movements.

2. Recording of saccadic eye movements

Images have been chosen from seven different topical classes (portraits, landscapes, technics, architecture, computer graphics, polygones, and textured polygons), in order to enable a differentiated analysis, and to avoid any bias on the relevant image features. Every class contained 7 images, hence a total number of 49 images has been investigated. All images were grey-level images of size 512 x 512.

Eleven paid subjects with a mean age of 27 years participated in the experiment. They were naive with respect to the objectives of the study. The subjects were instructed to view each image as careful as possible, to be able to perform subsequent tests with the images. The subjects were first shown a set of 10 randomly selected images, which were intended to serve as training images, and recordings were evaluated separately for these images. After a short pause and a recalibration, the main experiment was performed with the 25+24 images of the set. Each image was presented to the subject for five seconds, and eye movements were recorded during the whole presentation.

Apparatus: The experiments took place in a laboratory room providing an ambient illumination of approximately $0.1 cd/m^2$. Stimuli were presented on a 21" video monitor (CONRAC 7550 C21) with 1024x768 pixels spatial resoultion at a frame rate of 100 Hz. Screen background luminance was set to $3cd/m^2$; the luminance of the stimuli was $25cd/m^2$. Subjects viewed the screen binocularly from a distance of 80 cm. The total image size was 18x18 deg. Head movements were restricted by a biteboard and a forehead rest. Eye movements were measured with a SRI Generation 5.5 Purkinje-image eyetracker [4] sampled at 400 Hz. Its frequency response is better than 250 Hz with a noise level equivalent to about 20 arc sec rms. The eyetracker can follow saccades of 15 degrees or more without losing the eye. Each session started with a calibration procedure: the subject sequentially fixated 9 positions arranged on a circular array of 8 degrees radius. The eyetracker behaved linearly within this range. Static accuracy of the eyetracker was better than 0.1 degrees. Experiments were controlled by a 486 PC, which also performed automatic off-line analysis of the eye movement data in which saccadic latencies and saccade start and end positions were determined.

3. Statistical investigation of eye movements

From Fig. 2 it becomes apparent that the locations of saccadic eye movements are not randomly distributed across natural scenes. However, it is also evident that even if the saccadic selection process would be completely deterministic, there is a high degree of variability in the data. Because of this, and since we assume that statistical redundancies of the images have an important influence on the regions being selected for fixation, we have investigated the statistical properties of these fixated regions. For our analysis, we used the 7 images shown in Fig. 2.

3.1 Analysis of spatial variance

In a first step, we measured the signal variability of fixated image regions and compared it to the variance of regions, which had been chosen by a random number generator. Variance is computed as $\sum_{\vec{x}_{select}} \sum_{\vec{x} \in W} [u(\vec{x}_{select} - \vec{x}) - \bar{u}]^2$ where $\bar{u}$ is the mean within a local window W which in our analysis spanned $\frac{1}{8}$ of the total image size. Our analysis revealed that for most images of Fig. 2 the regions selected by eye movements had a significantly higher amount of signal variability (Tab. 1). For the whole set of images we obtained a variance ratio of $\sigma^2_{eye}/\sigma^2_{rand} = 1.35$.

Image	Monaco	Moscow	Lake	Linna	Girl	Poly	Tears	Σ
N_{fix}	125	131	159	141	136	140	139	971
σ^2_{eye}	1879	2065	930	2586	1598	716	2640	1753
σ^2_{rand}	1400	1812	957	2002	1438	512	1059	1299
$\frac{\sigma^2_{eye}}{\sigma^2_{rand}}$	1.34	1.14	0.97	1.29	1.11	1.40	2.49	1.35

Table 1 Analysis of spatial variance

3.2 Second-order statistics

The variance analyzed in the preceding section gives only a very crude approximation of the statistics within a local window. A more detailed characterization may be obtained by autocorrelation functions or their counterpart in the frequency domain, namely powerspectra. In Fig. 3 we compare the powerspectra of fixated and randomly chosen image regions. It turns out that the powerspectra obtained in both cases do not substantially differ in their basic shape. Similar results have

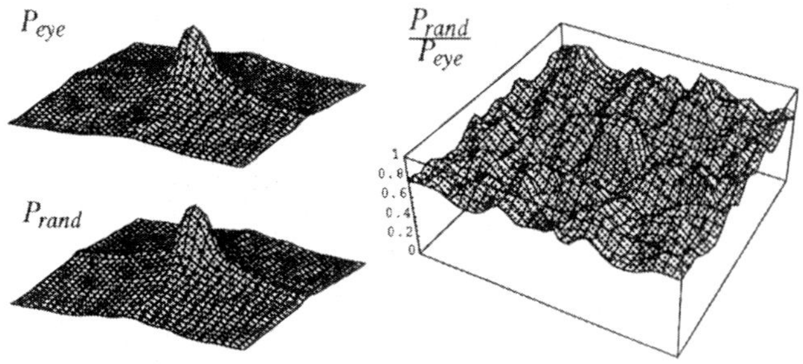

Figure 3 Powerspectra of image regions selected by eye fixations (upper left) and by a random number generator (lower left). The two powerspectra do not substantially differ in their basic shape. Their ratio (right) is consistently ≤ 1, which is in agreement with our analysis of the variance. For low frequencies the ratio is close to one, thus indicating an approximately equal contribution in both cases. For higher spatial frequencies fixated image regions contribute more power than randomly chosen regions.

been obtained in [14] Hence, second-order statistics are insufficient to obtain a detailed statistical description of fixated image regions. However, since the powerful selection process in saccadic scene analysis cannot be reduced to a mere variance selection, we have considered the higher-order statistics of the fixated regions.

3.3 Higher-order statistics

One measure for such higher-order dependencies is provided by the third-order cumulant c_3^u, which basically calculates the expectation of an image which has been multiplied by two shifted copies of itself. For a zero-mean stationary random process $\{[u(x,y)]\}=\{u(\vec{x})\}$ the third-order cumulant c_3^u can be defined as:

$$c_3^u(\vec{x}_1,\vec{x}_2) = \mathbf{E}[u(\vec{x}) \cdot u(\vec{x}+\vec{x}_1) \cdot u(\vec{x}+\vec{x}_2)] \qquad (1)$$

As for autocorrelations, it is also for higher-order statistics often more convenient to investigate the signal statistics in the frequency domain. The Fourier transform of c_3^u is known as the bispectrum C_3^U. Alternatively, the Fourier-Stieltjes representation of the process $\{u(\vec{x})\}$ offers the possibility to express the bispectrum directly in terms of the frequency components $dU(\vec{f})$ [11]:

$$\mathbf{E}[dU(\vec{f}_1) \cdot dU(\vec{f}_2) \cdot dU^*(\vec{f}_3)] = \begin{cases} C_3^U(\vec{f}_1,\vec{f}_2) \cdot d\vec{f}_1 d\vec{f}_2 & \text{if } \vec{f}_3 = \vec{f}_1 + \vec{f}_2 \\ 0 & \text{else} \end{cases}$$

From this equation it becomes apparent that the bispectrum measures the statistical dependencies between three frequency components, the sum of which equals zero. A direct computation in the frequency domain can also be derived from this notation [11].

In Fig. 4 we compare the bispectra of fixated image regions to those of randomly chosen areas. It becomes apparent that unlike the second-order spectra of Fig. 3 the two bispectra differ in their basic shape: For the randomly chosen regions we obtain a star-like shape which indicates strong statistical dependencies between frequency components which are aligned to each other with respect to orientation. In contrast, the bispectra of fixated regions are more circular, thus indicating strong simultaneous contributions from frequency components of different orientations within the local windows.

4. What are the relevant features?

4.1 Interpretation of the statistical results

How can we interpret these results? First, the saccadic selection system seems to avoid the fixation of image regions with little structural content (low variance), such as regions with approximately constant luminance. This fits with the basic operation of lateral inhibition in low level vision, since neurons with a center-surround antagonism suppress the responses to constant regions and enhance the response to variating signals like edges, lines etc.

Since such a basic distinction of structure vs. non-structure cannot differentiate between different types of structure, we first applied the standard second-order analysis via powerspectra, but could not obtain any reliable difference between fixated and randomly selected regions. However, we were able to find significant differences in the higher-order statistics. While the statistical structure of the average image regions is dominated by the presence of statistical dependencies between frequency components with aligned orientations, which is consistent with the frequent occurence of straight edges or lines, the bispectra of the fixated image regions show a different pattern. Here the bispectral "energy" is not concentrated on frequency components with aligned orientations, but is more or less evenly distributed on the various combinations. This implies, that the saccadic selection system avoids image regions, which are clearly dominated by a single oriented structure, like a straight line or edge, but selects instead regions which contain different orientations, like occlusions, corners, etc.

4.2 The concept of intrinsic dimensionality

These statistical effects fit well with the idea of an information-theoretic hierarchy in terms of *intrinsic dimensionality* [21; 22; 23]. The concept of intrinsic dimensionality relates the degrees of freedom provided by a signal domain to the degress of freedom actually used by a given signal and thus provides a hierarchy of local image signals in terms of different degrees of redundancy:

i0D-signals are constant, i. e. $u(x,y) = const.$ within a local window.

i1D-signals can locally be approximated by a function of only one variable, i. e. $u(x,y) = u(ax+by)$. Examples are straight lines and edges. Sinusoidal gratings, the eigenfunctions of linear systems, are also a member of this class.

i2D-signals are neither *i0D* nor *i1D*. Examples are corners, junctions, curved lines and edges, etc.

This hierarchy seems also well suited for the description of saccadic eye movements. First, *i0D*-signals are least frequently fixated, as indicated by the results on the higher variance of the fixated regions. Second, *i2D*-signals are more attractive targets than *i1D*-signals. An early report [10] suggested that the "gaze selects informative details", and that the most informative details seem to be curved contours (*i2D*-signals). Similar results have been obtained by [3] and [24] for fixations on simple polygons. In such figures the corners (*i2D*-signals) are more often fixated than the straight contour

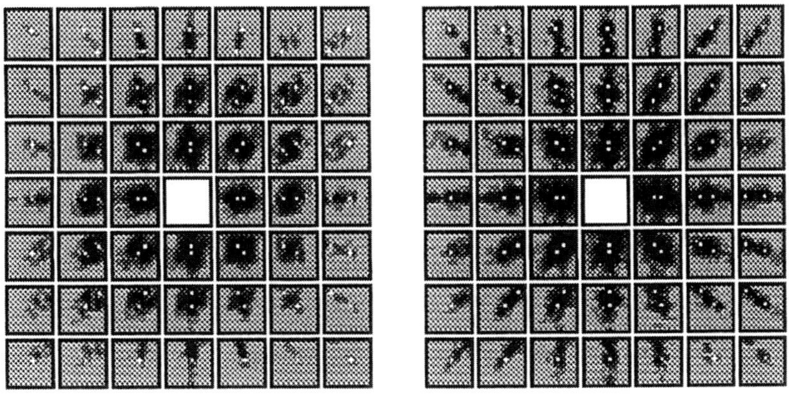

Figure 4 Bispectra of fixated image regions (left) and randomly selected image regions (right). Shown are several sections of the bispectral magnitude $|C_3([f_{x1}, f_{y1}], [f_{x2}, f_{y2}])|$ with $f_{x1} = const.$ and $f_{y1} = const.$ (cf. the coordinate axes depicted in Fig. 5). Note the different structure of the two bispectra, which is especially pronounced for contributions from higher spatial frequencies. In the random sample (right), the bispectral 'energy' is concentrated in those regions where the frequency components are aligned to each other with respect to orientation. This is consistent with our systematic investigations of the bipsectra of a large set of natural images [8]. In the bispectra of the fixated image regions the 'energy' distribution is more circular around the void points $[f_{x1}, f_{y1}] = [0,0]$, $[f_{x2}, f_{y2}] = [0,0]$ and $[f_{x1} + f_{x2}, f_{y1} + f_{y2}] = [0,0]$. This indicates that the eye (as opposed to the random number generator) selects more often image regions with strong statistical dependencies between frequency components of different orientation. In order to illustrate the distinctive structure, the two bispectra have been scaled to the same maximum. Actually the bispectral 'energy' for the fixations is higher by a factor of about four.

lines (*i1D*- signals), or the interior regions (*i0D*-signals). Attneave has emphasized the special importance of *i2D*-signals in the context of redundancy reduction and object recognition [1]. Furthermore, the least redundant *i2D*- information is basically sufficient for a reconstruction of the original image signal [2].

The hierarchy of intrinsic dimensionality seems to be reflected in the neural operations in early vision, which can be seen as the result of an efficient adaptation to the statistical properties of the natural environment. Evolution caused not only the development of linear isotropic and orientation-selective neurons, but also of nonlinear *i2D*-selective cells known as 'hypercomplex' or 'end-stopped'. In primates, about half of the neurons in area V2 of the visual cortex are selective to *i2D*-stimuli [13].

All this evidence implies that *i2D*-signals provide significant information in complex scenes, and that *i2D*-selective detectors may be important members of the class of "saliency detectors". However, unlike classical filters such operators are highly nonlinear devices, and require special techniques for their modelling [21; 22; 7].

4.3 *i2D*-selective operators

Starting from the Volterra-Wiener series expansion of nonlinear functionals, a quite general class of *i2D*- selective operators can be derived [7]. The Volterra series relates the input $u_1(\vec{x})$ of a nonlinear, shift invariant system to its output $u_2(\vec{x})$ in the following way:

$$u_2(\vec{x}) = h_0 + \int h_1(\vec{x}_1) \cdot u_1(\vec{x} - \vec{x}_1) \cdot d\vec{x}_1 + \quad (2)$$

$$+ \iint h_2(\vec{x}_1, \vec{x}_2) \cdot u_1(\vec{x} - \vec{x}_1) \cdot u_1(\vec{x} - \vec{x}_2) \cdot d\vec{x}_1 d\vec{x}_2 + \ldots$$

The quadratic part of Eq. (2) may be expressed in the frequency domain as

$$U_2(\vec{f}) = \int H_2(\vec{f}_1, \vec{f} - \vec{f}_1) \cdot U_1(\vec{f}_1) \cdot U_1(\vec{f} - \vec{f}_1) \cdot d\vec{f}_1 \quad (3)$$

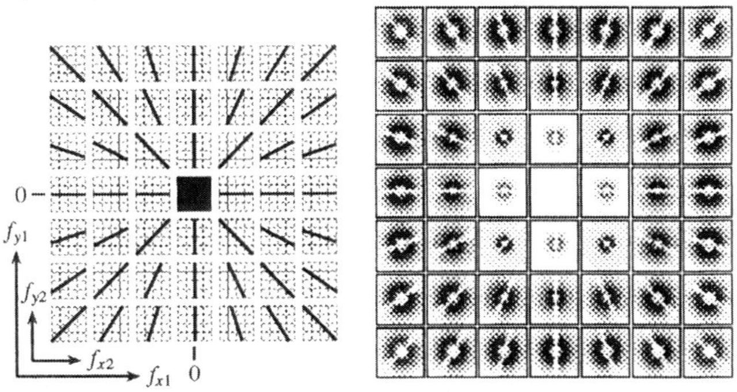

Figure 5 (Left) Illustration of the forbidden zones (indicated as black lines) for an *i2D*-selective quadratic Volterra kernel $H_2([f_{x1}, f_{y1}], [f_{x2}, f_{y2}])$ in the frequency-domain. A system whose symmetric kernel vanishes at the forbidden zones may be regarded as blocking frequency components of equal orientation [7]. (Right) Prototypical example of the kernel of an *i2D*- selective quadratic Volterra operator. Since the forbidden zones are taken into account, the operator will only respond to *i2D*-signals.

where H_2 is the Fourier transform of the second-order Volterra kernel h_2. Note that Eq. (3) may be regarded as the weighting of an AND-like conjunction between frequency components. A necessary and sufficient condition for a quadratic Volterra operator to be insensitive to *i0D*- and *i1D*- signals is given by [7]:

$$H_2([f_{x1}, f_{y1}], [f_{x2}, f_{y2}]) = 0 \quad \forall \quad f_{x1} \cdot f_{y2} = f_{y1} \cdot f_{x2}$$

The operation of systems adhering to this condition can be regarded as a blocking of aligned frequency components (see Fig. 5, an example is provided in Fig. 6).

4.4 Limits of data-driven approaches

By use of a comprehensive set of "saliency detectors" it may well be possible to identify all potentially interesting locations within a complex scene, but it would not be an optimal strategy to scan them all in a systematic fashion. Rather, the final selection must have the possibility to select a further subset out of this "preselection", and it must also be possible to

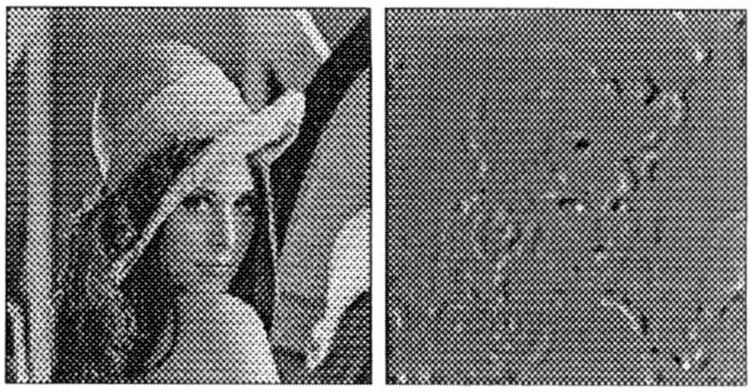

Figure 6 Application of *i2D*-selective operators to a natural image. The sparsely distributed regions selected by such operators carry substantial information about the image and are in good qualitative agreement with the saccadic fixations.

select points outside the preselected set (note, however, that untrained subjects have great difficulties in making an intentional saccade to a certain position on an unstructered smooth surface). This implies that any model which is based solely on botttom-up processes, can only provide an approximate statistical prediction of the fixations by human observers, and indeed it has been shown that top-down influences like the expectation of the observer, or the task to be performed will have a substantial influence on the pattern of saccadic fixations [20; 9]. We hence also investigate, how the bottom-up/low-level feature extraction process can be combined with top-down influences from a higher-level knowledge-based system.

5. High-level knowledge-based system

The high-level cognitive component realized by a knowledge-based system determines a sequence of saccadic eye movements to derive conclusions about a current scene on the basis of a learned sensorimotor representation of scenes. The system is able to learn links between eye movements and scenes, builds up knowledge about already analyzed scenes, and calculates the eye movements required to reach a maximum of information about the scene. The information gain is calculated with a parallel inference strategy called IBIG (inference by information gain) [15; 16] which is suitable for adaptive reasoning. The information gain strategy needs as knowledge-base a set of triples consisting of a datum, a hypothesis and an uncertainty value, which says to which degree the datum supports the hypothesis. Usually the datum corresponds to a symbolic proposition. To connect the symbolic reasoning with the low-level feature extraction processes in this system the datum is linked to sensor-data derived from the application of linear and nonlinear filters to the scene (see sect. 4. and [21; 22; 23]. Thus, the system combines results of signal processing with knowledge-based reasoning.

The representation of a scene is in two ways dynamic. It changes with new incoming scenes and it uses not only the sensor data, but also the motoric action required to obtain these data. A common view in computational vision is that objects are represented purely by afferent features. In contrast we use a *sensorimotor* representation which combines sensoric data and motoric actions, namely saccadic eye movements. Each scene[1] is represented by a set of eye movements, but not in a deterministic way by identifying a scene with a fixed sequence of eye movements [12], but in a qualitative way in that each eye movement supports a scene only with a certain evidence. The data-structure for the eye movement consists of a triple (first feature, relative position, second feature). Features correspond to classified data extracted from a multi-filter decomposition of the image. Since we have already shown a strong correlation between *i2D*-signals and fixation points, the classification is for the moment restricted to *i2D*-locations.

The eye movements and their uncertainty values associated with a scene are learned by the system, which is described below.

5.1 Learning the link between eye movements and objects

The link between eye movements and objects is represented by the uncertainty values in the knowledge-base. A hypothesis in the knoweldge-base corresponds to one object. The hypotheses space $\mathcal{H}$ is a strict hierarchy and therefore each non-leaf element may be interpreted also as a class. The set of possible objects O are the leaves of the hierarchy. Each scene is presented together with a label, which corresponds to a hypothesis $H \in \mathcal{H}$. This label is necessary for the "supervision" of the learning of the uncertainty values.

After analyzing the scene, the long-term memory of the system, i.e. the information over all analyzed scenes, is updated, using the content of the short-term memory, i.e. all eye movements generated on this scene, and a new knowledge base is generated (see bold arrows in Fig. 7). The long-term memory consists of a matrix $M = \{x_{ij}\}$, where each row i is associated with an eye movement E_i and each column j is associated with a possible hypothesis H_j.

For each row i, i.e. for each eye movement E_i, in the matrix $M = \{x_{ij}\}$ of the long-term-memory we compute a function

$$r_i(H_j) = \frac{x_{ij}}{\sum_k x_{ik}}, \forall H_j \in \mathcal{H}. \quad (4)$$

The ratio $r_i(H_j)$ tells, how many times a certain eye movement E_i had been done when an object with the label H_j was shown, divided by the number of all occurences of the eye movement E_i in all scenes shown.

It must be distinguished between "an object with label H was shown" and "an object of class H was shown".

If we want to know the minimum of how often an eye movement E_i was done, when an object of class H_j was shown, we must add all ratios which are assigned to subclasses of H_j.

$$\forall H_j \in \mathcal{H}: \; R_i(H_j) = \sum_{H_k \text{ subclass of } H_j} r_i(H_k) \quad (5)$$

This is only a lower limit. If an object of a superclass of H_j was shown, we do not know if it was of class H_j or not. We can also compute an upper limit U_i. At least, when an object of an inconsistent class to H_j was shown, we know that it could not have been an object which belongs to H_j.

It is not possible to compute from r_i exactly, how often an eye movement E_i was done, when an object of class H_j was

[1] In the current implementation of the system each scene shows one object and therefore there is an identity between scene and object. In following versions there will also be relations between objects.

shown. But it can be computed an interval $[R_i(H_j), U_i(H_j]$. Thus, r_i denotes the label ratio and $R_i(H_j)$ and $U_i(H_j)$ refer to the class ratio. Somewhere in the interval $[R_i(H_j), U_i(H_j]$ lies the class ratio.

The evidence which a certain eye movement E_i yields, can be expressed by the function r_i. The higher the value $r_i(H_j)$ the more supports this eye movement an object with label H_j. Thus, the more often in the past an object with label H_j led to the eye movement E_i, the higher is the direct support of the eye movement E_i for the label H_j. We are not so much interested in the direct support, but in the summarized support, that an object is H_j. Thus, we are interested in the value $R_i(H_j)$.

It can be shown, that the label ratio $r : \mathcal{H} \to [0,1]$ corresponds to a bpa $m : 2^O \to [0,1]$, and the lower class ratio R to a belief function $bel : 2^O \to [0,1]$ of the Dempster/Shafer-Theory [17]. Therefore Dempster-Rule of Combination can be applied for the combination of the evidences and the combined evidence of eye movements E_1 and E_2 is expressed by:

$$\forall H \subseteq \mathcal{H} : r_{12}(H) = \sum_{H_i \cap H_j = H} r_1(H_i) * r_2(H_j), \quad (6)$$

where $H_i \cap H_j$ refers to the class consisting of the intersection of the elements of the classes H_i and H_j.

To avoid contradictions which may be caused by Dempster-Rule the open-world-assumption [18] is assumed.This means, the label H must be interpreted as:"The shown object is a H or it is something which cannot be denoted by elements of $\mathcal{H}$." Thus, each label also includes all the unknown hypotheses, and the empty set denotes just this set of unknown hypotheses [19].

But as a final result for the belief situation we are interested in ratios with respect to the hypotheses space and therefore a normalization is applied to the ratios which lead to the original Dempster Rule of Combination with the closed-world assumption:

$$\forall H \subseteq \mathcal{H} \begin{cases} r_{12}(H) = \dfrac{\sum_{H_i \cap H_j = H} r_1(H_i) * r_2(H_j)}{1 - \sum_{H_i \cap H_j = \emptyset} r_1(H_i) * r_2(H_j)} & H \neq \emptyset \\ r_{12}(\emptyset) = 0 & H = \emptyset. \end{cases}$$

5.2 The inference strategy IBIG

IBIG [15; 16] is based on the Dempster/Shafer Theory applied on strict hierarchies [6]. In this scheme the concept of confirming and non-confirming belief is introduced. In the scene-analysis system confirming belief corresponds to lower class ratios computed from positive labels and non-confirming belief to lower class ratios computed from negative labels[2].

The Dempster Rule of Combination is used in order to calculate actual and potential belief situations from the confirming and non-confirming belief. In each step IBIG computes from the knowledge-base the actual belief situation and for each datum it computes the potential belief situation[3], i.e. the belief situation, which would emerge, if this datum could be confirmed. Thus, the influence of the potential belief on the actual belief situation is analyzed. The information gain is defined as the difference between the actual belief situation and the potential belief situation (for a detailed derivation see [15]).

The information increment strategy acts in parallel on the represented knowledge, while it organizes itself depending on the belief distribution as induced by the available data. The reasoning behaviour is determined by the permanently changing relationship between the knowledge structure and the growing corpus of actually available data. Simulations for a number of prototypical ambiguous data situations have shown that the adaptive behaviour of the strategy leads in fewer steps to a solution than common partitioning strategies [15].

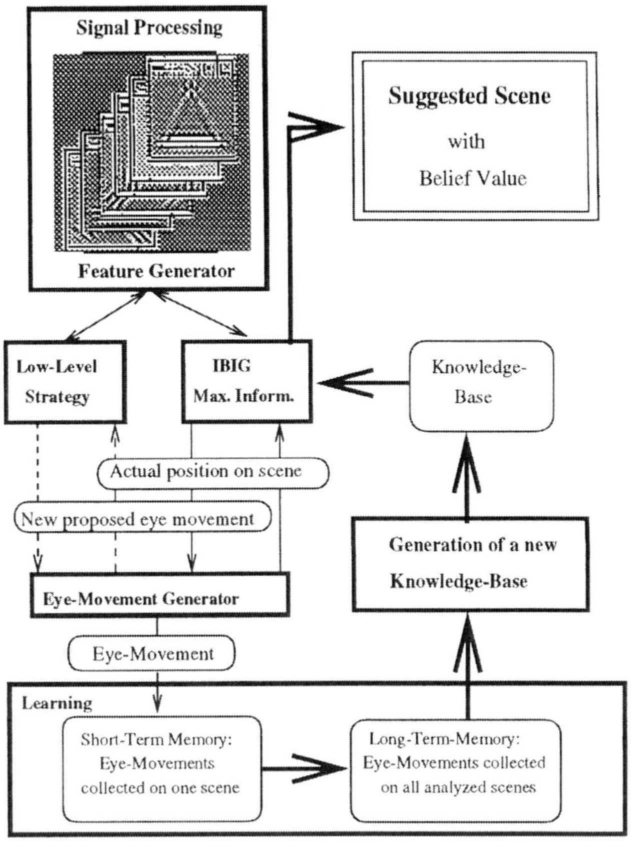

Figure 7 Architecture of the Scene-Analysis System.

5.3 The architecture

The system operates with two strategies - a low-level strategy and a high-level strategy (Fig. 7). The input for the two strategies are feature location pairs as gained by the feature generator from the filtered image. The low-level strategy takes control at the first fixation of the scene and in cases in which the high-level strategy yields no satisfactory result. It then generates eye movements based only on the actual scene and the actual fixation position. In contrast, the dominant, high-level strategy which contains IBIG uses also the knowledge-base generated from the long-term memory (Fig. 7).

IBIG computes in each step, depending on the label ratios which link the eye movements with the hypotheses, the eye movement which would yield the maximum change in

[2] Actually, there is no one-to-one correspondence between lower class ratios and confirming or disconfirming belief, but the lower class ratios have first to be decomposed, because a restricted subclass of belief functions is used.

[3] In contrast to that, in the original IBIG the potential belief is calculated for each node.

the belief situation. If the maximum change reaches a certain threshold, IBIG proposes this eye movement. This proposed eye movement (f_1, f_2, p) is combined with the coarse-grained peripheral representation of the feature classes to determine the final motor command for the generation of the eye movement.

6. Conclusion

We have reported first results of a multi-disciplinary research project on the analysis and modelling of the sensorimotor loop of saccadic scene analysis. We first performed experiments in which we recorded the saccadic eye movements of human observers on artificial and natural test images. These experimental data were used for an analysis of the statistical properties of those image regions which were selected by the saccadic eye movements as compared to that of an average set image regions which have been obtained by a purely random selection procedure. As a first result, our analysis indicated that regions with higher variance have a higher probability of being selected by saccadic fixations. However, we were not able to find further significant differences beyond these variance effects in our analysis of the power spectra. This is in agreement with similar results reported recently by [14]. We hence proceeded to a comparison of the two sets in terms of higher-order statistics, as reflected in the bispectral density. On this level, we obtained clear structural differences between the two sets, which indicate that image regions with purely *i1D*-signals (straight edges, for example) are less interesting than regions with a more complicated and statistically less redundant structure, as provided by *i2D*-signals like occlusions, corner points, and line endings, for example. This is in agreement with earlier results on visual information processing [1], and with informal descriptions of the "informative features" which have been deduced from recent experiments on saccadic eye movements (e.g. [10; 3; 24]). Simulations with nonlinear *i2D*-selective operators indicated a qualitive agreement between the operator activation map and the experimentally obtained fixation points.

Since it is well known, that saccadic fixations are not solely determined by bottom-up data processing but also by top-down (cognitive) influences [20], we considered an combination of the linear and nonlinear feature extraction processes of early vision with higher-level cognitive processing stages into one integrated architecture. In this architecture, the cognitive processing is modelled by an evidence-based reasoning system which is operating on a distributed object representation in which the objects are defined in terms of saccadic eye movements between some of their local spatial features. In such a system, the fixation of a local region of the scene corresponds to asking the most informative question about the relation between the scene and the systems current knowledge about it.

Acknowledgement: Supported by Deutsche Forschungsgemeinschaft, SFB 462 (B4, B5, and B6). We thank E. Pöppel, I. Rentschler, G. Hauske, and W. Prinz for their constant encouragement of our interdisciplinary research activities.

References

[1] F. Attneave. Some informational aspects of visual perception. *Psychological Review*, 61:183–193, 1954.

[2] E. Barth, T. Caelli, and C. Zetzsche. Image encoding, labelling and reconstruction from differential geometry. *CVGIP:GRAPHICAL MODELS AND IMAGE PROCESSING*, 55(6):428–446, 1993.

[3] V. Bozkov, Z. Bohdanecky, T. Radil-Weiss, L. Mitrani, and N. Yakimoff. Scanning open and closed polygons. *Vision Research*, 22:721–727, 1982.

[4] H. Crane and C. Steele. Generation-V dual-Purkinje-image eyetracker. *J. Opt. Soc. Am.*, 24:527–537, 1985.

[5] J. E. Dowling. *The Retina: An approachable part of the brain.* Harward University Press, Cambridge, 1987.

[6] J. Gordon and E. Shortliffe. A method for managing evidential reasoning in a hierarchical hypothesis space. *Artificial Intelligence*, 26:323–357, 1985.

[7] G. Krieger and C. Zetzsche. Nonlinear image operators for the evaluation of local intrinsic dimensionality. *[Special Issue on Nonlinear Image Processing] - IEEE Trans. Image Processing*, 5(6):1026–1042, 1996.

[8] G. Krieger, C. Zetzsche, and E. Barth. Higher-order statistics of natural images and their exploitation by operators selective to intrinsic dimensionality. In *Proc. IEEE Signal Processing Workshop on Higher-Order Statistics*, volume PR08005, pages 147–151. IEEE Computer Society, Los Alamitos, CA, July 1997.

[9] M. Land and S. Furneaux. The knowledge base of the oculomotor system. *Phil. Trans. R. Soc. Lond. B*, 352:1231–1239, 1997.

[10] N. H. Mackworth and A. J. Morandi. The gaze selects informative details within pictures. *Perception and Psychophysics*, 2:547–552, 1967.

[11] C. L. Nikias and A. P. Petropulu. *Higher-Order Spectral Analysis: A Nonlinear Signal Processing Framework.* Prentice-Hall, Englewood Cliffs, NJ, 1993.

[12] D. Noton and L. Stark. Scanpaths in saccadic eye movements while reviewing and recognizing patterns. *Vision Res.*, 11:929–942, 1971.

[13] G. A. Orban. *Neuronal operations in the visual cortex.* Springer, Heidelberg, 1984.

[14] P. Reinagel. Image statistics in the foveal visual field. In *Workshop on Natural Scene Statistics*, Jiminy Peak, MA, September 1997.

[15] K. Schill. Decision support systems with adaptive reasoning strategies. In *Foundations of Computer Science, Lecture Notes in Computer Science*, pages 417–427. Springer, Berlin, 1997.

[16] K. Schill, E. Pöppel, and C. Zetzsche. Completing knowledge by competing hierarchies. In B. D'Ambrosia, P. Smets, and P. Bonisonne, editors, *Proc. 7th conference on Uncertainty in Artfificial Intelligence*, pages 348–352. Morgan Kauffman Publ., San Mateo, CA, 1991.

[17] G. Shafer. *A Mathematical Theory of Evidence.* Princeton University Press, Princeton, 1976.

[18] P. Smets. Belief functions. In *Non-Standard Logics for Automated Reasoning.* Academic Press, London, 1988.

[19] E. Umkehrer and K. Schill. General perspective on the formalization of uncertain knowledge. In B. Ayyub and M. Gupta, editors, *Uncertainty Analysis in Engineering and the Sciences*, pages 21–35. Academic Press, 1997.

[20] A. L. Yarbus. *Eye movements and vision.* Plenum Press, New York, 1967.

[21] C. Zetzsche and E. Barth. Fundamental limits of linear filters in the visual processing of two-dimensional signals. *Vision Research*, 30:1111–1117, 1990.

[22] C. Zetzsche and E. Barth. Image surface predicates and the neural encoding of two-dimensional signal variation. In B. Rogowitz, editor, *Human Vision and Electronic Imaging: Models, Methods, and Applications*, volume SPIE 1249, pages 160–177, 1990.

[23] C. Zetzsche, E. Barth, and B. Wegmann. The importance of intrinsically two-dimensional image features in biological vision and picture coding. In A. Watson, editor, *Digital images and human vision*, pages 109–138. MIT Press, Cambridge, MA, 1993.

[24] C. Zetzsche and H. Deubel. Prediction of saccadic fixations by nonlinear 2d-detectors. Unpublished pilot study, 1990.

ACTION SELECTION AND BEHAVIORAL SEQUENCES

Planification versus sensory-motor conditioning: what are the issues?

A. Revel and P. Gaussier and S. Lepretre*and J.P. Banquet**
* ETIS ENSEA/UCP, 6, Av. du Ponceau 95014 Cergy CEDEX
{revel,gaussier,lepretre}@ensea.fr
** INSERM CREARE - l'institut des neurosciences - Paris VI

Abstract

This paper deals with the problem of linking a planification level to a sensory-motor level. We discuss the interest of conditioning when the environment is not predictable enough for motivated planification to work properly. We then present a neural implementation of the planification and we propose an architecture integrating both sensory-motor and planification level. We particularly discuss the interaction between the two levels.

1. Introduction

The motivation of our research group is to create a generic autonomous control system which could allow an autonomous robot, using a CCD camera as main source of information, to learn how to realize several complex tasks during the same "life" ("animat approach", see (McFarland, 1994; Meyer and Wilson, 1991)). In order to guide our research, we take inspiration from psychology and neurobiology expecting it could give us good hints to develop "well-fitted" architectures. Yet, according to an engineering point of view, inspiration from psychology can only be interesting if it allows the design of architectures that can solve problems which cannot be solved simply (or not solved at all) by classical approaches and that are not too expensive in terms of computing resources. Our claim is that the a priori knowledge coming from psychology and neurobiology we inject into our architectures allows to use "for free" structure and regularities of the environment without restricting the system adaptability. The main key for keeping adaptability is allowing the system to learn by itself how to react facing natural scenes in the environment. Thus, we focus our attention on the way animals can learn things from the environment in order to transpose it to our animats.

In psychology, two major learning theories confront each other. Behaviorism was introduced by Pavlov in the 30's (Pavlov, 1927) and suggests that the main part of learning capabilities can be explained by conditioning processes. Cognitivism, introduced by Tolman a bit latter after Pavlov (Tolman, 1932; Tolman, 1948), does not deny conditioning capabilities. Yet, cognitivists suggest that, to explain given complex skills such as maze path-finding, it is necessary to consider that animals can construct "cognitive maps" of their environment (a definition will be given in section 3) and use them for planification.

Artificial intelligence has obviously tried to echo these psychological results. The first approach has been to reduce the "intelligence" and especially, the human one, to its "cognitive" aspect. Many systems have thus been developed to allow the manipulation of internal symbols and the planification of "actions". Yet, on being confronted with reality, these systems have encountered difficulties. In fact, Harnad has emphasized the difficulty in grounding such systems in the real world (symbol grounding problem - (Harnad, 1990)). This problem comes from the fact that symbols are manipulated without regard to the sense they refer to (cf. the Chinese room problem — (Searle, 1987)).

Due to these difficulties, a "new" approach of artificial intelligence has emerged several years ago. The main claim of this approach is that intelligence "emerges" from the interaction of different levels of perception-action loops directly constructed from sensors and effectors. The first step has been to realize simple "reflex" behaviors directly connecting the sensors to the effectors with no more computation (phototaxis and obstacle avoidance behaviors - (Braittenberg, 1984)). In order to allow more flexible behaviors it has been necessary to introduce learning capabilities. At this point, a conditioning paradigm has been used as inspiration to build learning rules (Verschure et al., 1995; Pfeifer and Verschure, 1994; Pfeifer and Scheier, 1996). Those two levels are similar to the two first steps of the generic sumsumption architecture proposed by Brooks (Brooks, 1981). The next step would be to learn how to plan. Yet, this approach seems confronted with the inverse problem of the one encountered by the "cognitive" approach: it is very hard to build internal representations in order to plan actions without imposing to the system many a priori information on the environment.

In this paper, we try to see how the behaviorists and the cognitivists approaches may be reconciled. For that purpose, we try answering some questions on the interactions between conditioning and planification: Does planification always work ? How can complex behaviors be achieved using only conditioning ? Are conditioning and planification separate processes or are they linked together ?

In fact, in the first section of this paper, we will show it is possible to exhibit a maze experiment for which the use of planification does not work and that it is still possible to solve the problem using a simple reinforcement rule.

We then propose in the second section, a simple architecture that allows the learning of a cognitive map of the topological relationships between given "situations" (whatever they may be) so as to use them for planification. We propose using this algorithm to solve the problem of action selection in a maze and in an open environment where several motivations can appear.

In the third section, we claim it is necessary for the planification system to lay on a sensory-motor system in order to be grounded to the "real world". Yet, we show it is necessary to take into account a dynamical representation of the situations for the cognitive map and the sensory-motor system to work together.

In conclusion, we present unification perspectives of our approach, and we propose to validate our model with biological data.

2. Conditioning to acquire complex behaviors

Is it possible to exhibit an example in which actions planification is impossible ? In fact, as soon as encountered situations are regular enough, planification is a good strategy. The best example is obviously the maze problem which consists in finding the exit of a maze. Indeed, a cognitive map is the easiest way to remember where the exit is and to infer shortest ways to reach it. Yet, as soon as the maze is no more predictable for instance, the animat is always moved from maze to maze or the maze presents swinging doors), planification cannot be useful anymore. Conversely, conditioning can be used supposing that T-junctions are always marked with visual cues indicating where the exit is (e.g. turn arrows). Indeed, in behaviorism, each behavior is supposed to be the result of a single conditioning or a series of conditioning. The theory of behaviorism lays on the fact that a given sensorial stimulus (or set of stimuli) can be associated to an action when is it presented at the same time as an unconditional stimulus (classical conditioning — (Pavlov, 1927)) or because the association is reinforced (instrumental conditioning — (Skinner, 1953)). In the animat context instrumental conditioning is a good means of making an animat learn a behavior giving it only simple information on its success such as "true" or "false".

For the maze problem, the difficulty consists in solving the sensory-motor association problem while the reinforcement signal not only rewards a single action but the series of actions that has lead the animat to the goal. For instance, we can imagine using the maze represented figure 1. If pattern/movement associations are made properly, they allow the exit of the maze to be reached and a reward to be obtained.

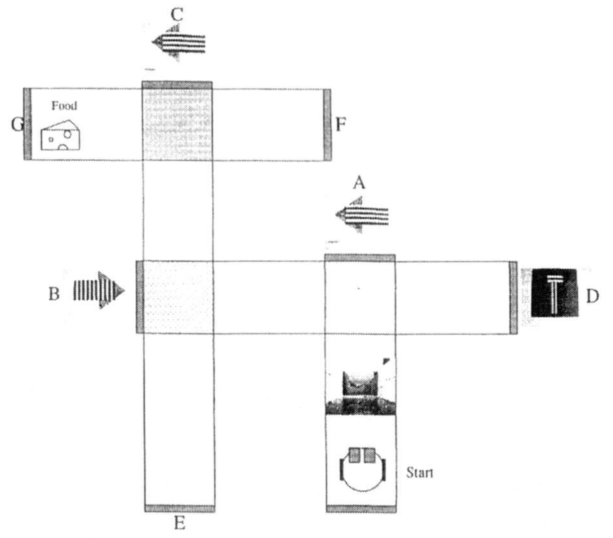

Figure 1 The maze used and corresponding patterns perceived by the robot (obtained with the robot CCD camera).

In order to tackle this association problem we had to develop our own conditioning algorithm. In fact, we have identified a problem which is usually totally forgotten. In classical sensory-motor association learning rules, such as Sutton and Barto's (Barto et al., 1983), noise is added to the output of the neuron. When the robot tests a set of sensory-motor associations and receives a punishment which is applied to all the stimulus/action pairs the robot has made, whatever their appearance frequency is. Thus, the learning rule calls into question the entire set of associations and there are as many chances of changing an association linked to a frequent situation (a corridor situation for instance) as of changing an association corresponding to a rare one (like a T-junction). Yet, as frequent situations appear more often (that is a pleonasm !), there are more chances this modification affects the animat behavior. One of the interesting features of the mechanism that we propose is that it allows us to take into account in the same way situations having different occurrence frequencies.

Another difficulty happens if the robot has to realize several different tasks. In an autonomous robotic context, the issue of allowing the robot to learn different behaviors endowed with the same architecture, learning rules and "memory", is an interesting one. If a task has been learned at a given moment, learning a new task should not interact with this previous acquisition unless the realization of the new task would lead to contradict associations already made. This memory effect is very important if the robot must keep in mind learned behaviors while going on exploring its environment. Besides, if only parts of the stimuli are new but the rest stays unchanged, it can help learning new behavior more rapidly. Furthermore, simple associations previously learned could help learning more difficult associations (this concept is referred as "shaping" or "teaching" (Kaelbling et al., pear)).

In order to integrate those constraints, we have proposed a neural learning rule modeling hypotheses testing capabilities. The idea is to allow the use of an hypothesis during a time long enough to test the consequences it could have and if it needs changing. In order to indicate the existence of an association between a particular recognition and an action, a simple binary value of the synaptic weights is enough. Yet, as the addition of noise at the neuron output can lead to uncertainties, we have decided to control diversity generation at the synaptic level. Our solution is to introduce a confidence measurement associated to the binary weight. It corresponds to the confidence in the input/output association the weight codes (see Fig. 2). When a reinforcement signal occurs, only the probability term is changed. Besides, a random draw is done in order to change weights whose confidence term is low. If there is no reinforcement variation neither the probabilities nor the weights are modified. However, information about the correlation between the input and the output of the weight goes on being stored. Such a mechanism gets the robot to behave as if it was testing different sets of hypotheses without calling into question hypotheses already tested.

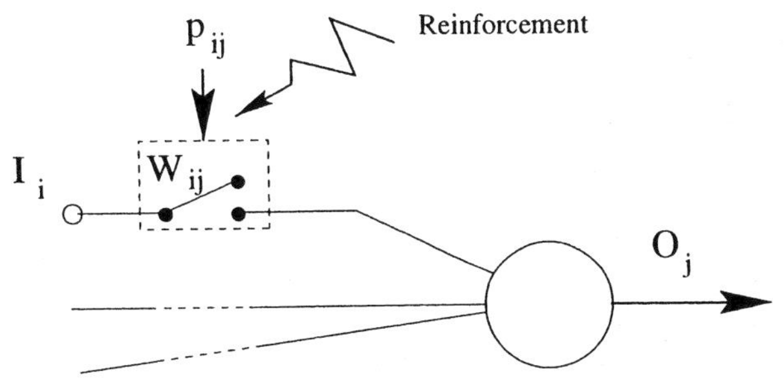

Figure 2 The PCR model

PCR is an algorithm which can be used in many applications depending on the "quality" of the reinforcement signal which is provided. If the reinforcement is sparse and crude, PCR is equivalent to a random search (yet, it is better because it only considers situations it really needs using). If the reinforcement signal varies according to a gradient, the algorithm uses this "slope" information and finds the solution quicker.

The maze problem has been actually realized on our Koala robot using the PerAc architecture (see (Gaussier et al., 1997b) — (Gaussier et al., 1996) for application

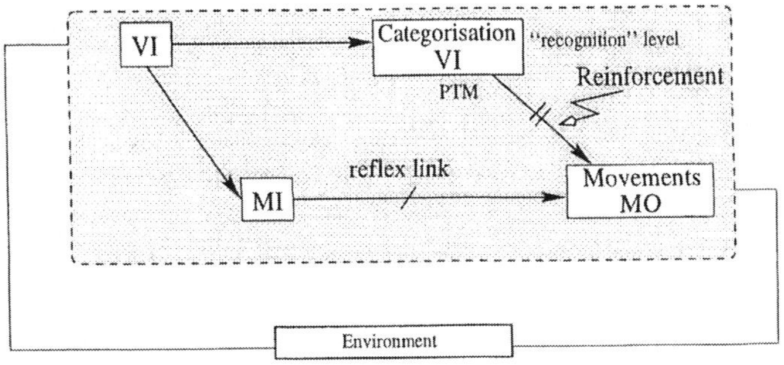

Figure 3 The PerAc architecture

on to object recognition and place learning). The main idea of this architecture is that both categorization and sensory-motor association must be controlled by the effect of the movements on the environment. Particularly, if the movement that has been performed at a given moment is reinforced, it probably means this situation is important and should be both categorized and linked or unlinked with this movement.

The PerAc architecture (Gaussier and Zrehen, 1994a)is made of 4 blocks (see figure 3): the Visual Input, the association map (or Visual Output - VO), the Motor Input and the Motor Output. The visual input is a representation of visual information the robot perceives through its camera. The association map is dedicated to the recognition of input patterns. In our experiment we use the Probabilistic Topological Map (PTM) described in (Gaussier and Zrehen, 1994b). The first interest of this self-organized map is to preserve topology (two similar situations are coded on close neurons) and to allow immediate learning. As learning is unsupervised, it is necessary to be able to select the salient information that must be coded. This is made possible thanks to a vigilance term equivalent to the one described in ART (Adaptive Resonance Theory - (Grossberg, 1976)). This map is a precious tool for the robot to be able to build rapidly an internal representation of the environment.

The Motor Input (MI) are reflex links already acquired or genetically given. In the context of classical conditioning, this block which would drive Unconditioned Responses (in our application, the MI group is a reflex keeping the robot in the middle of corridors). The Motor Output group (MO) selects which movement must be performed by the robot. The finally performed action corresponds to the most activated neuron. A motor movement is then proposed by the WTA.

The general behavior of this architecture depends strongly on the way links between the recognition map (VO) and the motor output group are learned (assuming that input stimuli are categorized by the PTM map and that WTA neurons use this learning process to create sensory-motor associations). Yet, it should be noted that this prior "categorization" of the stimuli contradicts the behaviorists' belief that conditioning needs no internal representation. The PCR algorithm is used to learn the links between VO and MO.

The main problem comes from the difficulty in categorizing the perceived situations in order to associate them with movements. To solve this problem, we have chosen to control the categorization according to the result of the action (see (Gaussier et al., 1996) for further information). Thus, the system may adapt the categorization to its use in the environment. In fact, using as often as possible the regularities of the environment is a major concern in our approach. This is made possible due to:

- the topology preservation, which exploits the environment continuity (as well at the perceptual level as at the action level).
- the generalization capabilities of the recognition map, which allows to learn only when it is necessary

It must also be noted that the sensory-motor association system lays on a reflex system which implements a priori information on the environment but not on the task to be performed. It can be thought as an adaption of the animat to the environment it lives in.

Along this section we have shown it was possible to build architectures with learning rules inspired from behaviorism, which allowed animat to realize a complex behavior. Yet, it can be noticed that at any level the robot can "choose" its behavior. In fact, the mechanism that we have developed allows the robot to learn fixed actions or sequences of actions.

We thus wonder about the capacity a robot must have to be able to select one action among several according to an internal "drive" it must reduce. We have developed an experiment in which a robot was able to go to two different places according to the level of two internal variables (comparable with "food level" and "water level"). When the robot is "hungry" (its "food level" variable is under a given threshold), it must go to the place it learned it corresponded to the place it can find "food". This can be done very easily learning how to go to those place (using the "place learning" algorithm) and modulating the activity of neurons coding the places according to the level of internal variables (inhibiting the place coding "food" when the corresponding variable is low and vice versa). But this mechanism is too simple to allow a real planification. For that purpose, we have been interested by the concept of internal representation as a "cognitive map" introduced by the psychologist Tolman.

3. Motivated planification

In this section, what we are interested in is to design a neural architecture allowing the robot to plan its actions to reach a goal. In the 30's, Tolman realized a maze experiment with rats, that could not be explained by behaviorists. The experiment consisted in comparing the results of two groups of rats: the first group was always rewarded while the second group was only rewarded after 11 days. According to behaviorism rats of the first group should have much better results than rats of the second group because they have been rewarded longer. In fact, results of the rats of the second group were equivalent. Tolman concluded that, although they was not rewarded, rats have learned a representation of the maze (what he called "latent learning"). This representation is called a "cognitive map" (Tolman, 1948). According to Gallistel (Gallistel, 1993):

> "A cognitive map is a record in the central nervous system of macroscopic geometric relations among surfaces in the environment used to plan movements through the environment..."

This definition supposes that such maps exist in the brain (which is still contested today) and that they take into account only macroscopic relationship between places in the environment. This definition makes no hypothesis of the nature of the information:

> "...A map in ordinary acception encodes relative **metric** positions. The relative metric position of a point is given by its **coordinates** (its vector position). Each coordinate of a point (each dimension of the position vector) specifies the distance of the point from an axis or **origin** of the system of coordinates."

It seems that Gallistel proposes that a cognitive map would be coded in Cartesian or polar absolute coordinates with metrics.

> "The thesis I will argue is that the intuitive belief that the cognitive maps of *lower* animals are weaker than our own is not well founded... There is experimental evidence that even insect maps are metric maps."

Yet, models not using the cognitive map concept have been proposed to explain the navigation of insects (they are based on a mechanism very similar to the one we used to learn "places" — see the interesting review on navigational systems (Trullier et al., 1997)). With Schmajuk, we prefer a more moderate definition of the cognitive map (Schmajuk and Thieme, 1992):

> "...The cognitive map [...] is a topological map, i.e, it represents only the adjacency, but not distances or directions, between places... The cognitive map allows the combination of information about spatially adjacent places, and thereby the inference of the connections to remote places."

To elaborate such a representation, we can consider a map of neurons fully interconnected. When a situation is recognized, it is coded on a neuron N_1, which is activated during a time T (see figure 4-a). When a new situation is coded on N_2, a simple Hebbian learning allows the learning of the time relationship (and thus the topological relationship) between those two situations (see figure 4-b).

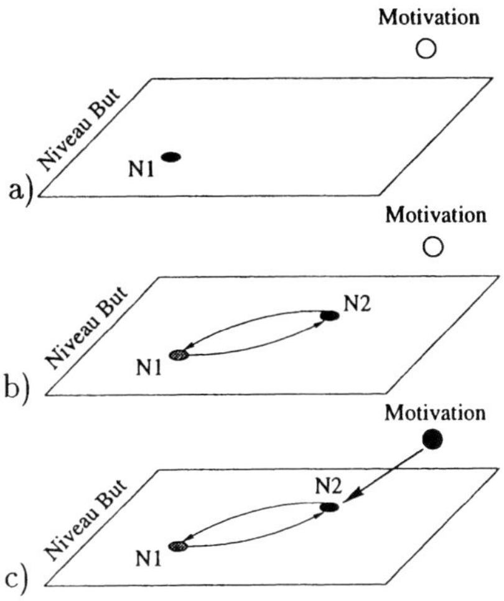

Figure 4 a) Neuron N_1 is activated. b) Neuron N_2 is activated while N_1 is still activated. Links $W_{N_1N_2}$ and $W_{N_2N_1}$ can be learned. c) The motivational neuron is activated. The link between this neuron and the neuron coding the last situation encountered is learned.

Indeed, if X_{N_i} is neuron N_i activity and $\overline{X}_{N_i}$ its corresponding short term memorization with:

$$\overline{X}_{N_i}(t+1) = \frac{T \cdot \overline{X}_{N_i}(t) + X_{N_i}}{T+1}$$

The Hebbian learning rule is given by :

$$dW_{N_1N2} = \epsilon \cdot X_{N_1} \cdot \overline{X}_{N_2}$$
$$dW_{N_2N1} = \epsilon \cdot \overline{X}_{N_1} \cdot X_{N_2}$$

The topological link between two neurons can thus be learned. By generalizing the mechanism to all the situations encountered, a graph of the spatial relationships between these different situations can be constructed. The last step to realize the planification, is to learn the link between the recognition of a situation and the satisfaction of a motivation. For that purpose, it must be considered that the recognition of this situation activates a given "motivational" neuron. Then, a simple Hebbian rule allows reinforcement of the link between the motivational neuron and the recognition of the situation (see figure 4-c).

In order to perform planification process, a solution can consist, as for resistive grids (Bugmann et al., 1995), in propagating backward the activation of the motivational neuron. In fact, the activity of a neuron in the cognitive map must be a function of its topological distance to the goal (in term of number of situations).

We propose the activation law given below:

$$y_i = \max_j (W_{ij} \cdot y_j)$$

As the neuron activity must be bounded by 1, connection weights must be bounded between 0 and 1. Indeed, if all weights value are 1, all neurons activity is 1 and it is impossible to decide which is the shortest path. The value of weights must thus be bounded with a value W which is less than 1. It must be noted that there must be several iterations before stabilization of the result (the minimal number of iterations is the number of intermediate situations to reach the goal).

The algorithm can thus be written:

1. Initialization :
   - N_{i_0} is the motivational neuron
   - $x_{i_0}(0) \leftarrow 1$
   - $x_i(0) \leftarrow 0, \forall i \neq i_0$
2. Do:
   - $\forall i,\ x_j(t+1) \leftarrow \max(W_{ij} \cdot x_i(t))$

   While the net is not stable (max distance between the current node and the goal in connections number)

We have proved that this algorithm is formally equivalent to Bellman's shortest graph distance algorithm (Bellman, 1958).

To fix the ideas, let us take the example of the navigation in the maze represented on figure 5. When a movement is possible between two "places" (two adjacent squares of the discretized environment), a link at a fixed value ($W = 0.9$) is created. After exploration, when the robot is brought back to the starting point A, it should perform sequence $A \rightarrow B \rightarrow C \rightarrow D$, instead of $A \rightarrow B \rightarrow E \rightarrow F \rightarrow G \rightarrow C \rightarrow D$.

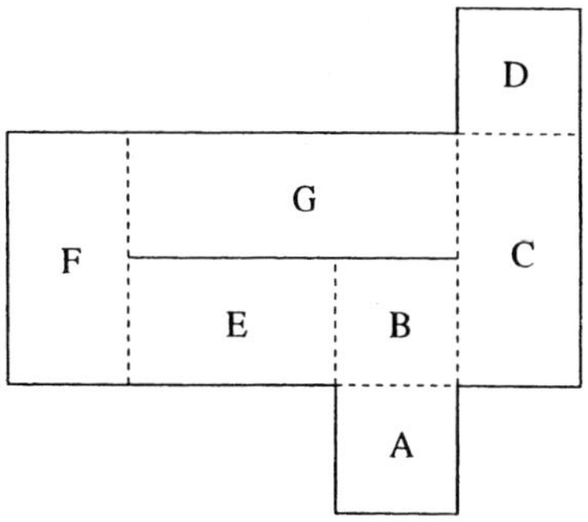

Figure 5 Complex maze with two pathways of different size leading from A to D.

If neuron coding D activity is forced to 1 due to its link with motivation, (see figure 6), according to neuron activity equation $y_i = \max_j (W \cdot y_j)$, and if $W = 0.9$, C and G activity is 0.9. F activity is $0.9^2 = 0.81$, and E is $0.9^3 = 0.729$. Besides, as B activity is computed according to the max value, its activity is $0.9^2 = 0.81$ instead of $0.9^4 = 0.6561$. Its activity thus directly depends on its distance to the goal (see figure 6).

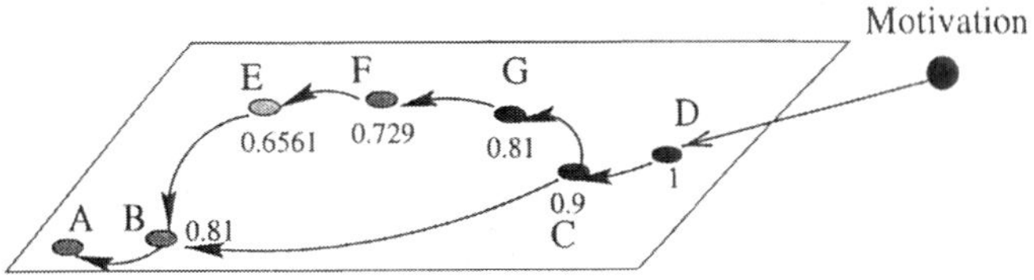

Figure 6 The motivational neuron activates the recognition of D. Due to the links learned during the exploration phase, there is a backpropagation of the activity. The neurons activity is directly a measurement of the distance to the goal.

We simulated the behavior of an animat evolving in a complex maze in which it could find different "food" sources (figure 7). The robot starts moving at random and gradually discovers its environment (walls, "food sources"). When a "food" source is reached, the neuron associated with the current position is also associated with the "hunger" motivation. In this implementation, it is considered that each time the robot moves, it consumes an internal resource associated to an "energy level" variable. When this variable level is too low (according to a given threshold), the robot begins the planification. If it has already discovered one or several "food" sources, it quickly reaches the nearest.

Moreover, in our tests, we have added the possibility of learning to avoid areas in order to test if the system was able to avoid a given area of the maze even if crossing this area it would reach the "food" source quicker. In fact, it has been possible only associating a lower weight value (0.7) to the link that would make the robot enter a "to avoid" area.

We have also simulated the behavior of an animat placed in an open environment. Thanks to a mechanism allowing "place learning" (merging the recognition – "what" – and the position – "where" – of landmarks (Gaussier et al., 1997a)), the robot can learn where "food" and "water" sources are when it can find them. Moreover, when it explores the environment, it can learn new places as soon as they are different enough from places it has already learned before (according to a "recognition" threshold) and it learns to link them with the last place encountered. From time to time, the animat builds a topological representation of its environment it can use to plan its actions. Besides, two internal variables corresponding to "food level" and "water level" are dynamically updated according to differential equations. When their level is too low, it activates a motivational neuron associated with the corresponding motivation ("hunger" resp. "thirst") which makes the robot plan its action to reach the nearest place where it can satisfy its motivation. We have tested this architecture on many different simulated open environments, and it seems to work properly even if:

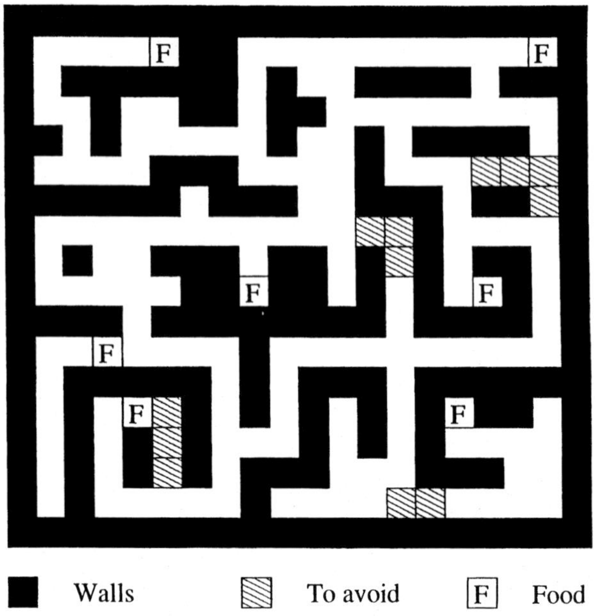

Figure 7 Complex maze used to test our planification algorithm efficiency.

- there are several places where the animat can be rewarded.
- the "food" or "water" sources disappear.
- the number of motivations is increased (e.g. adding a "homing" behavior)

In each case, the system is able to solve both shortest path finding and action selection problem.

Yet, in those simulations, we have supposed that it was not a problem to recognize situations and the movement was selected following the gradient of the cognitive map neurons activity. The problem is that, for an experiment in a real environment, we should be able to recognize situations and to know to which movements they can be linked. In fact, the planification algorithm should interact with a sensory-motor association system similar to the architecture we described in section 2.

4. Linking the SM level and the planification level

Usually, classical plan generation mechanisms are treated by resolution systems based on formal logic. The principle consists in representing knowledge as a priori rules

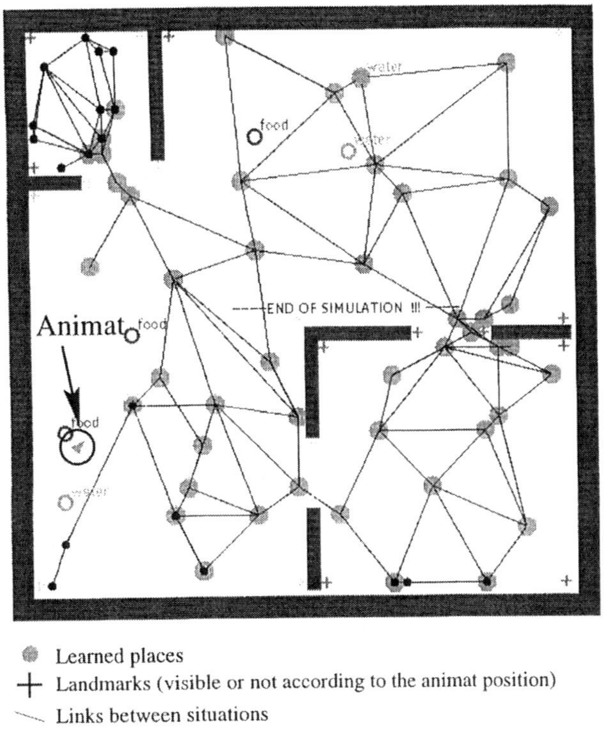

Figure 8 Simulation of an animat which plans its movements in an open environment in which both "food" and "water" sources can be found.

whose preconditions depend on the perceived situation and whose results consist in performing particular actions. Plan generation systems make the formal correspondence which allows to define, starting from the current situation, the set of rules to chain to reach the goal. Yet, in order to "ground" to the real world, interfaces between symbols corresponding to the rule preconditions and the situations they correspond to in the real world must be defined.

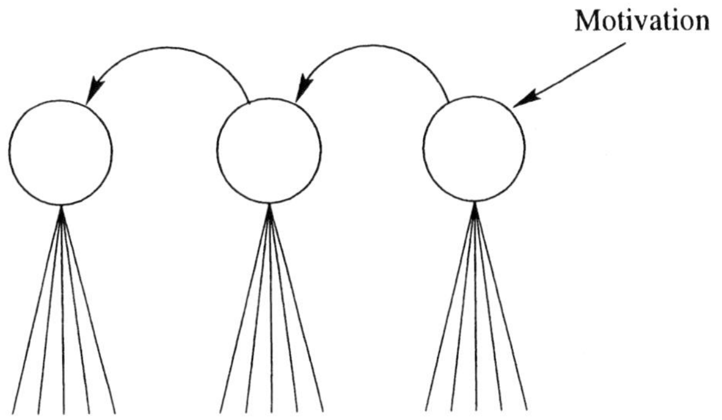

Figure 9 The informational flow corresponding to the recognition and to the planification are summed at the same level.

In our opinion, we think it is necessary to integrate within the architecture dedicated to planification a mechanism which allows the manipulation of internal representations elaborated from the perception of the environment. Usually, it is neglected that the internal representation must be constructed dynamically, while explorating the environment. It must be noted that planification can only be done on already known situations. There cannot be both learning and use of the information at the same time. It is thus necessary to integrate two working modes: an exploration mode and a planification mode. During exploration, the system must be able to learn different places (if these are perceptually different enough), to learn the topological relationship between those places, to learn the movement allowing to go from one place to another and to learn the places which can be associated to a goal. During planification, information learned during exploration can be used to reach the goal linked to the appearance of the corresponding internal motivation. In fact, as new situations may appear during plan execution, exploration and planification are not separated.

Then, why not use a mixed version of the sensory-motor learning architecture presented in section 2 and the planification architecture presented in section 3 ? Unfortunately, it is impossible to mix planification and recognition on the same neuron. Indeed, situation recognition and motivation backpropagation correspond to two distinct informational flows. The entire set of information cannot be treated by the same neuron map. If a single neuron must both decide which situation is recognized and propagate the motivational information from goal to sub-goals, there is obviously an ambiguity in the reason why the neuron is activated (see figure 9)[1].

At least two levels are thus necessary so as to separate the informational flows: one corresponding to the "goal" level, another corresponding to the recognition of the current situation and its association to a movement ("sensory-motor" level). As the information coded by a neuron must take into account both aspects of the information it codes ("goal" or "recognition"), corresponding neurons of the two levels must be linked together. Updating the activity of the neuron depends on the working mode in which the system is. During learning, the information must go "bottom-up" from the recognition level to the "goal" level so as to allow the cognitive map learning. Conversely, during planification, the information coming from the "goal" level must go "top-down" in order to select the action to do (see figure 11). This working mechanism corresponds to the neurobiological model of cortical columns proposed by Burnod (Burnod, 1989). The higher level of the column corresponding to the cortical level and the lower level to the thalamic one.

We can wonder if we can directly superimpose the planification level to the sensory-motor architecture described in 2.During the exploration phase, the system

[1] Another solution could be to dynamically deactivate the influence of a particular flow when necessary (this could be what happens in mammals' hippocampus)

could learn perceived situation and how to link them with actions. Besides, due to the superimposition of the "goal" level, a "cognitive map" could be learned according to categorized situations.

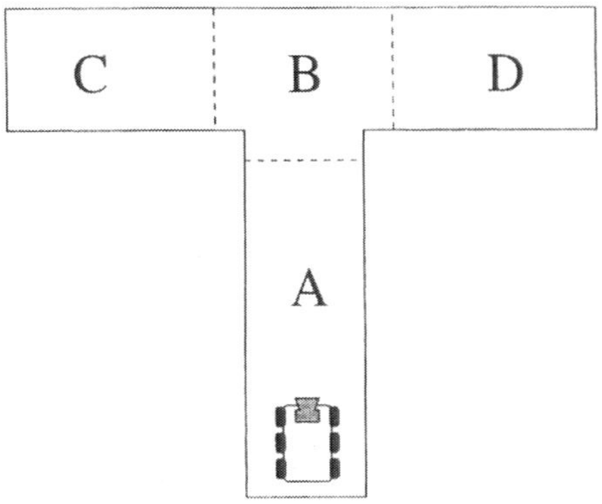

Figure 10 Maze example for planification

For instance, in the maze represented in figure 10, situation A is associated with "go ahead", B both with "turn left" and "turn right", C with the satisfaction of the motivation, D is a dead-end.

During planification, the system must use the information coming from the knowledge the goal is in C to perform the action sequence allowing to reach this goal starting from A. The problem is that when the animat arrives in B there is no way to choose which one must be performed (B is linked with two movements !).

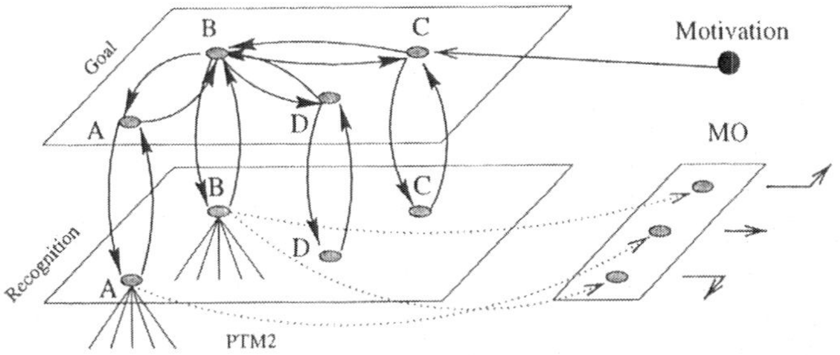

Figure 11 It is impossible to plan using only situation recognition. Indeed a situation can be linked to two different movements, and it is thus impossible to decide which action must be selected.

In fact, there should be two distinct representations each associated with one movement. Besides, this representation must take into account both the starting and finishing situations. A solution consists in building a representation of the transition between two situations (see figure 12). Let AB, the internal representation of the transition between A and B. The associated action is the movement allowing to go from A to B and is learned using PCR algorithm. The idea of this representation has been inspired by a model developed by (Banquet et al., 1998). As the internal representation is elaborated on transitions recognition and not only on scenes recognition, an internal representation is necessarily linked with only a single movement. In situation B, for instance, the animat may turn left (arrives in C) or turn right (arrives in D). Transition BC and BD will be created and respectively associated with "turn left" and "turn right" movements.

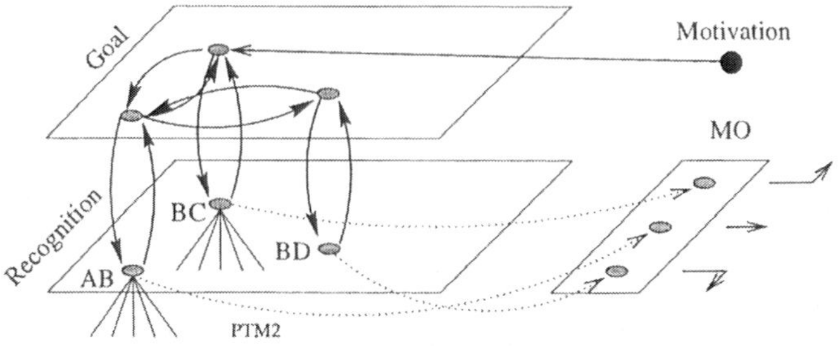

Figure 12 Use of the transitions for actions planification.

During exploration, the "recognition" level creates an internal representation for each transition between scenes and associates the movement allowing to go from one to the other. Besides, at the "goal" level, connections between representations are learned so as to create a graph of the topological relationships between scenes. Furthermore, when the animat reaches the goal, it learns the association between the recognition of the last transition and the motivation satisfaction. It is very important to notice that to create representation AB the animat must already be in B.

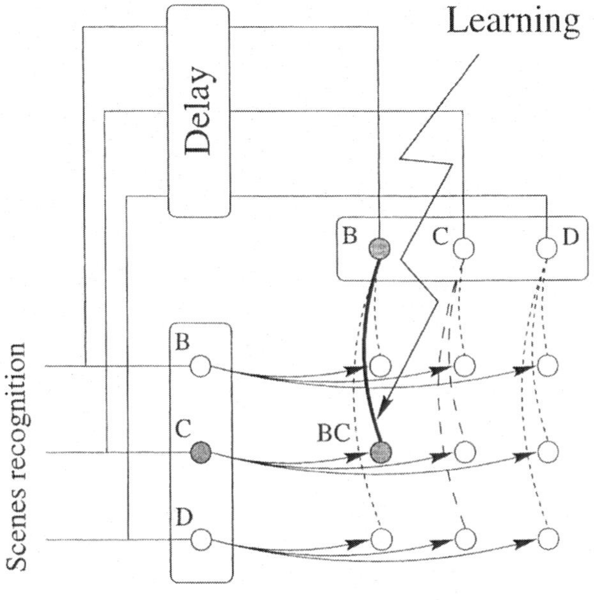

Figure 13 Fusion mechanism: direct input is fused with the delayed input in order to build a representation of the transition.

During planification, the motivation backpropagation towards the current state allows the activation of the graph nodes, thereby indicating the movement that it is necessary to perform in order to reach the goal. Conversely with the exploration phase, what is important now is to decide what movement to perform to reach the goal when recognizing a given place. It is thus necessary to build a system allowing a prediction of the scene(s) which can be reached from the current scene. This mechanism, linked to the motivation backpropagation algorithm should be able to select between different possible movements which is the best to reach the goal more rapidly. For instance, in B, the system must be able to predict both BC and BD transitions and make BC win in order to perform the corresponding movement (turn left).

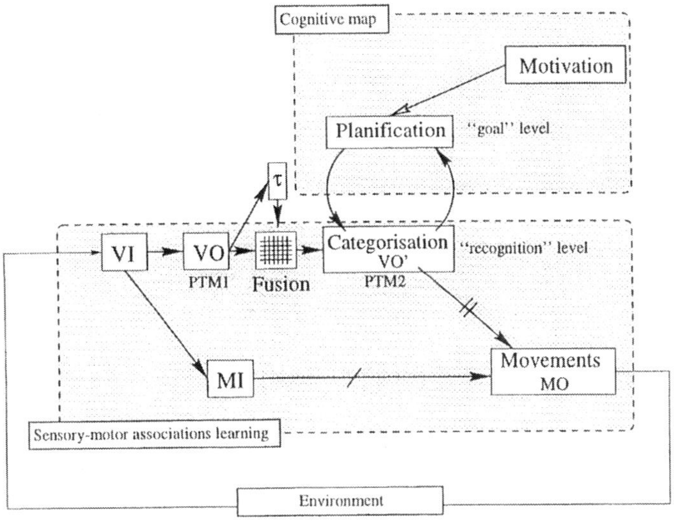

Figure 14 Starting from PTM1 (scenes recognition), a fusion mechanism of the input and the delayed input allows the elaboration of the transition recognition (PTM2). The activity of the "goal" level which depends on the distance from the goal is used to bias the recognition level activity so as to select the best action to reach the goal.

The transition learning mechanism is decomposed in two steps (see figure 13). The first step of the data treatment consists in learning perceptual scenes (PTM1). The second step is then to make the fusion of two successive situations in order to recognize it.

It is essential to rigorously sequence learning and use phases in order to make the architecture work properly for exploration and planification. The system architecture must thus take into account requirements imposed by both working phases.

The global architecture proposed is represented figure 14. For the moment, the solution we propose has only been tested in simulation. It is now being implemented for real. But we expect that we will be confronted with some severe problems. Indeed, while it has been shown that places can be learned very simply (Gaussier et al., 1997a), the fusion of two places in a transition requires important efforts on novelty detection and information compression.

5. Conclusion

In this article, we have proposed an architecture which allows the building of a cognitive map from the sensory-motor level (which includes in fact two sub-levels: reflex and conditioning). Our claim is that this superimposition of these levels is necessary for being able to integrate and control the different information flows. Yet, links between the different levels must take into account the system dynamic, merging information appearing at different time steps.

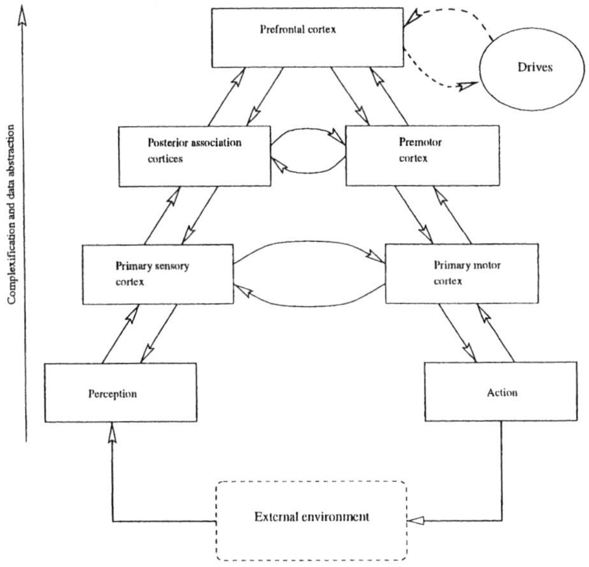

Figure 15 Schematic representation of the information integration in the brain.

In fact, we would like to draw a parallel between our architecture and biological data concerning the way the brain gradually integrates perceptual data to be able to perform more and more complex behaviors (see figure 15). At the top level, we can find specific integrative structures such as the pre-frontal cortex and the hippocampus.

At the lower level, the reflex system represents an a priori given by the genetic history of the species. In the robotic context, this a priori could be any knowledge on the structure of the task the robot must learn. Yet, our aim is not to constrain learning too much so that the robot is able to construct its knowledge by itself thanks to learning rules. In our opinion, the reflex system should only be used to facilitate the robot control by providing general knowledge on the environment. That is the reason why we are currently trying to implement our algorithm using dynamic field theory in order to make control smoother and more general (Schöner et al., 1995).

Our long term goal is to integrate in a same approach, robotic control and artificial intelligence. The architecture we propose is a first step to reach the "symbolic" level defended by classical AI, but we must go beyond. In particular, our architecture does not deal with the problem of planification structuration: our system can plan but it cannot plan to plan ! (see (Donnart and Meyer, 1996)) This problem is still very important if we consider the scaling problem. Indeed, we have shown that, at the sensory-motor level, the complexity directly depends on the number of possible associations there might be (number of input/output pairs). At the planification level, a focalisation system, based on the animat motivations, could allow selecting given sensory-motor schemes among all the possible ones. It would thus allow the solution of the problem of complexity explosion.

6. Acknowledgments

This research program is supported by a french GIS on cognitive science.

References

Banquet, J., Gaussier, P., Dreger, J., Joulain, C., Revel, A., and Grunther, W. (1998). *Cognitive science perspectives on personality and emotion*, chapter Space-time, order and hierarchy in fronto-hippocampal system: A neural basis of personality. Elsevier Science.

Barto, A., Sutton, R., and Anderson, C. (1983). Neuronlike adaptive elements that can solve difficult control problems. *IEEE transactions on system, man and cybernetics*, SMC-13(5):834–846.

Bellman, R. (1958). On a routing problem. *Quarterly of Applied Mathematics*, (16):87–90.

Braittenberg, V. (1984). *Vehicles : Experiments in Synthetic Psychology*. Bradford Books, Cambridge.

Brooks, R. (1981). A robust layered control system for a mobile robot. *IEEE Journal of Robotics and Automation*, 40:201–211.

Bugmann, G., Taylor, J., and Denham, M. (1995). *Neural Networks*, chapter Route finding by neural nets, pages 217–230. Alfred Waller Ltd., Henley-on-Thames.

Burnod, Y. (1989). *An adaptive neural network: the cerebral cortex*. Masson.

Donnart, J. and Meyer, J. (1996). Hierarchical-map building and self-positioning with monalysa. *Adaptive Behavior*, 5(1).

Gallistel, C. (1993). *The organization of learning*. MIT Press.

Gaussier, P., Joulain, C., Revel, A., and Banquet, J. (1996). Are shaping techniques the correct answer for the control of an autonomous robot ? In *UKACC International Conference on Control'96*, University of Exeter. IEE.

Gaussier, P., Joulain, C., Zrehen, S., Banquet, J., and Revel, A. (1997a). Visual navigation in an open environement without map. In *International Conference on Intelligent Robots and Systems - IROS'97*, pages 545–550, Grenoble, France. IEEE/RSJ.

Gaussier, P., Revel, A., Joulain, C., and Zrehen, S. (1997b). Living in a partially structured environment: How to bypass the limitation of classical reinforcement techniques. *Robotics and Autonomous Systems*, 20:225–250.

Gaussier, P. and Zrehen, S. (1994a). Complex architectures for autonomous agents. In Gaussier, P. and Nicoud, J., editors, *PerAc*, pages 278–290, Lausanne. IEEE Press.

Gaussier, P. and Zrehen, S. (1994b). A topological map for on-line learning : Emergence of obstacle avoidance. In *From Animals to Animats: SAB'94*, pages 282–290, Brighton. MIT Press.

Grossberg, S. (1976). Adaptive pattern classification and universal recoding, ii : Feedback, expectation, olfaction, and illusions. *Biological Cybernetics*, 23:187–202.

Harnad, S. (1990). The symbol grounding problem. *Physica D*, 42:335–346.

Kaelbling, L. P., Littman, M., and Moore, A. (to appear). Reinforcement learning: A survey. *Journal of Artificial Intelligence Research*.

McFarland, D. (1994). Animal robotics - from self-sufficiency to autonomy. In Gaussier, P. and Nicoud, J., editors, *From Perception to Action*, Lausanne, Switzerland. IEEE, IEEE Computer Society Press.

Meyer, J. and Wilson, S. (1991). From animals to animats. In Press, M., editor, *First International Conference on Simulation of Adaptive Behavior*. Bardford Books.

Pavlov, I. (1927). *Conditioned Reflexes*. Oxford University Press.

Pfeifer, R. and Scheier, C. (1996). Sensory-motor coordination: the metaphor and beyond. *Robotics and Autonomous Systems*.

Pfeifer, R. and Verschure, P. (1994). *The Artificial Life Route to Artificial Intelligence*, chapter The challenge of autonomous systems: Pitfalls and how to avoid them. MIT Press, Cambridge, MA.

Schmajuk, N. and Thieme, A. (1992). Purposive behavior and cognitive mapping: a neural network model. *Biological Cybernetic*, 67:165–174.

Schöner, G., Dose, M., and Engels, C. (1995). Dynamics of behavior: theory and applications for autonomous robot architectures. *Robotics and Autonomous System*, 16(2-4):213–245.

Searle, J. (1987). *Du cerveau au savoir*. Hermann.

Skinner, B. (1953). *Science and human behavior*. Macmillan Company, New York.

Tolman, E. (1932). *Purposive behavior of animals and men*. Irvington, New York.

Tolman, E. (1948). Cognitive maps in rats and men. *The Psychological Review*, 55(4).

Trullier, O., Wiener, S., Bethoz, A., and Meyer, J. (1997). Biologically based artificial navigation systems: review and prospects. *Progress in Neurobiology*, 51:483–544.

Verschure, F., Wray, J., Sporns, O., Tononi, G., and Edelman, G. (1995). Multilevel analysis of a behaving real world artifact: an illustration of synthetic neural modeling. *Robotics and Autonomous Systems*, 16(2-4).

Evolving Action Selection and Selective Attention Without Actions, Attention, or Selection.

Anil K Seth
Centre for Computational Neuroscience and Robotics
and School of Cognitive and Computing Sciences,
University of Sussex, Brighton BN1 9SB, UK
anils@cogs.susx.ac.uk

Abstract

A minimal animat architecture, consisting only of a set of autonomous, direct, and continuously active sensorimotor links, is shown to support a full range of 'action selection' phenomena. A genetic algorithm is used to engineer the activation functions supported by these links. No 'actions' are 'selected' in this model, and the use of artificial evolution means that there is no artificial separation of the problems of 'link design' from 'link fusion'.

Implications are drawn for how the concepts of 'action selection' and 'selective attention' may relate to the idea of coherence between sensorimotor processes.

1. Introduction

In order to survive and prosper in a dynamic and complex world, animals and animats need to engage in a variety of different behaviours at different times, at least some of which will be mutually incompatible. The problem of how, at any given time, they come to be doing one thing rather than another, is traditionally known as the *action selection* problem (hereafter referred to as the AS problem). Approaches to this problem, both in ethology and in autonomous agents research, generally assume the need for internal arbitration mechanisms to select the most appropriate action or behaviour according to internal and external conditions (Tinbergen, 1950), (Tyrell, 1993).

This paper argues that internal arbitration between behaviours is *not* the right way to view the AS problem. Instead, a scheme based on the coherence between continuous and concurrent sensorimotor processes (or links) is recommended. This scheme is inspired by (Braitenberg, 1984), and has already been shown to be viable by (Lambrinos and Scheier, 1995), see also (Scheier and Pfeifer, 1995), (Steels, 1994).

A novel methodology is adopted in this paper, in which artificial evolution is used to shape the activation function employed by the sensorimotor links. This methodology allows a minimal internal architecture to support a full range of AS phenomena, and avoids some potential problems associated with hand-design (discussed in Section 2.3).

Section 2 rehearses the theoretical background for this approach. This is based on two shortcomings common to many AS models. First, a confusion between behavioural and mechanistic levels of description, and second, a disregard for perception.

Section 3 describes the animat model. The architecture of this animat consists merely of a set of evolved, direct, sensorimotor links, without any interconnections. The animat exists in a continuous simulated world in which it must avoid 'traps', whilst periodically visiting 'food' and 'water' sources. Section 3 also presents the results from this model,which are discussed in Section 4, with special emphasis laid on how the AS problem and the 'selective attention' problem are related. Finally, Sections 5 and 6 describe further research in this direction, including robotic experiments, much of which is underway, and set out the conclusions to be drawn from the present work.

2. Theoretical Background

2.1 Behaviour

This paper considers a behaviour to be a joint product of agent, environment, and observer. For example, 'swimming' requires more than 'limb movement' and 'immersion in water'. It also requires an observer to classify these interactions as 'swimming', since a different observer may classify the same set of interactions in a different way ('underwater yoga', perhaps).

2.2 Behaviour and Mechanism in Ethology and Animats

Ethology is the study of animal behaviour *in situ*, and therefore one widely held ethological perspective accepts that behaviours are the product of continuous agent-environment interactions. This can be a very valuable perspective, but its value, in ethology, is tempered by the way in which much ethological theory generally assumes that entire behaviours have internal correlates. For example (Tinbergen, 1950), (see also (Lorenz, 1985)), proposed an early solution to the AS problem in terms of a hierarchy of 'behaviour centres', animated by the flow of some unspecified 'activation energy'. In each case, these 'behaviours', such as feeding, fighting, and so on, were clearly assumed to have direct internal correlates. However, behaviours are, as argued in the previous section, the products of environment-agent-observer interactions and so the agent-side (internal) component of behaviour cannot be identical to the behaviour itself[1].

Many recent models of AS employing animats employ the same premises, and concentrate on designing internal architectures for animats which arbitrate between internal correlates of entire behaviours. For example, (Tyrell, 1993) extends Tinbergen's fixed, winner-take-all hierarchy of behaviour centres to a 'free-flow' hierarchy (one that does not obey a strict winner-take-all rule), demonstrating an improvement in the efficacy of the mediation between behaviours of the animat concerned. However, the constituents of his hierarchy are still entire behaviours, such as 'avoid hazard' and 'find food'.

In a different example, (Maes, 1991) argues that 'behaviour selection emerges in a distributed fashion by parallel local interactions among the behaviours and between behaviours and environment'. Whilst Maes brings to light the importance of agent-environment interactions, she too works from the assumption that behaviours can be *parts* of agents, instead of agent-environment-observer interactions.

Behaviours, and any apparent selection between behaviours, should therefore *not* be explained purely at a behavioural level, but at a more mechanistic level.

2.3 Perceptual Deprivation

As (Maes, 1994) points out, AS models in general 'have a narrow-minded conception of the relationship between perception and action'. This situation persists despite a now general acceptance that action and perception should be considered jointly, as a part of an ongoing cycle of agent-environment interaction, (Ballard, 1991). What an agent is perceiving, or 'attending to', is likely to strongly influence what behaviour is appropriate. And similarly, the kind of behaviour currently being engaged in, will itself influence what features of the environment are worthy of perception and attention (the relationship between AS and selective attention is discussed further in Section 4).

Ethological study, taking place in natural environments, is not ideal for investigating the role of perception in behavioural control. Animat perception, on the other hand, is readily accessible. Nevertheless, most AS models do not extend to a satisfying incorporation of perception. In (Maes, 1991) perceptual variables are constrained to hold only predicate values. And (Tyrell, 1993) writes :

> It was decided that the small amount of added realism [from considering the perception of the animat] was not worth the large amount of extra time it would have taken to include[2].

This paper contends that the role of perception is essential in any explanation of AS that goes beyond assuming internal arbitration between behaviours. It contends that a mechanistic level of explanation should be grounded in terms of sensorimotor interactions with an environment, which are then observationally classified into behaviours, and into 'selection' between behaviours.

2.4 Action Selection through Sensorimotor Coherence

The position taken in this paper is as follows. The control of multiple behaviours is best understood in terms of the coherence of autonomous, and continuously active, sensorimotor links, or processes[3]. Action selection (and selective attention) are misleading interpretations of the general problem of coherence between these processes.

(Lambrinos and Scheier, 1995) propose a similar perspective, and describe an experiment in which a Khepera mobile robot is able to achieve a successful balance between collecting pegs, bringing the pegs to a 'home base', and periodically visiting a 'charging area'. This balance is achieved through coordination between several hand-coded, continuously active, sensorimotor processes. This work is a valuable step forward, and has the virtue of being demonstrable in the real world.

The present research differs in the following ways. First, the task environment, although not real-world, is sufficiently flexible to allow the investigation of performance according to a list of AS *desiderata* (fully described in Section 3). Second, the use of artificial evolution to design the sensorimotor processes allows the distinction between behaviour and mechanism to be rendered completely explicit. In (Lambrinos and Scheier, 1995), the reliance on hand-coding requires the artificial separation of the problems of process design (what each individual process should do) and process fusion (how to combine these processes to deliver coherent behaviour). If these problems are separated, then each sensorimotor process can potentially be considered in terms

[1] Of course, ethologists in the field have no direct access to the internal state of the organisms they study, and so, in the past, have had little alternative but to impute arbitrary behavioural primitives (see (McFarland and Sibly, 1975)). Nevertheless, ethological theories certainly have great predictive and comparative value, and have largely underpinned our understanding of behavioural *function*.

[2] Indeed, Tyrell's simulated environment is a discrete 'grid' world. This immediately constrains any possible role for perception to the rather trivial.

[3] The difference between a 'link' and a 'process', as used here, is that a 'process' is employed as a general term for a sensorimotor connection which may include intermediate stages, and a 'link' is specific to the present model, in which the connection consists of just one single transfer function.

of an eventual behaviour to be subsequently 'fused', and indeed their processes carry the names 'go-to-station', 'home' and so on. Although there is no explicit arbitration between these processes (they are all autonomous and concurrently active) there is still a certain tension between their affirmation that behavioural and mechanistic levels should not be confused, and this potential influence of behavioural descriptions on the design of the internal mechanism. The use of artificial evolution avoids this problem by allowing process design and process fusion to proceed concurrently, providing no scope for any influence of behavioural description over internal mechanism.

3. An Animat Model

The animat consists solely of a set of direct, autonomous, sensorimotor links, with no interconnections or 'artificial neurons'. Artificial evolution is used to specify the activation functions supported by the links. All of these links are active all the time.

The purpose of this model is to illustrate how such a minimal architecture can satisfy the requirements generally expected from an AS mechanism. To assess the quality of the model, we consider a list of such requirements, originally collated by (Tyrell, 1993), and expanded by (Werner, 1994) (more comprehensive versions of this list, which is admittedly somewhat arbitrary, can be found in either of these publications). Explicit tests for these properties are performed wherever possible.

An effective AS mechanism should:

- Prioritise behaviour according to current needs: for example, head for food if hungry, but don't fall into traps on the way.
- Allow contiguous action sequences to be strung together.
- Exhibit opportunism; for example by diverting to a nearby food source even if water is needed more.
- Balance dithering and persistence; for example, by drinking until full and then eating until full instead of oscillating between food and drink.
- Interrupt current behaviour; for example, if a trap suddenly appears, the animat should change course to avoid it.
- Deal with all subproblems; the AS mechanism should cope effectively in all situations.
- Prefer consummatory over appetitive actions.
- Use all available information (both internal and external).
- Support real-valued sensors and produce directly usable outputs.
- Be extensible and support learning.
- Allow 'parallel actions' to be executed (for example, walking and talking).

3.1 The Animat and Environment

The architecture of this animat was inspired by (Braitenberg, 1984), and is illustrated in Fig 1. It possesses two wheels, and three types of sensor, with one of each sensor type each side (giving a total of six sensors). There are three links from each sensor to the wheel on the same side; thus a total of nine links on each side.

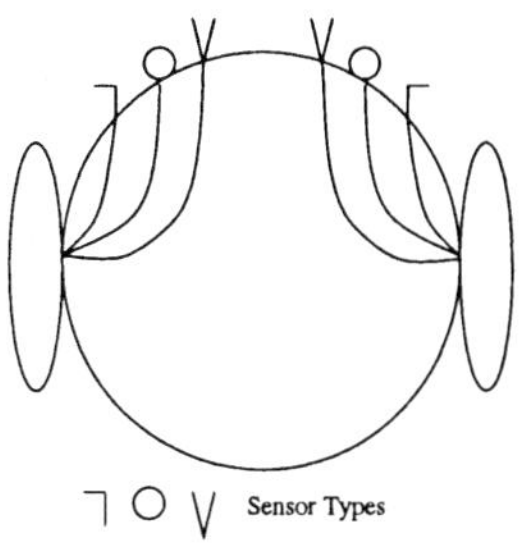

Figure 1 *Architecture of the animat: there are three sensor types, with each sensor connected directly to the wheel on the same side. Each illustrated connection actually comprises three genetically specified links, so the animat consists of 18 concurrent and direct transformations of sensory input into motor output.*

The animat also has two internal batteries which it must maintain at as high levels as possible. These levels diminish at a steady rate as long as the animat exists, and if both reach zero the animat 'dies'.

The links between the sensors and the motors simply transform the sensor input signal (ranged from 0-100) into a given output signal (ranged from -1 to 1) according to a transfer function (which can also be modified by battery level; see Section 3.2). The outputs from the links are combined at the wheels; the first 9 link outputs are summed, passed through a sigmoid function, and then scaled from -10 to 10 to set the left wheel speed. The links numbered from 9 to 18 are treated in a similar way to determine the right wheel speed.

The simulated environment is a continuous (in space, but not in time) unbounded area within which three types of object can exist, in addition to the animat. Each object is a circle, 16 units in radius (the animat is 5 units in radius, and all objects appear within a 200 by 200 unit area of the environment). The first object type (let us call this 'food') replenishes the animat's first battery level; the second type ('water') replenishes the second battery; and the third type ('trap') is dangerous, and will 'kill' the animat, if encountered. There are three instances each of food and water, and nine traps.

The three types of sensor respond to the distance from the nearest object of each type to the animat, with each sensor linearly ranging from 100 (at the source), to 0 (200 units or more distant). If an object is to the left of the animat, the relevant left sensor will respond with 20 percent greater activation (and *vice versa* if the object is to the right).

At the start of each trial, the objects and the animat are placed randomly within the environment (within a 200 by 200 unit range). The movement of the animat is then calculated on the basis of the changing wheel speeds (if both wheel speeds are set to +10, then the animat moves forward at a maximum speed of 2.8 units per time step). Each battery has a maximum (and intial) level of 200, which decreases by 1 each time step. If the animat encounters a food or water object, the appropriate battery level is restored to 200, and the object is replaced at a different random location. An encounter with a trap ends the trial.

3.2 Genetic Encoding Scheme

Each genotype consists of 83 integers in the range 0-99; thus 9 integers for each of the 9 links, and 1 each for the left wheel and right wheel sigmoid threshold values. Only 9 links need to be genetically specified since left/right symmetry is enforced. Figure 2 illustrates how the 9 integers for each link specify the shape of the activation function. The offset and thresholds are set by scaling[4] the 0-99 integer onto the range -100 to +100, with the restriction that the second threshold must follow the first. The gradients are set by scaling the integer to range from $-\pi/2$ to $\pi/2$, and then taking the tangent. The sigmoid thresholds are set by scaling on to a range from -3.0 to 3.0.

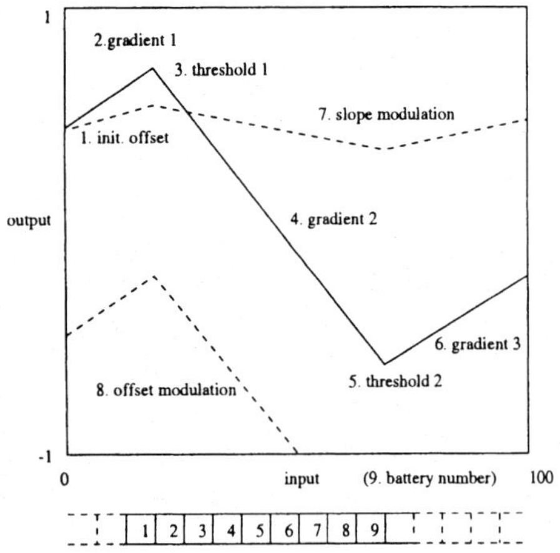

Figure 2 *Genetic encoding scheme: each sensorimotor link requires 9 integers to specify the various link parameters. The first 6 integers specify the basic shape of the activation function transforming the sensor input into an output signal, and the final 3 integers specify how this shape can be modified by a battery level. Each link can only be modified by a single battery (specified by integer 9), and the level of this battery can then influence both the overall gradient of the function (to a degree specified by integer 7), and the intial offset (to a degree specified by integer 8).*

The influence of the battery is slightly more complicated. If the ninth integer is even, then the function can be influenced by battery 1; if odd, then by battery 2. The relevant battery level ($\mathcal{B} \in (0:200)$) can affect two aspects of a function. First, through offset modulation ($\mathcal{O} \in (-1:1)$)):

$$out = out + ((\mathcal{B}/2) * \mathcal{O})$$

and second, through slope modulation ($\mathcal{S} \in (0:1)$) of the function:

$$out = out + (out * (((\mathcal{B} - 100)/100) * \mathcal{S}))$$

Note that there is no requirement that any given function, connecting a particular sensor type to a wheel, should be influenced by the battery corresponding to the object type that its sensor responds to.

3.3 The Genetic Algorithm

A distributed GA was employed, with a crossover rate of 0.5, and a per-bit mutation rate of 0.04. The population size was 100, and the fitness function simply rewarded high average battery level, calculated incrementally at each time step:

$$\mathcal{F} = (\mathcal{B}1 + \mathcal{B}2)/400.0$$

This function rewards animats that live long (by keeping at least one battery level above zero and avoiding traps), and that visit food and water sources as often as possible, up to a maximum lifespan of 800 time steps. It does *not* specify anything about what the animat should do in any particular situation.

Fit individuals evolved in about 200 generations (taking about 2 hours on a single user 143MHz Sun SparcStation). The results described in the next section are, however, drawn from the fittest individual of the 430th generation.

3.4 Results

The evolved animats displayed very effective AS behaviour. A sample trajectory is shown in Fig 3. The animat begins by navigating effectively through a series of food and water sources. At point 1, the threat of nearby traps is dealt with by backtracking and turning right towards some other food and water sources. At point 2, the animat displays opportunistic behaviour by visiting another nearby food source even though it had only just left a previous one. The animat then passes through another series of food and water sources before reaching its maximum lifetime of 800 timesteps.

Recalling the list of AS *desiderata*, the above example demonstrates the 'effective prioritisation of behaviour according to current needs', by generally heading towards the most needed type of source, without falling into any

[4] All scaling is linear.

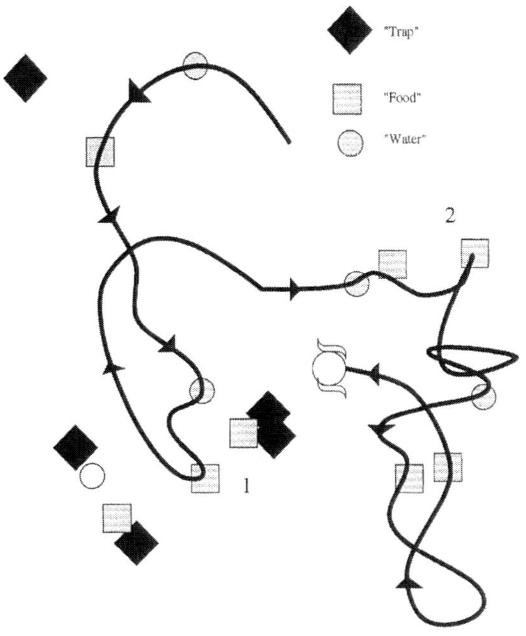

Figure 3 *A sample trajectory; animat passes through a series of food and water sources, backtracking to avoid traps at point 1, and displaying opportunistic behaviour at point 2.*

traps. The performance of 'contiguous action sequences' is illustrated by the way the animat moves smoothly from visiting one source to visiting another, and 'opportunistic behaviour' is also illustrated at point 2.

The following tests illustrate, more formally, that the evolved animat performs very well with respect to the requirements for effective AS:

A balance between dithering and persistence. In general, the animat would briefly slow down if equidistant between equally needed sources, before unequivocally plunging one way or the other. Fig 4 is a screen shot from a contrived situation in which the animat was placed equidistant from a cluster of 3 food sources (on the left), and a cluster of three water sources (on the right). Both battery levels were set at 100. The animat goes for the food first, and collects all three before returning for the water. The animat is therefore *persisting* at each task for an appropriate duration, rather than *dithering* between food and water collection.

Interrupts current behaviour if necessary. The animat was positioned at (0.0,0.0), with a food source at (100.0,0.0). Both battery levels were set at 150. As the animat approached the food source, and passed (30.0,0.0), a trap was suddenly introduced at (65.0,0.0). Fig 6a shows the trajectory of the animat as it changes course to successfully avoid the trap whilst still reaching the food. Fig 6b show the simple straight line trajectory traced by the animat if no trap is introduced, and Figs 6c and 6d illustrate the trajectory immediately before the trap appears, and immediately afterwards. The animat is clearly interrupting its direct navigation towards the food in order to avoid the trap.

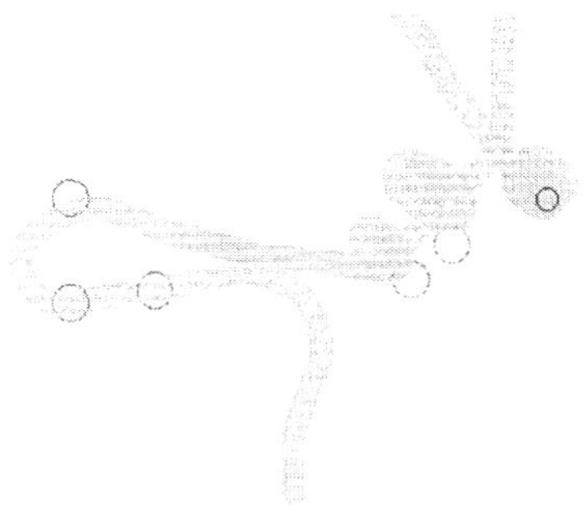

Figure 4 *A balance between persistence and dithering; animat collects all 3 food sources (on left) before collecting all three water sources (on right). The animat does not 'dither' between the food and water sources.*

Furthermore, the degree of this interruption is dependent on the battery level. Twenty trials were performed at each of 10 starting battery levels, between 100 and 200. Fig 5 illustrates that with low battery levels, the animat could not avoid the trap, but as the battery level increased, the animat avoided the trap more often.[5] This phenomenon has been termed 'varying attention' (see (Maes, 1991) for an animat example); when needing food, the animat appears to 'pay less attention' to traps.

Opportunism. In addition to the instance of opportunism shown in Fig 3, a further controlled experiment was performed. The animat was positioned at (-100.0,0.0), with a food source at (-90.0,0.0), and a drink source at (30.0,0.0). An additional food source was placed with an x position of -20.0, and a y position varying from 0.0 to 30.0.

After collecting the first food source, the animat is more in need of water. However, should another food source be *en route*, the animat will opportunistically visit it (Fig 6e). If, however, the second food source is too far out of the way, it will be ignored (Fig 6f).

Twenty trials were performed at each of 8 different y positions of the second food source, and Fig 5 illustrates that as this source becomes increasingly out of the way, the animat diverts to it less frequently.

Prioritisation. The sample trajectory (Fig 3) illustrates that, in general, the animat will always head for the most needed source type. Figs 6g and 6h illustrates this more formally, with the animat starting between two sources, facing away from both. The animat has one battery level set to 150, and the other set to 100. If the sources are equidistant (Fig 6g), the animat will go for the most needed (the source to the left). But if that source is

[5] This test was especially difficult for the animat, there being only one food source in a direct line with a trap. In normal situations (see Fig 3), encounters with traps occurred rarely, if at all.

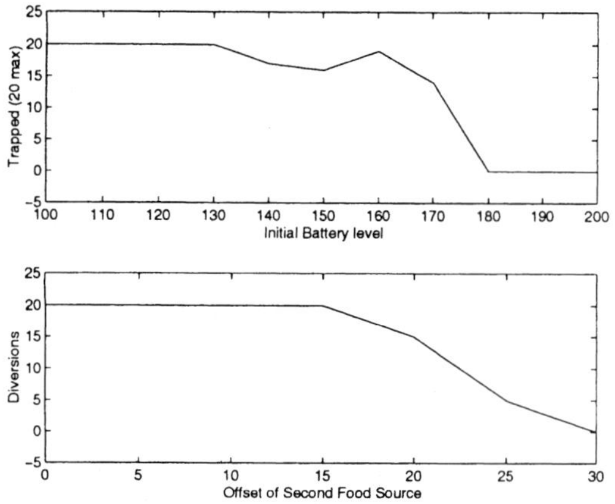

Figure 5 *Varying attention (top); the animat is more likely to fall into traps if it has a lower battery level. Opportunism (bottom); the animat will divert to collect a second food source en route to a water source, but does so less often as this second food source is placed increasingly out of the way.*

moved too far away (Fig 6h), then the animat will visit the source on the right instead. Thus the animat is striking a balance between prioritisation and opportunism.

Other Requirements. The remaining requirements for effective AS were not all applicable to this model, but are briefly discussed below:

- *Deals with all subproblems:* no particular subproblems arose other than evading traps, which the animat did very well.
- *Prefers consummatory over appetitive actions:* no distinction was made between the two in this model.
- *Uses all available information:* this is indeed the case, since all sensorimotor links are always influencing the motor output to some extent.
- *Has real valued sensors and produces directly usable outputs:* the animat consist of sensorimotor links and nothing else, so this is indeed true.
- *Allows parallel actions:* 'parallel actions' are undefined in this model.
- *Is easily extensible (perhaps through learning):* there is no reason why more sensors and effector types could not be added (perhaps to support 'parallel actions'), and 'learning' links could certainly be introduced.

4. Discussion

This model has illustrated that nothing more is required (under the conditions of this model) for effective AS behaviour than a set of independent sensorimotor links, and the influence of some internal state. In this model, a joint consideration of perception and action is paramount, and there is no arbitration between internally represented behaviours.

Fig 7 illustrates the shapes of the evolved activation functions for the links. All the links are influenced by one or other of the battery levels, even the links from the 'trap' sensors to the wheels. It is interesting to note that each set of three links is influenced by *both* batteries. That is, the 'food' sensorimotor links are not only influenced by the need for food, but also by the need for water. And the same applies for the 'water' sensorimotor links. This suggests the absence of any simple mapping from the behaviours 'get food', 'get water', and 'avoid trap' onto the sets of links. Indeed, there would be no reason to expect this to be the case. Thanks to the artificial evolution methodology, no distinction is made between 'link design' and 'link fusion', and so the internal mechanism need bear no relation to what may seem to be a sensible design strategy from an observer's point of view.

This work therefore advocates the position that action selection is not only not dependent upon internal arbitration, but also that it is not dependent on any kind of internalisation of behaviour at all. This has clear implications both for ethology and for animat research: theories of 'action selection' that assume the internalisation of behaviours may not be necessary, (and, as argued in Section 2, do suffer from some conceptual problems). Alternative perspectives that uphold the distinction between behaviour and mechanism, and that respect the inseparability of perception and action, are viable, and should be preferred. The relationship between the coherence of sensorimotor processes, and the observation of behaviours and behavioural 'selection' is one that invites further exploration in animat research.

It is also of interest that the overall behaviour of the animat could just as easily be described in terms of selective attention as in terms of AS. The animat appears to 'pay attention' to different features of the environment at different times, and indeed the term 'varying attention' given to the phenomenon recorded in Section 3.4 is perhaps a subtle recognition that the two problems are not different. Of course, in the present case the two problems are necessarily identical, since the presence of just a set of direct sensorimotor links precludes any distinctions being made between perception and action. All this is noteworthy since a substantial amount of work in cognitive science still considers selective attention in terms of a 'spotlight' designed to mediate between an external reality and an internal world model, with little consideration of the role of action (see, for example, (Treisman and Gelade, 1980))[6].

[6] This argument is in no way meant to invalidate this corpus of work. There is clearly an enormous gulf between the sensory apparatus of the present animat and the primate visual cortex, and psychology and neurology continue to contribute greatly to our understanding of the mechanisms involved in primate perception. All that is suggested here is that investigations into selective attention, in whatever

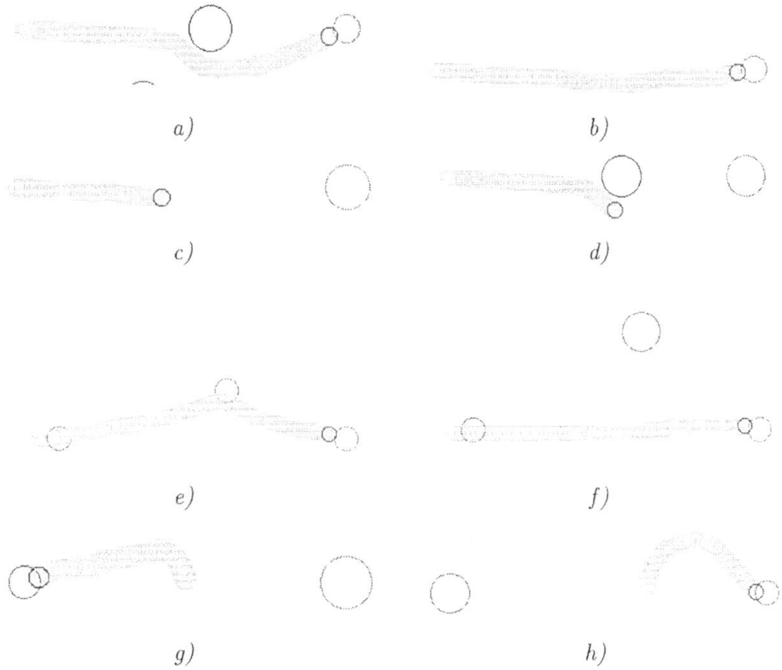

Figure 6 *Action selection tests; consumed sources appear as smaller circles; a) a trap suddenly appears, and the animat interrupts its progress towards the food in order to avoid it. b) with no trap, the animat progresses directly to the food source. c) trajectory of the animat at the point just before the trap appears... d)..and just after the trap appears. e) animat passes through a first food source, and will divert to a second food source (en route to water) if the diversion is not too far. f) if the second food source is too far away, the animat will proceed from the first food source to the water source. g) animat starts between two equidistant sources, and will head towards one needed most. h) if the 'most needed source' is too far away, then the animat will proceed to the other source.*

5. Further Work

A current project is to extend this methodology to real robots in real situations, with sensors that can no longer respond directly to the distance from a given object type. Artificial evolution will still be used to design the internal mechanism of the robots, since as the robots and the task demands become increasingly complex, the actual nature of appropriate sensorimotor processes is likely to be of considerable interest, as a way of understanding more about the relationship between behaviour and mechanism. If the sensorimotor processes are hand-designed, then this is harder to study.

6. Conclusions

This work has shown that a full range of action selection phenomena can be elicited from a minimal internal architecture, with neither explicit arbitration between internally represented behaviours, nor the hand-design of sensorimotor processes. The model is designed to make this theoretical point, and is not touted as the 'best' way of controlling behaviour in autonomous agents, even though it performs more than adequately in its context.

The novel methodology, of evolving the shape of each activation function, allows the design of each sensorimotor link to proceed concurrently with the fusion of all the processes into a coherent global behaviour. This avoids the potential influences of behavioural descriptions on internal mechanism that may arise in hand-designed architectures, where process design and process fusion are considered separately. Therefore it also allows the relationship between behaviour and mechanism to be the object of study.

Action selection behaviour is shown to emerge from coherence between autonomous, direct, and continuously active sensorimotor links. It is argued that just as the term 'action selection' is misleading through the disregard of perception, so the term 'selective attention' is misleading through the neglect of action.

domain, could benefit from a more detailed consideration of the role of action.

Acknowledgements

Thanks to Matt Quinn, Tom Smith, Andy Philippides, Emmet Spier, Phil Husbands, Hilary Buxton, and my anonymous reviewers. Financial support was provided by the EPSRC award no. 96308700.

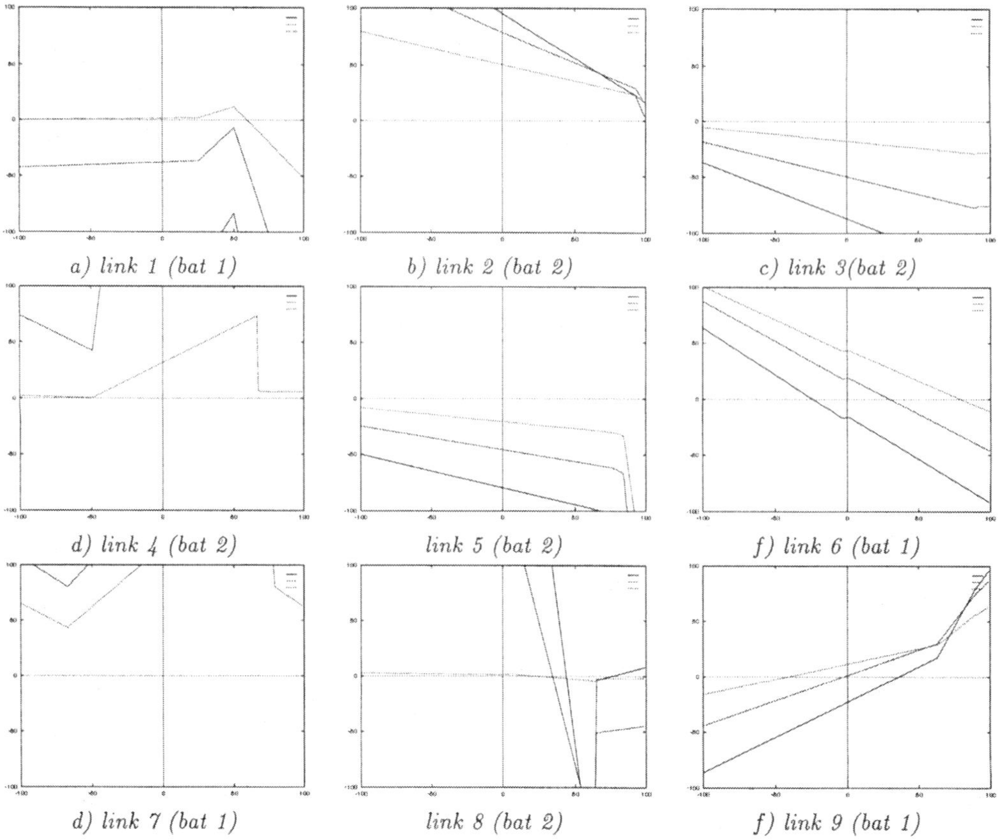

Figure 7 *Evolved activation functions; links 1-3 connect the 'food' sensors to the wheel, links 4-6 the 'water' sensors, and links 7-9 the 'trap' sensors. All axes are scaled from -100 to 100 (input along the x axis, and output along the y axis). The three lines on each graph represent the activation function at battery levels of 200,100 and 0, and the battery which influences the function is shown in parentheses. All these links are active all the time. Note that each set of three links is influenced by BOTH batteries, and contains a mix of excitatory and inhibitory outputs.*

References

Ballard, D. (1991). Animate vision. *Artificial Intelligence*, 48:57–86.

Braitenberg, V. (1984). *Vehicles: experiments in synthetic psychology.* MIT Press.

Lambrinos, D. and Scheier, C. (1995). Extended braitenberg architectures. Technical Report AI Lab no. 95.10, Computer Science Dept., University of Zurich.

Lorenz, K. (1985). *Foundations of ethology.* Springer-Verlag.

Maes, P. (1991). A bottom-up mechanism for behaviour-selection in an artificial creature. In Meyer, J. and Wilson, S., editors, *Proceedings of the first international conference on simulation of adaptive behaviour.*

Maes, P. (1994). Modelling adaptive autonomous agents. *Artificial Life*, 1:135–162.

McFarland, D. and Sibly, R. (1975). The behavioural final common path. *Philosophical Transactions of the Royal Society of London: B. Biological Sciences*, 270(907):265–293.

Scheier, C. and Pfeifer, R. (1995). Classification as sensory-motor coordination. In *Advance in Artificial Life: Proc. 3rd european conference on artificial life*, pages 656–667.

Steels, L. (1994). A case study in the behaviour-oriented design of autonomous agents. In Cliff, D., Husbands, P., Meyer, J., and Wilson, S., editors, *Proceedings of the 3rd International Conference on Simulation of Adaptive Behaviour*, pages 445–452. MIT Press.

Tinbergen, N. (1950). The hierarchical organisation of nervous mechanisms underlying instinctive behaviour. *Symposium for the society for experimental biology*, 4(305-312).

Treisman, A. and Gelade, G. (1980). A feature-integration theory of attention. *Cognitive Psychology*, 12:97–136.

Tyrell, T. (1993). *Computational mechanisms for action selection.* PhD thesis, University of Edinburgh.

Werner, G. (1994). Using second order neural connections for motivation of behavioural choice. In Cliff, D., Husbands, P., Meyer, J., and Wilson, S., editors, *Proceedings of the 3rd International Conference on Simulation of Adaptive Behaviour*, pages 154–164. MIT Press.

Dynamical Systems for the Behavioral Organization of an Anthropomorphic Mobile Robot

Axel Steinhage and Thomas Bergener
Institut für Neuroinformatik, Ruhr-Universität Bochum
44780 Bochum, Germany
Tel: +49 234 700 7969
Fax: +49 234 709 4209
Axel.Steinhage@neuroinformatik.ruhr-uni-bochum.de
Thomas.Bergener@neuroinformatik.ruhr-uni-bochum.de

Abstract

We present a generative approach to behavioral organization which uses continuous dynamical systems for switching between multiple actions of an anthropomorphic robot. The logical context of the possible actions is coded in matrices of parameters for a system of continuous differential equations. The sensor context is represented by a set of variable parameters depending on the sensor inputs. The switching between the behaviors is the result of nonlinear phase transitions in the solution of the underlying dynamical system. The stability of the overall system is guaranteed even for many different behaviors by keeping coupled behaviors on separated timescales. This is demonstrated on the anthropomorphic robot ARNOLD for the example of approaching a door. The robot searches for the door visually, recognizes it, points at it, and approaches it. The dynamical systems approach allows for a flexible change of a running behavioral sequence after unexpected perturbations brought about by dynamic changes in the environment.

1. Introduction

Autonomy basically means to generate behavior which is appropriate for the situation the system is currently in. Behavioral stability in this context means that from a set of possible behaviors that one is selected which is most adequate for the current situation and that a sufficient dynamic change of the environment leads to a categorical change of the selected behavior. In general, these environmental changes lead to changing sensor information and therefore the action selection mechanism must be controlled by the continuous sensor information. For example the target acquisition behavior requires that the target is detected by the corresponding sensors and if changes in the scenery cause an occlusion of the target, the target acquisition behavior must be replaced by a behavior which tries to find the target again. Sometimes there are situations in which not only the sensor information alone is sufficient to decide whether a certain behavior must be activated or not. This is the case in situations, in which not only a single behavior is needed to fulfill a task, but multiple parallel behaviors or even a whole sequence of behaviors must be acted out in a logical order. In addition to sensor information, the activation of a certain behavior must be controlled by a mechanism which selects the correct behavior in the sequence depending on the preceding behaviors in that case. This mechanism must also decide which behaviors must run in parallel and which are mutually exclusive.

In this paper we present such a mechanism. It is based on the *dynamical systems approach to behavior generation* (see (Schöner et al., 1995) for a review, (Steinhage, 1998) for an extended application). In this approach, behaviors are generated by evolving the solution of a nonlinear dynamical system in time. The behavioral state of the system is projected onto so called *behavioral variables* which must be selected according to some basic design principles. One of these principles is that the tasks to be fulfilled must be expressible as stable fixed points or *attractors* of the dynamical system and the time scale of the dynamics must be selected such that the behavioral variables are in these attractors at all points in time. While the feasibility of this approach for generating single sensor driven behaviors has been demonstrated in many applications before (e.g. (Schöner and Dose, 1992)(Bicho and Schöner, 1997)), we want to show how the concept can be scaled up to the generation of behavioral sequences and to action selection. As an example case we apply the approach to the problem of detecting, identifying and approaching a door, implemented on an anthropomorphic mobile robot.

2. The anthropomorphic robot ARNOLD

Arnold is an autonomous anthropomorphic robot for human environments (Fig. 1) (see (Bergener et al., 1997) for detailed description of the robot and its behavioral capabilities). Equipped with a mobile platform, a three degree of freedom stereo-vision head and a seven DoF robot arm Arnold is able to perform simple service tasks in human indoor environments. Grasping objects and navigating, as well as pressing light switches or door handles are examples for basic tasks that can be combined to perform complex jobs.

The robot's only sensor is a double stereo camera system that is mounted on a head with pan, tilt and vergence. Two pairs of CCD-cameras are used for foveal color vision and monochrome periphery sensing. The arm is anthropomorphic in its movability with wrist, elbow and shoulder and lifts weights up to one kilogram in a distance of 60 centimeter in front of the body. Two industrial Pentium-boards running the real-time operating system QNX control the entire robot. Rechargeable battery packs make robot autarkic for more than two hours.

Arnold's control architecture is based on the works of Brooks (Brooks, 1986) and Braitenberg (Braitenberg, 1984). A network of distributed control processes establishes paths of subsymbolic data flow from the sensors to the different actor devices. Most of the involved behavioral modules are designed or coupled following the *dynamic approach* as described in (Schöner et al., 1995) (see (Bergener and Dahm, 1997) for a sample application).

Figure 1 Robot Arnold

3. The Behaviors

To realize a reactive sequence to approach a certain door a set of behavioral modules and the necessary sensor capabilities had to be developed.

Since we did not want to realize a global navigation module that includes the target's position the target door must be recognized from image features. This means we have to detect the target using two basic perceptual capabilities:

- Find any door in the visual field and estimate its position relative to the robot.
- Classify a door in the visual field as 'target' or 'some other door'.

The first is solved using the spatial geometry of doors, i.e. a pair of vertical door posts with a certain distance. Therefore we acquire stereo image pairs with a horizontal view direction and search for nearly vertical lines in the images using an Adaptive Hough Transformation. A stereo matching of these lines in each image pair gives each door post's position relative to the robot. Pairs of vertical structures that have a distance of 80 ± 10 centimeters are candidates for the further check. To check if the detected door is the target we use an image classification module as described in (Kreutz et al., 1996). Autocorrelation features are extracted from a Laplace image pyramid and coded in a feature vector. A linear Bayes classificator is trained with feature vectors calculated from image sets of the target and other doors in the robots environment. The images are acquired from different view directions and distances to make the classifiction invariant with respect to these parameters.

3.1 Visual search and target tracking

The visual search of the target door and tracking the target are two basic behaviors (B_1 and B_2) for the choosen task. The first behavior B_1 means that the robot's head is rotated in steps of 20 degree counterclockwise. Stereo images are acquired at every position and sent to the door-detection and classification modules.

The targeting behavior B_2 directs the head towards the nearest door in the visual field und acquires new pairs of stereo images repetitively. By this the head will fixate the target even when the robot moves.

3.2 Pointing

Since the robot has no display or speech device we want it to point at a detected door before it proceeds. Hence we define the behaviors *pointing* B_3 (the arm moves up and points at the door), *parking* B_4(the arm moves to the parking position close to the body), and *safe* B_5 (the arm is locked in the parking position).

3.3 Target acquisition

Target acquisition should be the final behavior in a successful sequence of actions. The robot is moved towards the target and stopped in front of the door (behavior *local navigation* B_6).

3.4 Door recognition

The image classification module was trained with images of several doors in our lab - each door representing one class. Since we want to switch the classification module on or off whenever it is needed, we denote this capability as behavior B_7. To interact with other behaviors we formulate the recognition of the target door as behavior B_8. For the case that a scene is classified as some other door we define the behavior B_9. A behavior B_{10} denotes the internal state that the classificator is switched off or the classification result is not the target door.

4. Integrating the behaviors

To generate appropriate overall behavior, the described sub-behaviors must be organized in a sequence such that the required task can be fulfilled (Steinhage and Schöner, 1997a). While the active sub-behaviors are controlled by sensor inputs, the behavioral sequence depends additionally on logical conditions. Approaching the door for instance makes only sense if the door has already been detected by the visual search. The classical approach to incorporate logical dependencies between a number of instances is to assign a symbol to each of the instances and to design a symbolic algorithm which switches the instances on and off (Poole, 1995). The logical dependencies are encoded in the program structure of the symbolic algorithm. The most prominent example of a discrete symbolic algorithm controlling sub-behaviors consisting of continuous dynamical systems are the so called HYBRID SYSTEMS (Lemmon et al., 1993)(Brockett, 1993). For our purpose, this approach has some disadvantages (see (Steinhage, 1998) for a criticism): Scaling up the system to additional sub-behaviors requires a complete reprogramming of the symbolic algorithm as the logical interaction with the existing parts of the program structure must be captured. More importantly the sub-behaviors are controlled by continuous dynamical systems their organization through a discrete symbolic algorithm induces stability problems. As we have already explained, stable behavior is generated if the control variables of the sub-behaviors are in stable states at all points in time (Steinhage and Schöner, 1997b). These stable states are attractor states of the underlying continuous differential equation. In a symbolic algorithm the sub-behaviors are represented by symbols. The behavioral sequences are generated through switching the sub-behaviors on and off by manipulating the symbols in the algorithm. Unless the algorithm simulates a continuous dynamical system, the symbols are switched instantanously between their on and off states. As behavioral organisation implies that the sub-behaviors are activated depending on the activation states of the other sub-behaviors, the instantanous switching means a discontinuity for the dynamical system which controls the activation of the sub-behaviors. For the differential equation this may cause instabilities.

To avoid these complications, we developed a mathematical concept which organizes the behaviors through a sub-symbolic continous dynamical system. If the sub-behaviors are likewise described by continuous dynamical systems, the overall system which integrates and organizes all sub-behaviors can be entirely expressed by a system of continuous differential equations. This avoids the difficulties brought about by the combination of continuous dynamical systems and discrete symbolic algorithms.

4.1 The competitive dynamics

To make the problem of behavioral organization accessible for continuous dynamical systems, we assign a continuous variable n_i to every sub-behavior. This variable describes the state of activation of the corresponding behavior B_i: $n_i = 0$ means the corresponding behavior is not active, while $n_i = 1$ stands for an activated behavior B_i. A continuous dynamical system which controls this variable n_i therefore has to have two stable states: one for a deactivation and one for activation of behavior B_i. A possible choice is the following differential equation:

$$\tau \dot{n}_i = \alpha_i n_i - |\alpha_i| n_i^3 - \sum_{j \neq i} \gamma_{i,j} n_j^2 n_i + \xi_t \qquad (1)$$

Here, ξ_t is a gaussian white noise term, preventing the system from remaining in unstable states. To understand the characteristics of (1), we neglect the competitive term $\sum \gamma_{i,j} n_j^2 n_i$ at first. Depending on the so called *competitive advantage* α_i, the dynamical system relaxes into the corresponding stable fixed point on the timescale τ: for $\alpha_i > 0$, the dynamics relaxes to $|n_i| = 1$ and for $\alpha < 0$ the system relaxes to $n_i = 0$. Through this mechanism, the sub-behaviors can be switched on and off.

4.2 The competition matrix

Some of the described sub-behaviors can not run simultaneously. For example, the behaviors *tracking* and *visual search* are mutually exclusive: *tracking* should only be activated after the *visual search* is finished. Other behaviors can run simultaneously, like for instance *tracking* and *recognition.* To incorporate these logical conditions in our concept a *competition matrix* $\gamma_{i,j} \in \{0, 1\}$ is defined. If $\gamma_{i,j} = 1$ and $|n_j| = 1$, the state n_i remains in the stable fixed point $n_i = 0$ even for an $\alpha_i > 0$. This means, an active behavior B_j inhibits the activation of behavior B_i. Only when $\alpha_j < 0$ switches off behavior B_j, behavior B_i is free to be activated. If $\gamma_{i,j} = 0$, behavior B_i can be activated independently of behavior B_j.

4.3 The competitive advantage

The competitive advantage α has the following form:

$$\alpha_i = 2\rho_i \Psi_i - 1 \qquad (2)$$

$$\dot{\rho}_i = \frac{(1-\rho_i)\Psi_i}{\tau_{\rho,2}} + \frac{\rho_i(\Psi_i - 1)}{\tau_{\rho,1}} \qquad (3)$$

$$\tau_{\rho,1} << \tau_{\rho,2} \qquad (4)$$

Here, the variable $\Psi_i \in [0, 1]$ is the *input function* and ρ_i is the state of a *refractory dynamics.* The characteristics of these are the following: While (3) is in the stable fixed point $\rho_i = 1$, an input $\Psi_i \simeq 1$ leads to a positive competitive advantage $\alpha_i \simeq 1$ and behavior B_i can be switched on. If the input is switched off $\Psi_i \to 0$, the refractory dynamics relaxes to the fixed point $\rho_i = 0$ on the fast time scale $\tau_{\rho,1}$. A re-activation of behavior B_i through a $\Psi_i \to 1$ is now prevented for the time $\tau_{\rho,2}$ which it takes for the refractory dynamics (3) to relax back to the fixed point $\rho_i = 1$. By this, additional behavioral stability is generated for the behavioral organization: a behavior can be switched off instantaneously on a fast time scale but behavioral oscillations are prevented. Such oscillations occur, for instance, if a noisy sensor input is transmitted directly into the competitive advantage without filtering it with the low-pass activation behavior of the refractory dynamics (3).

4.4 The input function

The input function Ψ_i consists of two parts: the *sensor context* $C_i \in [0,1]$ and the *activation matrix* $A_{i,j} \in \{1,0\}$:

$$\Psi_i = C_i \prod_j \left(1 + A_{i,j}\frac{\tanh(c\,m_j) - 1}{2}\right), \quad c >> 1 \quad (5)$$

$$\dot{m}_i = \frac{(1-m_i)n_i^2}{\tau_{m,1}} + \frac{(1+m_i)(n_i^2-1)}{\tau_{m,2}} \quad (6)$$

$$\tau_{m,1} << \tau_{m,2} \quad (7)$$

The sensor context is a function which defines the necessary sensor condition for an activation of behavior B_i. The behavior *tracking*, for instance, requires that the cameras deliver the image of a door. Therefore the door-recognition algorithm sets the sensor context C_{tracking} corresponding to the confidence with which a door is recognized. If a behavior is not dependent on any sensor input at all, its sensor context is set to $C_i = 1$. While C_i characterizes the necessary *sensorial* context, the activation matrix $A_{i,j}$ defines the necessary *logical* context for the activation of behavior B_i. An element $A_{i,j} = 1$ means, behavior B_j must be active to allow for an activation of behavior B_i. For instance the *local navigation* to approach the door will only be activated if the arm is safe and the door is tracked and recognized. Through the multiplication in (5), the input Ψ_i will only be set, if both, the sensor context and all required sub-behaviors are active. The so called *memory dynamics* (6) defines a short term memory for the activation state n_i: if $|n_i| = 1$, the memory state m_i relaxes to the stable fixed point $m_i = 1$ on the fast time scale $\tau_{m,1}$ following the activation of behavior i immediately. If $n_i \to 0$, however, the memory m_i conserves the active state $m_i = 1$ for a time $\tau_{m,2}$ before it relaxes to the fixed point $m_i = -1$ which stands for a deactivated behavior B_i. This memory dynamics has two effects:

(1) The switch between two behaviors is made more stable. If behavior B_i requires that behavior B_j was active before but must not be active simultaneously with behavior B_i now, the behavioral state n_j may already be switched off while its memory image m_j is still active. Therefore the required condition of an active behavior B_j is still fulfilled in the input function (5) while in the competitive dynamics (1) no inhibition of behavior B_i through behavior B_j occurs. In practice this may be the case for the behavior *arm safe*: the behavior *local navigation* should only start after the arm has been parked but the arm should not be parked while the local navigation is active.

(2) As $A_{i,j}$ and also $\gamma_{i,j}$ are in general not symmetric, a causal direction is invented. Evolving the competitive dynamics (1) in time leads to a sequence of activation and deactivation of the sub-behaviors. The logics of this sequence is coded in the matrices $A_{i,j}$ and $\gamma_{i,j}$. Although these matrices are fixed, the resulting sequences depend on the sensor context $C_{i,j}$ and are therefore a function of the changing environment the robot moves in. A sequence may even be interrupted by an unexpected event, like the collision of the robot with a moving object. In this case, the memory dynamics (6) allows for a fast switch off of the currently active behavior and an alternative sequence can be started. Therefore the presented scheme allows for a reactive change in the generated behavioral sequences.

An important aspect makes this approach particularly useful for practical implementations: the whole architecture has a *local* character as only pairwise logical interactions between the sub-behaviors must be taken into account. This implies, that the integration of a new behavior does not require a complete reconfiguration of the dynamics: the existing structure, encoded in the matrices $\gamma_{i,j}$, $A_{i,j}$ and C_i, remains fixed while being extended by one row or column only. This additional row/column describes the interaction of the new behavior with the already existing ones.

The transformation of the activation state n_i into commands for the robot effectors is handled by a specific control architecture.

4.5 The PLANET*-architecture*

The implementation of the framework described above is based on PLANET (PLAtform for NETwork communication), our software for distributed robot control (Alberts Et Al, 1997). PLANET provides a way to concurrently run behavioral modules in a computer network connected by transparent communication channels. A configuration file defines the modules involved and the process interconnection scheme for an application. For this work one module represents the arbitrator that evaluates the differential equations as given in chapter 4.. The current state of the n-vector is send to all behavioral modules repetitively. In the same way the behaviors send their own *behavioral context* variable to the arbitrator module. The modules interpret their own activity n_i by switching their output connections on or off.

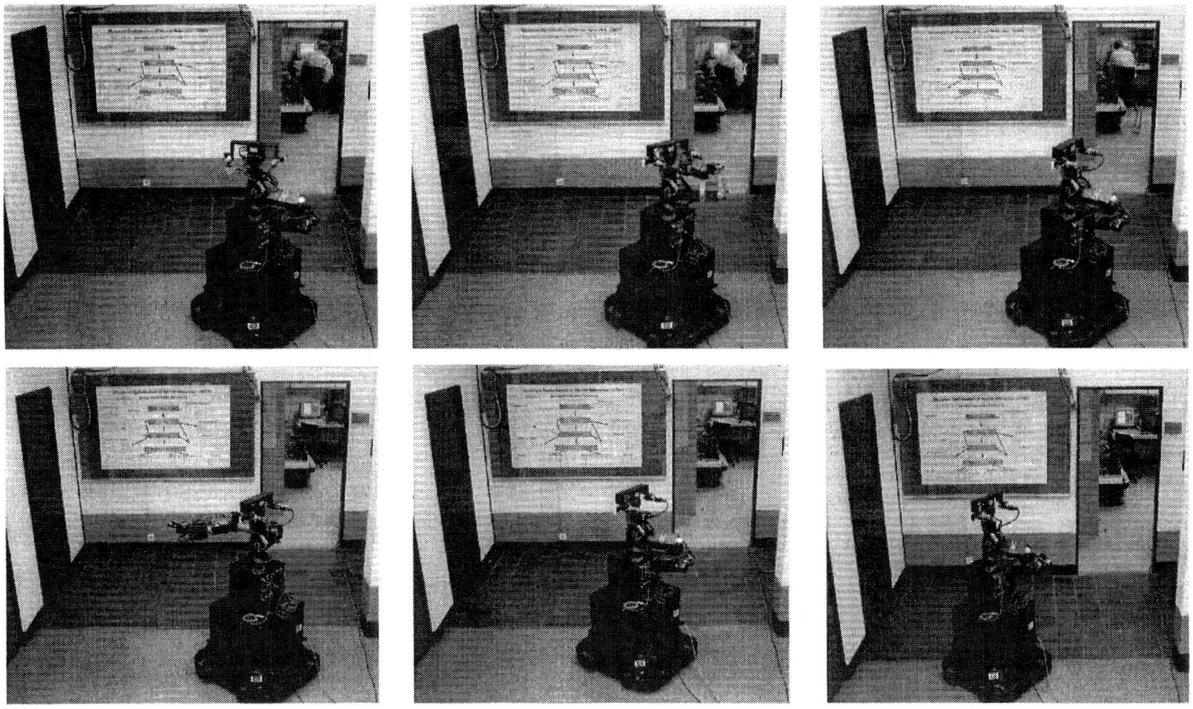

Figure 2 A sample run of the implementation: The target door is on the left in the images. Image 1: The robot detects the door on the right and starts the tracking behavior (compare Fig. 3, $t = 15s$). Image 2: The arm points at the detected door which is classified as *door 2* ($t = 45$) so the robot stops *tracking* it and starts *searching* again. Image 3: The robot detects the target door ($t = 77s$). ARNOLD points at it ($t = 82s$, image 4), classifies it as *door 1* ($t = 97s$), parks the arm (image 5) and starts to move towards the target after reaching the *arm safe* state ($t = 125s$, image 6).

5. Results

To generate the desired behavioral sequences the matrices γ and A were set up as follows:

$$\gamma = \begin{pmatrix} 0&1&1&0&0&1&0&0&0&0\\ 0&0&0&0&0&0&0&0&1&0\\ 0&0&0&1&0&1&0&0&0&0\\ 0&0&0&0&1&1&0&0&0&0\\ 0&0&0&1&0&0&0&0&0&0\\ 0&0&0&0&0&0&0&0&0&0\\ 0&0&0&0&0&0&0&0&0&0\\ 0&0&0&0&0&0&0&0&0&0\\ 0&0&0&0&0&0&0&0&0&0\\ 0&0&0&0&0&0&0&0&0&0 \end{pmatrix} \quad A = \begin{pmatrix} 0&0&0&0&0&0&0&0&0&0\\ 0&0&0&0&0&0&0&0&0&0\\ 1&0&0&0&1&0&0&0&0&0\\ 0&0&0&0&0&0&0&0&0&0\\ 0&0&0&0&0&0&0&0&0&0\\ 0&1&0&0&1&0&0&1&0&0\\ 0&1&0&0&1&0&0&0&0&0\\ 0&0&0&0&0&0&0&0&0&0\\ 0&0&0&0&0&0&0&0&0&0\\ 0&0&0&0&0&0&0&0&0&0 \end{pmatrix}$$

$\gamma_{1,2}$ e.g. means that behavior B_2 (tracking a door) suppresses behavior B_1 (searching a door). Since we only want to start the target acquisition if the robot is fixating the target door we set $A_{6,8} = 1$ and so on.

This definition of mutual exclusions and presuppositions produced in all cases reasonable sequences of actions: Even though the sensoric processes (door detection and recognition) produced noisy data and the robot was in many experiments disturbed by persons passing the scene and occluding the doors the dynamics in all cases relaxes to 'safe' states. Contrary to most other robot architectures this approach controls combinations of concurrently running behaviors. This coordination of two or more actions improves the robot's performance: Note that e.g. the behaviors *visual search* and *parking* can overlap in time.

Extensions of our sample application appeared fairly easy since new behaviors only add new rows and columns in A and γ and the interaction with existing behaviors has to be defined which is obvious in most cases.

Fig. 2 shows a sample run of the implementation. The corresponding behavioral states are shown in Fig. 3. The robot first detects a door which is not the target door and switches back from tracking to searching. When it detects the target door it points at it and - after parking the arm - moves towards the target.

Note that some sensor contexts (C_i) are set to one at all times, e.g. C_7 (see Fig. 3). The switching of these behaviors depends only on the states of the other behaviors and A and γ respectively. Most other context variables depend on aspects of the sensor information. The sensor context variables for 'tracking' and 'visual search' e.g. specify whether the door detection module found a door in the acquired images ($C_1 = 0$, $C_2 = 1$) or not ($C_1 = 1$, $C_2 = 0$).

To maximize this approach's robustness it is crucial to define one ore more fallback behaviors: In our example the behavior *visual search* has no presuppositions and

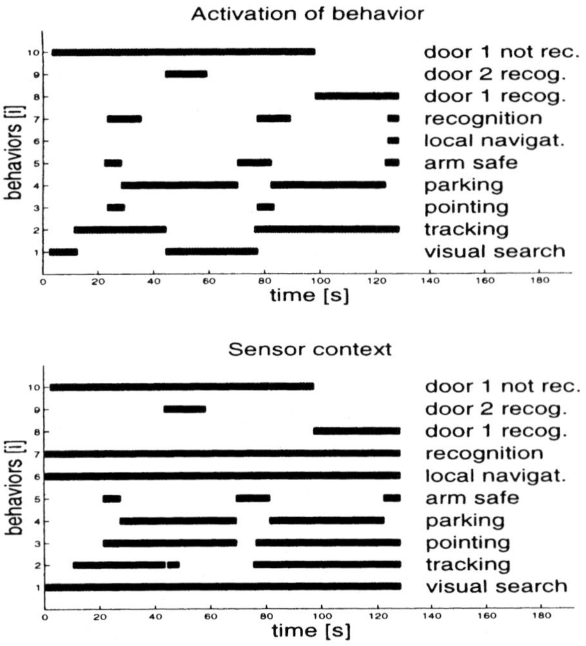

Figure 3 The plots show the behavioral avtivity (n_i) and sensor context (C_i) for a sample run of the implementation. The black bars mark periods of time at which a behavior is switched on ($|n_i| \simeq 1$) or a sensor context is present ($C_i \simeq 1$).

its sensor context is constantly set to one. Thus if the robot is strongly perturbed or the sensor processes can not produce reliable information the system will switch off all behaviors, activate the *visual search* and try to find the target again.

6. Conclusion

We presented a solution for the problem of behavioral organization, which is based entirely on continuous dynamical systems. By scaling up the *dynamical systems approach to behavior generation*, we achieved a stable integration of multiple sub-behaviors in a complex overall system. Although multiple nonlinear differential equations are coupled, the system remains stable and is easy to expand. The stability and flexibility of the system was demonstrated on the anthropomorphic robot ARNOLD for the example of detecting, identifying and approaching a door[1]. The generative aspect of the approach becomes evident by the fact that for a new application only the parameters of the dynamical systems, encoded in simple matrices, must be adapted.

7. Future work

Future work will focus on generation, adaptation and learning of the matrices A and γ. Since our approach gives an easy way to observe and interpret the current state of the robot in its environment, A and γ can be adapted to certain tasks or parts of the matrices could even be learned from scratch.

[1] This work was supported by the BMBF,NEUROS-project, grant number FKZ 01IN504 F

References

Alberts Et Al (1997). Entwicklung Einer Kommunikationsplattform Zur Steuerung Eines Autonomen Mobilen Roboters. Technical report, Fachbereich Informatik, Universität Dortmund.

Bergener, T., Bruckhoff, C., Dahm, P., Janßen, H., Joublin, F., and Menzner, R. (1997). Arnold: An anthropomorphic autonomous robot for human environments. In *Soave'97, Selbstorganisation Von Adaptivem Verhalten.*

Bergener, T. and Dahm, P. (1997). A Framework For Dynamic Man-Machine Interaction Implemented On An Autonomous Mobile Robot. In *Isie'97, Ieee International Symposium On Industrial Electronics.*

Bicho, E. and Schöner, G. (1997). Dynamic approach to autonomous robotics demonstrated on a low level-vehicle platform. *Robotics and Autonomous Systems, (in press).*

Braitenberg, V. (1984). *Vehicles. Experiments in Synthetic Psychology.* MIT Press, Cambridge, Mass.

Brockett, R. W. (1993). Hybrid models for motion control systems. In Trentelman, H. and Willems, J. C., editors, *Perspectives in Control*, pages 29–54. Birkhauser, Boston.

Brooks, R. (1986). A robust layered control system for a mobile robot. *IEEE Journal of Robotics and Automation, RA-2.(1).*

Kreutz, M., Völpel, B., and Janßen, H. (1996). Scale-invariant image recognition based on higher order autocorrelation features. *Pattern Recognition*, 29(1):Pp. 19–26.

Lemmon, M., Stiver, J. A., and Antsaklis, P. J. (1993). Event identification and intelligent hybrid control. In Grossman, R. L., editor, *Hybrid Systems*, pages 269–296. Springer, Berlin, viii edition.

Poole, D. (1995). Logic programming for robot control. In *Proceedings of the 14th International Joint Conference on AI (IJCAI-95), Montreal, Canada.*

Schöner, G. and Dose, M. (1992). A dynamical systems approach to task-level system integration used to plan and control autonomous vehicle motion. *Robotics and Autonomous Systems*, 10:253–267.

Schöner, G., Dose, M., and Engels, C. (1995). Dynamics of behavior: Theory and applications for autonomous robot architectures. *Robotics and Autonomous Systems*, 16:213–245.

Steinhage, A. (1998). *Dynamical Systems Generate Navigation Behavior (Ph.D. thesis).* Number ISBN 3-8265-3508-1 in Berichte aus der Physik. SHAKER-Verlag, Aachen, Germany.

Steinhage, A. and Schöner, G. (1997a). The dynamic approach to autonomous robot navigation. In *Proceedings of the IEEE International Symposium on Industrial Electronics, ISIE'97*, pages SS7–SS12. IEEE-publications.

Steinhage, A. and Schöner, G. (1997b). Self-calibration based on invariant view recognition: Dynamic approach to navigation. *Robotics and Autonomous Systems*, 20:133–156.

An extension of Maes' Action Selection Mechanism for Animats

VINCENT DECUGIS*, JACQUES FERBER**
* Groupe d'Etudes Sous-Marine de l'Atlantique (GESMA),
BP 42 , 29240 BREST NAVAL, FRANCE
decugis@gesma.fr
** Laboratoire d'Informatique, de Robotique et de Microélectronique de Montpellier,
161 , rue Ada, 34392 Montpellier Cedex 5, FRANCE
ferber@lirmm.fr

Abstract

Action selection is a central issue in animat design. Maes, among many others, has proposed an elegant solution to this problem (9),(10). Despite very interesting properties, this mechanism exhibits some problems, such as the one pointed out by Tyrrell in (15). We propose a solution to bypass these limitations by extending and deeply reorganizing this action selection mechanism to meet the requirements of an animat embedded in real world conditions.

1. Introduction

A large number of techniques have been proposed to solve the problem of action selection in animats. Tyrrell (14) gives an extensive review of these action selection mechanisms, or ASMs. Their roles are to choose over time the actions of an animat to optimize its chances of survival. The problem of surviving corresponds to the satisfaction of a set of viability constraints : maintaining one own temperature, getting food, drinking water ... These constraints are well suited for animats viewed as models of animals. Many ASMs have been proposed in this ethological context : Tinbergen (13), Lorenz (8), Baerends (1) or Halperin and Hallam ((7), (6),(5),(4)). For these mechanisms, motivation is the concept that enables the choice of the actions.

Some other ASMs tend to introduce the notion of goal in the action selection. Goals are a necessity for designing complex animats whose aims are not only to merely survive, but also to accomplish useful missions. Goals are often associated with the idea of planning. Classical planning, because of the heavy modelization efforts and computations it requires, has proved to be unadequate to design animats. Nethertheless, some ASMs exhibit planning properties. We studied one of them, introduced by Maes in (9) and (10), that we will call further the Maes' Action Selection Mechanism, or MASM. However, MASM lacks some requirements of real world applications, where goal-orientedness of ASM choices becomes a need. In an ethological simulation of Tyrrell, tt also exhibits problems with conflicting goals which are bound to reappear in our applicative context. Therefore we propose an extension of MASM for solving these problems. We illustrate the capacities of this new ASM with a series of experience in a realistic simulation of the Khepera robot.

2. The Maes' Action Selection Mechanism

2.1 Quick description of the ASM

The initial mechanism is extensively described and explained in its original presentation ((9),(10)). We do not aim at substituting to this description which fills a long paper on its own. We just remind here the key features necessary to understand the improvements introduced by our extension of the ASM.

The basic components of the MASM are the action components among which the ASM has to choose. Each action component is provided a structure precising its relations with the other components :

- a list of conditions that should be true for choosing the action.
- a list of positive consequences of the action.
- a list of negative consequences of the action.
- an "activation level" that indicates the relevance of the action to the current situation.

The conditions and consequences are facts or propositions about the situation of the animat in its environment. Some of these facts are considered to be a goal of the animat, i.e. the ASM will do whatever possible to make them becoming and remaining true. All these features are given a priori as a model of the animat action by its designer. The key principle of the ASM is the possibility of sharing activation between action components linked by causality. A fact that is both consequence of an action A and a condition of B is equivalent to a link. This link enables the diffusion of some activation proportional to A's level into B at each time step. As there is a constant input of activation into actions whose conditionning facts are true, there is a direct propagation which develops the consequences of the current situation. The goals input also activation in actions whom they are consequences. Retropropagation through a link, i.e. flow from B to A, is possible and diffuses the bias introduced by the goal activation input. There is thus a complex flow of activation between the action components, which has been constated empirically to converge to an equilibrium. The distribution of activation represents the repartition of relevance of the actions to the current situation for achieving the goals given to the ASM. The principle of the ASM is to choose out of this relevance distribution the best suited action to the current situation for the achievement of the goal. The execution of this action will modify the truth values of the the facts, and therefore leads to a new distribution, which leads to a new choice of action, and so on until one of the goal is reached.

2.2 Important features

This algorithm has very interesting properties:

- *Fault tolerance.* The links between the actions are not absolute rules. It can happen that an action has not all the expected consequences (action fault), or that a fact that was considered as true was in fact false (perception fault). The algorithm will take the real situation into account at the next choice.
- *Reactive planning.* The ASM of Maes has the unique ability of mixing planning and reactivity via the double direction of activation propagation. Direct propagation vehiculates the impact of the current situation, and favors the reactivity of the animat. Backward propagation introduces a bias in the choice of action towards the goals, and thus enables a special kind of planning.
- *adaptivity of links.* In the original mechanism, all links where supposed to have equivalent importance. For introducing adaptation in the selection of action, Maes proposed in (11) to use links with an associated strength. The modification of these strength according to a specific algorithm enables an adaptation of action selection.

From a strict planning viewpoint, it can be said that Maes' algorithm is a really good starpoint for designing an anytime fault-tolerant and reactive planning algorithm.

2.3 Known Problems

All these good properties are nethertheless balanced by some problems that restrict a lot its usage:

- *Deadlock in case of several goals.* This phenomenon has been pointed out by Tyrrell in (15) in an experiment where several contradictory goals where involved. A simulated animal endowed with Maes' ASM is instructed to fulfill both goals of eating and drinking. Experience has proven that when the animat nearly reach a pool of water, its need of food becomes stronger than its need of water. It then searches food, and when it nearly reaches a source of food, the drinking drive becomes reciprocally stronger, leading quickly the poor animat to death both by starvation and lack of water.
- *Synchronicity of action and perception.* Action selection obeys to a synchronous loop: checking truth value of facts, convergence of the activation levels, choice of an action, and then its execution. The problem in real world application is that there is usually no way to know whether an action is finished with success, failure, or whether it is still running. There is also no reason to believe that the truth value of facts will wait until the end of an action to change. Waiting this end may be a serious hinder to reactivity if slow actions are undertaken.

3. Proposed extension

These two major problems leaded us to introduce some important changes in the original Maes' ASM. We propose to give first the global idea of the transformations, and then to describe the new ASM more precisely, focusing by the way on the differences with the original ASM.

3.1 New architecture

To introduce an asynchronous process in the selection method, we consider that action selection should be done whenever there is a change in the truth value of the facts, i.e. whenever a redistribution of activation levels between the actions should happen. This ensures an optimal reactivity to the evolution of the environment, but has the drawback of making any action interruptible at any time. This is unavoidable as soon as we consider real world actions that are not instantaneous in their execution. But this will induce an action change only if the new event makes the previous action unappropriate.

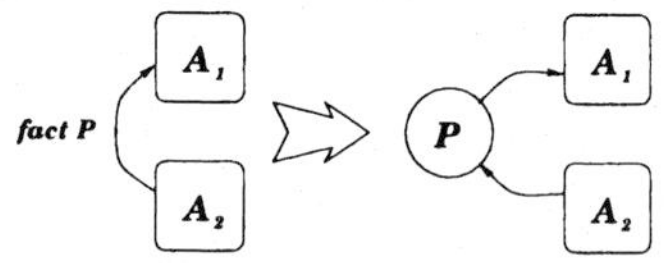

Figure 1 *Modification of the ASM components. In the original mechanism, facts where associated to links. In our proposed adaptation, we consider them as full components of the system, therefore doubling the number of links.*

We increase in this way the autonomy and the role of facts in the ASM. We found it interesting to give them corrolary a better formal position. In Maes' version, actions were "components" or "nodes" of a network, whereas the facts where just the links between them. We choose to consider the facts as real components of this network, that we will call further "perception components" (cf figure 1). This change gives a new signification to the links:

- a *perception to action link* is representing the relevance of the fact encapsulated in the perception for choosing the action.
- an *action to perception link* represents the positive consequence of the action on the evolution of the truth value of the fact in the perception.

The inhibition mechanism presented in Maes has in this formalism the drawback of introducing negative weights for the links. We use a different approach to take into account the negative effect of an action on a perception. Each fact P can be either true or false. We create two perceptions out of P, one being true when P is, and one being false, which is noted $\bar{P}$. A negative effect of action A on P will be instanciated by a link from A to $\bar{P}$ in the network.

At last, the Tyrrell's problem has to be taken into account. His analysis of the problem shows clearly that the problem comes from the presence of several goals in a single Maes ASM, and that it is impossible to solve it without changing deeply the propagation algorithm. On the other hand, Tyrrell reviews a large panel of other action selection mechanisms in (14). A good proportion of them are organized as a hierarchy of actions, especially ASMs inspired from ethology: Tinbergen (13), Lorenz (8) and Baerends (1). Each node of these hierarchies at a given level n is considered to be a complex action that is achieved by the help of its "children" actions, i.e. nodes of level $n-1$. Tyrrell points out the fact that this principle is too general to be implemented, because nothing is said on how to use this children actions to achieve the nth level action. This is precisely the kind of problem that a Maes ASM with a single goal can solve. We are therefore going to use the ethological-like hierarchical organization to distribute the different goals of our animat. At each node of this hierarchy, the action will be in fact equivalent to the goal of the underlying Maes ASM. Terminal nodes (nodes without child) of the hierarchy will be thus basic actions or behaviors of the animat.

3.2 Finer description of a node ASM

Let us note $\mathcal{P}$ and $\mathcal{A}$ the sets of perception and action components in the ASM. We suppose there are n_P elements in $\mathcal{P}$ and n_A elements in $\mathcal{A}$. We note α_{P_i} the activation level of the ith perception, and α_{A_j} the one of the jth action, and α_G the activation introduced by the goal. $f_{P_iA_j}$ is the strength of the link from P_i to A_j, and $f_{A_jP_i}$ is the strength of the link from A_j to P_j. v_{P_i} is the truth value of the fact included in P_i. The flow of activation through the links follow the following rules :

- link from P to A :
  - direct sense $\Delta\alpha_A = \theta_d f_{PA}\alpha_P$
  - indirect sense $\Delta\alpha_P = \theta_i f_{PA}\alpha_A$
- link from A to P :
  - direct sense $\Delta\alpha_P = \theta_d f_{AP}\alpha_A$
  - indirect sense $\Delta\alpha_A = \theta_i f_{AP}\alpha_P$

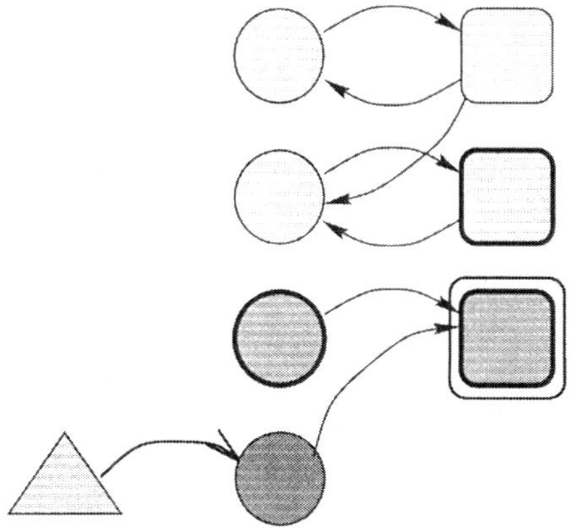

Figure 2 *Graphical synthetic representation of the flat action selection network. Perception and behavior components are placed in two facing columns. Links are represented between the components. The goal is represented by a triangle with a link to its aim. The gray level in the components indicates their activation level. A bold border for a perception component means that its truth value is true. A bold border for a behavior component indicates that all its mandatory conditions are true, and that it is thus choosable by the ASM. A double bold border for a behavior component means this is the chosen behavior of the ASM.*

Considering the evolution of the flow of activation over time, and aggregating all the activation diffusion phenomena, we can write :

$$\begin{cases} \alpha_{P_j}^{t+1} & = \alpha_{P_j}^t + \theta_d \sum_{A_i \in \mathcal{A}} f_{A_iP_j}\alpha_{A_i}^t \\ & +\theta_i \sum_{A_k \in \mathcal{A}} f_{P_jA_k}\alpha_{A_k}^t \\ & +\delta_{P_jG}\alpha_G + v_{P_i}(s^t) \\ \alpha_{A_j}^{t+1} & = \alpha_{P_j}^t + \theta_d \sum_{P_i \in \mathcal{P}} f_{P_iA_j}\alpha_{P_i}^t \\ & +\theta_i \sum_{P_k \in \mathcal{P}} f_{A_jP_k}\alpha_{P_k}^t \\ & +\delta_{A_jG}\alpha_G \end{cases}$$

At this point, it should be noted that the discrete time dynamical system regulating the activation level is diverging. Activation has to be normalized at each time step to ensure a constant global activation level present in the network.

This mathematical formulation has the advantage to be easily manipulated. We used it to prove that the normalized dynamical system is converging quickly to an equilibrium (3). In Maes' formulation the dynamics of activation was, as stated in (9), far more difficult to prove.

3.3 Modification of the link strength

As we put it forward, links have a different interpretation in our ASM. The adaptation mechanism proposed by Maes has therefore to be adapted.

The mechanism proposed in (11) is in fact evaluating the strength of a link from action A using the fact P as the probability that the selection of A will induce P becoming true. This probability is evaluated by the simpliest way: the Law of Large Numbers that enables out of a large set of facts to draw the probability of an event. More precisely, if we use n_{use} times the action A, and if it leads n_{succ} times to P being true, then the estimation of the strength of the link will be :

$$f_{AP} = \frac{n_{succ}}{n_{use}}$$

This estimation has a speed of convergence of $1/\sqrt{n_{use}}$, which is not intrinsically very quick, but proved to be sufficient in the experimentations. For our ASM, this estimation of probability corresponds to the strength of action to perception links. These strengths can therefore be reestimated after each use of the corresponding action, enabling an on-line learning.

Nothing is said on the perception to action links, which are always of strength 0 or 1 in Maes[1]. These links have been interpreted as showing the importance of the truth of their source fact for choosing the target action. The more important it is, the stronger the link should be. Without adaptation, this value must be hardwired at design time. But reinforcement learning seems to be an interesting technique for on-line evaluation of the strength of these links. We could for instance associate a reinforcement signal to the goal of the ASM. It would give a positive signal when the last action chosen by the ASM has reduced the "distance" for reaching the goal, and a negative one in the opposite case. Perception that were true at the time of choice of the action, and that gived activation through some links to the action helped it to having been chosen. The corresponding links have to be reevaluated after the end of this action. If the result was positive, these links should be strengthened, and weakened if it was negative.

4. Experimental test on a simulated Khepera

We use for our experiments the simulator of Khepera robot written by O. Michel. We implanted in the Khepera a series of basic reflex actions, most of them as small sensory-motor coupling of two neurons inspired from Braitenberg Vehicles (2), as shown in figure 3. We

[1] 1 when the fact is a condition of the action, 0 or no link if the fact is not.

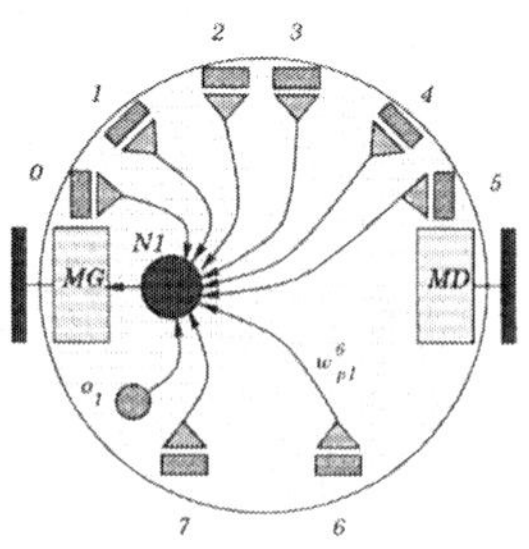

Figure 3 *The Khepera robot with a reflex architecture. The two motors (MG and MD) and the eight sensors (numbered from 0 to 7) are represented. One of the two symmetrical neuron is represented (N1). It receives input from all the sensors with a weight w^n_{pk} (p for proximity, k is the neuron number, n is the IR sensor number), and a constant input o_k.*

add a taxis behavior towards a special point in the environment. This behavior enables the Khepera to servo its heading towards this point, which can be thought of as a kind of beacon.

The first task after the deep changes we introduced to Maes ASM was to verify experimentally if the good properties of her algorithm were maintained. We therefore build a very simple ASM combining two behaviors of obstacle avoidance and taxis towards the center of the environment. Figure 4 shows that the result fulfills our hopes.

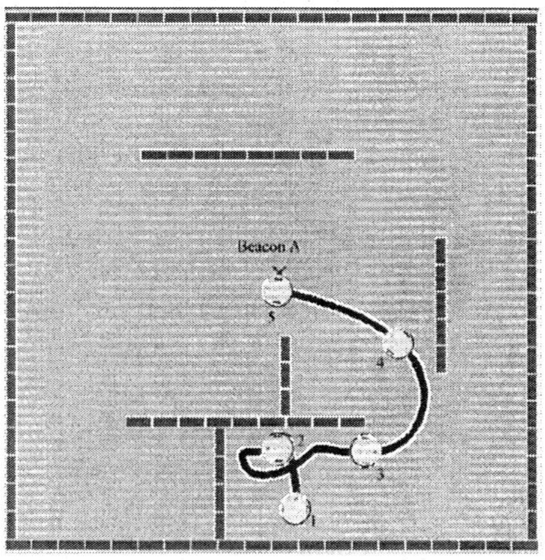

Figure 4 *Simple experience of action selection with two behaviors. The Khepera selects appropriate behavior to avoid obstacles when necessary. The numbers precise the different phases of its movement.*

We now have to verify that the Tyrrell's problem can

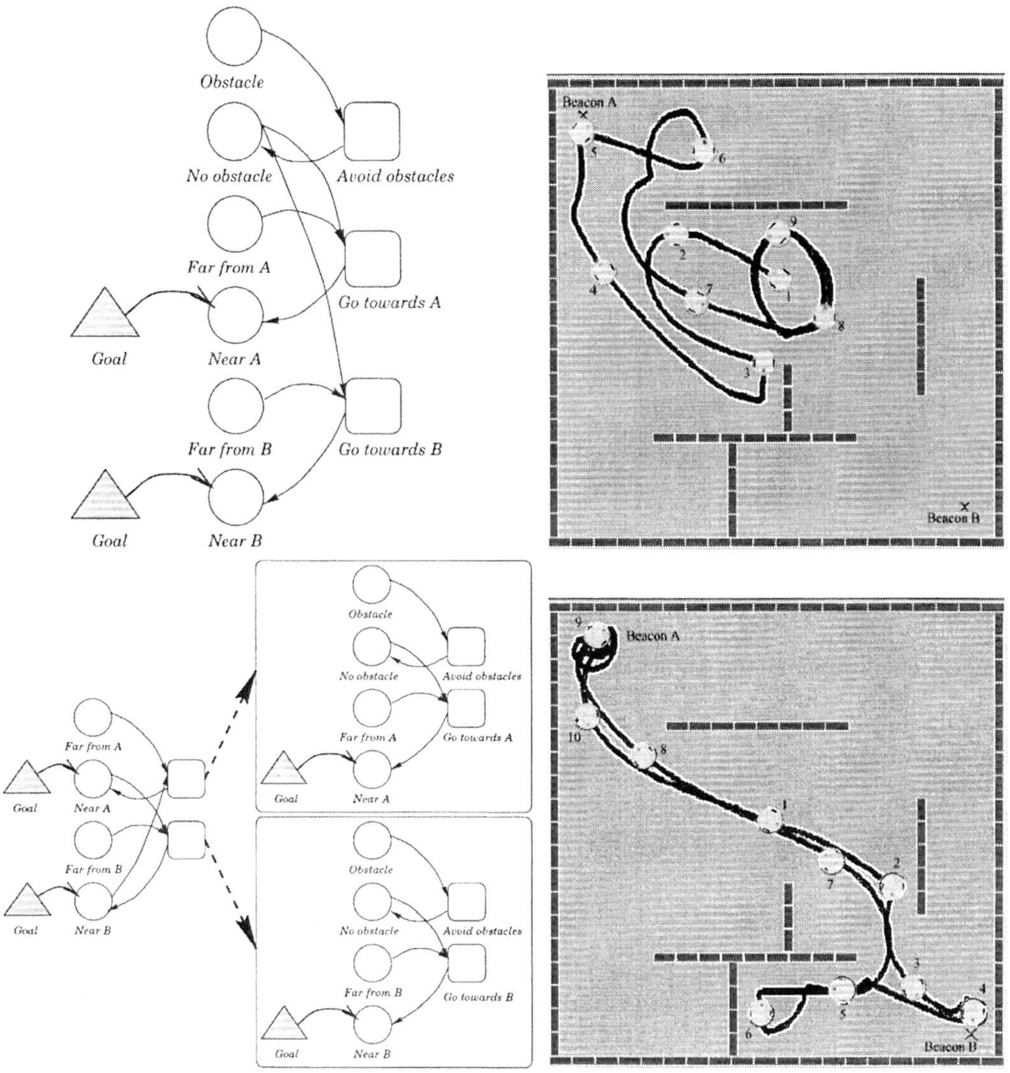

Figure 5 *Passing the double goal test without and with a hierarchical organization of the ASM. Upper part shows the non-hierarchical ASM and an example of the obtained behavior. The Khepera wanders most of time in the middle of the environment because it changes nearly at each action selection a new goal. Lower part shows the hierarchical ASM and the result of the experience. The Khepera alternatively go to the beacon A and to the beacon B. Once it has chosen its goal, it does not change it until it is reached.*

be solved with the help of a hierarchical organization of small ASMs with a single goal. In figure 5, we present the comparison between the hierarchical and non hierarchical solution for two conflicting goals. The Khepera is instructed to fulfill simultaneously the two goals : *go to the upper left corner*, and *go to the lower right corner*. To enable this, two different beacons have been put in the environment, one at each considered corner. The hierarchical ASM demonstrates a good stability in goal choice, and it can therefore be considered to have passed the test.

We tested also the adaptation of links in the khepera framework. We just verified the good transposition of the statistical learning of action to perception links, and did not yet implement our reinforcement learning suggestion.

5. Discussion

Now that we have shown the potentiality of the proposed ASM, it could be interesting to compare it with Tyrrell's proposal at the end of (14). After the comparison of a large number of different ASMs, he put forward an adaptation of Rosenblatt and Payton (12) architecture, called the free-flow hierarchy architecture. In this ASM, actions are situated at the bottom of a hierarchy of motivations from which they receive preferences. Actions and motivations also receive influences from sensory stimuli. This

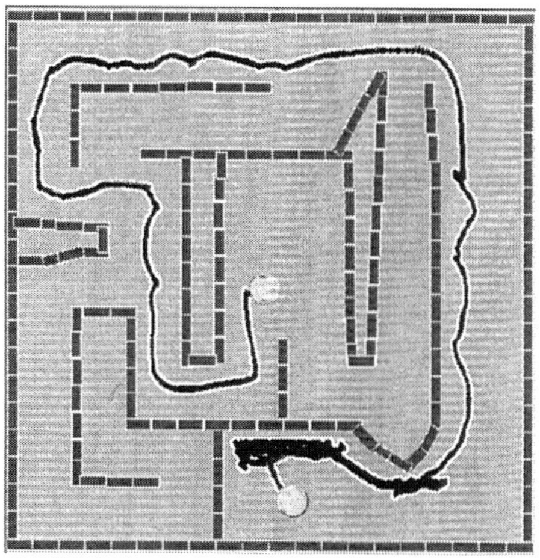

Figure 6 *Learning the consequences of actions. The picture shows the evolution of a Khepera in a maze. The goal was to reach the center. In the initial network, we only specified the perception to action rules, which are given in Maes' formalism by the conditions of action nodes. The robot has an erratic behavior at beginning, because it does not anticipate the results of its actions (no action to perception links). After a learning phase, it manages to explore the maze and reach its aim. This shows that learning from scratch is possible with this adaptation technique.*

ASM is argued to be superior to Maes' one, because ther is no choice at each node of the hierarchy. We proved in this work that choices at nodes is not a limitation and enables to solve the biggest problem of MASM. Moreover, the Rosenblatt architecture is very loose in the specification of the mechanism. This means that it is very difficult with this principle to design an ASM to solve a specific task, because there will be a very large number of possibilities to implement it. On the opposite, the adaptation property of MASM makes it very easy to design a specific solution. At last, Rosenblatt's ASM is able to combine several actions. It is an advantage if rules for mixing the actions exists, which is rather seldom according to our experience.

Another argument against MASM in (15) was that it takes just "digital" information into account. Our proposed modification shows clearly that analogic sensors can be compatible with MASM, since the basic selected actions can be sensory-motor coupling or other kind of reflexes. The point is not to be digital or not. It is to be easily computable from sensor raw data. Our experiments with Khepera used perception components obtained by a mere thresholding of sensors. These perception, though digital, are situated since there is an explicit way to compute them.

Finally, Rosenblatt's ASM does not provide to an animat any kind of planning, and is a purely reactive ASM. This is why we found it interesting to adapt Maes' ASM, which we found more suited to complex animats design, rather than Rosenblatt's, sufficient for ethological modelling purpose.

Our future work on the adapted MASM aims at integrating it in a real robot global control architecture, and at enhancing the adaptivity of the action selection mechanism.

References

G. Baerends. The functional organisation of behaviour. *Animal Behaviour*, 24:726–735, 1976.

V. Braitenberg. Vehicles: Experiments in synthetic psychology. MIT Press, Cambridge, Massachusetts, USA, 1984.

V. Decugis and B. Beauzamy. Convergence of a reactive planning algorithm. *submitted to Journal of Mathematical Analysis and Applications*, 1998.

B. Hallam. Fast robot learning using a biological model. Technical report, University of Edinburgh, 1994.

B. Hallam, J.R.P. Halperin, and J.C.T. Hallam. An ethological model of learning and motivation for implementation in mobile robots. Technical Report 629, University of Edinburgh, 1994.

B. Hallam and G. Hayes. Comparing robot and animal behavior. Technical report, University of Edinburgh, 1993.

J.R.P. Halperin. *A Connectionist Neural Network Model of Agression.* PhD thesis, Dpt of Ethology, Toronto University, Canada, 1990.

K. Lorenz. The comparative method in studying innate behavior patterns. *Symposia of the society of Experimental Biology*, 4:221–268, 1950.

P. Maes. How to do the right thing. *Connection Science Journal*, 1(3), February 1990.

P. Maes. Situated agents can have goals. *Robotics and Autonomous Systems*, 6:49–70, 1990.

P. Maes. Learning behavior networks from experience. In F. Varela and P. Bourgine, editors, *Towards a Practice of Autonomous Systems*, pages 48–57, 1991.

J.K. Rosenblatt and D.W. Payton. A fine-grained alternative to the subsumption architecture for mobile robot control, 1988.

N. Tinbergen. The hierarchical organization of mechanisms underlying instinctive behaviour. *Experimental Biology*, 4:305–312, 1950.

T. Tyrrell. *Computational Mechanisms for Action Selection.* PhD thesis, University of Edinburgh, 1993.

T. Tyrrell. An evaluation of Maes's bottom-up mechanism for action selection. *Adaptive Behavior*, 2(4):307–348, 1994.

Behaviors Coordination Using Restless Bandits Allocation Indexes

Yassine Faihe and Jean-Pierre Müller
IIIA - University of Neuchâtel
Emile Argand 11
CH-2007 Neuchâtel - Switzerland
Phone: +41 32 718 27 {27,07}
Fax:+41 32 718 27 01
{yassine.faihe, jean-pierre.muller}@info.unine.ch

Abstract

In order to remain viable and to reproduce an animal has to continuously deal with the problem of choosing the right behavior among several others (e.g. obtaining food, obtaining water, avoiding predators, ...) at the right time. In robotics this problem arises when we want to synthesize a complex behavior from elementary behaviors. Within the reinforcement learning framework we review the behaviors coordination methods proposed so far. Then we discuss their limitations and propose a new coordination method based on the restless bandits theory. Restless bandits allocation indexes are an extension of the Gittins indexes and are borrowed from the field of optimal scheduling. They concern problems involving the sharing of limited resources between several projects which are being pursued. The performance of the proposed method is illustrated through the postman robot problem and compared to the Hierarchical Q-learning (Lin, 1993).

1. Introduction

In order to remain viable and to reproduce an animal has to continuously deal with the problem of choosing the right behavior among several others (e.g. obtaining food, obtaining water, avoiding predators,...) at the right time. In robotics this problem arises when we want to synthesize a complex behavior from elementary behaviors. Such a problem is known as the behaviors coordination problem. In this article we will investigate it within the reinforcement learning framework.

Reinforcement learning is concerned with the class of learning problems in which an agent operating in an environment learns how to achieve a given task by trial and error. To do so it adjusts its decision process on the basis of a form of feedback commonly called reinforcement signal. This feedback is a scalar which reflects an internal measure of the robot's performance. The robot is rewarded (or punished) at each time step according to how close (or how far) it is from the target behavior. Thus the optimal behavior would be the one maximizing rewards and minimising punishments.

Reinforcement learning is very attractive because it raises the level of abstraction at which the designer intervenes (i.e., the robot is told what to do using the reinforcement function and not how to do it) and requires little programming effort (i.e., most of the work is done by autonomous training). Nevertheless and despite its mathematical foundations reinforcement learning can not be used as it is to make robots exhibit complex behaviors.

Such limitations are mainly due to the fact that both the complexity of the reinforcement function and the size of the state space with which the robot has to deal increase with the complexity of the behavior. Thus the learning process becomes slow and a wrong behavior may be generated. It is well known that breaking down the target behavior into simple behaviors which can work with simpler reinforcement functions and smaller state spaces, and then coordinating them should give better results.

We are currently developing a methodology (Faihe and Müller, 1998) which aims at helping the designer to build efficient control architectures for autonomous robots. The methodology setup a conceptual framework in which a number of methods regarding different stages of the analysis and design processes are to be defined. These stages includes the specification and the decomposition of the main behavior and the coordination of the sub-behaviors. In this paper we focus on the coordination problem and propose a method based on the restless bandits theory to solve it. The performance of the proposed method is illustrated through the postman robot problem which involves the coordination of concurrent and interfering behaviors. This problem is analogous to the optimal foraging problem which is usually faced by animals (Stephens and Krebs, 1986).

The remainder of this article is organized as follows. In the next section we expose the coordination problem and discuss some of its facets. Then we introduce the restless bandits theory. Next we explain how the experiments were carried out and show the obtained results.

Finally we conclude and give the outlines for future research.

2. The behaviors coordination problem

2.1 Formulation

Consider a collection of sensorimotor loops organized in a hierarchical structure (figure 1) in which the sensorimotor loops at a given level have a *direct causal influence* on the sensorimotor loops of the level below. In such a hierarchy decision making and learning occur at different levels but the interaction with the environment can only take place at the lowest level. Finally, each sensorimotor loop has its own internal state depending on the level at which it intervenes and on the task it has to solve as well.

Generally stated, the coordination problem within a hierarchy of sensory-motor loops consists in activating at each time step one sensory-motor loop at each level in order to generate the global expected behavior. A generic solution to this problem may consist in allocating an index to each sensory-motor loop and then activating the one with the highest index. The open question would then concern the meaning of such indexes and how to compute them.

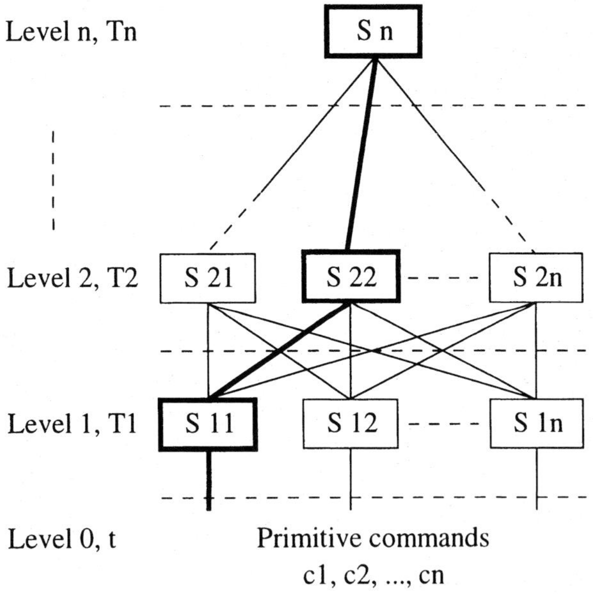

Figure 1 A hierarchy of sensorimotor loops. The path of active sensorimotor loops on a given time step is represented in bold.

2.2 Hierarchical reinforcement learning

It has been recognized that hierarchical methods have improved the performance of reinforcement learning. Such approaches allow a better exploration of the search space, the reuse of previously learned skills at a given level to acquire a new skill at the above level and speed up the overall learning process.

When we think about hierarchical methods the first idea that comes to mind is to treat the sensory-motor loops as macro-actions or closed-loop policies. This approach has been first introduced by Mahadevan and Connell (Mahadevan and Connell, 1992). In their work a global behavior consisting in box-pushing was decomposed into elementary sub-behaviors (finder, pusher, unwedger) which were learned independently using reinforcement learning. A hand-coded arbiter switches between the sub-behaviors according to their applicability condition and precedence. Lin (Lin, 1993) went beyond and proposed a system performing a complex task (finding a battery charger in an office environment and connecting to it) in which both the sub-behaviors and the arbiter were learned using Q-learning. In this case the collection of sub-behaviors (following walls on the robot's left/right hand side, passing a door, docking on the charger) $Sb_1, ..., Sb_n$ is used as a action set by the arbiter which learns $Q(state, Sb_i)$. This approach has been extended to the multi-level case by Dayan and Hinton (Dayan and Hinton, 1993) and more recently generalized by Dietterich (Dietterich, 1997).

Another idea consists in using sensorimotor loops as gating mechanisms. Their role would be to control the flow of action from bottom to top of the hierarchy. Approaches based on this scheme are called modular Q-learning. They were proposed by Whitehead *et al.* (Whitehead et al., 1993), Asada *et al.* (Asada et al., 1994) and Humphrys (Humphrys, 1997). In these action selection methods each module (sensorimotor loops) suggests the actions it wants to be executed and an arbiter chooses the one maximizing a given criterion. This criterion may be the sum of the different modules Q-values (greatest mass, maximize collective happiness) or simply the Q-value itself (nearest neighbor, maximize best happiness). Other criteria have been defined by Humphrys (Humphrys, 1997). However these methods were tested with hierarchies of two levels only and they sometimes require that the modules share the same actions.

2.3 Preemptive vs. non-preemptive schemes

In this section we discuss the differences between the two approaches exposed above which we call *behavior selection scheme* and *action selection scheme* and which are respectively non-preemptive and preemptive.

In the behavior selection scheme the learning process is uniform through the levels of the hierarchy and the problem is solved at different levels of abstraction. However from the second level of the hierarchy the action selection is replaced by the behavior selection and the time scale for decision making rises from t to T_1. This means that once a behavior is selected it will keep the control of the robot until its completion[1]. Hence learning will occur only at the termination of this behavior.

[1] A behavior is completed when it reaches a state which is a goal or in which it is not applicable anymore.

Because a selected behavior cannot be interrupted this scheme may have some drawbacks in problems involving the satisfaction of multiple and concurrent objectives. On the other hand exploration is improved because the state space is covered using big steps (Dietterich, 1997).

In the action selection scheme only one behavior will remain: it is the one which is produced by the overall system. It may be analysed at various levels as consisting of streams of actions ranging from reactive to planning operations. At each time step the system learns and makes decisions at each level of the hierarchy. Therefore any behaviors at any levels may interrupt each other. Such a continual interruption leads to inefficient exploration because it reduces the probability of reaching sensible states.

Let's illustrate these two schemes by the traditional ethological example of an animal having to satisfy both hunger and thirst drives. We assume that food and water are in different locations and that there are several levels of thirst and hunger. Suppose that the animal is hungry and that this activates the behavior leading it towards the food. If the thirst level becomes higher than the hunger one and the animal cannot interrupt the selected behavior it might die of dehydration en-route towards the food. On the other hand if it can interrupt its behaviors at any time and the levels of thirst and hunger become alternatively higher and lower one relative to the other, it may die of starvation or dehydration somewhere between the two locations.

These two approaches seem to be extremist but complementary. To find a compromise we can either introduce a model of fatigue to the action selection scheme or allow the interruption of behaviors in the behavior selection scheme. We opt for the second method because it is state based rather than the first one which is time based because we do not want to switch between behaviors on the basis of time sharing. Therefore we have to define intermediate states in which a behavior can be interrupted. Such states represent points in the trajectories of behaviors (i.e., from their activation to the their completion) where it is worth making a new decision. These states may be discovered by the robot. In addition, to interrupt an active behavior, the candidate behavior (i.e., the one with the highest index among the inactive behaviors) must not only have a greater index than the active behavior index but must also exceed it by a given constant $w/2$. The constant w is the width of the hysteris loop representing the behavior switching between active and passive phases (figure 2). This scheme seems to be the one implemented in the brain by the Basal Ganglia[2].

3. The restless bandits

The restless bandits are an extension of the multi-armed bandit problem and have been studied by Whittle (Whittle, 1988), and Weber and Weiss (Weber and Weiss, 1989). The initial problem concerns n projects, the state of project i at time t being denoted by $x_i(t)$. At each time step t only one project has to be operated. If the operated project is i then it will generate a reward $r_i(t)$ and make a transition $x_i(t) \to x_i(t+1)$ according to its transition probabilities P_i. The other $n-1$ projects remain frozen, i.e., neither produce reward nor change state. A project is said to be in an *active* or *passive* phase depending upon whether it is selected or not. Gittins (Gittins, 1989) has shown that an index policy is optimal for this problem. Such an index is denoted $I_i(x_i)$ and is a function of the project i as well as its state x_i:

$$I_i(x_i) = \max_{\tau>0} \frac{E\left[\sum_{t=0}^{\tau-1} \gamma^t r_i(t)\right]}{E\left[\sum_{t=0}^{\tau-1} \gamma^t\right]}.$$

This index can be interpreted as the maximal value of the reward density relative to the stopping time τ. The optimal policy will simply be to select the project with the greatest index.

Unfortunately this method cannot be directly applied to solve the coordination problem (the projects[3] being replaced by sensorimotor loops) because the fundamental assumption (i.e., the unselected projects remain frozen) is not valid anymore. This happens in many cases and especially in mobile robotics because the states of the sensorimotor loops are built from the same robot's perceptions and these perceptions evolve whatever the selected sensorimotor loop is.

To treat the restless bandits problem we will introduce the following notation:

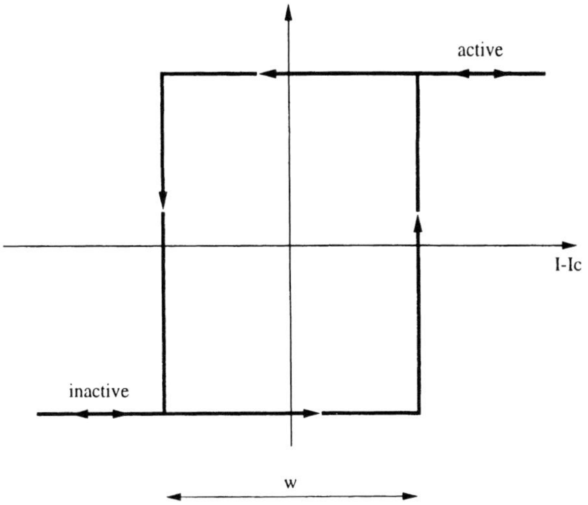

Figure 2 The hysteresis loop representing the behavior switching between the active and passive phases. Ic is the index of the candidate behavior and w is the width of the hysteresis.

[2] Personal communication from Tony Prescott

[3] To introduce the theory we will use the term project instead of sensorimotor loop.

- X_i is the set of states of project i;
- $P_i(x, y, k)$ is the probability that project i moves from state x to state y when it is in phase k, where $k = 1$ or 2 for respectively the active or passive phase;
- $r_{ik}(t)$ is the reward produced at time t by project i in phase k.

If we want to maximize the discounted sum of reward over an infinite horizon for a single project i we just have to solve the following optimality equation

$$V_i(x) = \max_{k=1,2} \left\{ r_{ik} + \gamma \sum_{y \in X_i} P_i(x, y, k) V_i(y) \right\},$$

where $V_i(x)$ is the value function of project i in state x. To do so we will compute the Q-values $Q_i(x, k) = E\left[r_{ik} + \gamma \max_{b=1,2} Q_i(x, b)\right]$ and then decide to activate or freeze the project according its Q-values.

Consider now the multi-project case. We are essentially interested in maximizing $\sum_t \gamma^t \sum_i r_{ik}(t)$ subject to $\sum_i l_i(t) = n - 1$ where $l_i(t) = 1$ if project i is passive at time t and $l_i(t) = 0$ otherwise (it means that at each time step only one project has to be active). Such a maximization amounts to maximizing $\sum_t(\gamma^t \sum_i r_{ik}(t) + \nu(t) \sum_i l_i(t))$ where ν is a Lagrangian multiplier. The new optimality equation to solve would then be

$$V_i(x) = \max_{k=1,2} \left\{ r_{ik} + \nu l_i + \gamma \sum_{y \in X_i} P_i(x, y, k) V_i(y) \right\},$$

or more compactly

$$V_i(x) = \max \left[L_{i1}, \nu + L_{i2} \right],$$

where $L_{ik} = r_{ik} + \gamma \sum_{y \in X_i} P_i(x, y, k) V_i(y)$. Whittle said that ν can be seen by an economist as a *'subsidy for passivity'* tuned at a level which guarantees that only one project is active at a time. The index of a project i in state x_i is then defined as being the value $\nu_i(x_i)$ of ν which makes $L_{i1} = \nu + L_{i2}$. It can be computed by using the Q-values of a project. We have

$$\nu_i(x_i) = Q(x_i, 1) - Q(x_i, 2).$$

Intuitively we can see that the index actually reflects the need of a project to be active with respect to the exploration and exploitation criteria. Indeed the value of ν increases if

- $Q(x_i, 1)$ increases which means that the project needs to be active (exploitation phase), or
- $Q(x_i, 2)$ decrease which means that the project does not want to be passive (exploration of the effects of its activation). This condition holds as far as the project is deteriorating during its passive phase (i.e., receiving negative rewards).

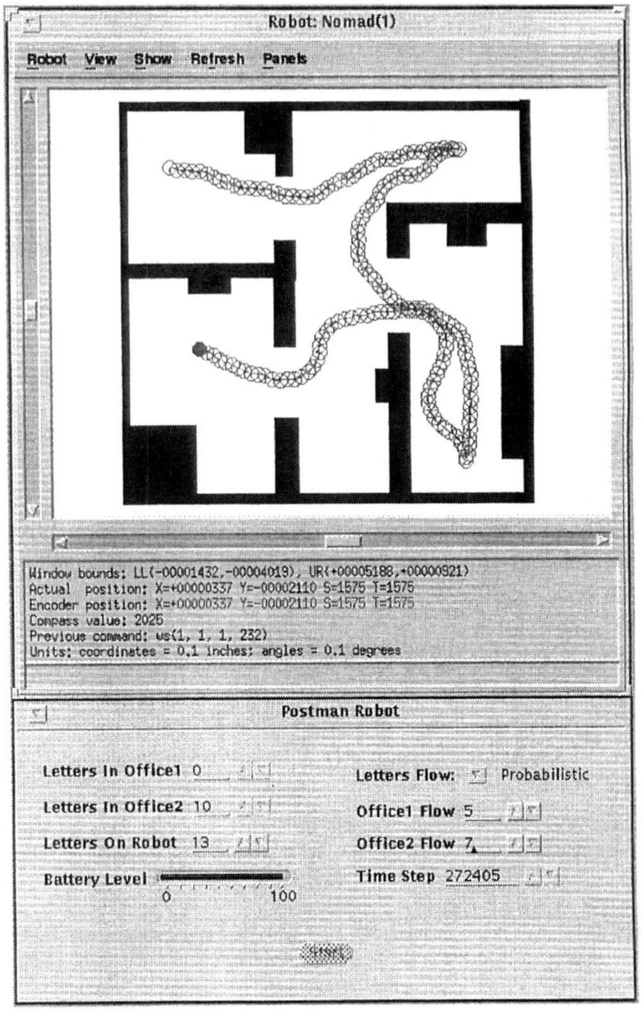

Figure 3 The Nomad 200 and the letters flow simulators. The robot learned how to move from one place to another. It needs about 70 steps to achieve a moving between two places.

4. Experiments

4.1 The postman robot problem

The postman robot is given a set of parallel and conflicting objectives and must satisfy them as best as it can. The robot acts in an office environment composed from offices, a battery charger and a mailbox. Its task is to collect letters from the offices and drop them in the mailbox. While achieving its postman's task as efficiently as possible the robot has to avoid collisions with obstacles and recharge its batteries to prevent break-downs. The following assumptions are made:

- The robot can sense the number of letters it holds, its battery level and the number of letters in each office;
- The robot gets the letters once it is in an office, posts the letters once it is near the mailbox, and recharges its battery once it is near the charger (because the real robot we have does not have any grasping or recharging devices).

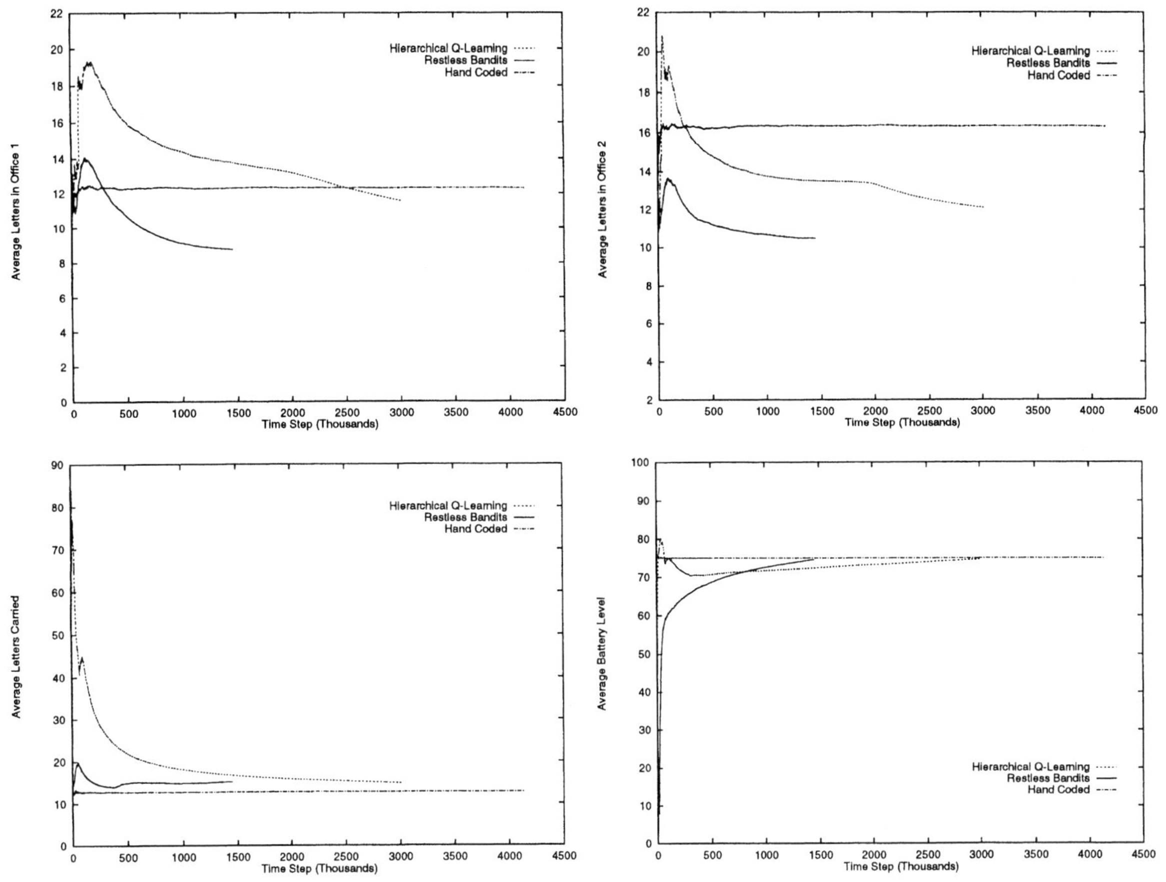

Figure 4 The performances of the postman robot. A comparison between the learning and the hand-coded robots.

4.2 Setting

The postman robot behavior was tested in an environment with two offices, one mailbox and one battery charger. The environment's size is approximately 13 m × 13 m. The experiments were carried out in the Nomad 200 simulator to which a letters flow simulator was added (figure 3). We had at our disposal a set of four elementary behaviors which had been previously learned (Faihe and Müller, 1998) using a connexionist Q-learning algorithm (Rummery, 1997). The configuration of the neural network was saved so that we could reuse it when needed. These behaviors consist in: moving to the first office, moving to the second office, moving to the mailbox and moving to the battery charger. Three coordination algorithms with the behavior selection scheme were tested: Hierarchical Q-learning (Lin, 1993), restless bandits coordination, and a hand coded coordination.

Hierarchical Q-learning The coordination is learned using connexionist Q-learning algorithm (Rummery, 1997). The reward function is defined as a gradient $r(t) = f(t) - f(t-1)$ of the function

$$f(t) = L_{o1}(t) + L_{o2}(t) + \lambda_1.L_r(t) + \lambda_2(t).(B_l(t) - Bth)$$

where L_{o1} and L_{o2} are respectively the number of letters in office1 and office 2, L_r the number of letters carried by the robot, B_l the battery level, and B_{th} a battery safety threshold. The function f can be seen as a Hamiltonian function in which $L_{o1} + L_{o2} + \lambda_1.L_r$ represents the objective function, $(B_l - Bth)$ the battery level constraint and $\lambda_2(t)$ the Lagrangian multiplier such as $\lambda_2(t) = 0$ if the constraint is satisfied and $\lambda_2(t) = \lambda_b$ otherwise. The positive constant λ_b weights the strength of the penalty.

The Q-values were approximated by feed-forward neural networks. We used one neural network for each of the elementary behaviors in order to reduce interferences between them. The input pattern of the neural networks is a 32 bits vector. It is constituted of a binary codification of the number of letters in each offices as well as those transported by the robot and the battery level (4 × 7 bits). The robot's location (in which place it is) is coded in the remaining 4 bits.

Restless bandits coordination The above function is now linearly decomposed into four functions: one for each elementary behavior. We have $f_1(t) = L_{o1}(t)$, $f_2(t) = L_{o2}(t)$, $f_3(t) = \lambda_1.L_r(t)$, $f_4(t) = \lambda_2(t).(B_l(t) - Bth)$. The reward function is also computed as a gradient: $r_i(t) = f_i(t) - f_i(t)$. The same Q-learning connexionist algorithm with the same neural networks and the same input pattern as in Hierarchical Q-learning was used.

However we needed two neural networks for each elementary behavior: one to approximate the Q-values of each phase (passive or active) of a behavior.

Hand coded coordination The algorithm used for this kind of coordination is the same as in (Faihe and Müller, 1998).

4.3 Results

We have run simulations for trials of 40000 steps. At each time step an elementary behavior is selected and performed. Best results were obtained with $\lambda_1 = 0.75$, $\lambda_b = 0.80$, and two neurons in the hidden layer. To evaluate the performances of the different kind of algorithms we defined the following metrics: the average letters in standby in the offices, the average letters carried by the robots, and the average battery level. These values were updated at each robot move (the lowest temporal resolution) and not at each behavior selection in order to guaranty a uniformity in the measures. A good postman robot would be the one which minimizes both the amount of letters in standby in the offices and the carried letters while keeping its battery level above the fixed threshold.

Figure 4 shows the obtained results with a probabilistic letters flow of 5% for office 1 and 7% for office2. The first remark concerns the superiority of the learned coordinations relative to the hand coded one. The main reason is that the learning robot implicitly takes into account some parameters like the distance between the different places or the letters flow. The hand coded algorithm would become complicated if these parameters would be explicitly used. We also notice that the restless bandits coordination gives better results than the Hierarchical Q-learning. With the former method there are in average 4.4 letters in standby in the offices less than with the latter method whereas the average letters carried by robot rises by only 0.3 letters. The energy management is almost the same for both methods. Moreover with the restless bandits coordination, the robot covers almost half of the distance that the one with Hierarchical Q-learning does. It means that sometimes the robot decides to stay at the same place because it is worth nothing to move to another place (e.g. there are not enough letters to pick up or to post). Finally we can justify this improvement by the fact that the reinforcement function is decomposed and that the index policy used allows a good balancing between exploration and exploitation.

5. Conclusion

A selection scheme as well as an index computation method for behaviors coordination were proposed in this paper and will be integrated to the design methodology we are developing. Hence complex behaviors could be acquired from elementary ones. The coordination scheme allows the robot to interrupt its ongoing behavior in case of urgency and to select another one. The index computation takes into account both the need for a behavior to be and not to be activated, so the exploration process is improved. The effectiveness of this method was demonstrated through the simulation of the postman robot problem.

We are currently validating the proposed behavior coordination scheme. Our future work will be devoted to (i) the autonomous discovery of preemption situations in the proposed selection scheme, (ii) the investigation of a systematic (then autonomous) method for behavior decomposition, (iii) testing the postman robot behavior on the real robot.

Acknowledgment

The authors would like to thank Paul Bourgine for fruitful discussions about the Gittins index theory and Tony Prescott for its comments on this paper. This work is supported by Swiss National Science Foundation grant 2000-046916.

References

Asada, M., Uchibe, E., Noda, S., S.Tawaratsumida, and Hosoda, K. (1994). Coordination of multiple behaviors acquired by vision-based reinforcement learning. In *Proceedings of the 1994 IEEE/RSJ International Conference on Intelligent Robots and Systems*.

Dayan, P. and Hinton, G. E. (1993). Feudal reinforcement learning. In *Advances in Neural Information Processing Systems 5*.

Dietterich, T. G. (1997). Hierarchical reinforcement learning with the MAXQ value function decomposition. Technical report, Oregon State University.

Faihe, Y. and Müller, J.-P. (1998). Behavior analysis and design: Towards a methodology. In Birk, A. and Demiris, J., editors, *Proceedings of the Sixth European Workshop on Learning Robots (EWLR6)*, Lecture Notes in Artificial Intelligence. Springer-Verlag. To appear.

Gittins, J. C. (1989). *Multi-armed Bandit Allocation Indices*. Willey.

Humphrys, M. (1997). *Action Selection methods using Reinforcement Learning*. PhD thesis, University of Cambridge.

Lin, L. J. (1993). Hierarchical learning of robot skills by reinforcement. In *Proceedings of the IEEE International Conference on Neural Networks*.

Mahadevan, S. and Connell, J. (1992). Automatic programming of behavior-based robots using reinforcement learning. *Artificial Intelligence*, 55:311–365.

Rummery, G. A. (1997). *Problem Solving With Reinforcement Learning*. PhD thesis, University of Cambridge.

Stephens, D. W. and Krebs, J. R. (1986). *Foraging Theory*. Princeton University Press.

Weber, R. R. and Weiss, G. (1989). On an index policy for restless bandits. *Journal of Applied Probability*, 27.

Whitehead, S., Karlsson, J., and Tenenberg, J. (1993). *Learning Multiple Goal Behavior via Task Decomposition and Dynamic Policy Merging*. Kluwer Academic Publishers.

Whittle, P. (1988). Restless bandits : Activity allocation in a changing world. *Journal of Applied Probability*, 25.

Behavior-Based Primitives for Articulated Control

Maja J Matarić*, Matthew Williamson†, John Demiris‡, Aswath Mohan*

*Computer Science Department, University of Southern California, Los Angeles, CA 90089-0781, mataric@cs.usc.edu
†MIT Artificial Intelligence Lab, 545 Technology Square, Cambridge, MA 02139, USA, matt@ai.mit.edu
‡Department of Artificial Intelligence, University of Edinburgh, Edinburgh, EH1 2QL, UK, johnde@dai.ed.ac.uk

Abstract

Our previous work has demonstrated how biologically-inspired behaviors can serve as an effective substrate for control, representation, and learning in mobile robots and multi-robot systems, in order to generate adaptive individual and group behavior. In this paper, we expand the behavior-based approach to the domain of manipulator control, and demonstrate first results in applying biological inspiration to a 20 degree-of-freedom humanoid dynamical torso simulation.

1. Introduction

Behavior-based systems take inspiration from biology, ethology, and neuroscience, in order to construct controllers for agents situated in noisy and dynamic environments (Matarić 1997). Behaviors impose a distributed, bottom-up approach to control, which eliminates bottlenecks and enables emergent functionality (Steels 1994). Aside from its pragmatic role in achieving distributed, real-time control in robotics, the behavior-based framework has also been used to model and implement simulations of biological systems. In this paper we apply behavior-based control to a novel domain, that of biologically-inspired 3D humanoid articulated motor control.

The vast majority of behavior-based systems to date have been applied to planar navigation problems. We present a generalization of behavior-based control to the domain of manipulator control, which has been largely under the umbrella of traditional robotics and geometric reasoning (Paul 1981, Brady, Hollerbach, Johnson, Lozano-Perez & Mason 1982). In contrast, we take inspiration from neuroscience and describe an approach toward implementing a general articulated motion controller based on distributed behaviors. We demonstrate the first stages of our approach on a 20 degree-of-freedom humanoid torso simulation with dynamics. Rather than use a single control scheme, which would be hard to design given the complexity of the problem, we decompose the control into a number of complementary strategies, choosing the most appropriate scheme for a given movement task.

2. Motivation

Biological motion provides a challenging model for control of articulated robots and graphical agents. Our work, both in navigation and now in manipulation, is inspired by a specific principle derived from evidence in neuroscience, which proposes an elegant organization of the frog and rat motor systems. Giszter, Mussa-Ivaldi & Bizzi (1993), Mussa-Ivaldi, Giszter & Bizzi (1994), and related work on spinalized frogs and rats suggests the existence of force-field motor primitives that produce complete high-level motor behaviors such as reaching and wiping. This evidence has been demonstrated by stimulating sections of the spinal cord, and measuring the resultant forces in the foot, which converge to a single equilibrium point, i.e., come to rest in the same position. Additionally, Mussa-Ivaldi et al. (1994) show that stimulating two or more regions in the frog spinal cord results in either a linear superposition of the corresponding force fields or a "winner-take-all" response that suppresses the other force fields.

This data suggest an elegant organizational principle for motor control, in which entire behaviors are coded with low-level force-fields, and can be combined into higher-level, more complex behaviors. The idea of supplying an agent with a collection of *basis behaviors* or *primitives* to generate motion, and combining those into a general repertoire for complex motion, is very appealing. Our previous work (Matarić 1995, Matarić 1997), inspired by the same biological results, has already successfully applied the idea to control of planar mobile robots. This paper extends the notion to agents with more DOFs, not by implementing a careful model of the frog force-fields, but rather by continuing to use the biological evidence as inspiration.

3. Experimental Testbed

We used a dynamic anthropomorphic simulation environment for implementing and testing our ideas. Adonis is a rigid-body simulation of a human torso, without an articulated spine, and with static legs (Figure 1).[1] The simulation consists of eight rigid parts or links: the head,

[1] Adonis was developed by Jessica Hodgins at Georgia Tech, using SD/Fast and models purchased from Viewpoint labs.

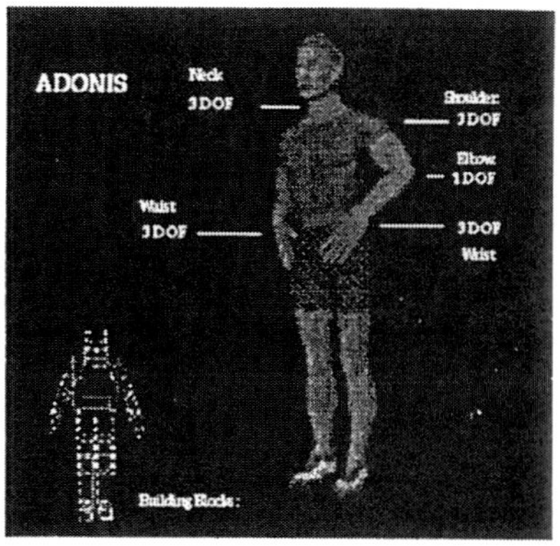

Figure 1 The 20 degree-of-freedom (DOF) physics-based torso simulation testbed. In the experiments presented here, 14 DOFs were used.

torso, left upper arm, right upper arm, left lower arm, right lower arm, left hand and right hand. The head is connected to the torso with a rotary three degree of freedom (DOF) joint, as is the torso to the waist. In the experiments described here, we used only the DOFs in the arms, consisting of three DOF rotary joints at the shoulders and wrists, and one DOF pin joints at the elbows. Thus, Adonis has a total of 20 DOFs of which we used 14. The testbed presents a very complex control problem and introduces novel challenges for the behavior-based framework. Much like a physical robot, the simulation software provides no built-in collision detection and avoidance capabilities, so the controller must resolve the many issues involved with self and other collisions.

4. The Behavior Primitive Approach

Designing a basis behavior set is a difficult problem, especially in a high-dimensional space. We drew inspiration and constraints from the human motor psychophysics literature, and selected a small initial set of general primitives with the goal of exploring their effectiveness, and subsequently using learning techniques to tune their parameters and generate their combinations. The following primitives were used:

1. **move-to-point:** Move the end-point to a Cartesian goal position, implemented using impedance control.
2. **get-posture:** Achieve an arm posture by adjusting the angles of the joints manually.
3. **avoid:** Avoid collisions, implemented based on generalized potential fields.

The implementation of each is described in turn.

4.1 Move-to-Point by Impedance Control

The general principle of impedance control (Hogan 1985) is to modulate the mechanical impedance of the end-point of an arm, by altering the torques at the arm's joints. Mechanical impedance for an object is defined as the relationship between an imposed disturbance and a generated force. For example, a compressed spring exerts a force proportional to the displacement. The impedance of such a system is constant and equal to the stiffness of the spring. For a more complicated mechanism like a robot arm, the mechanical impedance is determined by the control at the joint level. A mechanical arm can be made to appear as if a virtual spring and damper are connected to some equilibrium point; moving the point will drag the arm around, and the arm will automatically return to its equilibrium position if disturbed. Arranging the control of the arm in this way has advantages in terms of stability, especially when interacting with different environments (Colgate & Hogan 1989).

We implemented an impedance-based controller as follows. The force F from the "virtual" spring and damper is:

$$F = K(x_e - x) - B\dot{x} \tag{1}$$

where x is the 6-D vector defining the position and orientation of the end-point (hand) in space, x_e is the 6-D vector of desired positions and orientations, and K and B are stiffness and damping matrices. This desired force is implemented by applying torques (τ) at the joints, using the following simple relation (Craig 1989):

$$\tau = J^T F \tag{2}$$

The method has several advantages over position control techniques using inverse kinematics (Baker & Wampler II 1988). It is computationally simple, requiring only the forward kinematics and the Jacobian (Whitney 1982), and it is stable both when moving freely, and during contact with surfaces (Hogan 1985). It also fits well into the primitive-based approach; rather than use inverse kinematics for all joints, which would be very computationally intensive, a mixture of simpler control schemes can be used. For example, if **move-to-point** impedance control is used on the arms, and **get-posture** joint control on the body, the hands will remain in the same positions as the torso is moved.

To improve the robustness of the system, two additional terms were implemented. First, a term was added to the torque calculation to compensate for the effect of gravity on the limbs, so improving the tracking accuracy of the hand. Second, damping was applied at the shoulder to reduce redundant arm motions. With a small amount of damping at the shoulder joints, the arm moves smoothly without additional internal movements. The addition of these terms results in the following torque equation:

$$\tau = J^T F + G + B_s \tag{3}$$

where G is the gravity compensation, and B_s the added damping term.

The impedance controller was used to effectively implement the **move-to-point** behavior. The user simply specifies the desired position and orientation of the hand in Cartesian coordinates, and the impedance controller calculates the torques to move the arm to that position.

4.2 Get-Posture by Joint Angle Control

Many articulated movements involve achieving (and holding) postures, such as gestural communication, dance, etc., which are most easily defined in terms of joint angles. This capability is implemented in the **get-posture** behavior. PD servo control is used at the joints, and the desired angles are set manually by the user. The latter is a drawback of the method, however it allows the user to specify whole-arm position (rather than just the end-point as in **move-to-point**). Although we did not attempt to implement the motor control that might be used by frogs, **get-posture** is a reasonable approximation to a frog spinal field. The shape of the frog's force fields remains mostly unchanged when the animal is deafferented (Loeb, Giszter, Borghesani & Bizzi 1993), indicating that the field is largely generated by the spring-like muscles in the leg, acting through the leg kinematics. In Adonis, the analogue of the muscle springs are the PD servos of **get-posture**, which also generate a restoring force field at the hand.

Note that **move-to-point** and **get-posture** behaviors are duals of each other, since every posture involves a specific Cartesian location of the end-point, and every movement to a point results in a set of angles for the wrist, elbow and shoulder. Since the controller steps are sequential, and only one of these two controllers is used at a time, potential local minima problems are avoided. Task specifications typically vary between the two; some tasks are naturally presented in Cartesian space, while others are best described in joint angles (see Section 4.4).

4.3 Avoid by Using Virtual Forces

Detecting and avoiding collisions is a crucial part of control, especially in systems where self-collisions are possible. To prevent such collisions, we applied repulsive forces from obstacles to each moving arm. Unlike the other two behaviors, *avoid* runs in parallel with the other behaviors, and the repulsive forces become large only when a collision is imminent. The arms are constructed out of three point-mass building blocks, so the force is applied from the center of mass of the obstacle along a straight line to the centers of mass of the three blocks of the arm. Manz, Liscano & Green (1991) used repulsive force inversely proportional to the computed distance and directly proportional to the velocity of the approaching object. We adapted this approach, because our task required reaching targets that were frequently very close to the obstacle in 3D space, so the repulsive force had to be reduced after a certain proximity was achieved. This can be visualized as a force-field with a peak at a particular distance from the obstacle. Formally:

$$F = \frac{V\,S}{e^{D^2}} \tag{4}$$

where F is the applied repulsive force, V is the velocity, D is the distance, and S is the scaling factor. The force was computed for and applied to each component of the arm, as described above. Note that this force is added to the internally generated forces of Adonis' dynamical simulation, and serves to deflect the arms' trajectories away from obstacles and other body parts (the head being the most relevant in our task).

Avoid is successful when the initial distance between the body part and the obstacle is sufficiently large. When combined with **get-posture** and **move-to-point**, it can generate movement trajectories around obstacles and towards goals. However, the approach is not sufficient for generating smooth trajectories around obstacles. In fact, while direct reaching in biological systems is known to typically proceed along a straight Cartesian path (Flash & Hogan 1985), little is know about trajectory generation around obstacles. Most theories involve the use of carefully chosen intermediate via points (Miyamoto, Schaal, Gandolfo, Gomi, Koike, Osu, Nakano, Wada & Kawato 1996) which produce a smooth path around the obstacle. We are working on a similar implementation of a generic "curved-trajectory-generator".

4.4 The Task and Results

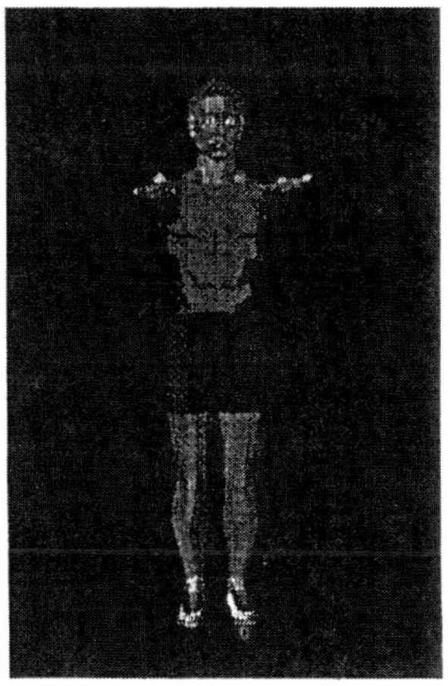

Figure 2 The start position for achieving sub-task five.

For the first demonstration of our approach, we chose a well-known motor task, the popular dance "Macarena", which involves a sequence of upper-body movements, whose order is well defined and success of achievement is easily evaluated. The goal behind our approach is to present a high-level, abstract description of the task, and have the motor behaviors execute the appropriate primitives to reach each of the sub-tasks. We used the following verbal specification of the Macarena to define the sub-tasks (sub-goals) for the motor primitives:

1. Extend right arm out, palm down
2. Extend left arm out, palm down
3. Rotate right hand (palm up)
4. Rotate left hand (palm up)

Figure 3 An example of an illegal solution to sub-task 5, based only on **get-posture** behavior.

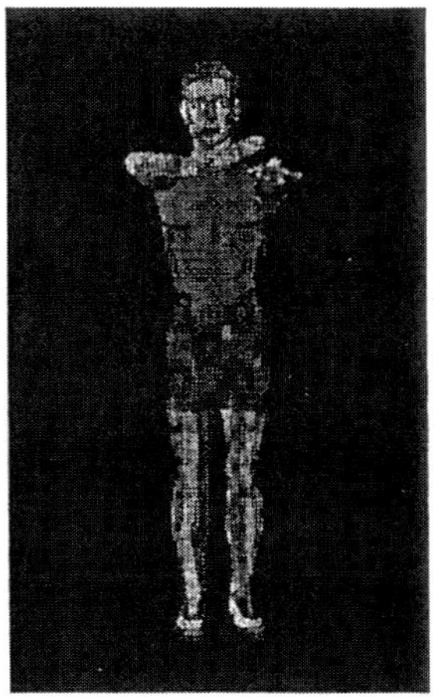

Figure 4 Successful achievement of subtask five by blending **get-posture** and **avoid** behaviors.

5. Touch right hand to top of left shoulder
6. Touch left hand to top of right shoulder
7. Touch right hand to back of the head
8. Touch left hand to back of the head
9. Touch right hand to the left side of the chest
10. Touch left hand to the right side of the chest
11. Put right hand to the right hip
12. Put left hand to the left hip

The Macarena task is sufficiently complex to test our approach; it involves the control of all of the 14 degrees of freedom in the two arms, and many of the movements involve the collision avoidance system. Furthermore, the task involves both Cartesian and joint-space specifications. The first four sub-tasks are best described in terms of postures (e.g., "extending the arm out" does not aim at a Cartesian end-point), while the remaining eight are best envisioned in Cartesian space.

A direct combination or blending of **avoid** with either of the other two behaviors (much like in our previous work on *flocking* (Matarić 1992)) produces legal and correct (i.e., goal-achieving) movements. However, the resulting trajectory is not optimal and may be quite unnatural. For example, the achievement of sub-task 5 requires the arm to move to the shoulder, which involves contacting the body but not colliding with it, while avoiding any contact with the head. A simple conservative addition of the two behaviors results in a trajectory that moves the arm first away from the head and toward the right shoulder, and then back around the face to the goal on the left shoulder. A small amount of parameter tuning results in a smoother trajectory. Unfortunately, the behavior tuning is specific to the shoulder-head configuration and does not generalize to other self-collisions. There are two ways of addressing this issue: assuming that self-collisions are critical enough to merit special-purpose fine-tuned avoidance strategies (as may be the case in biological movement (Arbib, Schweighofer & Thatch 1995)), or developing a higher-level smoothing behavior.

Figure 2 demonstrates the state of the system at the beginning of executing subtask 5; Figure 3 demonstrates an illegal solution, without the use of **avoid**; Figure 4 shows the same transition, with blending of the **avoid** behavior to generate a legal movement. Movies of Adonis performing the Macarena are available at http://www-robotics.usc.edu/~agents/macarena.html.

One of the benefits of the behavior decomposition of control systems is not only that there are different ways of structuring a given system (i.e., different collections of behaviors for solving the problem), but also that once a behavior decomposition is achieved, the specific behavior controllers can themselves vary, depending on the available sensors and effectors. To validate this point, we experimented with two different types of **move-to-point** behaviors: impedance control described above, and force-field control. The latter approach, described in detail in Matarić, Zordan & Mason (1998), applied forces to the end point of the arm to move it to the desired Cartesian location. The approach was limited since it imposed no constraints on the other DOFs of the arm, causing the resulting motion to resemble a rag doll; for example in sub-tasks one and two, the arm would reach the goal, but the elbow would swing about. Consequently, we opted for impedance control as a more general approach to generating movement in Cartesian space.

The current implementation of the Macarena does not address the question of sequencing behaviors; we assume the existence of a task-level plan. Our current research is addressing the production of such a plan automatically, directly from a continuous stream of movement data. Our preliminary experiments with human movement data have found salient cues based on the velocity profiles of the movements and zero-velocity breaks at the sub-goals. The issue of sequencing is addressed to the extent that **get-posture** and **move-to-point** are mutually exclusive in the Macarena task, and our implementation awaits termination of the current behavior before staring the next. Our task does not require concurrent execution of two behaviors on the different arms, but many more complex tasks would require this capability. The behavior-based framework we have implemented lends itself to scaling up to such coordination; it is not difficult to imagine how additional behaviors that constrain and guide coordinated two-arm movement can be added to the current system, and this is an active direction of our work.

5. Related Work

Manipulator control is a mature and active area of robotics, and a comprehensive review is well beyond the scope of this paper. Instead, we review only the most related approaches, both in behavior-based control in general, and in manipulation in specific. Manipulator work largely focuses on the difficult problems of inverse kinematics and dynamics (Paul 1981, Brady et al. 1982), and learning techniques are often employed to overcome the high dimensionality of the problem (Atkeson 1989, Schaal & Atkeson 1994). By using generic behaviors, we hope to develop "stereotyped" parametric solutions to common motor control problems, such as reaching to a point and achieving postures, so that a motor system can perform less run-time computation (i.e., avoid solving the find-path problem (Lozano-Pérez 1982) for each new movement task) without losing most of its generality. Previous work by Williamson (1996) is inspired by the same motivation; it demonstrates the control of a 6-DOF physical robot arm based on primitive postures. The control system uses four primitives (three reaching and one rest posture) combined using linear interpolation. This scheme was also used by Marjanović, Scassellati & Williamson (1996) to learn to reach toward a visual target. Their work and ours are both based on the same biological inspiration, but their system uses postural primitives while we use movement behaviors instead.

Behavior-based control has been used extensively in navigation; (Arkin 1987) introduced motor schemas, which are akin to behaviors, for mobile robot control, and related methods for combining them. Our **go-to-point** and **get-posture** behaviors are based on the same principle as the generic **homing** behavior used in Matarić (1995), which moves a mobile robot toward a specified 2D Cartesian goal, but are generalized here to the more complex 3D case. A simple special-purpose behavior-based arm controller was implemented by Connell (1988), for controlling a 3 DOF arm mounted on a mobile robot and used for soda can collection.

Sensorimotor primitives have been successfully applied to assembly tasks. Malcolm & Smithers (1990) used a hybrid system to generate a sequence of grasps in order to perform an assembly task with an ADEPT manipulator. Unlike theirs, our current system does not yet address the sequence generation process. Furthermore, while their execution module generates a specific path for each arm movement, our underlying behaviors are designed to be more general purpose. Morrow, Nelson & Khosla (1995) use primitives for translation, rotation, guarded moves, and grasping sequenced by a planner, and Nelson, Jägersand & Fuentes (1995) extend it to model "virtual tools" for grasping and assembly. The emphasis in this work is building skill libraries that can be robustly used by a planner. In contrast, our work is aimed at computationally simpler gross motion primitives.

6. Extensions and Conclusion

A key direction of our continuing work is in the application of the above primitives to the perceptual side of manipulation, in particular within an imitation learning framework (Kuniyoshi, Inaba & Inoue 1994, Demiris, Rougeaux, Hayes, Berthouze & Kuniyoshi 1997). Again inspired by biological data, which indicate that perceptual and motor systems may share some representational substrates, we have been working on adapting our motor primitives into sensori-motor behaviors, capable of not only generating but also recognizing typical movements. We endow the agent with a set of innate motor behaviors, and allow it to learn through imitation by observing other agents' movement. It does so by attempting to map the observed behaviors onto its own set of primitives, and internally (and externally, if needed) imitate them, thereby generating a learning signal.

The goal of the work described in this paper has been to take inspiration from biological motor primitives and apply it to the high-dimensional manipulator control problem. Based on our previous work with basis behaviors, we designed a set of motor primitives and used them to produce a non-trivial sequential motor task, the Macarena, using 14 DOFs of a dynamic torso simulation. We demonstrate how behavior-based techniques of behavior sequencing and blending can be directly applied to this new and more complex control domain, and continue to refine our approach to begin to tackle more sophisticated control problems.

Acknowledgments

This work was supported by the NSF Career Grant IRI-9624237 to M. Matarić. The authors thank Victor Zordan for help with the details of Adonis.

References

Arbib, M. A., Schweighofer, N. & Thatch, W. T. (1995), Modeling the Cerebellum: From Adaptation to Coordination, in Motor Control and Sensory-Motor Integration: Issues and Directions, *in* 'Motor Control and Sensory-Motor Integration: Issues and Directions', North-Holland Elsevier Science, pp. 11–36.

Arkin, R. C. (1987), Motor Schema Based Navigation for a Mobile Robot: An Approach to Programming by Behavior, *in* 'IEEE International Conference on Robotics and Automation', Raleigh, NC, pp. 264–271.

Atkeson, C. G. (1989), 'Learning Arm Kinematics and Dynamics', *Annual Review of Neuroscience* **12**, 157–183.

Baker, D. R. & Wampler II, C. W. (1988), 'On the Inverse Kinematics of Redundant Manipulators', *International Journal or Robotics Research* **7**(2), 3–21.

Brady, M., Hollerbach, J. M., Johnson, T. L., Lozano-Perez, T. & Mason, M. T., eds (1982), *Robot Motion: Planning and Control*, The MIT Press, Cambridge, Massachusetts.

Colgate, E. & Hogan, N. (1989), An analysis of Contact Instability in Terms of Passive Equivalents, *in* 'Proceedings of the 1989 IEEE International Conference on Robotics and Automation', pp. 404–409.

Connell, J. H. (1988), A Behavior-Based Arm Controller, MIT AI Memo 1025, MIT Artificial Intelligence Lab, Cambridge, MA.

Craig, J. J. (1989), *Introduction to Robotics: Mechanics and Control*, second edn, Addison-Wesley, Reading, Massachusetts.

Demiris, J., Rougeaux, S., Hayes, G. M., Berthouze, L. & Kuniyoshi, Y. (1997), Deferred Imitation of Human Head Movements by an Active Stereo Vision Head, *in* 'Proceedings of the 6th IEEE International Workshop on Robot Human Communication (RoMan97)', IEEE Press, Sendai, Japan.

Flash, T. & Hogan, N. (1985), 'The coordination of the arm movements: an experimentally confirmed mathematical model', *Journal of Neuroscience* **7**, 1688–1703.

Giszter, S. F., Mussa-Ivaldi, F. A. & Bizzi, E. (1993), 'Convergent force fields organized in the frog's spinal cord', *Journal of Neuroscience* **13**(2), 467–491.

Hogan, N. (1985), 'Impedance Control: An Approach to Manipulation', *Journal of Dynamic Systems, Measurement, and Control* **107**, 1–24.

Kuniyoshi, Y., Inaba, M. & Inoue, H. (1994), 'Learning by watching: extracting reusable task knowledge from visual observation of human performance', *IEEE Transactions on Robotics and Automation* **10**(6), 799–822.

Loeb, E. P., Giszter, S. F., Borghesani, P. & Bizzi, E. (1993), 'Effects of Dorsal Root Cut on the Forces Evoked by Spinal Microstimulation in the Spinalized Frog', *Somatosensory and Motor Research* **13**(2), 467–491.

Lozano-Pérez, T. (1982), Task Planning, *in* M. Brady, J. M. Hollerbach, T. L. Johnson, T. Lozano-Pérez & M. T. Mason, eds, 'Robot Motion: planning and control', MIT Press, Cambridge, MA, pp. 473–488.

Malcolm, C. & Smithers, T. (1990), Symbol grounding via a hybrid architecture in an autonomous assembly system, *in* P. Maes, ed., 'Designing Autonomous Agents', MIT Press, pp. 145–168.

Manz, A., Liscano, R. & Green, D. (1991), A comparison of real-time obstacle avoidance methods for mobile robots, *in* 'Proceedings, Experimental Robotics II, the 2nd International Symposium', Touluse, France.

Marjanović, M., Scassellati, B. & Williamson, M. (1996), Self-Taught Visually-Guided Pointing for a Humanoid Robot, *in* P. Maes, M. Matarić, J.-A. Meyer, J. Pollack & S. Wilson, eds, 'Fourth International Conference on Simulation of Adaptive Behavior', The MIT Press, Cape Cod, MA, pp. 35–44.

Matarić, M. J. (1992), Designing Emergent Behaviors: From Local Interactions to Collective Intelligence, *in* J.-A. Meyer, H. Roitblat & S. Wilson, eds, 'From Animals to Animats: International Conference on Simulation of Adaptive Behavior'.

Matarić, M. J. (1995), 'Designing and Understanding Adaptive Group Behavior', *Adaptive Behavior* **4**(1), 50–81.

Matarić, M. J. (1997), 'Behavior-Based Control: Examples from Navigation, Learning, and Group Behavior', *Journal of Experimental and Theoretical Artificial Intelligence* **9**(2–3), 323–336.

Matarić, M. J., Zordan, V. B. & Mason, Z. (1998), Movement Control Methods for Complex, Dynamically Simulated Agents: Adonis Dances the Macarena, *in* 'Autonomous Agents', Minneapolis, St. Paul, MI.

Miyamoto, H., Schaal, S., Gandolfo, F., Gomi, H., Koike, Y., Osu, R., Nakano, E., Wada, Y. & Kawato, M. (1996), 'A Kendama Learning Robot Based on Bi-Directional Theory', *Neural Networks*.

Morrow, J. D., Nelson, B. J. & Khosla, P. K. (1995), Vision and Force Driven Sensorimotor Primitives for Robotic Assembly Skills, *in* 'Proceedings of the IEEE International Workshop on Intelligent Robots and Systems (IROS-95)', Vol. 3, Pittsburg, PA, pp. 234–240.

Mussa-Ivaldi, F. A., Giszter, S. F. & Bizzi, E. (1994), 'Linear combinations of primitives in vertebrate motor control', *Proceedings of the National Academy of Sciences* **91**, 7534–7538.

Nelson, R. C., Jägersand, M. & Fuentes, O. (1995), Virtual Tools: A Framework for Simplifying Sensory-Motor Control in Complex Robotic Systems, *in* 'Proceedings of the IEEE International Workshop on Vision for Robots (IROS-95)', Pittsburg, PA, pp. 70–79.

Paul, R. P. (1981), *Robot Manipulators: Mathematics, Programming, and Control*, The MIT Press, Cambridge, Massachusetts.

Schaal, S. & Atkeson, C. C. (1994), 'Robot Juggling: An Implementation of Memory-Based Learning', *Control Systems Magazine* **14**, 57–71.

Steels, L. (1994), 'The Artificial Life Roots of Artificial Intelligence', *Artificial Life* **1**(1), 75–110.

Whitney, D. E. (1982), The mathematics of coordinated control of prosthetic arms and manipulators, *in* M. Brady, J. M. Hollerbach, T. L. Johnson, T. Lozano-Pérez & M. T. Mason, eds, 'Robot Motion: planning and control', MIT Press, pp. 287–304.

Williamson, M. (1996), Postural Primitives: Interactive Behavior for a Humanoid Robot Arm, *in* P. Maes, M. Matarić, J.-A. Meyer, J. Pollack & S. Wilson, eds, 'Fourth International Conference on Simulation of Adaptive Behavior', The MIT Press, Cape Cod, MA, pp. 124–131.

Selection Behaviour in Caddis Fly Larvae

Valentin A. Nepomnyashchikh
Institute for Biology of Inland Waters, Russian Academy of Sciences
152742 Borok, Yaroslavl obl., Russia
nepom@ibiw.yaroslavl.su

Abstract

I describe the selection of hard particles for case building in caddis fly larvae. The larvae prefer large and flat particles which are more profitable in terms of time and silk thread spent for building. When larva discover particles of various profitability in a random sequence, one may expect the sequence of larva's choices to be also random. However, the larvae tend to repeat acceptances and rejections of particles . This repetitive response sequence results in a partial choice of unprofitable particles. I developed a model of selection based on simple rules that real larvae use. The model qualitatively reproduces the preference for more profitable particles, as well as the repetition of responses and the partial choice. In the model, a random sequence of discovered particles is viewed upon as an external noise. The non-linear control mechanism transforms the noise into the repetitive behavioural sequence. Thus, the behaviour of model is the result of self-organisation in noise-driven system. I discuss how the self-organising order may interfere with adaptive behaviour.

1. Introduction

When the animals select some items in the environment, e.g., food or mates, they are widely regarded as optimising the selection. If two or more distinct types of items are discovered in the environment, an animal should use the most profitable type(s) only and ignore other(s). The profitability may equal, for instance, the energy or other gain obtained from the item divided by the amount of time required to handle it (McNamara and Houston, 1985; Stephens and Krebs, 1986). The optimal selection means that an animal should accept or reject items solely on the basis of their profitability. Therefore, if an animal discovers items of various profitability in a random order, than the resulting sequence of acceptances and rejections should be also random.

On the other hand, it is widely accepted that the animal behaviour is a self-organising process (see Cole, 1994 and Ferriere et al., 1996, for examples of chaotic dynamics of the behaviour). The output of self-organising systems tends to be time-correlated (and so is not random in this narrow sense), even under an influence of uncorrelated random input - white noise (Horsthemke and Lefever, 1984). It was indeed shown that the animals tend to respond or not to respond in the repetitive order to an apparently random sequence of items. For example, sticklebacks eat the discovered prey more readily if the previously discovered one was also eaten, and reject it more frequently after the previous one was rejected (Thomas, 1977). Larvae of caddis flies accept and reject items in the similar order when selecting the material for case building (Nepomnyashchikh, 1993a). One may speculate that the tendency to repeat responses may interfere with the optimal selection: some less profitable items may be accepted.

Here, I try to show how the non-random repetitive response sequences (RRS) emerges in the process of selection of randomly distributed items in caddis fly, *Chaetopteryx villosa* F., larvae, and how RRS are related with the partial choice of unprofitable items. I discuss also briefly the more general problem: how the self-organising behaviour relates to animal's adaptation to the environment.

2. Previous findings

The larvae inhabit creek bottoms and build their cases of sand and other hard particles they find at the bottom. If the case is partially destroyed, larvae must repair it quickly, as the case provides both camouflage and mechanical protection against predators (Johansson, 1991; Otto, Johansson, 1995). They do not use vision when testing a particle and test it with their legs. The duration of this test can be used as the measure of readiness to respond to a particle.

I monitored the selection of hard particles in 5th instar larvae when repairing partly destroyed case during 1 hr laboratory experiment (Nepomnyashchikh, 1993a). They readily used sand grains (G) and fragments of egg shell (F) for building and prefer large and flat Fs to small and round Gs. In the experiment, Fs and Gs were placed at the bottom of a Petry dish at random with no gaps between them. In the control, I provided larvae with Gs only. Larva came upon a particle ('discover' it, D), handled it and evaluated its shape and size ('tested' it, T) and then attached it (A) to the case with a 'silk' thread. A particle could be also rejected after T. The area of larvae's body surface covered with a single F was roughly eight times as large as the area covered with a single G, while the larvae spend for the attachment of F only twice as time as for G. Besides this, larvae spent less thread per area unit covered, if it used particles which have larger surface.

In the experiment, almost all discovered Gs were rejected after testing, provided Fs were abundant enough. On the other hand, more than 20% of all discovered Gs were attached in the control. It looks like the larvae tend to optimise their behaviour in terms of the acceleration of building and saving silk. However, most individuals never accepted all Fs, and never rejected all Gs in the experiment. The larvae used the following simple rules when selecting particles (Nepomnyashchikh, 1993a,b; Nepomnyashchikh and Podgorny, 1994):

1. Larva attached discovered F more readily, if the previous discovered particle was G, and it rejected F more frequently after discovering another F. On the other hand, G was rejected more frequently after discovering F, and attached more readily after another G. It means that the larvae compare the particle it is currently testing and the previous one.

2. Test duration for rejected sand grains varied. It was much longer if the grain was discovered immediately after attached shell fragment than after rejected grain. This suggests that the readiness to respond to a particle increases after the successful attachment.

3. If a single F was only provided for larva and the building started with attaching it, larva rejected more grains during the course of experiment in comparison with the control. Thus, the single "outstanding' particle is more important for caddis fly than a number of subsequently rejected ones.

3. The Model

I developed the model of larva behaviour. In order to simplify the model, I assume that all particles are flat, of the same shape, and differ only in the area of their surface. The larvae are assumed to measure the area instantly. They are also assumed to measure the area correctly, with the exception of extremely large or extremely small particles (see below). The other assumptions are based on above rules 1-3. Namely, I assume that the readiness $\boldsymbol{R_i}$ to respond to the particle i depend on 1) the relative difference between areas $\boldsymbol{q}$ of two particles found in succession, and 2) the readiness $\boldsymbol{R_{i-1}}$ caused by the previous particle:

$$\boldsymbol{R_i = aR_{i-1}} \exp [\boldsymbol{b(q_i - q_{i-1}) / q_{i-1}}],$$

where $\boldsymbol{a}$ and $\boldsymbol{b}$ are parameters. I assume also that 3) the ith particle is attached to the case, if $\boldsymbol{R_i}$ is equal to or exceeds a certain threshold $\boldsymbol{P}$ which is just the area of last particle attached, independently of the number $\boldsymbol{n}$ of other particles rejected since the last attachment: $\boldsymbol{R_i \geq P_{i-n}}$. If a particle was attached, then the new threshold is set up. The particle is rejected if $\boldsymbol{R_i < P_{i-n}}$.

In computer simulations, the distribution of particle's areas was Gaussian in the range 0.0 to 1.0 (mean size = 0.5, standard deviation = 0.2). I assume that it is hard for larvae to measure the area of extremely large or extremely small particles. For this reason, all areas were considered to be equal to 0.9 in case if they exceeded this value, and all areas smaller than 0.1 were considered to be equal 0.1. Particles of various area were discovered by "larva" in random order. The readiness ranged between 0.1 and 1.0. If the value of readiness exceeded these limits in the course of consecutive iterations, then I considered it to be equal to the nearest limit. Each run of simulations started at the lowest values of initial readiness and area of first particle discovered. The initial threshold was equal to 1.0. The run was aborted after either 12 particles had been attached, or number of particles discovered had exceeded 100. This matched roughly the speed of building by real larvae during the 1 hr experiment.

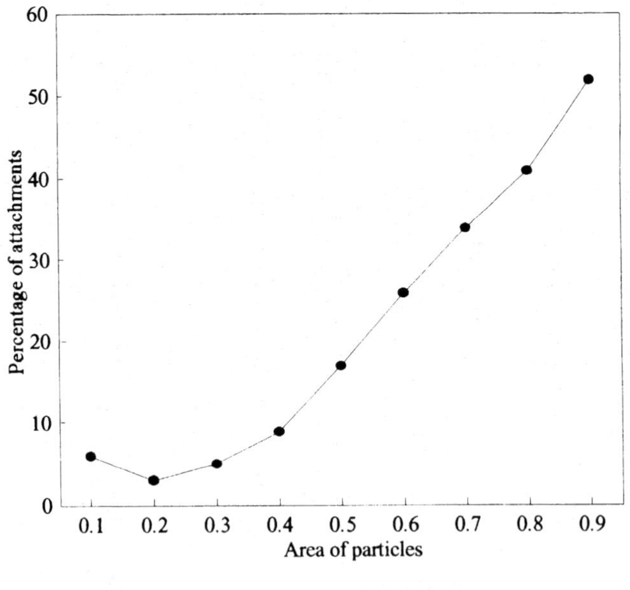

FIGURE 1: The dependence of percentage of attached particles on their area (arbitrary units) in the model (summed up data from 50 independent runs, a=0.89, b=0.83).

The results most similar to the behaviour of the real larvae were obtained using $\boldsymbol{a}$ = 0.80 to 0.90 and $\boldsymbol{b}$ = 0.75 to 0.85. Larger particles were preferred to smaller ones. On the other hand, some quantity of smaller particles was still attached, and a number of larger ones was rejected (Figure 1). If the size distribution of particles was made asymmetrical so that the encounter rate of larger particles increased, then greater amount of smaller particles were rejected. However, some number of small particles was always attached. Thus, the model reproduce the tendency to optimise the selection of particles, as well as the partial choice of less profitable particles and the partial rejection of more profitable ones observed in the experiment.

The repetition of attachments (A) and rejections (r) was observed in the majority of runs, for example (Wald-Wolfowitz test for binary sequences, $p<0.05$):

AAAArrrrrrrrrrrrrrrrArrrrrrArAAAAAA

despite the simulated sequence of particles was random. The series of repetitive attachments starts usually when a very large particle (i.e. close to 1.0) was discovered following a smaller particle resulting in a rise of the readiness and in the attachment. Then several smaller particles were also attached because of the high initial readiness. Thus, the model predicts that larvae would accept a less profitable particle after a more profitable one, if the difference between areas of the particles is not great and the readiness is high, and this would result in RRS.

4. Experiment

In order to verify the prediction, I provided larvae with large (2.0 to 2.6 square mm, L) and small (0.25 to 0.75 square mm, S) shell fragments in dishes. The ratio of encounter rates with Ls and Ss was 1:5. The average time D+T+A was equal to 200 s for attached S, and average time (D+T+A for attached L) + (D+T for all Ss rejected before L attachment) was equal to 370 s. The choice of Ls only under these conditions would have been optimal, because the average area of L was almost 5 times as large as that of S. I tested 26 individuals separately.

A significant preference of Ls to Ss was observed (Fisher exact test for each individual, $p<0.05$). However, both partial choice of Ss and partial rejection of Ls took place: in average, 20% and 40% of all discovered Ss and Ls respectively were attached to the case.

I derived three sequences S-S-attached L-S-S (notwithstanding were Ss attached or rejected) which occurred first in each individual record. Data from all individuals were combined for each of three sequences, and McNimara 'test for changes' was applied to compare frequencies of attachments of Ss. The frequency was greater after attached Ls than after rejected Ss in each of three sequences ($p<0.05$). Figure 2 shows data summed up for all 3 sequences. Significant tendency to repeat attachments and rejections was actually observed in 21 of 26 individuals tested, for example (Wald-Wolfowitz test, $p<0.05$):

AAArrrrrrrrrrArAAAAArrrrrrrrrrAAAAA

Series of attachments were usually initiated by the attachment of L after a number of rejected Ss, and then several Ss were also attached. One may conclude that the increased readiness really make the larvae to attach S after attached L, and this results in RRS.

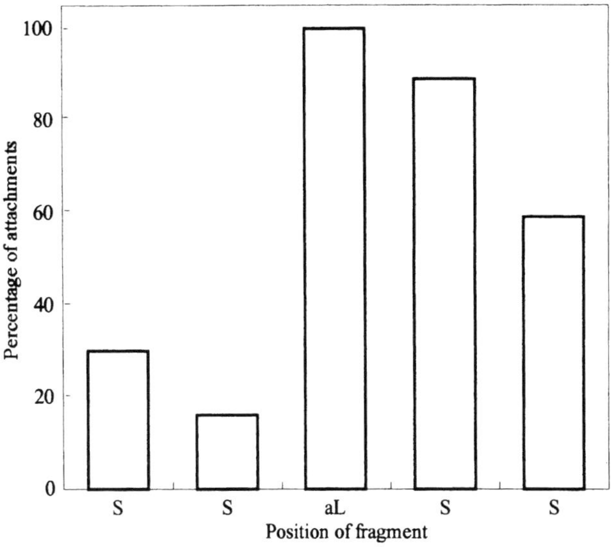

FIGURE 2: The dependence of percentage of attachments on the position of a small fragment in the sequence S - S - attached L-S-S.

Larvae tended to move to some other place in the dish after several consecutive rejections, but always stayed at the same spot and picked up the next particle after an attachment. This observation additionally suggests that the readiness to respond to a particle increased after an attachment.

5. Discussion

5.1. Origin of RRS

The model qualitatively matches the selection behaviour observed in the caddis fly larvae. It reproduces the preference for and partial rejection of more profitable particles. It correctly predicts RRS and the partial choice of less profitable particles. In the model, RRS and the partial choice results from variations of the readiness to respond to a particle. The model is extremely simplified, and various other causes of the partial choice may exist in the larvae's behaviour. In other animals, the partial choice always takes place (Werner and Hall, 1974; Krebs et al., 1977; Goss-Custard, 1977; Stephens, 1985), and various explanations for that were suggested (McNamara and Houston, 1987b). An animal may need sampling the environment to assess the relative profitability of different items, or it may be unable to identify them correctly (Krebs et al., 1977). If a considerable time is needed to recognise items, then it pays to accept some unprofitable ones (Hughes, 1979). However, the experiment (Section 4) showed that the larvae act in agreement with the model and really tend to accept an unprofitable item after they had accepted a profitable one.

I presume that the optimal selection in the larvae means that they try to maximise the speed of building and minimise the amount of silk spent for attachments. One may suggest another or additional explanations for the preference of large particles. For example, these particles may provide better camouflage or strengthen the case. However, the model explains the discrepancy between the preference and RRS independently of any gain the larvae maximise when they prefer large particles.

Larvae encountered particles of various size in the random sequence in our experiments, and one may consider the sequence to be a noise which influences the behaviour. In the model, the non-linear control system transforms the noise into RRS. The resulting non-random behaviour may be viewed upon as the case of self-organisation in the noise-driven system.

I based the model on three simple rules I had observed in the larvae. However, similar rules are likely to be used by other animals. The comparison of two particles by their relative difference in our model is just an another case of Weber's Law, and it was earlier shown that starlings compare amounts of consecutive food deliveries in accordance with the Law (Bateson and Kacelnik, 1995). The readiness of the animal to respond to an item was also shown to increase after the similar response was accomplished in the immediate past. Fish males, for instance, attack dummies of males more readily if they had a chance to attack an alive male shortly before (Heiligenberg, 1974). In the model, the threshold for an attachment depend on the last attached item, and does not depend on the whole sequence of attachments and rejections. The dependence on 'outstanding' events was also suggested earlier: The decision making in foraging starlings can be satisfactorily explained assuming that the birds only remember a minimum and maximum intervals between reinforcements (Brunner et al., 1996). If similar rules co-operate in the selection behaviour of many animal species, then one may expect RRS to be widespread.

5.2. RRS and the adaptive behaviour

Self-organising behaviour exhibits a non-random pattern in time and/or space. The pattern reflects basic principles of self-organisation, and may not aim at the optimal accomplishment of any task. If RRS matches the pattern of environment, it makes an animal well adapted to it. If not, then RRS would interfere with the adaptation.

Normally, caddis fly larvae crawl across a creek bottom seeking for food, appropriate site for moulting, etc. They don't test any particles before their case is damaged or need be enlarged after moulting, and the both events may occur in an unfamiliar area. For this reason, larvae have to start building using some general pre-existing strategy irrelevant to particular properties of the environment.

Building material is not distributed over the bottom randomly. Hard particles of various size, weight and shape accumulate in different patches because of heterogeneity in water velocity, turbulence and depth. If larva discover a particle of a certain type, then more particles of this type may be discovered nearby, and it is better to stay and pick up another particle instead of moving away. I pointed out (Section 4), that after an attachment larva tests another particle at the same place in the bottom, while may move to another place after rejections. This would help larva to allocate it's time more efficiently in the patched environment. Larva may also grasp the edge of a particle and identify initially its size and shape incorrectly; if particles of the same tipe are clamped together, then it pays to test a particle longer and more carefully after the previous one was found profitable. The larvae live in the world which is ordered in space and their behaviour sequence matches this order. In a sense, the self-organisation provides larvae with a simple deterministic model of their world and this may help them fairly efficiently.

The environment of some other animals is even more ordered. Flowers in an inflorescence produce nectar almost simultaneously, and bees tend to stay at an inflorescence if the first flower they visited contains nectar and leave to another inflorescence in the opposite case. As a result, the bee behaviour is almost optimal in terms of the rate of energy gain (Hodges, 1981). The similar movement pattern was also observed when artificial flowers were randomly distributed in space: the bees flew shorter distances following visits to rewarding flowers than after visits to empty flowers, as if rewarding flowers were clumped (Keasar et al., 1996). Any natural environment is ordered to some extent, and the repetition of similar events is a fairly good initial model to start the selection of items in a novel environment.

On the other hand, RRS interfere with the optimal selection. The larvae may assess the low profitability of a small shell fragment more correctly, when it tests it longer after a larger fragment. However, the larvae accept small fragments more frequently in this case (Figure 2). There is one more discrepancy between larva's decisions and the optimal behaviour. In models of the optimal behaviour, an animal should invest less time in unprofitable items, if the encounter rate of more profitable items becomes higher (McNamara and Houston, 1985). If an animal has limited memory, the invested time should depend on the recent experience (McNamara and Houston, 1987a). If larvae's memory is extremely limited and includes only one previous event, then one may expect the test duration of a sand grain to decrease after attaching a profitable shell fragment. However, the test duration increases (Section 2). It rather looks like the larvae expect similar events to repeat and so spend more time testing a sand grain after discovering a shell fragment. The animal, so to say, takes the model for the reality.

This means that self-organisation simultaneously helps animals to suit the environment and imposes constraints on the behavioural adaptation. The animals construct deterministic models of the world to foresee future events, while there are both determinism and randomness in the world. It was earlier pointed out that even the human brain is inclined to construct only deterministic models. The brain needs determinism and expects more order in the world than really exists. It even tries to find an order where there is no order at all, but the strategy of survival is based on such expectations (Turner and Pöppel, 1988).

5.3. Implications for the animat design

One may speculate that the principles of larvae's behaviour may be used as the prototype for the programming an autonomous agent. Consider the animat which is sent to a remote planet and select samples of some ore. Suppose that pieces of ore are spread over the surface of the planet and differ in profitability (in terms of size and metal content), but no information on the encounter rate and spatial distribution of pieces of different profitability is available. Suppose also that the animat has to collect maximal quantity of ore during predetermined period of time, which is too short to explore the environment before the start of selection. The animat will face a very fuzzy task. If the distribution of samples by size and ore content is not known, then no threshold, above which all samples should be considered good and collected, can be calculated. A handling of different pieces may take different time, and the animat must decide whether to keep processing a piece or pick up another one. The optimal decision is impossible if there is no information on the distribution of handling times (McNamara and Houston, 1985). The self-organising orderliness may help to accomplish the task in the absence of sufficient information on an environment. Caddis fly larvae accomplish the fuzzy task fairly efficiently, and so will do the animat, if it's behaviour is organised in similar way.

What seems more important, is that no animat can possibly be an optimal forager, even if it possesses the complete information on the environment and no constraints are imposed on it's memory, perception and learning abilities. Animat's behaviour is to be based on the principles of the animal behaviour (Maes, 1992). This implies that RRS and other possible manifestations of self-organising order should reveal themselves in it's behaviour and this would interfere with an optimal accomplishment of animat's task.

5.4. Future work

The larvae cannot possibly avoid errors when testing particles. The test takes a certain period of time, and the probability of errors should decrease as the test proceeds. The readiness may increase or decrease during the test depending on the sensory input. Under natural conditions, the larvae process particles: they clean them of a dirt and cut away soft parts. The readiness may depend on efforts invested in the processing. The density of particles may be low on a creek bottom, and a rather long time may be needed to discover a particle. The readiness may also depend on this time. I plan to investigate these issues experimentally and incorporate results into the model step by step. It may help to understand better how the self-organising system produces adaptive behaviour, as well as constraints it imposes on adaptation.

Acknowledgements

I thank Konstantin A. Podgorny, Pavel P. Umorin, Daniel Osorio and two anonimous reviewers for their helpful comments on the initial version of this paper. This work was supported in part by International Science Foundation (RLB000 and RLB300) and NATO (ENVIR.LG950428).

References

Bateson, M. and Kacelnik, A. Accuracy of the memory for amount in the foraging starling, *Sturnus vulgaris*. *Anim. Behav.*, 1995, 50, 431-443.

Brunner, D., Kacelnik, A. and Gibbon, J. Memory for inter-reinforcement interval variavility and patch departure decisions in the starling, *Sturnus vulgaris*. *Anim. Behav.*, 1996, 51, 1025-1045.

Cole, B. J. Chaos and behavior: the perspective of nonlinear dynamics. *Behavioral mechanisms in evolutionary ecology*, 1994, Ed. Real, L A. Univ. of Chicago Press, Chicago, 423-443.

Ferriere, R., Cazelles, B., Cezilly, F. and Desportes, J.-P. Predictability and chaos in bird vigilant behaviour. *Anim. Behav.* 1996, 52, 457-472.

Goss-Custard, J.D. Optimal foraging and selection of worms by redshank *Tringa totanus*. *Anim. Behav.*, 1977, 25, 10-29.

Heiligenberg, W. Processes governing behavioral states of readiness. *Advances in the Study of Behavior, Vol. 5*, 1974. Eds. Lehrmann, D.S., Rosenblatt, J. S., Hinde, R.A. and Shaw E. Academic Press, London, 173-200.

Hodges, C.M. Optimal foraging in bumblebees: hunting by expectation. *Anim. Behav.*, 1981, 29, 1166-1171.

Horsthemke, W. and Lefever, R. *Noise-induced phase transitions,* 1984. Springer-Verlag, Berlin.

Hughes, R.N. Optimal diets under energy maximization premise: the effects of recognition time and learning. *Amer. Naturalist*, 1979, 113, 209-221.

Johansson, A. Caddis larvae cases (Trichoptera, Limnephilidae) as anti-predatory devices against brown trout and sculpin. *Hydrobiologia*, 1991, 211, 185-194.

Keasar, T., Shmida, A. and Motro, U. Innate movement rules in foraging bees: flight distances are affected by recent rewards and are correlated with choice of flower type. *Discussion Paper # 120*, 1996. Center for Rationality and interactive Decision Theory, Jerusalem.

Krebs, J.R., Erichsen, J.T., Webber, M.I. and Charnov, E.L. Optimal prey selection in the great tit (*Parus major*). *Anim.Behav.*, 1977, 25, 30-38.

Maes, P. Behavior-based artificial intelligence. *Proc. SAB'92,* 1992. Eds. Mayer, J.-A., Roitblat, H.L. and Wilson, S.W. MIT Press, Massachusetts, 2-10.

McNamara, J. and Houston, A. A simple model of information use in the exploitation of patchily distributed food. *Anim. Behav.*, 1985, 33, 553-560.

McNamara, J. and Houston, A.I. Memory and the efficient use of information. *J. Theor. Biol.*, 1987a, 125, 385-395.

McNamara, J. and Houston, A.I. Partial preferences and foraging. *Anim. Behav.*, 1987b, 35, 1084-1099.

Nepomnyashchikh, V.A. The control of selection of particles for case building by larvae of *Chaetopteryx villosa* Fabr. (Trichoptera: Limnephilidae). *Proc. 7th Int. Symp. Trichoptera (Umea, Sweden, 1992),* 1993a. Ed. Otto, Ch. Universal Book Services, Oegstgeest, Netherlands, 263-265.

Nepomnyashchikh, V.A. Behaviour of caddisfly, *Chaetopteryx villosa* Fabr. (Limnephilidae: Trichoptera), larvae and optimal foraging theory. *Journal of General Biology*, 1993b, 54, 739-744 (in Russian).

Nepomnyashchikh, V.A. and Podgorny, K.A. The orderly sequence of behavior in the larvae of caddisfly, *Chaetopteryx villosa* Fabr. (Limnephilidae: Trichoptera) caused by random stimuli sequences. *Journal of General Biology*, 1994, 55, 328-336 (in Russian).

Otto, C. and Johansson, A. Why do some caddis larvae in running waters construct heavy, bulky cases? *Anim.Behav.*, 1995, 49, 473-478.

Stephens, D.W. How important are partial preferences? *Anim. Behav.*, 1985, 33, 667-669.

Stephens, D.W. and Krebs, J.R. *Foraging Theory,* 1986. Princeton University Press, Princeton, New Jersey.

Thomas, G. The ifluence of eating and rejecting prey items upon feeding and food searching behaviour in *Gasterosteus aculeatus* L. *Anim.Behav.*, 1977, 25, 52-66.

Turner, F. and Pöppel, E. The poetry, brain and time. *Beauty and the brain,* 1988. Eds. Rentschler, I., Herzberger, B. and Epstein, D. Birkhauser Verlag, Basel, 74-96.

Werner, E.E. and Hall, D.J. Optimal foraging and the size selection of prey by the bluegill sunfish (*Lepomis macrochirus*). *Ecology*, 1974, 55, 1042-1052.

Solving a Complex Prisoner's Dilemma with Self-Modifying Policies

Jieyu Zhao and Jürgen Schmidhuber
IDSIA, Corso Elvezia 36, CH-6900-Lugano, Switzerland
jieyu@idsia.ch, juergen@idsia.ch

Abstract

Self-modifying policies (SMPs) trained by the success-story algorithm (SSA) have been successfully applied to various difficult reinforcement learning tasks (Schmidhuber et al. 1997a, 1997b). Here we present new results on an application where two cooperating/competing animats have to solve a complex version of the prisoner's dilemma.

1 Overview

SMP/SSA. An animat's modifiable components that determine its behavior are called its policy. An algorithm that modifies the policy is called a learning algorithm. If the learning algorithm has modifiable components represented as part of the policy, then we speak of a self-modifying policy (SMP). SMPs can modify the way they modify themselves etc. They are of interest in situations where the initial learning algorithm itself can be improved by experience. How can we force some (stochastic) SMP to trigger better and better self-modifications? The *success-story algorithm* (SSA) addresses this question in a lifelong reinforcement learning context. During the learner's life-time, SSA is occasionally called at times computed according to SMP itself. SSA uses backtracking to undo those SMP-generated SMP-modifications that have not been empirically observed to trigger lifelong reward accelerations (measured up until the current SSA call — this evaluates the long-term effects of SMP-modifications setting the stage for later SMP-modifications). SMP-modifications that survive SSA represent a lifelong success history. This scheme has been used to solve several difficult reinforcement learning tasks unsolvable by more traditional RL methods tasks (Schmidhuber et al., 1997a, 1997b).

Outline. Basic principles of SMPs and SSA will be briefly reviewed in Section 2. Section 3 will describe an animat with a complex stochastic SMP, and Section 4 will present experiments with two such animats that need to learn to cooperate in a relatively complex environment to accelerate their reinforcement intake.

2 Fundamental Issues

2.1 A Single Animat

In what follows we will review basic concepts presented in (Schmidhuber et al., 1997a). Each animat lives in an unknown environment E from time 0 to unknown time T. Life is one-way: even if it is decomposable into numerous consecutive "learning trials", time will never be reset. The animat has an internal state S and a *self-modifying policy* SMP. Both S and SMP are variable dynamic data structures influencing probabilities of actions to be executed by the animat. Between time 0 and T, the animat repeats the following cycle over and over again ($\mathcal{A}$ denotes a set of possible actions):

> **REPEAT: select and execute $a \in \mathcal{A}$ with conditional probability $P(a \mid SMP, S)$.**[1]

Action a will consume time (Russell and Wefald, 1991; Boddy and Dean, 1994) and may change E, S, SMP.

Primitive learning algorithms (PLAs). Actions in $\mathcal{A}$ that modify SMP are called *primitive learning algorithms* (PLAs). To ensure non-vanishing exploration potential PLAs may not generate SMP-modifications that will let certain action probabilities vanish entirely. PLAs and other actions can be combined to form more complex (probabilistic) learning algorithms.

Checkpoints. The entire lifetime of the animat can be partitioned into time intervals separated by special times called *checkpoints*. In general, checkpoints are computed dynamically during the animat's life by certain actions in $\mathcal{A}$ executed according to SMP itself. The animat's k-th checkpoint is denoted c_k. Checkpoints obey the following rules: (1) $\forall k \;\; 0 < c_k < T$. (2) $\forall j < k \;\; c_j < c_k$. (3) Except for the first, checkpoints may not occur before at least one SMP-modification has been computed by some PLA since the previous checkpoint.

Sequences of SMP-modifications (SSMs). SSM_k denotes the sequence of SMP-modifications computed by

[1] Instead of using the expression *policy* for the conditional probability distribution P itself we reserve it for the animat's modifiable data structure SMP.

PLAs in between c_k and c_{k+1}. Since PLA execution probabilities depend on SMP, SMP influences the way it modifies itself.

Reward. Occasionally E provides real-valued reward. The cumulative reward obtained by the animat in between time 0 and time $t > 0$ is denoted $R(t)$, where $R(0) = 0$.

Animat goals. The animat's goal at checkpoint t is to use its experience to generate SMP-modifications accelerating reward intake: it wants to let $\frac{R(T)-R(t)}{T-t}$ exceed the current average speed of reward intake. To determine this speed it needs a previous point $t' < t$ to compute $\frac{R(t)-R(t')}{t-t'}$.

A problem. How can t' be specified in a general yet reasonable way? For instance, if life consists of many successive "trials" with non-deterministic outcomes (e.g., depending on other animats' behaviors), how many trials should the animat look back into time to evaluate its current performance? This question will be addressed by the success-story criterion introduced next.

2.2 Success-Story Algorithm

Success-story criterion (SSC). Let V denote the animat's time-varying set of past checkpoints that have been followed by long-term reward accelerations. Initially V is empty. v_k denotes the k-th element of V in ascending order. SSC is satisfied at time t if either V is empty (trivial case) or if

$$\frac{R(t) - R(0)}{t - 0} < \frac{R(t) - R(v_1)}{t - v_1} < \ldots < \frac{R(t) - R(v_{|V|})}{t - v_{|V|}}. \tag{1}$$

SSC demands that each checkpoint in V marks the beginning of a long-term reward acceleration measured up to the current time t.

Success-Story Algorithm (SSA). At every checkpoint we invoke the *success-story algorithm* (SSA):

1. **WHILE** SSC is not satisfied

   Undo all SMP modifications made since the most recent checkpoint in V.
   Remove that checkpoint from V.

2. Add the current checkpoint to V.

"Undoing" a modification means restoring the preceding SMP — this requires storing past values of SMP components on a stack prior to modification. (Components of SMP and elements of V can be stored on the same stack — see section 3.) Thus each SMP modification that survived SSA is part of a bias shift generated after a checkpoint marking a lifelong reward speed-up.

Implementing SSA. Using stack-based backtracking methods such as those described in section 3, one can *guarantee* that SSC will be satisfied after each checkpoint, despite interference from other animats. Although inequality (1) contains $|V|$ fractions, SSA can be implemented efficiently: only the two most recent still valid sequences of modifications (those "on top of the stack") need to be considered at a given time in an SSA call (see details in section 3). A *single* SSA call, however, may invalidate *many* sequences of modifications if necessary.

Timing SSA calls. Since the way of setting checkpoints depends on SMP, SMP can learn *when* to evaluate itself using SSA. This evaluation-timing ability of SMP is important in dealing with unknown reward delays.

Trials and "effective" trials. At any given checkpoint the animat has got just a *single* training example concerning the usefulness of any still existing SSM, namely, the reward per time since it occurred. All checkpoints in V represent starts of a yet unfinished "effective" trials (as opposed to externally defined trials with prewired starts and ends). In absence of empirical evidence to the contrary, each still valid SSM is assumed to have successfully layed the ground for later valid SSM. The older some surviving SSM, the more time will have passed to collect statistics concerning its long-term consequences, and the more stable it will be if it is indeed useful and not just there by chance.

Measuring effects of learning on later learning. Some SSM's success recursively depends on the success of later SSMs for which it sets the stage. Hence SSA (unlike previous RL methods) automatically measures earlier learning's influence on later learning. This provides a basis for "learning to learn" or "metalearning": SSA prefers SSMs making "good" future SSMs more likely. In other words, SSA prefers probabilistic learning algorithms leading to better probabilistic learning algorithms.

SMP/SSA's limitations. In general environments neither SMP/SSA nor any other scheme is guaranteed to find "optimal" policies that will lead to maximal cumulative reward for each animat (we will never know for sure which stocks will go up). SSA is only guaranteed to selectively undo those SSMs that were not empirically observed to lead to an overall speed-up of the animat's average reward intake (still, this is more than can be said about alternative RL algorithms). In particular, in "unfriendly" environments where the reward constantly decreases no matter which policy changes get executed, there is no way of justifying *permanent* SSMs by SSC (a trivial way of satisfying SSC is to never make a SSM).

Let us assume, however, that E, S and $\mathcal{A}$ (representing the system's initial bias) do indeed allow for SSMs triggering long-term reward accelerations. This is an instruction set-dependent assumption much weaker than the typical Markovian assumptions made in previous RL work, e.g., (Kumar and Varaiya, 1986; Sutton, 1988; Watkins and Dayan, 1992). Now, since we prevent all instruction probabilities from vanishing (see concrete implementation in section 3), SMP cannot help triggering

SSMs occasionally, and keeping those consistent with SSC. In this sense, the learner cannot help getting better.

3 A Self-Modifying Policy

Overview. In what follows, we will describe details of a single animat implementing the principles presented in section 2 (several such animats will interact in the experiment). Its self-modifying policy SMP is a set of variable probability distributions over a set of pre-designed instructions that include SMP-modifying and checkpoint-setting instructions.

Architecture. The architecture implementing the SMP is similar to Schmidhuber's (1995, 1997). There are m addressable *program cells* with addresses ranging from 0 to $m-1$. The variable, integer-valued contents of the program cell with address i are denoted d_i. An internal variable *Instruction Pointer* (*IP*) with range $\{0, \ldots, m-1\}$ always points to one of the program cells (initially to the first one). *IP*'s current position represents the internal state S (and can be used to disambiguate inputs). Instructions and their parameters are encoded by a fixed set I of n_{ops} integer values $\{0, \ldots, n_{ops}-1\}$. For each value j in I, there is an instruction B_j with N_j integer-valued parameters.

Stochastic, self-modifying policy. For each program cell i there is a variable probability distribution SMP_i over I. For every possible $j \in I$, $(0 \leq j \leq n_{ops}-1)$, SMP_{ij} specifies for cell i the conditional probability that, when pointed to by *IP*, its contents will be set to j and instruction B_j will be executed. The set of all current SMP_{ij}-values defines a probability matrix SMP (the learner's *current stochastic policy*) with columns SMP_i $(0 \leq i \leq m-1)$. If $IP = i$, then the contents of i, namely d_i, will be interpreted as instruction B_{d_i}, and the contents of cells immediately following i will be interpreted as B_{d_i}'s arguments, selected according to the corresponding SMP-values. See Figure 1.

Normal instructions. Normal instructions are those that do not change the SMP. An example is B_0 (later we will introduce many additional, problem-specific instructions for interaction with an environment):

B_0: *JumpHome* — *set* $IP = 0$ *(jump back to 1st program cell).*

Primitive learning algorithms (PLAs). By definition, a PLA is an instruction that modifies SMP. PLAs such as *IncProb* below and other instructions may be composed sequentially to form complex (probabilistic) learning algorithms. In what follows, the symbols $a1, a2, a3$ stand for instruction parameters in $\{0, 1, \ldots, n_{ops}-1\}$.

B_1: *IncProb(a1, a2, a3)* — Set $i := (a1 * n_{ops} + a2)/h$ (integer division); set $j := a3$. If $0 \leq i \leq m-1$, increase SMP_{ij} by α percent, and renormalize SMP_i (but prevent SMP-components from falling below a minimal value ϵ, to avoid near-determinism). We will use $\alpha = 15, \epsilon = 0.001$, $h = 3$.

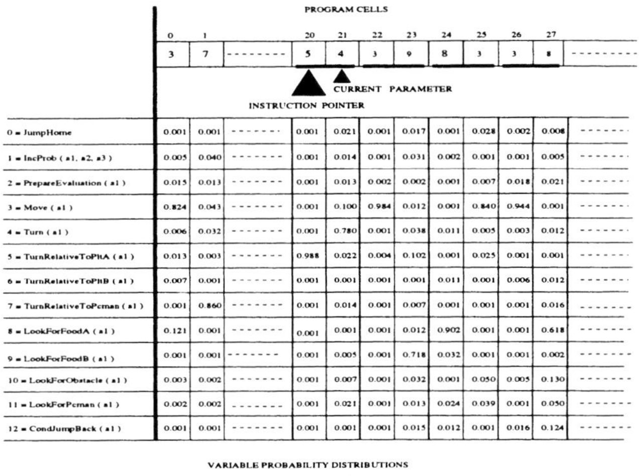

Figure 1: *Parts of a self-modifying policy SMP based on the parameterized instructions to be used in Experiment 3. IP currently points to program cell 20. The integer sequence 5 4 3 9 8 3 3 8 ..... is generated online according to SMP's current probability distributions. This immediately triggers execution of the instruction sequence* $B_5(4)B_3(9)B_8(3)B_3(8)$..... *with interpretation "Turn to plantation A; move 6 steps forward; look for food A, if it is within 4 steps in front then move 5 steps forward; otherwise, ..." (details in section 4).*

Although our PLAs do not allow for composing *arbitrary* probabilistic learning algorithms, they do allow for a wide variety of them. Later we will introduce another instruction: a call of Q-learning (Watkins and Dayan, 1992).

Evaluation instruction. By executing the instruction *"PrepareEvaluation"* the learner can influence the timing of the next checkpoint triggering an evaluation of all SMP-modifications that have survived so far:

B_2: *PrepareEvaluation(a1)* — Temporarily disable PLAs (e.g., *IncProb*) by preventing them from executing further policy modifications, until $5 \times (a1+1)$ (values chosen arbitrarily) additional, non-zero reinforcement signals have been received (this will define a checkpoint and trigger an SSA call — see basic cycle below).

Initialization. At time 0 (system birth) we initialize SMP with maximum entropy distributions (all SMP_{ij} equal), and set $IP = 0$. We introduce an initially empty stack $\mathcal{S}$ that allows for variable-sized stack entries, and the conventional *push* and *pop* operations. The integer variable $NumR$ (initially zero) is used to count down the remaining non-zero reward signals since the last execution of *"PrepareEvaluation"*. The Boolean variable SMP*changed* is initially false.

Basic cycle of operations. Starting at time 0, until time T (system death), the system repeats the following basic instruction cycle over and over again in real time:

1. Randomly generate an integer $j \in I$ according to matrix column SMP_{IP} (the probability distribution of the program cell pointed to by *IP*). Set program cell content $d_{IP} := j$. Translate j into the corresponding current instruction B_j. Look up the number N_j of cells required to store B_j's parameters. **IF** *IP*$> m - N_j - 2$, **THEN** set *IP* to 0, go to step **1.** **ELSE** generate instruction arguments for the N_j cells immediately following *IP*, according to their probability distributions SMP_{IP+1}, ..., SMP_{IP+N_j}, and set $IP \leftarrow IP + N_j + 1$.
2. **IF** B_j is a PLA, and not currently disabled by a previous *"PrepareEvaluation"* instruction, **THEN** push copies of those SMP_i to be modified by B_j onto $\mathcal{S}$.
3. Execute instruction B_j. This may change (1) environment, (2) *IP*, (3) SMP itself. In case (3), set SMP*changed* $= TRUE$. **IF** B_j is *"PrepareEvaluation"* and SMP*changed* $= TRUE$, **THEN** set $NumR$ equal to B_j's actual parameter plus one.
4. **IF** $NumR > 0$ and non-zero reward occurred during the current cycle, **THEN** decrement $NumR$. **IF** $NumR =$ 0, **THEN** call SSA to achieve the success-story criterion SSC by backtracking as follows:

SSA.1. Set variable t equal to current time (t is a checkpoint, see section 2).
IF there is no *"tag"* stored somewhere in $\mathcal{S}$ (tags are pairs of times and rewards pushed in earlier SSA calls — they reflect surviving checkpoints in V — see section 2), **THEN** push the tag (t, $R(t)$) onto $\mathcal{S}$, set SMP*changed* $= FALSE$, and go to **1** (this ends the SSA call).

SSA.2. Denote the topmost tag in $\mathcal{S}$ by (t', $R(t')$). Denote the one below by (t'', $R(t'')$) (if there isn't any tag below, set variable $t'' = 0$ — recall $R(t'') = R(0) = 0$).

SSA.3. **IF**

$$\frac{R(t) - R(t')}{t - t'} > \frac{R(t) - R(t'')}{t - t''}$$

THEN push (t, $R(t)$) as a new tag, set *SMPchanged* $= FALSE$, and go to **1.** This ends the SSA call.
ELSE pop off all stack entries above the one for tag (t', $R(t')$) (these entries will be former SMP-components saved during an earlier execution of step 2), and use them to restore SMP as it was before time t'. Then also pop off tag (t', $R(t')$). Go to **SSA.1**.

Each step above will consume various amounts of system life-time. Note a couple of differences to genetic programming (GP) (Cramer, 1985; Dickmanns et al., 1987): (1) no program populations, (2) self-modification ability, (3) stochastic policy, (4) instructions generated and executed online, (5) SSA calls evaluating the entire life so far and retrospectively adjusting effective trial starts.

Lifelong reward acceleration. After each call of SSA the history of still valid modifications is *guaranteed* to be a life-time success-story (in the worst case an empty one). There is no need to assume fully observable environments.

Speed? Due to the generality of the approach, no reasonable statements can be made about improvement *speed*, which indeed highly depends on the nature of E and $\mathcal{A}$. This lack of quantitative convergence results is shared by almost all other, less general RL and multiagent schemes.

4 A DIFFICULT POE TASK

The task in this section involves a complex partially observable environment (POE), and there are no externally defined trials. It may be regarded as a complex POE version of the prisoner's dilemma, e.g., (Axelrod, 1984).

Environment. There are two SMP/SSA-based animats A (black) and B (white) living in a 400×600 pixel POE with several obstacles (Figure 2). Two food types (food A and food B) are growing in the environment. Each piece of food occupies an area of 20×20 pixels. For some reason animat A can consume only food A (by moving to the food's location and touching it — then it will disappear), while animat B can consume only food B. During each basic cycle with probability 0.0001 random type food will grow at a random location (if outside of an obstacle). In two ***plantations*** with size 80×120 pixels, however, animats may also plant each other's food. During any given basic cycle, if animat B's center is in plantation A in the northwest corner, then it will replenish plantation A by planting up to $4 \times 4 = 16$ pieces of food A separated in a grid-like fashion (see Figure 2). Similarly, animat A may plant food B at plantation B in the southeast corner.

The situation is further complicated by a non-adaptive, prewired, circle-shaped pacman (with diameter of 50 pixels) trying to chase and catch the animats. The pacman's step size is 2 pixel widths. Normally during each basic cycle the pacman will (1) move one step in its current direction with probability 0.4, or (2) turn to face the closer animat and move one step forward with probability 0.2, or (3) randomly turn 45 degrees to the right or left with probability 0.4. In case the pacman collides with an obstacle, however, it will follow its boundary in the direction that best matches its current direction for the next 10 basic cycles or until it hits another obstacle. The pacman's strategy is sufficient to avoid being trapped by obstacles.

At birth (time 0) both animats are randomly rotated and placed near the northwest corner, the pacman near the southeast corner.

Reward. Reward/punishment is generated *only* if an animat encounters its food type (reward 1.0) or if it gets caught by the pacman (-10.0). Planting the other animat's food does not yield reward and seems like a waste of time at first glance: it requires the planter to move over to the other plantation where there is no food for it (except maybe the rare random food).

To significantly accelerate their reward intakes in the long run, however, both animats must learn to cooperate: they may not only greedily consume their own food but must also plant the other's food.

Animat details. Both agents share the same design. Each is equipped with limited "active" sight: by executing certain instructions, it can sense obstacles, food A and B, or the pacman, within up to 10 "steps" in front of it. The step size is 5 pixel widths. The agent can also move forward, turn around, turn relative to plantation A or Plantation B or the pacman. Each agent has $m = 52$ program cells, and $n_{ops} = 13$ instructions (including JumpHome/PLA/evaluation instructions B_0, B_1, B_2 from section 3):

B_3: *Move(a1)* — If $0 \leq a1 \leq 9$, then move $a1-3$ (if $3 \leq a1 \leq 9$) steps forward or $3 - a1$ (if $0 \leq n < 3$) steps backward in the current direction (Otherwise there is no action.)

B_4: *Turn(a1)* — If $0 \leq a1 \leq 7$ then change current direction D to $(D + \frac{a1}{4}\pi) mod(2\pi)$.

B_5: *TurnRelativeToPlantation A(a1)* — If $0 \leq a1 \leq 7$ then first turn towards the vector connecting the centers of animat and plantation A, then *Turn(a1)*.

B_6: *TurnRelativeToPlantation B(a1)* — (analogous to B_5).

B_7: *TurnRelativeToPacman(a1)* — (analogous to B_5).

B_8: *LookForFood A(a1)* — If $0 \leq a1 \leq 9$: if there is no food A within $a1+1$ steps in front of the animat then increase *IP* by 4 (this is a limited kind of conditional jump). If $IP > m - 1$ then set $IP = 0$.

B_9: *LookForFood B(a1)* — (analogous to B_8).

B_{10}: *LookForObstacle(a1)* — (analogous to B_8).

B_{11}: *LookForPacman(a1)* — (analogous to B_8).

B_{12}: *CondJumpBack(a1)* — If $a1 = 0$ and the animat is not in plantation A, or if $a1 = 1$ and the animat is not in plantation B, then reset animat's IP to the instruction next to last executed *CondJumpBack* (if there is no such instruction, reset $IP = 0$).

Results without learning. If we switch off the animats' self-modifying capabilities (*IncProb* has no effect), then the average reward intake speed is about -6.0×10^{-4} per instruction cycle and animat.

Results with SMP/SSA. Both animats die after 0.8×10^9 instruction cycles. By then, average reward intake per animat is 1.4×10^{-3} per instruction cycle (mean of 4 simulations). Performance is plotted against time in Figure 3.

Following the animats' behavior on the output screen we observe that both animats learn (1) to escape the pacman and (2) to cooperate with each other. A typical solution looks like this: together both animats go to one of the plantations. There one animat starts planting while the other is consuming. Soon the pacman will be attracted by this and approach the plantation. They let it get close but not too close. Then they both escape and move over to the other plantation. Since they can move faster than the pacman, this will buy them a little bit of time to cooperate again: the previous planter now consumes and the previous consumer plants. Then the cycle repeats itself in similar form.

Results without pacman. During the first 5×10^6 instruction cycles of runs without pacman the average reward intake speed per instruction cycle and animat is about 0.97×10^{-4}. After 0.8×10^9 instruction cycles it increases to 0.45×10^{-3}. Although the pacman greatly reduces initial reward intake speed it supports the animats' exploration process and thus helps to speed up learning.

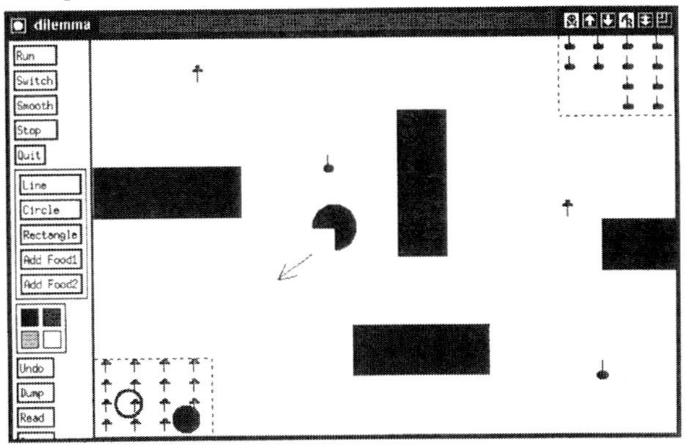

Figure 2: *Two interacting learning animats A and B in a POE with huge state space. After* 0.8×10^9 *basic cycles both animats know how to escape the pacman and how to cooperate by planting each other's food. In the snapshot above animat A is currently planting and B is consuming in plantation B (southwest corner) as the pacman is approaching.*

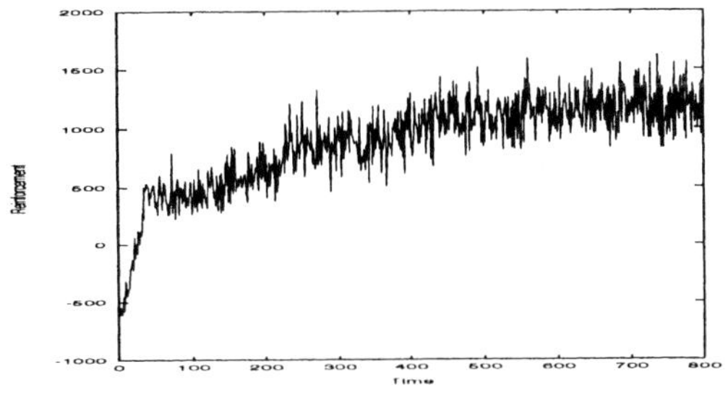

Figure 3: *Performance of a planting/consuming SMP/SSA-based animat. Reward intake is sampled at intervals of* 10^6 *instruction cycles (mean of 4 simulations).*

5 Conclusion

Unlike stochastic hill-climbing and other elitist methods such as genetic algorithms, SSA-based policy evaluation does not heavily depend on *a priori* knowledge about reasonable trial lengths. SSA never quits evaluating any previous policy change that has not yet been undone for lack of empirical evidence that it has contributed to a lifelong reward acceleration. Each invocation of SSA retrospectively establishes a success history of surviving self-modifications, each having been followed by more reward per time than all previous ones. We successfully exploited SSA's strength in dealing with uncertain policy evaluations in the context of multiagent learning with stochastic, self-modifying policies.

Future research. The primitive learning algorithms (PLAs) we plugged into SMP/SSA are quite simple. In future research we intend to plug in more powerful PLAs (such as recurrent neural net algorithms, Bayesian analysis, etc.). SSA will automatically take into account the time consumed by calls of various learning algorithms, and will focus on those leading to rewarding experiences.

6 ACKNOWLEDGMENTS

Jieyu Zhao is supported by SNF grant 2100-49'144.96 "Long Short-Term Memory".

References

Axelrod, R. (1984). *The Evolution of cooperation.* Basic Books, New York, NY.

Berry, D. A. and Fristedt, B. (1985). *Bandit Problems: Sequential Allocation of Experiments.* Chapman and Hall, London.

Boddy, M. and Dean, T. L. (1994). Deliberation scheduling for problem solving in time-constrained environments. *Artificial Intelligence*, 67:245–285.

Cramer, N. L. (1985). A representation for the adaptive generation of simple sequential programs. In Grefenstette, J., editor, *Proceedings of an International Conference on Genetic Algorithms and Their Applications*, Hillsdale NJ. Lawrence Erlbaum Associates.

Dickmanns, D., Schmidhuber, J., and Winklhofer, A. (1987). Der genetische Algorithmus: Eine Implementierung in Prolog. Fortgeschrittenenpraktikum, Institut für Informatik, Lehrstuhl Prof. Radig, Technische Universität München.

Gittins, J. C. (1989). *Multi-armed Bandit Allocation Indices.* Wiley-Interscience series in systems and optimization. Wiley, Chichester, NY.

Kumar, P. R. and Varaiya, P. (1986). *Stochastic Systems: Estimation, Identification, and Adaptive Control.* Prentice Hall.

Lin, L. (1993). *Reinforcement Learning for Robots Using Neural Networks.* PhD thesis, Carnegie Mellon University, Pittsburgh.

Russell, S. and Wefald, E. (1991). Principles of Metareasoning. *Artificial Intelligence*, 49:361–395.

Schmidhuber, J. (1995). Discovering solutions with low Kolmogorov complexity and high generalization capability. In Prieditis, A. and Russell, S., editors, *Machine Learning: Proceedings of the Twelfth International Conference*, pages 488–496. Morgan Kaufmann Publishers, San Francisco, CA.

Schmidhuber, J. (1997). Discovering neural nets with low Kolmogorov complexity and high generalization capability. *Neural Networks*, 10(5):857–873.

Schmidhuber, J., Zhao, J., and Schraudolph, N. (1997a). Reinforcement learning with self-modifying policies. In Thrun, S. and Pratt, L., editors, *Learning to learn*, pages 293–309. Kluwer.

Schmidhuber, J., Zhao, J., and Wiering, M. (1997b). Shifting inductive bias with success-story algorithm, adaptive Levin search, and incremental self-improvement. *Machine Learning*, 28:105–130.

Sutton, R. S. (1988). Learning to predict by the methods of temporal differences. *Machine Learning*, 3:9–44.

Watkins, C. (1989). *Learning from Delayed Rewards.* PhD thesis, King's College, Oxford.

Watkins, C. J. C. H. and Dayan, P. (1992). Q-learning. *Machine Learning*, 8:279–292.

Zhao, J. and Schmidhuber, J. (1996). Incremental self-improvement for life-time multi-agent reinforcement learning. In Maes, P., Mataric, M., Meyer, J.-A., Pollack, J., and Wilson, S. W., editors, *From Animals to Animats 4: Proceedings of the Fourth International Conference on Simulation of Adaptive Behavior, Cambridge, MA*, pages 516–525. MIT Press, Bradford Books.

NAVIGATION

Modeling Ant Navigation with an Autonomous Agent

Ralf Möller*, Dimitrios Lambrinos*, Rolf Pfeifer*
Thomas Labhart, Rüdiger Wehner****

* Department of Computer Science, University of Zurich
Winterthurerstr. 190, 8057 Zurich, Switzerland
(moeller,lambri,pfeifer)@ifi.unizh.ch

** Department of Zoology, University of Zurich
Winterthurerstr. 190, 8057 Zurich, Switzerland
(labhart,rwehner)@zool.unizh.ch

Abstract

The ability to navigate in a complex environment is crucial for both animals and robots. Many animals use a combination of strategies to return to significant locations in their environment. For example, the desert ant *Cataglyphis* is able to explore its desert habitat for hundreds of meters while foraging and return back to its nest precisely and on a straight line. The three main strategies that *Cataglyphis* is using to accomplish this task are path integration, visual piloting and systematic search (Wehner et al., 1996). In this study, we use the *autonomous agents* approach (Pfeifer, 1996) to gain additional insights into the navigation behavior of *Cataglyphis*. Inspired by the insect's navigation system we have constructed mechanisms for path integration and visual piloting that were successfully employed on the mobile robot Sahabot 2. The results of the navigation experiments indicate that a combination of these two mechanisms is capable to guide the agent precisely back to the target position.

1. Introduction

Bees and ants perform impressive navigation tasks in spite of their tiny ($< 1mm^3$) brains. For example, desert ants *Cataglyphis* individually perform foraging excursions that take them up to 200 m away from their nest. On finding a suitable prey, they return home unfailingly and in a straight line (Wehner, 1987).

More than two decades of field work (starting 1969, for a review see (Wehner, 1997)) have unraveled a lot of details about the behavioral repertoire that *Cataglyphis* employs when homing. The three main strategies used are path integration, visual piloting and systematic search (Wehner et al., 1996). Whereas in path integration compass information is primarily gained from the polarization pattern of the sky, geocentered information based on landmarks is also used in order to finally pinpoint its nest. Although there is a large number of behavioral data about the navigation behavior of *Cataglyphis*, and some mechanisms of peripheral signal processing have been unravelled, it is still largely unknown how the navigation system is implemented in the insect's nervous system. In this paper we use the *autonomous agents* approach to gain additional insights into the navigation behavior of insects. The goal of this approach is to develop an understanding of natural systems by building a robot that mimics some aspects of their behavior. The advantages of proceeding in this way are illustrated in our recent studies, in which an autonomous agent navigated by using directional information from skylight. A polarized light compass was developed that was based on neurophysiological evidence about the insect's neuronal circuit (Lambrinos et al., 1997).

In the current study we went one step further. First, we developed a path-integration system employing a polarized light compass. Second, based on biological evidence we implemented a landmark navigation system. Both systems were tested with robotic experiments.

2. Path integration

Central-place foragers such as bees and ants return to important places in their environment by primarily employing path integration, i.e., by means of a home vector that they continuously update by integrating all angles steered and all distances covered. To use this navigation mechanism, both distance information, but even more important, directional information must be available. Compass direction is gained from celestial cues, primarily from the polarization pattern of the blue sky (reviewed in (Wehner, 1994)).

2.1 Polarization vision in insects

Whereas in unpolarized light the e-vector oscillation occurs equally in all planes, in plane-polarized light there is a bias in favor of a certain oscillation plane (direction of polarization). Polarization in the sky occurs when the sunlight is scattered by atmospheric molecules. The degree of polarization is greatest for light scattered at an angle of 90° to the sunlight rays. The directions of polarization (e-vectors) form a regular pattern across the

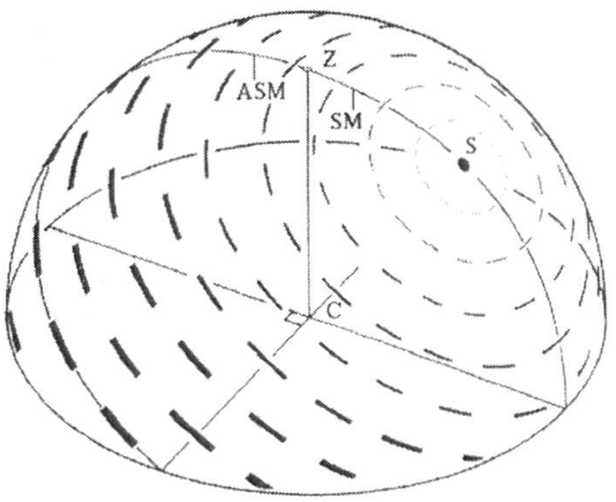

Figure 1 The pattern of skylight polarization in the sky as experienced by an observer in the center of the celestial hemisphere (C). Polarization in the sky occurs when the sunlight is scattered by atmospheric molecules. The directions of polarization (e-vector directions) form a regular pattern which is characterized by a symmetry plane formed by the solar (SM) and the anti-solar meridian (ASM). The solar meridian is the perpendicular from the zenith (Z) through the sun (S) down to the horizon. e-vector directions and degree of polarization are indicated by the orientation and the width of the black bars. In this figure, the solar elevation is 53° (from (Wehner, 1982)).

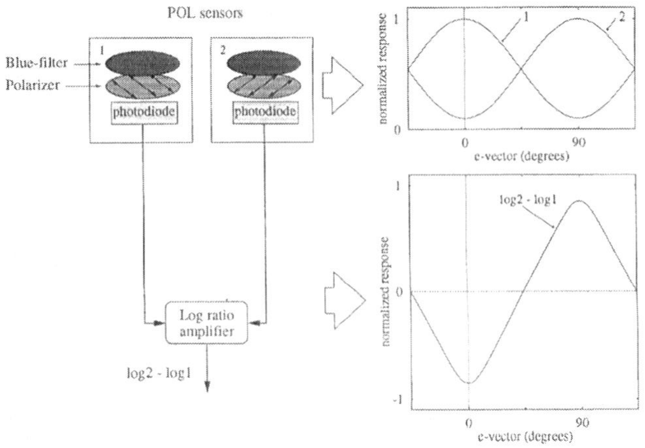

Figure 2 Diagrammatic description of a polarization-opponent unit (POL-OP unit). A POL-OP unit consists of a pair of POL-sensors and a log-ratio amplifier. The log-ratio amplifier receives input from the two POL-sensors and delivers the difference of their logarithmized signals. The e-vector responses of the POL-sensors (1, 2) follow a cos^2-function. Ordinates: normalized response, abscissa: e-vector direction.

entire celestial hemisphere (figure 1). The pattern of polarization has a mirror symmetry with respect to the plane defined by the solar and the anti-solar meridian. Due to the daily westward movement of the sun across the sky (by some 15° per hour), the symmetry plane, and with it the whole e-vector pattern, rotates about the zenith but retains two very important characteristics: its mirror symmetry, and the property that along the symmetry line the e-vectors are always perpendicular to the solar meridian.

Although the polarization pattern of the sky is invisible to humans, many insects exploit it by using the e-vector directions as a reference for compass orientation. Insects have compound eyes consisting of many ommatidia. Polarization vision is mediated by only a small group of specialized, upward-directed ommatidia situated at the dorsal rim of the eye (POL-area). Each ommatidium contains two sets of polarization-sensitive photoreceptors, which are tuned to orthogonal e-vectors (crossed-analyzer arrangement); for a review see (Wehner, 1994).

In insects, polarization-sensitive neurons (POL-neurons) have been found in two different neural centers, the medulla of the visual lobes (crickets: (Labhart, 1988)) and the central body (locusts: (Vitzthum, 1997)). The medullar neurons receive input from photoreceptors of the POL-area. The activity of these neurons is a sinusoidal function of e-vector orientation with an excitatory and an inhibitory part and with the maxima and minima separated by 90°, indicating that they receive antagonistic input from two polarization-sensitive channels with orthogonal e-vector tuning orientations. The advantages of such a crossed-analyzer configuration is that it enhances e-vector contrast sensitivity and that it makes the e-vector response insensitive to fluctuations of light intensity. There are three types of cricket POL-neurons that are tuned to different e-vector orientations, i.e., approximately 10°, 60° and 130° with respect to the body axis.

2.2 Polarization vision in a robot

Recent studies have shown that a polarized light compass implemented in analog hardware provides superior means for extracting compass information as compared to traditional methods based on other sources of directional information such as proprioception (Lambrinos et al., 1997). The basic component of the POL-compass are polarization opponent units (POL-OP units) that are functionally similar to the POL-neurons found in insects. Each POL-OP unit consists of a pair of *polarized light sensors* (POL-sensors) followed by a log-ratio amplifier (see figure 2). The POL-sensors are photodiodes with a linear polarizer and a blue transmitting filter on top. In each POL-OP unit the polarizing axis of one POL-sensor was adjusted 90° to the polarizing axis of the other sensor, thus mimicking the crossed-analyzer configuration in the POL-area of insect eyes. The signals of each pair of POL-sensors were fed into a log-ratio amplifier. Three pairs of POL-sensors were mounted on our mobile robot Sahabot 2 (see figures 3 and 4 (center)) and adjusted such that the polarizing axis of the positive channel was 0°, 60° and 120° with respect to the robot's body axis. The visual fields of the POL-OP units were about 60°

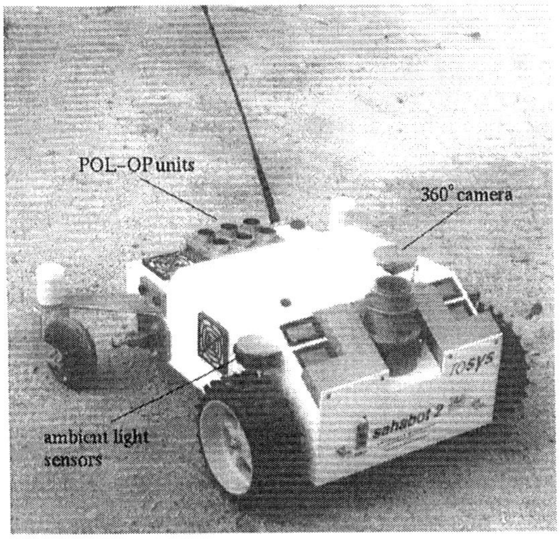

Figure 3 The mobile robot Sahabot 2.

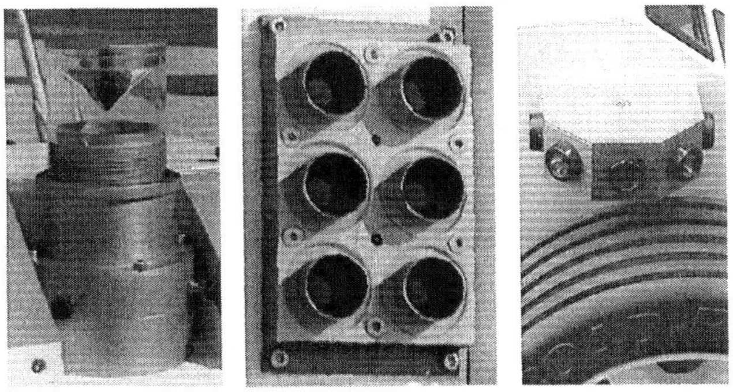

Figure 4 *Left:* Detailed view of the 360° camera system of the robot consisting of a camera with vertical optical axis (within the case) and a conical mirror aligned with the axis of the camera and mounted within an acrylic glass tube (top). Axial alignment and distance between mirror and camera can be adjusted. An infrared and a neutral density filter are mounted between camera and mirror. *Center:* Three pairs of POL-OP units were mounted on the rear part of the robot. They were tuned to 0°, 60° and 120° e-vector directions with respect to the robot's body axis. *Right:* One of the two ambient light sensor arrays at the sides of the robot.

and were centered around the zenith. The output of a POL sensor is described by the following equation:

$$s(\phi) = KI(1 + d\cos(2\phi - 2\phi_{max})). \qquad (1)$$

I is the total intensity $I = I_{max} + I_{min}$, with I_{max}, I_{min} being the maximum and the minimum intensity, respectively. d is the degree of polarization, ϕ_{max} is the value of ϕ that maximizes s (tuned e-vector direction), and K is a constant (Bernard and Wehner, 1977). The outputs of the POL-OP units are described by:

$$p_1(\phi) = \log\left(\frac{1 + d\cos(2\phi)}{1 - d\cos(2\phi)}\right)$$

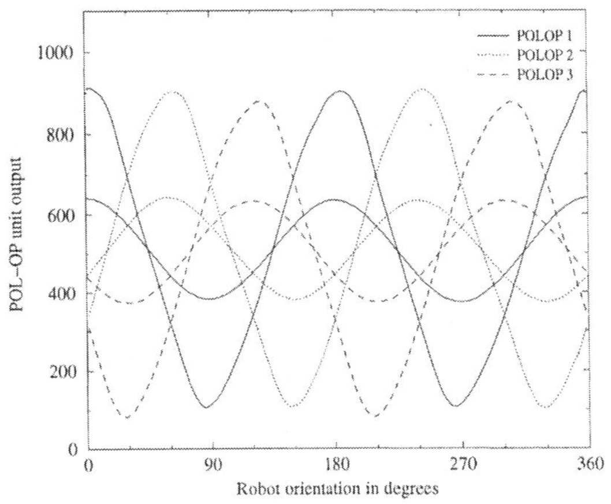

Figure 5 The outputs of the POL-OP units during a rotation of 360° at two different times of the day (i.e., with different solar elevation), late in the afternoon (large amplitude) and early in the afternoon (small amplitude).

$$p_2(\phi) = \log\left(\frac{1 + d\cos(2\phi - \frac{2\pi}{3})}{1 - d\cos(2\phi - \frac{2\pi}{3})}\right)$$

$$p_3(\phi) = \log\left(\frac{1 + d\cos(2\phi - \frac{4\pi}{3})}{1 - d\cos(2\phi - \frac{4\pi}{3})}\right) \qquad (2)$$

where $p_1(\phi)$, $p_2(\phi)$, $p_3(\phi)$ are the outputs of the POL-OP units tuned to 0°, 60° and 120° respectively, ϕ is the current orientation with respect to the solar meridian, and d is the degree of polarization.

There are two main models for using the POL-OP responses to derive compass direction: the *scanning model* and the *simultaneous model* (compare (Lambrinos et al., 1997)). With the *simultaneous model*, that is used in the present study, the heading direction can be determined continuously and no scanning movements are necessary during the journey. This can be done with a look-up table containing the output values of the POL-OP units and the corresponding orientation of the robot (Lambrinos et al., 1997).

Although the output signals of the POL-OP units are independent of the light intensity due to the cross-analyzer configuration, they are still dependent on the degree of polarization (see equations (2)). The amplitude of the POL-OP signals is proportional to the degree of polarization which changes during the day depending on the elevation of the sun and due to clouds. Figure 5 shows the outputs of the POL-OP units during a rotation of 360° at two different times of the day. Thus, the lookup table has to be updated during the course of the day.

One way to overcome this problem is to use a *scanning model* where only the maxima of the signals are evaluated (Lambrinos et al., 1997). For the *simultaneous model*, however, the change in degree of polarization during the day has to be taken into account. This can be done by normalizing the outputs of the POL-OP units in the following way. First, the POL-OP signal are delogarithmized by applying a sigmoid function:

$$\frac{1}{10^{p(\phi)}+1} = \bar{p}(\phi). \qquad (3)$$

Equations (2) then become:

$$1 - 2\bar{p}_1(\phi) = d\cos(2\phi) \qquad (4)$$

$$1 - 2\bar{p}_2(\phi) = d\cos(2\phi - \frac{2\pi}{3}) \qquad (5)$$

$$1 - 2\bar{p}_3(\phi) = d\cos(2\phi - \frac{4\pi}{3}). \qquad (6)$$

From the above we find:

$$\phi = \frac{1}{2} atan \left(\frac{2\bar{p}_2(\phi) + \bar{p}_1(\phi) - \frac{3}{2}}{\sqrt{3}\bar{p}_1(\phi) - \frac{1}{2}} \right) \qquad (7)$$

$$d = \frac{2\bar{p}_1(\phi) - 1}{\cos(2\phi)}. \qquad (8)$$

Since $p_1(\phi)$, $p_2(\phi)$, $p_3(\phi)$ are known (sensor readings), and $\bar{p}_1(\phi)$, $\bar{p}_2(\phi)$, $\bar{p}_3(\phi)$ are derived from these values, we can determine the values of ϕ and d. Notice that for this step only *two* POL-OP units are necessary (p_1 and p_2 in the above equations). Since the functions in equations (4–6) have a period of π, there will be two values of ϕ (ϕ and $\phi + \pi$) satisfying (4–6) for each triplet $\bar{p}_1(\phi)$, $\bar{p}_2(\phi)$, $\bar{p}_3(\phi)$. This ambiguity can not be resolved by referring to the POL-OP unit signals. A different sensory modality has to be involved. Disambiguation between the two candidate orientation values is done by employing the ambient light sensors of the robot. There are eight ambient light sensors arranged in two half-rings covering a visual field of 180° each. They are mounted in two metal cases in the front part of the robot (see figures 3 and 4, right). The values from the ambient light sensors are used to obtain a rough estimate (about 90°) of the robot's heading with respect to the sun. If the robot is oriented towards the sun then the ambient light sensor with the visual field enclosing the solar meridian will have a stronger response than the other sensors. This estimate is used to disambiguate between the two candidate orientations (ϕ, $\phi + \pi$) obtained from the POL-OP units.

We can now transform the current POL-OP readings to signals, that are independent of the time of the day. This is done by substituting the value of d from equation (8) in equations (4–6) instead of using equation (7). To optimize precision, the value of ϕ can be calculated by selecting the equation of the POL-OP unit for which the signal curve is steepest at that point, i.e., POL-OP unit 1 for 30°–60°, POL-OP unit 2 for 60°–90° and POL-OP unit 3 for 90°–120°.

Figure 6 shows the results of applying such a normalization process to the outputs of the POL-OP units shown in figure 5. The outputs of the POL-OP units during a rotation of 360° were recorded at two different times of the day, late and early in the afternoon. The signals

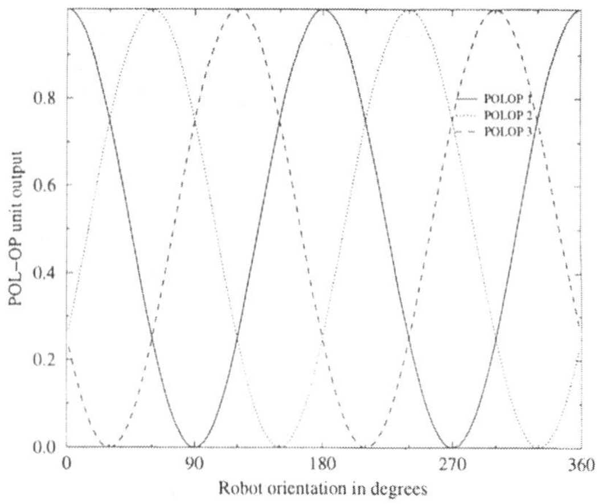

Figure 6 The outputs of the POL-OP units shown in figure 5 after normalization. The outputs of the corresponding units in the two different runs are completely overlapping in the figure.

become independent of the degree of polarization after the normalization process and are practically identical. Thus, they can be used to calculate the orientation of the robot with respect to the solar azimuth at any time of the day.

As we have seen until now, it is possible to derive the orientation of the agent with respect to the solar azimuth by using the polarized light compass. To use the POL-compass over an extended period of time, however, requires compensating for the daily sun movement. Complicating this task is that the rate of change of the sun azimuth is not constant. The sun azimuth changes slowly in the morning and faster around noon. Moreover, the ephemeris function, i.e., the function that describes the change of sun azimuth over time, depends on the season and the geographical latitude.

There is behavioral evidence that insects are equipped with such an ephemeris function (Lindauer, 1959). Moreover, recent studies in both bees (Dyer and Dickinson, 1994) and ants (Wehner and Müller, 1993) indicate that such an ephemeris function might be innate and that insects can refine its precision by experience. In our study we extended the POL-compass described above by including the ephemeris function valid for the location and season of our experiments.

2.3 Path-integration experiments

The POL-compass was used in a path-integration mechanism to keep an estimate of the robots position over time (for the ants' performances see (Müller and Wehner, 1988; Hartmann and Wehner, 1995)). The position of the robot was calculated as follows:

$$x(t + \Delta t) = x(t) + \cos(\theta(t))\, v(t)\, \Delta t \qquad (9)$$

$$y(t + \Delta t) = y(t) + \sin(\theta(t))\, v(t)\, \Delta t \qquad (10)$$

where $x(t + \Delta t)$, $y(t + \Delta t)$, $x(t)$ and $y(t)$ are the x and y coordinates of the robot at time $t + \Delta t$ and t, respectively.

Table 1 Overall performance of the robot with the two path-integration systems. Average error and standard deviations relate to deviations from the target position at the end of the experiments. n indicates the number of experiments.

	POL-compass	Proprioception
Average error [cm]	13.52	51.90
Standard deviation [cm]	6.29	28.83
Min [cm]	5	11.18
Max [cm]	27.46	108.85
n	9	9

$\theta(t)$ is the estimated orientation of the robot at time t, $v(t)$ is velocity of the robot as estimated from the wheel encoders, and Δt is the time step.

In order to compare the performance of the path-integration system employing the POL-compass (POL-compass system) with a system using proprioception (proprioceptive system), the same algorithm was used to implement another path-integration system that used the robot's wheel encoders to derive orientation information. In the experiments described below, both systems were running in parallel providing independent estimates of the robot's position over time.

Before each experiment, the robot was placed at the starting point and was aligned with the geographic north. During the experiments the robot had to follow a trajectory consisting of 15 "legs" of different orientations and different lengths by using one of the two path-integration systems, namely the POL-compass system or the proprioceptive system. During the experiments, the data from the sensors, i.e., the outputs of the POL-OP units, the light sensors and the wheel encoders were logged to the robot's on-board PC. At the end of each experiment, the final position of the robot was recorded. The distance between the final position of the robot and the starting position was used as an overall error measurement.

The data on overall robot performance are given in table 1 and will be summarized below. The robot performed well with both systems. The largest error recorded in these experiments was just 1 m. The average errors were 13.5 cm and 51.9 cm for the POL-compass system and the proprioceptive system, respectively. Figure 7 shows the results from the experiments with the 15-leg trajectories for both systems. The POL-compass system was clearly more precise than the proprioceptive system (U-test, $p > 0.0023$). Both the mean deviation from the target point and the standard deviation of the error were smaller.

During the experiments, the POL-compass system and the proprioceptive system produced different estimates of the position of the robot. Although the difference between these estimates was initially zero, it increased significantly over time. This can be seen more clearly in another experiment where the robot had to follow a trajectory consisting of 30 randomly generated legs. Figure 8 shows the trajectories as estimated from the POL-compass and the proprioceptive systems. During the experiment the robot was controlled by the POL-compass system, but estimation of the position was also derived from the proprioceptive system. Both estimates were logged in a file for later analysis. As it can be seen, the final position of the robot as estimated with the POL-compass system is very close to the real position of the robot which was measured with a meter stick at the end of the experiment. The final position of the robot as estimated from the proprioceptive system is about 3.39 m away from the real value. Figure 9 shows the deviation between the two estimates over time.

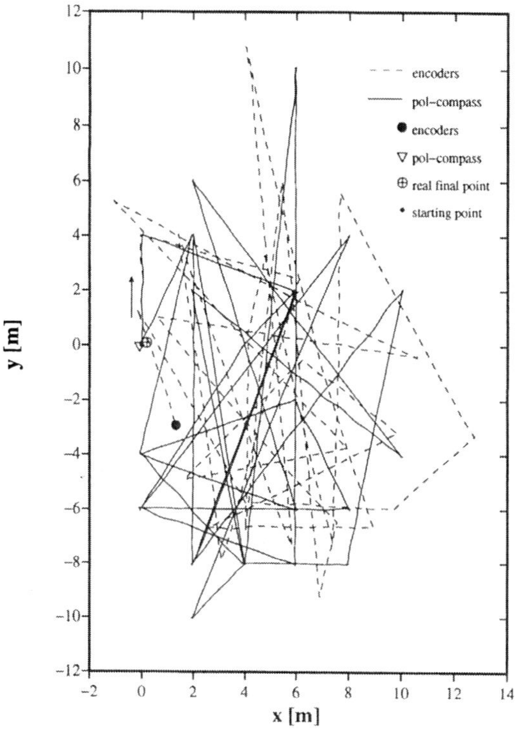

Figure 8 Trajectories estimated by both the POL-compass system and the proprioceptive system during an experiment with a 30 leg trajectory. The robot was controlled by the POL-compass system. The final positions of the robot as estimated by both systems and the real final position are also drawn. The arrow indicates the direction of the first leg.

From the results it is clear that a path-integration system employing a skylight compass will have an advantage over a system based on proprioception only. This general conclusion has been predicted, on theoretical grounds, by Benhamou et al. (1990). The polarization pattern of the sky provides a stable frame of reference that can be used to derive the most important piece of information which such a system requires, that of the compass information. Such a system is precise enough to guide the robot close to the target position so that another system based on visual landmark information can take over to finally pinpoint the target.

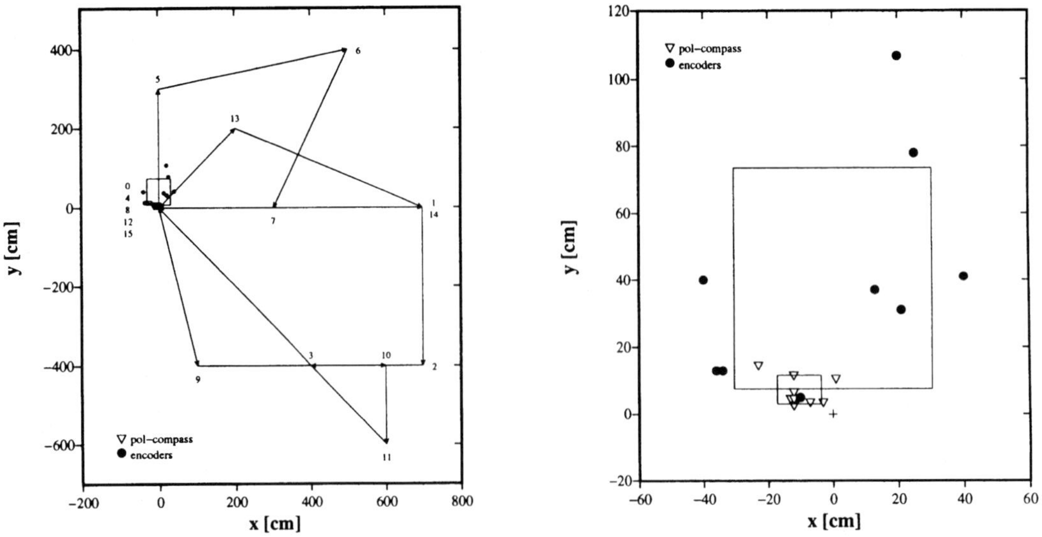

Figure 7 *Left:* Results of the experiments with a 15 leg trajectory. The trajectory is drawn starting from origin with the leg number printed at the end of each leg. Note that the trajectory passes three times through the origin. The final points of the robot during the experiments are drawn as filled circles (wheel encoder experiments) or triangles (POL-compass experiments). Mean and standard deviation within the data in the direction of the axes in both cases are indicated by boxes. *Right:* Zoom into the region around the starting position.

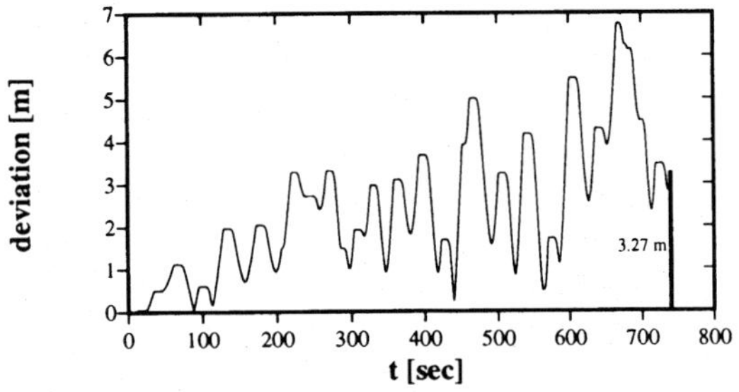

Figure 9 Deviation between the position estimates by the POLcompass system and by the proprioceptive system. At the beginning of the experiments both estimates are the same but the difference tends to increase over time.

3. Visual landmark navigation

3.1 Landmark navigation in insects

While path integration employing a skylight compass is the primary strategy that *Cataglyphis* ants are using to return to the vicinity of their nest, errors introduced by the path-integration process would result in a wrong estimate of the nest position (Wehner and Wehner, 1986). Since the nest entrance is an inconspicuous hole in the desert ground, which is invisible to the insect even from a small distance, alternative strategies have to be employed in order to finally locate the entrance. In the absence of visual landmarks, *Cataglyphis* will start a systematic search at the position where the nest is expected to be found after having reset its path-integration system (Wehner and Srinivasan, 1981). However, when landmark information is available, both bees (Cartwright and Collett, 1983) and ants (Wehner et al., 1996) will exploit it, and they will relocate the target position directly with a remarkable precision.

A number of experiments performed with bees (Cartwright and Collett, 1983) and ants (Wehner and Räber, 1979; Wehner et al., 1996) have unraveled important properties of the insect's landmark navigation system. The main conclusion from these experiments is that the animal learns a rather unprocessed visual snapshot of the scene around the goal position. By matching this snapshot to the current retinal image, the insect could derive the direction it has to move in order to relocate the target position where the snapshot was taken. There is evidence from experiments in which several parameters of the landmark configuration have been changed, that the snapshot matching strategies exploit differences in the bearing, apparent size and apparent height of the landmarks between snapshot and current retinal image.

3.2 Model of landmark navigation

Snapshot matching: A model for the matching of snapshot and current view was proposed by Cartwright and Collett (1983, 1987), which reproduces some aspects of the search behavior of bees; figure 10 (left) visualizes the matching process of this model. It assumes that the views on which the matching process is operating are horizontal images comprising dark and bright sectors corresponding to the landmarks and the gaps between them, respectively. A pairing between sectors in the two views is established by searching the closest sector of the same type (black or white) in the current view for each sector in the snapshot. Each pair of sectors contributes two vectors to the total home vector: a tangential vector point-

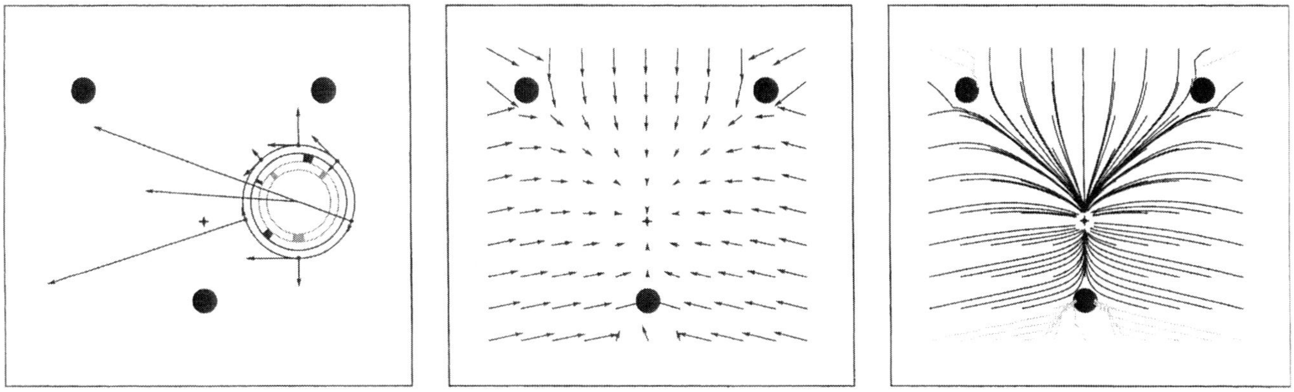

Figure 10 Simulation of the snapshot model (with proportional vector contributions) for a configuration of three landmarks (filled circles). *Left:* A snapshot (inner ring) taken at the nest position marked with a cross is matched with a current view (middle ring) taken at the position in the center of the circle. Each pair of matched sectors (both black and white) contributes with a radial and a tangential vector (attached to the outer ring in the diagram). The home vector (originating from the center) is obtained by averaging the contributing vectors. *Center:* A home vector can be assigned to each point of the plane. *Right:* Trajectories running tangentially to the home vectors. Dark courses end at the snapshot position, grey courses run into one of the landmarks.

ing from the snapshot sector towards the paired sector in the current view, and a radial vector, which points centrifugally, if the apparent size of the current view sector is smaller than the size of its counterpart in the snapshot, and vice versa. In principle, the model could also take into account differences in height, though this was not considered in the experiments described here. Averaging all contributions gives a home vector pointing approximately to the position of the target; a trajectory following these home vectors that are attached to each point of the plane will approach the location where the snapshot was taken (figure 10, center and right).

In the original model presented by Cartwright and Collett (1983), all contributing vectors have unit length. For the robot experiments described below, these unit vectors have been replaced by vectors with a length proportional to the difference in bearing, or apparent size, of the paired sectors. As shown in figure 11, this modification facilitates the use of the *length* of the home vector as a measure of discrepancy between current view and snapshot, called "disparity" here. This is not possible with the unit vector model, since the disparity is not decreased while approaching the snapshot position; it would only become zero if a perfect match between snapshot and current view is achieved. Using proportional contributions, the robot runs can be stopped when the disparity falls below a threshold. For the weighting of tangential and radial contributions the same ratio of 1:3 as given in (Cartwright and Collett, 1987) was used.

External compass reference: An important prerequisite of the matching procedure between snapshot and current view is the alignment of both views with respect to an external compass reference. Without proper align-

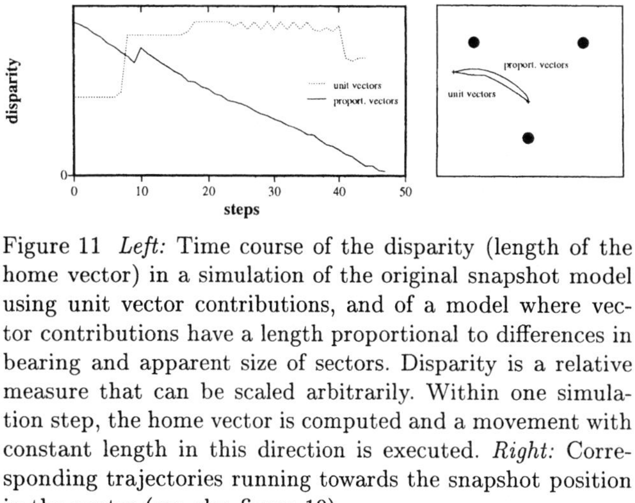

Figure 11 *Left:* Time course of the disparity (length of the home vector) in a simulation of the original snapshot model using unit vector contributions, and of a model where vector contributions have a length proportional to differences in bearing and apparent size of sectors. Disparity is a relative measure that can be scaled arbitrarily. Within one simulation step, the home vector is computed and a movement with constant length in this direction is executed. *Right:* Corresponding trajectories running towards the snapshot position in the center (see also figure 10).

ment, the number of mismatches between sectors — sectors in the two views that are paired, but do not correspond to the same landmark or gap — will increase significantly, resulting in an erroneous home vector. Bees might accomplish this alignment by maintaining a constant orientation of the body during learning and searching for a goal (Collett and Baron, 1994). For some species of *Cataglyphis*, body rotations interrupting the forward movement have been observed (Wehner, 1994), but it is not clear, if this behavior could also serve for the alignment of the views. An *internal* rotation of either snapshot or current view could be excluded for ants by experiments on intra- and inter-ocular transfer of information (Wehner et al., 1996). But even if the assumption of retinotopically fixed views holds and body rotation can be excluded, the alignment could be accomplished by storing a set of snapshots taken in different orientations at the same location and activating them selectively according to the current compass bearing (Cartwright and Collett, 1983).

Regardless of the method of alignment, ants could derive the required compass information from the polarization pattern (see section 2.2). In the robot experiments described here, we used the POL-compass to rotate the current view so that it is aligned with the snapshot.

3.3 Landmark navigation in a robot

Camera: The camera system mounted in the front part of the robot is shown in figure 4 (left). It consists of a CCD camera and a conical mirror in the vertical optical axis of the camera. The camera was mounted as low as possible on the robot within the construction constraints, which brings the mirror to a height of approximately 27 cm above the ground. The conical mirror provides a 360° horizontal view; the opening angle of the cone was determined so that the visual field extends ±10° around the horizon. In order to reduce the total light intensity, a neutral density filter is mounted between camera and mirror. An additional infrared filter was necessary to prevent an influence of thermal radiation on the camera image.

Image processing: Figure 12 illustrates the image processing steps from the camera image to a horizontal view with black and white sectors as required by the matching mechanism described in section 3.2. The same processing steps were applied to both snapshot and current views. In a first step, the camera image (obtained from a situation similar to the one shown in figure 14) is transformed into an azimuth-altitude representation. A mean grey value is determined from the transformed image and used to adjust the brightness parameter of the camera in a way, that the mean grey value of this image is kept constant over time. This is indispensable for the subsequent thresholding operation separating dark and bright regions. The final one-dimensional horizontal view is obtained by counting the number of black pixels within a part of each pixel column (limited by the two lines shown in the thresholded image) and applying a threshold of 50%.

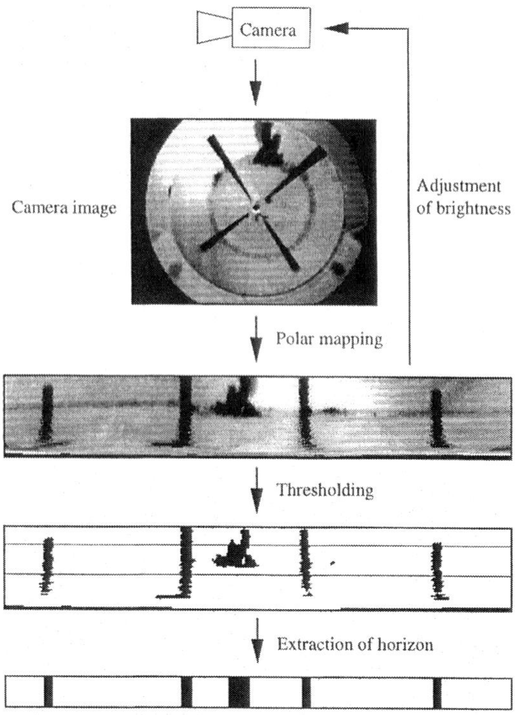

Figure 12 Image processing for landmark navigation. The image from the 360° camera (160 × 120 pixel) is transformed into a polar view (351 × 56). After applying a threshold to each pixel, the segmented horizon (351 × 1) is extracted from the area within the two lines; a pixel in the horizon will be black, if more than 50% of the pixel in the corresponding column are black. The landmark extracted in the center belongs to equipment in the vicinity, which was removed for the experiments.

3.4 Landmark navigation experiments

Figure 14 shows a typical setup for the robot experiments in the habitat of *Cataglyphis.* Black cylinders were used as landmarks; the same type of landmarks was also used for the ant experiments.

At the beginning of each experiment, the robot was placed at the starting position within the landmark array. The landmark array used for the experiments shown in figure 13 consists of 3 black cylinders with a height of 80 cm and a diameter of 30 cm, forming an equilateral triangle whose sides are 3 m long. The starting position was situated on a symmetry line of this triangle in a distance of 1 m from one of the landmarks. A same setup was used in the simulation described in figure 10.

Data from two experiments are visualized in figure 13. A snapshot was taken at the starting position, processed as described in section 3.3, rotated by the angle obtained from the POL-compass, and stored in the form of black and white sectors (figure 13, center). After taking the snapshot, the robot had to follow a certain direction for some distance (4 m for trajectory A, 2 m for trajectory B; see figure 13, left) corresponding to a short foraging journey (it could also have been displaced manually). At the end of this outward trip, control was handed over to the homing algorithm, which performed the extraction of a sectorized horizontal view from the camera image, the alignment of this view, and the computation of the home vector by matching the aligned view to the snapshot; transitions of the rotated current views are shown in the center of figure 13. The resulting home vector was used to set the direction of movement. As soon as the disparity became lower than a threshold, the experiment was stopped (figure 13, right).

Eight runs (including trajectories A and B in figure

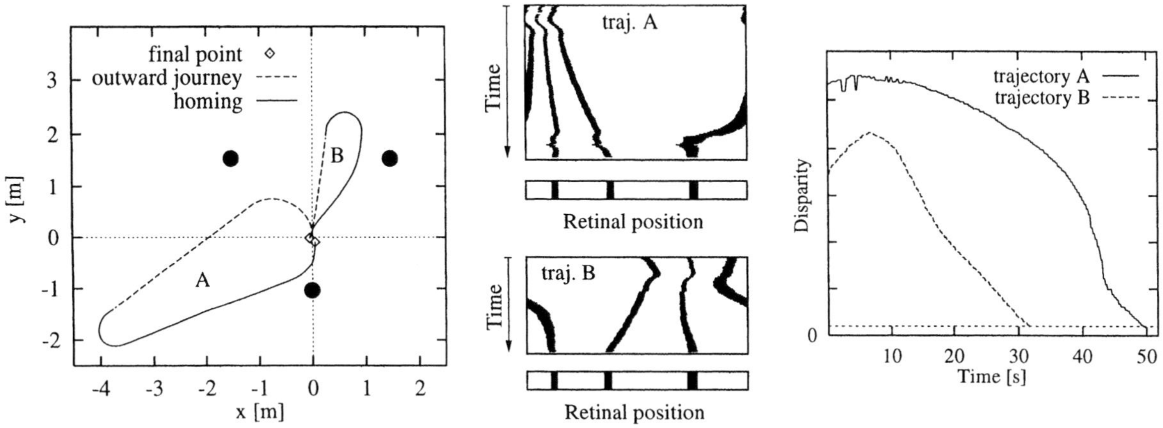

Figure 13 Landmark navigation experiment with the robot. *Left:* Two typical trajectories of the robot in an array of three landmarks (filled circles). From the end of the preprogrammed outward journey (dashed) towards the target position (0,0) the robot is guided by landmark navigation (solid line). *Center:* Transitions of the rotated current view over time towards the snapshots (images underneath each trace) for the two trajectories. *Right:* Time course of disparity between snapshot and current view for the two trajectories. The dashed line depicts the threshold used for terminating the runs.

Figure 14 Example of a landmark array used for the navigation experiments. The grid visible on the desert ground was used for the alignment of landmarks and robot, and for registration of the final points.

13) with different angles and distances of the outward journey were performed with this landmark configuration and with the same starting point; the final parts of the 8 trajectories are shown in figure 15 (left). The deviation between final and initial position of the camera axis was between 7 cm and 20 cm. This deviation does not reflect the precision of this homing mechanism, though, since the final points are locations where the disparity reached the termination threshold, but not locations with *minimal* disparity. For a visual evaluation of the data, those points on the trajectory are marked, where the disparity drops below 1.5 times the threshold. For comparison, the right part of figure 15 visualizes the disparity for points in the vicinity of the snapshot location in a simulation. Final points as well as the points where 1.5 $\times$ threshold was passed are in accordance with the disparity plot, since they can be thought to be located on roughly elliptic bands around the snapshot position, with all final positions on the inner band. Since these bands are not overlapping, noise induced to the disparity from the sensor signal seems to be in a range below the termination threshold. Therefore it can be assumed, that a reduction of the termination threshold will bring the final points closer to the snapshot location. This will require a specialized control algorithm for the final phase of the approach, that will allow the robot to locate the target point more precisely.

4. Conclusion

In this paper we used the *autonomous agents* approach to study the navigation behavior of insects. Inspired by the insects, we have developed navigation mechanisms that were successfully employed on the mobile robot Sahabot 2.

The goal of this study was to implement mechanisms of path integration and visual landmark navigation in an autonomous agent, and to test in the field whether a combination of these mechanisms is suitable to guide the agent precisely back to a target location. Both the path integration system using the POL-compass and the visual piloting mechanism exhibited a precision comparable to that of the insect. The model of landmark navigation proposed by Cartwright and Collett (1983), in conjunction with the POL-compass used for the alignment of views, could be shown to work on a robot and produce reasonable trajectories.

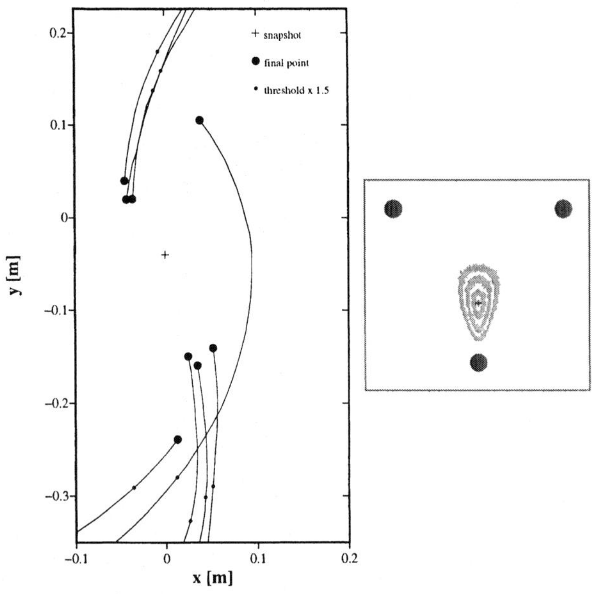

Figure 15 *Left:* Final parts of 8 homing trajectories from landmark navigation (for the configuration in figure 13, including trajectory A and B shown there). All points relate to the position of the camera axis. Small dots are placed on the trajectories at points, where the disparity dropped below a value of 1.5 times the termination threshold. *Right:* Disparity in the vicinity of the snapshot location. Grey regions are restricted by equidistant iso-disparity curves; disparity decreases when approaching the snapshot position.

Acknowlegdments

This work was supported by grants from the Swiss Federal Office for Education and Science (VIRGO TMR network), the Swiss National Science Foundation (grant 20-47257.96), and the Human Frontier Science Program. The Sahabot 2 was constructed in cooperation with Rosys AG, Hombrechtikon. Financial support by Rosys AG is kindly acknowledged. Many thanks to Christian Gfeller of Rosys for his skilled and patient cooperation, and to Marinus Maris, Hiroshi Kobayashi, and Lukas Lichtensteiger, who substantially contributed in the construction of the robot.

References

Bernard, G. D. and Wehner, R. (1977). Functional similarities between polarization vision and color vision. *Vision Res*, 17:1019–1028.

Cartwright, B. and Collett, T. (1983). Landmark learning in bees. *Journal of Comparative Physiology A*, 151:521–543.

Cartwright, B. and Collett, T. (1987). Landmark maps for honeybees. *Biological Cybernetics*, 57:85–93.

Collett, T. S. and Baron, J. (1994). Biological compasses and the coordinate frame of landmark memories in honeybees. *Nature*, 368:137–140.

Dyer, F. C. and Dickinson, J. A. (1994). Development of sun compensation by honey bees: How partially experienced bees estimate the sun's course. *Proc. Natn. Acad. Sci. U.S.A*, 91:4471–4474.

Hartmann, G. and Wehner, R. (1995). The ant's path integration system: a neural architecture. *Biological Cybernetics*, 73:483–497.

Labhart, T. (1988). Polarization-opponent interneurons in the insect visual system. *Nature*, 331:435–437.

Lambrinos, D., Maris, M., Kobayashi, H., Labhart, T., Pfeifer, R., and Wehner, R. (1997). An autonomous agent navigating with a polarized light compass. *Adaptive Behavior*, 6(1):175–206.

Lindauer, M. (1959). Angeborene und erlernte Komponenten in der Sonnenorientierung der Bienen. *Z. vergl. Physiol.*, 42:43–62.

Müller, M. and Wehner, R. (1988). Path integration in desert ants, *Cataglyphis fortis*. In *Proc. Natl. Acad. Sci. Neurobiology*, volume 85, pages 5287–5290, U.S.A.

Pfeifer, R. (1996). Building "Fungus Eaters": Design principles of autonomous agents. In *Proceedings of the Fourth International Conference on Simulation of Adaptive Behavior SAB96 (From Animals to Animats)*, pages 3–12, Cape Cod, Massachusetts, USA.

Vitzthum, H. (1997). *Der Zentralkomplex der Heuschrecke Schistocerca gregaria: Ein mögliches Zentrum des Polarisationssehsystems.* PhD thesis, Universität Regensburg.

Wehner, R. (1982). Himmelsnavigation bei Insekten. Neurophysiologie und Verhalten. *Neujahrsblatt der Naturforschenden Gesellschaft Zürich*, 184:1–132.

Wehner, R. (1987). Spatial organization of foraging behavior in individually searching desert ants, *Cataglyphis* (Sahara desert) and *Ocymyrmex* (Namib desert). In Pasteels, J. M. and Denebourg, J. L., editors, *From Individual to Collective Behavior in Social Incects*, pages 15–42. Birkhäuser, Basel.

Wehner, R. (1994). The polarization-vision project: championing organismic biology. In Schildberger, K. and Elsner, N., editors, *Neural Basis of Adaptive Behaviour*, pages 103–143. G. Fischer, Stuttgart.

Wehner, R. (1997). The ant's celestial compass system: spectral and polarization channels. In Lehrer, M., editor, *Orientation and Communication in Arthropods*, pages 145–185. Birkhäuser, Basel.

Wehner, R., Michel, B., and Antonsen, P. (1996). Visual navigation in insects: coupling of egocentric and geocentric information. *Journal of Experimental Biology*, 199:129–140.

Wehner, R. and Müller, M. (1993). How do ants acquire their celestial ephemeris function? *Naturwissenschaften*, 80:331–333.

Wehner, R. and Räber, F. (1979). Visual spatial memory in desert ants, Cataglyphis bicolor (Hymenoptera: Formicidae). *Experientia*, 35:1569–1571.

Wehner, R. and Srinivasan, M. (1981). Searching behavior of desert ants, genus cataglyphis (formicidae, hymenoptera). *J. comp. Physiol.*, 142:315–338.

Wehner, R. and Wehner, S. (1986). Path integration in desert ants: approaching a long standing puzzle in insect navigation. *Monitore Zool Ital (NS)*, 20:309–331.

Simulated and situated models of chemical trail following in ants

Titus Sharpe &[1] Barbara Webb

Department of Psychology, University of Nottingham,
Nottingham NG7 2RD, U.K.
email: Barbara.Webb@nottingham.ac.uk

Abstract

A simple neural network model for ant chemical trail following is proposed. It performs temporal differencing and comparison to control an osmotropotaxis response. The mechanism was first tested by implementing it in a virtual simulation, and using a genetic algorithm to find appropriate connection strengths. Resulting behavioural measures show strong similarity to data from ants and previous algorithmic models, including: non-linear effects of varying chemical strength; failure to follow trails crossed at large angles; worse trail following at faster speeds and better trail following with longer antennae. In a real-world implementation using chemical sensors on a robot following an alcohol-based trail, it was found necessary to use a somewhat different set of weightings to cope with the inherent unreliability of detecting chemical concentrations. It was still possible to show qualitatively similar behaviour under the same experimental conditions as the simulation model. We argue that these results may illustrate the nature of the agent-environment task space rather than proving the model 'correct'; but that the model nevertheless provides useful pointers to further investigation of this biological system.

1. Introduction

Ants navigate along chemical pheromone trails to reach a food source from the nest. This area of ant behaviour has been well studied and documented (e.g. Hangartner,1967; Calenbuhr & Deneubourg, 1992). However it is not known what is the effective neural mechanism which allows ants to follow trails by sensing the chemical. One way to investigate this is to propose and test plausible neural mechanisms through simulation (Webb, 1994). The results can contribute both specific hypotheses to aid neurophysiological investigations, and more general understanding of how perceptual systems work.

An important consideration in such an investigation is how to represent the interaction of the animal with its environment in sufficiently realistic terms. This is critical if the evaluation of the mechanism is to make any real contribution to biological understanding. On the other hand simplification and abstraction are necessary to make simulation viable and to understand how different factors interact.

Consequently this study takes a two-part approach to the problem. A hypothesised neural control mechanism is tested in a virtual simulation, and in a robot implementation. The simulation enabled us to prototype, evolve and test controllers rapidly; the robot allowed us to test results in the face of real world conditions. As it turned out, one was not a simply a test-base for the other but rather each gave us somewhat different insights into the problem of trail-following.

Our aim is to establish how well a relatively simple controller that utilises neural-style processing and exploits the real environment interaction that occurs in chemotaxis can reproduce the observed characteristics of trail-following in ants. Thus we will first describe the biological background and previous modelling efforts to establish a set of criteria for evaluating the model. We then propose a simple but plausible neural controller for the observed behaviour. This controller is implemented in a virtual simulation and adapted using a genetic algorithm, and also 'adapted' (hand-tuned) as the controller for a mobile robot with chemical sensors. Both are tested in a range of behavioural conditions. It is demonstrated that behaviour similar to the ant can be generated in the simulation and in the real world. Finally we discuss to what extent we can interpret the model as potentially the 'real' mechanism of trail-following in ants, and how this model may inform further behavioural and physiological investigation.

2. Background

Many animals use chemical trails as a means to communicate between members of a group of the same species. This involves one member of a group laying a specific mix of liquid chemicals along a useful path which will then be followed by other members. Species that show this behaviour include dogs, molluscs, termites, caterpillars and ants. Ants use trail following on an individual level which in turn leads to well organised patterns of behaviour on a nest wide scale. Ants use this form of behaviour for several tasks such as mass foraging, interconnection of nests and as a recruitment strategy for other ants. There is thus

[1] The hypotheses, simulations and experiments were devised and carried out by the first author under the supervision of the second author who produced this paper. Correspondence should be addressed to the second author.

likely to be strong selective pressure for effective and efficient trail-following.

The main use of chemical trail following is mass foraging for food. All ant species have to forage outside of the nest to collect food for the colony. This is often the most dangerous task for an ant as there are many predators outside of the nest; it probably also imposes the highest demand on navigational abilities. In the majority of ant colonies it is the oldest ants that forage. Different species of ant have different methods of foraging of which the two most common are random search and random search within in a specific orientation to the nest. Ants keep a knowledge of where they are in relationship to the nest and an approximate distance back to the nest. Therefore if they wish to return to the nest they can run in a straight line and they should arrive back at the nest. The way in which ants record their orientation and distance is by using the sun as an orientation cue, using optical flow and keeping track of their paces as a distance cue (i.e. path integration).

When the ants are returning to the nest certain species often lay chemical trails along the ground. The ants do this by dragging their rear glands and hindgut along the path. Six sub families of ant show this type of behaviour *Ponerinae, Dorylinea, Mymicinae, Aneuretinae, Dolichoderinae* and *Formicinae* (Holldobler, 1990). Depending on the species of ant it will lay a specific mix of pheromones with a specific selection of glands and hindgut. Twenty different ant pheromones have been found and there are likely to be many more yet to be identified. The pheromone mixture is species specific and in certain species nest specific.

This chemical trail then evaporates over a period of around ten minutes depending on the surface and pheromone type. This leaves a semi-ellipsoidal area (Bosset & Wilson, 1963) of gaseous pheromone around the trail. This is the active space in which an ant can sense the chemical. When the trail is laid to a nest entrance most ants encountering the trail respond immediately. The ants head out along the trail away from the nest.

2.2 Trail-following behaviour

The ants following the trail travel through the 'vapour tunnel' testing the air for the pheromone molecules, often moving their antennae about. Hangartner (1967) performed a pioneering experiment on ant trail following behaviour, see figure 1. This showed that an ant normally performs a sinusoidal walk along a trail. It can still orient itself along a trail with only one antenna or with its antennae crossed over: however the trail following becomes more inaccurate. Hangartner theorised that ants perform two types of chemical sensing, osmotropotaxis and klinotaxis. Osmotropotaxis (dual sensor mechanism) assumes the ant perceives the concentration of the pheromone with both antennae and computes the concentration difference to turn towards the higher concentration. Klinotaxis (unisensor mechanism) only provides information as to whether a change of direction is necessary. However it is not clear how these two mechanisms might interact in normal behaviour - in the current study we will focus on osmotropotaxis. The fact that the ant can still follow a trail with its antennae crossed suggests that the ant may use other sense mechanisms to orientate along the trail, i.e. it is a highly redundant system.

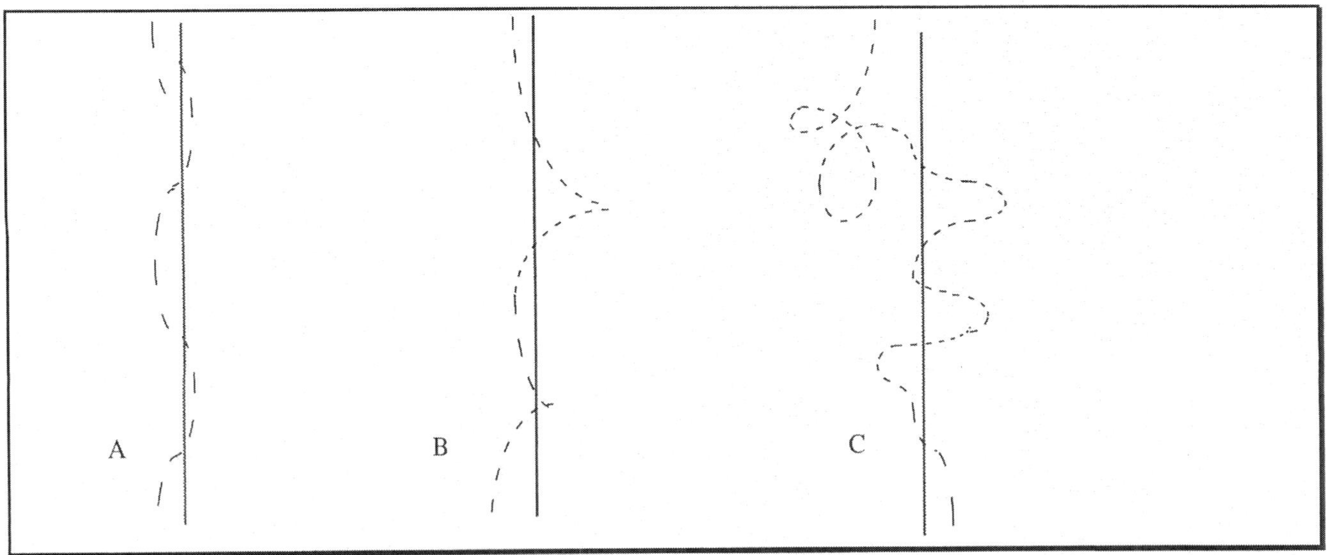

Figure 1: Hangartner's demonstration(1967) of worker ants orientating along a pheromone trail. A shows a normal worker ant following the trail sampling the chemical gradient. B shows a worker ant with its left antenna amputated. C shows a worker ant with its antennae crossed over.

There is evidence that trail orientation may be accompanied by guideline and phototatic orientation. It is important for the ants to know which way they are travelling along a trail (away from nest or towards the nest). It also helps explain how if the pheromone trail runs out the ant can still orientate itself back to its nest (Sudd & Franks, 1987). This suggests that ants use the normal orientation mechanism to keep track of their position and direction from the nest even when using a pheromone trail

A study by Pasteels *et al.* (1986) examined ants following circular trails and branched trails. The study found that an ant crossing a trail at angles close to the perpendicular of the trail didn't produce an orientation response. It was also found that the probability of an ant losing a trail remains constant with length and is independent of learning or fatigue - suggesting a reactive system (i.e. like a coin-toss, the length of run in which the trail has not been lost does not affect the current chances of losing the trail). Also, efficiency of orientation increases with concentration at low pheromone levels but at higher levels drops off. Calenbuhr & Deneubourg (1992) cite multiple studies that show "One of the characteristic experimental results is the peaked curve obtained when plotting the mean distance followed on the trail as a function of the amount of pheromone deposited"(p.368).

2.3 Algorithmic model of trail following

In the follow up to Pasteels *et al*'s study Calenbuhr & Deneubourg (1992) created a model for osmotropotactic behaviour. The algorithm they described managed to take into account all the findings of the primary study and held a close mathematical likeness to the way ants behave when trail following. The algorithm has the form shown in figure 2. Essentially the ant calculates the difference in concentration between the right and left antennae and changes direction towards the zone of higher concentration, with a non-linear relation between signal and response.

To back up their algorithm they did a second set of experiments that compared the theoretical and experimental situations. They varied several parameters in both conditions and found some strikingly similar results. These can be summarised as follows:

Parameter/ Phenomenon	Effect/ Result
v(speed)	Reducing speed leads to increased mean distance trail followed
a	Larger antennae span leads to increase in mean distance trail followed
$\ln(N/N_0)$	Probability of leaving the trail is independent of the distance followed
Proportion of non-orientations	Non-linear increase as a function of the amount of pheromone deposited
Mean distance vs. Amount pheromone	Peaked curve

Figure 3: Overview of similarity of theoretical and experimental results for Calenbuhr & Deneubourg (1992)

We are interested in how the behaviour described by these equations may actually be controlled by neural processes in the ant's brain. As the neural pathways have not been fully traced we have adopted the strategy of proposing and testing a simple plausible neural control mechanism to establish whether it can account for the variety of behavioural effects above when placed in realistic simulation environments and real world situations. This will provide a 'minimalist' starting hypothesis for further investigation.

Perception of the pheromone concentration with the tips of the antennae and determination of the concentration difference

$$\Delta C : C_{\text{left}} - C_{\text{right}}$$

Change direction according to the rule: $\Delta\theta = f(C_{left}, C_{right})\Delta t$

Where $f(C_{left}, C_{right}) = \text{sgn}(\Delta C)\dfrac{\theta_{\max}|\Delta C|^n}{|\Delta C|^n + REF^n}$

Movement of one unit step length:

$$\Delta x = (v\cos\theta)\Delta t$$
$$\Delta y = (v\sin\theta)\Delta t$$

Restart algorithm

Time for one perception of concentration to next $= \Delta t$. Change of angle of orientation from normal of trail $= \Delta\theta$. Difference between two antennae's concentrations $= \Delta C$. Speed of ant $= V$. Maximum turning rate $= \theta_{\max}$. Flexibility of response (real number 1-10) = n. Reference concentration that induces the half maximal turning rate = REF

Figure 2: Calenbuhr and Deneubourg algorithm for osmotropotactic orientation.

2.4 Neurophysiological evidence

Gronenberg (1996) notes that "with few exceptions we know almost nothing about the neuronal control of even simple behaviours in ants" (p.17). However, though there exists no layout of the neural controller of trail-following, there has been substantial investigation of the olfactory pathway in insects: the information below comes from studies on moths, bees and cockroaches as well as ants. Osorio, Getz and Rybak (1994) point out air turbulence makes it difficult to detect spatial and temporal differences in odour molecule concentration. Receptor cells on the antennae vary in sensitivity and specificity but there are pheromone specialist receptors. The responses made by the receptor cells to the concentration of odour is highly non-linear because of the dynamics of binding of odour molecules to receptors. Thus there are strong threshold and saturation effects. The firing of the receptor nerve cells is phasic-tonic.

Receptor axons make up the antennal nerve and transmit the signal to olfactory interneurons in the glomerular neuropil of the antennal lobes in the deutocerebrum (Masson & Mustaparta, 1990). This has been described as a labelled-line system, in that axons of specific pheromone receptors are grouped in the nerve and project to particular regions of the antennal lobe. Here there is convergence in the order of 1000:1 from receptor inputs to glomerular outputs. It is assumed this serves an integration function, to create reliable responses to minute concentrations of a pheromone (Duchamp-Viret *et al.* , 1989). Local interneurons in the antennal lobe overlap spatially with receptor projections and the few hundred output neurons, suggesting additional processing is involved in the temporally complex response properties of these outputs (Christensen et al 1993).

These projections from the deutocerebrum go to a number of sites in the protocerebrum including the mushroom bodies and lateral lobe. Pheromone processes appear to project to specific sites. The mushroom bodies have been associated with learning, multimodal integration and other 'higher' processes. Connections to descending neurons that may drive the motor response have not been mapped in ants (to our knowledge). In the moth, connections from olfactory protocerebral neurons to lateral accessory lobes (LAL) and thus to descending neurons has been shown. Interconnections between the LAL provide the earliest stage at which bilateral responses to olfactory stimuli are observed (Hildebrand, 1995).

2.5 Proposed model

We use model neurons typical of the neural representations given in connectionist or artificial neural network models. Each neuron has a simple activation rate and threshold value and the strength of the neural messages being passed between the neurones can be represented by a weight value for each connection. In our model each neuron has only one associated weight, and connects to only one or two other neurons, in a pre-specified architecture.

A simple initial model for osmotropotaxis can be seen Figure 4. Each sensor takes a reading of chemical concentration and turns it into a corresponding neural activity. Each sensor's neural activity then stimulates its own odour strength neurone and inhibits the opposite odour strength neurone. This means that which ever sensor is the more active, its odour strength neurone will be active, where as the opposite odour strength will not be activated at all. Therefore the ant's stimulated odour strength neurone will stimulate the opposite motor system to speed up, which in effect will cause the ant to head towards the greater concentration of gas (essentially, a Braitenberg vehicle). The strength of the speed neurone decides the general speed at which the ant travels and its speed of turning.

However, a number of important considerations suggest that this model needs to be elaborated so as to utilise temporal differences in chemical concentration:

- The pheromone signal is inherently unreliable due to air movement. Temporal averaging will provide a more accurate estimate of relative concentration.
- Gas sensors (used on the robot) do not provide exact values of chemical concentration but do respond to change in concentration. This is also a plausible interpretation of macroglomeruli output in insects.
- There is a small spatial chemical gradient between antennae. A larger temporal gradient can be seen when moving.
- Ants still manage to follow the chemical trail using klinotaxis with one antenna, i.e. temporal differences can be the cause of the orientation response.

Therefore a new neural layout was proposed that included both a temporal layer and a comparison layer. This can be seen in Figure 5 below. The response to chemical concentration (0) is, on each side, passed to two neurons (1 & 2). 1 connects to a "shunt" neuron (3), and the output of this is combined with the direct connection (2) by neurons 4 & 5. The output of this layer thus represents a simple temporal averaging or differencing (depending on the sign of the respective weights) of the signal: i.e. combining the direct signal from 2 and the signal delayed by one time step via 3. This then used to calculate a motor speed as before, although in this case the strength output (6) has been connected to the same side, so it should respond negatively when 4 is active and positively when the opposite 5 is active.

This mechanism was tested for control of osmotropotaxis in a virtual world, using a genetic algorithm to determine the weights, and also in the real world using a robot. The results were compared to ant behaviour and previous models by manipulation of the following environmental variables:

- The artificial ant's antennae size.
- The artificial ant's starting angle to the trail.
- The artificial ant's general speed.
- Amount of chemical laid down.

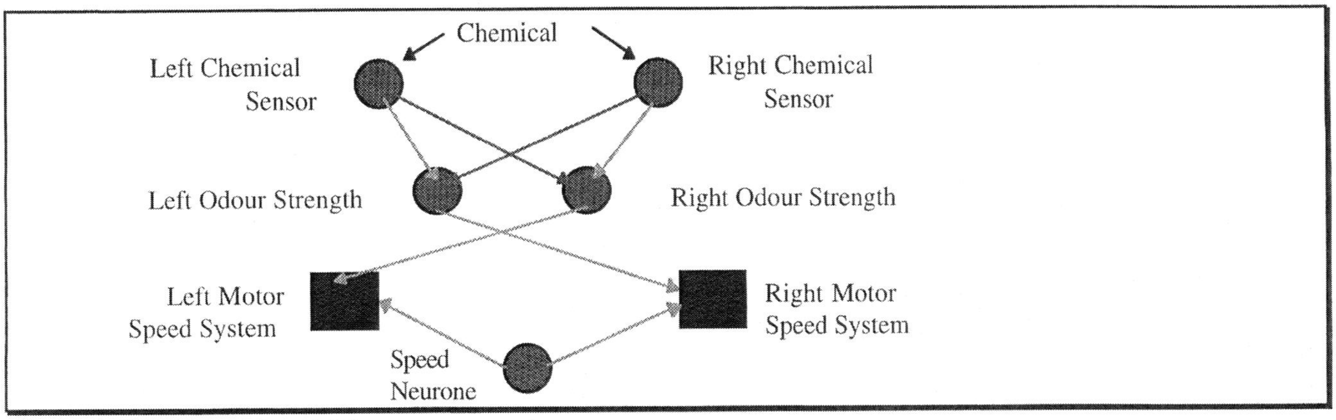

Figure 4: Simple neural model for osmotropotaxis.

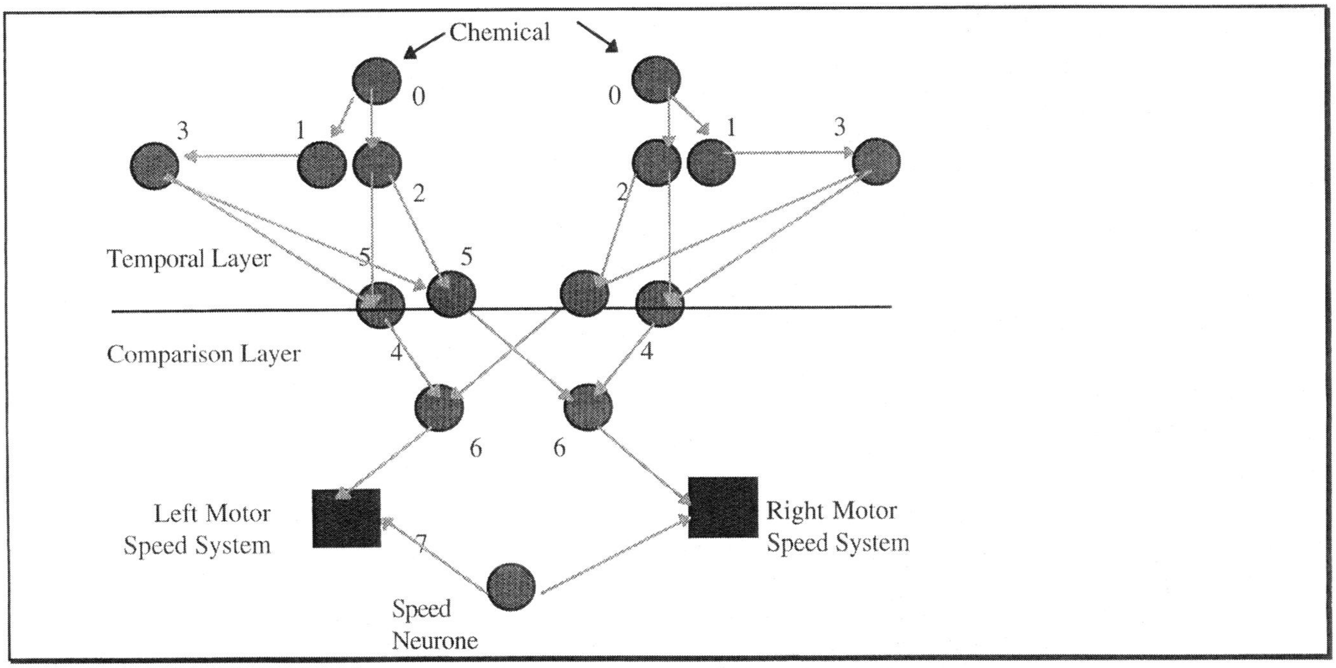

Figure 5: Proposed model for osmotropotaxis incorporating temporal layer (see text)

3. Virtual Simulation

The artificial ant was modelled using Java & VRML (Ames et al 1997) and a genetic algorithm used to adapt the weights in the network described above. The virtual world contains a flat plain with a straight line trail of chemical (this allows us to define an axis coincident with the trail, which simplifies calculations; it is also true for ants that trails tend to develop in straight line segments). The 3-d ant has to sample the chemical as it disperses from the trail, process the signal and then move appropriately (see figure 6). There are four functional units:

Left and right antennae position: The ant's position is represented as the Cartesian co-ordinates of its body centre; where the x-axis is given by the chemical trail, z is the horizontal distance and y the vertical distance.

The position of the (fixed) antennae then depends on the size and rotation of the ant. The distance of each sensor tip from the trail (x-axis) can be calculated by straight-forward trigonometry.

Calculation of gas concentration at the antennae positions: The following input is required for the equation to be calculated:

- amount of pheromone deposited (Q),
- the diffusion coefficient of the pheromone(D),
- the left or right antenna distance (r)
- length of time since trail deposited (t).

These values then can be calculated the equation:

$$Concentration = \frac{Q}{2\pi Dt}\exp\left\{-\frac{r^2}{4Dt}\right\}$$

The values used were taken from Calenbuhr & Deneubourg's model. An element of randomness of gas is created by multiplying the concentration with a random number and the randomness-of-gas variable.

Neural processing: this calculates the activation of each neuron in the network described in fig 5. The gas concentration is first amplified:

olfactory-output = gas-amplification *
gas-concentration / sensor-saturation-level;

This gives an output range between 1 and 0 for each gas level and this can be passed into the first neurone (neurone 0). Activation multiplies the input to each neuron by the appropriate weights, compares the sum with the neurons threshold, and produces an activation of 1 if above threshold or 0 if below. The sequence of activation calculation is actually carried out backwards, i.e. from neuron 7 to neuron 1 (rather than having activation propagate right through from 1 to 7) because it is important for the temporal encoding to have each layer of the neural network encoding temporally different changes in the chemical concentration. Symmetry was assumed i.e. the values of weights and thresholds for neurons on the right and left side were identical.

Movement: the weighted sum of activation from neurons 6 & 7 provides a motor speed strength for each side. If these are equal the ant moves an appropriate distance in a straight-line. If they are not equal, the difference between them is used to calculate a turning circle arc that can be used trigonometrically to update the position and heading of the ant.

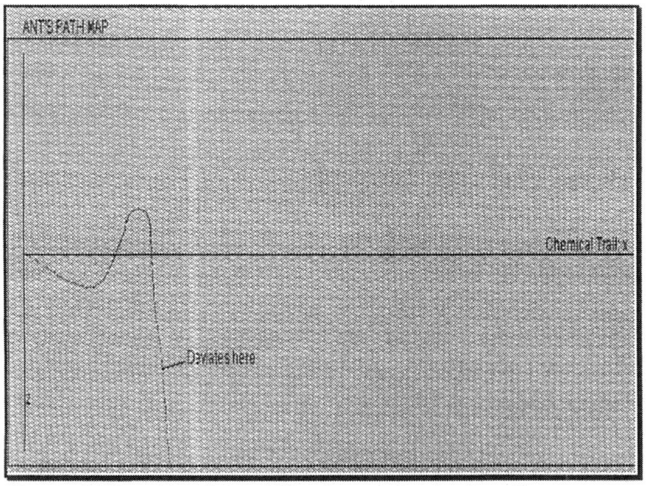

Figure 6 The virtual ant following a chemical trail

The weights and thresholds of the seven neurons and the gas amplification constant were adapted using a standard GA procedure. Weights were initially randomised, each ant attempted to follow the trail and its success was measured by the distance travelled within 15 units of the trail and the number of trail crossings. The best ants were mated using crossover and their offspring added to the next generation along with more randomly generated ants.[2]

3.2 Results

The neural layout used to produce the following results was the most successful individual in the fifth generation of the genetic algorithm run with a population of 30 ants, using the default values for all environmental variables (see below). We stopped when we had an ant capable of following the trail rather than waiting for average fitness to reach an asymptote, as we were more interested, here, in exploring the basic capabilities of the network then its evolutionary development (but see discussion).

The weight and threshold details of this ant are as follows:

	Weight	Threshold
Neurone 0 (to 1 & 2)	0.254	0.078
Neurone 1 (to 3)	0.057	0.006
Neurone 2 (to 4 & 5)	0.917	0.040
Neurone 3 (to 4 & 5)	0.852	0.005
Neurone 4 (to 6)	0.741	0.013
Neurone 5 (to 6 on opposite side)	-0.738	0.089
Neurone 6 (to motor)	-0.265	0.049
Neurone 7 (to motor)	0.867	0.005
Gas amplification	1.429	

It is interesting to note what these values imply about the mechanism, although the exact values are probably not critical. Neurones 2 and 3 both have positive weight which means the input is being time-averaged rather than time differenced. Neuron 6 has a negative weight so activation by 4 will slow speed, and activation by 5 will increase it.

There were four main tests used in the osmotropotactic evaluation. Each test was used to evaluate the four areas that Calenbuhr *et al.* (1992) used to test their model: amount of chemical, starting angle of the ant, speed of the ant and antennae span of the ant. For each test all the environmental variables were kept constant except the one being tested (the default values were: concentration 1E-9g/cm; antennae span 1; speed 0.5; starting angle 2 degrees). The variable being tested was tried over a range of values, with each value being tested twenty times. The repeated testing allowed for a reliable average to be calculated. The results gathered were the number of times the trail was crossed and the distance travelled along the trail inside a zone 15 units either side of the trail. Each trial lasted for two thousand loops of trail following behaviour. Both measures indicate the effectiveness of the trail following behaviour. Therefore they were multiplied together to give an overall "effectiveness" measure of trail following.

3.2.1 Effect of chemical concentration on trail following

The effect of chemical concentration was evaluated by varying the amount of chemical laid down. A wide range of chemical amounts were used from 1e-10 to 5e-6 [g cm^{-1}]. The results can be seen in Figure 7. This shows the non linear effect that the amount of chemical trail laid down has

[2] The simulation and GA interface can be found at http://www.psychology.nottingham.ac.uk/research/aarg/ant/project.html

on trail following. At low concentrations (below 1E-10), no chemical trail following occurs. Successful trail following occurs over a range from 3E-10 to 1E-9, an order of magnitude The peak in the graph (5E-10) indicates the amount of chemical/pheromone which induces the most effective trail following - interestingly this was not the value used during the evolution of the the circuit (1E-9).. Higher amounts of chemical cause a drop in chemical trail following ability.

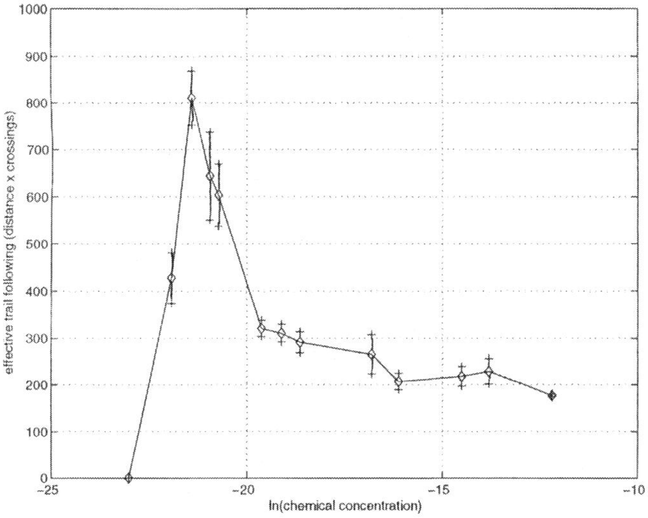

Figure 7: Effect of chemical concentration on effective trail following (measured as distance followed x number of crossings of the trail). Average score from 20 trials with standard error bars.

This graph shows a strong similarity to the effect already described for real ants: the characteristic peaked response curve. There is a threshold below which a negligible response occurs, a rise to a peak 'optimum' level of pheromone, and a drop-off after the peak (Calenbuhr et al 1992).

3.2.2. The effect of the starting angle of ant on chemical trail following

The effect of the starting angle was evaluated by varying the starting angles for the ant. Seven starting angles were used between from 0 to 45 degrees. The results can be seen in Figure 8 (top). The starting angle can clearly be seen to influence trail following. The graph indicates that starting angles of 0 to around 30 degrees can induce trail following. The graph also indicates the fact that the lower the starting angle of the ant the more effective the trail following is likely to be.

Real ant studies on how starting angle influences chemical trail following ability (Pasteels et al, 1986) show that ants starting with orientations less than thirty degrees to the trail are far more likely to follow that trail.

3.2.3 The effect of the ant's speed on chemical trail following

The effect that the speed of the ant had on chemical trail following was evaluated by multiplying both motor outputs by two and halving the number of cycles (2000 to 1000). The trail following from this new motor speed was examined in comparison to the normal motor speed trail following. The results can be seen in figure 8 (middle).

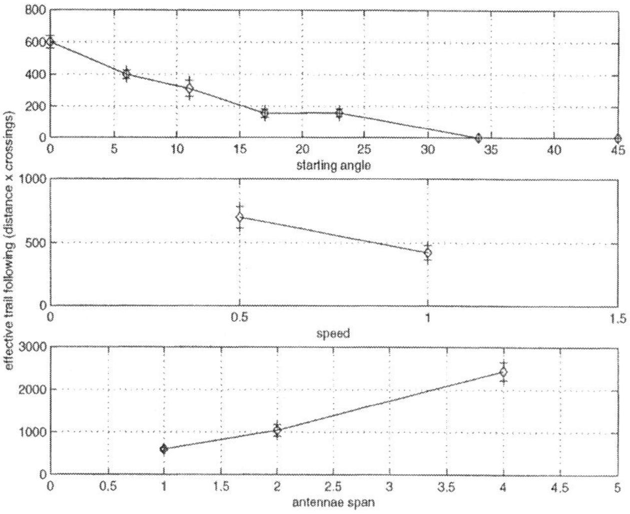

Figure 8: The effect of (top) the ant's starting angle (middle) the ants speed and (bottom) the ants antennae span on effective trail following. Average score from 20 trials with standard error bars.

The results indicate that ants at the slower speed follow trails more effectively than the faster ants. This trait is seen in real ants' chemical trail following: starved ants orientate along the trail at a slower rates and travel greater distances. Similar results were found in Calenbuhr & Denuebourg's (1992) model using two different speeds to orientate along the track.

3.2.4 The effect of the antennae span on chemical trail following

The effect of the antennae span on chemical trail following ability was evaluated by varying the distances between the antennae. Three sizes of antennae span were used 1, 2 and 4 [arbitrary units] . The results of the trail following can be seen in Figure 8 (bottom).

The graph shows that the larger antennal span is more effective at trail following. This same response is found in real ants with different antennae spans. In all species of ants their are different types of castes within the nest. The different types of castes often fulfil different roles in the nest and are therefore physically different from each other. In one species of ant, *P. pallidula*, there are two different castes of worker ants, major and minor. The major ants have a larger antennae span and the mean distance of trail they follow is twice as large as of that followed by the minor (Detrain et al., 1988). This indicates that larger antennal spans increase trail following ability in ants.

4. Robot model

The results with the simulation were interesting and encouraging - nevertheless there was the risk that the environment was insufficiently representative of the real world problem of trail-following, and the network described worked only because of this simplified task. Therefore the network was also tested in robot model that had to follow a trail of alcohol laid on the lab floor. This work was carried out in parallel with the simulation study (largely for practical reasons). Thus it should be regarded as an independent investigation of the proposed mechanism rather than an attempt to implement the results of the simulation.

4.1 Implementation

The robot was built from Lego bricks, which provided a sturdy but easily altered chassis. Two motors controlled the two drive wheels (left and right) and dragged two rear ground contacts that stabilised the robot. The processor board was a 68HC11 programmed in IC (Martin, 1992). The chemical sensors (Figaro SB-AQ1) detect alcohol (e.g. turpentine substitute) and their resistivity changes depending on the amount of the chemical in the air. They were interfaced to the board and the analogue voltage difference dependent on the resistance was converted into a 8-bit digital signal, which was inverted so that higher gas=lower resistivity=higher input. They were mounted on 'antennae' extending forwards and sidewards from the robot body. The default antennae span was 29cm.

Because the processor was relatively slow, instead of implementing the identical network used in simulation, a simplified version was adopted in which no thresholding was carried out. Appropriate values for the weights were determined by hand-tuning.

The gas sensor was slow at adjusting to decreases in the chemical- it could take over one minute to reach the original resting potential. Thus obtaining an accurate value of any chemical concentration in such a dynamic environment was difficult. Consequently the temporal layer was forced to do temporal differencing by making neurone 3 have a negative weight. (In later tests it was apparent that the temporal averaging that had evolved in the simulated ant was not viable to control the robot). Note that though neurons 4,5 & 6 here were all opposite in sign to the simulation this creates an equivalent behavioural effect i.e. slowing the motor to activation of 4 (same side) and increasing it for activation of 5 (opposite side).

The actual algorithm for the neural network was otherwise implemented in a similar manner as described for the virtual simulation.

There were other problems associated with the robotic system. There were differences between the left and right chemical sensors and the left and right motors. To overcome the bias problem of the chemical sensors, a sample concentration at both sensors is taken at the start. This starting concentration is used to gauge the difference between the chemical sensors. For this to work, the sensors must be equidistant from the chemical trail at the time of sampling. The motor bias is easily worked out by finding the ratio of motor speeds (left:right) that makes the robotic ant travel in a straight line. The output weights to the motors are multiplied accordingly. Other factors that could be manipulated were the number of chemical trail following cycles, the gas amplification and the speed neurone activation (neurone 7).

The criteria used to measure trail following effectiveness was similar to that done in the virtual simulation. The two measurements used were the number of times the trail was crossed and the distance the robot travelled along the track (max. 5.5 metres) without moving further than one metre from the track. The neural weights used for this test were as follows:

	Weight
Neurone 0	1.0
Neurone 1	1.0
Neurone 2	1.0
Neurone 3	-0.5
Neurone 4	-0.5
Neurone 5	1.5
Neurone 6	1.0
Neurone 7	1.0

The tests are not as extensive as for the simulation evaluation, largely because the robotic testing took far more time than the virtual testing. Therefore only five runs were performed on each test instead of the twenty done in the virtual evaluation. Also, changing the amounts of chemical proved difficult. The chemical used (turpentine substitute) was immiscible with water therefore could not be diluted. As a result twice the amount of chemical was laid on the trail in the hope that a different response would occur.

The same four tests as before were used in the robotic evaluation of osmotropotaxis: amount of chemical, starting angle of the ant, speed of the ant and antennae span of the ant.

4.2 Results

4.2.1 Effect of chemical concentration on trail following

The effect of the amount of chemical was evaluated by varying the amount of chemical laid down. A plastic bottle with a small hole punctured in the lid was used to distribute the chemical. To vary the amount of chemical laid on the trail a larger hole was punctured in the lid of the bottle. Therefore a large and a small amount of chemical of was laid for the two tests. The results of the trail following can be seen in Figure 9 (top).

This shows that trail following becomes less effective as the amount of chemical increases. There are not enough data points to compare to the curve from simulation or ant behaviour, but the results are consistent with these, in that a non-zero amount of chemical is necessary for trail following but increasing beyond some optimum point leads to reduction in trail following success.

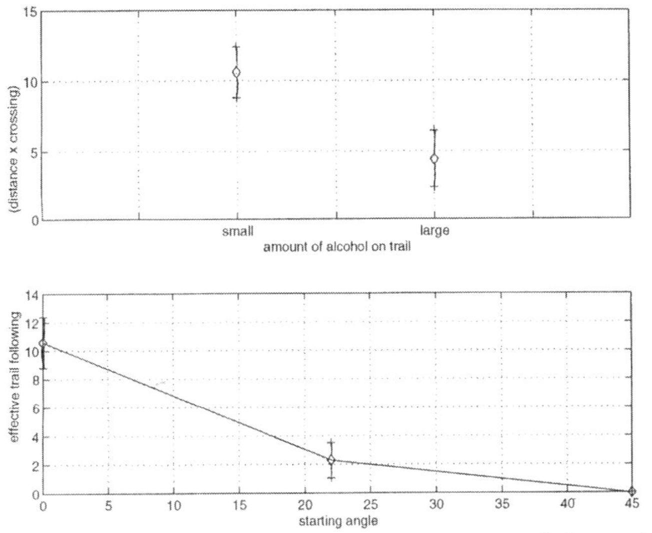

Figure 9: Effect of (top) chemical amount and (bottom) starting angle on effective trail following. Average from 5 trials with standard error bars.

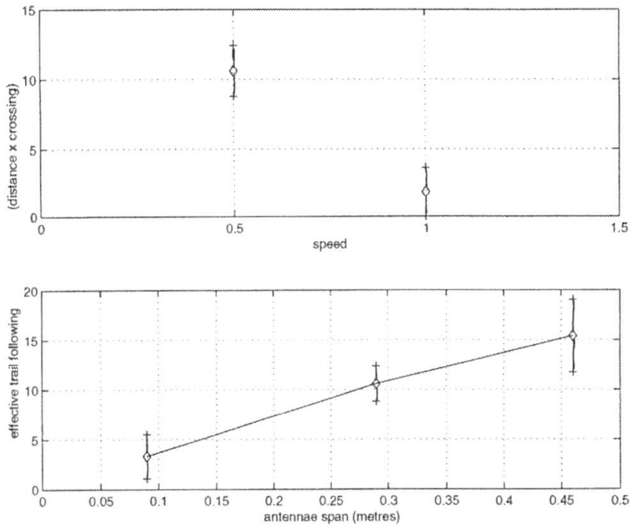

Figure 10 Effect on robot trail following of (top) speed and (bottom) antennae span. Average from 5 trials with standard error bars.

4.2.2 Effect of the ant's starting angle on trail following

The ant was started at either 0, 22 or 45 degrees to the direction of the trail. The results of the trail following can be seen in Figure 9(bottom).

The results show that the closer the starting angle is to the trail direction the greater the effective trail following becomes. This result is very similar to the virtual simulation response of the ant's starting angle. The result also remains similar to the result of Pasteels et al (1986) and Gerardy & Verharge's (1988) studies summarised in 3.2.2. Thus the robotic ant's starting angle affects chemical trail following in a similar manner to the effect in real ants.

4.2.3 Effect of the ant's speed on trail following

This effect was examined by measuring the robotic ant's trail following ability at normal speed and then at twice that speed (i.e. both left and right motor outputs were doubled in speed). The results of the trail following can be seen in the table in Figure 10 (top).

The results show that the slower speed of the robotic ant was more effective for trail following than the faster speed. This can also be seen in the simulated result. Both the results are similar to real results as covered in the virtual simulation evaluation.

4.2.4 Effect of antennae span on trail following

This effect was measured by changing the robots antennae width using different sized Lego bricks. The results of the trail following can be seen in are represented graphically in Figure 10 (bottom).

The results indicate that larger antennae span give better trail following. This result can also be seen in the virtual simulation results and in real ant data (see 3.2.4).

5. Discussion

The model we have tested in a virtual simulation and a robot implementation represents a first attempt to specify a neural control mechanism for osmotropotaxis in ant trail following. Although it is a very simple model it was possible to reproduce nearly all the observed behaviours of the ant, and of the mathematical model proposed by Calenbuhr & Denuebourg (1992).

Specifically we reproduced the following effects:

- The distance travelled on trails is affected in a non-linear fashion by the amount of pheromone on the trail
- The smaller the angle between the ant and the trail, the greater the chance that trail following will occur
- Increasing speed leads to an decrease in distance of trail followed
- Larger antennae span leads to an increase in distance of trail followed

The success of the model over this range and variety of behaviours is striking, and the common results over the two implementations suggest a reasonable robustness. However, rather than assume that we have hit on the exact neural layout of the ant, we believe that this suggests that these properties of the behaviour are more inherent in the nature of the sensor-environment interaction than dependent on details of the controller per se. Thus the fact that sensors saturate and have limited temporal resolution explains the shape of the response curve for different concentrations and the effect of speed. The relationship of spatial positioning of the sensors and the diffusion of chemicals from a trail can explain the angle and antennae span effects. By including these properties explicitly in the simulation, and by necessity reproducing them in the real-world implementation, it became relatively simple to find a control network that gave rise to ant-like behaviour.

However, though the network we implemented was initially derived from Braitenberg (1984) rather than ant physiology per se, it does provide some interesting suggestions regarding the further neuroethological investigation of this system. For example it provides a possible interpretation of the processing function within the glomeruli as temporal averaging or differencing (Christensen et al 1993). Similarly the fact that the two sides interact only at the level of motor output appears consistent with the observation that only at the level of descending neurons in the moth is there a bilateral olfactory response (Kanzaki et al 1991). It also shows that there need be no point in the processing of the stimuli where a 'difference' between the sensors is explicitly calculated (as it is in Calenbuhr & Deneubourg's model). As a minimal model it will likely prove too simplistic: but as with Webb's (1994) model of phonotaxis, there seems no compelling reason why a more complex form of control should be assumed, when this simple circuit can successfully reproduce many of the existing behavioural observations.

Critical tests to distinguish between this neural model and that of Calenbuhr & Denuebourg would probably require a more detailed, quantitative comparison that is not possible with current limitations of our implementations. Most of the real ant data comes from circular trails and we used straight trails (these were both easier to implement and are more realistic for natural trail following). Our models are not to scale, so distances, speeds and lengths are arbitrary. The diffusion co-efficient of the turpentine substitute used in our robot trials is probably different to ant pheromone. One other way to compare the models might be to look at how easily they can support additional behaviours such as klinotaxis.

It would be interesting to carry out a more systematic study of the weights evolving in the virtual simulation. For example, if run for longer, would a temporal differencing mechanism like that used in the robot emerge as a more successful solution than temporal averaging? It might be particularly interesting to examine the evolution while varying the sensor response properties to see whether there are parameter values that are particularly important in determining the success of certain controllers: e.g. can we specify the temporal character of sensors beyond which temporal differencing becomes necessary? Such results have potential application in analysing both the design and the development of the real neural mechanism underlying osmotropotaxis in ants.

References

Ames, A. L., Nadeau, D. R. & Moreland J.L. (1997) *VRML 2.0 sourcebook*. John Wiley & Sons, Inc. New York.

Bossert, W.H. & Wilson, E.O., (1963) The analysis of olfactory communication among animals. *Journal of theoretical biology* 5:443.

Braitenberg, V. (1984) Vehicles: Experiments in Synthetic Psychology, M.I.T. Press, Cambridge, Mass.

Calenbuhr, V., Chretien, L., Detrain, C. & Denuebourg, J.L. (1992) A model for osmotropotatic orientation (ii). *Journal of theoretical biology*, 158:395-407.

Calenbuhr, V. & Denuebourg, J.L. (1992) A model for osmotropotatic orientation (i). *Journal of theoretical biology*, 158:359-393.

Christensen, T.A., Waldrop, B.R., Harrow, I.D. & Hildebrand, J.G. (1993) Local interneurons and information processing in the glomeruli of the moth Manduca sexta *Journal of Comparative Physiology* A 173:385-399

Duchamp-Viret, P., Duchamp, A. & Vigouroux, M.; (1989) Amplifying role of convergence in olfactory system. A comparative study of receptor cell and second-order neurone sensitivities. *Journal of Neurophysiology*, 61, 1085-1094.

Gronenberg, W. (1996) Neuroethology of ants *Naturwissenschaften* 83:15-27

Hangartner, W. (1967). Spezifitat und Inaktivierung des Spurpheromons von *lasius fuliginious* Latr. und Orientierung der Arbeiterinnen im Duftfeld. *Zeitsschrift fur Vergleichende Physiology,* 57(2):103.

Hildebrand, J.G. (1995) Analysis of chemical signals by nervous systems *Proc. Nat. Acad. Sci. USA* 92:67-74

Holldobler, B. & Wilson, E.O., (1990) *The Ants*, Springer-Verlag, Berlin.

Kanzake, R., Arbas, E.A. & Hildebrand, J.G. (1991) Physiology and morphology of descending neurons in the pheromone processing pathways in the male moth Manduca sexta Journal of Comparative Physiology 169:1-14

Masson, C. & Mustaparta, H. (1990) Chemical information processing in the olfactory system of insects *Physiological Reviews* 70:199-245

Martin, F (1992) Interactive C Manual in Jones, J.L. The Mobile Robot Assembly Guide, A.K. Peters, Wellesley MA.

Michel, W. C., McClintock, T. S. & Ache B. W.; (1991) Inhibition of lobster olfactory receptor cells by an odour activated potassium conductance. *Journal of Neurophysiology*, 65, 446-453.

Osorio, D., Getz, W. M. & Rybak, J.; (1994) Insect vision and olfaction: different neural architectures for different kinds of sensory signal. Cliff, D. Husbands, P. Meyer, J. Wilson, S. (eds) *Animals to Animats 3*. MIT Press, Cambridge Mass.

Pasteels J.M.; Deneubourg, J.C.; Verhaeghe, J.L.; Boeve J.L. & Quintet Y. (1986) Orientation along terrestrial trails by ants. *Mechanisms in Insect Olfaction,* 14.

Sudd, J.H. & Franks, N.R. (1987)*The behavioural Ecology of Ants*. Blackie Publishing.

Traniello, J.F.A. & Robson, S.K. (1995) Trail and Territorial Communication in Social Insects. *Chemical Ecology of Insects* 2 - 7.

Webb, B.;(1994) Robotic Experiments in Cricket Phonotaxis. Cliff, D. Husbands, P. Meyer, J. Wilson, S. (eds) *Animals to Animats 3*. MIT Press, Cambridge Mass.

Reinforcement Landmark Learning

S. P. Toombs*, W. A. Phillips* and L. S. Smith**

* Department of Psychology
pat@cs.stir.ac.uk
wap@stir.ac.uk
** Department of Computer Science
University of Stirling,
Stirling, FK9 4LA. Scotland
lss@cs.stir.ac.uk

Abstract

Collett, Cartwright and Smith (1986) trained gerbils to find a hidden food reward at a fixed location relative to an arrangement of cylindrical landmarks. Having learnt where the food is, animals are tested by probe trials in which the food is absent, and their search pattern recorded. Testing with modified arrangements of the landmarks provides information about the computations underlying the animals behaviour. Experiments involving two and three landmarks are simulated using simple and reactive animats, embedded within a bounded, 2-D environment. Internal processing maps the sensory array, through a convolution network, to a topography preserving motor array that stochastically determines the direction of movement. Temporal difference reinforcement learning modifies the convolution network in response to a reinforcement signal received only at the goal location. These experiments are simulated with landmark distance coded as either a 1-D intensity array, or a 2-D vector array, plus a simple compass sense. Vector coding animats significantly outperform those using intensity coding and do so with fewer hidden units. More importantly, in the three landmark task, vector coding animats search behaviour closely matches that of gerbils when tested with modified landmark arrangements. This paper provides further evidence that complex spatial navigation behaviour need not be predicated on complex and navigation specific computations.

1. Introduction

Animal landmark learning experiments, such as those of Collett et al (1986), the focus of this paper, provide a rich source of behavioural data for computational modelling. Animals learn to move to a food source located relative to an arrangement of landmarks within an otherwise empty arena. After learning, animals are tested without food, and their search distribution measured. Search distributions in tests with modified landmark arrangements provide the data to compare with predictions of models of their behaviour.

Typically, models of the behaviour involve systems specific to spatial navigation, and internal representations of spatial location (eg. McNaughton et al, 1995; Redish and Touretzky, 1997). The simulations in this paper examine the extent to which behaviour similar to the animal's can be learned by simple animats whose internal processing consists of only a reactive neural network mapping a sensory array to a motor array. Animats have no navigation specific computation and learn from a non-specific reinforcement signal received only when they move within the goal region.

Section 2. outlines the gerbil landmark learning experiments of Collett et al (1986). Section 3. describes the sensory coding, internal processing and reinforcement learning algorithm. Section 4. examines animats after learning in terms of performance; how efficiently they have learned to move to the goal. More importantly, section 5. compares the search behaviour of animats with gerbils when tested with modified arrangements of landmarks.

2. Gerbil landmark learning

Collett, Cartwright and Smith (1986) trained gerbils to find a food reward at a fixed location relative to an arrangement of identical cylindrical landmarks. The environment for these experiments was a large circular arena painted black, and lit by a single bulb illuminating a central region and leaving the walls in darkness. As far as possible all sensory cues to the food location were removed, except for the visual stimuli provided by the landmarks. The food location itself was invisible. Between trials, the array of landmarks and food location was translated, but not rotated, and gerbils were released on each trial from random start locations within the arena.

After learning, gerbils were tested without food and their search distribution measured. Search distributions in tests with modified landmark arrangements provide the data to compare with the predictions of models of their behaviour.

2.1 Task 1

Collett et al (1986) found that gerbils are able to learn the invisible food location when it is specified only with respect to a single landmark. Because the landmark is a featureless cylinder, this implies that the gerbils have a non-visual direction sense.

2.2 Task 2

With the goal and two landmarks arranged as shown in fig. 1a, gerbils learned the food location as shown by their search pattern when tested with the training arrangement of landmarks (fig. 1b). When tested with only one landmark, gerbils spend most time searching in two locations, each at the same bearing and distance from one of the landmarks as was the goal during learning (fig. 1c). When tested with the distance between the landmarks doubled, gerbils search in the two internal of the four locations suggested by their single landmark search (fig. 1d).

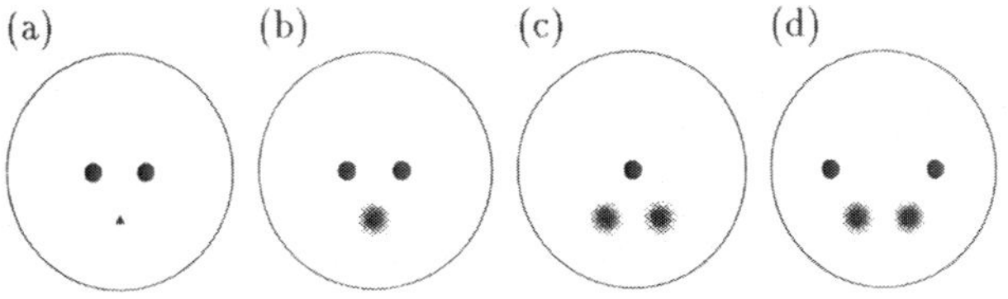

Figure 1 Task 2. (Highly schematic from Collett et al, 1986) (a) Learning arrangement: The 2 cylindrical landmarks are shown as black circles and the (invisible) food source as a triangle. (b)–(d) Search histograms. The dark patches show where gerbils, after learning, spent most time searching in test trials.

2.3 Task 3

Here, the food is located at the centre of an equilateral triangle (fig. 2a). Tested with this arrangement, gerbils focus their search at the goal location (fig. 2b). When tested with one of the landmarks moved away from the others, gerbils ignore the outlier and search in the location defined by the other two (fig. 2c). With a single landmark, gerbils search at the three locations with the same bearing and distance with respect to one of the landmarks as was the goal during learning (fig. 2d). Rotating the triangle of landmarks leads to the search behaviour shown in fig. 2e; gerbils direct most of their search to the centre of the rotated array, suggesting a knowledge of the relation between the three landmarks. In addition, they search to a lesser extent at three locations outside of the triangle. Each of these locations is at the same distance and bearing to two of the landmarks as was the goal in the learning arrangement. In fig. 2f, an extra landmark gives the gerbils a choice between landmarks in the training arrangement and their rotation. In this case, the animals search where the arrangement matches that experienced during learning.

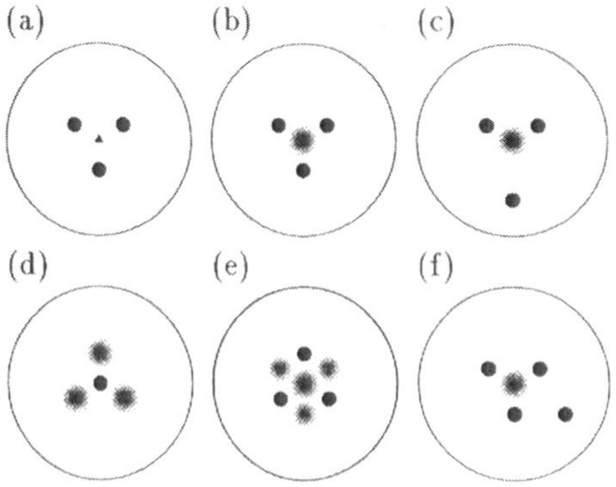

Figure 2 Task 3. (Highly schematic from Collett et al, 1986) (a) Learning arrangement. (b)–(e) Search histograms in test trials.

3. Simulations

Simulations of tasks 2 and 3 are reported here, since they allow tests with modified landmark arrangements that provide the important data for comparison with the gerbils. Animats are provided with a sensory array specifying the distance and relative bearing of landmarks, and a simple compass sense. The sensory input is reactively mapped to a topography preserving motor array that stochastically determines the direction of movement on each time step.

3.1 Sensory coding

Animats have a 360 degree view, provided by 60 evenly spaced receptor units, with non-overlapping receptive fields, each coding the distance of the nearest landmark falling within its receptive field. Landmarks only cause activity in the single receptor whose bearing they are closest to, and the activity of all other receptors is set to zero. Landmark distance in each direction is converted to a receptor activity by the function $f(d) = 2\sin^{-1}(10/d)$, capped at 1. This function is the angle in radians subtended by a circle of radius 10, and monotonically decreases from 1 to 0 as distance increases (fig. 3a).

Two methods of coding receptor activity are compared through simulation: intensity and vector coding.

Intensity coding. With intensity coding, a single continuous valued unit for each direction codes receptor activity, yielding a 1-D, 60 element, sensory array (fig. 3c, left column).

Vector coding. With vector coding, instead of having a single, continuous valued unit coding distance in each direction, a number of binary units are used, each coding for a non-overlapping range. At each landmark distance, a single one of these units is on and the rest off. In the simulations reported here, receptor activity (between 0 and 1) is split into 6 equally sized regions, corresponding to increasingly large distance ranges. Hence, the 2-D vector coded sensory array is of size 6×60 (fig. 3c, left column). Because only six binary units are used, vector coding of distance is a great deal coarser than the continuous valued intensity coding.

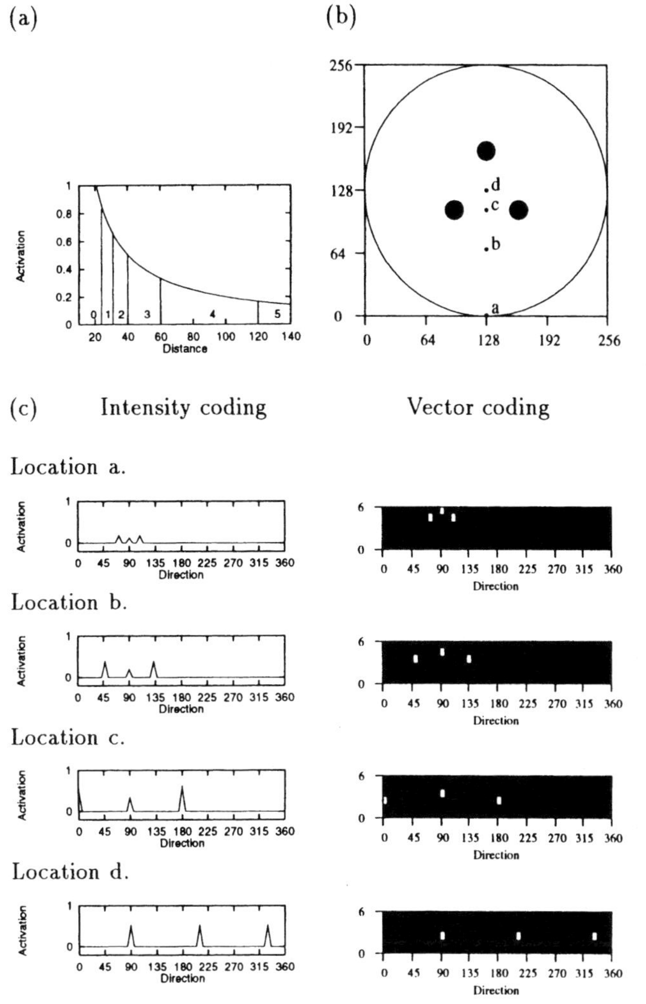

Figure 3 (a) Landmark distance in each direction is converted to a receptor activity between 0 and 1. With intensity coding, this value is coded by a single unit in each direction. With vector coding, each receptor activity is coded by 6 binary units with non-overlapping ranges. The vertical lines plot the boundaries between these ranges.
(c) plots the 1-D intensity coded sensory array (left column) and the 2-D vector coded array (right column) for locations a to d in (b).

3.2 Animat processing

The intensity, or vector coded sensory array is reactively mapped to a 15 element, topography preserving motor array, evenly covering 360 degrees, that stochastically determines the direction in which the animat moves a fixed distance. At each time step, the higher the value of a motor array element, the higher the probability that the animat will move in the corresponding direction.

Internal processing consists of convolution of the sensory array with a filter network—a standard feedforward neural network with a single output unit, and either a single layer of weights (direct), or hidden units and two layers of weights. To implement convolution, identical networks are centered at 15 evenly spaced directions and map their respective portions of the sensory array to a number between 0 and 1. Fig. 4 shows the internal processing in the vector coding condition.

Thus, a particular architecture within this framework is specified by the following parameters: dimensions and size of the sensory array, size of the 1-D motor array, filter network size, and filter network fan-in. In all the simulations reported here, network fan-in was set at 59 units, corresponding to 354 degrees.

Compass sense. In addition to the sensory input, networks also have a layer of compass units, providing allocentric direction information. In these simulations, 15 binary compass units are used giving a resolution of 24 degrees. The compass unit values are processed in the same way as the sensory input, so each hidden unit (or the output unit for direct networks) has weights to both sensory and compass units and activity is determined as the weighted sum across both sets of units. Compass units are binary with only one on and the rest off for each filter network position around the 360 degrees. Which compass unit is on depends upon the direction in which the network is pointing, as shown in fig. 4, with a different compass unit coding for each direction.

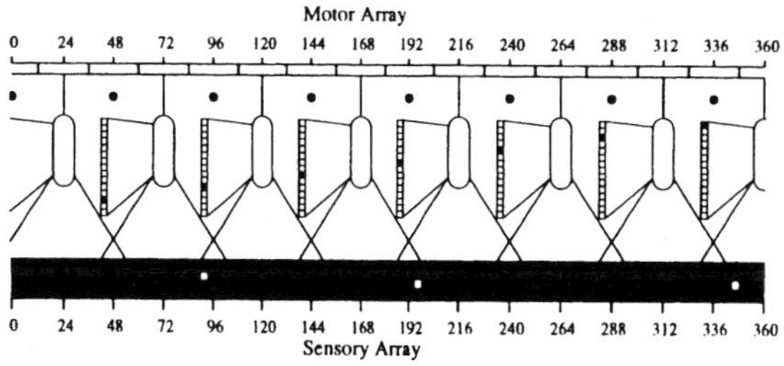

Figure 4 Animat sensorimotor system: The sensory array is convolved (with subsampling) with a filter network to produce a 15 element motor array. The sensory and motor arrays cover 360 degrees. A different binary compass unit is on for each direction in the convolution (black dots mark undrawn filter network positions)

3.3 Reinforcement Learning

The agent receives non-zero reinforcement only when it moves within the goal region. Having randomly stumbled into the goal region, the animat must use the non-specific reinforcement signal, received only having arrived at the goal, to modify its actions so as to steer more efficiently toward it in the following trials. Temporal difference learning (Sutton, 1988) is a method for solving this problem, based on the difference between temporally successive estimates of expected future reinforcement. Q learning (Watkins, 1989), is a form of temporal difference learning that maps state/action pairs to an estimate the number of steps to the goal for each action given the current state.

Motor array elements are taken as Q values, one for each direction in which the animat can move, q_i. These are converted to probabilities of moving in the corresponding direction, θ_i, by the function (Lin, 1992):

$$prob(\theta_i) = \frac{e^{q_i/T}}{\sum_k e^{q_k/T}} \tag{1}$$

where $0 < T \leq 1$ is a temperature parameter controlling the randomness of the action selection. With T near 0, the probability of selecting the direction with highest q is near 1, and the animats movements are near deterministic. With higher T, the direction favoured by the network has a lower probability of being selected, and the movement of the animat is more stochastic. A movement direction θ_m is randomly selected given these probabilities, and the animat moves a fixed distance in this direction.

The new location yields a new sensory array, mapped to a new motor array of Q values, q_i'. These are used to generate the error signal with respect to the direction actually moved, θ_m

$$error = \begin{cases} 1 - q_m & \text{if in the goal region} \\ \gamma \max\limits_k(q_k') - q_m & \text{otherwise} \end{cases} \tag{2}$$

This error is then backpropagated through the filter network so that weights are changed with respect to the input to the network in the direction in which the animat moved, θ_m. Standard backpropagation (Rumelhart et al, 1986) with momentum is used.

Parameters. All simulations use the same learning parameter values of: Temperature, T (in the stochastic action selection) = 0.02; Discount, $\gamma = 0.95$; backpropagation learning rate = 0.2; backpropagation momentum = 0.9.

3.4 Simulation Method

Animats were simulated for the above tasks in a bounded circular arena of radius 128 (distance units dropped). Because animats have no path integration, and are not sensitive to the arena walls, there is no point in translating the landmark arrangement between trials. Landmarks had a radius of 10, as did the circular goal region.

At the start of each trial, animats were placed at a random location, outside the goal region, and moved a distance of 10 in one of 15 evenly spaced directions on each time step, until they moved within the goal region. Animats received a reinforcement signal of zero on all time steps, except when moving into the goal region, when a reinforcement of 1.0 was received, and a new trial begun. If animats did not get to the goal within the maximum time of 500 steps, a new trial was begun. All simulations were replicated three times with different random initial weights.

The main focus of this work is on the behaviour of animats after learning, rather than on convergence speed. Hence, learning proceeded for 50K trials in all conditions, by which time all animats had easily converged; no signs of overlearning or learning instabilities were found.

After learning, animats were tested over 1000 trials from randomly chosen starting locations. Performance on each trial is measured as the minimum pathlength to the goal, divided by the actual pathlength. Thus, mean performance is between 0 and 1.0; with 1.0 corresponding to an animat taking the shortest route to the goal from every start location. A performance of 0.5 means that the animat took, on average, twice as many steps to get to the goal as the shortest route.

4. Performance on the learned task

4.1 Task 2

Fig. 5a plots mean performance, for both intensity and vector coding, as a function of filter network size. Intensity coding animats with 2, or less, hidden units perform at a random level, and with 4 or 8 hidden units perform around 0.5. Direct vector coding animats perform at a random level, but with 2 hidden units perform at 0.69; this does not increase with 4, or more, hidden units. The performance difference between 4 hidden unit, intensity coding animats, and 2 hidden unit, vector coding animats is significant (t-test: $t = 4.033$, $p < 0.02$).

4.2 Task 3

Fig. 5b plots mean performance for task 3. The performance of intensity coding animats increases with network size up to 4 hidden units, which perform at 0.50; around the same level as for intensity coding animats in task 2. Direct vector coding animats have a mean performance of 0.78; animats with hidden units do not do better. Direct vector coding animats significantly outperform 4 hidden unit, intensity coding animats (t-test: $t = 8.251$, $p<0.01$).

On both tasks, vector coding animats significantly outperform intensity coding animats, and do so with fewer hidden units.

5. Search behaviour

The important variable is the extent to which the search distribution of animats that have learned to move to the goal location matches that of the gerbils when tested with modifications of the landmark arrangement.

5.1 Task 2

Intensity coding. As shown above, intensity coding animats with 4, or more, hidden units, learn to move to the goal location. However, when tested with modified landmark arrangements, their behaviour is consistently very different to that of the gerbils. Tested with a single landmark, search is diffusely directed around it; tested

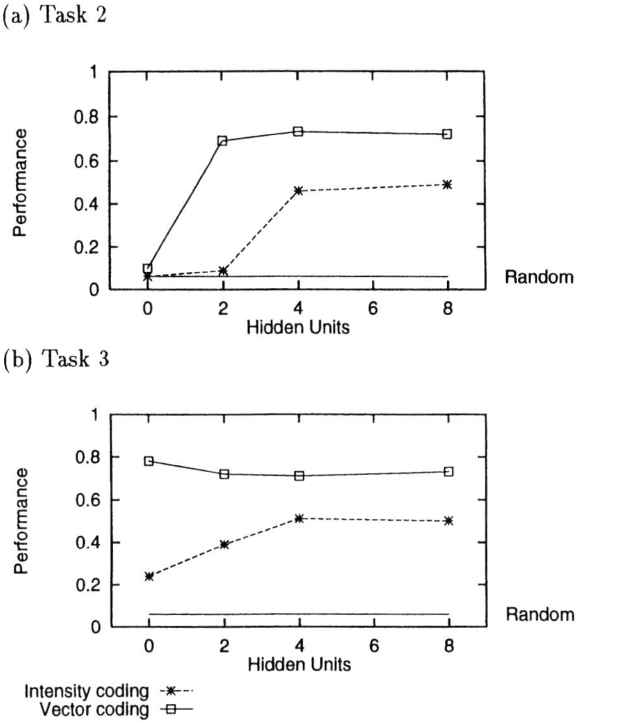

Figure 5 Mean performance after 50K learning trials as a function of sensory coding and filter network size for (a) task 2 and (b) task 3. Each data point is the mean of 3 animats tested over 1000 trials. All standard errors<0.05. Performance level of randomly behaving animat also shown.

with the distance between the two landmarks doubled, intensity coding animats consistently search in the single location where landmark bearing matches that during learning, unlike the gerbils (fig. 1).

Vector coding. Fig. 6 shows the behaviour and search histograms of a two hidden unit, vector coding animat performing at 0.70. This animat moves fairly directly toward the goal from anywhere in the arena except for the top left region (fig. 6a,b). Figs. 6c–e show search histograms and should be compared with those of the gerbils shown in fig. 1.

With the landmarks arranged as per learning, search is focussed upon the goal location (fig. 6c). Tested with a single landmark, the animat searches most at the location with the same distance and bearing from the landmark as was the goal from the lefthand landmark in the learning arrangement (fig. 6d). This suggests that the animat has only learned the goal location with respect to the lefthand landmark. However, when tested with the distance between the landmarks doubled, search is mostly confined to a single location between the landmarks (fig. 6e), rather than the two locations suggested by the single landmark search. This indicates that both landmarks influence this animat's search behaviour, although one is dominant. Note that this search behaviour is different to that shown by Collett et al's gerbils (fig. L.gerbs T2).

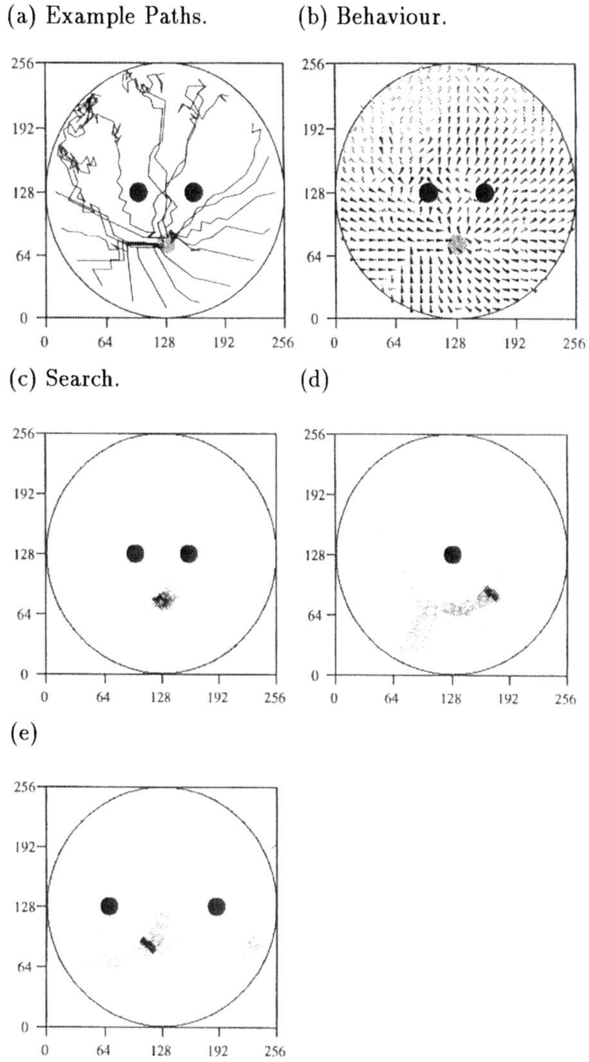

Figure 6 The behaviour and search patterns of a 2 hidden unit, vector coding animat on task 2 after learning. Details as in previous figures.
(c) Shows the search histogram when tested with the environment the same as during learning.
(d)–(e) show search histograms in modified environments.

5.2 Task 3

Intensity coding. As with task 2, intensity coding animats, despite having learned to move to the goal, show search consistently different to the gerbils when tested with modified landmark arrangements. With a single landmark, search is diffuse. With one landmark moved away from the others, intensity coding animats still search in the location specified by the remaining two landmarks, thus matching the gerbil's behaviour (fig. 2c). With the landmark array rotated, animats, unlike the gerbils, do not search in the center of the rotated landmark array.

Vector coding. Figs. 7 shows the behaviour and search histograms of a direct, vector coding animat performing at 0.78. This animat moves fairly directly to the goal

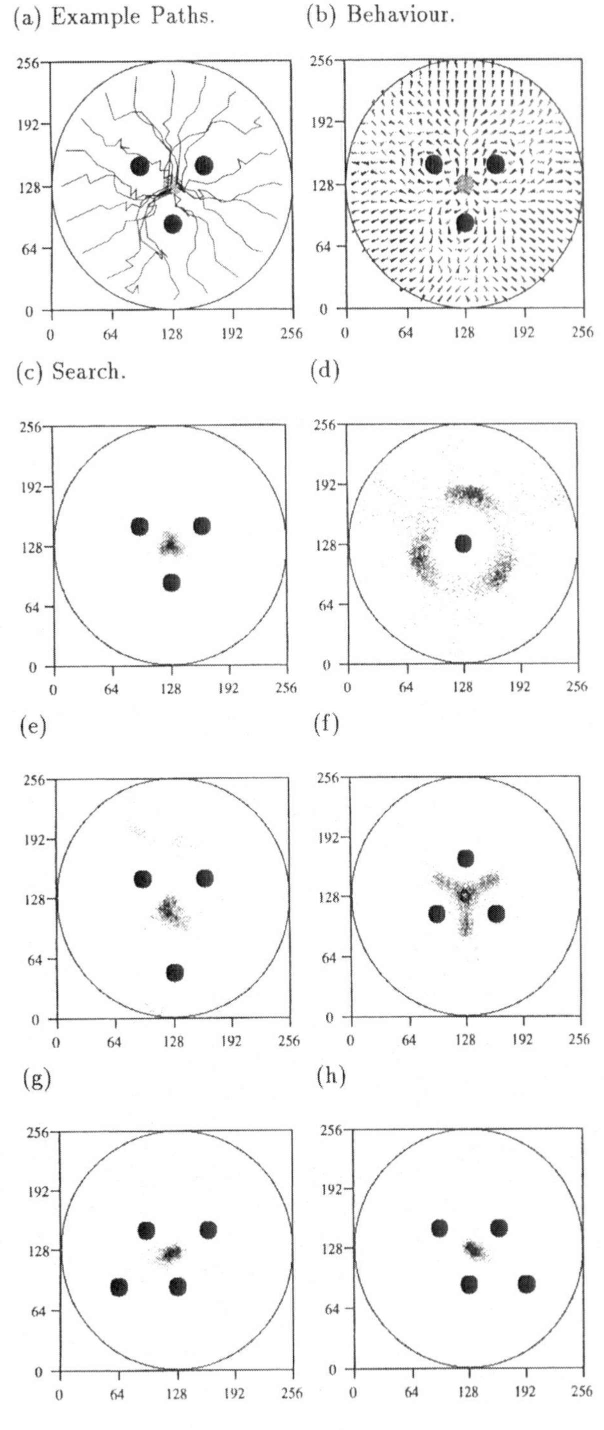

Figure 7 The behaviour and search patterns of a direct, vector coding animat on task 3 after learning. Details as in previous figures.
(c) Shows the search histogram when tested with the environment the same as during learning.
(d)–(h) show search histograms in modified environments.

from any start location (fig. 7a,b). Figs. 7c–h show search histograms and should be compared with those of the gerbils shown in fig. 2.

Tested with the learning arrangement, search is focussed at the goal location (fig. 7c). With a single landmark, the animat, like gerbils, searches mainly in three locations, each at the same bearing and distance from the landmark as was the goal in the learning arrangement (fig. 7d).

With one landmark displaced, the animat still searches around the region specified by the other two landmarks(fig. 7e). Like the gerbils, this animat ignores an outlying landmark when two out of the three are in the usual place.

With the triangle of landmarks rotated, search is greatest at the center of the rotated array; additionally, this animat searches between the center and three locations outside of the array (fig. 7f), as did the gerbils in Collett et al's (1986) study. Each of these locations is at the same bearing and distance from two of the landmarks as was the goal during learning. Despite the search distribution when presented with a rotated triangle of landmarks, this animat does not search at the center of the rotated triangle when given a choice between this and a location where both the relative and absolute bearings and distances match those experienced during learning (fig. 7g,h).

The search behaviour of this direct, vector coding animat matches that of the gerbils in Collett et al's (1986) study, for all the modified landmark arrangements shown in fig. 2. Approximately the same search distribution was shown by all three direct, vector coding animats.

Given Fig. 8 shows further search patterns for the direct vector coding animat from above. With one landmark removed, the animat searches mostly where the goal was (fig. 8a); this matches gerbil behaviour. Figs. 8b–d are behavioural predictions from the animat, and could be compared with gerbil search. In fig. 8b, the landmark arrangement is both rotated and stretched; the animat searches at the center of this array. With the size of the triangle of landmarks doubled, animats search at the center (fig. 8c), with less search at three outlying locations. With the arrangement size doubled and rotated, animat search has an interesting pattern around the center of the of the landmarks and again search at three outlying locations (fig. 8d).

6. Discussion

This paper shows that simple and reactive animats, adapting by reinforcement learning, are capable of learning tasks analogous to those learnt by Collett et al's (1986) gerbils. Given enough hidden units, both intensity and vector coding leads to animats that perform at a level far above chance. These results are obtained with a relatively coarse coding of the sensory input: land-

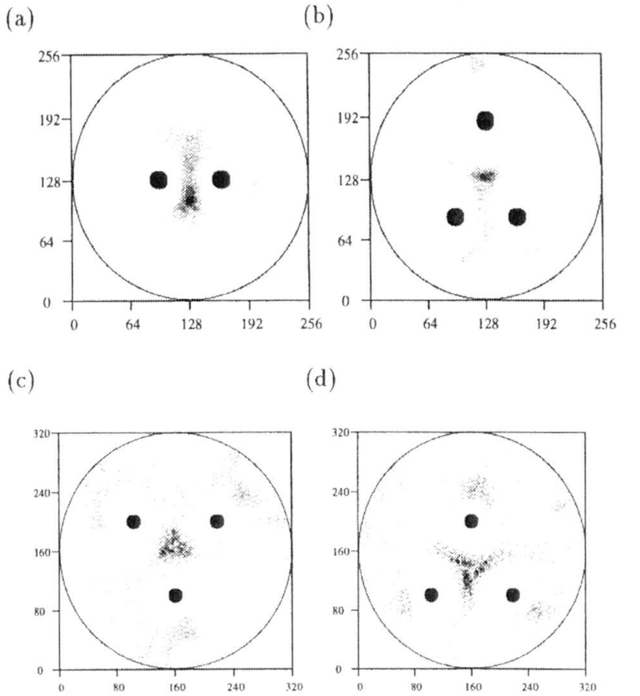

Figure 8 Further search patterns of the direct, vector coding animat of the previous figures. (a) matches the behaviour of gerbils, (b)–(d) are predictions.

mark bearing has a resolution of 6 degrees, the compass units of 24 degrees, and for vector coding, distance is split into one of just six bins.

Vector coding results in significantly higher performance, after learning, than intensity coding, and requires fewer hidden units.

In task 2, two hidden unit, vector coding animats, when tested with a single landmark, search mostly in a single location. When tested with the distance between landmarks doubled, animats search mostly in a single location within the array of landmarks, implying they are sensitive to the relation between landmarks in addition to their individual locations. However, gerbils search at two discrete locations out of the four where both distance and bearing match.

In task 3, direct vector coding animats search patterns closely matched that of the gerbils in all test arrangements. Consequently, these animats can be posited as a model of the gerbils behaviour, and falsifiable predictions of search patterns in further modified arrangements have been produced to compare with those of gerbils. The close match to gerbil behaviour is surprising because of the coarseness of coding, and the simplicity and generality of the animats. The results seem to suggest that Collet et al's (1986) results do not necessarily imply a computationally specific spatial navigation system.

Vector coding makes distance more explicit than does intensity coding, and this presumably underlies the superior performance of vector coding animats, and the smaller filter networks required to achieve peak performance. To respond to a landmark at a particular distance and relative bearing with vector coding is simply a matter of increasing a particular weight. With intensity coding, responding to the landmark at a particular bearing is as simple, but responding to a particular distance is a complex computation requiring, in general, two hidden units. Thus, intensity coding animats rely upon the relative bearing between landmarks since this is sufficient for competent performance and easily learned with intensity coding. Such a strategy is not an option for vector coding animats because each input unit codes both a relative bearing and a distance and so animats cannot help but become sensitive to both dimensions.

Analysis of the learned internal structure underlying the animats behaviour is lacking in this paper. The consistency of the behaviour of vector coding animats, and the small size of the filter networks required for efficient performance suggest that a detailed understanding of what these animats have learned to compute is possible.

The learned behaviour of these simple reactive animats provides further evidence that complex animal searching behaviour need not be predicated on complex and navigation specific computations.

Acknowledgements

Thanks to Paul Miller, Peter Hancock and Dario Floreano for discussion and improvements to previous drafts of this paper. This work was funded by EPSRC and Stirling University psychology department.

References

Clark, A. & Thornton, C. (1994). Trading Spaces: Computation, Representation and the Limits of Uninformed Learing. *Cognitive and Computing Sciences, University of Sussex, Technical Report*

Collett, T.S., Cartwright, B.A., & Smith, B.A. (1986) Landmark Learning and Visuo-spatial memory in gerbils. *J. Comp. Physiol. A.* 158, 835–851

Gallistel, C.R. (1990). *The Organisation of Learning.* Cambridge MA: MIT Press

Lin, L. (1992). Self-improving reactive agents based on reinforcement learning,planning and teaching. *Machine Learning* 8, 293–321.

Marr, D. (1982). *Vision* New York: W.H.Freeman and Company

McNaughton, B. L., Knierim, J. J., and Wilson, M. A. (1995). Vector Encoding and the Vestibular Foundations of Spatial Cognition: Neurophysiological and Computational Mecahnisims. In Gazzaniga, M. S. (Ed), *The Cognitive Neurosciences,* pp 1255–1267. MIT Press, Cambridge Mass.

Redish, A. D. and Touretzky, D. S. (1997). Navigating with landmarks: computing goal locations from place codes. In K. Ikeuchi and M. Veloso, (eds.), *Symbolic Visual Learning,* Oxford University Press.

Rumelhart, D.E., Hinton, G.E. & Williams, R.J. (1986). Learning internal representaions by error propagation. In: *Parallel distributed processing. Explorations in the microstructure of cognition. Vol 1.* Bradford Books/MIT press.

Sutton, R.S. (1988) Learning to predict by the methods of temporal difference. *Machine Learning* 3, 9–44.

Watkins, C.J.C.H. (1989) *Learning from delayed rewards.* PhD thesis, Kings college, Cambridge.

Animat Navigation Using a Cognitive Graph

Olivier Trullier[†‡], Jean-Arcady Meyer[†]
[†] AnimatLab, Département de Biologie, École Normale Supérieure,
46, rue d'Ulm, 75230 Paris Cedex 05, France
{trullier,meyer}@wotan.ens.fr
[‡] CNRS-Collège de France, LPPA, 11, place Marcelin Berthelot, 75005 Paris, France

Abstract

A model of the hippocampus as a "cognitive graph" is proposed. It essentially considers the hippocampus as an heteroassociative network that learns temporal sequences of visited places and stores a topological representation of the environment. Using place cells, head-direction cells, and "goal cells", we propose a biologically plausible way of exploiting such a spatial representation for navigation, which does not require complicated graph search algorithms. Simulations show that the resulting animat is able to navigate in continuous environments that contain obstacles. Furthermore, we make experimental predictions on simultaneous recordings of multiple cells in the rat hippocampus.

1. Introduction

The discovery of *place cells* in areas CA3 and CA1 of the rat hippocampus (O'Keefe and Dostrovsky, 1971) – cells that discharge selectively when the rat is in restricted regions of the environment (their *place fields*) – led to the idea that the hippocampus functions as a *cognitive map* of space (O'Keefe and Nadel, 1978). Cells that fire as a function of the animal's orientation in space, *head-direction cells*, have also been evidenced in various other regions of the rat brain (e.g. postsubiculum [Taube et al., 1990]). Thus, representations of position and orientation, necessary for any navigation system, are present in the rat brain. Here we propose a computational model of rat navigation that calls upon such representations.

The hippocampus is usually considered as an auto-associative network that learns, stores and recalls specific *episodes*, i.e. individual events or configurations of stimuli. According to such a view, the hippocampus learns the relationships between different stimuli that all characterize the same situation or event. At the time of recall, partial cues can be completed and the whole episode can be retrieved (Rolls, 1991). However, some people recently proposed that the hippocampus rather functions as an hetero-associative network, that learns, stores and recalls *sequences* of events (e.g. [Jensen et al., 1996]). According to this view, it learns the temporal relationships between successive, but different, configurations of stimuli. Thus, at the time of recall, the representation of a single event, associated with the intention of some action, leads to the *prediction* of the consequences of this action.

Several among the existing computational models of animal navigation (Trullier et al., 1997) also consider the hippocampus as an hetero-associative network in the spatial domain, that is, a network that learns the spatial relationships between neighboring, but distinct, places and learns how places are connected to one another. For instance, Schmajuk and Thieme (1992) proposed a model where the animat learns a transition matrix between any two distinct, non-overlapping places. In this model, the animat constantly compares where it is (active place cell), what places it predicts are neighbors (internal predictions) and what places are actually neighbors (views of adjacent places). It then modifies the synaptic weights between place cells to reflect the topological organization of places. Likewise, the model of Muller et al. (1996) learns what they called a "cognitive graph" from distinct but overlapping places. In this model, the animat moves around in the environment and reinforces the synaptic connections between simultaneously active place cells. These synaptic weights reflect the degree of overlap between place fields (the inverse of the distance between place field centers). However, in the former model, places have to be distinct, the animat has to "jump" from place to place, and the environment is discretized. In the latter model, learning is symmetric because it relies on simultaneously active place cells and there is no difference in the resulting representation between a movement from place A to place B from a movement from place B to place A. The model thus cannot be used to predict the animat's future positions from the current position.

Here we propose a mechanism by which the animat can learn another type of "cognitive graph". This graph represents the topological relations between overlapping places and it can be used to predict any place the animat will reach if it moves in a specific direction from its current position.

In contrast to the above mentioned models, and to

[†] Address for correspondence

other similar ones (see [Trullier et al., 1997]), we view exploration as a process by which an animat experiences a sequence of places. In other words, we don't only look at two neighboring places, like Schmajuk and Thieme (1992), and we don't only look at overlapping places, like Muller et al. (1996). However, our model, like others, will learn a topological representation of the environment, where neighboring places are represented by connected place cells in the hippocampus. The difference is that this learning strongly relies on sequences of physical moves in the environment, which are transformed into convenient spatial representations. This transformation process is inspired from the work of Jensen et al. (1996) who recently proposed a neural network model of short-term memory based on a biophysically plausible mechanism that can account for sequence learning. It exploits the superposition of high-frequency gamma oscillations (20-60 Hz) on top of a low-frequency theta oscillation (5-8 Hz) to segment time into cycles that are divided into discrete sub-cycles. Such a network is able to learn sequences of up to 7 ± 2 items. Neurons of this network, each representing a new event, maintain their activity through an intrinsic mechanism (a specific activity-dependent depolarizing current) and discharge within one gamma subcycle at each theta cycle. Subsequent events are represented by other neurons that discharge at *subsequent* gamma sub-cycles. Thus, within each theta cycle, successive gamma sub-cycles code for successive events and the whole sequence is stored in the correct temporal order. Thus, neurons that fire within a sub-cycle all correspond to the same particular event and ensembles of neurons that fire within successive sub-cycles correspond to two successive events. A simple Hebbian learning rule, applied to the connections from the first ensemble to the second one, enables the association of one event with the next in the sequence.

As an evidence that the hippocampus might actually function as a memory for sequences, Jensen and Lisman (1996) subsequently showed that their model could account for the *phase precession* of hippocampal place cells. The hippocampal EEG in the rat indeed exhibits the two types of oscillations needed in their model, that is, an overall theta rhythm – observed in the rat hippocampus when the animal is engaged in locomotor behaviors (O'Keefe and Recce, 1993; Skaggs et al., 1996) – and gamma oscillations (Bragin et al., 1995). Phase precession, or phase coding, refers to the specific temporal relationship between the discharge of place cells and the overall theta rhythm. As the animal traverses a place field, the corresponding place cell discharges first at a late phase and progressively at earlier phases of the theta cycle (Fig. 1). Thus, the discharge of a place cell indicates that the rat is within the corresponding place field, while the phase of the discharge indicates whether the rat is entering the place field (the place field is in front of the animal) or leaving the place field (the place field is behind the animal).

The model of Jensen and Lisman (1996) learns sequences of places and subsequently recalls them within a theta cycle. In other words, it predicts future positions from the current one. When the rat is at a certain position, the corresponding place cell discharges at the beginning of a theta cycle and, during the rest of the theta cycle, the hippocampus predicts the 7 places ahead of the animal that it will traverse if it keeps on moving. A step later, the animat is one place further and thus the places it predicted at the previous step are closer and the corresponding place cells fire earlier in the theta cycle (Fig. 2).

We will now describe a model that learns sequences of places to build a topological representation of the environment. This model uses putative *goal cells* (Burgess et al., 1994) downstream of such a spatial representation in order to navigate. We previously demonstrated that the corresponding animat could efficiently navigate in a grid-like environment with no obstacle (Trullier and Meyer, 1997). Here, we show that an improved model can cope with continuous environments and can move around obstacles.

2. The Model

The global architecture of our model (Fig. 3, top) is a simplified version of a previous model (Trullier and Meyer, 1997) that closely emulated the architecture of the rat hippocampus. Each of the 4 layers has a specific and distinct function but only the last two are simulated here.

2.1 Place cells

We first assume that place fields are uniformly distributed in the environment. As the animat moves in a continuous manner within the environment, only a subset of its place cells is active (typically the 5 place cells whose place fields are closest to the animat's position), in accordance with experimental data. Such a process, which keeps the total activity within the hippocampus very low, is thought to involve inhibitory interneurons, both through feedforward and feedback pathways, that are not explicitly modeled here (Marr, 1969; Sharp, 1991).

We have previously suggested that the role of dentate gyrus was to act as a short-term memory of the sequence of visited places and consequently to activate place cells in the CA3 region of the hippocampus through the strong mossy fiber synapses (Trullier and Meyer, 1997). Here, we do not explicitly simulate the dentate granular cells but we assume that place cells fire at specific phases with respect to the theta rhythm as the result of the activation by the dentate gyrus, that is, in the same temporal order as the sequence of visited places. In other words, if a place cell starts firing after a silent period (i.e. the animat enters the corresponding place field), we assign it the

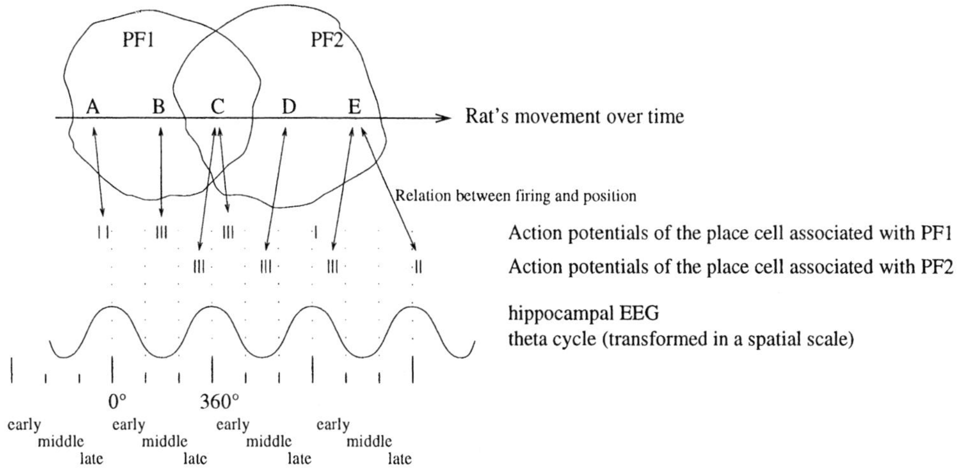

Figure 1 Phase shift in place cell firing during the rat's movement, as discovered by O'Keefe and Recce (1993). The horizontal axis represents both space (top, rat's movement) and time (bottom, theta cycle and spikes). When the rat is at A (it runs from left to right in this overhead view), the place cell corresponding to place field PF1 fires **late** in the theta cycle (spikes shown as vertical bars). In B, this cell fires at the **middle** phase and, when in C, at the **early** phase of the theta cycle. C also corresponds to the point of entry into place field PF2 with respect to the heading direction. The corresponding place cell fires **late** in the theta cycle. (Schematized on the basis of data from O'Keefe and Recce (1993) and from Skaggs et al. (1996))

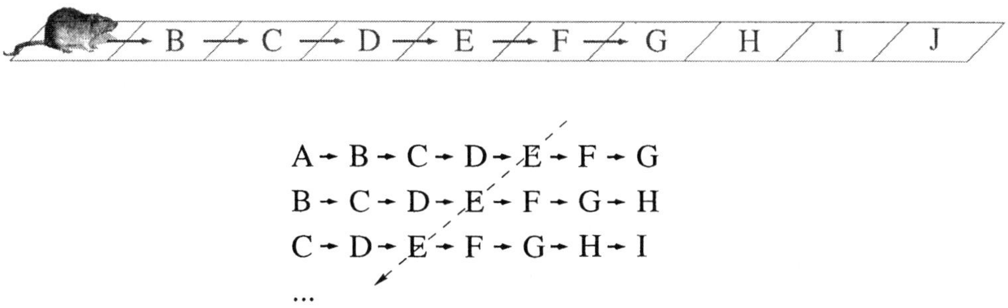

Figure 2 Phase precession explained by sequence learning. Each bin in this grid-like world corresponds to a unique place. The rat has learned the sequence of places from A to J.[1] It subsequently moves from A to J (top): when it is in A, it recalls the sequence from A to G (bottom); when it is in B, it recalls the sequence from B to H; and so on. Each movement and each prediction phase take a full theta cycle. Thus, the representation of the current place starts a new theta cycle and the prediction of place E comes earlier and earlier in the cycle (dotted arrow), i.e. the phase of firing of the place cell corresponding to E diminishes.

latest phase in the theta cycle. If it fires again at the next simulated time step (i.e. the animat hasn't moved out of the place field yet), its phase is decreased. This simulates phase precession and the fact that place cells fire within each theta cycle in the order of the experienced place sequence.

2.2 Topological representation

The CA3 region of the hippocampus has a dense recurrent connectivity with plastic synapses (Rolls, 1995). Like Schölkopf and Mallot (1994), we assume that these synapses can be gated by head-direction information and that the propagation of activity from the pre-synaptic cell to the post-synaptic cell, as well as synaptic modification of the connection, occur only if the animat is oriented in the direction corresponding to the gating head-direction information. Thus, as the animat explores its environment, synapses connecting place cells corresponding to place fields it traverses are modified. Synaptic weights are enhanced (Long-Term Potentiation or LTP)

[1] The model of Schmajuk and Thieme [1992] would have learned that A is next to B, B next to C, and so on. The model of Muller et al. [1996] would have learned nothing in this case because places are distinct. However, it would have learned that place fields PF1 and PF2 are overlapping in Figure 1.

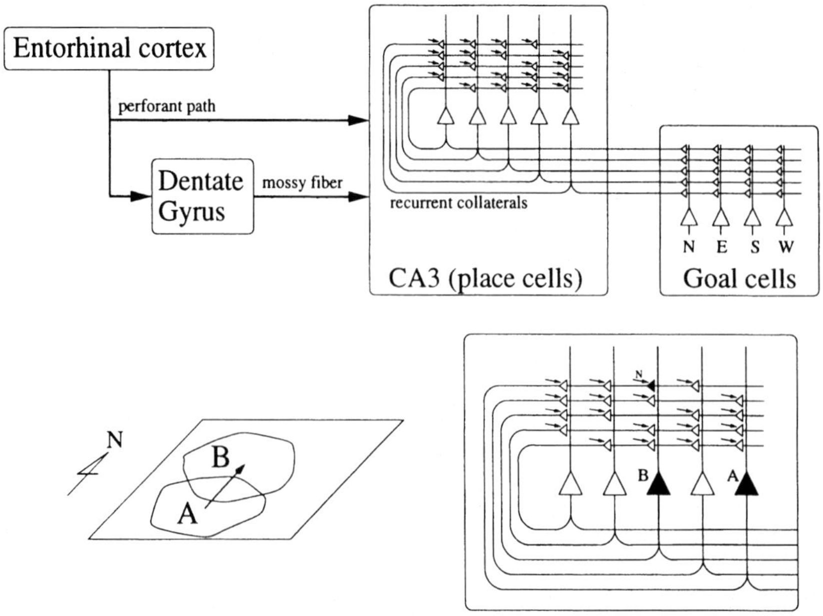

Figure 3 Architecture of the proposed model. (Top) We assume that the entorhinal cortex sends information about the spatial configuration of landmarks to the dentate gyrus and to the CA3 region through the perforant path. The dentate granular cells store this information as a short-term memory, in the correct temporal order corresponding to the temporal sequence of visited places (not simulated). This information is forced onto the CA3 pyramidal cells (large triangles) through the strong mossy fibers. These CA3 place cells then send projections (presumably through CA1 and subiculum) to putative goal cells. Small triangles are plastic synapses and small arrows on some synapses indicate that these synapses are gated by head-direction information. (Bottom) When the rat moves from place A to place B, as it is heading north (left), place cell A and place cell B fire in order (right, filled triangles). At the same time, the connection from place cell A to place cell B, gated by north head-direction information (small filled triangle with labeled arrow) is reinforced.

if both pre- and post-synaptic cells are active at successive phases (Fig. 3, bottom). They are depressed (Long-Term Depression or LTD) if either the pre- or the post-synaptic cell is active while the other is silent. They progressively stabilize at a value that reflects the overlap between place fields, if the place fields are oriented in the appropriate direction. For instance, the connection gated by the "northward" information between place cell a and place cell b in Figure 4 has been enhanced during exploration because the animat went from place field A, corresponding to a, to place field B, corresponding to b, as it was moving northward. Similarly, the connection gated by the "eastward" information between place cell a and place cell c has been enhanced. However, the connection gated by the "northward" information between a and c has almost not been enhanced (if it were, it would be because the animat traversed through the overlapping region between place fields A and C while facing north) and all connections between place cell a and place cell d have been depressed because place fields A and D are too far apart from each other.

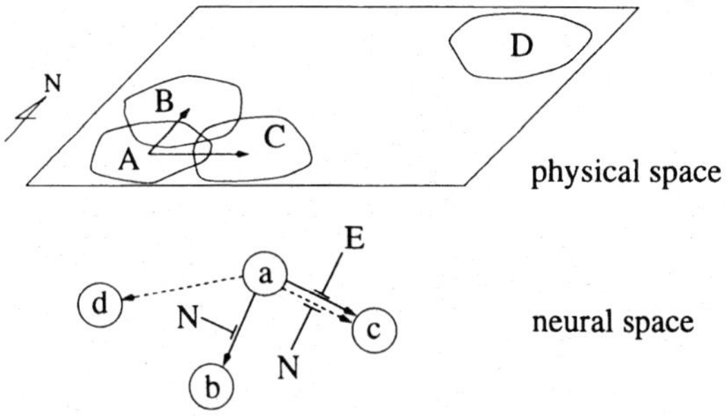

Figure 4 The modified connection between two place cells in neural space corresponds to the facts that the corresponding place fields are neighbors and that the post-synaptic one is in the direction corresponding to the head-direction that modulates the connection, with respect to the pre-synaptic one.

In the end, the population of CA3 place cells with their recurrent synapses can be viewed as a directed graph, where nodes are places and links represent the fact that nodes are adjacent – such links being labeled by head-direction information. In other words, it is a topological representation of the environment (Fig.5a).

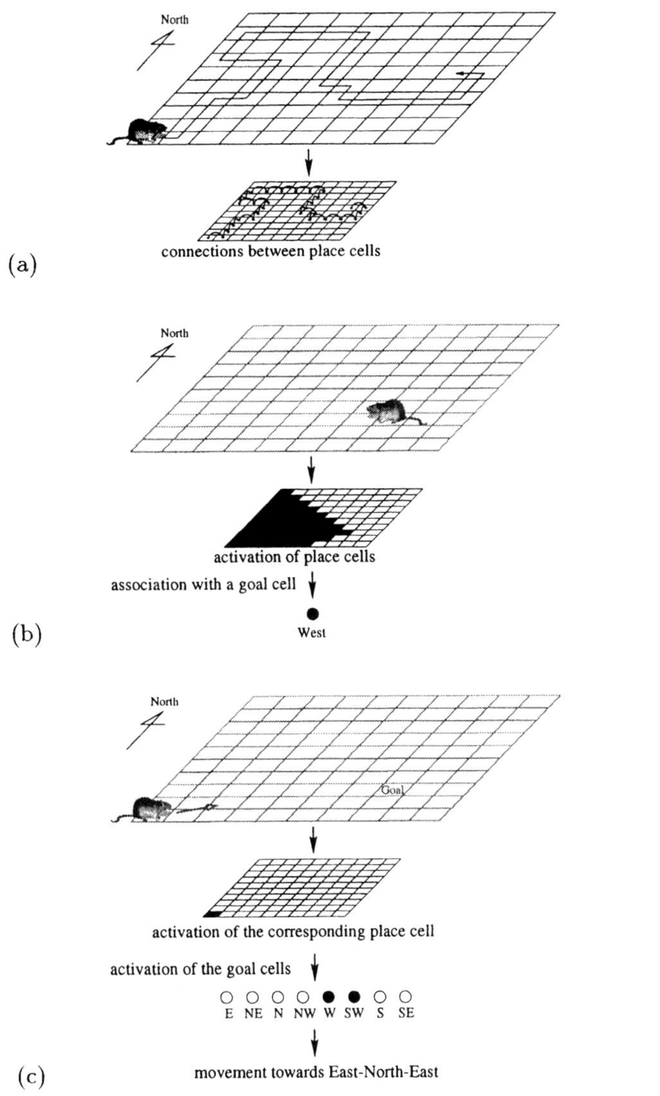

Figure 5 Summary of the principles of our model. The simulations are decomposed into three phases, corresponding to exploration (a), association with a goal location (b), and exploitation (c). (a) As the animat moves randomly, connections between place cells are modified to reflect the topological organization of the corresponding place fields. Direction information is also included. (b) At a goal, the information corresponding to the current place is propagated through the head-direction gated synapses. All the place cells lying in a specific direction with respect to the current goal location are triggered and are then associated with one goal cell. (c) Each place cell triggers the firing of the appropriate goal cells. These goal cells code, as a population, the direction to follow to return to the goal. (We used a grid-like world for illustration purposes. Also note that place cells are not topographically organized in the rat hippocampus. They are placed topographically here to ease the visibility of the learned connections.)

2.3 Goal representation

Such a topological representation can be used to predict the places that can be reached from the current place, when moving in a certain direction. Propagation of activity through a topological graph is a mechanism that has been used (e.g. [Matarić, 1991] and [Schmajuk and Thieme, 1992]) for path planning. However, in such approaches, either the signal propagates back from the goal, which implies that there's a mechanism triggering the place cell that represents the goal place, or the signal propagates forward from the current place, but there must be a mechanism to detect that the goal place has been reached by the prediction.

Alternatively, Burgess et al. (1994) hypothesized the existence of so-called goal cells, downstream of the hippocampus, that would code for the animat's position with respect to the goal, based on the information coming from the place cells. For instance, the "East" goal cell would fire when the animat is to the East of the goal. In the model they proposed, Burgess et al. used phase coding to reinforce the connections from specific place cells to the goal cells. To learn a specific location it will need to return to later on (a goal location), their animat, upon finding such a location, looks about in all directions. In each of these directions, modification of the connections between place cells and the corresponding goal cell is forced to occur at the end of the theta cycle. As a consequence, only the place cells that fire at a late phase, that is, place cells that have place fields lying in front of the animat are involved in the process. The "place field" of a goal cell can then be defined as the sum of the place fields of the associated place cells – cells from which the connections have been reinforced – and indeed lies in the direction the animat is facing, with respect to the goal location. However, each of these place cells that fire at a late phase must have their place fields containing the goal location (when theta rhythm is present, a place cell only fires if the animat is within the corresponding place field). In other words, the "place fields" of goal cells cannot extend at a great distance from the goal because the size of place fields are limited. As a consequence, navigation abilities are extremely dependent upon the size of the place fields and the environment has to be small enough. This is because the model of Burgess et al. does not encode any spatial relationship between places and cannot predict all the places that lie ahead of the animat. Furthermore, no mechanism for dealing with obstacles is proposed.

In this work, we propose to combine the mechanism of signal propagation with the existence of goal cells. As in the model of Burgess et al., the animat, upon finding a goal location, looks about in all directions. In each of these directions, the activity of the place cell representing the goal location – where the animat currently is – is propagated within the CA3 network, through the connections that are gated by the current head-direction information (Fig.5b). *All* the place cells that represent locations *in the direction the animat is facing* are thus

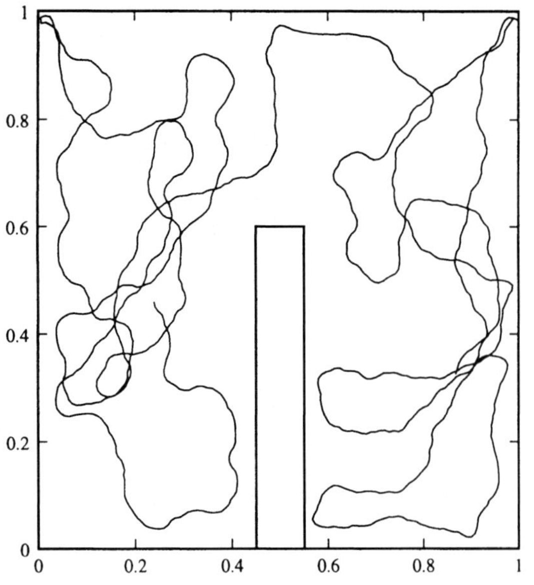

Figure 6 An example of the exploratory behavior.

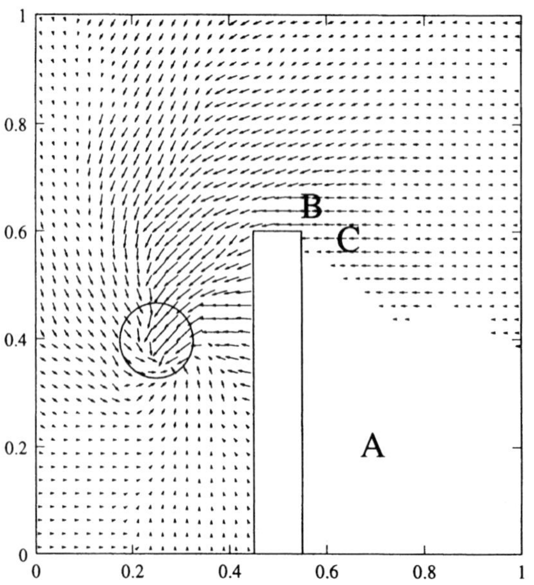

Figure 7 Vector field generated by the goal cells learned at the goal location. This vector field is generated by virtually putting the animat in each position within the environment and computing the direction it would follow by measuring the output of the goal cells. The limit between the region where there is information (arrows) and the region where there is none is not a straight line because place fields are scattered randomly, because place fields have a certain extent, and because synaptic weights have been learned with some noise. For instance, a place cell whose place field is at B received the information from the goal location when the animat was facing north-east but its place field extends behind the corner. As a consequence the place cell corresponding to place C was also activated and so on.

activated. They represent, as an ensemble, the corresponding region with respect to the goal (West in the figure), regardless of distance from the goal. The connections from these activated place cells to the respective goal cell are enhanced through a Hebbian learning rule. Consequently, any of the associated place cells becomes able to trigger the firing of the goal cell. In other words, the firing of the place cell corresponding to the location of the rat will activate the appropriate goal cells that will represent where the animat is with respect to the learned goal. For instance, Figure 5c shows a place cell that has been associated with, and thus activates, the "South" and the "South-West" goal cells. This means that the animat is at the South-South-West of the goal.

Rats, upon finding a reward, often look all around, as if to learn the configuration of cues visible from there. They might also propagate the information corresponding to where they are in the cognitive graph, in different directions. When the rat is not moving, hippocampal activity usually goes into another mode, called LIA (Large-amplitude Irregular Activity) where many more cells fire at the same time. This mode is usually associated with a "recall mode" (Buzsáki, 1989). Our animat is supposed to learn the connections between place cells and goal cells during such periods.

3. Navigating with a Cognitive Directed Graph

The animat moves in a continuous manner in a continuous environment containing obstacles. Its movement is specified by its constant velocity and the maximum angle by which it can rotate at each simulated time step. An obstacle avoidance mechanism based on the use of optical flow (Duchon, 1996) is also called upon when the animat is too close to a wall. Figure 6 shows an example of the exploratory trajectory of the animat.

Once the topological representation is acquired (the animat has explored for a fixed length of time), the animat looks for reward. When it finds one, it recruits a set of goal cells to learn the goal location. However, as can be deduced from the working principle of Figure 5, the information from a goal location propagates in a certain direction, in approximately straight lines. It doesn't skirt around obstacles. As a consequence, this set of goal cells cannot provide the information about the direction to the goal all over the environment. In the example of Figure 7, the animat would be lost if put at location A.

We thus introduce the notion of subgoal. When the animat is trying to reach the goal, it can be in two modes. If some goal cells are firing and provide the direction towards the goal, it follows this direction; if no goal cell fires, the animat starts moving randomly to look for information. At the location where it finds information, it recruits a new set of goal cells (subgoal cells) and learns the connections from the CA3 place cells to these subgoal cells in the same way it learned the connections from the place cells to the first set of goal cells. It then proceeds

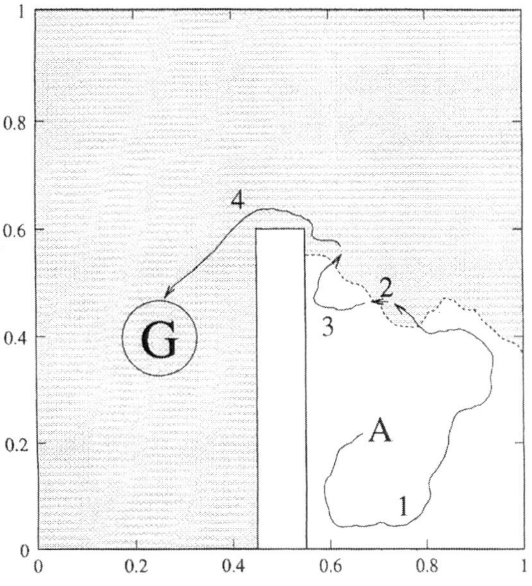

Figure 8 4 successive trajectories in the process of creating a subgoal. At the end of path 1, the animat recruits a set of subgoal cells but this turns out to be inefficient, since path 2 leads back the animat into the "silent" zone. The animat recruits another set of subgoal cells at the end of path 3, which proves successful, since path 4 leads to the goal (G).

and follows the direction coded by the first set of goal cells. If it succeeds in reaching the goal, it validates the set of subgoal cells. If it moves again into a region without information, it removes the newly recruited set and starts looking for information again. Figure 8 illustrates this process.

If the environment contains more obstacles, the animat will have to recruit other sets of subgoal cells. As a consequence, at any given position in the environment, more than one set of (sub)goal cells can be active, for the same goal. The animat has to choose which direction to follow. Thus, we introduce yet another mechanism. When the animat is lost and looking for information, it recruits a new set of subgoal cells as soon as one of the existing sets is active. We will say that the new set is associated with this active set. For instance, the first subgoal will be associated with the goal, the second subgoal with the first, etc. Some kind of distance information is implicitly coded in this association. The set of goal cells leads to the goal and is at distance 0 from the goal. The first set of subgoal cells leads to a position where the set of goal cells will be active, so the first subgoal is at (topological) distance 1 from the goal. The second subgoal is thus at (topological) distance 2 from the goal, and so on.

Figure 9 shows how our model performs in the example environment we've been using so far. Note that the trajectories are not straight to the goal. This is due to the scale of the place cell representation, to the fact that exploration is never quite exhaustive, and to the fact that all connections are not stabilized at their theoretical values.

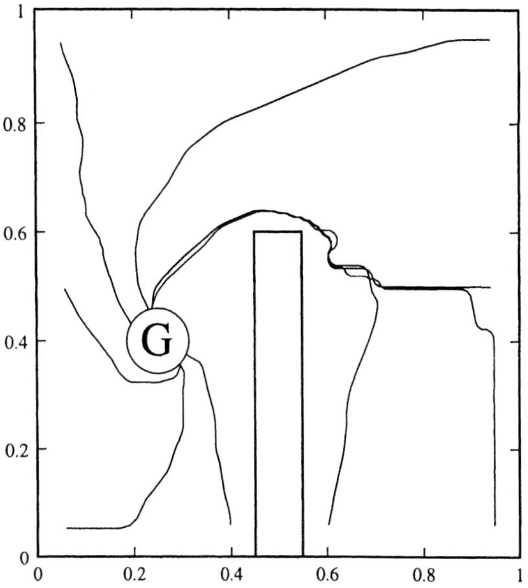

Figure 9 Resulting trajectory during exploitation.

Figure 10 shows other examples of environments that require several subgoals.

4. Discussion

There are, to our knowledge, only two other models based on hippocampal place cells that can cope with continuous environments with obstacles. Gerstner and Abbott (1996) proposed a model where the activity of place cells is modulated by the animat's position *and* by the goal location. This implies that each place cell gets information about where the goal is. This is not what is thought to happen in the rat hippocampus. Speakman and O'Keefe (1990) have indeed shown that place cell activity is independent of goal locations. By contrast, our model, like the one by Burgess et al. (1994), exhibits *latent learning.* It doesn't need the presence of any reinforcement to build a spatial representation of the environment, which is, by essence, independent of goal locations. Then, the animat can recruit as many sets of (sub)goal cells as needed to be able to return to as many goals as needed, without ever changing the underlying spatial representation.

The other model that is based on hippocampal place cells and can cope with continuous environments with obstacles is the one by Muller et al. (1996). Beside the already mentioned fact that their "cognitive graph" cannot be used to predict the animat's future positions, their model, in order for the animat to return to a goal while avoiding obstacles, relies on a biologically unrealistic graph search algorithm stemming from Artificial Intelligence research on path planning. Here we used sequence learning instead of correlational learning. This is what is thought to occur in the rat hippocampus. Indeed, phase coding compresses temporal sequences by storing events that can be hundreds of msec apart into successive

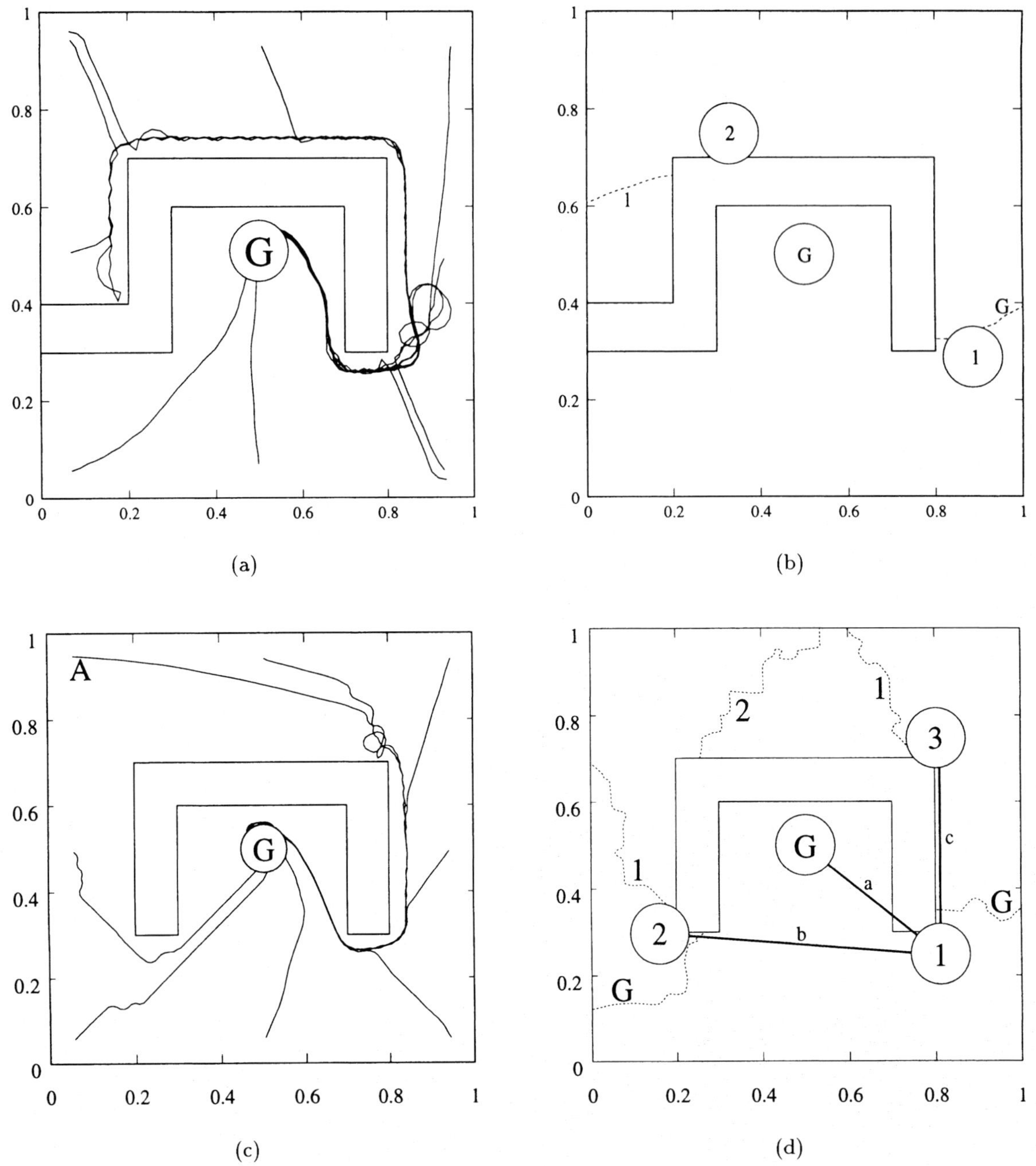

Figure 10 Trajectories during exploitation within two other environments (a,c) and the respective positions of the recruited subgoals (b,d). The dotted lines indicate where the corresponding set of (sub)goal cells (label near the line) stops being active.

gamma sub-cycles that are 10 to 20 msec apart. This latter time separation seems to be the ideal timing for learning (LTP) to occur (Skaggs et al., 1996). Moreover, phase coding maintains the correct ordering of place cell firing and thus enables asymmetric learning of the connections that can now take into account the orientation of the animat and how it moved from one place to the other. We also propose in our model a biologically-plausible mechanism to "read out" the cognitive graph, a mechanism that is modified and extended from the original idea of Burgess et al. (1994).

Finally, our model relates more closely to the rat hippocampal architecture and to physiological mechanisms than previous models. As we previously suggested (Trullier and Meyer, 1997), we can make several experimental predictions. (1) We propose that phase precession in the CA3 region is forced by dentate gyrus and enables sequence learning. This hypothesis implies that phase precession also occurs in dentate gyrus, which it does (Skaggs et al., 1996). But it also implies that, if we simultaneously record several dentate granular cells and several CA3 pyramidal cells, we should be able to see that place cells whose place fields overlap those of the dentate granular cells fire just after the granular cells, a correlation supporting the idea that the recorded granular cells force the discharge of the recorded place cells. (2) We propose that the learning of the goal location occurs during another state of hippocampal activity, namely LIA (Large-amplitude Irregular Activity), observed in particular during awake immobility of the rat, and which is usually associated with "recall" by the hippocampus of previously learned episodes (Buzsáki, 1989). During such a state, a large number of pyramidal cells in the CA3 region are activated (but far from all). We thus propose that this state is the result of a signal propagation within the CA3 region that activates all the place cells whose place fields lie ahead of the animat. If we simultaneously record a lot of CA3 (not CA1) place cells during LIA, we should be able to see a tendency of the activation to be limited to specific place cells.

Apart from place cells that code for locations in an absolute reference frame, our model doesn't include representations for objects or specific places. In particular, obstacles are *not* represented internally but obstacle avoidance emerges from the fact that some pairs of place cells are not connected because of the obstacles. The positions of subgoals are also not represented explicitly. The animat seems to move from subgoal to subgoal, closer and closer to the goal, but doesn't actually "recognize" that a subgoal has been reached. This is the result of the competition between sets of goal cells. In fact, our model doesn't actually perform "path planning": it doesn't have to choose among a certain number of different paths. Consequently, our model rather belongs to the category of navigation models that we called "Place recognition-triggered response" (Trullier et al., 1997).

Our current model is not meant to cope with dynamically-changing environments. Once a set of goal cells is learned, the animat exploits the information this set provides, without comparing with new sensory information. Real animals, on the contrary, constantly update their internal representations. In particular, if obstacles move, disappear or appear, they restart a new phase of exploration. We are currently looking into this kind of process to improve our model.

5. Conclusion

We have proposed a new model of the rat hippocampus based on several ideas from different models previously published by others in the literature. We have been able to show that such a model can successfully navigate in continuous environments containing obstacles. We also have shown that our model is based on biologically-plausible architecture and mechanisms. Furthermore, we made several experimental predictions based on our new way of interpreting the roles of the different structures in the rat hippocampal formation.

We are still working however on ways to include other mechanisms, so that our model can cope with dynamically-changing environments.

Acknowledgments

We would like to thank members of the "groupe modèle" at LPPA (S.Wiener, J.Droulez, N.Brunel & A.Berthoz) for helpful discussions. Work supported by the CNRS GIS "Sciences de la Cognition". O.T. supported by a CNRS BDI fellowship.

References

Bragin, A., Jandó, G., Nádasdy, Z., Hetke, J., Wise, K., and Buzsáki, G. (1995). Gamma (40-100Hz) oscillation in the hippocampus of the behaving rat. *Journal of Neuroscience*, 15(1):47–60.

Burgess, N., Recce, M., and O'Keefe, J. (1994). A model of hippocampal function. *Neural Networks*, 7(6/7):1065–1081.

Buzsáki, G. (1989). Two-stage model of memory trace formation : A role for 'noisy' brain states. *Journal of Neuroscience*, 31:551–570.

Duchon, A. P. (1996). Maze navigation using optical flow. In Maes, P., Mataric, M. J., Meyer, J.-A., Pollack, J., and Wilson, S. W., editors, *From Animals to Animats 4 : Proceedings of the Fourth International Conference on Adaptive Behavior.* The MIT Press/Bradford Book.

Gerstner, W. and Abbott, L. F. (1996). Learning navigational maps through potentiation and modulation of hippocampal place cells. *Journal of Computational Neuroscience*, 4:79–94.

Jensen, O., Idiart, M. A. P., and Lisman, J. E. (1996). Physiologically realistic formation of autoassociative memory in networks with theta/gamma oscillations : Role of fast NMDA channels. *Learning and Memory*, 3(2/3):243–256.

Jensen, O. and Lisman, J. E. (1996). Hippocampal CA3 region predicts memory sequences : Accounting for the phase precession of place cells. *Learning and Memory*, 3(2/3):279–287.

Marr, D. (1969). A theory of cerebellar cortex. *Journal of Physiology*, 202:437–470.

Matarić, M. J. (1991). Navigating with a rat brain : A neurobiologically-inspired model for robot spatial representation. In Meyer, J.-A. and Wilson, S. W., editors, *Proceedings of the First International Conference on Adaptive Behavior : from Animals to Animats.* The MIT Press/Bradford Book.

Muller, R. U., Stead, M., and Pach, J. (1996). The hippocampus as a cognitive graph. *Journal of General Physiology*, 107:663–694.

O'Keefe, J. and Dostrovsky, J. (1971). The hippocampus as a spatial map. Preliminary evidence from unit activity in the freely moving rat. *Experimental Brain Research*, 34:171–175.

O'Keefe, J. and Nadel, L. (1978). *The Hippocampus as a Cognitive Map*. Clarendon Press, Oxford.

O'Keefe, J. and Recce, M. (1993). Phase relationship between hippocampal place units and the EEG theta rythm. *Hippocampus*, 3(3):317–330.

Rolls, E. T. (1991). Functions of the primate hippocampus in spatial and nonspatial memory. *Hippocampus*, 1(3):258–261.

Rolls, E. T. (1995). A model of the operation of the hippocampus and entorhinal cortex in memory. *International Journal of Neural Systems*, 6(Supp. 1995):51–70.

Schmajuk, N. A. and Thieme, A. D. (1992). Purposive behavior and cognitive mapping : A neural network model. *Biological Cybernetics*, 67:165–174.

Schölkopf, B. and Mallot, H. A. (1994). View-based cognitive mapping and path planning. Technical Report TR-7, Max Planck Institut für biologische Kybernetik.

Sharp, P. E. (1991). Computer simulation of hippocampal place cells. *Psychobiology*, 19(2):103–115.

Skaggs, W. E., McNaughton, B. L., Wilson, M. A., and Barnes, C. A. (1996). Theta phase precession in hippocampal neuronal populations and the compression of temporal sequences. *Hipocampus*, 6(2):149–173.

Speakman, A. and O'Keefe, J. (1990). Hippocampal complex spike cells do not change their place fields if the goal is moved within a cue controlled environment. *European Journal of Neuroscience*, 2:544–555.

Taube, J. S., Muller, R. U., and Ranck, J. B. (1990). Head-direction cells recorded from the postsubiculum in freely moving rats : Description and quantitative analysis. *Journal of Neuroscience*, 10(2):420–435.

Trullier, O. and Meyer, J.-A. (1997). Place sequence learning for navigation. In Gerstner, W., Germond, A., Hasler, M., and Nicoud, J.-D., editors, *Artificial Neural Networks - ICANN'97*, pages 757–762. Springer-Verlag.

Trullier, O., Wiener, S. I., Berthoz, A., and Meyer, J.-A. (1997). Biologically-based artificial navigation systems : Review and prospects. *Progress in Neurobiology*, 51:483–544.

Efficient Model-Based Exploration

Marco Wiering, Jürgen Schmidhuber
IDSIA
Corso Elvezia 36, Lugano, Switzerland
marco@idsia.ch
juergen@idsia.ch

Abstract

Model-Based Reinforcement Learning (MBRL) can greatly profit from using world models for estimating the consequences of selecting particular actions: an animat can construct such a model from its experiences and use it for computing rewarding behavior. We study the problem of collecting *useful* experiences through exploration in stochastic environments. Towards this end we use MBRL to maximize exploration rewards (in addition to environmental rewards) for visits of states that promise information gain. We also combine MBRL and the Interval Estimation algorithm (Kaelbling, 1993). Experimental results demonstrate the advantages of our approaches.

1. Introduction

In reinforcement learning (RL) problems an animat repeatedly receives input from the environment, selects and executes an action, and may receive reinforcement. At a given time the animat's goal is to select the action leading to maximal future cumulative reward. RL-based methods essentially learn how much long-term reward the animat will receive on average for selecting a particular action in a particular state. The expected future cumulative reward intakes are stored in a value function which builds the basis for generating rewarding behaviors. Direct RL methods like Q-learning learn the value function from experimenting with the current policy without using a world-model. These methods have been shown to learn significantly slower in discrete state spaces than *model-based* RL (Moore and Atkeson, 1993), which first estimates the transition probabilities between states and the immediate rewards for these transitions, and then computes a policy by applying dynamic programming-like techniques (Bellman, 1961) to the estimated model.

The problem. Since some RL animat is only able to learn from what it has experienced, the success of its policy heavily depends on the utility of its experiences. Optimal experimental design (Fedorov, 1972; Dodge et al., 1988) tries to gather those experiences which are most useful for computing good approximate solutions. In RL the problem of selecting alternative non-greedy actions is called exploration or dual control (Dayan and Hinton, 1993). Deviating from the current greedy policy (which always selects the action with the highest Q-value), however, usually causes some loss of immediate reinforcement intake. The animat faces the problem of spending as little time as possible on exploration without affecting its ability to find optimal policies.

Previous work. Schmidhuber (1991a) and Thrun (1992) present comparisons between *directed* and *undirected* exploration methods. *Directed* exploration methods use special exploration-specific knowledge to guide the search through alternative policies. *Undirected* exploration methods use randomized action selection methods to guess useful experiences. Previous research has shown significant benefits of using directed exploration (e.g., Schmidhuber 1991a; Thrun 1992; Storck, Hochreiter and Schmidhuber, 1995, Wilson 1996, Schmidhuber 1997). Koenig and Simmons (1996) shows how undirected exploration techniques can be improved by using the action-penalty rule which makes unexplored actions look more promising — this decreases the advantage of directed exploration.

The Interval Estimation (IE) algorithm (Kaelbling, 1993) uses second order statistics to detect whether certain actions have a potential of belonging to the optimal policy. IE computes confidence intervals of Q-values and always selects the action with largest upper interval boundary. Previous results (Kaelbling, 1993) show that IE works well for action selection in bandit problems (Berry and Fristedt, 1985).

Model-based exploration. Directed exploration methods learn an *exploration value function* in the same way standard RL methods learn a problem-oriented value function. Therefore we may simply define an exploration reward function determining immediate exploration rewards and let the selected RL method learn *exploration Q-values*. Most previous methods use Q-learning for learning where to explore, e.g., (Schmidhuber, 1991a; Thrun, 1992; Storck et al., 1995). We extend this work by using model-based RL (MBRL) to learn exploration policies. Since MBRL can outperform its direct RL counterpart (Sutton, 1990; Moore and Atkeson, 1993) we expect that it can also improve learning where to explore: model-based exploration is promising because it allows for immediately selecting actions based on global (as opposed to local) expected information gain.

Objective. Finding optimal solutions is considered a hard and unprosperous activity in artificial intelligence. We will focus on a more modest goal: given some RL problem, how can we learn with minimal experiences a policy whose performance is not more than ϵ percent below the optimum? Kearns and Singh (1998) proved that this goal can be achieved in polynomial time for general Markov decision problems. We will use a slightly adapted version of prioritized sweeping (PS) (Moore and Atkeson, 1993) to learn both an exploration policy and a problem-oriented policy, and combine this approach with (a) frequency-based and recency-based exploration reward functions, and (b) our novel Model-Based Interval Estimation (MBIE) update rule, which combines IE and MBRL.

Outline. Section 2 describes MBRL and introduces our version of prioritized sweeping. Section 3 addresses exploration issues in RL and mentions the exploration reward rules used in the experiments. Section 4 addresses issues related to MBRL for learning exploration policies and introduces MBIE. Section 5 describes experimental results on a task involving a 50 × 50 maze with one optimal goal and two suboptimal ones. Section 6 concludes.

2. Model-Based Reinforcement Learning

2.1 Markov decision problems

We consider discrete time steps $t = 1, 2, 3, \ldots$, a finite set of states $S = \{S_1, S_2, S_3, \ldots, S_n\}$ and a finite set of actions A. Let s_t denote the state at time t, and $a_t = \Pi(s_t)$ the action, where Π represents the learner's policy mapping states to actions. The transition function P with elements $P_{ij}(a) := p(s_{t+1} = j | s_t = i, a_t = a)$ for $i, j \in S$ defines the transition probability to the next state s_{t+1} given s_t and a_t. A reward function R maps state/action pairs (SAPs) $(i, a, j) \in S \times A \times S$ to scalar reinforcement signals $R(i, a, j) \in I\!R$. The reward at time t is denoted by r_t. A discount factor $\gamma \in [0, 1]$ discounts later against immediate rewards. The controller's goal is to select actions which maximize the expected long-term cumulative discounted reinforcement, given an arbitrary initial state $\in S$. The value $V^{\Pi}(s)$ is a prediction of the expected discounted cumulative reward to be received in the future, given that the process is currently in state s and policy Π will be used in the future:

$$V^{\Pi}(i) = E(\sum_{k=0}^{\infty} \gamma^k R(s_k, \Pi(s_k), s_{k+1}) | s_0 = i)$$

Action evaluation functions (Q-functions) $Q^{\Pi}(i, a)$ return the expected future discounted reward for current state i, current action a, and subsequently executed policy Π:

$$Q^{\Pi}(i, a) = \sum_j P_{ij}(a)(R(i, a, j) + \gamma V^{\Pi}(j))$$

2.2 Estimating a model

Inducing a model from experiences can simply be done by counting the frequency of observed experiences. Towards this end our animat uses the following variables:

$C_{ij}(a)$:= nr. of transitions from state i to j after executing action a.

$C_i(a)$:= number of times the animat has executed action a in state i.

$R_{ij}(a)$:= sum over all immediate rewards received after executing action a in state i and stepping to state j.

A maximum likelihood model (MLM) is computed as follows:

$$\hat{P}_{ij}(a) := \frac{C_{ij}(a)}{C_i(a)} \text{ and } \hat{R}(i, a, j) := \frac{R_{ij}(a)}{C_{ij}(a)} \quad (1)$$

After each experience the variables are adjusted and the MLM is updated. In deterministic environments one experience per SAP is sufficient to infer the true underlying model. In stochastic environments, however, we need resampling. For resampling a good exploration strategy is essential.

2.3 Prioritized Sweeping (PS)

Dynamic programming (DP) techniques could immediately be applied to the estimated model, but online DP tends to be computationally very expensive. To speed up DP algorithms, some sort of efficient update-step management should be performed. This can be done by prioritized sweeping (PS) (Moore and Atkeson, 1993) which assigns priorities to updating the Q-values of different state/action pairs (SAPs) according to a heuristic estimate of the update sizes. PS keeps track of a ***backward model*** relating states to predecessor SAPs. Following the update of a state-value, the state's predecessors are inserted in a priority queue. Then the priority queue is used for updating Q-values of actions of those states with highest priority.

Our PS. Moore and Atkeson's PS (M+A's PS) calculates the priority of some state by checking all transitions to updated successor states and identifying the one whose update contribution is largest. Our variant allows for computing the *exact* size of updates of state values since they have been used for updating the Q-values of their predecessors, and yields more appropriate priorities. Unlike our PS, M+A's PS cannot detect large state-value changes due to many small update steps, and will forget to process the corresponding states.

Our implementation uses a set of predecessor lists *Preds*(j) containing all predecessor states of state j. We denote the priority of state i by $|\Delta(i)|$, where the value $\Delta(i)$ equals the change of $V(i)$ since the last time it was processed by the priority queue. To calculate it, we constantly update all Q-values of predecessor states of currently processed states, and track changes of $V(i)$. The details are as follows:

Our Prioritized Sweeping:

1) Promote the most recent state k to the top of the priority queue
2) $\forall$ a do:
   3 $Q(k,a) := \sum_j \hat{P}_{kj}(a)(\hat{R}(k,a,j) + \gamma V(j))$
4) While $n < U_{max}$ AND the queue is not empty
   5 Remove the top state s from the queue
   6 $\Delta(s) := 0$
   7 $\forall$ Predecessor states i of s do:
      8 $V'(i) := V(i)$
      9 $\forall$ a do:
         10 $Q(i,a) := \sum_j \hat{P}_{ij}(a)(\hat{R}(i,a,j) + \gamma V(j))$
      11 $V(i) := \max_a Q(i,a)$
      12 $\Delta(i) := \Delta(i) + V(i) - V'(i)$
      13 If $|\Delta(i)| > \epsilon$
         14 Promote i to priority $|\Delta(i)|$
   15 $n := n + 1$
16) Empty priority queue, but keep the $\Delta(i)$ values

Here U_{max} is the maximal number of updates to be performed per update-sweep. The parameter $\epsilon \in I\!R^+$ controls update accuracy. Note that another difference to M+A's PS is that we remove all entries from the queue after having processed the maximal number of states.

3. Exploration

3.1 Undirected Exploration

Undirected exploration methods rely on pseudo-random generators, e.g., the *Max-random* and *Boltzmann* exploration rules. We will use Max-random which outperformed Boltzmann in a number of experiments (Thrun, 1992; Caironi and Dorigo, 1994).

Max-random exploration rule. The Max-random rule, also known as ϵ-greedy (Sutton, 1996) or pseudo-stochastic (Caironi and Dorigo, 1994), is the simplest one. It uses a single parameter P_{max} denoting the probability of selecting the action with highest Q-value, and selects a random action otherwise. In case there are multiple actions with highest Q-value, we select one of them stochastically.

3.2 Directed Exploration

Directed exploration techniques use knowledge of what an animat has learned to direct the exploration behavior to the most interesting parts of the state space. All they require is a local reward function determining which experience is interesting (e.g. Schmidhuber, 1991a,b, 1997). It takes the place of the standard MDP reward function. The MDP transitions and the experiences stay the same, but now we learn two Q-functions: the *exploration Q-function* and the *exploitation Q-function*. Here we look at two types of exploration: recency-based, and frequency-based.

Recency-based. Select the action which has been selected least *recently*. The local reward function for exploring SAP (s_t, a_t) is:

$$R^E(s_t, a_t, *) := \frac{-t}{K_T}, \tag{2}$$

where K_T is a scaling constant and t the current time step. The asterisk stands for the don't-care symbol — the exploration reward is *local* because it depends on the current state/action pair only.

Frequency-based. Explore actions which have been executed least *frequently*. The local reward function is simply:

$$R^E(s_t, a_t, *) := -\frac{C_{s_t}(a_t)}{K_C}, \tag{3}$$

where K_C is a scaling constant.

4. Learning Exploration Models

We can use all RL methods to learn exploration value functions. While most previous approaches used standard Q-learning for learning it (Schmidhuber, 1991a; Thrun, 1992; Storck et al., 1995), we prefer to use prioritized sweeping (PS) instead. PS allows for quickly learning exploration models which may be useful for learning Q-values estimating global information gain, taking into account yet unexplored regions of the state-space.

Replacing reward. The nice thing about using MBRL is that all explorative transition rewards can be based entirely on the current reward. Suppose that an animat selects an action which has already been executed several times. Computing the local exploration reward by averaging over all previous transition rewards would not result in the desired reward measure. For instance, with frequency-based exploration rewards the exploration model's estimate $\hat{R}^E(i, a, *)$ would be the average over $1, 2, \ldots, C_i(a)$ instead of just $C_i(a)$. Direct RL cannot circumvent this problem, but with MBRL we can just replace the estimated reward $\hat{R}(i,a,j)$ by $R^E(i,a,j)$ for all j with $\hat{P}_{ij}(a) > 0$, that is, we update all rewards for outgoing transitions from the current state/action pair to take the latest available information into account.

Never-ending exploration. The exploration utilities continually change; there is no such thing as a stable, optimal exploration function. This is not a fundamental problem — in fact the goal of exploration is to search for alternative paths in order to find better and better policies; therefore the exploration policy should never converge. In particular applications, however, e.g., in limited life-time scenarios (Berry and Fristedt, 1985; Thrun, 1992; Schmidhuber et al., 1997), we want to increase or switch to exploitation after some time.

Interval Estimation. To explore efficiently, an animat should not repeatedly try out actions that certainly cannot belong to the optimal policy. To reduce the set of optimal action candidates we extend the interval estimation algorithm (IE) (Kaelbling, 1993) to make it amenable for MBRL.

Standard IE selects the action with the largest upper bound for its Q-value. To compute upper bounds it keeps track of the means and standard deviations of all Q-values.

MBIE, however, uses the model to compute the upper bound of Q-values. Given a set of outgoing transitions from SAP (i, a), MBIE increases the probability of the best transition (the one which maximizes $\gamma V(j) + R(i, a, j)$), depending on its standard deviation. Then MBIE renormalizes the transition probabilities and uses the result for computing the Q-values. The following algorithm can be substituted for line 3 and 10 in our PS algorithm:

Model Based Interval Estimation:

a $m := Argmax_{j:\hat{P}_{ij}(a)>0}\{R(i,a,j) + \gamma V(j)\}$
b $n := C_i(a)$
c $P := \hat{P}_{im}(a)$
d $P^+_{im}(a) := (P + \frac{z^2_\alpha}{2n} + \frac{z_\alpha}{\sqrt{n}}\sqrt{P(1-P) + \frac{z^2_\alpha}{4n}})/(1 + \frac{z^2_\alpha}{n})$
e $\Delta_P := P^+_{im}(a) - \hat{P}_{im}(a)$
f $\forall j \neq m$
  g $P^+_{ij}(a) := \hat{P}_{ij}(a) - \frac{\Delta_P C_{ij}(a)}{C_i(a) - C_{im}(a)}$
h $Q(i,a) := \sum_j P^+_{ij}(a)(\hat{R}(i,a,j) + \gamma V(j))$

Here z_α is a variable which determines the size of the confidence bounds. Step 8d elaborates on the commonly used $z_\alpha\sqrt{P(1-P)/n}$, and is designed to give better results for small values of n — see (Kaelbling, 1993) for details.

MBIE hybrids. Although IE seems promising it does not clearly outperform Q-learning with Boltzmann exploration due to problems of estimating the variance of a changing Q-function in the beginning of the learning phase (Kaelbling, 1993). Since MBIE also relies on initial statistics we propose to circumvent such problems by starting out with some other exploration method and switching to IE once some appropriate condition holds.

This is done as follows: we start with frequency-based exploration and keep tracking the cumulative change of the problem-oriented value function. Once the average update of the V^Π function (computed over the most recent N time-steps) falls below $\eta \in \mathbb{R}^+$, we (I) copy the rewards and Q-values from the problem-oriented model to the exploration model, and (II) switch to IE: we apply asynchronous value iteration (Bellman, 1961) to the model; the iteration procedure calls MBIE for computing Q-values and ends once the maximal change of some state value is less than $\epsilon \in \mathbb{R}^+$.

Simultaneous policy learning. The model-based learner simultaneously learns both exploration policy and problem-oriented policy. After each experience we update the model and use PS to recompute the value functions.

5. Experiments

5.1 Maze-tasks

To compare the different RL methods discussed in previous chapters we use a 50 × 50 maze shown in Figure 1. It consists of about 20% blocked states and 20% punishing states (these are inserted randomly). Blocked states are represented by black fields, penalty states by grey fields, free states by white fields. In each state the animat can select one of four actions: *go north, go east, go south, go west.* The starting state (S) is located 1 field north/east of the south-west corner. There are three absorbing goal states, two of them are suboptimal. The optimal goal state (G) is located 1 field south/west of the north-east corner, the suboptimal goal states (F) are located in the north-west and south-east corners. Once the animat hits the goal it is reset to its initial position. Selected actions are replaced by random actions with 10% probability.

Reward function. Actions leading to a blocked state are not executed and punished by a reward of −2. Steps leading to free states are punished by a reward of −1. Actions leading to a penalty state yield a reward of -10. If the animat finds the optimal goal state it will receive a reward of 1000. For finding a suboptimal goal state it gets a reward of 500. The discount factor γ is 0.99.

Comparison. We compare the following exploration methods: Max-random, directed model-based exploration techniques using frequency-based and recency-based reward rules, and MBIE. The latter starts out with model-based exploration using the frequency-based reward rule, and switches to IE once the value function hardly changes any more.

The goal is to learn good policies as quickly as possible. We computed an optimal policy using value iteration (Bellman, 1961) and tested this optimal policy by executing it for 1,000 steps. We computed its average reinforcement intake by testing it 10,000 times, which resulted in 7590 ± 2.

For each method we conduct 20 runs of 100,000 learning steps. During each run we measure how quickly and how often the animat's policy collects 95%, 99% and 99.8% of what the optimal policy collects. This is done by averaging the results of 1000 test runs conducted every 2000 learning steps — each test run consists of executing the greedy policies (always selecting actions with maximal Q-value) for 1000 steps.

Parameters. We set the accuracy parameter $\epsilon := 0.1$, and $U_{max} := 100$ for both learning the problem-oriented and exploration value functions. The explo-

Figure 1 *The 50 × 50 maze used in the experiments. Black squares denote blocked fields, grey squares penalty fields. The animat starts at S and searches for a minimally punishing path to the optimal goal G. Good exploration is required to avoid focusing on suboptimal goal states (F).*

ration reward's discount factor γ is set to 0.99 for frequency-based and to 0.95 for recency-based exploration. The constant K_C (used by the frequency-based reward rules) is set to 50; K_T (used by the recency-based reward rule) is set to 1000. We used two values for P_{max} (for Max-random exploration): 0.2 and 0.4. The value of z_α (for MBIE) is set to 1.96 (which corresponds to a confidence interval of 95%). The combination of MBIE and model-based exploration switches to MBIE once the value function has not changed by more than 0.1 (η) on average within the 1000 (U) most recent steps.

Exploration Rule	95% (freq)	99% (freq)	99.8% (freq)
MBIE	25K (20)	42K (19)	66K (18)
Frequency-based	24K (20)	50K (16)	66K (10)
Recency-based	27K (19)	55K (18)	69K (9)
Max-random 0.2	43K (4)	52K (4)	68K (4)
Max-random 0.4	— (0)	— (0)	— (0)

Table 1 *The number of steps required by several exploration methods for obtaining p-optimal policies (and how many runs found them at all) (K stands for 1000).*

5.2 Results

Table 1 shows significant improvements achieved by learning an exploration model. The undirected exploration methods focus too much on suboptimal goals (which are closer and therefore easier to find). This often prevents them from discovering near-optimal policies. On the other hand, exploration model-based learning does favor paths leading to the optimal goal. Using the frequency-based reward rule by itself, the animat always finds the optimal goal although it fails to always find 99.8% optimal policies. Figure 2 shows a large difference between learning performances of exploration models and max-random exploration. It does not clearly show the distinction between frequency-based or recency-based exploration and the extension with MBIE. Switching to MBIE after some time (between 35,000 and 55,000 steps), however, significantly improves matters. First of all, Table 1 shows that this strategy finds optimal or near-optimal policies in 90% of the cases, whereas the others fail in at least 50% of the cases. The second improvement with MBIE is shown in Table 2. MBIE collects much more reward during training than all other exploration methods, thereby effectively addressing the exploration/exploitation dilemma. In fact it is the only exploration rule leading to a positive cumulative reward score.

Exploration Rule	Best run result	Training Performance
MBIE	7.57K ± 0.05K	350K ± 40K
Frequency-based	7.55K ± 0.06K	-45K ± 9K
Recency-based	7.54K ± 0.11K	-120K ± 10K
Max-random 0.2	4.8K ± 1.4K	-190K ± 16K
Max-random 0.4	4.1K ± 0.3K	-62K ± 19K

Table 2 *Average and standard deviation of the best test result during a run and the total cumulative reward during training.*

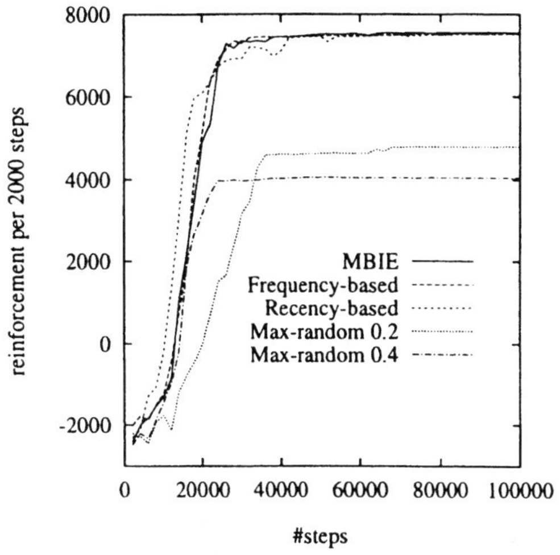

Figure 2 *Cumulative reward during test runs, averaged over 20 simulations.*

6. Discussion

Undirected exploration relies on a random generator for selecting actions. When applied to tasks with multiple absorbing goal states, it faces major difficulties in finding the optimal one. Our exploration models, however, allow for globally maximizing exploration rewards. This allows for discovering good policies circumventing suboptimal goal states. We compared two types of exploration rewards: frequency-based, and recency-based. Although frequency-based exploration performed slightly better in our stationary environments, recency-based reward rules may make more sense in non-stationary ones.

Optimal exploration. When it comes to finding optimal policies, methods using external MDP rewards for focusing on actions with large probability of being optimal will often save some SAP visits over methods based solely on local exploration reward rules. Furthermore, cumulative external MDP rewards obtained during training should be taken into account in attempts at approaching the exploit/explore dilemma. That is why we introduced MBIE, a method combining Kaelbling's interval estimation (IE) algorithm (Kaelbling, 1993) and model-based reinforcement learning (MBRL). Since MBIE heavily relies on initial statistics, we switch it on only after an initial phase during which an exploration model is learned (according to, say, the frequency-based local exploration reward rule). In our experiments this approach almost always led to 99.8%-optimal policies.

Future work: function approximators. Our ideas have been presented in the context of discrete state spaces. We can extend them to function approximators in the same way we can extend model-based RL methods to learning in continuous state spaces. This is an interesting topic for future research.

7. Acknowledgments

Many thanks to Rafał Sałustowicz, Nic Schraudolph, and Jieyu Zhao for helpful discussions. We are thankful to the Swiss Center for Scientific Computing (CSCS/SCSC) for providing additional computing power. This work was supported in part by SNF grant 2100-49'144.96 "Long Short-Term Memory".

References

Bellman, R. (1961). *Adaptive Control Processes.* Princeton University Press.

Berry, D. and Fristedt, B. (1985). *Bandit Problems: sequential allocation of experiments.* Chapman and Hall, London/New York.

Caironi, P. V. C. and Dorigo, M. (1994). Training Q-agents. Technical Report IRIDIA-94-14, Université Libre de Bruxelles.

Dayan, P. and Hinton, G. (1993). Feudal reinforcement learning. In Lippman, D. S., Moody, J. E., and Touretzky, D. S., editors, *Advances in Neural Information Processing Systems 5*, pages 271–278. San Mateo, CA: Morgan Kaufmann.

Dodge, Y., Fedorov, V., and Wynn, H., editors (1988). *Optimal Design and Analysis of Experiments: Proceedings of First International Conference on Optimal Design and Analysis of Experiments.* Elsevier Publishers.

Fedorov, V. V. (1972). *Theory of optimal experiments.* Academic Press.

Kaelbling, L. (1993). *Learning in Embedded Systems.* MIT Press.

Kearns, M. and Singh, S. (1998). Near-optimal performance for reinforcement learning in polynomial time. Retrievable from http://www.research.att.com/~mkearns/.

Koenig, S. and Simmons, R. G. (1996). The effect of representation and knowledge on goal-directed exploration with reinforcement-learning algorithms. *Machine Learning*, 22:228–250.

Moore, A. and Atkeson, C. G. (1993). Prioritized sweeping: Reinforcement learning with less data and less time. *Machine Learning*, 13:103–130.

Schmidhuber, J. (1991a). Curious model-building control systems. In *Proc. International Joint Conference on Neural Networks, Singapore*, volume 2, pages 1458–1463. IEEE.

Schmidhuber, J. (1991b). A possibility for implementing curiosity and boredom in model-building neural controllers. In Meyer, J. A. and Wilson, S. W., editors, *Proc. of the International Conference on Simulation of Adaptive Behavior: From Animals to Animats*, pages 222–227. MIT Press/Bradford Books.

Schmidhuber, J. (1997). What's interesting? Technical Report IDSIA-35-97, IDSIA.

Schmidhuber, J., Zhao, J., and Wiering, M. (1997). Shifting inductive bias with success-story algorithm, adaptive Levin search, and incremental self-improvement. *Machine Learning*, 28:1:105–130.

Storck, J., Hochreiter, S., and Schmidhuber, J. (1995). Reinforcement driven information acquisition in nondeterministic environments. In *Proceedings of the International Conference on Artificial Neural Networks*, volume 2, pages 159–164. EC2 & Cie, Paris.

Sutton, R. S. (1990). Integrated architectures for learning, planning and reacting based on dynamic programming. In *Machine Learning: Proceedings of the Seventh International Workshop.*

Sutton, R. S. (1996). Generalization in reinforcement learning: Successful examples using sparse coarse coding. In D. S. Touretzky, M. C. M. and Hasselmo, M. E., editors, *Advances in Neural Information Processing Systems 8*, pages 1038–1045. MIT Press, Cambridge MA.

Thrun, S. B. (1992). The role of exploration in learning control. In *Handbook of Intelligent Control: Neural, Fuzzy and Adaptive Approaches.* Florence, Kentucky 41022: Van Nostrand Reinhold.

Wilson, S. W. (1996). Explore/exploit strategies in autonomy. In Meyer, J. A. and Wilson, S. W., editors, *Proc of the Fourth International Conference on Simulation of Adaptive Behavior: From Animals to Animats 4*, pages 325–332. MIT Press/Bradford Books.

Autonomous Decision Making in Local Navigation

Percy Dahm[1] **and Carsten Bruckhoff**[2]
Institut für Neuroinformatik, Ruhr-Universität Bochum
44780 Bochum, Germany
Tel: +49 234 700 [1]7975 / [2]5570
Fax: +49 234 709 4209
Percy.Dahm@neuroinformatik.ruhr-uni-bochum.de
Carsten.Bruckhoff@neuroinformatik.ruhr-uni-bochum.de

Abstract

We present a neural field approach towards local navigation that is capable of endowing a mobile robot with features of autonomy. In particular we refer to the ability (and appetency) of making decisions and keeping them stable in a varying environment. The quality of a decision is represented by the strength of an activation over the neural field. We employ this strength for influencing the robot's behavior. The theoretical framework is validated by an implementation on our mobile service robot called 'ARNOLD'. Since its only sensor is an active stereo camera system we highlight the importance of gaze control and low-level short-term memory for the stability of the overall behavior.

1. Introduction

In the narrower sense autonomy means to be independent. Since it is quite difficult to implement this universal definition on a mobile robot, we will focus on one of the most important aspects of autonomy: the ability of decision making.

Regarding the underlying decision making process one has to consider at least three basic properties: First, decisions are always goal-directed. Second, for reasons of stability a decision has to be represented as an internal state of the robot. Finally, the robot must be able to acquire actively all the information which is necessary for the decision.

As an example we investigate the problem of local navigation. Here the robot's objective is to reach a target under the constraint of avoiding obstacles. Thus, a decision corresponds to a certain direction of movement. Considering a noisy sensor and a previously unknown or dynamically varying environment such a decision has to be stabilised internally. Finally, we are convinced that, whenever a robot is endowed with more than one skill, attention-based information acquisition is necessary. Therefore, the robot has to extract autonomously the most relevant information with respect to its current behavior. In our case this leads to the need for controlling the gaze direction. In the course of our studies we will show that this is an inherent property of the model we developed.

For the purpose of giving a more universal concept we employ the 'neural field' introduced by Amari (Amari, 1977). It can be described by a competitive, nonlinear dynamic over a field of neurons. Although Amari's original intention was to model cortical neurophysiology, in the past years several authors considered it to be suitable for applications in the field of mobile robotics (Engels and Schöner, 1995; G. Schöner, 1995; Bicho and Schöner, 1997).

Our work regarding autonomous decision making is related to that of Engels (Engels and Schöner, 1995), who uses neural fields in the context of memory representation, making 'the local path planning more global'. The level where Engels determines the robot's movement direction is substituted by an approximation described in (Amari, 1977). Since we are especially interested in the dynamics of this level we keep the original Amari-field. Instead of introducing a separate level of memory we endow the stimulus field with an exponentially decaying dynamic.

Finally, we show that the strength of activation over the neural field can be used as an important parameter of the robot's overall behavior.

2. The neural field

The neural field introduced by Amari (Amari, 1977) is described by an integro-differential equation:

$$\tau \dot{u}(\varphi, t) = -u(\varphi, t) + \int_0^{2\pi} w(\varphi, \varphi')\sigma(u(\varphi'))d\varphi' + h + s(\varphi, t) \tag{1}$$

In our case $u(\varphi)$ is the activation with respect to the direction φ. h is a global constant that fixes the threshold for the localisation interaction defined by the sigmoid function $\sigma(u)$:

$$\sigma(u) = \frac{1}{e^{-Cu} + 1} \tag{2}$$

$s(\varphi)$ represents the input with respect to $u(\varphi)$, and τ determines the time scale of the dynamics. The aspects of decision making mentioned above are strongly related to the integral term in eq.(1). It contains the symmetric interaction kernel mediating the coupling between the neurons voting for φ and φ', respectively. The interaction is realised by a short-ranged excitative term and a global inhibitive constant H_0:

$$w(\varphi, \varphi') = k \cdot e^{-C'(\varphi-\varphi')^2} - H_0 \tag{3}$$

It can analytically be shown that eq.(1) has three basic types of solutions. The homogeneous solution $u = h$ (ϕ-solution), an instable localised solution (a_1-solution) and a stable localised solution (a_2-solution). Further details can be found in (Kishimoto and Amari, 1979).

From here it becomes obvious that eq.(1) is particularly appropriate for modeling a decision making process:

- a localised peak can be interpreted as a decision
- in particular, if this peak is of an a_2-type, a decision is self-stabilising
- the uniqueness of a decision can be realised by choosing a global inhibition (see eq.(3))

Furthermore, we are able to measure the quality of a decision which is proportional to the strength of the corresponding peak. This will be important in ambiguous situations where the stimulus field is multi-modal.

3. Application to local navigation

With respect to mobility most robots possess two degrees of freedom: one for rotation and one for translation. Consequently, keeping this separation is a natural choice. The neural field is used to decide in which direction the robot is moving. Hence, it encodes angular directions from 0 to 2π. Since the field itself is periodic, the interaction kernel in eq.(3) has to be periodic, too. The stimulus field $s(\varphi)$ contains the angular obstacle information and the direction of the target. The latter one is represented by a wide ranged, bell shaped function $t(\varphi)$, which is greater than zero over the whole field. The smallest value of $t(\varphi)$ has to be at least as great as necessary to evoke an activation, i.e., it must be greater than $|h|$ in eq.(1). The target stimulus $t(\varphi)$ has its maximum in target direction and its minimum in the opposite direction. Obstacles are understood as forbidden movement directions. Hence, the corresponding stimuli $o(\varphi)$ are always negative. Moreover, the absolute values of the stimuli reflect the distance of the robot from the corresponding obstacle:

$$o(\varphi) = -A \cdot e^{-\delta \cdot d(\varphi)^2} \tag{4}$$

The constant A denotes an amplitude and $d(\varphi)$ the distance measured by the sensor. (We set $d(\varphi)$ to infinity if the corresponding direction φ represents a free path). δ fixes the shape of the range limiting transfer function.

In order to reach a stable decision the neural field needs continuous information from the stimulus field. However, as our robot's only sensor is an active stereo camera system we have to cope with the problem of a restricted view angle of the sensor input. For compensating this 'lack' of information we introduce a short-term memory with respect to $s(\varphi)$:

$$\begin{aligned}\dot{s}(\varphi) = & \frac{1-\theta(\beta/2 - |\varphi-\alpha|)}{\tau_1} \cdot (t(\varphi) - s(\varphi)) \\ & + \frac{\theta(\beta/2 - |\varphi-\alpha|)}{\tau_2} \cdot (t(\varphi) + o(\varphi) - s(\varphi))\end{aligned} \tag{5}$$

where α is the current gaze direction and β the angle of view. θ is a step function:

$$\theta(\beta/2 - |\varphi - \alpha|) = \begin{cases} 1 & : \beta/2 - |\varphi-\alpha| > 0 \\ 0 & : \beta/2 - |\varphi-\alpha| \le 0 \end{cases}, \tag{6}$$

and $\tau_2 \ll \tau_1$. Thus, if φ lies within the current view angle interval, $s(\varphi)$ relaxes on a fast time scale to $t(\varphi) + o(\varphi)$, i.e., to the current sensor input. Out of the view angle interval $s(\varphi)$ relaxes on a much slower time scale to the current target stimulus $t(\varphi)$. Therefore, eq.(5) enables the robot to store information about its environment for a certain time, that can be adjusted by the choice of τ_1.

Of course, with respect to translatory movements this information is not updated correctly. However, this is not necessary: First, the information is only used as a coarse and temporarily limited description of the environment. Second, the direction of the gaze is controlled by the maximum activation of the neural field. Therefore, the most relevant information, namely that, which corresponds to the current movement direction, is updated almost immediately.

Furthermore, it is important to note that we apply a coordinate system which is robot centered, but globally fixed with respect to its orientation. This choice helps to keep both, the information stored in the stimulus field and the fields activation, as continuous as possible.

Now, it is clear that the gaze direction, i.e., our active vision head, plays an important role in the decision making process. Hence, the corresponding degree of freedom has to follow the field's activation on a fast time scale. By contrast, it is reasonable that the robot's rotatory and translatory velocity must be slowed down if the quality of a decision is not sufficient. This can easily be achieved by setting the corresponding velocity components proportional to the thresholded maximum activation of the field:

$$\begin{aligned} v'_R &= v_R \cdot \sigma(u_{max}, t_R) \\ v'_T &= v_T \cdot \sigma(u_{max}, t_T), \end{aligned} \tag{7}$$

with

$$\sigma(u_{max}, t_{R/T}) = \frac{1}{e^{-\gamma(u_{max} - t_{R/T})} + 1} \tag{8}$$

$v_{R/T}$ denotes the original velocity and γ the slope of the sigmoidal function $\sigma(u_{max})$. t_R and t_T are the thresholds of activation that are necessary to evoke a rotatory or translatory movement.

4. Experimental setup

Figure 1 Arnold

We have tested our model for decision making on our autonomous mobile robot 'ARNOLD'. It is based on a TRC Labmate platform with two degrees of freedom (DoF). Arnold is equipped with a 7 DoF arm and a double stereo camera head with 3 DoF as its only sensor. A detailed description of Arnold can be found in (Bergener et al., 1997).

The control architecture of Arnold is based on the behavior robotics paradigm introduced by V. Braitenberg (Braitenberg, 1984) and R. Brooks (Brooks, 1986). Most of Arnold's behaviors are based on nonlinear dynamical systems. A first application, the tracking of a human hand using all its degrees of freedom, is described in (Bergener and Dahm, 1997).

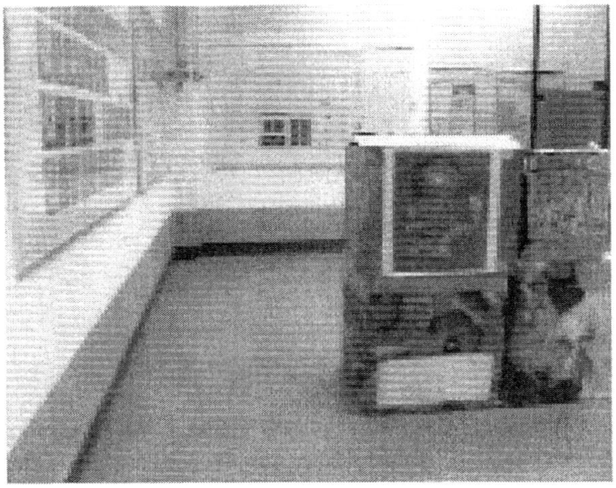

Figure 2 · Experimental setup

The sensor module for obstacle avoidance is based on the 'Inverse perspective mapping' (IPM) (Mallot et al., 1991). Here the image of the right camera is mapped onto the image of the left camera under the assumption that the robot moves on a horizontal plane and the parameters of the cameras are known. A difference image is calculated and the pixels which exceed a certain threshold represent an obstacle in the field of view.

Figure 3 A person blocks Arnold's way

The neural field approach for decision making in local navigation has been applied successfully in several experiments. In the following we will only describe one example in more detail. Arnold has to move along the wall and reach a target position laying 5.5 meters straight ahead. While approaching the gap, which is only slightly wider than the robot itself, a person is blocking the way. Arnold has to decide where to go to reach the target position. Figures 2 and 3 show the two situations. Figure 2 shows the obstacle free path to reach the target position and figure 3 shows the person blocking the path.

5. Experimental results

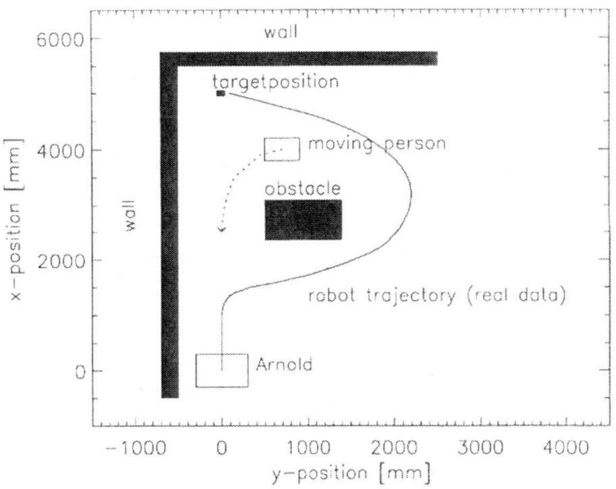

Figure 4 Robot trajectory

Arnold performs the experiment in real time, it moves at a maximum speed of $250\frac{mm}{s}$. Here the limiting factor is the module which calculates the obstacles, it needs about $300ms$ to evaluate one image pair. In order to perform the given task the robot needs about one minute and the data has been sampled every second.

The trajectory of the robot in measured odometry coordinates is shown in figure 4. Arnold shows the desired behavior. It moves straight in target direction, stops when the person blocks the path, gathers new information to decide where to go, passes the obstacle and changes to target direction behind the obstacle.

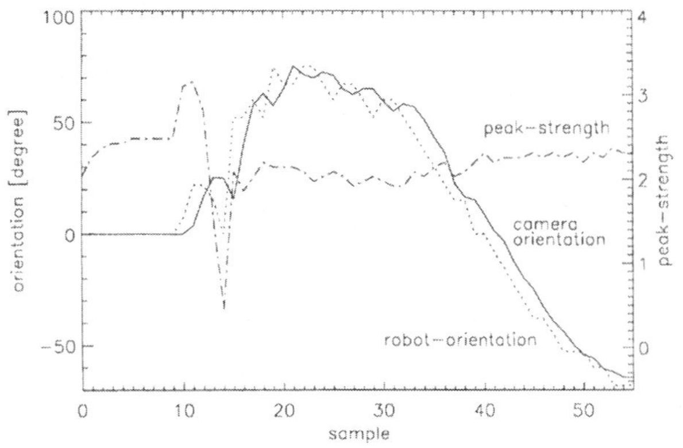

Figure 5 Orientation and peak strength

Figure 5 shows the orientation of the robot and the camera head, as well as the peak's strength in the neural field representing the quality of the current decision.

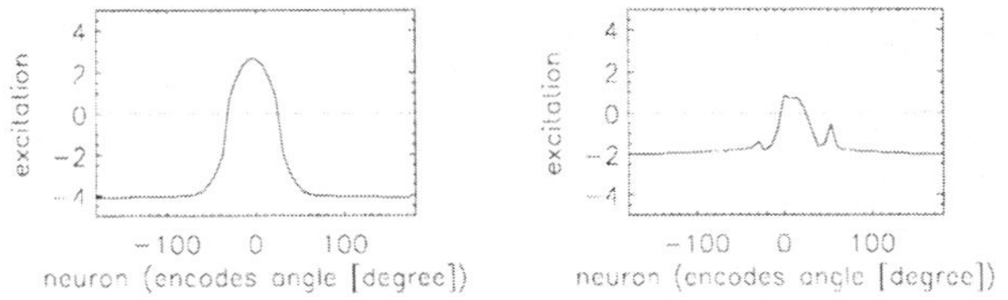

(a) Step 7 (b) Step 15

Figure 6 Neural field (1)

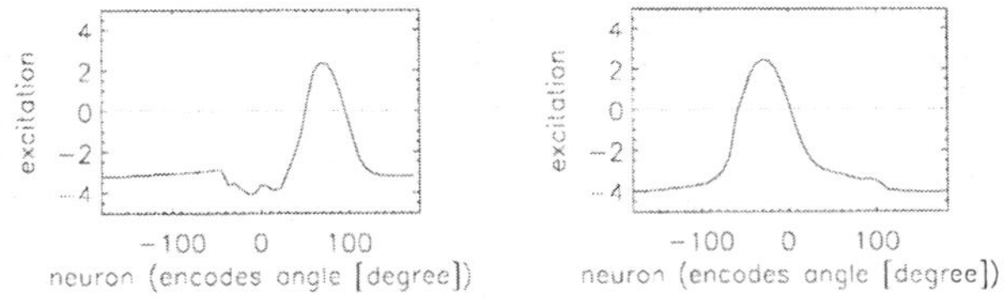

(a) Step 21 (b) Step 45

Figure 7 Neural field (2)

The short increase of strength at about sample 9 is caused by the fact that the peak is self-stabilising, i.e., of a_2-type. Hence, being surrounded by obstacles, the peak's height initially increases compensating its decreasing width. However, due to the competition the 'weakened' peak brakes down, and a new peak arises voting for another direction of movement.

Figures 6,7 and 8, 9 show four steps of the neural field (fig. 6, 7) and the short-term memory (stimulus) (fig. 8, 9).

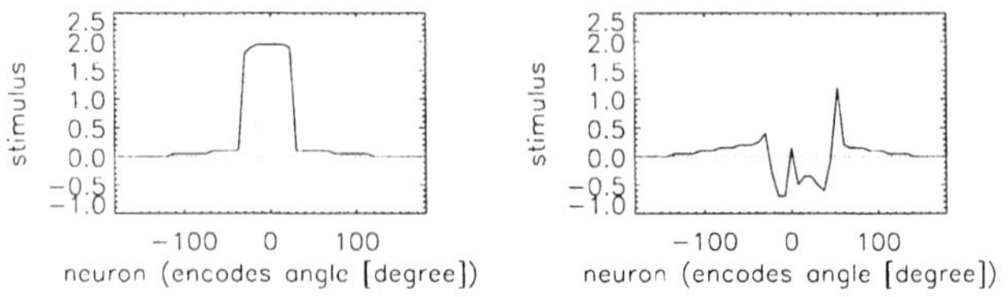

(a) Step 7 (b) Step 15

Figure 8 Short-term memory (stimulus) (1)

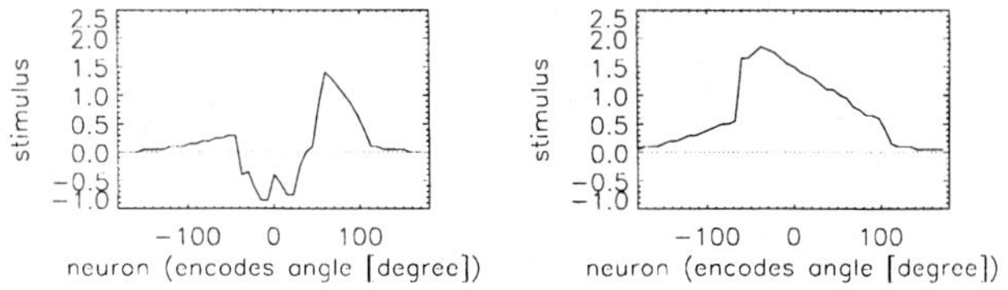

(a) Step 21 (b) Step 45

Figure 9 Short-term memory (stimulus) (2)

As an example we have picked four situations of the stimulus and the neural field to discuss the decision making process. Figure 8 (a) shows the stimulus while the robot is moving in target direction. In the field of view the short-term memory is updated on a fast time scale, outside the field of view it is updated on a slow time scale. The peak in the neural field (fig. 6 (a)) relaxes to target direction with large strength, so that the robot moves at its fastest velocity. The person blocks the robot's way and the stimulus in the field of view is updated on a fast time scale to negative values. Therefore, the peak in the neural field collapses (fig. 6 (b)) and Arnold stops.

After Arnold has estimated the obstacle's position on its right, the short-term memory shows no further obstacles in the field of view (fig. 9 (a)) and the memorised obstacles in front of the robot. A new peak arises in the neural field at about 90° (fig. 9 (a)) regarding the globally fixed coordinate system concerning the orientation. Therefore, Arnold begins to move along the obstacle.

The peak slowly moves to target direction and Arnold looks on its left side to decide whether it has reached the end of the obstacle or not. As long as the obstacle to its left is still there, the obstacle position is updated in the short-term memory. The resulting small oscillations can be seen in figure 5, sample (19-33).

After the robot has reached the end of the obstacle, the short-term memory shows no obstacle (fig. 9 (b)) and the direction to the target position. The peak in the neural field has a stable position in the same direction (fig. 7 (b)) and the robot is approaching the given target position.

6. Concluding remarks and future work

We presented a model for autonomous decision making which is exclusively based on a nonlinear competitive dynamic system, the so-called neural field. As an example we investigated the problem of local navigation. Regarding this application we were able to show that our model contains basic properties of autonomous decision making:

- decisions are goal-directed
- decisions are self-stabilising
- decisions can be made in a unique fashion
- the quality of a decision is an important behavioral parameter
- information acquisition is organized by the decision making process itself

We are convinced that our results can be transfered to general effector control problems. Hence, our recent work is dedicated to the field of manipulators, where we are faced with a higher dimensional task space. However, this only makes the underlying mathematics regarding the neural field more sophisticated, not the concept itself. Moreover, the uniform language carries the advantage of coordinating mobility and manipulation in an easy fashion.

7. Acknowledgments

We thank W. von Seelen and C. Engels for inspiring discussions. The technical assistance of M. Neef and R. Menzer is acknowledged. The project 'NEUROS' is supported under grant 01IN504F from the German Ministry of Research.

References

Amari, S.-i. (1977). Dynamics of pattern formation in lateral-inhibition type neural fields. *Biological Cybernetics*, (27):77–87.

Bergener, T., Bruckhoff, C., Dahm, P., Janßen, H., Joublin, F., and Menzner, R. (1997). Arnold: An anthropomorphic autonomous robot for human environments. In *SOAVE'97, Selbstorganisation von adaptivem Verhalten.*

Bergener, T. and Dahm, P. (1997). A framework for dynamic man-machine interaction implemented on an autonomous mobile robot. In *ISIE'97, IEEE International Symposium on Industrial Electronics.*

Bicho, E. and Schöner, G. (1997). Target position estimation, target acquisistion and obstacle avoidance. In *Proceedings of the IEEE International Symposium on Industrial Electronics, ISIE'97.*

Braitenberg, V. (1984). *Vehicles: Experiments in Synthetic Psychology.* The MIT Press.

Brooks, R. A. (1986). A robust layered control system for a mobile robot. *IEEE Journal of Robotics and Automation*, 2(1):14–23.

Engels, C. and Schöner, G. (1995). Dynamic fields endow behavior-based robots with representations. *Robotics and Autonomous Systems*, 14:55–77.

G. Schöner, M. Dose, C. E. (1995). Robotics and autonomous systems: Dynamics of behavior: theory and applications for autonomous robot architectures. *Robotics and Autonomous Systems*, (16).

Kishimoto, K. and Amari, S. (1979). Existence and stability of local excitations in homogeneous neural fields. *Journal of Mathematical Biology, Springer*, 7:303–318.

Mallot, H. A., Bülthoff, H. H., Little, J. J., and Bohrer, S. (1991). Inverse perspective mapping simplifies optical flow computation and obstacle detection. *Biological Cybernetics*, 64:177–185.

Motivated Animat navigation: a visually guided approach

P. Gaussier, C. Joulain, J.P. Banquet

* Neurocybernetic group, ETIS, Ecole Nationale Superieure d'Electronique et de ses Application - Cergy Pontoise University
6 av du Ponceau, 95014 Cergy Pontoise Cedex, France
gaussier, joulain@ensea.fr
** CREARE, university Paris VI

Abstract

In this paper, we describe how a mobile robot only controlled by visual information can retrieve particular goal locations in an open environment according to the level of associated motivations and to its current location. The interest of our control architecture relies in its simplicity and its robustness. Our robot does not need a precise map nor to learn all the possible positions in the environment to be able to navigate correctly. The neural control architecture used is inspired from neurobiological studies about the brain visual processes and the integration of those information for navigation behaviors. We emphasize the interest to go back and forth between robotics and biology. At last we show neurobiological modelization of the hippocampus has helped us to imagine a solution for scaling our architecture to more complex problems.

1. Introduction

In our previous works (Gaussier and Zrehen, 1994; Gaussier and Zrehen, 1995; Gaussier et al., 1996) we proposed a neural architecture inspired from neurobiological data to control autonomous robots. Our architecture allows a visually guided robot to discover, learn and retrieve autonomously a particular place in an open environment (both in simulation and in real environments). This model explains how animals use landmarks (recognition of snapshot images) to navigate from 2D perceived images (Bachelder and Waxman, 1994). Our main goal was just to show animals did not need to have or to learn a large number of positions in an open environment to be able to return to a learned place (for instance the platform in the Morris swimming pool (Morris, 1981)). Moreover we showed there is no need of an internal allocentric representation of the world, like a cartesian map which is used in a lot of models (for a large review of existing models including ours see (Trullier et al., 1997)). Our navigation model only uses a cascade of two simple PerAc (Perception-Action) neural architectures (Gaussier and Zrehen, 1995) to control the robot (from the visual information extracted from the CCD camera to the motor output).

In mammals, like rats, the achievement of navigation tasks involves a particular brain structure called the hippocampus. Electro-physiological recordings of hippocampus cells performed in CA3 and/or in CA1 regions have shown the existence of cells that maximally respond when the animal (a rat for instance) is at a particular place in its environment (O'Keefe, 1989). These cells have been called "place cells".

In the first part of the paper, we show a simple navigation model that allows an animat to choose and to reach a particular goal according to the animat location and to its motivations. As others we will use Hs to merge information coming from the parietal and temporal visual associative areas at the level of the CA3 region (processes that we study in the frame of a neurobiological project (Banquet et al., 1997)). The main difference of our algorithm is its implicit use of the topological properties of an Euclidian space which allows our algorithm to be very simple and robust. Successful transposition to robotic experiments will reinforce this proposition. We will also propose an explanation about why primates do not show "place cell" activity but only some kind of "view cell" activity as reported by (Rolls and S.M.O'Mara, 1995).

2. Landmark-based navigation

Our navigation model shows it is possible to return to a learned goal position if a minimum of 3 locations have been learned in the immediate neighborhood of the goal. If the animal or the robot (the animat) is able to associate the recognition of these places with the movement needed to reach the goal, then it will join the goal whatever its starting point. Using this strategy, there is no need for a robot or an animal to have a "cognitive map" of the environment to navigate in an open environment. The open environment is defined as a surface surrounded by the landmarks and without obstacles to avoid.

The PerAc architecture is a neural computation architecture proposed to solve a wide variety of control problems requiring learning capabilities (in contrast to adaptation capabilities[1]). It consists in an action level (a

[1] Learning is something that can be performed quickly and

hardwired pathway able to play the role of a reflex mechanism) and a perception level trying to recognize particular situations and to associate them with the correct action through an associative or a reinforcement learning rule. The perception level allows the robot to react to a situation even if it is too complex to allow the action pathway to propose an answer (the goal is not in sight for instance) by generalization of previously learned situations (the landmark configuration in the navigation task). Moreover, if the action pathway induces a negative reward, the links between the recognition of the perceived situation and the current robot action can be inhibited and a link with an action avoiding a negative reward can be learned. This kind of architecture is not in competition with Brook's subsumption (Brooks and Stein, 1994) and other robot controller architectures (Steels, 1995) because it continues in the same direction under a neural formalism (an obstacle overtaking reflex is directly implemented on the low level layer of our architecture for instance).

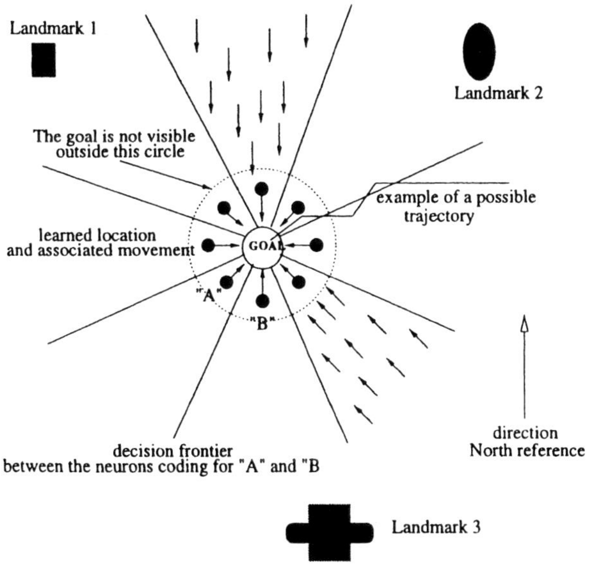

Figure 1 Example of a goal retrieval with an absolute direction.

At the beginning of the exploration phase, we suppose our robot (Prometheus) moves randomly, looking for something interesting. When it finds its goal, it moves around it in order to explore various positions in its proximity. At these places, it learns both the landmark configuration (represented by a set of local views and their angles) and the direction leading to the goal. Later, when the robot wants to find the goal, it considers the information of the place cells (i.e. the cells which react to a specific set of local views associated to their azimuth) and moves in the direction associated to the most activated place cell (competitive mechanism) to reach the goal. Thus at each step, the distance to the target is reduced (Fig.1) and the robot returns inevitably to the learned position. A complete description of a neural implementation of the learning process can be found in (Gaussier and Zrehen, 1995). The PerAc architecture for place learning realizes an approximation of a potential field function (Arkin, 1987) without the cost of knowing the goal and the robot location in a cartesian space.

When the goal is in sight (goal recognition), a neuron corresponding to its angular position relative to the robot's facing position is activated in the Target Azimuth group (we suppose that the robot has previously learned what the goal looks like (Joulain et al., 1997)). A shifting mechanism activates a neuron in the Direction of Movement Proposal (DMP) group by adding an angle corresponding to the angle between the robot and the north direction. The inverse shifting mechanism is applied to the output of the Direction of Movement group, by subtracting the same angle. Thus the robot movement is correct whatever the robot orientation is (conservation of the sensory topology in the whole neural network architecture).

First to recognize a place, the robot must be able to isolate a local view (using the focus of attention mechanism that will be presented later). Next, information on landmark recognition and on the associated angle are merged to produce a unified representation that can be easily learned and matched with previously learned representations. This merging process is represented by the matrix product of the information corresponding to "what" (Landmark) and "where" (Azimuth) the objects are (Fig. 2 Landmarks Azimuths group). A time integration process allows to suppress the sequential aspect of the scene exploration (spatio-temporal merging). With this representation, there is no need to "recognize" specifically what the landmarks are (a fridge, a chair...), it is only important to distinguish them and to know their angular position. Even if a landmark is missing (for instance if the fridge is removed or if the "image" of the scene (Landmarks Azimuths) is too noisy), the other landmarks can allow a good recognition (we have shown that several landmarks can be removed, hidden or displaced without disturbing the global recognition of the scene (Zrehen, 1995)).

The place representation (LA) is learned by the Place Cells group of neurons. If the robot recognizes the goal, it moves in that direction (reflex link between Target Azimuth and Direction of Movement on Fig. 2). Otherwise, learned association between Place Cells and Direction of Movement allows an efficient generalization to the whole

involves a variation in the structure of the control architecture whereas the adaptation consists in a slower variation of the structure parameter. Adaptation can induce learning if there is a non linear modification of the adapted parameters that induce a non linear variation of the system response.

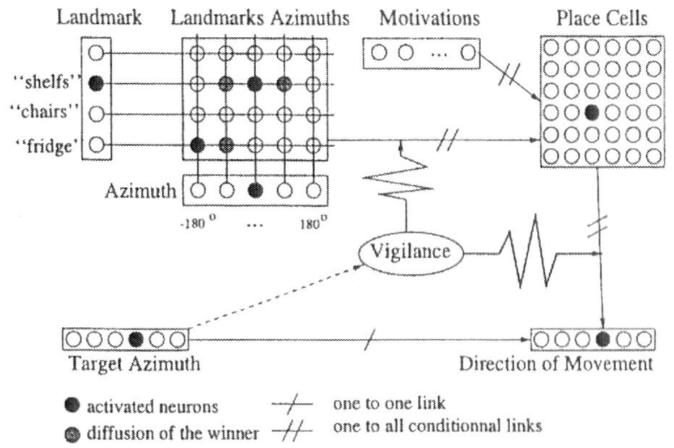

Figure 2 The navigation neural network. Target azimuth and Direction of movement are WTA. The Landmarks azimuths group emphasizes merging of visual and motor flow.

environment. Consequently, this process creates basin of attraction in which the robot always moves in order to come closer to the goal. Moreover, it has been mathematically proved that there was no local minimum induced by the competition between the action neurons within the domain bounded by the set of landmarks (Gaussier and Zrehen, 1995).

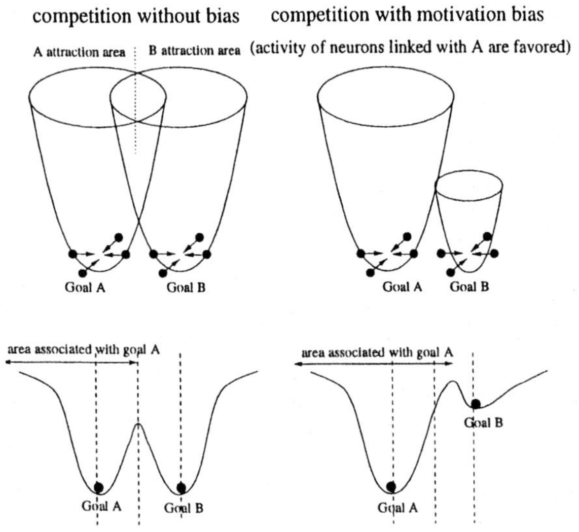

Figure 3 Bias introduced by motivations in goal selection.

The motivations allow the robot to bias competition between potential goals (Fig. 3) and then to reach the desired goal (Zrehen and Gaussier, 1997) if it belongs to the same visual environment than the current robot location. Motivations are controlled by two internal variables associated to the perception of something similar to the food and water levels. As soon as the robot is on a supply location, the level of the associated ressource is immediately increased of a constant quantity.

2.1 An efficient robotic implementation

The visual input comes from a 384×288 gray-scale CCD camera. Its field of view spans on about 70 degrees. To build the panoramic view a servo-motor is used to pan the camera. The robot takes 24 images with a 7.5 degrees rotation between each capture. The central vertical bands of each image (30×288 pixels) are merged to constitute the global panoramic view (the central band is only used because the camera distorts the image sides). A 1246×288 pixels panorama is obtained. The resulting field of view is about 250 degrees. It is not a complete 360 degrees image but it is enough for our application.

The test environments are our every-day working rooms with the chairs, tables, shelfs, workstations, dustbins... The environments are used as they are without any change. In addition, the more complex the environment is, the more numerous landmarks can be, and thus the more efficient our algorithm is. The experiments were done in a room of about $7.2m \times 5.4m$ (the size of our Koala robot is about $30cm \times 30cm$).

Figure 4 16 examples of 32×32 local views learned as landmarks and their azimuth in the panoramic image.

The first step of the navigation algorithm is to find where the possible landmarks are. All the image columns are averaged and weighted with a heavier weight for the points near to the center of a column. The resulting one dimensional signal is differentiated and the local maxima and minima are used to point out what will be the center of local views. For each selected point, a 32×32 pixels local view is built by averaging the 148×288 pixels of the corresponding panoramic image (an example of local views and their positions is shown on Fig. 4). The **y** axis is just scaled whereas a logarithmic transformation is used for the **x** axis. Each local view is then compared with each learned local view. This comparison is a simple difference between pixels of the local views. Views which are sufficiently activated are used as landmarks, i.e. their positions in the image are compared with the ones in the learned panorama. The sum of the absolute angles between these positions gives us the similarity between panoramas. So the movement associated to the best corresponding learned location is performed.

2.2 Experimental results

The algorithm has been tested in different rooms (Gaussier et al., 1996; Gaussier et al., 1997a; Gaussier et al., 1997b). Figure 5 represents the map of one of these rooms. In these experiments the robot reaches the goal and stays in its vicinity with a precision of 5cm (each move was about 20cm and the robot size is about 30cm). Even if the starting point of the robot can be very far from the goal, the robot is able to reach the goal with great precision (the precision can be increased with smaller steps).

A compass sensor[2] is used to allow a correct movement decision whatever the robot's orientation. This compass sensor gives an indication on the camera direction, we only have to subtract this rotation angle to the angle found between viewed landmarks and learned landmarks. After this shift, the comparison is made with the learned views and the best corresponding movement is performed (according to the body orientation).

A first limitation is the coarseness of the movements. So, we combine the different responses of each "place cell" in function of their activity. We have to apply a transformation on each activity to enhance differences because activities are very close. The contrast enhancement of the movement direction is performed as follows: $S_i' = \left(\frac{S_i}{S_{max}}\right)^n$ and $\mathbf{V} = \sum_i S_i' \cdot \mathbf{V_i}$ where S_i is the value of the place cell i, $\mathbf{V_i}$ its associated movement, S_{max} the value of the most activated place cell and $\mathbf{V}$ the robot movement vector. n is chosen high enough to avoid that place cells with a low activation (place cells coding for a distant location) imply a motor contribution (here $n = 10$). Conversely, this equation allows to sum the contributions of place cells which are the most activated and which have almost the same value.

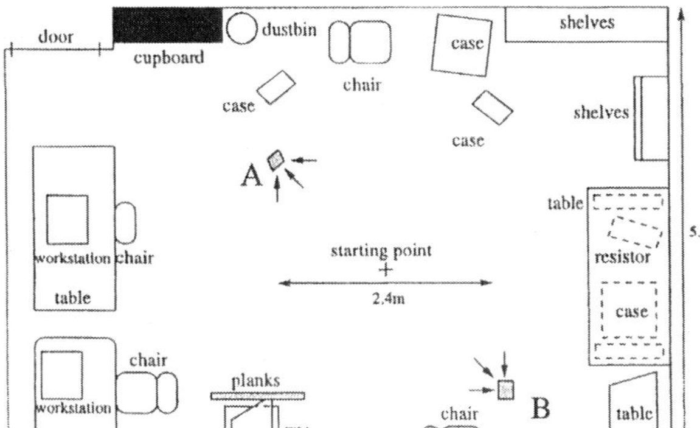

Figure 5 Map of a room used for simple motivated navigation. We can see the two goals A and B. The 3 arrows around each goals represent the learned positions and the associated movements. The distance between each learned position and the center of the goal is only 30 cm.

In the following experiments the aim is to simulate a system with two different goals. We have introduced in our environment two goals named A and B (see Fig.5,6,7). They are made of metallic paper sticked on the floor that special sensors of our robot can detect. When the robot is on a goal, the corresponding motivation is solved (its intensity decreases) and so the robot is attracted by the other goal. We made several experiments on the robot (see Fig.8,9), with motivations evolving with time. As we can see the robot is always able to reach the goals, even if the size of the goal is small and if the robot is far from it. In fact, in those experiments, goals are two rectangles ($16cm \times 14cm$ and $16cm \times 20cm$) and the distance between goals is 3.5m. It is important to compare this distance with the distance between learned positions and goals which is only 30cm. Figure 8 and 9 represent the robot trajectories, goals positions (A left-up and B right-down) and the position of an obstacle added during the experiment (not present at that place during learning). In fact those trajectories are recorded from the robot odometer. Odometry is only used to give us information on trajectories and it is not used by the system, the system only uses visual cues and the compass (and infra-red sensors to avoid obstacles). The robot does not use cartesian information and has no memory of its previous locations. Though the goal can be visible (the metallic paper on the black floor) the robot does

Figure 6 Panoramic image constructed by our system. We can see the experimental room with the two goals A and B (see plan Fig.5).

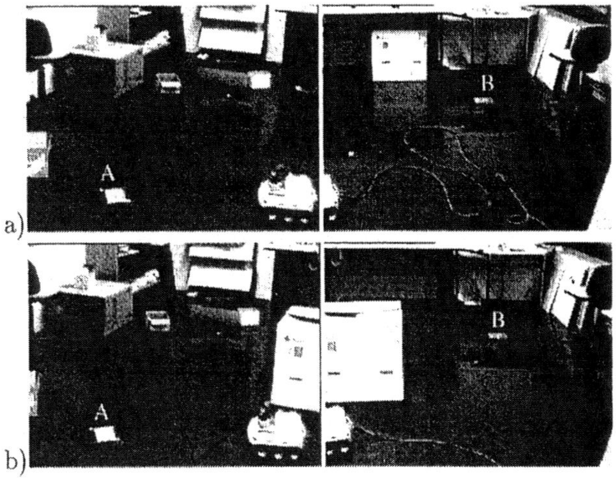

Figure 7 The experimental room, we can see the robot and the two goals A and B. On the image b we can see the introduced obstacle which is a landmark moved in the center of the experimental area.

[2] Pewatron sensor model 6070 for example

not use this information. As we can see on Figure 11 a) when the robot is far from the goal (3.5m) this one is invisible. Moreover the shortest distance to see the goal is 80cm (Fig.10 a), closer (40 cm Fig.10 b) the goal is no more visible due to the camera position on the robot body (and we have verified it has not been learned as a landmark).

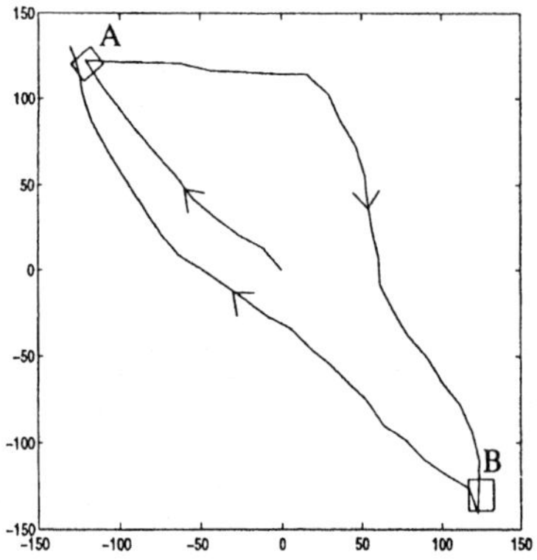

Figure 8 Path followed by the robot attracted by two kinds of supply locations (linked with two different motivations). The robot starts at the center of the room and it is more motivated by goal A than B. When it reaches A the corresponding motivation is decreased and the robot is attracted by B. Finally at B the motivation besomes once again A.

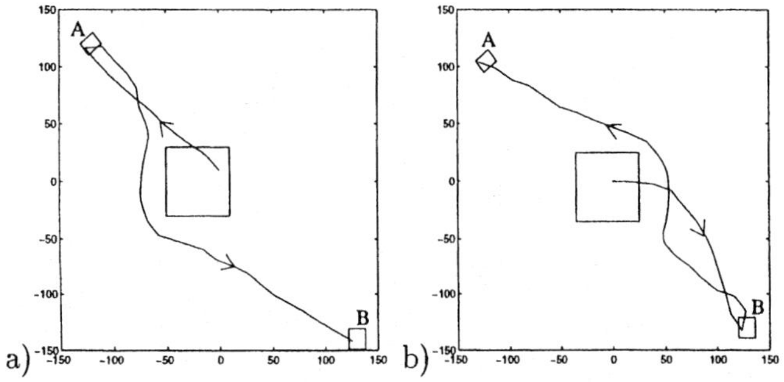

Figure 9 In these 2 examples the robot starts from the center of the room without obstacle. Then, when it reaches the goal associated to the most activated motivation, an obstacle is put at the center (the large box) and the robot haves to reach the other goal. In a) the first goal is A and in b) it is B.

To show that our system actually does not need to see the goal and can merge goal seeking and obstacle avoidance behaviors, we have introduced a huge obstacle (a $60cm \times 60cm \times 60cm$ box) in the center of the experiment area. Those experiments are shown in Figures 9 and 11, we can see that the goal is totally occluded by the obstacle. The learned positions are always the same, the robot starts without obstacle and when it reaches the first goal we move the obstacle at the center of the room. There the robot uses an obstacle overtaking strategy combined with our navigation model (like a Braittenberg vehicle (Braitenberg, 1984)). In fact the robot quits the goal reaching strategy and runs the obstacle following strategy until it succeeds in retrieving the desired orientation, given by the goal reaching strategy. Hence we avoid the local minima problems that appear when two contradictory behaviors must be performed at the same time. The figures 8 and 9 show that trajectories do not go to the goal directly but always reach it. We have done several other experiments with different starting points and starting orientations without any problem. If we want the robot to succeed from any position in the room the robot just haves to learn some positions all around each goal (in fact 6 positions around each goal are enough to have good performances, 3 positions all around can be sufficient).

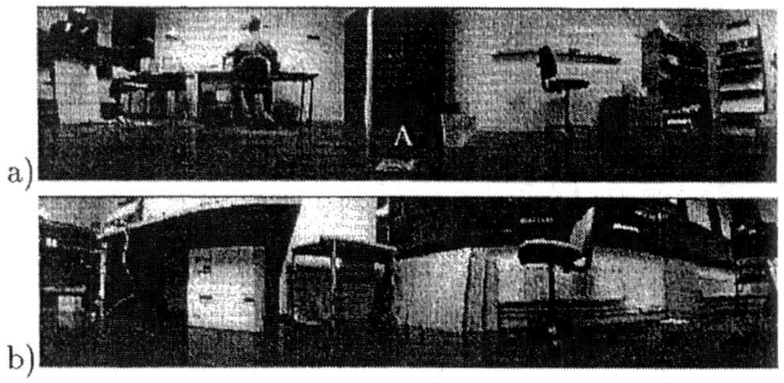

Figure 10 The robot is close to a goal. In image a) the robot is about 80 cm from the goal A, we can see it on the panoramic view. In image b) the robot is about 40 cm from the goal B, it is too close, we can not see the goal.

Figure 11 The robot is close to the goal. In image a) the robot is about 80 cm from the goal A, we can see it on the panoramic view. In image b) the robot is near the introduced obstacle which occults some of the learned landmarks.

3. Conclusion

First concerning the navigation algorithm, we can notice that none of the navigation experiments we have modelized and analyzed in this paper need a cartesian map of the environment or metric information. This makes it necessary for the system to have a mechanism to maintain an internal direction (proprioceptive information).

We think our level of modelization cannot account for the neurobiological details of each brain structure.

Nevertheless, it allows to test for the coherence of global brain models and to verify if two functional boxes can really be connected. Going back and forth between robotics and biology is full of promise. For instance, our robotic system will soon be improved by a better model of the hippocampus. Indeed, most of the time is spent in taking images to build a panoramic view (14 seconds to take the images and less than 1s to compute the movement). This problem can be easily solved if the merged representation of "what" and "where" information was not reset after each move. The robot could then only turn its head to verify old information in order to update it. Moreover, the scalability problem which appears when merging thousands of landmarks recognition with hundreds of possible azimuths, can also be solved by using a more realistic CA3 model. If the merging matrix is no more a hardwired "AND" operator but a limited set of neurons able to learn possible situations, the size of the matrix will no more directly depend on the possible landmark shapes. The computation complexity could then be reduced. At last, if the field of view of the robot is reduced to 180 degrees, the neurons that code the merged representantion of azimuths and landmarks (local snapshot) identification stop firing like "place cells" but have a "view cells" like activity (see Fig. 12) as shown in monkeys by Rolls (Rolls and S.M.O'Mara, 1995). Biological inspiration allows to realize very efficient robot controller and pattern recognition systems and at the same time robotic experimentation appears to be a very promising tool to test biological models in more real life conditions.

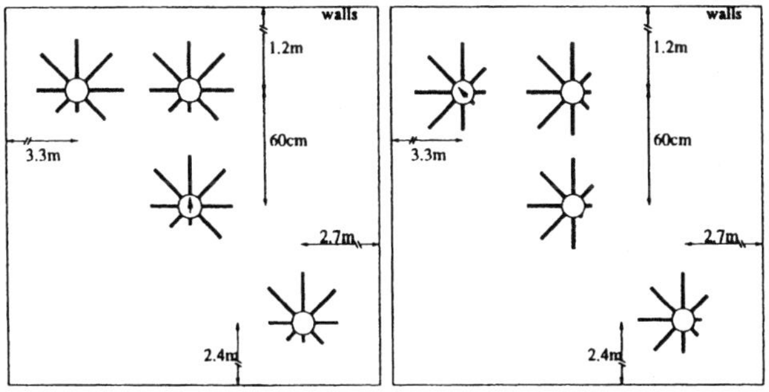

Figure 12 When the robot visual field is reduced "place cell" like activity turns into "view cell". Fig. a) and b) shows the activities of 2 views cells according to the robot location represented by circles. The arrow in one of the circles represents the position and the orientation used for the view learning. The directional vectors represent the neuron activity when the robot is facing a particular view.

References

Arkin, R. (1987). Motor schema-based mobile robot navigation. *International Journal of Robotics Research*, pages 92–112.

Bachelder, I. and Waxman, A. (1994). Mobile robot visual mapping and localization: A view-based neurocomputational architecture that emulates hippocampal place learning. *Neural Networks*, 7(6/7):1083–1099.

Banquet, J., Gaussier, P., Dreher, J., Joulain, C., and Revel, A. (1997). *Cognitive Science Perspectives on Personality and Emotion*, chapter Space-Time, Order and Hierarchy in Fronto-Hippocamal System: A Neural Basis of Personality. Elsevier Science BV Amsterdam.

Braitenberg, V. (1984). *Vehicles : Experiments in Synthetic Psychology*. Bradford Books, Cambridge.

Brooks, R. and Stein, L. (1994). Building brains for bodies. *Autonomous Robots*, 1:7-25.

Gaussier, P., Joulain, C., Moga, S., Quoy, M., and Revel, A. (1997a). Autonomous robot learning: What can we take for free ? In *International Symposium on Industrial Electronics - ISIE'97*, pages SS1–SS6, Gumarães, Portugal. IEEE.

Gaussier, P., Joulain, C., Revel, A., and Banquet, J. (1996). Are shaping techniques the correct answer for the control of visually guided autonomous robots ? In *UKACC International Conference on Control'96*, pages 1248–1253, University of Exeter. IEE.

Gaussier, P., Joulain, C., Zrehen, S., Banquet, J., and Revel, A. (1997b). Visual navigation in an open environement without map. In *International Conference on Intelligent Robots and Systems - IROS'97*, pages 545–550, Grenoble, France. IEEE/RSJ.

Gaussier, P. and Zrehen, S. (1994). Complex architectures for autonomous agents. In Gaussier, P. and Nicoud, J., editors, *PerAc*, pages 278–290, Lausanne. IEEE Press.

Gaussier, P. and Zrehen, S. (1995). Perac: A neural architecture to control artificial animals. *Robotics and Autonomous System*, 16(2-4):291–320.

Joulain, C., Gaussier, P., and Revel, A. (1997). Learning to build categories from perception-action associations. In *International Conference on Intelligent Robots and Systems - IROS'97*, pages 857–864, Grenoble, France. IEEE/RSJ.

Morris, R. (1981). Spatial localization does not require the presence of local cues. *Learning and motivation*, (12):239–260.

O'Keefe, J. (1989). *Neural Connections, Mental Computation*, chapter Computations the hippocampus might perform, pages 225–284. MIT Press.

Rolls, E. and S.M.O'Mara (1995). View-responsive neurons in the primate hippocampal complex. *Hippocampus*, (5):409–424.

Steels, L. (1995). A selectionist mechanism for autonomous behavior acquisition. *Robotics and Autonomous Systems*, 15.

Trullier, O., Wiener, S., Bethoz, A., and Meyer, J. (1997). Biologically based artificial navigation systems: review and prospects. *Progress in Neurobiology*, 51:483–544.

Zrehen, S. (1995). *Elements of Brain Design for Autonomous Agents*. PhD thesis, Swiss Federal Institute of Technology Lausane.

Zrehen, S. and Gaussier, P. (1997). Building grounded symbols for localization using motivations. In Husbands, P. and Harvey, I., editors, *Proceedings of the Fourth European Conference on Artificial Life ECAL97*, pages 299–308, Brighton, UK.

Landmark-Based Navigation for a Mobile Robot

Carl Owen and Ulrich Nehmzow
University of Manchester
Department of Computer Science
Manchester M13 9PL
United Kingdom
owenc,ulrich@cs.man.ac.uk

Abstract

In this paper we present a landmark based navigation mechanism for a mobile robot. The system uses a self-organising mechanism to map the environment as the robot is led around that environment by an operator. Detected landmarks, and their relative position towards each other, are recorded in a map that can subsequently be used to plan and execute paths from the robot's current location to a 'goal' location specified by the user.

The main motivation for the research described in this paper is to develop a mobile robot navigation system that is robust (through the use of perceptual landmarks), and allows the robot to plan *arbitrary* paths within its known environment. The system presented here achieves these objectives.

1. Introduction

Research in mobile robotics has produced two main approaches to modeling environments: metric and topological. In the first scheme, an accurate metric map of the robot's environment is either constructed by the robot or supplied by a human designer (Kampman and Schmidt, 1991; Knieriemen and von Puttkamer, 1991). In the latter approach the environment is modeled as a graph containing nodes representing distinct locations, pathways between locations are denoted by arcs connecting the appropriate nodes (Kurz, 1996; Yamauchi and Beer, 1996; Zimmer, 1996; Mataric, 1992).

A metric map has the advantage of being a simple and natural representation for human users. However, due to the amount of detail contained in such a representation, these maps are time consuming to construct, require large amounts of memory, and are often over specified for the task of general navigation.

Conversely, the topological approach gives a compact representation since only distinctive places within the environment are encoded. In addition, this type of map is well suited for use with the various path planning algorithms that have been developed within the field of artificial intelligence (e.g. A^*, Best First Search). One of the main problems with this method is *perceptual aliasing* i.e. distinct locations within the environment appearing identical to the robot's sensors.

Various approaches to the problem of perceptual aliasing have been utilized within the topological mapping paradigm. One method is to effectively increase the robot's perceptual resolution by adding and combining additional information from differing sensor modalities (sensor fusion). In (Kortencamp and Weymouth, 1994), for example, vision is used to augment the sonar data of the robot. However, this type of approach cannot be guaranteed to disambiguate all situations, and is more useful as a tool for reducing, rather than eliminating, perceptual ambiguity.

Other systems use positional information to disambiguate perceptually similar but physically distinct locations (Kurz, 1996; Zimmer, 1996; Mataric, 1992). However, since positional information based on the robot's internal odometry is subject to drift effects, some means of correction is required if the robot is to map anything other than small scale environments. In (Kurz, 1996), for example, an extended Kalman-filter is used for drift compensation.

The system described in this paper constructs a topological map of the environment based on a process of self-organisation of the robot's sensory data. The approach of self-organisation to landmark detection was chosen for two reasons. Firstly, interpreting the world using the robot's relatively impoverished sensors is difficult for the human designer. Thus, user defined landmarks tend to be rather simplistic. The type of environment that can be categorised using this method is therefore restricted. Secondly, since the clustering techniques used in self-organisation enable a generalisation over perceptions, this approach gives a robust, noise tolerant, method of landmark detection.

Global positioning information does not form part of the representation of the system described here, thus obviating the need for drift compensation. An exploration strategy is used to resolve perceptual ambiguity.

The paper is structured as follows: Section 2. deals with the navigation mechanism. In this section the implementation details of the system are discussed. Section 3. sets out the results of several experiments conducted within the robotics laboratory of Manchester University. This section also details a performance metric that was used to measure the robot's navigational ability. Section 4. draws conclusions and discusses further work.

2. The navigation system

The navigation mechanism consists of two main processes:

1. Mapbuilding - construction of a topological vector map.
2. Map interpretation - path planning and execution.

These two processes are performed consecutively, i.e. the mapbuilding process is fully completed before map interpretation begins.

The robot used for our experiments was a Nomad 200 mobile robot (see figure 1). This robot is equipped with sixteen ultrasonic range finding sensors (range up to 6.5 m), sixteen infrared (IR) sensors (range up to 60 cm), twenty tactile sensors, a compass and a monochrome CCD camera. In the experiments described here, only the sonar sensors and compass were used.

Figure 1 The Nomad 200 mobile robot.

2.1 Mapbuilding

The concept of the vector map is depicted in figure 2.

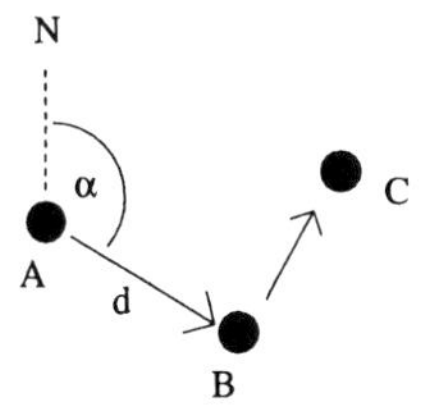

Figure 2 The vector map. Three points 'A', 'B' and 'C' represent particular perceptual landmarks within the environment, the arcs connecting the nodes record the compass direction, α, and distance, d, between these landmarks.

By employing a vector mapping of this type we restrict the use of odometry to the measurement required between locations, thus limiting the accumulation of odometry error.

In previous work, a Self-Organising Feature Map (SOFM) (Kohonen, 1988) has been used to map the robot's environment. Using this method, mapbuilding is achieved by performing an unsupervised clustering of the sensory data of the robot as it moves through the environment (Nehmzow and Smithers, 1991; Owen and Nehmzow, 1996). When the network has settled, regions on its surface represent 'perceptual landmarks' within the robot's environment.

However, the requirement for a settlement period is relevant to the issue of 'lifelong learning'. If the environment were to change after training then a further period of settlement would be required before vector links could be added. In addition, the size of the network needs to be pre-defined before use. Determination of optimal network size is difficult since no indication of the number of landmarks that the robot will perceive can be gained by simply observing the environment in which the robot will operate. Choosing too small a network would result in saturation, and a highly unstable network. On the other hand, the larger the network the more expensive the algorithm becomes at lookup time.

Other network methods that have been used in the context of landmark detection include the Restricted Coulomb Energy (RCE) classifier (Reilly et al., 1982), used, for example, by Kurz (Kurz, 1996), and the Adaptive Resonance Theory (ART) classifier (Carpenter and Grossberg, 1987), used, for instance, by Duckett and Nehmzow (Duckett and Nehmzow, 1997). Neither of the latter methods require a settlement period to achieve a stable clustering of the input space (thus allowing immediate placement of vector links), and both are able to grow to accommodate new classifications.

Since, as discussed, pre-determination of network size is difficult, and lifelong learning is seen as a future development for this system, it was decided that one of the growing networks would be used as the basis of the mapping mechanism. The RCE model was chosen in preference to the ART classifier since the latter requires a large number of parameters to be set before use, the values of which can only be determined experimentally.

Although the SOFM is better able to generalise on noisy inputs, smoothing of the input vector can reduce noise (see below), and in practice the RCE network was found to produce a stable clustering of the robot's perceptual space.

The clustering mechanism The RCE-Classifier uses a method of classification based on self-organisation. Each class is represented by a *representation vector* (R-vector). Training the RCE-Classifier involves determining the R-vectors.

When a pattern is presented to the classifier, the input is compared to each of the existing R-vectors, using some form of similarity measure (e.g. dot product) in order to determine a winner (i.e. the R-vector of highest similarity). If the similarity between the input pattern and the winning R-vector is within a predetermined threshold then the input pattern belongs to the class of this winning R-vector. If the similarity is outside the threshold then the input pattern becomes a new R-vector. Thus the boundaries of classes are determined by the nearest neighbour law. Figure 3 shows an example for a two-dimensional input vector.

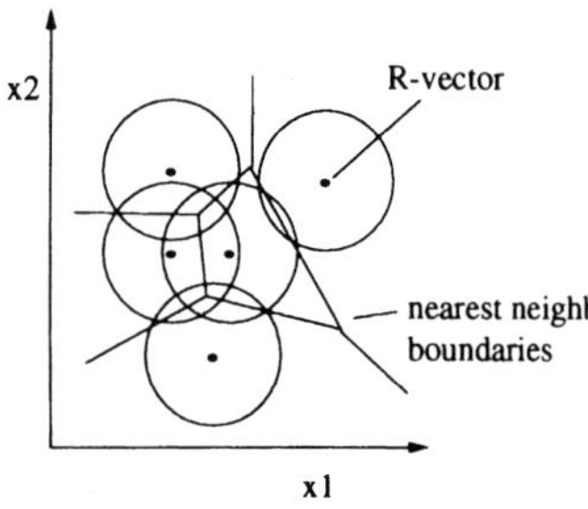

Figure 3 RCE-Classifier, two-dimensional example. Each 'dot' in the diagram represents an R-vector. The circles surrounding each R-vector denote the 'threshold area' within which an input pattern must fall in order to belong to the corresponding R-vector's class. In the case of patterns falling within more than one threshold area, the nearest neighbour law applies.

The input vector The input vector in this application consists of the sixteen readings of the robot's sonar sensors. This input vector is normalised and the dot product of the input and R-vector is used as the measure of similarity for the RCE-classifier. The threshold for adding new R-vectors was fixed at 0.9.

In using the RCE-classifier to process the input vector, perceptual features within the environment are determined autonomously by the robot; each class generated by this method can be viewed as a 'perceptual landmark' within the environment.

Since landmarks in this scheme are determined by the sensory patterns obtained by the robot, perception will be affected by the direction in which the sensors are facing. This means that a location within the environment could be perceived differently when approached from different directions. In order to overcome this problem the robot's on board compass is used to align the turret (which contains the sensors) to compass north at all times. Thus, the input vector generated by the system will depend on position alone, regardless of steering orientation.

Since sonar sensor signals are prone to noise, some form of preprocessing is desirable. In these experiments a smoothing procedure is applied by delaying any large change in the input data by a pre-defined number of time steps, with the reading in question simply taking on its old value for the duration of this period. If the new data value still holds on completion of the allotted interval, it is then allowed to pass through to the classifier as part of the input vector. The purpose of this mechanism was to remove "spikes" from the input data. Both the amount of change defined as large, θ, and the number of time steps to skip, γ, were determined experimentally by observing data time series from several locations within the computer science building.

Building the vector map The vector map is built by the robot as it is led around the environment by the user. Movement of the robot is restricted to forward or turn (these commands cannot be carried out simultaneously), and the robot travels at a constant speed of 4 inches/sec.

As the robot moves forward it continuously takes in all 16 sonar readings and processes them using the RCE-classifier, with preprocessing as detailed above, to generate the 'perceptual landmarks'. Each time the perception changes the robot takes note of the distance travelled since the start of the last perception. As some of these landmarks can occupy very small areas, and as such can be easily missed on subsequent visits to the same location, only landmarks that persist for a pre-determined minimum distance are considered for the purpose of navigation. In the results detailed here, only landmarks that are visible over a travelling distance of over 4 inches are added to the vector map.

Each place node of the vector map consists of a perception number and a list of the places connected to that node. For each connected place, details of compass direction, distance between the centres of the nodes, and size of the place node's influence are stored.

Each time a place node is allocated it is added to the list of connections for the previous node. In addition, this previous node's details are added to the current node's list (i.e. each link is recorded bi-directionally). Figure 4 shows an example of three connected place nodes and the corresponding vector map.

Storing the size of the place node's influence area allows the robot to travel to the centre of this node once found. Note that this size can be multiply defined for one particular place node, this is necessary since the size of a node's influence will depend on the direction from which it is approached (see figure 5). The distance measurement gives an upper bound on distance to travel in searching for the connected place.

Perceptual aliasing The system, as proposed so far, takes no account of perceptual aliasing (identical sensory perceptions at different locations). As such, distinct locations within the environment that have similar perceptual signatures may excite the same place node. One solution to this problem would be to allow the robot to explore the environment on finding a location whose perception has previously been encountered. Exploration in this instance would involve following the links given for each place node that shares the current perception. If,

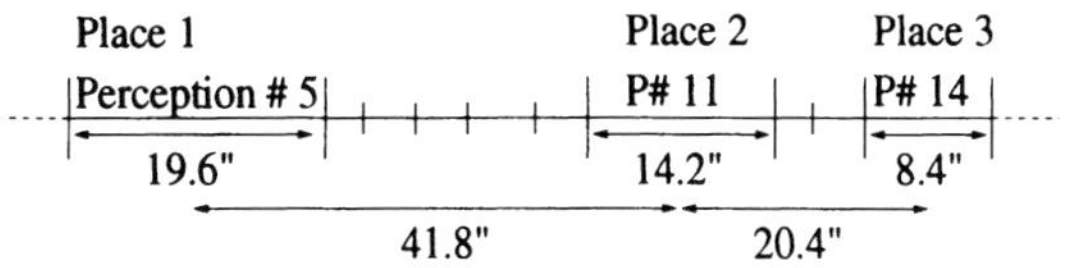

Place 1	Place 2	Place 3
Perception number: 5	Perception number: 11	Perception number: 14
Number of links: 2	Number of links: 2	Number of links: 2
Link 0:	Link 0:	Link 0:
Place: 0	Place: 1	Place: 2
Direction: 31	Direction: 29	Direction: 26
Distance: 338	Distance: 418	Distance: 204
Size: 224	Size: 196	Size: 142
Link 1:	Link 1:	Link 1:
Place: 2	Place: 3	Place: 4
Direction: 211	Direction: 209	Direction: 207
Distance: 418	Distance: 204	Distance: 150
Size: 142	Size: 84	Size: 144

Figure 4 Three connected place nodes and the corresponding vector map. Start and end points for each perception (as given by the RCE-classifier) are denoted by vertical lines along the path. Here only three 'landmarks' (labelled *5*, *11* and *14*) are persistent enough to warrant place node generation, these landmarks have been given the corresponding place labels, *1*, *2* and *3* in the vector map. Distance and size in the vector map are measured in tenths of inches.

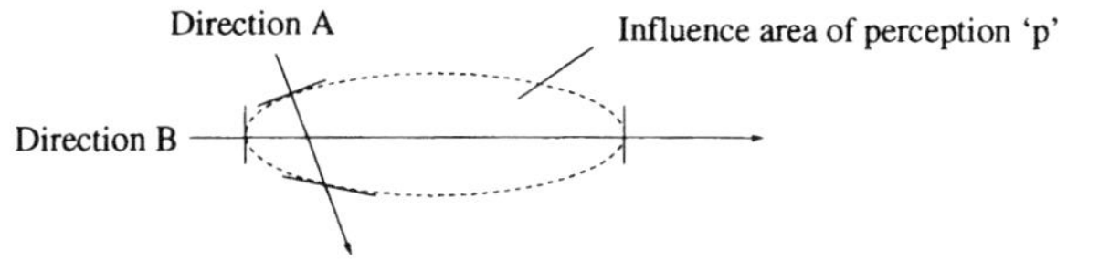

Figure 5 An example of multiply defined place size. The dotted ellipse indicates the influence area of perception 'p'. The arrows indicate two different directions from which this perception was encountered, the 'size' of 'p' measured for direction *A* will be smaller than that for *B*.

for a given place node, the correct perception is encountered along each of the links, then the system assumes that this place has been re-encountered and adds new links accordingly. Otherwise, a new place node is created. An exception to this behaviour is made when the robot moves from a previously encountered place node along a link already contained in that node's link list; if the place found along this link is as expected then no exploration is performed (i.e. the robot uses prediction when traversing previously followed paths). We call this strategy ***incremental mapbuilding***, and this is the mechanism used in the experiments described here.

2.2 Map interpretation

The object of the map interpretation phase is to plan a route from the current location to an arbitrary, externally specified goal location using the vector map. The basis of the planning mechanism is the 'best-first search' algorithm, which is used in this instance to determine the *shortest* known path between the current location and the goal location.

Path following Once a path to the goal has been generated the robot's path following task can begin. For this task the robot takes each node in turn and attempts to find the next node along the path by taking the trajectory indicated on that path (the goal and starting location being provided by the user).

If the robot fails to find a given node then it reverses to the centre of the node on the path that was last identified. From here a new path is calculated with this node as the start position.

The goal location is only assumed to have been found when it appears in the right context according to the planned path (i.e. if the robot happens to encounter a place with the same perception as the goal before it appears on the planned path, then it is assumed *not* to be the goal location).

3. Experimental results

In this section we present the results of testing carried out with the system in a simple environment set up within the robotics laboratory at Manchester. Figure 6 shows the environment used in the experiments, the route along which the robot was guided through the environment (chosen arbitrarily), and the corresponding map generated by the system.

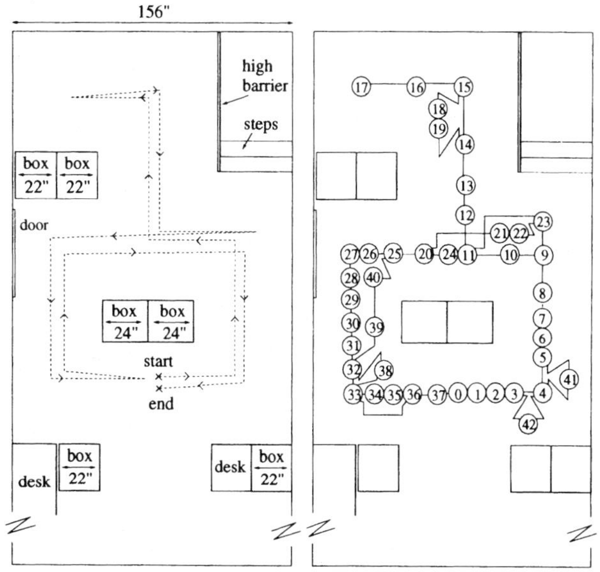

Figure 6 The left diagram shows the environment used in the experiments, the dotted line depicts the path along which the robot was guided. The corresponding map generated by the system is shown on the right (each circled number indicates a place node on the map).

For this experiment four different goal and start positions were chosen from the environment so as to cover the area traversed by the robot during mapbuilding. The results for each of these trials are given below. In each diagram the shortest route from start to goal is shown, along with any alternative routes taken by the robot. Each of the four trials was completed ten times by the robot.

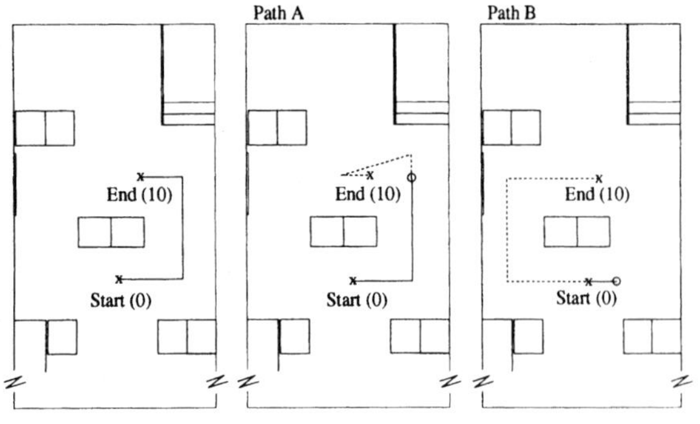

Figure 7 Trial 1. The figure on the left shows the shortest path from start location to goal location. In 8 out of 10 attempts the robot traversed this path successfully. The other two figures show two alternative routes taken by the robot due to failure to locate place nodes along the shortest path. The point at which failure occurred is indicated by a circle, the dotted line denotes the new path.

Trial 1 (figure 7): In 8 out of the 10 attempts made by the robot, the shortest route was traversed successfully. For the remaining two attempts, the robot failed to locate one of the place nodes on each occasion, thus initiating a new path planning phase. In both cases the robot was able to follow the new path successfully to the goal location (see figure 7).

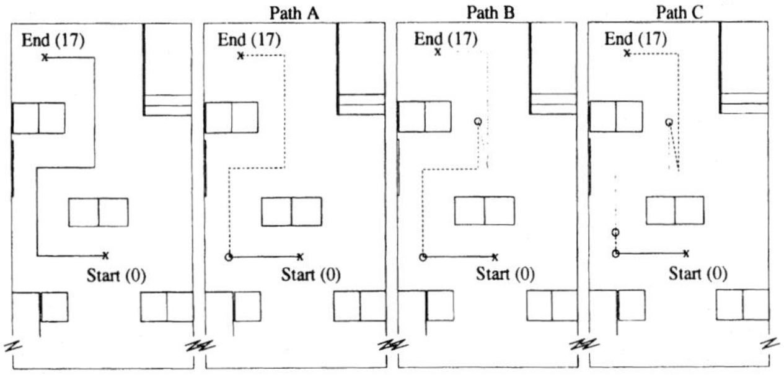

Figure 8 Trial 2. The figure on the left shows the shortest path from start location to goal location. In 6 out of 10 attempts the robot traversed this path successfully. In two of the remaining cases only one new path needed to be generated (see path A). Path A looks identical to the shortest route due to the fact that re-planning was possible via alternative nodes along this route. Of the remaining two attempts, one required two new paths to be generated (path B), whilst the other required three (path C). Each new path is indicated by a dotted line with a circle at the start point.

Trial 2 (figure 8): In 6 out of the 10 attempts for this trial, the shortest route was traversed successfully. Of the remaining four attempts, two required one new path generating phase in order for the robot to locate the goal, one required two and the other required three (see figure 8).

Trial 3 (figure 9): In this trial, for all 10 attempts the robot successfully traversed the shortest path.

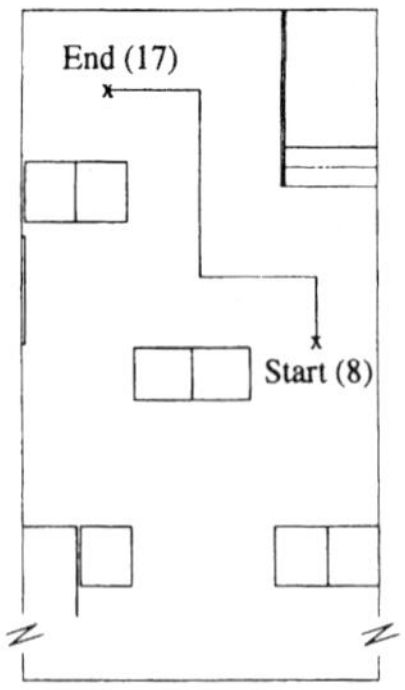

Figure 9 Trial 3. This figure indicates the shortest path from start to goal for this trial. In all 10 attempts this path was traversed successfully.

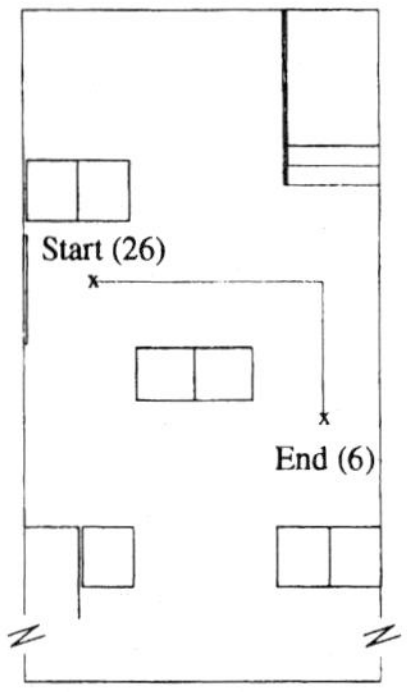

Figure 10 Trial 4. This figure indicates the shortest path from start to goal for this trial. In all 10 attempts this path was traversed successfully.

Trial 4 (figure 10): As in trial 3, for all 10 attempts the robot successfully traversed the shortest path.

As can be seen from trials 1 and 2, the robot is not always able to find a place at the location given by the map. This phenomenon can be particularly prominent at locations near to junctions or cluttered areas where even small deviations in the robot's position and/or orientation can substantially alter the robot's perception. However, the effects can be alleviated somewhat by extending the training period, thus giving opportunity for acquiring alternative perceptions for locations where perception is problematic. By taking the robot repeatedly to the same physical location, path planning is facilitated using alternative perceptions associated with that location.

3.1 A performance metric

In order to gain some quantitative measure of the robot's performance in its navigation task, we defined a metric, *distance efficiency.* To obtain this metric, the length of a particular run made by the robot is measured as a percentage of the shortest possible distance between the start and goal locations. The results for the distance efficiency metric are shown in figure 11.

From this graph we can see that trials 1 and 2 were problematic for the robot. As discussed in section 3., this

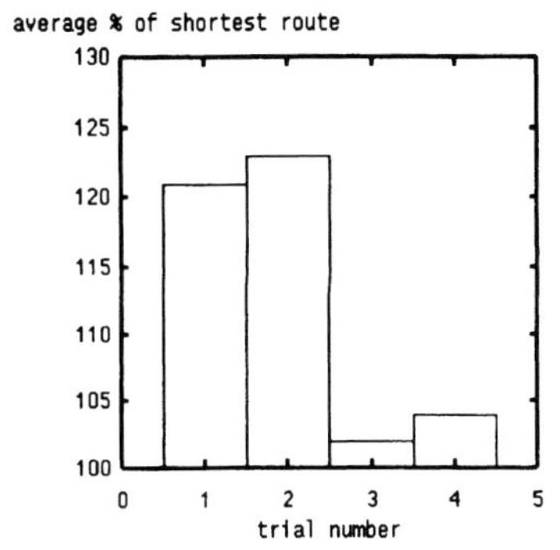

Figure 11 Distance efficiency. For each trial the total shown is an average over 10 traversals from start to goal location.

was due to the robot's inability to locate certain nodes along the generated path, thus resulting in new path planning and route traversal phases.

A particularly poor performance in trial 1 is due to the fact that, on one occasion, an alternative route along the shortest path was not available, thus requiring a lengthy detour in order to reach the goal (see path B, figure 7). In trial 2, replanning was required in four out of the ten attempts, with certain nodes proving consistently difficult to detect (see figure 8).

4. Conclusions and further work

In this paper we have presented a navigation system for a mobile robot, and tested it experimentally on a Nomad 200 mobile robot. The system uses a self-organising mechanism to build a representation of the environment based on landmark recognition. The representation takes the form of a vector map whereby recognised locations within the environment are linked together by arcs denoting direction and distance between them. Exploration is used both to disambiguate perceptually similar locations, and to recognise places that have been previously visited. Global positional information is not required. A metric was introduced (Distance Efficiency) and used to measure the robot's navigational behaviour.

In order for a robot navigation system to be useful it must operate in real world situations (Owen and Nehmzow, 1997). Much of the work to date in the field of mobile robot navigation has been conducted in small scale laboratory environments - the question arises of how these systems will scale up when faced with more complex environments covering much larger distances. Large scale experiments with the navigation mechanism described here are currently underway at Manchester University.

Our results show that the robot is not always able to find a place at the location indicated by the map. Rather than re-plan an alternative route immediately on failure to find a place node, a better approach would be to make a more concentrated effort to locate the lost node. To do this, one could simply re-attempt to locate the node along the trajectory on which failure occurred, or perform a more systematic search strategy. In doing this, lengthy detours such as the one depicted in figure 7 (path B) might be avoided. In addition, this behaviour gives the robot the opportunity to reach the goal where no alternative routes are available.

Although not tested in these experiments, one can assume that the system has some ability to cope with *dynamic environments.* If, for example, a path becomes blocked then the perception near the blocked position will change, thus forcing the robot to plan an alternative route (the perception will not be as expected). This property needs to be tested experimentally. In addition, some way to store information regarding such environmental change would be useful.

With the present system, mapbuilding and navigation are carried out as two separate phases. For dynamic environments this approach is unsuitable since changes to the robot's surroundings will introduce errors into the map. A more appropriate strategy would be to intersperse mapbuilding with map interpretation (lifelong learning). With this approach the robot would continue to map new information as it attempts to follow a path.

References

aisb97 (1997). *AISB-97 Workshop on Spatial Reasoning in Mobile Robots and Animals*, Technical Report Series UMCS-97-4-1, ISSN 1361-6161, http://www.cs.man.ac.uk/csonly/cstechrep/titles97.html. University of Manchester.

Carpenter, G. and Grossberg, S. (1987). Art2: Self-organisation of stable category recognition codes for analog input patterns. *Applied Optics*, 26(23):4919–4930.

Duckett, T. and Nehmzow, U. (1997). Experiments in evidence-based localisation for a mobile robot. In (aisb97, 1997).

Kampman, P. and Schmidt, G. (1991). Indoor navigation of mobile robots by use of learned maps. In (Schmidt, 1991).

Knieriemen, T. and von Puttkamer, E. (1991). Real-time control in an autonomous mobile robot. In (Schmidt, 1991).

Kohonen, T. (1988). *Self Organisation and Associative Memory.* Springer Verlag, Berlin, Heidelberg.

Kortencamp, D. and Weymouth, T. (1994). Topological mapping for mobile robots using a combination of sonar and vision sensing. In *Proceedings of the AAAI.*

Kurz, A. (1996). Constructing maps for mobile robot navigation based on ultrasonic range data. *IEEE Transactions on Systems, Man, and Cybernetics–Part B: Cybernetics*, 26(2):233–242.

Mataric, M. J. (1992). Integration of representation into goal-driven behaviour-based robots. *IEEE Transactions on Robotics and Automation*, 8(3).

Nehmzow, U. and Smithers, T. (1991). Mapbuilding using self-organising networks. In Arcady Meyer, J. and Wilson, S., editors, *From Animals to Animats*, Proceedings of SAB90 Paris, pages 152–159. MIT Press, Cambridge Mass. and London, England.

Owen, C. and Nehmzow, U. (1996). Route learning in mobile robots through self-organisation. In *Euromicro workshop on advanced mobile robots.* IEEE Computer Society, ISBN 0-8186-7695-7.

Owen, C. and Nehmzow, U. (1997). Middle scale robot navigation - a case study. In (aisb97, 1997).

Reilly, D. L., Cooper, L. N., and Erlbaum, C. (1982). A neural model for category learning. *Biological Cybernetics 45*, pages 35–41.

Schmidt, G., editor (1991). *Information Processing in Autonomous Mobile Robots.* Springer Verlag, Berlin, Heidelberg, New York.

Yamauchi, B. and Beer, R. (1996). Spatial learning for navigation in dynamic environments. *IEEE Transactions on Systems, Man, and Cybernetics–Part B: Cybernetics*, 26(3):496–505.

Zimmer, U. R. (1996). Robust world modelling and navigation in a real world. *Neurocomputing*, 13(2-4):247–260.

A Subsymbolic+Symbolic Model for Learning Sequential Navigation

Ron Sun, Todd Peterson
University of Alabama
Tuscaloosa, AL 35486, USA
{rsun, todd}@cs.ua.edu

Abstract

To deal with reactive sequential decision tasks, we present a learning model CLARION, which is a hybrid connectionist model consisting of both localist and distributed representations, based on the two-level approach proposed in Sun (1995). The model learns and utilizes procedural and declarative knowledge, tapping into the synergy of the two types of processes. It unifies neural, reinforcement, and symbolic methods to perform on-line, bottom-up learning. Experiments in various situations are reported that shed light on the working of the model.

1. Introduction

This paper presents a hybrid model that unifies neural, symbolic, and reinforcement learning into an integrated architecture. It addresses the following three issues: (1) It deals with concurrent on-line learning: It allows a situated agent to learn continuously from ongoing experience in the world, without the use of preconstructed data sets or preconceived concepts. (2) The model learns not only low-level specific skills but also high-level generic (declarative) knowledge (which is beyond traditional reinforcement learning algorithms as will be discussed later). (3) The learning is bottom-up: generic knowledge is acquired from an agent's experience interacting with the world through the mediation of low-level skills. This differs from top-down learning in which low-level knowledge is acquired through "compiling" mostly externally given high-level knowledge (Anderson 1983). Although the essential motivation for this model is cognitive modeling, this paper will focus only on computational experiments.

Reactive sequential decision tasks (Sun and Peterson 1995) involve selecting and performing a sequence of actions, step by step, on the basis of the current state, or the moment-to-moment perceptual information (hence the term "reactive"). At certain points, the agent may receive *payoffs* or *reinforcements* for their actions performed at or prior to the current state. The agent may want to maximize the total payoffs. Thus, the agent may need to perform *credit assignment*, to attribute the payoffs/reinforcements to actions at various points in time (the temporal credit assignment problem), in accordance with various aspects of a state (the structural credit assignment problem). There is in general no teacher input. The agent starts with little or no a priori knowledge. One example involves learning to navigate through mines (see Figure 1).

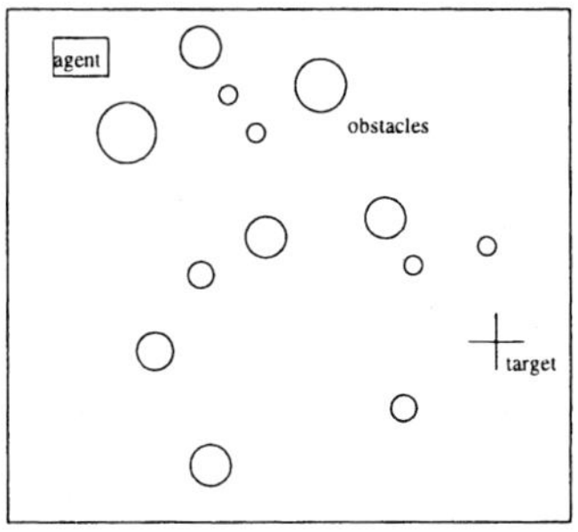

Figure 1 Navigating Through A Minefield

To acquire low-level specific skills in these tasks, there are some existing methods available. Chief among them is the temporal difference method (Sutton 1990), a type of reinforcement learning that learn through exploiting the difference in evaluating actions in successive steps and thus handling sequences in an incremental manner. This approach has been applied to learning in mazes, navigation tasks, and robot control (Sutton 1990, Lin 1992). But they do not learn generic knowledge (generic rules). This approach can be extended to dynamic programming and partially observable Markov decision process models; however, these models often require a domain model to begin with.

In terms of learning generic knowledge or generic rules for such tasks, however, the characteristics of the task render most existing rule learning algorithms inapplicable. This is because they require either preconstructed exemplar sets (thus learning is not online; Michalski 1983, Quinlan 1986), incrementally given *consistent* instances (Mitchell 1982, Fisher 1986, Utgoff 1989), or complex manipulations of learned structures when inconsistency is discovered (which is typically more complex than the limited time a reactive agent may have; Hirsh 1994). Non-stationary ("drifting") worlds are more than noise and inconsistency as considered by some learning algorithms because they involve changes over time and lead to radical changes in learned knowledge. This often requires extra dedicated mechanisms. Above all, most of the rule learning algorithms do not handle the learning of sequences which necessarily involves temporal credit assignment.

2. Hybrid Models

How can an agent develop a set of skills that are highly specific (geared towards particular situations) and thus highly efficient but, at the same time, acquire sufficiently general knowledge that can be readily applied to a variety of different situations and be communicated to others? Although humans seem to possess such abilities and be able to achieve an appropriate balance between the two sides, existing systems fall short. What appears to be missing is the duality and coexistence of both procedural and declarative knowledge (or both specific skills and generic knowledge). There has been a great deal of work demonstrating the difference between procedural knowledge and declarative knowledge: for example, Anderson (1983), Keil (1989), and Sun (1994). It is believed that a balance between the two is essential to the development of complex cognitive agents. This is based on two lines of argument. First, there are ample psychological data that support the distinction between procedural and declarative knowledge and the need for both. Anderson (1983) initially proposed the distinction based on such data; Fitts and Posner (1967), Keil (1989), and Sun (1994) subsequently made similar points also based on psychological data. Second, there are many philosophical arguments for making this distinction and achieving an appropriate balance between the two. Dreyfus and Dreyfus (1987) proposed the distinction between analytical and intuitive thinking; Smolensky (1988) proposed the distinction between conceptual (publicly accessible) and subconceptual processing; in addition, the distinction between conscious and subconscious processes, although controversial, is well known (cf. James 1890). We are not aiming to capture all of the above dichotomies. Denoting more or less the same thing, these dichotomies serve as justifications for our main distinction, between (conceptual) declarative knowledge and (subconceptual) procedural skills, which is distinguished by the difference in accessability (declarative knowledge is easily accessable and linguistically expressible, procedural skills are not; Anderson 1983, Sun 1995).

Generic declarative knowledge has some advantages which make it indispensable to a learning agent despite the fact that procedural skills are more efficient or easier to learn. These advantages are:

- It helps to guide the exploration of new situations, and reduces the time necessary to develop specific skills in new situations. In other words, it helps the transfer of learned skill (as shown psychologically by Willingham et al 1989).
- Generic knowledge can help to speed up learning. If properly used, generic knowledge that is extracted on-line during skill learning can help to facilitate the learning process itself (Willingham et al 1989).
- Generic knowledge can also help in communicating learned knowledge and skills to other agents.

A two-level hybrid model provides the needed framework for representing both types of knowledge. CLARION is similar to the model in Sun (1995) but is specifically designed for reactive sequential decision making. It consists of two levels: The bottom level contains procedural knowledge (Anderson 1983) and the top level contains declarative knowledge in the form of propositional rules. An overall pseudo-code algorithm that describes the operation of CLARION is as follows:

1. Observe the current state x.
2. Compute in the bottom level the Q-values of x associated with each of all the possible actions a_i's: $Q(x, a_1)$, $Q(x, a_2)$, ..., $Q(x, a_n)$.
3. Find out all the possible actions (b_1, b_2, ..., b_m) at the top level, based on the input x and the rules in place.
4. Compare the values of a_i's with those of b_j's, and choose an appropriate action b
5. Perform the action b, and observe the next state y and (possibly) the reinforcement r.
6. Update the bottom level in accordance with *Q-Learning.*
7. Update the top level with *Rule-Extraction-Revision.*
8. Go back to Step 1.

2.1 The Bottom Level

Reinforcement learning. A Q-value is an evaluation of the "quality" of an action in a given state: $Q(x, a)$ indicates how desirable action a is in state x (which consists of sensory input). To acquire the Q-values, we use the *Q-learning* algorithm (Watkins 1989; a temporal difference reinforcement learning algorithm). In the algorithm, $Q(x, a)$ estimates the maximum discounted cumulative reinforcement that the agent will receive from the current state x on. The updating of $Q(x, a)$ is based on minimizing $r + \gamma e(y) - Q(x, a)$, where γ is a discount factor, $e(y) = \max_a Q(y, a)$, and y is the new state resulting from action a. Thus, the updating is based on the *temporal difference* in evaluating the current state and the action chosen. Through successive updates of the Q function, the agent can learn to take into account future steps in longer and longer sequences notably without explicit planning (Watkins 1989).

Implementation. To implement Q-learning, we chose to use a four-layered network, in which the first three layers form a backpropagation network for computing Q-values and the fourth layer (with only one node) performs stochastic decision making. The output of the third layer (i.e., the output layer of the backpropagation network) indicates the Q-value of each action (represented by an individual node), and the node in the fourth layer determines probabilistically the action to be performed based on a Boltzmann distribution (Watkins 1989). The lookup table implementation of Q-learning is out of question here because of the (likely) continuous input space and when discretized, the resulting huge state space (e.g., in the navigation task there are more than 10^{12} states). Some kind of function approximator has to be used (Sutton 1990, Lin 1992, Tesauro 1992) and as a result the convergence of learning is not guaranteed.

2.2 The Top Level

Rule encoding. Declarative knowledge is captured in a simple propositional rule form. Although we can directly use a symbolic rule representation, to facilitate correspondence with the bottom level and to encourage uniformity and integration, we chose to use a localist connectionist model instead. Basically, we connect the nodes representing the conditions of a rule to the node representing the conclusion (Sun 1992, Towell and Shavlik 1993). That is, we directly translate the structure of a rule set to that of a network. For more complex rule forms including predicate rules and variable binding, see Sun (1992).

Rule learning. We devised a novel rule learning algorithm because of the differing characteristics of our tasks: concurrent (on-line) learning, reactivity, lack of teacher input and a priori knowledge, "drifting" environments, and requisite revisions. The basic idea is as follows: we perform rule learning (extraction and subsequent revision) at each step, which is associated with the following information: (x, y, r, a), where x is the state before action a is performed, y is the new state entered after an action a is performed, and r is the reinforcement received after action a. If some action decided by the bottom level is successful then the agent extracts a rule that corresponds to the decision and adds the rule to the rule network. Then, in subsequent interactions with the world, the agent verifies the extracted rule by considering the outcome of applying the rule: if the outcome is not successful, then the rule should be made more specific and exclusive of the current case; if the outcome is successful, the agent may try to generalize the rule to make it more universal. Rules are in the following form: *conditions* $\longrightarrow$ *action*, where the lefthand side is a conjunction of individual conditions each of which refers to a primitive: a value range or a value in a dimension of the (sensory) input state.

Let us look into a refined version of the overall algorithm that fleshes out some rule learning details (in steps 8 and 9).

Do the following for each decision step:
1. Observe the current state x.
2. Calculate the Q-values of all the actions in the current state: $Q(x, a_1), Q(x, a_2), ..., Q(x, a_n)$
3. Find the set of rules that matches the current state. Choose a rule based on specificity (choose at random if more than one rule has the same highest specificity). (The action suggested by the chosen rule has an activation value 1.)
4. Combine the outcomes of the previous two steps: Based on the values of actions from the two levels, select an action b to be performed using either the *percentage combination* or the *stochastic combination*.
5. Perform the action b.
6. Enter the new state y resulting from the action. Possibly receive a reinforcement r.
7. Calculate all the Q-values for the new state: $Q(y, a_1), Q(y, a_2), ..., Q(y, a_n)$
8. Check the current criterion for rule extraction and revision. Update statistics (with respect to the action performed).
8.1. If the result is good according to the current criterion, and there is no rule matching that state and that action, then perform *extraction* of a new rule: state $\longrightarrow$ action. Add the extracted rule to the rule network.
8.2. If the result is bad according to the current criterion, revise all the matching rules using *shrinking* and *deletion*.
8.2.1. Remove the matching rules from the rule network.
8.2.2. Add the revised versions of the rules into the rule network.
8.3. If the result is good according to the current criterion, then generalize the matching rules.
8.3.1. Create new rules using *expansion*.
8.3.2. Add the expanded rules to the rule network to replace the original rules.
9. Perform *merge* to combine existing rules.

At each step, we update the following statistics for each rule condition and each of its minor variations (i.e., the rule condition plus/minus one value), with regard to the action a performed: that is, PM_a (i.e., Positive Match) and NM_a (i.e., Negative Match). Here, positivity/negativity is determined by the following inequality: $\max_b Q(y, b) - Q(x, a) + r > threshold$, which indicates whether or not the action is reasonably good. Based on these statistics, we calculate the information gain measure; that is,

$$IG(A, B) = log_2 \frac{PM_a(A) + 1}{PM_a(A) + NM_a(A) + 2} - log_2 \frac{PM_a(B) + 1}{PM_a(B) + NM_a(B) + 2}$$

where A and B are two different conditions that lead to the same action a. The measure compares essentially the percentage of positive matches under different conditions A and B (with the Laplace estimator; Lavrac and Dzeroski 1994). If A can improve the percentage to a certain degree over B, then A is considered better than B. In the algorithm, if a rule is better compared with the match-all rule (i.e, the rule with the condition that matches all inputs), then the rule is considered successful (for the purpose of deciding on expansion or shrinking operations).

We decide on whether or not to extract a rule based on a simple success criterion which is fully determined by the current step (x, y, r, a):

- *Extraction*: if $r + \gamma e(y) - Q(x, a) > threshold$, where a is the action performed in state x and y is the resulting new state (that is, if the current step is successful), and if there is no rule that covers this step in the top level, set up a rule $C \longrightarrow a$, where C specifies the values of all the input dimensions exactly as in x.

The criterion for applying the *expansion* and *shrinking* operators, on the other hand, is based on the aforementioned statistical test. Expansion amounts to adding an additional value to one input dimension in the condition of a rule, so that the rule will have more opportunities of matching inputs, and shrinking amounts to removing one value from one input dimension in the condition of a rule, so that it will have less opportunities of matching inputs. Here are the detailed descriptions of these operators:

- *Expansion*: if $IG(C, all) > threshold1$ and $\max_{C'} IG(C', C) \geq 0$, where C is the current condition of a matching rule, *all* refers to no condition at all (with regard to the same action specified by the rule), and C' is a modified condition such that $C' = C$ plus one value (i.e., C' has one more value in one of the input dimensions) (that is, if the current rule is successful and the expanded condition is potentially better), then set $C'' = argmax_{C'} IG(C', C)$ as the new (expanded) condition of the rule. Reset all the rule statistics. Any rule covered by the expanded rule will be placed in its children list. [1]
- *Shrinking*: if $IG(C, all) < threshold2$ and $\max_{C'} IG(C', C) > 0$, where C is the current condition of a matching rule, *all* refers to no condition at all (with regard to the same action specified by the rule), and C' is a modified condition such that $C' = C$ minus one value (i.e., C' has one less value in one of the input dimensions) (that is, if the current rule is unsuccessful, but the shrunk condition is better), then set $C'' = argmax_{C'} IG(C', C)$ as the new (shrunk) condition of the rule. Reset all the rule statistics. Restore those rules in the children list of the original rule that are not covered by the shrunk rule. If shrinking the condition makes it impossible for a rule to match any input state, delete the rule.
- *Deletion*: The same as *Shrinking*.
- *Merge*: when the conditions of two rules are close enough, the two rules may be combined. If one rule is covered completely by another, it is put on the children list of the other. If one rule is covered by another except for one dimension, produce a new rule that covers both.

Note that although the accumulation of statistics is gradual, the acquisition and revision of rules is one-shot and all-or-nothing.

2.3 The Whole Model

In the overall algorithm, Step 4 is for making the final decision of which action to take by incorporating outcomes from both levels. We combine the corresponding values for an action from the two levels by a weighted sum; that is, if the top level indicates that action a has an activation value v (which should be 0 or 1 as rules are binary) and the bottom level indicates that a has an activation value q (the Q-value), then the final outcome is $w_1 * v + w_2 * q$. Stochastic decision making with Boltzmann distribution based on the weighted sums is then performed to select an action out of all the possible actions. Relative weights or percentages of the two levels are automatically set based on the relative performance of the two levels.

2.4 Some Analyses

We can contrast the characteristics of the two levels. The top level is discrete, all-or-nothing, rigorously verified (through experience), and without random exploration, and it learns through trial-and-error in a one-shot fashion. The bottom level is continuous, graded, not rigorously verified, and with random exploration, and it learns in a gradual and cumulative fashion. Thus they complement each other. Note also that the generalization of rules complements the generalization in the bottom level: While the bottom level generalization is continuous/graded, rule generalization at the top level is discrete/crisp (in terms of results, not the process), thus capturing different kinds of regularities. Because they possess different characteristics, each level tends to learn differentially; thus a combination of the two, through stochastic "averaging", is likely to result in improved performance (similar in a way to "stacking" and "bagging"; Breiman 1996).

The temporal difference in Q-values that determines rule learning encourages the exploration of those steps (state/action pairs) that are worth exploring, i.e., leading to an improvement (a new state with a higher Q-value). The criterion encourages the increase of Q-values, since the action and the updating that are based on large temporal differences lead to the increase of the Q-value for the state and the action.

When learning is well under way, the actions leading to large temporal differences (≥ 0) tend to be the best or near best actions in given states, because a large

[1] The children list of a rule is created to keep aside and make inactive those rules that are more specific (thus fully covered) by the current rule. It is useful because if later on the rule is deleted or shrunk, some or all of those rules on its children list may be reactivated if they are no longer covered.

temporal difference means a new state with an above average Q-value compared with other possible new states (each Q-value tends to be the average of all the possible outcomes from the corresponding action and is also "averaged" by neighboring Q-values due to network generalization/approximation).

3. Experiments

We tested CLARION on the simulated navigation task as shown in Figure 1. The agent has to navigate an underwater vessel through a minefield to reach a target location. The agent receives information only from a number of instruments. As shown in Figure 2, the sonar gauge shows how close the mines are in 7 equal areas that range from 45 degrees to the left of the agent to 45 degrees to the right. The fuel gauge shows how much time is left before fuel runs out. The bearing gauge shows the direction of the target from the present direction of the agent. The range gauge shows how far the target is from the current location. Based only on such information, the agent decides on (1) how to turn and (2) how fast to move. The agent, within an allotted time period, can either (a) reach the target (a success), (b) hit a mine (a failure), or (c) run out of fuel (a failure again). A random mine layout is generated for each episode. The time allotted to the agent for each episode is 200 steps. There are 41 inputs and thus more than 10^{12} states.

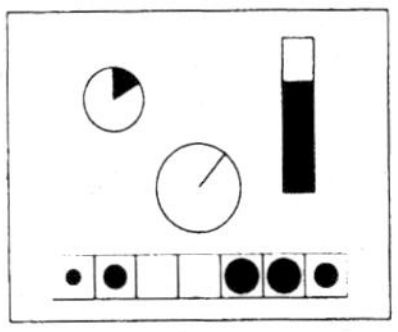

Figure 2 The Navigation Input

The display at the upper left corner is the fuel gauge; the vertical one at the upper right corner is the range gauge; the round one in the middle is the bearing gauge; the 7 sonar gauges are at the bottom.

Mine Density During Training: 10

	Train 10	B+T 10	B 10	T 10	B+T 30	B 30	T 30	B+T 60	B 60	T 60
CLARION	840.6	87.0	82.5	19.0	57.5	61.5	5.0	34.5	26.0	1.5
	49.9	19.6	8.7	19.8	23.0	17.5	11.8	17.8	9.4	2.3
Q	836.9		67.5			47.0			30.0	
	66.0		36.6			31.7			23.5	

Figure 3 Learning/transfer from 10-mine minefields.

B+T, B, and T denote the performance of the whole system, the bottom level, and the top level, respectively, at indicated mine densities (after the two levels are trained together). The first horizontal group contains the data generated by CLARION. The second horizontal group contains the data of the bottom level used alone with Q-learning as the sole learning method. In each group, the first row contains success rates and the second row contains corresponding standard deviations.

Learning speed. The "train" columns of Figures 3, 4, and 5 show the difference between CLARION and the bottom level alone (trained with pure Q-learning) in terms of learning effectiveness, which is measured by the number of successful episodes out of a total of 1000 episodes during training (averaged over 10 runs). In the figures, the first horizontal group of data, in which the first row contains success rates and the second row contains corresponding standard deviations, is generated by CLARION. The second horizontal group contains data from the bottom level used alone with Q-learning as the sole learning method. It appears that the higher the mine density (that is, the more difficult the task), the better CLARION is compared with the bottom level alone. The performance is undifferentiable in the 10-mine case. However, in the 60-mine case the superiority of CLARION (over the bottom level alone with Q-learning only) is statistically significant (with t tests, $p < 0.05$).

Transfer. The right three blocks of Figures 3, 4, and 5 show the transfer data, where transfer is measured by the percentage of successful episodes in the new setting by the trained models (each trained model is applied to minefields that contain a different number of minefields for a total of 20 episodes; the data is averaged over 10 runs). The data generally follows the pattern that the higher the mine density is, the lower the success rate is, no matter what the training setting is (whether it is at 10, 30, or 60 mine density). Moreover, the transfer performance of the models trained at different mine densities is very comparable when tested in minefields of the same density; for example, the models trained at 10-mine and 30-mine minefields perform almost indistinguishably in 60-mine minefields. As also indicated by the tables, CLARION outperforms the bottom level alone (trained with Q-learning) in transfer; the higher the mine density, the more pronounced the difference. The differences are statistically significant in the 30-mine and 60-mine cases (using t tests, $p < 0.05$). Finally, comparing the transfer performance of the top level, the bottom level, and the whole system, we notice that the whole system always performs at least as well as the bottom level, and sometimes significantly better (such as for those models trained at the 60-mine density), in which case there seems to be a synergy between the two levels (in the sense that the whole system performs better than either levels) and rules help to improve the transfer performance. But in those cases where the performance is comparable, does that mean we do not really need the top level? The answer is no, because if we compare the transfer performance of the bottom level alone after it is trained only with Q-learning with the transfer performance of the bottom level after it is trained together with the top level using the CLARION learning algorithm, we see the performance of the former is generally worse.

Mine Density During Training: 30

	Train	B+T	B	T	B+T	B	T	B+T	B	T
	30	10	10	10	30	30	30	60	60	60
CLARION	683.5	70.0	69.5	21.5	56.0	68.5	17.0	30.0	17.5	2.0
	51.3	21.3	25.3	31.6	27.0	17.0	25.6	11.2	14.3	6.0
Q	563.0		52.0			33.5			8.5	
	292.7		40.0			30.9			11.2	

Figure 4 Learning/transfer from 30-mine minefields.

Trained performance. The right three blocks of Figures 3, 4, and 5 also contain the trained performance data, which is defined to be the percentage of successful episodes in the same setting as used in training by the trained models (each trained model is applied to the minefields for a total of 20 episodes; the data is averaged over 10 runs). Comparing the trained performance of the top level, the bottom level, and the whole system, we again notice that the whole system sometimes performs better than the bottom level and the top level alone, which suggests a synergy in these cases. The bottom level after being trained as part of CLARION generally performs better than the bottom level after being trained alone using Q-learning only, which again suggests the advantage of CLARION.

Mine Density During Training: 60

	Train	B+T	B	T	B+T	B	T	B+T	B	T
	60	10	10	10	30	30	30	60	60	60
CLARION	372.6	88.5	64.0	30.0	70.5	60.0	14.5	42.0	42.5	13.0
	138.0	11.2	30.2	32.0	17.7	31.6	29.2	17.9	22.8	24.1
Q	80.0		53.0			19.0			6.5	
	54.0		28.8			17.1			6.3	

Figure 5 Learning/transfer from 60-mine minefields.

We have also done extensive experiments in a maze domain (Sun and Peterson 1995) and the same conclusions hold.

Acknowledgements: This work is supported in part by Office of Naval Research grant N00014-95-1-0440. The simulator for the navigation task was provided by Naval Research Lab (D. Gordon, A. Schultz, J. Ballas).

References

J. R. Anderson, (1983). *The Architecture of Cognition*, Harvard University Press, Cambridge, MA

L. Breiman, (1996). Bagging predictors. *Machine Learning*, Vol.24, No.2, pp.123-140.

H. Dreyfus and S. Dreyfus, (1987). *Mind Over Machine*, The Free Press, New York, NY.

D. Fisher, (1987). Knowledge acquisition via incremental conceptual clustering. *Machine Learning*. 2, 139-172.

P. Fitts and M. Posner, (1967). *Human Performance*. Brooks/Cole, Monterey, CA.

H. Hirsh, (1994). Generalizing version spaces. *Machine Learning*, 17, 5-46.

W. James, (1890). *The Principles of Psychology*. Dover, New York.

F. Keil, (1989). *Concepts, Kinds, and Cognitive Development*. MIT Press. Cambridge, MA.

N. Lavrac and S. Dzeroski, (1994). *Inductive Logic Programming*. Ellis Horword, New York.

L. Lin, (1992). Self-improving reactive agents based on reinforcement learning, planning, and teaching. *Machine Learning*. Vol.8, pp.293-321.

R. Michalski, (1983). A theory and methodology of inductive learning. *Artificial Intelligence*. Vol.20, pp.111-161.

T. Mitchell, (1982). Generalization as search. *Artificial Intelligence*, 18, 203-226.

R. Quinlan, (1986). Inductive learning of decision trees. *Machine Learning*. 1, 81-106.

P. Rosenbloom, J. Laird, A. Newell, and R. McCarl, (1991). A preliminary analysis of the SOAR architecture as a basis for general intelligence. *Artificial Intelligence*. 47 (1-3), 289-325.

P. Smolensky, (1988). On the proper treatment of connectionism. *Behavioral and Brain Sciences*, 11(1):1–74.

R. Sun, (1992). On Variable Binding in Connectionist Networks, *Connection Science*, Vol.4, No.2, pp.93-124.

R. Sun, (1995). Robust reasoning: integrating rule-based and similarity-based reasoning. *Artificial Intelligence*. 75, 2. 241-296.

R. Sun, (1997). Learning, action, and consciousness: a hybrid approach towards modeling consciousness. *Neural Networks*, special issue on consciousness. 10 (7), pp.1317-1331.

R. Sun and T. Peterson, (1995). A hybrid learning model of reactive sequential decision making. In: R. Sun and F. Alexandre, (eds.) *The Working Notes of The IJCAI Workshop on Connectionist-Symbolic Integration*.

R. Sun and T. Peterson, (1998). Some experiments with a hybrid model for learning sequential decision making. *Information Sciences*. in press.

R. Sutton, (1990). Integrated architectures for learning, planning, and reacting based on approximating dynamic programming, *Proc.of Seventh International Conference on Machine Learning*. Morgan Kaufmann. San Mateo, CA.

T. Tesauro, (1992). Practical issues in temporal difference learning. *Machine Learning*. Vol.8, 257-277.

G. Towell and J. Shavlik, (1993). Extracting Refined Rules from Knowledge-Based Neural Networks, *Machine Learning*. 13 (1), 71-101.

P. Utgoff (1989). Incremental induction of decision trees. *Machine Learning*. Vol.4, 161-186.

C. Watkins, (1989). *Learning with Delayed Rewards*. Ph.D Thesis, Cambridge University, Cambridge, UK.

D. Willingham, M. Nissen, and P. Bullemer, (1989). On the development of procedural knowledge. *Journal of Experimental Psychology: Learning, Memory, and Cognition*. 15, 1047-1060.

EMERGENT STRUCTURES AND BEHAVIORS

Risk Sensitive and Ambiguity Reducing Dominance Interactions in a Virtual Laboratory.

Charlotte K. Hemelrijk

AI Lab, Department of Computer Science, University of Zürich, Switzerland, Fax 0041-1-363 00 35, Email: hemelrij@ifi.unizh.ch

Abstract

In this paper the effects of dominance interactions on social-spatial structure (i.e. centrality of dominants) and social interaction patterns are studied in a simple virtual world: artificial entities inhabiting it are just grouping and performing dominance interactions in which the effects of winning and losing are self-reinforcing. I compare three strategies of attack. The first is based on the contention that aggression stops once individuals recognize the status of others ('the Ambiguity Reducing strategy'). The second represents another popular view among ethologists, that attack depends on risks involved. The third version is a kind of control system. Here, entities attack obligatory whomever they encounter. In all three systems a dominance hierarchy develops and an increase in intensity of aggression leads to clearer differentiation of ranks, which has a cascade of effects on frequency of aggression, stability of ranks and spatial structure. Furthermore, a spatial structure with dominants in the center emerges in the control and risk sensitive strategy. In the risk sensitive system, bidirectionality of attack is decreased at higher intensities of aggression. Since the latter is observed in certain primate species too, other patterns that emerged for the risk sensitive strategy may be used as research hypotheses for these species.

1 Introduction

Dominance interactions form an important aspect of social behavior in many animal species (e.g. Bonabeau et al, 1996; Rowell, 1974), but contradictory views abound on the subject. As to its origin, dominance is seen as an inherent quality of an individual (Ellis, 1994) or - oppositely - as a relational quality (Bernstein, 1981; Francis, 1988) that may develop by chance and the self-reinforcing effects of winning and losing (Rowell, 1974; Chase et al. 1994).

It is generally assumed that a dominance hierarchy functions to reduce aggression by dissolving ambiguity of relationships. Yet at the same time it is maintained that animals should continuously strive to obtain as high a rank as possible (Datta and Beauchamp, 1991) and should do so 'economically', i.e. while taking risks into account (Parker, 1974). The suggested benefits of high rank are manifold. Apart from priority of access to food and sexual partners, they may also include acquisition of a safe location, which according to the theory of the 'selfish herd', should be in the center of the group (Hamilton, 1971). For some primate species, such as Japanese macaques, dominant individuals are indeed reported to occupy the center whereas subordinates remain at the periphery (e.g. Itani, 1954, Yamada, 1966). In other species, such as baboons, however, Altmann (1979) reported individual positioning to be random, whereas Rhine and Westland (1981) state males to be peripheral.

These contradictory views are partly due to the close intertwinement of aggression and dominance (e.g. Francis, 1988), which in turn may also be interconnected with other domains of behavior as emphasized by Thierry (1985ab,1990). He made plausible how species differences in characteristics of dominance interactions, such as intensity of aggression, have implications for other aspects of behavior, such as infant handling and frequency of counter-attack (i.e. bidirectionality).

Such a systems theoretical perspective to dominance as adopted by Thierry for primates presents empirical scientists with a number of difficulties. Keeping all variables constant but one, the practice of experimentalists, is no longer meaningful. Especially when feedback processes come with non-linearity, unexpected macro-patterns may arise by self-structuring. These cannot be described as linear combinations of independent factors and therefore not be detected by an analysis of variance. Nor can these patterns be understood from observations on its end-state. The best way to study such a system is, therefore, not by top-down analysis, but by using a bottom-up approach, i.e. by following how the dynamics from interactions among entities unfold into macro-patterns (and in turn, these macro-patterns affect the behavior of the entities). By covering the complete ontogenetic process of a system, its end-patterns are thus more easily understood (Braitenberg, 1987).

Since such a procedure is problematic for natural systems, studying artificial analogons becomes important. Here, individual oriented models may serve as a kind of 'virtual laboratories' (Epstein and Axtel, 1996). In these models, agents are implemented in an artificial world and their behavior is studied in the same way ethologists study real animals. Various explanations for certain phenomena can be represented as variations in the rule sets of the artificial agents. If some of the behavioral consequences are similar to findings on real animals, additional social interaction patterns emerging in the artificial world may function as a 'search image' for further data-analysis on the species under concern. In this way te Boekhorst and Hogeweg (1994) studied the grouping behavior of orang-utans by comparing field data with patterns in an individual oriented model on artificial apes.

Here, I will follow this approach regarding specifics of dominance interactions. I use an individual oriented model inspired by Hogeweg (1988). It consists of simple entities that do nothing else than aggregate and perform dominance interactions in which the effects of winning and losing are self-reinforcing. Entities are cognitively simple in that they perceive dominance ranks of others directly and do not recognize others individually. I will show that a dominance hierarchy and a spatial structure with dominants in the center develop and explain how a close connection between spatial- and rank-distance originates. I will vary intensity of aggression and compare three strategies of attack that represent popular views among ethologists. In the first attack strategy, aggression is ambiguity reducing, thus entities attack partners more often the smaller the rank-distance with their partners. The second one is a risk sensitive strategy, in that entities are more likely to perform a dominance interaction with those from win they are more likely to win. The third one is a kind of control experiment, in which entities always fight upon meeting others ('Obligate attack'). I will compare these strategies as to their socio-spatial structure (i.e. the centrality of dominants), stability of dominance hierarchy and to some other patterns of social interaction that are often studied in animals (such as the overall frequencies of aggression and counter-attack). The results will be evaluated in order to solve some of the contradictions mentioned above. More specifically, I will show that the Risk Sensitive strategy delivers patterns of counter-attack that are similar to those observed in certain primate species (Thierry, 1985a). On the basis of these and other outcomes of the model, I will present some suggestions for further data analysis on these primate species.

2 Methods

In this section I will present a description of the model and outline how spatial structure, bidirectionality of attack and stability of a hierarchy is measured.

2.1 The Model

The model is individual-oriented and event-driven (see Hogeweg & Hesper 1979; Hogeweg, 1988; Villa, 1992; Judson, 1994). The modeling environment (written in object-Pascal, Borland Pascal 7.0) consists of three parts:

* the 'world' (toroid) with its interacting agents,

* its visualization,

* special entities that collect and analyze data on what happens in the 'world' (cf. the 'recorders' and 'reporters' of Hogeweg, 1988).

Unlike in former lattice-based models (Hemelrijk, 1996,1997), the 'world' presented here consists of a continuous space of 200 by 200 units. Therefore, agents are able to move in any direction. They have an angle of vision of 120 degrees and their maximum perception distance (MaxView) is 50 units. I will confine myself to a small ensemble of 8 entities. Agents group and perform dominance interactions according to the sets of rules described below (figure 1).

2.1.1 Grouping rules

In the literature, two opposing forces affecting group structure are often postulated: on the one hand animals are attracted to one another, because being in a group provides safety. On the other hand, aggregation implies competition for resources and this drives individuals apart.

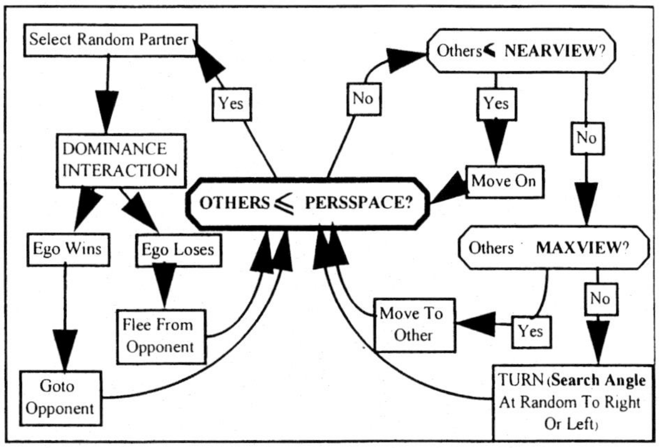

Figure 1. Flow chart for the behavioral rules of the entities

The forces leading to aggregation and spacing are realized in the model by the following set of rules (cf. Hogeweg, 1988):

- If an agent sees another within a critical distance (parameter PerSpace), it performs a dominance interaction with that entity. In case several agents are within PerSpace, the interaction partner is chosen at random. If the agent wins the interaction, it moves towards its opponent, otherwise it makes a full turn and moves away from its rival.

- If nobody is in its PerSpace, but an agent perceives others within a distance of NearView, it continues moving in its original direction.

- If an agent detects it nearest neighbors outside NearView, but within its maximum range of vision (= MaxView), it moves towards them.

- If an agent does not perceive any other agent within MaxView, it searches for group members by making a turn over an angle (= SearchAngle) of 90 degrees at random to the right or left.

2.1.2 Dominance types

Conventionally, rank-acquisition is attributed to a (possibly inherited) quality of an individual (Ellis, 1994). However, experimental results from various animal species (for a listing see Bonabeau et al, 1996) have shown chance and self-reinforcement determine that winning: once an animal has won, it has a larger chance of winning again in a subsequent fight (the so-called 'winner-effect', see Chase et al, 1994). Because this study emphasizes the self-organizing properties of interactions, I confine myself to the self-enhancing effects of winning/losing by starting with completely identical entities.

About the perception of rank by others, a number of hypotheses are entertained by various authors (e.g. Barnard and Burk, 1979; Hemelrijk, 1996,1997). Either individuals recognize others individually and estimate rank of partners based on their experience with them, or the capacity to win is directly perceived from external cues, such as pheromones in social insects (e.g. van Honk and Hogeweg, 1981) and body postures in crustaceans (e.g. Copp, 1986). The simplest case of direct perception is studied here.

2.1.3 Dominance interactions.

Interactions between agents with direct perception of dominance ranks (i.e. Perceivers) are modeled after Hogeweg & Hesper (1983) as follows:

1. Each entity has a variable DOM (representing the capacity to win a hierarchical interaction).

2. After meeting one another in their PerSpace, entities display and observe each other's DOM. This represents an active display and only through such a display the partner obtains information about the DOM value of its opponent. Subsequent winning and losing is determined as follows by chance and values of DOM:

$$w_i = \begin{cases} 1 & \dfrac{DOM_i}{DOM_i + DOM_j} > RND(0,1) \\ 0 & \text{else} \end{cases} \qquad (1)$$

Where w_i is the outcome of a dominance interaction initiated by agent i (1=winning, 0=losing). In other words, if the relative dominance value of the interacting agents is larger than a random number (drawn from a uniform distribution), then agent i wins, else it looses. Thus, the probability of winning is larger for the one who is higher in rank and this is proportional to the relative DOM-value with its partner.

3.Updating of the dominance values is done by increasing the dominance value of the winner and decreasing that of the loser:

$$\begin{aligned} DOM_i &:= DOM_i + \left(w_i - \frac{DOM_i}{DOM_i + DOM_j} \right) * STEPDOM \\ DOM_j &:= DOM_j - \left(w_i - \frac{DOM_i}{DOM_i + DOM_j} \right) * STEPDOM \end{aligned} \qquad (2)$$

The consequence of this system is that it behaves as a damped positive feedback: winning by the higher ranking agent reinforces their relative DOM-values only slightly, whereas winning by the lower ranking gives rise to a relatively large change in DOM. To keep DOM values positive, their minimum value was arbitrarily put at 0.01. STEPDOM is a scaling factor that varies between 0 and 1 and is analogous to intensity of aggression. High values imply a large change in DOM-value when updating it, and thus indicate that single interactions may strongly influence future outcomes of conflicts. Conversely, low STEPDOM-values represent low impact. Unless stated otherwise, STEPDOM is set at 0.5.

4. Winning includes chasing the opponent one unit distance and then turning randomly 45 degrees to the right or left in order to reduce the chance of repeated interactions between the same partners. The loser responds by fleeing under a small random angle over a predefined FleeingDistance.

From now on, the initiation of a dominance interaction will also be referred to as 'attack' for short.

2.1.4. Probability of attack.

The probability to attack has been suggested to depend on, among others, potential risk of defeat or injury (Parker, 1974;Datta and Beauchamp, 1991). This 'risk-sensitive attack' system was implemented as follows. Upon meeting another and observing its DOM-value, an entity may or may not attack. This depends on the outcome of a 'mental' battle, which follows the rules of a DODOM interaction. If ego looses the mental interaction it will refrain from any action (thus displaying 'non-aggressive' proximity). If it wins the mental battle it will start a dominance interaction. After winning is decided, DOM-values of both partners are updated. This implementation bears similarity to the SKINNY's of Hogeweg (1988). However, SKINNIES are 'cognitively' more sophisticated in the sense that they recognize others individually and have a mental representation of the rank of partners.

I will compare centrality of dominants to two other attack systems, described before (Hemelrijk, in press). The first one is the 'Obligate attack system', which can be viewed as a kind of 'default' strategy (Hemelrijk, 1996, 1997) since activated entities always engaged in dominance interactions when encountering others nearby. In the other one, the probability of attacking a partner linearly decreases with the rank-distance to that partner (the Ambiguity Reducing attack system, Hemelrijk, in press). This Ambiguity Reducing attack system is therefore partly the counterpart of the Risk Sensitive strategy.

2.1.5. Timing regime

Since parallel simulations cannot be run on most computers, a timing regime regulating the sequence of the activation, has to be included. The type of timing regime influences the results of a simulation. A biologically plausible timing regime must be locally controlled, i.e. by other entities and not by a monitor (e.g. Goss & Deneubourg, 1988). In the timing regime used here, each entity draws a random waiting time from a uniform distribution. The entity with the shortest waiting time is activated first. The decay of waiting time is the same for each entity. However, if a dominance interaction occurs within NearView of an agent, the waiting time of this agent is reduced stronger.

2.2 Experimental setup and Data collection

To compare results of the Risk-Sensitive System to the other two attack systems, it was studied for the same parameter setting (n=8, persSpace=4, nearView=24, searchAngle=90 degrees, FleeingDistance=2 units).
For each of three values of STEPDOM (0.1, 0.5 and 1.0) 10 runs were conducted.
During a run, every change in spatial position and heading direction of each entity was recorded. Every time step (consisting of 160 activation) the distance between agents was calculated. Dominance interactions were continuously monitored by recording: 1) the identity of the attacker and its opponent; 2) the winner/loser; 3) the updated DOM-values of these entities.

2.3 Measures of spatial centrality of dominants and hierarchical stability and bidirectionality of attack

The degree with which dominants occupied the center was measured in two ways, by the spatial directions of others around ego and by the average distances of partners towards ego. Using circular statistics (Mardia, 1972), for each scan the centrality of each individual was calculated by drawing a unit circle around it and projecting the direction of other group members (as seen by ego) as points on the circumference of that circle. Connecting these points with the origin gives vectors. The length of the mean vector represents the degree in which the position of group members relative to ego is clumped; longer mean vectors reflect more direction and indicate lower centrality (i.e. 'encirclement'). Thus stronger centrality of higher-ranking entities is reflected in a larger negative correlation between rank and encirclement.
The second measure is the Kendall rank correlation between dominance value and the average distance of Ego to others. Again, centrality of dominants is represented by a negative correlation. Both measures appeared to be strongly correlated and only the rank-encirclement correlation will be used in the analysis.

The stability of the dominance hierarchy was expressed as the relative number of rank reversals. This was established by calculating the Kendall rank correlation between the dominance ranks of entities at successive intervals of two time steps (320 activation).

Bidirectionality of attack is calculated as a τ_{Kr}-correlation between an actor and receiver matrix of attack (Hemelrijk, 1990). This statistic measures the correlation between the corresponding rows of two social interaction matrices and the method reckons with the statistical dependency due to recurrent observations on the same individual (Hubert, 1987). A higher τ_{Kr}-value corresponds to a higher degree of bidirectionality of attack.

3 Results

In this section, I will first summarize the results of the Obligate and Ambiguity Reducing attack system which I described elsewhere (Hemelrijk, in press). Next, these outcomes will be compared to those of the Risk Sensitive attack system.

3.1. Obligate and Ambiguity Reducing Attack System

In the Obligatory attack system, a higher STEPDOM strengthens social-spatial structure, whereas for Ambiguity Reducing attack rates the opposite was found (Figure 2).
To explain these results, note that the higher the value of STEPDOM the stronger a single event of winning (or losing) influences the outcome of future interactions. Thus, starting from completely identical entities, ranks differentiate faster at higher STEPDOM. In the system with Obligatory attack, this results in larger average distances among entities: by being defeated again and again, losers move away further and further from others. Consequently, the frequency of aggression drops and this lowers the probability of a rank-reversal. The thus induced higher stability and larger differentiation of ranks enhance social-spatial structure as follows. The larger the rank-distance, the higher the probability that a dominant will win from and chase away a subordinate. Thus, the physical distance between subordinates and dominants reflects their rank-distance. This correspondence develops through the outcomes of interactions. If during this process entities reverse ranks (i.e. the hierarchy is unstable), there is insufficient time for this correspondence to materialize.

In the Ambiguity Reducing attack system entities also reduce their aggression frequency as ranks differentiate, but in this case they increasingly tolerate partners more distant in rank. Obviously, this impairs spatial structure so that the majority of these correlations were non-significant. Thus, I conclude that a spatial structure with true centrality of dominants does not develop. However, the spatial structure is not reversed (i.e. into a configuration with dominants at the periphery and subordinates in the center). This is due to certain similarities shared by the Obligate attack and the Ambiguity Reducing attack system: in both, dominants win more often the larger their rank-distance with subordinates and aggressive interactions especially occur among rank-near entities. Note that in the Ambiguity Reducing attack system prevalence of interactions among rank-near entities is implemented a priori as a behavioral rule, whereas in the Obligate attack system it is due to the emergent proximity of rank-near entities.

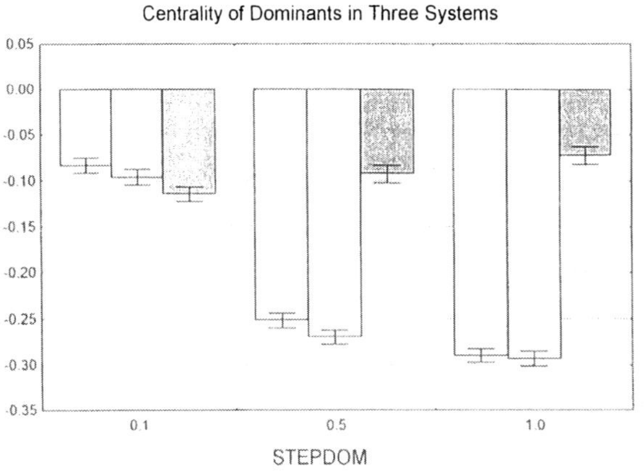

Figure 2. Rank-encirclement correlation for various Attack Systems and intensities of aggression. White bars: Risk-Sensitive attack. Shaded bars: Obligate Attack. Striped bars: Rank-Distance decreased Attack system.

3.2. Risk Sensitive Attack system

How does Risk Sensitive Attack relate to the other two systems? As in 'Obligate Attack', centrality of dominants increases with STEPDOM (figure 2). This is not surprising, because what happens during a run is similar to the processes that take place in the Obligate Attack system (see figure 3 for an overview). First, differentiation of dominance ranks results in larger inter-individual distances (figure 4) and second, because of this, the frequency of attack reduces (correlation between mean distance and frequency of attack for N=130, at STEPDOM0.1 –0.424; SD0.5 –0.335; SD1.0 –0.280). Third, lowered aggression strengthens the stability of the hierarchy (correlation between frequency of attack and stability of hierarchy for N=128, at SD0.1 –0.253; SD0.5 –0.325; SD1.0 –0.421). This enforces spatial structure in the way explained for the Obligate attack system.

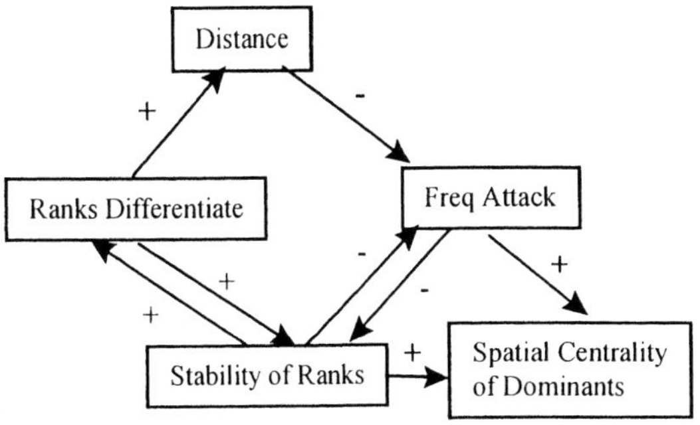

Figure 3. Summary of results for the Risk Sensitive attack system. Arrows indicate direction of the effect.

Resemblance with the Obligatory attack system comes about because rank-same entities stay close in both systems. In case of Obligatory attack this is so because rank-same entities chase away each other with the same probability and therefore stay on average in the same neighborhood. In the Risk Sensitive system closeness is due to this too, but also the fact that rank-same entities are likely not to attack one another.

Both systems differ, however, in the degree of stabilization of ranks. Since in the Risk Sensitive system, attack is less challenging, hierarchies remain undifferentiated longer than in Obligate attack system (Wilcoxon Matched Pairs Signed Ranks test for N=128, SD0.1 z=2.09 p<0.037; SD0.5 z=4.35, p<0.000014; SD1.0 z= 2.24, p<0.025). In turn, this lower stability may explain why spatial structure of the Risk Sensitive system falls somewhat short of the Obligate attack system for all values of STEPDOM (figure 2).

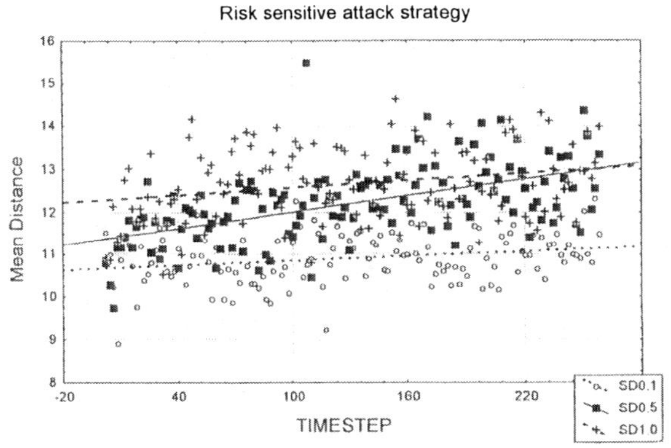

Figure 4. Mean distance among 8 Risk Sensitive entities over time. SD=stepdom; one time step equals 160 activation.

3.3 Some other effects of spatial structure

When looking at consequences for spatial structure, there appears hardly any difference between the Obligate and the Risk Sensitive attack system. There are, however, clear differences with respect to certain other social patterns.

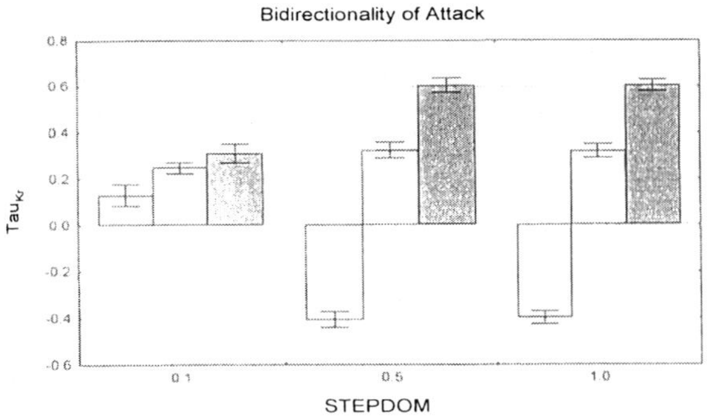

Figure 5. Bidirectionality of attack in three Attack systems for varying intensities of aggression (mean τ_{Kr} plus/minus S. E.) White bars: Risk-Sensitive attack. Shaded bars: Obligate Attack. Striped bars: Rank-Distance decreased Attack system.

First, at higher STEPDOMs, bidirectionality of attack decreases in the Risk Sensitive system, whereas in the other systems the opposite is found (figure 5). In the Risk Sensitive system entities are able to distinguish more clearly from whom they might win and therefore become more selective in whom they attack at higher STEPDOMs. Note that this selectivity is asymmetric: entities attack rank-distant partners more often only if they are lower ranking than themselves. Due to clearer rank differentiation, attack becomes less bidirectional at high STEPDOM. In the Ambiguity Reducing attack system, selectivity concerns only rank-distance, irrespective whether partners are higher or lower ranking than self is. This symmetry implies that rank-near entities will mutually attack each other more often at high STEPDOM. This induces enhanced bidirectionality.

Bidirectionality also increases in the Obligate attack system at higher STEPDOM. Here, it is an emergent property of conspicuousness and stability of spatial structure: by mutually encountering each other in a statistically stable neighborhood, (rank-near) entities attack each other bidirectionally.

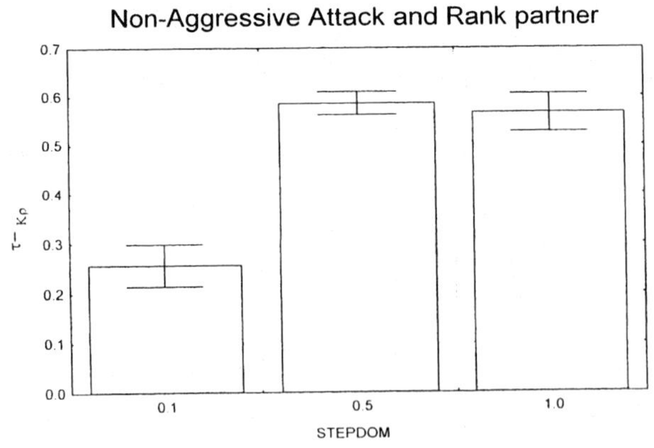

Figure 6. Initiation of Non-Aggressive Proximity Rank of Partners and STEPDOM.

Furthermore, in the Risk Sensitive system and particularly at high STEPDOMs, entities approach non-aggressively more often partners of higher rank (figure 6). (Obviously, this is complementary to their attacking subordinates more often the lower their rank.) In contrast, in the Ambiguity Reducing system, such a correlation is absent, because entities approach high as well as low-ranking partners more often the larger the rank-distance.

4 Discussion

It is generally believed that the function of a dominance hierarchy is the reduction of aggression. This contention is based, among other things, on the observation that aggression subsides some time-period after unacquainted individuals have been put together (e.g. chickens: Guhl, 1968; primates: Kummer, 1974). In other words, once the rank position of others has been acknowledged, no further aggression is needed. By making the probability of attack negatively dependent on rank distance, this perspective was implemented as the Ambiguity Reducing strategy in this study.

At first sight, an ambiguity reducing strategy seems to be at odds with another the popular view, namely that individuals are continuously striving for higher rank positions and do so while taking risks into account. Yet, the model demonstrates that aggression reduction takes place also in the Risk Sensitive and the Obligate attack system, although this is not obvious from the rule-set. In these two systems, decline of attack is an emergent property. As in the Ambiguity Reducing strategy in these two systems, stronger differentiation of rank increases stabilization of a hierarchy. However, in the Ambiguity Reducing strategy this leads to a strongly diminished attack-rate among rank-distant entities; herewith the communal attraction of entities is no longer countered and hence they 'huddle' closer together. In the other two systems, however, stronger differentiation, via repeated winning of certain entities from others, drives entities further apart. Because of this increased inter-individual distance, encounter frequency drops and this leads to a decline in aggression in the Obligate and Risk Sensitive strategies. Thus, aggression reduction is a consequence of spatial structure, and it is in this way that the apparent contradiction between the Ambiguity Reducing system on the one hand and the Obligate and Risk Sensitive strategy on the other hand is dissolved.

Not only does spatial structure reduce aggression, it also sustains its own robustness. Because individuals are spatially sorted out according to rank classes, they tend to have rank-similar entities as their nearest neighbors. Possible rank-reversals, therefore, mostly take place within the same rank category, which hardly disturbs the spatial structure.

Spatial structure, in this model, is thus an interwoven, essential part of the dynamics of the complete system. It is the consideration of spatial dynamics that leads to parsimonious results. Dominance centered spatial structure is a point in case. Conventionally, centrality of dominants is viewed as an independent feature that is only linked by virtue of the benefits it accrues to dominants.

Hence the separate adaptive explanation is that individuals have been evolved to seek a central location in order to be protected against predators ('selfish herd' theory from Hamilton, 1971). Such an explanation is clearly superfluous in this model.

Another strength of these models is that they may provide guidance for investigation of empirical data. If some patterns of the model correspond to findings on real animals, other model outcomes may be used as a search image for further analysis (te Boekhorst and Hogeweg, 1994). For instance, Thierry (1985ab) showed how bidirectionality of attack in primates depends on the species-specific intensity of aggression. In the rhesus monkey (Macaca mulatta), a species that is characterized by more intense forms of aggression, such as biting, bidirectionality was low, whereas in the mildly aggressive Celebes monkeys (Macaca tonkeana) counter-attack was more frequent. The same association between aggression intensity and bidirectionality was found in the Risk Sensitive artificial entities.

If dominance processes in macaques indeed reflect a Risk Sensitive strategy, it is worthwhile to look for the other patterns of the model in these monkeys as well. Therefore, we are presently studying whether rhesus monkeys non-aggressively approach higher-ranking group-members more often and display the dominance centered spatial configuration more clearly than Celebes monkeys.

Acknowledgments

I like to thank Rolf Pfeifer and Bob Martin for continuous support and René te Boekhorst for improving former versions of this paper. This work is supported by the Swiss National Science Foundation by a grant from the Marie Heim-Voegtlin Foundation.

References

Altmann, S. A. (1979). Baboon progressions. Order or chaos ? A study of one-dimensional group geometry. Anim. Behav., 27, 46-80.

Barnard, C. J., & Burk, T. E. (1979). Dominance hierarchies and the evolution of 'individual recognition'. J. theor. Biol., 81, 65-73.

Bernstein, I. S. (1981). Dominance: the baby and the bathwater. Behavioral and brain sciences, 4, 419-457.

Boekhorst, I. J. A. te, & Hogeweg, P. (1994). Effects of tree size on travelband formation in orang-utans:data-analysis suggested by a model. In R. A. Brooks & P. Maes (Eds.), Artificial life (pp. 119-129). Cambridge, Massachusetts: The MIT Press.

Bonabeau, E., Theraulaz, G., & Deneubourg, J.-L. (1996). Mathematical models of self-organizing hierarchies in animal societies. Bulletin of mathematical biology, 58(4), 661-717.

Braitenberg, V. (1987). Vehicles. Experiments in synthetic psychology. Cambridge, London: The MIT Press.

Chase, I. D., Bartelomeo, C., & Dugatkin, L. A. (1994). Aggressive interactions and inter-contest interval: how long do winners keep winning ? Anim. Behav., 48, 393-400.

Copp, N. H. (1986). Dominance hierarchies in the crayfish Procambarus clarkii (Girard, 1852) and the question of learned individual recognition (Decapoda, Astacidea). Crustaceana, 51(1), 9-24.

Datta, S. B., & Beauchamp, G. (1991). Effects of group demography on dominance relationships among female primates. I. Mother-daughter and sister-sister relations. American Naturalist, 138, 201-226.

Ellis, L. (Ed.). (1994). Reproductive and interpersonal aspects of dominance and status (1 ed.). Westport: Greenwood publishing group.

Epstein, J. M., Axtell, R. (1996). Growing artificial societies. Social science from bottom up. Cambridge: The MIT Press.

Francis, R. C. (1988). On the relationship between aggression and social dominance. Ethology, 78, 223-237.

Goss, S., & Deneubourg, J. L. (1988). Autocatalysis as a source of synchronised rhytmical activity in social insects. Insectes Sociaux, 35(3), 310-315.

Guhl, A. M. (1968). Social inertia and social stability in chickens. Anim. behav., 16, 219-232.

Hamilton, W. D. (1971). Geometry for the Selfish Herd. J. theor. Biol., 1971(31), 295-311.

Hemelrijk, C. K. (1990). Models of, tests for, reciprocity, unidirectionality and other social interaction patterns at a group level. Anim. Behav., 39, 1013-1029.

Hemelrijk, C. K. (1996). Dominance interactions, spatial dynamics and emergent reciprocity in a virtual world. In P. Maes, M. J. Mataric, J-A Meyer, J Pollack, & S. W. Wilson (Eds.), From Animals to Animats 4: Proceedings of the fourth international conference on simulation of adaptive behavior (pp. 545-552). Cambridge, MA: The MIT Press/Bradford Books.

Hemelrijk, C. K. (1997). Cooperation without genes, games or cognition. In P. Husbands, I. Harvey (Eds.), 4th European conference on artificial life, . (pp511-520). Brighton, UK: MIT-Press.

Hemelrijk, C. K. in press. Spatial centrality of dominants without positional preference. Sixth International Conference on Artificial Life. Eds. C. Adami, R. Belew, H. Kitano, C.Taylor.

Hogeweg, P. (1988). MIRROR beyond MIRROR, Puddles of LIFE. In C. Langton (Eds.), Artificial life, SFI studies in the sciences of complexity (pp. 297-316). Redwood City, California: Adisson-Wesley Publishing Company.

Hogeweg, P., & Hesper, B. (1979). Heterarchical, selfstructuring simulation systems: concepts and applications in biology. In B. P. e. a. Zeigler (Eds.), Methodologies in systems modeling and simulation (pp. 221-231). Amsterdam: North-Holland Publ. Co.

Hogeweg, P., & Hesper, B. (1983). The ontogeny of interaction structure in bumblebee colonies: a MIRROR model. Behav. Ecol. Sociobiol., 12, 271-283.

Honk, C. van, & Hogeweg, P. (1981). The ontogeny of the social structure in a captive Bombus terestris colony. Behav. Ecol. Sociobiol., 9, 111-119.

Hubert, L. J. (1987). Assignment methods in combinatorial data analysis (first ed.). New York and Basel: Marcel Dekker, inc.

Itani, J. (1954). The Monkeys of Mt Takasaki. Tokyo: Kobunsha.

Judson, O. P. (1994). The rise of the individual-based model in ecology. Trends. Ecol. Evol., 9, 9-14.

Kummer, H. (1974). Rules of dyad and group formation among captive baboons (Theropithecus gelada). In Symposium 5th Congress Int'l. Primat. Soc., (pp. 129-160). S. Karger, Basel.

Mardia, K. V. (1972). Statistics of directional data. London and New York: Academic Press.

Parker, G. (1974). Assessment strategy and evolution of fighting behavior. J. theor. Biol., 47, 223-243.

Rhine, R. J., & Westland, B. J. (1981). Adult male positioning in baboon progressions: order and chaos revisited. Folia Primatol., 35, 77-116.

Rowell, T. E. (1974). The concept of social dominance. Behav. Biol., 11, 131-154.

Thierry, B. (1985a). Patterns of agonistic interactions in three species of macaque (Macaca mulatta, M. fascicularis, M. tonkeana). Aggressive Behavior, 11, 223-233.

Thierry, B. (1985b). Social development in three species of macaque (Macaca mulatta, M. fascicularis, M. tonkeana). Behavioural processes, 11, 89-95.

Thierry, B. (1990). The state of equilibrium among agonistic behavior patterns in a group of Japanese macaques (Macaca fuscata). C. R. Acad. Sci. Paris, 310(3), 35-40.

Villa, F. (1992). New computer architectures as tools for ecological thought. Trends Ecol. Evol., 7, 179-183.

Yamada, M. (1966). Five natural troops of Japanese monkeys of Shodoshima Island (I): distribution and social organization. Primates, 7, 315-362.

Structural Coupling of Cognitive Memories Through Adaptive Language Games

Luc Steels
Sony Computer Science Lab
6 Rue Amyot 75005 - Paris
VUB AI Lab - Brussels
E-mail: steels@arti.vub.ac.be

Abstract

The paper investigates how a group of distributed agents may develop congruent cognitive memories in the form of networks of prototypes. Memories are congruent if they structure reality in a sufficiently similar way for the agents to cooperate and communicate. Each agent develops his own memory independently of the others and based on his own history of experiences. Agent memories are weakly coupled when the agents are in the same environment. They encounter the same sorts of objects and therefore make similar generalisations. Memories can also be more strongly coupled if the agents communicate and use their cognitive memories to structure and conceptualise reality for communication. The paper explores both types of coupling. Experiments are reported where agents build autonomously internal cognitive memories and develop through self-organisation a shared lexicon that makes their memory more congruent.

1 Introduction

Even the simplest cognitive agent needs a memory to store experienced situations. There is a wide consensus in the cognitive sciences that (1) such a memory is *hierarchical*, i.e. that there are structures recording more specific and more general situations, and (2) the nodes in these possibly tangled hierarchies correspond to *prototypes*, as opposed to concepts defined in terms of necessary and sufficient features (Lakoff, 1987). The representation of a prototype is generally agreed to take the form of a *schema*, which contains the typical features of the prototype, possibly enriched with other information such as demons, defaults, associative links to other prototypes, etc.

An agent must autonomously develop his own memory based on a stream of situations presented by the environment. This raises the issue how a group of autonomous agents - which cannot inspect each other's brainstates directly and have each only limited experience with the environment - may arrive at a network of prototypes which is sufficiently shared that communication and cooperation between the agents is possible. This paper hypothesises that sharing may be achieved by *structural coupling*. Structural coupling is a concept introduced by Maturana and Varela for explaining how independent systems may nevertheless develop a mutual congruence. When the activities of each impacts indirectly the other through the environment, they consequently become coordinated - without a coordinator (Maturana and Varela, 1972).

Structural coupling is present, but in a relatively weak form, when different agents are presented with similar situations emanating from the same environment. A stronger form of coupling occurs when the structure of cognitive memory plays a role in communication. Rather than inventing new names for every possible situation that may occur, the agent conceptualises reality in terms of the prototypes in his memory, and verbalises this conceptualisation. The other agent is necessarily forced to adopt or hypothesise similar conceptualisations and hence structure his memory in a similar way. This way a co-evolution of language and cognitive memory is seen: The language lexicalises prototypes developed by the agents, and the agents adopt prototypes lexicalised by the language.

There is clear evidence for the presence of this phenomenon in human natural languages. A language lexicalises prototypes at different level of a prototype hierarchy. For example, English has words for animal, mammal, horse, and mare. Each of these words indicates progressively more specific prototypes for a female horse. There are differences between languages for which prototypes are lexicalised. For example, English does not have a single word for female elephant - some other language might. Also the inheritance relation between prototypes is not always simple.

This paper explores strong structural coupling of cognitive memories through an emergent lexicon. The next part of the paper presents a mechanism by which a single agent may build up his cognitive memory in the form of a tangled hierarchy of prototypes. Then a mechanism is presented whereby a group of agents may build up and acquire a lexicon for describing situations in terms of words associated with prototypes. The next part of the paper then studies results of simulation experiments in which a group of agents progressively build up their cognitive memories and their lexicons. It is seen that congruence indeed arises. The results reported here are taken from software simulations but they will be ported to physical robots - as we have done with similar experiments in lexicon formation coupled to discrimination tasks (Steels and Vogt, 1997).

The research discussed in this paper fits within a larger research effort to understand the origins of language and meaning (Hurford, 1998). Several researchers have adopted the viewpoint that language is a complex adaptive system and consequently that mechanisms from 'artificial life' and 'simulation of adaptive behavior' can be fruitfully applied to study the emergence and maintenance of language systems (see the overview in (Steels, 1997)).

2 An Adaptive Cognitive Memory

The main role of memory is to compactly store experienced situations in such a way that they can be retrieved later. A particular situation or a prototype of a situation is represented as a *schema*. Each schema contains a *description-set* which contains a set of descriptions. A schema is called a *situation-schema* when it describes a particular situation, it is a *prototype-schema* when it describes a class of situations. There is no formal difference between the two types of schemata.

A description contains a *content* in the form of a feature, such as '(colour red)'. A description also contains *justifications*. Each *justification* decomposes into an indication of the *strength* of a description and a *reason* for its adoption. The reason indicates the operation that was used to create the description as well as other descriptions used in this operation. The strength specifies how crucial the description is in the schema, in other words how certain it is that the description must be part of the schema. The strength can also be negative, implying that the description is known not to be part of the schema.

In the case of situation schemata, the sensory channels and sensory processing routines provide an indication of certainty and hence strength. In case of prototype schemata, the strength is progressively learned from the use and success of a schema in use. When a description has more than one justification, the one with the highest strength is assumed to be the main justification and determines the overall strength of the description.

Schemas can contain *slots* which are filled by other schemas. When such a link exists, all the descriptions associated with the filler are inherited by the schema. They consequently do not have to be stored with the inheriting schema. Thus there could be a prototype-schema for 'horse' and a schema for 'mammal' and a slot in the horse schema filled by the mammal schema so that all descriptions associated with the mammal prototype are also associated with the horse prototype even though they are not stored with the horse prototype. A schema can have several slots. For example, there can be a prototype-schema for 'female horse' which inherits both from the prototype for 'female' and for that of 'horse'. The slots clearly implement specialisation/generalisation hierarchies widely used in AI and object-oriented programming. The justification mechanism relies on well-established techniques from truth-maintenance systems.

We now face two technical problems: (1) How can an agent retrieve the minimal set of prototype schemas that covers a situation, and (2) How can an agent organise his memory so as to represent in the most compact way the situations he has encountered so far, in other words how can he build up his cognitive memory.

2.1 Retrieving prototype schemata

The retrieval is proposed to take place in two steps. In the first step all schemas are activated whose descriptions match, even partially. Then the schemas are ordered based on their score and the best schema-set, i.e. the minimal set with the highest score is chosen.

A description d_s, called the source-description, matches with another description d_t called the target-description if its contents are equal. The match also results in a *score*, which is designed to accomodate flexible matching of schemata, compatible with the notion of a prototype. The score $< l, g >$ decomposes into a *loss* l and a *gain* g. Let u_s be the strength of the source description and u_t that of the target description, and $m = 1$ if the content of d_s matches with the content of d_t and $m = 0$ otherwise. The score is equal to $m \times u_t \times u_s$. When the score is negative, loss is equal to the absolute value of the score, otherwise gain is equal to the score.

The result of matching a source-schema s_s with a target-schema s_t consists of pairs relating each target description with each source description. A global score is calculated by taking the average loss and the average gain of all matches. There are possibly descriptions in the target-schema s_t which do not have any matching equivalent in the source-schema (in which case the loss is 1.0 for that description) and vice-versa.

In a first pass, a new situation-schema (the source) is matched with every prototype-schema (the target) resulting in a *match-matrix* for every prototype-schema,

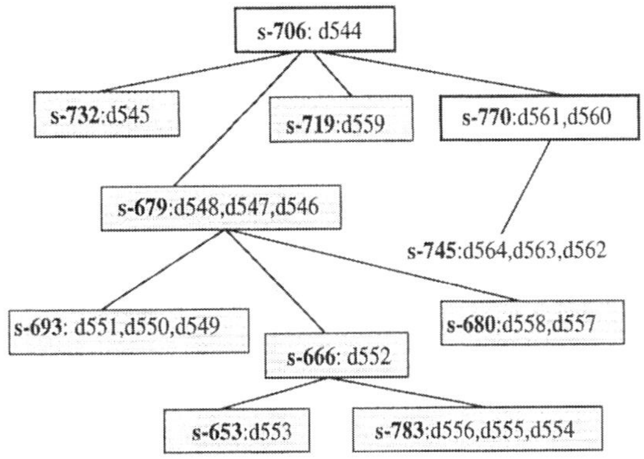

Figure 1: Partial prototype hierarchy in one agent. Each node specifies the name of the schema and the descriptions in it. For example, schema 719 has the description d559 in its description set. The links indicate inheritance relations between schemas. For example, s-719 inherits from s-706.

which indicates for every description what the outcome has been. Some of the descriptions in the target-schema are slots filled by other schemata. These descriptions can only be matched when an examination of the corresponding prototype schemata has been performed. So in a consecutive series of passes, match-matrices are further completed, causing the results of matching to propagate downwards in the schema hierarchy.

Consider for example the prototype hierarchy in figure 1, and a new situation schema with the following description set:

```
s-496: d544, d561, d560, d566, d565
```

There are only two prototypes that completely match: s770 and s706. All the prototypes that appear shaded partially match (because they all inherit directly or indirectly from s-706). Prototypes that inherit from s706 but have no other matching descriptions will progressively have less and less gains and higher losses. d566 and d565 do not match anywhere. The match-matrix for s770 is as follows. Total score is $< 0.0, 1.0 >$.

Schema	Description	Loss	Gain
s770	[slot s-706]	0.0	1.0
s770	d561	0.0	1.0
s770	d560	0.0	1.0

Given match-matrices for every schema in memory, the next step is to find the best match. This is the schema with the highest total score and the one that covers most of the descriptions in the source. In the present case, this is s-770.

2.2 Reorganising memory

When there is a set of prototype schemata that completely cover the situation schema with a perfect match, then the new situation can be *assimilated* by the memory. No changes are necessary and the new situation is just another case of what has already been seen before.

A more complex case arises when there are no prototype schemata covering the new situation. We then say that the memory needs to *accomodate* the new situation, in the sense that it has to adapt the repertoire of available schemas and their interrelationships. The terms assimilation and accomodation have obviously been taken from Piaget (Piaget, 1965). He was the first to see memory as an adaptive complex dynamical system with assimilation and accomodation as primary operators. Concretely, there are two cases.

1. When no matching target schema at all could be found, we are dealing with a completely new case. Hence the memory is expanded with a new prototype that contains the descriptions of the new situation.

2. The second case is more complex. A prototype (or chain of prototypes - see below) has been found but it is only a partial match. In this case a *lift operator* is used to repair memory.

Consider a source and target schema as in figure 2. There are some descriptions that match, there are some descriptions in the target that do not match with any in the source, and there are some descriptions in the source that do not match with any in the target. The lift-operator constructs a new prototype which contains all the descriptions common to the source and the target. It then constructs slots between the source and the target stripped off from the common descriptions now put in the new prototype.

There are two optimisations possible. First of all it can be that there are no descriptions in the source left after all common descriptions have been absorbed into the new prototype. The source schema covering the situation is then equal to the new prototype (see figure 3a). Alternatively, it could be that all the descriptions in the target occured in the source, so that the existing target schema acts as the new prototype and only a slot filled by the target must be established in the source schema (figure 3b).

The lift-operator needs to be applied recursively when repairs are made to different nodes in the hierarchy. If a schema inherits from another schema that has been changed, then this requires another lifting operation, and equally so for all downward prototypes in the hierarchy. Also in this recursive application optimizations are possible.

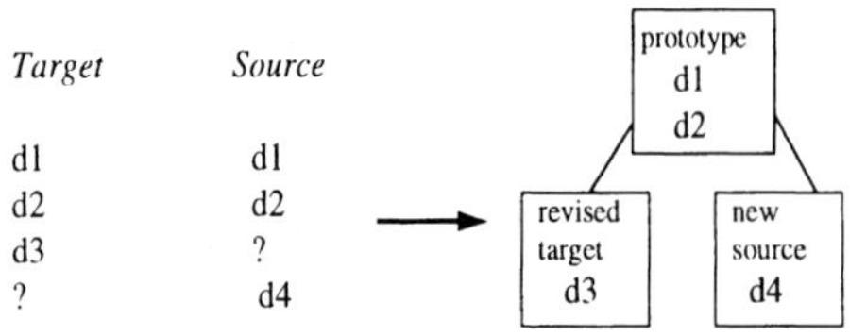

Figure 2: The lift-operator abstracts out the common descriptions from source and target and creates slots filled by links to the newly formed prototype.

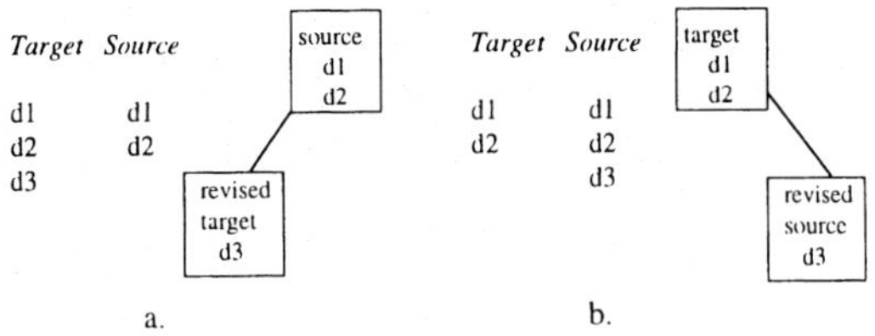

Figure 3: A new prototype need not always be constructed. 3a (left) occurs when all descriptions in the source are in the target schema. 3b occurs when all descriptions in the target were in the source.

2.3 Experiments

In testing the cognitive memory, a particular structure is first generated by the experimenter, for example a tree of prototypes or a tangled hierarchy. Then an environment is constructed which consists of a set of possible situations satisfying this structure. For example a hierarchical tree of certain dimensions is generated randomly and then a set of situations is generated to conform with this tree. The structure is known to the experimenter but not to any one of the agents. This turns out to be a good way to test the strength and limitations of the mechanism before using it on real world data for which the possible structure is also unknown to the experimenter.

Next situations from the environment are presented to the agent and the experimenter can follow in how far the agent manages to reflect the hidden structure introduced by the experimenter. The evolution of a cognitive memory with respect to a given hidden structure can be followed in a graph such as the one in figure 4. The hidden structure for this experiment contains 35 nodes organised in a strict hierarchy. The graph shows how the network grows when new situations are encountered. It shows clearly that the repair operations proposed above are remarkably efficient in extracting the hidden structure. Even though the situations are presented in an arbitrary order, the network reflects completely the hidden structure imposed by the experimenter, but unknown to the agent, as soon as all situations have been seen at least once. The network is also maximally efficient because it does not contain any prototypes that are not relevant to cover at least one situation. In the case of tangled hierarchies with multiple inheritance the performance will not be that good. There are occasionally different ways to accomodate a new situation and the network may perform generalisations which are not born out later.

The cognitive memories of different agents might be different because they may have seen the situations in a different order and may have consequently constructed different abstract prototypes. This is particularly true for very large hidden structures, where different agents may have seen only parts of the set of possible situations, or when there are multiple possibilities for structuring memory.

3 Lexicon Formation

Agents encounter situations through their normal interactions with the environment. We now consider what happens when the agents also engage in communicative exchanges describing to each other which situations they encounter. They do not create a new name for every new situation but instead structure the situation in terms of their cognitive memory and then use names associated with the prototypes that assimilate the situation as best

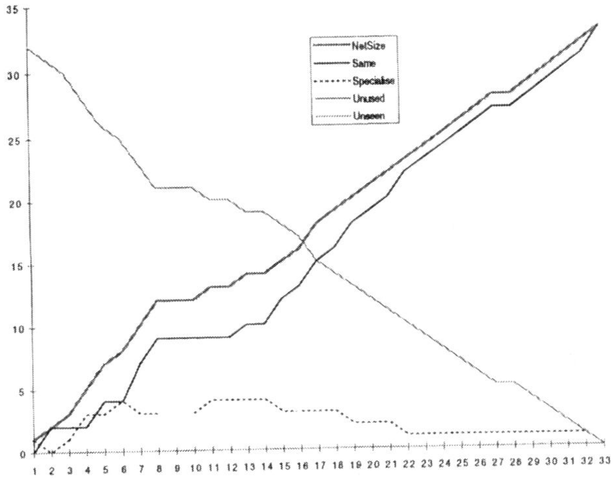

Figure 4: Evolution of cognitive memory for a single agent. *Net-size* plots the size of the network. *Same* measures how much the internal structure of the memory is equal to the hidden structure imposed by the experimenter. *Specialise* measures how many prototypes still need to be generalised or specialised. *Unseen* indicates how many situations are left that have not been seen yet. *Unused* is the number of prototypes that are superfluous. There are none in this example.

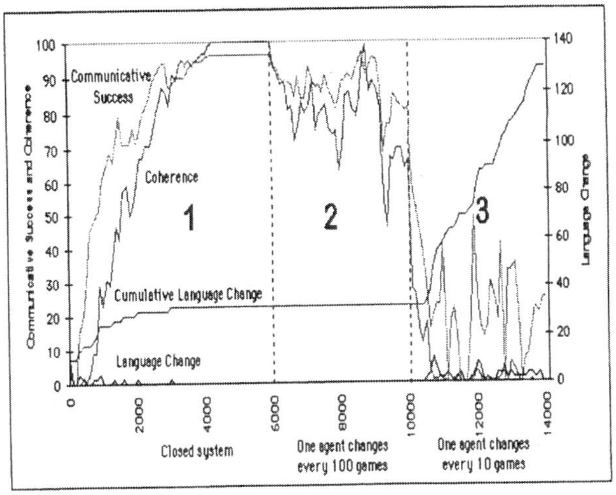

Figure 5: Results of experiments with a group of 20 agents who name 10 different items. A language quickly forms itself (phase 1) and resists an in- and outflow of its population (phase 2). When the flow is too high (phase 3) the language disintegrates.

as possible.

The lexicon formation process used in the experiments has been extensively described and researched in other papers (see for example (Steels 1996)), and is here only summarised. Each agent is assumed to have a lexicon L, which consists of a set of *associations* in the form of triples: <form, meaning, score>. The form is equal to a randomly generated string of letters. The meaning is equal to a prototype schema. The score $0.0 \leq u \leq 1.0$ indicates the strength of the association. There can be more than one association for the same form (homonymy) and more than one form for the same meaning (synonymy).

It has been shown in earlier work that a group of agents will arrive at a shared lexicon when a positive feedback loop is introduced between score and use. The score is based on success in a communication. The more an association is successful, the higher its score and thus the more agents use it. This causes success even more to increase and a winner-take-all situation arises where one form starts to dominate for every meaning. We have also been able to show that the lexicon continues to grow if new meanings are added and is resilient to a change in the population. This is illustrated in figure 5, which shows the results of experiments with naming single items in a changing population of agents.

3.1 Expressing conceptualisations

An utterance will consist of several words naming the different prototypes involved. Of course, it is possible to define a unique name for every prototype but it is more desirable to re-use as much as possible existing words. For example, if the two prototypes from which a new prototype p inherits are lexicalised, there is no need to introduce a new word for p. At the same time, it is also desirable to have a minimum number of words in the message itself. Often used prototypes should therefore be lexicalised even if the prototypes from which they inherit have their own names.

The minimal set of words covering a situation-schema is called its *cover-set*. To discuss the principles for constructing this set, it is useful to represent the set of prototypes assimilating a given situation-schema in a tree, further called the *coverage tree*, with the most specific prototypes near the top of the tree. This tree represents the relevant parts of the prototype network, but upside down compared to figure 1.

Let n be the top-node of a coverage tree, then the words contained in an utterance, the *cover-set*, are assembled by performing a top-down walk through the tree starting from n. If n has been lexicalised, then its associated word is the only word in the cover-set, otherwise the cover-set is equal to the union of the cover-sets of its sisternodes (see figure 6).

3.2 Constructing a Lexicon

When an agent is playing the role of speaker and there are nodes in the coverage tree which are not lexicalised, then new words may be created with a certain proba-

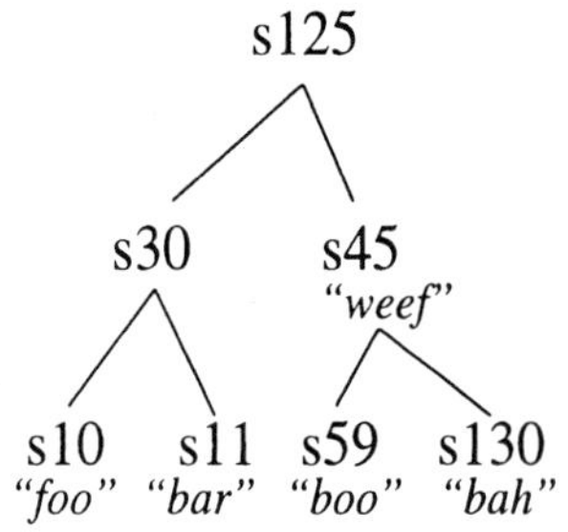

Figure 6: The coverage tree represents the hierarchies of prototypes that assimilate a given situation-schema, from the more specific to the more general. For this particular tree, the cover-set, which is the minimal set of words covering s125, is equal to {foo, bar, weef}

bility the word creation probability w_c. The agent will create the minimum set of words. For example, if there are no words at all, the top node of the coverage-tree is lexicalised. If some sister-nodes are lexicalised the non-lexicalised other sister-nodes are. At this point there are no provisions yet for lexicalising often used nodes, although this could be achieved by making lexicalisation dependent on the frequency of use of a prototype.

3.3 Acquiring a Lexicon

When the lexicon of the hearer fails to cover all the words transmitted by the speaker, repair actions need to be performed, possibly both to the memory and to the lexicon. In general, changes are only made when the hearer can make reasonably educated guesses. Otherwise it is better to wait for a situation that has less uncertainty. More specifically the following cases can be seen:

1. If the word is completely unknown, the hearer stores a new association, with a probability w_a, the word absorption probability, between the most specific prototype schema and the word.

2. If there are several words, some of which are known and the other ones unknown, an attempt is made to find a conceptualisation that would map onto the words. Occasionally it will be necessary to perform new lift operations to create prototype schemas compatible with new words. For example, if words for "female" and "horse" are used by the speaker, the hearer is forced to decompose his prototype for "female-horse".

4 Congruence of Memory

We are now in a position to consider the growth of cognitive memories independently constructed by individual agents but coupled through a shared environment *and* through the adaptive language games described in the previous section. The first step is to develop a measure to identify whether memories are indeed congruent. This is not easy because each agent develops different schemas, and an agent may have seen parts that other agents did not see yet.

We say that two prototype schemas are congruent iff *all* their descriptions (including the ones they inherit) match. This means that two prototypes are congruent if they cover the same set of situations. The schemas are not necessarily equal because the prototypes from which they inherit may be structured differently. Next, we pair the test network to the memory of an agent, by constructing a mapping from all nodes in the test network to their congruent schema in the agent, if there is one. The coverage of a network is equal to the percentage of nodes that have a congruent schema. The coverage for a group of agents is equal to the number of nodes in the test network for which at least one agent has a congruent schema. The group congruence of a test network schema s is equal to the percentage of agents that have a congruent prototype schema for s.

Figure 7 shows the results of experiments with a group of 10 agents. The test network contains 43 nodes with a hierarchy of maximum depth 5 and maximum width 5. The description-set in each prototype varies from 0 to 5 descriptions. The agents start with no cognitive memory and no shared lexicon. They are within the same environment (partly explaining similar generalisations), but also engage in language games in which they conceptualise and then verbalise situations. We see clearly that congruence is progressively arising.

Figure 8 shows the build up of the lexicon of the agents for the same experiment. We see that the lexicon slowly builds up in parallel with the increased congruence in the memories. The experiment runs for 5000 games, after which a high success rate is reached.

5 Conclusions

The paper has investigated how the cognitive memories of distributed autonomous agents may become congruent based on structural coupling. Weak structural coupling happens already when agents are within the same environment and encounter similar situations. A stronger form occurs when coupling takes place through language. The agents use the structure of their memories to conceptualise situations and transmit the lexicalisation of these conceptualisations rather than unique names for every situation. The organisation of memory therefore influ-

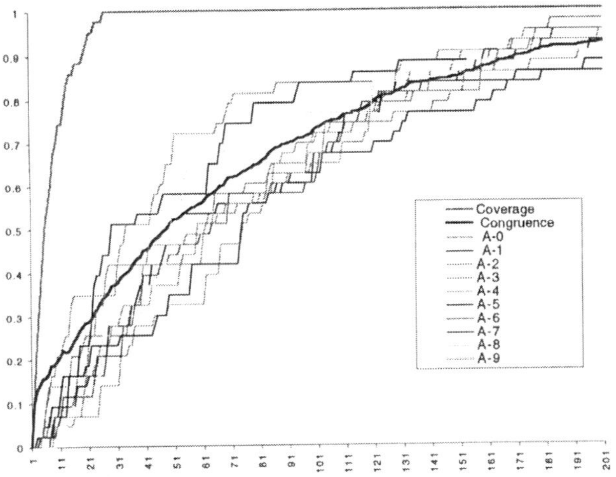

Figure 7: This diagram shows the evolution towards memory congruence, per cycle of 10 games. The thick line going up fastest is the coverage of the group. Very quickly, at least one agent has a congruent node for every schema in the test network. The second thick line corresponds to the average congruence for all agents. The other lines indicate the congruence between the memory of a single agent and the memory of the total target network.

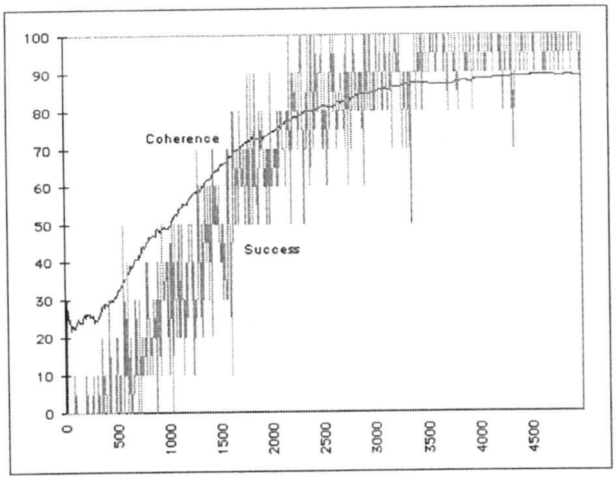

Figure 8: This diagram shows the game success reaching almost total success once the memories have stabilised. Language coherence builds up alongside communicative success and memory congruence.

ences the lexicon in order to economise on the number of words.

These experiments are a step towards understanding how shared language and meaning may evolve in groups of distributed autonomous agents located in open, changing environments. We are at present working on cognitive memories with more complex representations for a prototype (including variables so that unification can be used for matching prototypes with situations), and with a category formation system that constrains the possible fillers of a slot and thus makes the matching process more efficient.

Acknowledgement

The research described in this paper was financed and conducted at the Sony Computer Science Laboratory in Paris. The simulations presented have been built on top of the BABEL toolkit developed by Angus McIntyre (McIntyre, 1997) of Sony CSL. Without this superb toolkit, it would not have been possible to perform the required investigations within the time available. Discussions with Frederic Kaplan and Joris Van Looveren have contributed to the ideas expressed in this paper.

References

Lakoff, G. (1987) Women, Fire and Dangerous Things: What Categories Reveal about the Mind. University of Chicago Press, Chicago.

McIntyre, A. (1997) The Babel Toolkit for Studying Language Formation and Evolution. Submitted for Publication.

Hurford, J., M. Studdert-Kennedy and C. Winter (ed.) (1998) Evolution of Human Language. Edinburgh Univ. Press. Edinburgh.

Maturana, H.R. and F.J. Varela (1972) Autopoiesis and Cognition. The Realization of the Living. D. Reidel Pub. Cy. Dordrecht.

Piaget, J. (1965) L''equilibration des structures cognitives. PUF, Paris.

Steels, L. (1996) Self-organizing vocabularies. In: Langton, C. (ed.) (1996) Proceedings of Alife V, Nara Japan. The MIT Press, Cambridge Ma.

Steels, L. (1997a) The synthetic modeling of language origins. *Evolution of Communication*, 1(1):1-35. Walter Benjamins, Amsterdam.

Steels, L. and P. Vogt (1997) Grounding adaptive language games in robotic agents. In: Harvey, P. and P. Husbands (eds.) (1997) Proceedings of European Conference on Artificial Life. The MIT Press, Cambridge MA.

Learning to Perceive the World as Articulated: An Approach for Hierarchical Learning in Sensory-Motor Systems

Jun Tani[*], **Stefano Nolfi**[**]

[*] Sony Computer Science Lab.
Takanawa Muse Building, 3-14-13 Higashi-gotanda, Shinagawa-ku,Tokyo , 141 JAPAN
tani@csl.sony.co.jp

[**] Institute of Psychology, National Research Council
15, Viale Marx - 00187, Rome, Italy
stefano@kant.irmkant.rm.cnr.it
(During stay at Sony CSL)

Abstract

This paper describes how the internal representation of the world can be self-organized in modular and hierarchical ways in a neural network architecture for sensory-motor systems. We develop an on-line learning scheme – the so-called mixture of recurrent neural net (RNN) experts – in which a set of RNN modules becomes self-organized as experts in multiple levels in order to account for the different categories of sensory-motor flow which the robot experiences. The proposed scheme was examined through simulation experiments involving the navigation learning problem, in which a robot equipped with range sensors traveled around rooms of different shape. It was shown that representative building blocks or "concepts" corresponding to turning right and left at corners, going straight along corridors and encountering junctions are self-organized in their respective modules in the lower level network. In the higher level network, the "concepts" corresponding to traveling in different rooms are self-organized by combining the "concepts" obtained in the lower level into sequences. The robot succeeded in learning to perceive the world as articulated at multiple levels through its recursive interactions.

1. Introduction

How can sensory-motor systems attain an internal representation of the world in structurally organized ways? The consensus in cognitive science and artificial intelligence is that complex worlds can be represented efficiently utilizing modular and hierarchical structures of symbol systems (Newell1980). However, it is still not well understood how such modular and hierarchical representation, if they existed, becomes self-organized in analog neural systems by means of their iterative sensory-motor interactions.

The difficulty lies in the question of how the continuous sensory-motor flow can be perceived as being articulated into sequences of meaningful representative modules. Kuniyoshi (Kuniyoshi, Inaba, & Inoue1994) addressed this articulation problem in the robot learning context. In his experiment with an assembling robot, the robot recognizes the various task performances by decomposing them into sequences of modular representations. Subsequently, the robot is able to learn various tasks in terms of compositions of the reusable modular representations obtained. For attaining such a modular representation, the task performance was temporally segmented by means of detecting "meaningful changes" in the observed sensory flow. The problem, however, is that the definitions of these "meaningful changes" were predetermined by designers. Our investigation is of how a robot can define "meaningful changes" by itself, the use of which allows a task performance to be segmented into reusable modules.

Robot navigation learning, which has a quite long research history, faces the same type of problem. There are basically two types of approach. One is the neural network learning approach. Krose (Krose & Eecen1994), Zimmer (Zimmer1996) and Nehmzou (Nehmzou1996) showed that for relatively simple workspaces, localization problems for robots can be solved using the topological preserving map scheme (Kohonen1982). It is, however, difficult to scale-up using this scheme since the very plain representation by a single neural network hardly organizes the modular and hierarchical structure of the learned contents at all. The other approach is the machine learning approach, which is used in landmark-based navigation (Kuipers1987; Mataric1992). In this approach, the travel of the robot is temporally segmented by means of landmarks such as turning at corners, encountering junctions, or going straight along corridors. This temporal segmentation enables the abstraction of robot experiences into a simple chain representation of these landmark types. The scheme can be scaled-up much more readily than the neural network learning approach since the landmarks play the roles of the representative modules. However, the problem is that the landmark types, which are defined by designers, are not necessarily intrinsic to the perceptions of a robot. The representative modules such as corners, junctions, or corridors, if necessary to the problem's solution, ought to be generated from the robot's experiences.

In this paper, we introduce a novel scheme based on the dynamical systems approach (Pollack1991; Beer1995; Schoner, Dose, & Engels1995; Smith & Thelen1994; van Gelder) whereby the problems of articulation and structural formation of modules and hierarchy are explained solely by the dynamical systems terms such as self-organization, coherence and phase transition. The scheme has been developed as inspired by a modular and hierarchical learning method using neural nets, namely the mixture of experts proposed by Jacobs and Jordan (Jacobs & Jordan1991). We have extended this original architecture dramatically such that it can cope with the learning of not only spatial patterns but also spatio-temporal patterns which sensory-motor systems are inevitably involved with. The reader will see that the sensory-motor flow is articulated in an autonomous manner in the form of modules, and that the modules' hierarchy becomes self-organized in our proposed architecture.

The paper introduces robot navigation learning as a prototype problem; our simulation experiments will illustrate how a set of primitive representative building blocks or "concepts" emerge and how they lead to the construction of "concepts" in the higher level in a dynamic fashion. Our hierarchical learning approach is developed in combination with the prediction learning scheme which is described in the next section.

2. Prediction Learning Using Sensory-Motor Flow

Learning to predict the next sensation means that the system must acquire some analogical model of the target observed. Elman (Elman1990) was the first to show that a recurrent neural network (RNN) can learn to predict word sequences by extracting regularity hidden in example sentences. Tani (Tani1996) applied RNN prediction learning to the navigation learning problem. In this scheme, a robot learns to predict how its future sensory sequences will depend on the action sequences taken in a given workspace. It was shown that a real mobile robot with a range sensor learned structure hidden in an obstacle workspace from the sensory-motor flow experienced. The structure of the environment was embedded in an attractor self-organized in the RNN by means of prediction learning. It was further shown that the obtained internal model by means of RNN dynamics was utilized to generate mental plans for goal-directed behavior. However, a crucial criticism of this scheme is that the prediction of sensory input is made in reality in a discrete temporal manner by means of the predefined branching mechanism. Branching plays the role of landmarks and invokes the temporal segmentation of the sensory-motor flow. Our new experiment is to attempt to eliminate these types of predefined mechanisms for temporal segmentation in the hope that the robot itself will find them.

One possible way to implement temporal segmentation of the sensory-motor flow is to focus on the magnitude of its change with time (Billard1996). For example, while a robot travels by following a straight wall using the range image, the image will be almost invariant. However, the sensory-motor state will change dramatically when the robot encounters a corner and starts turning left or right. This rapid change can be used as a signal for segmentation between the two behaviors of following straight wall and turning at a corner. However, the difficulty in this scheme is that the cornering behavior can be segmented several times since the sensory-motor state probably changes rapidly all through the cornering process. It is clear that the change of the sensory-motor state at a single moment provides only partial information about the on-going behavioral process. A specific mechanism is required whereby a meaningful time interval of the behavioral process, such as a cornering behavior, can be recognized as a unique event through extracting its specific spatio-temporal structure from the sensory-motor flow.

3. New Scheme

Our new proposal in this paper is to use multiple module RNNs, each of which competes to become an expert at predicting the sensory-motor flow for a specific behavior. The experts achieve their status through learning processes. For example, one module RNN would win in predicting the sensory-motor flow while traveling around a corner; another would win while following a straight wall. The switching between the winning RNN modules actually corresponds to the temporal segmentation of the sensory-motor flow. The essential point in this scenario is that the segmentations take place by means of pronounced changes in the observed dynamical structure in the sensory-motor flow, rather than just in temporal differences in the sensory-motor state. These highly pronounced changes correspond to switching between the dynamical functions, each of which is embedded in an RNN through having learned the specific sensory-motor flow. One might ask how each RNN can choose to learn its corresponding sensory-motor flow. The speciality of each module is determined during the processes of on-line learning. The competition between the modules during the simultaneous processes of recognition and learning result in generating their specialties. The next section will introduce a new architecture called the mixture of RNN experts which has been extended from the original idea of the mixture of experts first expounded by Jacobs and Jordan (Jacobs & Jordan1991).

3.1 Architecture

Figure 1 shows the proposed architecture for the mixture of RNN experts (MRE) which is used for the prediction-learning of the sensory-motor flow. Fig. 1(a) shows a hierarchical architecture consisting of two levels; more levels are possible in general.

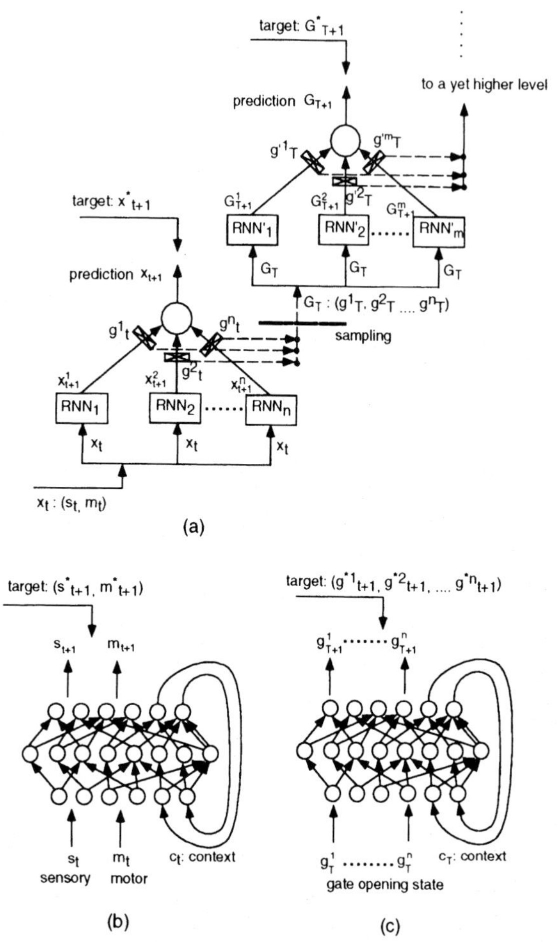

Figure 1 The complete architecture of the mixture of RNN experts for prediction learning. (a) hierarchical learning architecture, (b) details of each RNN module for learning the sensory-motor flow in the lower level, and (c) RNN module for learning the gate opening dynamics in the higher level.

Each RNN module in the lower level receives the sensory-motor inputs, $X_t : (s_t, m_t)$, and outputs the prediction of the sensory-motor inputs at a time Δt afterwards in the form $X_{t+1} : (s_{t+1}, m_{t+1})$, as shown in Fig. 1(b). The total output of the network is obtained from the weighted average of each output with its associated value of gate opening at the time g_t^i for all modules. The gate opening is computed dynamically with time using the prediction errors of each module, which are obtained from the difference between the prediction (s_{t+1}, m_{t+1}) and the outcome (s_{t+1}^*, m_{t+1}^*). The gate opens more if its module produces a relatively lower prediction error than the other modules. The module with the lowest error over a suitable time interval becomes the winner. The original work on the mixture of experts (Jacobs & Jordan1991) used a gating network which selected the module with the closest correspondence to the inputs. In our architecture, without using a gating network the module is activated autonomously as the result of dynamical competition between all modules over some time interval, utilizing on-line monitoring of the prediction errors. The winning module changes from one module to another as the profile of the sensory-motor flow changes with time.

The higher level network learns the gate opening dynamics of the lower level network. More specifically, each RNN module in the higher level samples the gate opening state of the lower level in the current time step $G_T : (g_T^1, g_T^2 .... g_T^n)$ and makes a prediction for the next time step G_{T+1}, as shown in Fig. 1(c). T denotes the time step in the higher level; the higher level sampling interval ΔT is much larger than that in the lower level. The modules in the higher level compete for gate opening $g_T'^i$, in the same way as shown for the lower level, and the resultant gate opening can be sent to yet higher levels in a recursive manner. The higher level network observes the lower level activities by means of perceiving its gate opening dynamics while the lower level network perceives the sensory-motor flow. In this manner, the signal is "bottom-up" as abstracted from one level to the next.

3.2 Algorithm

This subsection describes the mathematical formulae for the proposed scheme of the MRE. Suppose a single level network consists of n RNN modules, where x_t^i, y_{t+1}^i, y_{t+1}^{*i} and g_t^i are the inputs, the outputs, the target outputs for teaching and the gate opening of the i-th module RNN, respectively. x_t and y_{t+1} correspond to the sensory-motor state or the gate opening state depending on the levels of the network.

The "soft-max" activation function is used to represent the i-th gate opening g_t^i given by:

$$g_t^i = \frac{e^{s_t^i}}{\sum_{j=1}^{n} e^{s_t^j}} \tag{1}$$

where s_t^i is the current internal value of the i-th gate opening. The total output of the network is y_{t+1}, given by:

$$y_{t+1} = \sum_{i=1}^{n} g_t^i \cdot y_{t+1}^i \tag{2}$$

We define the following likelihood function L which is maximized for prediction learning: it has been obtained by modifying the original definition of Jacobs and Jordan (Jacobs & Jordan1991).

$$ln\, L = ln \sum_{j=1}^{n} g_t^i \cdot e^{\frac{-1}{2\sigma^2}\|y_{t+1}^* - y_{t+1}^i\|^2} \quad (3)$$

σ denotes a scaling parameter.

Both the weight of each RNN and the gate opening are updated simultaneously such that the likelihood function is maximized. This point is essential for the on-line learning scheme. In order to obtain the update rules for these two processes, we consider the partial derivatives of the logarithm of the likelihood function with respect to the internal value s^i and with respect to the output of the i-th RNN y^i given by:

$$\frac{\partial ln\, L}{\partial s_t^i} = g(i \,|\, x_t, y_{t+1}^*) - g_t^i \quad (4)$$

$$\frac{\partial ln\, L}{\partial y_t^i} = g(i \,|\, x_t, y_{t+1}^*) \frac{(y_{t+1}^* - y_{t+1}^i)}{\sigma^2} \quad (5)$$

where $g(i \,|\, x_t, y_{t+1}^*)$ is the *a posteriori* probability that the i-th module RNN generated the target vector y_{t+1}^*, in terms of x_t. Explicitly, this is given by

$$g(i \,|\, x_t, y_{t+1}^*) = \frac{g_t^i \cdot e^{\frac{-1}{2\sigma^2}\|y_{t+1}^* - y_{t+1}^i\|^2}}{\sum_{j=1}^{n} g_t^j \cdot e^{\frac{-1}{2\sigma^2}\|y_{t+1}^* - y_{t+1}^j\|^2}} \quad (6)$$

where $\|y_{t+1}^* - y_{t+1}^j\|^2$ represents the square of the error of the current prediction. Eq. (4) denotes the direction of update for the internal gate opening value s_t^i. The differentiation of $ln\, L$ with respect to y_{t+1}^i involves the error term $y_{t+1}^* - y_{t+1}^i$ weighted by the *a posteriori* probability associated with the i-th module RNN as shown in Eq. (5). Thus the connective weights of the RNN are adjusted to correct the error between the output of the i-th RNN and the global target vector, but only in proportion to the *a posteriori* probabilities. By this means, the individual RNN which is the expert for the on-going input sequence tends to learn exclusively. The error distributed to each module RNN is:

$$error_{t+1}^i = g(i \,|\, x_t, y_{t+1}^*) \cdot (y_{t+1}^* - y_{t+1}^i) \quad (7)$$

The details of the derivation of Eq. (4) to Eq. (7) are given in Ref. (Jacobs & Jordan1991).

Upon obtaining the mathematical formulae, the actual update of the gate opening and the connective weights for each RNN are computed through the use of the back-propagation through time (BPTT) algorithm (Rumelhart, Hinton, & Williams1986). In this computation, the sequence of the sensory-motor inputs as well as the gate internal states for the previous l steps are stored temporally in the window memory. When new sensory-motor inputs are received, the window memory is shifted one step forward, the forward and backward computation by means of the BPTT are iterated for *nepochs* times, and finally the sequence of l steps of the gate internal states as well as the connective weights for each RNN module are updated. The update for s_k^i, which is the ith gate internal state in the kth step in the window memory, is obtained as being:

$$\Delta s_k^i = \epsilon_g \cdot \frac{\partial ln\, L}{\partial s_k^i} - \eta_g \cdot (s_k^i - s_{k-1}^i) \quad (8)$$

The first term in the right-hand side of the equation represents the direction of the update obtained in Eq. (4); the second term represents the damping term which suppresses abrupt changes in the gate opening; ϵ_g and η_g are parameters. This update is computed in the forward direction in the window memory from k=1 to k=l. The error obtained from Eq. (7) is back-propagated (Rumelhart, Hinton, & Williams1986) through the window memory for each RNN; the update of the connective weights is obtained by means of the steepest descent method utilizing parameters for the learning rate ϵ and for the momentum α.

4. Simulation Experiments

4.1 The environment

The scheme proposed above was investigated in the context of the navigation learning problem by simulation. We assumed a mobile robot with a sensor belt on its forward side holding 20 laser range sensors. The robot, upon perceiving the range image of its surrounding environment, maneuvers in a collision-free manner using a variant of the potential method (Khatib1986). (For further details of this maneuvering scheme, see Ref. (Tani1996).)

For our simulations, we adopted two different rooms, namely Room A and Room B connected by a door, as shown in Fig. 2 (a). Fig. 2 (b) shows an example of the sensory-motor flow which corresponds to the robot travel indicated by the dotted line in Fig. 2 (a). In this workspace, the robot travels around one room three times, then enters the other room going through the newly opened door and travels around the other room three times. The on-line learning experiment was conducted while the robot moved between rooms for a total of 5 room encounters. The entire travel of the robot in this simulation took about 2100 Δt steps. The lower level network, which consists of 5 RNN modules each of which has 6 inputs, 6 outputs, 4 hidden units and 2 context units, learns to predict the sensory-motor state in the next step. The higher level network, which consists of 5 RNN modules each of which has 5 inputs, 5 outputs, 4 hidden units and 2 context units, learns to predict the gate opening state in the lower level network in the next step. Other parameter settings for the networks are $\epsilon = 0.002$, $\alpha = 0.9$, $\epsilon_g = 0.007$, $\eta_g = 0.02$. These settings are the same for both levels. The sampling interval in the higher level is 10 times longer than that in the lower level ($\Delta T = 10 \cdot \Delta t$). We observed how modules become

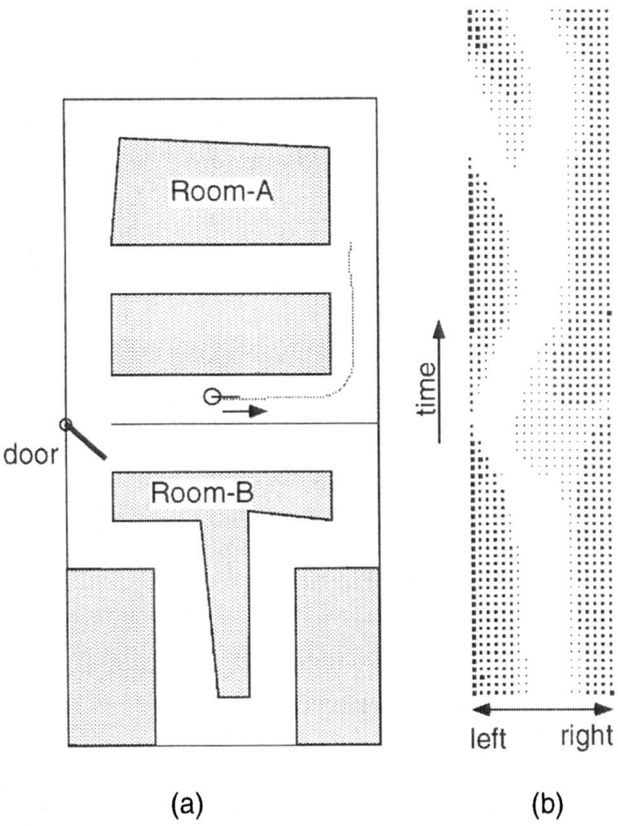

Figure 2 (a) Simulation workspace consisting of two rooms connected by a door. (b) the time development of the simulated range image while the robot traveled.

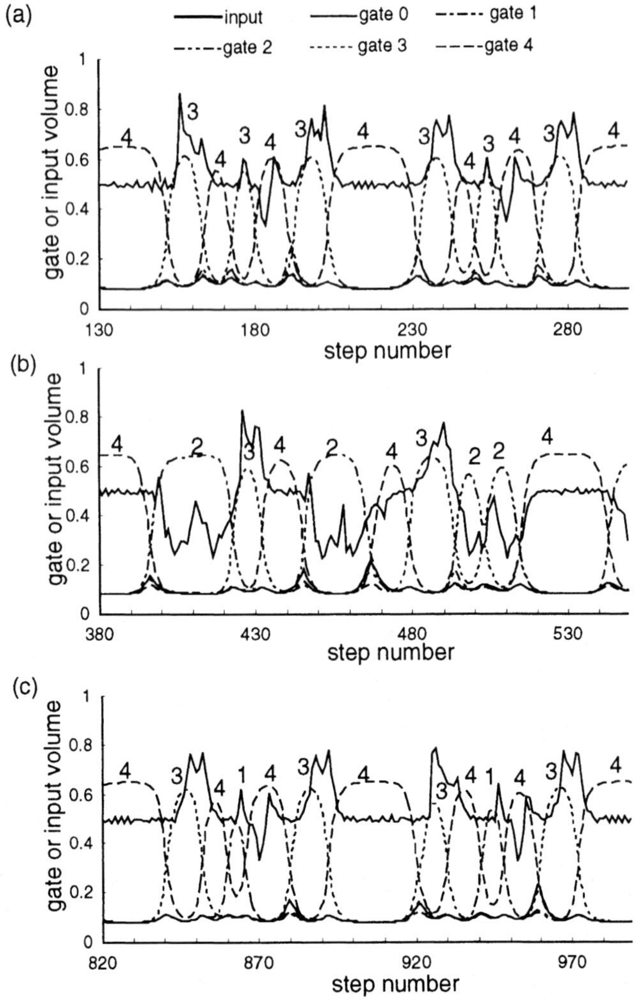

Figure 3 Time development of the opening of 5 gates and of a motor input in the lower level network for three different periods. The number near the data denotes the current winning gate.

self-organized in a hierarchical manner by looking at the gate opening dynamics taking place during the prediction learning of the two levels.

4.2 Results

We recorded gate opening dynamics both in the lower and in the higher levels during the entire learning process. First, let us consider the gate opening processes in the lower level network. Figure 3 shows the time development of each gate opening state and of the motor input in the lower level for three different periods. Figure 4 illustrates when and which module wins in the lower level network along the course travelled for each of the three different periods. Fig. 3 (a) shows the profiles for the period from step 130 to step 300 while the robot travelled in Room A for the first time. It can be seen that gate4 and gate3 open in turn as the profile of the motor command changes. In Fig. 4 (a), it is seen that the opening of gate4 corresponds to following a straight wall, while the opening of gate3 corresponds to both a left turn at a corner and to passing a T-junction. Fig. 3 (b) shows the profiles for the period from step 380 to step 550, when the robot experienced Room B for the first time. One can see that gate4, gate2 and gate3 open in turn. Fig. 4 (b) shows that these opening events corresponded to following a straight wall, making a right turn at a corner and making a left turn at a corner, respectively. Fig. 3 (c) shows the profiles for the period from step 820 to step 990, when the robot travelled around Room A for the second time. A remarkable finding is that the gate opening dynamics for this period differ from those observed during the first encounter with Room A. From Fig. 4 (c), one can see that the opening of gate3, which corresponded to both making a left turn at a corner and passing a T-junction in the previous encounter, now corresponds only to making a left turn at a corner, and that the opening of gate1 now corresponds to passing a T-junction. After this period, the learning processes in the network appeared to have stabilized and no further dramatic changes in the

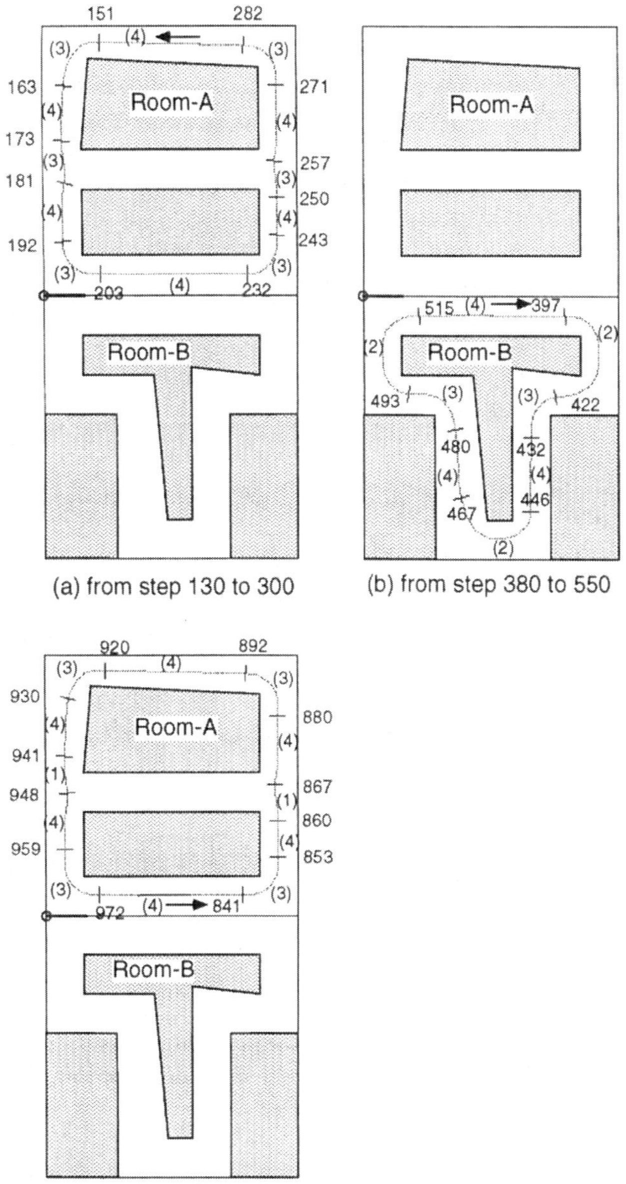

Figure 4 Along the trajectory of the robot travel indicated by the dotted lines, the number in parenthese indicates which module wins and the number without parenthese denotes the step number when the module switching took place.

correspondence of the gate openings were found. By the end of the simulation, four types of meaningful concepts were generated using 4 RNN modules out of the 5 modules available to the lower level network. An important observation is that the process of generating concepts is totally dynamic in the sense that the correspondence between the RNN modules and their associated behavior is not static during the on-line learning process.

Next, we describe the gate opening dynamics in the higher level network. Figure 5 shows the opening of the 5 gates for the whole period of on-line learning. (The step number in this graph denotes the sensory-motor step number in the lower level, for clarity.) One can see that

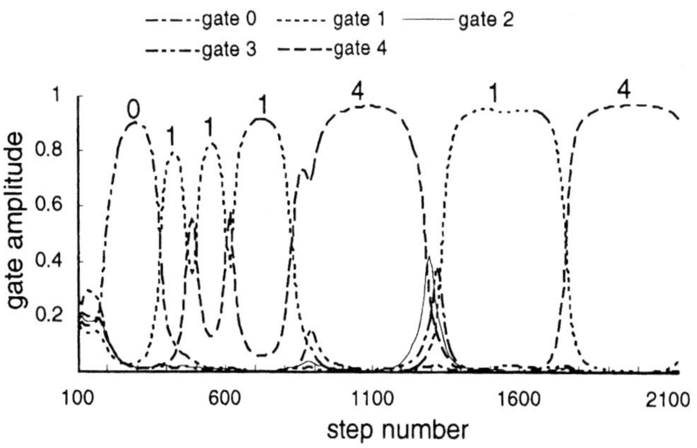

Figure 5 Gate opening dynamics in the higher level network during the whole process of learning.

the stable switching of the gate opening between gate4 and gate1 takes place after 800 steps. This switching actually corresponds to the movement between rooms during the travel, where the open state of gate4 and gate1 correspond to travel in Room A and in Room B, respectively. We observe that gate0 opened only in the beginning while the robot traveled in Room A for the first time. The dynamic replacement of module0 by module4 for the representation of Room A evidently took place because the module representation in the lower level network also changed, as noted above. It is readily understood that the dynamics in higher level network can be stabilized only after stabilization occurs in the lower level network.

From so far we have obtained results in the simulation experiments, it can be concluded that the proposed MRE architecture was successful in learning about the environment in a hierarchical way through the sensory-motor interactions of the robot. The lower level network learned to predict the row profile of the sensory-motor flow by organizing the modular representation of specific behavior. The higher level network did likewise for the sequences of segmented behavior by creating the higher concept of a room. Therefore, it can be said that the robot is not just perceiving the current sensory-motor flow but that it is also recognizing the background context of its behavior and situation.

We repeated this learning experiment for five times with different initial conditions including the starting position of the robot in either Room A or Room B and with different random initial connective weights of the networks. By looking at structures self-organized in the higher level network in these five experiments, equivalent module structures to those in the previous results, representing Room A and Room B, were found in four cases out of the five. Following this, we observed the lower level structures for these four cases and found that equivalent module structures to the previous result appeared in three cases, while the structures were different in one case. In the case where we did not observe clear module structure corresponding to two separate rooms in the higher level, it was observed that the lower level structures continued to change gradually which prevented the higher level structures from stabilizing. The stability in the higher level depends substantially on that in the lower level. These results reveal that the self-organization processes do not always arrive at one optimal solution. They can generate unstable and non-optimal structures by chance. Further studies are required for detailed understanding of "stability and diversity" problems in the hierarchical learning scheme.

5. Analysis and Discussion

5.1 On the dynamic mechanism for articulation

We have seen that building blocks for representing specific sensory-motor structures are self-organized in the lower level; the building blocks in the higher level are constructed by combining those in the lower level. The results may be interpreted as being the emergence of internal "symbols". However, the definition of our "symbols" is quite different to that used in traditional cognitive science studies (Newell & Simon1976; Newell1980). The "symbols" in our scheme are articulated not by the external designer's views but by the view intrinsic to the robot through its own experiences. In fact, the articulation emerges through the interactions between the system and its environment. For the purpose of putting forward this argument, we will now attempt to explain the qualitative mechanism of articulation from the dynamical systems perspective.

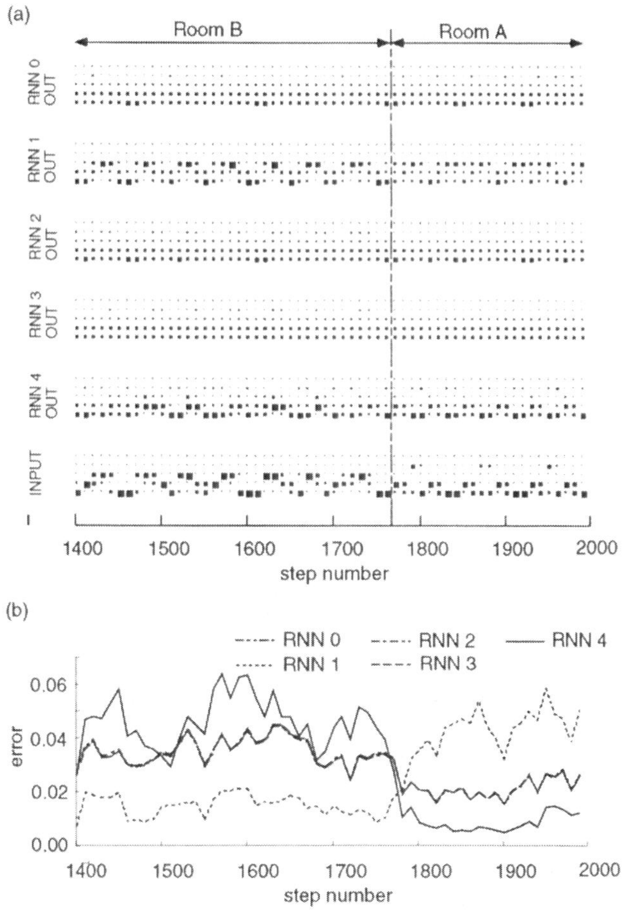

Figure 6 (a) shows the sequence of inputs (shown in the bottom row) and the corresponding prediction outputs by each RNN module (shown in the upper five rows) in the higher level network. (b) indicates the time dependence of the prediction error by each RNN module.

In order to understand how the switching of modules takes place dynamically as corresponding to room entering, we examined the time series of prediction outputs by each RNN module in the higher level network. Figure 6 shows the time series of prediction outputs and the corresponding error for each RNN module in the higher level network recorded from step 1400 to step 2000 during which period the robot moved from Room B to Room A. In Fig. 6 (a), the upper five rows represent the sequence of prediction outputs by the five RNN modules. The five squares aligned vertically in each five column row represent the values predicted for the five coming inputs (the sampling of the five gate openings in the lower level network) by their size. The largest square area corresponds to the value of 1.0 and zero area corresponds to a value of 0.0. The bottom row represents the sequence of five inputs. The robot moved from Room B to Room A at around step 1770 (denoted by a hatched line in the figure). Fig. 6 (b) shows the time series of the prediction error for each module. By looking at Fig. 6 (b), it is observed that the prediction error by RNN 1 is continuously minimized relative to the other RNNs until step 1770. During this period, it is seen, from Fig. 6 (a), that the output sequence by RNN 1 keeps coherence with the input sequence to a certain extent; RNN 0, RNN 2 and RNN 3 are not activated at all, while RNN 4 is activated but is incoherent with the input sequence. On the other hand, the output sequence by RNN 4 becomes coherent with the input sequence after step 1770 through minimizing its error, while the output sequence of RNN 1 loses coherence with the input. This switching of the winning RNN modules takes place rather quickly within several iteration steps of the RNNs. We speculate that the robot recognizes its surrounding in a particular room by means of achieving temporal coherence between the dynamics of a particular RNN module and the environmental dynamics.

In order to investigate the essential dynamical structure of the organization of the coherence and the incoherence appearing in the coupling between the internal and the environmental dynamical systems, we compared the phase trajectory of each RNN module for its intrinsic dynamics and for its coupled dynamics with Room A and with Room B. Figure 7 shows such a comparison. The five plots in the upper row in this Figure show

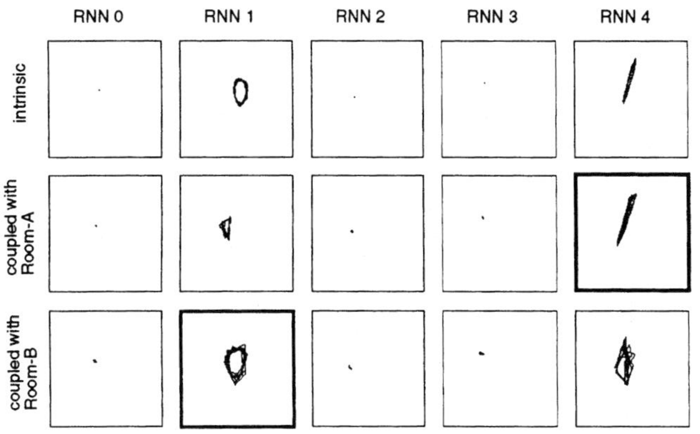

Figure 7 The phase trajectory for each RNN module in the higher level is shown for its intrinsic dynamics and for its coupled dynamics with Room A and Room B, in the upper row, middle row and bottom row, respectively.

the phase trajectory for the intrinsic dynamics of each RNN. Each RNN can be iterated autonomously without the input sequences by installing feed-back loops from the prediction outputs to the inputs. (This autonomous iteration of each RNN corresponds to mental rehearsing of the learned temporal patterns.) We plotted the activation trajectory for two context units in X and Y during the autonomous iteration using fixed connection weights for each RNN which had been obtained in step 2000 in the previous experiment. It is observed that the trajectories of RNN 0, RNN 2 and RNN 3 converge to a fixed point and that those of RNN 1 and RNN 4 do not converge to a point but generate invariant closures in the phase space. Further analysis showed that the dynamical structure of RNN 0, RNN 2 and RNN3 is characterized by a fixed point attractor, while the dynamical structure of RNN 1 and RNN 4 is characterized by a quasi-periodic attractor. The five plots in the middle row and the bottom row show the phase trajectories of the coupled RNN dynamics with the Room A and Room B environments, respectively. The phase trajectories are plotted for the context units' activation states in response to receiving the input sequences while the robot travelled around the two rooms. In these plots, it is seen that RNN 0, RNN 2 and RNN 3 each generated the same fixed point attractor from their intrinsic dynamics regardless of whether the RNN was coupled with Room A or with Room B. The plots for RNN 1 and RNN 4 are different. RNN 1 exhibits a similar phase trajectory to its intrinsic one when coupled with Room B; RNN 4 does likewise when coupled with Room A. This observation coincides with the previous observation that the prediction sequence by RNN 1 and by RNN 4 became coherent with the input sequence while travelling around Room B and Room A, respectively.

In our previous work (Tani1996), we studied how the RNN prediction process can be *situated* in the environment through sensory-motor experiences. Our analysis (Tani1996) showed that the prediction process is successful when coherence is achieved between the internal RNN dynamics and the environmental dynamics. The entrainment (Endo & Mori1978; Beer1995) of the RNN dynamics by the input flow can take place when the RNN learns to share the dynamical structure of the environment (Tani1996). The same mechanism can explain the autonomous switching of modules in the current study: one module is activated in a manner inhibitory to the others by achieving coherence between its intrinsic dynamics and with a specific dynamical structure hidden in the input flow. When the essential dynamical structure in the input flow changes, the currently activated module loses coherence with the flow, while another module is activated through gaining coherence with the flow. This winning module switching takes place rather quickly as a phase transition of the dynamical state from one stable attractor to another. By means of switching of coherence among the repertoire of the intrinsic dynamics, this phase transition actually results in the articulation which the system perceives internally for the structural changes in the sensory-motor flow (see the schematic diagram for this explanation in Figure 8.)

5.2 Recursive chains from state to function

Another important aspect which should be discussed is the relationship between state and function in the hierarchical learning. The direct observation of the sensory-motor state provides only non-robust information about its present process since the state can evolve in many ways. What should be focussed on is rather the spatio-temporal structure hidden in the time development of the state, since such structures could be similar in many cases even when the state is changing quantitatively. The RNN, which is basically an adaptive type dynamical function, is used for capturing such similar structure from the observed time development of the state. This time development of the state is, eventually, represented by one of the RNN functions. The higher level observes that which RNN function is currently activated in the lower level in terms of its gate opening state. This gate opening state can vary as the result of structural change in the lower level. The resultant time development of the

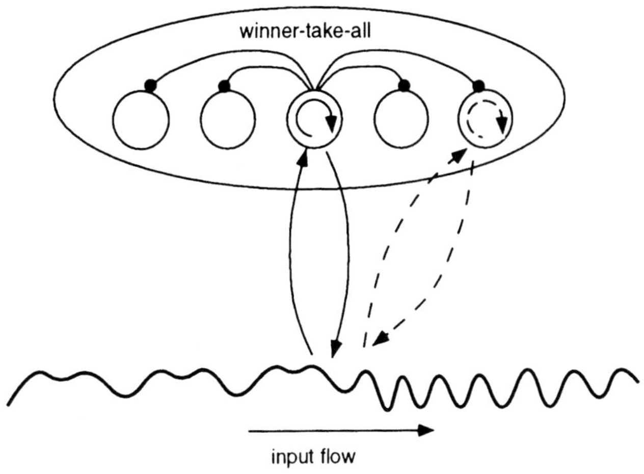

Figure 8 Dynamic mechanism of articulation. The input flow is articulated when coherence with the flow is switched among the repertoire of the intrinsic dynamics contained in multiple modules.

gate opening state is again captured by the RNN functions in the higher level. Here, we see that the aim of the hierarchical learning is to organize such recursive chains from the state to the function, and from the function to the state, through the level of abstraction.

5.3 Related works

The current study was inspired by other adaptive agent research based on the dynamical systems approach by, for example, Beer (Beer1995), Yamauchi and Beer (Yamauchi & Beer1994) and Smithers (Smithers1996). However, it is notable that the dynamical systems approach, which especially emphasizes coherence and structural coupling (Varela, Thompson, & Rosch1991) of the internal system with its environmental dynamical systems, has explained only the low level behavior mechanisms—for instance, walking by a legged agent (Beer1995), or simple visually guided behavior (Beer1996). In the current research, we attempted to apply the dynamical systems approach to the question of higher-order cognition by introducing mental capacities for anticipation in the agents. Indeed, it has been found that anticipation plays an essential role for the cognitive mechanism of articulation. As we have illustrated in the preceding sections, the articulation of the input flow is triggered when the prediction of the next future event fails, causing temporal incoherence in the structural coupling. It is plausible that the higher-order cognition is achieved by utilizing not only the coherence but also the incoherence appearing in the interaction between the internal and the environmental dynamics (Tani, Yamamoto, & Nishi1997). So far, one candidate alternative dynamical systems approach for achieving higher-order cognition is the theory of neural dynamic fields proposed by Schoner (Schoner, Dose, & Engels1995), which aims to achieve a cognitive function from the lower level to the higher level, such as in going collision-free maneuvering to achieving goal-directed planning in the navigation problem. It is, however, not clear if this theory can be extended to address the issues of articulation and hierarchical learning.

Moreover, we should relate our scheme to others using the architecture of the mixture of experts (Jacobs & Jordan1991). Gomi and Kawato (Gomi & Kawato1993) showed that modular representation is self-organized through sensory-motor interactions for manipulated objects in the mixture of expert networks with a gating network. Although this study shares a similar concept that a certain internal model can be acquired in modules, it has not been extended to deal with the problems of articulation and hierarchy. Jordan and Jacobs (Jordan & Jacobs1994) developed the original architecture of the mixture of experts further in order to incorporate some hierarchical structures. The "hierarchy" in their study (Jordan & Jacobs1994) refers to the structure of the recursive function approximation within multiple layers, which differs from our study of how articulation takes place across multiple levels as a function of time.

6. Conclusion

In this paper, we have proposed a novel scheme of hierarchical learning for sensory-motor systems using the mixture of RNN experts. The scheme was examined though simulation experiments concerning on-line navigation learning. The results indicate that the robot learns to articulate a continuous sensory-motor flow dynamically while modular and hierarchical structure are self-organized internally in a recursive manner across multiple levels. We explained the observed mechanism of articulation qualitatively by using the dynamical systems language.

The proposed approach can be developed in many ways in the future. The example considered in this paper was limited to the prediction learning of the sensory-motor flow. One future development could be the inclusion of motor or action-learning mechanisms. The scheme should be extended to cover both prediction-learning and behavior-learning to ensure that "concepts" can also be self-organized for the purpose of action generation. Then after, goal-directed planning can be carried out by utilizing the acquired internal representation embedded in the intrinsic dynamics of the RNNs, as had been described in Ref. (Tani1996). One limitation of the present study is that all the interactive processes were undertaken using only the bottom-up pathway in the architecture. A more general model is one in which top-down processes interact with bottom-up processes such that a module can be activated by manas of of bi-directional interactive dynamics between the top-down prediction from the higher level network and the bottom-up signals from the lower level network. In future research, we will study how goal-directed behavior can be generated in the extension of the proposed scheme by considering the shortcoming mentioned above.

References

Beer, R. 1995. A dynamical systems perspective on agent-environment interaction. *Artificial Intelligence* 72(1):173–215.

Beer, R. 1996. Toward the evolution of dynamical neural networks for minimally cognitive behavior. In *Proc. of the Fourth Int. Conf. on Simulation of Adaptive Behavior*, 421–429.

Billard, A. 1996. Do you follow me? or learning to speak through imitation for social robots. Master's thesis, Dept. of AI, University of Edinburgh.

Elman, J. 1990. Finding structure in time. *Cognitive Science* 14:179–211.

Endo, T., and Mori, S. 1978. Mode analysis of a ring of a large number of mutually coupled van der Pol oscillators. *IEEE Trans. Circuits Syst.* CAS-25(1):7–18.

Gomi, H., and Kawato, M. 1993. Recognition of Manipulated Objects by Motor Learning. *Neural Networks* 6:485–497.

Jacobs, R., and Jordan, M. 1991. Adaptive mixtures of local experts. *Neural Computation* 3(1):79–87.

Jordan, M., and Jacobs, R. 1994. Hierarchical mixtures of experts and the EM algorithm. *Neural Computation* 6(2):181–214.

Khatib, O. 1986. Real-time obstacle avoidance for manipulators and mobile robots. *The Int. J. of Robotics Research* 5(1):90–98.

Kohonen, T. 1982. Self-organized formation of topographically correct feature maps. *Biological Cybernetics* 43:59–69.

Krose, B., and Eecen, M. 1994. A self-organizing representation of sensor space for mobile robot navigation. In *Proc. of the Int. Conf. on Intelligent Robotics and Systems '94*, 257–263.

Kuipers, B. 1987. A qualitative approach to robot exploration and map learning. In *AAAI Workshop Spatial Reasoning and Multi-Sensor Fusion (Chicago)*, 774–779.

Kuniyoshi, Y.; Inaba, M.; and Inoue, H. 1994. Learning by Watching: Extracting Reusable Task Knowledge from Visual Observation of Human Performance. *IEEE. Trans. on Robotics and Automation* 10(6):799–822.

Mataric, M. 1992. Integration of representation into goal-driven behavior-based robot. *IEEE Trans. Robotics and Automation* 8(3):304–312.

Nehmzou, U. 1996. Flexible control of mobile robots through autonomous competence acquisition. *Measurement and Control* 28(2):48–54.

Newell, A., and Simon, H. 1976. Computer science as emprical inquiry: symbols and search. *Communication of the ACM* 19:113–126.

Newell, A. 1980. Physical symbol systems. *Cognitive Science* 4:135–183.

Pollack, J. 1991. The induction of dynamical recognizers. *Machine Learning* 7:227–252.

Rumelhart, D.; Hinton, G.; and Williams, R. 1986. Learning internal representations by error propagation. In Rumelhart, D., and McClelland, J., eds., *Parallel Distributed Processing*. Cambridge, MA: MIT Press.

Schoner, G.; Dose, M.; and Engels, C. 1995. Dynamics of behavior: Theory and applications for autonomous robot architectures. *Robotics and Autonomous Systems* 16:213–245.

Smith, L., and Thelen, E. 1994. *A dynamic systems approach to the development of cognition and action*. Cambridge, MA: MIT Press.

Smithers, T. 1996. On what embodyment might have to do with cognition. In *AAAI Fall Symposium, Embodied Cognition and Action*, 113–118.

Tani, J.; Yamamoto, J.; and Nishi, H. 1997. Dynamical Interactions between Learning, Visual Attention, and Behavior. In Husbands, P., and Harvey, I., eds., *Proc. of the Fourth European Conf. of Artificial Life (ECAL'97)*. Cambridge, MA: MIT press.

Tani, J. 1996. Model-Based Learning for Mobile Robot Navigation from the Dynamical Systems Perspective. *IEEE Trans. System, Man and Cybernetics Part B, Special issue on learning autonomous robots* 26(3):421–436.

van Gelder, T. The dynamical hypothesis in cognitive science. *Behavior and Brain Sciences*.

Varela, F.; Thompson, E.; and Rosch, E. 1991. *The embodied mind*. Cambridge, Mass.: MIT Press.

Yamauchi, B. M., and Beer, R. D. 1994. Sequential behavior and learning in evolved dynamical neural networks. *Adaptive Behavior* 2(3):219–246.

Zimmer, U. R. 1996. Robust world-modelling and navigation in a real world. *NeuroComputing* 13:247–260.

The evolution of subtle manoeuvres in simulated hockey

Alan D. Blair* & **Elizabeth Sklar****
* Dept. of Computer Science & Electrical Engineering
University of Queensland
4072, Australia
blair@cs.uq.edu.au
** Dept. of Computer Science
Brandeis University
Waltham, MA 02254, USA
sklar@cs.brandeis.edu

Abstract

A simulated hockey environment is introduced as a test bed for studying adaptive behavior and evolution of robot controllers. A near-frictionless playing surface is employed, partially mimicking zero gravity conditions. We show how a neural network using a simple evolutionary algorithm can develop nimble strategies for moving about the rink and scoring goals quickly and effectively.

1. Introduction

In recent years, there has been growing interest in using machine learning and evolutionary techniques for developing software controllers that play competitive or co-operative games in some kind of physical environment, either real or simulated. These have included: avoiding obstacles (Xiao et al., 1997; Lee et al., 1996; Brooks, 1986), foraging for "food" (Werger & Mataric, 1996), playing pursuit/evasion games (Miller & Cliff, 1994), discriminating objects (Beer, 1996), fighting for control of a cube (Sims, 1995) and RoboCup soccer (Kitano et al., 1995).

Following in these traditions, we have developed a simulated hockey game called *Shock* which is rather like "table-top" or "air" hockey. One or more players engage in games with the aim of shooting a puck into the opposing goal during an allocated time period. One player may be controlled by a human user; all other players are controlled by the simulation software.

Shock is similar in style to robot soccer simulators (Kitano et al., 1995; Salustowicz et al., 1998) and has much in common with (Balch, 1997), also being implemented in Java. However, Shock differs from previous systems in some significant ways. The near-frictionless environment in which Shock players operate provides special challenges distinct from those of typical playing surfaces, and the compactness of the rink distinguishes it from games played on an open plane (Miller & Cliff, 1994; Funes et al., 1998). While most work in robotics is concerned with collision *avoidance*, the Shock environment naturally gives rise to a strategy of collision *management*. Indeed, it is often advantageous to make strategic use of collisions in order to arrive at a desired position or orientation. The Shock players are rectangular in shape affording development of strategies involving internal movement and co-ordination different from those of circular players. In addition, Shock players are equipped with two-dimensional *holonomic* actuators which enable them to move with more grace and dexterity than standard one-dimensional wheels would allow.

We note that the near-frictionless Shock environment mimics to some extent the conditions found in outer space. Lately, hardware designs that will sustain a robot operating in zero gravity have been investigated. The Zero-G robot (Raibert et al., 1989) was designed to operate in a weightless environment such as a space station, and relies on a system of parallel rebound surfaces in order to control its movement. In work at CMU, the Self Mobile Space Manipulator (SM2) is being used to develop basic locomotion and manipulation capabilities for use on the international Space Station Freedom, with the ultimate goal of assisting astronauts during extra-vehicular activity and replacing astronauts in performing simple, dangerous or routine tasks (Nechyba & Xu, 1994).

2. The Simulated Hockey Domain

Shock is played in a rectangular rink, 1m wide by 1.5m long, with a goal at each end 150mm wide (Figure 1). Each player has a rectangular body 150mm by 50mm and a mass of 500 grams. The puck is circular in shape with a radius of 25mm and a mass of 100 grams.

Collisions between the players, puck and walls are calculated by the "spring" method of collision handling (Keller et al., 1993) and are totally elastic. This means that each experiences a restoring force in the normal direction, proportional to the depth of penetration. The

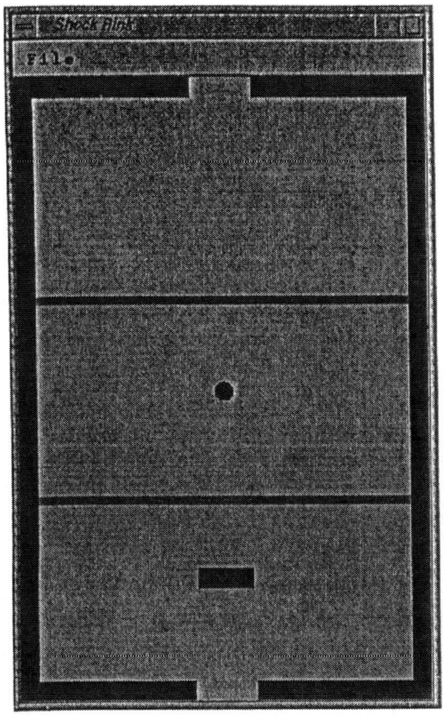

Figure 1 The Shock Rink.

puck collides frictionlessly with both players and walls, and therefore never acquires any spin. Each player experiences sliding friction when it collides with another player or a wall (i.e. a force in the opposite direction to the relative tangential motion and proportional to the restoring force). The players and puck also experience a frictional force as they slide on the rink that is proportional to their velocity and in the opposite direction.

We imagine each player as having a *skate* at either end of its body with which it can push on the rink in any direction (Figure 2). Actuators allowing 2-dimensional opposing movement in this way are difficult to implement in real hardware, but for the purpose of simulation we felt such a design might provide more interesting opportunities to study adaptive behavior.

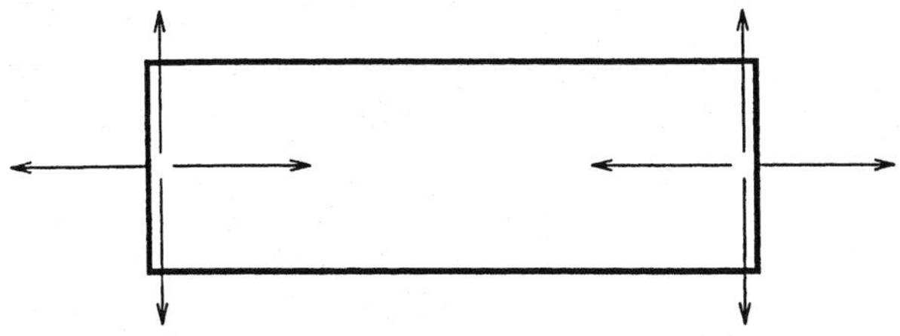

Figure 2 The Shock Player.

The output of the simulator's controller is in the form (x_L, y_L, x_R, y_R) where (x_L, y_L) and (x_R, y_R) represent the forces exerted at the left and right skate, respectively. We impose a restriction on the magnitude of the applied forces as follows:

$$x_L^2 + y_L^2 + x_R^2 + y_R^2 < \mathrm{F}_{\max}^2,$$

where $\mathrm{F}_{\max} = 0.2\mathrm{N}$. There is some redundancy in the format of the output since each player really only has three degrees of freedom available to it.

Each player is given complete information expressed in global coordinates, in the spirit of robot soccer where an overhead camera is often positioned above the playing field. In future work, we plan to develop players who instead receive their input from simulated local sensors.

The *game initial condition* (GIC) is a vector specifying a starting location and orientation for each player, and a starting location for the puck. All values are chosen at random, with the restriction that each player begins somewhere in the third of the rink nearest its own goal (see Figure 1). If only one player is present, the puck begins somewhere in the remaining two thirds of the rink; otherwise it begins somewhere in the middle third.

The simulation is updated in time intervals of 0.01 seconds, which runs in approximately real-time when the graphics window is open. When the graphics are switched off, the simulation runs faster, completing a game of 10 simulated seconds' duration in about half a second. Graphics mode is provided for several reasons: (1) to illustrate that the physics we are simulating appears natural, (2) to allow us to peek at the progress of the evolving players and (3) to provide an interface for the human player.

3. Network Architecture

Shock was designed so that players could be controlled by an expert system, fuzzy controller or any other kind of intelligent software. One of our objectives in building the system was to create an environment that could be used for comparing various machine learning techniques. The present work focuses on the following architectures:

1. simple recurrent (Elman, 1990) neural networks with 4, 6 and 8 hidden units
2. feed-forward networks with 6 and 8 hidden units
3. linear controller

All these architectures use 12 inputs and 4 outputs. Two of the inputs specify the location of the puck in a global co-ordinate system for which the origin $(0, 0)$ is at the center of the rink and the width of the rink is exactly 2.0 units. Four additional inputs indicate the locations of the player's left and right skate in the same co-ordinate system. The other 6 inputs convey the *observed velocity* of the puck and the left and right skate – i.e. the difference between current position and previous position, divided by the time interval.

The networks are fully connected and use a hyperbolic tangent function at the hidden layer. The four-dimensional output $\mathbf{z}$ of the network is converted to an applied force vector as follows:

$$(x_L, y_L, x_R, y_R) = \mathrm{F}_{\max} \tanh\left(||\mathbf{z}||\right) \frac{\mathbf{z}}{||\mathbf{z}||}.$$

This preserves the direction of the output vector, altering only its magnitude, and also allows the network to sometimes apply a very small force by choosing (x_L, y_L) and (x_R, y_R) to be nearly opposites of each other.

4. Evolutionary Paradigm

One long-term aim is to *co-evolve* different players to play games with each other lasting one minute or more of simulated time. However, before beginning multi-player simulations, we need to determine an architecture and a learning paradigm that will enable individual players to develop the basic skills necessary for the task. We therefore begin by studying a single-player 10-second version of Shock in which one player, beginning from a random GIC, moves about the rink until either (a) the puck goes into one of the goals or (b) the allotted time of 10 seconds elapses. In each game, a score of +1 is awarded for hitting the puck into the enemy goal, −1 for hitting it into the *friendly* goal (i.e. the player's own goal) and 0 for failing to hit it into either goal. Note that, even with no opponent, the need to avoid hitting the puck into the friendly goal is an important part of the task, since it prevents the player from winning by simply hitting the puck around randomly until a goal is scored.

We use an evolutionary hill-climbing algorithm in which a *champ* neural network plays two games with random GIC's and is then challenged by a series of *mutant* networks until one is found to do "better" (see below) than the champ; the champ's weights are then adjusted in the direction of the mutant:

1. mutant ← champ + gaussian noise[1]
2. mutant plays 2 games with same GIC's as champ
3. if mutant does better than champ,
   champ ← 0.9 * champ + 0.1 * mutant

If 100 mutants fail to do better than the champ with these GIC's, they are discarded and 2 new GIC's are generated. After the champ's weights have been adjusted, two new GIC's are chosen for the new champ and the process continues. The practice of making only a small adjustment in the direction of the mutant was introduced in previous work on backgammon (Pollack & Blair, 1998) on the assumption that most of the strategies of the (well-tested) champion would be preserved, with only limited influence from the mutant – since two games are not enough to determine for sure whether the mutant is really better, or just a lucky novice.

[1] with standard deviation of 0.05 for each weight

A total of 8 preliminary runs were conducted in order to refine various features of this algorithm. In the first five preliminary runs, only one game was used to compare the mutant and the champ, but the resulting players tended to degrade in performance after initial improvement – indicating that one game is not enough to make a good comparison.

Exposing the player to a variety of initial conditions is important for developing robust strategies (Tesauro, 1992). In the first two preliminary runs, the puck was constrained to always start in the middle third of the rink. The evolved players generally learned to hit the puck near to the enemy goal, but then failed to move up and hit the puck again – thus missing an opportunity to score. In subsequent runs, the puck was allowed to start anywhere in the upper two-thirds of the rink, which seems to provide an advantage because it enables players to learn the task "backwards from the end game" as is common in reinforcement learning (Tesauro, 1992).

What does it mean to do "better" than the champ? In theory, it should mean that the mutant is more likely to score goals than the champ. However, in the early stages of evolution the players are very unlikely to score at all, so it may become necessary to jump start the process. One way to do this is by introducing *partial credit* adjustments in the fitness measure which reward the player for moving closer to the puck, or for moving the puck closer to the enemy goal and away from the friendly goal. On the other hand, such adjustments run the risk of introducing artifacts which may distract the player from its main task (Angeline & Pollack, 1992). For example, a player seeking partial credit might push the puck close to the goal without bothering to score, or it might hover close to the puck without touching it, for fear of moving it in the wrong direction.

We attempt to balance these considerations by dividing the evolution into two phases: Phase 1 is the partial credit phase in which the mutant is considered to be better than the champ in the following circumstances:

1. the mutant scores in enemy goal but not the champ
2. the champ scores in friendly goal but not the mutant
3. both score in enemy goal, but mutant does it faster
4. both score in friendly goal, but mutant does it slower
5. neither scores a goal, but mutant moves the puck to a place where the *distance ratio* is smaller [2]
6. neither player ever hits the puck, but mutant is closer to puck at the end of the game than the champ was

In Phase 2, the fitness is evaluated purely on the basis of goals scored. Several GIC's are tested until two are found for which the champ scores either 0 or −1 (i.e. fails to hit the puck into the enemy goal in the allotted

[2] We define the *distance ratio* to be $d_1/(d_1 + d_0)$ where d_1 and d_0 are the distance from the puck to the enemy goal and the friendly goal, respectively.

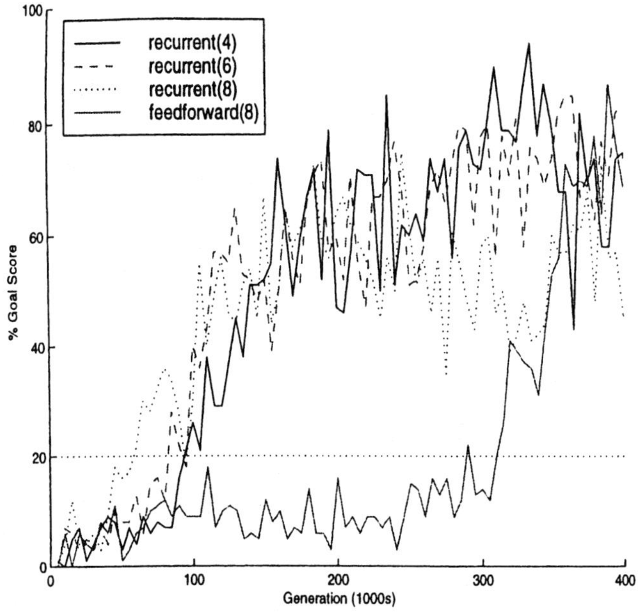

Figure 3 Score out of 100 games every 5000th generation for recurrent networks with 4, 6 and 8 hidden units and feed-forward network with 8 hidden units, annealed at 20%.

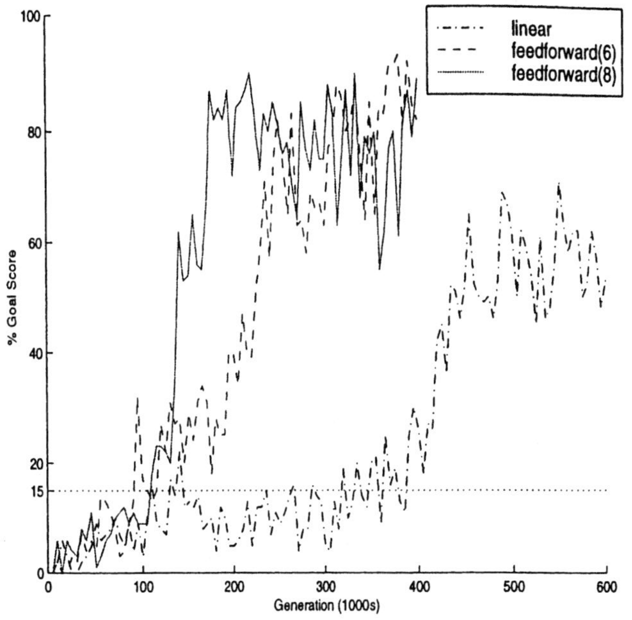

Figure 4 Score out of 100 games every 5000th generation for linear controller and feed-forward neural networks with 6 and 8 hidden units, annealed at 15%.

time). Then the mutant is considered to be better if it achieves a higher score than the champ for one of these GIC's, and an equal or higher score than the champ for the other GIC. Phase 2 was introduced because the eight preliminary runs, which used only the Phase 1 scheme, were found to produce inferior results.

5. Results

Following the 8 preliminary runs described in Section 4, a further 9 experimental runs were conducted in order to test the algorithm and compare different architectures and sizes for the neural network controller. Due to limited resources, each run was conducted once; ideally we would have executed each run several times and averaged the results.

In earlier work on phased learning (Pollack & Blair, 1998), we found that it is better to base the timing of the transition or *annealing* from one phase to another on internal diagnostics rather than at a pre-determined generation. To this end, we test our player every 5000 generations by having it play 100 games from different GIC's and using the total score for these games as a measure of its performance. When the performance according to this measure reaches some pre-determined *annealing threshold*, we switch from Phase 1 to Phase 2. The choice of an appropriate annealing threshold is a delicate and important issue, particularly when different architectures are being compared. If the threshold is set too low, causing the evolution to anneal prematurely, then the new evaluation criteria may prove so difficult that no mutant ever prevails. If it is set too high, the threshold may never be reached, denying us the potential benefits of subsequent phases. A threshold chosen for one kind of architecture may turn out to be inappropriate for another.

We initially chose 20% as our annealing threshold because all three recurrent networks reached this level within 100K (100,000) generations, after which their performance (without annealing) started to level off. With annealing implemented, their performance rose steadily to around 60% by generation 180K (Figure 3). On the other hand, the linear controller and feed-forward neural networks failed to reach this 20% threshold within 200K generations, which initially led us to suspect that these non-recurrent architectures might be inadequate for the task. However, when these runs were continued, the 8-unit feed-forward network finally reached the 20% level at generation 290K and, after annealing, its performance increased to 70% by generation 360K. We then repeated these runs with the annealing threshold set to 15% instead of 20% and found (Figure 4) that the 8-unit (respectively, 6-unit) network now annealed at 110K (resp. 95K) and reached 80% performance by 180K (resp. 370K). The linear controller annealed at 110K but took around 450K generations to reach 50% performance.

Figures 5 to 8 illustrate the steady advance in skill of the evolving recurrent network with 4 hidden units. The puck and player are shown in their initial configuration (solid) and at one-second intervals (outlined) while the dotted lines trace their trajectories throughout the game.

At generation 10K (Figure 5) the player approaches the puck but doesn't hit it. By generation 15K (not

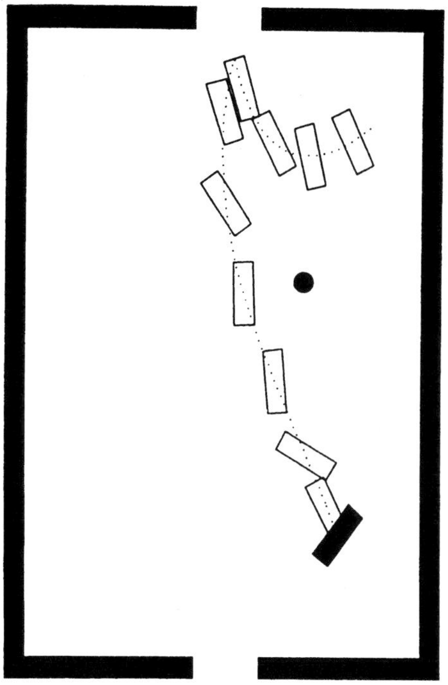

Figure 5 10K player, open shot.

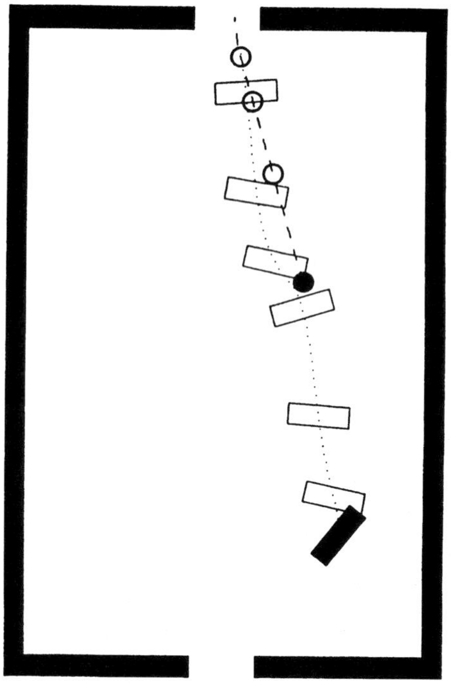

Figure 6 100K player, open shot.

Figure 7 100K player, difficult shot.

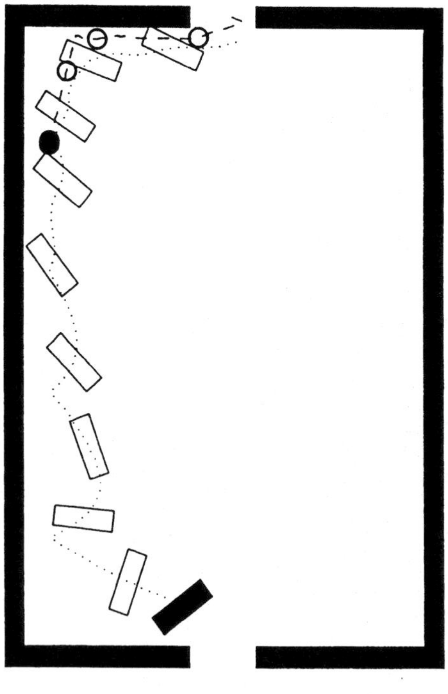

Figure 8 300K player, difficult shot.

shown) it can generally hit the puck once, not necessarily in the right direction. At generation 100K if the player has a fairly open shot at the goal (Figure 6) it hits the puck in the right direction and follows up in case a second shot should be required. If the same (100K) player is presented with a more challenging initial condition (Figure 7) it is able to nudge the puck away from the side wall, but upon hitting the back wall the puck fumbles from its grasp. By generation 300K the player has gained enough dexterity to see this situation through to a successful conclusion (Figure 8).

The other networks displayed similar kinds of behavior, but each developed its own particular style. The 6-unit recurrent network tends to maintain a greater distance from the puck, bouncing it once or twice off the back wall before scoring. Sometimes, when dribbling the puck along the back wall as in Figure 8, it will overshoot the goal, but then chase the puck and start dribbling it back in the other direction to take a second shot. The 8-unit feed-forward network tends to stay very close to the puck, almost sliding it along the rink, and moves very slowly when it gets near to the goal. The linear controller hits the puck once, then backs up to the middle of the rink before approaching the puck again to take a second shot. It has difficulty scoring unless there is an open shot at the goal. This, together with its slow learning rate, suggest that a linear control mechanism may be inadequate for the task.

In further work, we hope to gain a better understanding of the relative strengths and weaknesses of the various types of networks, both on their own and against opponents.

6. Conclusions and Further Work

We have developed a simulated hockey environment and demonstrated how neural network controllers, learning to play on their own by means of an evolutionary algorithm, can develop reliable skills for scoring goals quickly and effectively. Our ultimate "goal" is to build teams of players for contests lasting a minute or more of simulated time.

We are currently applying co-evolutionary methods to this task, and also investigating on-line learning algorithms, including supervised and reinforcement learning, which may reduce the amount of computation currently required. Co-evolution and interaction with other players will undoubtedly bring a whole new set of complexities and challenges to this task (Salustowicz et al., 1998; Haynes et al., 1995). Nevertheless, given the nimble strategies we have been able to develop in the single-player scenario, it is not unreasonable to suppose that equally subtle manoeuvres appropriate for interaction with co-evolving opponents might be developed in the multi-player case.

In parallel work, we are building a user interface for an interactive version of Shock that will enable human players to test their skill against our various computer opponents. Check our web site for this coming attraction:

`http://www.demo.cs.brandeis.edu`

7. Acknowledgements

Thanks to Gordon Wyeth, David Blair, Janet Wiles and Jordan Pollack for helpful comments, and to Macquarie University for hosting the first author during the writing of this paper. This work was supported primarily by a University of Queensland Postdoctoral Fellowship.

References

Angeline, P. & Pollack, J. (1992). Evolutionary induction of subroutines. In *Proc. 14th Annual Conference of the Cognitive Science Society*, pp. 236–241.

Balch, T. (1997). Learning roles: Behavioral diversity in robot teams. *AAAI97 Workshop on Multiagent Learning.*

Beer, R. (1996). Toward the evolution of dynamical neural networks for minimally cognitive behavior. In (Maes et al., 1996), pp. 421–429.

Brooks, R. (1986). A robust layered control system for a mobile robot. *IEEE Transactions on Robotics and Automation*, 2:14–23.

Elman, J. (1990). Finding structure in time. *Cognitive Science*, 14:179–211.

Funes, P., Sklar, E., Juillé, H., & Pollack, J. (1998). Animal-animat coevolution: Using the animal population as fitness function. In *Proceedings of SAB-5.*

Haynes, T., Sen, S., Schoenefeld, D., & Wainwright, R. (1995). Evolving a team. In *AAAI 1995 Fall Symposium on Genetic Programming.*

Keller, H., Stolz, H., Ziegler, A., & Bräunl, T. (1993). Virtual mechanics – simulation and animation of rigid body systems. Rept. Comp. Sci. Dept. 8/93, U. Stuttgart.

Kitano, H., Asada, M., Kuniyoshi, Y., Noda, I., & Osawa, E. (1995). Robocup: The robot world cup initiative. In *IJCAI-95 Workshop on Entertainment and AI/ALife.*

Lee, W.-P., Hallam, J., & Lund, H. (1996). A hybrid gp/ga approach for co-evolving controllers and robot bodies to achieve fitness-specified tasks. In *Proc. 1996 IEEE Int'l Conf. Evolutionary Computation*, pp. 384–389.

Maes, P., Mataric, M., Meyer, J.-A., Pollack, J., & Wilson, S. (Eds.) (1996). *From Animals to Animats 4: Proceedings of the Fourth International Conference on Simulation of Adaptive Behavior.* MIT Press.

Miller, G. & Cliff, D. (1994). Protean behavior in dynamic games: Arguments for the co-evolution of pursuit-evasion tactics. In *Proceedings of the Third International Conference on Simulation of Adaptive Behavior*, pp. 411–420.

Nechyba, M. & Xu, Y. (1994). Sm2 for new space station: autonomous locomotion and teleoperation control. *Proc. IEEE Int'l Conf. Robotics and Automation*, 2:1765–1771.

Pollack, J. & Blair, A. (1998). Co-evolution in the successful learning of backgammon strategy. *Machine Learning (to appear).*

Raibert, M., H.B. Brown, J., Chepponis, M., Hodgins, J., Koechling, J., Dustman, D., Brennan, W. K., Barrett, D., Thompson, C., Hebert, J., Lee, W., & Borvansky, L. (1989). Dynamically stable legged locomotion. AI technical report 1179, MIT.

Salustowicz, R., Wiering, M., & Schmidhuber, J. (1998). Learning team strategies: soccer case studies. *Machine Learning (to appear).*

Sims, K. (1995). Evolving 3d morphology and behavior by competition. *Proc. Artificial Life 4*, pp. 28–39.

Tesauro, G. (1992). Practical issues in temporal difference learning. *Machine Learning*, 8:257-277.

Werger, B. & Mataric, M. (1996). Robotic "food" chains: Externalization of state and program for minimal-agent foraging. In (Maes et al., 1996), pp. 625–634.

Xiao, J., Michalewicz, Z., Zhang, L., & Trojanowski, K. (1997). Adaptive evolutionary planner/navigator for mobile robots. *IEEE Trans. Evol. Computation*, 1(1):18–28.

Multifunctionality: a fundamental property of behavior mechanisms based on dynamical systems

Herbert Jaeger
GMD, Sankt Augustin, Germany
herbert.jaeger@gmd.de

Abstract

The mechanisms supporting robot behaviors are increasingly often designed as dynamical systems. Such mechanisms are inherently multifunctional. This means that they can exhibit different qualitative behavior in different circumstances. Multifunctionality makes system design difficult. On the other hand, it may be beneficial for adaptiveness, since it allows qualitative changes in a robot's behaviors without changing the supporting mechanisms.

1. Introduction

The "behavior based" approach to designing mobile robots (Brooks, 1991) has been very fertile. However, a formal theoretical foundation, comparable to the role of logics in traditional AI, has been lacking in the beginning. A promising road toward such a theoretical foundation is to view agents as dynamical systems (Beer, 1997)(Steinhage and Schöner, 1997). This perspective is also being developed in cognitive science (van Gelder and Port, 1995). In these approaches, agents (animals, humans, or robots) are described with formalisms which lend themselves to an analysis in terms of dynamical systems theory. Such formalisms include differential equations (e.g., (Jaeger and Christaller, 1998)(Large et al., 1997)), recurrent neural networks (e.g. (Beer, 1995)), and stochastic automata (e.g. (Baasye et al., 1995)). The most common kind of formal analyis concerns the description of attractor structures emerging from learning or evolution (e.g. (Robertson et al., 1993)(Smithers, 1995)).

In the present article I wish to point out a basic property of dynamical-systems based robots, namely, the inherent multifunctionality of input-driven dynamical systems. I shall demonstrate the ubiquity of this property, and discuss the difficulties as well as the chances that it might afford.

First, a clarification of terminology. By the term *functionality,* I refer to a combination of a *mechanism* with a particular *purpose.* In the context of this article, I deal with computational mechanisms. Examples are all kinds of functions and algorithms, dynamical systems, neural networks, etc., when they are physically implemented on a robot. There, they (usually) serve a particular purpose, e.g. extracting features from a sensor signal, or producing a base oscillation for a leg controller, etc. *Multifunctionality* refers to the fact that a particular mechanism can serve many purposes.

This fact is well known. It has been referred to under various names, e.g. "relative systems" (Shimizu, 1993), "functional indeterminacy", and "multifunctionality" (Pasemann, 1994).

The multifunctionality described by those authors results from bifurcations. In this article, I wish to point out that multifunctionality even arises without bifurcations, due to nonlinearities and exchanges in relative timescales of system inputs.

In order to illustrate this fact, I shall describe a "pushing boxes" behavior of a robot, which was programmed as a dynamical system according to the "dual dynamics" design scheme (section 2). The behavior is ruled by simple differential equations. One of them is inspected in section 3, where I describe how the differential equation serves a particular purpose in the robot. In section 4 I will show that the same equations may support different functionalities when fed with different input. In the final discussion section I argue that this kind of multifunctionality may turn out to be beneficial for adaptiveness, although it presents difficulties for design.

2. Background: dual dynamics and the Black Knight

I start this section with a brief introduction to the "dual dynamics" (DD) scheme. A more detailed account can be found in (Jaeger and Christaller, 1998).

DD is a formal scheme for designing behavior control systems for mobile robots. The scheme prescribes how to specify a collection of behaviors as a comprehensive dynamical system. It consists of many coupled subsystems, each of which is responsible for controlling a particular behavior. These behavior subsystems are specified through ordinary differential equations.

Like in many behavior-oriented control architectures, behaviors are ordered in levels in the DD scheme, too. At the bottom level of a DD behavior hierarchy, one finds *elementary* behaviors. These are sensomotoric coordinations with direct access to external sensor data and actuators. Typical examples are `move_forward` and `turn_left`. Higher levels are constituted by increasingly comprehensive behaviors. They also have access to sensoric information but cannot directly activate actuators. Typical examples are `work` and `replenish_energy`.

Elementary behaviors are different from higher-level behaviors in that they are made from two subsystems, which serve different purposes. This has given the approach its name, "dual dynamics".

The first of these subsystems is called the *target dynamics*. It calculates target trajectories for all actuators which are relevant for the particular behavior. Its output consists of as many variables as there are degrees of freedom to be controlled.

A requirement for the target dynamics is that this system must not undergo bifurcations. This is what makes elementary behaviors elementary. For instance, the target trajectories of `turn_left` in a simple 2-wheeled robot, which moves on a flat surface, are likely to remain qualitatively unchanged in different instances of the maneuver. Thus, `turn_left` would be a good candidate for an elementary behavior in such a simple robot. By contrast, in a walking machine which has to cope with different surfaces, it is likely that there will be qualitatively different maneuvers for turning left in different circumstances. Each of them would thus yield a separate elementary behavior.

The other subsystem of an elementary behavior is its *activation dynamics*. It regulates a single variable, the behavior's *activation*. The equation ruling this variable should be written in a way that the variable displays a dynamic range between 0 and 1. A value of 1 means that the behavior is fully active, whereas 0 means that it is completely inhibited.

The activation dynamics is allowed to undergo bifurcations. The control parameters which induce these bifurcations are the activation variables of higher-level behaviors. This is the core idea behind DD. The ways of how, exactly, these bifurcations are induced by changing control parameters, are highly constrained by the DD formalism. These constraints and the fact that bifurcations are confined to the one-dimensional activation dynamics subsystems, warrants the transparency of the DD design scheme.

I would like to emphasise that an elementary behavior is not "called to execute" from higher levels. The level of elementary behaviors is fully operative on its own and would continue to work even if the higher levels were cut off. The effect of higher levels is not to "select actions", but to change the overall characteristics of the elementary level, by inducing bifurcations in that level.

This concludes the introduction to DD. In the remainder of the section, I sketch the Black Knight (BK) robot and its `pushbox` behavior.

BK (now deceased) was a typical 2-df robot made from Lego bricks. It was built and maintained at the VUB AI Lab[1] by Peter Stuer and Dany Vereertbrugghen. Programming was done using the PDL programming language, which was also developed at the VUB AI Lab. PDL is well suited to implement control systems written in differential equations[2].

BK's basic set of tasks is to wander around in an arena limited by plywood walls; avoiding obstacles while wandering; knocking into light-emitting "pushboxes"; and recharging at a power station when the batteries run low. Fig. 2 shows a view of this setup.

Figure 1 The Black Knight approaching the charging station (brightly lit construction). A pushbox emitting a thin horizontal line of red light is seen at the right.

One of BK's behaviors is to "work" by reapeatedly knocking into one of several cylindrical pushboxes scattered in the arena. The pushboxes emit a modulated red light. BK can distinguish pushboxes from other light-sources due to this modulation. When a pushbox is knocked repeatedly, it gradually dims its light until it cannot longer be recognised by BK. A typical `pushbox` episode proceeds in three phases: (i) the robot gets into recognition distance of a light-emitting pushbox, and steers toward it, (ii) then drives head-on into it (a "knock"), retracts a bit, and repeats a few times until (iii) the pushbox is almost completely dimmed, thereby becoming an ordinary obstacle, from which the robot turns away.

3. A closer look at the pushing boxes behavior

In this section I explain the activation dynamics of the `pushbox` behavior, a simple differential equation which

[1] http://arti.vub.ac.be/~cyrano/robot_home.html

[2] The implementation was done for teaching purposes and is richly documented. It is obtainable at http://www.gmd.de/People/Herbert.Jaeger/Resources.html

successfully served its intended purpose. However, in the next section it will turn out that this simple equation might support quite different functionalities, too.

The activation dynamics has two kinds of inputs. One kind of input enables the pushbox behavior to detect the pushboxes by virtue of the modulated light they emit. More specifically, the readings of the narrow-angled and wide-angled photodiodes are preprocessed to yield two signals M_l and M_r, which roughly correspond to "modulated red light is seen at the left (right, respectively) before the robot". Furthermore, while M_l and M_r rise sharply with modulated light influx, they decay exponentially, providing a kind of "afterimage" or "short-term memory" effect.

M_l and M_r are used to alert the pushbox behavior by switching on its activation, when even only a little modulated light is seen. Technically, in the activation dynamics (1) the sum $M_l + M_r$ is fed into a steep sigmoid σ which essentially returns 1 when $M_l + M_r$ rises over a certain small threshold, and returns 0 otherwise.

$$\dot{A}_{pbx} = 20 \cdot A_{work} \cdot (\sigma(M_l + M_r) - A_{pbx}) + decayterm \qquad (1)$$

The other input into the activation dynamics is the activation A_{work} of the work behavior. Mathematically, it appears as a factor. If the work behavior is not active, then $A_{work} = 0$, and (1) is dominated by the decay term. Intuitively, that means that when BK is not in working mood, it will not be tempted by pushboxes. If, however, the activation of work is high (i.e., about 1), then the dynamics of (1) is dominated by the main term $(\sigma(M_l + M_r) - A_{pbx})$. This term makes A_{pbx} follow $\sigma(M_l + M_r)$. Since the latter term basically can attain the values 0 and 1 due to the sigmoid, the resulting overall dynamics of (1), when work is activated, is a soft "toggling" between 0 and 1. This toggling switch is triggered when $M_l + M_r$ rises/falls over/under the small threshold of 0.03, which is the switching point of the sigmoid σ.

Figure 2 depicts a typical pushbox episode, recorded when BK was in working "mood" (i.e., when A_{work} was about 1). The variable R_{pbx} is the signal issued to the right motor by the target dynamics of the pushbox behavior. It is given in the figure to indicate BK's motor action during that pushbox episode, which consisted in an initial right curve for the first approach until the first knock, followed by a back-and-forth maneuver for the second knock, and a final retraction from the pushbox after it became invisible. The activation A_{pbx} quickly went to 1 when M_r started to climb, and quickly fell to 0 after modulated light "afterimages" had died out. In sum, the dynamics specified in (1) behaved in the way it was supposed to behave, namely, as a "toggle switch" for A_{pbx}.

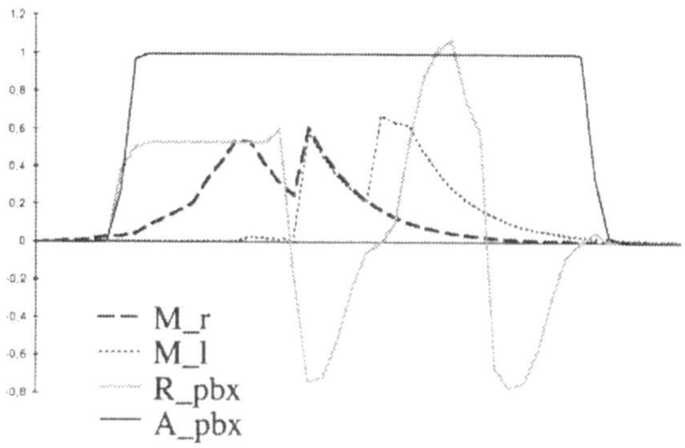

Figure 2 A trace of a pushbox episode comrpising two knocks. Duration is approximately 8 seconds. For variables see text.

4. Other functionalities supported by the mechanism

Now I shall re-consider equation (1) and list some functionalities which it might assume, other than the toggling functionality just described. In each case, I briefly state what characteristics the input into (1) must have, in order to make (1) assume that functionality.

Since the names of variables in (1) refer to the functionality within BK, it is better to rename them, in order to avoid confusion. For the sake of simplicity, I also join $M_l + M_r$ into a single variable. This gives the following condensed version of (1):

$$\dot{x} = 20 \cdot a \cdot (\sigma(b) - x) + decayterm \qquad (2)$$

With a little experience in analysing differential equations, one will find it easy to detect many other functionalities (2) might acquire, besides the "toggling switch" characteristics. Here is a list of some of them:

Low-pass filter. Assume again that a remains fixed at 1, and that b is a signal of mixed frequencies with a small amplitude averaging at a value of 0.03, i.e. the threshold of σ.

In these circumstances, (2) works as a low-pass filter. Roughly, frequencies with a period greater than 1/20 will be suppressed, while lower frequencies will be followed by x.

Adaptable low-pass filter. Just like before, assume that input b is a signal of mixed frequencies. However, this time allow a to take various values between 0.1 and 10.

Then, (2) works as a low-pass filter like before, but with cutoff frequencies set by a in a period range from 1/2 to 1/200.

Signal amplifier. Again, let a be fixed at 1. Assume that b is a weak signal which slowly (relative to time increments of 1/20) varies in a range of 0.03 ± 0.001. In this context, the mechanism (2) will appear as an amplifier, with x following the input signal in a range of $0.5 \pm 0.001 \times S$, where S is the steepness of σ.

Leaky spike integration. Assume that $a = b$, and that b is some kind of "spiking" signal, i.e. it consists of short pulses, separated by refractory periods, which are long compared to spike duration, and where the signal is zero. Assume further that spike duration is shorter by at least an order of magnitude than 1/20 time units, and that the length of the refractory period is of the same order of magnitude as the time constant of the decay term in (2).
Under these conditions, (2) works as a leaky integrator of spikes. More concretely, the current value of x will measure the average number of spikes which occured in the past, with more recent spikes weighted stronger. The measurement is highly nonlinear, saturating at a value of 1.

Two factors are crucial for multifunctionality in these examples, namely, nonlinearities (represented by the sigmoid in (2)), and changes in relative timescales of a vs. b vs. the decay term vs. the overal system time scale fixed by the time constant 20.

Note that through all these examples, (2) does not bifurcate: the phase portraits of the system are characterised by a single point attractor through all variations of input, in all examples.

5. Discussion

The preceding sections illustrate that even simple dynamical systems can support different functionalities. These differences in functionalities need not result from bifurcations, but can arise from different characteristics of system input only. If bifurcations were included into the picture, multifunctionality would become even more pervasive.

Multifunctionality obviously makes a dynamical-systems oriented design more difficult, compared to design schemes which rely on functions in the mathematical sense, like e.g. typical C procedures or feedforward neural networks. Likewise, recurrent neural networks (which are dynamical systems) are notoriously more difficult to train than feedforward networks. A designer, who wishes to obtain a specific input-output trajectory from a differential equation, has to take into account the system's internal state evolution, which is connected to the desired IO-characteristics in no generally treatable way. Furthermore, the designer has to be careful about the exact dynamical properties and timescales of inputs, in order to fish a particular functionality from the reservoir of potential functionalites of a dynamical system.

Considering these problems, what reasons are there for using dynamical systems? One obvious advantage of dynamical systems is that they are intrinsically temporal formalisms (detailed discussion in (van Gelder, 1998)). This renders them natural for achieving functionalities which are time-sensitive, in particular, for motor control; and renders them charming for researchers who believe that cognition grows out of bodily activity. However, I will not further this aspect here, but instead point out another possible advantage of dynamical systems.

Multifunctionality of mechanisms affords a robot with a reservoir of novel ways of behaving. As I hope to have demonstrated, this reservoir is a rich one. A strategy for a robot to exploit multifunctionality might consist of the following three steps: (i) get into novel situations, e.g. by playing or exploration, (ii) discover that the behavior-generating mechanisms respond in novel ways due to multifunctionality, and (iii) reinforce and memorise novel responses if they are beneficial, and replay them when similar situations occur. I would call this a *discover-and-memorise* strategy for generating adaptive novel behaviors.

Obtaining new functionalities by a discover-and-memorise strategy should be compared to the more familiar way of obtaining novel functionalities, namely, by evolutionary strategies. I use here the term "evolutionary" in a broad sense, to denote any scheme which first randomly *modifies* an existing system and then *tests* the result by evaluating the resulting fitness. Evolutionary schemes for finding new functionalities do not necessarily require many generations of agents, but can work inside a single agent (e.g., (Steels, 1997)).

A big difference between discover-and-memorise vs. modify-and-test strategies is that in the former, an already existing (but dormant) reservoir of novelty is tapped, whereas in the latter, novel functionalities are added to the system. Conversely, this entails that discover-and-memorise leaves existing functionalities intact, whereas modify-and-test may be destructive. Thus, the former seems intrinsically safer than the latter. It might also be faster since new functionalities are harvested together with the situations in which they arise. By contrast, modify-and-test is apt to produce undetectable functionalities for situations which will never occur.

One price to be payed for the benefits of multifunctionality is that it may be difficult to keep subsystems within the operative range of a particular functionality. Keeping multifunctionality within bounds may be one of the reasons why homeostasis is so important in animals.

To my knowledge, a discover-and-memorise strategy capitalising on multifunctionality has not yet been realized on a robot. A prerequisite for investigations in that direction is to be able to program robots on a dynamical systems basis in a convenient fashion. The dual dynamics design scheme is a step toward that goal.

Acknowledgments. I wish to express my deep gratitude to Luc Steels, Peter Stuer and Dany Vereertbrughen at the VUB AI Lab. Thanks to Tim van Gelder for insightful discussions, and an anonymous reviewer for very helpful comments. The work described in this article was mostly carried out under a postdoc grant from GMD, Sankt Augustin.

References

Baasye, K., Dean, T., and Kaelbling, L. (1995). Learning dynamics: System identification for perceptually challenged agents. *Artificial Intelligence*, 72:139–171.

Beer, R. (1995). A dynamical systems perspective on agent-environment interaction. *Artificial Intelligence*, 72(1/2):173–216.

Beer, R. (1997). The dynamics of adaptive behavior: a research program. *Robotics and Autonomous Systems*, 20:257–289.

Brooks, R. A. (1991). Intelligence without reason. A.i. memo 1293, ftp'able at http://www.ai.mit.edu/, MIT AI Lab.

Jaeger, H. and Christaller, T. (1998). Dual dynamics: Designing behavior systems for autonomous robots. *To appear in Artificial Life and Robotics.*

Large, E., Christensen, H., and Bajcsy, R. (1997). Scaling the dynamic approach to autonomous path planning: planning horizon dynamics. In Pollack, M., editor, *Proceedings of IJCAI-97, Vol. 2*, pages 1360–1365. Morgan Kaufmann.

Pasemann, F. (1994). Neuromodules: A dynamical systems approach to brain modelling. In Herrman, H., Wolf, D., and Pöppel, E., editors, *Proceedings of the Workshop: Supercomputing in Brain Research - from Tomography to Neural Networks,*, pages 331–348. HLRZ, KFA Jülich, Germany, World Scientific.

Robertson, S., Cohen, A., and Mayer-Kress, G. (1993). Behavioral chaos: Behind the metaphor. In Smith, L. and Thelen, E., editors, *A Dynamic Systems Approach to Development: Applications*, pages 119–150. Bradford/MIT Press, Cambridge, Mass.

Shimizu, H. (1993). Biological autonomy: The self-creation of constraints. *Applied Mathematics and Computation*, 56:177–201.

Smithers, T. (1995). On quantitative performance measures of robot behavior. *Robotics and Autonomous Systems*, 15(1/2):107–134.

Steels, L. (1997). Synthesising the origins of language and meaning using co-evolution, self-organization, and level formation. In Hurford, J., editor, *Evolution of Human Language*. Edinburgh University Press, Edinburgh.

Steinhage, A. and Schöner, G. (1997). Self-calibration based on invariant view recognition: Dynamic approach to navigation. *Robotics & Autonomous Systems*, 20(2-4):133–156.

van Gelder, T. (1998). The dynamical hypothesis in cognitive science. *To appear in Behavioural and Brain Sciences.*

van Gelder, T. and Port, R., editors (1995). *Mind as Motion: Explorations in the Dynamics of Cognition.* Bradford/MIT Press.

Species Abundance Adapted to the Energy Flow in Ecosystem Simulation

Tokufumi Yamamoto, Michael Kirley & Xiaodong Li
School of Environmental and Information Sciences
Charles Sturt University
PO Box 789 Albury NSW 2460 Australia
e-mail: tyamamoto@csu.edu.au, mkirley@csu.edu.au, xli@csu.edu.au

Topic areas: ecological simulation, emergent properties, collective and social behavior.

Abstract

The numerical models featuring energy flow between *soil, plant* and *herbivore* are examined. The goal of this work is to observe the relationship between body size and species abundance of the *herbivore*, resulting from adaptive behavior of population. The species in the simulations are defined by the parameter of energy flux between organisms and replication strategies. Our result shows that the individual behavior builds up the distribution of species abundance resembling that of real ecosystem under the various conditions. This means that the amount of available resources affects not only the number of organisms, but also the adaptive behavior of the population in a local ecosystem. The observed adaptation in the present simulation study therefore indicates that there are more sustainable strategies than one under a given condition. The modelling technique presented here can also be applied to more detailed ecosystem simulations.

1. Introduction

A number of ecological models have been proposed in literature to examine the relation between species which have different energy intake and replication strategy. Some models, however, ignore the effect of spatial interaction between organisms and restriction imposed by the spatial dimension. In this work, we propose the numerical simulation based on logistic equation which is held by individuals located on a 2-D landscape to investigate adaptive behavior of ecosystem. The aim of this work is to examine a sustainable set of parameters that maintains necessary energy for ecosystem, rather than searching a unique solution for organisms which dominate under a given condition. So the focus of the study is not the individual behavior, but the collective behavior.

Various computer simulations have been used to examine the spatial distribution of organisms in an ecosystem. The 'forest game' proposed by Solé and Manrubia (1995) simulates the growth of trees in a cellular automaton format (see Wolfram 1984, 1994, for cellular automaton). In the forest game, the organisms (trees) are allocated in a 2-D grid, as they interact with each other. As a result, a self-organized critical state is found in the distribution of canopy-gaps in the simulated forest, while $1/f$ noise is found in the time series of the biomass of the forest (see Bak & Chen (1991) and Bak, Tang & Wiesenfeld (1987) for self-organized critical state and $1/f$ noise). A simulation like forest game shows emergent structures of ecosystem, namely rainforest, which develops and maintains a typical distribution of organisms. Although such a simulation does not predict a particular behavior of an organism in the ecosystem (e.g. predict when a particular tree falls and so forth), it provides us an overall picture of the dynamics of the ecosystem concerned.

The research on the energy (nutrient) flow between organisms has also been recognised as of importance in describing the distribution and the dynamics of ecosystem. But an analytical work on energy flow in ecosystem is limited to relatively simple models. In other words, an analytic method using difference equations to describe the dynamical behavior may be used when there are a few agents in the ecosystem or many, but identical agents in the system (for a general analysis from systems science point of view see Hirata & Ulanowicz, 1984; Ulanowicz & Norden, 1990). That is, the diversity and heterogeneity in the system hinder the analytical method from describing ecosystems.

The diversity among agents is realised in the 'Echo' model proposed by Forrest and Jones (1994). In this model, one agent involves in a number of activities, including *trade*, *mating* and *combat*. The agents have resource to *trade* with other agents in the Echo, while the consequences of *mating* and *combat* are decided by the values of the agents' resources or other parameters inherited by the agents. By implementing those characteristics into the individuals in the model, the macroscopic patterns of Echo is found to resemble those of real ecosystem, especially species abundance. A similar pattern in species abundance has been found in the form of Preston's "canonical" distribution (Preston 1948, 1962a, 1962b), which has a relation between population size and abundance.

Although the above ecosystem simulations do not concern the details of behavior in the ecosystem, the individual characteristics of agents (organisms) has to be implemented. As Forrest and Jones (*ibid.*) point out, the simulation that demonstrates the macroscopic pattern, with the expense of microscopic detail, is needed for ecological study which deals with a high degree of complexity. The results from the above mentioned simulations suggest that it is necessary to change our point of view towards understanding the heterogeneous system like ecosystem. What we ought to do is to demonstrate the dynamical behavior of the system, rather than predicting the details of individual activity.

Since the prime goal of our numerical model is to demonstrate the dynamical pattern of ecosystem, some important points are reviewed in the following section, concerning the abundance of species and energy flow within an ecosystem. The processes in our model are defined in Section 3, followed by the results. In the final two sections, the energy flow in the real and artificial ecosystem is discussed to evaluate the present simulation study, and to further improve the numerical modelling concerning ecosystem.

2. Body size and abundance in animal population

The relation between body size and abundance has been investigated by many authors (e.g. Damuth, 1981). In most cases, there are more small organisms, whereas large organisms are found less. Such observations are formalised in an inverse relation between the abundance of animal species and the body size (metabolic requirements). Logarithmically transformed data of body size and abundance of animal population generally form a straight line with a negative slope in a double-logarithm scale (Damuth, *ibid.*).

This correlation indicates that the energy consumed by organisms in a unit area is independent of the body size of animal (Damuth, 1981, 1991). This correlation is commonly explained by the effect of competition for resources, since individual metabolic requirements scale proportionally to the body size. Some authors call this relationship 'energetic equivalence rule' (Nee *et al.*, 1991).

Marquet, Navarrette and Castilla (1990) assembled a number of data of marine communities, which showed an inverse relation between body size and abundance. But it is difficult to claim that the inverse relation under consideration results from only inter-specific competition for resources, because different communities belong to different environments and are subject to the different conditions. Besides, Marquet, Navarrette and Castilla (*ibid.*) point out that the relation is highly susceptible to sampling bias.

One exception is the population of birds, which does not show clear relation between body mass and abundance. But a common view on this is that the bird's three-dimensional move allows more flexible strategy than "two dimensionally moving" creatures (Cotgreave & Harvey, 1994). The following model does not include three dimensional moves. A general review of the inverse relation between abundance and body size is summarised by Cotgreave (1993).

3. The model

The organisms in the simulation are represented by a set of parameters, since the flux of energy is the focus of investigation. The energy in the model is carried by three parts, namely *soil*, *plant* and *herbivore*. The *soil* does not have any parameter to do with metabolism because it is not an organism. Then the *soil* simply reserves nutrient at each site to support the *plants* and *herbivores*. Therefore the only parameter that the *soil* has is the current amount of energy (nutrient) held at the site.

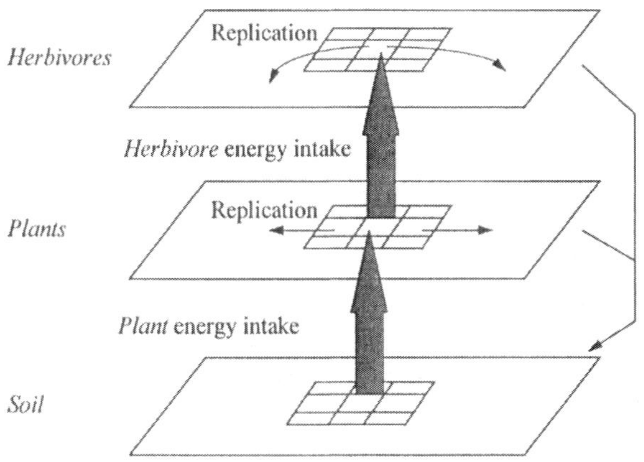

Figure 1: Diagram of the energy flow between the layers of *soil*, *plant* and *herbivore*. The thick arrows represent the energy flow associated with metabolism. Thin arrows within a layer represent the energy flow by means of replication. When the organisms die, the energy held by organisms goes back to the *soil*.

A 2-D grid is used to accommodate *soil*, *plant* and *herbivore*. The size of system is 97 by 97 (which has 9409 sites). A periodic boundary condition is used. Therefore, the organisms (*plant* and *herbivore*) at the edge of the landscape can interact with the organisms at the other side. This boundary condition prevents any asymmetric interaction at the edge of the landscape.

3.1 Energy flow (metabolism)

The *plant* and *herbivore* have two parameters to determine the amount of energy to absorb from the layer underneath. One parameter is *growth rate* (r), while the other is *capacity* (C). The transfer of energy per step (represented by t) is defined by the value of *vitality* (V), using the traditional logistic equation:

$$\frac{dV}{dt} = rV(1 - \frac{V}{C}) . \qquad \text{(eqn. 1)}$$

The maximum *vitality* (size) of an organism is therefore determined by C. The values of C and r are held by individual agents so as to develop a heterogeneous population. The above logistic equation is one of the rudimentary type. However, since the form of the equation is the same for all organism in the simulation, our discussion below does not have to include a scrutiny of the type of equation (for the different types of logistic equation see Olson *et al.*, 1992).

The value dV/dt is regarded as an energy intake of *plant* and *herbivore*. Therefore, if the *plant* cannot obtain the amount, dV/dt from the *soil* at the site concerned (in other words the *soil* has the energy less than dV/dt), then the *plant* dies. The same applies to the *herbivore*. That is, when the *herbivore* cannot gain dV/dt from the *plant* beneath the *herbivore* under consideration, then the *herbivore* dies.

A maximum value of the *capacity* for the *plants* (denoted by C_P) is set to 10.0, while the maximum value of *capacity* for *herbivores* (C_H) is set to 5.0, because the plants have to be abundant to support consumers in many ecosystems. The value of *growth ratio* ranges 0 to 1, because it represents, by equation 1, how much gap between the *capacity* and the current value of *vitality* is fulfilled. If the value of r exceeds 1.0, then the *vitality* may exceed the value of C, which is designated for the maximum "size" of the organism. Therefore in this simulation, the value of *vitality* is always below the value of C.

3.2 Replication

The *herbivores* can replicate when it is more than T_M steps old (when they become "matured"), while *herbivores* are expected to "live" T_H steps. In our experiments, the values T_M=11 and T_H=15 are used. Therefore, the value of available energy allowing, the *herbivores* can replicate themselves during the "age" of 12 to 15 step old. The *herbivores* may replicate more than once at a single step. But the number of replication is limited less than N_R, which is held by individual herbivore. The cost of replication (S) is defined by

$$S = \frac{C_H}{N_R} \qquad \text{(eqn 2)}$$

per offspring. The value of S is defined so as to provide a trade-off relation between the number of replications and the initial *vitality* of offspring. In other words, the *herbivore* can reproduce many offspring if they are small, whereas they can reproduce only a few if they are big. Also this formula conserves the value of *vitality*. However, there is no relation between C defined above and N_R. Therefore the organisms can have any combination of C and N_R. But N_R is defined to be an integer less than 16 for a technical reason associated with data gathering on the computer.

To replicate, the *herbivores* must have enough *vitality*. For this reason, the value of *vitality* is checked before the replication process. If the *vitality* is less than the product of S and the number of the replications, then the *herbivore* is not allowed to replicate at the time step concerned. When the replication takes place, the new *herbivores* are allocated at the vacant sites (chosen randomly) in the vicinity of the "parent" cell, and given the same parameters (*capacity*, *growth rate* and N_R) as those of the parent cell. The simulation does not concern sexual reproduction. The mating process is therefore not included in this simulation.

The reproduction of the *plants* in this simulation is designed simpler than that of *herbivores*. The *plants* do not replicate more than once at a time step. But if a *plant*-free site is adjoined by the site occupied by *plant*, then the unoccupied site will receive a half of the *vitality* of the adjoined *plant*, while the "parent" loses a half of its *vitality*. If more occupied sites than one are adjoined by the unoccupied site, then the parent *plant* is chosen randomly.

3.3 Mutation

Mutation affects all the parameters described above, but their occurrences are independent. Each time step, 0.1% of the total sites occupied by either *plant* or *herbivore* are subjected to the mutation of C (capacity), r (growth rate) or N_R (number of replications). After the site is chosen, the value that shifts the original values of either C, r or N_R is determined by random function. At the same time, whether increase or decrease the value is also determined randomly. If the altered value is outside the range of parameters ($0.0 \le C_H \le 5.0$, $0.0 \le C_P \le 10.0$, $0.0 \le r \le 1.0$ and $1 \le N_R \le 15$), then the organism is regarded as dead. Intuitively, the mutation that brought the parameter out of the range is treated as a "lethal" mutation.

The initial values of all the parameters are randomly distributed within the ranges described above. The initial value of *vitality* is mentioned in the result below. This initial value does not mean that all organisms have the same initial value at the start. Uniformly distributed random values less than the initial-*vitality* are allocated to the *soil*, *plants* and *herbivores* at the first step. Thus the high and low initial-*vitality* imply the "rich" and "poor" environment, respectively.

4. Result

The model is run under the condition which can sustain enough number (>90%) of producers (*plants*) covering the landscape. But the present study focuses on the dynamical behavior of consumers (*herbivores*). Therefore we compare the results of the environments with poor and rich conditions, both of which still have enough producers. Even in this simple format of the model, the *herbivores* holding different *capacity* can form colonies. Figure 2 shows a typical landscape, which displays different "species" of *herbivores* (it might be inaccurate to call them *species*, but see Section 5 below).

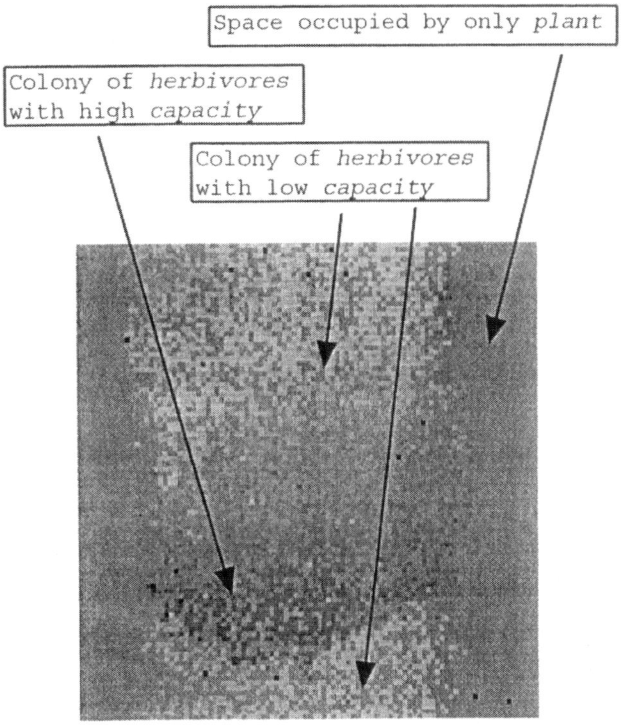

Figure 2: A typical landscape of the simulation. Different organisms are represented by different hue of dots. Initial-*vitality*=30.0, at 5500th step.

It is interesting to see that the landscape shows a gradual change in *capacity* upwards on Figure 2, whereas there is a sharp change between the high-*capacity herbivores* and their immediate southern neighbours in the landscape. These gradation and sharp change in *capacity* of *herbivores* in the landscape are often seen in the landscape. But at the moment, the cause of such geometrical variation of *capacity* is not clear.

Regarding the relationship between body size and abundance of organisms, the data of *capacities* are assembled below. The parameter *capacity* does not exactly correspond to the body size in a real ecosystem, but the *capacity* is a maximum value of *vitality* in the model. Thus, the value of *capacity* is used to represent the body size of organisms in the diagram.

When the landscape is filled with much *vitality* (nutrient), the relationship between abundance and body size in the diagram is seen in Figure 3. In this situation, *herbivores* with a large *capacity* (body size) can survive, thanks to the plenty of nutrients. It is worth noting that the gradient of regression line (not shown) of the plot in Figure3 is about -0.98, which is just under -0.75, which is theoretically suggested value (Nee *et al.*, 1991).

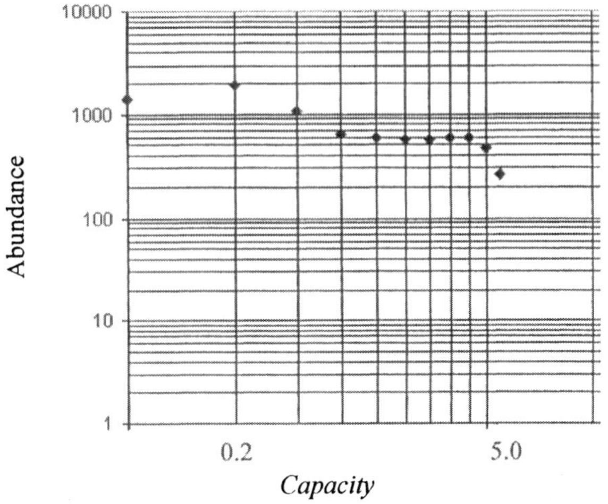

Figure 3: Double-logarithm diagram of *capacity* and *abundance* for a "rich" environment. Initial-*vitality*=30.0, at 10,000th step. The plot may be estimated by a regression line of a shallow slope.

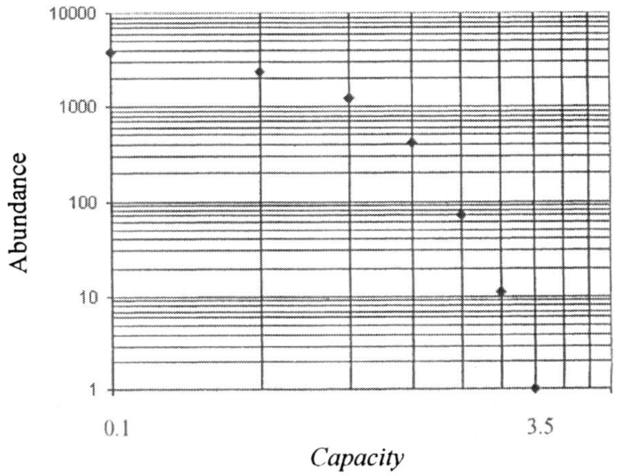

Figure 4: Double-logarithm diagram of *capacity* and *abundance* in the landscape with "poor" condition. Initial-*vitality*=9.0, at 7,000th step. *Herbivores* with more than 3.5 of *capacity* do not appear.

When the landscape provides poor nutrition, the relationship between body size (*capacity*) and abundance shows a steep slope, as seen in Figure 4. Although the relationship between body size and abundance could be estimated by a regression line, the gradient is about -3.8, which is well below the reported value in the past studies (e.g. Damuth, 1981, 1991; Cotgreave, 1993). Besides, in a poor condition, larger organisms cannot survive.

Various combinations of parameters are yet to examine. Therefore, in the future research, we expect that the different patterns in the relationship between body size and abundance can appear under the different conditions. But here, the amount of resources available to the organisms clearly affects the pattern of population abundance. It is worth noting that, even in a poor condition, comparatively large organisms (*capacity*) still survive despite of their small population.

5. Adaptation appearing in population dynamics

It is encouraging that the results of our model show a variety of patterns in the relationship between body size and abundance at this early stage of research, while real ecosystems show diverse patterns of species abundance too.

As a number of authors suspect (e.g., Cotgreave & Harvey, 1994), various causes determine the distribution of species abundance. One strong factor is competition for resources in a local ecosystem. The present simulation study confirms that inter-specific competition affects the species abundance. Comparison between poor and rich environments suggests that the distribution of species abundance is significantly affected by the overall flow of nutrients in the landscape. In a rich environment, both of small and large organisms can maintain a reasonable size of population, because there is no severe competition in terms of resources (competition for space does exist). Contrary, in a poor environment, large organisms fail to obtain sufficient nutrient to survive.

Another cause of the species abundance is phylogeny. The effect of phylogeny is regarded as a major reason why the relationship between body size and abundance in bird community does not often show the inverse covariation (Cotgreave & Harvey, 1994). However, our computational model does not include genotype or phenotype. Then we can discuss the result of the model in the view of energy flow alone.

5.1 Energetic equivalence rule

The most convincing and general reason for the inverse correlation between body size and abundance is "energetic equivalence rule" (Odum, 1971). This rule states that the population abundance follows the distribution which provides equal energy consumption within a unit area. Therefore the large species appear less frequently, whereas small species are found more.

If we accept an assumption that the species abundance is solely determined by the "energetic equivalence rule," then the gradient of the correlations between body size and abundance must follow the curve of gradient -0.75, since the individual metabolic requirements increase proportionally to the body size with the power of +0.75 (Odum, *ibid.*). This argument of course is too simple, because the organisms involve more activities than the role of energy flow. However, interestingly, the present simulation study shows that a variety of distributions can emerge, even if the population dynamics depends on mainly energy flow. Besides, the metabolic rate in our model is not directly defined by the power of body size (see eqn 1). The result of the model thus tell us that the energetic equivalence rule can result from the metabolism rate which is only roughly correlated to the body size.

5.2 Validation of the model

There are two questions concerning the relation between the present simulation study and real ecosystem. One is the classification of species. The other is whether or not the model can describe the intrinsic dynamics of population fluctuation.

Regarding the first question, the classification of species is not defined by genotype or phenotype in the present simulation, because our aim is to examine the relationship between body size and metabolic rate, while different phenotypes or genotypes happen to have the same metabolic rate. If the model has a definite phylogeny (e.g., like Echo model by Forest & Jones, 1994), we could divide the species according to the genetic linage. But such classification is not necessary in our model, because all that the organisms do in our model is to consume resources, and die when exhausted. The model presented here is therefore constructed by the flows of energy associated with feeding. That is, the classification in our model relies on the body size, not on the genetic characteristics of the species. Metaphysically speaking, the classification may be done in various ways including classification by weight of the organism.

Concerning the second question, some authors point out the intrinsic problem of computational models, especially density dependent model of population dynamics (Cousens, 1995). Although the present study of population dynamics differs from conventional models in that the logistic equation (definition of energy flow) is held by individuals not by a population as a whole, the problem of density dependent model needs attention.

A major concern in relation to the computational model of population dynamics is how to distinguish the cause of fluctuation in population resulting from the intrinsic dynamics of the population and the cause from environmental noise. This point of view, however, relies on the assumption that the intrinsic behavior, which is ideally deduced from the mechanism of population dynamics, can be separated from environmental noise. The assumption that the effect of noise and intrinsic dynamics are separable is not usually mentioned in literature explicitly. However, the question how we describe the intrinsic dynamics of population cannot be answered without assuming that the intrinsic dynamics can be separated from noise in the first place.

Many believe that if the model is realistic enough, then the model can display the intrinsic dynamics of population and so forth. But of course, being realistic/unrealistic and intrinsic/extrinsic must not be confused. In a broad view including a general study of complex system, it is important to examine the form of power law in relation to the energy flow in the landscape. In the previous studies, one of the authors claimed that the emergent phenomena, especially power law (like the relationship between body size and abundance of animal species) and $1/f$ noise may well be caused by the interaction between irreducible processes (Yamamoto 1995, 1997), using formalistic models. In a comparatively concrete model, the present study also shows that the collective behavior of ecosystem adapt to a sustainable species distribution. It is worth noting that even if each individual does not seem to make any efforts to adapt to the given environment, the whole dynamics leads the population to a state, which has been observed in a real ecosystem.

6. Conclusion

The ecological model focusing on energy flow between producer and consumer showed that the energetic equivalence rule can be resulted from the metabolic rate which is roughly related to the body size. The amount of resources in the landscape also contributes to the distribution of the species abundance. This result will explain the relationship between body size and abundance in a specific environment.

In the future research, we will see more detailed relationship between individual metabolism and species abundance. It is also important to see how foraging strategy affects the species abundance. The species abundance and individual strategy are at the opposite side of ecological scale, which are difficult to examine simultaneously in a real ecological system, but a simulation study like in this work will help link those two subjects. The theoretical implication of the model will provide the insight into directing the field research also.

Acknowledgments

One of the authors (T.Y.) has been supported by Charles Sturt University Writing-Up Award. The computer model is implemented on the CM5 at the South Australian Centre for High Performance Computing.

References

Bak, P. & Chen, K. (1991). Self-Organized Criticality. *Sci. Am.* **264**, 46-53.

Bak, P., Tang, C. & Wiesenfeld, K. (1987). Self-organized criticality: an explanation of $1/f$ noise. *Phy. Rev. Lett.* **59**, no. 4, 381-384.

Cotgreave, P. (1993). The Relationship between Body Size and Population Abundance in Animals. *Trends in Ecology and Evolution.* **8** (7) 244-248.

Cotgreave, P. & Harvey, P. H. (1994). Phylogeny and the relationship between body size and abundance in bird communities. *Functional Ecology.* **8**. 219-228.

Cousens, R. (1995). Can we determine the intrinsic dynamics of real plant populations? *Functional Ecology.* **9**. 15-20.

Damuth, J. (1981). Population density and body size in mammals. *Nature.* **290**. 699-700.

Damuth, J. (1991). Of size and abundance. *Nature.* **351**. 268-269.

Forrest, S. & Jones, T. (1994). Modeling complex adaptive system with Echo. In "*Complex Systems: Mechanisms of Adaptation.*" 3-21. R. J. Stonier and X. H. Yu (eds.). IOS Press, Amsterdam.

Hirata, H. & Ulanowicz, R. E. (1984). Information Theoretical Analysis of Ecological Networks. *Int. J. Sys. Sci.* **15**, no. 3, 261-270.

Marquet, P. A., Navarrete, S. A. & Castilla, J. C. (1990). Scaling Population Density to Body Size in Rocky Intertidal Communities. *Science.* **250**. 1125-1127.

Nee, S., Read, A. F., Greenwood, J. J. D. & Harvey, P. H. (1991). The relationship between abundance and body size in British birds. *Nature.* **351**. 312-313.

Odum, E. P. (1971). *Fundamentals of Ecology.* 3rd ed. Sanders, Philadelphia.

Olson, M. H., Watokinson, A., Mackenzie A., Young, T. P., Berryman, A. A. & Ginzburg, L. (1992). Intuition and the Logistic Equation. *Trends in Ecology and Evolution.* **7** (9) 314-337.

Preston, F. W. (1948). The commonness, and rarity, of species. *Ecology.* **29** (3) 254-283.

Preston, F. W. (1962a). The canonical distribution of commonness and rarity: Part I. *Ecology.* **43** (2) 185-215.

Preston, F. W. (1962b). The canonical distribution of commonness and rarity: Part II. *Ecology.* **43** (2) 410-432.

Solé, R. V. & Manrubia, S. C. (1995). Are rainforests self-organized in a critical state? *J. Theor. Biol.* **173**, 31-40.

Ulanowicz, R. C. & Norden, J. S. (1990). Symmetrical Overhead in Flow Networks. *Int. Systems Sci.* **21**, no. 2, 429-437.

Wolfram, S. (1984). Cellular automata as models of complexity. *Nature*, **311**, 419-424.

Wolfram, S. (1994). *Cellular Automata and Complexity: Collected papers.* Addison-Wesley.

Yamamoto, T. (1995). Critical state between local and global interaction. *Appl. Math. Comp.* **73**, 153-180.

Yamamoto, T. (1997). Emergent Property Caused by Interaction between Heterogeneous Measurement Processes in Complex Systems. *15th IMACS World Congress on Scientific Computation, Modelling and Applied Mathematics.* vol. 4, 671-676.

Morpho-Functional Machine: Robotics Approach of Amoeba Model based on Vibrating Potential Method

Hiroshi Yokoi, Wenwei Yu, Jun Hakura, and Yukinori Kakazu

* Laboratory of Autonomous Systems Engineering, Graduate School of Engineering, Hokkaido University
Kita3, Nishi-8, Kita-ku, Sapporo,Hokkaido 060, JAPAN
{ yokoi, hakura, yu, kakazu}@complex.eng.hokudai.ac.jp

Abstract

Soft-mechatronics is a new research area for a developmental robotics approach. This paper reports on the methodology and results of our approach to soft-mechatronics via the imitating of mollusc-like characteristics in through both hardware and software fusion technology. The proposed approach for soft-mechatronics here is developing a Morpho-Functional Machine by imitating amoeba motion and structure. This paper describes the Vibrating Potential Method to control the Morpho-Functional Machine, that consists of a new field technique called the Vibrating Potential Field plus a new parameter tuning method inspired from thermodynamics. The field model creates self-organising, gathering behaviour, through physical interaction with potential fields. The computer simulation shows typical characteristics of swarm intelligence, such as, energy gathering, thermotaxis, and obstacle avoidance. Three types of robots are built with these three characteristics. This paper shows the possibility of various behaviours produced by Morpho-Functional Machine.

1. Introduction

Hardware design of adaptive systems is one of the most significant research areas, in Real World Computing. The challenge is to design a physical system that behaves adaptively, flexibly, intelligently, and with emergence. Accordingly, some specialised communities have begun to discuss evolvable hardware, distributed autonomous robots, evolutionary robotics, and animat approaches.

These new approaches have evolved come out an Animat Robotics. "Morpho-Functional Machine" is a coined term that means a machine behaves like a creature with adaptive and flexible deformation. This developmental approach started as a research project within the Research Project for the Future of JSPS, the Science University of Tokyo and Hokkaido University corroborated on the project which focused on mollusc imitation. This paper reports on methodology and results of our approach for soft-mechatronics mollusc imitation.

Specifically, we have focused on the flexible motion which emerges from the competitive life game of amoeba cells. Artificial (pseudo) cells are designed to adapt to new environment into energy scrambling phenomena. These artificial cells are called unit. The unit consists of sensor(s), controller(s), indicator(s) and actuator(s). Finding and selecting set s posed biggest challenge of this project. In this paper, a mathematical model of the mutual action of units and a parameter tuning method for each unit is described. To approximate dynamics of cell interaction, sensing and indications are mediated by vibrating potential fields (Yokoi, 1992). Actuations are described as kinematic equations. The strength, stiffness, and density of sensors are described by system parameters. These parameters are tuned so that each unit receives increasingly more energy than the neighbouring units. This paper shows hardware designs and control rules extracted from living NC4 (the most researched species of amoeba) toward flexible motion.

2. Related works

2.1 Flexible Motion

The spatial movement of amoeba cells is driven by cAMP. The detail of cell motion is explained (Varnum 1987). Deformation of pseudopod and oscillation of cell expansion results in flexible motion. Such flexible motion is controlled by a spatial pattern of cAMP waves.

2.2 Information in Biological Medium

The morphogenesys of amoeba is adaptive for environment based on chemotaxis and thermotaxis that as shown in (Siegert, 1992) and (Steinbock, 1993). The adaptive behavior of amoebas results in adaptive grouping mechanism which are self-organised with a declining degree of freedom (D.O.F.) from unicellular state with higher D.O.F. to higher ordered multi-cellular state. The objective application to create obstacle avoidance with amoeba like behavior.

2.3 Computer Simulation

The amoeba-like motion consists of the displacement of the gravity centre and the flexible deformation of cell body. Such motion realises complex searching behavior including global and local interactions that describes grouping behavior of population depending spatial patterns (Agarwal, 1994), (Shaw, 1995), (Mackay, 1978). These amoebic characteristics are related to the dynamics of the autonomous tuning character in the amoebas body. Specifically, the stored energy and heteronomous movement of body parts depends on the oscillation and density information propagated from its own living field (Abe, 1994). The proposed model consists of structure and interaction. The structure and the interaction are mathematically described as unit group structure using communication and interactive communication through wavy information.

2.4 Unit-based modelling approach

The unit-based modelling approach taken here also can be called "cell-based approach". Cell-based approaches in modelling the systematic multi-celled organisms are based on the cell theory developed by Schleiden, M.J. That is, "Cells are units of the structure and functionality of all organisms, and, in a sense, are first-order elements of organisms." (Fleischer,1994) made developmental models that can represent some characteristic behaviours of cells.

For the field technique, the theory of cellular automata (CA) is a basic model to generate or calculate field patterns from local interactions among cells. Turing's morphogen model is also a more basic model of the field technique. Ueda's work, which concerned Physarum intelligence model concluded that self-organising the chemical patterns at chemical molecular level cause the intelligent behaviour (Ueda, 1993). As for the systems with emergent functionality, (Steels, 1991) theoretically discussed in the form of dynamic systems.

The Unit-based modelling consists of sensor(s), controller(s), indicator(s) and actuator(s). All information propagated from the indicator is reflected by suitable fields, and picked up by the units sensors. This simulation technique of propagating information is called "field technique". The field technique and movable phenomena of particles, is called VPM (vibrating potential method) proposed by (Yokoi 1996). VPM is used in cell movement imitation and other engineering problem.

2.5 Hardware design of Amoeba model

To imitate amoeba characteristics, both hardware design and theoretical approaches are necessary. In other word, question is such as "how to measure", "how to produce", and "how to control" of amoeba arise. For the function of amoeba-like motion, large structural deformation will be required in the hardware design, and also a large degree of freedom to control motion.

To realise this, we defined Morpho-Rate to fix the quantity. Here, the morpho-rate **MR** consists of terms dealing with structure, measurement, control, and actuation. The **x1** means the possibility of deformation that is defined as a proportional ratio between Max(D) and Min(D). Max(D) which means the maximum diameter when the body is expanded, and Min(D) means the minimum diameter when the body is shrunk. Fig. 1 (a) shows a draft image of the possibility of deformation. **x2** is a measure of plasticity that is defined as a rate of motive force and maximum local stress, where, p(r) is local stress as shown in Fig. 1 (b) and (c). **x3** is a density of the sensor, and **x4** is the density of the actuator. Where n(s) is the total number of sensors, n(a) is the total number of actuators, and V is the volume of the machine.

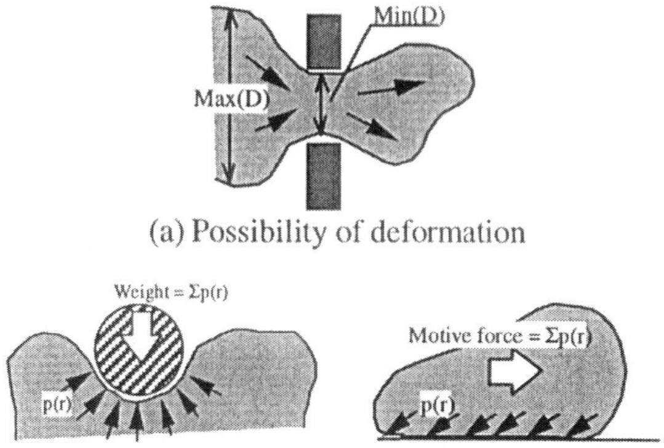

Figure.1. Concept of Morpho-rate

$$\mathbf{MR} = \{\mathrm{sign}(\mathbf{x1}) + \mathrm{sign}(\mathbf{x2}) + \mathrm{sign}(\mathbf{x3}) + \mathrm{sign}(\mathbf{x4})\}/4 \quad (1)$$

Where, $\mathbf{x1}=\mathrm{Max}\{D\} / \mathrm{Min}\{D\}$, $\mathbf{x2}=|\Sigma p(r)| / \mathrm{Max}\{\, |p(r)|\, \}$, $\mathbf{x3}=n(s) / V$, $\mathbf{x4}= n(a) / V$

The hardware design of the Amoeba model requires high MR. Fig. 2 shows one ideal concept model of hardware design. To increase x1 and x2 as many control points as possible should be defined. To increase x3 and x4, as many distributed sensors, indicator and actuators as possible should be assigned around the edge of unit. Sensors and Indicators should communicate through spatial media. Therefore, an Unit-based approach and field techniques are suitable for this purpose.

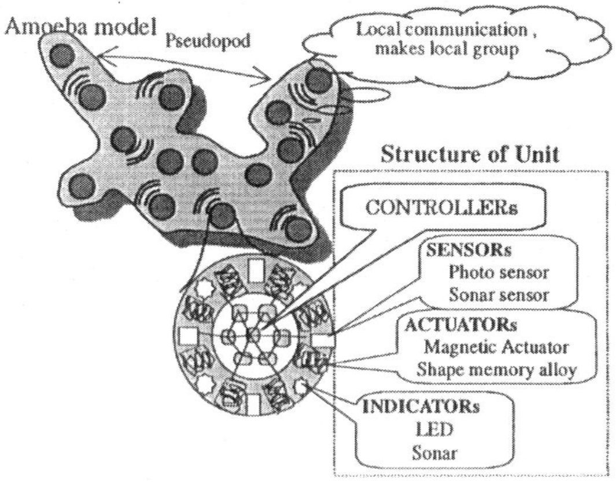

Figure 2. Concept of Amoeba model

3. Amoeba model

The mathematical design of the amoeba model consists of Sensor(s), Actuator(s), Indicator(s), and Controller(s). A sensor's function is to detect information. A indicator's function is broadcast information. Those two functions are described "Field" and "Potential functions". An actuator's function is to physically interact with the environment, that as is i.e. "Unit Motion". A controller function is to make decisions based on information from the three other functional elements. The controller also tunes System parameters in a mechanism called "Adaptation".

{ **H** } =	{ **Force** } +	{ **Signal** } +	{ **Energy** }
Field	Physical interaction	Information	Reward

{ **P** } =	{ O(**H**) }	{**Parameter**}	{ **Energy** }
Actuation	Sensor	System Parameter	Consumption

{ **K** } =	Opt{ **Internal Entropy** }
Parameter	Optimisation of System Parameter

3.1 Field (Sensor and Indication)

One virtual field is suggested the model of biological medium. The whole field, **H** called VPF, consists of the accumulated individual fields that are generated by each units. The multi-body's interactions among units can be obtained by the examine interaction between one unit and VPF, i.e. as if it were the interaction between two body's. Two body's interaction offers special feature that a unit observes the status of other units without knowledge of relationship among them. The VPF mathematically is described as an equation using potential functions (h, w, E, hl) on individual fields (ψ, χ, τ, φ). In the equation, the information is defined only as propagating a wavy signal, and described using potential functions on spatial intervals and time.

$$\mathbf{H}(r, \phi, t) = \sum_i^{\text{Unit number}} \{\mathbf{h}_i(r,t)\psi(\phi) + \mathbf{w}_i(r,t)\chi(\phi) + \sum_i^{\text{Unit number}} \mathbf{hl}_{ij}(r,t)\varphi_i(\phi)\varphi_j(f)\} + \mathbf{E}(r,t)\tau(\phi) \quad (2)$$

Where, $\mathbf{H}(r,\phi,t)$ is the vibrating potential field (whole field), ϕ is total co-ordinate axis, i is the unit number, $\mathbf{E}(r,t)$ is the energy function of the environment, $\mathbf{h}_i(r,t)$, $\mathbf{w}_i(r,t)$, $\mathbf{hl}_{ij}(r,t)$ are information propagated from unit i (potential function), r is the position vector. From $H(\boldsymbol{r}, \phi, t)$, a unit interacts with other units on $\psi(\phi)$, $\chi(\phi)$, $\tau(\phi)$, $\varphi(\phi)$ according to the amplitude of each individual field. This spreading characters of field is mathematically described using the partial differential equation u=u(r,t) for each field parameter {α,β,γ,δ, ε} as Eq.(3).

$$\alpha\frac{d^2u}{dt^2} + \beta\frac{du}{dt} + \gamma\nabla^2 u + \delta\nabla u + \varepsilon u = 0 \quad (3)$$

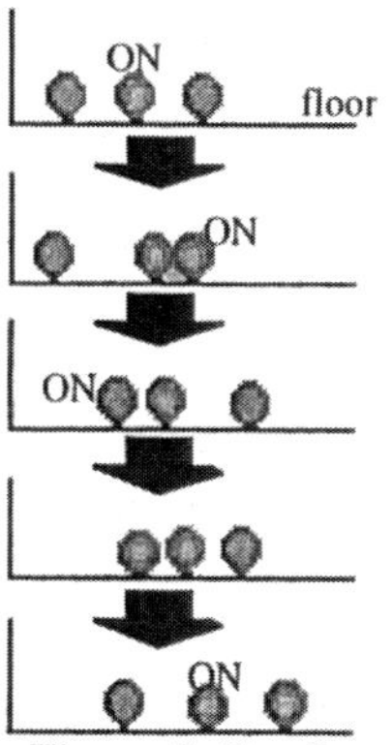

Movement of the unit group is driven by the phase shift of the attractive potential function. The gravity center of a unit group is advanced. Where, "ON" is active unit of attractive/repulsive potential functions. Friction of floor is set at 0.1 (where unit velocity equals 0), and 0.3 (where velocity does not equal 0)

Figure 3. Fundamental movement of unit group

Unit communication is reflected by the multiple VPF **H** in Eq. (1). The Sensor and the Indicator are described as a wave equation shown in Eq. (4). Selection of the type of Sensor and Indicator is defined by the System Parameter vector K(t)=(Kψ Kχ, Kτ, Kφ). Boundary conditions should be different for each property to avoid confusion.

$$\frac{\mathbf{K}(t)}{2}\frac{d^2\zeta(\phi)}{d\phi^2} + E_{env}(\boldsymbol{r}_i, t)\zeta(\phi) = 0$$

$$(\zeta(\phi) = \psi(\phi), \chi(\phi), \varphi(\phi), \tau(\phi)) \quad (4)$$

Where, $\psi(\phi)$, $\chi(\phi)$, $\varphi(\phi)$, $\tau(\phi)$ are unit co-ordinate axis, $\mathbf{E}_{env}(r_i,t)$ is interacting energy in Eq. (5).

$$dE_{env} = \sum_i^{\text{Unit number}} E_{out}(ri,t)\, dt \quad (5)$$

3.2 Unit Motion (Actuation).

The unit motion is defined according to the field information and the unit's stored energy, and is realised as a result of transitions by the external stimulus on unit internal parameters. In Eq. (6), unit motion p(t) is derived by Lagrange equations of motion where $\mathbf{H}(r,\phi,t)$, E_{out}, K(t) are potential function. Where integral $\mathbf{H}(r,\phi,t)$ means observation of field information, E_{out} is outflowing energy from the unit to the environment, and K(t) is the System Parameter. Fig. 4 shows one example of unit motion, where the sign "ON" means active input from the field. The computation shows that the reciprocal activation input creates movement of the gravity centers of the units through attractive/repulsive potential functions.

$$\dot{p}(t) = M \cdot \nabla \oint \mathbf{H}(r, \phi, t)(\zeta(\phi)+\varphi(\phi))d\phi\; E_{out}(t)\,\mathbf{K}(t) \quad (6)$$

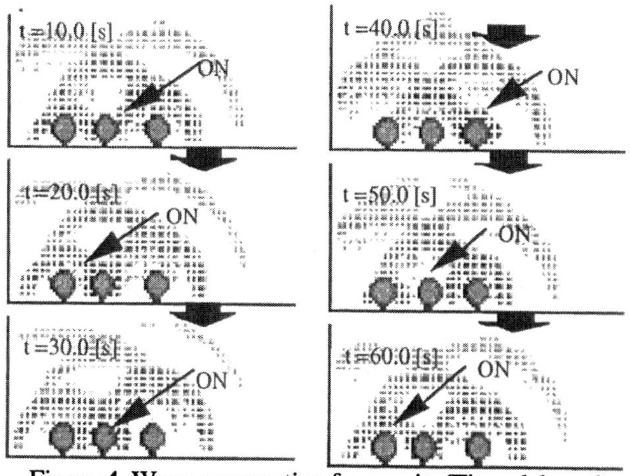

Figure 4. Wave propagation from unit. (Time delay of activation drives reciprocal action of wavy field)

The propagating wave pattern is produced from a Speaker in the Indicator that is deformed by the tuned delay timing of each unit.

3.3 Unit Information (Potential Functions)

The unit information in this paper deals with the amplitude of the propagating information in a field. Two types of information is applied. One is a density type that simply diffuses with the declining along with a strength of received information that works attractive and repulsive potential function. The other is a wavy type that propagates out while gradually declining, and suffers resonance, reflection and refraction depending on the boundary condition of the field. The unit produces such information, and influences surrounding units according to the distance among them. This information's are realised by setting values on the suitable individual fields, and also it is defined by Strength, Delay, Frequency , and Direction for each set information $\mathbf{h}i(r,t)$, $\mathbf{w}i(r,t)$, $\mathbf{hl}ij(r,t)$, and $\mathbf{E}(r,t)$. Fig. 4 shows reciprocal activation according to the wave input for neighbouring units.

3.4 Objective function

The adaptive behavior of units using local information requires an objective function for decentralised control. An ideal objective function of a decentralised system leads the units to an orderly configuration depending on the environment. In this section, the objective function for the system parameter $\mathbf{K}(t)$ is set in accordance with adaptive mechanics of living creatures. According to the field of statistical mechanics, each unit conserves order by minimising the differential calculated gain of entropy ds. The ds is derived by the input-output relation of in the energy environment. Stated values on each unit i are inflowing heat $\mathbf{E}in=\mathbf{E}in(ri,t)$, stored energy $\mathbf{Q}=\mathbf{Q}i(t)$, work $\mathbf{W}=\mathbf{W}i(t)$, $\mathbf{E}out=\mathbf{E}out(ri,t)$, and spatial heat environment $\mathbf{E}env=\mathbf{E}env(r,t)$. The inflowing heat $\mathbf{E}in$ is input energy from the environment that is transmitted through wave propagation minus threshold TH as shown in Eq. (7)

$$E_{in} = \oint \mathbf{H}(r, \phi, t)\, \chi(\phi)\, d\phi - TH \qquad (7)$$

The stored energy of each unit $\mathbf{Q}$ shown Eq. (8).

$$d\mathbf{Q} = (\mathbf{E}in - \mathbf{E}out - \mathbf{W})\, dt \qquad (8)$$

$\mathbf{W}$ is given energy from the unit to the environment, and is defined by the velocity $\mathbf{v}(t)$ of each unit in Eq. (9).

$$\mathbf{W} = \frac{1}{2} M\, v(t)^2 \qquad (9)$$

The outflowing heat $\mathbf{E}out$ from each unit is a function of $\mathbf{Q}$ and $\mathbf{E}env$ in Eq. (10).

$$d\mathbf{E}out = (Q - \mathbf{E}env)\, dt \qquad (10)$$

Since such stated values are connected with other stated values, numerical computation is applied to obtain a results. Now, the differential calculated gain of entropy ds is define by Eq. (11) using stated values the $\mathbf{E}in$, $\mathbf{E}out$, and $\mathbf{E}env$.

$$ds = \frac{d\mathbf{E}out - d\mathbf{E}in}{T(\mathbf{E}env)} \qquad (11)$$

The system parameter $\mathbf{K}(t)$ is derived by Eq. (12) in order to minimise ds.

$$\frac{\partial \mathbf{K}(t)}{\partial t} = -\frac{\partial\, ds}{\partial \mathbf{K}(t)} \qquad (12)$$

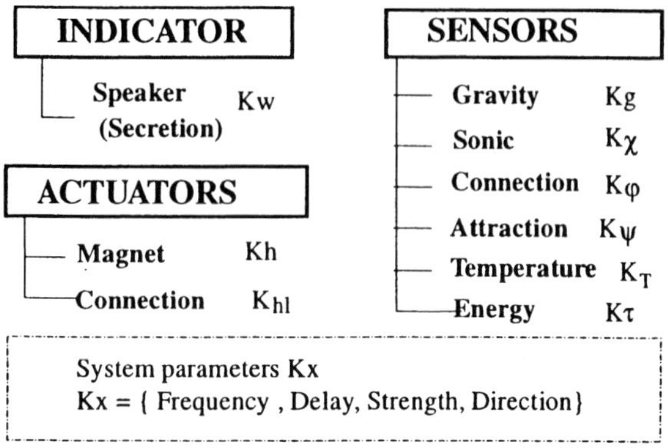

Figure 5. System parameters K(t)

Where, the system parameter K(t) consists of a Sensor part, an Indicator part and an Actuator part as shown in Fig. 5. The sensor part has $K\psi$ $K\chi$, $K\tau$, $K\varphi$ (boundary conditions of individual field), Kg (gravity), and Kt (temperature). The indicator part has Kw (Speaker). The actuator part has Kh (magnetic potential) and Khl (elastic potential). Each parameter has 4 degrees of freedom {Strength, Delay, Frequency, Direction}.

Using Eq. (12), the adaptive behavior of each unit group are shown in Fig. 6. The number of unit in each group is 30, 15, 9, 6, 3 and 2. All unit groups acquired goal searching behavior except the 2 unit group.

4. Experimental results of amoeba model

This section shows the Organisation of Research Groups for the amoeba model

1. Extraction of Amoeba characteristics by Sequential Imaging and Pattern Recognition.
2. Hardware design of Amoeba model.

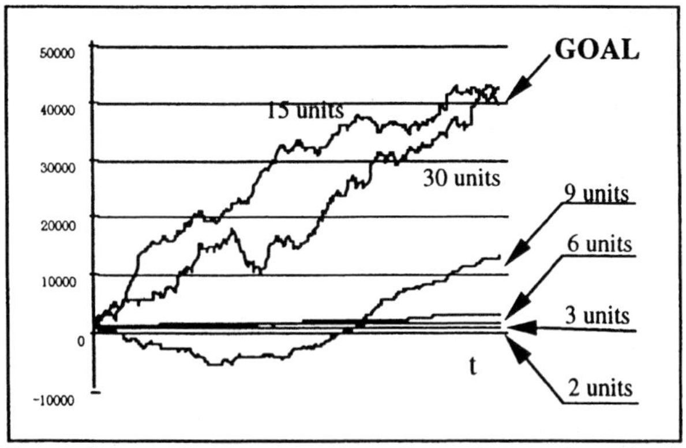

Figure 6. Moving distance depends on the number of units

4.1 Extraction of Amoeba characteristics by Sequential Imaging and Pattern Recognition

Microscopic images of NC4 show flexible motion and swarming states of cells. we identified the potential functions of unit group to be fit for the shape of cell membranes (Sato 1997). Our experiments show that by using our approach the external shape of cell motion and the period of oscillation reflected original amoeba. The characteristics of cell motion can be derived based on the position and shape of potential function using artificial cell method as shown in Fig. 7.

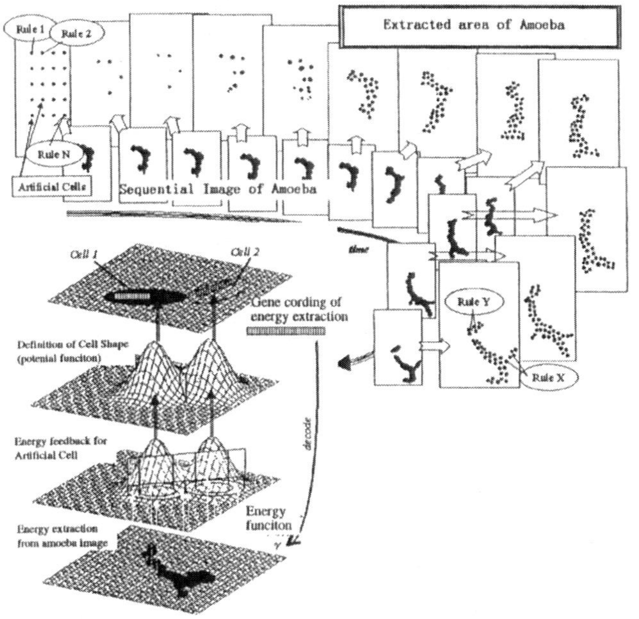

Figure.7. Autonomous tracking of amoeba sequential image

4.2 Hardware design of amoeba model.

The flexible function of the amoeba model is applied to the robotics design. The conceptual design of the amoeba model works based on collective states of unit groups interacting with physical environment shown in Fig. 2.

Three types of robots are proposed, such as, Distributed Autonomous Swimming robot, Star Fish type robot, and SMA Net. The Star Fish type is a structured networks of a SMA and a skeleton shown in Figure 8. The each legs of this robot has a CPU and sensors, and is actuated by a bias type SMA through thermal control. The multi-agent programming method is proposed as a controller of a goal acquisition mechanism for the Star Fish type, and also reciprocal adaptation of legs control creeping, rotating and walking behaviours. The SMA Net type consists of lattice structure shown in Fig. 9. This robot established looper like creeping motion through the mutual action of SMA net.

Figure 8. Star Fish type

Figure 9. SMA Net type

The next figure shows swimming units. The real amoeba will not swim, however, we are applied to the control problem of distributed autonomous robot to show the powerful capability of the amoebe model. Fig. 12 shows the adaptation process for goal acquisition. Which indicate trial number of attempts to get maximum energy its goal. Fig. 10 shows best motion, in which the robot swims to and holds its goal for within 45 steps. Fig. 13 indicates the average amount of energy which each unit received.

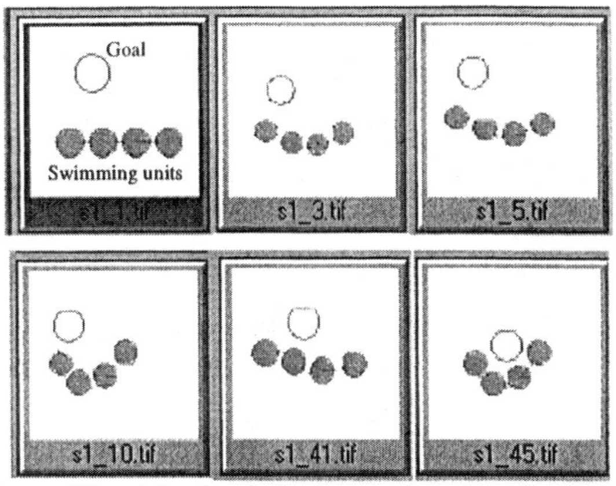

Figure 10. Adaptive behavior produced by distributed autonomous swimming robot.

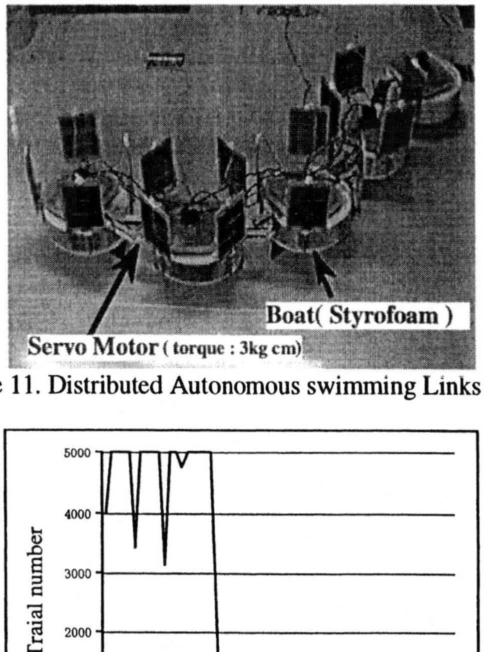

Figure 11. Distributed Autonomous swimming Links

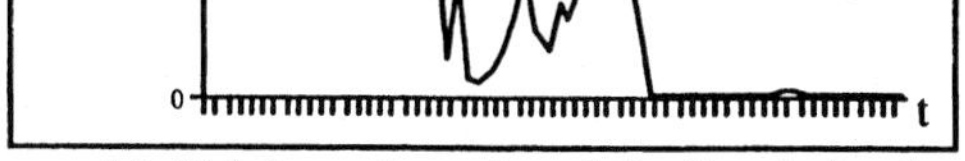

Figure 12. Trial number of goal finding behavior of distributed autonomous swimming robot

The Distributed Autonomous Links has CPU, sensors (4 solar panels), and 1 servo motor for each link unit as shown in Fig. 11. Each link unit has a learning mechanism of sensor-actuator mapping that tunes according to energy consumption. Through a competitive learning process, the group of link unit obtain adaptive behaviours to approaching energy sources.

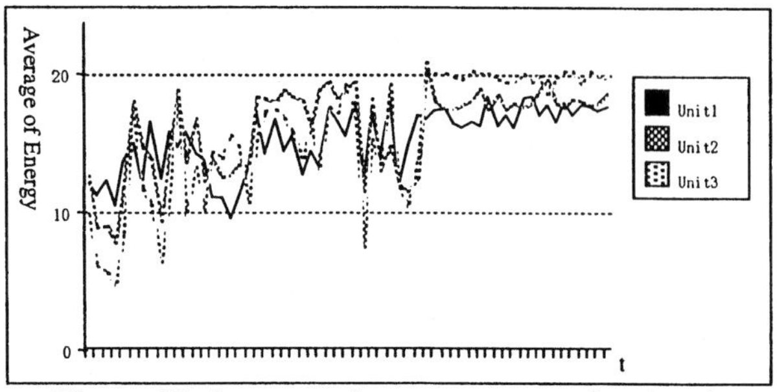

Figure 13. Average of energy storage of each unit

These amoeba robots are structured as collective units groups that realise distributed autonomous systems. The unit has a controller, a sensor, and an actuator in its own body. The units communicate to other unit through assyncronized wavy radiation. The results show competitive learning reflex reciprocal adaptation, and autonomous synchronisation of phase delay produce a creeping motion and walking behavior.

Fig. 14 shows the Morpho-rate MR of each mechanism. The MR is the average of each line. NC4 means real amoebas, and model means amoeba model produced by computer simulations. The number enclosed in parentheses is number of gathering units. As a result, this figure shows the real amoeba have a high Morpho-rate, and the amoeba model with 2 gathering units obtained comparable Morpho-rate.

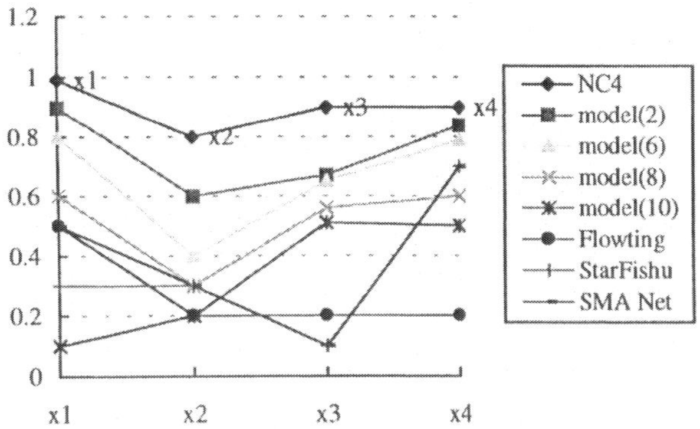

Figure 14. Morpho-rate of proposed model and robots

5. Conclusion

This study tries to develop the morpho-functional machine to achieve flexible deformable robot design. For this purpose, amoeba robots are proposed based on the amoeba model structure of collective unit groups. The amoeba model describes competitive adaptation mechanism through mutual action of units using the vibrating potential method. The results show competitive learning reflex reciprocal adaptation, and autonomous synchronisation of phase delay reflex creeping motion and walking behavior.

This research is supported by Science Research Foundation.

Reference

Abe, T., Early, A., Siegert, F., Weijer, C., and Williams, J. (1994). Patterns of Cell Movement within the Dictyostelium Slug Revealed by Cell Type-Specific, Surface Labeling of Living Cells, Cell Press, Cell, Vol.77, pp.687-699.

Agarwal, P. (1995), "The Cell Programming Language," *Artificial Life*, 2:1, pp.37-77.

Fleischer, K. and H. A. Barr (1994), A Simulation Test bed for the Study of Multicellular Development: The Multiple Mechanisms of Morphogenesis, *Artificial Life III*, Reading, MA: Addison-Wesley, pp.389-416.

Kudo, M., Yokoi, H., and Kakazu, Y. (1997). Constructing Amoeba-Like Behavior System:Mobile Automata in Vibrating Field, MIT Press, Fourth European Conference of Artificial Life, pp.284-289.

Mackay, S. A. (1978). Computer Simulation of Aggregation in Dictyostelium Discoideum, Company of Biologists Limited, T. Cell Sci., 33, pp.1-16.

Sato, Y., Yokoi, H., and Kakazu, Y. (1997), Extraction of Flexible Motion from Sequential Amoeba Image., ASME Press, Intelligent Engineering Systems Through Artificial Neural Networks, Vol. 7, pp.421-426.

Steinbock, O., Siegert, F., Muller, S.C., and Weijer, C. J. (1993). Three-dimensional waves of exitation during Dictyostelium morphogenesis, Biophysics, Proc.Natl.Acad.Sci.USA, Vol.90, pp.7332-7335.

Steels, L. (1991). Towards a Theory of Emergent Functionality, *From* Animals to Animats: Proceedings of the First International Conference on the Simulation of Adaptive Behavior, Cambridge, MA: MIT Press, pp.451-61.

Shaw, M.W. (1995). Simulation of population expansion and spatial pattern when individual dispersal distributions do not decline exponentially with distance,The Royal Society, Proc. R. Soc. Lond. B259, pp.243-248.

Siegert, F., and Weijer, C.J. (1992). Three-dimensional scroll waves organize Dictyostelium slugs, Developmental Biology, Proc. Natl. Acad. Sci. USA, Vol. 89, pp. 6433-6437.

Ueda, T. (1993). Controlling Perception and Behavior of the Cell, Nikkei Science, pp.32-39.

Varnum-Finney, B., Voss E., and Soll, D.R. (1987). Frequency and Orientation of Pseudopod Formation of Dictyostelium Discoideum Amebae Chemotaxing in a Spatial Gradient: Further Evidence for a Temporal Mechanism, Alan R. Liss, Inc., Cell Motility and the Cytoskeleton, Vol. 8, pp.18-26.

Yokoi, H., and Kakazu, Y. (1992). An Approach to the TSP by a Bionic Model, HEURISTICS, The Journal of Knowledge Engineering, pp.13-27.

Yokoi, H., Mizuno, T., Takita, M., Hakura, J., and Kakazu, Y. (1996). Amoeba Like Self-Organisation Model using Vibrating Potential Field, Proc. A-Life V, pp.32-39.

Modeling of Emergent Ecology for Simulating Adaptive Behavior of Universal Computer Programs

Shinichiro Yoshii, Satoshi Ohashi, and Yukinori Kakazu
Complex Systems Engineering, Hokkaido University
N-13, W-8, Kita-ku, Sapporo 060, Japan
{yoshii, ohhashi, kakazu}@complex.eng.hokudai.ac.jp

Abstract

To further our understanding of mechanisms that exhibit life-like behavior, attention should be given to the emergent dynamics by which a new action or agent, as the representation of that action, is brought forth. The problem is how to model the intuitive notion of the emergence of an action, without involving arbitrariness in the modeling process. In considering how to implement such a notion onto a computer, it could be interpreted as the generation of a new algorithm, or the generation of a new code establishing a mapping between a symbol and an information. This paper proposes an ecological model, named PROTEAN, consisting of Turing machines as a platform on which to simulate the emergent dynamics of computational procedures. Although the system is not provided with any particular mechanism for self-reproduction, the constituents of the ecosystem adapt themselves to improve their utilization of the environment for survival. This paper demonstrates and discusses how adaptive behaviors of algorithms may be achieved with PROTEAN through computer simulations.

1. Introduction

Simulation is the term applied to the process of imitating a given phenomenon through a different medium, for example, in particular, a computer capable of simulating any computable situation. Although the definition of life still remains open to question, much work has been devoted to achieving a full simulation or realization of life-like behaviors on a computer in order to derive its essential scheme. What underlies such research is the concept that life itself can be understood as a complex series of computable processes. The discovery of DNA, for example, argues in favor of the analogy of life as being computably programmable.

Then, how should we simulate life-like behavior with a computer? In other words, what kind of model is appropriate for the simulation of life? Even if we don't have a strict definition of life, we can observe life-like properties in various phenomena, such as increase of complexity, adaptiveness, and emergence. In fact, the notion of emergence itself is also difficult to define in a way that distinguishes simply quantitative changes from those cases where something really new emerges. For the purpose of this paper, we should formalize it as the appearance of a new action or agent as the realization of that action.

In this context, most simulation models for life-like behavior are unable to realize the emergence of new actions. Rather than modeling a process whereby something really new emerges, those models seem to focus on the implementation and simulation of particular aspects of life-like adaptive behavior. This arises from their modeling process to meet a specific purpose. For example, evolutionary computing techniques such as Genetic Algorithms are modeled on an analogy of Darwinian natural selection. In their general modeling, phenotypic behaviors are encoded, or, computationally programmed in genotypes. Although their evaluating scheme based on their own fitness function may achieve a hill-climbing by which the average level of adaptivity increases, it only serves to satisfy a specific criterion, resulting in the convergence of the population. No matter how those genotypes are recombined, there is little chance of anything completely new emerging.

It seems usual in most conventional models for adaptive behavior to be obtained according to the criteria defined by their designers. It is this embedded arbitrariness in conventional modeling that results in restricting its capability to allow new structures to emerge. From this viewpoint, one might say that these conventional models highlight the adaptive acquisition of data built in in advance, rather than an emergent process whereby new action may emerge.

The reason, therefore, why Tierra developed by T. Ray (Ray, 1991) is superior to these traditional approaches, is its evolutionary process characterized by prototype agents operating free of a fixed set of possibilities given in advance by a designer. The essence of the Tierra system lies in the representation of the modeling, called Tierran language, similar to assembly language for a computer. However, from our viewpoint, even the Tierran language may involve some arbitrariness with respect to the giving of semantics to each atomic operation and the whole system that contains some operations to control its dynamics.

This paper is concerned with the modeling of the emergence of new actions resulting in life-like adaptive behavior. In implementing this onto a computer, it could be interpreted as the generation of a new algorithm within a computer program and the generation of a new code establishing a mapping between a symbol and an information. To further our understanding of mechanisms that exhibit truly adaptive behavior, attention should be given to the modeling of an emergent process whereby new actions may be brought forth. Adaptive behavior should be understood as the result of complex relations brought about by the emergence of new actions, rather than beginning with the application of a specific scheme designed to simulate a certain phenomenon.

This paper proposes an ecological model, named PROTEAN, consisting of computer programs described by Turing machines. A Turing machine is a mathematical model of computing or algorithms, enabling any computational procedure that we know today to be described. In particular, we use the notion of a universal Turing machine (UTM), which is really a programmable Turing machine that, depending on its program, can simulate any other Turing machine's operation. In the proposed ecological model, computer programs are decoded onto the UTM, where they interact with each other by performing their various functions. Through computer simulations, this paper discusses the evolutionary dynamics behind the adaptive behavior of these computer programs.

2. Proposed Model: PROTEAN

PROTEAN (Platform for Recursive Ontogenetic Turing-machine Ecosystem for Autopoietic Networks) is an emergent system aimed at simulating autopoietic behaviors of computer programs described in the form of a Turing machine. In PROTEAN, survival games take place on the UTM, with dynamically changing ecological resources being competed for. The essential features of PROTEAN are the representation of a UTM, the interaction procedure between constituents, and the constitution of the system as a whole. For more details about PROTEAN, refer to (Yoshii and Kakazu, 1998) and also our web site at `http://junji.complex.eng.hokudai.ac.jp/export/yoshii/WWW/index.html` where its source code is available.

2.1 Universal Encoding of Algorithms for UTM

The basic model of a Turing machine has a finite control and a tape head that scans one cell of the input tape at a time. Formally, a Turing machine is denoted as follows:

$$T = (Q, \Sigma, \Gamma, \delta, q_0, B, F), \tag{1}$$

where

Q : the finite set of states,

Γ : the finite set of allowable tape symbols,

B : the symbol of Γ, is the blank,

Σ : the set of input symbols

δ : the next move function, a mapping from $Q \times \Gamma$ to $Q \times \Gamma \times \{L, R\}$,

q_0 : the initial state ($q_0 \in Q$),

F : the set of final states ($F \subseteq Q$).

To make the definition of PROTEAN simple, the input symbol is set to $\Sigma = \{0, 1\}$, and we assume that *Left*, *Right*, and *Stop* are the operations of the Turing machine and every Turing machine begins with the internal state 0. Thus, our Turing machine is defined as follows:

$$T = (Q, \{0, 1\}, \Gamma, \delta, 0, B, F), \tag{2}$$

where $Q = \{q_1, \ldots, q_r\}$ and $\Gamma = \{a_1, \ldots, a_s\}$. It should be noted that this simplification does not impose any restrictions on the possible functions of such Turing machines.

Let $\Delta_T = \{\alpha_1, \ldots, \alpha_t\}$ be an encoding of Turing machine T. T can be identified with the following set:

$$\Delta_T = \{(q_i, a_j, q_k, a_l, d) \in Q \times \Gamma \times Q \times \Gamma \times \{L, R, Stop\} \mid \delta(q_i, a_j) = (q_k, a_l, d)\}. \tag{3}$$

In this context, a code of each element of Δ_T can be defined as follows:

$$\overline{\alpha_t} = \overline{(q_k, a_l, d)}. \tag{4}$$

Suppose a UTM reads a description tape where *Left*, *Right*, and *Stop* are encoded as 110, 1110, 11110, and the input symbols 0 and 1 are encoded as 0 and 10. Our Turing machine is able to know what is written on the tape by means of counting the number of subsequent 1s and detecting 0. One can encode this Turing machine without using any mark to signify the end of a one move description, since it is possible to discriminate one move from another by the 110 included in *Left*, *Right*, and *Stop*. Thus, the code $\overline{T}$ of Turing machine T is described as follows:

$$\overline{T} = \overline{\alpha_1}\,\overline{\alpha_2}, \ldots, \overline{\alpha_t}. \tag{5}$$

This above procedure offers us a universal method of encoding any computational procedure into a simple bit string†. Conceptually speaking, PROTEAN has the capability of describing and producing any kind of algorithm computable by Turing machine. A tape for the UTM is nothing more than the coded version of a Turing machine that performs the task desired of the UTM. Thus, we can regard a tape as a genotype that encodes the Turing machine's functionality, while a decoded Turing machine on the UTM is called a phenotype because it is dictated completely by that tape.

2.2 Ecological Resources

Figure 1 shows the ecosystem which consists of the description tapes of the Turing machines. Unlike a Turing machine which can make use of an almost infinite amount of memory as long as the real computer environment can allow, the total amount of ecological resources in this ecosystem is static and limited by the amount of memory. In the ecosystem, each Turing machine occupies its own memory block, dependent on its description length.

† Ideally, a Turing machine can read and write out an infinite tape during execution. Unfortunately, we cannot provide our "real" computer with an infinite memory like a "virtual" computer. Thus, in our implementation, in addition to the above encoding, a method is used to encode a Turing machine as a bit string, in order to decrease the actual memory space required by the real computer, without loss of generality. For more details, see (Yoshii and Kakazu, 1998).

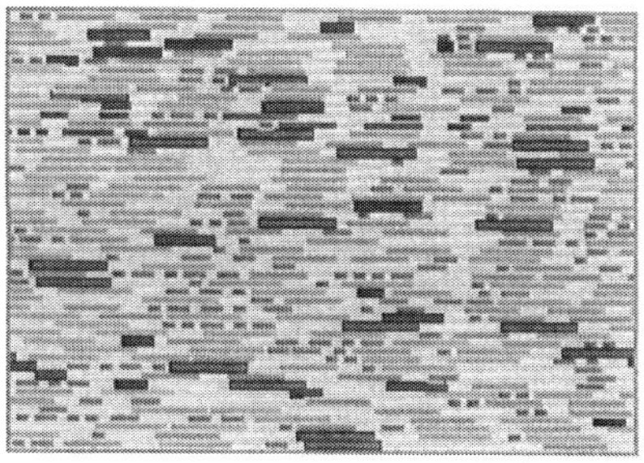

Figure 1 Ecological Resource.

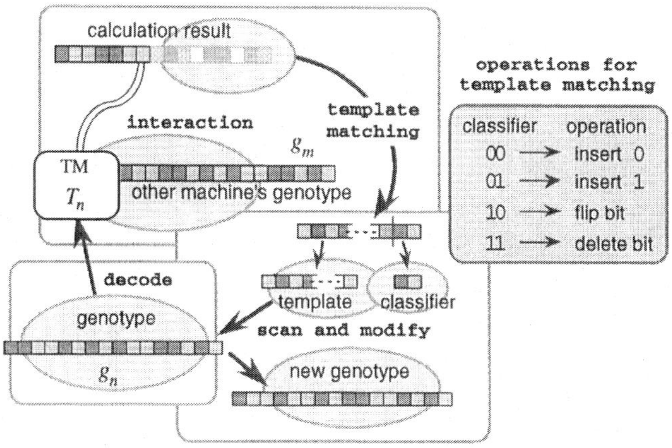

Figure 2 Interaction between Turing machines.

2.3 Interaction between Turing Machines

Figure 2 illustrates an overview of an interaction. First, a genotype g_n is decoded into a phenotype (i.e., a specific Turing machine T_n), which is then interpreted by the UTM. Next, the decoded Turing machine, T_n, reads and operates on another genotypic machine description, g_m. If this tape is accepted, the genotype of that machine is reproduced and applied to the following template matching procedure: to begin with, the output is divided into two parts – a latter two-bit part, and a former part. The latter two-bit and the former part are defined as a classifier and a template, respectively. Then, the template starts to scan its corresponding patterns on that genotype, and the operation, according to the classifier, is applied to each bit just behind the patterns matching those of the template. In this context, a Turing machine, which can accept a tape while generating an output template which doesn't match its own genotype, can easily self-reproduce. The newly generated genotype will dictate a new Turing machine, or in other words, a new algorithm. This means that such an interaction can realize a process whereby a program can directly operate on the description of another program to generate a new algorithm.

2.4 Parallel Processing against Halting Problem

The way interaction takes place in PROTEAN is based on the technical assumption that an algorithm is a definite step-by-step procedure guaranteed to terminate after a finite number of steps. However, in fact, there is no way to tell in advance whether a Turing machine will accept its input tape or not. In order to avoid this insoluble problem, known as the halting problem, PROTEAN adopts parallel processing and compulsory termination: Each Turing machine reads, in parallel, the other machines' descriptions chosen at random. When a Turing machine accepts a tape and some partition is vacant, the parent Turing machine allocates a memory block for its child there. On the other hand, if another machine or machines have already occupied that memory block, the parent can terminate their interaction process and rewrite the memory. As a result, those machines that face the halting problem are eliminated from the ecosystem.

2.5 Characteristics of the Model

Some of the properties of our model may be regarded as being similar to those of Tierra (Ray, 1991) or other evolutionary models which use bit strings (Banzhaf, 1994) (Ikegami and Hashimoto, 1995). However, the following points make PROTEAN stand out from former works.

First, there are no arbitrary rules for self-organization or self-reproduction, as well as no special metaphor. Even how an interaction is interpreted, and what kind of result will be obtained, are not predictable beforehand, but deterministically dependent on the interaction between Turing machines. Intrinsically, the PROTEAN modeling does not restrict the possibilities for algorithms: any kind of algorithm can be produced. Second, there are no explicit parameters to control the dynamics, such as, a parameter which would reap constituent machines in the system if the memory filled up to some specified level. The sole condition for a Turing machine to survive is its being able to accept its local resource more effectively than the others, and self-reproduce, where possible. Consequently, the fitness of each machine is dependent dynamically on its relation to the others within the system and the consequent global environment thereof. Thus, the problem of ill-defined machines becoming stuck is also left up to the system's dynamics, by which they will be eliminated from the ecosystem.

3. Computer Simulation

This section describes the self-organizing dynamics that resulted in the adaptive behavior of the Turing machines.

3.1 Emergence of Self-reproductive Networks

The first computer simulation started from randomly generated well-defined Turing machines whose descriptions were initialized to be in the range from 0 through 100000 in decimal (i.e., base 10) notation. The system was filled to 90% with these descriptions for the initial configuration.

All the Turing machines started to interact with each other and generate new description tapes immediately after the computer simulation began. Even if the description generated was a genotype dictating an ill-defined Turing machine, it could be used as a datum for the subsequent generation of new machines. As the simulation time steps proceeded, the computer programs diversified through subsequent interactions with second or third products. Figure 3 illustrates the reaction pathways between them. These figures are abbreviations as their total size doesn't allow full inclusion here. Each number in this figure is the description tape of a Turing machine described in decimal notation. Figure 3 a) shows new Turing machines being generated from interactions between T_{11107} and others. It should be noted here that the Turing machines shown in shaded regions are those that didn't originally exist in the initial configuration, but were newly generated through interactions.

In fact, these newly generated machines performed important duties for subsequent interactions: defining $T_l(m) = T_n$ as the interaction where Turing machine T_l read genotype m which dictated Turing machine T_m and generated a new genotypic description for Turing machine T_n. The following interactions were observed:

$$T_{11107}(8491) = T_{11107}, \quad (6)$$

$$T_{11107}(84163) = T_{11107}, \quad (7)$$

$$T_{11107}(10595) = T_{11107}, \quad (8)$$

$$T_{11107}(11107) = T_{10595}. \quad (9)$$

Thus, Turing machine T_{11107} could reproduce itself by interacting with the resources generated, such as Turing machine T_{8491}, for example, as shown in Figure 3 c). It made use of the initial resource, consisting of genotypic descriptions for T_{12083} and so on, and, then, generated the description tape 8491. Because description tape 8491 dictated an ill-defined machine, as shown in Figure 4, it merely behaved as a tape for providing data for the self-reproduction of T_{11107}. As a whole, T_{11107} achieved self-reproductive behavior.

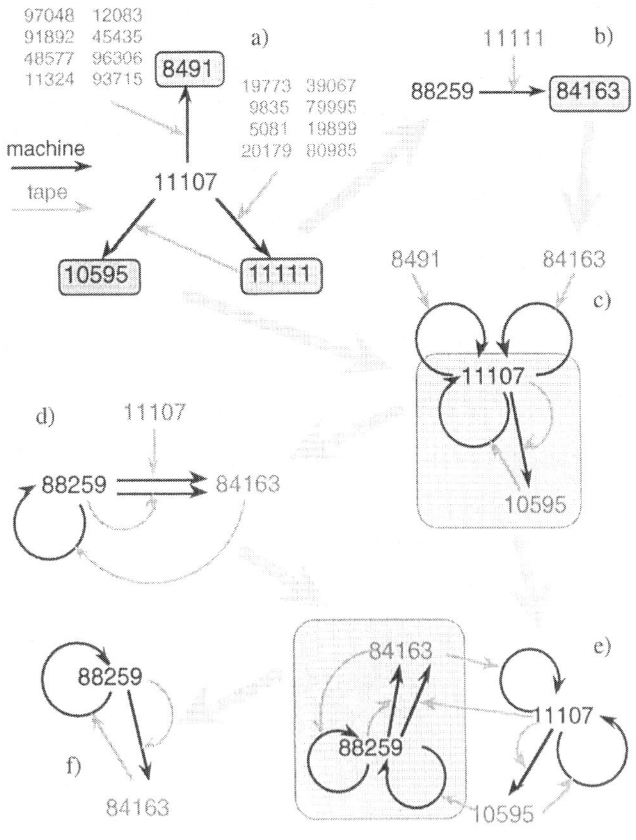

Figure 3 Reaction network between Turing machines.

```
Q×Γ    Q×Γ×{L,R}
0_0 -> 0_0 R
0_1 -> 69_1 STOP
```

Figure 4 Ill-defined Turing machine 8491.

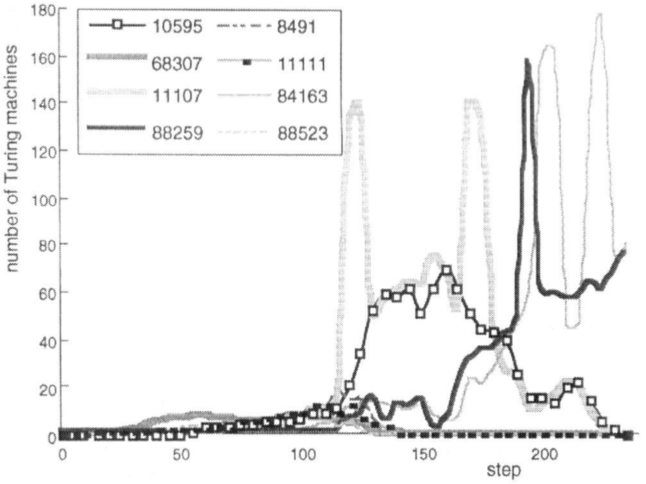

Figure 5 Change in number of representative Turing machines.

On the other hand, it is important to note the following series of interactions: T_{11107} could self-reproduce exponentially by reading description tape 10595, which was generated by the interaction where T_{11107} operated on its own description tape. As a result, the reaction network between Turing machine T_{11107} and description tape 10595, shown in the shaded region of Figure 3 c), drove the other machines out of the ecosystem, due to its vigorous growth. Figure 5 indicates a change in the number of certain specific representative Turing machines, though we cannot follow all the complex reaction pathways. You can see their number increase very quickly around the 120th step.

However, this reaction network did not completely dominate the entire ecological resource. If we look at the reaction network shown in Figure 3 d), we can see that Turing machine T_{88259} generated tape 84163 from the rich resource of T_{11107}, and that it could reproduce itself by interaction with tape 84163. Consequently, this network formed a parasitic network, as shown in Figure 3 e). Because the reaction network between Turing machine T_{88259} and tape 84163 could utilize the entire ecological resource more effectively than the self-reproducing network of T_{11107}, this network superseded the other Turing machines, as can be seen from Figure 3 f) and Figure 5.

3.2 *Altruistic Behavior toward Symbiosis*

This second simulation demonstrates an altruistic dynamics toward symbiosis in more complex Turing machines than those performing only simple functions in the first simulation above. It started from a configuration with the five kinds of description tape described in Figure 6, and adopted a base 100 notation system as a short-hand notation for the

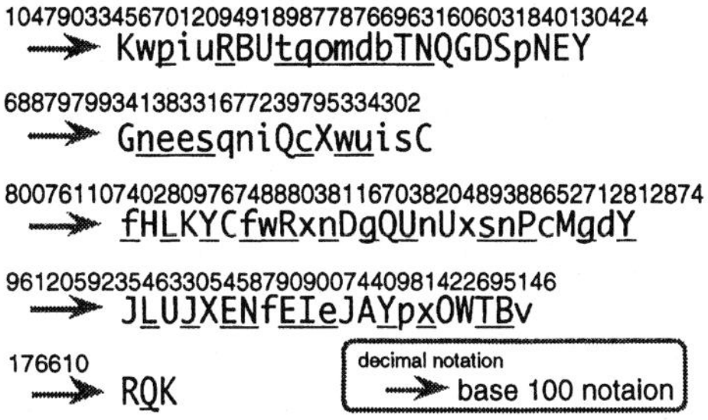

Figure 6 Turing machines in 100-adic notation in simulation.

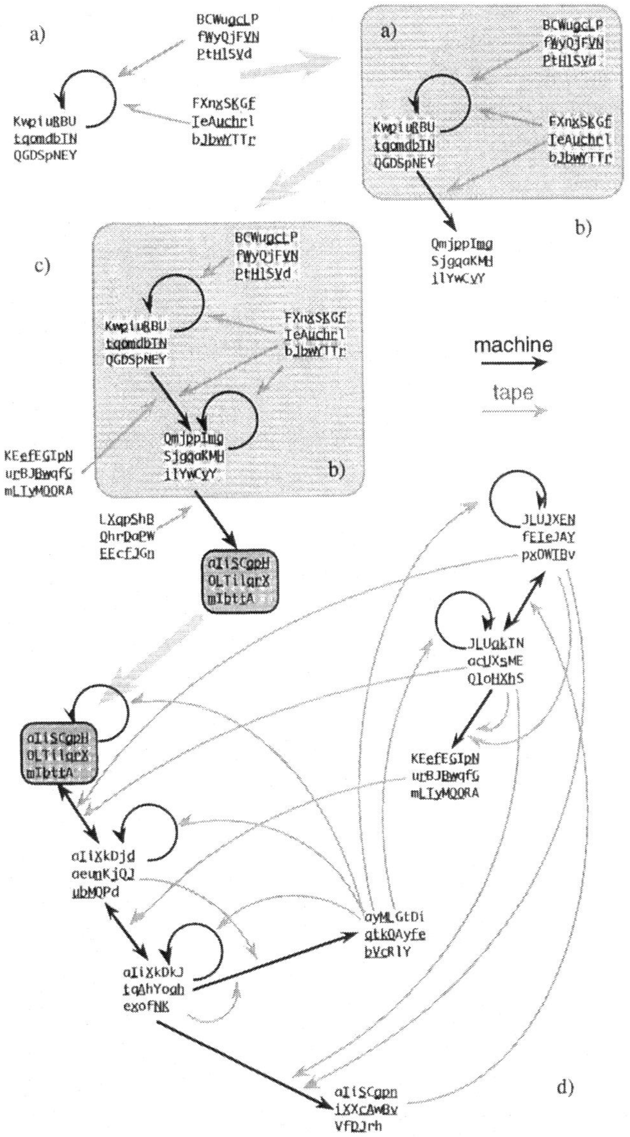

Figure 7 Symbiosis of self-reproductive networks.

long descriptions, using 100 characters such as "A, ..., Y, a, ..., y, A, ..., Y, a, ..., y", as their length didn't allow representation in decimal notation.

Look at the reaction networks, such as Figure 7 a), b), and c). These denote the growth of altruistic networks, where the initial Turing machines generated new Turing machines, one after another, through their interactions. "Altruistic", here, means that the growth of these networks was of

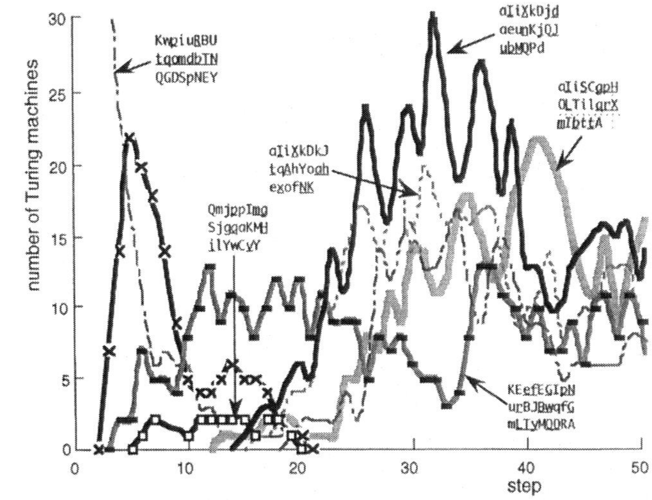

Figure 8 Change in number of Turing machines in symbiosis.

no benefit to their generators. From the exoscopic viewpoint, they seem to transform themselves into these subsequent networks by their accepting the changing ecological resources. Thus, Turing machine $T_{\mathrm{KwpiuRBUtqom\,dbTNQGDSpNEY}}$, for example, disappeared because of its accepting the resource and generating other machines that could utilize the resource more effectively, as seen from Figure 8. Figure 8 also shows a change in the number of certain specific representative Turing machines, concentrating on the earlier dynamics of the ecosystem where more interesting behavior was observed.

Up to about the 20th step, various Turing machines appeared and disappeared. Finally, the reaction network shown in Figure 7 d) emerged in the ecosystem. This reaction network is very interesting in respect to the following points: first, it consists of various kinds of self-reproductive network; and, second, the constituent Turing machines also behaved as tapes that were made use of by other interactions. Thus, in total, these networks form a large and complicated hypercycle (Eigen and Schuster, 1977) consisting of various Turing machines. This symbiotic hypercycle network seems to be stable against a certain degree of fluctuation, though we have not investigated this in detail, yet.

4. Discussion

These Turing machines behaved as real computer programs, processing information, or, as data, providing information. Although their description tapes as data showed altruistic behavior, they could survive when a parasitic relation, in which they were produced from other reactions, was possible. However, in the ecosystem where there is no static fitness function, such a relation is actually fragile, as their fitness is dynamically dependent on their relations to others and the consequent environment thereof.

Fitness is a property related to adaptivity of a system constituent that determines the possibility that that constituent will survive through reproducing or being produced. It refers to the capability to cope with and control over environmental resources. Adaptivity of computer programs in

PROTEAN means the capability of accepting surrounding data, and then reproducing themselves. In this respect, the generated Turing machines are highly adaptive compared to the ones eliminated from the ecosystem.

PROTEAN can make new types of constituent emerge through their interactions without a mutative operation unlike Tierra. Each of those new species gives rise to a new environmental condition, in which these species all depend on each other. The appearance of new species gives an opportunity for the appearance of parasites and makes less fit competitors go extinct. The emergent symbiotic network consisting of self-reproducing Turing machines is similar to "core network" reported in (Ikegami and Hashimoto, 1995). However, the number of possibilities is infinite in PROTEAN, that is conceptually capable of dealing with any kind of algorithm computable by a UTM. This leaves evolution unpredictable and open-ended rather than being restricted by the model.

Evolution in general tends to grow fitness, and, in addition, the dependencies or other linkages among constituents become more complex in the ecosystem as shown in the simulations above. Furthermore, those constituents seem to produce increases of their functional complexity through the evolutionary process. It is interesting to see the simulated behavior, from the viewpoint of the complexity of an algorithm. Fortunately, it is easy to define the degree of complexity of a Turing machine by measuring the amount of resources required to execute its function, even though this is based on the assumption that a difficult computation will require more resources than a less difficult one.

According to one definition (Brookshear, 1989), *time complexity* means the amount of time required to perform a computation, which will correspond to the steps of a head moving in the case of a Turing machine. On the other hand, *space complexity* refers to the amount of storage space required by a computation. The space complexity of a Turing machine is defined as the number of tape cells required. Thus, Figure 9 shows the transition of both complexities, with respect to the average number of constituent Turing machines in the second simulation. This figure indicates that the complexity of algorithms self-organizes to a certain attractor. Note that the complexity becomes greater compared to the initial complexity.

Evolution is a largely unpredictable and contingent of events, where some fluctuations may lead to a drastic change that will affect the course of future development. However,

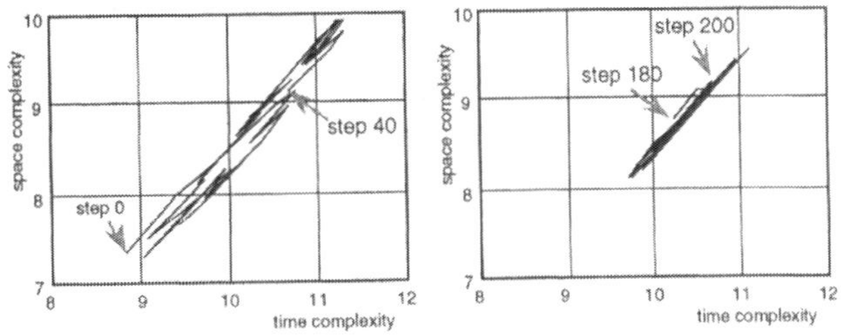

Figure 9 Self-organized complexity of constituent machines.

the trajectory in the transition of the computational complexity helps us to predict to what kind of point the system will reach. While our system produces an increase in algorithmic complexity, it eventually reaches an attractor where the requisite variety of constituents is in balance with their environmental constraints.

Since we didn't employ any mutative operation in the simulations above, the final version of this self-reproductive network shows no further change after reaching the attractor. However, it might be possible to realize boundless evolution, if a mutative process is employed. In this paper, we didn't examine such a situation as our objective was the modeling of an emergent system consisting of computer programs representing the step-by-step deterministic procedure of an algorithm.

5. Conclusion

For the further understanding of life-like adaptive behavior, it is necessary to highlight the emergent dynamics by which new actions may come forth. The important thing in PROTEAN is to be able to achieve a modeling that can process any kind of information.

With PROTEAN, we were able to observe some interesting behavior even though it has no special metaphors for self-organization. Adaptive behavior seems to be an intrinsic property of complex systems, consisting of many constituents capable of processing information and reconfiguring themselves. The Turing machines in PROTEAN read, stored, and interpreted information immanent in their environment, and then, exhibited autopoietic dynamics. Macroscopically, "autopoietic" means those dynamics that result from the transformation of self-reproductive networks. Those self-assembling aspects are of central importance in understanding simulated emergent adaptive behavior.

References

Banzhaf, W. (1994). Self-organization in a System of Binary Strings. Proc. of Artificial Life IV, pp. 109-118. The MIT Press.

Brookshear, J. G. (1989). *Theory of Computation: Formal Languages, Automata, and Complexity*. The Benjamin/Cummings Publishing.

Eigen, M. and Schuster, P. (1977). *The Hypercycle: A Principle of Natural Self-Organization*. Naturwissenschaften 64.

Ikegami, T and Hashimoto, T. (1995). Coevolution of Machines and Tapes. *Advances in Artificial Life*. Proc. of Third European Conference on Artificial Life, pp. 234-245, Springer.

Ray, T. S. (1991). An Approach to the Synthesis of Life. *Artificial Life II*, Santa Fe Institute Studies in the Sciences of Complexity, Proc. Vol. X, pp. 371-408. Addison-Wesley.

Yoshii, S. and Kakazu, Y. (1998). Self-Assembling Networks of Computer Programs. Proc. of Intelligent Autonomous Systems 5. (to be appeared)

LEARNING

Role of Behavior in the Development of Complex Neuronal Properties

Nikolaus Almássy and Olaf Sporns
The Neurosciences Institute, 10640 John J. Hopkins Drive, San Diego, California 92121, USA
almassy@nsi.edu   sporns@nsi.edu

Abstract

It has long been recognized that changes in neural activity within an organism's brain may cause specific changes in its behavior. However, much less is known about the specific role of self-generated movement and of behavioral interactions with an environment in the development of complex neuronal properties. To address this question, we designed Darwin V, which consists of a simulated neuronal model embedded in a real world device capable of autonomous behavior. The model has four major components: a visual system incorporating several key properties of mammalian visual cortex, a taste system based on conductance, sets of motor neurons capable of triggering behavior, and a diffuse ascending (value) system. As Darwin V encounters stimulus objects in its environment, synaptic modifications within its visual maps produce pattern-selective and translation invariant neuronal responses. Due to value-dependent synaptic modifications in sensorimotor connections, visual neurons become linked to motor neurons triggering appropriate behaviors. Closer analysis reveals that the acquisition of translation invariance depends critically upon continuous self-generated movements of Darwin V. In addition, the allocation of map areas selective for particular stimulus classes depends upon frequency and historical sequence of stimulus encounters. In summary, the analysis of Darwin V provides several examples for the role of environment and behavior in the development of complex neuronal responses.

1 Introduction

Traditionally, theoretical frameworks of pattern classification have approached the problem by attempting to devise strategies for partitioning often high-dimensional input spaces into distinct regions corresponding to categories. These approaches do not typically take into account several key characteristics of how biological systems achieve their superior performance in categorizing natural environments. It has been argued that biological perceptual categorization is not the result of disembodied information processing or of (even neurally implemented) algorithmic procedures (Edelman, 1987). Rather, such categorization necessarily involves exploration and active movement resulting in sensory exposure and sampling. This gives rise to the hypothesis that the construction of perceptual categories and the development of their underlying neuronal components depends critically on actual experience. The sampling of statistical patterns in sensory inputs depends upon factors related to body structure (including sensory and motor organs, i.e., the "phenotype"), saliency, and the history of interactions in the real world. Our goal here is to analyze specific examples for ways in which behavioral experience modifies visual neuronal responses related to categorization.

In actual biological organisms, the interactions between brain activity, behavior and environment are often difficult to analyze, due to the fact that their simultaneous observation over prolonged periods of time is all but impossible to achieve. To overcome these limitations, we constructed a series of computational models (Reeke et al., 1990a,b, Edelman et al., 1992, Reeke and Sporns, 1993) with structural and dynamic properties realistic enough to allow direct comparison to biological organisms. Each consisted of a simulated nervous system, a phenotype with sensory and motor organs, and an environment containing various stimuli. The design of these systems, called synthetic neural models, allows the experimenter to correlate behavioral and neural events occurring at multiple levels of organization, from synaptic modifications to changes in overall behavior.

In a previous simulation, Darwin III, categorization was based on a combination of visual and tactile sensory inputs and showed emerging characteristics such as generalization and ranking of stimuli according to their typicality. The simulated nervous system of Darwin IV (Edelman et al., 1992) was embedded in a mobile real world device capable of autonomous behavior. Objects could be discriminated according to their color, and "taste" (mediated by conductivity of objects in the environment) served as a secondary sensory modality, triggering unconditioned reflexes and neuronal responses of a value system. Value responses signaled the occurrence of salient events in the environment and were broadcast to widespread regions of the nervous system via diffuse projections. Value system responses enabled the selective modification of sensorimotor connections linking visual responses to motor behaviors.

Perceptual categorization within the visual modality provides an appealing test case to study the role of behavioral and environmental interactions on properties intrinsic to the nervous system. Many studies in a variety of species have identified possible neural correlates of categorization. The mammalian visual system contains neurons in the inferior temporal cortex that respond to visual stimuli with complex patterns, shapes and forms (Gross, 1992, Tanaka, 1996). These neurons have large receptive fields giving rise to responses that are invariant with respect to stimulus translation across the visual field (Gross and Mishkin, 1977, Tovee et al., 1994). Evidence suggests that tuning properties of visual cortical neurons are subject to significant postnatal modification and refinement (Rodman, 1994), at least in part as a result of exposure to actual stimuli. Experience-dependent cortical plasticity has been extensively studied in modalities other than vision, throughout postnatal development and adulthood (Merzenich et al., 1988, Kaas, 1991).

In the present paper, an extension of an earlier and more complete study (Almássy et al., 1998), we study Darwin V, composed of two main parts: (a) its simulated nervous system contains cortical and subcortical networks, sets of motor neurons capable of generating various behaviors, and a diffuse ascending (value) system; (b) this simulation is embedded in a mobile real world device capable of autonomous behavior in an environment. As Darwin V moves and interacts with objects, pattern selective and translation invariant neuronal responses develop. The sensing of "taste" (a highly salient sensory signal) leads to responses in a neural value system which are used adaptively to link visual to motor centers controlling behavior.

Our present analysis provides evidence for the importance of self-generated translational movements for the emergence of pattern selective and translation invariant neuronal properties within its simulated nervous system. We also report a strong correlation in the amount of cortical space ("map area") allocated to neurons selectively responding to particular object classes according to their overall frequency of encounter. Furthermore, early imbalances in object encounters lead to lasting deficits that are difficult to compensate by later increases in exposure to stimuli that have been encountered only rarely during early experience. Darwin V provides a working example of a biologically-based model of cortical development and plasticity, allowing their study in relation to factors extrinsic to the nervous system, such as phenotype, bodily motion, and history of stimulus encounters.

2 Implementation

The robotic device used in these experiments (called NOMAD, for Neurally Organized Multiply Adaptive Device; see figure 1a) is constructed on a circular mobile platform with three steerable wheels that permit independent translational and rotational motion. Sensory input is provided by a miniature color CCD camera and a rigid electromagnetic "snout" at the front of the platform to sense the conductiveness of the surface of the test objects. This binary conductivity signal provides a "taste" input to the nervous system. Eight infrared (IR) sensors are mounted at 60° intervals around the mobile platform. The IR sensors are responsive to the boundaries of the environment and are used to trigger avoidance reflexes.

The on-board interface between the sensors, effectors, and the simulated nervous system does not initiate or control any behavior. Communication between NOMAD and the simulated nervous system takes place over a serial line that allows an asynchronous and continuous interaction between the simulated nervous system and NOMAD. All effectors operate under open loop control, i.e., movements, once initiated, can only be interrupted by the occurrence of new sensory events (Verschure et al., 1995).

2.1 Device

Input to the visual system consists of a 64×64 pixel array derived by spatial averaging a full-resolution video image. Visual input (sampled at a maximum rate of 30 f/sec) is either a wide-angle field of view derived from a 218×218 subregion of the original video image (corresponding to a viewing angle of 58.3 degrees, with a final resolution of 0.9 degrees per pixel; Figure 1b) or a narrow-angle field of view (97×97 subregion with a viewing angle of 24.7 degrees, with a final resolution of 0.4 degrees per pixel; Figure 1c). The wide-angle field of view is used during exploratory search; a switch to the narrow-angle field of view occurs when a visual target (defined by its luminance contrast) is detected in NOMAD's vicinity. When vision signals the presence of an object directly in front of the snout, the electromagnet in the snout becomes activated and, following contact, conductivity is registered across its multiple exposed electric contacts.

2.2 Environment

The environment consists of an enclosed area (3 by 2.4 m) having black cloth-covered walls (1.5 m high) with a floor covered by opaque black plastic panels, on which there are 6 to 9 metal cubes (10 cm on each side, subtending a maximal visual angle of about 7.5 degrees). The top surfaces of these cubes are covered with various removable 10-by-10 cm black-and-white patterns on cardboard. All other surfaces of the cubes are featureless and black. All experiments with Darwin V reported in this paper were carried out with multiple exemplars of two basic designs: "blobs" (several white patches about 2.5 cm in diameter), and "stripes" (thickness about 5 mm, evenly spaced). Note that "stripes" can be viewed in either a horizontal or vertical orientation, yielding together with "blobs" a total of three classes of visual patterns to be discriminated. The metal cubes come in two varieties: the vertical surfaces of one kind are exposed metal and conductive, while the surfaces of the other kind are covered with transparent plastic rendering them non-conductive. These two kinds of cubes are visually indiscriminable by Darwin V. In the experiments described in this paper, all objects with "blobs" were conductive, while all with "stripes" (horizontal or vertical) were non-conductive. Cubes were initially distributed randomly throughout the environment and were manipulated only by NOMAD's magnetic snout. Because NOMAD's re-

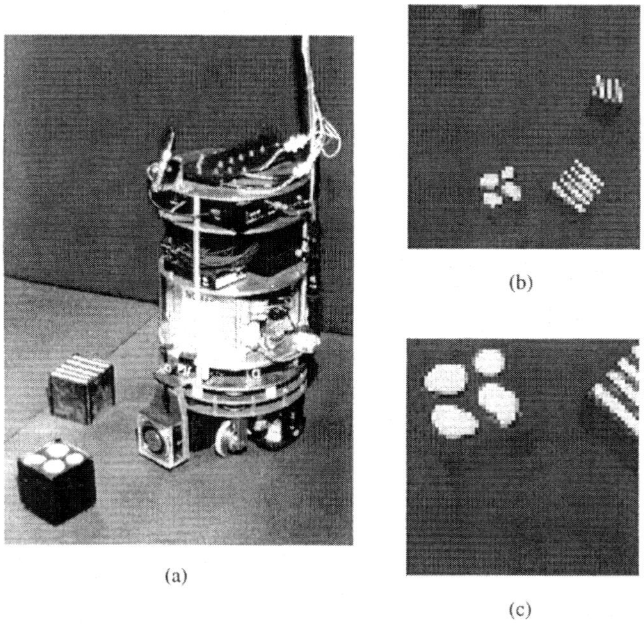

Figure 1: (a) NOMAD is equipped with a magnetic snout, a conductivity sensor, eight infrared sensors and a color CCD camera. An appetitive behavioral response consists of prolonged gripping and sensing, a turn to the left, followed by releasing the object and turning back to the original position; an aversive behavioral response consists of an immediate release of the object and turning away. (b) Spatially averaged wide-angle field of view, 64×64 pixels, 58.3 degrees approximate viewing angle. (c) Spatially averaged narrow-angle field of view (same situation), 64×64 pixels, 24.7 degrees approximate viewing angle.

flexes do not permit the manipulation of blocks when IR sensors are triggered, occasionally blocks had to be moved away from the walls by the experimenter.

2.3 *Nervous System*

Darwin V's nervous system is modeled using ASE, developed in C at the Neurosciences Institute. The model consists of a number of areas representing different brain regions containing multiple neuronal units of different types. Such units are taken to represent local populations of neurons that are organized into neuronal groups (Edelman, 1987). The simulated nervous system contains a total of 14,994 neurons and 380,890 synaptic connections. It consists of four major components: a visual system containing subcortical and cortical networks, a taste system, sets of motor neurons capable of triggering motion, and a value system (Figure 2). Values for anatomical, cell activity, and synaptic parameters are given in Table 1, Table 2, and Table 3, respectively.

Visual input (a 64×64 pixel greyscale image) is relayed to a receiving area R ("retina"). R emits topographic excitatory connections to area VA_p ("primary visual cortical area"). The ordered spatial arrangement of these connections results in responses that are selective for particular local features of stimuli. There are three interleaved maps of units within VA_p that are selective for visual "blobs," or for horizontal and vertical "stripes" (called VA_p-B, VA_p-H and VA_p-V, respectively). Units in VA_p-H and VA_p-V are mutually connected by local inhibitory connections, while both are locally inhibited by units in VA_p-B. Afferent connectivity and mutual inhibition produce VA_p responses that are selective for "blobs," or for horizontal or vertical "stripes" within the input image. Each of the interleaved maps of units in VA_p sends excitatory projections to excitatory units in VA_s (higher-order or "secondary visual cortical area"). This projection is sparse (each unit in VA_s receives input from about 2% of units in VA_p) but the spatial extent (arborization width) of input connections corresponds to about two thirds the width of VA_p. Within VA_s, excitatory units connect among themselves in a local neighborhood (3×3 neuronal units) and to inhibitory units over longer distances (11×11 units); inhibitory units (VA_{si}) project in turn to excitatory VA_s units within a 3×3 region. Excitatory efferents from VA_s connect to units in motor centers, which trigger appetitive (M_{app}) and aversive (M_{ave}) behaviors. The motor centers also receive sensory input directly from taste sensing units. These connections are arranged such that taste units responding to an appetitive stimulus (T_{app}) connect only to the motor center triggering the appetitive motor response (M_{app}), while aversive taste (T_{ave}) activates only the aversive motor center (M_{ave}). Both appetitive and aversive taste sensors send excitatory connections to a value system (S; Figure 2), which reflects salience of inputs and incorporates several key properties of diffuse ascending systems of the brain. In particular, inputs to the value system signal salient sensory events and result in its activation. Its widespread output projections modulate synaptic changes in cortical areas (see Equation 2 below). In the model, the sensing of taste results in the activation of S, affecting synaptic changes in connections linking higher sensory to motor areas.

Area R ("retina") also emits topographic projections to an area C ("colliculus") consisting of two interleaved maps of excitatory (C_e) and inhibitory (C_i) units. C_e units connect to each other over a short distance (3×3 units) and to inhibitory units over longer distances (11×11 units). Inhibition extends over an annulus-shaped region of width 8 and diameter 11. This intrinsic connectivity within area C leads to activity patterns that disambiguate possible multiple targets within the visual field by favoring targets that are brighter than others. Note that collicular responses are only used for visual approach behavior and do not directly bias cortical visual responses in VA_s related to pattern discrimination. Area C_e projects to motor units driving rotational (M_{rot}) and translational (M_{tra}) movement of NOMAD. To accelerate the development of pattern discrimination and facilitate behavior in the present experiments, the connection strengths within these projections were not subject to gradual synaptic modification. Instead, they were assigned initial values resembling distributions obtained in earlier work by value-dependent synap-

Table 1: Properties of anatomical projections (connection types) in Darwin V. Arborization shape is indicated as either square (□), circular (○) or ring (⊙). Values for initial connections strengths ($c_{ij}(0)$) represent means with negative numbers indicating inhibitory connections. See text for definitions of areas.

Projection	Arborization shape and size		$c_{ij}(0)$
R → VAp-B	○	7×7	0.03
R → VAp-H,V	□	9×1, 1×9	0.037
VAp-B → VAp-H,V	○	5×5	-0.2
VAp-H ↔ VAp-V	□	7×7	-0.027
VAp → VAs	○	33×33	0.03
VAs → VAsi	○	11×11	0.015
VAsi → VAs	○	3×3	-0.155
VAs → VAs	○	3×3	0.205
VAs → Mapp,ave	□	11×11	0.0019
Mapp ↔ Mave	□	5×5	-0.1250
Tapp,ave → S, Mapp,ave	□	7×7	0.0035
R → Ce	○	3×3	0.155
Ce → Ce	○	3×3	0.1
Ce → Ci	○	11×11	0.015
Ci → Ce	⊙	11×8	-0.3250
Ce → Mrot,tra	□	11×3, 3×11	0.07

tic modification during sensorimotor training (Edelman et al., 1992).

2.4 *Cell Activation and Synaptic Rules*

At each cycle (iteration) t of the computer simulation, the activities $s_i(t)$ of all neuronal units are recomputed. In experiments conducted in a real environment with a freely behaving device, each cycle corresponds on average to about 400-600 milliseconds of real time. As in previous models (e.g. Reeke et al., 1990b), this activity level represents the average firing rate of a local population of neurons. Synaptic connections from unit j to unit i are designated c_{ij} and are subject to activity-dependent modification. The total synaptic input $A_i(t)$ to unit i is given by

$$A_i(t) = \sum_{l=1}^{M}\sum_{j=1}^{N_l} c_{ij}\, s_j(t)$$

where M is the number of different anatomically defined connection types and N_l is the number of connections per type M projecting to unit i. Negative values for c_{ij} correspond to inhibitory connections. The activity level of unit i is given by

$$s_i(t+1) = \phi(\tanh(g_i[A_i(t) + \omega s_i(t)])) \quad (1)$$

$$\phi(s) = \begin{cases} 0 & \text{if } s < \sigma_i \\ s & \text{otherwise} \end{cases}$$

where ω determines the persistence of unit activity from one cycle to the next, σ_i is a unit specific firing threshold and g_i is a scale factor.

In the model, several of the connection types (Figure 2) are subject to activity-dependent modification according to a synaptic rule. This rule incorporates dependency on pre- and postsynaptic activity as well as on value V and allows for synaptic potentiation and depression:

$$\Delta c_{ij}(t) = \epsilon(c_{ij}(0) - c_{ij}(t)) + \eta\, s_j(t)\, F(s_i(t))\, V \quad (2)$$

$$V = \begin{cases} v(t) & \text{if } c_{ij} \text{ is value-dependent} \\ 1 & \text{otherwise} \end{cases}$$

$$F(s) = \begin{cases} 0 & \text{if } s < \Theta_1 \\ k_1(\Theta_1 - s) & \text{if } \Theta_1 \le s < (\Theta_1 + \Theta_2)/2 \\ k_1(s - \Theta_2) & \text{if } (\Theta_1 + \Theta_2)/2 \le s < \Theta_2 \\ k_2(s - \Theta_2) & \text{otherwise} \end{cases}$$

$s_i(t)$ and $s_j(t)$ are activities of post- and presynaptic units, respectively, η is a fixed learning rate, ϵ is a decay constant ($\epsilon \ll 1$), $c_{ij}(0)$ is the initial ($t = 0$) weight of connection c_{ij}.

Synaptic changes depend upon both pre- and postsynaptic activity and can result in both strengthening and weakening of connections. Which of these alternatives is the case is determined by the function F which sets limits and rates for synaptic potentiation and depression depending upon the postsynaptic activity (Bienenstock et al., 1982). F is a piecewise linear function defined by two thresholds ($0 < \Theta_1 < \Theta_2 < 1$) and two inclinations ($k_1, k_2$); a sketch of this function is shown in Figure 2 (insert on upper right).

Darwin V's nervous system contains an area resembling a diffuse ascending value system (S; see Figure 2) whose responses signal the occurrence of salient events. The outputs of S modulate synaptic changes in connections linking VAs to Mapp and Mave. In Equation 2 the influence of value is designated by the term V. For connections that are subject to value-dependent modification, the value term V is given by the instantaneous activity $v(t)$ of the value system S.

Table 2: Values of parameters defining properties of neuronal units in Darwin V (see Equation 1).

Area	g	σ	ω
R	1.29	0.5	0.0
VA_p-B	1.33	0.6	0.5
VA_p-H,V	1.50	0.5	0.5
VA_S	1.35	0.04	0.03
VA_{si}	1.35	0.02	0.15
$M_{app,ave}$, $T_{app,ave}$	2.00	0.0	0.3
S	2.00	0.1	0.13
C_e, M_{rot}, M_{tra}	1.4	0.05	0.24
C_i	1.35	0.05	0.09

Experimental evidence suggests (Kirkwood et al., 1995) that key parameters of neural plasticity may vary over the course of postnatal development. Since our model was intended to cover early as well as late stages of development, we attempted to replicate observed trends in postnatal visual cortical plasticity, effectively modeling a transition from an earlier "critical period" to a later "adult period." Connections linking VA_P to VA_S undergo several stages of plasticity, characterized by progressive changes in learning rate η, synaptic decay ϵ and parameters of the postsynaptic function F (Table 3). These changes, from earlier to later stages of development, are qualitatively as follows: synaptic plasticity decreases in magnitude, synaptic weights become more resistant to spontaneous decay back to baseline, and synaptic potentiation requires higher levels of postsynaptic activity. In the set of experiments described in (Almássy et al., 1998) changes in plasticity occurred at specified time points within a run. To facilitate comparison of multiple individual experiments with different stimulation histories, in the present experiments these changes were dependent on the cumulative neuronal activity $\overline{a}$ (see Table 3) in area VA_S, evaluated as the sum of activity levels since the start of a given run. Cumulative neuronal activity is intended to reflect overall levels of sensory exposure. The values chosen for $\overline{a}$ result in switching at times comparable to those described in (Almássy et al., 1998).

3 Results

3.1 *Basic Behavior of Darwin V*

Our approach allows continuous on-line recording of many neuronal and behavioral variables over time, enabling their correlated analysis. First, we describe Darwin V's behavior before and after "normal" visual development.

Before development, Darwin V indiscriminately approaches visual objects until they are located within the visual fovea and grips them to allow sensing of their conductivity ("taste"). In the present experiments, all "blobs" were conductive, while all "stripes," irrespective of orientation, were non-conductive. Sensing conductivity produces activation of the value system and the motor centers which trigger appetitive or aversive behaviors. Figure 3 (left) shows examples of activity traces of visual, taste, motor, and value networks before extensive experience. For all objects encountered, visual networks are activated first, followed by taste, which in turn immediately activates behavioral motion and value. Sensing of taste and activation of the value system are used to condition behavioral responses. Taste serves as the unconditioned stimulus (US) triggering an appetitive or aversive unconditioned response (UR). Value signals are used to modify visuomotor connections.

In the present experiments, the developmental period is completed after an average number of 673.4 (+/− 134.2) seconds of autonomous behavior (corresponding to 1685.1, +/− 141.4, computational cycles). During each run on average 46.0 (8.9) objects are encountered (N=9). After development (Figure 3, right), Darwin V continues to explore the environment and approach all bright objects. By this time, through conditioning, visual signals have become capable of triggering motor responses independent of taste. Visual input from "stripes" leads to immediate activation of M_{app}, followed by the execution of an appetitive response, which is accompanied by registration of taste and activation of the value system. In contrast, visual input from "blobs" produces an immediate aversive response, triggered by visual input to the appropriate motor center (M_{ave}). This preempts activation of grip action, tasting, or value responses. Conditioning, involving value-dependent modification of visuomotor connections, results in a behavioral change: after development, visual signals serve as conditioned stimuli (CS) capable of triggering appetitive or aversive conditioned responses (CR). After conditioning is completed, on average more than 90% of the behavioral responses are triggered by vision.

3.2 *Translation Invariance and the Continuity of Movement*

Several studies of the primate visual system have indicated that invariant and selective object-related responses arise only at "higher" hierarchical levels and depend for their construction on the progressive elaboration of visual input from "lower" to "higher" areas via convergent/divergent feedforward connectivity (for a review see Rolls, 1992). The present model approximates such a multi-stage process with two simulated visual cortical areas (VA_P and VA_S), linked by widely convergent/divergent modifiable connections.

The development of translation invariance in Darwin V depends on a combination of several neural and behavioral constraints: (a) Visual areas are hierarchically arranged, with a "lower" area (VA_P) responsive to elementary features of the stimulus, and a "higher" area (VA_S) that acquires responses to more global features. (b) "Lower" and "higher" areas are linked by modifiable convergent/divergent connectivity. (c) Neuronal units in "higher" areas show temporally sustained responses, due to their extended receptive fields and the local interactions between excitatory and inhibitory cell populations. (d) Signals from objects are displaced continuously

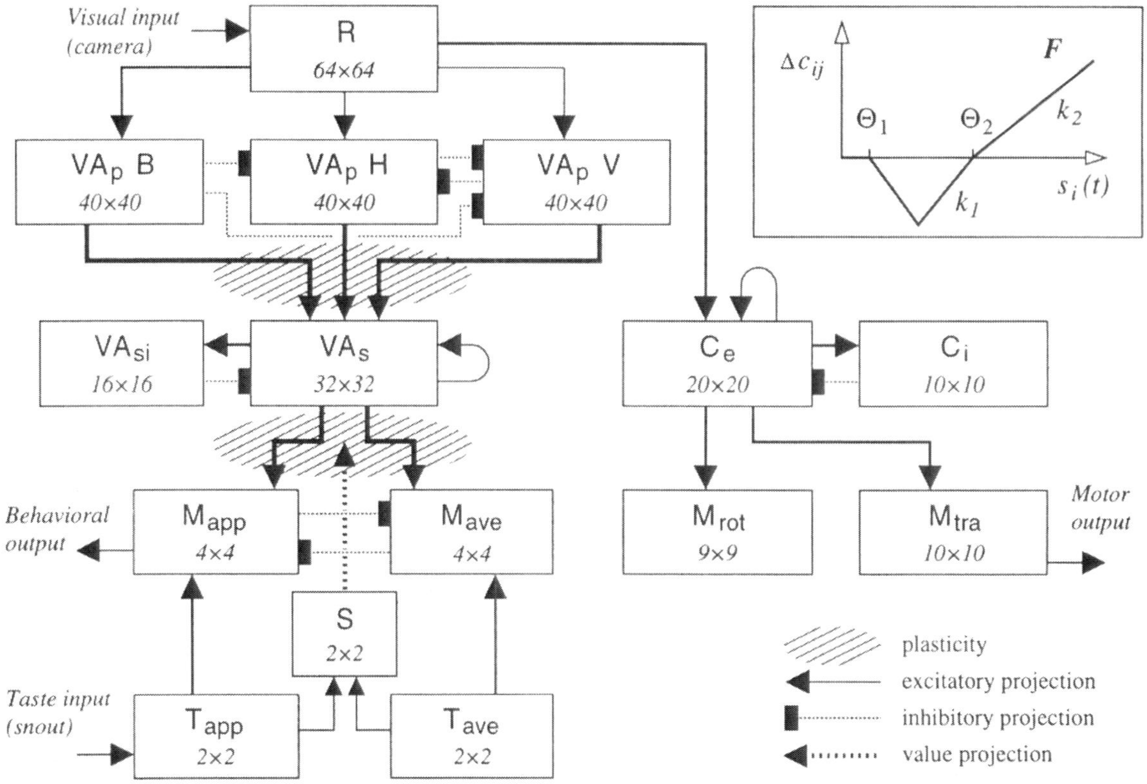

Figure 2: Schematic of the anatomy of DARWIN V's nervous system (for areas see text; for detailed settings of anatomical and physiological parameters see tables 1 and 2). Each box represents a modeled area containing neuronal units, numbers within boxes indicate dimensions of neuronal array. Arrows represent excitatory (solid line) and inhibitory (dotted line) projections between areas; arrow width indicates the approximate degree of convergence/divergence between areas. Modifiable projections are indicated by cross-hatching. Value projections are represented by a stippled arrow and area of termination indicates projections that are subject to value-dependent modification. Insert on upper right is a sketch of function F (see Equation 2 for details).

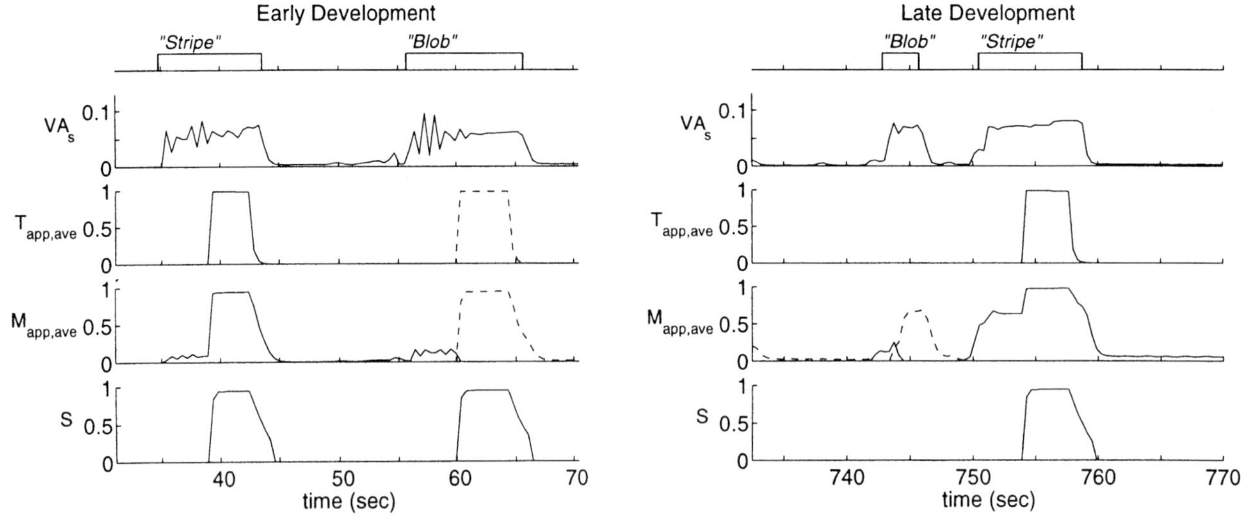

Figure 3: Comparison of average group activity in Darwin V during early (left) and late (right) visual development. The type and duration of the object encounter is indicated on top and labeled "Stripe," and "Blob." Individual plots show (top to bottom) activities of average VA_S visual area, taste areas (T_{app}, solid line; T_{ave} stippled line), average motor areas (M_{app}, solid line; M_{ave}, stippled line) and the value system (S).

Table 3: Values of parameters governing synaptic change (see Equation 2). For connections between VA_P and VA_S, switching from "early" to "middle" to "late" development is dependent upon the cumulative activity of excitatory units in VA_S. (see text)

Projection	Cumulative activity $\bar{a}$	η	ϵ	Θ_1	Θ_2	k_1	k_2
$VA_P \rightarrow VA_S$	$0 < \bar{a} < 19.2$	0.044	0.0006	0.14	0.27	0.45	0.1
	$19.2 < \bar{a} \leq 38.4$	0.035	0.0005	0.14	0.31	0.45	0.1
	$\bar{a} > 38.4$	0.026	0.0004	0.14	0.35	0.45	0.1
$VA_S \rightarrow M_{app,ave}$		0.3	0.002	0.01	0.25	0.10	0.035

over the visual field, due to smooth self-generated movements during visual exploration and approach.

Crucial for the development of translation invariance in our model are sustained (long-lasting) neural responses at a time scale commensurate with stimulus translation across the visual field. Sustained inferior temporal neural responses lasting about 300-800 milliseconds have been demonstrated in several studies (Nakamura et al., 1994, Rolls and Tovee, 1995). In the present model, sustained activity in VA_S is partly due to intrinsic connections which tend to produce foci of activation surrounded by lateral inhibition. As objects are encountered, repeated sustained co-activation of neighboring units tends to produce similar patterns of synaptic change in their afferent and efferent connectivity.

To evaluate translation invariance of neuronal responses within area VA_S, we tested neuronal responses to visual objects presented at different locations within the visual field. Examples of VA_S response patterns for two individual presentations of stimuli at different locations (separated by about 15 degrees of visual angle) are shown in figure 4. We calculate their mutual overlaps the angle ϕ between the two activation vectors $\mathbf{a}$ and $\mathbf{b}$, according to

$$\cos(\phi) = \frac{\mathbf{ab}}{|\mathbf{a}| \cdot |\mathbf{b}|}. \tag{3}$$

Before development, VA_S activity patterns are characterized by variable foci due to local excitation and lateral inhibition. Patterns recorded for widely separated stimuli show little or no overlap (Figure 4). After "normal" development, activity patterns are more sharply tuned and patterns recorded for objects presented at different locations show significant overlap (Figure 4), i.e., neurons respond in manner that is invariant with respect to lateral stimulus displacement (translation). To test the role of movement in the development of this property, we immobilized Darwin V and presented it with sequences of stationary images of visual patterns flashed at random positions within its field of view ("discontinuous" development), while maintaining the same overall coverage, intensity and frequency of stimulation as was used during "normal" development. After "discontinuous" development, widely separated stimuli elicit activity patterns that had little or no overlap in VA_S. Shapes and sizes of receptive fields for the vast majority of neuronal units in VA_S were virtually unchanged from before the onset of visual experience (data not shown).

3.3 Experience-Dependent Map Formation and Plasticity

By comparing VA_S activity patterns before and after "normal" development, we found that most neuronal units within VA_S were selective for one, or at most two different patterns of the training set (Figure 5). To allow comparison for presentations of objects evenly distributed throughout the visual field, we recorded activity patterns for brief stationary presentations of objects in an evenly spaced 32×32 grid of locations. From this we obtained average maps of neural activity, which discount location-dependent effects. While average maps obtained before development tended to be flat and featureless (due to the high degree of variability of neuronal responses; data not shown), maps obtained after "normal" development consistently showed numerous localized populations of neurons (neuronal groups, Edelman, 1987) responding selectively to particular stimulus patterns.

Each behavioral experiment with Darwin V produced an individual pattern of neuronal groups that varied in extent and spatial distribution. This prompted us to investigate if behavioral factors such as frequency or sequence of encounter have systematic effects on these patterns. To allow quantitative comparison of multiple runs, we estimated the overall area in map VA_S allocated to specific responses for a given stimulus as the sum of neuronal units responding higher than a threshold μ. Figure 6a shows summaries obtained from a total of 9 runs of similar length. Average map areas for each of the presented patterns were calculated at the end of each experiment, and are plotted together with the total number of objects encountered during the entire length of the experiment. There is a significant correlation between the total number of stimulus encounters and the number of neuronal units within VA_S that are selectively responding to that stimulus.

All data shown in Figure 6a was acquired from runs that did not show any significant variations in object frequencies from early to late stages of development. If, however, certain object classes are less frequently encountered during early experience, the map area allocated to these objects will be reduced. This reduction persists even if frequencies of encounter are balanced during later developmental stages (Figure 6b). This indicates that sensory maps are more resistant to changes in selectivity during late development, an effect due in part to reduced levels of plasticity (Equation 2 and Table 3) and competition from already well established neuronal groups. Historical fluctuations in stimulus encounters ("non-stationarity" of

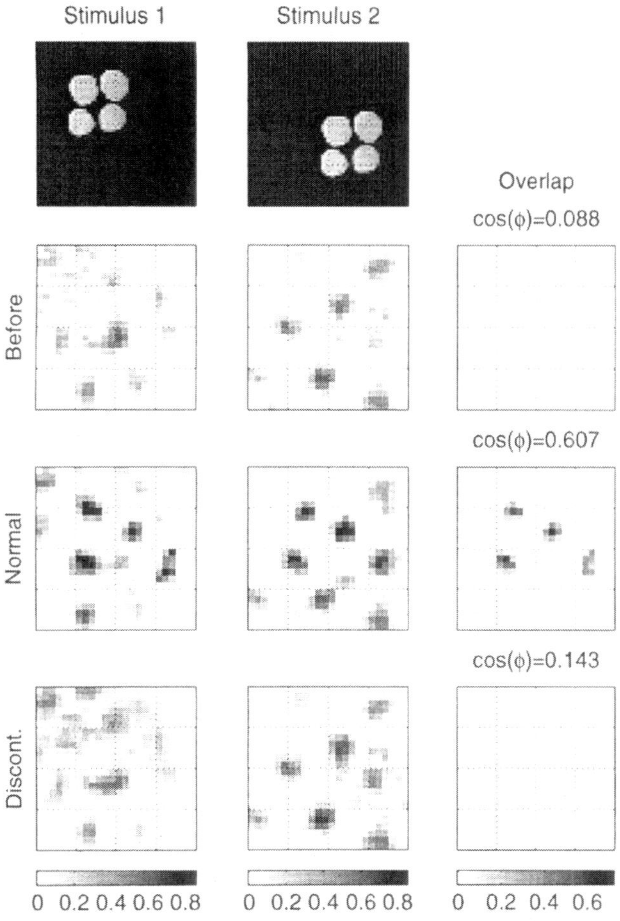

Figure 4: Activity patterns of area VA_S recorded before, after "normal", and after "discontinuous" development. Individual stimuli as seen through NOMAD's camera are shown at the top. Panels labeled 'Overlap' display the element-by-element product of the two activity pattern and values for $cos(\phi)$.

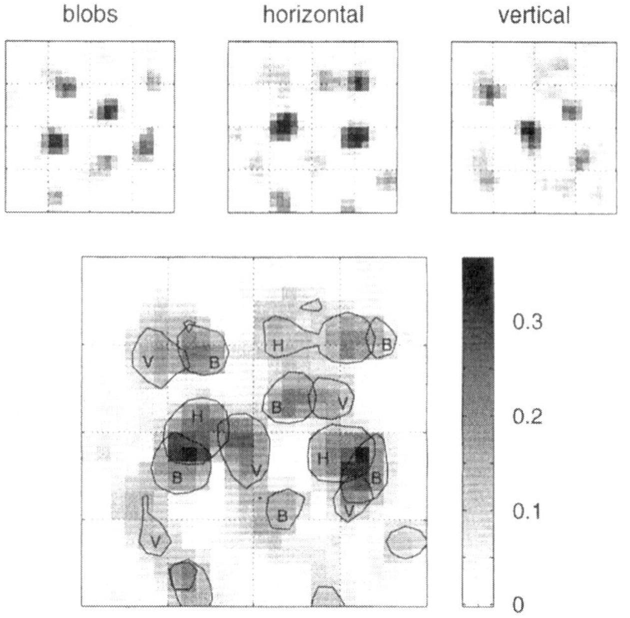

Figure 5: Average (location-independent) activity patterns in area VA_S obtained after "normal" visual development. Maps showing responses to "blobs", "horizontal" and "vertical" stripes are shown at the top, the corresponding composite map is displayed at the bottom.

environments) or behavioral patterns that increase or reduce exposure to particular stimuli can therefore have significant effects on the formation of sensory maps.

4 Discussion

While much is known about how neurons control behavior, much less attention has focused on the consequences of behavioral interactions within an environment on the distinct characteristics of neuronal responses. Darwin V provides an example of a simulated model of cerebral visual cortex embedded in a real world device that can be simultaneously analyzed at the neural and behavioral level. We have shown that the emergence of pattern selective and translation invariant neuronal properties is dependent upon self-generated movement and exploration. Furthermore, simulated cortical maps capable of plastic changes will develop and reorganize according to the frequency and sequence of stimulus encounters experienced.

While the neuronal bases of perceptual invariances in real animals are still largely unexplored (but see e.g. Blakemore and van Sluyters, 1975, for an early study on the development of orientation preference), several authors have invoked developmental (learning) processes in the generation of translation invariance. Some computational models of inferior temporal cortex (Gochin, 1994, Wallis and Rolls, 1997) have been developed previously. Specifically, it is proposed that the translation invariance of inferior temporal neuronal responses can be accounted for by a Hebbian-type learning mechanism (called the trace rule) that allows linkage of successive stages in the translational movement of objects. In contrast, translation invariance in Darwin V develops in the course of unrestricted, non-canonical stimulus displacements, without using a synaptic rule that explicitly incorporates time delays. While biological implementations for such rules may exist (i.e., slow receptors or calcium dynamics), other factors such as sustained neuronal discharge may play a similar role. Most importantly, however, the timing of these effects must be matched by an organism's sampling of statistical properties of its environment, largely dependent upon movements of the sensory apparatus. Viewed from this perspective, which invariances will develop is not determined by the nervous system alone, but by how its body and sensory systems move and interact with the world. Bodily motion and behavior thus become crucial mediators in how brains extract invariant statistical patterns from highly variable environments.

A wealth of evidence suggests that key properties of sen-

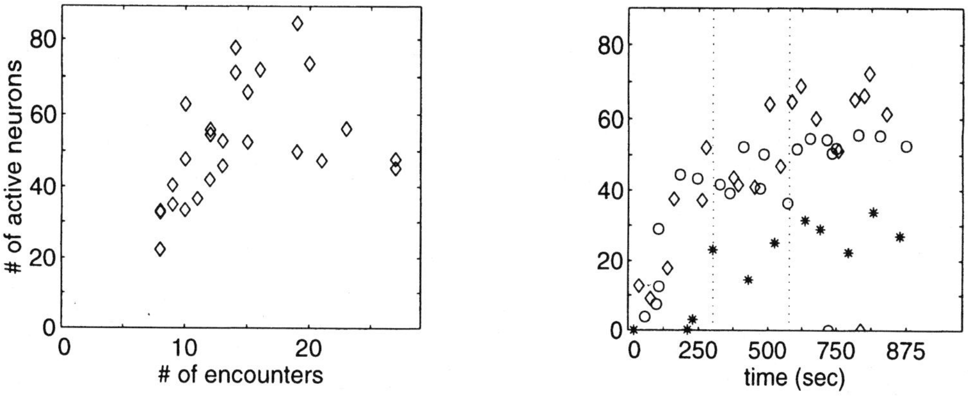

(a) (b)

Figure 6: (a) Scatter plot of map areas and total number of encounters obtained at the end of 9 individual experiments (r_{xy} = 0.4344). (b) Development of neuronal responses over a single representative experiment. Objects from one class ("blobs", circles) were less frequently encountered during early development (0-600 seconds), but appear at a frequency equal to that for "horizontals" (asterisks) and "verticals" (diamonds) during later development (>600 seconds). Each data point indicates the number of VA_S responding higher than $\mu = 0.5$ during an individual block encounter.

sory and motor maps, such as the areal extent and distribution of neuronal responses selective for particular stimuli, are experience-dependent (Merzenich et al., 1988, Kaas, 1991, Nudo et al., 1996). In studying the development of sensory maps within Darwin V's nervous system we attempted to relate the initial formation of these maps as well as their dynamic adjustments to ongoing behavioral interactions. In agreement with experimental results (Nudo et al., 1996), and previous simulation studies (Pearson et al., 1987), the number of neuronal units responding selectively to particular object classes is related to the overall amount of sensory exposure. To evaluate the relative impact of variations in experience during early versus late visual development we incorporated progressive changes in cortical plasticity similar to those reported in experimental studies (Fox, 1995). We found that an early bias towards particular stimulus classes results in significant shifts in neuronal specificity. These shifts are relatively stable and tend to persist even if this bias is removed during later experience. This is due in part to lower levels of plasticity and to competition from already formed neuronal groups. It is important to realize that in natural environments frequencies of stimulus encounter are rarely constant ("stationary") over extended time periods and are often a function of behavioral patterns and preferences by the organism. Future studies of behaving neural systems in rich environments are needed to further investigate the intricate interplay of behavior, stimulus selection and neural function.

Recent work in areas such as machine vision (Ballard, 1991, Bajcsy, 1988, active perception) and robotics (Brooks, 1991, Pfeifer and Scheier, 1997) has provided insights into how solutions to seemingly complex tasks and problems can arise from the action of simple behaving systems in an environment. Our approach aims at embodying anatomically and physiologically detailed nervous systems within real world devices to study the interaction of neural and behavioral variables. Understanding these relationships will be essential in formulating neurally based theories of brain and cognitive function.

It has been recognized previously that working with devices in real environments has several advantages, including increased realism (e.g., natural "noise") and a large repertoire of configurations of stimuli (not all of which are anticipated in advance). The results reported here support an even more fundamental rationale for why neurobiological models of behavior (in particular those focusing on learning or development) should be implemented as "embodied" systems (interfaced to real world devices and operating in real time) rather than studied as pure simulations. The interplay between neural and behavioral variables can only be adequately captured if ongoing neural changes are allowed to have consequences in behavior that can in turn modify neuronal processes. The results of our work with Darwin V indicate that modes of sensory sampling as well as changes in the statistical distribution of sensory input patterns have important roles in the development of properties thought to be intrinsic to the nervous system.

Acknowledgments

This work was carried out as part of the theoretical neurobiology program at The Neurosciences Institute, which is supported by Neurosciences Research Foundation.

References

Almássy, N., Edelman, G. M., and Sporns, O. (1998). Behavioral constraints in the development of neuronal properties: A cortical model embedded in a real world device. *Cereb. Cortex* (in press).

Bajcsy, R. (1988). Active perception. *Proceedings of the IEEE*, 76(8):996–1006.

Ballard, D. (1991). Animate vision. *Artificial Intelligence*, 48:57–86.

Bienenstock, E., Cooper, L., and Munro, P. (1982). Theory for the development of neuron selectivity: Orientation selectivity and binocular interaction in visual cortex. *J. Neurosci.*, 2:23–48.

Blakemore, C. and van Sluyters, R. C. (1975). Innate and environmental factors in the development of the kitten's visual cortex. *J. Physiol.*, 248:663–716.

Brooks, R. A. (1991). Intelligence without representation. *Artificial Intelligence*, 47:139–159.

Edelman, G. (1987). *Neural Darwinism*. Basic Books, New York.

Edelman, G., Reeke, Jr., G., Gall, W., Tononi, G., Williams, D., and Sporns, O. (1992). Synthetic neural modeling applied to a real-world artifact. *Proc. Natl. Acad. Sci. USA*, 89:7267–7271.

Fox, K. (1995). The critical period for long-term potentiation in primary sensory cortex. *Neuron*, 15:485–488.

Gochin, P. M. (1994). Properties of simulated neurons from a model of primate inferior temporal cortex. *Cereb. Cortex*, 4(5):532–543.

Gross, C. and Mishkin, M. (1977). The neural basis of stimulus equivalence across retinal translation. In Harnad, S., Doty, R., Goldstein, L., Jaynes, J., and Krauthammer, G., editors, *Lateralization in the Nervous system*, pages 109–122. Academic Press, New York.

Gross, C. G. (1992). Representation of visual stimuli in inferior temporal cortex. *Phil. Trans. R. Soc. Lond. B*, 335:3–10.

Kaas, J. (1991). Plasticity of sensory and motor maps in adult mammals. *Annu. Rev. Neurosci.*, 14:137–167.

Kirkwood, A., Lee, H.-K., and Bear, M. F. (1995). Co-regulation of long-term potentiation and experience-dependent synaptic plasticity in visual cortex by age and experience. *Nature*, 375:328–331.

Merzenich, M., Recanzone, G., Jenkins, W., Allard, T., and Nudo, R. (1988). Cortical representational plasticity. In Rakic, P. and Singer, W., editors, *Neurobiol. of Neocortex*, pages 41–67, New York. Wiley.

Nakamura, K., Matsumoto, K., Mikami, A., and Kubota, K. (1994). Visual response properties of single neurons in the temporal pole of behaving monkeys. *J. Neurophysiol.*, 71(3):1206–1221.

Nudo, R., Milliken, G., Jenkins, W., and Merzenich, M. (1996). Use-dependent alterations of movement representations in primary motor cortex of adult squirrel monkeys. *J. Neurosci.*, 16(2):785–807.

Pearson, J., Finkel, L., and Edelman, G. (1987). Plasticity in the organization of adult cortical maps: A computer simulation based on neuronal group selection. *J. Neurosci.*, 7:4209–4223.

Pfeifer, R. and Scheier, C. (1997). Sensory motor coordination: The metaphor and beyond. *Robotics and Autonomous systems*, 20:157–178.

Reeke, Jr., G., Finkel, L., Sporns, O., and Edelman, G. (1990a). Synthetic neural modeling: A multilevel approach to the analysis of brain complexity. In Edelman, G., Gall, W., and Cowan, W., editors, *Signal and Sense: Local and Global Order in Perceptual Maps*, pages 607–707. Wiley, New York.

Reeke, Jr., G. and Sporns, O. (1993). Behaviorally based modeling and computational approaches to neuroscience. *Annu. Rev. Neurosci.*, 16:597–623.

Reeke, Jr., G., Sporns, O., and Edelman, G. (1990b). Synthetic neural modeling: The "Darwin" series of recognition automata. *Proc. IEEE*, 78:1498–1530.

Rodman, H. (1994). Development of inferior temporal cortex in the monkey. *Cereb. Cortex*, 5:484–498.

Rolls, E. T. (1992). Neurophysiological mechanisms underlying face processing within and beyond the temporal cortical visual areas. *Phil. Trans. R. Soc. Lond. B*, 335:11–21.

Rolls, E. T. and Tovee, M. J. (1995). The responses of single neurons in the temporal visual cortical areas of the macaque when more than one stimulus is present in the receptive field. *Exp. Brain Res.*, 103:409–420.

Tanaka, K. (1996). Inferotemporal cortex and object vision. *Annu. Rev. Neurosci.*, 19:109–139.

Tovee, M. J., Rolls, E. T., and Azzopardi, P. (1994). Translation invariance in the responses to faces of single neurons in the temporal visual cortical areas of the alert macaque. *J. Neurophysiol.*, 72(3):1049–1060.

Verschure, P., Wray, J., Sporns, O., Tononi, G., and Edelman, G. M. (1995). Multilevel analysis of a behaving real world artifact: an illustration of synthetic neural modeling. *Robotics and Autonomous Systems*, 16:247–265.

Wallis, G. and Rolls, E. T. (1997). A model of invariant object recognition in the visual system. *Progress in Neurobiol.*, 51:167–194.

Learning Hierarchical Control Structures for Multiple Tasks and Changing Environments

Bruce L. Digney*
Robotics Institute
Carnegie Mellon University
Pittsburgh, PA, 15232, USA
Phone (412) 268-7084
bdigney@ri.cmu.edu

Abstract

While the need for hierarchies within control systems is apparent, it is also clear to many researchers that such hierarchies should be learned. Learning both the structure and the component behaviors is a difficult task. The benefit of learning the hierarchical structures of behaviors is that the decomposition of the control structure into smaller transportable chunks allows previously learned knowledge to be applied to new but related tasks. Presented in this paper are improvements to Nested Q-learning (NQL) that allow more realistic learning of control hierarchies in reinforcement environments. Also presented is a simulation of a simple robot performing a series of related tasks that is used to compare both hierarchical and non-hierarchal learning techniques.

1. Introduction

The need for hierarchical structures within learning control systems is clear. Without some form of hierarchy, a learning system would be bogged down in the innumerable details of the lowest level of control. Many researchers have recognized this and have imposed hand crafted hierarchies. Although the imposed hierarchies result in more tractable problems, they also impose the designers preconceived notions on the control system. It is desirable that the control system able to generate its own hierarchical structure. Recently, work on Nested Q-learning (NQL) (Digney, 1996) has shown that a hierarchical structure can be learned in a reinforcement learning environment. Although this method generated hierarchical structures, it was considered to be seriously handicapped by the need to classify every distinct sensory state as a feature. With each distinct feature becoming the termination point for a new behavior the control system quickly became overwhelmed with choices of behaviors, of which the vase majority were irrelevant. In this paper, extensions able to remove that handicap are introduced. The solution is to have only useful features emerge. To do this two emergence criteria are introduced; high frequency of state occurrence and high reinforcement gradients. As only features with a high probability of relevance emerge learning will be much faster. A hierarchical learning control system is developed and tested against a non-hierarchical learning system.

2. Background

An issue closely linked to robot learning is the architectures in which the control strategies are implemented (Tyrrel, 1992). The two main ideologies are flat and hierarchical. Figure 1 shows the flat and the hierarchical architectures schematically. Flat architectures have direct connections from sensors to actions, through a single level of control. When some situation is perceived all behaviors compete for control with the strongest response winning. Hierarchical architectures have a more indirect coupling of perceptions to actions through a hierarchical control structure. When a situation is perceived, some high level behavior becomes active and issues commands to one or more lower level behavior(s). In turn, these lower level behaviors control yet lower level behaviors until primitive actions are activated and some physical action(s) is performed. Both architectures have benefits and drawbacks and have been used to implement learning techniques. One of the major benefits of hierarchical structures in robot learning is that learning difficult tasks can be made more tractable by clever design at the hierarchical structure. By abstracting away many details of a complex world to lower level behaviors, learning can more easily be implemented in the higher levels.

Many researchers have recognized the need for hierarchical structures in learning control systems (Maes and Brooks, 1990) (Long-Ji, 1993) (Dayan and Hinton, 1993) (Singh, 1992). These approaches are all similar in the respect that the structures are hand designed and individual components are learned or vise versa. Previous work in Nested Q-learning (NQL) (Digney, 1996) demonstrated the autonomous construction of a control hierarchy able to learn both the structure and the behaviors using a reinforcement learning technique based on Q-learning (Barto and Watkins, 1989). When the agent initially started out, no information about structure or

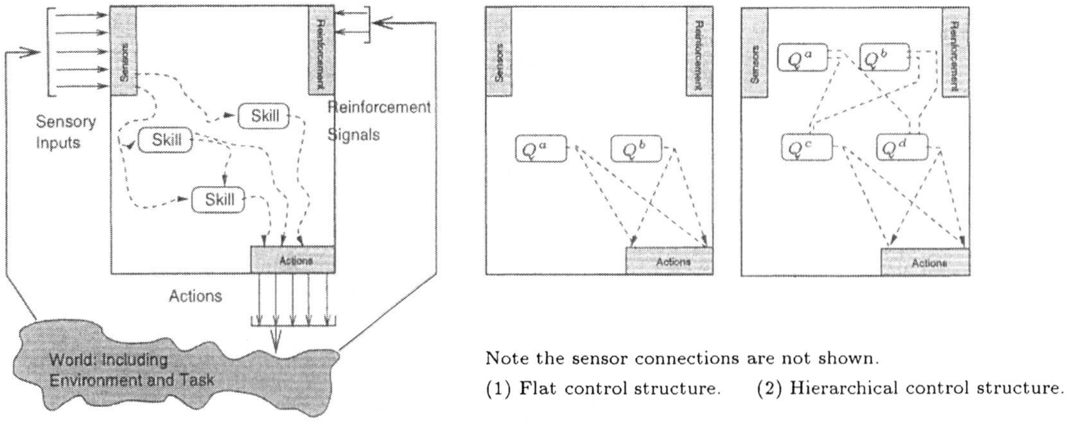

(a) Control schematic.

(b) Flat and Hierachical control structures.

Figure 1 Schematic of control architectures: (a) sensory, action and reinforcement signal configuration for a learning control system and (b) control architectures: (1) flat structure and (2) hierarchical structure. Note: $Q^?$ represents behaviors and the sensory connections in (b) have been removed for clarity.

useful behavior was given to it. The control system discovered distinct recognizable features in its world and learned the relationship between different features, its environment and its tasks. These relationships formed behaviors which were usually a hierarchical assembly of other behaviors and primitive actions. The structure that emerged was of as many levels as the task required. Using NQL, the control system also learned bottom-up reactive/opportunistic functions for each behavior which served to 1) provide a high level command source at the top of the hierarchy and 2) facilitate the invocation of previously learned beneficial behaviors in new situations without the necessity of relearning them.

One of the simplifying assumptions made during the initial work on NQL was that all perceivable states in the environment were valid termination points for possible behaviors. Useful behaviors would then emerge out of this pool of candidate behaviors. This assumption works for only very small problem spaces. When the robot is faced with problems of any realistic size, the pool of candidate behaviors grows too large very quickly. In this paper, a method for removing this assumption and allowing only behaviors with a high probability of relevance to emerge, leaving all other behaviors undiscovered is presented.

3. Nested Q-Learning with Emergent Behaviors

Consider an animat placed in some world and expected to perform some task defined by delayed rewards. As the animat moves about in its world, it experiences reinforcements from various sources. High negative reinforcements result from damaging actions, nominally negative reinforcements represent energy expended and positive reinforcement signals represent successful performance of some task(s). It is also able to perceives changes in its state as it moves around. A typical reinforcement learning problem is to learn the state, x, action, u, evaluation function $Q(x, u)$ such that the total positive reinforcement received by the animat is maximized. This results in a single monolithic evaluation function $Q(x, u)$ that, although able to solve the current problem, the information gained, albeit likely gained at less cost, will be useless in new tasks.

Nested Q-learning decomposes the monolithic evaluation function into a hierarchical structure of smaller evaluation functions, each representing a behavior which can later be invoked and reused in a new task or environment. To achieve this, the NQL algorithm allows the selection of behaviors in addition to primitive actions. These behaviors are defined as learned sequences that bring about some distinct termination state. Originally, these states were defined as any distinct state. This was a severely limiting and unnatural assumption. In this paper, only behaviors that prove useful emerge while all other possible behaviors remain undiscovered. These emergent behaviors will be selected based on two criteria. First, high changes in reinforcement gradient and secondly, a high state occurrence rate or frequency of visits to a particular state.

As the animat moves around, it perceives itself in various states and entering most of these states results in nothing more than normally occurring reinforcement signals. However, some states will represent dramatic changes in reinforcement. It is reasonable to assume that high reinforcement signal gradients might represent a non-typical and likely useful location in the state space. The high reinforcement signal gradients will be used to differentiate useful features from irrelevant ones. The animat's state space, as perceived by sensors will be discrete and represented by x, with $x \in \{ 0, 1, \cdots x \cdots x_{max} \}$ where x_{max} is the maximum number of distinct states currently existing but can increase as new states are discovered and x is the animat's current state. As the agent arrives at state x it experiences some total reinforcement signal, R_{TOTAL}. This reinforcement signal can be high negative in situations such as collisions with obstacles, but is usually some nominal value indicating the energy expended in reaching the current state. In addition to driving the adaptive mechanism, this signal can be used to learn a squared external reinforcement signal gradient mapping function $m(x)$ which learns the areas of non-typical reinforcement in the state space. This mapping will be used to generate potentially useful features in the state space and is determined as follows:

$$e_m = m(x) - \left(\frac{\partial R_{TOTAL}}{\partial t} \right)^2 \tag{1}$$

or

$$e_m = m(x) - \left(\frac{\Delta R_{TOTAL}}{\Delta t} \right)^2 \tag{2}$$

and

$$m(x) \longleftarrow m(x) - \eta e_m \tag{3}$$

where e_B is the environment, x is the current state, η is the learning rate and $\longleftarrow$ represents learning by some type of incremental parametric storage device such as a neural network. High changes in the gradient of the reinforcement signal during the transition from state to state, learned as a high value in $m(x)$, indicates a non-typical location in state space and a potentially useful feature. Eventually, as $m(x)$ converges, potentially useful features emerge using a simple threshold operator

$$x = \begin{cases} \text{feature} & \text{if } m(x) > T_1 \\ \text{non-feature} & \text{otherwise.} \end{cases} \tag{4}$$

where T_1 is the minimal squared external reinforcement gradient for a state to be considered as a feature.

Another criteria for useful behavior emergence is the occurrence or frequency of visits to particular state space locations. During operation, important states will be visited more often than less important ones. However, the animat must already be successfully performing one or more tasks for these features to emerge. At first, the reason for doing this might seem a bit unclear. Why generate new behaviors once the animat is operating successfully? The reason is that features useful in one or more tasks are likely to be useful in other, yet to be learned tasks. By learning the decomposition of the currently well learned tasks, portions of what has been learned can be used in new tasks. To learn this decomposition, a occurrence function $c(x)$ is defined. The strength of this function is increased by some small amount upon each visit to each state,

$$c(x) \longleftarrow c(x) + \epsilon \tag{5}$$

where ϵ is some incremental amount. This function is also periodically decayed by some decay factor,

$$c(x) \longleftarrow \lambda c(x) \text{for all } x \tag{6}$$

where λ is the decay factor, $0 < \lambda < 1$ and x denotes all currently discovered states. The result is the emergence of states that are more commonly visited or high occurrence. These commonly visited states can then be extracted as potentially useful features and used to generate behaviors. For the animat to extract such useful features from the occurred function $c(x)$, a threshold function similar to Equation 4 is used

$$x = \begin{cases} \text{feature} & \text{if } c(x) > T_2 \\ \text{non-feature} & \text{otherwise.} \end{cases} \tag{7}$$

where T_2 is the criteria for determining whether or not a state is visited often enough to warrant it emerging as a feature. When a single task is being performed, key states from the state trajectories are extracted. That is, the key states of the action state sequence which must be visited for successful outcomes. When more than one task is being performed, these equations decompose the action sequences of the tasks into sub-trajectories (sub-behaviors) that the tasks have in common.

Now that the two criteria for useful feature emergence have been developed, the extension to NQL can be discussed. The core of the NQL algorithm remains very much as previously described (Digney, 1996). However, instead of having all possible behaviors discovered and available for learning and use very soon after the animat begins operation, features emerge and are learned and used continuously throughout the life of the animat. This continual emergence of features causes added difficulties with learning and must be taken into consideration. This is because behaviors that emerge early will not be able to benefit from behaviors that emerge later. Again, consider an animat initially placed within a world without any form of previous experience or knowledge of its environment or task. The animat has some primitive actions, a, from which it can choose to physically act within its world. These are $a \in \{ a_0, a_1, \cdots a_j \cdots a_J \}$ where a_j is a primitive action and J is the total number of primitive actions. Initially, without any behaviors having emerged, the possible actions available to the animat, u, are only primitive actions, $u \in \{ Q^{a_0}, Q^{a_1}, \cdots Q^{a_j} \cdots Q^{a_J} \}$ where Q^{a_j} is a non-adaptive behavior representing the performance of primitive action a_j and should not be confused with the adaptive behaviors Q^{f_i} and u is the action to be taken by the animat. As the agent gains experience within its task and environment, potentially useful features will emerge according to the previously described criteria. These features, f_i, are labeled $f \in \{ f_0, f_1, \cdots f_i \cdots f_I \}$ where f_i is a discovered feature and f_I is currently the last discovered feature. The total number of action (both primitive actions and behaviors) that may be taken by the agent now includes both the primitive actions and the emergent behaviors, $u \in \{ Q^{a_0}, \cdots Q^{a_j} \cdots Q^{a_J}, Q^{f_0}, \cdots Q^{f_i}, \cdots Q^{f_I} \}$. The total number of possible actions is the sum of all primitive actions and all currently possible behaviors, $u_{total} = J + I$. The state of the agent is established by the state of all the incoming sensors and is represented by a discrete value x_l which ranges from x_1 through x_{max}. The evaluation function for each feature is a function of the robot's current state and all possible actions, $Q^{f_i} = f(x, u)$. The evaluation function for each behavior now becomes, $Q^{f_i} = f(x_1, \cdots x_{max}, Q^{a_1} \cdots Q^{a_J}, Q^{f_1}, \cdots Q^{f_I})$, where both the number of states, x_{max}, and the number of behaviors, f_I, are open ended and subject to initial discovery and then to increases or decreases due to ongoing changes. It is seen that the learning algorithm described above becomes nested and possibly recursive. That is, the evaluation function, $Q^{f_{desired}}$, can invoke other behaviors including itself while attempting to reach the feature, $f_{desired}$. It is this nested nature that will allow hierarchical control structures to emerge.

As the agent interacts with the environment it receives an external reinforcement signal(s), r_{EXT}. It is through this signal that the agent is driven to perform tasks of external benefit. The reinforcement signal is defined as

$$r_{EXT} = \begin{cases} 0 & \text{if external task is achieved} \\ -R_{EXT} & \text{otherwise} \end{cases} \tag{8}$$

where r_{EXT} is an external reinforcement signal and R_{EXT} is a positive constant. In addition, there are various internal reinforcement signals, r_{INT}. These drive the agent to perform tasks of internal benefit such as *avoid danger* and *find fuel.* In the nested Q-learning algorithm there is also a reinforcement signal that effects only the currently active behavior(s). The reinforcement signal drives the action of the agent to reach the desired feature

$$r_{FEAT} = \begin{cases} 0 & \text{if } x = f_{desired} \\ -R_{FEAT} & \text{if } x \neq f_{desired} \end{cases} \tag{9}$$

where r_{FEAT} is the feature's reinforcement signal and R_{FEAT} is a positive constant. Upon performing a particular selected action, u^*, be it a primitive action or a behavior, the robot advances from state x_v to the next state x_w and incurs a total reinforcement signal, r_{TOTAL}. Included in this total reinforcement signal is the cost of performing the selected action. This cost includes the physical cost of performing the action and possibly the mental (computational) cost of choosing the action. These costs are designated as r_{LOW}, with

$$r_{LOW} = \begin{cases} -C \\ \quad \text{if } u^* \text{ is a primitive action} \\ \sum_{k=0}^{K} r_{TOTAL}^{u*}(k) \\ \quad \text{if } u^* \text{ is an adaptive behavior} \end{cases} \tag{10}$$

where $\sum_{k=0}^{K} r_{TOTAL}^{u*}(k)$ is the total reinforcement signal from the invoked behavior summed over the number of steps, K, required to perform the behavior, u^*, and C is a constant that reflects the cost of performing the primitive action. Of course, if many lower levels of behaviors are employed, this process becomes nested at r_{LOW} which represents the experiences (physical and mental) of all lower levels. The total reinforcement signal for the invoking behavior becomes

$$r_{TOTAL} = r_{EXT} + r_{INT} + r_{FEAT} + r_{LOW} \tag{11}$$

where r_{TOTAL} is the sum of all reinforcement signal sources. This total reinforcement is used to construct the Q functions of expected reinforcement, from which useful top-down control strategies will emerge. The error, e_Q, is defined to be,

$$e_Q = \gamma \cdot \max_u \{Q_{x_w,u}\} - Q_{x_v,u^*} + r_{TOTAL} \tag{12}$$

where γ is the temporal discount factor $0 < \gamma < 1$ and $\max_u\{Q_{x_w,u}\}$ is the current prediction of the maximum total future reinforcement remaining when the agent leaves state x_w. This error is used to adapt the evaluation functions,

$$Q_{x_v,u=u^*}(k+1) = Q_{x_v,u=u^*}(k) + \eta_Q \cdot e_Q \tag{13}$$

$$Q_{x_v,u\neq u^*}(k+1) = Q_{x_v,u\neq u^*}(k) \tag{14}$$

where η_Q is the rate of adaptation and k is the index of adaptation.

Features may emerge at any time during operation and their corresponding behaviors are added to the evolving behavior pool. Incorporating these newly introduced behaviors into the action selection mechanism is required. This makes action selection more difficult than in the action selection mechanism described in previous work (Digney, 1996). For the top-down goal directed action/behavior selection, a random based exploration policy is used. Although this is not the most effective exploration policy, it is easily implemented and other more efficient forms of exploration have been studied in depth elsewhere (Thurn, 1992). For the currently invoked behavior, $Q^{invoked}$, action/behavior selection is determined according to

$$u^* = \begin{cases} argmax_u\{Q^{ALL} + E_{invoked}\} \\ \quad \text{if } Q^{invoked} = Q^{f_i} (\text{a behavior}) \\ a_j \\ \quad \text{if } Q^{invoked} = Q^{a_j} (\text{a primitive action}) \end{cases} \tag{15}$$

where argmax_u is a maximum function taken over all primitive actions and existing behaviors, $E_{invoked}$ is the exploration policy and Q^{ALL} is all possible choices including both other behaviors and primitive actions. Note, that there can be more that one behavior invoked at any time. In fact, it is usual that many behaviors will be active at the same time in response to either top-down or bottom-up directives. The nested nature is seen again in Equation 15 where the currently active behavior is able to select another behavior or a primitive action. If another behavior is selected that behavior may also go on and select yet another behavior and so on. If a primitive action is selected, Q^{a_j}, then the physical actuator action a_j is performed. The exploration policy for behavior $Q^{invoked}$ that is discovered at time d_i is

$$E_{inv} = RAND(k_{inv}STEP(t^{inv} - d^{inv}_{dis})e^{-\tau(t^{inv}-d^{inv}_{dis})} + k_{other} \sum_{i=0,\ i \neq inv}^{I} STEP(t^i - d^i_{dis})e^{-\tau(t^i - d^i_{dis})} \quad (16)$$

where t_i is the behavior local time measured relative to each behavior (and is effectively proportional to the number of times each behavior has been invoked), d^{inv}_{dis} is the time of discovery for the invoked behavior, d^i_{dis} is the discovery time for behavior, i, and t^{inv} and t^i are the behavior local time for the invoked, $Q^{invoked}$, and the other behaviors Q^i. The time constants, τ, control the rate at which the exploration contribution to action selection is reduced. The constants k_{inv} and k_{other} controls how much exploration is due to the invoked behavior and how much is due to other behaviors that may have been discovered later. The $STEP(w)$ is the a step function,

$$STEP(w) = \begin{cases} 1 & \text{if } w > 0. \\ 0 & \text{if } w < 0. \end{cases} \quad (17)$$

The first term directly following k_{inv} in Equation 16 represents the contribution to exploration due to the discovery of the currently invoked behavior itself. That is, the behavior will have an initially high exploration component, decaying towards pure exploitation. The second term of Equation 16 represents the contribution to exploration due to the discovery of other behaviors. If this second term were zero then behaviors that are discovered after earlier behaviors have converged would never be tried by those converged behaviors. Without partially refreshing the exploration component of the earlier behaviors, the earlier learned behaviors may never learn to benefit from the new behaviors as they emerge.

The preceding derivation describes a NQL technique through which a top-down action selection mechanism will generate a hierarchical control structure and cascade goal seeking commands downward through the structure as behaviors are discovered continuously throughout the life of the animat. Each behavior, whenever invoked, will in turn invoke other behaviors and/or primitive actions in an attempt to fulfill the desired goals of higher behaviors. The bottom-up or sensory based action selection mechanism and how it fits with the top-down action selection to result in a flexible reactive control hierarchy is described elsewhere (Digney, 1996) and is not explored in this paper.

4. Simulation

To evaluate the NQL algorithm and it's capabilities, (the emergent features and the subsequent learning of a hierarchical structure of behaviors), an animat and the two dimensional environment of Figure 2 were used. The animat was placed in a grid world 6 by 10 units in size as shown in Figure 2(b). This grid world was bounded by an impassable barrier and additional barriers could be erected within the world in any configuration desired. Different from related work (Digney, 1996) in which the animat had a number of different sensors, the animat in this study was only capable of perceiving its location within the grid world. In this study, the animat's primitive actions were capable of moving it *left*, *right*, *up*, *down* or *no motion* shown in Figure 2(a). These actions were non-deterministic and when invoked they only achieved their desired outcome with a probability of 0.75. Otherwise another action was taken randomly. Figure 2(b) also shows the initial configuration of the grid world. The animat always started from the indicated starting region located on the left most side. There were three goal locations labeled A, B and C, placed as indicated in Figure 2(b). Although their positions are indicated, the goals were not necessarily always active or visible to the animat. The goals were activated or made visible to the animat as desired to test the capabilities of the control system. An internal barrier was placed as shown in Figure 2(b), forcing the animat to pass through a narrow passage en route to the goals. In previous work, the animat learned to associate various goals with their appropriate external stimuli (i.e. blue overhead light caused animat to move toward blue floor panel). In this paper, such associations were not studied and the animat was instructed to seek all goals. This was implemented as a random selection over all goals that had been discovered up to that time.

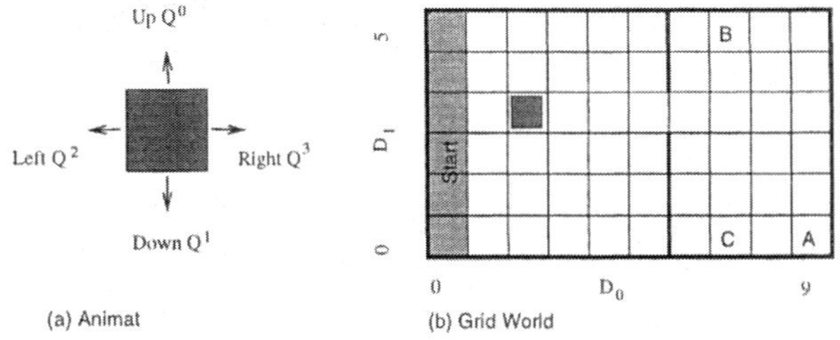

Figure 2 The animat and its grid world. Note the five primitive actions (up down, left, right and stay), the start locations, the initial barrier and the location of the three possible goals.

4.1 Emergent Features, Behaviors and Structures

The world was initially configured with barriers as shown in Figure 2(b). Two goals, A and B, were activated and goal C was left undetectable. The control system initially had no behaviors available to it as shown in Figure 3(a) and its actions were thoroughly random and undirected. This can be contrasted to previous work (Digney, 1996) in which the control system was quickly flooded with potential useful behaviors and then had to sort out the useful from the many irrelevant. Once placed in its world, the animat randomly moved about undirected. After a short period of time it learned that a high reinforcement signal gradient occurred near active goal locations (the animat first discovered goal A before goal by chance). This constituted the discovery of a feature, f_0, and the behavior Q^5 emerged. The subsequent expansion of the control system at the emergence of Q^5 is shown in Figure 3(b). Next, the second goal location (goal B) feature, f_1, was discovered and the control system expanded to include behavior Q^6 as shown in Figure 3(c).

Note that at this point only two goals are currently active and two behaviors in existence. These have resulted from two reinforcement gradient based features emerging from $m(x)$. As the animat performed these two behaviors, two other features, based on high occurrence peaks in function $c(x)$ were discovered: f_2 at spatial location $D_0 = 5$ and $D_1 = 3$ and f_3 at $D_0 = 6$ and $D_1 = 3$. They emerged as the locations adjacent to the mouth of the barrier; clearly important locations in the state space. The hierarchical control system expanded to include behaviors Q^7 and Q^8, as shown in Figure 3(d). The two existing behaviors quickly integrated the new behaviors into their action sequences. The hierarchical structure of the control system is further shown in the new action sequence when behavior behavior Q^6 was activated is shown in Table 2 and Figure 4. This shows that the sub-behaviors, Q^7 and Q^8 are evident in the state-action-behavior trajectories of the higher level behaviors Q^5 and Q^6 (not shown). For convenience, the behaviors, features and locations discussed are summarized in Table 1.

At time ≈ 150, the third goal (goal C) was activated and made detectable to the animat. Shortly thereafter, its squared reinforcement gradient was learned and another feature, f_4 emerged at location $D_0 = 9$ and $D_1 = 0$. The control system expanded to incorporate the new behavior, Q^9 as shown in Figure 3(e). From the behavior sequence of Table 3 and Figure 5, it is seen that behavior Q^9 quickly exploited the existing behaviors Q^5, Q^6 and to a greater extent Q^7 and Q^8. The performance of the hierarchical control system and its component behaviors are shown in Figure 8(b). The primitive actions have a constant performance and are not shown. The events previously described are shown between time = 0 and time ≈ 150. For graphing purposes, any undiscovered behaviors were given a dummy performance value of -200 until their time of discovery. From this plot, it is seen that the learning of behavior Q^9 was considerably quicker than

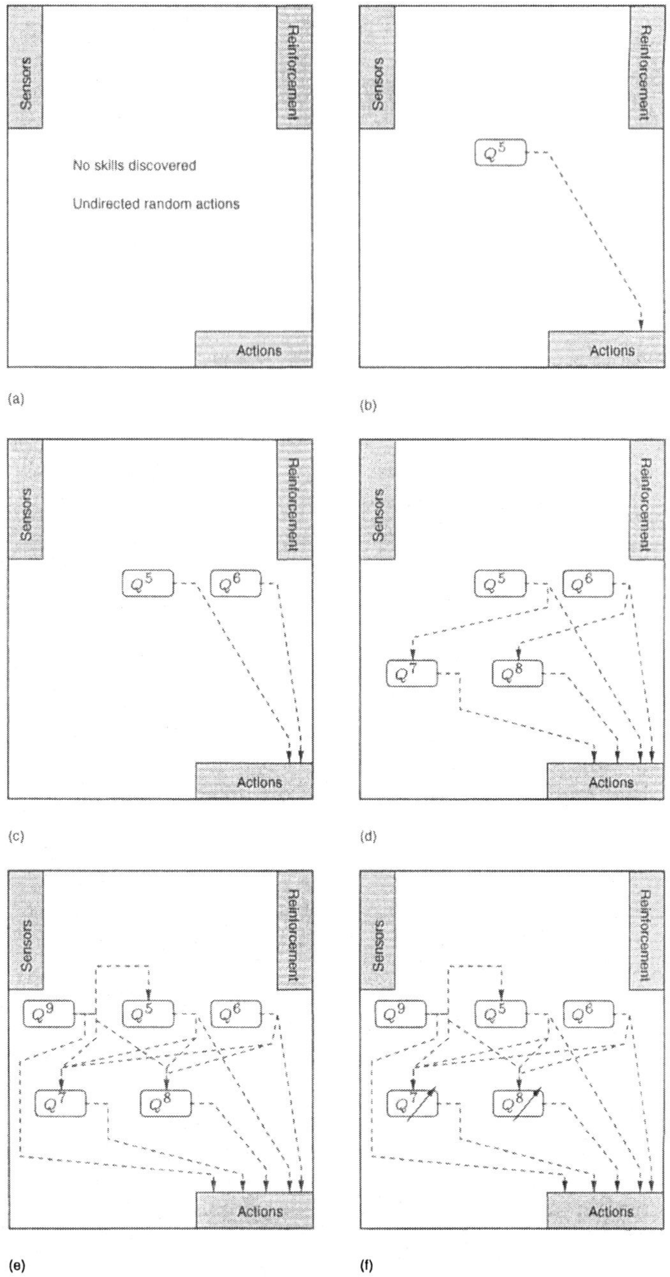

Figure 3 Emergent behaviors and structures.

Feature	Behavior	Location D_0	D_1	Discovery Time	Comments
NA	Q^0	NA	NA	NA	Primitive action
NA	Q^1	NA	NA	NA	Primitive action
NA	Q^2	NA	NA	NA	Primitive action
NA	Q^3	NA	NA	NA	Primitive action
NA	Q^4	NA	NA	NA	Primitive action
f_0	Q^5	9	0	≈ 50	First goal (A)
f_1	Q^6	7	5	≈ 50	Second goal (B)
f_2	Q^7	5	3	≈ 100	Door Entrance
f_3	Q^8	6	3	≈ 110	Door Exit
f_4	Q^9	7	0	≈ 150	Third goal (C)

Table 1 Behaviors, Features and Locations for the Hierarchical Control System.

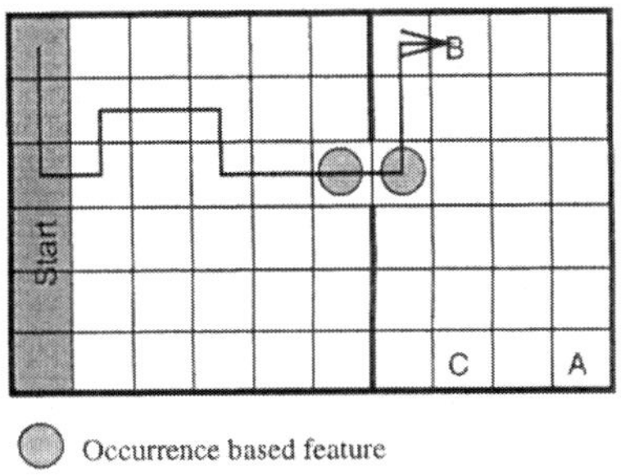

Figure 4 Animats's path for behavior Q^6.

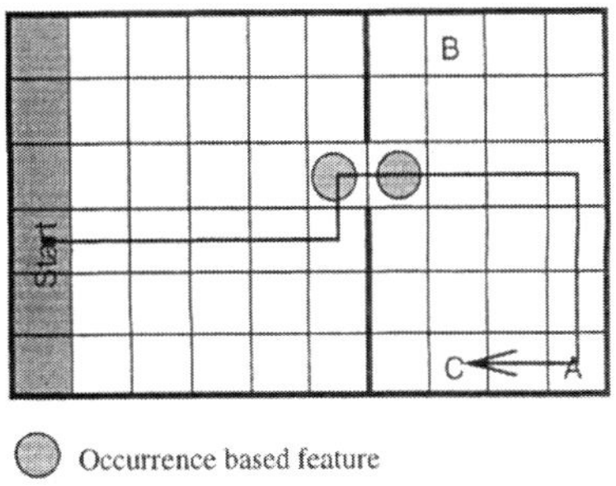

Figure 5 Animats's path for behavior Q^9.

Depth	Behavior	D_0	D_1	Comments
1	$- > Q^6$	0	5	Start (Go to B)
2	$- - > Q^8$	0	5	Go to door exit
3	$- - - > Q^1$	0	5	Primitive action
3	$- - - > Q^3$	0	4	"
3	$- - - > Q^3$	0	3	"
3	$- - - > Q^0$	1	3	"
3	$- - - > Q^7$	1	4	Go to door entrance
4	$- - - - > Q^3$	1	4	Primitive action
4	$- - - - > Q^3$	1	4	"
4	$- - - - > Q^3$	2	4	"
4	$- - - - > Q^1$	3	4	"
4	$- - - - > Q^3$	3	3	"
4	$- - - - > Q^3$	4	3	"
3	$- - - > Q^3$	5	3	At door entrance
2	$- - > Q^0$	6	3	At door exit
2	$- - > Q^3$	6	4	Primitive action
2	$- - > Q^0$	7	4	At goal(B)

Table 2 Behavior Q^6 before the new barrier. See Figure 4 for path. Note the arrow length represents the depth into the structure.

Depth	Behavior	D_0	D_1	Comments
1	$- > Q^9$	0	2	Start(Go to C)
2	$- - > Q^3$	0	2	Primitive action
2	$- - > Q^7$	1	2	Go to door entrance
3	$- - - > Q^3$	1	2	Primitive action
3	$- - - > Q^3$	2	2	"
3	$- - - > Q^3$	3	2	"
3	$- - - > Q^3$	4	2	"
3	$- - - > Q^0$	5	2	"
3	$- - - > Q^3$	4	2	"
3	$- - - > Q^3$	4	2	"
3	$- - - > Q^0$	5	2	At door entrance
2	$- - > Q^5$	5	3	Go to A
3	$- - - > Q^3$	5	3	Primitive action
3	$- - - > Q^3$	6	3	"
3	$- - - > Q^3$	7	3	"
3	$- - - > Q^3$	8	3	"
3	$- - - > Q^1$	9	3	"
3	$- - - > Q^1$	9	2	"
3	$- - - > Q^1$	9	1	"
2	$- - > Q^2$	9	0	At A
2	$- - > Q^2$	8	0	At goal (C)

Table 3 Behavior Q^9 before the new barrier. See Figure 5 for path. Note the arrow length represents the depth into the structure.

the learning of the two previous reinforcement gradient based features, Q^5 and Q^6. This new behavior, Q^9, was learned more quickly because it was able to exploit the previously learned behaviors, Q^7 and Q^8, and to a lesser extent Q^5 and Q^6. The initial effort expended in learning to decompose the two behaviors Q^5 and Q^6 into Q^7 and Q^8 paid off by speeding the learning of the new behavior Q^9.

4.2 Recovery From Changes

Next the capabilities of the Nested Q-learning technique to adapt to changes were evaluated. After the third goal was introduced an additional barrier was introduced at and the animat was required to recover. At time ≈ 200, the configuration of the world was changed to the configuration shown in Figure 6. This configuration was almost identical to the initial one, except for the addition of a new barrier at $D_0 = 2$. This barrier effectively forced the animat to move to the lower edge of its world when traveling from the start location towards the goal locations on the right side. Behavior performance as displayed in Figure 8 (b) shows that after the introduction of the new barrier, the animat recovered and was able to relearn its control strategies.

From the resulting state-action-behavior trajectories for the relearned behavior Q^6, it is seen that most of the adaptation was confined to the lower level behaviors. This trajectory is shown in Figure 7 and Table 5. Figure 3(f) shows the final hierarchical structure with the lower levels adapted to the new barrier.

4.3 Comparison with Non-hierarchical Control Systems

An obvious criticism of this approach is that the complexities added by introducing intermediate behaviors in addition to the primitive actions will increase the number of choices available to the animat, thus making the learning more difficult. For a single task this is indeed correct, but it is the improvement in the animat's ability to learn

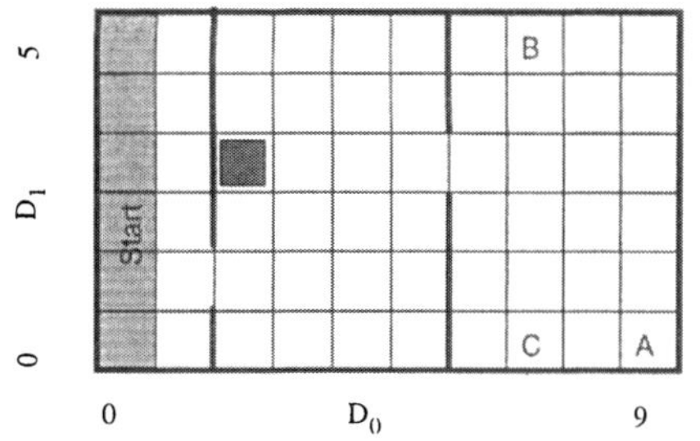

Figure 6 Animat's gridworld with new barrier introduced.

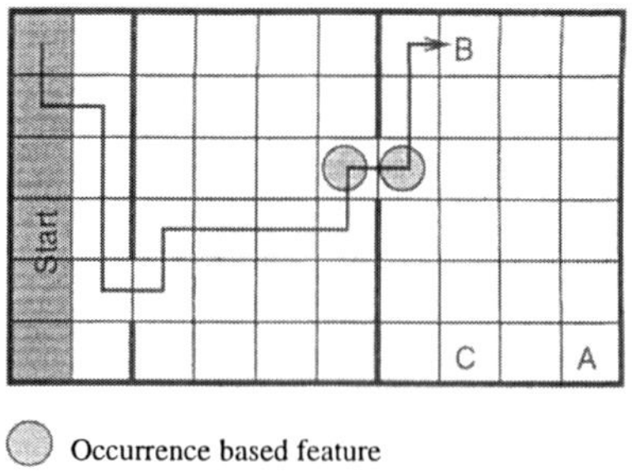

Figure 7 Animats's path for behavior Q^6 after new barrier.

Table 4 Behaviors, Features and Locations for the Non-Hierarchical Control System.

Feature	Behavior	Location D_0	Location D_1	Discovery Time	Comments
NA	Q^0	NA	NA	NA	Primitive action
NA	Q^1	NA	NA	NA	Primitive action
NA	Q^2	NA	NA	NA	Primitive action
NA	Q^3	NA	NA	NA	Primitive action
NA	Q^4	NA	NA	NA	Primitive action
f_0	Q^5	9	0	≈ 50	First goal (A)
f_1	Q^6	7	5	≈ 50	Second feature (B)
f_2	Q^7	7	0	≈ 150	Third feature (C)

Depth	Behavior	D_0	D_1	Comments
1	$- > Q^6$	0	5	Start (Go to B)
2	$-- > Q^8$	0	5	Go to door exit
3	$--- > Q^1$	0	5	Primitive action
3	$--- > Q^1$	0	4	"
3	$--- > Q^1$	1	4	"
3	$--- > Q^1$	1	3	"
3	$--- > Q^1$	1	2	"
3	$--- > Q^7$	1	1	Go to door entrance
4	$---- > Q^3$	1	1	"
4	$---- > Q^0$	2	1	"
4	$---- > Q^3$	2	2	"
4	$---- > Q^3$	3	2	"
4	$---- > Q^3$	4	2	"
4	$---- > Q^0$	5	2	At door entrance
3	$--- > Q^3$	5	3	At door exit
2	$-- > Q^0$	6	3	Primitive action
2	$-- > Q^3$	6	4	"
2	$-- > Q^3$	6	5	At goal (B)

Table 5 Behavior Q^6 after the new barrier. See Figure 7 for path. Note the arrow length represents the depth into the structure.

many tasks over its lifetime that is argued to be worth the poorer performance at learning a single, isolated task. To compare the NQL techniques of emergent hierarchical control with non-hierarchical control, an identical set of tasks was learned by a non-hierarchical control system. The non-hierarchical control system is presented with the same animat/grid world/barrier/multi-task scenario as was described and presented to the hierarchical control system in the previous sections.

The non-hierarchical control system had identical learning parameters to the hierarchical system, except that it was not allowed to generate hierarchies. In fact, the identical simulation code was used, with the hierarchy generating capabilities disabled. The non-hierarchical control system was presented with an identical sequence of events. That is, initially only two goals were detectable, then, later, a third goal and then a new barricade were introduced. The performance of all these tasks, for both the non-hierarchical and hierarchical control systems are shown in Figure 8. It was seen that both hierarchical and non-hierarchical control systems performed poorly at learning their first tasks, namely Q^5 and Q^6. However, at a later time when a new task was introduced, the non-hierarchical system had to learn it from scratch whereas the hierarchical system could use previously learned behaviors whenever beneficial. Similarly, whenever changes occurred (e.g. the new barrier) the non-hierarchical control system had to adapt each behavior individually while in the hierarchical adaption was confined to the the elemental behaviors effected and minimized the effect on the other behaviors. The performance at the third task (locating goal C) for the hierarchical and non-hierarchical control systems is shown in Figure 9(a) and (b) respectively. Note that the third task in the hierarchical control system is Q^9 and in the non-hierarchical control system it is Q^7. This encapsulation of control strategies into behaviors allowed the hierarchical control system to clearly out-perform the non-hierarchical control system, especially when new tasks were introduced and changes in the world occurred.

To further quantify this improvement, a measure of the overall performance was introduced. This measure, is the usage weighted average performance at any given time of all the behaviors that currently existed within both the non-hierarchical and hierarchical control systems and is plotted in Figure 10(a) and (b), respectively. The rate of learning is similar on the first task, but as new tasks and changes are introduced, the non-hierarchical control system is clearly out-performed by the self-generating hierarchical control system.

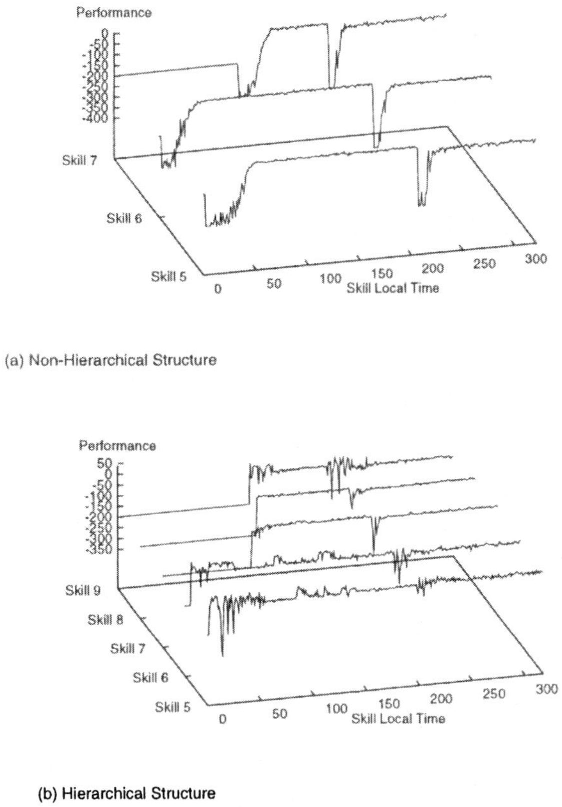

Figure 8 Comparison of Performance: (a) Non-Hierarchical (b) Hierarchical.

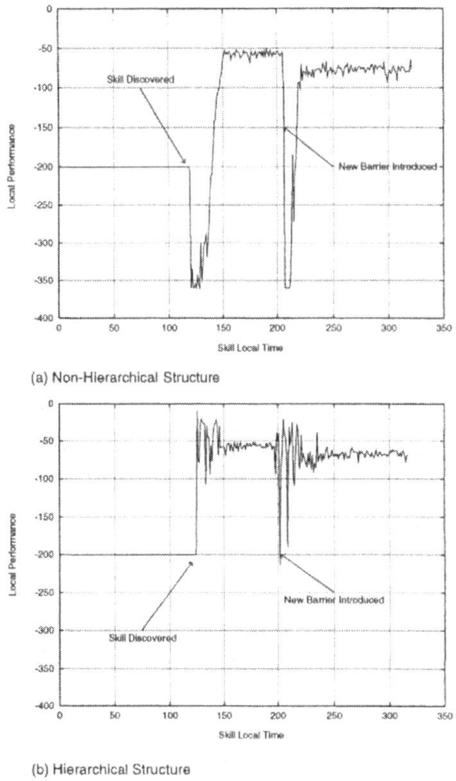

Figure 9
Performance of a new behavior for (a)non-hierarchical(Q^7) and (b)hierarchical (Q^9) control systems. Note that the time of discovery and barrier addition are indicated.

5. Discussion

In the preceding simulation studies two criteria were used to discover useful features. The squared external reinforcement signal gradient and the recurrence of states. The squared external reinforcement gradient criteria led to the discovery of goal locations within the state space. Although not explicitly simulated in this paper, had there existed areas of high negative reinforcement signals, these too would have been discovered as useful features. Although they would not be directly useful they would be indirectly practical, representing areas of the state space to be avoided (behaviors that are learned to **NOT** be active). The occurrence criteria led to the discovery of important locations of the state space that were used during task performance. These common locations became the point for decomposition of the tasks into smaller sub-tasks (or sub-behaviors).

By decomposing its behaviors into smaller sub-behaviors, the animat's control system demonstrated a capability to reuse existing knowledge and improve its performance at learning new but related tasks. This was evident when a new behavior (Q^9) was requested once the control system had learned the first two tasks (Q^5 and Q^6) and decomposed them into their elemental components (behaviors Q^7 and Q^8). As the animat began to learn behavior Q^9, it invoked behavior Q^7 to move the animat to the mouth of the barrier in one step. At this time behavior Q^7 was operating very well and this allowed a single command from the infant behavior Q^9 to move all the way from the barrier mouth. Behavior Q^5 was then invoked to bring the animat to within two steps of its goal. The invocation of Q^5 resulted in a somewhat inefficient overall path to the goal, but when one thinks about this, it is how animals and humans often behave. They often prefer to use established behaviors and routines at the expense of efficiency.

Another benefit of the decomposition of behaviors into a hierarchical structure is the resistance to change. High level behaviors and any new behaviors that are to be learned are ultimately composed of some number of lower level behaviors. When changes occur will often be confined to the lower level behaviors that are affected and the higher level behaviors will remain intact. In these simulations, when a new barrier was placed in between the start location and the initial barrier, most of the disruption was confined to the behaviors that moved the animat from the start to the initial barrier's mouth, namely behaviors Q^7 and Q^8. Some disturbance was shown in all behaviors, but this was to be expected because lower level behaviors during their period of recovery effected the higher level behaviors by briefly feeding back higher

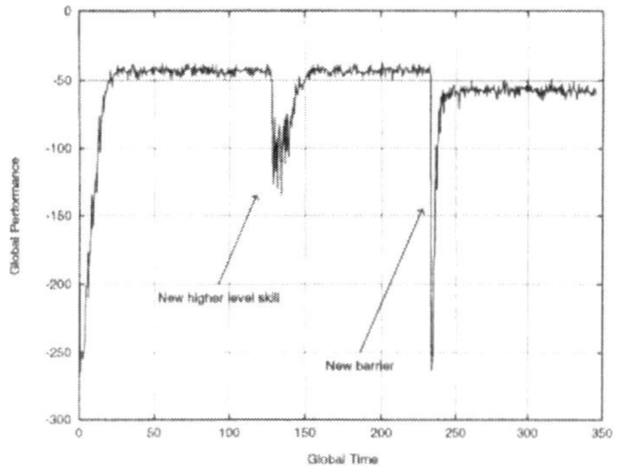

(a) Non-Hierarchical Structure

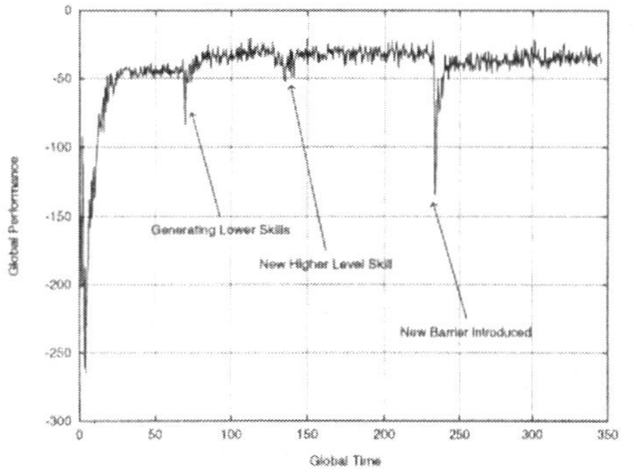

(b) Hierarchical Structure

Figure 10 Overall behavior performance for (a)non-hierarchical and (b) hierarchical control systems.

negative reinforcements. Once the brief disturbance settled, the lower level behavior had learned how to navigate through the additional barrier and the higher level behaviors had remained the same. If the simulation were to be continued, additional recurrence based features would be discovered near the doorway of the new barrier.

To compare the benefits of the self-generating hierarchical structure with non-hierarchical control systems, an identical animat was used in an identical situation. From the results, it is seen that the non-hierarchical structure was forced to learn everything from scratch whenever a new task was attempted. Also, when changes occurred, each behavior had to be adapted in its entirety, resulting in much poorer performance.

It is clear that for this example the self-generating hierarchical control structure of NQL can outperform non-hierarchical control structures. However, there is much more work that must be done before such systems can be used in real applications. Most notably, the technique described in this paper still uses discrete spaces, states, actions and behaviors and generalization within these states, actions and behaviors is not possible. Operation within continuous space and generalization methods for NQL are currently being pursued (Digney, 1998) . This should allow the operation of applications in more realistic situations with real sensors and actuators. Also, the results presented indicated that the animat was beginning to view its world at a rudimentary level of abstraction. This abstraction will be necessary for learning at higher levels where seeing the world with a high level of detail recognition would be unnecessary and impeding. Current work will further concentrate on techniques for generating such abstracted features in realistic sensory systems.

6. Conclusions

A Nested Q-learning technique was developed for generating hierarchical control structures for a simple animat. Unlike previous work, this technique employed some simple selective criteria that allowed only those features with a good probability of becoming useful to emerge. This resulted in faster learning of the hierarchical control structure. The structure itself was shown to represent the abstracted higher level behaviors and their decomposed elemental sub-behaviors. Decomposition resulted in information easily being transferred to new but related tasks and a higher resilience to change as changes were confined to the behavior level which they affected, leaving most other behaviors untouched. The self-generating hierarchical control system was shown to outperform a non-hierarchical control system whenever new tasks were added or when change occurred. Although not yet ready for realistic application, the steps necessary for advancement have been outlined.

References

Barto, A.G. Sutton, R. and Watkins, C. (1989). Learning and sequential decision making. In *COINS Technical Report.*

Dayan, P. and Hinton, G. (1993). Feudal reinforcement learning. In *Advances in Neural Information Processing Systems 5*, San Mateco, CA. Morgan Kaufman.

Digney, B. (1996). Emergent hierarchical control structures: Learning reactive/hierarchical relationships in reinforcement environments. In Materic, editor, *From animals to animats 4: SAB 96*, pages 363–373, Cape Cod, USA. MIT Press-Bradford Books.

Digney, B. (1998). Learning in continuous domains with delayed rewards. In *SAB 98 (submitted).*

Long-Ji, L. (1993). Hierarchical learning of robot skills. In *IEEE International Conference on Neural Networks*, pages 181–186, San Francisco. IEEE Press.

Maes, P. and Brooks, R. (1990). Learning to coordinate behaviors. In *Eighth National Conference on Artificial Intelligence*, pages 796–802.

Singh, P. (1992). Transfer of learning by composing solutions of elemental sequential tasks. *Machine Learning*, 8(3/4):323–339.

Thurn, S. (1992). Efficient exploration in reinforcement learning. In *Technical Report CMU-CS-92-102.* Carnegie Mellon University, School of Computer Science.

Tyrrel, T. (1992). The use of hierarchies for action selection. In *From animals to animats 2: SAB 92*, pages 138–148, Massachussets. MIT Press-Bradford Books.

Improving the DAC Architecture by Using Proprioceptive Sensors

Ralf Salomon
AI Lab, Department of Computer Science, University of Zurich
Winterthurerstrasse 190, 8057 Zurich, Switzerland
salomon@ifi.unizh.ch

Abstract

Previous research has proposed the distributed adaptive control (DAC) architecture as a self-organizing controller for autonomous mobile robots. Despite its successes, practical experience has indicated several options for improvement. This paper proposes a new control architecture for mobile robots that can be seen as an extension of the DAC architecture. The control architecture consists of a controller and a value system. The controller uses the activity patterns of visual sensors to determine the motor commands, whereas the value system receives stimuli from *proprioceptive* sensors. This design decision is justified by the following arguments: (1) both components are significantly decoupled by using different sensory modalities; (2) the feedback of proprioceptive sensory patterns is omnipresent in biological systems and has been widely neglected in control systems; and (3) proprioceptive sensors operate more reliably *and* can be used more efficiently than visual sensors, such as pixels in a CCD camera. Practical experiments with the Khepera robot show that by using proprioceptive sensors, the control architecture considerably better adapts to changing environments *and* that it yields very robust behavior with respect to sensor failures.

1 Introduction

One goal of research on autonomous agents is to build robots that are able to behave without human control in dynamically changing, partially unknown environments. Research on evolutionary robotics (Cliff *et al.*, 1992; Floreano and Mondada, 1996a, 1996b; Harvey *et al.*, 1996; Nolfi and Parisi, 1995; Salomon, 1996), for example, employs different evolutionary algorithms (Baeck and Schwefel, 1993), such as genetic algorithms (Goldberg, 1989) or evolution strategies (Rechenberg, 1973; Schwefel, 1995), to evolve complete control architectures. These systems, however, are mostly limited to adaptation on an evolutionary time scale, since most of the evolved controllers do not feature any learning or self-organizing processes. Exceptions are controllers presented in Floreano and Mondada (1996a) and (Nolfi and Parisi (1995), which feature learning to a limited extend.

For the goal of achieving robust behavior in dynamically changing environments, the employment of self-organizing processes is another, very promising option. The distributed adaptive control (DAC) architecture (Verschure *et al.*, 1992) is a good example of such efforts. Section 2 reviews the DAC architecture and discusses some of its limitations. Even though the DAC architecture proposes a powerful self-organization scheme, real world experiments indicate several options for improvement, especially for mobile robots that behave in the real world rather than in simulation.

Existing neuronal control architectures, e.g., DAC, employ learning and self-organization for only those connections that connect to visual sensors. Surprisingly little attention has been devoted to feeding back proprioceptive activity patterns. Proprioceptive patterns refer to stimuli arising within an organism. Section 3 proposes a new control architecture, called self-organizing-through-propriocetion (STP), that can be seen as an extension of the DAC architecture. The STP architecture uses proprioceptive data in order to better evaluate chosen actions with respect to visual data mediated by the environment, i.e., the controller uses different modalities. Section 4 summarizes the test environments as well as all parameter settings. All experiments were done with the Khepera robot. Khepera's body is 32 mm high and 55 mm in diameter, and it is equipped with eight infrared and eight ambient light sensors, as well as two motors. The robot and the approximate location of the sensors can be seen in Figure 1. It should be noted that Khepera's sensor readings are very noisy. A detailed descrip-

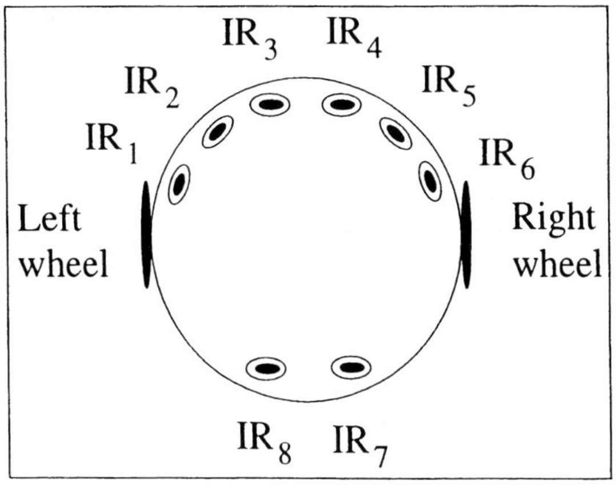

Figure 1 The Khepera robot and the approximate location of the infrared sensors $IR_1 \dots IR_8$.

tion of the robot and its electrical parts can be found in the *Khepera Users Manual*. Section 5 presents some results obtained with the proposed control architecture, and finally, Section 6 concludes with a detailed discussion.

2 Background

The background of this paper is mainly given by the DAC architecture, which is described in detail in Subsection 2.1. The DAC architecture features self-organization and life-long learning without distinguishing between a learning and a performance phase. Due to its design, however, the DAC architecture exhibits a few problems that are discussed in Subsection 2.2.

2.1 The DAC Architecture

The distributed adaptive control (DAC) architecture has already been proposed in Verschure *et al.* (1992). As can be seen from Figure 2, the DAC architecture consists of four unit layers and the four connection layers ML_1, ML_2, AV and AP. The motor layer consists of several units that execute certain reflexes, such as move forward, turn right, back up, etc. The command units are activated in a winner-take-all manner, where the move-forward-unit is activated by default. The robot's action space is completely defined by the motor command layer, since no other output unit is present in the architecture.

Each command unit can be triggered by a unit of the avoidance or approach layer. These two layers are connected to the motor units via hard-wired ML_1 and ML_2 connections. The approach layer receives its input data from a target detector (a nose), and tries to steer the robot toward a target. If, for example, the robot senses a food source on the right-hand-side, the approach layer triggers a turn-right reflex.

Units in the avoidance layer are triggered by a collision detector. If, for example, a collision occurs on the left-hand-side, the avoidance layer might trigger a command unit that performs a back-up-and-turn-right reflex. In addition, the avoidance layer inhibits the approach layer in case of a collision. This form of a dynamic suppression of another module has been proposed in Brooks (1986).

The combination consisting of the the collision and target detectors, the avoidance and approach layers, as well as the motor command layer provides the basic control system. This subsystem is sufficient to move the robot around and find food while recognizing obstacles. The collision detector and the target detector together provide a *value system*, since these detectors provide feedback about the quality of the robot's behavior.

The avoidance layer and the approach layer are both connected to a sensory layer via fully connected, modifiable connections located in the AV and AP layer, respectively. The sensory layer receives its input from a range finder, which can be implemented by infrared proximity or ambient light sensors. The activation of a unit in the avoidance or approach layer not only triggers a specific command unit but also triggers the execution of a learning step.

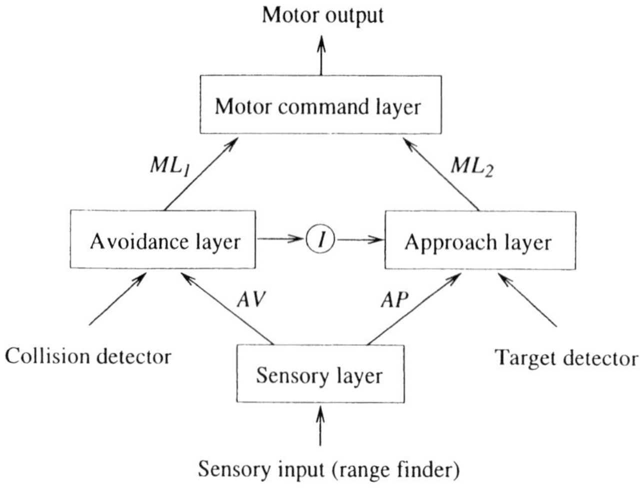

Figure 2 The distributed adaptive control (DAC) architecture.

In mathematical terms, the system is defined as follows. The activation of an approach or avoidance unit s_i^λ is given by

$$s_i^\lambda = f\left(c_i^\lambda + \sum_{j=1}^{N} s_j K_{ij}^\lambda\right) , \qquad (1)$$

where λ denotes either the approach or the avoidance layer, c_i^λ is the input from the corresponding target or collision detector, s_j is the activation of the range-finder unit j, K_{ij}^λ represent the weights of the AV and AP connections, and $f(\cdot)$ is a threshold function. In a learning step, the weights K_{ij}^λ are updated as follows

$$K_{ij}^\lambda \leftarrow K_{ij}^\lambda + \frac{1}{N}\left(\eta^\lambda s_i^\lambda s_j - \epsilon \bar{s}^\lambda K_{ij}^\lambda\right) , \qquad (2)$$

where N is the number of the range-finder units, η^λ is the learning rate, ϵ is the decay rate, and $\bar{s}^\lambda$ is the average activation of layer λ. The learning rule (2) consists of a Hebbian term (η^λ) and an active forgetting term (ϵ).

With the pre-programmed part of the DAC architecture, the robot can move around. In case it experiences a collision or locates a food source, the corresponding detectors initiate the execution of an appropriate reflex and a learning step. Over time, the DAC architecture learns to associate range finder data with the collision and target detector. Once the connections K_{ij}^λ have sufficient strength, the range finder can trigger units in the avoidance layer initiating certain reflexes executed by the corresponding command units before any unit in the collision or approach layer is activated. An extensive analysis of many successful experiments done in simulation can be found in Verschure *et al.* (1992).

2.2 Limitations of the DAC Architecture

Many successful DAC simulation experiments are reported in Verschure *et al.* (1992). These simulations reveal, however, that a fair amount of the DAC architecture has to be determined by the designer. As is evident from equations (1) and (2), learning takes place only in layers AV and AP, and thus, self-organization is limited to these two layers. In addition, several parameters, such as ϵ, the two learning rates η^λ, and various thresholds for $f(\cdot)$ have to be specified. In the simulation experiments presented in Verschure *et al.* (1992), these parameters have been varied considerably, indicating that they have a significant influence on the robot's behavior.

In addition to simulation, the DAC architecture has also been implemented[1] on the Khepera robot. As expected from simulation, the control architecture works fine for environments with a low density of obstacles. As in the simulation (Verschure *et al.*, 1992), Khepera does not have any collision sensor, and therefore, the collision sensors are derived from the infrared proximity sensors. After normalization, each proximity sensor provides values in the range [0..1], and a value of 1 is used to indicate a collision. More formally, the collision detector is derived as follows:

$$c_i^\lambda = g(s_j, \theta_c) \; , \qquad (3)$$

where $g(\cdot)$ denotes a threshold function with threshold θ_c. Since the implementation uses Khepera's six frontal sensors, the collision detector consists of six units, which are all hard-wired to appropriate command units.

Since both the infrared and the collision detectors are highly correlated, they actually represent only one modality and learning becomes very easy. It is reported in Verschure *et al.* (1992) that the robot did not experience any further collision after 600 learning steps have been performed. For a collision free behavior, it is sufficient to set all K_{ii}^λ connections above function threshold. In other words, setting $K_{ii} > \theta_c$ in equation (3) leads to an avoidance behavior without activating any collision unit.

The high correlation between both detectors can also be observed in the AV layer after a sufficiently large number of learning steps. Each learning step predominantly increases the connections K_{ii}^λ that connect corresponding collision and infrared sensors. Due to real-world correlations, however, each learning step also reinforces connections to neighboring sensors. Essentially, when setting all parameters appropriately, the robot learns to move around without hitting any obstacle by establishing a matrix $\boldsymbol{K}$ with strong diagonal elements and some significant connections around the diagonal.

The collision detector is part of the value system, and thus plays an important role in the entire system. One problem is that the infrared sensors send very imprecise data. For a constant agent-object distance, the actual sensor reading depends on the ambient light, the material the sensed object is made of, the sensor's technical characteristic, and additional noise. Hence the value system is very sensitive with respect to these sensors.

Furthermore, this sensitivity requires a very high reliability of the collision sensors. If deriving many collision detectors from a high-resolution infrared sensor or a CCD camera, these sensors must be very reliable, which is very unlikely in the real world. For example, if the reliability of each pixel of a 512×512 CCD camera is "only" $p = 0.999999$, DAC constantly assumes a collision with a probability $p = 1 - 0.999999^{512 \times 512} = 0.23$. It should be noted that a reliability of $p = 0.999999$ is hard to meet in real-world environments. For instance, if the robot is working in a sewerage system, it is indeed very likely that the camera is exposed to some dirt and that it is consequently very likely that some of the pixels will be sending erroneous information.

In addition, the derivation of the collision sensors from the proximity sensors has the effect that the DAC architecture *assumes* a collision if a proximity sensor exceeds a certain threshold. The control architecture cannot *determine* whether or not a collision has occurred; the decision can be false in both ways, i.e., a collision and a non-collision can be mistakenly assumed.

By deriving the collision sensors from another sensory input device, the DAC architecture implements a significant amount of system-environment assumptions, which are not sufficiently grounded in the sensory-motor couplings; the environment *indirectly* feeds back the motor activities to the control architecture. It can happen that the control architecture assumes a collision even though the robot has sufficient distance to any obstacle, or it might happen that the range finder provides sensor readings below collision threshold, even though the robot is blocked by an obstacle. The first problem can be observed in the following experiment.

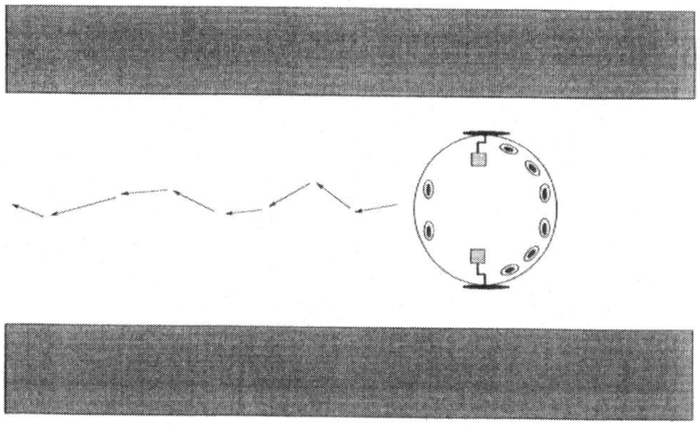

Figure 3 In a narrow aisle, the DAC architecture constantly backs up a robot.

In Figure 3, the robot is located in a small aisle. Due

[1] The DAC implementation for the Khepera robot can be obtained by sending an email to lambri@ifi.unizh.ch.

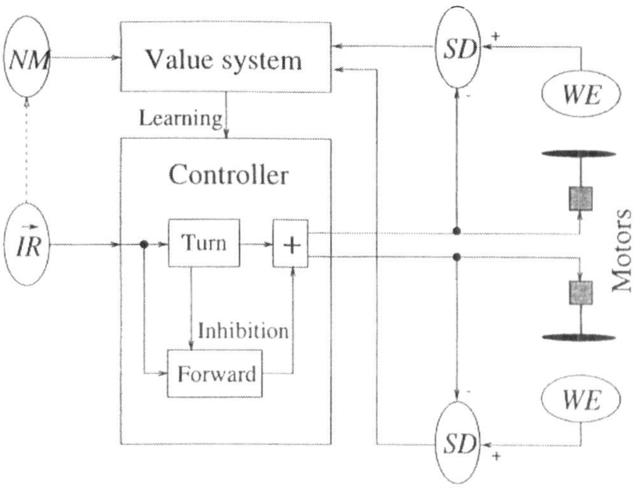

Figure 4 The self-organization-through-proprioception (STP) architecture. For details, see text.

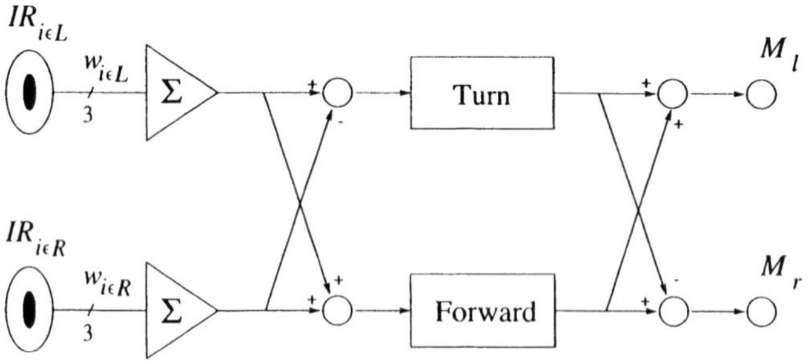

Figure 5 The implementation of the STP controller module. For details, see text.

to technical characteristics, the DAC architecture (constantly) receives sensor readings indicating a collision on either lateral side, even though the robot does not touch a wall. This effect does not occur in simulation, since the range finder provides precise data, and thus reliable information about collisions. Since the DAC architecture constantly assumes a collision in this narrow aisle, it constantly triggers certain reflexes, which constantly moves the robot backwards.

The reason for the robot behavior presented in Figure 3 is that too many assumptions were used in the design process of the control architecture. If a robot is to exhibit robust behavior in a dynamically changing, partially unknown environment, the control architecture should not assume specific sensor values for particular situations. By contrast, the control architecture must be self-adaptive and should tune as many parameters as possible by appropriate mechanisms. Furthermore, it can happen that a real-world sensor fails. A collision sensor, for example, can always send a value of 0 or 1. A robust control architecture should be capable of dynamically dealing with such a situation.

Besides the limitations discussed in this subsection, the DAC architecture offers a very valuable approach to self-organization. The self-organizing process utilizes a value system in order to correlate different sensors for choosing different actions. The following section, uses some of the ideas introduced in the DAC architecture to propose a new self-organizing control architecture that can be seen as an extension of the DAC architecture.

3 The STP architecture

This section describes the self-organization-through-proprioception (STP) architecture in full detail. It is helpful to first describe the STP architecture on a functional level and to then consider implementation details.

As can be seen from Figure 4, the STP architecture consists of a controller, a value system, and several sensors. The controller is inspired by the subsumption architecture (Brooks, 1986) and consists of two modules, which are often called processes. Each module determines its individual contribution to the motor activation by using the activity of the infrared proximity sensors. The contributions are then combined and sent to the motors. The "turn" module can inhibit the activity of the "forward" module.

The value system uses *proprioceptive* data originating from the motor side, which is a different sensory modality than is used by the actual controller. At the motor side, speed difference *SD* sensors calculate the difference between the desired and the actual speed, which is measured by attached wheel encoders *WE*. Depending on the *SD* sensor readings, the value system decides whether or not a collision has occurred. The idea is that in case of a collision, the surface-to-wheel friction significantly slows down the wheels. The effect, however, depends on the surface, the wheels, and the electrical characteristics of the motors. When using Khepera on a wooden surface, for example, the wheels are mostly stopped by the environment, but sometimes, they are spinning. To deal with this problem, the value system uses an additional non-moving sensor *NM*, which is described below.

Let us now discuss the STP architecture on the implementation level, which is shown in Figure 5. The implementation seems different from the functional description as presented in Figure 4, since both modules use the same connections **w**. The number of parameters is kept small as to avoid tuning by too many degrees of freedom. As can be seen from Figure 5, the weighted sum $IR_i w_i$ is separately calculated for the left-hand-side $L = \{1, 2, 3\}$ and the right-hand-side $R = \{4, 5, 6\}$. The sum of both values is fed into the forward unit, whereas the difference is used by the turn unit. Each unit determines its contribution, and the sum and difference are fed to the left and right motor, respectively. The activation of the forward unit s_f is calculated as

$$s_f = 1 - 1.2 \tanh\left(\sum_{i=1}^{6} IR_i w_i\right), \qquad (4)$$

which yields values in the range $s_f \in [-0.2, 2.2]$. That is, the forward unit yields any speed between fast forward and slow backwards. The activation of the turn unit s_t is given by

$$s_t = \tanh\left(\sum_{i=1}^{3} IR_i w_i - \sum_{i=4}^{6} IR_i w_i\right) . \quad (5)$$

The left and right motor activations M_l, M_r are finally

$$M_l = s_f + s_t,\ M_r = s_f - s_t . \quad (6)$$

It is easy to see that both motor activations are in the range $M_l, M_r \in [-1.2, 3.2]$.

As already mentioned, the value system uses an additional non-moving sensor NM. This sensor is derived from the infrared sensors in the following way. For each frontal sensor, the difference between the current and previous time step is maintained $\Delta IR_i^t = IR_i^t - IR_i^{t-1}$, where the superscript indicates the time. The vector length $nm = \sqrt{\sum_i (\Delta IR_i^t)^2}$ is then used to update the sensor in a set-to-peak, decay fashion:

$$NM \leftarrow \begin{cases} nm & \text{if } nm \geq NM \\ \tau_{nm} NM & \text{if } nm < NM \end{cases} \quad (7)$$

Essentially, the NM sensor is a *flow* sensor as is omnipresent in nature. This sensor monitors the amount by which the activation of the infrared sensors are changing, since high changes indicate movements. The NM sensor value decays with the time constant τ_{nm}, and is set to the current change value, if the current change value exceeds the current sensor value. If the NM sensor value falls below a certain threshold, the value system assumes that the robot is not moving. As is shown in Section 5, the use of the NM sensor improves the robustness in cases where the SD sensors fail. The implementation uses the NM sensor, since Khepera does not feature other motion sensors, such as a speed measure device/wheel.

Prior to each calculation of the motor activations, the connection weights w_i are updated depending on whether or not the value system indicates a collision. In case of a collision, the weights are increased according to

$$w_i \leftarrow w_i + \eta IR_i , \quad (8)$$

where η is the learning rate. In case of a non-collision, the connection strength is reduced by a decay factor ϵ

$$w_i \leftarrow \epsilon w_i + w_i(1 - \epsilon)(1 - IR_i) . \quad (9)$$

The motivation of the learning rules is as follows. In case of a collision, learning rule (8) increases each weight w_i proportional to the activity of its corresponding sensor IR_i. This increase leads to a stronger inhibition of the forward unit s_f and a stronger excitation of the turn unit s_t. Depending on the obstacle, the ambient light, the robot's morphology, etc., the robot is eventually able to move away from the obstacle. In the non-collision case, learning rule (9) decreases each connection strength in order to avoid over generalization. This decrease depends on the connection strength w_i *and* the sensor reading IR_i. If sensor i does not sense any obstacle, i.e., $IR_i \approx 0$, the weight w_i remains unchanged. If sensing an object $IR_i \gg 0$, however, the weight is decreased, since the high sensor reading does not lead to a collision. It should be noted that by applying learning rule (9), the connection strength w_i cannot increase, since $0 \leq IR_i \leq 1$. The intention of both learning rules (8) and (9) is to project the collision-sensor correlation onto the weights $\mathbf{w}$.

Even though the STP architecture is inspired by the DAC architecture, significant differences can be identified. First, the STP architecture does not feature an output layer, in which each unit represents a fixed, predefined reflex. Rather, different STP modules/processes generate continuous activations, which are combined at the motor side. Second, both the STP controller and the STP value system use different sensory modalities; the value system makes its decision upon proprioceptive activity patterns. This design decision is justified by the assumption that such proprioceptive sensors are *more reliable* than visual sensors, such as a CCD camera or infrared proximity sensors. Third, since always one of the learning rules is applied in each step, the weights continuously change, which increases the chance of chosing appropriate actions in erroneous situations, such as sensor faults. The main goal of the STP architecture is to yield adaptive and robust behavior, and Section 5 shows to which extend this goal has been achieved.

4 Methods

The proposed STP architecture was implemented on the Khepera robot and was tested in different environments. The first environment is an arena that is shown in Figure 6. The arena is of size 60x45 cm and the walls are made from wood with a height of 3 cm. This arena has narrow corridors at many locations. As already explained in Figure 3, the DAC architecture does not exhibit any useful behavior in this environment, since these corridors trigger backup-and-turn reflexs most of the time.

The second environment consists of an arena of size 100x150 cm with white walls of height 30 cm. Here, 10 obstacles of size 10x10 cm were randomly distributed.

Unless otherwise stated, the following parameter settings were used. The learning rate and decay factor were set to $\eta = 0.1$ and $\epsilon = 0.95$. All weights were initially set to $w_{1 \leq i \leq 6} = 0.5$. Prior to their usage, all infrared sensor readings, which are in the range $[0, 1023]$, are normalized by the interface such that they are in the range $[0, 1]$.

Since Khepera's speed values are valid in the range $[-40, 40]$, the interface maintains a speed scale factor s_s, by which the motor activities M_l and M_r are multiplied prior to setting the motor speeds. In all experiments,

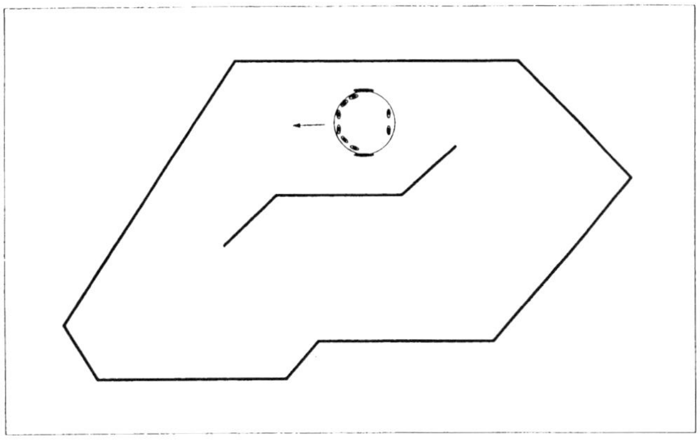

Figure 6 The first test environment.

the speed scale factor was set to $s_s = 3.0$. Similarly, the incoming wheel encoder values are divided by the same factor.

Since Khepera's speed values are limited to integer values, a floating-point-to-integer conversion is required. In order to avoid artifacts and round-off errors, the interface adds [0,1]-equal-distributed random numbers to the motor speeds at each time step. By this means, a controller output of, for example, 0.5 is converted into a "0-1-..." sequence. Without this special conversion, a continuous control output would be mapped onto a step function, which can cause further problems.

The value system assumes a collision, if one of the speed difference sensors SD reports a value greater than 0.8. The non-moving sensor uses a decay factor $\tau_{nm} = 0.95$. The value system also assumes a collision, if the NM sensor yields a value smaller than 0.15 *and* if the averaged infrared sensor activity is above 0.2.

In the presence of noise, it can be very useful to low pass input signals. A low-pass filter can be implemented by $x \leftarrow x(1-t_c)+t_cI$, where x is the low-passed signal, t_c is a time constant[2], and I is the input signal.

In the experiments reported in this paper, only the speed difference sensors SD were low-pass filtered with a time constant $t_{sd} = 0.7$. In the presence of ambient light produced by electrical bulbs, low-pass filtering of the infrared sensors would be of particular interest. However, the 100 Hz noise cannot be appropriately filtered, since sensors can be read only once within a 100 millisecond time interval. If low-pass filtering of the infrared sensors were applied, the controller would become to slow for yielding reasonable behavior. Therefore, low-pass filtering of the infrared signals was not considered.

5 Results

The STP architecture was tested in different environments under various environmental conditions, and exhibited the expected adaptive behavior. This section discusses some of these experiments in full detail and explains some of its adaptive behavior properties.

The first test environment is an arena of size 60x45 cm as shown in Figure 6. This arena has several small aisles in which the DAC architecture does not exhibit reasonable behavior, since the robot constantly backs up as already discussed in Section 2.2. In this arena, the STP architecture has no problem to move around. Normally, it moves straight ahead and avoids obstacles. It sometimes collides with a wall, and after some steps, moves away continuing with moving around. In all runs, the robot never got stuck. Collisions occur infrequently, with each collision taking approximately 8-10 steps.

Figure 7 shows how the left and right weights $w_{1\le i\le 3}$, $w_{4\le i\le 6}$ develop over 1000 time steps, which are arbitrarily chosen from a long run. Figure 7 clearly shows that the value system recognizes a collision at around time step 120, 260, 390, 770, and 890. It is obvious that a collision causes a significant increase of those weights w_i that receive high activity from their associated infrared sensors IR_i (see also learning rule (8)). Between two subsequent collisions, each weight is slowly reduced proportional to the activity of its associated infrared sensor.

From Figure 7 several further observations can be made. First, the decrease resembles a $\exp(-cx)$ function, which can be expected from learning rule (9). Second, horizontal line segments indicate that in these time steps, the associated infrared sensor had no activity. Third, most of the time, the lateral connections w_1 and w_6 have the smallest values. Mostly, these two weights are quickly decreased, since the high activity of a lateral sensor is not correlated with a collision. For example, while following a wall, the lateral sensor has a high activity close to the value 1, and the weight is decreased, since the value system triggers learning rule (9). Conversely, the frontal weights are slowly decreased, since the frontal sensors send much less activity. Forth, Figure 7 shows that the robot experienced more collisions on the left-hand-side, which indicates that in this particular time series just by chance, the robot was mainly circling around clockwise.

The next example consists of the same experimental setup, including all parameter settings, but with significant ambient light produced by electrical light bulbs. As is explained in Section 4, ambient light significantly modifies the sensor readings for an unchanged agent-object distance. Different sensor readings obviously lead to a modified control characteristic given by equations (4) and (5). Figure 8 shows that consequently, the STP architecture evolves different connection strengths.

The way how the STP architecture adapts to changed sensor readings can be seen from Figure 9, which shows the connection strength averaged over the shown 1000 time steps of all six weights in a dark environment (left bars) and under ambient illumination (right bars). From

[2] Actually, t_c represents the reciprocal of a time constant, i.e., t_c is the first derivative of $dx(t)/dt$ of a continuous signal.

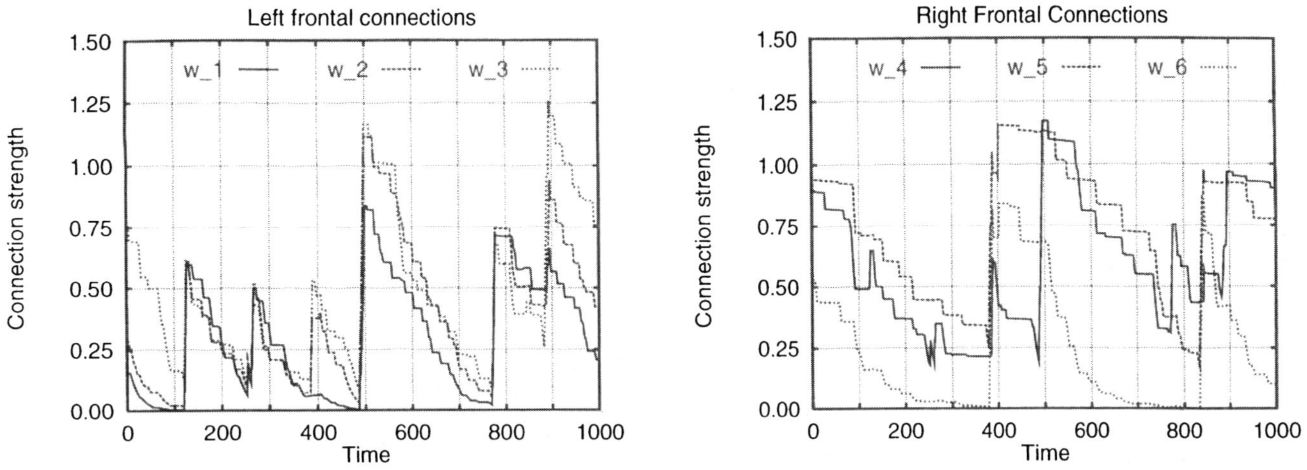

Figure 7 The evolution of the weights $w_{1\leq i\leq 6}$ in the first test environment *without* any ambient light.

Figure 9, it can be seen that ambient illumination leads to a higher average connection strength for the frontal sensors and a lower average connection strength for the lateral sensors. In a dark environment, the infrared sensor readings are generally lower. A lower sensor reading would lead to a smaller distance to obstacles, which in turn, increases the probability of a collision. Consequently, the connections remain at higher values, since learning is triggered earlier. The frontal connections are only slightly affected, since obstacles do no appear that often in front of the agent. It should be noted that the observer can hardly distinguish the behavior under different illumination.

In the second environment, the STP architecture easily controls the robot. Most of the time, all infrared sensors yield very low activation that is on the noise level. Consequently, the weights w_i remain constant over longer periods of time. In other words, the much smaller density of obstacles is represented in higher average values of the connections.

The last experiment studies the behavior of the STP architecture when the surface-wheel friction is not high enough to sufficiently slow down the wheels in case of a collision. This situation is depicted in Figure 10, which plots the activity of the non-moving sensor *NM*, and the two speed contributions s_f and s_t. At time step 8, the robot collides with an obstacle. In the presented situation, the *SD* sensors do not have significant activity, since the wheels are freely spinning. The robot, however, is not moving, and consequently, the infrared sensors send constant values. Therefore, the activity of the *NM* sensor falls below threshold, and at time step 44, the value system recognizes the collision. The STP architecture then behaves as already discussed.

A significant amount of experimentation was devoted to sensor changes/failures. To this end, the sensor readings of selected sensors were artificially multiplied by a constant $c \in [0.1, 10]$ or were even set to the constant values 0 or 1. In all cases, the STP architecture exhibited reasonable behavior. When setting a sensor to the constant value 1, the STP architecture had the most problems. In this case, the robot was spinning after a collision for a few steps until "normal" behavior could be observed. In all other cases, the observable behavior was hardly changed.

Finally, this section discusses some parameter settings mentioned in Section 4. The value system recognizes a collision, if the *SD* sensors values exceed a threshold normally set to the value 0.8. This parameter setting is not critical. When using a value greater than 0.8, the value system requires some more steps to recognize a collision. This extra amount of time is required to further weaken the forward unit's inhibition (equation (4)), which is due to the weight decay caused by learning rule (9). Similarly, changing the values for the learning rate η and decay factor ϵ leads to slight changes in the observable behavior. A decay factor ϵ closer to 1, would stretch the time between two subsequent collisions. However, a smaller decay factor allows for faster adaptation to new environments.

6 Discussion

This paper has proposed the self-organizing-through-proprioception (STP) architecture that is significantly inspired by the DAC architecture already presented in Verschure *et al.* (1992). From the DAC architecture, the STP architecture adopts the general idea of using a value system for training the actual controller. The other components have been added or modified as follows. The controller of the STP architecture employs different modules that send continuous values to the motors, rather than employing a set of (action) reflexes that are triggered in an on/off mode. Furthermore, the value system has been decoupled from the controller by using different sensory modalities; the value system uses *proprioceptive* activity

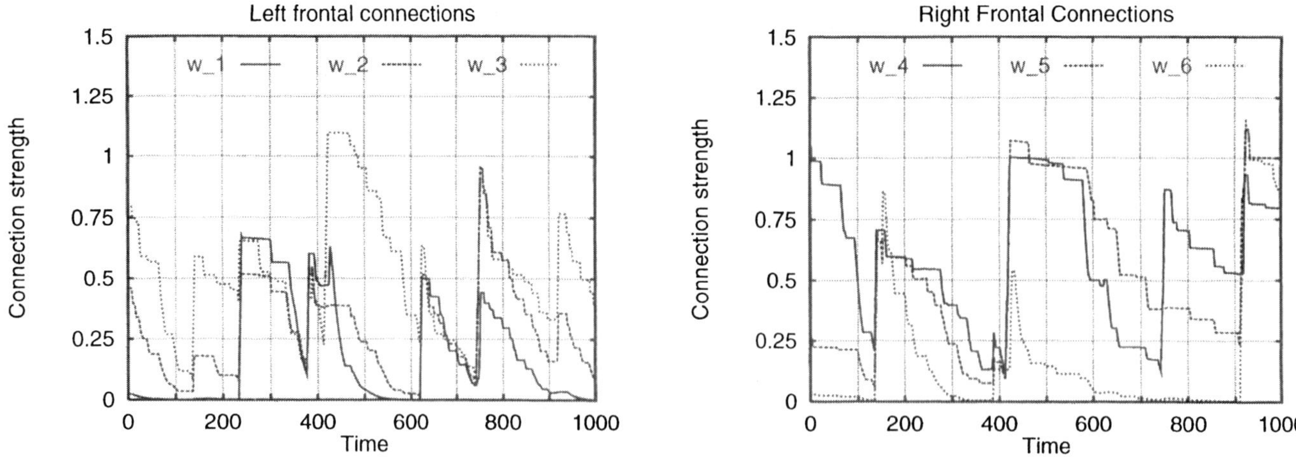

Figure 8 The evolution of the weights $w_{1\leq i\leq 6}$ in the first test environment *with* ambient light.

patterns originating from the motor side. By employing *additional* proprioceptive feedback, the STP architecture is able to monitor and evaluate its own controller. In other architectures, such as DAC, the effect of the control (motor) actions can be observed only indirectly through the environment and visual sensors. The experimental results presented in this paper have shown that the STP architecture adapts to different environmental conditions, and that it exhibits very robust behavior with respect to sensor changes/failures.

In order to maintain focus on the general ideas, the implementation of the STP architecture was not tailored for the Khepera robot. The first example that concerns this point is given by Khepera's on-board PID controllers, which help the designer to operate the robot in a forward mode, since the PID controllers (try to) regulate the motors such that their speeds are as close as possible to the desired speed settings. In other words, the designer can decide, which motor speed might be appropriate for a given situation, and the desired speed is probably maintained by the on-board PID controllers. This *forward* strategy, however, hides a large amount of information, which makes it harder to detect a collision and impossible to detect whether the robot moves uphill or downhill. Further control architectures to be developed by future research might implement these PID controllers themselves in order to make additional sources of feedback signals available. These additional feedback signals would allow for tuning further system characteristics that depend on the agent-environment interaction.

As a second example, the speed difference *SD* sensors could have been tailored for Khepera. It is theoretically conceivable that during a collision, the weights might be increased such that the robot turns and tries to move backwards into the obstacle. In this situation, the *SD* sensors would still indicated a collision, and the robot would be trapped. This problem can be overcome by tak-

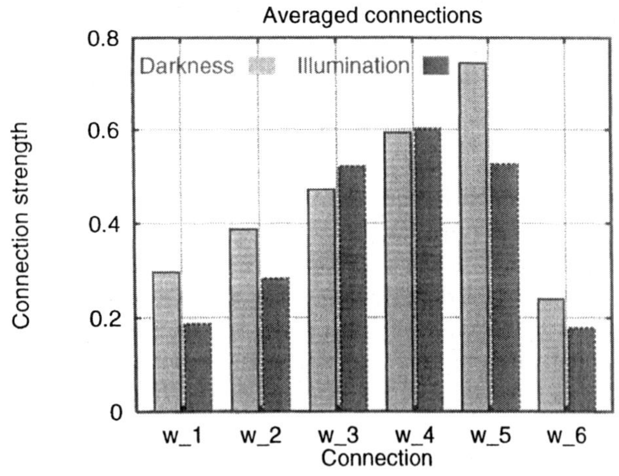

Figure 9 Connection weights averaged over 1000 time steps in a dark (light grey bars) and an illuminated (dark grey bars) environments.

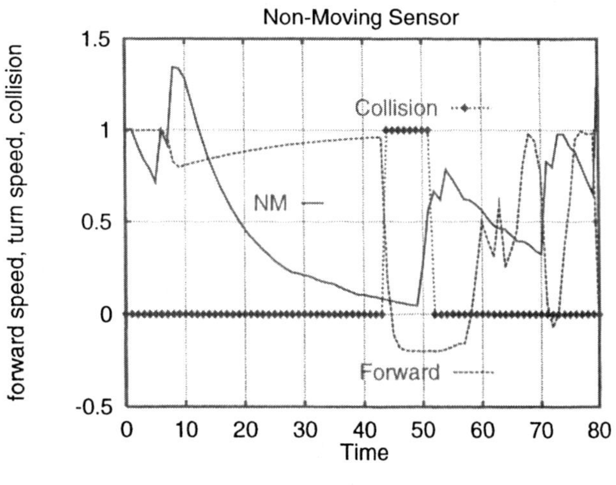

Figure 10 Recognizing a collision by the non-moving sensor *NM*.

ing the *SD* sensors into account only if the desired speed is forward directed. In the first place, this problem arises from the robot, and was therefore not considered in the previous sections, since it can be easily fixed as already outlined.

The stability of the integrated agent-control system is of major concern. For this goal, some important time constraints have to be considered. For stability of such feedback control systems, the time constants of outer loops should be higher than those of inner loops. For example, the PID controllers should be slower than the actual motors, and the *SD* sensor readings should be low-passed such that they change slower than the motor speeds; otherwise a temporarily speed lag could be considered a collision. When adding further control loops to self-adapt further parameters, such as the learning and decay rates η and ϵ, the possible changes should be much smaller than the average weight modifications. Otherwise instability could emerge due to oscillations. A possible option for self-adaptation of η and ϵ could be the definition of an additional need or an additional performance measure, such as the overall energy consumption of the robot.

Further research will be dedicated to improve the value system. Besides adding some of the features discussed above, the main focus will be on the learning rules. The weakest point of the value system is that it triggers *either* learning rule (8) *or* (9). It seems desirable to develop a value system that activates each learning rule to a certain degree. As a result, all learning rules would establish a dynamic equilibrium, rather than "jumping" between discrete stages. It seems that in this way, many more learning components could be simultaneously considered, since the value system would not need to select a particular learning rule in a winner-take-all style. Another important topic for future research will be the integration of a CCD camera. Since the value system is (besides the non-moving sensor *NM*) entirely based on motor feedback, it can be expected that this integration does not cause further difficulties.

Acknowledgements

This work was supported in part by a Human Capital and Mobility fellowship of the European Union, grant number ERBCHBICT941266. Thanks to Christopher Lueg and Peter Eggenberger for helpful discussion. Special thanks are due to Dimitrios Lambrinos for providing the DAC implementation on Khepera.

References

Bäck, T. and Schwefel, H.-P. (1993). An Overview of Evolutionary Algorithms for Parameter Optimization. *Evolutionary Computation*, **1**(1):1-23.

Braitenberg, V. (1986). *VEHICLES, Experiments in synthetic psychology.* MIT Press, Cambridge, MA.

Brooks, R. (1986). A robust layered control system for a mobile robot. *IEEE Journal of Robotics and Automation*, **2**:14-23.

Cliff, D., Husbands, P., and Harvey, I. (1992). Evolving Visually Guided Robots. In Meyer, J.-A., Roitblat, H.L., and Wilson, S.W., editors, *From animals to animats 2*, pages 374-383. MIT Press, Bradford Books, Cambridge, MA.

Floreano, D. and Mondada, F. (1996a). Evolution of Homing Navigation in a Real Mobile Robot. *IEEE Transactions on Systems, Man, and Cybernetics-Part B*, **26**(3):396-407.

Floreano, D. and Mondada, F. (1996b). Evolution of Plastic Neurocontrollers for Situated Agents. In Maes, P., Mataric, M., Meyer, J.-A., Pollack, J., Wilson, S.W., editors, *From Animals to Animats 4: Proceedings of the Fourth International Conference on Simulation of Adaptive Behavior*, pages 402-410. MIT Press, Bradford Books, Cambridge, MA.

Goldberg, D.E. (1989). *Genetic Algorithms in Search, Optimization and Machine Learning.* Addison-Wesley Publishing Company.

Harvey, I., Husbands, P., Cliff, D., Thompson, A., and Jakobi, N. (1997). Evolutionary Robotics: the Sussex Approach. *Robotics and Autonomous Systems, Special Issues on "Practice and Future of Autonomous Agents"*, Pfeifer, R., and Brooks, R., editors, **20**(2-4):205-224.

Khepera Users Manual, Laboratoire de microinformatique, Swiss Federal Institute of Technology (EPFL), 1015 Lausanne, Switzerland.

Nolfi, S. and Parisi, D. (1997). Learning to Adapt to Changing Environments in Evolving Neural Networks. *Adaptive Behavior*, **5**(3):343-363.

Pfeifer, R. (1996). Building "Fungus Eaters": design principles of autonomous agents. In Maes, P., Mataric, M., Meyer, J.-A., Pollack, J., Wilson, S.W., editors, *From Animals to Animats 4: Proceedings of the Fourth International Conference on Simulation of Adaptive Behavior*, pages 3-12. MIT Press, Bradford Books, Cambridge, MA.

Rechenberg, I. (1973). *Evolutionsstrategie.* Frommann-Holzboog, Stuttgart, Germany.

Salomon, R. (1996). Increasing Adaptivity through Evolution Strategies. In Maes, P., Mataric, M., Meyer, J.-A., Pollack, J., Wilson, S.W., editors, *From Animals to Animats 4: Proceedings of the Fourth International Conference on Simulation of Adaptive Behavior*, pages 411-420. MIT Press, Bradford Books, Cambridge, MA.

Schwefel, H.-P. (1995). *Evolution and Optimum Seeking.* John Wiley and Sons, Inc, New York, NY.

Verschure, P., Kröse, B. and Pfeifer, R. (1992). Distributed Adaptive Control: The Self-Organization of Behavior. *Robotics and Autonomous Systems*, **9**:181-196.

Temporal Pattern Learning in a Spiking Neuron Chain

Matt Southall (ICL Institute of IT), **Tom Scutt** (ICL Institute of IT), **Barbara Webb** (AI Group, Dept. of Psychology)*

University of Nottingham
Nottingham NG7 2RD
UK
+44 115 9515151
Email and Extension Numbers
Matt Southall: mjs@psychology.nottingham.ac.uk, Ext. 18311
Tom Scutt: Tom.Scutt@nottingham.ac.uk
Barbara Webb: Barbara.Webb@nottingham.ac.uk Ext 15295

Abstract

A simple neural network demonstrates the ability to learn Morse code-like temporal patterns, presented via a microphone. A neuronal model, which provides facility for 'spiking' neurons, is used as the basis for this network. The network consists mainly of a chain of neurons, connected so as to fire in sequence. After a period of unsupervised learning, the chain is capable of repeating a temporal pattern upon which it has been trained. The system is particularly robust with respect to noise in the input pattern. Suggestions are made as to how this system could be extended, by replacing the simple chain structure with complex trees and interconnected 'blobs' of neurons. The example of birdsong as a biological system of temporal pattern learning is used as inspiration for further work in this area.

1. Introduction

This paper describes a first step towards a temporal pattern learning system, built using a 'spiking' neural network model. It aims to avoid some of the criticisms levelled at much of the 'classic' connectionist movement, not least of which is the problem of neurobiological implausibility. This is accomplished in two ways. Firstly, the microstructure of the model draws heavily on findings from the field of neurophysiology and is thus more plausible (as a model) than many neural networks. Secondly, the task performed by the network is drawn from real animal temporal learning systems, particularly the song-learning sub-system of certain bird species. As yet the macroscopic structure of the system presented is not intended as a direct model of the neurophysiology of this system but is rather a network built using approximately biologically plausible neurons and designed to accomplish a specific task. Nevertheless, we hope ultimately to gain insights into the actual mechanisms behind the biological example. Crick wrote, "...within most modellers a frustrated mathematician is trying to unfold his wings." (Crick, 1989). This paper is more the opposite case, a working system has been built using inspiration from a 'frustrated neurobiologist' trying to work out why tiny networks of slow, noisy neurons still outperform pattern recognition algorithms implemented in precise hardware performing millions of calculations per second.

1.1 Spiking Networks

A common criticism of classic connectionist neural network models is their insistence that it is sufficient to encode the spike rate of a neuron as a simple level of activation. Data from real neuronal systems show that individual spikes are significant in the amount of information they carry. Not only can we show that behavioural decisions are often taken within time windows that would only allow a neuron to spike once or twice, but application of Shannon's

* Matt Southall devised and implemented the model based on Tom Scutt's spiking neural network model. The paper was written with the help and under the supervision of both Tom Scutt and Barabara Webb.

information theory to spike trains shows that, contrary perhaps to intuition, individual spikes can represent more than one 'bit' of information (Rieke et al., 1997). Furthermore, classic neural networks are 'clocked', i.e. all the nodes in the network update simultaneously and the network enters a new state. This process repeats itself with a fixed period. This makes it difficult to imagine such networks displaying phenomena such as phase locking and comparison in auditory neurons. This is a mechanism which allows many animals (including humans) to locate sounds with low frequency (higher sounds are locatable by the volume difference due to the acoustic shadow of the head). Two neurons lock spiking to the phase of the wave from each ear respectively. A third neuron compares spike timing differences between the two auditory neurons and so provides directional information from the phase of the wave at each ear. There are many more such examples.

The assumption that the wealth of data carried by spike timings and patterns can be reduced to a simple 'spikes per second' measurement is not a reasonable one. Individual spike timing is highly important in neural encodings and surely any neural network concerned with the processing of temporal data will benefit from the use of a spiking network model. In fact, most non-spiking neural network models deal only with static patterns. Those that do learn and produce temporal patterns typically accomplish this by adding some sort of short-term memory to 'classic' network architectures. For example, context sensitivity can be added to a standard neural network by having time delayed input nodes (e.g. through recurrent feedback), which allow the network to have some knowledge of its previous states. This enables the network to become context sensitive. What it does not allow is the learning of absolute temporal patterns. Such a network can learn that B *will* follow A but cannot necessarily learn *exactly when* B will follow A. These models could learn to tell jokes but would never gain a sense of comic timing.

1.2 Birdsong Learning

The behaviour of songbirds is of particular interest when studying temporal pattern learning in neural networks. There is a great deal of literature concerning the phenomenon of birdsong, including behavioural, physiological and neurobiological. The work presented here represents a preliminary step towards a temporal pattern learning system that borrows from biological examples. Nothing here is intended as a model of the bird song-learning system, rather it is an artificial system that borrows from biology when it seems useful to do so and possibly the seed of what could become a model of birdsong neurophysiology

Although there is of course some genetic influence on different birdsong (not least morphology), it is accepted that much of a bird's singing is learned from its neighbouring conspecifics. Birds living in close proximity will typically have learned their song types from the same few birds and so will share song types (Beech et al., 1996), whereas geographically separate communities of the same species may well exhibit vastly different repertoires (Baker, 1996), albeit within the same species specific syntax.

There is evidence to suggest that, in some species, the ability to recognise and select for a particular song occurs in a particular early stage of development. This stage is distinct from later stages where the bird uses auditory feedback to 'crystallise' its own song, implying that the song is somehow stored in memory for later comparison when attempting to reproduce that song (e.g., Konishi 1985, Marler 1987). It is this developmental process, which proceeds in distinct stages, that is of interest here. In particular, this model is concerned with possible mechanisms via which a temporal pattern may be learned and stored in a spiking neural network. This type of temporal pattern learning is different in character to that exhibited in humans (e.g., learning of musical patterns). In contrast to the perception and production of music, birds sing their highly stereotyped songs at a fixed tempo. What is important here is the absolute rather than the relative timing of events. To the best of our knowledge, birds do not possess the ability to sing their song at significantly differing tempos (it is worth pointing out here that bird *calls* are a completely different matter and may be repeated with varying frequencies depending on circumstance).

1.3 Spiking Neuron Model

As mentioned above, a spiking neural network model has special relevance to temporal pattern learning. The work presented here uses such a model (based on that of Scutt and Damper, 1991). As in real nervous systems, signals are transmitted as discrete spikes, the frequency of which typically encodes the signal intensity. This model is designed such that the behaviour (i.e. membrane potential) of individual neurons is similar to that of real neurons.

The model uses an approximate state-system for simulating the changes in each neuron's membrane potential; the exact nature of this state-system depending on the type of neuron involved. A 'normal' neuron is treated as being in one (or, occasionally, more than one) of six states (see Figure 1). At each time cycle, for each neuron, the membrane potential (MP) is adjusted as follows:

1. Use the current MP to determine the state of the neuron and carry out appropriate operations:

A (*MP above resting, below threshold*):

Subtract TIME-CONSTANT from MP
B (*MP above threshold, below peak (FIRED flag off)*) :
Add ATTACK to MP & send potential to synapse(s)
C(*MP above peak*):
Set MP = HIGH; set FIRED flag to on; send potential to synapse(s)
D(*MP between peak and minimum, FIRED flag on*):
Subtract DECAY from MP & if MP > BaseMP then send potential to synapse(s)
E(*MP below minimum*):
Set MP = LOW; Set FIRED flag to off
F(*MP above minimum, below resting*):
Set MP = MP + (BaseMP - MP)/1.5

2. Add any incoming synaptic potential to MP; and if the neuron is noisy, randomly add or subtract a small amount of potential.

The constants ATTACK and DECAY, used in states B and D, represent the rapid increase and decrease in potential associated with a spike; and the constants HIGH and LOW, used in states C and E represent the maximum and minimum potential. The TIME-CONSTANT in state A can be used to give the neuron properties of leaky integration. The function in state F simply describes a smooth recovery curve from minimum to resting potential.

During states B, C, and D while the neuron is above threshold it will send potential to all neurons with which it synapses. The amount depends on the weight of the synaptic connection, and how far the sending neuron's MP is above BaseMP. The sign (positive or negative) of the synapse's weight indicates whether the effect on the postsynaptic neuron is excitatory or inhibitory.

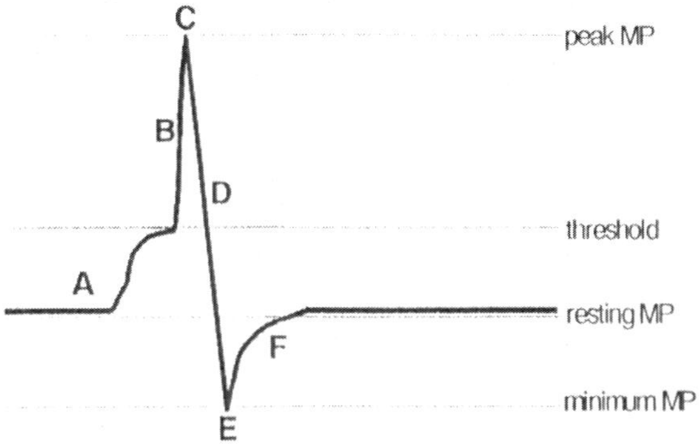

Figure 1. *The states passed through by the model neuron during spiking. A= Above resting, below threshold; B= Above threshold, below peak MP; C= Above peak MP; D= Post firing; E= Below minimum MP; F= Above minimum, below resting.*

The parameters for each neuron are variable, to provide different thresholds for firing, different time constants (rate of decay of membrane potential) etc. This leads to a second important difference between this and more typical neural networks: Parameters can be adjusted so that the artificial neurons imitate different types of real neurons such as bursting, plateau and pacemaker neurons. For example, giving a neuron a small negative time constant (i.e., a positive decay), will lead to a neuron that slowly builds membrane potential until it reaches threshold, upon which it will fire, hyperpolarise and repeat the cycle. This is essentially a pacemaker neuron, whose period can be varied by changing the magnitude of the negative time constant.

The model also incorporates a range of different synapse types, including conditioning, habituating and sensitising synapses. Habituating synapses are simply connections which reduce their weight by some function every time they fire. This will have the effect of gradually desensitising a network to repeated stimuli. The connection weight will slowly return to its normal value over a period of time. The sensitising and conditioning connections are 'synapse on synapse' connections i.e., They connect a neuron to another synapse, allowing that neuron to change the weight of that synapse. In the case of the sensitising synapse, this simply involves adding a large amount to the weight, making the connection sensitive again after habituation, to re-sensitise the network to a stimulus.

Conditioning synapses are the most complex forms of connection. These synapses change the weight of another synapse according to a function of the difference between the last firing times of both the conditioning neuron and the connection it is synapsing on. The function provides a roughly bell shaped curve with a long tail. The peak of the curve occurs at the optimal 'inter-stimulus interval', i.e., the time period since the connection's last firing when the conditioning neuron fires. If the conditioning neuron fires too soon or too late after the conditionable synapse, the weight increase will be small or zero. Simply put, a conditioning synapse changes the weight of another synapse when it fires but only if the other synapse has recently fired (recent being defined with respect to the 'optimal inter-stimulus interval').

Specifically, whenever a neuron fires, all its axonal synapses are time-stamped with the current system time. Whenever one of these synapses is of the type 'CONDITIONING', it will affect the weight of the synapse to which it connects according to the following functions:-

$$isi = \frac{t_j - t_i}{\rho}$$

$$\Delta w_j = w_i \times (isi \times e^{-isi})$$

Where:-

- t_i and w_i are the time stamp and weight, respectively, of the conditioning synapse.

- t_j and w_j are the time stamp and weight, respectively, of the synapse being conditioned.

- isi is the 'inter-stimulus interval'

- ρ is a constant used to set the optimal inter-stimulus interval, i.e., the isi at which Δw_j is maximal. For the purposes of this study ρ was set such that this maxima occurred at around 300 milliseconds.

These conditioning synapses are useful in that they enable the network to 'associate' two stimuli if they persistently occur together. Furthermore, this method of learning is based on real neuronal mechanisms (e.g., Bartha and Thompson, 1995).

The reader is referred to Scutt and Damper (1991) and Scutt (1994) for further description of this model and examples of its use.

2. Implementation

The software used was based on Scutt's Pascal version but the code has been heavily modified and ported to C/C++ and now runs on several systems including Rug Warrior and Khepera Robots. The actual codebase used for these experiments consists of the main network simulator which runs on a Linux host, the robot communications package also running on the Linux host and the other end of the communications package which runs on board a Rug Warrior robot, providing the host with regular sensor updates and using the output of any 'actuator neurons' to control the robot's motors, beeper, etc. By adjusting the software parameters, a neural model can be built that provides facility for excitatory and inhibitory interactions on the millisecond scale, in common with much other neural modelling (Arbib, 1995, pp5). It should be noted that although only the robot's microphone was made use of here, there are several other sensors and actuators on a Rug Warrior robot. The spiking neural network software has been designed to take advantage of all of these and so can be used as a more general robot control system.

2.1 A Pattern Learning Neuron Chain

This model uses a chain of neurons; each connected to the next neuron in the sequence via a connection of sufficient weight that a single action potential will propagate along the chain. Due to the inherent action potential propagation delay in the spiking neural network model, this process will take a certain amount of time. Once the first neuron in the chain fires, the relative firing times of the rest of the neurons will be approximately fixed. When a subset of these neurons are connected to some other (pattern output) neuron, they can cause a fixed temporal pattern of spike activity in that neuron. For example, if every odd numbered neuron in the chain is connected to an 'LED' neuron on a robot, the LED will flash intermittently with a fixed period. If every fourth neuron is connected to the LED neuron, it will flash intermittently with double this period. Using long chains and various patterns of connection, complex temporal patterns can be reproduced.

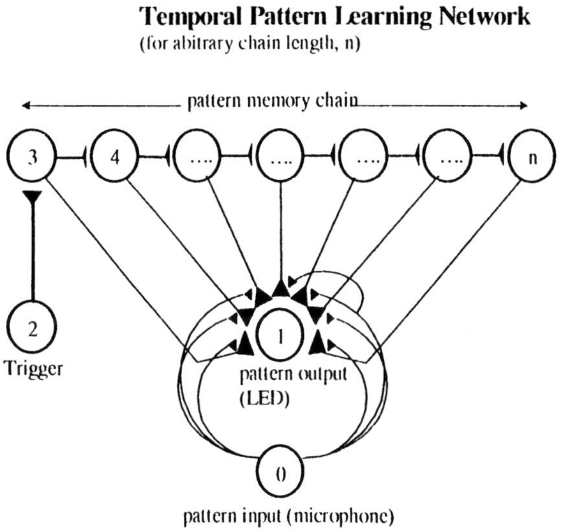

*Figure 2. An example of the pattern learning chain structure. This example shows a chain of Length 'n'. The actual chain used in these experiments was 27 neurons long. The numbered neurons are as follows:- (0) is the input neuron connected to the microphone. (1) is the output neuron, connected to an LED. (2) is the trigger neuron which starts the activation moving along the chain. (3) and (4) are the first two neurons in the chain which continues up to (n). As can be seen, each chain neuron synapses on the output neuron (1), and each of **these** connections is synapsed on by a conditioning synapse from the pattern input neuron (0).*

2.2 The Learning Process

In order to learn a temporal pattern, the connections between the chained neurons and the output neuron must be formed in response to the training signal. This model achieves this by initialising the network with each chain neuron having a zero weight connection to the pattern output neuron. The pattern input neuron then forms a conditioning 'synapse on synapse' on each of these connections (see Figure 2). The conditioning algorithm causes an increase in weight of each of these connections whenever the pattern input neuron fires. The weight increase is governed by the conditioning function (see above) and is usually zero. If the connection from chain to pattern output has recently fired then the change in weight will be non-zero (with the largest weight increases occurring if the difference between the firing times of the pattern input and the chain neuron is equal to the optimal inter-stimulus interval). The result of this process is that any neuron in the chain that fired at approximately the same time as the input neuron will become more strongly connected to the output neuron. As the neurons in the chain fire in a fixed sequence, this pattern of connections will represent a temporal pattern; specifically, the pattern presented at the input neuron during learning.

The chain can be of arbitrary length with the only practical limit being computational power. The chain used in these preliminary experiments was relatively short, consisting of only 27 neurons. The input pattern neuron was stimulated by manually by tapping the microphone in some particular rhythm. The program was forced to operate at less than its maximum speed in order to make the action potential propagate down the chain slowly enough that a human could tap a recognisable and repeatable rhythm during this time. The trigger for the chain (i.e. the trigger for sending action potentials along it) was a simple switch on board the robot. Ideally, the trigger for the chain should be in the same sensory modality as the pattern input neuron, in this case the microphone. For the purposes of this trial, the microphone was not used (due to initial problems with 'neural echoes'). The conditioning parameters were set such that several (around ten) repetitions were required for the network to fully replicate the pattern. Due to the nature of the (manually generated) stimulus pattern, the rhythm would vary to a small degree during these repetitions. For the purposes of this experiment, the output neuron was connected to an LED.

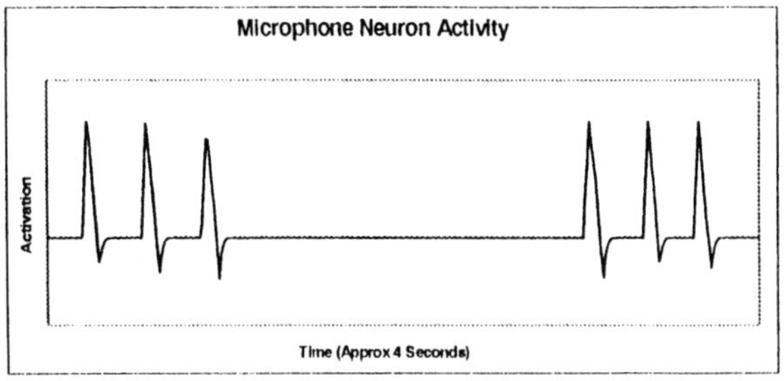

Figure 3. *Close up of microphone (input neuron) activity*

3. Results

3.1 Input Patterns

During the course of several trials, the activity of all neurons was logged. Additionally, spike times and weight changes were recorded. Although logged data was time-stamped to an accuracy of 100 μs, the specific timing of spikes is far less important than the relative timing. The network's 'perceived' time can be arbitrarily altered by varying a delay loop within the program and so the time scales in the graphs presented here are provided merely to

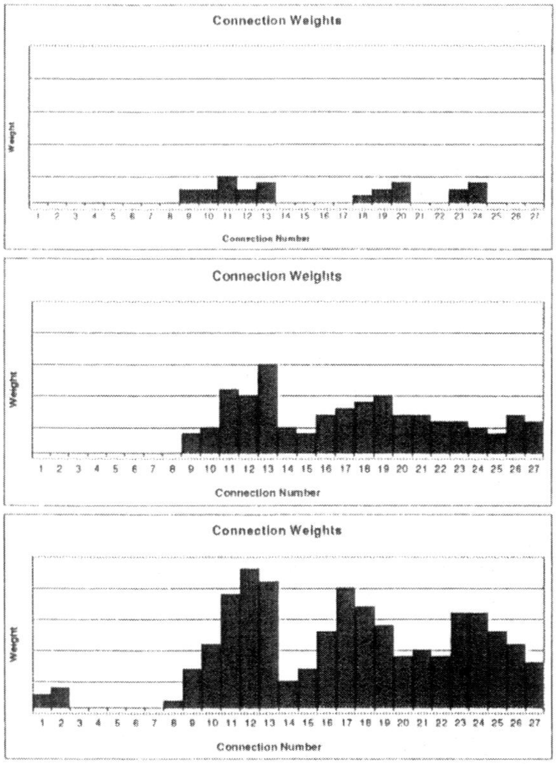

Figure 4. *Shows the weight of a connection from a chain neuron to the pattern output neuron, plotted against the connection number (ie. the number of the neuron in the chain from which the connection is made). The three graphs are taken from different points in the trial and are in temporal order, top to bottom. The three peaks corresponding to the three microphone spikes (see Figure 3) are clearly visible.*

give an idea of the sort of rhythms being learned. The network could be run at speeds far greater than these.

Figure 3 shows a typical activation pattern for the input neuron during stimulation. Each repetition of this pattern was caused by three short taps on the microphone. Before each repetition, a signal was sent to the first neuron in the chain, via the trigger neuron (Figure 4 shows that, by the end of the trial, the microphone had picked up the sound of the trigger switch opening again and begun to condition the first few neurons in the chain to stimulate the output neuron).

3.2 Output Patterns

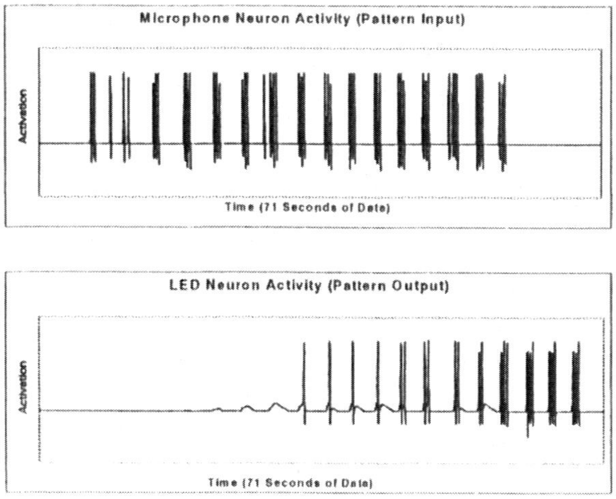

Figure 5. *Shows the activation of the input and output neuron during the same time period. During this time, the output neuron begins to fire until finally it repeats the entire pattern in the absence of the input.*

Figure 5 shows how after a few repeats of the stimulus pattern, the activation in the output neuron begins to mimic that of the input neuron, first with a single spike and finally with a set of three spikes. During the process of learning the network has begun to 'join in' the pattern repetitions until, towards the end of the trial, triggering the chain (with no microphone input) is sufficient to generate the entire pattern. A close up of the activation of the output neuron is shown in Figure 6.

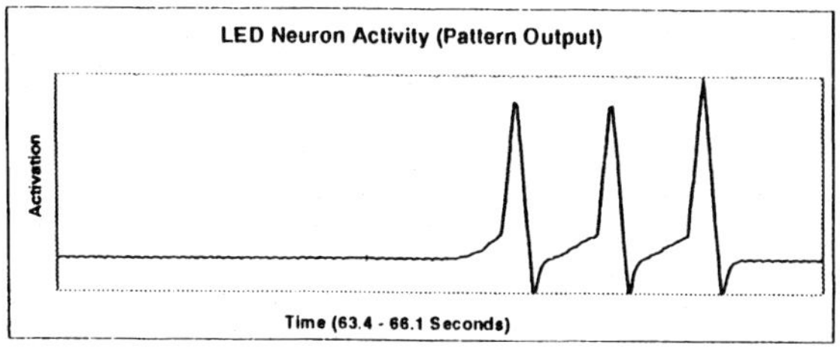

Figure 6. *A close up view of the activation of the pattern output neuron during a repetition of the pattern. Notice how the gradient of the slope up to threshold is quite shallow. The neuron is temporally summating multiple input spikes.*

3.3 Learning

The activation level of the output neuron is a lot slower to reach threshold and spike than that of the input neuron. The very fast 'attack' of the input neuron is coded into the model for practical reasons (not least the fact that weak stimulation of the input neuron by the microphone makes for unsociably loud experiments). The rather slower attack of the output neuron is explained by the variability of the input pattern. It is theoretically possible that an identical input pattern on each occasion could cause a single neuron in the chain to gain a heavily weighted connection to the output neuron. In practice however (especially with manually generated signals), the weight change that occurs for each particular stimulus in the input pattern is spread over a group of connections from adjacent neurons. Some of this spread is due to the weight change function of the inter-stimulus interval but, with the parameters used for these trials, this spread should have been limited to two neurons either side of the peak. Figure 4 shows weight changes in connections up to four neurons away from the peak. It is clear from this that some of the spread is due to slight variations in the input pattern.

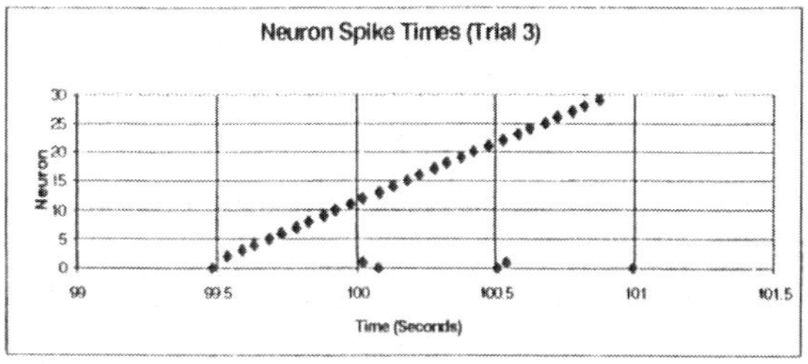

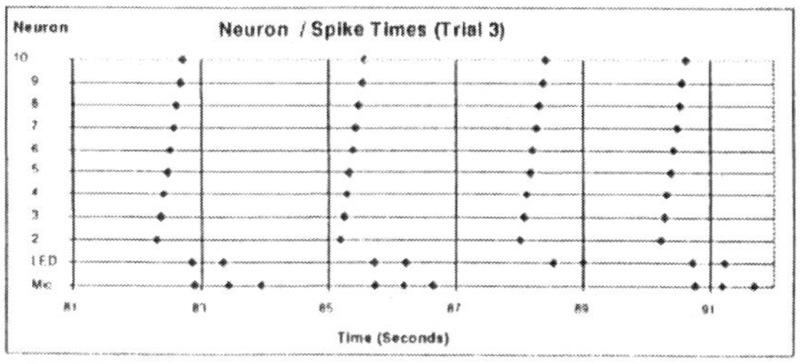

Figure 7. *Two plots of neuron number against neuron spike time. Each point represents one spike. The chain is visible as a line of points on each graph, with only the first few members of the chain visible on the bottom graph. The bottom neuron and next neuron up on each graph are the input and output neurons, respectively.*

These slight variations are visible in Figure 7, which shows neuron spike time plotted against neuron number. The neuron chain is clearly visible as an ascending line of neurons (with the first neuron in the line being the trigger).

At the bottom of the graph the input neuron (0) can be seen providing the input pattern. The output neuron (1) is simultaneously 'singing along' but there are slight discrepancies between the input on each occasion. The match is, however, approximately correct.

3.3 Temporal Summation

As suggested above, each spike in the output neuron is the result of several neurons in the chain having low strength connections to the pattern output neuron, rather than a one to one, high strength connection from a single chain neuron. This is made possible by the fact that the neuronal model used here has a 'time constant'. This is the rate at which the 'activation' (analogous to membrane potential) decays back to its basal state. If this rate is sufficiently low, then the input from temporally distributed spikes is summated and can cause a neuron to fire. An examination of Figure 6 shows clearly how the activation of the pattern output neuron is not being driven suddenly over its threshold, but rather is summing the incoming spikes from several neurons and slowly building up its activation.

4. Discussion and Further Work

4.1 More Chains

In these trials, a delay loop was introduced to slow the program down to the point where patterns of a 'tempo' perceptible to (and producable by) humans could be used. Experiments involving up to one hundred neurons have since been conducted and, on initial inspection, perform similarly to the shorter example presented here. This is exactly as would be expected, longer patterns can be learnt. By increasing the length of the neuron chain, decreasing the size of the delay loop and the optimal inter-stimulus interval, patterns of similar length but greater precision can also be learnt.

Although in this experiment it was required that the 'trainer' triggered the learning chain (a possibly unrealistic expectation for a robot required to learn in the real world), the input neuron could, in principle, be used to trigger the chain. The connection from the input neuron to the first chain neuron in this case would have to be a massively habituating connection that would not retrigger the chain while a spike was propagating along it. Failure to inhibit retriggering during learning would result in 'echoes' and eventual saturation of the output neuron. Future versions of this network will also ensure that the connections from the neuron chain to the output neuron will gradually return to zero weight over time. The stimulus pattern will have to be constantly represented to the network in order to stop the chain from 'forgetting' it. This will also mean that the network no longer runs the risk of saturation if it spends too long attempting to learn any particular pattern along with the inherent noise.

Preliminary experiments with different conditioning parameters have shown that it will be possible to build chains that can learn a pattern from a single exposure to it, although then noise becomes a serious problem, as any 'rogue' spikes in the input (an inevitability in a real world environment) become part of the output pattern. The advantage of learning over several repetitions is the averaging out of noise.

4.2 Trees

Although these initial networks are fairly small and the patterns uncomplicated, it is easy to imagine more complex networks involving larger ranges of input neuron types, each one conditioning the connections to output neurons for different types of actuation, e.g., frequencies, leg movements etc. When conditioned by a suitable complementary set of input neurons, the temporal patterns learnt would not seem quite so simplistic.

This idea can be further extended by imagining a 'tree' of such chains. Here, the output neuron from one chain could simply be the triggering neuron for another chain. This would allow sequencing of temporal patterns with sub-patterns such as for example syllables in birdsong. It would be possible to teach such trees the 'syllables' from a pattern and then teach the sequence of syllables, provided a sensory network could recognise both the pattern elements constituting a syllable and then the syllables of which a 'song' consists.

4.3 Blobs

All these models so far require a great deal of predetermined structure for anything useful to occur during the learning process. One final possible network type could simply be a 'blob' of neurons with random interconnections. An input to the network should cause activation patterns to 'resonate' around the network for a short while. Parameters for the neurons, such as their recovery time from hyperpolarisation (refactory period) and the connection weights and types between them would have to be carefully adjusted to prevent the blob 'self-oscillating'. This blob could contain loops, branches etc. and so learn and produce very complex patterns. The problem with this sort of network is basically one of blame assignment. If the network learns the wrong pattern, for example, if a neuron in a loop becomes conditioned to excite an actuator neuron creating an echo not present in the input, it is difficult to see how the network could correct this sort of mistake. One possible method would be to borrow some ideas from the separate stages of birdsong development. The system could initially learn the pattern using the simple chains or trees described above. Once the pattern is learnt it could later move into the 'production' phase where a neuron blob resonates and produces output, with the chain memorised pattern being used to condition and adjust parameters for neurons and connections within the blob.

4.3 Feedback and Blame Assignment

This idea raises the problem of feedback. In order to effectively solve the blame assignment problem during unsupervised crystallisation of a memorised song, the system must be able to analyse its own output in some way and compare it to the input. There is evidence for such feedback pathways within the bird song system. It is also suggested that this feedback pathway has little effect on full adult song, once learnt, but is important in the development of this song.

The task of making feedback useful in determining the correctness of the output, will be facilitated by the use of specialised neurons, selective for particular temporal patterns within the output. Such neurons have already been identified in the bird forebrain nucleus, where neurons have been recorded that are selective for the bird's own song but respond less to the song with the syllables reversed and minimally to the entire song reversed (Lewicki 1996).

5. Conclusion

This paper has set out an example of how unsupervised learning of arbitrarily complex absolute temporal patterns can occur in a spiking neural network. The system was simple, effective and robust with respect to noise. These properties are due in no small part to the choice of neural model. For example, temporal summation of incoming spikes proved to be an essential ingredient in allowing the network to learn a signal which was not necessarily exactly the same on each presentation. Also the spiking network allowed us to encode (in the weights) the absolute temporal structure of the Morse-like signal, as opposed to the relative ordering of it's components.

Future work will attempt to extend this system to include multiple input and output neurons and more complex structures. It is hoped that this will allow the learning of more complex patterns such as those found in birdsong. Also, blending chain, tree and, possibly, 'blob' based neural networks with what is known about the learning of birdsong, particularly song templates and feedback pathways, should contribute usefully to the study of unsupervised temporal pattern learning systems.

References

Arbib, M.A. (Ed.) (1995) The handbook of brain theory and neural networks. *MIT Press.*

Baker, M.C. (1996) Depauperate meme pool of vocal signals in an island population of singing honeyeaters. *Anim. Behav.* 51:853-858.

Bartha, G.T. Thompson, R.F. (1995) Cerebellum and Conditioning. *In Arbib, M.A. (ed) The handbook of brain theory and neural networks. MIT Press.*

Beecher M.D, Stoddard P.K, Campbell S.E, Horning C.L. (1996) Repertoire matching between neighbouring song sparrows. *Anim. Behav.* 51:917-923.

Crick, F. (1989) The recent excitement about neural networks, *Nature, 337, pp129-132.*

Konishi, M. (1985). Birdsong: from behaviour to neuron. *Ann. Rev. Neurosci.* 8:125-70.

Lewicki, M.S. (1996) Intracellular characterization of song-specific neurons in the zebra finch auditory forebrain.

Marler, P. (1987). Sensitive periods and the roles of specific and general sensory stimulation in birdsong learning. *In: Imprinting and cortical plasticity (Rauschecker, J.P. Marler, P. eds), pp99-135. New York Academic.*

Reike, F. Warland, D. de Ruyter van Steveninck, R. Bialek, W. (1997) Spikes: Exploring the Neural Code. *The MIT Press, Cambridge, MA.*

Scutt, T.W. and Damper, R.I. (1991). Computation modelling of learning and behaviour in small neuronal systems. *Proceedings of International Joint Conference on Neural Networks, Singapore*, 430-435.

Scutt, T.W. (1994) The five neuron trick: using classical conditioning to learn how to seek light/ *in Cliff, D.T. et al (eds) From animals to animats 3: Third International Conference on Simulation of Adaptive Behaviour (pp87-96). MIT Press.*

Computational Models of Classical Conditioning: A Comparative Study

Christian Balkenius, Jan Morén
Lund University Cognitive Science
Kungshuset, Lundagård
S-222 22 Lund, Sweden
christian.balkenius@fil.lu.se
jan.moren@fil.lu.se

Abstract

We describe computer simulations of a number of associative models of classical conditioning in an attempt to assess the strengths and weaknesses of each . The behavior of the Sutton-Barto model, the TD model, the Klopf model, the Balkenius model and the Schmajuk-DiCarlo model are investigated. All models are shown to have problems explaining some of the available experimental data. The ISI curves for trace and delay conditioning for all the models are presented together with simulations of acquisition and extinction, reacquisition, blocking, conditioned inhibition, secondary conditioning and facilitation by an extra stimulus. We also present cases where some of the models show an unexpected behavior.

1. Introduction

Although traditionally seen as a very simple phenomenon, classical conditioning has offered unexpected resistance to theoreticians. Still, almost a hundred years after Pavlov's initial experiments (Pavlov 1927), there exists no model capable of explaining the full range of phenomena under this heading. In fact, classical conditioning contains some of the hardest problems in learning today. It is most unsatisfactory that the general opinion is that conditioning is simple to model and has been solved a long time ago.

We believe that the main reason for this situation is the limited amount of detailed studies that have been made of the different models. This is partly because they can be very hard to implement correctly. Very seldom are the models specified in sufficient detail to allow other researchers to use the models. When mathematical descriptions are given, they are often incomplete or confusing.

In this paper, we compare a number of models of conditioning. We have chosen a number of popular associative models. The models were selected as they are fairly similar and make similar claims about their abilities.

2. The Models

In this section, we describe the models tested in our investigation. The models are the Sutton-Barto model proposed by Sutton and Barto (1981), the more recent development of this model, the TD-model (Sutton and Barto 1987), the model described by Klopf (1982, 1988), the model developed by Balkenius (1995, 1996, 1998), and the model proposed by Schmajuk and DiCarlo (1992).

2.1 The Sutton-Barto Model

The Sutton-Barto (SB) model (Sutton and Barto 1981), is an early time-derivative model of reinforcement in conditioning. It works in real-time in contrast to the earlier Rescorla-Wagner model (Rescorla and Wagner 1972), from which it is descended. This model is a precursor to many later computational models, including the TD model, the SBD model (Blazis *et al.* 1986) and the Klopf model (Klopf 1988). The reinforcement in the SB model is the time derivative of the sum of the stimulus strengths:

$$\dot{Y}_t = Y_t - Y_{t-\Delta t},\ Y = \sum_i CS_i V_i + V_{US} US, \qquad \text{(SB1)}$$

where the V_i is the stimulus strength of CS_i, and V_{US} is the strength of US. With suitably chosen constants, these equations will account for all predictions of the Rescorla-Wagner model. An eligibility trace mechanism is added to account for temporal relationships:

$$\overline{X}_{i,t+1} = (1-\delta)\overline{X}_{i,t} + \delta X_{i,t},$$
$$X_{i,t} = \begin{cases} 1, & CS_i \text{ present} \\ 0, & CS_i \text{ absent} \end{cases} \qquad \text{(SB2)}$$

The rate of decay δ is in the range $0<\delta<1$. The reinforcement signal and the eligibility trace are combined as:

$$\Delta V_i = \beta \dot{Y} \times \alpha_i X_i \qquad \text{(SB3)}$$

with α_i and β_i positive constants as in the Rescorla-Wagner model. During the simulations reported below, the constants chosen were α_i=0.1, β=1, δ=0.2; these are the values used by Sutton and Barto.

2.2 The TD Model

The TD (temporal difference) model (Sutton and Barto 1987) was introduced as an extension of the earlier Sutton-Barto model (Sutton and Barto 1981). According to the TD-model, the goal of conditioning is to predict the temporally discounted value of all future rewards. At time t, this prediction is called $\overline{V}_i$ and is calculated as the sum of all weights V_i for the CS_i:s present at time t:

$$\overline{V}_t = \max(\sum V_i CS_i, 0). \qquad \text{(TD1)}$$

Here, the weight V_i represents the contribution from CS_i to the total prediction. During learning, these weights are updated as:

$$\Delta V_i = \beta\left(\lambda_{t+1} + \gamma \overline{V}_{t+1} - \overline{V}_t\right) \times \alpha_i \overline{X}_i, \qquad \text{(TD2)}$$

where λ_{t+1} represents the strength of the US at time t+1, γ is the discount factor, α_i and β are learning rate constants, and $\overline{X}_i$ is a trace of stimulus CS_i. Thus, the first term of (TD2) can be seen as the expected change in US from t to t+1 (called reinforcement by Sutton), while the second term is a function of temporal proximity to the CS. The definition of $\overline{X}_i$ is identical to (SB2) above. Using our implementation, we successfully reproduce the simulation data as presented in Sutton and Barto (1990).

2.3 The Klopf Model

The Klopf model was introduced in 1982 as the Drive-Reinforcement (DR) model (Klopf 1982), but we present the model as described in Klopf (1988). The output, y(t), is calculated as the sum of all CS representations x_i multiplied with their corresponding weights w_i.

$$y(t) = \sum_{i=1}^{n} w_i(t)x_i(t). \tag{K1}$$

During conditioning the weights change as,

$$\Delta w_i(t) = \Delta y(t)\sum_{j=1}^{\tau} c_j |w_i(t-j)| \Delta x_i(t-j), \tag{K2}$$

where c_j are learning constants, $|w_j|$ are the magnitude of the individual weights, and Δx_i the change in the CS representation. It is stated in Klopf (1988) that only *positive* changes should be considered in the equation above. It should also be noted that there are both excitatory and inhibitory weights in the model. These are treated separately and are constrained to stay on the positive and negative side respectively. It is also required that all weights must be larger than 0; we use 0.1.

Equation (K2) illustrates a fundamental assumption of the Klopf model: that changes in the output should be correlated with changes in the inputs to determine whether learning should occur. When both the input and the output changes, the weights should increase. To allow non-zero ISI:s, each CS is assumed to leave an eligibility trace in the system. This is represented by the sum in equation (K2) which is used as a memory that extends backward in time.

2.4 The Balkenius Model

Like the Klopf model, the model presented in Balkenius (1995, 1996, 1998) is based on a neural interpretation of the conditioning mechanism. But contrary to the Klopf model, the Balkenius model uses a network rather than a single node. Like the Klopf model, the Balkenius model separates inhibitory and excitatory learning, but this is made explicit in the formulation of the model. Here, only the positive side of the network is described. The equations for the negative side are identical. The output of the model is given by,

$$CR(t) = x^+(t) - x^-(t), \tag{B1}$$

$$x^+(t) = \sum_{i=1}^{n}\sum_{j=0}^{\tau} w_{ij}^+(t)\left[\Delta CS_i(t-j)\right]^+. \tag{B2}$$

The extra index *j* in equation (B2) corresponds to a tapped delay-line for each CS with length τ. This approach differs from the eligibility traces used in the other models but have a similar role. The representation is usually called a multiple-element stimulus trace and has the advantage that it can support more complex associations that a single eligibility trace. On the other hand, it requires many more variables since the number of weights must equal the length of the stimulus trace.

The weight change is calculated as,

$$R^+ = \left[US + \delta\left(x_{t+1}^+ - x_{t+1}^-\right) - \left(x_t^+ - x_t^-\right)\right]^+. \tag{B3}$$

In this equation, δ is a discount factor responsible for secondary conditioning. We used the value 0.9 in the simuations below. This equation can be compared to equation (TD2) for the temporal difference model. In that model, the absolute values are used in the learning equation. In this model, it is instead the *changes* that contribute to the reinforcement. This is consistent with the idea that the model tries to predict the *level* of the US rather than the integral over it as the TD model.

Like in the Klopf model, changes in the US level are correlated with changes in the CS levels and during conditioning, the weights change according to the equation,

$$\Delta w_{ij}^+(t) = \gamma^+ R^+\left[\Delta CS_i(t-j)\right]^+, \tag{B4}$$

where γ^+ is the learning rate.

2.5 The Schmajuk-DiCarlo Model

The Schmajuk-DiCarlo (SD) model was introduced in 1992 and was shown to model a number of classical conditioning phenomena (Schmajuk and DiCarlo 1992). Especially interesting is its ability to model the effect of various types of configurational stimuli and the effects of hippocampal lesion on conditioning. Here, however, we will only investigate the more fundamental abilities of the model. The complete characterization of the model can be found in Schmajuk and DiCarlo (1992) and we will only describe the equations responsible for the dynamics found in the simulations. A stimulus, CS_i, gives rise to a short-term memory trace X_i, described by,

$$\frac{dX_i}{dt} = -K_1 X_i + K_2\left(K_3 - X_i\right)CS_i. \tag{SD1}$$

K_1, K_2 and K_3 are constants that determine the passive decay of the trace, the rate of increase, and the maximum level of the trace, respectively. The associative strength, VS_i changes according to the equation,

$$\frac{dVS_i}{dt} = K_5 K_6\left(1 - |VS_i|\right)EO, \tag{SD2}$$

where K_5 is an output level constant and K_6 is the learning rate constant. EO describes the error in the prediction of the model. This error is calculated as the difference between the US level and the sum of all stimulus traces multiplied with their respective weights,

$$EO = US - \sum_i K_5 X_i VS_i. \tag{SD3}$$

In the simulations, we followed the complete description of the model given in Schmajuk and DiCarlo (1992), and the equations above should serve only as an indication of the dynamics of the model and not as a complete description.

3. The Experiments

We have tested the models above on a representative set of classical conditioning experiments All experiments were selected on the grounds that at least one of the models claim to reproduce the effect. Some of the models handle much more complicated cases than the ones presented, but our main concern here is to see to what extent the different models can account for a variety of basic learning effects.

3.1 Acquisition

Acquisition is the fundamental process in classical conditioning. In acquisition, an association is established when a CS is paired with a US.

$$CS + US \Rightarrow CS \rightarrow CR$$

First, a CS is followed by the US for a number of presentations. Second, a CS will elicit a CR when presented on its own. All models tested can easily model acquisition.

A feature of some empirical studies of acquisition of associations is that the response level forms an S-shaped curve similar to a sigmoid function (Schneiderman, Fuentes and Gormezano 1962). Of the models tested in this paper only the Klopf model, and to some extent, the Balkenius model show this effect. Interestingly, the Klopf model loses this feature during reacquisition and for the Balkenius model, it only appears when the ISI is larger than 1 tick.

3.2 Inter-stimulus interval effects

The effect of the inter-stimulus interval on the response level is one of the primary real-time effects of classical conditioning. The rate of conditioning is optimal at a certain ISI. For example, in classical eye-lid conditioning in rabbit, the optimal ISI is approximately 250 ms and decays exponentially with time (Smith, Coleman and Gormezano 1969, Schneiderman and Gormezano 1964). Also, independently of the ISI, the CR tends to appear slightly before the US (Desmond 1990).

Empirical studies have been done both with trace conditioning, where the CS terminates before the US is presented, and with delay conditioning, where the CS offset occurs at the onset of the US, or later (see figure 1).

We have tested the various models with three types of timing (figure 1). In *trace* conditioning, the CS and US have fixed lengths and only the ISI changes. In trace conditioning, the ISI can be both positive and negative. In the first type of delay conditioning, called *delay A*, the CS terminates at the onset of the US. In the second type, *delay B*, the CS is present until the termination of the US.

An empirically determined profile, according to Schneiderman (1966) and Smith et al. (1969) can be found in figure 2a. All models tested in this paper exhibit ISI timing effect, albeit for very different reasons.

The inter–stimulus interval effects are poorly modelled by the SB model. For example, the model shows inhibitory conditioning with ISI close to zero during trace conditioning (figure 2b). This seems to be inconsistent with empirical data (Smith *et al.* 1969). The model also fails to account for ISI effects during delay conditioning, with no decay of asymptotic learning with longer ISI:s, and with overall learning level dependent on proximity between CS_{off} and US_{off}, as seen in figure 2b. As Sutton and Barto notes (Sutton and Barto 1990), this model does not, and is not intended to, account for these effects.

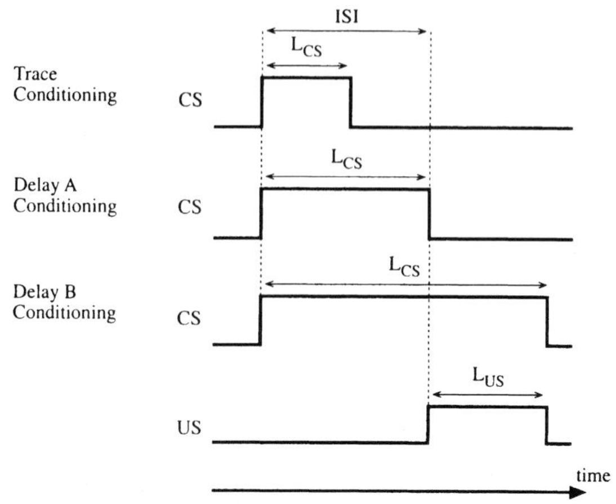

Figure 1. Delay and trace conditioning. For trace conditioning, the CS terminates before the onset of the US. In delay conditioning of type A, the length of the CS (L_{CS}) is equal to the inter-stimulus interval (ISI), for delay conditioning of type B, the length of the CS is equal to the ISI plus the length of the US, that is the CS and the US terminates simultaneously.

The TD model reproduces the ISI-dependency shown in empirical studies. Figure 2c shows the asymptotic weights as a function of the ISI for trace and delay conditioning. The ISI-curve is an emergent effect of the learning equations and handles a variety of different ISI:s and stimulus lengths.

In the Klopf model, the learning constants are explicitly chosen to reflect the ISI-curve for delay conditioning (Klopf 1988). Figure 2d shows the simulation result for the ISI experiments. Although the Klopf model has the ability to produce a range of ISI-curves depending on the learning rate constants, it has almost no predictive value in this area since the curve is essentially rigged by the constants c_i. Another limitation is the fixed length memory buffer used for the eligibility trace. The use of a memory buffer limits the temporal resolution of the model and the length of the ISI.

The Balkenius model differs from the other models in the way the ISI effect is modelled since it results from secondary conditioning within the stimulus trace. Figure 2e shows the ISI curve for the model. The curve directly reflects the discount factor as the asymptotic value of the weights converges to δ^{i-1} for an inter-stimulus interval, $i \geq 0$. However, the model does not distinguish between delay and trace conditioning and like the Klopf mode, it uses a fixed length memory trace.

The ISI curves for the Schmajuk-DiCarlo model are shown in figure 2f. Together with the TD model, the SD model is the only one to model the difference between the trace and delay conditioning in a qualitatively correct way.

3.3 Extinction

In an extinction experiment, a previously acquired response disappears as the CS is presented without a subsequent US. The training procedure can be summarized as,

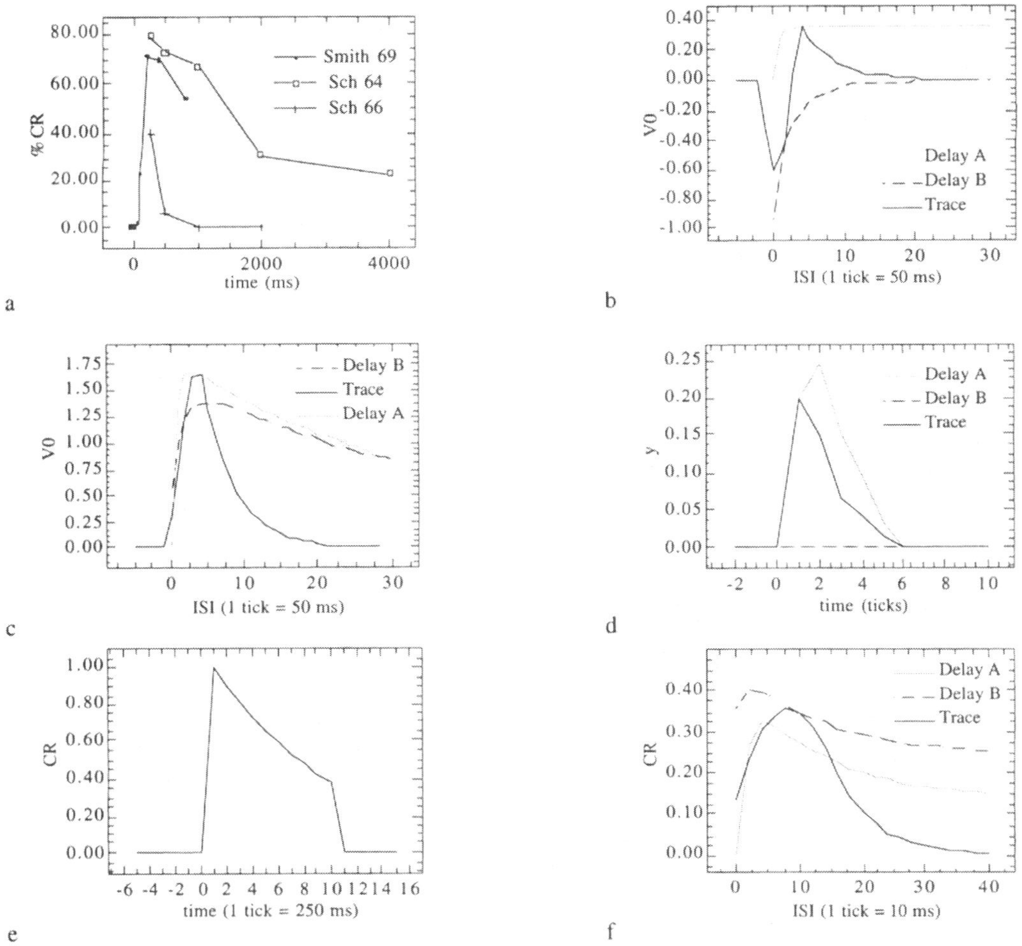

Figure 2. (a) The effect of the inter-stimulus interval in classical nictitating membrane conditioning. Delay B conditioning after Scheniderman and Gormezano 1964). Trace conditioning after Smith et al. (1969) and Schneiderman (1966). (b) The SB model. (c) The TD model. (d) The Klopf model. (e) The Balkenius model. (f) The SD-model.

$CS_1 \Rightarrow CS_1 \rightarrow$ no CR

This is again a very fundamental ability of any conditioning models as it shows that learning is reversible. When the CS is presented on its own, it no longer predicts the US and should no longer produce a CR. All the models tested can handle this situation, which is not very surprising.

3.4 Reacquisition Effects (Savings effects)

The reacquisition, or savings, effect appears when an animal relearns an association that has previously been extinguished. Learning the second time is faster than during the original learning phase, and is faster still the third time (Pavlov 1927).

We tested the reacquisition effect by running the association–extinction cycle four times. Of the models described here, only the Klopf and the Schmajuk-DiCarlo models show any reacquisition effects, and for the Klopf model, it only appears for the first relearning phase.

Although Klopf (1988) presents a simulation where the reacquisition effect is clearly visible, this effect only occurs for the *first* reacquisition. All reacquisitions after the first one are identical for the parameters used in Klopf (1988). This was not illustrated in the original article.

3.5 Blocking

Blocking is one of the best known features of classical conditioning (Kamin 1968, Rescorla and Wagner 1972). The regime for blocking is:

$CS_1 + US \Rightarrow CS_1 \rightarrow CR$

$(CS_1{+}CS_2) + US \Rightarrow CS_2 \rightarrow$ no CR

The parenthesis is used to indicate that CS_1 and CS_2 are presented simultaneously. When CS_1 already predicts the US, there will be no conditioning of CS_2 to CR; CS_1 blocks CS_2. All models tested perform well on tests for this effect. The blocking experiment shows that acquisition of an association is dependent on earlier learning and all stimuli present influence learning in a non-local way.

3.6 Conditioned Inhibition

In Pavlov's account for extinction, he describes it as the acquisition of an inhibitory association that removes the effect of the excitatory association from the CS (Pavlov 1927). That is, extinction is considered an active process rather than a passive decay of an association. The existence of an inhibitory association can be shown in a conditioned inhibition experiment. In such an experiment, the animal is first conditioned with a test stimulus CS_2. In a second phase, trials occur where the CS_0 is paired with the US, and trials where CS_0 together with CS_1 do not predict the US. Finally, CS_1 is paired with CS_2. If CS_1 has acquired inhibitory properties, it will be able to inhibit the CR that CS_2 would otherwise produce:

Phase I	CS_2 + US
Phase II	CS_0 + US
	(CS_0 + CS_1) + no US
Test	(CS_1 + CS_2) $\rightarrow$ no CR

It is assumed that CS_1 takes on its inhibitory properties during the trials where it is presented together with CS_0 that on its own would predict the US. All models tested exhibit conditioned inhibition.

3.7 Secondary Conditioning

When CS_1 has been associated with an US, CS_1 acquires reinforcing properties and can itself be used to reinforce a CS_2. This is called secondary conditioning. The learning regime for this effect is:

CS_1 + US => CS_1 -> CR

CS_2 + CS_1 => CS_2 -> CR

The effect is typically weak and very dependent on the exact timing of CS_1 and CS_2, as CS_1 will extinguish at the same time that CS_2 is reinforced. All models except the Schmajuk-DiCarlo model can handle secondary conditioning effects. This effect is important if one wants to use classical conditioning as a part of an instrumental learning system (Barto, Sutton and Watkins, 1990, Klopf, Morgan and Weaver 1993, Balkenius 1995). In this case, the chaining of responses is often described as secondary conditioning.

For the SB model, secondary conditioning with no overlap between CS_0-US or CS_1-CS_0 works as expected; CS_1 is reinforced by CS_0, albeit weakly, as CS_0 is simultaneously extinguished. However, when the inter-stimulus intervals are short enough to make the stimuli overlap, an erroneous inhibitory conditioning is produced. Briefly, the reason is that when CS_0 and US overlap, V_0 is initially small, making $\dot{Y}$ small. Thus, change in X is the primary factor in changing V_0. The greatest change in X is for CS_{off}, and when overlapped with US, this negative change will influence V_0, producing the effect. The analogous effect is achieved for phase two of secondary conditioning, as V_0 is large and V_1 is small.

In the TD model the presence of secondary conditioning depends on the stimulus length. Interestingly, the TD model suffers from a similar problem as the Sutton-Barto model for secondary conditioning: overlapping CSs produce inhibition rather than excitation, as shown in figure 3. The reason this effect is not apparent for first-order acquisition, is that the US level is not taken into account when computing $\overline{V}_i$, in contrast to the Sutton-Barto model.

Klopf (1988) reports simulations of secondary conditioning which we have reproduced. It is interesting to note that this experiment results in the inhibitory effect illustrated above when run with the TD model.

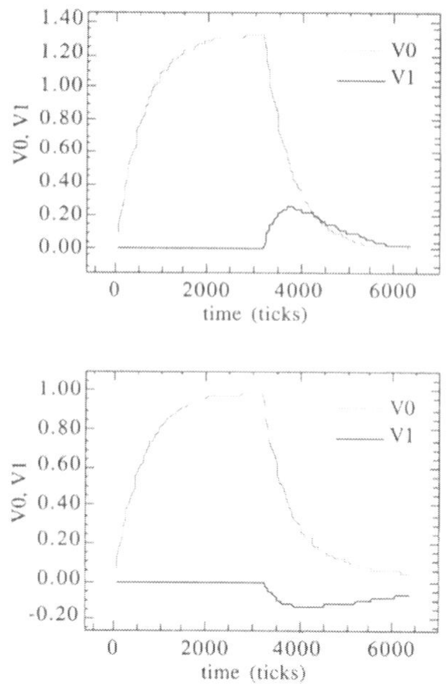

Figure 3. Secondary conditioning in the TD model with and without overlapping stimuli. The upper curves show V_0 and the lower curves show V_1. (top) With non-overlapping CS_0 and CS_1, the secondary conditioning effect is clearly visible. (bottom) With overlapping conditioned stimuli, V_1 shows inhibition rather than excitation.

Secondary conditioning is also handled by the Balkenius model. In fact, this model is the only to give robust secondary conditioning independently of timing. This however, is done at the cost of ignoring the difference between trace and delay conditioning.

The SD model does not allow for secondary conditioning. Since the change in associative strength depends on the difference between the US and the aggregate prediction given by the sum in equation (SD3), there is no room for secondary conditioning.

3.8 Facilitation

Facilitation of acquisition can be seen when an extra stimulus CS_2 is introduced inbetween CS_1 and the US. If the conditioning to CS_1 is weak due to a long inter-stimulus interval, the extra CS_2 will facilitate conditioning to CS_1.

Normal	CS_1 + US => CS_1 -> weak CR
Facilitated	CS_1 + CS_2 + US => CS_1 -> strong CR

Note that the interval between CS_1 and the US is the same in both cases. One explanation for the facilitation effect is to consider it a form of secondary conditioning. It is therefore interesting to see that the Schmajuk-DiCarlo model that does not demonstrate secondary conditioning still shows the facilitation effect as do all the models tested.

4 Discussion

The models in this study have different objectives and it is thus not surprising that the performance differs dramatically for various experiments. None of the models claim to provide a model of all of classical conditioning phenomena; indeed, they are intended to provide explanations of only a few specific experimental situations. The experiments themselves were chosen for their appearance as examples in the texts describing the various models, or their widespread use as examples of classical conditioning.

The results should not be interpreted as any kind of ranking of the models, but instead be seen as an aid to understanding of the strengths and weaknesses of each model, and to gain a better understanding of the field of classical conditioning.

Conspiciously absent are statistical models and experiments such as Gallistel (1992). Nor have we included attentional models (Grossberg 1975). We intend to extend these simulations with many more models and experiments.

Acknowledgments

The support from The Swedish Council for Research in the Humanities and Social Sciences and the Swedish Foundation for Strategic Research is gratefully acknowledged. Other papers pertaining to this subject can be found at http://lucs.fil.lu.se/Projects/Conditioning.Habituation/.

References

Balkenius, C. (1995). *Natural intelligence in artificial creatures*. Lund University Cognitive Studies 37.

Balkenius, C. (1996). Generalization in instrumental learning. In Maes, P., Mataric, M. J., Meyer, J.-A., Pollack, J., Wilson, S. W. (Eds.) *From Animals to Animats 4: Proceedings of the Fourth International Conference on Simulation of Adaptive Behavior* (pp. 305–314). Cambridge, MA: The MIT Press/Bradford Books.

Balkenius, C. (1998). *A neural network model of classical conditioning I: The dynamics of learning*. Lund University Cognitive Studies, 68.

Barto, A. G., Sutton, R. S., Watkins, C. J. C. H. (1990). Learning and sequential decision making. In Gabriel, M., Moore, J. (Eds.) *Learning and computational neuroscience: foundations of adaptive networks* (pp. 539-602). Cambridge, MA: MIT Press.

Blazis, D. E. J., Desmond, J. E., Moore, J. W., Berthier, N. E. (1986). Simulation of the classically conditioned nictitating response of a neuron-like adaptive element: a real-time variant of the Sutton-Barto model. *Proceedings of the Eigth Annual Conference of the Cognitive Science Society* . : Erlbaum.

Desmond, J. E. (1990). Temporally adaptive responses in neural models: the stimulus trace. In Gabriel, M., Moore, J. (Eds.) *Learning and computational neuroscience: foundations of adaptive networks* (pp. 421-456). Cambridge, MA: MIT Press.

Gallistel, C. R. (1992). Classical conditioning as an adaptive specialization: a computational model. In Medin, D. L. (Ed.) *The psychology of learning and motivation: advances in research and theory* (pp. 35–67). New York: Academic Press.

Garcia, J., Koelling, R. A. (1966). Relation of cue to consequences in avoidance learning. *Psychonomic Science,* 4, 123-124.

Grossberg, S. (1975). A neural model of attention, reinforcement, and discrimination learning. *International Review of Neurobiology,* 18, 263–327.

Kamin, L. J. (1968). Attention–like processes in classical conditioning. In Jones, M. R. (Ed.) *Miami symposium on the prediction of behavior: aversive stimulation* (pp. 9–31). Miami: University of Miami Press.

Klopf, A. H. (1982). *The hedonistic neuron: a theory of memory, learning and intelligence*. Washington, D. C.: Hemisphere.

Klopf, A. H. (1988). A neuronal model of classical conditioning. 16, 2, 85–125.

Klopf, A. H., Morgan, J. S., Weaver, S. E. (1993). A hierarchical network of control systems that learn: modeling nervous system function during classical and instrumental conditioning. 1, 3, 263–319.

Pavlov, I. P. (1927). *Conditioned reflexes*. Oxford: Oxford University Press.

Rescorla, R. A., Wagner, A. R. (1972). A theory of Pavlovian conditioning: variations in the effectiveness of reinforcement and nonreinforcement. In Black, A. H., Prokasy, W. F. (Eds.) *Classical conditioning II: current research and theory* (pp. 64–99). New York: Appleton-Century-Crofts.

Schmajuk, N. A., DiCarlo, J. J. (1992). Stimulis configuration, classical conditioning, and hippocampal function. *Psychological Review,* 99, 268–305.

Schneiderman, N. (1966). Interstimulus interval function of the nictitating membrane response of the rabbit under delay versus trace conditioning. *Journal of comparative and physiological psychology,* 62, 397-402.

Schneiderman, N., Fuentes, I., Gormezano, I. (1962). Acquisition and extinction of the classicaly conditioned eyelid response in the albino rabbit. *Science,* 136, 650–652.

Schneiderman, N., Gormezano, I. (1964). Conditioning of the nictitating membrane of the rabbit as a function as the CS–US interval. *Journal of comparative and physiological psychology,* 57, 188-195.

Smith, M. C., Coleman, S. R., Gormezano, I. (1969). Classical conditioning of the rabbit's nictitating membrane response at backward, simultaneous, and forward CS-US interval. 69, 226-31.

Sutton, R. S., Barto, A. G. (1981). Toward a modern theory of adaptive networks: expectation and prediction. *Psychological Review,* 88, 135-170.

Sutton, R. S., Barto, A. G. (1987). A temporal-difference model of classical conditioning. *Proceedings of the ninth conference of the cognitive science society* . : Erlbaum.

Sutton, R. S., Barto, A. G. (1990). Time-derivative models of Pavlovian reinforcement. In Gabriel, M., Moore, J. (Eds.) *Learning and computational neuroscience: foundations of adaptive networks* (pp. 497-538). Cambridge, MA: MIT Press.

Emotion-driven Learning for Animat Control

Sandra Clara Gadanho, John Hallam
Department of Artificial Intelligence
University of Edinburgh
5, Forrest Hill
Edinburgh EH1 2QL
sandrag@dai.ed.ac.uk, john@dai.ed.ac.uk

Abstract

Models of emotion are often suggested as a way of providing an evaluation of the current behaviour of an agent. In this work, we investigate whether emotions can actually provide suitable reinforcement signals for a Q-learning system to learn adaptive policies. For this purpose a recurrent network model of emotion consistent with the somatic somatic marker hypothesis of Damásio was developed. Experimental work was done in a realistic mobile robot simulator in a simple foraging-like task. Experiments revealed that having emotions providing a context evaluation for direct use as a reinforcement signal does not work, but using them as modifiers for learning system parameters could be fruitful.

1. Introduction

The fact that emotions are considered to be essential to human reasoning suggests that they might play an important role in artificial creatures as well. This paper reports work done in trying to use emotions in autonomous robot control. Emotions seem to have solved in mammals many cognitive problems involved in autonomous systems with multiple goals and uncertain environments (Oatley, 1987). It has been argued before in the field of Artificial Intelligence that emotions are the ultimate source of intelligence and might provide robots with the autonomy they need (Toda, 1994); doubts have even been posed on whether machines can exhibit intelligence without any emotions (Minsky, 1986).

In robotics, emotions are often used in a social role as a sort of communication mechanism that allows the robot to report its internal state to others (Shibata et al., 1996) or makes it capable of generating empathy emotions in people, by creating an illusion of life in a believable character (Bates, 1994). Yet, the focus of the present research will not be the external demonstration of emotions, but its use in the control of a learning robot.

The work was done under an animat philosophy (Wilson, 1991), by building bottom-up a biologically inspired complete agent by synthesis. A robot was equipped with a recurrent network model of "emotions" which incorporates the important computational features of Damásio's somatic marker hypothesis (Damásio, 1994). Yet, the developed model is based on a simplified hormone system and is far from the complexity of true emotions experienced by humans. Experiments were carried out on a simulated Khepera robot in an animal-like adaptation task to assess the usefulness of emotions as a source of context value in a reinforcement learning task.

In the following sections, sections 2 and 3, a short description of the emotion model developed and of the adaptive controller used is presented. Next, in section 4, a description of the experiments is given and the results are discussed. A more detailed description of the system and experiments made can be found in the extended version of this paper (Gadanho and Hallam, 1998). Finally, section 5 concludes that emotions modelled using our realisation of Damásio's ideas do not provide as good a reinforcement as expected, but may be used successfully in the modulation of learning.

2. Emotions

It is well established that emotions play an important role in reasoning, even in terms of basic mechanisms like memory and remembering (Schwartz and Reisberg, 1991), although most neurologists have usually given them a more disruptive than helpful function.

People like to have a Cartesian approach to life in thinking that all their reasoning is purely logical. In fact, individuals do not always make rational choices (Grossberg and Gutowski, 1987) and pure logical reasoning shows serious faults when used to model human reasoning in the field of Artificial Intelligence (Dreyfus, 1992). Recent neurophysiological research suggests that our thinking is not so detached and ungrounded. With the help of the emotions, the feelings provided by our body play an important role in reasoning. This is the central claim of the somatic-marker hypothesis[1] (Damásio, 1994).

[1] Marker because it marks an internal image and somatic because it is the body that does it.

Damásio makes a clear distinction between the concepts of emotion and feeling that will be used in the current work. Feeling designates the process of monitoring the body. Feelings offer us the cognition of our visceral and musculoskeletal state. Emotion is a combination of a mental evaluative process with dispositional responses to that process, mostly toward the body proper but also toward the brain itself. Emotions all generate feelings, but only some feelings generate emotions. If feelings are associated with emotions then the body signals will move from the background to the foreground of our attention.

Somatic markers are special instances of body feelings (visceral and non-visceral sensations) generated by emotions, that are acquired by experience based on internal preference systems and external events and which help to predict future outcomes of certain scenarios. They will force attention on the negative or positive outcome of certain options, that can be immediately defeated leaving fewer alternatives or can be immediately followed. This way, the somatic markers provide humans with a reasoning system that is free from many of the faults of formal logic, namely the need for much computational and memory power for having every option thoroughly evaluated.

Many emotions theorists agree that emotions are most helpful on focusing attention in the relevant features of the problem at hand (De Sousa, 1987; Tomkins, 1984; Plutchick, 1984; Scherer, 1984; Panksepp, 1982).

Inspired by the ideas that have been presented, an emotion model was developed that is described next.

2.1 Model

A large subset of theories of emotions is based on cognitive appraisal theories (Lazarus, 1982; Power and Dalgleish, 1997), although some evidence exists to suggest that emotions can be aroused without cognition (Zajonc, 1984).

Following the psychologists' main stream, most AI models of emotions are based on an analytic approach (Sloman et al., 1994; Pfeifer, 1982; Bates et al., 1992) that tries to endow the model with the full complexity of human emotions as perceived from an observer's point of view. In opposition to this kind of approach, a bottom-up approach was followed here. In both ontological development and evolution the full richness of emotions is only achieved at a final stage. In the first stages of these processes only certain basic emotions are present and, later, other more complex emotions develop on top of these.

It is the purpose of this model to have an emotion system that provides the decision making process of the robot with a value judgement based on what the robot feels. The model that was developed — figure 1 — is based on four basic emotions: Happiness, Sadness, Fear and Anger. The intensity of each emotion is based on the robot's current internal feelings. These feelings are: Hunger, Pain, Temperature, Restlessness and Eating. Each

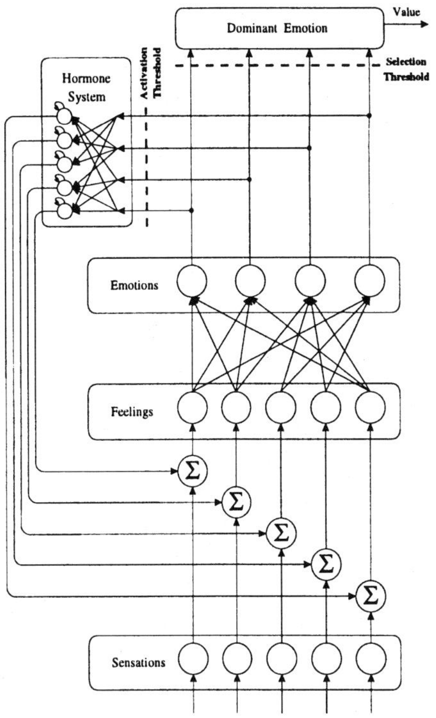

Figure 1 Emotion model.

emotion is defined by a set of constant feeling dependencies and a bias. The values of the dependencies were carefully chosen to provide adequate emotions for the possible body states. For example, the sadness intensity will be high if hunger and restlessness are high and the robot is not eating.

Furthermore, based on what was suggested in (Damásio, 1994), the emotion state should also influence the way the robot feels. In general, the body reactions aroused by emotions also give rise to the emotions that create them. In this model each emotion tries to influence the body state in such a way that the result body state matches the state that gives rise to that particular emotion.

When an emotion is active, *i.e.* its intensity value is significantly large, then it will influence the body through a hormone system, by producing appropriate hormones.

The hormone system in the model is a very simplified one. It consists in having one hormone associated with each feeling. A feeling intensity is not a value directly obtained from the value of the body sensation that gives rise to it, but from the sum of the sensation and hormone value. The hormone values can be (positively or negatively) high enough to totally hide the real sensations from the robot's perception of its body.

The hormone quantities produced by each emotion are directly related to its intensity and its dependencies on the associated feelings. The stronger the dependency on a certain feeling, the greater quantity of the associated hormone is produced by an emotion.

On the one hand, the hormone mechanism provides a sort of fight between the emotions to gain control over the body which is ultimately what selects which emotion will be dominant. On the other hand, what the robot feels — its body image — is not only dependent on its sensations but is also dependent on its emotional state.

The hormones' values can increase quite rapidly, allowing for the quick build up of a new emotional state, and decrease slowly allowing for the persistence of an emotional state even when the cause that has given rise to it is gone, another of the characteristic features of emotions.

The dominant emotion is the emotion with the highest intensity, unless no emotion intensity exceeds a selection threshold. In this case, there will not be a dominant emotion and emotional state will be neutral. Emotions were divided into two categories: positive and negative. The ones that are considered "good" are positive (only Happiness, in the set of emotions used), the others are considered negative. This way a value judgement can easily be obtained from the emotion system by considering the intensity of the current dominant emotion and whether it is positive or negative.

In summary, the model of emotions described provides not only a value judgement about the current situation, but also influences the body perception. In order to evaluate the role of emotions in reasoning, this value should be used for the actual control of the robot, determining its behaviour (Albus, 1990; Wright, 1996; Moffat et al., 1993). The next section describes an adaptive controller that will make use of such a value.

3. Adaptive Action Selection Controller

It is widely accepted that robot learning is beneficial, because it is often difficult or even impossible for the designer to anticipate all possible scenarios the robot will be confronted with. In the present work an adaptive controller based on reinforcement learning was used. The controller developed — figure 2 — tries to maximise the evaluation received by selecting one of six possible discrete actions, taking into account the current sensor readings and robot feelings, and the previously received evaluations.

The controller has two separate modules:

ASSOCIATIVE MEMORY MODULE — This plastic module uses feed-forward networks to associate the sensor readings and feelings with the current expected value of each one of the six robot actions (slow move forward, fast move forward, turn left, turn right, stop, and move slowly backwards with a slight twist to the right). Q-learning (Watkins, 1989) was used in an implementation very similar to the one reported by Lin (Lin, 1992). The networks were used to learn by backpropagation utility functions that model $util(x, a)$:

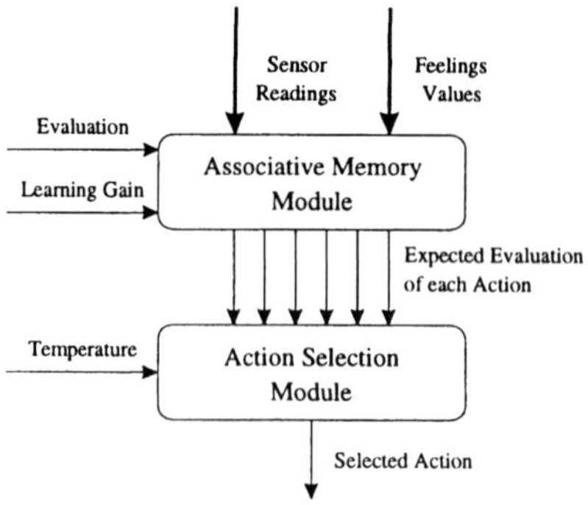

Figure 2 Adaptive action selection controller.

$$util(x, a) = R + \gamma eval(y) \quad (1)$$

The value R is the immediate reinforcement. The function $eval(y)$ is the expected cumulative discounted reinforcement starting from state y reached by doing action a in state x. The discount factor (γ) was set to 0.9. For each iteration, the target value u will be given to the network whose action was used in the previous iteration:

$$u = R + \gamma \text{Max}\{util(y, k) \mid k \in actions\} \quad (2)$$

ACTION SELECTION MODULE — Based on the information provided by the previous module, this module makes a stochastic selection of the action to take at each step. For a temperature T, the probability of selecting action a is:

$$P(a) = \frac{e^{\frac{v_a}{T}}}{\sum_{i=1}^{6} e^{\frac{v_i}{T}}} \qquad v_i = \text{Value of action } i \quad (3)$$

In order to have more coherent behaviour, the selection of a new action is not made every cycle: there is a certain inertia of the current action that is directly correlated with its probability.

4. Experiments

4.1 Experimental setup

Experiments were carried out in a Khepera simulated robot (Michel, 1996) within the environment shown in figure 3. This is a closed environment with some walls and three lights surrounded by bricks.

The emotion-based robot controller was tested in the context of an animal-like creature with self-maintenance needs, making it easier to conceptually ground the emotion system. Simulated feeding needs were therefore added to the robot. The robot is always losing energy, the more it uses its motors the more energy is used up. It can recover its energy from light. Apart from feeding itself by standing next to the lights, the robot is supposed

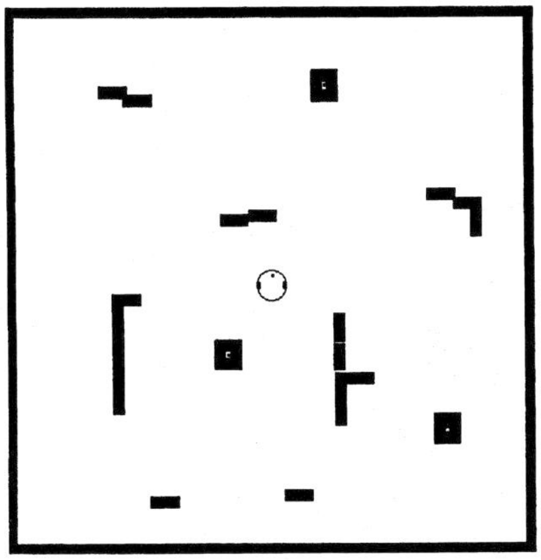

Figure 3 The robot and its environment.

to wander in order to avoid getting restless and avoid walls because bumping into the walls causes it pain.

All experiments[2] consisted in having the robot learn for a total of 120,000 steps. The robot was evaluated in every 2,000 steps. The robot's evaluation was based on the reinforcement values it received in a 2,000 step testing period with learning turned off. There were two evaluations, each one based on a different reinforcement function. One was the mean of the emotion-dependent reinforcements and the other was the mean of the sensation-dependent reinforcements obtained during its test period (a detailed description of the reinforcement functions will be given in the next section). The higher the value, the better is the evaluation. Good robot behaviour is usually associated with positive reinforcement and bad behaviour with negative reinforcement.

For each experiment, fifty different robots were used, each placed in a randomly selected starting position. The experimental data that will be shown in the results graphs are the means of the reinforcement values obtainded by each of the fifty robot during the testing periods. The error bars show the 95% confidence interval.

4.2 Results

The most usual role attributed to emotions in the functionality of a complete working agent is that of providing an evaluation of the state of the world[3], *e.g.*, (Albus, 1990; Wright, 1996). It is often assumed that human decision making consists in the maximisation of positive emotions and minimisation of negative emotions, *e.g.*(Tomkins, 1984). Therefore we expect that *an emotion-based evaluation of the context should be adequate as a reinforcement signal for policy acquisition by Q-learning.*

To test this hypothesis an experiment was done, consisting in giving the controller described previously an emotion-dependent reinforcement value. The reinforcement was set to be the intensity of the current dominant emotion or zero if there was no dominant emotion. If the dominant emotion was a negative one then its (positive intensity) value would be negated.

Results are shown in figure 4. The left graph shows the values for the reinforcement function used in this experiment. The right graph shows the reinforcement function based on direct sensations and it is there only for direct comparison with the results of the controller presented next. The analysis of the results suggested the existence of two quite differentiated populations, one that manages to learn the task and another that does not. For this reason, in the presentation of the results two populations were distinguished, based on the emotion-dependent reinforcement obtained at the last ten evaluation points. The trials that had quite negative reinforcement for these testing points formed one population which amounted to 38% of the total number of trials. The remaining trials were included in the "successful" population.

It should be noticed that the robot's adaptation task must be achieved in a limited amount of time. If the robot takes too long to adapt, the reinforcement will lose meaning and the task will become impossible. The reason for this is that if the robot does not learn to feed itself, it will get increasingly hungry. It will eventually arrive in a state where it will keep getting low reinforcement on account of its hunger, independently of what it does. This is why two very different populations can co-exist: only one managed to learn the task on time.

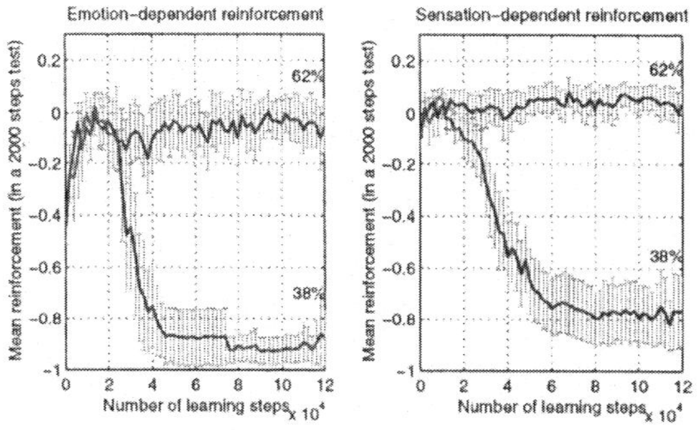

Figure 4 Learning controller with emotion-dependent reinforcement.

At this point, the role of emotions was re-evaluated. The reason for the poor results obtained appears to be that the controller is not receiving the kind of reinforcement it needs. The controller needs a good evaluation of the situation as it stands at the moment and not the misleading mixed evaluation of present and recent past situations that the emotions provide.

[2] A detailed account of all parameters used and their values is given in (Gadanho and Hallam, 1997).

[3] There are also some researchers, *e.g.*, (Michaud et al., 1996; Shibata et al., 1996) who give emotions the somehow more sophisticated role of monitoring the robot's performance so that the robot's plans or actions can be changed if necessary.

It looks as if *the reason for the emotion-dependent evaluation failure is the recurrent and lateral influences of emotions in the model.* To test this hypothesis, a control test was made that consisted in giving the controller a putatively more adequate reinforcement signal. This new signal was based on the previous one but without any temporal or lateral side effects. The sensations were used instead of the feelings to calculate the value of each emotion and the highest of these values was selected to be the reinforcement value (see figure 5). As in the previous experiment, the value of "bad" emotions was negated. This procedure provides a more traditional reinforcement that directly reflects the immediate situation. The results obtained in this experiment are shown in figure 6. They are a considerable improvement on the results obtained with the previous experiment which used emotion-dependent reinforcement.

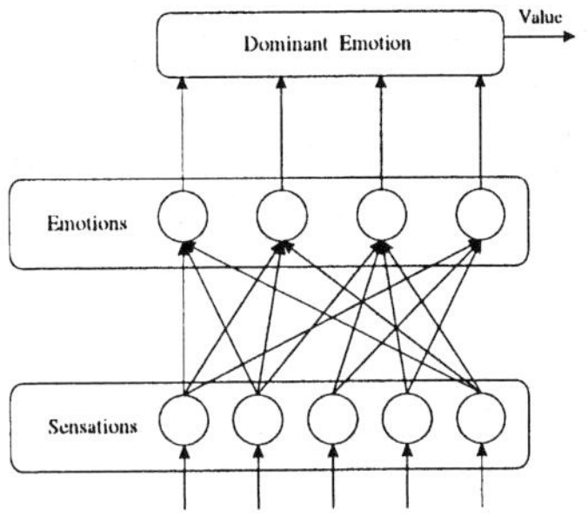

Figure 5 Truncated emotion model used to obtain the sensation-dependent reinforcement.

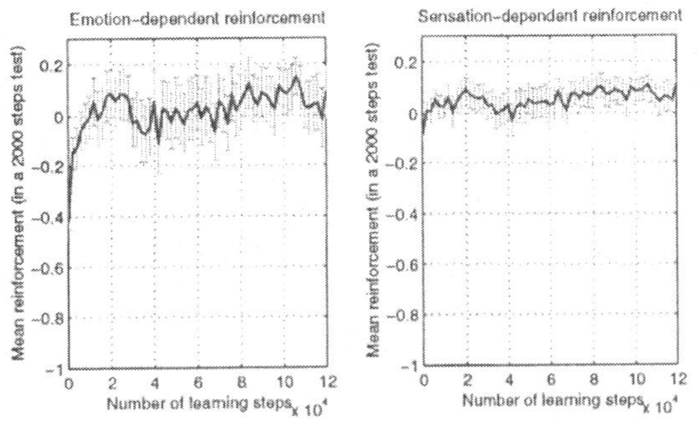

Figure 6 Learning controller with sensation-dependent reinforcement.

The results reported so far do not seem very impressive in terms of final reinforcement obtained. Even the most successful do not seem to do much more than to keep their reinforcement. Yet, it should be clear that even just keeping the reinforcement is quite good. A controller selecting randomly between all available actions will actually have decreasing reinforcement (because of increasing hunger) through out the entire experiment.

To determine how much reinforcement a reasonably successful behaviour might receive, a behaviour was designed to take full advantage of its environment and achieve maximum reinforcement. New experiments were made with the robot starting off with this human designed behaviour. With no learning this behaviour ends up receiving better reinforcements than the previous experiments. When the robot is allowed to learn either with sensation-dependent or emotion-dependent reinforcement, it maintains the initial behaviour and keeps similar reinforcement values. So, although such successful behaviour was not achieved by the learning system alone, this behaviour is stable under the learning algorithm. It should be noticed that when designing the behaviour, a slightly different set of actions had to be used and that the same behaviour cannot be reproduced with the set of actions normally used. Nevertheless, slightly more sophisticated behaviours were learned in the previous experiments that achieved the same kind of reinforcement for long periods of time.

Experiments		Emotion reinforcement	Sensation reinforcement
Human designed	no learning	0.54	0.30
	with learning	0.53	0.29
Sensation learning		0.07	0.08
Emotion learning	group 1	−0.04	0.04
	group 2	−0.90	−0.78
Random		−0.81	−0.32

Table 1 Mean reinforcement values in the last ten testing points of each experiment.

Table 1 summarises the results obtained. Given the results presented in this paper and further experiments made to comprove their correctess (Gadanho and Hallam, 1998), one has to conclude that the emotions are quite unsuccessful in providing a good reinforcement value.

Another issue relevant to the model used is whether using feelings or sensations for the robot's body perception makes a difference in terms of its final performance. In general, the values given as input to the neural networks are not the robot's raw sensations, but feelings that are influenced by the emotions through the hormone system. In the particular task tested, the influence of emotions on body perception is unnoticeable in terms of final behaviour.

5. Conclusion

The model of emotions behaved appropriately when tested on the animat, in the sense that the robot consistently displays plausible contextual emotional states during the process of interacting with the environment. Furthermore, because its emotions are grounded in body feelings, and not direct sensory input, it manages to avoid sudden changes of emotional state, from one extreme emotion to a completely different one. The more different the emotions are, the more difficult it is to change from one to the other. The physiological arousal caused by emotions has been repeatedly left out of cognitive theories of emotions[4], because it is not considered cognitively interesting, yet without it emotions lack their characteristic inertia (Moffat et al., 1993).

Unfortunately, the experiments done failed to show that emotions can be used for evaluation in robot learning. More importantly, the results do show that the role of emotions is more intricate than that often assumed and that a simple approach to the use of emotions as context judgement values suitable for direct use as reinforcement is not very successful when a more than usually realistic emotion model is used. The emotions do not really provide a good evaluation of what is going on at any one moment, but are a sort of mixed evaluation the robot acquired from its past experiences. This may be good to modulate its behaviour, but should not to be taken at face value when trying to predict the outcome of each one of its actions. Preliminary results (Gadanho and Hallam, 1998) suggest that emotions can be successfully used in modulating the learning rate and the exploration versus exploitation ratio.

Acknowledgements

The first author is supported by a scholarship from the program PRAXIS XXI. Facilities for this work were provided by the University of Edinburgh.

References

Albus, J. S. (1990). The role of world modeling and value judgement in perception. In *Proceedings of the 5th IEEE International Symposium on Intelligent Control*, Los Alamitos, California. IEEE Computer Society Press.

Bates, J. (1994). The role of emotions in believable agents. Technical Report CMU-CS-94-136, Carnegie Mellon University, School of Computer Science.

Bates, J., Loyall, A. B., and Reilly, W. S. (1992). Integrating reactivity, goals, and emotion in a broad agent. CMU-CS 92-142, Carnegie Mellon University, School of Computer Science.

Damásio, A. R. (1994). *Descartes' error — Emotion, reason and human brain.* Picador.

De Sousa, R. (1987). *The rationality of emotion.* The MIT Press.

Dreyfus, H. L. (1992). *What computers still can't do: A critique of artificial reason.* The MIT Press.

Gadanho, S. C. and Hallam, J. (1997). Exploring autonomy mechanisms: the role of emotions. DAI Research Paper 851, University of Edinburgh.

Gadanho, S. C. and Hallam, J. (1998). Emotion-driven learning for animat control. DAI Research Paper 881, University of Edinburgh.

Grossberg, S. and Gutowski, W. (1987). Neural dynamics of decision making under risk: Affective balance and cognitive-emotional interactions. *Psychological Review*, 94(3):300–318.

[4] The FEELER system (Pfeifer, 1982) can be considered an exception.

Lazarus, R. S. (1982). Thoughts on the relations between emotion and cognition. *American Psychologist*, 37:1019–1024.

Lin, L.-J. (1992). Self-improving reactive agents based on reinforcement learning planning and teaching. *Machine Learning*, 8:293–321.

Michaud, F., Lachiver, G., and Dinh, C. T. L. (1996). A new control architecture combining reactivity, planning, deliberation and motivation for a situated autonomous agent. In *From animals to animats 4*. The MIT Press.

Michel, O. (1996). Khepera simulator version 2.0. Robot simulator written at the University of Nice Sophia–Antipolis. Downloadable from the World Wide Web at `http://wwwi3s.unice.fr/~om/khep-sim.html`.

Minsky, M. (1986). *The Society of Mind.* Simon and Schuster, New York.

Moffat, D., Frijda, N., and Phaf., R. (1993). Analysis of a model of emotions. In et al., A. S., editor, *Prospects for Artificial Intelligence*, pages 219–228. IOS Press.

Oatley, K. (1987). Editorial: Cognitive science and the understanding of emotions. *Cognition and Emotion*, 1(3):209–216.

Panksepp, J. (1982). Toward a general psychobiological theory of emotions. *The behavioural and brain sciences.*

Pfeifer, R. (1982). Cognition and emotion: An information process approach. CIP working paper 436, Department of Psychology, Carnegie-Mellon University.

Plutchick, R. (1984). Emotions: A general psychoevolutionary theory. In Scherer, K. R. and Ekman, P., editors, *Approaches to emotion.* Lawrence Erlbaum, London.

Power, M. and Dalgleish, T. (1997). *Cognition and Emotion.* Psychology Press.

Scherer, K. R. (1984). On the nature and function of emotion: A component process approach. In *Approaches to emotion.* Lawrence Erlbaum, London.

Schwartz, B. and Reisberg, D. (1991). *Learning and memory.* W.W. Norton and Company, Inc.

Shibata, T., Ohkawa, K., and Tanie, K. (1996). Spontaneous behavior of robots for cooperation - emotionally intelligent robot system. In *Proceedings 1996 IEEE International conference on Robotics and Automation.*

Sloman, A., Beaudoin, L., and Wright, I. (1994). Computational modeling of motive-management processes. In *Proc. of the Conf. of the Int. Society for Research in Emotions*, Cambridge. ISRE Publications. Poster.

Toda, M. (1994). The urge theory of emotion and cognition: Chapter 2 basic structure of the urge operations. SCCS technical report 94-1-01, Chuyko University.

Tomkins, S. S. (1984). Affect theory. In *Approaches to emotion.* Lawrence Erlbaum, London.

Watkins, C. (1989). *Learning from delayed rewards.* PhD thesis, King's College.

Wilson, S. W. (1991). The animat path to AI. In *From animals to animats.* MIT Press.

Wright, I. (1996). Reinforcement learning and animat emotions. In *From animals to animats 4.* The MIT Press.

Zajonc, R. B. (1984). On primacy of affect. In Scherer, K. R. and Ekman, P., editors, *Approaches to emotion.* Lawrence Erlbaum, London.

Skill Acquisition and Self-Improvement for Environmental Change Adaptation of Mobile Robot

Takashi Minato and Minoru Asada
Dept. of Adaptive Machine Systems
Graduate School of Engineering
Osaka University
Suita, Osaka 565, Japan
minato@er.ams.eng.osaka-u.ac.jp
asada@ams.eng.osaka-u.ac.jp

Abstract

Learning and development are essential processes for an animat to adapt itself to environmental changes so as to accomplish a given task. This paper proposes a single mechanism for learning and self-improvement that results in learning curves similar to the "U-shape" phenomena observed in several psychological experiments concerning the human learning process such as in language acquisition. The basic idea is that (1) the animat monitors its success rate in goal achievement so as to perceive environmental changes instead of relying on signals from a teacher, and (2) in order to reuse acquired knowledge and accelerate reinforcement learning, the animat does not memorize the action values but transfers only the learned policy. The resultant policy (a state transition map where transitions indicate the best actions) may not be optimal in any given environment but it may be able to better handle differences between environments. We apply this model to a mobile robot navigation problem for which the task is to reach the target while avoiding obstacles by means of uninterpreted sonar and visual information. Our experimental results demonstrate the validity of the model.

1. Introduction

Learning and development are essential processes for biological and artificial systems alike. Robots, which may be required to adapt themselves to different environments, provide a typical example. Conventional methods achieve goals in different environments by employing a different module to cope with each kind of environment. However, such methods are limited by the capacity of processing modules. How do biological systems overcome this problem and how could adaptive animats do the same?

In psychology, several experiments indicate the "U-shape" phenomena in the learning of various kinds of skills in humans (Elman et al., 1996). First learning improves monotonically, then performance drops, and finally it rises up again. A typical example can be observed in children who are learning the past tense, that is, to conjugate both regular and irregular verbs correctly (Rumelhart and McClelland, 1986). Researchers had disputed whether a single or a dual mechanism was implicated. An simple solution involves one mechanism for regular verbs and another for irregular. Instead, Rumelhart and McClelland (Rumelhart and McClelland, 1986) showed that a single mechanism plus a carefully designed learning schedule could give the same U-shaped results. Despite much debate and criticism (Pinker and Prince, 1988; Plunkett and Marchman, 1991; Marcus et al., 1992), micro U-shaped curves have been observed in child learning processes for vocabulary development, past tenses of English verbs, physical event cognition, and so on. An artificial neural network has produced similar results (Plunkett et al., 1992), indicating that a single mechanism could cope with different tasks. New tasks were introduced after fixing the policy for the tasks learned so far. That way, the policy that had already been learned was not unlearned while new skills were being acquired.

In this paper, we propose a single mechanism for learning and self-improvement in a mobile robot. The robot must overcome a navigation problem in different environments. The robot learner continually monitors its success rate in achieving the goal in order to perceive changes in the environment when it encounters them. Thus, we distinguish our system from models assumed in several psychological experiments and in artificial neural networks as applied to supervised learning since these models make use of explicit signals from a teacher. The learner does not need to find changes in the environment unless its success rate worsens.

A task domain similar to ours has been dealt with in the lifelong learning area of machine learning (Thrun and Mitchell, 1996). This approach reuses learned policies as *a priori* knowledge to accelerate improvement. Tanaka and Yamamura (Tanaka and Yamamura, 1997) applied a similar idea to a simple grid-world navigational task using a method which combined reinforcement learning with stochastic gradient ascent. We distinguish our method from these because

1. although they accelerate learning, the robot must have a different policy for each kind of environment; our model has only a single policy.
2. more importantly, to learn multiple policies, the robot is explicitly informed of a change in environment; the robot in our model detects a change only when necessary, because the success rate worsens.

To realize a single policy mechanism and accelerate learning, the robot does not keep the action values obtained by reinforcement learning in the previous environment. It only uses the policy (state transition map where transitions indicate the best actions) for action selection.

We also adopt the "learning from easy missions" (LEM) paradigm (Asada et al., 1996) by which the initial positions of the robot are controlled to accelerate the learning. Since LEM is basically considered as a technique for learning in a single environment, we do not deal with it in details here.

The resultant policy obtained by our model does not seem optimal in each individual environment, but may absorb the differences between multiple environments. The remainder of this article is structured as follows. First, the method is explained in details with a brief summary of reinforcement learning, especially Q-learning. Next, the task and some assumptions are given. Finally, we examine the experimental results to test the validity of the model and consider future work.

2. Learning and Self-Improvement

2.1 Basics of Reinforcement Learning

Before getting into the details of our system, we briefly review the basics of Q-learning (Kaelbling, 1993). For a more through treatment, see (Watkins and Dayan, 1992).

We assume that the robot can discriminate the set $\boldsymbol{S}$ of distinct world states, and can take the set $\boldsymbol{A}$ of actions on the world. The world is modeled as a Markov process, making stochastic transitions based on its current state and the action taken by the robot. Let $T(s, a, s')$ be the probability that the system will transit to the next state s' from the current state-action pair (s, a). For each state-action pair (s, a), the *reward* $r(s, a)$ is defined.

Without initial knowledge on T and r, we construct incremental estimates of the action values called Q values on line. Starting with $Q(s, a)$ at any value, usually 0, every time an action is taken update the Q value as follows:

$$Q(s,a) \Leftarrow (1-\alpha)Q(s,a) + \alpha(r(s,a) + \gamma \max_{a' \in A} Q(s',a')). \quad (1)$$

where r is the actual reward value received for taking action a in a situation s, s' is the next state, and α is a learning rate (between 0 and 1).

2.2 Algorithm

As described in **1**, the basic ideas of our method are:

1. by monitoring its success rate, the robot can decide when to restart Q-learning, regardless of the actual change in the environment it encounters (this approach is quite different from existing methods); and,
2. in order to accelerate learning, the action values obtained by Q-learning in the previous environment are not reused for Q-learning in the current environment, but only the policy (action selection) is used. Actually, we have attempted to reuse the action values, but we have often observed that they prevented the robot from learning a new policy.

The algorithm is as follows:

1. Quantize the state space as S.
2. Apply Q-learning to the initial environment, and obtain the policy $P : \boldsymbol{S} \rightarrow \boldsymbol{A}(\boldsymbol{A}$:action set) with the success rate R_s
3. Apply P to any environments unless R_s decrease.
4. If R_s decreases, then find states $\boldsymbol{S_r} \subset \boldsymbol{S}$ where P fails to achieve the goal, and modify P for such states by applying Q-learning as follows until R_s recovers to pre-specified adaptability rate β.
   (a) Apply Q-learning to $\boldsymbol{S_r} \subset \boldsymbol{S}$. Action selection during the learning is as follows:
      if $s \in \boldsymbol{S_r} \cup \boldsymbol{S_n}$ follow the normal action selection in Q-learning, where $\boldsymbol{S_n} \subset \boldsymbol{S}$ denotes inexperienced states.
      else follow P
   (b)
5. Go to 3 with the obtained policy P.

The adaptability rate β determines the extent to which re-learning occurs, that is:

$$R_{sd} = R_{sc} + \beta(R_{sp} - R_{sc}), \quad (2)$$

where R_{sp}, R_{sc}, R_{sd} denotes the success rates in the previous environment, in the current one, and the desired one, respectively.

Based on the selected state vector, we apply the algorithm to the given task with the following specifications:

- the learning rate $\alpha = 0.25$, and the discounting factor $\gamma = 0.9$.
- If the robot reaches the target, the positive reward 1 is given. Otherwise 0.
- One trial terminates if the robot reaches the target, makes a collision with any obstacles, or the given time limit expires.

3. Task, Robot, and Assumptions

3.1 Our Robot

Our robot has a Power Wheeled Steering (hereafter PWS) system driven by two motors. We can send commands to each motor independently. In our experiment, we quantized each motor command $\omega_{l(r)}$ into three levels which correspond to forward, stop, and backward, respectively. Totally, the robot has 9 actions.

The robot is equipped with a ring of 12 ultrasonic range sensors (ranging from 0.0 to 300 cm), which have high accuracy for incident angles of less than 15° from the surface normal. The robot is also equipped with a CCD camera. These sensors have their inherent characteristics as follows:

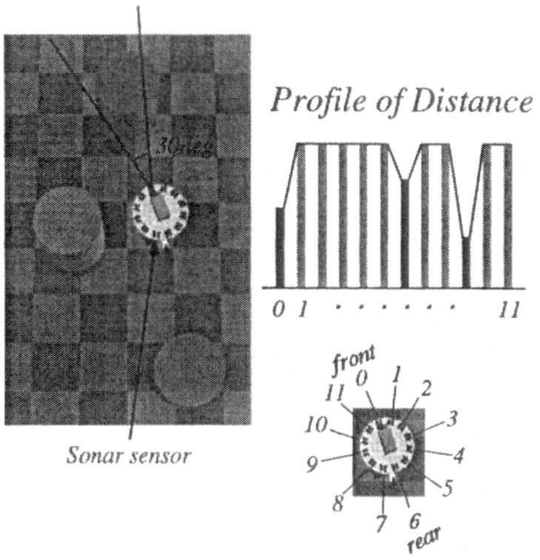

(a) sonar

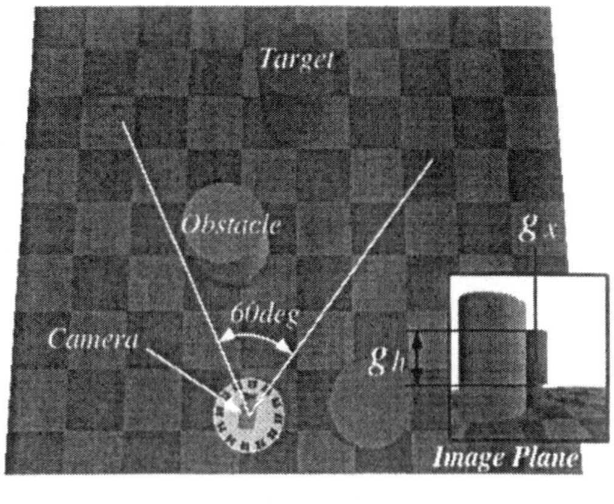

(b) vision

Figure 1 Sensory information

- **Sonar**
  Using 12 sonar sensors, our robot can sense its surrounding environment in robot centered polar coordinates as a profile of the distance D_i $(i = 0 \sim 11)$ as shown in Figure 1(a). Each sonar sensor in the ring has a field view of roughly 30°. Sonar sensors cannot identify the object (the target or something else).
- **Vision**
  Image processing provides the position and size of the target in the image, even if the object's sensory projections are deformed by occlusion (see Figure 1(b)). However, it is not given information on how to detect obstacles and, therefore, cannot detect them.

3.2 Task

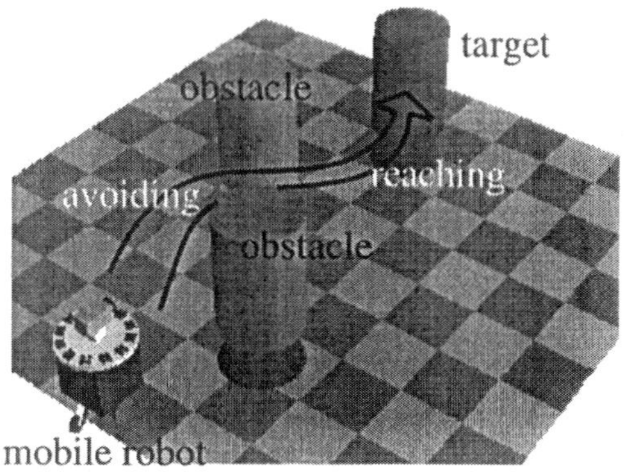

Figure 2 Task and environment

The task of the robot is to reach the target while avoiding obstacles as shown in Figure 2. As mentioned earlier, there are two difficulties with this task:

- visual and sonar information has not been preassigned to specified roles in order to accomplish the given task; therefore, the robot has to learn what kind of information is to be used in which situation. In other words, the sensory data has not been interpreted for the robot, and
- both target reaching and obstacle avoidance tasks have to be achieved simultaneously, through the learning process.

Nakamura et al. (Nakamura et al., 1996) have devised a system that has a limited ability to cope with the above problems in a single isolated environment, but it suffered from the curse of dimensionality: a huge state space. In addition, the environment may change in a few ways here:

1. target and obstacles configuration may change, and
2. the number of obstacles may also change.

Therefore, a learned policy obtained in a single environment may not be applicable to different environments, and usually it takes an enormous amount of time if the robot learns from scratch.

3.3 State Vector Selection

As mentioned above, the state space construction problem is one of the most serious issues in reinforcement learning even in a single isolated environment. Since our robot has a considerably large sensor space, we have to build a reduced-size state space from the original sensor space. As primitive features, we have selected the center position g_x and the height g_h of the target image from vision, and the following from sonar profile (see Figure 3).

- d_{min}: minimum range value
- d_{max}: maximum range value observed
- d_{mean}: mean range value observed
- d_{diff}:$d_{max} - d_{min}$
- θ_{min}: direction of d_{min}
- θ_{max}: direction of d_{max}
- θ_{mean}: mean between θ_{min} and θ_{max}
- θ_{diff}: width between θ_{min} and θ_{max}

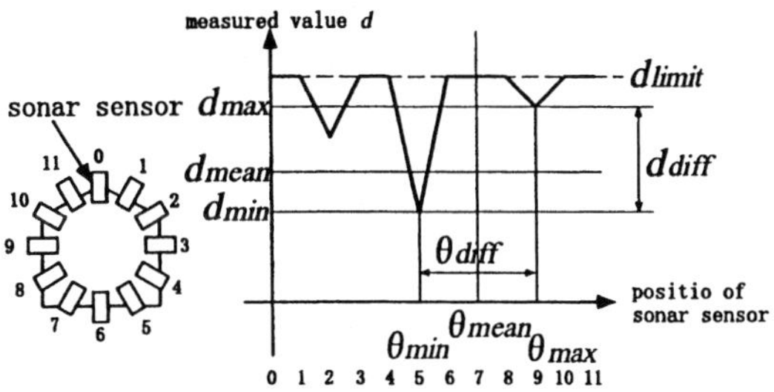

Figure 3 Primitive features from sonar profile

Since these features still constitute a large feature space, we have checked all combinations of state vectors under the constraints of memory space and limited learning time in a single environment, and selected the following state vector $\boldsymbol{x}$ for the task. To focus on the skill acquisition and self-improvement, we skip the details of this procedure, which have been published elsewhere (Minato and Asada, 1998).

$$\boldsymbol{x} = \begin{pmatrix} x_1 \\ x_2 \\ x_3 \\ x_4 \end{pmatrix} = \begin{pmatrix} g_x \\ g_h \\ \theta_{min} \\ d_{min} \end{pmatrix} \quad (3)$$

4. Experimental Results

In order to show the validity of the proposed method, we have applied three kinds of methods to a series of five different environments called A, B, C, D, and E of which top views are shown in Figures 4 and 5 where a solid black circle and gray circles indicate the target and obstacles, respectively. As we can see, the configuration of the goal and obstacles, and/or the number of obstacles are different from each other.

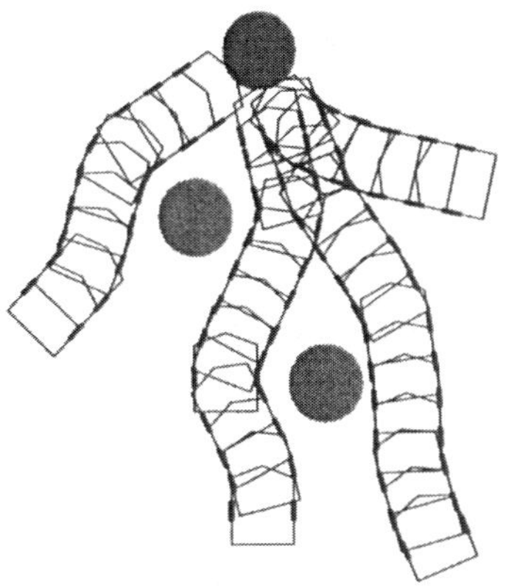

Figure 4 Environment A and successful trajectories

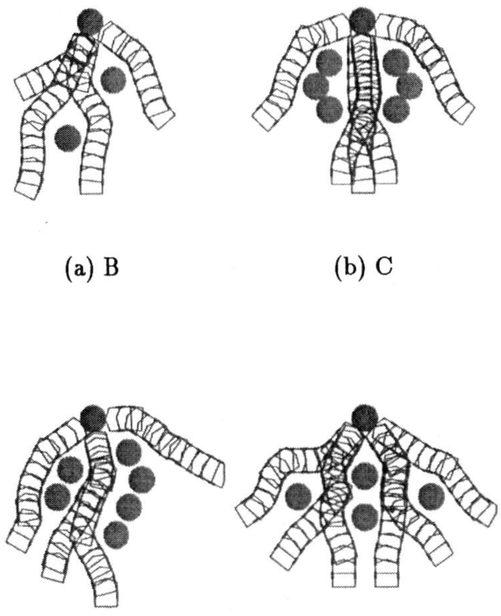

(a) B (b) C

(c) D (d) E

Figure 5 Four more environments and successful trajectories

Three methods are:

1. Q-learning with multiple policies: the robot is informed when the environments changes, and all action values are initialized for new learning
2. The proposed method:
3. Another method similar to the proposed one: instead of policy transfer for action selection, all the action values are always retained during the learning.

Each trial terminates when

- the robot reaches the goal,
- the robot makes a collision with any obstacles, or
- the pre-specified period (300 time steps) is expired.

Figure 6 shows the change of the success rate of the Q-learning with multiple policies. The horizontal axis indicates the number of trials. The success rate is measured every 500 trials, the first 200 of which are for normal Q-learning, and the remaining 300 of which are for success rate measuring based on the current policy (action selection is fixed). If the success rate stably achieves the pre-specified one (here 90%), then the policy is fixed and all 500 trials are for success rate measuring. As mentioned in **1**, we applied the LEM (Learning from Easy Missions) paradigm, in this case, three learning stages (easy, moderate, and hard ones) are prepared, therefore two sudden drops can be seen until around the 35,000th trial where the environments changes from A to B, and all the action values are reset to all zeros for new learning in B. Similarly, the robot encounters C, D, and E. Totally, about 145,000 trials are needed for the robot to adapt itself to different environments.

Figure 7 shows the change of the success rate of the proposed method. The pre-specified success rate is 90%, the same as the above, and we set the adaptability rate β as 0.8. The shape of the curve until around the 35,000th trial is completely the same as in Figure 6, but hereafter, the curve has different shape from that by the first method. The changes of the environments have not been informed, but the robot has perceived these changes by monitoring the success rate. Comparing with Figure 6, the total learning time is almost half, and when it encounters A again, the robot has not perceived any changes because the success rate has not dropped down. This implies that the resultant policy is capable of not simply adaptation but also generalization, too.

As the acquired knowledge, the proposed method transfers only the current policy. Further, we may expect to use the action values to accelerate new learning. Figure 8 shows the result of this attempt where it has taken much longer time to adapt itself to B and C, and what's worsen is no convergence can be seen for D until the 200,000th trial. The main reason seems that the state transitions can be different and sometimes opposite, therefore it may take much more time to obtain the correct action values than in case of resetting all action values to zeros.

Examples of the successful trials are shown in Figures 4 and 5 as robot trajectories, some of which do not seem optimal due to a single policy mechanism.

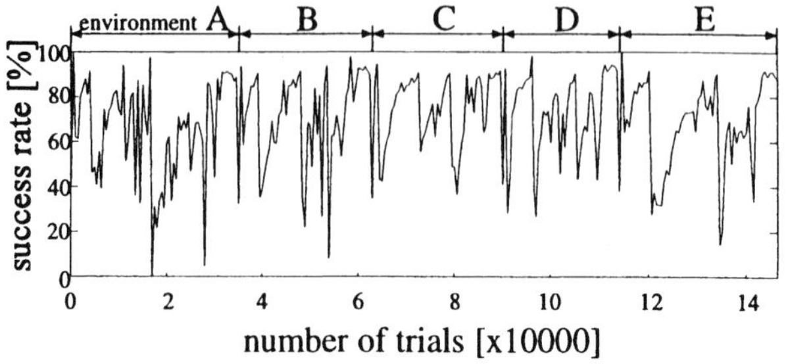

Figure 6 Q-learning with multiple policies

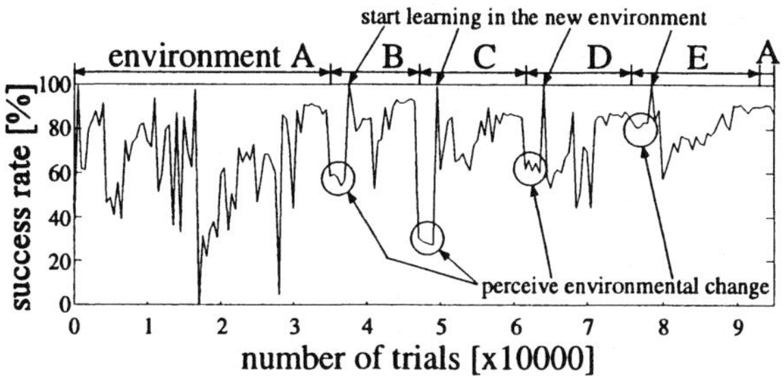

Figure 7 The proposed method

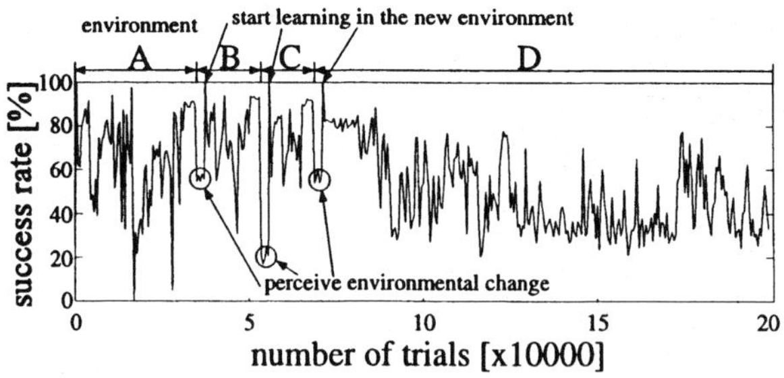

Figure 8 Another method

5. Discussion

We have proposed the model of skill acquisition and self-improvement for a mobile robot to adapt itself to different environments. The learned policy have shown the similar curves in human child learning process often seen in psychological experiments, that is, micro U-shapes. Similar to the method in (Rumelhart and McClelland,

1986), that is, carefully designed input schedule, we have implemented the LEM paradigm to control the order of the situations the robot encounters.

There are many issues to be considered: 1) the definition of the task class in which the robot can gradually skill up the learned policy, 2) state vector selection which is currently off-line process but should be included in online learning process, and 3) real robot experiments.

Acknowledgement

We appreciate Dr. Karl MacDorman, Department of Systems and Human Science, Osaka University for his valuable comments and suggestions.

References

Asada, M., Noda, S., Tawaratumida, S., and Hosoda, K. (1996). Purposive behavior acquisition for a real robot by vision-based reinforcement learning. *Machine Learning*, 23:279–303.

Elman, J. L., Bates, E. A., Johnson, M. H., Karmiloff-Smith, A., Parisi, D., and Plunkett, K. (1996). *Rethinking Innateness.* The MIT Press, Cambridge, Massachusetts.

Kaelbling, L. P. (1993). "Learning to achieve goals". In *Proc. of IJCAI-93*, pages 1094–1098.

Marcus, G., Ullman, M., Pinker, S., Hollander, M., and T. J. Rosen, F. X. (1992). *Overregularization in language acquisition.* Monographs of the society of for research in Child Development, 57.

Minato, T. and Asada, M. (1998). Environmental change adaptation for mobile robot navigation. In *Proc. of IEEE/RSJ International Conference on Intelligent Robots and Systems 1998 (IROS98)*, page (submitted).

Nakamura, T., Morimoto, J., and Asada, M. (1996). Direct coupling of multisensor information and actions for mobile robot behavior acquisition. In *Proc. of 1996 IEEE/SICE/RSJ International Conference on Multisensor Fusion and Integration*, pages 139–144.

Pinker, S. and Prince, A. (1988). On language and connectionism: Analysis of a parallel distributed processing model of language acquisition. *Cognition*, 28:73–193.

Plunkett, K. and Marchman, V. (1991). U-shaped learning and frequency effects in a multi-layered perception: Implications for child language acquisition. *Cognition*, 38:43–102.

Plunkett, K., Sinha, C., MØller, M. F., and Strandsby, O. (1992). Symbol grounding of the emergence of symbols? vocabulary growth in children and a connectionist net. *Conection Science*, 4(3-4):293–312.

Rumelhart, D. E. and McClelland, J. L. (1986). On leaening the past tenses of english verbs. In Rumelhart, D. E. and McClelland, J. L., editors, *Paralle ditributed processing: Exploration in the micro structure of cognition. Volume 2. Psychological and biological models.* The MIT Press, Cambridge.

Tanaka, F. and Yamamura, M. (1997). An approach to lifelong reinforcement learning through multiple environments. In *6th European Workshop on Learning Robots*, pages 93–99.

Thrun, S. and Mitchell, T. (1996). Lifelong robot learning. Technical Report IAI-TR-93-7, University of Bonn, Dept. of CS III.

Watkins, C. J. C. H. and Dayan, P. (1992). "Technical note: Q-learning". *Machine Learning*, 8:279–292.

Dynamic Action Sequences in Reinforcement Learning

Jan Morén
Lund University Cognitive Science
Kungshuset, Lundagård
S–222 22 Lund, Sweden
jan.moren@fil.lu.se

Abstract

Reinforcement learning is a popular method for learning in autonomous dynamical systems. One of the most popular reinforcement learning methods is Q-learning, where the evaluation function and action selection function is combined in one data structure. However, Q-learning suffers from poor scalability and slow convergence, problems typically addressed by clustering of states or by using a hierarchical action system. Hierarchical Q-learning, presented in this paper, provides a simple mechanism for dynamical creation of hierarchical action sequences, that solves the scalability problems of regular Q-learning, while retaining its simplicity. By creating dynamical action sequences and using them to generalize over the state-space, the model is able to increase the learning speed without prior assumptions about the structure of the state-space.

1. Introduction

The reinforcement learning paradigm is commonly used for many applications in machine learning, such as navigation and motor control. Its attractiveness stems from the ability for unsupervised learning, where all that is needed is a reinforcing signal from the world. The learning system is conceptually divided into a *value function*, that estimates the desirability of the current state, and a *policy function* that determines the optimal action to take from the current state. For a general introduction to reinforcement learning, see (Kaelbling et al., 1996).

Q-learning, developed by Watkins (Watkins, 1989), is a popular method for implementing reinforcement learning. It consists of a function Q (for "Quality") over all states and actions, a learning rule that updates an estimate of Q, and a response model using Q as a basis for a policy. The learning rule is strongly reminiscent of the TD model (Sutton and Barto, 1990).

Q-learning has several advantages compared to other learning methods. While other reinforcement learning methods separates the value function from the policy function, Q-learning brings them together into one function with a single learning rule, making it both easier to implement and easier to understand.

There are two problems with Q-learning, however. First, Q-Learning does not scale well; the learning time grows rapidly with the size of the state-space, making Q-learning unsuitable for real-world applications without extensive preprocessing. Q-learning is thus not suitable as a model of animal learning.

Second, Q-learning does not generalize skills. Any knowledge gained in one part of the state-space is ignored in all other parts. As a state-space often exhibits some regularity, it would be efficient to utilize this information.

In these respects, the Q-learning method can be seen as a base upon which to build more comprehensive learning methods.

Below, I propose a learning method that speeds up learning by taking advantage of any regularity of the state-space. It is a hierarchical chunking system situated on top of ordinary Q-learning and will allow dynamic creation and use of hierarchical action sequences. This results in faster learning and transparent generalization of action sequences. This is accomplished by a simple and consistent chunking system and a straightforward means for applying the resulting control architecture. This idea was first explored in (Morén, 1995).

2. Reinforcement Learning

Q-learning is a straightforward reinforcement learning model and has become very popular, but it suffers from poor scalability and lack of generalization. As the state space grows, the learning time grows exponentially, which is disastrous for any real-world applications. A number of approaches to solve this problem will be presented below.

To be able to have a meaningful discussion about learning speed, we must agree on what we mean. As a tentative definition, consider a discrete state space with a start state and a goal state. The minimal learning time for a system in such a space is defined as the number of runs (or trials) from start to goal required to transfer information about the goal state to the start state. In practice the need for exploration will increase learning time above this lower limit. This definition factors

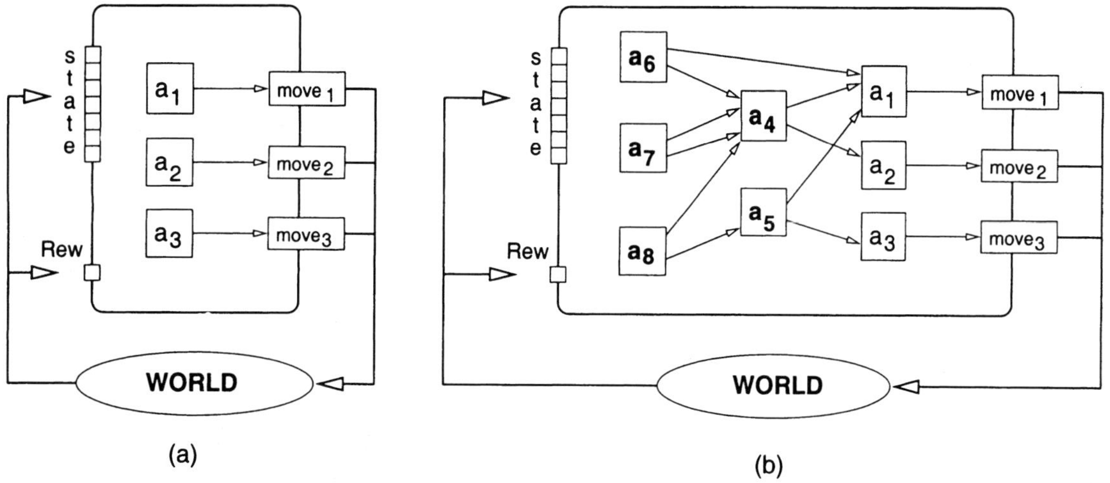

Figure 1 (a): A reinforcement learning system. An action selection mechanism chooses an action a_i that will execute a physical movement. The result will be a new state and a reinforcing signal. (b): A hierarchical reinforcement system. If the chosen action $a = \langle a_i, a_j \rangle$, it will execute a_i and a_j in turn, until the level of physical action is reached.

out performance of the reinforcement system from the behavior of the exploratory system (which is very interesting in itself but outside the scope of this paper).

2.1 Related Work

Due to its appealing features, there have been a number of attempts to speed up the learning time of Q-learning. These models can be categorized into three commonly used techniques: employing a world simulation, clustering of states, and creating action sequences. These three techniques are not mutually exclusive, but can profitably complement each other.

Internal world This is a class of techniques where actions can be generated and tested without spending time or taking risks in the real world. Dyna-Q (Sutton, 1990) is an example of this approach: the Q function is used in internal simulations of the environment where an uncertainty factor (based of the amount of time since the state-action pair was last used) influences the reward function.

Such a model will speed up learning and enhance *re*learning in a changing environment. But, of course, the internal world model will have the same problems of scalability as the original Q-model.

Clustering states Another class of methods used to speed up Q-learning and to allow for generalization is to somehow reduce the perceived number of states by treating similar states as if they were the same. This can be done by approximating the Q-function with a neural network (see cf. (Lin, 1992)), thus generalizing over several states. This can also be accomplished by explicit clustering of "similar" states.

The problem with approximating Q functions is that the convergence results for Q no longer holds (Watkins and Dayan, 1992). This can result in slower learning than in regular Q-learning. Note, though, that this approximation generally is done to allow for generalization of actions, not to speed up Q-learning. If the state-space is predictable so that the approximations are close to the optimal Q-function, this approach will tend to work well.

The number of states can also be conceptually reduced by explicit clustering of similar states. A method using a CMAC is found in (Sutton, 1996). A method employing hierarchical clustering of states is (Balkenius, 1996). Here, each state is simultaneously a part of several larger-scale states, and learning occurs at each scale. In absence of any information at a lower scale, the information at higher scales is used.

A hybrid approach using classifier systems to dynamically create clusters of states is described in (Munos and Patinel, 1994).

Action sequences Instead of combining states, it is feasible to combine actions into sequences. This is typically done using a hierarchical system consisting of only two levels (a 'world' level and a 'detail' level), and the structure is predefined and static. Examples of such methods are (Maes and Brooks, 1990) and (Singh, 1992).

It would be beneficial if the system could create the structures dynamically as it learns the state-space. This would increase the adaptability of the system, as the structures are created in response to the demands of the state-space. It does result in a more complicated model; indeed, the mechanisms to create these dynamical structures can become more complex than the reinforcement model itself.

One such system – called Nested Q-Learning – has been proposed by (Digney, 1996), where structures on many levels are constructed by the system without supervision. This method builds hierarchical action sequences in a top-down fashion, with a complex action learning what actions to perform to achieve its goal. It is able to gracefully adapt to a changing environment, creating only those action sequences needed to navigate in the state-space. However, the learned sequences are usable only at the state where it was defined, leading to poor generalization of action sequences. To compensate for this, Nested Q-Learning requires a second, independent action selection mechanism working bottom-up to exploit serendipitous events during exploration. The present model, in contrast, needs only one mechanism to create action sequences. As all sequences are available everywhere, no special subsystem is needed to generalize action sequences over the state-space.

2.2 Q-learning

Q-learning lends itself well as a foundation upon which to build more complex reinforcement learning systems and is used as the basis for the present model. A brief description follows; for a more complete treatment, see cf. (Watkins, 1989).

There is a set of primitive actions

$$A = \{a_1 \dots a_N\},$$

and a set of states

$$S = \{s_1 \dots s_M\}.$$

For Q-Learning, the state-action space has to be Markovian to have guaranteed convergence, i.e. the states have to be uniquely identifiable and finite in number, and the actions have to be deterministic (Watkins and Dayan, 1992).

A function Q_{sa} is defined over all actions and states, that is the expected future reward for every pair of states and actions. It is used both as a basis for an evaluation function and as data for a policy.

The update function for Q-learning when moving from state x to y by action a is

$$Q_{xa} = Q_{xa} + \alpha E_{xa}.$$

α is the learning rate. E_{xa} is the error in the current estimate of Q_{xa}, defined as

$$E_{xa} = (R_y + \gamma V_y - Q_{xa}),$$

where γ is a discount factor for future rewards. V_s is the value of the state:

$$V_s = \max_a(Q_{sa})$$

i.e. V_s is the maximum expected reward that can be obtained at state s.

We also need a policy or action selection mechanism over Q. Any reasonable policy will do; ϵ-greedy[1] is a common choice.

[1] Use the optimal action (as defined by Q) at the current state with probability $P = (1 - \epsilon)$, otherwise select an action at random. With a suitable value of ϵ (typically 0.05 to 0.01), a reasonable balance between exploration and exploitation will be struck. For a discussion of this aspect of reinforcement learning, see (Kaelbling et al., 1996).

2.3 Hierarchical Q-learning

The model is an extension of Q-learning that uses hierarchical action sequences to improve the learning rate and generalize learned skills. As discussed above, it is desirable to create the sequences dynamically.

We want to create new compound actions as in fig 1(b), effectively increasing the number of actions available to the system. Now, the actions a_1 to a_N are the "primitive" actions, in the sense that they actually change the state of the system when used. Under conditions defined below, Hierarchical Q-learning will create new actions $a_c = \langle a_i, a_j \rangle$, $c > i, j$ and define the Q function for the new action over all states.

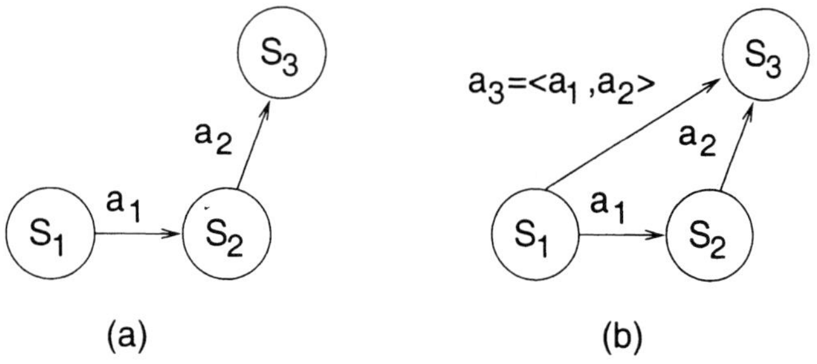

Figure 2 (a): Three states S_1 to S_3 with primitive actions a_1 and a_2 leading between them. (b): A new action a_3 with component actions a_1 and a_2 has been created. This action can be seen as shortening the path between S_1 and S_3 from two steps to one.

Action sequences are represented as binary trees. This will make it conceptually easy to define new sequences in terms of old ones, and the representation also makes it straightforward to update the Q-values for every subsequence as they are executed.

A new action will be constructed whenever:

- two consecutive actions a_i and a_j from state s_1 and state s_2, respectively, both have a positive error $E_a > k, k > 0$, and
- $a_i = \max_a(Q_{s_1 a})$ and $a_j = \max_a(Q_{s_2 a})$. This second requirement, that each action be the highest valued action from its state, is to prevent local loops which would slow down the learning.
  The k parameter above is a threshold regulating the creation of new action sequences. It is used to achieve a tradeoff between Q-table size and ease of learning. The new action $a_c = \langle a_i, a_j \rangle$ will be defined over all states S, with

$$Q_{xa_c} = \begin{cases} \alpha(R + \gamma V_y), & x = s_i \\ Q_{xa_i}, & \text{otherwise} \end{cases}$$

In other words, set Q_{xa_c} to the expected future reward of the final state of the new action, and allow the action a to be tried out at other states by setting its Q value equal to the Q of its first action component. This implements the heuristic that a good a_i makes it more probable that the new a_c will also be good.

The error function E will be redefined for non-primitive actions as the sum of errors over the two lower-level actions. Thus it is changed to the recursive definition

$$E_a = \begin{cases} e_{a_p} + e_{a_q}, & a = \langle a_p, a_q \rangle \\ (R_{s_j} + \gamma V_{s_j} - Q_{s_i,a}), & \text{otherwise} \end{cases}$$

The update function for Q is defined in the same way as for regular Q-learning:

$$Q_{xa} = Q_{xa} + \alpha E_{xa}.$$

The action sequences thus created are not indivisible. Every action, whether primitive or compound, will update the Q-table irrespective of it is part of a larger sequence or not. Similarly, a new action sequence can be created when the system is in the middle of executing one.

To be able to define a new action$a_c = \langle a_i, a_j \rangle$, a small fixed memory holding the last two actions are needed. This memory must be recursively updated when executing an action sequence, enabling the system to define new sequences even when in the middle of running an existing one.

If a primitive action fails for any reason, that failure will be propagated up to all action sequences involved in executing that action, making sure that none of them will be tried again from that state.

2.4 Example

To see what happens when learning to navigate through a state-space, let's look at a simplified maze environment. In figure 3, we see a very simple environment consisting of seven states and one possible action at every state. In this environment, the system can thus only do the optimal thing. What is interesting is instead how new action sequences would be built and how information about the goal state would be propagated.

Figure 3 An idealized "Maze" with seven states including the start state to the left and the goal state to the right. At every state, there is only one possible action.

Let us first look at the behavior of Q-learning in this environment. At the first trial, Q-Learning would move to the goal state G, where Q_5 would be updated with information about the whereabouts of the goal state. At the next trial, the information would propagate to state 4. In all, it would take six trials for any information about the goal state to spread to the start state. This is clearly an optimal scenario for Q-Learning, and shows that at best, Q-learning can learn a route through a state-space at a rate proportional to the minimal path length.

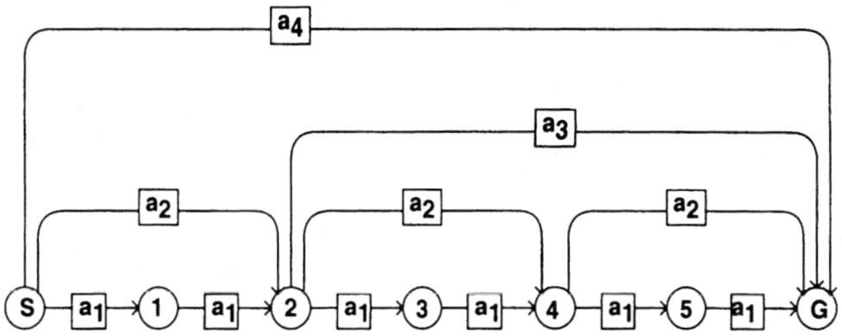

Figure 4 The maze from fig. 3 with new action sequences as defined by the hierarchical Q-learning model. In contrast to regular Q-learning, the number of actions available at every state increases as new sequences are created.

During the first trial, hierarchical Q-learning would respond in an identical way to Q-learning. When the system enters the goal state at the second trial, however, the system will – with reasonable constants – have fulfilled the conditions for creating a new action sequence $a_2 = \langle a_1, a_1 \rangle$ from state 4 to state G. Action a_2 will also be defined over all other states. If the system chooses a_2 at state 2 and 4, respectively, the conditions are again met for a new action sequence: $a_3 = \langle a_2, a_2 \rangle$ – see picture 4. At the next trial again the sequence $a_4 = \langle a_2, a_3 \rangle$ will be defined. Hierarchical Q-learning can thus spread information from the goal state at a rate of $\log_2(n)$ for n equal to the minimal path length.

3. Simulations

To verify that the system increases performance as expected, a series of simulations have been run. The system ran a number of maze-learning tasks in a very simple grid-world. Though the results are preliminary, they nevertheless show that chunking of hierarchical action sequences clearly improves on learning speed.

3.1 Setup

There are three primitive actions $a_1 \ldots a_3$ that move forward, turn 90° clockwise and turn 90° counterclockwise. The state information is the spatial location and the direction faced by the system.

The actions are relative to the current position and direction of the system, so each cell in the maze is coded as four states – one for each possible direction the system can face. A graphical representation of the mazes is given in figure 5.

This type of environment – while rather unrealistic – is easy to implement and the simplicity makes it easy to understand the dynamics of a simulation. Also, a discrete, deterministic, environment makes it feasible to isolate different capabilities and limitations of a system.

For the purpose of these simulations, a learning rate of $\alpha = 0.2$, error threshold of $k = 0.2$ and a discount factor of $\gamma = 0.95$ were chosen.

The simulations were run 20 times on the mazes shown above both with and without the hierarchical extension. The mazes were varied in size from $3*3*4 = 36$

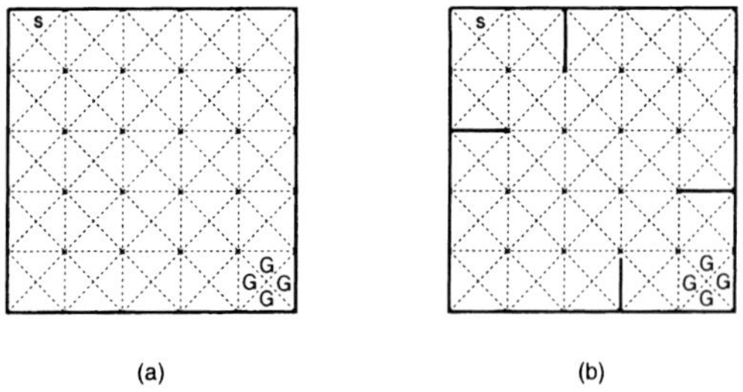

Figure 5 In (a) an empty maze with 5*5*4=100 states. Every location consists of four states, one for each direction faced by the system. In (b) a more complex maze with the same number of states as in (a).

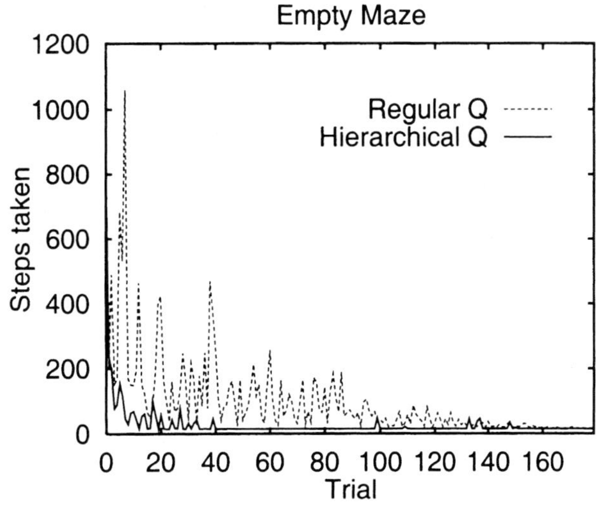

Figure 6 Steps taken to the goal states for Q and HQ during one trial of the empty 8*8*4 maze. It is clear that HQ rapidly exploits the regularity of the state-space.

to $8*8*4 = 256$ states. For every size, maze and method, the average learning time (in number of trials) and standard deviation were calculated.

3.2 Results

The results for an empty maze and for the minimally difficult maze is given in figure 7 and 8. These figures show the learning speed of Q and Hierarchical Q respectively as a function of the number of states. It is clear that Hierarchical Q is able to utilize information about the state-space much more efficiently than Q, by developing appropriate action sequences. The number of new actions for a 256 state maze is typically around 25; this depends on the value of the k parameter, as discussed in section 2.3.

Another aspect of the learning process is the behavior of the models during learning. In figure 6, the number of steps taken as a function of trials are plotted for Q-learning and Hierarchical Q-learning. Both models converge, but Q-learning only slowly approaches the optimal path, taking 155 learning trials to settle down on the path. Hierarchical Q-learning, in contrast, efficiently exploits the regularity of the state-space and rapidly

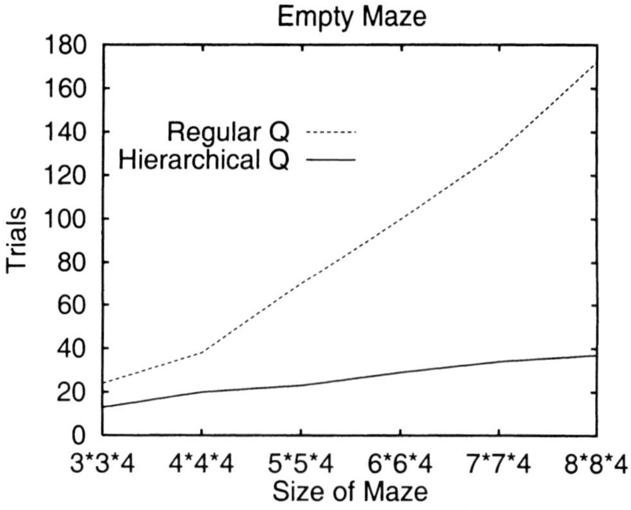

Figure 7 Number of trials to learn the optimal path in the empty maze as shown in figure 5 (a). The size of the maze ranges from $3*3*4 = 36$ to $8*8*4 = 256$ states. The data for each size is an average of 20 runs.

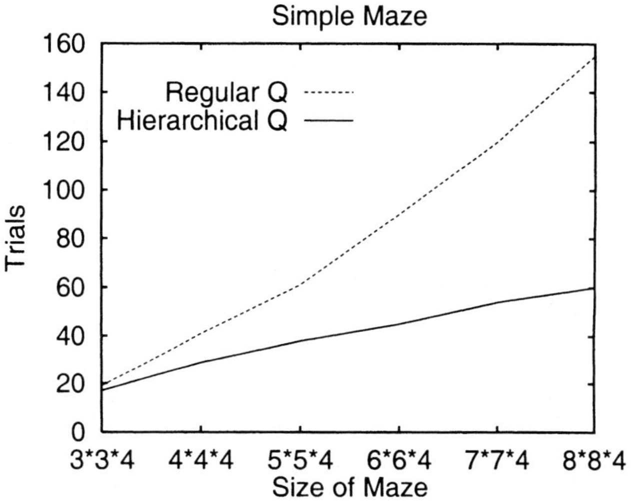

Figure 8 Number of trials to learn the optimal path in the complex maze as shown in figure 5 (b). The size of the maze ranges from $3*3*4 = 36$ to $8*8*4 = 256$ states. As above, the data is an average of 20 runs.

spreads information about good and bad paths alike, achieving convergence after around 40 learning trials.

These simulations are by no means complete. However, they do show that there is real and substantial improvement on the learning speed compared to regular Q-learning.

4. Discussion

There are two aspects of reinforcement that have not yet been adequately tried in simulation: relearning in a changing environment and the handling of non-unique states.

4.1 Relearning

When a Q-learning system encounters an unexpected failure such as a moved/removed goal state or a disallowed action, this information is transmitted backwards as slowly as positive information, as discussed earlier. Depending on the structure of the state space, this can mean using an even larger number of trials learning a new path than it took to learn the original one. In this respect, Q-learning is not very flexible.

The present model will transfer the negative information even faster than it did the positive information. When learning a path through the state space, the action sequences had to be created and tested. When an unexpected failure occurs, the entire structure of action sequences are already available to transfer the change backwards to earlier states in an efficient manner.

4.2 Non-unique states

The problem is one of disambiguing identical states using information gained about the structure of the state space. One way is to encode what action the system used to arrive at the present state. Thus, every state would be encoded in the Q-table as a tuple $\langle s, a \rangle$.

Using this strategy with regular Q does not work, as encoding the last action gives very little additional information. With only a few actions leading to the state, failures are bound to be common.

With action sequences as defined by hierarchical Q-learning, there need never be a shortage of actions to disambiguate states. Every state will be represented in a tuple with every action sequence; if a failure occurs due to a mix up of states, those tuples that are too short to disambiguate the states will be punished, and those tuples are less likely to be referenced again. Also, the system will not disambiguate identical states until it has to, effectively partitioning identical states into equivalence classes. This is very preliminary, though, and should be taken as work in progress.

5. Conclusion

Presented here is a model of hierarchical reinforcement learning built on Q-Learning by Watkins (Watkins, 1989). It dynamically creates action sequences by hierarchically combining actions whenever two conditions are met: both actions have to be "good" actions in the sense that they improve the systems ability to reach the goal state; and they have to be the best action at the given state, i.e. they have to have the highest Q-value of the actions from that state.

Simulations of the hierarchical system in a simple deterministic grid world have been run. The simulations indicate that the hierarchical system can utilize any regularities of the state-space to create action sequences that dramatically decrease learning time.

References

Balkenius, C. (1996). Generalization in instrumental learning. In Maes, P., Mataric, M. J., Meyer, J.-A., Pollack, J., and Wilson, S. W., editors, *From animals to animats 4*, pages 305–314, Cambridge, MA. MIT Press.

Digney, Bruce, L. (1996). Emergent hierarchical control structures: Learning reactive/hierarchical relationships in reinforcement environments. In Maes, P., Mataric, M. J., Meyer, J.-A., Pollack, J., and Wilson, S. W., editors, *From animals to animats 4*, pages 363–372, Cambridge, MA. MIT Press.

Kaelbling, L. P., Littman, M. L., and Moore, A. W. (1996). Reinforcement learning: A survey. *Journal of Artificial Intelligence Research*, 4:237–285.

Lin, L. J. (1992). Self-improving reactive agents based on reinforcement learning, planning and teaching. *Machine Learning*, (8):293–321.

Maes, P. (1994). Modelling adaptive autonomous agents. *Artificial Life Journal*, 1(1&2).

Maes, P. and Brooks, R. A. (1990). Learning to coordinate behaviours. In *Eight National Conference on Artificial Intelligence*, pages 796–802.

Morén, J. (1995). Hierarchical chunking improves q-learning. Master's thesis, Department of Computer Science, Lund University.

Munos, R. and Patinel, J. (1994). Reinforcement learning with dynamic covering of state-action space : Partitioning q-learning. In Cliff, D., Husbands, P., Meyer, J.-A., and Wilson, S. W., editors, *From animals to animats 3*, pages 354-363, Cambridge, MA. MIT Press.

Singh, P. S. (1992). Transfer of learning by composing solutions of elemental sequential tasks. *Machine Learning*, (8):323–339.

Sutton, R. S. (1990). Integrated architectures for learning, planning and reacting based on approximating dynamic programming. In *Proceedings of the Seventh Int. Conf. on Machine Learning*, pages 216–224. Morgan Kaufmann.

Sutton, R. S. (1996). Generalization in reinforcement learning: Successful examples using sparse coarse coding. *Advances in Neural Information Processing Systems*, (8):1038–1044.

Sutton, R. S. and Barto, A. G. (1990). Time-derivative models of pavlovian reinforcement. In Gabriel, M. and Moore, J., editors, *Learning and computational neuroscience: foundations of adaptive networks*, chapter 12, pages 497–537. MIT Press, Cambridge, MA.

Watkins, C. J. C. H. (1989). *Learning from Delayed Rewards*. PhD thesis, Kings College, Cambridge.

Watkins, C. J. C. H. and Dayan, P. (1992). Q-learning. *Machine Learning*, (8):293–321.

Learning Concepts in Software Agents

Uma Ramamurthy[*], **Stan Franklin**[**], **Aregahegn Negatu**[***]
[*]St. Jude Children's Research Hospital[1]/University of Memphis
332 North Lauderdale, Memphis, TN 38105, U.S.A.
urmmrthy@memphis.edu
[**]University of Memphis[2]
Department of Mathematical Sciences, Memphis, TN 38152, U.S.A.
stan.franklin@memphis.edu
[***]University of Memphis
Department of Mathematical Sciences, Memphis, TN 38152, U.S.A.
negatua@msci.memphis.edu

Abstract

This concept-paper explores issues related to learning new concepts in software agents which inhabit dynamic domains. We argue that agents learn based on what they already know and agents solve new problems which they encounter by making analogies to previously solved problems of similar type. We explore these issues within the scope of the Cognitive Agent Architecture and Theory (CAAT) [Franklin, 1997]. The architecture of a multi-agent system, CMattie which is based on the CAAT strategy and inhabits an email-based dynamic domain, is described. CMattie gathers information from humans, composes announcements of next week's seminars, maintains a mailing list, mails the weekly seminar announcement to members of that mailing list, and learns new variations to seminars and other such events which have to be announced to the members of the mailing list as her domain changes with new types of seminar-like events. Learning mechanisms being implemented in this system which enable CMattie to adapt to her changing domain are described. Through her learning, CMattie acquires new domain-specific concepts, thus adapting to her dynamic domain.

1. Introduction

Computer networks provide an infrastructure for an efficient information exchange and processing; as a result they improve individual and group performances. E-mail is one of the most used tools that help people transfer and manipulate information. In our academic setting, e-mail service is so important that many standard activities are based on it. Introducing intelligent agents into computer systems can automate some simple and not so simple routine organizational tasks so that human workers will be freed from such routine but tedious work to concentrate on some other higher-level and non-routine tasks.

1.1 Cognitive Agent Architecture and Theory

We have been participating in an intelligent system research using the Cognitive Agent Architecture and Theory (CAAT) strategy [Franklin, 1997]. Basic cognitive processes include learning, memory (long term & short term), concept formation, perception, attention, problem solving, decision-making, thinking, etc. An autonomous agent that exhibits many or most of these cognitive processes (functions) can be referred to as a cognitive agent [Franklin, 1996] [Franklin, 1997]. Agent architectures may be characterized according to their capabilities (learning, planning, etc.), their properties (learning: deliberative, reflexive, monatomic, non-monatomic, etc.) and the nature of environments (static, dynamic, etc.) and domains for which they are designed. In general, agent architecture can be stated as a subset of a system that contains and manages the underlying and primitive resources of an agent. A methodical study of these resources and their management will assist in determining the necessary, sufficient and optimal distribution of resources for the development of agents that exhibit some type of intelligence.

In the design space [Sloman, 1995], we develop specifications (what capabilities and what properties/characteristics in each capability are needed for a specific domain) for a control structure which in turn has some type of underlying theory to support it. So, in CAAT strategy, architecture design entails the creation of theories of cognition to reinforce or modify existing theories or to serve as a hypothesis for a new theory. Conversely, any

[1] First author supported in part by Cancer Center Support CORE Grant, P30 CA 21765 and by American Lebanese Syrian Associated Charities (ALSAC).

[2] Second author supported in part by NSF grant SBR-9720314 and by ONR grant N00014-98-1-0332

improvements in theory should lead to possible enhancements of the architecture. This synergy between theory and architecture should lead to productive experimental work in artificial intelligence, cognitive science and other related fields.

1.2 VMattie

VMattie is a software agent which gathers information from humans, composes announcements of next week's seminars, maintains a mailing list and mails the weekly seminar announcement to members of that mailing list, all without human supervision. VMattie "lives" in a UNIX-based system. It is an intelligent software agent living in a complex dynamic environment, with multiple high-level perceptions and actions satisfying many drives.

VMattie is a multiagent system with all internal actions accomplished by microagents called **codelets**. VMattie's architecture is based on multi-layer, feed-forward neural network based categorization mechanism and Behavior Networks [Maes, 1989] [Maes, 1990]. The Behavior Networks component of VMattie is actually an extension of the classic Behavior Networks. Using the Behavior Networks component, VMattie deals with message Input Processing Workspace along with the associated codelets -- form the message understanding module of VMattie.

Several distinct drives operate in parallel in this system. The salient ones amongst these drives are:

1. To understand every incoming mail message,
2. To acknowledge every incoming mail message,
3. To maintain complete information on every ongoing seminar (in its Tracking Knowledge Base),
4. To keep the mailing list current, and
5. To mail out the weekly seminar announcements on time.

The urgency of these drives varies with incoming mail messages and the nearing of the time for mailing the weekly seminar announcements. This variation in drive-urgency is an enhancement over the classic Behavior Network architecture. Drives provide activation to behaviors.

The behavior net consists of various behaviors with their associated links: successor links, predecessor links and conflictor links [Maes, 1989]. In VMattie, behaviors get instantiated, thus supporting variables. For example, the add-address-to-list behavior asks which address to add to the list. The behaviors consist of preconditions, activation, add-list and delete-list accommodating variables in the

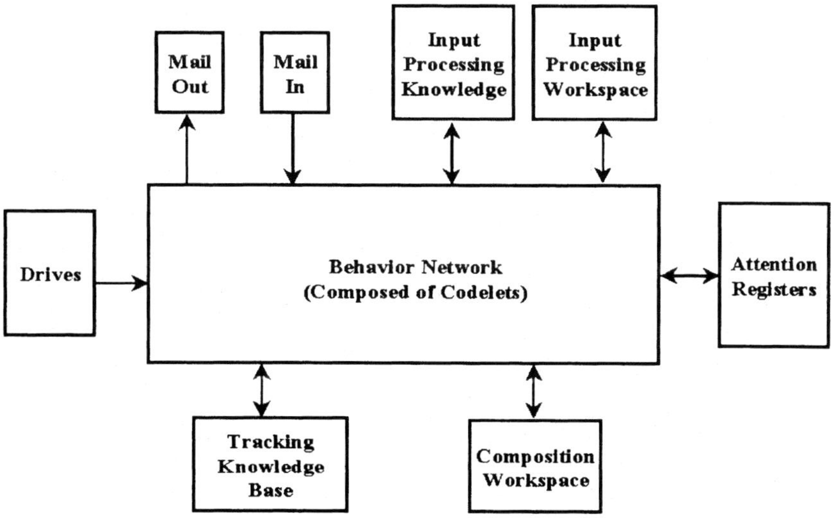

Figure 1: VMattie's Architecture

variables, with variable drives and with activation from internal states, thus providing informal planning and action selection in this system. The message understanding module of VMattie is composed of a feed-forward neural network mechanism working with codelets to achieve the natural language understanding within a narrow-domain when VMattie perceives its input, namely incoming mail messages.

In VMattie's architecture shown in Figure 1, the three units -- Drives, Behavior Network and Attention Registers -- form the action-selection module with important extensions. Each behavior is implemented by a collection of codelets. The two units -- Input Processing Knowledge and contents of any of these. Activation spreads through the instantiated behaviors.

Attention Registers hold information extracted from an incoming mail message. Each register holds the content of a field, so that codelets which need it can get it from the Attention Registers. Such fields include name-of-the-organizer, email-address, seminar-name, etc. When filled, Attention Registers provide activation to behaviors that can use their contents.

The Input Processing Knowledge holds the declarative knowledge needed to understand the incoming mail messages. For example, it knows the various forms of "Wednesday" -- Wed., wed., Wednesday, etc., and the names of the buildings where seminars are held. Input Processing Workspace holds the contents of an incoming mail message while codelets associated with the Workspace attempt to understand the mail message and intermittent results. In regular consultation with the Input Processing Knowledge, inferences are made. The most significant inference of all is the type of the incoming mail message. Once an incoming mail message is understood -- every significant phrase or word is given a field name and the type of the incoming mail message is inferred, then the primitive codelets transfer this information to the Attention Registers. Primitive codelets are not directly connected to any behavior or neural network node, and perform housekeeping functions in the VMattie system.

The Tracking Knowledge Base unit holds information used in composing outgoing mail messages. Examples of such information are default data for ongoing seminars as regards the name of the seminar, the organizer of the seminar, the usual seminar time and place. Behaviors update the default data. This unit maintains the mailing list for announcements with help from behaviors. The Tracking Knowledge Base also holds the templates for the different types of outgoing mail messages from the VMattie system.

The outgoing mail messages are composed in the Composition Workspace. In this unit, choosing of the template and filling the appropriate fields takes place. The information for the composition comes from the Tracking Knowledge Base and the Attention Registers.

2. CMattie

CMattie is the next experiment under the CAAT strategy. CMattie "lives" in a UNIX-based system. She is a software agent that gathers information from humans, composes announcements of the next week's seminars, maintains a mailing list and mails the weekly seminar announcements to members of that mailing list, all without human supervision.

In CMattie, new mechanisms have been added to the VMattie's architecture. This has resulted in a modified, more complex architecture for CMattie (shown in Figure 2). CMattie has a limited capacity Global Workspace [Baars, 1988] which is capable of broadcasting to all the processes. An attention mechanism, the Spotlight, controls access to the Global Workspace of coalition of processes.

CMattie has memories based on Sparse Distributed Memory architecture [Kanerva, 1988] and case-based memory [Kolodner, 1993]. CMattie has an emotional mechanism that enables her to be emotional when appropriate. CMattie has a sense of self-preservation which enables her to be concerned with her environment, namely her resource needs and the status of the UNIX-based system that she "lives" in. Most important of all, CMattie learns. She learns new concepts and new behaviors related to those newly learnt concepts in her dynamic domain.

CMattie has an episodic memory which acts as an intermediate memory. This is a case-based memory system. In her episodic memory, she stores sequences of mail messages that form episodes in her domain. Using this episodic memory, she can relate to new events which are similar to past events in her domain and which she is capable of understanding from her built-in domain knowledge. This memory acts as intermediate memory as information stored here is used in learning domain knowledge and if found relevant to the overall drives of the system, the information is passed on to the long-term memory of the system, the Sparse Distributed Memory.

Due to the very nature of her domain, namely, interaction through mail messages, reinforcement learning is best suited for many of CMattie's learning mechanisms. Reinforcement learning [Kaelbling, 1996] is a trial and error approach to learning. The agent learns by adjusting the

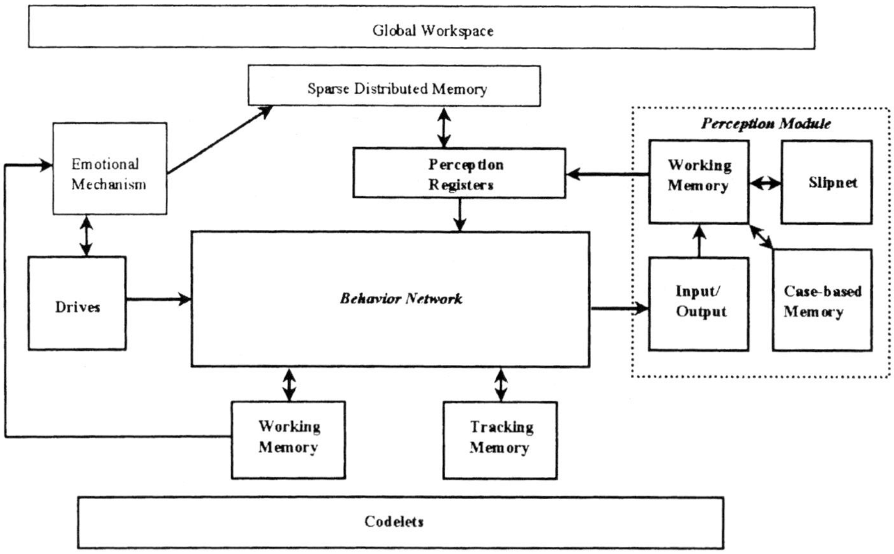

Figure 2: CMattie's Architecture

mapping from her domain states to her actions based on positive and or negative feedbacks. CMattie has built-in domain knowledge and built-in behaviors. When her domain changes, she attempts to interpret the new events in her domain in terms of her built-in knowledge and behaviors. This elicits some feedback from the members on her mailing list who send her mail messages. Based on that feedback, CMattie learns that she needs to modify some of the concepts in her domain knowledge, thus creating new concepts and further, new behaviors.

The various aspects of learning in CMattie are outlined below:

1. Associative learning in the codelets or processes based on Pandemonium Theory [Jackson, 1987].
2. Learning occurs automatically in the Sparse Distributed Memory architecture which is being used as long-term memory in CMattie.
3. The Slipnet in CMattie's Perception Module forms part of her built-in domain knowledge. CMattie learns new Slipnet nodes and fits them into her existing Slipnet.
4. For the new Slipnet nodes, CMattie learns associated new codelets.
5. When a change occurs in CMattie's domain knowledge, most often it calls for new behaviors. Based on the changes in her domain knowledge and feedback from her environment, she modifies her existing behaviors to learn new behaviors for her action-selection.

The last three mechanisms of learning in the CMattie system listed above are some form of reinforcement learning. In this paper, we focus on learning mechanisms for items (3) and (4) above.

3. Perceptual Learning

The Perception Module of CMattie is based on the Copycat architecture [Mitchell, 1993]. The Copycat system is based on the thesis that analogy-making lies at the core of understanding and analogy-making is itself a process of high-level perception. The Copycat system interprets and makes analogies between situations in an idealized microworld involving letter-string analogy problems. The basic objects in this microworld are the 26 letters of the alphabet. A typical analogy problem in this world is as below:

$$\textbf{abc} \rightarrow \textbf{abd}$$
$$\textbf{ijk} \rightarrow \ ?$$

i.e., one is given the change from string **abc** to the string **abd** and asked to come up with a similar change for the target string **ijk**.

Copycat's architecture consists of 4 parts: (1) The **Slipnet**, (2) **Workspace**, (3) **Pool of Codelets**, and (4) **Temperature**. The **Slipnet** is a network of nodes and links which represent the permanent concepts of the system. Copycat builds perceptual structures in the **Workspace** to represent the system's current understanding of the given analogy problem. The **Pool of Codelets** consists of various perceptual and higher-level structuring agents called **Codelets**, waiting to run. A **codelet** is a small piece of code that carries out some local task. These agents build and sometimes destroy the perceptual structures in the **Workspace**. The **Temperature** measures the amount of disorganization or entropy in the system's understanding of the given analogy problem and controls the degree of randomness used in making decisions.

The domain of the Copycat system is predefined and fixed. Hence there is no learning in this system. On the contrary, CMattie "lives" in a dynamic domain. As her domain changes, to enable her to perceive this dynamism, she learns new concepts in her Perception Module.

3.1 Concepts and Features

Concepts are objects or logical units of ideas that are used to represent entities. In the domain of CMattie, Seminar, Seminar Organizer, E-mail Address, Periodicity, Location, Title of Talk, etc. are considered concepts. Such concepts need to be represented in such a way that cognitive processes relevant in the domain will manipulate these concepts effectively.

In CMattie, concepts are manipulated continuously. After sensing raw data from the environment, a perceptual cognitive activity takes place. The perception involves building of instances of known concepts, detection and creation of new concepts, making appropriate relations among concepts. In our design, we like to have a simple conceptual representation, yet powerful enough to be used in a relatively complex domain. The basic characteristics of a concept are represented by its features or attributes. In a given situation, a feature of a concept has a specific measure called "value". For instance, a "seminar" concept has a feature called "name" whose values could be "Complex System" or "Cognitive Sciences". A concept may have more than one feature. The same "seminar" concept can have another feature called "organizer" whose values could be "S. Franklin" or "M. Garzon". A concept can be a feature for another concept. For instance, "organizer" which is a feature of the "seminar" concept is by itself a concept which has a feature called "e-mail address" with a value, say "franklin@msci.memphis.edu."

In CMattie, concepts are defined not necessarily by a single node in the Slipnet of her Perception Module. A concept has a core and a set of features. Each of these may be individual nodes or group of nodes. The various nodes of the Slipnet are connected to one another through weighted links between them. One of the concepts in CMattie's built-in domain knowledge is the **Seminar**. The **Seminar** concept has the following features:

- **Name** of the seminar
- **Organizer** of the seminar
- **Location** where the seminar will be held
- **Date** of the seminar
- **Day** of the seminar
- **Time** at which the seminar will be held
- **Speaker** of the seminar
- **Title of talk** for the seminar

- **Periodicity** of the seminar

In Figure 3, a small segment of the Slipnet from the Perception Module of CMattie is shown. The direction of the links indicates the possible spread of activation between the nodes.

When a Seminar-Organizer sends her a message announcing a seminar with a seminar name that she has never seen before, she attempts to treat such a message as similar to seminars that she already knows. The learning mechanism here is based on the premise that **the agent learns based on what it already knows**. When the message understanding mechanism in the Perception Module attempts understanding this message, the system recognizes that it is a **initiate-seminar-message** for a seminar, but the **name** of this seminar is not part of the built-in domain knowledge. In such a situation, CMattie has built-in mechanisms to converse with the sender of the message to determine if the sender wishes to initialize a new seminar. She sends an acknowledgement to the sender of the message stating that a new seminar with that seminar name will be initialized in the system, with the sender as the organizer for that seminar. Reinforcement is provided to CMattie by the response she might or might not get for such an acknowledgement. Based on such a feedback, she creates a new node for this seminar name and links it to the **name** node (which is also a feature of the seminar concept) in the Slipnet. When a new node is generated, the underlying codelets for that node are generated as well. This process is quite straight forward, as the new codelets are based on existing similar codelets for other seminar name nodes. Once this process is complete, CMattie has understood the incoming message and the Perception Module sends the relevant fields from the understood message to the Perception Registers.

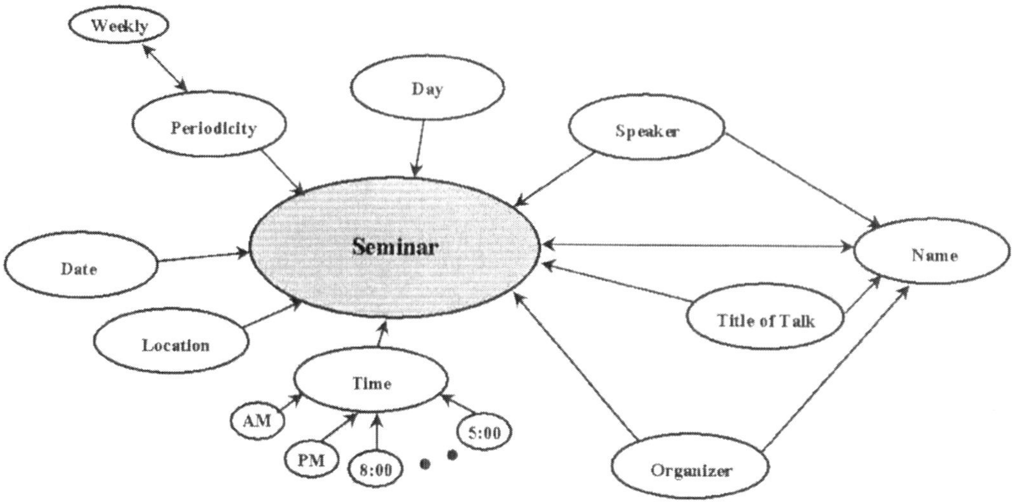

Figure 3: Some nodes and weighted links in CMattie's Slipnet

As can be seen from the above figure, **Time** is a feature of the **Seminar** concept. But **Time** by itself is also a concept with a set of its own features. The concept **Seminar** is much deeper and has a higher **depth value** than the concept **Time**. The links are weighted and these weights aid in reinforcement learning in the Slipnet.

Next question we need to ask is "how does CMattie learn new concepts ?"

3.2 Learning new Concepts

CMattie's senses are the mail messages she receives. In her Perception Module, the Slipnet contains most of her domain knowledge. Using this domain knowledge, she understands the input, namely mail messages, sent to the CMattie system. This understanding process occurs in the Perceptual Working Memory through limited amount of natural language understanding. Through her built-in domain knowledge, CMattie is capable of understanding different types of messages related to the mailing list she maintains and messages related to seminars.

CMattie has a limited number of seminars already defined in her Slipnet. CMattie "knows" about these seminars through the built-in **seminar** concept and its features: She "knows" --

- a seminar is held once a week;
- it has an organizer and a name;
- each week, it might have a different speaker;
- it has a different title-of-talk;
- it is usually held at the same location, on the same day of the week and at the same time, unless stated otherwise.

The second type of learning which takes place in the Perception Module is when CMattie learns concepts which are not completely identical to the built-in seminar concept, but slightly different from it. In CMattie's domain, the Colloquia, Dissertation Defenses, Dissertation Committee Meetings, Faculty Meetings fall in the category of concepts which are not totally identical to the seminar concept. This learning mechanism is based on the premise that **every new**

situation/problem is viewed by the agent in terms of a previously solved problem (analogy-making). When CMattie receives a message of such a non-seminar event, say a Colloquium, she treats it as a **speaker-topic message** for a seminar. She sends an acknowledgement to the sender stating that she is initializing a new seminar by the name "Colloquium Seminar" in the system and the sender is the organizer for this seminar. This misunderstanding effects one or more of the following events dependent on the sender:

- The wrong acknowledgement might elicit a negative response from the sender, starting an episode. The "conversation" between CMattie and the sender is stored in the case-based memory system of the Perception Module as an episode. This episode provides the reinforcement and along with the limited natural language understanding based on this episode, CMattie learns that Colloquium is similar to the seminar concept with slightly different features. In the case of colloquium, the **periodicity** feature has a different value. CMattie learns that through reinforcement from her domain.
- The sender might ignore this wrong acknowledgement and CMattie includes the Colloquium Seminar in her weekly seminar announcement.
- CMattie sends out the weekly seminar announcement with the Colloquium Seminar listed in it. That announcement might elicit a negative message from the sender of the original Colloquium message, starting a "conversation" between CMattie and the sender. This episode is stored in CMattie's case-based memory to aid in her learning what a colloquium is.
- The sender of the original Colloquium message might ignore the weekly announcement with the Colloquium Seminar listed in it. But he/she might respond to the reminder sent by CMattie the following week, when CMattie doesn't receive a **speaker-topic message** for the Colloquium Seminar. This, again, generates an episode aiding CMattie to learn about a colloquium.
- The sender of the original Colloquium message ignores all the reminders. This acts as a feedback to the CMattie system to modify the **periodicity** feature of the seminar concept, giving rise to a new concept which is similar to the seminar concept with a modified set of feature-values.

Irrespective of which of the above events occur, CMattie eventually learns the new concept called Colloquium which is closely related to the seminar concept. This concept is of the same depth as the seminar concept with a modified set of features. CMattie also learns related new codelets for the colloquium concept, by copying and modifying the codelets for the built-in seminar concept, to be able to correctly perceive and understand a **Colloquium message** when she receives one the next time.

4. Conclusions

In this paper, we briefly described the architecture of a software agent, CMattie, whose design considerations are based on the CAAT strategy. We also showed how CMattie learns new concepts to adapt to her dynamic domain. Our focus here was to discuss perceptual learning to detect changes in her domain, generate new concepts in CMattie's domain knowledge so as to enable her to respond appropriately to novel situations in her domain. The learning mechanisms are based on the premises: (1) Agents learn from what they already know, and (2) Every new situation/problem is viewed in terms of a previously solved problem. At the time of writing this paper, the VMattie system has completed preliminary testing and is starting on a beta test and the CMattie system is being implemented. We plan to focus on issues related to unlearning in the future stages of this project.

References

Bernard J. Baars. (1988). "A Cognitive Theory of Consciousness", Cambridge University Press.

Stan Franklin. (1995). "Artificial Minds", MIT Press.

Stan Franklin and Art Graesser. (1996). "Is it an Agent, or just a Program?: A Taxonomy for Autonomous Agents", Proceedings of the Third International Workshop on Agent Theories, Architectures and Languages, Springer-Verlag.

Stan Franklin. (1997). "Autonomous Agents as Embodied AI," Cybernetics and Systems, special issue on Epistemological Issues in Embodied AI, 28 p. 499-520.

John V. Jackson. (1987). "Idea for a Mind", SIGART Newsletter, No. 181, July 1987, p. 23-26.

Leslie Pack Kaelbling, Michael L. Littman, Andrew W. Moore. (1996). "Reinforcement Learning: A Survey", Journal of Artificial Intelligence Research 4 p. 237-285.

Pentti Kanerva. (1988). "Sparse Distributed Memory", MIT Press.

Janet Kolodner. (1993). "Case-based Reasoning", Morgan Kaufmann Publishers.

Pattie Maes. (1989). "How to do the right thing", Connection Science Journal, Vol. 1, No. 3, March 1989.

Pattie Maes and Rodney A. Brooks. (1990). "Learning to Coordinate Behaviors", Proceedings of the Eighth National Conference on Artificial Intelligence, AAAI Press/MIT Press, p. 796-802.

Melanie Mitchell. (1993). "Analogy-Making as Perception", MIT Press.

Aaron Sloman. (1995). "Exploring Design Space and Niche Space", Proceedings of the Fifth Scandinavian Conference on AI, Trondheim, May 1995, Amsterdam: IOS Press.

Hongjun Song and Stan Franklin (forthcoming), Action Selection using Behavior Instantiation.

MOTOR LEARNING

MMC - a recurrent neural network which can be used as manipulable body model

Holk Cruse, Ulrich Steinkühler, Christian Burkamp

Fac. of Biology, University of Bielefeld, Postfach 100131, D-33501 Bielefeld, Germany
e-mail: holk.cruse@biologie.uni-bielefeld.de

Abstract

A recurrent network is proposed which can be used as a manipulable body model to solve different kinematic tasks as the inverse kinematic problem, the direct kinematic problem or any mixed problem. The model may be used for planning a movement, or „thinking“, by being uncoupled from the motor output, or it may be used for direct motor control. The network is based on a new type of neuronal network called MMC net which is similar to but shows some essential differences to the Hopfield type network. These are (1) no symmetrical weights are necessary in the MMC net. (2) Furthermore, no clipping functions are necessary which allows for real valued outputs. (3) No limited number of discrete attractors, but an infinite number of attractors which form a continuum are possible in the MMC network.

1 Introduction

All behaving systems - whether living or artificial - may be classified into one of two types. One type comprises strictly sensory or data- driven, feedforward systems, such as simple reflex machines. In these cases the extant input determines the output. Although the properties of such systems might be changed by learning, the system at any moment can be regarded as having a static world model. The other type of systems has a form of world model which can be used to make predictions concerning future states of the world. Such a system may possess an internal world model which is "manipulable" in the sense that the internal world model can be cut off from the sensory input and the motor output and then used to mentally "play" with different possibilities. Such a model could be used for thinking, for reflecting and - when connected to a value system - for decision making. A system with such a model is often called a cognitive system. These two types may not form two distinct classes but rather form the extreme ends of a continuum. There are, for example, intermediate systems, namely systems with non-manipulable world models which nevertheless can make predictions and therefore do not slavishly rely on the extant sensory input. Important biological examples of such models are the form of central pattern generators more appropriately called central oscillators, elements which can produce rhythmic changes in activity in the absence of sensory or other external influences. Such a central oscillator, for example the circadian clock, provides a prediction of future situations which can be employed to control actions and also to control and integrate sensory input.

It has often been argued (e.g. Brooks 1991) that systems using manipulable, or explicit world models may represent unnecessarily complicated solutions to a given problem. Indeed many problems can be solved by astonishingly simple reactive systems by exploiting the properties of the real world and "use the world as its best own model" (Brooks 1991). There are even cases where reactive systems are not only simpler but even offer solutions to a problem which may not be solved by an explicit model (e.g. Cruse et al. 1996). Although, however, such reactive systems can be unexpectedly adaptive, their capacity is limited insofar as, apart from trial and error learning, no invention of qualitatively new behavior seems to be possible in the life time of such a system. In other words, "thinking" may not be possible for such a reactive system.

On the other hand, by intuition we have the impression that human beings, at least, have some kind of world model which can be used for planning and possibly also for controlling behavior. This is supported by the following observations. The conscious part of our world model, what we "see of the world" is not a direct projection of the sensory input. When looking at a picture, for example, we are not aware of the effects of the eye movements, but have the impression of a stable sceenary. This internal image is not a direct copy of the sensory input, but seems to be constructed such that it is only updated here and there by using the sensory input (Simons and Levin 1997, Neumann 1990). A neuronal system which may be described as a manipulable world model is also required for the capability of humans and other animals to "mentally rotate" visually given objects. Again in this case, more than a direct representation of the sensory input appears to be necessary to solve this task. Therefore, we seem to use such manipulable world models which leads to the question of how could such a manipulable model be represented in neuronal terms?

In this paper, we will approach this question by describing an ANN model which can be interpreted as a model of the own body. We begin with a body model because we feel that in both ontogenetic as well as phylogenetic terms, the body model may form the basis of any developed world model that we use for internal manipulation. For the brain, the own body is the most immediate part of the world.

What are the requirements to be fulfilled by such a "manipulable" body model? To be a general purpose model it first of all should allow to represent a system with redundant (extra) degrees of freedom. This is already the case when an arm with three joints is free to move in a two-dimensional plane. In this case the classical problem of inverse kinematics, namely to find three angle values such that the hand points to a given position in the two-dimensional plane, allows for many different solutions and the model should be able to select one of these solutions. To this end, the system should allow to represent, in principle, all geometrically possible solutions. In order to adapt the model to the geometrical conditions of a really given hardware system, it should also be able to cope with built-in constraints like, for example, joint limits. Furthermore, such a model should be capable of dealing with external constraints like for example holding the hand at a fixed angle.

When solving these questions such a system may be confronted with two problems. First, as is well known from robotics, in particular when solving the inverse kinematic problem, manipulator positions may appear which produce singularities because they require a division by a very small value or even by zero. Therefore, one question is how to avoid singularities. Second, the system has to deal with the problem of sensor fusion, when different signals coming from different sense organs are used to describe the same values. For example, the position of the tip of the hand can be measured by visual sensors. At the same time, mechanosensors may measure the joint angle values. By application of coordinate transformations the latter values can also be used to compute the position of the tip of the hand. However, due to measurement and computational errors both results usually do not agree exactly. Therefore, the question arises how to deal with this ambiguous data set. Apart from another problem which will not considered here further, namely to deal with constraints given by obstacles in the workspace, all these problems will be solved by the model explained below.

In order to allow for an explanation as simple as possible, we restrict the description to a model of only a part of the body, namely a three-joint arm moving in a two-dimensional plane. It thus fulfills the condition of having an extra degree of freedom. It should, however, be stated already here that this model can be easily expanded to arbitrarily complicated geometries in 3D space.

A mechanical version of such a body model has been discussed by Mussa Ivaldi et al. (1988). These authors pointed out that the inverse kinematic problem could be easily solved by a mechanical model of the manipulator provided with springs to simulate the muscles, in the following way. By simply moving the tip of the (arbitrarily complicated) manipulator in the direction of the position of the intended target point, the joint angles of this mechanical model automatically find their appropriate values. The angle values of this model could then be used to control the real arm. Mussa Ivaldi et al. called this the "passive motion paradigm". Here, we will propose a way of how this idea could be implemented in neuronal hardware.

In general terms, the model consists of a recurrent network which relaxes to adopt a stable state corresponding to a geometrically correct solution, even when the input does not fully constrain the solution. The underlying idea is that several geometrical relationships are used to calculate the same value (e.g. the shoulder angle) several times in parallel. The final output is then obtained by calculation of the mean value of these multiple computations (MMC). This is then fed back for the next iteration. Any special constraints of the variables (e.g. joint limits) can be introduced after the mean value has been computed. In this way, the MMC-type network can be used as a kinematic inverse model which allows for very different solutions of the redundancy problem, and which can easily cope with limitations of the joint space or the workspace.

The model will be shown in two steps. First, we introduce a very simple, linear model which, however, is only applicable to manipulators with variable segment lengths, and, in the 3D case, with ball and socket joints. Second, we will expand this system by some nonlinear control structures. In this version, the system can be used to describe manipulators with fixed segment lengths, and the joints may be constrained to being simple hinge joints.

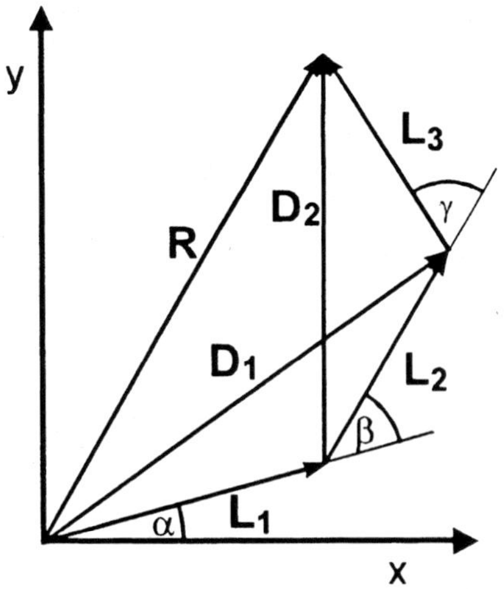

Fig 1. An arm consisting of three segments described by vectors $\mathbf{L_1}$, $\mathbf{L_2}$, and $\mathbf{L_3}$, which are connected by the three joints. The joint angles are α, β, and γ. The position of the endeffector is described by vector $\mathbf{R}$. Furthermore, two additional vectors $\mathbf{D_1}$ and $\mathbf{D_2}$, describing the diagonals, are shown.

2 The linear model

To introduce a simple linear MMC model, we will use a manipulator with three joints in a 2D space (Fig. 1) as an example. The net could, however, be directly extended to the 3D case and an arbitrary number of segments. The coordinates of the joints can be described in a Cartesian coordinate system x, y, the origin of which is situated in the "shoulder" of the manipulator. We describe the segments as vectors L_1, L_2, L_3. For later use, we will also introduce the three angles α, β, and γ which describe the relative position of the segments, as shown in Fig. 1. The vector R pointing to the endeffector, and the vectors D_1, D_2 describing the diagonals, are additionally introduced (Fig. 1). For these vectors, the following seven equations hold:

$$\begin{array}{l} L_3 + L_2 + L_1 - R = 0 \\ L_2 + R - D_2 - D_1 = 0 \\ -L_3 + L_1 + D_2 - D_1 = 0 \\ L_2 + L_1 - D_1 = 0 \\ L_3 + L_2 - D_2 = 0 \\ L_3 - R + D_1 = 0 \\ L_1 - R + D_2 = 0 \end{array} \qquad (1)$$

In this system, each value L_1, L_2, L_3,R, D_1, and D_2 appears four times. Therefore, for each value four of these equations can be used to calculate this value. For example, L_3 could be determined by

$$\begin{array}{rl} L_3 & = -L_2 - L_1 + R \\ & = L_1 + D_2 - D_1 \\ & = -L_2 + D_2 \\ & = R - D_1 \end{array} \qquad (2)$$

According to the MMC principle, all four values L_3 can be summed, and this sum divided by four to obtain the mean value L_3 m (for a graphical illustration see Fig. 8). The same result could be obtained by adding the four equations:

$$\begin{array}{rl} L_3m & = 1/4\,(-2\,L_2 + 2\,R + 2\,D_2 - 2\,D_1) \\ & = -0.5\,L_2 + 0.5\,R + 0.5\,D_2 - 0.5\,D_1. \end{array}$$

Correspondingly, for the other values, the following equations can be obtained:

$$\begin{array}{l} L_2m = 1/4\,(-2\,L_3 - 2\,L_1 + 2\,D_2 + 2\,D_1) \\ L_1m = 1/4\,(-2\,L_2 + 2\,R - 2\,D_2 + 2\,D_1) \\ Rm = 1/4\,(2\,L_3 + 2\,L_1 + 2\,D_2 + 2\,D_1) \\ D_2m = 1/4\,(2\,L_3 + 2\,L_2 - 2\,L_1 + 2\,R) \\ D_1m = 1/4\,(-2\,L_3 + 2\,L_2 + 2\,L_1 + 2\,R) \end{array} \qquad (3)$$

Thus, the coefficients of the system of these six equations (2) take values of 0.5, -0.5 or zero.

The next step is to decompose the vectors into the x and y components. This leads to two sets of corresponding linear equations, where each vector is replaced by its x- or its y-component, respectively. These equations are graphically illustrated by the network shown in Fig. 2. The left column of values shows the inputs, the horizontal line the output values of the two equation systems. The connection weights are either 0.5 (full circles), -0.5 (open circles), or zero (no circle). The two networks, one for each component, are plotted together, the net for the x components is shown by continuous lines, that for the y components as broken ones. (For an extension to the 3D case, a third parallel net for the z-component had simply to be introduced). Each subnet corresponds to the system of equations (3) shown above.

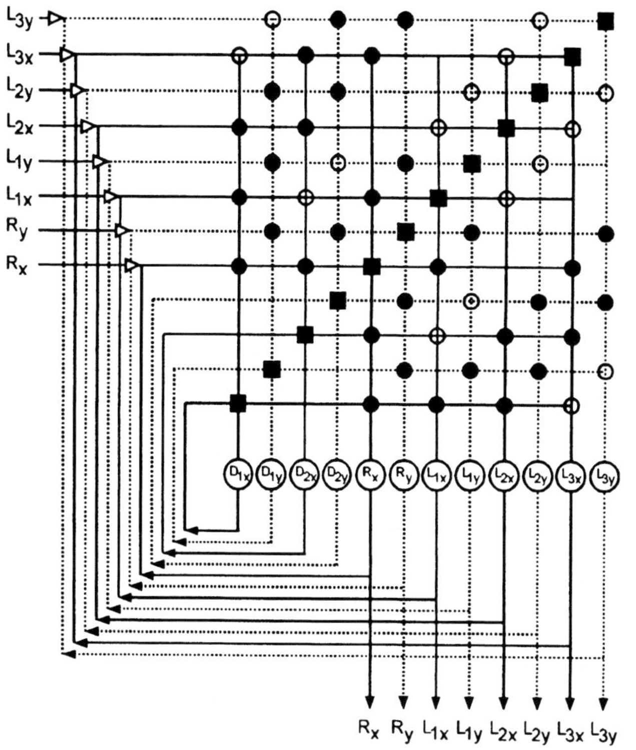

Fig. 2. The linear version of the MMC net. The output values represent the components of the vectors **R**, $\mathbf{L_1}$, $\mathbf{L_2}$ and $\mathbf{L_3}$ (lower line). Continuous lines represent the subnet for the x components, the broken lines that for the y components of the five vectors. To calculate the inverse kinematics, the components of **R** have to be provided as input for the net. However, any other combination of input values could be chosen instead. The possible input values are shown on the vertical column at the left. For the chosen input, the recurrent line is suppressed, and only the input value is fed into the net, symbolized by the open arrowheads.

According to the MMC principle, the mean values can be used as output values but are at the same time fed back as input for the next iteration. This forms the simple recurrent network of Fig. 2. To solve the inverse kinematic problem, R_x and R_y, the components describing the desired position of the endeffector, are given as input to the complete network. This means that the recurrent connections of R_x and R_y are suppressed by the external input. This is symbolized by the open arrowheads in Fig. 2. During a number of iterations the network relaxes, and the output values then represent a geometrically correct position of the manipulator pointing to the desired endeffector position R_x, R_y.

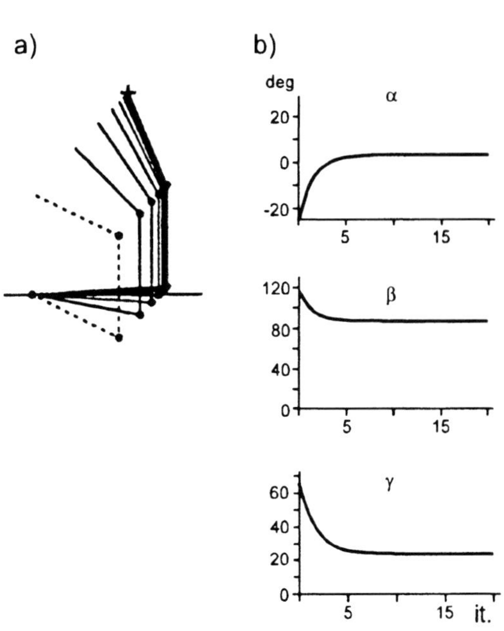

Fig. 3. Relaxation/movement of the linear version of the three-joint arm net (Fig. 2). (a) The arm moves from a starting configuration (dotted lines) to the prescribed position of the endeffector (cross). As can be seen, the basal segment $\mathbf{L_1}$ is lengthened during the relaxation. (b) The temporal changes of the three-joint angles α, β, and γ during the relaxation. Abscissa is no. of iteration cycles.

In the system of equations (3), all weights at the diagonal are zero, which means that there are no self-excitations of the individual units. If such couplings are introduced as shown for the case of L_3 :

$L_3m = 1/(4+d)\ (-2\,L_2\ + 2\,R + 2\,D_2 - 2\,D_1 + d\,L_3)$,

the dynamic behavior of the net is changed such that the relaxation becomes the smoother the higher the "damping" values d at the diagonal. The introduction of these values has the advantage that oscillations, which sometimes occur during the relaxation, can be successfully damped. Thus, some kind of low-pass property is introduced into the system. For the calculations in the examples shown here, we used d = 5. Therefore, the weights in Fig. 2 change from 0.5, -0.5, and zero to 2/9, -2/9, and zero, respectively. The weights d arranged along the diagonal are shown as squares in Fig. 2.

Fig. 3a,b shows an example of how the system behaves when moving from a given starting configuration to an end position defined by the coordinate of the endeffector. For this purpose, the desired endeffector coordinates R_x, R_y have to be given as input values for the net. The system always finds a solution for the inverse problem but, as the segment lengths are not fixed, they might change during relaxation. In Fig. 3a this is obvious for the basal segment L_1.

By application of a graph theoretical approach it could be shown for this system (and the general 3D version) that from any given starting position, it always converges to a stable and geometrically correct final state (Steinkühler and Cruse 1998).

3 The nonlinear model

We will now expand the model by a nonlinear part which allows us to keep the segment lengths constant and, furthermore, to provide, as external input, not only the Cartesian coordinates of the endeffector, but also the values of the joint angles α, β, and γ. For this purpose, the latter have to be calculated using the Cartesian coordinate values provided by the linear net shown in Fig. 2. This is schematically shown in Fig. 4 by boxes P and T (see below). These values could then be used as output values and could also be fed back as recurrent input to the net for the next iteration cycle. Before this is possible, the corresponding inverse transformations have to be performed which are symbolized by boxes Q and T (Fig. 4). Transformation P simply represents the arctan function to determine the angle value from the components L_x and L_y. For the second and subsequent segments, this is only possible after a rotation of the coordinate system. This is done by the corresponding circuit T. Circuit Q corresponds to the inverse of P and represents the calculation of the Cartesian components from the segment length L and the rotation angle α, β, or γ. The inverse of T is obtained by using the negative value of the rotation angle, as indicated by -1 in Fig. 4 (for details see Steinkühler and Cruse 1998).

In this version, we now have the possibility to deliberately influence the joint angles at the input side and also use these values at the output. For a manipulator with fixed segment lengths, these values L_1, L_2, L_3 will always be fixed at the input and therefore need not be fed back. Important variables are the Cartesian coordinates of the endeffector R_x and R_y, and the three joint angles α, β, and γ. Just as in an earlier MMC version (Cruse and Steinkühler 1993), an arbitrary input vector can now be chosen, and, after relaxation, the system provides a geometrically correct and complete output vector. The system calculates a complete set of state variables, i.e., in our example the three joint angles and the endeffector coordinates when only a subset of these five variables α, β, γ, R_x and R_y is given. Thus, the MMC net, simply by varying the input values of the net, can be used to calculate the inverse kinematics (input R_x, R_y), the direct kinematics (input α, β, γ), or any mixed problem (input e.g. α, R_x) including underconstrained cases. This furthermore means that the task can be deliberately changed from one moment to the next. Giving an example for the underconstrained inverse kinematic case, Fig. 5 shows the movement of the manipulator using the same starting configuration and endpoint coordinates as in Fig. 3. Fig. 6 shows an example for a mixed problem. In this case the final position is determined by a given wrist angle α and a given value R_x for the x component of the endeffector position. The net selects a geometrically possible position which suffices these conditions.

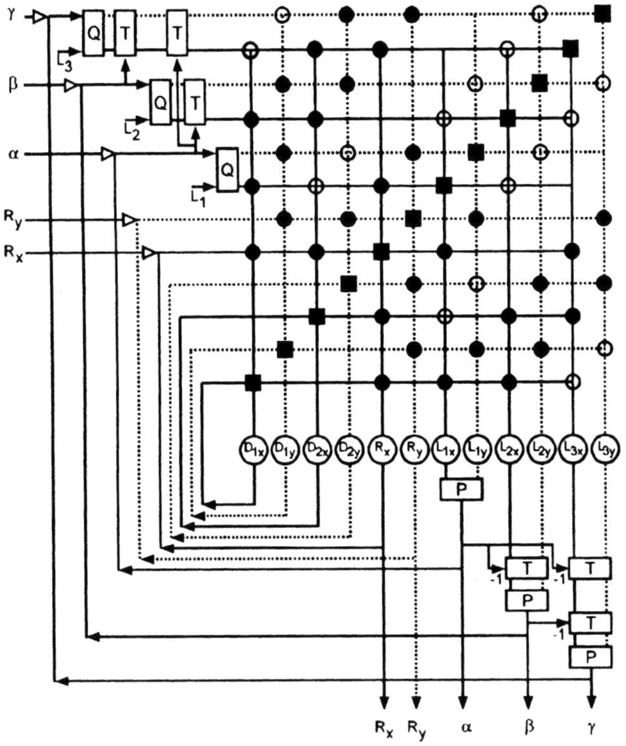

Fig. 4. The MMC net as shown in Fig. 2, but with nonlinear subsystems in the feedback loops. These nonlinear systems P, T, and Q are explained in the text. The squares represent the weights, d, responsible for the damping of the oscillations.

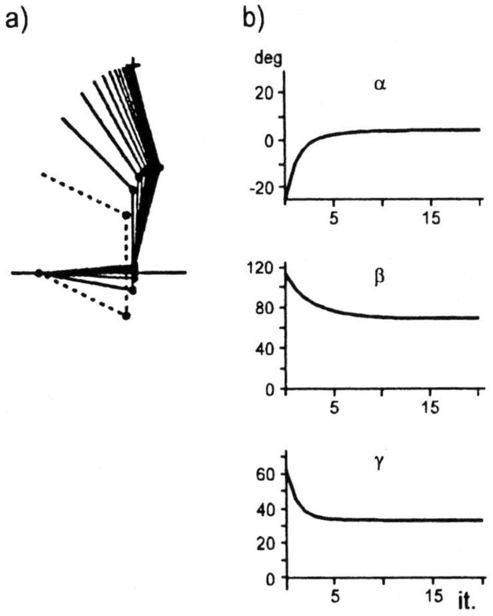

Fig. 5. Relaxation/movement of the nonlinear version of the three-joint arm net (Fig. 4). (a) The arm moves from a starting configuration (dotted lines) to the prescribed position of the endeffector (cross). As can be seen, the segments now maintain their length during relaxation. (b) The temporal changes of the three joint angles α, β, and γ during relaxation. Abscissa is no. of iteration cycles.

A special case is presented by the example in Fig. 7. Here, the coordinates of the endpoint cannot be reached by the manipulator because they specify a point outside its workspace. Nevertheless, the network tries to follow the desired coordinates as well as possible. The manipulator finally points in the direction of the desired endpoint coordinates.

The MMC principle can be easily expanded to three dimensional cases by adding a third net for the z component parallel to the already explained nets for the x and y components (see Figs. 2 and 4). This has been applied for a humanoid 7 degree of freedom arm by Steinkühler and Cruse (1998) and to a 18 degree of freedom, six legged walking system (Cruse et al. 1996).

An important problem in classical systems for calculating the inverse kinematics is the possible occurrence of singularities. These occur when, for some special positions, a division by zero becomes necessary. As a division by zero is not possible in our network, there are no singularities. Even when the tip of the arm has to be moved to the margin of the workspace, or even to targets outside the workspace (Fig. 7), no such singularities occur.

The system can be further extended by adding constraints as, for example, limited angle ranges for individual joint angles. This is possible by the introduction of such limits into the nonlinear element P (Fig. 4). Soft constraints, e.g. cost functions which attribute to some angle values higher costs than to others, can also be introduced as shown by Steinkühler and Cruse (1998)

Up to now, it has not been possible to prove the convergent behavior of this nonlinear system in general. Only for selected cases, namely the case of the direct kinematic and some cases of variable, but limited segment lengths, i.e., for the case of a manipulator with prismatic joints, this proof was possible (Steinkühler 1994).

4 Asymmetrical weights

The construction of the network proposed above necessarily leads to a net with symmetrical weights. This means that the weight values are symmetrical with respect to the symmetry axis given by the diagonal of the weight matrix (see Figs. 2 and 4). However, versions with asymmetrical weights may also show the property to relax to stable states. Asymmetrical weight distributions can easily be obtained in the actual version, when, for the calculation of the mean value, the four equations (2) are weighted differently. As this could be done independently for the calculation of each of the six mean values, the equations (3) could show very different, in particular asymmetrical weight

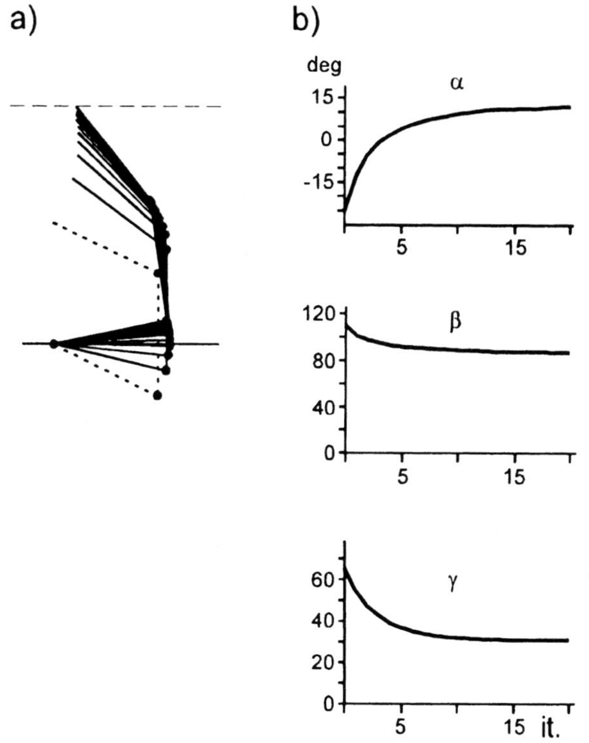

Fig. 6. Relaxation/movement of the three-joint arm network when the final position is determined by a given wrist angle α and a given value R_y for the y component of the endeffector position. (b) The temporal changes of the three joint angles α, β, and γ during relaxation. Abscissa is no. of iteration cycles.

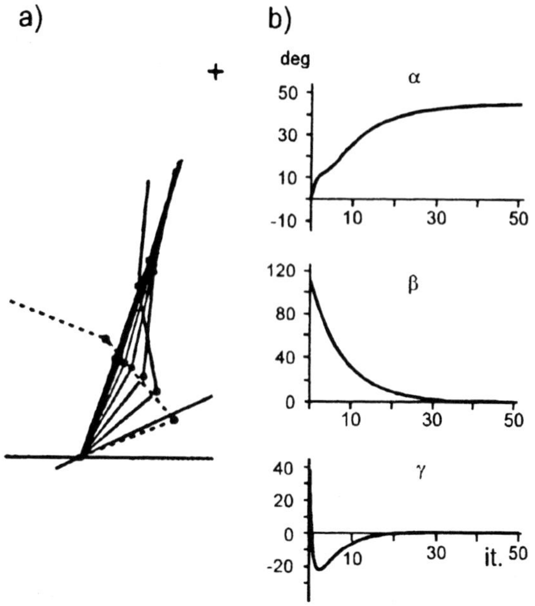

Fig. 7. Relaxation of the net when the desired endpoint coordinates (cross) lie outside its workspace. Only the arm configuration of every fifth cycle is shown in (a). (b) The temporal changes of the three joint angles α, β, and γ during relaxation. Abscissa is no. of iteration cycles.

distributions. This is illustrated in Fig. 8 showing the network for only one, say the x component, but representing all the 4 x 6 = 24 individual equations as separate neuronal units. The input weights (closed or open circles) have to be kept fixed in order to maintain the mathematical correctness of the 24 equations. Always four units are grouped to calculate the mean value $\bar{x}$. Before this calculation, the output of each unit can be multiplied by an arbitrary "base" weight, shown by filled triangles in Fig. 8. It can be shown that an infinite number of asymmetrical, but only one symmetrical form of weight distributions exist for a given geometrical system (for a proof see Burkamp 1996). These base weights have to be taken into account for the calculation of the (weighted) mean. In contrast to the lumped versions shown in Figs. 2 and 4, Fig. 8 also illustrates the massively parallel structure which is the basis of the error tolerance of the MMC system. The 24 base weights can be changed arbitrarily. We used genetic algorithms (100 individuals per generation, cross over, mutation) and applied an evaluation function which tried to minimize the relaxation time. One example for such a weight distribution is shown in Tab. 1 in the format of the lumped version (see Figs. 2, 4).

Tab. 1. An asymmetrical weight distribution producing faster relaxation compared to the symmetrical version. The weights are given in the same format as in Figs. 2 and 4. Values d represent the damping factors.

	D1	D2	R	L1	L2	L3
L3	-1.0	0.23	1.	-0.04	-0.44	d
L2	0.	0.22	1.	-0.39	d	-0.36
L1	0.09	-0.78	1.	d	-0.26	-0.17
R	0.91	0.77	d	0.65	-0.3	0.80
D2	0.09	d	0.	-0.61	0.74	0.20
D1	d	0.0	0.	0.35	0.56	-0.64

5 Discussion

The MMC net proposed here can be considered as a model which can be used to manipulate knowledge concerning the own body when searching for solutions for a given kinematic task. In other words, it can be used as a basis for "mentally moving" geometrically specified objects. This is similar to the task of mental rotation (Shepard and Metzler 1971), but more difficult to the extent that in our case the geometrical relationships between the body parts are not fixed, but, in general have to be changed to find a solution. Furthermore, the task, in general, is underconstrained which means that a

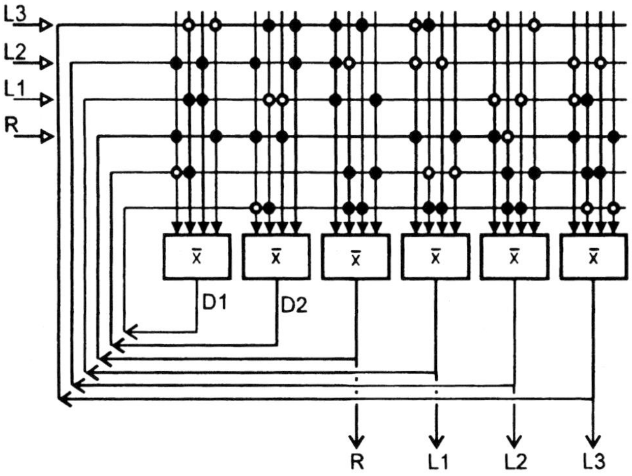

Fig. 8. The MMC net for the three-joint planar arm. Only the net for one spatial component is shown. Weights are 1 (closed circles) or -1 (open circles). The black triangles represent arbitrary weights.

decision between a number of possible solutions has to be made. Using the MMC net, this is possible for arbitrarily complicated body geometries.

To some extent the MMC principle is similar to the Hopfield net (Hopfield 1982, 1984), but there are some essential differences, too. The MMC network as described in sects. 1 - 3 uses symmetrical weights as does the Hopfield net. For Hopfield nets this is a necessary condition. In contrast, the MMC principle also allows for asymmetric weigths. As in the Hopfield net, an attractor can be reached, even if some of the input variables do not have the appropriate values. However, the Hopfield net permits only a limited number of discrete attractors, whereas the MMC net provides an infinite number of attractors which form a continuum. In the case of the Hopfield net, weights typically are introduced by application of the Hebb rule, leading to a symmetrical weight distribution. For the MMC net, the weights of a unit simply have to represent the coefficients of a mathematically correct equation. Nevertheless, appropriate weights might be found by application of learning methods like genetic algorithm (see above) or backpropagation. For the latter, the input vectors have to be compared with the output vector obtained after relaxation. The differences for each component could be used as error signals to train the weights of the nets shown in Figs. 2 or 4.

Another basic difference between the Hopfield net and the MMC net is the following. A crucial property for a Hopfield net to converge is that all units are provided with a clipping function. This leads to stable states where the activation of all units is either approximately 1 or -1. For the MMC such clipping functions are, however, not necessary. Conversely, an important property of the MMC net is that it can provide real valued outputs representing, for example, an angle value. The MMC net converges although it lacks the clipping functions. The reason for this appears to be that the calculation of the mean values avoids an unlimited increase in the activation of the units. The convergence proof holds only for the linear symmetrical case, but simulation experiments have shown that the nonlinear symmetrical version (Fig. 4) and versions with asymmetrical weight distributions (Tab. 1) show stable behavior as well.

Like the Hopfield net, the MMC net acts as an autoassociator which completes an incompletely specified input vector. In the case of a non-redundant system (Steinkühler and Cruse 1998, Fig. 3), the MMC net simply performs a coordinate transformation. For the redundant case, the system also selects one out of many possible solutions. The selection depends on the earlier state of the system, but we cannot explicitly formulate the underlying cost function.

The MMC net provides a solution for the problem of sensor fusion, occurring when different sensory inputs represent the same value. As an example, in the case of the three-joint arm, the endeffector coordinates R_x and R_y could be given by visual input, the angles α, β, and γ by mechanoreceptor inputs. These different values are integrated within this holistic system. In this way, the MMC net concatenates both the visual and the mechanoreceptively given spaces to form a common, dynamic representation.

One might argue that a system like the MMC net is of no direct relevance for the neurobiologist insofar as there is no immediate idea how to test whether or not such a network exists. Although this is probably true - similarly one might argue that conditions necessary for Hopfield nets may usually not be fulfilled in brains - we would like to argue that both types of networks are of interest because they can provide ideas on possible properties of recurrent systems. Because the MMC net does not require symmetrical weights and as it allows for graded output values, this net is biologigally more plausible and may, therefore, offer a new view onto the abundancy of recurrent connections in the brain. One aspect is that the MMC principle shows an alternative to the well known cortical maps to represent geometrical information. Contrary to such topological representations, the MMC net requires an extremely small number of units. Nevertheless, it offers a high flexibility by being able to represent all geometrically possible positions, thereby being able to exploit the redundancy of the mechanical system.

What might be the functional sense of such an internal representation of the body? Such a representation may not be of interest for a system with an extremely simple body as for example a robot of cylindric form and only two active wheels. In such cases the motor output usually consists of two values, one for each motor. However, the control of a complex mechanical system with a high number of degrees of freedom may require different control structures. One possibility is to apply a MMC-like structure because it is able to find solutions for animals or animats of arbitrary complicated body geometry. The capability to select between a number of possible solutions is an important property for a system showing adaptive behavior on the level of motor control.

One could, however, speculate that this principle might also be applied to higher-level tasks. This will be explained in the following two, admittedly very speculative, considerations.

Application of brain imaging techniques has shown that during the mental imagination of a body movement nearly the same cortical areas are active as when performing real movements. These areas include motor areas as well as parietal areas (Crammond 1997). This finding suggests that the neuronal substrate responsible for imagining a body movement contributes to the control of an actual movement. When a movement is only imagined, the system may be uncoupled from the lower-level motor system. This imagination of an action may be regarded as to form the basis of „thinking". What might be the neuronal basis of this representation of the body? Here we propose that an MMC-like system might be considered as a quantitative hypothesis for a neuronal realization of a mental body model: this system can represent all geometrical positions and it can be applied for direct control of movement or for movement planning. For further arguments why such a mental model may be helpful or even necessary, see Kawato (1996).

Another aspect shows how such a recurrent network may help to guide the interpretation of neurophysiological results. As summarized by Fuster (1995, pp. 134-137) there was a discussion whether cells found in area 7 of a monkey which are excited when the monkey was looking at or was reaching to a given target should be termed motor or sensory neurons. Application of the MMC principle might serve as a way to solve this apparent ambiguity. In a MMC type system, the internal connections cannot be labeled as either motor or sensory elements. This distinction is only possible for those channels which connect the MMC system to the external world. Therefore, a classification of such neurons as more sensory or more motor might not be sensible.

In this context, the MMC principle might, furthermore, be considered as a hypothesis showing how the two memory sections described by Fuster (1995), namely the motor memory, largely distributed in the precentral cortex, and the perceptual memory, localized in the postcentral cortex, might be concatenated to form a common system. In this way, the MMC principle may provide a quantitative basis for the phenomena called "Gestaltkreis" by V. v. Weizsäcker (1950) or "action-perception cycle" by Arbib (1981).

Internal world models are used in GOFAI (Brooks 1997): these Good Old Fashioned Artificial Intelligence systems are based on reasoning about their world model, they use symbols and explicit goals. Self-adaptation is usually not possible. Therefore, the application of world models is not in accord with the philosophy of most representatives in the field of adaptive behavior. However, an orthodox rejection of internal world models might overlook that there are ways in between these GOFAI type models and the pure behavior-based approaches. We argue that this is the case for our model: it represents a world model without explicit rules; there is no processing of symbols; the system can self-adapt or calibrate by interaction with the world (see above); the solutions result from the dynamics of the system. These are all properties typical for the so called behavior-based systems. Therefore, the MMC net may be considered an inherently behavior-based system being „abused" as a world model. The MMC system can help to find solutions for a number of tasks listed in Brooks (1997, sect. 2.3) as are, for example, coordinate body, head and eye motions, and/or select a grasping strategy, a task which requires the coordination of many subsystems.

Very interesting related ideas have been proposed by Morasso and Sanguineti (1994) and by Kawato (1996). However, in both cases the realization is different from our approach. Kawato uses separate representations of the direct and the inverse kinematic solution. Both subsystems are then coupled to form a „bidirectional" recurrent system. Morasso and Sanguineti propose a system consisting of three parallel topological maps representing the external space, the joint space and the muscle space. Again there are recurrent connections between these layers. This architecture based on spatial coding necessarily requires more neural units then does the MMC approach which in the version proposed here is based on intensity coding. Both architectures appear to be somewhat less „holistic" than the MMC principle. Another completey different approach using production rules to form a mental model has been proposed by Donnart and Meyer (1996). Their MonaLysa can build a cognitive map of its environment which can be applied for planning trajectories to find a goal while avoiding obstacles. In this way the system develops an internal model of the world. For this it does however not require a detailed model of its own body.

Finally, it should be stated that there are, of course, a number of open questions. Apart from providing a convergence proof for the nonlinear MMC version a selection of particular asymmetric weights in order to gain special overall properties might be of interest. As another example, we want to mention the question of how to simplify large MMC nets as in the case of modelling the 19 body insect by attempting to modularize the MMC net, so that only a smaller number of connections is necessary.

Acknowledgements

This work was supported by the Körber Foundation, BMFT (grant no. 01 IN 104 B/1), and DFG (grant no. Cr 58/8-2).

References

Arbib, M. (1981): Perceptual structures and distributed motor control. In: Handbook of Physiology: Nervous systems 2, V.B. Brooks (ed.). pp. 1448-1480. Bethesda MD. Amercan Physiol. Soc

Brooks, R. A. (1991): Intelligence without reason. IJCAI-91, Sydney, Australia, 569-595

Brooks, R.A. (1997): From earwigs to humans. Robotics and Autonomous Systems 20, 291-304

Burkamp, Ch. (1996) Untersuchungen zum Lernen in rekurrenten künstlichen Netzen. Diplomarbeit Universität Bielefeld

Crammond, D. J. (1997): Motor imagery: never in your wildest dreams. Trends in Neurosciences 20, 54-57

Cruse,H., Steinkühler, U. (1993): Solution of the direct and inverse kinematic problems by a common algorithm based on the mean of multiple computations. Biol. Cybernetics, 69, 345-351

Cruse, H., Bartling, C., Dean,J., Kindermann, T., Schmitz, J., Schumm, M., Wagner, H. (1996): Simple solutions to complex problems by exploitation of the physical properties. In: Maes, P., Mataric,M.J., Meyer, J.-A., Pollack,J., Wilson, S.W. (eds.): From animals to animats 4. p. 84-93, MIT Press, Cambridge MA

Donnart, J-Y, Meyer, J-A (1996) Spatial exploration, map learning, and self-positioning with MonaLysa. In: Maes, P., Mataric,M.J., Meyer, J.-A., Pollack,J., Wilson, S.W. (eds.): From animals to animats 4. p. 204-213, MIT Press, Cambridge MA

Fuster, J. M. Memory in the Cerebral Cortex. MIT Press, Cambridge MA 1995

Hopfield,J.J. (1982): Neural networks and physical systems with emergent collective computational abilities. Proc.Natl.Acad.Sci.79,2554-2558

Hopfield,J.J. (1984): Neurons with graded response have collective computational properties like those of two-state neurons. Proc. Natl. Acad. Sci. 81, 3088-3092

Kawato, M. (1996): Bidirectional theory approach to integration. In: Inui, T., McClelland,J. (eds.) Attention and Performance XVI. p. 335-367. MIT Press

Mussa Ivaldi, F.A., Morasso, P., Zaccaria, R. (1988): Kinematic networks - a distributed model for representing and regularizing motor redundancy. Biol. Cybern. 60, 1-16

Morasso, P., Sanguineti, V. (1994): Self-organizing body-schema for motor planning. J. Motor Behavior 27, 52-66

Neumann, O. (1990): Direct parameter specification and the concept of perception. Psychol. Res. 52, 207-215

Shepard, R.N., Metzler, J. (1971): Mental Rotation of Three-Dimensional Objects. Science 171, 701

Simons, D.J., Levin, D.T. (1997): Change blindness. Trends in Cognitive Sciences 1, 261-267

Steinkühler, U. (1994): MMC-Modelle zur Lösung kinematischer Aufgabenstellungen eines redundanten Manipulators. Dissertation Universität Bielefeld

Steinkühler, U., Cruse, H. (1998): A holistic model for an internal representation to control the movement of a manipulator with redundant degrees of freedom. Biol. Cybernetics (in press)

v. Weizsäcker, V. Der Gestaltkreis. Thieme, Stuttgart 1950

http://www.uni-bielefeld.de/biologie/Kybernetik

From lampreys to salamanders: evolving neural controllers for swimming and walking

Auke Jan Ijspeert*, John Hallam*, David Willshaw**

* Dept. of Artificial Intelligence, U. of Edinburgh
5 Forrest Hill, Edinburgh EH1 2Q, UK
{aukei, john}@dai.ed.ac.uk
** Centre for Cognitive Science, U. of Edinburgh
2 Buccleuch Place, Edinburgh EH8 9LW, UK
david@cns.ed.ac.uk

Abstract

This paper presents how neural swimming controllers for a lamprey can be adapted for controlling both the swimming and the walking of a salamander-like animat. Using a Genetic Algorithm (GA), we extend a connectionist model of the biological Central Pattern Generator (CPG) controlling the swimming of a lamprey (Ekeberg, 1993) to control the locomotion of a 2D mechanical simulation of a salamander.

We first summarize experiments on the evolution of alternative swimming controllers for the lamprey (Ijspeert et al., 1998). The aim of that work was to study whether there exists other neural configurations than that found in the lamprey which could control swimming with the same efficiency and to develop a method for developing neural locomotion controllers using a GA. We then present how that method, namely a staged evolution of the neural configuration of a connectionist model, can be used to extend swimming controllers to control both swimming and walking. Controllers which, similarly to CPGs of animals, can produce complex oscillations when receiving simple excitatory signals are thus developed. In particular, we generate a controller which can switch from swimming to walking and produce different speeds of motion depending on its excitation.

1. Introduction

Although there exists a variety of different locomotor movements among vertebrates, their control systems appear relatively similar across different species (Grillner et al., 1991). Locomotion is controlled by networks of interneurons located in the spinal cord which produce the complex patterns of oscillations necessary for motion. As demonstrated in the 1960s by Orlovsky's classical experiment on the cat (as reported in (Grillner, 1996)), these networks, called Central Pattern Generators, need only simple excitatory signals from the brain to produce the coordinated oscillations necessary for motion.

Our research focusses on biologically inspired neural locomotion controllers for autonomous agents and how to develop them using a Genetic Algorithm (GA). In particular, we are interested in which kinds of neural configuration can control the anguiliform swimming observed in lampreys and the swimming and walking of salamanders. Both lampreys and salamanders project themselves in water by undulation of the body, without use of fins or limbs. The undulation is a traveling wave propagating from head to tail. When the salamander switches from swimming to walking, its trunk ceases to propagate an undulation and, instead, performs an S shaped standing wave with the nodes at the level of the girdles (Frolich and Biewener, 1991). The bending of the trunk helps the salamander to increase the reach of its limbs which are attached laterally to the trunk.

The work presented here follows experiments in which we used a GA to evolve swimming controllers for a simulated lamprey with alternative neural configurations to the observed biological configuration (Ijspeert et al., 1998). The work was inspired by a biological connectionist model of the lamprey's CPG (Ekeberg, 1993). We developed artificial controllers using an incremental approach where increasingly complex control mechanisms are developed using elements of the previous evolutionary stage. Here, as the method proved to be a useful design technique, we apply it again to study whether controllers for swimming can be extended to control both swimming and walking of a salamander-like animat. Walking controllers based on oscillatory networks of the swimming controllers are generated for the control of a 2-dimensional mechanical simulation of the salamander.

We are interested in the development of control mechanisms which, similarly to the CPGs of animals, can produce complex oscillations when receiving simple input signals. A potential application of these controllers could be the control of locomotion of an amphibian robot. Another motivation is to gain some insights on how CPGs function in real animals. Note that the evolution from swimming to walking controllers and its application to robotics has been studied before (Lewis, 1996), but at a more abstract level than the connectionist approach taken here.[1]

In this paper, we firstly present, in Section 2, Ekeberg's connectionist model of the lamprey's CPG (Ekeberg, 1993). We then summarize our work on the incremental evolution of alternative swimming controllers based on neuron models similar to those of Ekeberg (Ijspeert et al., 1998) (Section 3). An example of how the same method could potentially be used as a tool for developing biological models is also given (Section 3.4). Finally, in a further evolutionary stage, we show how the controllers can be extended to control both the swimming and the walking of a salamander-like animat (Section 4).

2. Connectionist model of the lamprey's Central Pattern Generator

The swimming controller of the lamprey has been extensively studied and is probably one of the best known vertebrate locomotion controllers. The controller is a Central Pattern Generator which can produce the patterns of oscillation necessary for locomotion without oscillating input either from the brain or from sensory feedback. The complete CPG is located along the spinal cord of the lamprey and, as small parts of the spinal cord can be made to oscillate when isolated, is thought to be composed of interconnected segmental oscillators. During swimming, each segment oscillates with a small phase lag compared to its rostral neighbour which leads to a wave of neural activity, and hence an undulation of the body, to travel from head to tail. The lamprey usually maintains a constant wavelength corresponding to approximately the length of the body and changes speed of swimming by changing the frequency of oscillation.

Physiological observations have allowed neurobiologists to make a model of the neural circuitry of the CPG (Buchanan and Grillner, 1987; Grillner et al., 1988) whose capacity to produce most of the observed patterns of oscillations has been demonstrated through simulations both at a neurobiological (Ekeberg et al., 1991; Wallén et al., 1992) and a connectionist level (Buchanan, 1992; Williams, 1992; Ekeberg, 1993). The biological circuitry connects four type of neurons: motoneurons MN, excitatory interneurons (EIN), contralateral inhibitory interneurons (CIN) and lateral inhibitory interneurons (LIN) (Figure 1).

We have reproduced Ekeberg's connectionist model of the biological controller (Ekeberg, 1993). In this model, populations of functionally similar neurons are represented by single neuron units which are modeled as leaky integrators with a saturing transfer function. The output u of a neuron corresponds to the mean firing frequency of the population it represents ($\in [0, 1]$) and is calculated as follows:

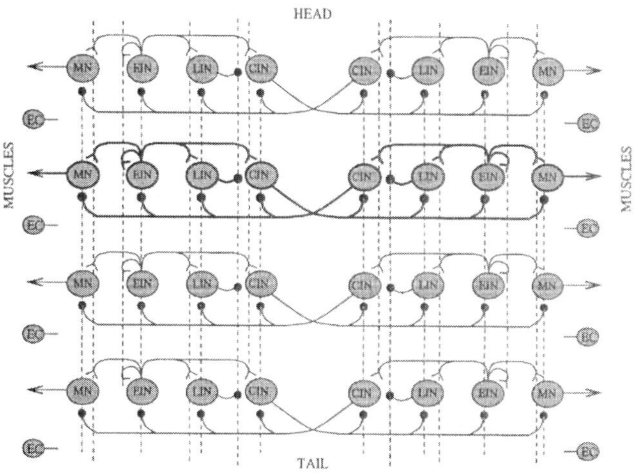

Figure 1 Model of the lamprey's CPG (based on (Buchanan and Grillner, 1987)). The controller is made of interconnected segmental oscillators (only 4 segments shown). Each segmental oscillator is composed of populations of neurons of four different types: 3 types of interneurons EIN, CIN and LIN and the motoneurons MN (one neuron unit represents a whole population of similar neurons in the lamprey). Inhibitory and excitatory connections are ended by • and ∨ respectively. The controller receives feedback from the stretch sensitive edge cells EC (connections not shown).

$$\dot{\xi_+} = \frac{1}{\tau_D}(\sum_{i \in \Psi_+} u_i w_i - \xi_+) \tag{1}$$

$$\dot{\xi_-} = \frac{1}{\tau_D}(\sum_{i \in \Psi_-} u_i w_i - \xi_-) \tag{2}$$

$$\dot{\vartheta} = \frac{1}{\tau_A}(u - \vartheta) \tag{3}$$

$$u = \begin{cases} 1 - \exp\{(\Theta - \xi_+)\Gamma\} - \xi_- - \mu\vartheta & (u > 0) \\ 0 & (u \leq 0) \end{cases} \tag{4}$$

where w_i are the synaptic weights, Ψ_+ and Ψ_- represent the groups of pre-synaptic excitatory and inhibitory neurons respectively, ξ_+ and ξ_- are the delayed 'reactions' to excitatory and inhibitory input and ϑ represents the frequency adaptation observed in some real neurons.

The controller is made of 100 interconnected segmental oscillators. The segmental oscillators have the observed biological connectivity as shown in Figure 1 and the interconnections between segments are obtained through extending the connections from one neuron to another in one segment to the corresponding postsynaptic neuron in neighbouring segments. Because the interconnections in the lamprey are not well known, Ekeberg has simplified them by chosing a symmetric exten-

[1] Lewis evolved abstract structures called Ring Rules representing explicitly the frequency and the phase of different oscillators and applied them to control the swimming of a simulated lamprey and the walking of a robot with a one-joint flexible spine.

sion (in the rostral and caudal direction) for all connections (but not necessarily with the same extent) except for the extensions from CIN which project more caudally.

The model is able to produce patterns of oscillations which are very similar with those observed in the lamprey. When the segmental networks receive adequate excitation from the brainstem, they oscillate regularly with the left and right neurons out of phase. The frequency of oscillations increases with the level of excitation. The interconnections between segments lead to traveling waves of oscillations when the most rostral segments receive extra excitation compared to the others. The lag between segments relative to the period of oscillation is then constant over the whole spinal cord, except for the first and last ten segments at the extremities where it decreases towards zero. The value of the lag can be varied with the amount of extra excitation, and the higher the extra excitation, the larger the lag (or the shorter the wavelength). Interestingly, the frequency of oscillation and the wavelength of the undulation can be changed nearly independently. The model can therefore reproduce the capacity of the real lamprey to cover a whole range of different frequencies of oscillations while keeping the wavelength of the undulation constant.

We have also reproduced Ekeberg's mechanical model of the lamprey's body interacting with water (Ekeberg, 1993).[2] The body is made of 10 rigid links connected through one-degree of freedom joints. Muscles are connected to each link and are modeled as a combination of springs and dampers whose spring constants are determined by the motoneurons' signals (from every tenth neural segment). Propulsion through water is obtained when the traveling waves of neural activity are transformed into traveling undulations of the body. The mechanical simulation allows a direct evaluation of the efficiency of the CPG to control swimming (in terms of speed of swimming, for instance). It also allows study of the effect of sensory feedback from stretch sensitive cells (the edge cells) located on both sides of the spinal cord (Viana Di Prisco et al., 1990).

Our simulations produce frequencies of oscillations between 1.6 and 5.5 Hz, phase lags per segment and relative to the oscillation period between 0.0% and 2.4% and speeds of swimming between 0.0 and 0.50 m/s. Because of the correction of some mechanical parameters, we obtain speeds of swimming which are approximately 40% smaller for the two quantitative examples given in (Ekeberg, 1993).

3. Evolution of swimming controllers

In (Ijspeert et al., 1998), we developed alternative swimming controllers by using a GA to define suitable connections and synaptic weights between neurons similar to those of Ekeberg's model. Our motivations were to evaluate how efficient GAs could be for the design of neural controllers and to investigate whether there exist alternative neural configurations to the biological one found in the lamprey. The controllers were evolved in three stages. First, segmental oscillators were evolved; then multi-segmental controllers were generated by evolving the couplings between copies of a chosen segmental oscillator; and, finally, connections providing sensory feedback from stretch sensitive cells were added. A similar real number GA was used for the three stages and will be described in Section 4.2. We briefly summarize here the results of the experiments.

3.1 Evolution of segmental oscillators

Segmental oscillators were generated by evolving both the connectivity between the 8 neurons composing the biological segment and the *sign* (excitatory and inhibitory) of each interneuron. The evaluations were based on the neural activity of the oscillators, and the fitness function was defined to reward the following characteristics:

1. The production of regular oscillations of the motoneurons, with one peak of activity per period and with the left and right neurons out of phase.
2. A frequency of oscillation which can be varied and which increases monotonically with the level of external excitation.
3. A minimal set of connections.

We evolved 10 populations of 100 chromosomes for 500 generations. All runs converged to final populations composed of networks producing regular oscillations with variable frequency. Although the fittest oscillator of one of the runs had a similar structure to the biological one, a variety of different segmental oscillators were produced which differed not only in their connectivity, but also in the sign of the interneurons and in the number of neurons active in the creation of oscillations (a majority of solutions use only four interneurons). All fittest oscillators covered a larger range of frequency than Ekeberg's segmental network (largest range: between 1.1 and 11.3 Hz).

3.2 Evolution of the coupling between oscillators

We generated complete CPGs by evolving the interconnections (i.e. the extents in the caudal and rostral direction of the projections of each segmental connection) between 100 copies of a segmental network. The fittest segmental oscillators of each evolution of the previous evolutionary stage were chosen. The evaluation was based on simulations of both the neural activity and the mechanical movements of the body. In order to obtain the same swimming characteristics as the biological controller, we defined the fitness function to reward:

[2] Our simulation is identical to Ekeberg's except for the correction of some mechanical parameters. See Table 2 in Appendix A.

1. generation of stable oscillations in the 100 segments with coordinated phase differences for the creation of traveling undulations of the body,
2. ability to change the speed of swimming by changing either the frequency of oscillation or the wavelength of the undulation,
3. ability to change the frequency and the wavelength independently, by changing, respectively, the global excitation level and the amount of extra excitation on the most rostral segments.

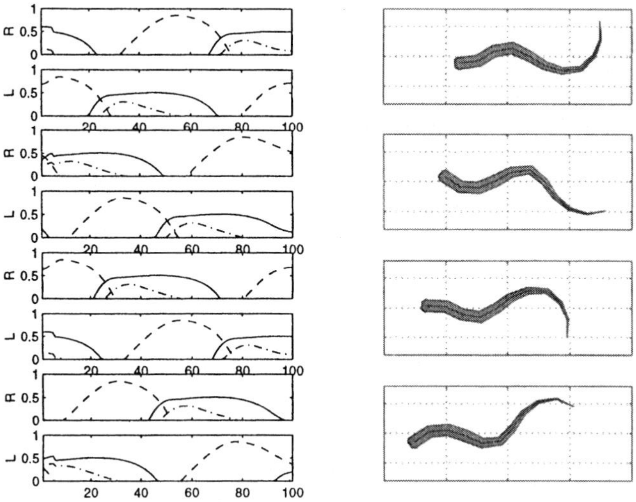

Figure 2 Simulation of one of the evolved artificial swimming controllers. The frequency of oscillations is 4.0 Hz, the phase lag between segments is 1.1% of a cycle duration and the speed of swimming is 0.41 m/s. *Left:* Neural simulation, the continuous line correspond to the MN, the dashed line corresponds to the CIN and dotted-dashed line to the EIN. *Right:* Mechanical simulation, the dotted vertical lines are separated by 100 mm, and there are 60 ms between each snapshot.

We evolved 10 populations of 40 chromosomes (each population based on one of the fittest evolved segmental oscillators). All runs developed couplings between the segmental oscillators which produce stable oscillations with the necessary phase lags for swimming (Figure 2 shows the simulation of one of them). Similary to the biological model, swimming at different speeds and with different wavelengths of undulations could be produced by varying simple external signals (the excitation applied to the CPG). Although it was not rewarded by the fitness function, turning could also be induced when one side of the controller receives more excitation than the other. The evolved controllers could generally cover larger ranges of frequency, phase and speed (maximum speed 0.60 m/s) than Ekeberg's model but with less independence in the control of the frequency and the phase between segments.

3.3 Evolution of sensory feedback from stretch sensitive cells

The CPGs evolved in the second stage did not receive feedback from the mechanical simulation. In the real lamprey, sensory feedback is provided to the CPG by stretch sensitive cells (the edge cells) situated on both sides of the spinal cord. There are both inhibitory and excitatory edge cells and they emit signals proportional to their elongation (Viana Di Prisco et al., 1990). Ekeberg showed (Ekeberg et al., 1995) that this sensory feedback is necessary for crossing a barrier of water with local speed opposite to the direction of swimming. We carried out a preliminary experiment on how to evolve this sensory feedback for the artificial controllers (i.e. the synaptic weights of excitatory and inhibitory connections from the stretch sensitive cells to the neurons of the segmental networks) in order to allow them to cross a similar speed barrier. The barrier was 150 mm wide (half the length of the simulated lamprey) and the speed of the water flow opposite to direction of swimming was set to be 40% higher than the speed of swimming of the lamprey. The fitness function was defined to reward:

1. progression through the barrier in the direction of swimming,
2. minimal deviation in the direction,
3. minimal difference of speed between swimming with and without sensory feedback.

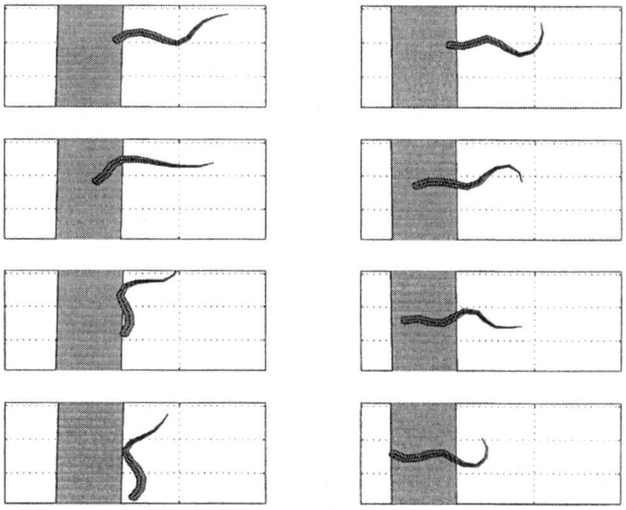

Figure 3 Effect of a speed barrier on swimming without (left) and with (right) evolved sensory feedback.

We carried out 5 evolutions with populations of 40 chromosomes using one of the fittest controller of the previous evolutionary stage. Without sensory feedback, the lamprey was not able to cross the speed barrier (Figure 3 *left*). The increase in local forces at the entrance to the barrier disturbs the undulation of the body and leads to extra bending of the most rostral part of the body which forces the lamprey to change its direction of swimming. The lamprey eventually swims perpendicularly to its initial direction. After 80 generations, all runs developed controllers with sensory feedback allowing them to cross the barrier (Figure 3 *right*). The feedback enables the crossing by sending excitatory signals to the motoneurons leading to a stiffer body and by acting as a burst terminator which tends to switch the neural activity from one side to the other when the latter is too much extended.

3.4 GA as a tool for neurobiological modeling

We believe that a GA can be a useful tool for neurobiologists to develop models, as it can be used to automatically generate instantiations of variables which are difficult to measure. As an illustration of this, we repeated the evolutions and restricted the space of possible solutions to controllers which preserve the segmental connectivity and the sign of neurons observed in the real lamprey. The evolutionary process generated thus a significant part (the synaptic weights and the interconnections) of the connectionist model that Ekeberg set up by hand.

We used the same evolutionary stages as for the artificial controllers. We firstly evolved segmental oscillators by evolving the synaptic weights of a network with a fixed configuration corresponding to the biological segmental connectivity. We found that Ekeberg's set of synaptic weights could be modified in order to optimize the frequency range. Frequency ranges much closer to those observed in real lampreys (between 0.25 and 10 Hz) could be obtained (the best solution covered a range between 0.9 and 11.0 Hz). The main observation is that the increase of the frequency range is obtained through an increase of the strength of some inhibitory connections (the contralateral CIN to LIN and ispilateral LIN to CIN connections).

In the second evolutionary stage, we evolved the couplings between segmental oscillators corresponding to those of Ekeberg's biological model. As these coupling are not perfectly known in the real lamprey, the idea was to study which kind of interconnections can produce phase lags between segments which are constant over the spinal cord and independent of the frequency of oscillation. There are opposing views on the origin of these phase lags in the real lamprey, in particular whether they are due to differences in the intrinsic frequencies of the oscillators or to the nature of the coupling, and, in the latter case, whether the dominant coupling is ascending or descending (see for instance (Matsushima and Grillner, 1992; Wadden et al., 1997)). We evolved interconnections with the same fitness function as for the artificial controllers and developed couplings which could produce the kind of phase lags observed in the real lamprey. The evolved solutions cover large ranges of lags and can reach slightly higher speeds than with Ekeberg's coupling. The evolved couplings in the five different runs we carried out (populations of 40 chromosomes) were all very similar and significantly different from the simplified coupling of Ekeberg's model. We find that the evolved CPGs seem to have no dominant coupling (for instance, no systematic asymmetry) and they present both rostral and caudal asymmetries of projection depending on the segmental connection. As for Ekeberg's coupling, the CPGs rely both on the nature of the couplings and on differences of intrinsic frequencies for creating travelling waves. The coupling is optimized for waves to travel from head to tail and the extra excitation of the most rostral segments determines the exact phase lag (giving more excitation to those segments amounts to increasing their intrinsic frequency).

Finally, we also evolved sensory feedback to Ekeberg's controller which enabled the simulated lamprey to cross a speed barrier. Interestingly, the evolved connections correspond very closely to those observed in the real lamprey, with all connections from excitatory and inhibitory stretch sensitive cells established in the real lamprey (Viana Di Prisco et al., 1990) being also present with the same sign in the evolved controllers. Crossing a speed barrier may therefore be a good example of the situation for which sensory feedback has been developed through natural evolution for the real lamprey.

4. Evolving a locomotion controller for a salamander-like animat

We present next how swimming controllers can be extended to control the walking of a salamander-like animat. Keeping the idea of staged evolution in which increasingly complex control systems are incrementally built using elements of the previous stage, this can be seen as a fourth evolutionary stage following the evolution of the swimming controllers. As little is known about the neuronal circuits underlying the locomotion of salamanders (the locomotion of salamanders has only been investigated using kinematic and electromyographic studies, see for instance (Delvolvé et al., 1997)), this evolutionary stage can not be compared with neurobiological data. We evolve walking controllers for a simulated salamander based on two interconnected oscillatory networks which are copies of the segmental oscillators of the swimming controller. Two experiments are realised, with walking controllers extended from the biological controller of the lamprey (Ekeberg's model) in the first one, and from one of the best evolved artificial swimming controllers in the second one. The aim is to develop controllers which, when receiving simple excitatory signals, can switch from swimming to walking and control locomotion at different speeds.

4.1 Mechanical simulation

We have developed a mechanical simulation of a salamander-like animat by extending the simulation of the lamprey. To the ten rigid links representing the spinal cord of the lamprey, are added four rigid links corresponding to the limbs (Figure 4). All links are connected through one degree of freedom joints. Muscles, modeled as a combination of springs and dampers, are connected in parallel to each joint and can be contracted through an increase of their spring constant by the corresponding motoneuron. There are a total of 26 muscles, 18 muscles for the 9 joints of the spinal cord and one extensor and one flexor muscle for each limb.

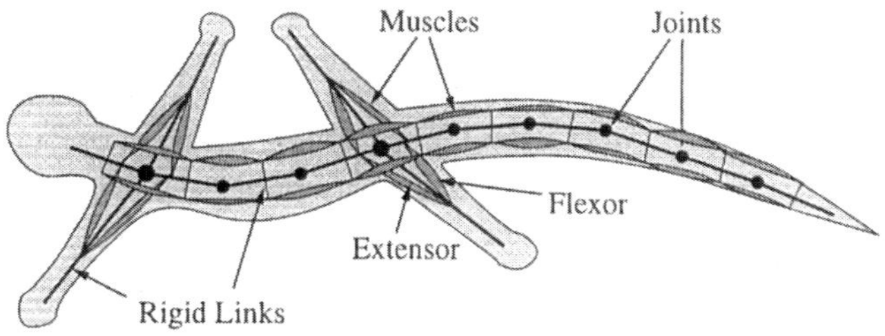

Figure 4 Mechanical configuration of the salamander-like animat.

The accelerations of the links are due to three forces: the torques due to the muscles, inner forces linked with the mechanical constraints and the forces due to the environment. The forces due to the mechanical constraints ensure that the segments stay connected at all times. The forces due to the environment depend on whether the salamander is in water or on the ground. In water, it is assumed that each link (limbs included) is subjected to inertial forces due to the water. On ground, all trunk links are subjected to a friction force, representing the fact that the trunk of the salamander slides on the ground when the salamander is walking. As only the accelerations in the horizontal plane are calculated, we represent the contact of a limb with the ground as an extra mechanical constraint. We assume that the contact in itself is determined by the signals sent to the horizontal extensor and flexor muscle of the limb. The limb is assumed to be in contact with the ground when the signal of the flexor is larger than that of the extensor, and in the air otherwise. The motoneurons for the flexor and extensor therefore not only determine the torque of the limb, but also its stand and swing phase. Once a foot is on the ground, a mechanical constraint is added which ensures that the foot does not move until it is lifted (i.e. the foot is temporarily 'pinned' to the ground). For more details see Appendix A.

4.2 Evolution of walking controllers

Similarly to the evolved swimming controllers, walking controllers are developed by evolving the connections and synaptic weights between neurons modeled by equations 1-4.

We evolve walking controllers under the following assumptions:

1. the walking controllers consist of two interconnected segmental oscillators which are copies of those of the swimming CPG,
2. the segments send signals to the muscles of the limbs and of the trunk, with the anterior and posterior segment projecting to the anterior and posterior muscles respectively,
3. the walking and the swimming controller receive excitation from the brain stem through different pathways.

Our aim is to evolve controllers which can switch between swimming and walking depending on which part of the controller is excited (Figure 5), and which can produce a speed of locomotion proportional to the level of excitation.

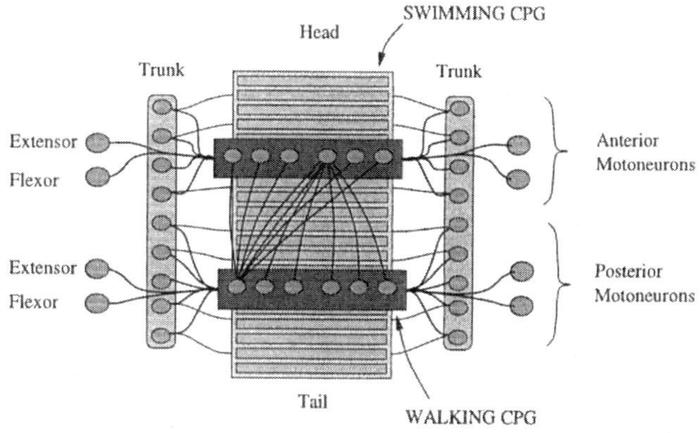

Figure 5 Neural configuration of the locomotion controller for the salamander-like animat. The controller is composed of two parts, the swimming and the walking CPG. All neurons receive an input from the brain stem (connections not shown), and swimming or walking patterns are generated depending on which CPG is excited.

Encoding A 'chromosome' encodes the synaptic weights of: the projections from the neurons of the 2 segments to the motoneurons, the connections from the brain stem to the motoneurons, and the connections between the 2 segments (see Figure 5). The connections within a segment and those from the brain stem to the segments are fixed and correspond to those of the swimming controller. We assume a left/right symmetry. The chromosome is made of 39 genes $\in [0, 1]$, with each gene encoding a synaptic weight and corresponding (via a linear transformation) to a real value between -5 and 0 if the pre-synaptic neuron is inhibitory or between 0 and 15 otherwise.

Genetic Algorithm The algorithm used is a variation of the standard GA (see for instance (Goldberg, 1989)) with the usual binary encoding being replaced by a real

number encoding. The algorithm starts with a randomly generated initial population. At each generation, a fixed number of children are generated using two operators, *crossover*, and *mutation*. The *crossover* operator chooses couples of parents, with a rank-based probability, for breeding and creates couples of children either by two-point crossover (probability *Prob_Xover*) or by simply copying the two parents. The *mutation* operator mutates each gene of the children with a probability *Prob_mut* and the mutation consists of adding or subtracting a small random number within a mutation range:
$new\_value = old\_value + Mut\_Range \cdot rand$
where $rand$ is a random number $\in [-0.5, 0.5]$. The children are then evaluated, and the size of the population is kept constant by rejecting, at each generation, the worst solutions of the increased population (old population plus children). The evolutions are carried out with populations of 40 chromosomes and 12 children at each generation. The crossover probability is 0.5, the mutation probability is 0.4 and the mutation range is 0.2.

Evaluation Solutions are rewarded for their capacity:

1. to perform fast walking in a straight line,
2. to cover a large range of different speeds when the excitation of the CPG is varied, with the speed increasing with the excitation, and
3. to use four limbs.

The last point was added to prevent premature convergence to solutions using only 2 limbs. Initial tests (without the third point) showed that this often happened (especially with these small populations), because using 4 limbs handicaps progression unless the limbs are perfectly coordinated (in which case higher speeds can be reached).

The mathematical definition of the fitness function is the following:
$Fitness = fit\_max\_speed \cdot fit\_speed\_range \cdot fit\_limbs$
$\in [(0.05)^6, 1]$, where

- *fit_max_speed* varies linearly between (and is limited to) 0.05 and 1 when the maximum speed of the salamander varies between 0.0 and 0.5 m/s respectively.
- *fit_speed_range* varies linearly between (and is limited to) 0.05 and 1 when the speed range (difference between the highest and the lowest speed) varies between 0.0 and 1.0 m/s respectively. The speed range is measured by making a set of simulations at different levels of excitation, and only ranges in which the speed increases monotonically with the excitation level are rewarded.
- *fit_limbs* is made of four factors, one for each limb, which vary linearly between (and are limited to) 0.05 and 1 when the time on the ground of each limb varies between 0 and 25% of the simulation duration, respectively.

The simulations are carried out for a fixed duration (1000ms) with asymmetric initial conditions for the neural activity (left interneurons of the anterior segment excited) and with an initial straight body with the limbs perpendicular to the body.

Results We carried out two sets of 5 runs (populations of 40 chromosomes), runs 1-5 are evolutions of walking controllers based on the biological controller of the lamprey (Ekeberg's model), and runs 6-10 are based on one of the fittest evolved artificial swimming controllers.

After 40 generations, all runs converged to controllers able to produce a walking gait with the speed of motion dependent on the level of excitation. The fitness and the maximum speed of the fittest solutions of each run are given in Table 1. All evolved controllers use all four limbs for locomotion, except the controller of run 4 which uses only the hindlimbs (which explains its low fitness value). There seem to be no significant differences in the type of walking developed by the two sets of runs, except that the runs based on the biological controller reach on average slightly higher speeds.

Table 1 Fitness and maximum speed of the evolved walking controllers

Run	1	2	3	4	5	6	7	8	9	10
Fit	0.38	0.71	0.37	0.00	0.62	0.27	0.15	0.14	0.48	0.60
Spd m/s	0.42	0.73	0.42	0.37	0.62	0.36	0.25	0.24	0.48	0.58

While all solutions move their left and right limbs strictly out of phase, the phase relation between fore- and hindlimbs varies among the nine controllers using four limbs. Five controllers have developed a strictly out of phase relation between fore- and hindlimbs (runs 1,2,3,5 and 9), two have a relation which is almost out of phase with one oscillator leading slightly the other (runs 7 and 8) and two have have fore- and hindlimbs which are in phase which means that they have the limbs on the same side being lifted simultaneously (runs 6 and 10).[3] All five solutions which, similarly to the salamander, move their fore- and hindlimbs out of phase have developed connections to the anterior trunk motoneurons such that one side of the anterior part of the trunk contracts simultaneously with the flexor on the same side leading to an increased reach of the extended and lifted limb of the other side.

Figure 6 (up) shows the simulation of walking controlled by the evolved solution of run9. The walking mode is obtained when the walking CPG receives excitation from the brain stem. Increasing the level of excitation leads to an increase of the frequency of the oscillators and

[3] These solutions may not be stable in a three-dimensional simulation (risk of rolling).

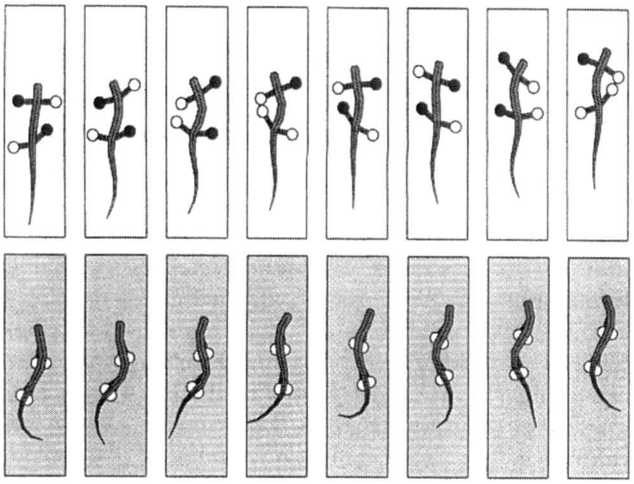

Figure 6 Walking (*Up*) and swimming (*Bottom*) salamander (fittest of run 9). A filled circle indicates that the limb is on the ground.

therefore to an increase of the speed of locomotion. Although it is not rewarded in the fitness function, for some solutions a limited turning can be induced when asymmetrical excitation is applied. When the swimming CPG is excited and excitation is given to the flexor muscles of the limbs (in order to stick the limbs to the trunk), swimming can still be produced (Figure 6, bottom). However, because of the extra forces due to the limbs (inertia and forces due to the water), the speed of swimming obtained is approximately 35% lower than that of the lamprey.

5. Discussion

Inspired by the CPG of the lamprey, we have developed connectionist models for the control of undulatory swimming and of salamander-like walking. The evolution of swimming controllers for the lamprey alternative to Ekeberg's biological has demonstrated that there exists a variety of possible solutions which can control swimming with the same efficiency. The solutions differ from the biological model in terms of which neurons are interconnected, of the excitatory or inhibitory influence of the interneurons and even of the number of neurons involved in the rhythmogenesis. Similarly to Ekeberg's model, the evolved solutions can modulate the frequency and the wavelength of the undulation, hence the speed of swimming, by varying the global level of excitation of the CPG and the extra excitation given to the most rostral segments. Turning can also be induced when asymmetrical excitation is given.

Walking CPGs have been successfully developed using oscillators of the swimming controllers. Most evolved CPGs produce a walking gait (or trot to be more accurate) similar to that of salamanders with the fore- and hindlimbs out of phase and contractions of the upper part of the trunk which increase the reach of the limbs. The speed of the walking increases with the excitation level. As only small changes of directions can be induced with asymmetrical excitation, further evolutions should include a factor in the fitness function rewarding the capacity to turn.

We have developed walking controllers based on oscillators of swimming controllers which have specialised to control the limbs. It seems reasonable to think that a similar specialisation has occurred in vertebrates through natural evolution. It is well established that evolution from swimming gaits to legged gaits has seen morphological changes of the bones and the musculature of some segments of the trunk to become fins and then limbs. It is therefore quite probable that the oscillators for the limbs have followed a corresponding specialisation from the trunk segmental networks (see (Cohen, 1988) for a discussion). It is less clear, however, how the interconnections between the segments of the trunk have evolved to allow both travelling waves for swimming and standing waves for walking. As far as we know, the neural circuitry responsible for the control of locomotion in the salamander is not well understood yet, although (Delvolvé et al., 1997) presents an interesting hypothesis of how limb oscillators may influence a trunk CPG similar to that of the lamprey. Here we have assumed (without any biological evidence) that the walking CPG has direct connections to the muscles of the trunk (without influencing the interneurons of the swimming CPG), with the anterior segment projecting to the upper part of the trunk and the posterior segments projecting to lower part, i.e. the tail. The swimming or walking gaits are then obtained depending on whether the segments of the trunk (the swimming CPG) or the segments of the walking CPG receive excitation from the brainstem.

Note that our mechanical simulation is only a crude description of the body of the salamander, especially in the way the limbs are simulated and because only accelerations in the horizontal plane are calculated. Salamanders have three-jointed limbs which therefore need much more complex control than the rigid links of our 2-dimensional simulation. Our interest, however, was mainly to study the neural mechanisms for switching from travelling to standing waves, and we could imagine that extending the controller to a more realistic 3D model could be done by coordinating oscillators for the different muscles of a limb with the signals of the corresponding principal oscillator that we have evolved.

Controllers were evolved using a GA with a staged-evolution approach. Although these swimming and walking controllers have strong evolutionary links, the GA was used as a design tool and not as a simulation of natural evolution. The method of staged evolution has allowed us to incrementally build controllers with each stage using elements of the previous stages. Such a decomposition has the advantage of reducing the time needed to evolve a complete controller by structuring the evolution and avoiding evaluation of whole controllers whose segmental oscillators do not oscillate correctly even when isolated. It presents the risk, however, of developing sub-optimal controllers as the fittest solutions of one stage may not necessarily be the best starting elements of the next stage. Although the decomposition and the simple encoding scheme (for instance, fixed neuron models and fixed maximum number of neurons) have strictly limited the type of controllers that could be evolved, the multidimensional search space was still large enough to generate a variety of interesting CPGs.

The main advantage of using a GA for the design of neural controllers is its ease of implementation and use. There are, for instance, no mathematical restrictions on the fitness function. Unlike most error functions minimised by learning algorithms for dynamical neural networks (like variations of the backpropagation algorithm), the fitness function does not need to be differentiable or even continuous. There is also no need to produce a specific oscillation (limit cycle) that the network should learn. This has allowed us to define the desired oscillations indirectly, either by qualitative aspects (as in the first evolutionary stage) or by the performance of the mechanical simulation. Another interesting aspect of the GA is the parallel search of the parameter space it performs (each chromosome is in a different place in that space), reducing the risk of convergence to local maxima. This parallel search and the stochastic nature of GAs makes them particularly suitable for visiting the space of possible solutions and designing, potentially, a variety of solutions.

It is interesting to see that a GA could also be used by neurobiologists to develop biological models. As we demonstrated for the lamprey's swimming controller, a significant part of a connectionist model —the synaptic weights— can be generated by the GA when biological observations are used to constrain the encodings and to define the fitness functions. Although it is possible, in the case of the lamprey, to define satisfactory values by hand (as Ekeberg did), the GA proved useful for searching the different possibilities and to optimise the unknown values for specific criteria. In particular, we showed that Ekeberg's model could be improved to cover a larger range of frequencies, much closer to that observed in real lampreys. From a general point of view, note that, although the GA can be very useful for demonstrating that a model can produce a specific behaviour (by finding efficient sets of unknown variables), it is less useful for invalidating a hypothetical model as an inability to find successful variable instantiations may be due to failings of the model or to problems with the GA set up, or both.

6. Conclusion

This paper has described how a Genetic Algorithm can be used to develop connectionist Central Pattern Generators for the control of the swimming and the walking of a simulated salamander-like animat. Inspired by the swimming controller of the lamprey, we first generated, in three evolutionary stages, controllers which can produce the patterns of oscillation necessary for anguiliform swimming. Walking controllers were then developed based on two oscillators of the swimming controllers. Similarly to biological CPGs, the resulting controllers can produce complex patterns of oscillation when receiving simple excitatory signals. In particular, the evolved controllers can switch between swimming and walking gaits and produce locomotion at different speeds depending on the excitation.

Acknowledgements: Auke Jan Ijspeert is supported by grants from the Swiss National Research Fund and from the European Commission (Marie Curie Fellowship). Facilities were provided by the University of Edinburgh.

A Mechanical simulation of the salamander

Newton's law of motion The acceleration of each link i (trunk and limb) depends on the torques due to the muscles T, the inner forces due to the constraints F_{in} and the forces due to the environment F_{env}:

$$m_i\ddot{x_i} = F_{env,i,x} + \sum F_{in,i,x}$$
$$m_i\ddot{y_i} = F_{env,i,y} + \sum F_{in,i,y}$$
$$I_i\ddot{\varphi_i} = \sum T_i - \sum F_{in,i,x}\tfrac{l_i}{2}\sin\varphi_i + \sum F_{in,i,y}\tfrac{l_i}{2}\cos\varphi_i$$

where x_i and y_i are the position of the middle of the link and φ_i is its angle.

Geometry The trunk of the salamander is 300 mm long (10 links of 30 mm). Each link has an elliptical cross section with height of 30 mm and width of 20 mm for the four first links and then decreasing towards 0 at the tail. The limbs are modeled as cylinders of 50 mm long and 10 mm width which are attached to the first and the fourth joints of the trunk. The masses and moments of inertia of the links are calculated by assuming the density of the salamander to be constant and equal to that of water (Table A).

Torques due to the muscles The torque acting at a particular joint is determined by the motoneuron activities (M_1 and M_2) of the corresponding parallel muscles:

$$T = \alpha(M_1 - M_2) + \beta(M_1 + M_2 + \gamma)\Delta\varphi + \delta\Delta\dot{\varphi}$$

where $\Delta\varphi$ is the difference bewteen the actual angle of the joint and the default angle (when $M_1 = M_2 = 0.0$). The parameters of the spinal cord muscles are: $\alpha = 9.4$ [N mm], $\beta = 0.94$ [N mm], $\gamma = 10$ and $\delta = 94$ [N mm ms]; and those of the limb muscles: $\alpha = 30.0$ [N mm], $\beta = 3.0$ [N mm], $\gamma = 10$ and $\delta = 300$ [N mm ms].

Table 2 Parameters for the mechanical simulation.

link	w_i [mm]	m_i [g]	I_i [g mm^2]	$\lambda_\perp$ [Ns^2/m^2]	$\lambda_\parallel$ [Ns^2/m^2]
1	20.0	14.1	1414	0.45	0.3
2	20.0	14.1	1414	0.45	0.2
3	20.0	14.1	1414	0.45	0.1
4	20.0	14.1	1414	0.45	0.0
5	17.2	12.2	1137	0.45	0.0
6	15.0	10.6	944	0.45	0.0
7	11.7	8.3	691	0.45	0.0
8	8.3	5.9	465	0.45	0.0
9	5.0	3.5	271	0.45	0.0
10	1.7	1.2	90	0.45	0.0
Limbs	1.0	4.0	850	0.25	0.1

The parameters of link 1-10 correspond to those of the lamprey simulation. Note that we have corrected some of Ekeberg's parameters which did not correspond to the assumptions made in the text (see (Ijspeert et al., 1998)).

Forces due to the environment In water, it is assumed that the forces due to the water are mainly inertial (high Reynolds number). It is also assumed that the water is stationary and that the parallel and perpendicular components of that force on each segment can be calculated separately. The components of the force can therefore be calculated as: $F_{env\parallel} = \lambda_\parallel v_\parallel^2$ and $F_{env\perp} = \lambda_\perp v_\perp^2$, where $v_\parallel$ and $v_\perp$ are the components of the speed of the link relative to the water and $\lambda_\parallel$ and $\lambda_\perp$ are coefficients which depend on the density of the fluid, the area perpendicular to the movement and the drag coefficient C dependent on the shape of the link (here $C_\perp = 1$ and $C_\parallel = 0$ for all links except those close to the head. See Table A).

On the ground, the links of the trunk are subjected to a friction force opposite to their motion and proportional to their weight (gm): $F_{env} = \mu g m$. In the simulations presented here, μ has been fixed to 0.8.

Inner forces due to the mechanical constraints Inner forces arise from the constraints on the movements by the joints which force the rigid links to stay connected. On the ground, new constraints are added as soon as a limb is assumed to be on the ground (flexor activity greater than extensor activity). The end position of the limb (the foot) is then fixed and inner forces are created to keep that position constant until the foot is lifted.

References

Buchanan, J. (1992). Neural network simulations of coupled locomotor oscillators in the lamprey spinal cord. *Biological Cybernetics*, 66:367–374.

Buchanan, J. and Grillner, S. (1987). Newly identified 'glutamate interneurons' and their role in locomotion in the lamprey spinal cord. *Science*, 236:312–314.

Cohen, A. (1988). Evolution of the vertebrate central pattern generator for locomotion. In Cohen, A. H., Rossignol, S., and Grillner, S., editors, *Neural control of rhythmic movements in vertebrates*. Jon Wiley & Sons.

Delvolvé, I., Bem, T., and Cabelguen, J.-M. (1997). Epaxial and limb muscle activity during swimming and terrestrial stepping in the adult newt, *pleurodeles waltl. Journal of Neurophysiology*, 78:638–650.

Ekeberg, Ö. (1993). A combined neuronal and mechanical model of fish swimming. *Biological Cybernetics*, 69:363–374.

Ekeberg, Ö., Lansner, A., and Grillner, S. (1995). The neural control of fish swimming studied through numerical simulations. *Adaptive Behavior*, 3(4):363–384.

Ekeberg, Ö., Wallén, P., Brodin, L., and Lansner, A. (1991). A computer-based model for realistic simulations of neural networks i: The single neuron and synaptic interaction. *Biological Cybernetics*, 65:81–90.

Frolich, L. and Biewener, A. (1991). Kinematic and electromyographic analysis of the functional role of the body axis during terrestrial and aquatic locomotion in the salamander *ambystoma tigrinum. Journal of Experimental Biology*, 62:107–130.

Goldberg, D. E. (1989). *Genetic Algorithms in Search, Optimization, and Machine Learning.* Addison-Wesley.

Grillner, S. (1996). Neural networks for vertebrate locomotion. *Scientific American*, pages 48–53.

Grillner, S., J.T., B., Wallén, P., and Brodin, L. (1988). Neural control of locomotion in lower vertebrates. In Cohen, A. H., Rossignol, S., and Grillner, S., editors, *Neural control of rhythmic movements in vertebrates*, pages 1–40. Jon Wiley & Sons.

Grillner, S., Wallén, P., and Brodin, L. (1991). Neuronal network generating locomotor behavior in lamprey: Circuitry, transmitters, membrane properties, and simulation. *Annu. Rev. Neurosci.*, 14:169–199.

Ijspeert, A., Hallam, J., and Willshaw, D. (1998). Evolving swimming controllers for a simulated lamprey with inspiration from neurobiology. *Dept. of Artificial Intelligence Research Paper 876, University of Edinburgh.* A shorter version of this paper has been submitted to Adaptive Behavior.

Lewis, M. (1996). *Self-organization of locomotory controllers in robots and animals.* PhD thesis, Faculty of the Graduate School, University of Southern California.

Matsushima, T. and Grillner, S. (1992). Neural mechanisms of intersegmental coordination in lamprey: local excitability changes modify the phase coupling along the spinal cord. *J. of Neurophysiology*, 67:373–388.

Viana Di Prisco, G., Wallén, P., and Grillner, S. (1990). Synaptic effects of intraspinal stretch receptor neurons mediating movement-related feedback during locomotion. *Brain Research*, 530:161–166.

Wadden, T., Hellgren, J., Lansner, A., and Grillner, S. (1997). Intersegmental coordination in the lamprey: simulations using a network model without segmental boundaries. *Biological Cybernetics*, 76:1–9.

Wallén, P., Ekeberg, Ö., Lansner, A., Brodin, L., Traven, H., and Grillner, S. (1992). A computer-based model for realistic simulations of neural networks ii: The segmental network generating locomotor rhythmicity in the lamprey. *J. of Neurophysiology*, 68:1939–1950.

Williams, T. (1992). Phase coupling by synaptic spread in chains of coupled neuronal oscillators. *Science*, 258:662–665.

BISAM: From small mammals to a four-legged walking machine

W. Ilg*, K. Berns*, H. Jedele*, J. Albiez*, R. Dillmann*
M. Fischer**, H. Witte**, J. Biltzinger**, R. Lehmann**, N. Schilling**

* Gruppe Interaktive Diagnose- und Servicesysteme
Forschungszentrum Informatik Karlsruhe, Germany
ilg@fzi.de

** Institut für Spezielle Zoologie und Evolutionsbiologie
Friedrich-Schiller-Universität Jena, Germany
b5fima@rz.uni-jena.de

Abstract

In this paper we apply principles of mammalian locomotion to the development of an autonomous robot. The relevance of biological analyses to the fields of kinematics, construction and locomotor control is discussed. The control architecture presented here is based on coupled neuro-oscillators and models the locomotor behaviour of the machine on different levels. The locomotion of animals is controlled by only a small set of parameters.

1. Introduction

The dynamic stability of quadrupedal mammalian locomotion relies on a coordinative system responding even to standardized conditions in a highly flexible way. Intra- and interlimb kinematics are the obvious expression of activities of central and peripheral control systems. According to a hypothesis of R. Full (Full, 1998) "The locomotor apparatus has a mind of its own", control is neural but ruled by mechanics. Musculature as a peripheral system acts in positioning the elements rather independantly of complex neural control but is self-stabilizing.

In this study we try to transfer biological solutions of quadrupedal locomotion to quadrupedal walking machines.

Control systems of walking machines have to move the center of gravity of the machine along a defined trajectory in an adequate gait in a stable manner. The desired foot positions are usually calculated according to the gait equations and to the foot trajectory planning system in the body-related coordinate system. From the foot positions the corresponding joint angles are calculated using inverse kinematics. The control hardware compares the measured joint angles to the desired value realizing a closed-loop control. This method is costly in terms of computational effort and not robust against disturbances. Compared to animal locomotion, kinematic constraints are not considered to make the control more easy and efficient.

In our interdisciplinary work we transfer biological control strategies and their functional principles to the control of a quadrupedal animat, which is more efficient and robust than the devices available. For this purpose, we have to describe the adaptivity and its hierarchical structure in specific natural constructions. The limbs of mammals are built of a chain of four bars linked by three joints (Fischer, 1994). Limit circles are realized by neuro-oscillators (Kolb et al., 1998), and the influence of their possible bundling onto the robustness of the system. Our quadrupedal animat should be able to perform different gaits in the future. At first we concentrate on in-phase gaits (gallop, hopping, half-bound and bound). Symmetrical gaits (walk and trot) may cause greater problems induced by resonance. One question is how far the use of elastic elements may be helpful (Cavagna, 1970), (Blickhan, 1989), (Witte, 1996). The reflux of knowledge into the biological sciences may be helpful to further the understanding of the principles of vertebrate locomotion.

1.1 Related Work

In the past decade, much work has been invested into the analysis of the locomotion and its control in six-legged insects (Cruse, 1990). The results have been transferred to six-legged walking machines (Buehrle and Cordes, 1996). In contrast to animals and animats with two or four legs, the mechanical stabilization of hexapods is more simple, since two intrinsically stable tripod states alternately can be used for a continuous stabilization by statical means.

The modelling of mammal-like locomotion is essentialy restricted to interlimb coordination, in which several approaches to realize gait patterns are presented (Collins and Richmond, 1994),(Pribe et al., 1997). Intralimb coordination is investigated in bipedal models, where different approaches based on neuro-oscillators have been developed (Taga, 1995).

Berkemeier reported one approach to an elastic mammal-like leg for a quadrupedal walking machine (Berkemeier and Desai, 1996), (Berkemeier and Fearing, 1994).

2. Derivation of models

2.1 Material and methods

Our generalization of the limbs of quadrupedal mammals is the result of a comparison of metric and kinematic parameters in 10 different small mammals (*Dasyuroides byrnei, Monodelphis domestica, Procavia capensis, Tupaia glis, Ochotona rufescens, Rattus rattus, Tragulus javanicus, Galea musteloides, Saguinus oedipus, Eulemur fulvus*). The selection comprises, for example, carnivorous marsupials, small ruminants, primates, ranging in mass from 150 g to over 3 kg and in size from 20 cm to more than 60 cm crown-rump-length.

Individuals of the named species were habituated to run at different speeds on a horizontal treadmill. Uniplanar cineradiographic recordings were taken at the Institut für den wissenschaftlichen Film (IWF, Göttingen) using a Arritechno R35-150 camera at different speeds: 24, 48, 96, 150 frames/s. Films were copied on a videotape and then analysed using a computer with a frame grabber. Motion analysis was done by digitalization of coordinates, correction of aberrations and calculation of metric and kinematic parameters.

2.2 Results

All the mammals studied in this group plus published data (e.g. on cats) show a strikingly similar pattern in the kinematics and basically the same strategy to gain speed. Higher speeds correlate with a reduction of stance duration (= higher frequency), while swing duration is rather constant, and to an increase of stride length by introducing a suspension period in which longer distances are covered. Both metric changes have only a slight impact on angular movement schemes but might be combined with different inter- (trailing vs. leading limb, cf. (Hildebrand, 1977)) and intralimb coordinations whenever higher speeds imply changes in gaits - which is not necessarily the case.

2.3 Models

In plantigrade and semi-digitigrade animals (ground contact is performed by the "hand") the forelimb acts as a four-bar three-link chain, whereas in digitigrade animals (ground contact via the "finger tips") only three elements (composite of lower arm and hand, upper arm, shoulder blade) with two joints are markedly used (figure 1). The hindlimb is composed of three or four elements (lower leg and foot, femur, and a composite of lumbar spine and pelvis) depending on the gait. Propulsion is mainly executed in the highest possible fulcrum: the superior border of the scapula (= shoulder blade), and the hip joint and/or lumbar spine + pelvis. Scapular rotation and translation accounts for more than three quarter of forelimb stride length; the rest is shared by the shoulder and elbow joint. The wrist joint and other carpal and phalangeal joints do not primarily serve for propulsion but ameliorate the ground contact with their telescoping abilities. Under a propulsive aspect the hand is considered as a mere prolongation of the lower arm.

In symmetrical gaits (walk, trot), back limb drive is mainly due to hip joint movements, while in in-phase gaits (gallop, bound) sagittal lumbar spine movements are added. Like in the forelimb, the more distal the joints are, the lesser they actively contribute to propulsion and the more they serve as elastic springs and shock absorbers. In contrast to the forelimb, transition from symmetrical to inphase gaits implies a different kinematic behaviour of the contralateral limbs (trailing and leading limb).

The geometry of the limbs like the length of elements and their positioning is determined by locomotory and non-locomotory factors. Here, we consider only the first aspect.

The length of the elements and, thus, of the limbs is determined by only a few factors. The location of the forelimb's foot down lies always very close or just perpendicular under the eye. Therefore, the length of the forelimb can be defined as: three rectified elements must cover the distance from the scapular fulcrum to the point on the ground right below the eye. The length of the hindlimb has to fulfill the conditon that three rectified elements (symmetrical gait) or four (inphase gait) reach a location at foot down which lies closest to the foot up point of the forelimb or - at the best - fits right into it (follow-the-leader-principle). Herewith instabilitites due to irregular grounds are reduced up to 50 %.

During locomotion the intralimb positions of the different elements are held rigidly. The more or less rectangular position of neigboured elements and the parallel orientation of shoulder blade / lower arm or pelvis / lower leg and femur / foot at foot down and foot up is a standard parameter of the locomotion of smaller mammals (figure 2). The zig-zag configuration of the limb is responsible for the adjustment angles at key events of each cycle: foot down and foot up.

Intralimb kinematics have to describe the timing of defined events (foot down and foot up) as well as the displacements of the elements and of the joints during the whole cycle. As principal movements are executed in high pivots, the distal elements are just carried along. Flexions in all limb joints are strongly synchronized with the two key events of each cycle. At foot down the joints take up load, at foot up the limb has to be elevated. Extensions are observed in the second third of stance and swing phases. In contrast to this biphasic pattern of the joint movement, the propulsive elements (scapula, lumbar spine, femur) show a monophasic behaviour of the angular scheme. The posterior and anterior displacements start already before touch down and lift off; e.g. the scapula reaches its minimum at the late swing and is backwards accelerated expending already 20% of angular movements before foot down.

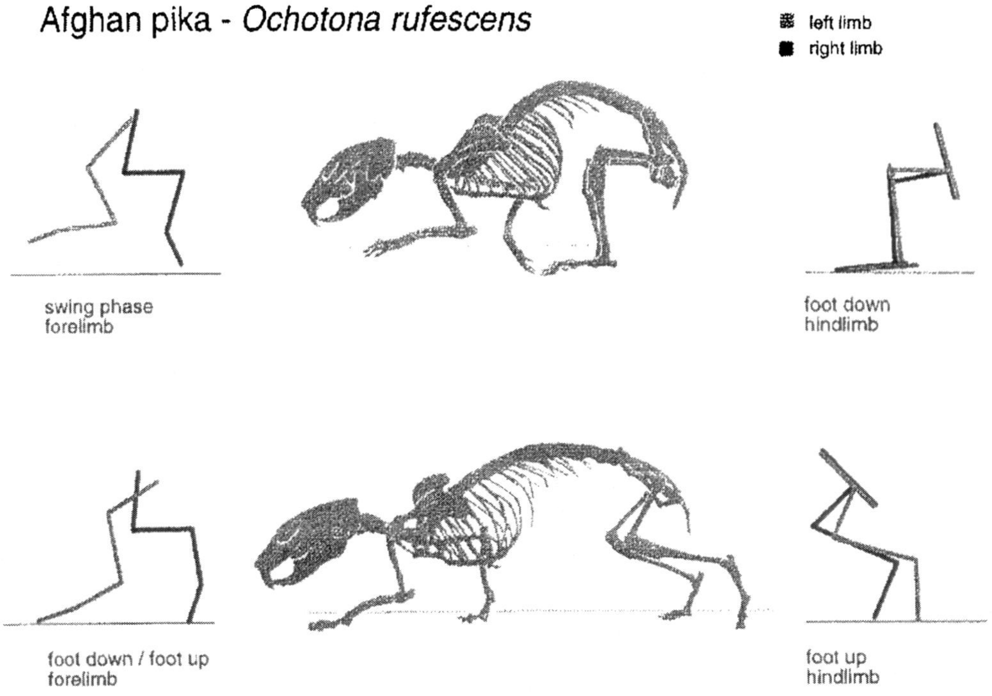

Figure 1 Drawing of the skeleton of a small mammal (*Ochotona rufescens*) during in-phase locomotion (top) and symmetrical locomotion (bottom). Also shown are stick figures derived from cineradiography for the forelimb (left) and hindlimb (right).

2.4 General strategy

The lower variability within the time schemes of the angular orientations of the elements in comparison to the joint angles indicates realization of a hierarchical structure: primary control parameters are the displacements of the elements in the global frame, joint movements have to be derived secondarily and serve as tuning mechanisms.

3. Transfer to technical walking

It is essential to analyze the basic locomotory behaviour in animals and animates within the same conceptual framework.

The movements recorded from tupaia and the other small mammals mentioned can not be transferred easily to the control of a quadrupedal walking machine. Therefore, we use a biologically inspired neuro-oscillator based on control architecture and integrate it with the knowledge about the characteristics of the rhythmic behaviour of cyclic locomotion.

3.1 The quadrupedal walking machine BISAM

The four-legged walking machine BISAM (Biologically InSpired wAlking Machine) consists of a trunk, four identical legs and the head. The trunk carries the microcontrollers, the processing units, the on-board energy supply and the stereo-head.

Each leg consists of four segments, which are interconnected by three parallel revolute joints. The legs are attached to the trunk by means of a fourth revolute joint (figure 3).

The trunk of the walking machine is segmented into four parts. The shoulder and hip segments are connected with two active degrees of freedom to the two middle segments. The latter are linked by an additional joint. Legs are moved by horizontal rotations of shoulders and hip (figure 4). This architecture simulates the basic features of the mammalian design.

The hardware control architecture consists of seven microcontrollers C 167 and one PC-card. There is one C 167 for each leg, two for the body and one for the control of the camera head. The microcontroller processes the lower control levels and sensor information. Higher control levels are executed on the PC-card. For a detailed description of the hardware architecture see (Cordes et al., 1997).

3.2 Locomotion Control

We introduce the neuro-oscillator model reported in (Matsuoka, 1987) and (Taga, 1995) for the representation of rhythmic movements. Using these neuro-oscillators we build structures to model the movement of tupaia and the quadrupedal walking machine BISAM. First experi-

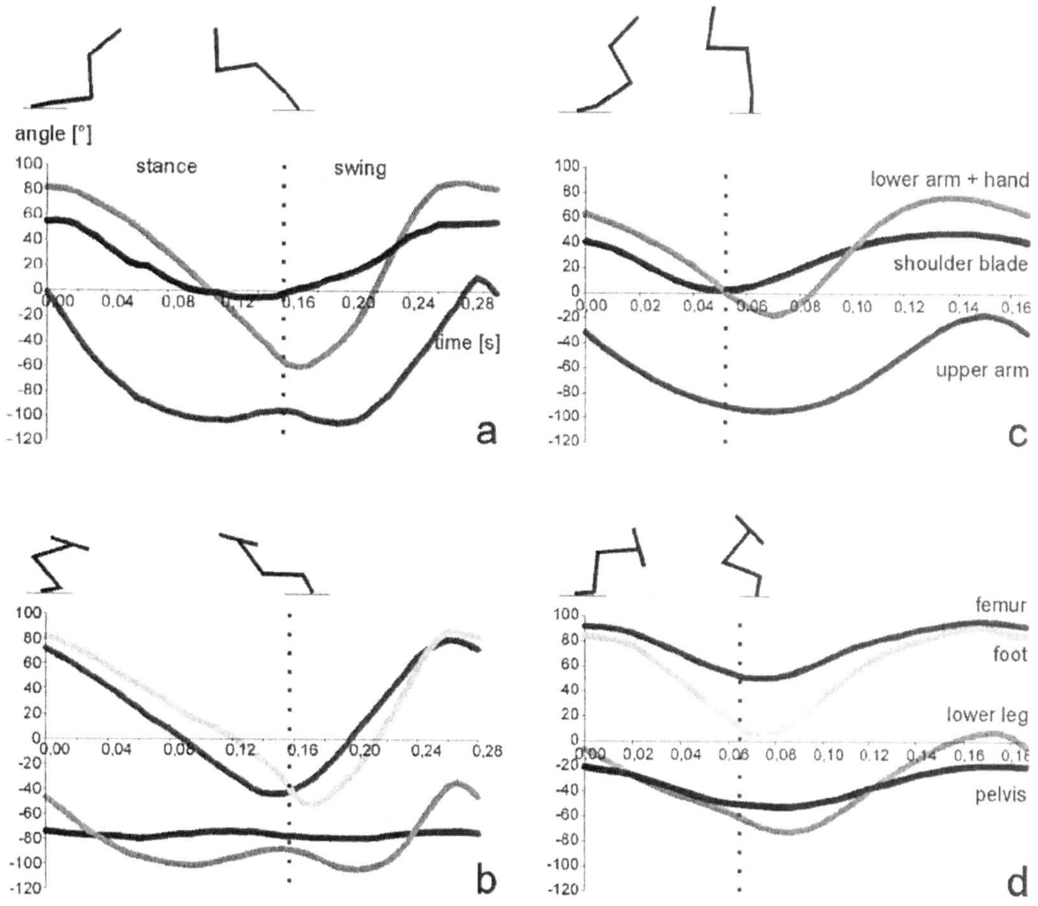

Figure 2 Angular orientation of forelimbs (top) and hindlimbs (bottom) of *Tupaia glis* (left) and *Ochotona rufescens* (right) during a one step cycle of in-phase locomotion. Time is scaled in percent of cycle duration.

ments show that the integration of sensor information is essential for a suitable modeling of the movements.

Neuro-oscillators One of the most simple neural networks models, which generates oscillatory behaviour, consists of two tonically excited neurons, with self-inhibitory effects and inhibitory links. This model was originally proposed by Brown (Brown, 1914) to describe the alternating activation of flexors and extensors of a cat limb during walking. This model can be mathematically represented as follows (Matsuoka, 1987)

$$T_r \delta x_i/\delta t + x_i = -\sum_{j=1}^{n} a_{ij} y_i + u_i - b f_i \qquad (1)$$

$$T_a \delta f_i/\delta t + f_i = y_i \qquad y = max(0, x) \qquad (2)$$

where x_i is the membrane potential of the neuron i, u_i the external excitation, f_i describes the degree of self-inhibition. The inhibitory effect is realized by the weighted activities included in equation 1. The time constants T_r and T_a change the frequency, the amplitude of the oscillator can be influenced by the external excitation u_i.

The capability to integrate sensor information S_i is shown in equation 3.

$$T_r \delta x_i/\delta t + x_i = -\sum_{j=1}^{n} a_{ij} y_i + u_i - b f_i + S_i \qquad (3)$$

In figure 5 one simple neuro-oscillator to model the underlying extensor-flexor-principle with sensor feedback is shown.

Modelling of elementary locomotion The neuro-oscillators described are used to simulate the movement of the forelimb of tupaia (figure 6). The coupled neuro-oscillators generate a typical trajectory, a limit cycle. This attractor behaviour ensures a robust control behaviour according to disturbances. In our neuro-oscillator architecture we realize local couplings only between directly connected segments. This kind of coupling supports the applicability of the model to whole animals or walking machines.

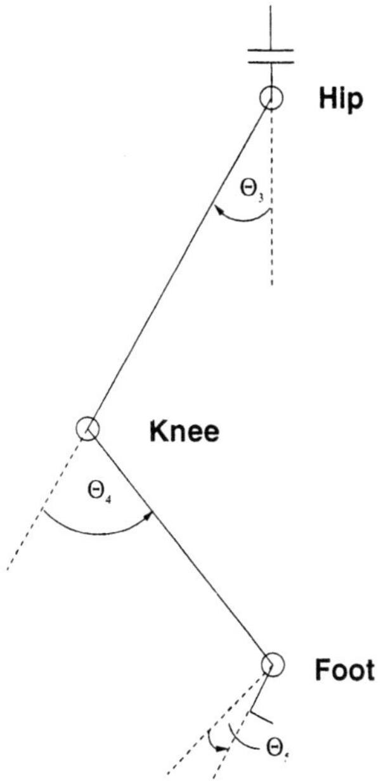

Figure 3 Kinematic description of an animate leg with three parallel joints for movements in the sagittal plane.

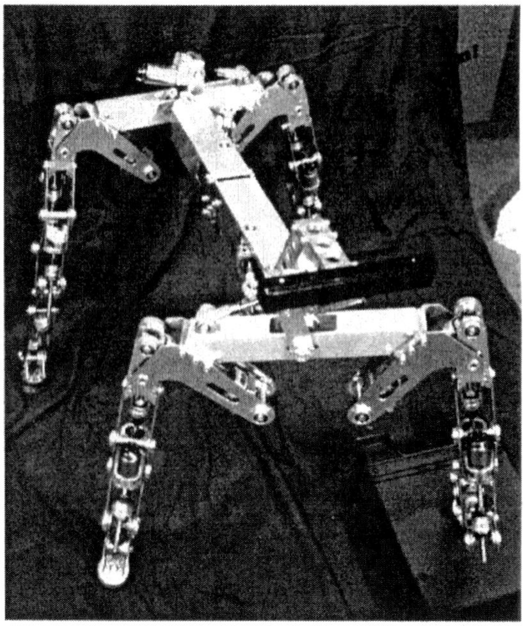

Figure 4 Photograph of the quadrupedal walking machine BISAM in mammal-like position. Due to the five active degrees of freedom in the trunk and the ability to rotate the shoulder and pelvis, the machine realizes key elements of mammal-like locomotion.

In figure 7 experiments to model the basic movements of tupaia are shown for the forelimb. First, the orientation sequences of the several segments are modelled by horizontal angle sequences. The joint angle movements can be generated by the superpositioning of the horizontal angle sequences. The segment orientations can be modeled in amplitude and phase in such a way, that the resulting joint angle sequences produce the characteristics of the sequences recorded in cineradiography. An example of a time scheme is shown for the joint angle of the shoulder (figure 7 bottom). The high positive peak represents the touch-down of the foot at the end of the swing phase. The low positive peak represents the end of the stance phase. Between those peaks are different negative peaks, which can be modelled by the neuro-oscillators. These high-frequent oscillations are provoked by translation of the scapula. The modelling of the upper arm is sloppy (figure 7). This is caused by the differences between the kinematics of tupaia and the simplified model (figure 8), which has no degree of freedom in the forehand.

Figure 8 illustrates the modelling of tupaia in walk in the positions described above.

In our future work further neuro-oscillators will be integrated to generate three dimensional walking behaviour.

Some experiments for robust control First results show robust behaviour of the neuro-oscillators in situations which differ from the modelled elementary behaviour. In these cases the integration of sensor information, the attractor behaviour of the neuro-oscillators, and the couplings between the different extensor-flexor-oscillators

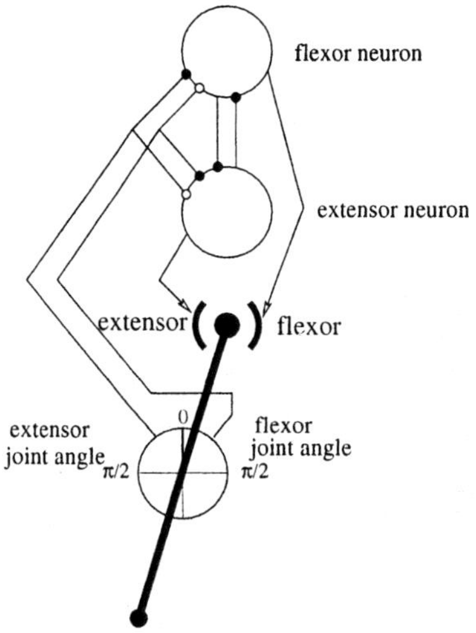

Figure 5 A basic neuro-oscillator to model the movement of one segment of a limb, using the extensor-flexor principle with sensory feedback.

are important to generate flexible and robust movements.

If the footsensor signals ground contact before the swing phase of the standard trajectory is finished, the movements of the limb segments must be corrected. For this purpose, the sensor signal must be integrated into the neuro-oscillator architecture in a suitable manner. In figure 9 the behaviour generated for the upper arm segment is shown, when the sensor signals ground contact. The swing phase is interrupted and the segment is traced back for the stance phase. Because of the attrac-

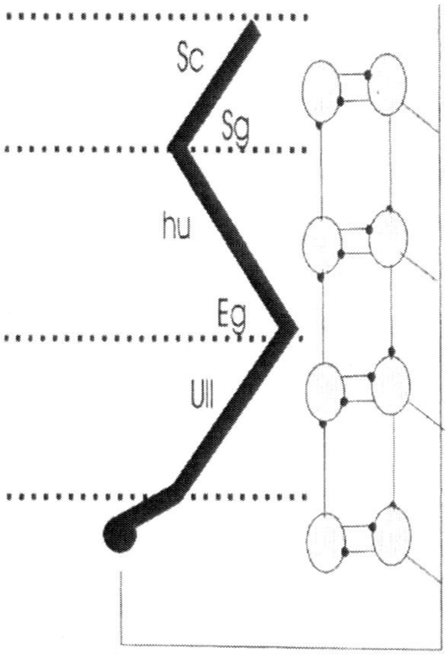

Figure 6 Left: Schematic drawing of the kinematic construction of a forelimb of *Tupaia glis*. Relevant angles are also shown: shoulder joint angle (Sg), elbow joint angle (Eg), angles of scapula (sc), humerus (hu) and ulna (ul) relative to horizontal plane. Right: Drawing of the neuro-oscillator architecture used for the modelling of the forelimb movement of *Tupaia glis*. There is one elementary two-neuron-oscillator for each segment and a local inhibitory coupling between connected segments. The signals of the foot sensor is transmitted to all segments.

tor behaviour the normal trajectory is initiated in the next step. The sensor signal is modeled in equation 3 as negative S_i (The amount correlates with T_r).

If the ground is not reached when the swing phase of the standard trajectory terminates, the step has to be lengthened. This is modelled by an inclinometer which signals the slanting of the trunk. The behaviour is exemplarily shown at the upper arm segment and the joint angle of the shoulder in figure 10. Both angles have to be increased for a lengthening of the swing phase and an additional stretching of the limb. Increasing amplitude of an angle is an result of a positive sensor feedback S_i in equation 3 to increase $\dot{x}$.

The movements of the whole leg in both cases of extended and interrupted swing phase is shown in figure 11 Due to the integration of foot sensor and inclinometer information in the coupled neuro-oscillators (see equation 1) the architecture is able to react to uneven terrain in an adequate way. After passing the obstacles, the oscillator returns to its standard trajectory.

Experimental Learning of Senso-Motoric Interactions For the experimental learning of the senso-motoric interactions a Reinforcement Learning method based on the AHC-approach (Barto et al., 1983),(Ilg et al., 1997) is used. This algorithm consists of a critic element which delivers an internal evaluation of the actual state and action elements which determine the next control values.

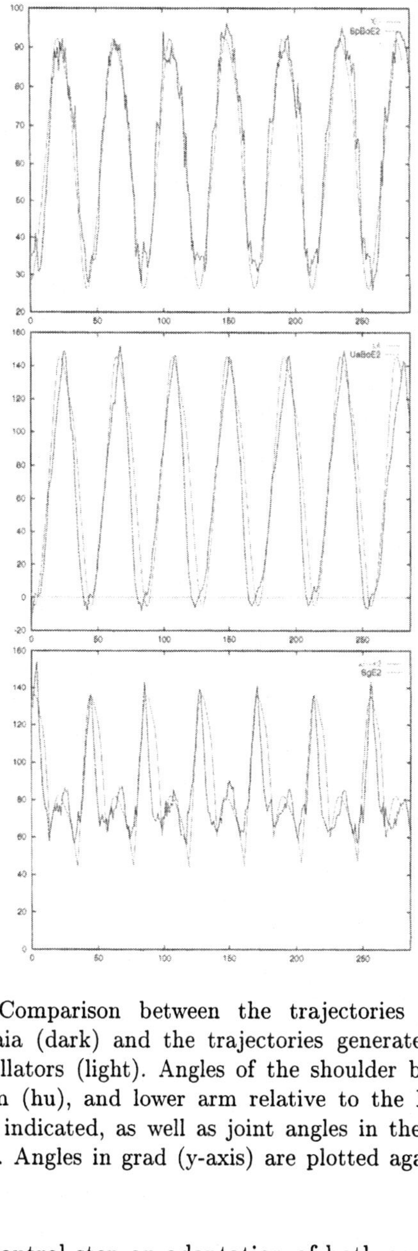

Figure 7 Comparison between the trajectories recording from tupaia (dark) and the trajectories generated by the neuro-oscillators (light). Angles of the shoulder blade (sc), upper arm (hu), and lower arm relative to the horizontal plane are indicated, as well as joint angles in the shoulder joint (Sg). Angles in grad (y-axis) are plotted against time (x-axis)

In each control step an adaptation of both components with the TD(λ)-algorithm takes place.

The state space representation used by the learning method is incrementally constructed using self-organizing RBF-networks. The RBF-net builds localized receptive fields which divide the input space into regions of limited size thus allowing localized learning of a function within the boundaries of such a region. This property makes RBFs a suitable tool for online function approximation. In (Ilg et al., 1997) a method is described by which the topology of the RBF network can be constructed according to the learning task.

The integration of the Reinforcement Learning component into the adaptive control architecture can be seen in figure 12. The sensor information of the inclinometer is put into the RBF-network, which determines actions to influence the weight of the current inclinometer input to

Figure 8 Snapshot of the artificial tupaia controlled by neuro-oscillators.

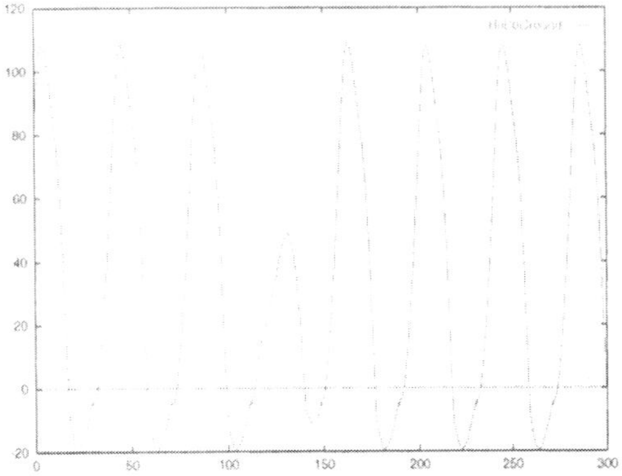

Figure 9 Angular orientation sequence of the upper arm during an experiment including the foot sensor signalling ground contact. As soon as ground contact is transmitted, the swing phase is interrupted.

the neuro-oscillators. The evalution of the generated actions is based on the sequence of inclinometer responses. The adaptation of the RBF-network is done according to the evaluation sequence. If the inclinometer shows the possible overturning of the walking machine the recent actions have to be adapted.

4. Conclusion

In this paper we describe interdisciplinary investigations using the results of biological research for the control of quadrupedal walking machines. For this purpose data recorded from small mammals are analysed to identify principles which can be used to control a robot. One of the major tools is the characterization of kinematic schemes of the mechanical coupling and consecutive superposition of the segment movements.

The movements of Tupaia and of our quadrupedal walking machine are modelled by a biologically inspired control architecture based on neuro-oscillators. A robust and flexible control is reached by the integration of sensor information and a specially coupled neuro-oscillator topology. This is shown by several experiments using the forelimb of Tupaia.

The final objective is to develop an adaptive control architecture based on the neuro-oscillators of a whole animate. The adaptive component should be used to learn the coordination of the elementary oscillating behaviours of the segments or joints in an experimental way for generating complex movements.

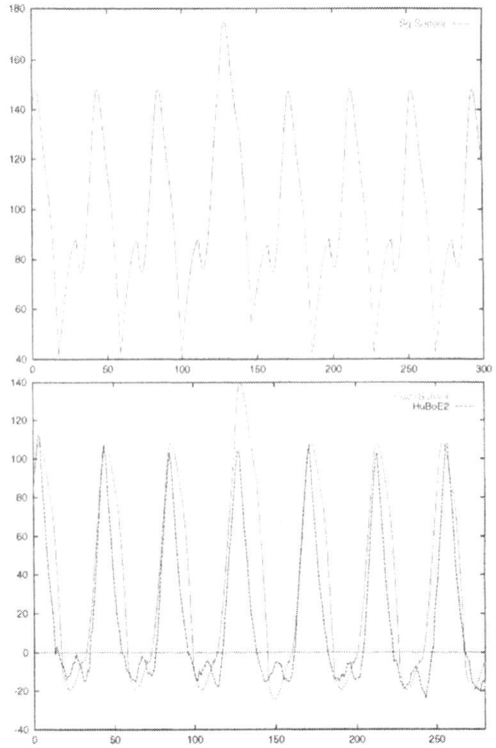

Figure 10 Extension of the swing phase illustrated by the example of the shoulder joint (top) and the upper arm segment (bottom).

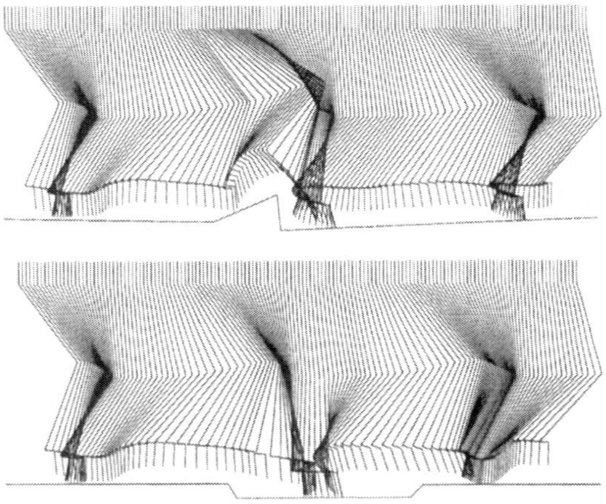

Figure 11 Stick figures of walking experiments in uneven terrain with interrupted (top) and extended (bottom) swing phase.

Acknowledgments

This research is funded by the Deutsche Forschungsgemeinschaft (DFG), grants DI-10-1 and FI-410-4-1.

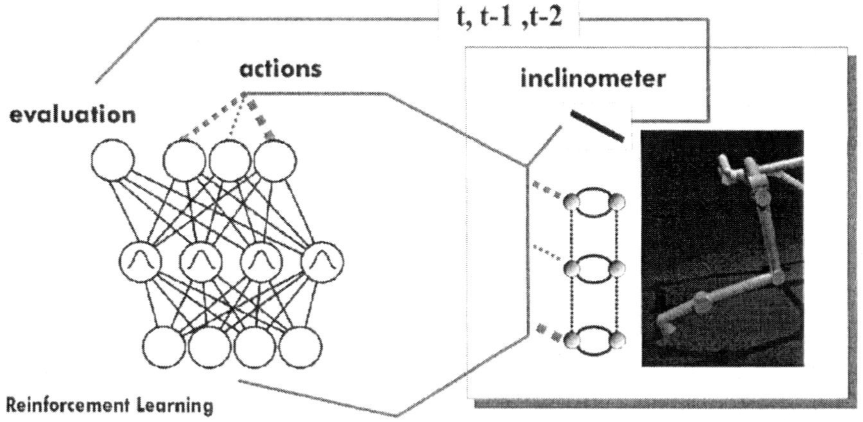

Figure 12 Adaptive control architecture for learning senso-motoric interactions.

References

Barto, A., Sutton, R. S., and Anderson, C. (1983). Neuron-like adaptive elements that can solve difficult learning control problems. *IEEE Transactions on Systems, Man and Cybernetics*, SMC-13:834–846.

Berkemeier, M. D. and Desai, K. V. (1996). Design of a robot leg with elastic energy storage, comparison to biology, and preliminary experimental results. In *IEEE International Conference on Robotics and Automation.*

Berkemeier, M. D. and Fearing, R. (1994). Control experiments on an underactuated robot with application to legged locomotion. In *IEEE International Conference on Robotics and Automation*, volume 1, pages 149–154, San Diego, CA. IEEE Computer Society Press.

Blickhan, R. (1989). The spring-mass model for running and hopping. *Journal of Biomechanics*, 22:1217–1227.

Brown, T. G. (1914). On the nature of the fundemantal activity of the nervous centeres. *Journal of Physiologie.*

Buehrle, P. and Cordes, S. (1996). Modeling, simulation and realization of an autonomous six legged walking machine. In ASME Design Engineering Technical Conferences.

Cavagna, G. (1970). Elastic bounce of the body. *Journal of applied Physiology*, 29(3):279–282.

Collins, J. J. and Richmond, S. A. (1994). Hard-wired central pattern generators for quadrupedal locomotion. *Biological Cybernetics*, 71:375–385.

Cordes, S., Berns, K., Eberl, M., Ilg, W., Schoenung, F., and Baeker, A. (1997). On the design of a four-legged walking machine. In *International Conference on Advanced Robotics ICAR'97.*

Cruse, H. (1990). Coordination of leg movement in walking animals. In Meyer, J.-A. and Wilson, S. W., editors, *From Animals to Animates: Proceedings of the First International Conference on Simulation of Adaptive Behavior*, pages 105–119, Paris, France. The MIT Press, Cambridge, Massachusetts.

Fischer, M. S. (1994). Crouched posture and high fulcrum, a principle in the locomotion of small mammals: The example of the rock hyrax (procavia capensis)(mammalia hyracoidea). *Journal of Human Evolution*, 26:501–524.

Full, R. (1998). Opening lecture of the 1st Congress on Biomechanics Jena, germany. BIONA Report

Hildebrand, M. (1977). Analysis of symmetrical gaits. *Journal of Mammology*, 58:131–158.

Ilg, W., Mühlfriedel, T., Berns, K. and Dillmann, R. (1997). Hybrid Learning Concepts based on Self-organising Neural Networks for Adaptive Control of Walking Machines. *Robotics and Autonomous Systems*, 22(3):317–329. Special Issue: Robotot Learning - The new Wave.

Kolb, T., Ilg, W., and Wille, J. (1998). Using time-discrete recurrent neural networks as nonlinear oscillators. In *IEEE World Congress on Computational Intelligience.*

Matsuoka, K. (1987). Mechanisms of frequency and pattern control in the neural rhythm generators. *Biological Cybernetics*, 56:345–353.

Pribe, C., Grossberg, S., and Cohen, M. A. (1997). Biped and quadruped gaits and bifurcations. *Biological Cybernetics*, 77:141–152.

Taga, G. (1995). A model of the neuro-musculo-skeletal system for human locomotion II: Real-time adaptability under various constraints. *Biological Cybernetics*, 73(2):113–121.

Witte, H. (1996). Beiträge zur Anatomie und Biomechanik elastischer Elemente in Bewegungsapparaten. Ruhr-Universität Bochum. Medizinische Fakultät Bochum. Habilitationsschrift.

Adaptive Behavior of Quadruped Walking Robot under Periodic Perturbation

Satoshi ITO[*], Hideo YUASA[**], Zhi-wei LUO[*], Masami ITO[*], Dai YANAGIHARA[***]

[*] Bio-Mimetic Control Research Center (RIKEN)
Anagahora, Shimo-shidami, Moriyama-ku, Nagoya 463-0003, Japan
satoshi@bmc.riken.go.jp luo@bmc.riken.go.jp itom@bmc.riken.go.jp

[**] Graduate School of Engineering, Nagoya University
Furo-cho, Chikusa-ku, Nagoya 464-8603, Japan
yuasa@ishidalab.nuem.nagoya-u.ac.jp

[***] Brain Science Institute (RIKEN)
Hirosawa 2-1, Wako, Saitama 351-0198, Japan
dai@postman.riken.go.jp

Abstract

Adaptive behavior of animals can be observed in the locomotion, as the variations of gait patterns. Physiological experiments demonstrated that decerebrate cats can adapt to periodic perturbation applied by the treadmill, and change their gait pattern with respect to the environmental changes. Based on this fact, we regard the adaptation as the process that adjusts memorized motion patterns to be more appropriate for the changed environments. Following this idea, we describe the adaptive behavior of decerebrate cat locomotion using a central pattern generator (CPG) model. We also take into account of the coupling of oscillators and limbs dynamics, and propose an adaptive control approach for the limb movements.

1. Introduction

Adaptation is one of the most interesting abilities in animal behavior. It was difficult to realize it by machines artificially. Usually, machines were designed to move under the specific conditions in the restricted environments. If these conditions vary, machines can not always act in the desired manner. Conversely, animals can adapt to the variations of environments or task requirements, and achieve their intended tasks successfully. An example of such adaptive behaviors can be found in animal locomotion. For example, if a animal lost the function of one leg, it can acquire a new walking pattern under this constraint and finally become to walk automatically with the new pattern. However, in the case of conventional walking machines, the locomotion might be impossible unless we teach the new method to walk.

Throughout this paper, we consider the adaptation in animal locomotion from the viewpoint of the gait variations, i.e., the changes of the rhythms for the leg movements. These types of the rhythms are reproducible. Thus, it is considered that locomotion rhythms are memorized as programs in the rhythm generators, i.e., central pattern generators (CPG) (Grillner, 1975; Delcomyn, 1980), and the CPG generate the rhythms in a feedforward manner. In our opinion, the adaptation in locomotion is equal to the change of memorized rhythm patterns with respect to the environmental variations.

The experimental paradigm for adaptive locomotion satisfying the above conditions has been proposed (Yanagihara and Kondo, 1996). According to their reports, cats were gradually acquiring a new gait pattern with repeating the locomotion on the treadmill, although the cats could not walk with the steady gait patterns initially (see Sect. 2). In this paper, we aim at realizing the above adaptive behavior of decerebrate cats by machine.

2. Adaptive locomotion of cats

Concerning the adaptive locomotion of quadrupedal animals, the following experimental results were reported by Yanagihara et al. (Yanagihara and Kondo, 1996) who elucidated that a neural transmitter nitric oxide (NO) plays an essential role in learning mechanism of cerebellum. They designed such a special treadmill as shown in Fig. 1. This treadmill consists of three moving belts, each of which can be driven independently at the different speed. The left forelimb (LF) and left hindlimb (LH) of the cat are separately placed on the different belts, whereas the right forelimb (RF) and right hindlimb (RH) are placed together on the other belt. If the speed of each treadmill are different, the movement of the limb is periodically disturbed whenever the cat places the limbs on the treadmill belt.

Firstly, they drove all treadmill belts normally with the same low speed (normal condition), and trained de-

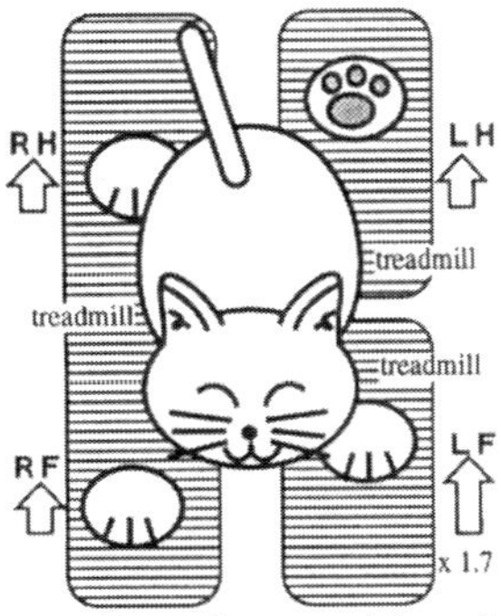

Figure 1 Experiment with a decerebrate cat.

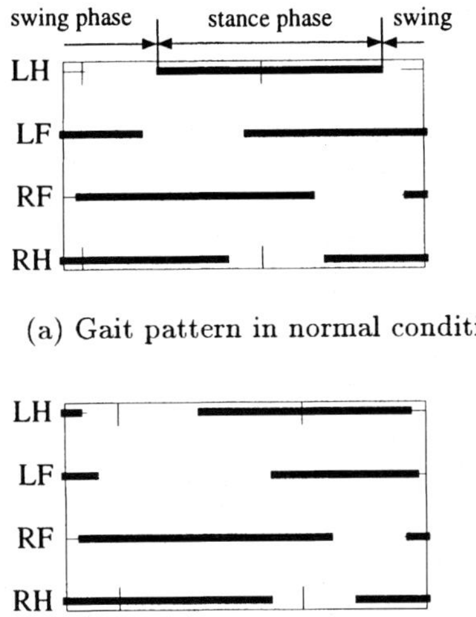

(a) Gait pattern in normal condition.

(b) Adapted gait pattern in perturbed condition.

Figure 2 Gait diagrams of cat locomotion. Thick lines denote support phase, and only one step cycle is expressed. The gait pattern changes according to adaptation.

cerebrate cats[1] to walk on such a treadmill. Then, decerebrate cats can walk on this treadmill with the same gait pattern 'walk' (Fig. 2(a)) as the intact cats'.

Next, they changed the speed of one treadmill belt, on which the cat places its left forelimb, 1.7 times faster than the others (perturbed condition), and observed the cat's behavior through three trials. Each trial includes 60 to 100 steps A few minutes of intervals were taken between each trial. At the first trial, any steady gait patterns can not be observed. However, at the second trial, the decerebrate cat obtained a new gait pattern (Fig. 2(b)) after some steps. At the last trial, the cat showed the new pattern acquired at the second trial from the beginning of the locomotion experiment.

[1] The decerebrate cat is a cat whose high nerve systems such as cerebral cortex are surgically cut at the midbrain. The locomotion of decerebrate animals means that the high nerve systems do not influence the control and the rhythm generation in locomotion.

3. Mathematical modeling of adaptive locomotion

3.1 Adaptation mechanism

We interpret the experimental results by Yanagihara et al. as follows: Firstly, the cat had memorized the walk gait for the normal condition. When the speed of the treadmill belt changed, the walk gait had become less appropriate to the changed environment. Accordingly, the cat adjusted the memorized gait pattern 'walk' into a new one by repeating locomotion. The fact that a new gait pattern appeared at the beginning of the third trial in the perturbed condition implies that the cat had memorized this new locomotion pattern.

In order to explain our concept of adaptation mechanism schematically, we use the potential function in Fig. 3 for a while (see also (Ito et al., 1997)). Although each limb movement is dynamic and periodic, the relative phases of four limb movement will be constant if the gait pattern is stationary. Therefore, the gait pattern can be treated as fixed point, i.e., the minimum point of the potential in a relative phase space. Note that gradient expresses the forces which effect to realize the memorized motion pattern. Accordingly, at the minimum point, any forces do not work since the gradient of the potential vanishes.

In the case of decerebrate cats, the walk gait initially becomes the minimum point (Fig. 3(a)). From the characteristic of potential function, the memorized motion pattern is stable, in other words, corresponds to an attractor. Even though locomotion is perturbed in an impulsive-like manner, it will soon regain its original gait pattern.

However, if the perturbation becomes periodic, the next perturbation will disturb the motion pattern again before the motion pattern is completely regained. Consequently, the perturbation and gradient force of potential will balance each other, as shown in Fig. 3(b). As mentioned above, the gradient force is force which realizes the memorized motion pattern. In Fig. 3(b), the memorized motion pattern is not realized even though this force is always working, which means that the memorized motion pattern is not appropriate to the changed environment.

Therefore, the memorized pattern, i.e., an attractor in motion pattern should be adjusted so that the gradient force can achieve it. This is equivalent to adjusting the minimum point of the potential in order to decrease the gradient force, as shown in Fig. 3(c). We consider this adjustment process as 'adaptation'.

3.2 Rhythm generator dynamics

Figure 4 shows a model of CPG. It is composed from four coupled oscillators. We assume that the oscillator phase θ_i $(i = 0, 1, 2, 3)$ directly represents the phase of the periodic limb movement. The coupling in Fig. 4 may

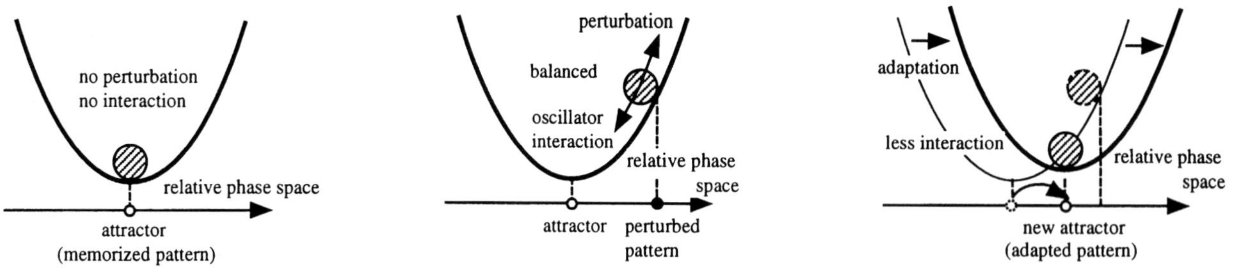

(a) Normal locomotion.

(b) Perturbed locomotion without adaptation.

(c) Perturbed locomotion with adaptation.

Figure 3 Adaptation mechanism in perturbed locomotion.

not always coincide with the actual oscillator connections in animals. This is only for mathematical convenience. A different coupling may also generate the same rhythm pattern mathematically.

Each limb moves in either swing phase or support phase. The character of limbs movements is essentially different in each phase. Before defining a rhythm generator dynamics, we divide the phase space of oscillators into two regions and assign them to the swing phase and the support phase, respectively. As shown in Fig. 5, we set the support phase to the region $\cos\theta_i \geq \gamma$, while the swing phase to the region $\cos\theta_i < \gamma$. Here, we determine γ by

$$\gamma = \cos \pi\beta. \tag{1}$$

where β is duty factor that denotes the proportion of the support phase in one step cycle. From Fig. 2(a), we set $\beta \cong 2/3$.

In the support phase, limbs are always in contact with the treadmill and cannot move freely. Thus, we describe the dynamics of the supporting limbs as

$$\dot{\theta}_i = \rho_i \quad (i = 0, 1, 2, 3), \tag{2}$$

where ρ_i $(i = 0, 1, 2, 3)$ is a variable representing the speed of the treadmill belt. Equation (2) means that the limb movement is forced by the treadmill.

In the swing phase, limbs can move freely. Thus, it is possible to adjust the phase of limb movement according to interactions among oscillators.

$$\dot{\theta}_i = \omega_i + f_i \quad (i = 0, 1, 2, 3). \tag{3}$$

Here ω_i $(i = 0, 1, 2, 3)$ denotes a natural angular velocity of oscillator. As the limbs shows the same movements in normal condition, we initially set them as follows:

$$\omega_0 = \omega_1 = \omega_2 = \omega_3 = \omega. \tag{4}$$

In addition, f_i $(i = 0, 1, 2, 3)$ denotes the interaction term. According to Yuasa and Ito (Yuasa and Ito, 1990),

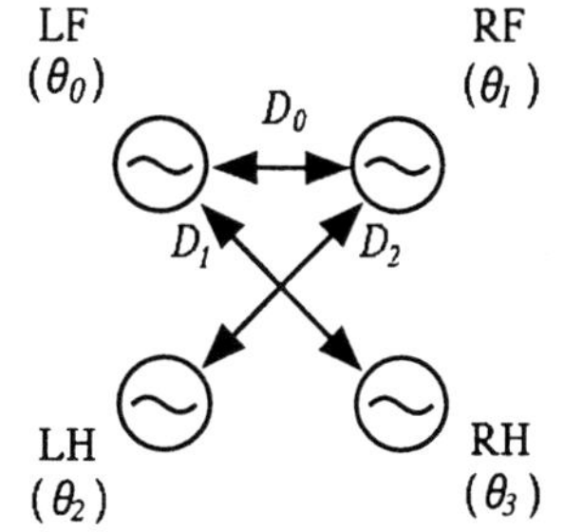

Figure 4 Connection of oscillators in our CPG model (LF: left forelimb, RF: right forelimb, LH: left hindlimb, RH: right hindlimb). D_0, D_1, D_2 are the desired values of each relative phase.

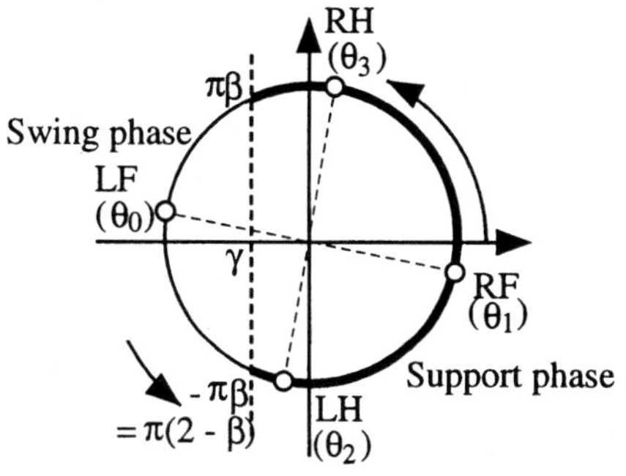

Figure 5 Swing phase and support phase. In this case, only the LF (θ_0) is in the swing phase and the others (LH, RF, RH) are in the support phase.

we can control relative phases to any values with a potential function in relative phase space. Using this method, f_i $(i = 0, 1, 2, 3)$ is given as follows:[2]

$$f_0 = \tau_\theta(\theta_1 + \theta_3 - 2\theta_0 - D_0 - D_1) \tag{5}$$

$$f_1 = \tau_\theta(\theta_0 + \theta_2 - 2\theta_1 + D_0 - D_2) \tag{6}$$

$$f_2 = \tau_\theta(\theta_1 - \theta_2 + D_2) \tag{7}$$

[2] These oscillator interactions can be calculated with local information, in other words, each oscillator should know only the phase of coupled oscillators. Therefore, phases of all the oscillators, i.e., global information are not necessary.

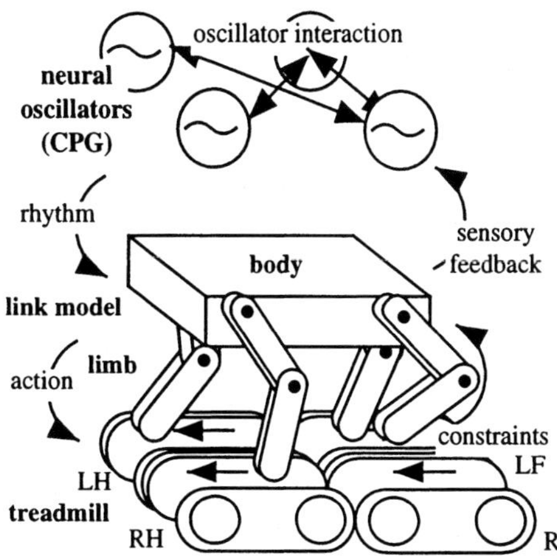

Figure 6 A model of adaptive behavior in the cat locomotion.

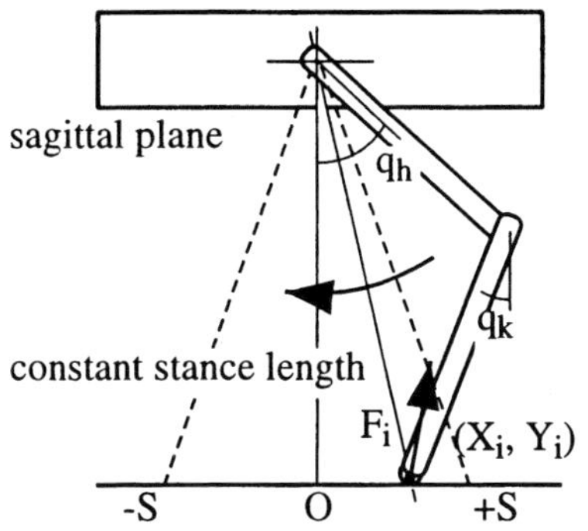

Figure 7 The structure of the limb. Each limb has 2 DOF: hip joint (q_f) and knee joint (q_k). F_i is a ground reacting force.

$$f_3 = \tau_\theta(\theta_0 - \theta_3 + D_1), \tag{8}$$

Here, τ_θ is a constant parameter which determines the magnitude of oscillator interaction, D_0, D_1 and D_2 are desired value of relative phases. For example, if we want to generate the walk gait in Fig. 2 (a), we should set them as

$$D_0 = \pi, D_1 = \frac{3}{2}\pi, D_2 = -\frac{1}{2}\pi. \tag{9}$$

We show the potential function in Appendix A.

3.3 Adaptation dynamics

The parameters which determine the minimum point of the potential function are ω_i $(i = 0,1,2,3)$ and D_j $(j = 0,1,2)$. Adaptation dynamics change them to decrease the oscillator interactions. As the cost functions we choose the integrated value of interaction f_i for one cycle,

$$F_i = \int_T f_i dt \quad (i = 0,1,2,3), \tag{10}$$

and the integration of the squared sum of interaction f_i

$$V_D = \int_T \sum_{i=0}^{3} \left\{ \frac{1}{2\tau_\theta} f_i^2 \right\} dt. \tag{11}$$

The resultant adjustment rules are given as

$$\omega_i^{(n+1)} = \omega_i^{(n)} + \tau_\omega \int_T f_i dt \quad (i = 0,1,2,3), \tag{12}$$

$$D_0^{(n+1)} = D_0^{(n)} + \tau_D \int_T (f_0 - f_1) dt, \tag{13}$$

$$D_1^{(n+1)} = D_1^{(n)} + \tau_D \int_T (f_0 - f_3) dt, \tag{14}$$

$$D_2^{(n+1)} = D_2^{(n)} + \tau_D \int_T (f_1 - f_2) dt, \tag{15}$$

where n denotes the number of step cycles, T is the duration of one step cycle, f_i $(i = 0,1,2,3)$ is the force given by eqs. (5)–(8), and τ_ω and τ_D are parameters that influence the convergence of ω_i and D_j, respectively. We show in Appendices B and C the derivation of this dynamics from the cost functions. This dynamics should be slower than that of locomotion.

4. Control of limb movements

4.1 Assumptions

We extend the mathematical expression in the previous section to include the limb dynamics. Then, the coupling of the oscillator and limb dynamics, or the design of the interaction between two dynamics, becomes an important problem. Figure 6 shows a sketch of our quadruped model.

For the simplicity, we make the following assumptions to focus on the coupling of the two dynamics.

- The walking motion is restricted within the sagittal plane.
- The balance of the body is not considered, i.e., the body is supported at the fixed position on the treadmill.
- As shown in Fig. 7, each of four limbs has the same structure with 2 degrees of freedom of the motion: the hip joint (q_h), and the knee joints (q_k).
- The stance length for each limb is constant from $-S$ and $+S$ with no relation to the walking speed.
- The ground reacting force from the treadmill is detectable.

From the second assumption, we can treat the motion equations of the limbs separately in each four limbs. Then we can obtain four decoupled motion equations of two link systems.

4.2 Control in support phase

In the support phase, the oscillator dynamics should follow the limbs dynamics. It is because the limb movements are under constraint of the treadmill movements and are reflected to the oscillator phases. The limbs have to generate the large force along the gravitational direction, while following the movement of treadmill along the horizontal direction. Accordingly, we apply the

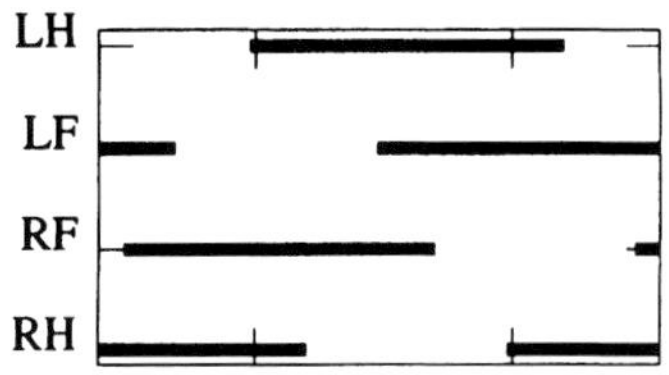

(a) Gait pattern in normal condition.

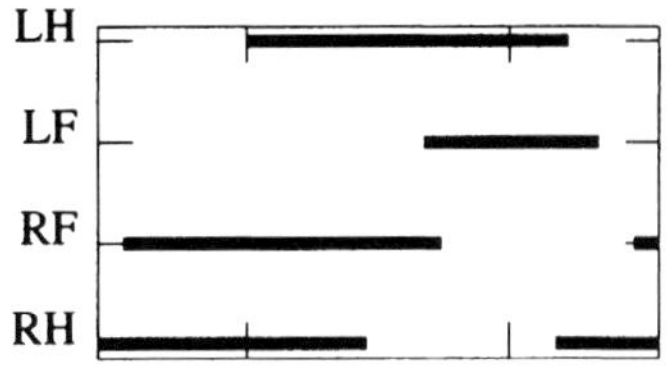

(b) Adapted gait pattern in perturbed condition.

Figure 8 Simulation result of gait change by adaptation dynamics.

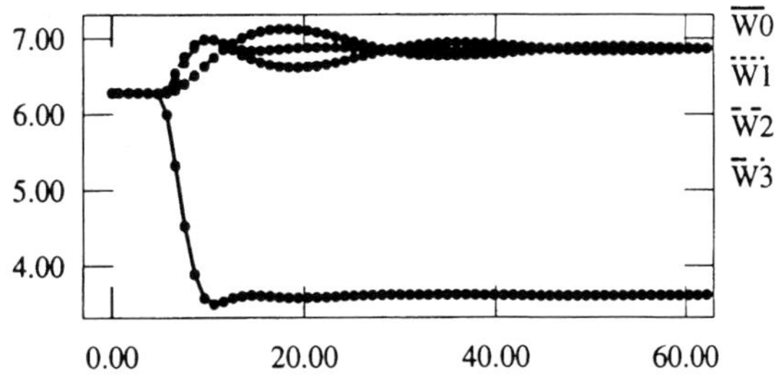

Figure 9 Change of ω_i $(i = 0, 1, 2, 3)$ due to adaptation.

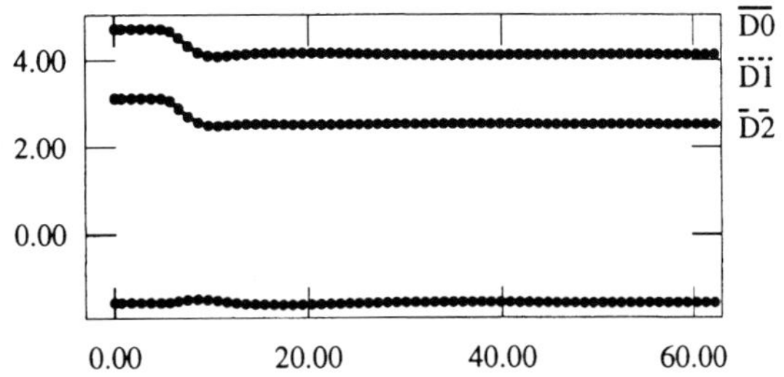

Figure 10 Change of D_j $(j = 0, 1, 2)$ due to adaptation.

impedance control such that high-impedance in the gravitational direction and low-impedance in the horizontal direction. We can compute the joint torque τ_i so that the response to the ground reacting force F_i becomes

$$M_d \ddot{X}_i + D_d(\dot{X}_i - \dot{X}_{di}) + K_d(X_i - X_{di}) = F_i \tag{16}$$

where $X_i = [x_i, y_i]$ is the position of the toe in the absolute coordinate, X_{di} is its desired position, M_d, D_d and K_d are matrices denoting the desired inertia, viscosity and elasticity. We design X_{d_i} to follow the treadmill belt as

$$x_i = \hat{\omega}_i t, \quad y_i = H_{sp}. \tag{17}$$

Here, $\hat{\omega}_i$ is an estimated speed of treadmill, and H_{sp} is a desired height of the toe in the support phase. To contact with the treadmill at any time, we set $H_{sp} < 0$.

In the support phase we change the oscillator phase according to the position of the toe. The toe moves along the horizontal direction from $+S$ to $-S$ (Fig. 7), while phase evolves from $-\pi\beta$ to $\pi\beta$ (Fig. 5). If we assume that the phase evolves at the constant rate when the toe moves at the constant horizontal speed, we can compose the following continuous mapping from the toe position to the oscillator phase,

$$\theta_i = -\frac{\pi\beta}{S} X_i. \tag{18}$$

The oscillator dynamics becomes

$$\dot{\theta}_i = -\frac{\pi\beta}{S} \dot{X}_i. \tag{19}$$

This equation must replace eq. (2) if we consider the limb dynamics.

4.3 Control in swing phase

In the swing phase, on the other hand, the oscillator dynamics govern the limb movement dynamics. Followed by the oscillator rhythms, the timing of limb movements, i.e., the relative phases are adjusted.

At first, we calculate the desired trajectory of toe from the oscillator phase θ_i. In the swing phase, phase evolves from $\pi\beta$ to $\pi(2 - \beta)$ (Fig. 5), while toe moves along the horizontal direction from S to $+S$ (Fig. 7). Under the similar condition in the support phase, we can generate the desired toe position as follows:

$$x_{di} = \frac{S}{\pi(1 - \beta)}(\theta - \pi), y_{di} = H_{sw}. \tag{20}$$

Here, H_{sw} is a desired height of the toe in the swing phase. Then, we determine the joint torques,

$$\tau_i = J^T(q_i)[D(\dot{X}_{di} - \dot{X}) + K(X_{di} - X)] \tag{21}$$

to follow this trajectory, where D and K are respectively the velocity and position feedback gain. Note that, in swing phase, there is no reaction force form the treadmill, i.e., $F_i = 0$

5. Simulations

We executed the computer simulation including the limbs dynamics. At 5.0(s) after the simulation started, we switch the treadmill speed 1.7 times faster. We set time constants as $\tau_\theta = 2.0$, $\tau_\omega = 0.25$ and $\tau_D = 0.045$.

Fig. 8 shows the simulated gait diagram, which is very similar to the experimental result of the decerebrate cat of Fig. 2. Figure 9 and Fig. 10 show the adjustment of angular velocity ω_i and the desired relative phase D_j. These show that the minimum point of potential function, i.e., memorized gait pattern is gradually changing.

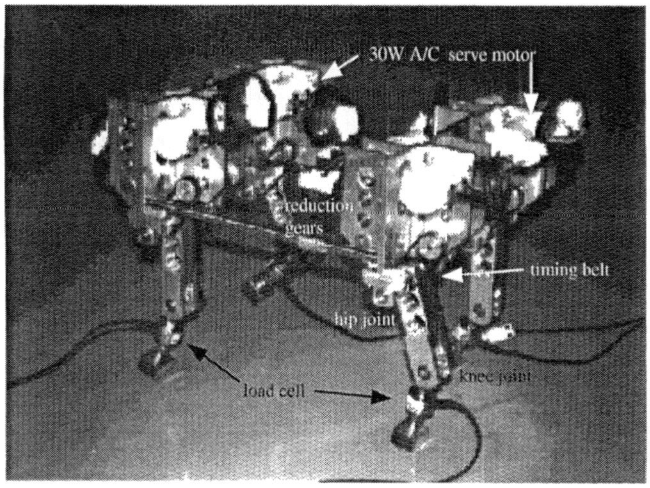

Figure 11 A quadrupedal walking robot.

6. Conclusion

We have successfully simulated the gait change with the treadmill speed. As the future works, we will experimentally examine it with a quadrupedal walking robots shown in Fig. 11. The experimental results of robot will be presented at the conference.

Appendices

A Dynamics of relative phases

Using eqs. (3) to (8), we can calculate the dynamics of relative phases, $\phi_0 = \theta_1 - \theta_0$, $\phi_1 = \theta_3 - \theta_0$, $\phi_2 = \theta_2 - \theta_1$ as:

$$\begin{aligned}\dot{\phi}_0 &= \omega_1 - \omega_0 + \tau_\theta(-\phi_0 + \phi_2 + D_0 - D_2) \\ &\quad -\tau_\theta(\phi_0 + \phi_1 - D_0 - D_1), \end{aligned} \tag{22}$$

$$\dot{\phi}_1 = \omega_3 - \omega_0 + \tau_\theta(-\phi_1 + D_1) - \tau_\theta(\phi_0 + \phi_1 - D_0 - D_1), \tag{23}$$

$$\dot{\phi}_2 = \omega_2 - \omega_1 + \tau_\theta(-\phi_2 + D_2) - \tau_\theta(-\phi_0 + \phi_2 + D_0 - D_2). \tag{24}$$

It is obvious that they are gradient dynamics with the potential function given as follows:

$$\begin{aligned} V &= \frac{1}{2}\tau_\theta[(\phi_0 + \phi_1 - D_0 - D_1 + \frac{\omega_0}{\tau_\theta})^2 \\ &\quad +(-\phi_0 + \phi_2 + D_0 - D_2 + \frac{\omega_1}{\tau_\theta})^2 \\ &\quad +(-\phi_2 + D_2 + \frac{\omega_2}{\tau_\theta})^2 \\ &\quad +(-\phi_1 + D_1 + \frac{\omega_3}{\tau_\theta})^2)]. \end{aligned} \tag{25}$$

In fact, eq. (22) to (24) can be derived from $\dot{\phi}_0 = -\partial V/\partial\phi_0$, $\dot{\phi}_1 = -\partial V/\partial\phi_1$ and $\dot{\phi}_2 = -\partial V/\partial\phi_2$.

B Adjustment of angular velocity in the swing phase

The dynamics of the oscillator in the swing phase are given by

$$\dot{\theta}_i = \omega_i + f_i \quad (i = 0, 1, 2, 3). \tag{26}$$

Integrating them during the swing phase, we obtain

$$\Delta\theta_i = \theta_i(t_f) - \theta_(t_s) = \omega_i T_{sw} + F_i, \tag{27}$$

where

$$F_i = \int_{T_{sw}} f_i dt, \tag{28}$$

and $T_{sw} = t_f - t_s$ is the duration of the swing phase. If $F_i > 0$ (or < 0), then θ_i is accelerated (or decelerated). In order to reduce the interaction F_i, we adjust ω_i in proportion to F_i as

$$\omega_i^{(n+1)} = \omega_i^{(n)} + \tau_\omega F_i \quad (i = 0, 1, 2, 3). \tag{29}$$

If $F_i = 0$, then we do not change ω_i.

Since interactions do not work in a support phase, we can change eq. (28) as follows:

$$F_i = \int_T f_i dt \quad (i = 0, 1, 2, 3). \tag{30}$$

C Modification of desired relative phase

When the cost function is given by eq. (11), we can adjust the minimum of the potential function D_j as

$$\frac{dD_j}{dt} = -\tau_{D'}\frac{\partial V_D}{\partial D_j} \quad (j = 0, 1, 2). \tag{31}$$

It can be expressed in the discrete form as

$$\begin{aligned} D^{(n+1)} &= D^{(n)} + \tau\frac{dD_j^{(n)}}{dt} \\ &= D^{(n)} - \tau_D\frac{\partial V_D}{\partial D_j}. \end{aligned} \tag{32}$$

Thus, eqs. (13)–(15) can be derived from eqs. (5) – (8).

References

Delcomyn, F. (1980). Neural basis of rhythmic behavior in animals. *Science*, 210:492–498.

Grillner, S. (1975). Locomotion in vertebrate:central mechanisms and reflex interaction. *Physiological Reviews*, 55:247–304.

Ito, S., Yuasa, H., wei Luo, Z., Ito, M., and Yanagihara, D. (1997). A mathematical model of adaptation in rhythmic motion to environmental changes. In *Proc. of IEEE SMC'97*, pages 275–280. IEEE.

Yanagihara, D. and Kondo, I. (1996). Nitric oxide plays a key role in adaptive control of locomotion in cat. *Pro. Natl. Acad. Sci. USA*, 93:13292–13297.

Yuasa, H. and Ito, M. (1990). Coordination of many oscillators and generation of locomotory patterns. *Biological Cybernetics*, 63:177–184.

COLLECTIVE BEHAVIOR

Market Trading Interactions as Collective Adaptive Behavior

Dave Cliff
Artificial Intelligence Laboratory
Massachusetts Institute of Technology
CAMBRIDGE MA 02139, U.S.A.
davec@ai.mit.edu

Janet Bruten
Hewlett-Packard Laboratories Bristol
Filton Road
BRISTOL BS12 6QZ, U.K.
jlb@hplb.hpl.hp.com

Abstract

We argue that human economic interactions, particularly bargaining and trading in market environments, can be considered as adaptive behaviors. Moreover, the tools and techniques of adaptive behavior research could be profitably employed to build predictive models of existing or planned market systems. In addition to applications in economic modeling, "trading animats" could find use in market-based resource-allocation and control, and in internet-based commerce. After a brief overview of core concepts in experimental economics (where human trading behavior is studied under laboratory conditions), we propose that such experiments could and should be used as 'benchmarks' for evaluating and comparing different architectures and strategies for trading animats. We then present empirical results from simulations that invite a Braitenberg-style eliminative materialism perspective on the dynamics of experimental retail markets. In these experiments, an elementary machine learning technique endows simple software animats with the capability to adapt while interacting via price-bargaining in market environments. The environments are based on artificial retail markets used in experimental economics research. We demonstrate that groups of simple agents can exhibit human-like collective market behaviors, and discuss the implications.

1 Introduction

The majority of research in adaptive behavior has concentrated on developing artificial autonomous agents (i.e., animats) that exhibit behaviors common to many species of animals. Typical examples are spatiotemporal behavior patterns such as obstacle-avoidance, wall-following, and navigation to goal locations. Very few papers in the adaptive behavior literature have examined behaviors that are exhibited exclusively by humans.

In this paper we argue that human economic interactions, particularly bargaining and trading in market environments, can be considered as adaptive behavior despite being uniquely human. One strength of adaptive behavior research is its equal emphasis on synthesis and analysis, both in explaining behaviors of animals and in designing animal-like artifacts. We argue that this approach can be applied to human economic activity, and we describe experiments where simple adaptive animats interact within environments based on experimental retail markets used to evaluate human trading behavior. We show that the collective behavior of the trading animats is similar to that of humans, and that explanations of the animat markets could have significance for the way in which comparable human activity is explained.

If groups of simple artificial agents interact to exhibit market-level phenomena that are similar to those of human markets, explanations of how the phenomena arise in the artificial system may be viewed as candidate explanations for the same phenomena in human markets. If adaptive behavior techniques can be used to build explanatory models of existing or planned market systems, it should also be possible to use such models for predictive purposes. We doubt that current agent technology and human performance data could be combined to create genuinely novel explanatory or predictive models, but we see this as a worthwhile and challenging aim for future work. However, in addition to economic modeling (either theoretical or applied), there are two other significant uses in which animats with human-like bargaining or trading behaviors could be employed: internet-based commerce, where software agents could autonomously seek out and purchase items on behalf of a human user, possibly entering into bargaining (or "haggling") interactions with vendor-agents; and market-based control (MBC), where economics is used as a source of inspiration, metaphors, and terminology for developing solutions to problems in distributed resource-allocation and control [5]. Thus, there is a need for 'bargaining' mechanisms that allow software agents to agree on prices for transactions in decentralized market systems.

Section 2 explains our arguments in more detail. We start with a brief review of some seminal work in experimental economics, where theoretically predictable market dynamics emerge as a result of the collective behavior of human trading agents. Arguments for the development of adaptive behavior models of such market activity are given in Section 2.2. In Section 2.3 we note the near-total lack of relevant work in the published literature on adaptive behavior and the related field of artificial life. We describe our experimental methods and results in Section 3, and discuss the implications in Section 4.

2 Economics and Adaptive Behavior

In this paper we assume familiarity with the basic microeconomics of price-determination and equilibration in markets: most text-books on (micro)economics discuss interactions between market supply and demand in some detail: we provide a summary in [9]. The fundamental observation is that the point of intersection of the supply and demand curves for a free market determines that market's *equilibrium price*, which we denote by P_0, and its *equilibrium quantity*, which we denote by Q_0.

Conditions under which a market can attain equilibrium from a given set of initial conditions, and the nature of the approach to equilibrium, have been the subject of intense research. Many models of equilibration assume *perfect competition*, where homogeneous indivisible units of a commodity are traded by large numbers of buyers and sellers, none of whom are sufficiently powerful to individually have any impact on market price, and all of whom are aiming to maximize utility (i.e., sellers maximizing profits, buyers minimizing costs). The nature or organization of some markets makes perfect competition unlikely or impossible, yet price determination can still occur. However, as the number of individuals in the market falls, the likelihood of collusion and the formation of rings or cartels increases: as the number of sellers is reduced, the market approaches an *oligopoly*, where the behavior of individual sellers in the market is highly dependent on the likely initiatives or responses of the other sellers. In such cases, the actions of individual traders can have a significant impact on the market price, depending on the organization of the market.

Market organization concerns the regulations governing the information available to the agents in the market and the agents' *opportunity sets* (i.e., the possible actions they can perform). Together, these affect the method by which a market price is determined. In most markets, prices are determined via a particular style of *auction*. One particular style of auction that has received significant attention in the literature is the *continuous double auction*, or CDA. In a CDA, a group of buyers and a group of sellers simultaneously and asynchronously announce *bids* (e.g., "buy at \$1.40") and *offers* (e.g., "sell at \$5.20"). Traders quote prices continuously, typically with both groups of traders altering their quote-prices (increasing bids and decreasing offers) in an attempt at securing a deal. At any time, a seller is free to accept the bid of a buyer, and a buyer is free to accept the offer of a seller. One practical reason for the interest in CDAs is that they have for many years been the basis of trading in major international financial markets.

While there are many other styles of auction, much work has been concentrated on CDAs, motivated by the fact that they are often very fast and efficient, despite (or possibly because of) the volumes of information exchanged. This interest has resulted in a research literature discussing CDAs (see, e.g., [13] and [11, pp.125–172] for reviews). The motivation for such work comes from a desire to better understand *how* the organization of the market, and the behavior of the traders in that market, affects the speed and efficiency of the market. Solving this problem, i.e. developing a theory of how the transaction prices of human traders converge on the theoretical equilibrium price, requires data gathered under controlled conditions. Such data has been generated by work in experimental economics, discussed next.

2.1 Experimental Economics

The use of 'laboratory methods' in economics, conducting controlled experiments to test theoretical hypotheses and predictions, has been of interest since at least the 1930's. In a typical simple experiment, a group of human subjects are each given the means to buy one unit of an arbitrary commodity; while another group are each given one unit of the commodity to sell. Each buyer will be a given a *limit price*, the maximum that buyer should pay for a unit of the commodity; and each seller will be given a minimum limit price below which the seller's unit should not be sold. Each trader's assignment of limit prices is private (i.e., not known by any other trader). Typically, different buyers will be given different limit prices, as will different sellers. The distribution of limit prices determines the supply and demand, and hence the values of P_0 and Q_0, for the experimental market. The subjects are then allowed to buy and sell within a particular market mechanism. Theoretical hypotheses concerning the effects of a market's structure on its dynamics can be tested, and the implications explored by varying parameters such as supply and demand, the traders' opportunity sets, or the amount and quality of information available to each trader. Factors of interest may include the temporal dynamics of transaction prices, the amount of potentially-available profit that is extracted from the market by the sellers, etc.

In a seminal paper, Smith [16] reported on experiments conducted from 1955 to 1961. All of Smith's experimental methods are similar to that described above: a cohort of human subjects are divided into a group of sellers and a group of buyers, and the two groups then trade within some specified market mechanism.

Each experiment is run as a sequence of distinct trading periods or 'days'. During each day, any trader entitled to quote can do so at any time. The traders interact by quoting prices until one is accepted. When this happens, the buyer and seller are considered to have entered into a binding transaction: for both traders, the number of units they are entitled to trade in is reduced by one. When a trader's entitlement reaches zero that trader drops out of the market for the remainder of that day. This process continues until quotes no longer lead to transactions, or when some predetermined time-limit (typically 5–10 minutes) is reached, at which point the day ends. If there are more days to run, the entitlements of all traders are then restored to their start-of-day values and the market reopens for another day's trading.

In a typical experiment, trading in the first day is characterized by early transactions taking place at prices that differ significantly from the P_0 value: as the day progresses, transaction prices approach P_0. On subsequent days, transaction prices are initially nearer P_0, and approach it faster. In most of the experiments described in [16], only the prices of agreed transactions were recorded.

To better characterize the convergence of transaction prices, Smith introduced a "coefficient of convergence", α, that is computed at the end of each day's trading. This is defined as $\alpha = 100\sigma_0/P_0$, where, for a set of n transaction prices denoted by $P_{i:i\in\{1,\ldots,n\}}$, $\sigma_0 = (\frac{1}{n}\sum_{i=1}^{n}(P_i - P_0)^2)^{0.5}$. In most of Smith's experiments, α tends to decline from one trading day to the next. Smith also monitored the *allocative efficiency* of the market. This is defined as the total profit actually earned by all the traders, expressed as a percentage of the maximum possible total profit that could have been earned by all the traders. Typically, after one or two trading periods, human traders achieve allocative efficiency very close to 100%.

Smith's [16] results were some of the first to demonstrate that markets consisting of small numbers of traders could still exhibit equilibration to values predictable from classical microeconomic theory. To appreciate why this is significant, it is necessary to consider the underlying supply and demand curves in more detail.

Consider, for example, five sellers (denoted by the letters A to E) and five buyers (F to J) each with an entitlement to trade one unit. Possible supply and demand curves are illustrated in Fig. 1. From the figure, it is clear that P_0 is \$2.50, and Q_0 is three. As is also evident, the supply and demand curves for this experiment are stepped: this is because the commodity is dealt in indivisible discrete units, and there are only a small number of units available in the market. Thus, supply and demand in this simple market differs appreciably from the smoothly sloping curves traditionally discussed in economics textbooks, where the step-changes involved can be treated as infinitesimally small.

Furthermore, in most classical theories of price determination and equilibration that assume or require a large number of traders, if an individual trader drops out of the market, the supply and demand curves remain essentially the same. But in markets with few participants, this is not the case. Consider the simple market illustrated in Fig. 1: if the first trade in the market is between seller C and buyer I (at a price, say, of \$2.65: giving profits of \$0.15 for C and \$0.35 for I) then these two traders drop out of the market, and the resultant supply and demand curves intersect to give values of P_0 and Q_0 that are significantly different.

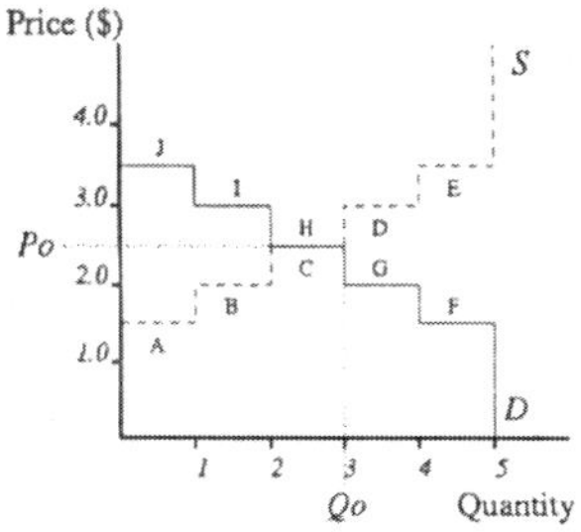

Figure 1: Supply and demand curves for ten traders. The supply curve is shown as a dashed line for extra clarity. There are five sellers (A to E), each with one unit to sell, and five buyers (F to J), each with the right to buy one unit. The step-changes in the quantities supplied and demanded are dependent on the limit prices of the individual traders, as indicated by the labels A to J. The intersection gives equilibrium values P_0=\$2.50 and Q_0=3.

Matters are further complicated when the difference between the *underlying* and *apparent* supply and demand curves is considered. Each trader in the market will be trying to make a profit, so buyers will be quoting prices lower than their individual limit prices, and sellers will be quoting prices higher than their limit prices. Because the limit price of each trader is private, the prices quoted by the traders give only a weak indication of the *underlying* supply and demand determined by the limit prices: the *apparent* supply and demand, based on the actual prices quoted, may be significantly different. In the literature, the apparent supply and demand are also referred to as the *seller's offer array* and *buyer's bid array*: or collectively as the *bid-and-offer arrays* [11, p.300]. Note that, in the example above, although there is a difference between the underlying supply and demand before and after the trade between C and I, an observer of (or participant in) the market does not have access to this information: the underlying schedules can only be guessed at by the traders, and the only clues they have are the quote-prices observed in the market.

To illustrate the difference, assume that each trader aims for a particular 'profit margin', expressed as a percentage of that trader's limit price. Say that we take the market of Fig. 1 and assign each trader a randomly chosen profit margin in the range 0% to 50%. Then a demand curve constructed from the bid array (i.e., the apparent demand) will be lower than the underlying market demand curve, and the apparent supply curve resulting from the offer array will be higher than the underlying market supply curve. The regular stepped appearance of the underlying curves is likely to be distorted, possibly so much that the rank-ordering of traders' prices also alters. These shifts are highly likely to yield values for the apparent equilibrium price and quantity that differ significantly from the underlying values of P_0 and Q_0. For graphical illustrations of this, see [6, 9].

Finally, it should be noted that the bid and offer arrays

are dynamic, and can alter rapidly. Because no trader has full knowledge of the underlying supply and demand, traders might base their profit levels on an initial guess that is then refined on the basis of the prices subsequently quoted by the competition (other members of the trader's group) and opposition (traders in the other group), and on the basis of which quotes lead to transactions and which are ignored. In a CDA, such information arrives in a continuous asynchronous stream. Despite this, the humans in Smith's experiments rapidly approach the competitive equilibrium predicted from theory. Fig. 2 shows one of Smith's experimental results. As is clear on the right-hand side trace, transaction prices converge toward the predicted P_0 over the 4 trading days, and the number of transactions per day varies from 5–8.

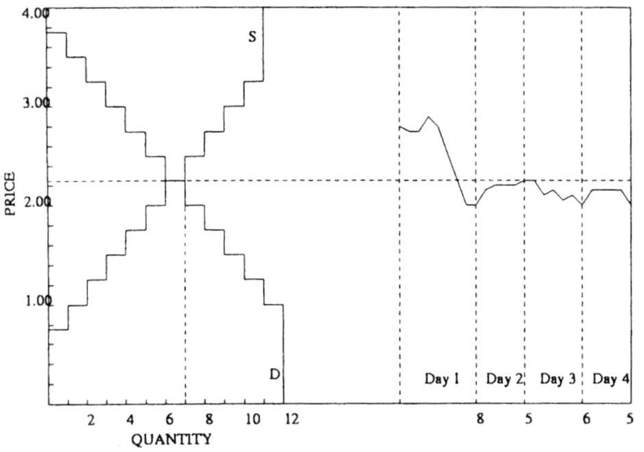

Figure 2: Left-hand side shows the market supply and demand, with theoretical equilibrium values P_0=\$2.25 and Q_0=7. Right-hand side shows time-series of transaction prices from four 'days' when human traders interact via a one-sided auction (offer-only) 'retail' market. Numbers on the horizontal axis of the right-hand figure indicate market volume (quantity of transactions) for each day. After Chart 8 of Smith [16].

Human beings are notoriously smart creatures: the question of just how much intelligence is required of an agent to achieve human-level performance is an intriguing one. This question was addressed by Gode & Sunder [14] who described a set of experiments similar in style to Smith's, but in which human traders are replaced by "zero intelligence" (ZI) programs that submit randomly-priced bids and offers in electronic CDA markets. Gode & Sunder reported that, when the ZI traders were constrained to prevent them entering into loss-making transactions, their market dynamics (in terms of Smith's α measure, and in terms of allocative efficiency) were very similar to, or even indistinguishable from, those of human traders in (almost) identical market experiments. Thus, Gode & Sunder concluded that the market's attractive dynamics, i.e., low α and high allocative efficiency, is not much affected by the intelligence, rationality, learning, or memory capabilities of the traders. Rather, it is due to the structure of the market (i.e., traders interacting via a CDA and prevented from loss-making transactions). It appears that, so long as the market is structured correctly, traders with *no intelligence at all* can give human-like market dynamics.

Gode & Sunder's paper has had significant impact in the experimental economics literature (in [9] we provide a list of 30 texts that approvingly cite Gode & Sunder's work). However, we have recently presented analytic and empirical results which conclusively demonstrate that Gode & Sunder's results only occur in special circumstances, and that in general the ZI traders fail to give human-like market performance [6, 8]. Thus, more than zero 'intelligence' is required of artificial trading agents to give human-like collective behavior and market dynamics. And because of this, methods developed in adaptive behavior research should be relevant.

2.2 *Trading as Adaptive Behavior*

Given that nonzero 'intelligence' is required of artificial traders to give human-like market dynamics, one possibility is to use traditional artificial intelligence (AI) techniques, based on deliberative reasoning with explicit representations of 'knowledge' or 'beliefs', as the foundation for building artificial trading agents. However, we believe that the assumptions underlying adaptive behavior research imply that it is a more attractive approach.

Although precise definitions are rarely articulated, much adaptive behavior research is consistent with the definition that a behavior is adaptive if, when an agent exhibits that behavior, it increases its chances of survival (or reduces its chances of ceasing to exist). This requires that the behavior is well-matched to the prevailing environmental conditions, and this carries an implicit assumption that the environment with which the agent interacts is nontrivial: i.e. that it is, to significant extents: complex, dynamic, unknown, unpredictable, and unforgiving of mistakes. Because of this, the agents studied in adaptive behavior are fundamentally *situated*, in the sense that their (inter)actions cannot be premeditated because it is impossible to predict with any accuracy all possible future conditions, events, or outcomes. Such situatedness requires that the agents are responsible for co-ordinating perception and action for extended periods of time without human intervention (i.e., that they are *autonomous*). Finally, adaptive behavior research typically places strong emphasis on the resource-limited nature of real-world interactions: agents in real environments do not have unlimited computational power or memory space or time in which to decide on appropriate actions. The stark contrast between these assumptions and those of the vast majority of work in AI is manifest.

Compare this explicit characterization of adaptive behavior with the situation faced by human traders in a real CDA market or in one of Smith's experiments: traders need to coordinate perception and action in the sense that they must accumulate and integrate data from diverse asynchronous sources, and act upon that data in good time (the value or integrity of much market data decays rapidly with time). In a competitive market, the environment is clearly dynamic and unforgiving. Relevant information (such as other traders' profit margins, or the information held by other traders) is rarely known or predictable, and it is unlikely that any trader will experience acts of kindness or selfless altruism. In real markets, traders that consistently fail to make profits will not last long. To summarize: if you're a trading agent, you can't be sure of anything except that the world is out to get you and that if you snooze you lose. For these reasons, trading agents need to be situated and autonomous. And, as adaptive behavior research is fundamentally concerned with the analysis or design of situated autonomous agents, there should be some use for adaptive behavior tools and techniques in studying or creating microeconomic systems. For instance, much work in adaptive behavior has explored the use of adaptation methods such as reinforcement learning (based on attempting to maximize "reward") and artificial evolution (based on attempting to maximize "fitness"). When trading agents interact within a market environment, measures of reward or fitness can be clearly and unambiguously equated with profit or utility.

Much work in microeconomics has been devoted to analytic studies that are based on a number of simplifying assumptions necessary to maintain tractability. Simulations of the sort common in adaptive behavior research could be built to explore models that are too complex for formal analytic treatment but sufficiently simple (relative to real systems) that they have useful predictive or explanatory power.

Given these arguments, it seems reasonable to expect that some prior work in the adaptive behavior literature will have explored issues in the collective behavior of agents that bargain and trade in market-based environments. Yet, surprisingly, the field seems to have almost totally ignored such issues.

2.3 (The Lack Of) Related Work

There are 256 papers published in the proceedings of the first four SAB conferences, and (at time of writing), a little over 60 papers published in the *Adaptive Behavior* journal. Given the arguments of the previous section, it would seem reasonable to expect that a fair proportion of papers in the adaptive behavior literature should deal with economic behaviors.

However, as far as we are aware, there is only *one* paper that deals explicitly with economic activity: this is Beltratti & Margarita's study of the evolution of trading strategies among heterogeneous artificial economic agents [2]. Beltratti & Margarita worked with artificial 'stock markets' populated by trading agents with neural network controllers that determined their trading strategies. While this work is novel and interesting, it is of very little relevance to price-determination or bargaining behaviors, because whenever two agents entered into a transaction, the price was (arbitrarily) set at the average of the predictions of the two agents [2, p.495]. Thus, it appears that no work at all published in the adaptive behavior conference proceedings or journals addresses the issues we are discussing here.

Adaptive behavior research is often characterized as related to work in *artificial life* (a-life). Many papers in the a-life literature deal with systems that exhibit complex coherent global behavior arising from the interaction of groups of components which are individually simple in comparison to their global behavior. Given this focus, the price-equilibration of groups of traders operating in market-based environments would seem to be a natural candidate for a-life research.

However, the a-life journal and conference proceedings show a distinct lack of such research. While there is a small core of work on the *iterated prisoner's dilemma*, a classic game-theory problem in which the emergence of cooperative behavior among non-altruistic agents can be explored, and which is of direct relevance to oligopolistic markets, we know of only two papers published in the a-life literature that explicitly study market trading strategies: [15] and [12]. Both of these papers report on the application of elementary evolutionary adaptation methods to optimize simple trading strategies for speculative markets, and both directly set all transaction-prices to be equal to P_0 via a centralized process that collects quote prices from all individuals, determines the supply and demand, and then calculates P_0: [15, p.191], [12, p.326]. For this reason, neither of these two papers are relevant to the study of equilibration or bargaining behaviors.

The apparent lack of work in a-life on agents with bargaining behaviors for market-based environments is confirmed in a recent review paper by Tesfatsion, [18]. The paper presents a summary overview of aspects of a-life especially relevant for the study of decentralized market economies. The two main areas of research activity discussed are: the combination of evolutionary game theory (e.g., iterated prisoner's dilemma) with preferential partner selection (i.e., the ability to choose or refuse particular opponents in the game); and an extension of this, where trade networks can form and evolve, using the prisoner's dilemma game to model risky trades between individuals. Thus, there is no emphasis on bargaining mechanisms in this work, and the indications from [18] are that little or no work in a-life is comparable to the work on trading agents discussed here.[1]

[1]While relatively little work in economics is directly comparable, some recent work combines biologically-inspired computing techniques with artificial markets: see e.g. [1]. For discussion of other relevant work in economics, see [6, 9].

2.4 Summary

It is not clear to us why the published research in adaptive behavior, and also in artificial life, has largely ignored economic activity. We find it curious, given that there are strong reasons for considering economic activity as adaptive behavior. For surely, adaptive behavior research methods could (and should) be used to develop animats that engage in economic interactions. In addition to being used as potential scientific models of human activity, such animats could find use in several potential application areas including predictive simulations, market-based control, and internet-based commerce. The clear correlation between profitability (or utility-maximization) and 'fitness' or 'reward' should allow for many of the tools and techniques of adaptive behavior research to be readily transferred to autonomous agents that interact with market environments. Like many environments in adaptive behavior research, markets can be dynamic, uncertain, and unforgiving.

Given that the simplified and constrained market environments introduced by Smith have formed the basis of much work in experimental economics over the last 30 years, it seems wise to commence by attempting to develop "intelligent" trading agents with the bargaining capabilities necessary for small groups of traders to show price-equilibration in market scenarios similar to those studied with human subjects in experimental economics. Concentrating on a defined class of environments should facilitate sharing of techniques and comparison of data.

The full details of equilibration in markets of human traders are yet to be explained. The development of trader animats capable of price-determination as a collective behavior in market-based environments could advance our understand of existing microeconomic systems and open up new application areas. Thus, we anticipate a melding of research in adaptive behavior and market microeconomics that we expect to be highly productive, and hopefully also highly profitable. Next, we describe results from our ongoing work directed at this aim.

3 Animat-Trader 'Retail' Markets

Our current trading agents are given adaptive capabilities via elementary machine-learning techniques. Because our agents are intended to have minimal intelligence, but not zero intelligence, they are referred to as "ZIP" traders: an acronym for "zero-intelligence-plus".

In other publications [6, 7] we have shown that our ZIP traders do not suffer from the failings that afflict Gode & Sunder's ZI traders. Furthermore, we have noted that the collective behavior of groups of ZIP traders is *human-like*, by which we mean that ZIP traders in experimental CDA markets give scores on the standard metrics of market performance (such as the Smith's α measure of convergence [16], allocative efficiency, and also profit dispersion [14]) that are very similar to those given by human traders in the same markets. Here, we demonstrate that the market behavior of ZIP traders is human-like in another sense: ZIP traders fail to exhibit rapid equilibration in a particular style of non-CDA market, and their mode of failure is very similar to that of human traders in a similar non-CDA experiment reported by Smith [16].

Smith [16] reported results from an experimental model of common retail markets, where sellers announce prices and buyers either purchase at the offer price or ignore the offer, without giving any indication of what range of transaction prices they would be willing to bid. The results from Smith's experiment, and his explanation of those results, are intriguing. Smith's expectation was that transaction prices would settle at levels higher than the theoretical P_0, indicating that the structure of retail markets offers advantages to the sellers. But this did not happen: instead, transaction prices settled at levels significantly below P_0. Smith explained this as being due to buyers that never quite recovered from having been 'badly fleeced' in the early stages of the experiment, where transactions occurred at high prices before equilibration had driven them lower.

If groups of simple artificial agents interact to exhibit market-level phenomena that are similar to those of human markets, explanations of how the phenomena arise in the artificial system may be viewed as candidate explanations for the same phenomena in human markets. We illustrate these arguments by means of an example. We present results from experiments where ZIP traders adapt and interact via price-bargaining in market environments based on the artificial 'retail' market that Smith [16] used in his experimental economics research, and we demonstrate that groups of simple agents can exhibit human-like collective market behaviors. We then note that, while it is often tempting to offer explanations of human market behavior in terms of the mental states of the agents in the market, our agents are sufficiently simple that explanations of the human-like collective market behavior of our agents cannot be phrased in terms of mental states; thereby inviting comparisons with Braitenberg's influential "law of uphill analysis and downhill invention", with eliminative materialism in the philosophy of cognitive science, and with dynamical-systems-based analyses of adaptive behavior.

Section 3.1 introduces the mechanisms of adaptation in ZIP traders. Section 3.2 presents results showing that ZIP-trader 'retail' markets exhibit the same failure qualities as Smith's human-trader 'retail' markets. Given this similarity, we note that explanations of the animat markets could have significant impact on the way in which comparable human activity is explained. In particular, the simplicity of the ZIP trading mechanisms means that explanations of their failures cannot be phrased in terms of mental states. We further discuss the implications of this in Section 4.

3.1 ZIP Traders

The emphasis in our work is on creating simple software animats for bargaining in market-based environments. This emphasis on simplicity comes not only from a desire for computational efficiency (important in engineering applications if hundreds or thousands of animats are active on a network), but also in a speculative scientific attempt at sketching the minimum mechanistic complexity necessary and sufficient for explaining human bargaining behaviors in specific market environments.

Space restrictions prevent us from presenting a full discussion of the rationale for the current design of ZIP trader agents, and from presenting exhaustive results. The intention here is to briefly summarize key aspects of the design before presenting illustrative results. Cliff [6] gives a complete discussion of the design, shows results from many experiments in different types of market environment, and includes all the C source-code for the system. A recent thesis by van Montfort [21] replicated our CDA results, and explored the use of ZIP traders in spatially distributed markets where there may be potentially hundreds or thousands of traders, and as arbitrage agents capable of buying units of commodity in one market for subsequent re-sale into another market, exploiting differences in price between the two markets.

Each ZIP trader operates by maintaining a *profit margin* that it uses for calculating the price it 'quotes' (offers or bids) in the market: the profit margin determines the difference between the limit price for the unit of commodity the agent is trading and the price that agent quotes for that unit. The 'aim' of each ZIP agent is to maximize profit generated by trading in the market. If an agent's profit margin is set too low, it will miss out on potential profit when it makes a transaction with another agent, so all agents are constantly trying to increase their profit margins. But if an agent sets its profit margin too high, it may miss the opportunity to make transactions with other agents, because the price it offers is less attractive than the prices offered by competing agents. Clearly, what it means for the profit margin to be "too high" or "too low" is dependent on the context of the market conditions, and this varies dynamically. Thus, the problem of designing a trading agent can be considered as a combination of two issues: the *qualitative* issue of deciding *when* to increase or decrease the profit margin, and the *quantitative* issue of deciding *by how much* the margin should be altered.

For reasons discussed in detail in [6], each ZIP trader makes the qualitative decision of when to alter its margin on the basis of four factors. The first factor is whether the agent is *active* in the market: agents are active until they have sold or bought their full entitlement of units of the commodity. The remaining three factors concern the last quote by any agent in the market: we refer to this as $\mathcal{Q}$. Each ZIP trader notes whether $\mathcal{Q}$ was an offer or a bid, whether $\mathcal{Q}$ was accepted (i.e., led to a transaction) or rejected (ignored by the traders in the market), and whether $\mathcal{Q}$'s price, $q(t)$, is greater than or less than the price the ZIP trader would currently quote. We refer to the price a ZIP trader i would quote at time t as that trader's *quote-price*, $p_i(t)$, which is calculated from i's limit price $\lambda_{i,j}$ (for i's jth unit of commodity) and i's current profit coefficient $\mu_i(t)$ using $p_i(t) = \lambda_{i,j}(1 + \mu_i(t))$. Thus, a seller's profit margin is raised by increasing μ_i and lowered by decreasing μ_i, such that $\mu_i(t) \in [0, \infty); \forall t \forall i$. The situation is reversed for buyers: they raise their margin by decreasing μ_i and lower it by increasing μ_i, subject to $\mu_i(t) \in [-1, 0]; \forall t \forall i$.

A ZIP seller raises its profit margin whenever $\mathcal{Q}$ was accepted and $p_i(t) \leq q(t)$. It lowers its margin only if it is still active and $\mathcal{Q}$ was an offer with $p_i(t) \geq q(t)$, or if $\mathcal{Q}$ was a bid that was accepted and $p_i(t) \geq q(t)$. Similarly, a ZIP buyer raises its profit margin whenever $\mathcal{Q}$ was accepted and $p_i(t) \geq q(t)$, and it lowers its margin when it is active and either $\mathcal{Q}$ was a rejected bid with $p_i(t) \leq q(t)$ or $\mathcal{Q}$ was an accepted offer with $p_i(t) \leq q(t)$.

Note that the value $q(t)$ is simply the current quote-price of one of the traders in the market: if that trader is designated by the index j then $q(t) = p_j(t)$.

The quantitative issue of by how much the profit margin should be altered is addressed by using a simple machine-learning rule. Specifically, the learning rule we use is *Widrow-Hoff with momentum*, which also underlies back-propagation learning in neural networks. Briefly, this adjusts the actual output of a system toward some *target* output value, at a speed determined by a learning rate β, and with a simple 'memory' or 'momentum' parameter γ. In each ZIP trader the target-price is a stochastic perturbation of $q(t)$, and each trader i uses this in combination with β_i and γ_i to adjust its profit-coefficient $\mu_i(t)$. The profit-margin update rule is:

$$\mu_i(t+1) = (p_i(t) + \Gamma_i(t))/\lambda_{i,j} - 1$$

where $\Gamma_i(t)$ is the momentum term:

$$\Gamma_i(t) = \gamma_i \Gamma_i(t-1) + (1-\gamma_i)\beta_i(\mathcal{R}_i(t)q(t) + \mathcal{A}_i(t) - p_i(t))$$

for $t > 0$, and $\Gamma_i(0) = 0 : \forall i$.

The target-price is generated by multiplying $q(t)$ by a *relative* coefficient $\mathcal{R}_i(t)$ and then adding a small *absolute* perturbation $\mathcal{A}_i(t)$. The values for $\mathcal{R}_i(t)$ and $\mathcal{A}_i(t)$ are stochastically generated (IID) for each trader, every time $\Gamma_i(t)$ is calculated. When the trader's quote-price is being increased, $\mathcal{R}_i = \mathcal{U}(1.0, 1.0+c_\mathcal{R})$ and $\mathcal{A}_i = \mathcal{U}(0.0, c_\mathcal{A})$, where $\mathcal{U}(c_{\text{lo}}, c_{\text{hi}})$ denotes a uniformly distributed random real value over the range $[c_{\text{lo}}, c_{\text{hi}}]$. When the trader's quote-price is being decreased, $\mathcal{R}_i = \mathcal{U}(1 - c_\mathcal{R}, 1.0)$ and $\mathcal{A}_i = \mathcal{U}(-c_\mathcal{A}, 0.0)$. For further details of how learning is implemented in ZIP traders, see Cliff and Bruten [6, 7].

In the experiments reported in this paper the following parameter values were used. Each trader's value for

β_i was set randomly from $\mathcal{U}(0.1, 0.5)$, and each trader's value for γ_i was set randomly from $\mathcal{U}(0.0, 0.1)$. When calculating $\Gamma_i(t)$, all traders use $c_{\mathcal{R}} = 0.05$ and $c_{\mathcal{A}} = 0.05$ in the distributions for $\mathcal{R}_i$ and $\mathcal{A}_i$. The initial profit coefficients (i.e., $\mu_i(0)$) for the sellers were set randomly from $\mathcal{U}(0.05, 0.35)$ and for the buyers from $\mathcal{U}(-0.35, -0.05)$.

3.2 Experiments

In Smith's paper [16], the one non-CDA market that he examined involved a one-sided auction, where only sellers were allowed to quote offers: buyers were not allowed to quote bids, but passively observed the prices offered by the sellers. Each buyer therefore had the privilege of being able to ignore offer-prices that were too high and accept those that were within their range, without giving any indication of their limit prices. Smith proposed this as an approximation to an ordinary retail market, where sellers bear the responsibility of advertising their prices and buyers decide whether to buy or not without entering into any kind of bargaining or haggling process. Smith's results from this experiment were shown in Fig. 2. Smith's comments on his expectations and actual results for this experiment are significant:

> "Since sellers desire to sell at the highest prices they can get, one would expect the offer prices to be high, and, consequently, one might expect the exchange [i.e., transaction] prices to show a persistent tendency to remain above the predicted equilibrium. The result was in accordance with this crude expectation in the first market period [i.e., day] only.... Since sellers only were making offers, the prices tended to be very much above equilibrium. Five of these offers were accepted at prices ranging from \$2.69 to \$2.80... The competition of sellers pushed the offer prices lower and the remaining buyers made contracts at prices [of \$2.35, \$2.00, and \$2.00]. The early buyers in that first market period never quite recovered from having subsequently seen exchange prices fall much below the prices at which they had bought. Having been badly fleeced, through ignorance, in that first trading period, they refrained from accepting any high price offers in the remaining three periods of the test. This action, together with seller offer price competition, kept exchange prices at levels persistently below equilibrium for the remainder of [the experiment]." [16].

The ZIP traders can be used in a straightforward copy of Smith's experimental retail market. The supply and demand curves for the ZIP market are the same as those shown in Fig. 2. In Fig. 3 we show the mean daily transaction prices from 50 experiments where the ZIP traders operate in Smith's 'retail market'. As can be seen, the average transaction prices are typically less than \$2.00 (significantly below the P_0 value of of \$2.25). There also

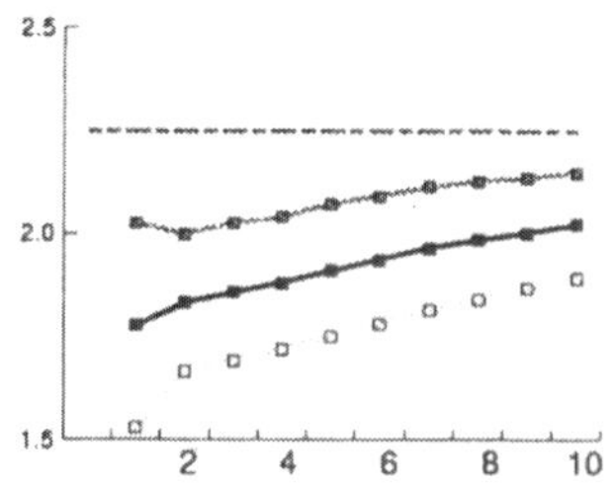

Figure 3: Mean ZIP transaction prices, averaged over 50 experiments, for 'retail-market' experiments with the supply and demand shown in Fig. 2. Horizontal line indicates P_0=\$2.25. Data points show mean plus and minus one standard deviation.

appears to be little or no convergence towards equilibrium, or reduction in variance as the experiment progresses. The apparent lack of convergence or reduction in variance can be better understood by examining individual price trajectories: Figures 4 to 7 show time-series of the transaction prices in four individual experiments using ZIP traders in the 'retail' market with supply and demand as illustrated in Fig. 2. As can be seen, in all four experiments the market converges to a fairly constant value for transaction prices by Day 4, but the value that is converged on varies: in Figures 4 to 6, all trades on Day 10 are within \$0.15 of P_0, while in Fig. 7 no trade is less than \$0.40 off the equilibrium price. As is clear in Fig. 3, the price converged on is, on average, significantly less than P_0. Thus, there is a strong qualitative agreement between results from our ZIP traders and Smith's [16] observations of human subjects in his experimental 'retail markets'.

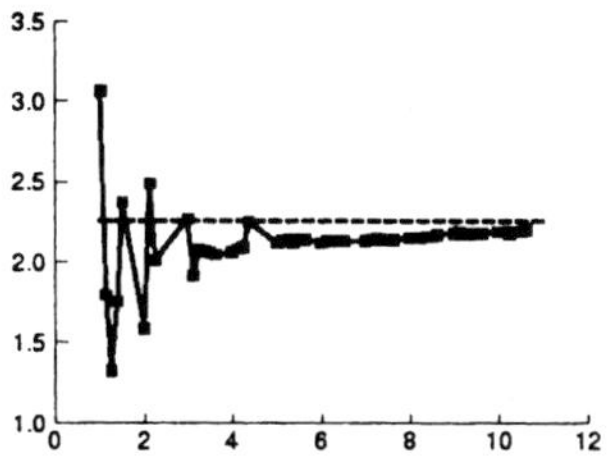

Figure 4: Transaction price time series for one 'retail market' experiment, seed=3453.

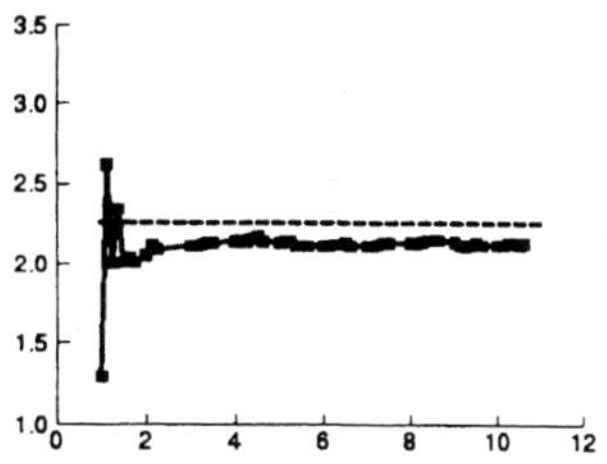

Figure 5: Transaction price time series for one 'retail market' experiment, seed=3522.

Of these four single experiments, the price series in Fig. 4 most closely resembles that of Smith's subjects: only three transactions occur at transaction price more than a few cents above the equilibrium price; while many more occur at prices lower than equilibrium, which is approached *very* slowly, from below. Smith's explanation was that this is due to early transactions at high prices preceding a series of low-price transactions that induce a resistance to higher prices in 'fleeced' traders.

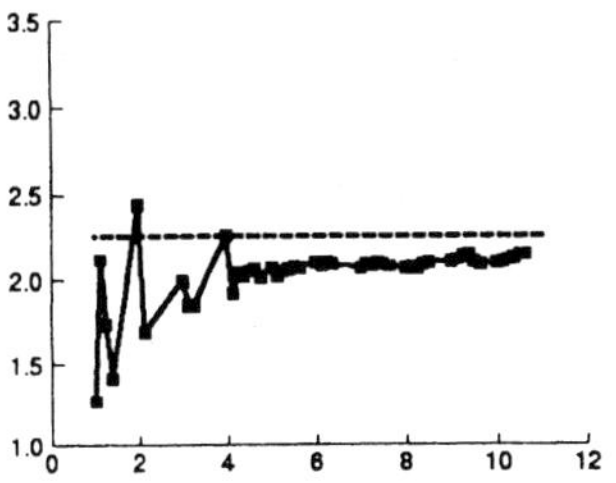

Figure 6: Transaction price time series for one 'retail market' experiment, seed=3553.

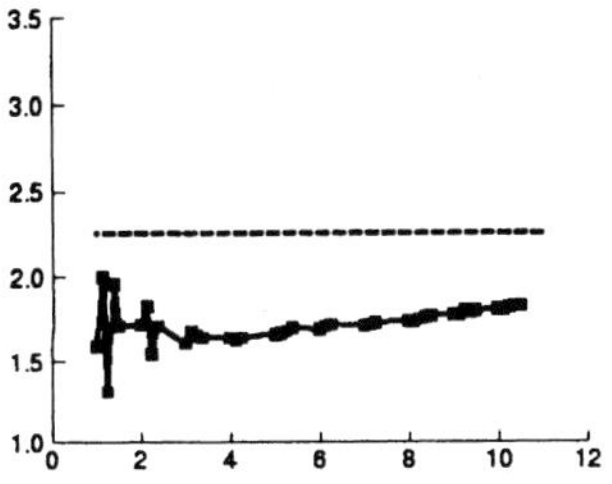

Figure 7: Transaction price time series for one 'retail market' experiment, seed=3591.

Thus, in addition to our demonstration in other publications [6, 7] that ZIP traders can give human-like collective behavior in CDA markets, the results presented here show that the dynamics and the modes of failure of ZIP traders are also similar to those of humans in Smith's [16] one-sided auction experimental model of retail markets.

4 Discussion

The similarity between our ZIP results and those from Smith's human subjects suggests a line of reasoning similar to that underlying much adaptive behavior research. This reasoning relies on noting that there is one key difference between our results and Smith's. Smith was working with human subjects, where there is a natural temptation to offer explanations in terms of mental states. In the passage quoted above, Smith talks of the human buyers "never quite recovering" from "having been badly fleeced". It is not clear from the original text whether this account is inventive conjecture on Smith's part, or the result of properly conducted post-experiment interviews. But even if these comments are the result of interviewing those subjects who ended up as 'fleeced' buyers, the danger of introspective *a posteriori* accounts of behavior are well known.

The crucial difference then, between Smith's work with humans and our work with ZIP traders, is that in the ZIP traders there are no place for such mentalistic descriptions of the behavior of the agents in the market. There is nothing, not even an evolved neural network, in which the ZIP agents could hide the mental states of 'never quite recovering' or noticing that they have been 'badly fleeced'. Any explanation of what causes the ZIP-agent markets to approach equilibrium slowly and from below is *forced* to be framed in terms of the interactions of the simple ZIP adaptation mechanisms, because there is *nothing else* in the system that could cause the observable phenomena.[2] Let us assume that a causal mechanistic explanation for how the collective behavior of ZIP traders gives rise to some market-level phenomena can be developed, and call it $\mathcal{E}$. Then $\mathcal{E}$ can also be considered a candidate explanation for the behavior of groups of human traders. Naturally, if it can be demonstrated that the ZIP traders are using adaptation mechanisms that could not be employed or implemented by humans, then $\mathcal{E}$ is a very weak explanation, or no explanation at all. But if such counter-arguments to $\mathcal{E}$ cannot be readily advanced, $\mathcal{E}$ should properly be considered as a putative explanation for the human behavior, which can be subjected to experimental evaluation or falsification. And, crucially, here $\mathcal{E}$ cannot be phrased in terms of mental or emotional states, because the ZIP traders have nothing that corresponds to such states.

The failure of ZIP traders to converge on a competitive equilibrium (i.e., a steady sequence of transaction prices at the P_0 value) in 'retail' markets is due simply to the fact that although the buyer and seller profit-margins are altered symmetrically when in a CDA, the prevention of bids in the one-sided 'retail' market introduces an asymmetry: although the traders raise their margins under symmetric conditions, an active buyer b lowers its margin only when $\mathcal{Q}$ was *accepted* at a price $q(t) \geq p_b(t)$, while an active seller s will lower its margin when $q(t) \leq p_s(t)$ regardless of whether $\mathcal{Q}$ was accepted or not. In essence, this demonstrates that, despite the good equilibration properties of CDA markets where both buyers and sellers are trading according to the ZIP strategy described in Section 3.1, the asymmetry of opportunity sets (i.e., the prevention of bids) in the 'retail' market prevents equilibration by ZIP traders because their trading strategies depend on the bilateral flow of information found in CDA markets. While it may be possible to alter the ZIP strategies to give good equilibration in retail markets, or even in both retail markets *and* CDA markets, the key issue here is that our explanation of ZIP traders' failure to reach a competitive equilibrium is not reliant on them having vague and difficult-to-define mental states such as 'never quite recovering from being badly fleeced'.

By specifying and observing simple synthetic trading agents, it is possible to demonstrate the same overall market behavior without relying on abstract or vague descriptions of the mental states of the participants in the market. In this sense, the work described here is similar to other work in adaptive behavior that is justified by the principle that it can be more fruitful and more parsimonious to attempt an understanding of how some behavior is generated by *synthesizing* an artificial system that exhibits that behavior, rather than by *analyzing* a natural system that exhibits the same behavior: a principle that Braitenberg [3] named the "law of uphill analysis and downhill invention". Although it is often difficult to resist the temptation to describe the cognitive behaviors of animals (and humans in particular) in terms of mental states, there is growing support for a counter-approach, where the intention is to explain observed behaviors in terms of the dynamics of causal mechanistic interactions, rendering the mental-states-based accounts obsolete. These ideas first gained credence in the phi-

[2] Assuming, of course, that the code for the system has no bugs.

losophy of mind, where they are most strongly associated with Churchland's *eliminative materialism* [4], and their relevance to work in artificial autonomous agents has been discussed by Smithers [17], van Gelder [19, 20], and Cliff & Noble [10]. Thus, our work here can be viewed as a step in the direction of adopting an eliminative materialism or dynamical systems perspective on the economic activity of human agents.

So, we have demonstrated here that ZIP traders can give results qualitatively similar to those of humans in 'retail market' experiments. In doing so, we have demonstrated a point of more general significance: that techniques common in adaptive behavior research can be used to cast new lines of inquiry on the human experimental economics data. Given that ZIP traders exhibit human-like behavior and have no mental states, of how much genuine use are mental states in the explanation of human market behavior?

5 Conclusion

The development of computational mechanisms that allow groups of software agents to exhibit bargaining behaviors in market-based environments satisfies a number of needs. In market-based control, simple mechanisms are required to give computationally efficient, robust, and truly distributed resource allocation and control. Such mechanisms could also be employed in the growing field of internet-based commerce. Moreover, such mechanisms act as mechanistically rigorous statements of potential models of human bargaining behaviors, although it is likely that more complex mechanisms would be required to further account for the many subtleties and nuances of human behavior: empirical work in experimental economics and human psychology would also be necessary to validate any models. Once validated, such model agents could be used for conveniently testing theories concerning the behavior of humans in different market structures and conditions.

The arguments we presented in earlier papers [6, 8, 9] indicate a need for bargaining mechanisms more complex than the constrained stochastic generation of bid and offer prices used by Gode & Sunder's ZI traders [14]. Our work on ZIP traders, reported here and in other papers [6, 7] should be viewed as a preliminary sketch of what forms such bargaining mechanisms might take. The ZIP traders are more complex than ZI traders, but only slightly, and in any case are manifestly much less complex than humans. Nevertheless, the results from the ZIP traders, both in terms of equilibration in CDA markets and failure to equilibrate in Smith's one-sided auction model of retail markets, are clearly closer to those from human experimental markets than are the results from ZI traders. It is reassuring to see that the ZIP mechanisms can give such human-like results, but there is much further work that could be done in exploring the behavior of ZIP traders in more complex market environments, and in attempting to extend the behavioral sophistication of such traders without unduly adding to their complexity.

References

[1] W. Arthur, J. Holland, B. LeBaron, *et al.* Asset pricing under endogenous expectations in an artificial stock market. In W. Arthur, S. Durlauf, & D. Lane, eds, *The Economy as Evolving Complex System II*, pp.15–44. Addison Wesley, 1997.

[2] A. Beltratti & S. Margarita. Evolution of trading strategies among heterogeneous artificial economic agents. In J.-A. Meyer, H. Roitblat, & S. Wilson, eds, *From Animals to Animats 2: Proc. SAB92*, pp.494–501. MIT Press, 1992.

[3] V. Braitenberg. *Vehicles.* MIT Press, 1984.

[4] P. M. Churchland. *A Neurocomputational Perspective.* MIT Press, 1989.

[5] S. Clearwater, ed. *Market-Based Control: A Paradigm for Distributed Resource Allocation.* World Scientific, 1996.

[6] D. Cliff. Minimal-intelligence agents for bargaining behaviors in market-based environments. Tech. Rep. HPL-97-91, Hewlett Packard Labs, 1997.

[7] D. Cliff & J. Bruten. Less than human: Simple adaptive trading agents for CDA markets. Tech. Rep. HPL-97-155, Hewlett Packard Labs, 1997.

[8] D. Cliff & J. Bruten. More than zero intelligence needed for continuous double-auction markets. Tech. Rep. HPL-97-157, Hewlett Packard Labs, 1997.

[9] D. Cliff & J. Bruten. Shop 'til you drop I: Market trading interactions as adaptive behavior. Tech. Rep. HPL-98-58, Hewlett Packard Labs, 1998.

[10] D. Cliff & J. Noble. Knowledge-based vision and simple visual machines. *Phil. Trans. Roy. Soc.: B*, 352:1165–1175, 1997.

[11] D. Davis & C. Holt. *Experimental Econ.* Princeton, 1993.

[12] M. de la Maza & D. Yuret. A futures market simulation with non-rational participants. In R. Brooks & P. Maes, eds, *Artificial Life IV*, pp.325–330. MIT Press, 1994.

[13] D. Friedman & J. Rust, eds. *The Double Auction Market: Institutions, Theories, and Evidence.* Addison-Wesley, 1992.

[14] D. Gode & S. Sunder. Allocative efficiency of markets with zero-intelligence traders. *J. Pol. Econ.*, 101(1):119–137, 1993.

[15] C. Nottola, F. Leroy, & F. Davalo. Dynamics of artificial markets. In F. Varela & P. Bourgine, eds, *Towards a Practice of Autonomous Systems*, pp.185–194. MIT Press, 1991.

[16] V. L. Smith. An experimental study of competitive market behavior. *Journal of Political Economy*, 70:111–137, 1962.

[17] T. Smithers. Taking eliminative materialism seriously. In F. Varela & P. Bourgine, eds, *Towards a Practice of Autonomous Systems: Proc. ECAL91*, pp.31–40. MIT Press, 1992.

[18] L. Tesfatsion. How economists can get Alife. In W. Arthur, S. Durlauf, & D. Lane, eds, *The Economy as an Evolving Complex System II*, pp.533–564. Addison Wesley, 1997.

[19] T. van Gelder. What might cognition be if not computation? Tech. Rep. 75, Indiana University Cognitive Sciences, 1992.

[20] T. van Gelder. The dynamical hypothesis in cognitive science. *Behavioral and Brain Sciences*, forthcoming, 1998.

[21] G. van Montfort. Economic agents for controlling complex systems. Master's thesis, Delft University of Technology Faculty of Technical Mathematics and Informatics, Sept. 1997.

Sewer Robotics

Joachim Hertzberg, Thomas Christaller, Frank Kirchner, Ulrich Licht, and **Erich Rome**

GMD - German National Research Center for Information Technology
Schloss Birlinghoven, 53754 Sankt Augustin, Germany
⟨first name⟩.⟨last name⟩@gmd.de

Abstract

We expose the application area of building and controlling autonomous service robots for performing inspection tasks in sewerage systems that are inaccessible for humans. The problem is explained, the physical and legal boundary conditions for operating sewer robots are described, the state of current sewer inspection technology is sketched, and the need for using advanced technology in this area motivated. Based on that, we identify AI research and technology topics that are crucial for sewer robotics and give pointers to related work, including work that we have been doing since the last four years.

What's in a Sewer

Most municipalities in the world run sewers. They may differ in age, making, extension, and elaboration, but in any case they incorporate a large infrastructure investment—for many towns their largest item in capital goods overall. The standards of sewage treatment and, consequently, the size of sewers differ greatly in different countries, but their overall length is enormous, to say the least. Total sewer pipe length in Japan, e.g., was 250,000 km by the end of 1994, with a yearly installation of additional 15,000 km (Fukui 1994). Communal sewers in Germany were about 360,000 km in length in 1990 (Bölke 1996, Ch. 1); sewers on private ground in Germany are estimated to be of twice that size.

There are two types of communal wastewater systems: Storm drains and sanitary sewers. Although not exactly healthy, water in storm drains is relatively clean and is sometimes lead directly into rivers and lakes. Water from sanitary sewers should obviously be treated in purification plants before being released. Many towns have a combined system for both purposes—the separate and the combined designs both have their advantages and disadvantages. We use the term *sewer* to unspecifically denote storm drains, sanitary sewers, and combined systems.

Quite a variety exists in pipe materials. Concrete, PVC, and glazed clay are typically used today; older pipes may also consist of cast iron, or be made of brick. The normal condition inside a used sewer pipe is: Dirty, damp, slippery, and around 15°C. The sewage level in a pipe depends on the type of sewer (storm, sanitary, combined) and on recent precipitation: A typical filling degree for a combined sewer at dry weather is 5 %, with higher average levels in sanitary sewers, and levels in storm drains varying greatly.

Pipe diameter, gradient and material are normally chosen such that the sewage flows fast enough to cleanse the pipe, but not so fast as to erode it. As a rule, the smaller the diameter, the faster the flow must be, and, hence, the greater the gradient. For a pipe of 20 cm diameter, the range of gradients is 1:15 - 1:300, typical values being 1:50 - 1:150; for a 80 cm pipe, 1:50 - 1:1000 is admissible (Stein & Niederehe 1992, Sec. 2.3.4). A typical flow is 0.5 - 2 m/sec, with peaks around 6 m/sec. To limit flow speed in steep areas, sewers contain vertical down-pipes, which may be some meters high.

If designed, built, and used properly, a sewer may just work for a long time: Keding, van Riesen, & Esch (1990) report that 27% of the communal sewers in western Germany were 50 years and older. However, it is mandatory to put effort into sewer maintenance to guarantee proper function, to protect the large investment capital that went into building them, and to prevent damage from the sewer itself, from the area adjacent to it, and from sewage treatment plants.

Maintenance has to begin with inspection. Larger pipes, i.e., those wider than 80 cm inner diameter, are considered accessible for humans and can be inspected by walking through. Only relatively few mains are accessible, the vast majority of sewer pipes is not, diameters going down to 15 - 10 cm towards inlets from single houses. The procedure for inspecting such inaccessible pipes is determined by the damages that must be found, of which there are two broad classes: clogs and leaks. Typical causes for clogs are encrusted sediments, laterals jutting out into the main due to ill construction, and tree roots grown in through leaks, such as in Fig. 1.

While clogs might impede proper working of the sewer and cause spectacular, sometimes expensive mal-

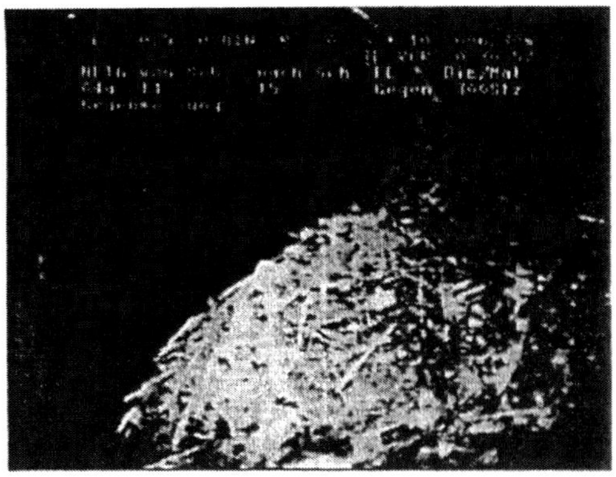

Figure 1: *Image from a conventional sewer inspection: Tree roots grown in through a leaking joint. (Image by courtesy of rhenag, Cologne)*

function, they are typically not as insidious as leaks are: Once clogs really matter, they expose themselves, and before that, they can easily be detected, often even by the archaic technique of peeping through the pipe from manhole to manhole. Leaks are more subtle. Most frequently, they come from corrosion of the gaskets around joining pipe segments, or from cracks in the pipe material itself as from construction faults or from strain on the pipe by shift of the embedding soil.

Leaks are harmful for two reasons: wastewater may leak out, and groundwater may leak in. Water leaking out from a sanitary or combined sewer may pollute soil and groundwater, which may cause environmental problems—in Germany, this may even be a statutory offense. Wastewater flowing out in larger quantities may wash away soil, potentially undermining streets, pavements, and foundations of buildings. Groundwater seeping in, particularly into sanitary sewers, may dilute the volume at the purification plant, with the danger of washing away the bacteria in the biological purification process—thereby literally killing this process and causing high cost to start it up again. Finally, as cracks in the pipe material grow and strain or corrosion joins the effect, there is a danger that pipes collapse.

In the rest of this paper, we argue that building autonomous mobile robots for inspecting inaccessible sewers is feasible, makes sense economically, and necessarily involves AI technology and research. The results are based on a joint feasibility study (Berns *et al.* 1996; 1997) and on some technical work, on which we will briefly report below. We start with a sketch of the state of the art in sewer inspection, based on which we explain the potential of inspection using autonomous mobile robots. After that, we identify the AI technologies that we have found relevant for the endeavour, pointing then to existing concrete work in that direction. We conclude by summing up the arguments.

State of the Art in Sewer Inspection

A standard, state-of-the-art technology of inspecting inaccessible sewers is emerging now over different countries. This section gives a sketch of the recent inspection practice; Bölke (1996) describes it in all detail—at least its German version.

As a boundary condition for technology and research in sewer inspection, there are large differences in the procedural regulations concerning if, when, and how to inspect. In the US, e.g., there is, to the best of our knowledge, no regulation that would urge or force a community to regularly inspect their sewers. In Germany, on the other hand, there are states in which inspection of the complete sewers of all communities are mandatory every ten years. The European norm EN 752-7 (CEN 1997) includes general inspection requirements as well as ample reference to national regulations. In any case, we assume that many municipalities all over the world have developed or will develop a regular sewer inspection pattern—motivated by laws or by own insight.

Physics and chemistry pose other boundary conditions for sewer inspection, which are non-relaxable. Although sewers are mostly filled with normal air, explosive or corroding gases can occur. Sewage may gush out of laterals at any time. In consequence, all sewer inspection equipment must be explosion-proof, corrosion-resistant, and water-protected.

State-of-the-art inspection of inaccessible sewers is *visual*. Fig. 2 shows the scheme: A human tele-operates a wheeled platform equipped with a colour, high-resolution video camera and a lighting system. The camera, which normally points along the pipe axis while driving, can be turned to inspect frontally every point on the pipe walls. To have small debris and light sediments removed, inspection is normally done shortly after flushing.

The operator's duty is to detect, localize, classify and protocol all features in the pipe, such as laterals and damages. The inspection is documented by a video tape of the complete procedure, by a graphical and textual protocol of all findings, and by photos of particular damages, like the one in Fig. 1. The protocol is sometimes requested in digital form suitable for input into sewer information systems, which are instances of geographical information systems (GIS) with sewer-relevant objects and relations pre-defined and with special features, such as for calculating flow and capacity.

The camera platform is tethered, the cable serving for energy supply, for transmission of control from the operator, for data transmission to the operator, for measuring the distance travelled, and for pulling in manually the platform in case of malfunction. Of obvious help, the cable is also a tremendous hindrance. While the platform drives along a pipe, the cable rubs along the pipe bottom. Depending on its weight and on the ruggedness of its surface and of the pipe, the friction increases with the distance travelled, effectively stopping the platform typically after 50 - 200 m. If bends are present, the range of operation shrinks again: Cable

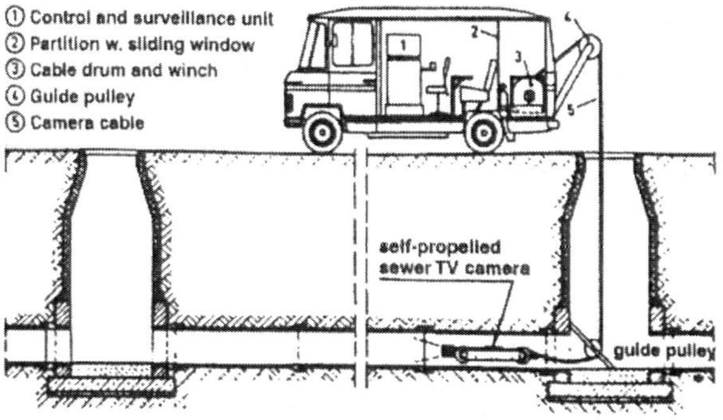

Figure 2: *Schema of current visual sewer inspection. From (ATV 1991), adapted.*

friction is severely increased by bends, and the design of the platform frames, which is typically longish and rigid, makes turning along narrow bends physically impossible. The same ist true for steps, which are frequent in sewers at places where laterals of smaller diameters join mains with larger diameters and their bottom level not meeting: It is currently considered unthinkable—or at least unheard-of—that an inspection platform climb into such a lateral.

This hindrance by the cable has its price—in terms of handling trouble and, consequently, in terms of inspection cost. The camera platform has to be hauled in frequently, which means: It has to go back, be lifted up (for which purpose an assistant has to climb down into the manhole), stowed away in the control van, the van has to be driven the short way to the manhole last seen by the camera, the platform be set through the manhole into the pipe (assistant climbs down again), and off it goes—until the next stop. Moreover, the driving speed of the camera platform in the pipe is normally restricted to enforce inspection accuracy. Specifications of maximum speeds and average inspection speeds vary; 15 cm/s and 2 cm/s, resp., are typical values (Bölke 1996, p. 197).

In consequence of all that, an inspection team of operator and assistant using operating equipment (van, control post, cable handling mechanism, camera platform) for over hundred thousand dollars can inspect just a few hundred meters per work day, 600 m being an expected value. A typical price for an inspection (camera operation, data analysis, documentation) these days in Germany is around 8.70 DM (approx. $5) per meter (Berns *et al.* 1996, Part III), which means that a complete inspection of the communal sewers in Germany would currently cost $1.8 Billion. The price, by the way, is in the end paid by the users of the sewer, i.e., citizens and industry.

If the cable causes trouble, why not do without? High-bandwidth wireless transmission through underground pipes suffers from high damping—so the video images do not arrive undisturbed. Odometry is difficult in a slippery sewer—so positions of detected features cannot be localized precisely. Lighting up a sewer for good sight consumes much energy—so a battery-powered vehicle can impossibly work all the time with full lights on. These points, of course, are also true for an autonomous platform, so we will revisit them below.

The main inspection sensor is the camera. Some sewer features are obvious from the video image, such as laterals or drastic damages, some require considerable expertise to detect. Used pipes typically show some degree of corrosion and are covered with a film of dried mud and algae, hiding smaller cracks to direct sight. Particularly hard to see is damage under the streamlet of sewage, which is always present in a sewer (except storm drains during a drought). Another field for expertise and training is to assess from the video image the degree of pipe deterioration, given a damage. This is the chief task for the inspection operator, as it determines if and how urgently action must be taken: It makes a difference whether a crack is just on the surface, or goes through the pipe material, or—as part of a crack web—puts at risk the stability of the pipe.

Recently, equipment can be added to the camera platforms, often as separate, disposable modules. Calibration tools allow the width of joints and cracks to be measured from the video image. Inclinometers along the main platform axis help measure and record pipe fall. Special units allow laterals to be inspected from the platform in the main pipe; typically, they are built as rods of a few meters length, on the tip of which sits a small fish-eye video camera and a separate lighting system. They can be reeled out and in from the platform, thereby being pushed into and pulled out of the lateral, respectively, allowing to literally push further tele-operated video inspection.

Potential Benefits of a new Technology

A radical change in sewer inspection technology, based on using autonomous mobile robots, could enhance inspection quality, yet reduce cost. Moreover, it could open up a perspective to view differently the whole process of running and maintaining a sewer. In this section, we describe, first, the benefits of substituting current inspection technology by autonomous sewer robots, and, second, sketch the vision of keeping autonomous robots permanently in the sewer. These results are abstracted from the study by Berns *et al.* (1996).

Substituting current technology

The challenge for sewer inspection is to cut the cable. If this were possible within the limits of a robust technology, then there would be a large potential for saving cost and for enhancing inspection quality.

The savings potential should be obvious from the previous description of handling current inspection platforms. For a more detailed analysis, consider the scenario that an autonomous inspection robot works in the sewer all day; the human operator can contact it by wireless transmission at selected shafts or manholes that it passes, transmitting inspection data to the operator and control information to the robot. The wireless connection can easily be established by hanging an antenna into the manhole; being close to the robot, transmission rates can get arbitrarily high, at low energy consumption. As the robot continues inspecting, the operator team can prepare the next contact point and cross-check recently transmitted data.

Assuming a net inspection speed of 10 cm/s for 6 hours per day on site, of which 1 hour is spent sitting for data transmission, expected inspection distance per day is 1,800 m, i.e., triple the distance usually achieved today, at comparable investment and operation cost.

As to inspection quality, there are recurring complaints about the state of the art—e.g., (Bölke 1996, p. 2)—that it be inherently subjective, qualitative, and error-prone. Autonomous inspection robots would not magically *solve* the problem, but *require* to use an algorithmic, maybe partly heuristic, but systematic procedure for feature detection. We will come back to this point below. In any case, developing sewer robots for substituting current inspection technology would push the existing need for "objective", standardized tools and procedures for detecting features in sewers.

Another quality issue comes together with the cost argument. As sewer inspection today is prohibitively expensive, many municipalities run their sewers not knowing what is really going on down there. For example, they would flush sewers regularly (in Germany about once a year, at current cost $\approx$4.50 DM/m), not knowing where that is required by actual sedimentation, and where the sewer is neat and clean by regular sewage flow. Moreover, municipalities have no practical chance to run sample inspections for checking whether the flushing job has been done well—be it by own staff or by contractors. It is suspected (Berns *et al.* 1996, Part III) that the sheer existence of cheaper inspection technology would raise the quality of maintenance work for sewers; in particular, more detailed knowledge about the sewer reality—which would come from more frequent inspection—would help reduce flushing to the occasions necessary.

The new technology could also be used in *private* sewers, which work the same way like public ones, but—at least in European countries—are currently subject to less rigid regulation. Probably not many house-owners wish to operate personal sewer robots, but there are huge private sewers, e.g., in the chemistry industry or the military, and some of these owners want to document high environmental standards. In all such cases where the sewer owner is also its user and has to pay for the maintenance, the cost argument is crucial, and it will become unavoidable one day when regulations for running private sewers get tighter.

The vision: Life-long sewer robots

The strategy of replacing current inspection technology by autonomous robots seems promising as a first step both for exploring the field and for familiarizing involved staff with a new technology. The full potential of sewer robots, though, exploited only in another, more visionary scenario: Robots permanently working in the sewer. To make this a reality, some organizational and technical problems must be solved.

To supply the sewer robots with energy and mission data, and to collect the data that they have recently acquired, docking stations must be installed (preferably in manholes). Energy is to be transmitted contact-free by inductive coupling, and data as above by short-distance, low-energy, high-bandwidth wireless transmission. Existing sewer information systems must be changed so as to allow on-line access by robots.

On the technical side, robots that work, in principle, life-long in the sewer have to have certain physical, sensor, and control features. They must be able to follow bends and steps to not be encaged in a single straight sewer section. They must be able to withdraw to a suitable place (such as a manhole) and shut up in case of sudden raise in sewage level, e.g., at heavy rain. Their mechanics and motion control as well as their sensor configuration and interpretation must be able to cope with the sewer non-flushed. Their positioning must be fool-proof. And—to mention the most obvious point last—there must be a rescue strategy and procedure in case a sewer robot breaks down in action.

Each and every of these points is a challenge in itself—not to mention their combination. It seems, though, as if they can be developed gradually from the previous scenario of replacing current technology, where new features can be added to the "conservative" sewer robots if, when, and as far as they yield in terms of cost or quality. Berns *et al.* (1996, Part IV) sketch a family of research projects over ten years, which in part depend on each other and could be steps towards realizing life-long sewer robots. To end this section, let us name their topics, showing both the research problems to be solved and some concrete application potential of sewer robotics: Each of these items would help make sewer maintenance cheaper or better in some respect, but are currently far beyond the feasible.

- Multi-segment, articulated sewer robot platform. There is some ongoing work about this, which will be described below.
- Sensor equipment and interpretation for detecting injection of certain illegal (poisonous, corroding) substances into sewers.
- On-site automatic characterization of the pipe state (geometry, corrosion, flow, condition of joints).
- Autonomous measuring of sewers and features in them.
- Sewer information system for providing robot access to all sewer-relevant data available, for managing the information provided by sewer robots, and for assisting humans in deciding about maintenance actions.
- Small (three to five) team of sewer robots and the required infrastructure for operating jointly in sewer for a limited amount of time.

An AI View on Sewer Robotics

Sewers are real world. There are three basic problems that make real world robot applications difficult—be it sewer robotics or any other class of robot application. In particular, they make it impossible to rely solely on classical control theory approaches to robot control, but require employing some AI technology.

First, the world is not completely known, i.e., no domain model is given in advance that includes all important and relevant domain features, so the robot must be prepared to adapt its world model when needed. This fact in sewers necessitates regular inspection in the first place: If everything of interest in a sewer were known, then there would be no reason to inspect. Not all of the *a priori* unknown sewer features directly affect robot control: Overlooking a leaking joint is an inspection fault, but does the robot no harm. Some features painfully matter for control, though. The robot *must* know an encrustation to avoid getting stuck in the pipe. It *must* know a lateral sticking into the main to avoid damaging its sensors. It *must* know a dislocation of the pipe bottom to avoid crashing down.

Second, the world is only partially observable, i.e., sensor data may be ambiguous or noisy and cannot be assumed to be correctly interpretable for deciding about and guiding action. Again, this is the case both for the very inspection part of sensing and for the sensing necessary to warrant the robot's integrity and orientation. (The sensor configurations may overlap that get used for these two observation tasks). The usual sensoric problems that exist even for lab robots are intensified for sewer robots by difficult and differing lighting conditions, dampness, and differences in pipe materials; drips and dirt may splash on and adhere to the sensors; and the whole system, including the sensors, must be sealed water-tight.

Third, actions can fail, i.e., they cannot be executed with guaranteed precision and success. Whoever walked a sewer knows it is impossible to position a robot precisely on the slippery ground. Stopping it does not mean it will stand still: It may drift towards the pipe bottom and along the pipe fall, pushed by the sewage current. The exact surface within the sewer is not known in advance: pipes that used to be circular when built years ago may have changed by encrustation, wear-out or damage; the exact geometry is downright unknown at shaft bottoms and at joining laterals. In sum, motion (and inertia, for that matter) has to be controlled based on sensory feedback, which can safely be assumed to be unreliable.

A sewer has obviously much structure that can be exploited for sewer robot control: After all, the robot works in a pipe system; it must be prepared to encounter many, but not arbitrary world features; it has a limited set of tasks to do; no humans physically interfere. So we need not solve all the problems of AI first to be successful in sewer robotics. Getting first experiences with sewer robot control is even possible in lab pipe networks. For instance, we have used a dry, non-operational concrete sewer pipe network (Fig. 3) to that end; using small indoor PVC pipe networks is a cheap and handy alternative.

On the other hand, the combination of incompleteness of knowledge, partial observability, and action imprecision challenges AI at heart. Fields like uncertainty handling, perception, planning, navigation, and various facets of machine learning have to contribute to sewer robot control. If the long-range vision of sewer robotics includes the background information infrastructure of off-line mission planning for sewer robots and of extracting knowledge from inspection data and from various other potentially relevant information sources available in a municipal administration, then reasoning, decision-making, and data-mining, are also involved.

In the rest of this section, we give some detail on the AI view of problems that are unavoidable in sewer robot control, even in the relatively conservative view of using such robots for substituting current technology. We also briefly state approaches to attacking these problems.

A brief remark before starting. Focusing on the AI challenges in sewer robotics does not imply everything else about such robots is easy—it is not! The trouble starts with designing and building a suitable mechanics that is small, robust, water-proof, dirt-resistant, flexible, efficient, affordable, what have you. Required engineering efforts include developing new, smaller, and better sensors with automatic cleaning devices or dirt rejecting covers, lighter batteries with higher capacities, as well as robust, long-life drives and transmissions. The non-AI challenge continues with designing the appropriate controller, computer and sensor hardware and operating systems. The final frontier is social: Convince the sewer maintenance people (and their standardization committees) to let a robot work on its own and believe what it finds. The point is: not AI alone is needed to make sewer robots practical—but it won't work without.

Motion control

To control motion sensibly—ultimately, to apply the right voltage to the right motor at the right time—a robot needs a kinematic model of itself in its current environment, how detailed or crude, how explicit or implicit that model may be. If the robot in question is a manipulator arm, say, in a car factory, the model is normally programmed or taught-in, and the robot kinematics (motor precision, stiffness) as well as the environment are designed such that the model fits. That does not work for sewer robots. Let us give some detail about two independent, but related points: Motion imprecision and sewer robot kinematics. We have found learning techniques helpful for attacking both problems, and we here point at the respective learning approaches used in our own sewer robotics studies.

Figure 3: *The LAOKOON-net, a dry sewer net not in operation, constructed by rhenag, at the GMD site in Sankt Augustin, Germany.*

Motion imprecision. The sewer has to be accepted as the environment of a sewer robot as is, with all its slip and sludge. Two motor commands that are identical in terms of forces and timing are likely to produce different motions at different places. This is the same as in office navigation robots, just more of it. In addition, robots in pipes are in permanent danger of toppling over, so permanent tilt control is mandatory.

As an example, consider KURT (Kirchner & Hertzberg 1997), a six-wheeled, shoe box sized sewer robot test platform built to study sewer robotics problems in a dry sewer—the LAOKOON-net (Fig. 3). Owing to its flexible wheel control, KURT is able to turn on the spot on flat terrain. Turning in a manhole in a sewer for entering a lateral, however, is more intricate. Due to the pipe geometry, the bottom of a manhole is dented. Turning in this dent, KURT does not always touch the ground with all its wheels: Only the front and back wheels would have permanent contact during a turn. Occasionally, the middle wheels also touch ground, causing unpredictably varying traction and making useless every practical model of rotational forces and timings.

In our study, we have solved this problem in the following way. The current junction is first classified as an X-, T-, or L-junction, using data of an ultrasound scan. (The result of this classification is also used for determining the current location mission control—see below.) For each of these classes, a sequence of subgoals was defined manually for each turning maneuver (right, straight, left), respecting the junction class geometry and KURT's sensor configuration. For example, turning right in an X-junction means to advance into the manhole until the open pipe is straight to the right; then turn right until the pipe is seen by the front-right sensor; then keep turning until it is seen by the front sensor; then check whether the junction can be correctly classified in KURT's new orientation; and then enter the pipe. While turning, the ridges at the bottom of the manhole are used as additional help for localization. The motion during each of the turning phases is best described as setting-back-and-forth-while-turning.

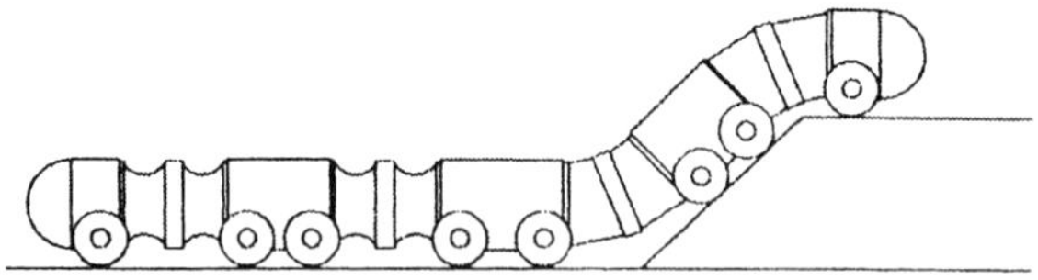

Figure 4: *Sketch of a multi-segment, articulated sewer robot (Berns* et al. *1996).*

Sewer robot kinematics. For sewer robotics, the kinematics of the robot itself is another challenge. Remember that pipes to inspect go down to 15 cm in diameter. Remember further that sewer robots have to climb small steps and turn at bends. Consider components necessary for such a robot: Motors, gears, axles; batteries; data buses, computer, storage, and controller hardware; a variety of sensors; data transmission components; and a housing around all that. Then chances are high that a rigid robot frame containing it all does not fit into the small pipes or is not maneuvrable enough to turn or climb.

An obvious solution is to make the frame flexible horizontally and vertically. An example design is sketched in Fig. 4, where the idea is to have rigid containers for the necessary components and couple them by movable joints of three degrees of freedom (pan, tilt, rotate). Depending on the geometry (number of containers, container and wheel diameter, ground clearance, container length, turning angles of joints), a robot with such a kinematic design can obviously provide the required maneuvrability, space, and small diameter.

This comes at a cost. Assume we have five segments. Each segment has its own propulsion on either side, making 22 DOF for the whole robot: 5×2 for propulsion, plus 4×3 for the segments. It would be a nuisance to formulate a kinematic model for a 22 DOF machine. Worse, if this model should be worth anything, then all mechanic parts would have to be so precise and stiff that they would probably weigh and/or cost too much for a practical sewer robot. Finally, whenever the robot slips, tilts, or twists on uneven and slippery pipe surface, forget any precise analytic model whatsoever. Other sewer robot designs than the one in Fig. 4 are possible, of course; the required maneuvrability together with the space constraints, however, necessitates a great number of DOF in any design.

Our studies suggest that a closed-loop, hierarchical version of Q-Learning—HQL (Kirchner 1997)—allows the problem to be tackled. We have done experiments using a 16 DOF, six-legged walking machine. The motion pattern to be learned in this case was stable walking towards a target, but the experiments seem to indicate that the learning results can also be applied to multi-DOF sewer robot motion coordination. The general idea is to define basic motion behaviors in terms of which the complex overall motion is then hierarchically assembled. In the walking machine, for instance, such basics are the swing and stance motions of individual legs. The hierarchical organization of the original, complex motion allows the original learning space to be reduced to a set of relatively handy sub-spaces—the original space is enormous, consisting of all possible time sequences of motor commands along the multitude of DOF.

Mission control

On a higher level, planning and plan execution monitoring are relevant AI techniques for sewer robotics. Given an initial position and a goal position on a digitized sewer map, the first task is to find a suitable path. This is graph search in the map, where the graph is typically close to a tree for real sewers, and often a desired path is even explicitly given, namely, the inspection route. If there is a challenge in sewer robotics, it is not in path planning.

Physical navigation along a given path can be more tricky. We assume it is not economic to plaster sewers with beacons and keep them clean and readable. As a fact, signals of the Global Positioning System (GPS) are inaccessible in sewers. In consequence, a sewer robot has to rely on on-board equipment to determine its position with respect to the map. No general statement is possible about how difficult this is in practice. Odometry is lost in a slippery sewer pipe. However, given an ideal map (which is correct, complete, and states the joining angles of laterals) and given a method to measure reliably turning angles of the robot (such as by using gyrometers or summing over the internal tilt angles in a multi-segment design as in Fig. 4), then navigation according to existing landmarks, such as manholes or joining laterals, is straightforward.

Now consider as an example the current sewer mapping situation in German municipalities. Traditionally, the only information available is a construction drawing of considerable age, typically coming on large rolls of paper, with drawings of more recent additions, coming on more large rolls of paper. Existing construction features are not sure to be in the plans: Sometimes features are changed yet not reported in the construction process, and illicit additions occur. As mentioned earlier, a number of cities have started to build up sewer information systems, which include digitized maps. The map information is provided by preparing the old drawings manually or by feeding in available information from recent state-of-the-art inspection protocols. In any case, currently existing sewer maps cannot in general be assumed to be correct and complete. Add to that possible inaccuracies in motion control (see above) and feature detection (see below), and there lurks a navigation problem.

Our studies in sewer robotics include work on navigation of KURT in the LAOKOON-net (Hertzberg & Kirchner 1996). The basic idea is to represent possible action and feature classification uncertainty in terms of partially observable Markov decision processes as has been done for office navigation robots (Simmons & Koenig 1995). In a sewer, the situation is simpler than in office buildings in that pipes prevent drifting past landmarks: the robot is sure to pass every landmark that is on its way. Moreover, manholes are that different from pipes in terms of sensor input that they can be assumed to be detected with certainty (but may be mis-classified). We have not attempted to correct the sewer map in this line of work, but allowed for relatively probable turning errors and landmark classification errors. Using more adequate sensors for turning control seems to give a chance to handle map flaws.

The visionary, long-term sewer robotics scenario would include many more mission control tasks that require AI technology. The first set consists of off-line mission planning tasks. Consider the idea of generating inspection plans or flushing plans for the immediate future automatically from all available sewer data, such as time of construction, materials, topology, and results of past inspections: an obvious application field for knowledge-based systems.

Yet more data is potentially relevant for sewer maintenance than is available today in sewer information systems. All that is known elsewhere in a municipal administration about recent construction work (water and energy supply, telecommunication, road-building) about types of industry located in different areas, about geologic conditions, even about vegetation (certain types of trees close to sewer pipes), can give hints about where to inspect or flush next. The challenge here is to join these data that vary in source, format, date, precision, and quality, to help in decision-making for sewer maintenance. Once it is possible to combine these various information sources, data-mining technology lends itself for extracting hidden regularities.

A second set of mission-control tasks in the long-term scenario comes from the robot autonomy that this scenario requires. There is no limit for decision-making, perception, learning, and cooperation methods that would help autonomous sewer robots organize their work well.

Feature detection

Finally, there is the problem of correctly interpreting sensor data of the sewer, or sewer *perception,* in AI terms. This subject spans a wide range of applications and problems, already sketched above: navigating, detecting damages, analyzing chemical substances, characterizing pipe state, identifying other robots, and measuring sewers.

General constraints on sensors for sewer robots. A number of constraints apply to the types and numbers of sensors that may be selected for the above tasks. Remember that a sewer robot has to fit into the cylindrical sewer pipes with their mostly small diameter. A reasonable place to mount the sensors for perceiving the environment are the two ends of the robot's housing. A fully autonomous robot, with its battery of limited capacity and its narrow housing, thus can be equipped only with a rather small number of sensors, compared to a large office floor robot, where the sensors may be distributed over almost the entire housing.

The sensors should be minimized in weight, size, energy consumption, and computing power needed to process their data. For instance, current off-the-shelf laser scanners are too large and too heavy to be mounted on the multi-segment robot of Fig. 4. There would be little or no space left for other sensors, and it would make the end segments of the robot too heavy to be lifted.

A camera can only be used if the sewer is lit up. Light consumes much more power than, e.g., an ultrasound sensor, and thus reduces the possible maximum range of the robot and its operation time, if used extensively. Depending on the algorithms used, continuous video image processing may also consume a considerable amount of computing power, which might be needed for other purposes, e.g., the complex motion control.

However, video cameras are the type of sensor which sewer operation people are used to, and which yield the richest data of all existing sensor types. It makes sense to use this wealth of data for specific tasks, which can hardly be solved using data of other sensors, like simple ranging devices. For instance, the robot test platform KURT classifies different types of manholes properly with high reliability, but it cannot distinguish individual manholes of the same type. If it were possible to identify individual manholes on the basis of image data, then position uncertainty could be significantly reduced.

Sensors for navigation. Cameras with a high light sensitivity are now available, allowing to reduce drastically the energy needs for lighting purposes as long as only landmark identification is concerned (damage detection might require different resolution and lighting). Moreover, if vision is limited to processing of still video images, then the power consumption for lighting gets uncritical, and so does computing time consumption. In consequence, navigation can rely on image analysis as a source of information. More intricate cases can also be handled: For instance, if ranging sensors detect an obstacle in a pipe, then vision may be needed to determine whether this obstacle can be crossed safely.

Above, landmark classification in KURT has already been reported as determining the type (X,T, or L) of the joins in a manhole. This shows that unexpensive sensing techniques can also successfully be used for sewer navigation: In fact, KURT uses only inclinometers and ultrasound sensors for this task. However, ultrasound sensors may fail in a real sewer as different pipe materials reflect ultrasound differently and thus may yield unreliable data in some cases.

Under the difficult sensor conditions in a sewer, more than one environment sensor type should be used at any time, i.e., the unescapable uncertainty should be countered by redundancy. It seems useful to determine the type of pipe material in order to decide which sensors to use best in the current pipe. To achieve this, sophisticated methods for texture analysis of still video images of the top—other parts may be too soiled—of the surrounding pipe are required. (Knowing the pipe material, by the way, may yield additional position information as it should be included in sewer maps.) In sum, sensor fusion is crucially needed.

Sensors for automatic generation of sewer state descriptions. Of course, navigation is only a means towards an end, namely, detection of faulty sewer conditions, which requires damage detection, sewer measurement, and sewage analysis. On the engineering side, for all of these applications either new sensor types have to be developed (damage detection, sewage analysis), or known sensor types have to be modified in order to make them suitable for use in an autonomous robot (sewer measurement, sewage analysis).

The complexity of sensor data processing for the applications varies. While a simple diagnosis system might suffice to process the data of a few electrochemical sensors for sewage analysis, damage detection and sewer measurement may be arbitrarily complicated. If vision is employed for extracting exact metrical information from images, then automatic offline camera calibration (Faugeras 1993) is a challenging problem that has to be solved. Other than that, the methods and techniques for processing the sensor data depend on the technical sensors, which are yet to be developed—so nothing concrete can be said here.

Existing Work in Sewer Robotics

As we have shown, sewer robotics poses a great variety of hard AI problems, many of which are known from other service robotics applications and are special in sewer robotics. The problem areas addressed above include robot architectures, robot control, robot navigation, sensor fusion, sensor interpretation, landmark detection, mapping, learning, and complex motion control; a more visionary sewer robotics scenario will definitely bring action planning, cooperation, and data mining into play. Given that we feel we have understood the relevant application problems fairly well but have only first approaches to their solutions, presenting potentially relevant AI technology comprehensively would amount to giving an overview about much of AI in general, which is obviously out of scope here.

As another relevant line of research, there is work that aims at creating more "intelligent" sewer inspection vehicles. Completed projects include KARO (Kuntze *et al.* 1995) and PIRAT (Campbell, Rogers, & Gilbert 1995). Both projects were aimed at producing "smart" vehicles—equipped with several different sensor devices—which are connected to a mobile control and surveillance unit by a cable. The research focus of both projects was developing methods for the automatic interpretation of sensor data to discover pipe damages mechanically and thus help automatize sewer inspection.

KARO employed fuzzy logics to fuse and interpret data from different types of sensors. In the PIRAT project, an expert system classified discovered pipe damages. In both systems, the main control routines were not running on on-board hardware, but on a computer in the mobile control unit. The KARO platform has been described by its creators as a still experimental prototype, whereas the makers of PIRAT report to have employed their system for inspecting five kilometers of sewers in Melbourne, Australia.

Some other projects related to sewer inspection are underway. Eiswirth *et al.* (1995) describe preliminary work for a project going into the KARO and PIRAT direction. Clarke (1995) uses optical triangulation for measuring sewer pipe profiles automatically, aiming at a detection of noticable pipe profile deformations.

The only work of which we are aware that deals explicitly with designing autonomous sewer robots in the sense of this paper is the one in the context or succession of the above-mentioned feasibility study LAOKOON (Berns *et al.* 1996). In particular, Kirchner & Hertzberg (1997) describe the sewer robot test platform KURT; Hertzberg & Kirchner (1996) give details about KURT's navigation under uncertainty; Cordes *et al.* (1997) report the concept of the multi-segment sewer robot platform; and Kirchner (1997) presents results about learning motion coordination.

Conclusion

Let us conclude the paper by summing up the line of argumentation. Sewers are an important and expensive part of communal infrastructure. Maintaining them requires considerable effort; in some countries, this effort is even required by law. The current technology for inspecting inaccessible sewer pipes causes relatively high cost, which is in the end paid by the citizens and industry of the communities. Running autonomous sewer robots for the task promises to reduce the inspection cost to about a third of the cost for the conventional procedure. AI technology is crucially needed for controlling a sewer robot, as sewer robotics involves a particular combination of the imprecision, lack of knowledge, and resource constraints that are typical for so many real world service robot applications. On the other hand, the problem is such that using current AI technology promises to yield usable results in reasonable time.

Sewer robotics is a typical instance of service robotics. So, in terms of AI technology, advanced methods in robot architectures, robot control, robot navigation, sensor fusion, sensor interpretation, landmark detection, mapping, learning, and complex motion control are required right away; more AI fields can contribute to solving problems in detail or to tackling the general vision of long-term sewer maintenance work by autonomous robots. Sewer robotics has a potential both for exploiting available AI technology and for launching interesting AI research problems. As in every instance of service robotics, the AI technology in use has to be integrated with a wealth of non-AI technologies, where sewer robotics poses interesting mechatronics and sensor problems as well.

Yet sewers have not been one of the main problem domains of AI. The reason is simple. Sewers are not easily accessible, sewer operations people dislike strangers playing around in the pipes, sewers are not the most convenient experimentation ground, robot experiments in sewers cost real money, they require to be real certain about the operation of the robots—and finding practical solutions for real problems is hard. Judging from our own domain experience within the last four years, however, the field is worth some effort: An interesting and rewarding area for applying AI technology and doing AI research opens up, once you get serious about doing sewer robotics.

Acknowledgements

This work has been supported by the German Federal Ministry of Education, Research, and Technology (BMBF) in the joint feasibility study LAOKOON (523-4001-01IW501 A) and the joint project MAKRO (02-WK9702/4), respectively, project partners being rhenag, FZI and Hitzel. This work is also partly funded by the European Commission's TMR project VIRGO under grant number ERBFMRXCT960049. This support and cooperation is gratefully acknowledged.

References

ATV (German Association for the Water Environment). 1991. *Inspection, Repair, Rehabilitation and Replacement of Sewers and Drains. Part 2: Optical Inspection (Advisory leaflet M 143, Part 2).*

Berns, K.; Cordes, S.; Eberl, M.; Grimm, H.-J.; Hertzberg, J.; Ilg, W.; Kemmann, M.; Kirchner, F.; Lagermann, B.; Licht, U.; Müller, B.; Prchal, P.; di Primio, F.; Rome, E.; Stapelfeldt, H.; and Suna, R. 1996. *BMBF-Machbarkeitsuntersuchung LAOKOON: Abschlußbericht.* rhenag, FZI, GMD. Final report to BMBF, in German.

Berns, K.; Christaller, T.; Dillmann, R.; Hertzberg, J.; Ilg, W.; Kemmann, M.; Rome, E.; and Stapelfeldt, H. 1997. LAOKOON – lernfähige autonome kooperierende Kanalroboter. *KI* 11(2):28–32.

Bölke, K.-P. 1996. *Kanalinspektion. Schäden erkennen und dokumentieren.* Berlin: Springer.

Campbell, G.; Rogers, K.; and Gilbert, J. 1995. PIRAT - A system for quantitative sewer assessment. In *Proceedings of the 12th International No-Dig Conference*, 455–462. Hamburg: Messe und Congress GmbH.

CEN (European Committee for Standardization). 1997. *Drain and sewer systems outside buildings –*

Part 7: Maintenance and operations. Final draft, Nov. 1997.

Clarke, T. 1995. The development of an optical triangulation pipe profiling instrument. In *Optical 3-D Measurement Techniques III – Applications in inspection, quality control and robotics*, 331–340. Karlsruhe: Wichmann.

Cordes, S.; Berns, K.; Eberl, M.; Ilg, W.; and Suna, R. 1997. Autonomous sewer inspection with a wheeled, multiarticulated robot. *Robotics and Autonomous Systems* 21:123–135.

Eiswirth, M.; Hötzl, H.; Kramp, J.; Lazar, C.; and Merkler, G.-P. 1995. Qualifying methods for leakage detection. In *Proceedings of the 12th International No-Dig Conference*, 195–208. Hamburg: Messe und Congress GmbH.

Faugeras, O. 1993. *Three-Dimensional Computer Vision*. Cambridge, MA: MIT Press.

Fukui, T. 1994. Current status of nondigging techniques applied to sewerage work in Japan. In *4th International Conference on Pipeline Construction*, volume Japanese Forum, 1–14. Hamburg, Germany, October 16 – 20, 1994.

Hertzberg, J., and Kirchner, F. 1996. Landmark-based autonomous navigation in sewerage pipes. In *Proceedings of the First Euromicro Workshop on Advanced Mobile Robots (EUROBOT '96)*, 68–73. Los Alamitos, CA: IEEE. (Kaiserslautern, Germany, October 9–11, 1996).

Keding, M.; van Riesen, S.; and Esch, B. 1990. Der Zustand der öffentlichen Kanalisation in der Bundesrepublik Deutschland. Ergebnisse der ATV-Umfrage 1990. *Korrespondenz Abwasser* 37:1148–1153.

Kirchner, F., and Hertzberg, J. 1997. A prototype study of an autonomous robot platform for sewerage system maintenance. *Journal of Autonomous Robots* 4(4):319–331.

Kirchner, F. 1997. Q-Learning of complex behaviours on a six-legged walking machine. In *Proceedings of the Second Euromicro Workshop on Advanced Mobile Robots (EUROBOT '97)*, 51–59. Los Alamitos, CA: IEEE. (Brescia, Italy, October 22–24, 1997).

Kuntze, H.; Schmidt, D.; Haffner, H.; and Loh, M. 1995. KARO - A flexible robot for smart sensor-based sewer inspection. In *Proceedings of the 12th International No-Dig conference*, 367–374. Hamburg: Messe und Congress GmbH.

Simmons, R., and Koenig, S. 1995. Probabilistic robot navigation in partially observable environments. In *Proceedings of the International Joint Conference on Artificial Intelligence (IJCAI '95)*, 1080–1087. San Mateo, CA: Morgan Kaufmann. (Montréal, Canada, August 20–25, 1995).

Stein, D., and Niederehe, W. 1992. *Instandhaltung von Kanalisationen*. Berlin: Ernst & Sohn, 2nd edition.

Learning Distributed Strategies for Traffic Control

David E. Moriarty, Simon Handley, and Pat Langley
Daimler-Benz Research and Technology Center
1510 Page Mill Road, Palo Alto, CA 94304
{moriarty,handley,langley}@rtna.daimlerbenz.com

Abstract

In this paper, we cast the problem of managing traffic flow in terms of a distributed collection of independent agents that adapt to their environment. We describe an evolutionary algorithm that learns strategies for lane selection, using local information, on a simulated highway that contains hundreds of agents. Experimental studies suggest that the learned controllers lead to better traffic flow than ones constructed manually, and that the learned controllers are robust with respect to to blocked lanes and changes in the number of lanes on the highway.

1. Introduction

In recent years, there has been growing interest in the distributed behavior of large populations of independent agents, and in the ability of such agents to adapt not only to their environment but to each other's behavior. Most research in this area has focused on agent populations designed to mimic those that occur in the natural world, such as colonies of ants and schools of fish. However, the artificial urban environment created by humans contains another important example of distributed agent behavior: automobile traffic on roads and highways.

The traffic domain has much to recommend it as a fertile source of research problems. Clearly, each agent (a driver and his vehicle) has independent control of its actions, but its behavior must take into account physical constraints, such as staying on the road and avoiding collisions, and the behavior of many other agents. Another advantage is that there exist clear criteria for evaluation, such as maximizing traffic flow and minimizing lane changes. In addition, the domain supports complex behaviors of the overall system, such as traffic jams, even though the individual agents are relatively simple. Also, most researchers have personal experience in traffic environments and, presumably, have good intuitions about reasonable strategies and behaviors. Finally, progress in this area could lead to improvements in actual traffic conditions and thus increase the quality of life for drivers.

In this paper, we describe research on distributed adaptation in the traffic domain. Following Moriarty and Langley (1998), we formulate the problem of traffic management from a distributed, car-centered perspective and the task of improving this process in terms of distributed machine learning. We assume that each agent receives local information about the vehicles that immediately surround it, including their location and speed, and that the agent determines which lane to drive in but not the vehicle's speed. The aim of learning, and thus our measure of performance, is not the behavior of individual traffic agents but rather the behavior of the traffic system as a whole. To this end, we want the learning module to develop a control strategy for lane selection that considers not only the maintenance of each car's desired speed, but that takes into account how its selection will affect speeds of other cars. For instance, we would like the cars to organize themselves into a cooperative system that lets the fast drivers pass through, while still letting the slow drivers maintain their speeds.

We begin by presenting our formulation of the traffic control task, and its associated learning problem, in more detail. We next describe the inputs and outputs of our reactive vehicle controllers, along with our genetic approach to learning distributed control strategies. After this, we characterize our simulated traffic environment and report experimental studies of the system under a variety of conditions. Elsewhere (Moriarty & Langley, 1998), we showed that our approach learns controllers that are robust to changes in the proportion of learned to 'selfish' cars on the highway and to changes in traffic density. Here we focus instead on the system's ability to generalize to situations that involve blocked lanes and different numbers of lanes from those used in training. We close the paper with a discussion of related work on distributed learning and our plans for future research.

2. Distributed Traffic Control

Our approach to traffic management involves a reformulation of the problem into a distributed artificial intelligence task, in which cars coordinate lane changes to maintain desired speeds and reduce total lane maneuvers. We make no assumptions about the level of automation of the cars. Lane selection information could be provided to the driver, who completes the maneuver, or to a regulation controller in an automated car (Pomerleau, 1995; Varaiya, 1993). Here we consider in more detail our formulation of the performance and learning tasks.

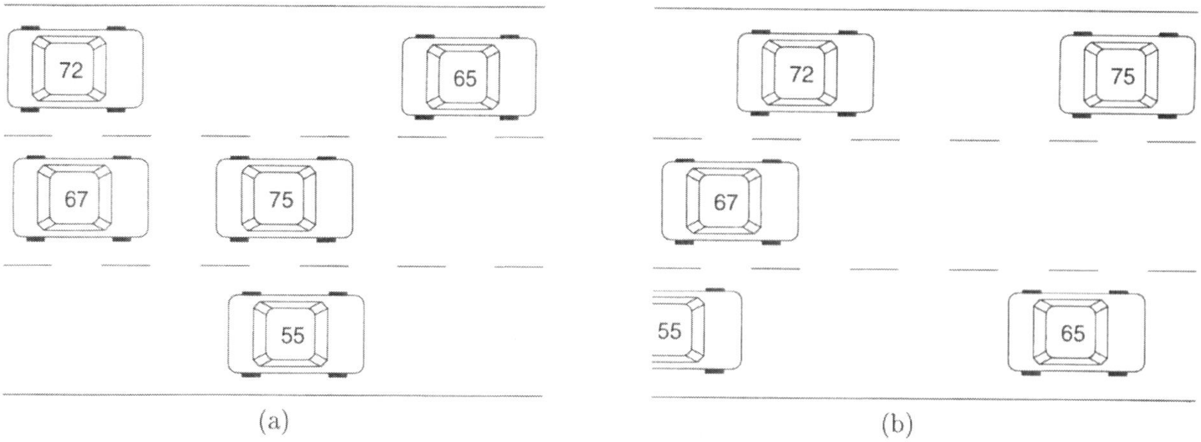

Figure 1 (a) An example traffic situation in which the traffic flows from left to right and the number on each car shows the car's speed. (b) Traffic after reorganization in which car 75 and 65 swap lanes, followed by another lane change by car 65, so that all cars can maintain their desired speeds.

2.1 Definition of the Problem

Most work on advanced traffic management views cars as tokens that follow simple, selfish rules of behavior. These management systems affect the flow of the car tokens by controlling external, fixed-position devices such as traffic signals, ramp meters, speed limits, and dynamic lanes. Surprisingly, little research has addressed how the cars themselves might sense and intelligently affect traffic dynamics.[1]

Our view is that cars are not blind tokens, but rather can sense their environment and act cooperatively to achieve desired global behavior. More specifically, cars can learn to organize themselves by traffic lanes to increase overall traffic throughput, reduce the average number of lane changes, and maintain the desired speeds of drivers. Car-centered control, specifically lane selection, should therefore complement existing traffic management efforts by providing better behavior between traffic signals.

Figure 1(a) illustrates a situation in which lane coordination is beneficial. The figure shows five cars along with their speeds, which we will use as identifiers. Car 72 is quickly approaching car 65 and will be unable to pass because of the position of car 67. Without reorganization, car 65 forces car 72 to reduce its speed and wait for car 67 to pass car 65, which will decrease traffic throughput and car 72's satisfaction. An efficient solution to this problem is for car 75 and car 65 to immediately swap lanes, followed by car 65 moving into the bottom lane, as shown in Figure 1(b). This maneuver ensures that no speeds are reduced and no throughput is lost.

We recast the traffic management task as a problem in distributed artificial intelligence, where each car represents an individual agent in a multi-agent system. Cars act on their world (the highway) by selecting appropriate lanes to drive in. They interact with other cars by competing for resources (the spaces or slots on the highway). Each action is local in nature, and may not produce any noticeable benefit to the car. Collectively, however, the local actions can improve the global performance of the traffic. For example, yielding a lane to a faster car does not produce any local benefit to the slower car, but does increase the overall traffic throughput and let the passing car maintain its desired speed.

Global traffic performance could be defined in many different ways. Governments want high traffic throughput, whereas drivers want to maintain desired speeds with few lane changes. We selected the driver-oriented metric, since drivers are likely to be the harshest critics of cooperative driving. The performance measure P for a set of cars C contains two terms, one that penalizes deviations in speed and one that penalizes lane changes,

$$P(C) = \frac{\sum_{t=1}^{T}\sum_{i=1}^{N}(S_{it}^{a} - S_{it}^{d})^2}{TN} + 4 \times 60 \times \frac{\sum_{i=1}^{N} L_i}{TN}, \quad (1)$$

where T is the total number of time steps (in seconds), N is the number of cars, S_{it}^{d} is the desired speed of car i at time t, S_{it}^{a} is the actual speed of car i at time t, and L_i is the total number of lane changes for car i over T time steps. The first constant, 4, is a weighting factor, whereas the second, 60, converts the lane changes per second into lane changes per minute. The goal is to minimize the difference between actual speeds and desired speeds, modulated by the number of lane changes, averaged over several time steps and over all learned cars on the road. Each speed difference is squared to penalize extreme behavior. For example, driving 60 m/h 90% of the time and 10 m/h 10% of the time gives an average of 55 m/h but is clearly less desirable than driving 56 m/h 50% and 54 m/h 50% of the time, which gives the same

[1] One exception is the work of Carrara and Morello in the DOMINC project.

average. Squaring the error from desired speed gives a higher evaluation to the more consistent strategy.

The problem is thus to find a lane-changing strategy that minimizes equation 1. A naive strategy for each car, which most traffic management systems assume, is to select the lane that lets it most consistently achieve its desired speed and only change lanes if a slower car is encountered. The disadvantage of such a strategy is that it does not take into account the global criteria of traffic performance. A slow car should not drive in the "fast" lane simply because it can maintain its desired speed. We will refer to cars that employ the naive strategy as *selfish*, since they maximize the local performance of their respective car. We are interested in *smart* strategies that maximize the aggregate performance of traffic through cooperative lane-selection strategies.

2.2 Communication and Coordination

The previous section defined the problem of car-centered traffic management, but left open some important issues about the level of communication between cars and the knowledge available about other cars' decisions and states. The multi-agent literature is often divided on these matters, and we feel that it is not central to the problem definition. Still, we should describe our assumptions about communication and state information.

We assume that cars have access to information on their own state, including knowledge of their current driving speed and the driver's desired speed. One could imagine a driver specifying desired speeds at the start of a trip, or the system could infer this information from the driver's historical behavior. We also assume that agents can perceive limited state information of surrounding cars, such as their relative speeds. The system could sense this information using radar or receive it directly from other cars via radio waves or the Internet. Agents should also sense which surrounding cars are cooperative and which are selfish. Again, the system could infer cooperation from driver behavior or direct communication.

Figure 2 illustrates the input for an agent in a specific traffic situation. The middle car receives as input its current speed, its desired speed, the relative speeds of surrounding traffic, and whether surrounding cars are cooperative or selfish. The range and granularity of the relative speed inputs could be adjusted to take into account both local traffic and distant traffic. For example, it may prove beneficial to receive not only relative speeds of individual cars in the immediate vicinity, but also relative speeds of groups of cars in farther ranges.

We assume that the controller's output consists of three options: stay in the current lane, change lanes to the left, or change lanes to the right. The output does not specify the best lane to drive in, but rather whether the lanes immediately to the left or immediately to the right are better than the current lane. This control provides flexibility, since it does not depend on the number of lanes or on knowledge of the current driving lane. Thus, controllers that learn on a three-lane highway should, at least in principle, generalize to greater or fewer lanes.

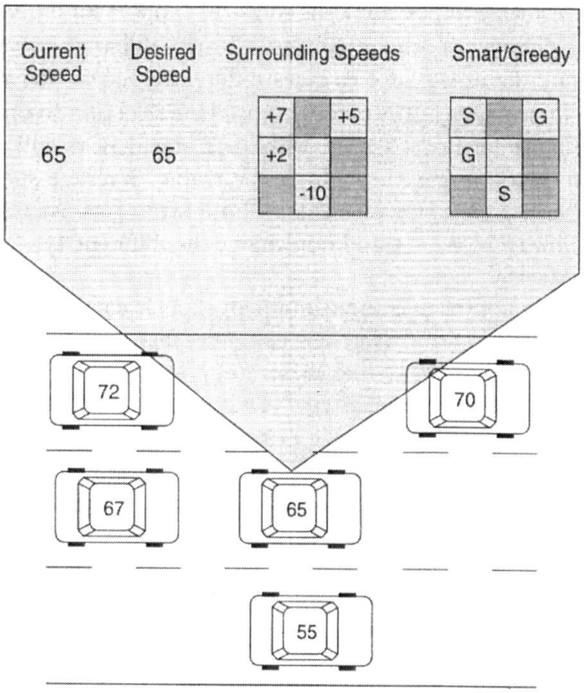

Figure 2 An illustration of the input to each agent. The shaded region shows the current input information for the middle car. The agent has access to its current speed, its desired speed, the relative speeds of surrounding traffic, and whether other cars are smart or selfish.

We assume that the controller's output represents a ranking of the three possible choices, with the highest ranked choice that is both valid and safe being selected as the car's next action. For a recommendation to be valid, there must be a lane available in the specified direction. For a recommendation to be safe, there must not be a car in the same longitudinal position in the new lane. It is always safe to remain in the current lane. The system could also incorporate other safety assurances, such as detecting whether a lane change produces an unsafe spacing between cars in the new lane. For example, one might specify that a slow car should not move in front of a fast car even if there is immediate space for it in the fast car's lane, since the fast car will likely close that space during the span of the lane change.

The higher-level safety and validation process relieves the controller of the overhead in deciding which lanes are safe and centers the control problem on lane selection. In other words, by removing the problem of validation and safety, the controller can focus on and more easily learn to rank lanes. This approach is analogous to separating the identification of legal moves from the selection of desirable moves in game playing.

Another important issue concerns support for individual differences among drivers. Clearly, different drivers should be able to select lanes differently. Slower drivers will normally (but not always) use lane selection to open up lanes for faster traffic, whereas faster drivers will select lanes to get through slower traffic. Average-speed drivers will employ elements of both strategies. At issue is how to represent and implement the different types of strategies.

One approach is to maintain an explicit control policy for each type of driver. For example, fast drivers would utilize a fast lane-selection strategy and slow drivers a slow lane-selection strategy. A disadvantage of this approach is that it requires *a priori* knowledge of the number of driver types and the boundaries that separate them. Also, it does not provide a smooth transition between styles of driving. A driver on a boundary would be forced into one of the two surrounding strategies instead of an interpolation between the two.

A better approach is to parameterize the driving style and use it as input to a single control policy. Each car would contain the same control policy, but since it receives driving style as input, it behaves differently for different types of drivers. In this case, driving style is simply the desired speed. No a priori decisions are necessary regarding the number of lane-selection strategies or their boundaries. Moreover, since the different strategies are keyed to a continuous input (desired speed), there can be smooth transitions and interpolations between different lane-selection strategies.

2.3 Approaches to Intelligent Lane Selection

Creating distributed lane-changing controllers by hand appears quite difficult. It is unclear whether experts exist in this domain and, even if they do, experts often find it difficult to verbalize complex control skills, which creates a *knowledge acquisition bottleneck.* Also, the innumerable traffic patterns and varying driving styles create a large problem space. Even with significant expert domain knowledge, hand crafting a controller that operates effectively in all areas of the problem space may not be feasible.

Another solution is to apply machine learning to develop intelligent controllers through direct experience with the domain. A learning algorithm would modify the controller based on good and bad experiences in the problem space. This approach frees us from the task of acquiring and encoding expert domain knowledge, since it discovers examples of good and bad decisions through direct experience. Moreover, the controllers are not necessarily fixed and could continue to learn and adapt with new experiences.

The lane-selection problem appears out of reach of the more standard, *supervised* machine learning methods (e.g., Quinlan, 1986; Rumelhart, Hinton, & Williams, 1986). In supervised learning, control policies are formed from examples of correct behavior. In the case of intelligent lane selection, supervised learning requires demonstrations of good and bad lane selections. Without expert domain knowledge, it is difficult to generate these examples. In some control problems, supervised learning is used to mimic the behavior of people (e.g., Pomerleau, 1992; Sammut, Hurst, Kedzier, & Michie, 1992). For intelligent lane selection, however, this is exactly what we do not want to model. We believe that most drivers do not select lanes intelligently, but are rather more selfish in nature. Thus, it seems misguided to use real driver behaviors as a basis for learning cooperative lane selection.

A more flexible machine learning approach that is capable of learning from general rewards instead of behavioral examples has been termed *reinforcement learning.* The rewards provide only a general measure of proficiency over the task and do not explicitly direct the learner toward any course of action. The learner adjusts its actions through trial and error interactions with the environment to maximize the reward signal. In the lane-selection problem, agents receive the rewards defined by equation 1 at specific time steps. In response, they adjust their lane-selection strategies using some reinforcement learning algorithm to maximize the reward function.

The literature includes two main types of approach to reinforcement learning. One class of methods learn through calculation of temporal differences (Sutton, 1988; Watkins & Dayan, 1992; Kaelbling, Littman, & Moore, 1996) over rewards, which lets them acquire mappings from state-action pairs onto expected values. Another class of methods search more directly through the space of control policies (Grefenstette, Ramsey, & Schultz, 1990; Holland & Reitman, 1978; Moriarty & Miikkulainen, 1996; Whitley, Dominic, Das, & Anderson, 1993; Wilson, 1994), often using evolutionary algorithms to this end. Our approach, to which we now turn, relies on an evolutionary algorithm as the primary mechanism for reinforcement learning, but it also incorporates a technique similar to temporal-difference learning to handle smaller strategy refinements.

2.4 Machine Learning for Lane Selection

Elsewhere (Moriarty & Langley, 1998), we have described our distributed learning system in detail, so here we only review its main components. The system represents its control knowledge as a feedforward neural network with one hidden layer, as depicted in Figure 3. This network includes 16 input units, 12 hidden units, and three output units, with full connections between adjacent levels. The input nodes correspond to information about the car's current and desired speeds, as well as the speeds of surrounding vehicles, whereas the output nodes specify whether to stay in the current lane, to move left, or to move right. On each time step, the controller uses the input values to compute activations for each output node, then selects the action with the highest activation.

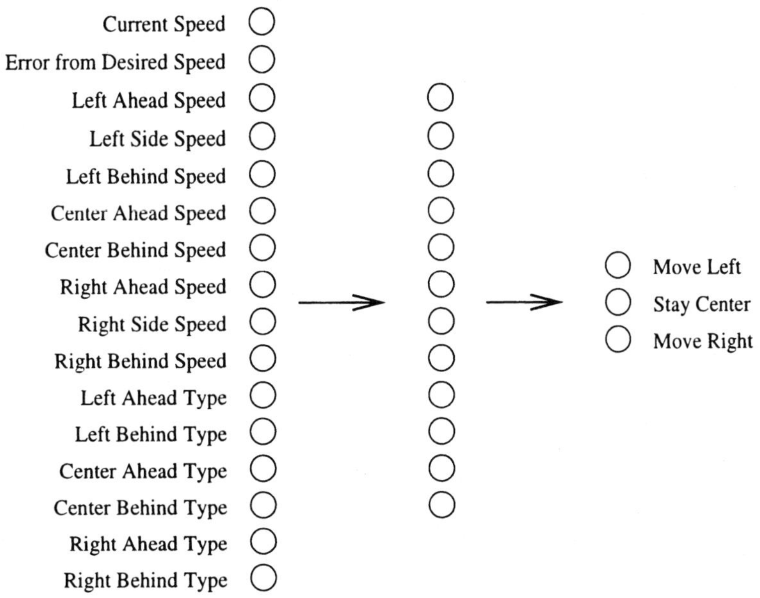

Figure 3 The input and outputs to the neural network for lane selection.

The learning system relies on three interrelated modules to determine the weights on the network's links, and thus to acquire robust controllers. The first component is SANE (Moriarty & Miikkulainen, 1996; Moriarty, 1997), which carries out genetic search through the space of feedforward networks, given a network architecture, by operating at two distinct levels. At one level, the module retains a population of complete controllers, each defined as a collection of hidden-layer neurons. The second-level population consists of these individual neurons, each of which specify the weights on their input and output links. Each member of the neuron population can appear in zero or more members of the controller population.

SANE assign fitness to each candidate controller by converting its encoding (stored as bit strings) into a complete neural network, then using that network in a simulated traffic environment for 400 seconds. The system determines the fitness of a given controller by applying equation 1 to traces of the traffic behavior. SANE also assigns fitness to each candidate neuron based on the fitnesses of the controllers to which it contributes. The algorithm selectively applies genetic operators, such as crossover, on members of each population to generate new members, repeating the evaluation and generation process many times.

The second learning module is responsible for seeding the initial populations. Rather that starting SANE with random populations of controllers and neurons, the system initializes them with candidates that are likely to be useful. The module accomplishes this feat by collecting behavioral traces of the hand-written 'polite' controller that we described earlier. These provide training cases that take the form of sensor-action pairs, which the system passes to a supervised algorithm (backpropagation) to learn weights that approximate the polite controller. The module repeats this process a number of times to generate a collection of controllers and neurons that form 25% of the two initial populations.

As noted above, the SANE module evaluates the fitness of candidate controllers by running them in the simulator for fixed time periods. Because the reward signals generated by this scheme are infrequent, we added a third learning module that relies on more immediate feedback. After every ten simulated seconds, this component checks to determine if the overall traffic performance has changed since the last measurement. If performance has improved substantially, it labels all actions taken during this period as desirable; if performance has worsened, it labels all invoked actions as undesirable. In either case, the module passes these actions (and their associated sensory inputs) to the backpropagation algorithm, which alters the controllers weights to either encourage or discourage their use.

Recall that the system's reward signal is based on equation 1, which assumes global information about overall traffic behavior. Clearly, such information is not currently available to actual cars or their drivers, but we predict this will change as vehicles come to include positioning devices and gain access to the Internet, which will let them report their position and speed to a central facility. For now, we have been forced to rely on a simulated traffic domain, which has also encouraged us to use an offline training regimen to collect accurate statistics over extended runs. However, the basic approach also lends itself to online learning, though we expect the learning rate would decrease in this scenario.

3. Experimental Evaluation

Our approach to distributed learning appears to offer a viable method for acquiring lane-selection strategies and thus improving overall traffic performance. However, whether the method works in practice is an empirical question, and in this section we report experimental studies of the system's adaptive behavior.

3.1 A Simulated Traffic Environment

To evaluate traffic management through intelligent lane selection, we developed a simulator to model traffic on a highway. For each car, the simulator updates the continuous values of position, velocity, and acceleration at one second intervals. The acceleration and deceleration functions were set by visualizing traffic performance under different conditions and represent our best estimate of the behavior of actual drivers. We adjust acceleration (A) using the equation $A(s) = 10s^{-0.5}$, where s represents the current speed in miles per hour (m/h).

Deceleration occurs at the rate of -2.0 m/h per second if the difference in speed from the immediate preceding car is greater than twice the number of seconds separating the two cars. In other words, if a car approaches a slower car, the deceleration point is proportional to the difference in speed and the distance between the cars. If there is a large difference in speed, cars will decelerate sooner than if the speed differences are small. If the gap closes to two seconds, the speed is matched instantaneously. The simulator allows lane changes only if the change maintains a two-second gap between leading and following cars.

The simulated roadway is 3.3 miles long, but the top of each lane "wraps around" toroidally to the bottom, creating an infinite stretch of roadway. We designed the simulator as a tool to efficiently evaluate different lane-selection strategies, and thus it makes several assumptions about traffic dynamics. The current model makes five primary assumptions:

- all cars are the same size;
- all cars use the same acceleration rules;
- cars accelerate to and maintain their desired speed if there are no slower cars directly ahead;
- lane changes are instantaneous; and
- there are no curves, hills, on ramps, or exit ramps.

Although none of these assumptions hold for real-world traffic, they do not appear crucial for evaluating the merits of intelligent lane selection, and removing them unnecessarily complicates the model. In future work, however, we hope to expand our experiments to more realistic simulators such as SmartPATH (Eskafi, 1996).

During training, the learning system uses the traffic simulator to evaluate candidate lane-selection strategies. Each evaluation or *trial* lasts 400 simulated seconds and begins with a random dispersement of 200 cars over three lanes on the 3.3 mile roadway. Desired speeds are selected randomly from a normal distribution with mean 60 m/h and standard deviation 8 m/h. In each trial, the percentage of smart cars is randomly selected from a uniform distribution with a minimum percentage of 5%. All other cars follow the selfish lane-selection strategy outlined in Section 2.1.

To simulate congestion caused by lane closures and merging, we blocked portions of either the far right or far left lanes during training. Lane closures last for one mile and only one closure exists at any given time. There is an equal probability that the far right or far left lane will be blocked. A lane-selection strategy perceives a blocked lane as a car with a speed of zero.

Each training run begins with a population of 75 random lane-selection strategies and 25 seeded strategies, which are modified by SANE and the local learning module over 30 simulated driving hours. SANE keeps track of the best strategy found so far based on its performance over a trial. When the system finds a better strategy, it is saved to a file for later testing. The saved strategies are each tested over ten 2000-second trials and the best is returned as the final strategy.

We developed two hand-written controllers for use as benchmarks and to provide the simulated environment. The *polite* controllers follow four rules:

- If your desired speed is 55 m/h or less and the right lane is open, then change lanes to the right.
- If you are in the left lane, a car behind you has a higher speed, and the right lane is open, then change lanes to the right.
- If a car in front of you has a slower current speed than your desired speed and the left lane is open, then change lanes to the left.
- In the previous situation, if the left lane was not open but the right lane is open, then change to the right.

We based these rules on our interpretation of the "slower traffic yield to the right" signs posted on the highways. The selfish strategy described earlier uses only the last two rules.

3.2 Evaluation of Intelligent Lane Selection

Our earlier studies (Moriarty & Langley, 1998) evaluated the learned controllers' behavior as we varied the density of traffic and the ratio of learned to selfish controllers. The results, which showed reasonable behavior over a wide range of densities and ratios, encouraged us to carry out additional additional studies to further test the adaptive nature of the learned controllers.

We designed our first experiment to evaluate the polite, selfish, and learned strategies in the presence of lane closures. Recall that, during training, we closed one mile of either the far left or far right lane and the location of the closure changed every 500 simulated seconds. We replicated lane closures in testing to determine each

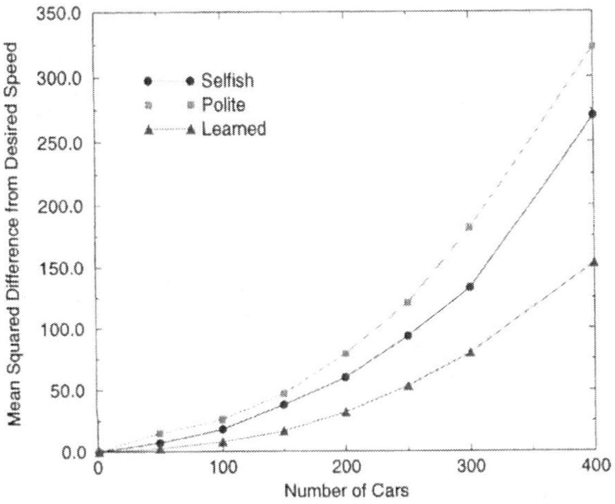

(a)

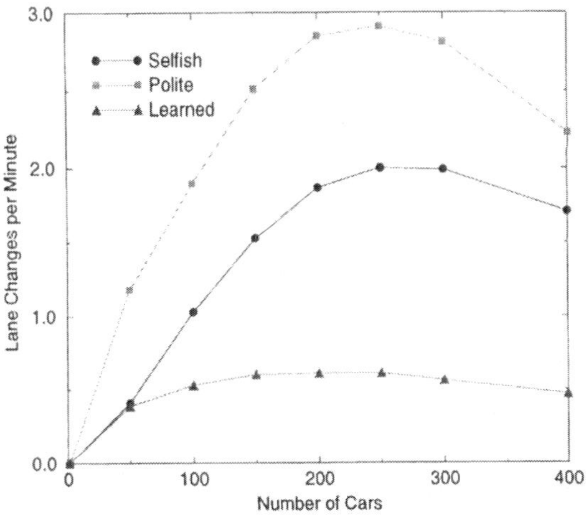

(b)

Figure 4 Traffic performance when portions of the lanes were blocked.

strategy's ability to handle high congestion areas created by the merging traffic. The degree of merging in these tests is extreme (a blocked lane every 13 miles), to fully test the robustness of the three strategies.

Figure 4 plots the mean squared error in desired speeds and the average number of lane changes with closed lanes. Surprisingly, the polite strategy performed worse than the selfish one when lanes were blocked, which differed from our earlier results with no closures. Figure 4(a) shows that, under a high degree of merging, it is better to act greedily than politely. The large errors that the polite strategy incurs come when portions of the rightmost lane are closed. Since the polite strategy directs all of its slow drivers into the right lane, it becomes difficult to merge them back into the two faster lanes when this lane is blocked. This difficulty causes large bottlenecks in the right lane and creates high errors in desired speed. Since the selfish strategy assigns no lane bias based on driving speed, it is less affected by right lane closures.

Although the learned strategy also directs its slower drivers to the right lane, its response to bottlenecks is even more robust than the selfish scheme. Under the learned strategy, faster drivers in the center and left lanes maneuver to let slower drivers merge more easily, which eases congestion. These seemingly altruistic behaviors were learned because reinforcement comes from the aggregate traffic performance. Additionally, the learned cars have relative speed sensors that can detect slow speeds in traffic ahead. Thus, the learned strategy can merge the cars much earlier than the polite strategy, which does not begin to merge until a slow car or closed lane forces it to decelerate. Overall, the learned strategy incurs substantially lower driving errors and performs only a fraction of the lane change maneuvers as the other two strategies.

The second experiment evaluated the learned strategy's ability to adapt to four-lane highways. As noted in Section 2.2, we designed the controller input to ignore the car's actual lane, and the output to reflect only whether the left or right lane is better than the current one. Thus, in principle an effective strategy learned only on a three-lane highway should perform well on four lanes. Since the learning system only experienced three-lane highways in training, this experiment serves as another test of the learned strategy's adaptability.

Figure 5 plots the error in driving speed and average number of lane changes using four lanes of traffic. Since there is more lane capacity, we need up to 600 cars in this study. The figure shows that the learned strategy achieves the same performance gains over the polite and selfish strategies in four lanes of traffic as it does in three lanes. In dense traffic, the learned strategy incurs one third to one quarter of the driving speed error for the selfish strategy and one half of the error for the polite strategy. As with three lanes of traffic, the polite and selfish strategies make substantially more lane change maneuvers than the learned controller.

Figure 6 provides a visualization of the lane utilization for the three strategies with four traffic lanes using 300 cars. This shows that the selfish strategy assigns no lane bias based on driving speed, whereas the polite strategy exhibits a sharp transition between slow and fast driving styles. The graph for the learned strategy is very similar to its three-lane counterpart, giving further evidence that it generalizes to more than three lanes.

With only three lanes, the learned strategy encumbers the right lane with many slow drivers and uses the other lanes to organize the faster drivers. Under four lanes of traffic, however, the fastest drivers are placed more consistently in the leftmost lane than under three lanes. For example, drivers with desired speeds of 80

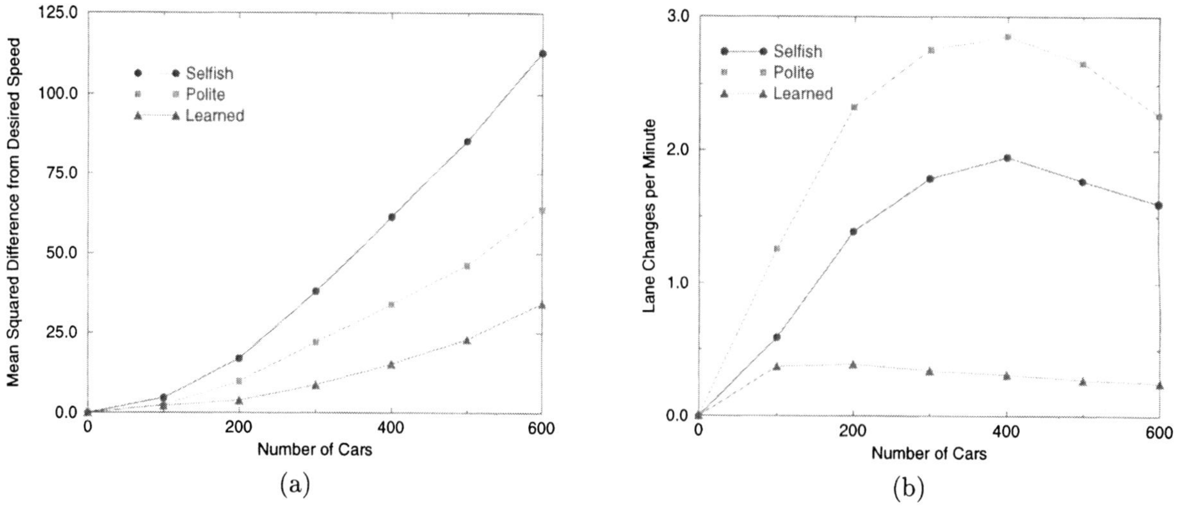

Figure 5 Traffic performance with four driving lanes; no lanes were blocked in this condition.

m/h drive in the left lane 83% of the time with four lanes of traffic, compared to only 57% of the time with three lanes. Most of the traffic organization occurs in the middle two lanes, with the middle-speed drivers. This strategy seems reasonable, since the middle-speed drivers make two different types of lane changes: passing slow cars and yielding to fast cars, and therefore must reorganize more frequently.

4. Related and Future Work

One can roughly divide research on communities of learning agents into two broad categories. The first, often called *multi-agent learning*, refers to situations in which the agents have shared goals and thus cooperate, either explicitly or implicitly, to achieve those goals. Examples of this approach include Schultz, Grefenstette, and Adams' (1996) work on multi-robot herding behavior, Matarić's (1994) efforts on foraging, in which four robots acquire social rules that reduce disruption, Tan's (1993) studies of reinforcement learning among predators cooperating to track down prey, and Sen and Sekaran's (1998) use of reinforcement learning to improve two-agent coordination in block pushing. Stone and Veloso (1997) present an extensive review of work on multi-agent learning, including their own results on soccer playing, so we will not try to be exhaustive here.

Another category focuses on situations that involve many agents, typically more than in multi-agent settings, each of which pursues its own goals. Research on such *distributed learning* seems less common than multi-agent work, but it also a bears a closer relation to our own approach. Perhaps the best-known effort of this sort revolves around Holland's (1996) ECHO, a simulation framework designed to study the behavior of complex biological systems, such as the interaction of plants, herbivores, and carnivores in an ecosystem (Schmitz & Booth, 1997). Schoonderwoerd, Holland, and Bruten (1997) use distributed agents to balance loads in telecommunications networks, but learning occurs only in the sense that agents lay down ant-like trails to improve performance. Grand, Cliff, and Malhotra's (1997) work on CREATURES is more akin to our own work, with independent agents that exist in a simulated environment, receive rewards, and change their behaviors with experience.

There does exist some work on machine learning for traffic control, but this has focused on learning for individual driving agents. For example, Sukthankar et al. (1996) use reinforcement learning to acquire control strategies for a vehicle that operates on a simulated highway among other cars controlled by hand-crafted strategies. Similarly, McCallum (1996) reports a system that uses reinforcement learning to acquire a single-agent controller for 'New York driving', which involves weaving around slower traffic. There also exists a substantial literature on more traditional approaches to traffic management that typically involve more centralized control, which we have reviewed in a separate paper (Moriarty & Langley, 1998).

Although we have found no other work on distributed learning in the traffic domain, we remain excited about its potential as a fertile research testbed. In future work, we plan to improve our traffic simulator to include durative lane changes, entrance ramps, and exit ramps. We believe these additions will produce more realistic congestion patterns and thus increase the need for intelligent lane selection. The revised simulator will also let the vehicles increase or decrease their speed, which should provide further improvements in traffic flow, at least for learned controllers.

In the longer term, we envision an extended simulator that supports a network of interconnected highways. Each car would be given a destination and, to the extent

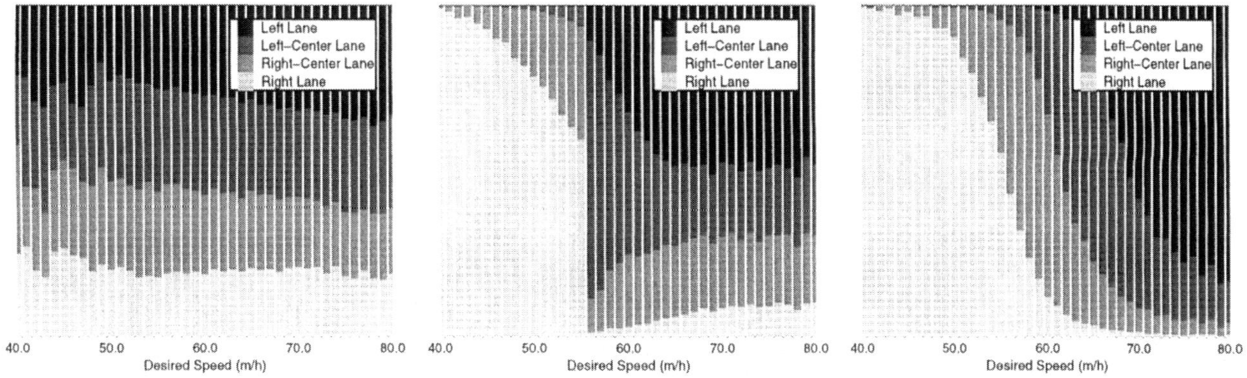

(a) (b) (c)

Figure 6 Utility of lanes with respect to desired speeds for the (a) selfish, (b) polite, and (c) learned strategies with four traffic lanes.

that multiple routes are available, the controller will select among routes just as it currently selects among lanes. This will require access to higher-level information about the distribution of cars on the various highways, so that the distributed controllers can select routes that avoid congestion. These variations on the task of distributed traffic management should provide a rich set of problems to drive our research in years to come.

However, our research methodology should remain much the same, in that we will construct systems that control traffic in a distributed manner and we will study those systems' adaptive behavior under a variety of experimental conditions. We invite other researchers to join us in our exploration of an intriguing domain that remains poorly understood despite its relevance to our everyday lives.

Acknowledgements

We would like to thank the anonymous reviewers for their helpful comments and Dan Shapiro for his evaluation of traffic simulators.

References

Carrara, M., & Morello, E. Advanced control strategies and methods for motorway of the future. In *The drive project DOMINC: New concepts and research under way.*

Eskafi, F. (1996). *Modeling and simulation of the automated highway system.* Ph.D. thesis, Department of Electrical Engineering and Computer Science, University of California, Berkeley.

Grand, S., Cliff, D., & Malhotra, A. (1997). CREATURES: Artificial life autonomous software agents for home entertainment. In *Proceedings of the First International Conference on Autonomous Agents*, pp. 22–29. New York: ACM Press.

Grefenstette, J. J., Ramsey, C. L., & Schultz, A. (1990). Learning sequential decision rules using simulation models and competition. *Machine Learning, 5*, 355–381.

Holland, J. H. (1996). *Hidden order: How adaptation builds complexity.* Reading, MA: Addison-Wesley.

Holland, J. H., & Reitman, J. S. (1978). Cognitive systems based on adaptive algorithms. In Waterman, D. A., & Hayes-Roth, F. (Eds.), *Pattern-directed inference systems.* New York: Academic Press.

Kaelbling, L. P., Littman, M. L., & Moore, A. W. (1996). Reinforcement learning: A survey. *Journal of Artificial Intelligence Research, 4*, 237–285.

Matarić, M. J. (1994). Learning to behave socially. In *Proceedings of the Third International Conference on Simulation of Adaptive Behavior*, pp. 453–462. Cambridge, MA: MIT Press.

McCallum, A. K. (1996). Learning to use selective attention and short-term memory in sequential tasks. In *Proceedings of Fourth International Conference on Simulation of Adaptive Behavior*, pp. 315–324 Cape Cod, MA.

Moriarty, D. E. (1997). *Symbiotic evolution of neural networks in sequential decision tasks.* Ph.D. thesis, Department of Computer Sciences, University of Texas at Austin.

Moriarty, D. E., & Langley, P. (1998). Learning cooperative lane selection strategies for highways. In *Proceedings of the Fifeenth National Conference on Artificial Intelligence* Menlo Park, CA: AAAI Press.

Moriarty, D. E., & Miikkulainen, R. (1996). Efficient reinforcement learning through symbiotic evolution. *Machine Learning, 22*, 11–32.

Pomerleau, D. (1995). RALPH: Rapidly adapting lateral position handler. In *Proceedings of the 1995 IEEE Symposium on Intelligent Vehicles*, pp. 506–511 Detroit, MI.

Pomerleau, D. A. (1992). *Neural network perception for mobile robot guidance.* Ph.D. thesis, School of Computer Science, Carnegie Mellon University, Pittsburgh, PA.

Quinlan, J. R. (1986). Induction of decision trees. *Machine Learning, 1*, 81–106.

Rumelhart, D. E., Hinton, G. E., & Williams, R. J. (1986). Learning internal representations by error propagation. In Rumelhart, D. E., & McClelland, J. L. (Eds.), *Parallel distributed processing: Explorations in the microstructure of cognition, Volume 1: Foundations.* Cambridge, MA: MIT Press.

Sammut, C., Hurst, S., Kedzier, D., & Michie, D. (1992). Learning to fly. In *Proceedings of the Ninth International Workshop on Machine Learning*, pp. 385–393. San Francisco: Morgan Kaufmann.

Schmitz, O., & Booth, G. (1997). Modelling food web complexity: The consequences of individual-based, spatially explicit behavioral ecology on trophic interactions. *Evolutionary Ecology, 11*, 379–398.

Schoonderwoerd, R., Holland, O., & Bruten, J. (1997). Ant-like agents for load balancing in telecommunications networks. In *Proceedings of the First International Conference on Autonomous Agents*, pp. 209–216. New York: ACM Press.

Schultz, A. C., Grefenstette, J. J., & Adams, W. (1996). RoboShepherd: Learning a complex behavior. In *Proceedings of RoboLearn-96: International Workshop for Learning in Autonomous Robots* Key West, FL.

Sen, S., & Sekaran, M. (1998). Individual learning of coordination knowledge. *Journal of Experimental & Theoretical Artificial Intelligence, 10.*

Stone, P., & Veloso, M. (1997). Multiagent systems: A survey from a machine learning perspective. Tech. rep. CMU-CS-97-193, School of Computer Science, Carnegie Mellon University.

Sukthankar, R., Hancock, J., Baluja, S., Pomerleau, D., & Thorpe, C. (1996). Adaptive intelligent vehicle modules for tactical driving. In *Proceedings of the AAAI-96 Workshop on Intelligent Adaptive Agents*, pp. 13–22 Portland, OR. Also available at `http://www.cs.cmu.edu/~rahuls/Shiva/`.

Sutton, R. S. (1988). Learning to predict by the methods of temporal differences. *Machine Learning, 3*, 9–44.

Tan, M. (1993). Multi-agent reinforcement learning: Independent vs. cooperative agents. In *Proceedings of the Tenth International Conference on Machine Learning*, pp. 330–337. San Francisco: Morgan Kaufmann.

Varaiya, P. (1993). Smart cars on smart roads: Problems of control. *IEEE Transactions on Automatic Control, 38*, 195–207.

Watkins, C. J. C. H., & Dayan, P. (1992). Q-learning. *Machine Learning, 8*, 279–292.

Whitley, D., Dominic, S., Das, R., & Anderson, C. W. (1993). Genetic reinforcement learning for neurocontrol problems. *Machine Learning, 13*, 259–284.

Wilson, S. W. (1994). ZCS: A zeroth level classifier system. *Evolutionary Computation, 2*, 1–18.

A Society of Agents in Environmental Monitoring

João Costa Seco*, Carlos Pinto-Ferreira, Luís Correia***

* Dept. de Informática, Universidade Nova de Lisboa
Quinta da Torre, 2825 Monte da Caparica, Portugal
{jcs,lc}@di.fct.unl.pt

** Instituto de Sistemas e Robótica, Instituto Superior Técnico
Av. Rovisco Pais, 1, 1096 Lisboa Codex, Portugal
cpf@isr.ist.utl.pt

Keywords: Agents, Agent societies, Autonomous robots, Simulation models, Pollutant monitoring

Abstract

The evaluation of pollutant levels is a key aspect on the issue of keeping a clean environment. Conventional techniques include the utilisation of a fixed setup incorporating pollutant sensors. However, these approaches are a very long way from an accurate monitoring. Thus, to improve pollutant monitoring on a power plant chimney, the use of robotic agent societies (mobile robots) is suggested. This suggestion is adequate in pollutant monitoring when the environment is hostile and/or the region to be sampled has large dimensions.

However, the implementation of a system incorporating robotic agents raises complex technological problems. Before a set of any kind of real robotic agents is implemented, an accurate evaluation must be performed. What this paper describes is a simulated application of small flying robotic agent societies (helicopter models) monitoring a pollutant cloud.

This simulation intends to show that an "intelligent" search method works better than a systematic or random procedure. In this kind of environment (dynamic and non-structured) and using mobile robotics to meet a goal such as this, a behavioural control architecture seems to meet the performance objectives.

The behaviours designed to control the agents are prepared to implement individual needs (survival and navigation) and social needs (follow or gather group). The agents as individuals are capable of performing such a mission, however, global results are enhanced by social strategies.

1. Introduction

The simulation of a robotic agent society monitoring a pollutant cloud, originated from a power plant chimney, is the major issue on this work and was the motivation for a project proposal (Seco, 1997) dedicated to reactive agent societies.

These agents have a simple mission: to go around a chimney and sample the pollutant cloud sending values to a central processing unit on the ground which builds a global map. This map is transformed into an image that holds information about cloud direction, pollutant concentration, etc., allowing decision makers to evaluate and change the burning conditions of the power plant.

Present cloud monitoring approaches utilise a set of land sensor stations that transmit their readings to a central processing unit and by some means of estimation, values are predicted and decisions are made.

The use of robotic agents to search and sample a polluted area permits direct assessment of the real values and a more accurate monitoring process (on the air and also on the ground or underwater).

However, the sampling of a large area is not an easy job. A systematic or random search process may either take a while or require a large number of agents. On the other hand, the use of "intelligent" agents reduces the number of agents needed to perform such a task in a reasonable period of time. With an interpolation process, good results are obtained if samples are taken in "interesting" locations.

This so called "intelligence" can be achieved by a reactive architecture based on behavioural modelling with dynamic decision making (Correia, 1995). This kind of architecture is designed to deal with multiple sensors, multiple goals and be robust in non-structured environments (Brooks, 1986).

To assess the efficiency of the social behaviour, a simulator creates a modelled pollutant cloud, launches a set of robotic agents that sample the cloud using several group behaviour strategies and the collected clouds may then be compared to the one that was monitored. This allows a decision on what is the best behaviour configuration for such a group of agents.

Group strategies were inspired on Kube and Zhang's work (Kube and Zhang, 1992) on minimal group behaviours. This work concerns a non-interference strategy as a base for more complex behaviours such as follow or gather.

The use of simulation gives the chance of testing this

kind of behaviours without building the real agents. However, this is only the iceberg's tip, the gap between simulation and reality represents a large amount of work. So, options on computational models were taken in order to keep it at a simple level as a way to concentrate attention on behaviours design.

The whole system is composed by a few parts: The world simulator, the agents, and the processing unit. The first one, is in charge of all computational models, from helicopters to pollutant and gives agents and the processing unit a three dimensional environment to work on. Agents produce actions based on sensor readings and the processing unit produces clouds based on samples taken by the agents.

2. Agent Society Description

2.1 The Proposed Architecture

In this application, agents are small helicopter models equipped with perception, actuation and computational devices. Based on this equipment, there is a high level control architecture that must be capable of performing a mission. This control is built according to an architecture which is a way to implement a social behaviour adequate to the mission goals.

There were options taken in all these topics. What the next section explains is the context in which they were made. The perception and actuation are basic parts of the agent and need to be described in detail. The other important part of the agent is its programming and its interaction with the previous parts.

The Vehicles: Agent sensors provide values resulting directly from the perception they have on the environment. The properties they were designed to provide are:

1. Short range perception: Sonars, infrared detector;
2. Relative location assessment: Altimeter, base direction detector;
3. Self status: Fuel level reader, climb state detector;
4. Long range distance perception: Device for measuring the minimal and maximal distance to other agents based on a ultrasound receiver;
5. Environmental conditions: Pollutant reader, pollutant gradient detector;
6. Data communication: Digital radio receiver, landing permission detector;
7. Mission status: Sample density evaluator within a neighbourhood (based on a communication with the central processing unit);
8. Group status: Gather group detector (radio signal detector);

When dealing with an embodied agent, there are actuators that influence the agents movement and others that actuate over a communication channel. The movement actuators behave by giving the standard helicopter moves: Bank (left and right), Pitch (forward and backward), Yaw (left and right), and Rotor speed (higher or lower). On the other hand there are communication actuators that provide: Data communication with other agents and processing unit (Radio emitter); and Long range presence announcement (Ultrasound emitter).

This kind of vehicles is very sensitive to actuator and environmental changes (e.g. the wind, the rain), thus, it is necessary to ensure that the vehicle stays on a stable position.

To meet this goal, the existence of a fast control layer was presupposed. This layer is called *Executor* and it is responsible for keeping the referred stable position by monitoring the sensors and operating directly over the actuators within a fast and efficient control cycle. It also receives commands from the high level control layer and applies them to the actuators. These commands are described in Figure 1.

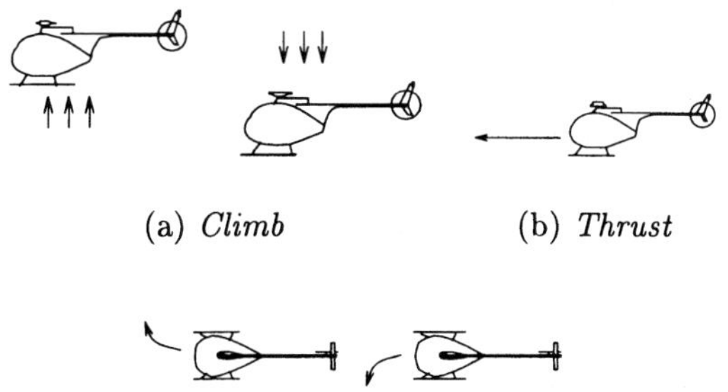

(a) *Climb* (b) *Thrust*

(c) *Yaw*

Figure 1 Movement commands to the executor

Besides movement commands the executor layer implements some other kind of predetermined actions: Sample submission (Sending pollutant readings according to pollutant sensor values); Landing base communication (Requests permission to land) and Communication with other agents (Sending them a predefined signal).

The Central Processing Unit: This component is a ground facility that receives all the readings from the agents and produces the expected global result. So, in this context there is no individual result for each agent, there are only social results. This set of samples is processed in order to produce useful information for the power plant decision makers.

Besides the sample values, the processing unit also registers the agents global location for each sample and the calculated cloud. This location can be obtained by a triangulation method and the global cloud is calculated by gathering the samples from all agents and performing data interpolation.

2.2 Programming

The structure that commands the executor works by receiving information from sensors and sending it decisions.

The approach taken is composed by a set of behaviours in a dynamic decision making architecture (Correia, 1995). Each behaviour has two outputs: an action signal and an activity signal. This activity signal varies from null activity to a fixed maximum according to the sensor stimuli. The decision is made comparing the activity output from each pair of conflicting behaviours. This kind of architecture seems to be more suited to unstructured environments than planning (classic, reactive or opportunistic).

Agent Programming: The agent high level control module is a programmable structure that is composed by a set of behaviours. This set of behaviours is built according to a programming method. The basic tasks assigned to this method are: information distribution; mission decomposition into a sequence of operations; individual behaviours design and category assignment.

The first step (information distribution) derives from the agent architecture. All the information that an agent possesses is built from its sensor values or given by outside entities that communicate with it. In this case, the values transmitted by outside entities are the number of samples taken in the agent's neighbourhood as well as the indication of some other agent gathering a group.

In the second step, a sequence of operations is produced. It includes all the major actions that the agent is intended to do. The sequence obtained for this mission is: Lift off and climb to a safe altitude; Navigate around the pollutant cloud in an "intelligent" way; Take samples according to the sensed pollutant; Refuel and proceed as many times as needed; Return home and land.

This sequence provides the necessary information for the next step. That designs individual behaviours to fulfill all the agents operations. Finally, these behaviours are assigned to a set of previously defined categories: survival, navigation, mission and group. Each one of these categories corresponds to a priority level. The survival category is the most important, the group category is the least important and the other two categories, navigation and mission, are of medium importance.

Additionally, mechanisms for coordination and communication between behaviours were applied allowing the implementation of a structured behaviour set.

A Set of Minimal Behaviours: The identified priority classes and the corresponding behaviours are as follows:

1. Survival behaviours:
   (a) Avoid collisions (horizontal direction);
   (b) Avoid collisions (vertical direction);
   (c) Watch fuel level;
   (d) Keep altitude range;
2. Navigation behaviours:
   (a) Wander;
   (b) Approach base;
   (c) Land;
3. Mission behaviours:
   (a) Maintain altitude;
   (b) Follow positive gradient;
   (c) Follow negative gradient;
   (d) Collect samples;
   (e) Avoid over-explored areas;
   (f) End mission;
4. Group behaviours:
   (a) Keep minimum distance from other agents;
   (b) Keep maximum distance from other agents;
   (c) Follow group;
   (d) Gather group;

The functionality of each behaviour module is such that the interaction of multiple modules fulfills the desired global behaviour. It is as follows:

1. Avoid collisions (horizontal direction): this behaviour is responsible for obstacle avoidance. It takes place at the agent present altitude. Some fixed rules were built in order to allow the agent to respond to all possible sonar snapshots.
2. Avoid collisions (vertical direction): it is a complement to the *Avoid collisions (horizontal direction)* functionality using a bottom sonar. The reason for this logical split is the need for a separate control in the landing process.
3. Watch fuel level: This behaviour increases the agents "will" to return home and land, according to the fuel sensor value, by coordinating *Approach base* and *Land* behaviours. If a lower limit is reached, this behaviour asks for landing permission to the landing base.
4. Keep altitude range: The purpose of this behaviour is to keep the agent within an acceptable range of altitudes. At the lower level, this behaviour prevents collisions with most common obstacles (e.g. buildings). On the other hand, the upper level prevents agents from going too high and loose radio control.
5. Wander: This behaviour influences actuators in a random fashion. However, it obeys to a set of probabilistic rules defined to provide a smooth wandering.
6. Approach base: This behaviour tries to guide the agent home. It tries to maximise the *base direction* sensor value by changing the agent direction. If the agent has no landing permission and it is located directly above the landing base (through the infrared detector value), the behaviour guides it away preventing accidents.
7. Land: This behaviour depends on the landing permission and on the infrared sensor that detects presence over the landing base. If conditions are met the behaviour lands the vehicle smoothly by inhibiting the *Avoid collisions (vertical direction)* and *Keep altitude range* behaviours.
8. Maintain altitude: The cloud reconstruction method as well as the agents stable position are based on

the horizontal position. So, this behaviour tries to keep the agent work divided in horizontal layers. The agents manage to go up or down when this behaviour suspends its activity.

9. Follow positive gradient: The navigation strategy is composed by two behaviours that follow the pollutant gradient. This one follows positive gradient and the other one, the negative gradient.
10. Follow negative gradient: It is complementary to the previous behaviour. The switching of the gradient following method causes the widening of the agent working area and consequently the collecting of a richer variety of samples.
11. Collect samples: Pollutant values are discretised in such a way that only values with significant changes are stored.
12. Avoid over-explored areas: The number of samples taken by the whole society in the present area is given by a sensor and it can be used to decide if this is an interesting area to explore. If that value is too high the agent should go away and explore other areas.
13. End mission: Agents should be able to decide when to finish their mission. This behaviour activates the landing procedure after verifying that the number collected samples is equally high for a very long time.
14. Keep minimum distances from other agents: Tries to maintain other agents outside a defined neighbourhood. This is achieved by maximising to an upper level the minimum distance value. This is a basic individual behaviour towards social behaviour.
15. Keep maximum distances from other agents: This behaviour, on the other hand, tries to maintain the other agents within a broader neighbourhood by minimising the maximum distance value. This behaviour results in a group maintenance strategy within the distance sensor range.
16. Follow group: Based on a communication mechanism with the *Keep maximum distances* behaviour, it leads the agent towards the others. If all agents activate this behaviour at the same time they form a more compact group. The behaviour is activated on the absence of pollutant or a call from another agent invoking this next behaviour.
17. Gather group: This behaviour does not change the agent movement but it tries to gather a group around it by sending a gather signal to all the others.

Emergent Individual Behaviour: The behaviour priority classes give the possibility of testing all aspects of the global behaviour.

The survival behaviours implement the lowest level of agents capabilities: "being alive". The *Avoid collisions (horizontal and vertical directions)*, *Watch fuel level* and *Keep altitude range* behaviours, keep the agent out of trouble and within a safe working altitude. The navigation behaviours (*Wander*, *Approach base* and *Land*) complement this first global behaviour providing some tools towards an interesting "life time" in the environment. The kind of a combined trajectory (navigation and survival) that can be obtained is shown on the first image in Figure 2.

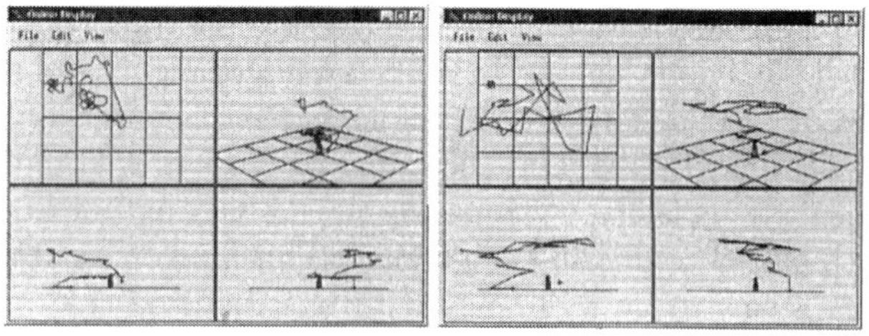

Figure 2 One agent trajectories (Three ortagonal views and a perspective view)

On the other hand, the mission behaviours (*Maintain altitude*, *Follow positive gradient*, *Follow negative gradient*, *Collect Samples*, *Avoid explored areas* and *End mission*) provide the agent the "intelligence" necessary to analyse and monitor the pollutant cloud. The resulting trajectories are of a "come and go" kind, from and to the chimney in order to explore the cloud from the centre to the edges and back as presented in the second image on Figure 2.

Emergent Group Behaviour: The group behaviours design was based on the group strategies presented in (Kube and Zhang, 1992) where several kinds of minimal cooperation based on simple communication protocol are used.

There are three major incremental group strategies which can be used separately. The simplest one is a *non-interference* strategy based on the known minimum distance to other agents (direct sensor reading). This assures a sparse distribution of the agents in the search space. The second group strategy is a *follow* feature that allows an agent to approach other agents in order to search for pollutant without external clues (assuming that other agents are inside the cloud). The most complex strategy is intended to improve other agents efficiency by calling them to interesting places (high level of pollutant values). It is called *Gather group* and it is based on both explicit communication and the *follow* feature.

The basic group behaviours are *Keep minimum distance* and *Keep maximum distance.* The maximising (minimising) procedure for the *Keep minimum distance* (*Keep maximum distance*) behaviour is a trial and error method that toggles between left and right yaw movements when the distance increases (decreases).

These behaviours maintain the referred distances in a range that is known and modifiable by other behaviours. In this way, other behaviours can influence the agents global group behaviour by incrementing or decrementing the maximum or minimum distances.

The non-interference strategy results directly from the use of *Keep minimum distance* behaviour with a standard goal distance for the minimum distance to maintain. This distance should be the radius for non-interference that the agent should try to maintain. This is a unilateral behaviour but it results rather well in societies because every agent is trying to maintain that distance as minimum. If one agent tries to decrement that distance far beyond the limit all others would run away.

The follow feature is implemented by the *Follow group* behaviour that decrements the maximum distance. This results on a unilateral approach behaviour that does not interfere with the previous strategies thus maintaining a constant minimum distance. The gather group strategy has a behaviour called *Gather group* that does not influence the agents movement. However, it does influence other agents group behaviour through a simple communication protocol. All the agents that receive the gather signal and do not have a "good" pollutant reading start their follow procedure and hopefully approach the calling agent. This could fail because they may not be located near that agent and start following some other agent that did not respond to the call. If this happens, this other agent must also have a good pollutant reading.

3. Pollutant Modelling

The pollutant processing has two separate procedures, an environment simulation for the original cloud generation and a sample processing and collected cloud generation.

3.1 Original Cloud Modelling

An analytical formula from the laminar jet dispersion theory (yun Kuo, 1987) was used in order to simulate a cloud produced by a power plant chimney. This creates a cloud like the one represented in the first image on Figure 3. This cloud has a paraboloid shape and surrounds completely the chimney area. This means that the higher value is always located at the chimney exit and along the chimney axis.

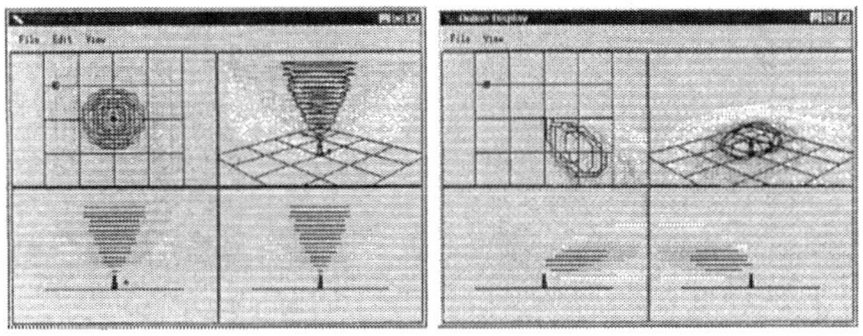

Figure 3 Original pollutant clouds

On the other hand, a more complex and harder to monitor model was experimented in order to verify the agents' robustness. It is a statistical gaussian dispersion model and it is based on a wind vector and a dispersion class. The pollutant values decrease uniformly along the wind vector as presented in the second image on Figure 3 (Ferreira and Câmara, 1996). On the simulator structure there is a separate module that feeds the pollutant reader sensor according to the agent locations and chosen pollutant model.

3.2 Sample Interpolation and Cloud Building Algorithm

All samples are gathered by the central unit with the information of the pollutant value, sampling time and sample location. Thus it is possible to draw a window on the time axis and follow a possible evolution of the cloud. In the present model this is not due to the static nature of the pollutant model.

The samples are used to calculate a three dimensional grid with a fixed step that contains all the samples taken. The interpolation used to obtain the value grid is a quadratic one and it is applied after a selection of some eligible points. The eligible point criterion is dictated by a maximum number of points ordered by proximity within a defined range[1].

This grid is then passed through a discretisation procedure into several levels of pollutant value. The visualisation is obtained by a contour extraction algorithm (chain-coding) applied on each horizontal layer of the discrete grid.

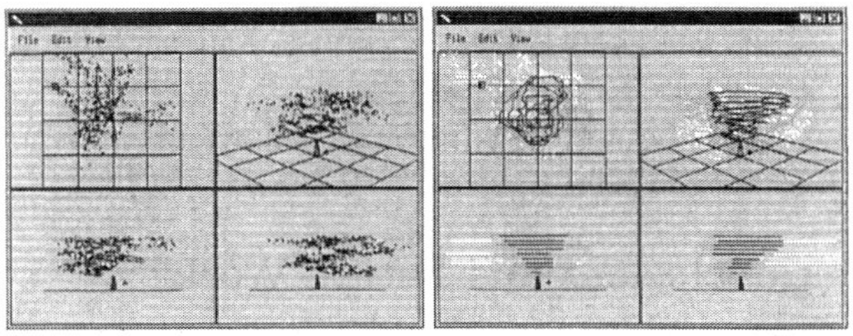

Figure 4 Agent Society Results

4. Results

Several agent configurations (subsets of all the predefined behaviours) were tested with the simpler pollutant model: Individual behaviours, non-intereference strategy, follow group strategy, and gather group strategy. A three agent population simulated for one hour using several behaviour configuration generates a large amount of clouds. These clouds were analysed and simple statistical results were obtained. The clouds were built from an average of 300 samples for an average 30000 points grid.

In the simulator interface, an example of collected

[1] The six closest points within a range of 200m are considered to be eligible for every point on the grid.

samples and corresponding cloud can be observed in Figure 4.

4.1 Cloud Evaluation

Using the same discretisation, applied to the original cloud, on the collected samples, it is possible to compare the two clouds and define an evaluation criterion based on similarity. This evaluation lacks quantification as it is based on a visual assessment. However, the evaluation resulting on *Good* or *Bad* clouds builds a success rate on a range of runs.

This criterion is defined by three rules. A cloud is considered *Good* if there are at least three distinct pollutant level values, and there are pollutant readings all around the chimney, and the two highest layers can be considered similar to the original ones (they are in paraboloid shape with approximately the same size for the laminar jet model), otherwise the cloud is considered to be *Bad.*

4.2 Group Results

Applying this criterion to the results obtained on all the runs, a percentage of *Good* clouds is obtained for each agent configuration. The behaviour configurations and correspondent results are:

1. Individual behaviours: All the survival and navigation behaviours – 40% of *Good* Clouds.
2. Non-interference: The previous configuration plus the *Keep minimum distance* behaviour – 60% of *Good* Clouds.
3. Follow Group: The previous configuration plus the *Keep maximum distances* and the *Follow Group* behaviours – 65% of *Good* Clouds.
4. Gather Group: The previous behaviours plus the *Gather Group* behaviour – 70% of *Good* Clouds.

5. Conclusions and Future Work

Pollutant monitoring based on robotic agents improved significantly when relying upon group strategies. All strategies implement a better search method than the previous one, spreading and coordinating all agents through the search space.

Individual behaviours designed to search and sample the environment do not guarantee that agents do not overlap work regions or even that they find the pollutant cloud. With the use of the non-interference strategy agents decrease the number of overlapping situations. With a unilateral follow group behaviour, agents manage to find the cloud assuming others have already found it. At last, the gather group strategy tries to improve even more the global efficiency.

Concerning the second pollutant model presented, the challenge is to find the cloud in the first place. So, another group behaviour was created (*Spread Group*) intending to build a wide net of agents and in this way increase the probability of one of them finding the cloud and then gather the group. The previous behaviours present good performance once the cloud is found, translating the success issue to the finding procedure. Although this configuration is not stable, there is an 85% success rate on finding the cloud in a reasonable amount of time.

References

Brooks, R. A. (1986). A robust layered control system for a mobile robot. *IEEE Journal of robotics and automation*, RA-2(1):14–23.

Correia, L. (1995). Veículos autónomos baseados em comportamentos - um modelo de controlo de decisão. PhD thesis, Universidade Nova de Lisboa, Faculdade de Ciências e Tecnologia.

Ferreira, F. C. and Câmara, A. S. (1996). Digital video air pollution monitoring. In *Proceedings of the 15th International Conference in Optical Sensing and Environmental Process Monitoring*, Dallas.

Kube, C. R. and Zhang, H. (1992). Collective robotic intelligence. In (Meyer et al., 1992), page 460.

Meyer, J.-A., Roitblat, H. L., and Wilson, S. W., editors (1992). *From animals to animats 2 - Proceedings of the second international conference on simulation of adaptive behavior.* MIT Press.

Seco, J. C. (1997). Monitorização de nuvens de poluente através de sociedades de agentes robóticos simulados. Master's thesis, Universidade Nova de Lisboa.

yun Kuo, K. K. (1987). *Principles of Combustion.* John Wiley & Sons, New York.

Assessing the role of social development in the evolution of cooperation

Ezequiel A. Di Paolo
School of Cognitive and Computing Sciences
University of Sussex
BRIGHTON BN1 9QH, U.K.
ezequiel@cogs.susx.ac.uk

Abstract

The role of space-time structuring processes has recently come into focus as an important ingredient in the understanding of cooperative interactions in different contexts (Boerlijst and Hogeweg, 1991; Hemelrijk, 1997; Di Paolo, 1997; Di Paolo, 1998). This paper explores a similar role for a different sort of contextual process: phenotypical development under social conditions. A minimal ontogenetic, reaction-network model maps newborn phenotypes into adult ones according to genetic constitution and parental influence. In this way three coupled processes determine the distribution of behaviors within a population: selection, development and spatial organization. The presence of dynamic "developmental structures" is shown to play an important role in stabilizing cooperative coordination even when individual costs are against it. Resulting coordination levels prove to be much more stable than with spatial distribution alone. Some of the resulting self-promoting and self-regulating developmental structures are discussed.

1. Introduction

A tacit supposition in many evolutionary models is that those factors that remain outside the influence of natural selection are either fixed or vary at random. The evolution of a trait is often studied under the assumption that the relevant environmental factors change so slowly that they may be considered to be static (e.g. the distribution of food patches) or their variation is dependent upon so many contingencies that, for large populations, they may be considered to change randomly (for instance, mating opportunities). These are fruitful assumptions under which the evolution of the particular trait of interest becomes, as a dynamic process, decoupled from a myriad of other processes which, if considered, would make the treatment of the problem a very difficult task[1]. However, it is apparent that in many cases this is an oversimplification, even for the localized understanding that is intended to be obtained from such models. The distribution of food patches will partly depend on the efficiency of foraging strategies, and mating opportunities will depend, for instance, on space-time constraints.

This is a criticism that has been repeated many times (Bohm, 1969; Lewontin, 1982; Odling-Smee, 1988, among others) and should be taken seriously if an understanding of evolutionary processes in general is to be achieved. On the other hand, it is clearly impractical to model a whole ecosystem if we intend to understand the changes of a single trait in a particular species. If we are to find a compromise between accuracy, realizability and explanatory value, the relevance of those contextual factors which are not completely independent of the evolutionary process itself should be fully justified if we wish to include them in our models.

The aim of this paper is to accomplish such a justification in the particular case where the evolution of *cooperative social coordination* is coupled with a process of *social development*. Since, at this stage, we will only be interested in abstracting some basic dynamic features we can afford to ignore much of the complexities involved by using a rather simple but justifiable model.

2. Social coordination and dynamic structures

We are interested in the evolution of communicative behaviors. At first instance, and for reasons fully justified elsewhere (Di Paolo, 1998; Maturana and Varela, 1980), these will be better understood in terms of behavioral patterns of coordination rather than in terms of signals and information exchange. At a purely operational level, communication involves the interaction between two or more autonomous organisms, whose actions become mutually oriented as a result. In particular we will consider those cases of behavioral coordination that involve some sort of cooperation among participants with conflicting interests, since this sort of behavior has always been problematic when viewed from a neo-Darwinian framework.

From a game-theoretic perspective cooperative communication can only be understood if certain conditions are met regarding the distribution of cost-benefit relations among individuals differing in some quality and

[1] In problems involving co-evolution assumptions like these are weakly relaxed so that the mutual influence of two or more "single" evolutionary processes can be accounted for.

degree of relatedness. The corresponding models are usually more sophisticated than simpler, "single trait" models, in that a distribution of possible strategies is assumed to exist and the beneficial properties of a given strategy are not independent of this particular distribution. In all other respects most of these models still assume either a static or a random variation of the rest of the relevant factors. For instance, they assume no structure in the interaction patterns of the individual organisms: *individuals play randomly with any partner in the population, and all individuals play all roles in the game with a fixed uniform frequency.*

Recently, the importance of a spatial distribution in the population of organisms that constrains both reproductive patterns and possible interactions has been put into focus. It is found in these cases that cooperation is stabilized even under circumstances where simpler models predict the contrary. This has been demonstrated both by mathematical considerations and computer models in the case of the Prisoner's Dilemma (Axelrod, 1984) and communication games (Ackley and Littman, 1994; Di Paolo, 1998). Depending on the particulars of the model there may be more than one possible explanation of this phenomenon. Axelrod argues that spatial clustering favours reciprocity if the playing strategy is cooperative (Axelrod, 1984, p. 65 - 69). Ackley and Litmann suggest, but fail to prove, that, since the areas where mating partners are chosen from and offspring allocated tend to coincide with areas where coparticipants are selected from, a process of kin-selection is responsible for stabilizing cooperation. In this way spatial distribution ensures that neighbours will be highly related so that cooperating with them will tend to increase inclusive fitness. Yet another possibility is *shown* to be at work in a model of an action-response game (Di Paolo, 1997; Di Paolo, 1998) and in a formally relevant model of distributed chemical catalytic interactions (Boerlijst and Hogeweg, 1991). In both these cases structured spatial entities and processes play a non-trivial role in the evolutionary process with which they are coupled. Di Paolo (1997; 1998) shows how agents playing an action-response game aggregate into structured clusters which influence the distribution of resources in the area occupied by them and modify the relative frequencies with which agents participate as players of the first role (action) and second role (response) in such a way that cooperative interactions result even for a range of individual costs against them.

These latter cases point to contextual processes or entities which are neither fixed nor randomly fluctuating, but depend either on the evolutionary process itself, or on some other non-decouplable aspect of the situatedness of organisms in their environment, (such as the distribution of resources). We may propose other processes or entities of this kind. In particular most gregarious animal species manifest some sort of social structure either in terms of phenotypical differentiation or behavioral patterns. The dynamics of such structures may constitute a very rich process in itself with far from obvious repercussions for the stabilization of cooperative interactions. For instance, Hemelrijk (1997) studies the effects on cooperative interactions of a self-reinforcing dominance hierarchy in spatially situated agents. A different sort of social process can be found when phenotypical development is affected by social influence. As an example we may think of parental imprinting of sexual preferences (Hess, 1973) and cultural processes in general. These processes will have an effect on the behavioral constitution of populations and consequently they will constrain the actual interaction patterns that become established within them. Since coordinated action is *not* an individual trait the relevance of these mechanisms should be evident.

The assertion that social structures can play an important role in the evolution of cooperative behavior is far from surprising. The point we wish to stress, though, is that the focus of attention should be on coupled dynamics, i.e. social structures not just as a static background against which organisms will try to optimize their behaviors, but as part of a generalized process in which organisms participate.

3. A model of social development

Our purpose is to make a *first order* assessment of the relevance that a socially influenced developmental process may have for understanding the evolution of cooperative action coordination, and also to try to identify what sort of dynamic constraints such a process may impose. This justifies the choice of an abstract, and indeed crude, model of development under social conditions. However, we will want our model to account for the following relevant aspects:

1. Phenotypes are the result of a dynamic process which is affected both by genetic constitution and coupling between the developing organism and its medium.
2. The nature of the process itself is not totally determined by genetic constitution nor by environmental coupling, (for instance, the number of possible attractors a developmental system may have depends on the specific transition dynamics and may not be easily changed during evolution).
3. In most natural cases of social development, the most relevant social coupling is between organism and parent(s).

The following scheme is an abstraction of these aspects. Let us suppose that an individual adult phenotype P_a^i is totally determined by its phenotype at birth P_b^i with the exception of certain minor changes which will depend both on P_b^i and the phenotypes of its parents $P_a^{p_1,p_2}$. P_b^i depends only on the genetic constitution of i. Parental influence is not arbitrary. We may define a *phenotypic metric* that, in principle, need not be the same as the genotypic metric in terms of Hamming distances, nor the same as a "behavioral" metric, but may take into account other traits besides those that are being studied. Under such a metric two phenotypes may be said to be close or apart. Parental influences will be taken into consideration only if the following condition holds:

$$||P_b^i - P_a^{p_j}|| \leq D, j = 1 \text{ or } 2,$$

for D a small distance; the influence will be such that:

$$P_a^i = Dev(P_b^i, P_a^{p_j}),$$

and

$$||P_b^i - P_a^i|| \leq D.$$

This means that a social influence on development will be manifested when parents are not very distant from the offspring's phenotype at birth (so that genetic constitution does not become irrelevant), and it will be such that the resulting adult phenotype will remain within the same distance. If the above condition is not fulfilled then $P_a^i = P_b^i$.

4. Computer model

The abstract reaction-network model described above can be easily implemented in the context of an action-response game similar to those presented in (Di Paolo, 1997; Di Paolo, 1998). A population of agents are distributed in a two dimensional spatial environment. During its lifetime an individual agent does not leave its finite local habitat which is defined as a circle of radius R_0 around a fixed position with real-valued coordinates, and where all the interactions involving that agent take place. At each time step agents are chosen asynchronously with equal probability to play the first role in the game. An agent selected in this way will try to randomly locate a partner from its neighbourhood with a probability of success proportional to the local density of agents. At the same time a food repository will be randomly selected from those located in the neighbourhood. The first player performs an action A and the second player performs a response R. If A and R are complementary in the sense described below, the two players are said to be coordinating their behavior. In such a case both agents share equally the payoff which is an amount of energy extracted from the local repository. If they do not coordinate the first agent receives a proportion c of the energy in the repository, where $0.5 < c \leq 1$ and the other agent receives nothing.

In this simplified version A and R are fixed for all circumstances in a given individual: they do not depend on food types nor on a history of interactions. They are represented by two integer numbers between 0 and 15. The condition for A and R to be complementary is simply that $parity(A) = 1 - parity(R)$ where $parity(.)$ returns 1 for even numbers and 0 for odd numbers. From this it is possible to distinguish four possible behavioral types depending on the combination of parities for A and R.

Although all possible phenotypes will fall into one of the four possible behaviors (Figure 1), in order not to trivialize the issue of development, we have worked with the larger number of 256 behaviorally (but not developmentally) neutral phenotypes. These are encoded in an 8-loci binary haploid genome which defines P_b^i using binary encoding[2] where A is encoded in the top 4 bits and R in the lower 4 bits. The distance of allowed influence D was set to 2 (modulo 256).

Agents store the energy gained from the games they play and use it for self-maintenance (a fixed energy cost is discounted at each time step), and for reproducing once a certain amount has been accumulated. Selection of mating partner is based only on locality constraints, and offspring are placed within the neighborhood of the first parent. Neighborhoods do not overlap completely so that gene flow is not prevented. The offspring's genotype results from a uniform crossover operation on the genotypes of the parents plus random point mutations with a fixed probability μ per loci.

The adult phenotype of newborn agents is obtained as a function of P_b^i and $P_a^{p_1,p_2}$, observing the conditions discussed in the previous section. For each value of P_b^i its developmental function $Dev(.)$ is represented by an integer *symmetric* matrix M^{P_b} of $(2D+1) \times (2D+1)$ which is defined at the beginning of the simulation run. The elements of this matrix are integers randomly selected from the interval $I = [P_b^i - D, P_b^i + D]$. The indexes of the rows and columns correspond to the same interval, so that the first element of the first row will have indexes $(P_b^i - D, P_b^i - D)$. If the adult phenotypes of both parents fall within I the adult phenotype for the offspring will be:

$$P_a^i = M^{P_b}(P_a^{p_1}, P_a^{p_2}).$$

If only one of the parental phenotype falls within I the remaining index is taken to be equal to P_b^i:

$$P_a^i = M^{P_b}(P_a^{p_1}, P_b^i).$$

and if none of the parental phenotypes falls within I the adult phenotype is taken to be equal to the phenotype at birth. M^{P_b} defines a *constrained random reaction-network*. Table 1 shows an example of M^{P_b} for $P_b^i = 50$ and $D = 2$. For instance, if the parental adult phenotypes are 51 and 52, the resulting adult phenotype will be 49.

[2] Tests performed using Gray encoding showed no significant difference in the results.

Table 1 Developmental matrix for $P_b^i = 50$.

	48	49	50	51	52
48	50	50	52	51	53
49	50	50	51	52	53
50	52	51	48	50	49
51	51	52	50	50	49
52	53	53	49	49	53

5. No development

How can we expect this model to behave? Based on previous experience and mathematical models we know how the model will behave if no social development is present, (for full details see Di Paolo, 1998):

No Space. Actions and responses may be odd (O) or even (E) so that there are four possible strategies related as shown in Figure 1, where arrows indicate a cooperative relationship between first player (arrow's starting point) and second player (arrow's end). Absent connections indicate non-cooperative relationships.

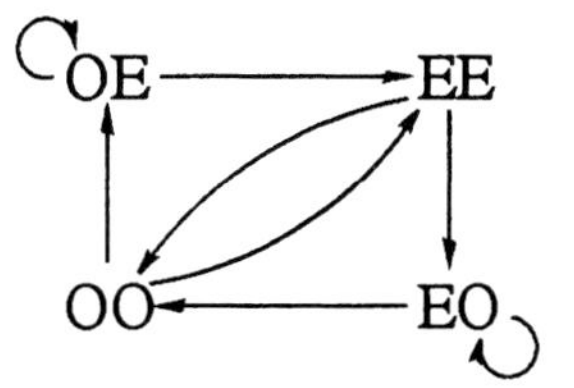

Figure 1 Cooperative relations between strategies. (EO) and (OE) are called "self-cooperating" or simply "cooperating" strategies.

Under these circumstances a dynamic game-theoretic analysis shows the presence of only two attractors. A fixed point stable attractor (ESS) where the population is divided equally between strategies (EE) and (OO) and so defining a baseline level of cooperation of 50 %, and also a limit cycle attractor, where each of the four types becomes dominant for a certain period, only to be followed by a different type in the order shown by the straight arrows in Figure 1. The basin of attraction for this second case is much larger than for the ESS. So, in the mixed medium situation the population will indefinitely oscillate between periods of cooperation and non-cooperation.

Space. When agents relate spatially results are different. In all cases a toroidal arena of 200 by 200 was used, the local neighbourhood radius R_0 was equal to 8 and the mutation rate to 0.002 per loci. 40,000 food resources were uniformly distributed across the arena, and energy related parameters were chosen so that population sizes varied between 700 and 1000 agents. The population tends to be distributed in relatively isolated,

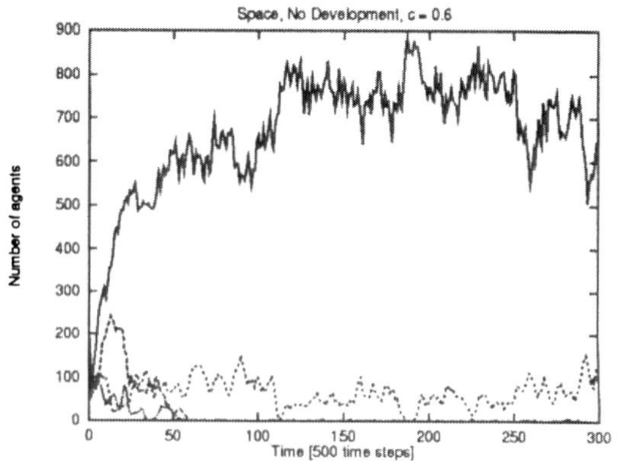

Figure 2 Population constitution for spatial game with no development, $c = 0.6$. Solid line: agents of type (EO), dotted line: agents of type (OO).

stable clusters. We do not find oscillations, and the resulting level of cooperation depends on parameter c, the payoff obtained by the first player in the case of non-coordination. For $c = 0.5$ (neutral cost) cooperation is established in the whole population (that is, the population becomes fixed at (OE) or (EO) or both of them, each in separated clusters). As c increases the level of cooperation decreases linearly until $c \cong 0.7$ where the baseline case of 50 % cooperation is reached. This shows that cooperation is stable even for some values of c greater that 0.5 (i.e. with individual costs against it), and that this level decreases gradually as c increases up to approximately 0.7. It has been observed that populations tend to be constituted largely by a mixture of strategy (EO) and a minor amount of (OO) or, conversely, by (OE) and a minor amount of (EE). Figure 2 shows an example of such a case, for a run of 1.5×10^5 time steps (roughly 400 generations). The population is rapidly invaded by a (self-)cooperating strategy (EO) and a small number of agents of type (OO) is maintained since they take advantage of the (EO) strategy, however the spatial dynamics prevents an invasion (other strategies disappeared early in the simulation). As c increases beyond 0.6 the constitution ceases to be stable and all strategies participate in a strong dynamics.

6. Results with social development

We find some interesting qualitative differences with the results discussed in the previous section when the evolutionary process incorporates development under social conditions. All simulations[3] were run using the same parameters and for various values of c between 0.6 and 0.95.

No Space. We mentioned that the imposition of spatial constraints changed the evolution of cooperation from an oscillatory pattern into a stable level. Here we

[3] Data and code are available from: `http:\\www.cogs.susx.ac.uk\users\ezequiel\dev\code.html`.

ask if a process of social development by itself could constrain the evolutionary dynamics in a similar way. The answer is no. The dynamics remain oscillatory, although qualitatively different in that strategies become phase-locked by pairs. The reason for this is not completely understood yet and further exploration is needed.

Space. Results are similar to the case of evolution with no development except that the range of c's where cooperation is stabilized is larger: $0.5 \leq c \leq 0.8$. More interestingly, almost for the totality of this interval, the population is constituted *purely* by a single cooperative strategy (either (OE) or (EO)), and only for $c \cong 0.8$ do mixed strategies appear. This means that the global level of cooperative coordination will not decrease smoothly as c increases, but will be maintained at near 100 % and then change rather abruptly for $c \cong 0.8$. Figure 3 shows a typical simulation run with $c = 0.6$.

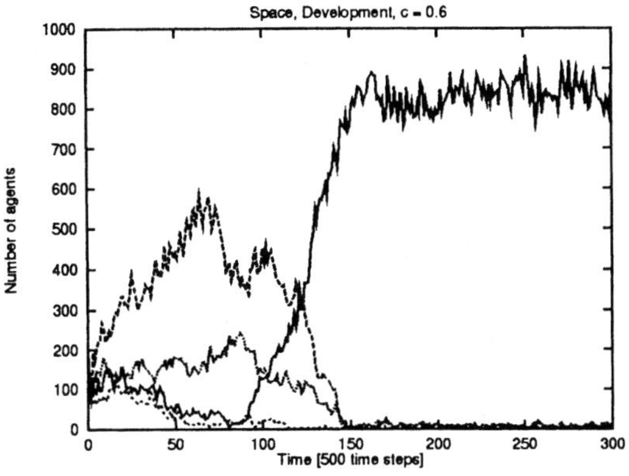

Figure 3 Population constitution for spatial game with development, $c = 0.6$. Each line represents one of the four strategies, solid line: (EO)

If we analyse the population constitution in detail we find that, within the range of c's where cooperation is stable, the population tends to be constituted mainly by a single phenotype either of type (EO) or (OE), with a developmental structure of the type shown in Figure 4.a. In this figure numbers in squares represent phenotypes at birth, numbers in ovals represent adult phenotypes, full arrows show a developmental relation and dashed arrows a parental influence on development. Figure 4.a shows that an agent i with $P_b^i = 238$ will develop into $P_a^i = 240$ if its parents happen to be $P_a^{p_j} = 240$.

By the term developmental structure we mean not only the logical relationship between offspring and parental phenotypes but the *actual* developmental relationship instantiated by the different types present in the population. The latter also depends on agent distribution and genetic relationships among other things.

As Figure 3 shows, there are no other strategies besides the main cooperating one (solid line). This marks a difference with the case without development (see Figure

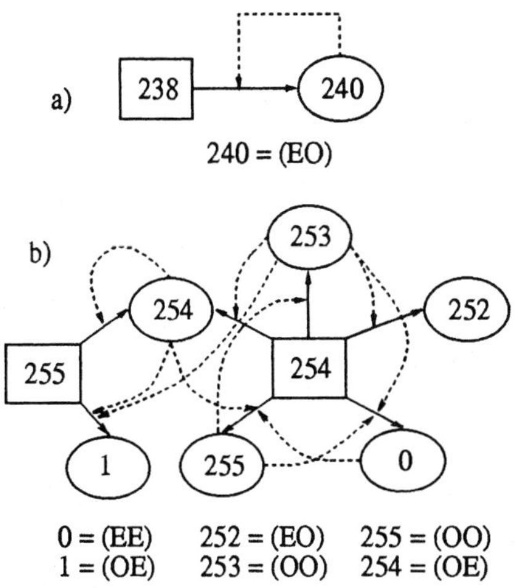

Figure 4 Developmental structures (see text) and corresponding behavioral types for a simulation with $c = 0.8$. a) Main self-promoting structure, b) secondary structure

2). This is due to the fact that the presence of a developmental structure promotes additional stability. While spatial organization seems to prevent an invasion by competing strategies, it cannot prevent their existence since they are just "one-mutation away" from the main component. With development the relationship between phenotypes of this main component and competing phenotypes is not so simple. It is not enough to be near in genetic space, but competing phenotypes must at the same time conform to the logic of the developmental space in such a way that the resulting behavioral relations still produce a competing interaction; a much more demanding constraint. As c increases there will be more pressure to meet this constraint. This is what we observe as we approach the limiting value of $c = 0.8$. Figure 5 shows the constitution of the population for such a value. Here we also see the main cooperating strategy constituting most of the population (solid line). The corresponding phenotype is also associated with the developmental structure shown in Figure 4.a. In addition, we observe that other strategies are not completely wiped out. By examining the population we discover that some agents relate through a more complex developmental structure (Figure 4.b).

The structures shown in Figure 4 originate from three different genotypes. The Hamming distance between '238' and '254', and between '254' and '255' is 1 meaning that uniform crossover will not be likely to introduce new genotypes. However, the two developmental structures are separate since developmental influences exist only if adult and newborn phenotypes differ in less that D[4]. From the behavioral perspective we see that two of the adult phenotypes produced in the second structure are of type (OO) which, under ideal circumstances,

[4] Tests were run with values of D between 2 and 5, with similar results.

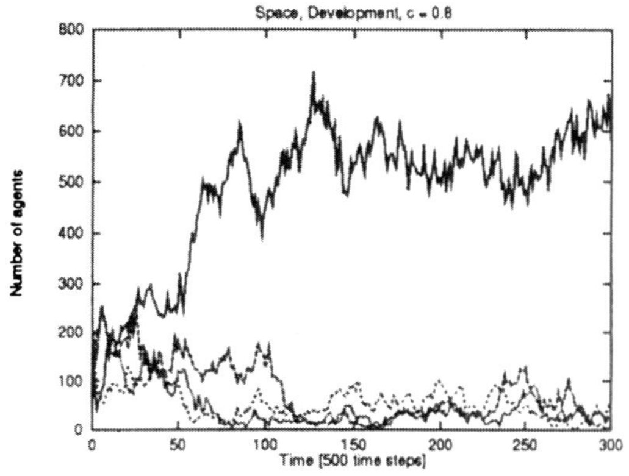

Figure 5 Population constitution for spatial game and development, $c = 0.8$

would invade a population constituted mainly of type (EO). However, as can be seen in Figure 4.b, these adult phenotypes are not "free" to grow but they are constrained by the amount of adult phenotypes that *favour* their development, some of which *compete* with them on the behavioral plane. In spite of what Figure 4.b may suggest all the phenotypes in this developmental structure remain bounded in number. In fact, if we study the time evolution of cooperating and non-cooperating phenotypes in this structure we find a very strong correlation in their variation ($\rho = 0.825$) as if the overall effect were behaviorally neutral which suggests that some complex self-regulation is taking place.

7. Discussion

As we said earlier our main purpose is purely exploratory, hence the abstractions made in this early model. The relevance of social development for understanding the evolution of cooperative coordination could well be supported independently by theoretical arguments. We chose to propose a minimal model where the nature of such influence could be explored.

We observe that in addition to the stabilization provided by spatial organization, cooperative coordination receives further stabilization by the presence of stable developmental structures which constrain the range of viable competing strategies. Further understanding of these structures could be obtained from other studies using reaction-networks. By the introduction of a chemical analogy, we may use notions such as catalysis, hypercycles, quasi-species, etc., (Eigen and Schuster, 1979), whereby developmental relations can be seen as favouring the production of certain group of phenotypes in a self-sustaining manner. But we must be careful, as the differences are also apparent, especially in terms of the other two "dimensions" in which these entities exist: the behavioral and the genetic.

We would like to emphasize that this study has looked at the issue of social development from an abstract perspective, that of an instantaneous process coupled with, but not completely governed by differential reproduction. This is a strong simplification of the issue of development under social conditions, especially if we are interested in its behavioral and dynamic aspects. Whatever the complexities of natural cases we can at least be certain that those features abstracted in our model have relevant consequences for the study of cooperative action coordination. We think this is enough justification for reconsidering the role played by natural selection as the main explanatory factor when social behaviors are involved.

Acknowledgements

Thanks to Inman Harvey, Phil Husbands and Jason Noble for valuable discussions. The author acknowledges the support received from the *Consejo de Investigaciones Científicas y Técnicas de la República Argentina* and an Overseas Research Student Award.

References

Ackley, D. H. and Littman, M. L. (1994). Altruism in the evolution of communication. In Brooks, R. A. and Maes, P., editors, *Artificial Life IV*. MIT Press.

Axelrod, R. (1984). *The Evolution of Cooperation*. Basic Books, New York.

Boerlijst, M. C. and Hogeweg, P. (1991). Spiral wave structure in pre-biotic evolution: hypercycles stable against parasites. *Physica D*, 48:17 – 28.

Bohm, D. (1969). Addendum on Order and neo-Darwinism and Some comments on Maynard-Smith's contributions. In Waddington, C. H., editor, *Towards a Theoretical Biology 2, Sketches*, pages 90 – 93 and 98 – 105. Edinburgh University Press.

Di Paolo, E. A. (1997). Social coordination and spatial organization: steps towards the evolution of communication. In Husbands, P. and Harvey, I., editors, *Proceedings of the 4th European Conference on Artificial Life*. MIT Press.

Di Paolo, E. A. (1998). An investigation into the evolution of communication. *Adaptive Behavior*, 6 - 2:285 – 324.

Eigen, M. and Schuster, P. (1979). *The Hypercycle: A principle of Self-organization*. Springer Verlag, Berlin.

Hemelrijk, C. K. (1997). Cooperation without genes, games or cognition. In Husbands, P. and Harvey, I., editors, *Proceedings of the 4th European Conference on Artificial Life*. MIT Press.

Hess, E. (1973). *Imprinting : early Experience and the Developmental Psychobiology of Attachment*. Van Nonstrand.

Lewontin, R. (1982). Organism and environment. In Plotkin, H. C., editor, *Learning, development and culture: essays in evolutionary epistemology*. Wiley Chichester.

Maturana, H. and Varela, F. J. (1980). *Autopoiesis and Cognition: The Realization of the Living*. D. Reidel Publishing, Dordrecht, Holland.

Odling-Smee, F. J. (1988). Niche-constructing phenotypes. In Plotkin, H. C., editor, *The Role of Behavior in Evolution*, pages 73–132. MIT Press, Cambridge, Mass.

Formation of Morphology and Morpho-Function in a Linear-Cluster Robotic System

Naoki Kawai and Fumio Hara
Department of Mechanical Engineering, Science University of Tokyo
1-3 Kagurazaka, Shinjuku-ku, Tokyo 162-8601, Japan.
e-mail: {kawai, hara}@hafu0103.me.kagu.sut.ac.jp

Abstract

This study focuses on the morphology formation and morpho-function in a linear-cluster robotic system which is composed of linearly connected autonomous mobile modules. The equations of motion of the linear-cluster robotic system are derived under dynamic interaction with its physical environment, and computer simulations are performed for two cases of physical environment condition: one is for two dimensional friction plane and the other for a maze-like arena surrounded with walls. When each module has simple rules of locomotion determined according to the direction of interaction force acting from a neighboring module, various morphologies have been formed. When adding intelligence of obstacle avoidance and locomotion towards the object and goal signals to the modules, the linear-cluster robotic system has achieved the task to find the object, enclose it and then carry the object to the goal destination by changing its morphology. Finally some of the future works are discussed from the view of balancing the complexities of morphology and intelligence in cluster robotic systems.

1 Introduction

Biological systems are extremely adaptive to the drastically changing physical conditions of the surrounding external environment and their mechanical functions are considered to have evolved so as to survive the hardships of nature. It is speculated that such mechanical functions emerge through changing their own morphology depending on the environment so as to perform their tasks more effectively.

From this viewpoint, we propose a robotic system which can create its mechanical function by changing its own morphology through the dynamic interaction between the system and external environment. One of these robotic systems is a so-called cluster-robotic system which is a distributed autonomous robotic system composed of mobile modules physically connected with each other. The cluster robotic system may change its own morphology by changing the connection between the modules and achieve the assigned task in various environments.

A few studies have been performed in the field of distributed autonomous robotic system based on this aspect of robotic system. Kokaji et al.(Kokaji, 1994) proposed the algorithm which allows a group of homogeneous mechanical units to form various shapes. In this algorithm the target shapes are prescribed in advance. Fukuda et al.(Fukuda, 1988) proposed a cellular robotic system (CEBOT) which consists of many functional units called "cells". CEBOT system has the ability of dynamic system reconfiguration depending on the task given. Yim(Yim, 1993) presented the typical locomotion modes of a modular type reconfigurable robot, for example, locomotions appeared in earth worm, caterpillar, and spider. For a study of snake-like robot, Hirose(Hirose, 1993) developed "active cord-mechanism" (ACM). These studies, however, have not focussed on the aspect of dynamic interactions associated with formation of robotic morphology.

This paper focuses on the formation of morphology and morpho-function in a cluster robotic system which is composed of linearly connected autonomous mobile modules. The equations of motion of the linear-cluster robotic system are derived by taking into account its dynamic interaction with physical environment, and computer simulations are performed for various interaction-physics parameters to investigate the characteristics of robotic morphology formation on a two-dimensional friction plane. Each module has simple rules of locomotion: 1) when a module is pulled by a neighboring module, the module generates a driving force in the pulling-force direction, and 2) when a module is pushed by a neighboring module, the module generates a driving force in the pushing-force direction. The computer simulations show various morphologies generated in the cluster robotic system which are majorly specified by the magnitude of friction force acting from the environment plane, and we find especially a critical friction-steering angle relationship for steady circular motion of arch-shaped morphology emerged in the robotic system.

When we add the intelligence of obstacle avoidance and locomotion towards the light signals representing an object and a goal to the modules, and set the robotic system in a maze-like arena containing a circular object (light A) and a goal (light B), the behavior of the linear cluster robotic system is investigated by computer simulations. The computer simulations demonstrate dynamic changes in the robotic morphology formation according to the physical conditions of the environment and the robotic system has achieved the task to find the object, enclose it and then carry the object to the goal destination. We demonstrate how physical interaction between a linear-cluster robotic system and its environment is important to create a certain form of the robotic system and how it can be used to perform an adaptive behavior in a given environment.

2 Linear-cluster robotic system

As a first step, we focus on a linear-cluster robotic system which consists of linearly connected autonomous mobile modules as shown in Fig.1. Each module can rotate with respect to a pivot joint, and the robotic morphology can be changed depending on the forces determined by mechanical interaction between modules and between the robotic system and the environment.

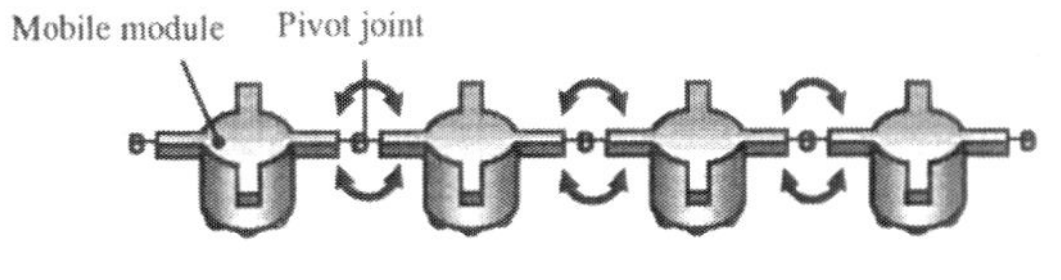

Fig.1 Linear-cluster robotic system consisted of linearly connected autonomous mobile modules

2.1 *Mechanical feature of mobile module*

Each module has four driving wheels which enable the module to translate and rotate (Fig.2). When a module moves on a flat plane, we take into account the friction force and friction torque acting between the flat plane surface and the wheels. Force sensors which can measure the magnitude and direction of forces acting on each connecting joint are set on each pivot joint. Each of four arms has two touch sensors (right side and left side of each arm end). At the top of the module an optical sensor is positioned to detect the direction of light signal from an object and a goal, respectively. Each module determines its own action autonomously according to simple rules of action (see section 3.2). Note that in this paper we do not discuss about the generation of these simple rules which should be definitely an important future work.

2.2 *Morpho-function of linear-cluster robotic system*

As the generation of robotic morphology in the linear-cluster robotic system is considered to be strongly associated with dynamic interaction between the robotic system and its environment, the robotic morphology formation is first focused on under a simple dynamic, but non-linear interaction through friction force and friction torque between the mobile modules and the flat plane. When considering the object(baggage) carrying problem in the linear-cluster robotic system, the mechanical function of carrying the object may be created by changing its own morphology according to physical environment conditions such as the shape and size of the object. Then the robotic system can achieve the task even when the shape and size of object(baggage) may be changed. For example, if the linear-cluster robotic system takes a wavy-line morphology, the system may carry a flat board, and if it takes a S-shaped morphology, the system may carry two kinds of objects not mixed each other.

3 Formulation of dynamics of a linear-cluster robotic system

3.1 *Equations of motion*

A linear-cluster robotic system can be modeled as a multibody dynamic system in which each body is linearly connected with each other by pivot joints. As shown in Fig.3, the equations of motion of the i-th module are described by the following set of equations:

for translation;

$$m_i\ddot{x}_i = -f_{(i-1)x} + f_{ix} + T_i\cos\phi_i + F_i\cos\theta_i + P_{ix} \quad (1)$$

$$m_i\ddot{y}_i = -f_{(i-1)y} + f_{iy} + T_i\sin\phi_i + F_i\sin\theta_i + P_{iy} \quad (2)$$

for rotation;

$$I_i\ddot{\alpha}_i = M_i - \frac{\dot{\alpha}_i}{|\dot{\alpha}_i|}N_i + \left(f_{ix} + f_{(i-1)x}\right) r\sin\alpha_i - \left(f_{iy} + f_{(i-1)y}\right) r\cos\alpha_i + Q_i \quad (3)$$

$$(f_0 = f_{10} = 0)$$

where

F = friction force acting on module (0~4.71N)
I = moment of inertia with respect to P (0.01kg·m^2)
M= driving torque acting on module
N = torque of friction (0~0.29N·m)
P= collision force
Q= torque due to collision force
T= driving force acting on module (0~5N)
f = forces acting on connecting points
i = the number of module (1...n)
m = mass of module (2kg)
r = length of each arm (0.15m)
x = x coordinate of center position (P) of module
y = y coordinate of center position (P) of module
α = angle of module

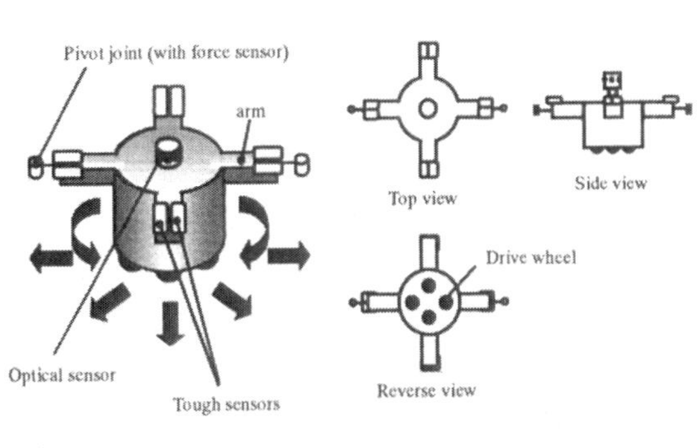

Fig.2 Schematic view of mobile robot module

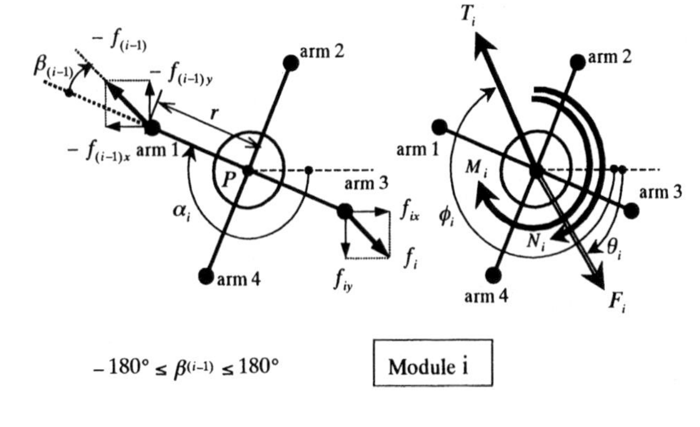

Fig.3 Physical quantities for the module i

β = angle of f with respect to arm 1
θ = angle of friction force
ϕ = angle of driving force
$[\ ]_x$ = x component of each quantity
$[\ ]_y$ = y component of each quantity
$[\ ]_i$ = quantity for i-th module

These equations are numerically solved by the Runge-Kutta(Williams, 1972) method for the case n=10, where the time step was carefully selected to obtain physically reasonable results.

3.2 *Rules of locomotion*

Each module has the following simple rules of locomotion which were determined through try and error procedure. When the modules 2 to 10 measure the magnitude and direction of the force acting on the connecting joint $-f_{(i-1)}$ and $\beta_{(i-1)}$, respectively, the driving force is determined according to Table 1 and Fig.4. The head module (module 1) moves forward with a steering angle γ_1 (see Fig.5). For the case when a mobile module collides with surrounding walls, the collision force P_i and torque Q_i are taken into account.

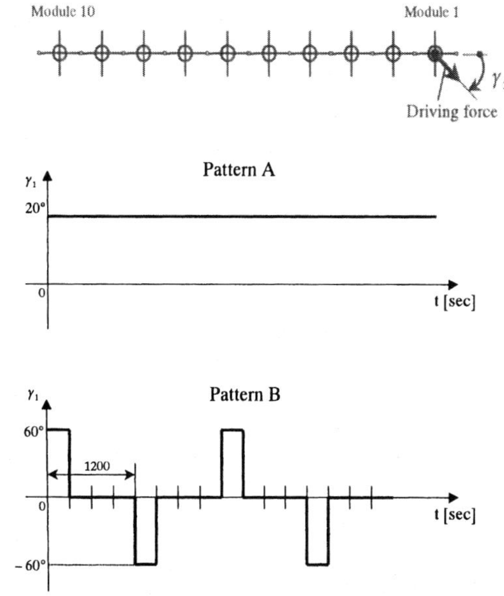

Fig.5 Steering angle of driving force of module 1

Table 1 Rules of locomotion action

Direction of $-f_{(i-1)}$	Direction of driving force
$-90° < \beta_{(i-1)} < 90°$	Forward (arm 1)
$90° < \beta_{(i-1)} \leq 180°$ or $-90° > \beta_{(i-1)} \geq -180°$	Backward (arm 3)

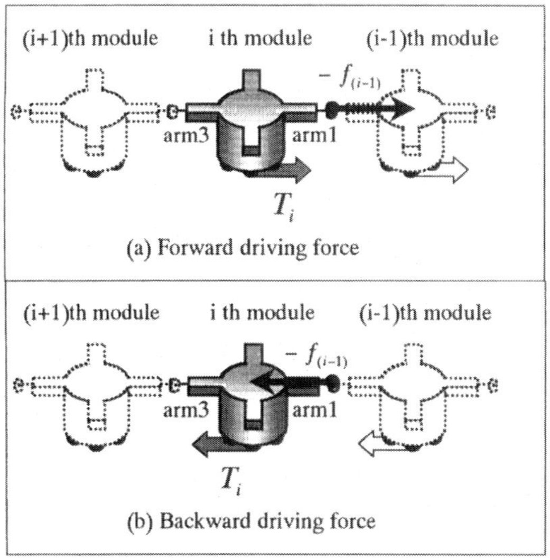

Fig.4 Rules of locomotion action

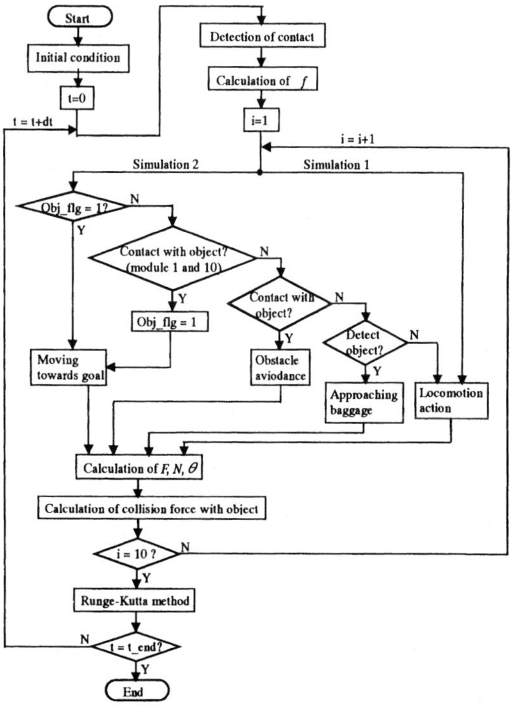

Fig.6 Simulation procedure (Simulation 1 and Simulation 2)

4 Simulation 1

4.1 *Simulation conditions*

We calculated the equations of motion of the linear-cluster robotic system to investigate the characteristics of robotic morphology formation to clarify how the morphology in a linear-cluster robotic system be created through the dynamic interaction between the robot modules and the friction plane.

Of course, the motion of the head module is considered to influence essentially on the robotic morphology formation and thus the head module is implemented with a different action rule from other modules, that is, the head module is assumed to move forward with various steering angles γ_1. Figure 5 shows two cases of action rule for the head module. In Pattern A, the head module generates a constant driving force in a certain direction (steering angle γ_1) with respect to its own body. In Pattern B, it generates an alternate driving force whose direction changes in three ways, i.e., rightward, forward, and leftward, successively. The value of the friction coefficient between the modules and the flat plane is changed to examine its effect on the characteristics of the formation of the cluster robotic morphology.

The simulation procedure is shown in Fig.6(see Simulation 1).

4.3 *Simulation Results*

Figure 7 shows the typical formation of robotic body shape, or morphology in the linear-cluster robotic system in pattern A. The cluster robotic system could form various morphologies depending on the magnitude of the friction coefficient. For example, the following body shapes after a long time elapse(say 2000 sec) were found: 1) J-shaped(μ =0.24, at 2000[sec]), 2) Waterdrop-shaped (μ =0.22, at 2000[sec]), 3) Horseshoe-shaped (μ =0.1725, at 1500[sec]), 4) Round-shaped (μ =0.1725, at 2000[sec]), 5) Arch-shaped (μ =0.17, at 2000[sec]).

Only the arch-shaped morphology seems to circulate steadily on the friction arena while maintaining the arch-shape. Figure 8 shows the critical relationship between the steering angle γ_1 and friction coefficient μ for the steady circulation of the arch-shaped cluster robotic system. Figure 9 shows the various shapes of wavy-line morphology in the case of Pattern B. This figure suggests that the linear-cluster robotic system may have a possibility to move forward with taking a S-shaped form which emerged through its dynamic interaction with the flat friction plane and the time-wise change in the steering angle. It is here noted that the cluster robotic system does not know the target shape and each module does not intend to form a certain shape as a whole body, but the physical interaction law can essentially determine the morphology in the cluster robotic system.

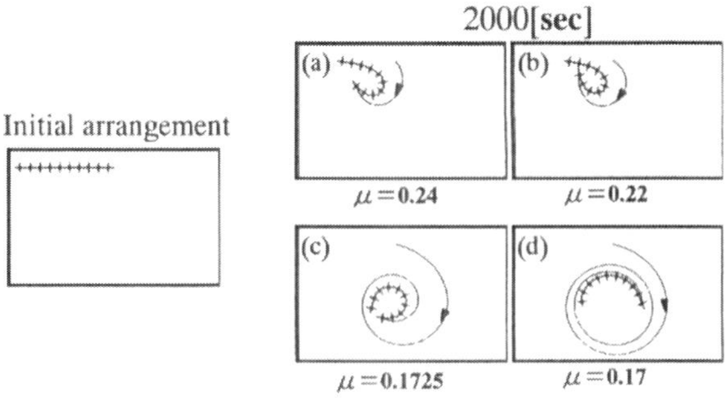

Fig.7 Morphology examples obtained in Pattern A
μ= coefficient of friction, Thin solid line =trajectory of center position of the head module

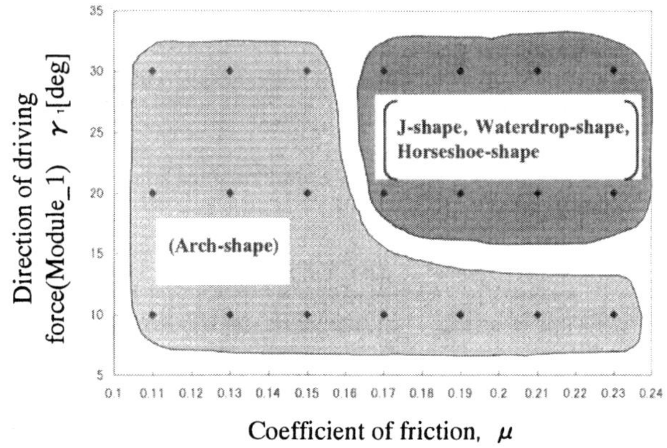

Fig.8 Map of morphology
◆ = calculation points

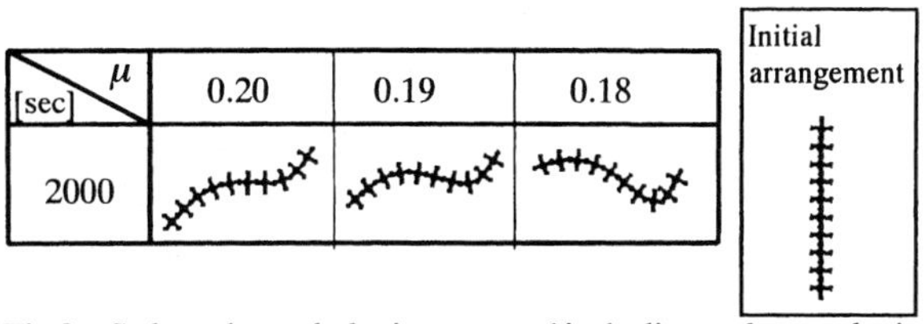

Fig.9 S-shaped morphologies appeared in the linear-cluster robotic system (Pattern B, μ= coefficient of friction)

5 Simulation 2

5.1 *Baggage carrying problem*

We consider the object (baggage) carrying problem in a linear-cluster robotic system. In addition to the above rules of locomotion, the intelligence of obstacle avoidance, approaching baggage, and moving towards goal is incorporated. Note here that the intelligence is, in this paper, assumed to be a priori given in the cluster robotic system, and we do not discuss the issue of how to evolve the intelligence. However, when the intelligence is given, our present interest is placed in demonstration of the ability to achieve the task of baggage carrying, which emerges through change in the robotic morphology.

To the cluster robotic system we assign the task to pass through a narrow and winding corridor, approach and enclose the baggage, and carry it to the goal. The following intelligence has been generated by hand in try and error manner:

Obstacle avoidance: When a module touches any object(wall, baggage, and other modules), it takes the action of obstacle avoidance. Figure 10 shows the details of action rules used for obstacle avoidance.

Approaching baggage: As shown in Fig.11, when a module detects the light signal A emitted from the baggage, it takes the action to steer itself according to the detected light direction. In this action each module does not intend to capture the baggage, only to move towards it. However as a result of both actions of obstacle avoidance and approaching baggage, the entire system happens to enclose the baggage. (The action selection priority among these actions is described in section 5.2).

Moving towards goal: Once the head module and module 10, (tail module) touch the baggage simultaneously, each module changes its action from the obstacle avoidance and approaching the baggage to the action of moving towards the goal (light B). Note that the baggage and the goal were assumed to emit the light signals A and B, respectively and the direction of these light signals to be detected with 8 circumferential divisions by appropriate optical sensors installed on each module as shown in Fig.11.

5.2 *Action selection priority*

The action selection priority among these action rules is shown in Table 2. Note again that the action selection priority has been determined through try and error manner in this paper. However this kind of sensor-motor coordination, or action selection should be intensively studied in future from the viewpoint of evolving and balancing the robotic morphology and intelligence under the given tasks and environmental conditions.

Basically, each module, except for the head module, moves according to the rules of locomotion and the head module moves forward with $\gamma_1 = 0°$ until finding the baggage. When the modules detect the baggage, they move towards it according to the action rules shown in Fig.11. If the modules touch the baggage, they move away from it according to the action of obstacle avoidance. After the touch sensor is off, the module approaches the baggage again. While modules stay in the area where the baggage light signal A can be detected, they take the actions of obstacle avoidance and trying to approach the baggage over and over again.

Each module was assumed to be able to know whether the entire system has enclosed the baggage or not from the communication that the head and tail modules have touched the baggage. After enclosing the baggage, the modules move towards the goal according to the action of moving towards the goal. Once this action is taken, the modes of obstacle avoidance and approaching baggage are not activated.

5.3 *Simulation Results*

Figure 12 shows the time trajectory of the linear-cluster robotic system in the course of passing a narrow corridor, enclosing the baggage and carrying it to the goal. In Fig.12, we found 3 kinds of robotic morphologies, first one is S-shaped and appears during passage through the narrow, winding corridor(2), second one is an arch-shape which appears in passing the upper right hand corner edge of 90° degree(3), and the final one is the circular shape to enclose the baggage(4). This figure suggests that the linear-cluster robotic system formed various morphologies on the friction arena according to the dynamic interaction with the environment and to its intelligence implemented beforehand, and finally the robotic system succeeded in finding the baggage, and carrying it to the goal by changing its morphology.

6 Discussion

In Simulation 1, though each robot module except the head one was implemented with the same rule of locomotion, and the differences in the values of friction coefficient of the flat plane caused the system to form the various morphologies. This result implies that the mechanical interaction with the external environment plays an important role in the formation of morphology in a cluster robotic system under a given intelligence and there exists certain physical laws which are essential to the formation of robotic morphology.

In Simulation 2, it was demonstrated that the cluster robotic system achieved a task to pass through the narrow and winding

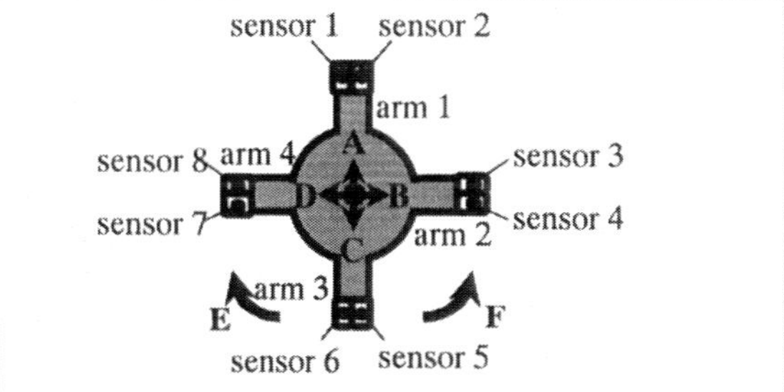

Touch sensor No.								Direction of driving force
1	2	3	4	5	6	7	8	
1	0	0	0	0	0	0	0	B
0	1	0	0	0	0	0	0	D
1	1	0	0	0	0	0	0	B or D (random)
0	0	■	■	0	0	0	0	D
0	0	0	0	■	■	0	0	A
0	0	0	0	0	0	■	■	B
■	■	■	■	0	0	0	0	F
■	■	0	0	0	0	■	■	E

1 = sensor on, 0 = sensor off

■ means [1,0],[0,1],or[1,1].

Fig.10 Action rules for obstacle avoidance

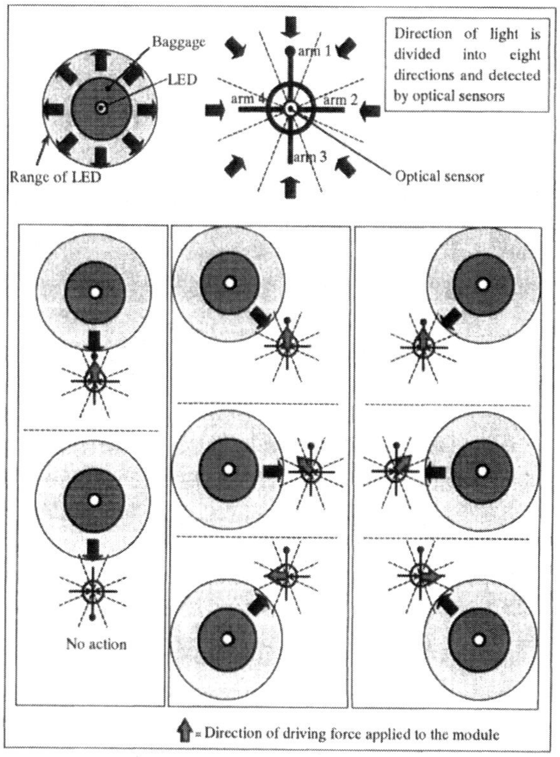

Fig.11 Action rules for approaching baggage

Table 2 Action selection priority for baggage carrying problem

Priority	Action	Conditions for each action			
		Force sensor (Pivot joint)	Light sensor	Touch sensor	Touch sensors of Module 1 and 10
Low	Rules of locomotion action	On	Off	Off	
↓	Approaching baggage	*	On	Off	
	Obstacle avoidance	*	*	On	
High	Moving towards goal	*	*	*	On

* = Do not care.

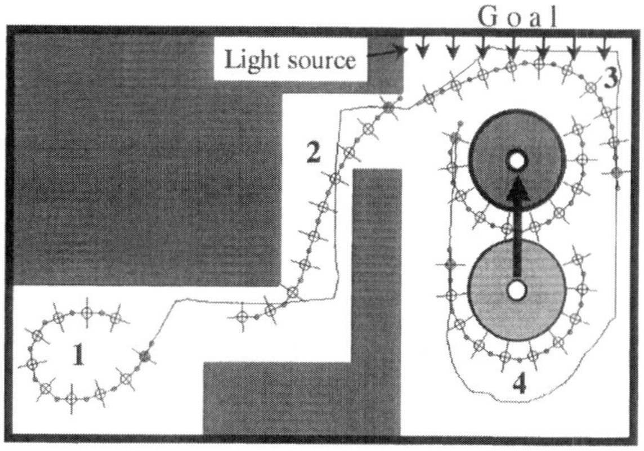

Fig.12 Trajectory of linear-cluster robotic system in baggage carrying problem

corridor, to enclose it and to carry the baggage to the goal by changing its morphology. This may be called as morpho-function emerged in the cluster robotic system. But the intelligence needed for generating the proper morphology in the cluster robotic system was implemented by hand in this paper. The study of evolution of intelligence through dynamic interaction with external environments is considered as one of the most important issues. For this evolution of intelligence, a learning mechanism seems to be one solution, but surely we need more elaboration to this problem.

Secondarily the balance of complexity in robotic morphology and intelligence is considered an important design issue for realizing a cheap robotic system (Pfeifer, 1996). And we need to develop the measure of complexity in the morphology and intelligence for clarifying this design issue. However it is still not clear how to balance them to obtain a cheap design.

Finally the multibody dynamic model and environmental conditions used in this paper seem to be simplistic for discussing the robotic morphology and morpho-function in real environments, and more realistic model and environmental conditions such as mechanical nonlinearity and uncertainty in friction coefficient must be taken into account to clarify the robustness of morpho-function in real worlds. Of course experimental studies using hardware cluster robotic systems and real environments must be carried out.

7 Conclusions

We proposed the concept of morpho-function in a cluster robotic system which emerges through changing robotic morphology according to dynamic interaction with external environments.

We demonstrated the importance of physical interaction in the formation of robotic morphology and morpho-function by showing the emergence of various morphologies in a linear-cluster robotic system. The robotic system was composed of linearly connected autonomous mobile modules and each module moved according to simple rules of action and to the dynamic interaction between the modules and the environment. Furthermore, the baggage carrying problem showed that the cluster robotic system can achieve a certain task by changing its own morphology according to the environment conditions by adding intelligence of obstacle avoidance and of moving towards the light signals representing the baggage and the goal.

Finally we discussed the future works to be done, some of which are as follows:
1) more realistic computer simulations to clarify the robustness of morpho-function, 2) evolution of robotic intelligence through dynamic interaction with external environments, and 3) experimental study using hardware cluster robotic systems in real world.

Acknowledgment

This research work was done as a part of the Research for the Future Project sponsored by the JSPS in 1996~1997 and the authers express their thanks to those involved in the program.

References

Kokaji,S,et al.(1994). Self Organization of a Mechanical System, *Distributed Autonomous Robotic Systems.* Springer-Verlag, pages 237-242.

Fukuda,T,et al.(1988). Dynamically Reconfigurable Robotic Systems, *Proc. Int. IEEE Conf. Robotics and Automation.* pages 1581-1586.

Yim,M.(1993). A Reconfigurable Modular Robot with Many Modes of Locomotion, *Proc.of the JSME Int. Conf. on Advanced Mechatronics.* pages 283-288.

Hirose,S.(1993). *Biologically Inspired Robots (Snake-like Locomotor and Manipulator).* Oxford University Press.

Williams,P.W.(1972). *Numerical computation.* Nelson.

Pfeifer, R.(1996). Building "Fungus Eaters":Design Principls of Autonomous Agents, *From animals to animat 4. Proc. SAB'96,* pages 3-12.

Collective sorting and segregation in robots with minimal sensing

Chris Melhuish, Owen Holland, and Steve Hoddell
Intelligent Autonomous Systems Engineering Laboratory
Faculty of Engineering, University of the West of England,
Coldharbour Lane, Frenchay, Bristol BS16 1QY
o-hollan, cr-melhu@uwe.ac.uk, sej-hoddell@wpg.uwe.ac.uk

Abstract

Many ants sort their brood so that brood items at similar stages of development are grouped together, and separated from items at different stages of development. The only model proposed to date assumes that ants engaged in brood sorting can sense both the type of element they are carrying, and the local spatial density of that type of element; the model was demonstrated in simulation using two types of objects. This paper describes a system of simple homogeneous autonomous mobile robots which are able to segregate or sort two types of physically identical objects differing only in colour, yet can sense only the colour of the object they are carrying, and have no capacity for spatial orientation or memory. This shows that this sorting problem can be solved by agents simpler than was previously supposed possible.

1. Introduction

In their seminal paper, Deneubourg *et al* (1991) presented a simulation showing that simple mobile agents could achieve the clustering of scattered objects of a single type, and the grouping and sorting of objects of two different types. For sorting, the agents needed to be able to sense the local densities of the different types of brood items, which was achieved by using a short-term memory, and also needed to know the type of any brood item they were carrying. Clustering was the result of the mechanism operating on only a single type of item.

Beckers, Holland, and Deneubourg (1994), achieved clustering with an even simpler mechanism, using physical robots which were unable to detect whether or not they were moving any objects, which had no memory, and which could sense the local density of objects only as being below or above a fixed threshold. This paper reports a further robot study which investigates whether this technique can be extended to the sorting problem to achieve the task in a simple way.

Spatial sorting is an informal notion; in order to structure this investigation, some formalisation is required. We divide spatial sorting into four basic types:

- Clustering: grouping a class of objects within a continuous area that is a small fraction of the area of the available environment
- Segregation: grouping two or more classes of objects so that each occupies a continuous area of the environment which is not occupied by members of any of the others
- Patch sorting: grouping two or more classes of objects so that each is both clustered and segregated, and each lies outside the boundary of the other
- Annular sorting: forming a cluster of one class of objects, and surrounding it with annular bands of the other classes, each band containing objects of only one type

2. Sorting in social insects

A type of patch sorting carried out by some social insects was described by Deneubourg (1991): "...the eggs are arranged in a pile next to a pile of larvae and a further pile of cocoons, or else categories are placed in entirely different parts of the nest...if you tip the contents of a nest out onto a surface, very rapidly the workers will gather the brood into a place of shelter and then sort it into piles as before." However, sorting need not be into piles; Franks and Sendova-Franks (1992) describe the annular brood sorting behaviour of *Leptothorax unifasciatus,* which forms concentric rings of progressively more widely spaced brood items at different stages of development.

The mechanisms of sorting in ants are not known, but Deneubourg's (1991) model certainly captures many features of ant behaviour. However, Franks and Sendova-Franks (1992) note that "it is not, in its present form, sufficient to create concentric brood clusters" and speculate that other factors may be involved, such as self-sorting by size (Barker and Grimson, 1990), and pheromones determining the space given to each type of brood item.

3. Sorting in robots

There do not appear to be any accounts of sorting analogous to brood sorting being carried out by physical robots. The related task of clustering using minimally capable robots was demonstrated by Beckers, Holland, and Deneubourg (1994), and Maris and te Boekhurst (1996). Most accounts in the literature deal with foraging tasks where the objects must be brought back to a designated location which is found by some means of navigation; see Cao et al (1995) for a useful bibliography and review.

4. Materials and methods

The robots and experimental environment used in this study were designed to investigate a range of social insect behaviour. The design of the robot system was inspired by the work of Franks and his collaborators on certain species of *Leptothorax* (Franks and Sendova-Franks, 1992; Franks *et al*, 1992). These ants live in small colonies in cracks in rock, and are behaviourally adapted to life in two dimensions. It is therefore possible to study their behaviour by providing them with a particular two dimensional habitat - the space between two glass slides - which allows unrivalled opportunities for observation and recording.

Leptothorax is also peculiarly suitable as a model for robotic investigations of some collective behaviours, because current robots, which are wheeled, only operate well in two dimensions. *Leptothorax* building behaviours tend to involve the movement of single lumps of material which are placed next to other lumps rather than piled on top of them. Similar functionality can be achieved with robots by pushing and pulling building blocks around the floor, rather than lifting them and piling them up.

The robots, known as U-bots, were designed and built in our laboratory to provide a flexible and capable platform for a range of collective robot experiments. (See Melhuish *et al*, 1998, for a full technical description). For this set of experiments, the robots were equipped with a gripper, which was designed to sense, grip, retain, and release frisbees (Figure 1) The frisbees, which are either red or yellow, are placed on the floor concave side up. If a robot moves towards a frisbee which is directly in its path, the frisbee will fit neatly inside the semicircular part of the gripper. As the gripper moves to enclose the frisbee, the raised edge of the frisbee contacts the lower parts of two small weighted 'barbels' mounted above the gripper which are free to swing fore and aft, but cannot move sideways. As the gripper continues to move forward, the barbels ride over the edge of the frisbee and dangle freely inside the concave area. The function of the barbels is to retain the frisbee when the robot turns on the spot. When the frisbee is right inside the gripper, an optical sensor detects its colour. (In order to ensure correct identification under all conditions of lighting, the red frisbees have been painted with black paint, with a white central portion to enable accurate visual location).

The functionality of the gripper is augmented by two further mechanisms. The first is a pin mounted at the rear of the gripper, which can be lowered by a small electric motor so that it projects down inside the concave rim of the frisbee. If the pin is in the raised position, and the robot reverses, the frisbee will be left in position. However, if it is

Figure 1: A U-bot with frisbees

in the lowered position, it will engage the rim, and the frisbee will be pulled backwards.

The second mechanism is similar to that used by Beckers *et al* (1994): the whole gripper is suspended so that, when pushing a frisbee, a backwards-acting force greater than some threshold value can move the gripper and trigger a microswitch. Most importantly, the force can act on the ends of the gripper jaws (when a misaligned frisbee is struck) as well as on the area within the jaws. The preset value is adjusted so that the microswitch is not triggered when the gripper encounters a single frisbee, but is always triggered by two or more.

The robots are run in a large octagonal arena with 4 metre sides (Figure 2); a video camera mounted 6 metres above the centre of the arena is linked to a PC based video capture system which automatically records an image at preset intervals.

Figure 2: The robot arena

5. Experiments and results

5.1 Experiment 1: Clustering

Our first task was to check that the new robot and gripper arrangement, combined with a suitable algorithm, could reproduce the clustering observed by Beckers *et al* (1994). The robots were programmed with a rule set corresponding to that used by Beckers, but with a slight difference imposed by the nature of the arena boundary. Beckers' arena had a deformable boundary wall which enabled robots to push their grippers past pucks trapped on the boundary, and to scoop them up. The new boundary is made of rigid plastic pipe, and we found that it was extremely difficult to adjust the infra red obstacle sensors to the exact distance to enable frisbees left on the boundary to be retrieved. The modified rule set, in decreasing order of priority, became:

Rule 1:
if (gripper pressed & *Object* ahead)
then make random turn away from object
Rule 2:
if (gripper pressed & no *Object* ahead)
then reverse small distance
make random turn left or right
Rule 3:
go forward

The rules work as follows. A U-bot approaching another U-bot or the boundary will detect it as an obstacle, but will continue moving until it hits it with its gripper. The combination of the gripper being pressed and an obstacle being detected will then trigger a random turn away from the side on which the obstacle was detected. If the U-bot was pushing a frisbee, it will retain the frisbee during the turn due to the action of the barbels. If a U-bot's gripper is pressed in the absence of an obstacle, as happens when it attempts to push more than one frisbee, the U-bot will reverse for a short distance (just over half a frisbee radius) and make a random turn; the effect of this is to leave the frisbee at the location where the gripper was triggered. If neither of these conditions is met, the U-bot moves forwards in a straight line.

If a frisbee is touching the boundary, a U-bot heading straight for it will receive it into its gripper, which will immediately be triggered; since the boundary wall will have been detected, the U-bot will obey Rule 1 and turn away from the boundary, keeping the frisbee in its gripper. It will then set off in a straight line (Rule 3) in the new direction, taking the frisbee with it. The U-bots can thus remove single frisbees from the boundary. However, if a frisbee lies so that it is touching a frisbee which is in turn touching the boundary, a U-bot which runs into the first frisbee will have its gripper triggered, but will be too far away from the boundary to sense it. It will therefore obey Rule 2, and reverse, leaving the frisbee.

22 red frisbees were placed in the arena at a uniform and regular spacing, and 6 robots were released. The video system was set to record a frame every 5 minutes. The red frisbees behaved in the expected way, first aggregating in small clusters, then forming larger ones, and finally, after 1 hour 50 minutes, forming a cluster of size 20, which we had decided was an acceptable criterion for completion. However, we noticed that in the middle phase of the experiment the robots had great difficulty 'stealing' frisbees from the intermediate sized clusters. The problem seemed to be geometric in origin: even in a small cluster it was rare for a robot to strike the cluster at a point which would allow a frisbee to be removed without triggering the gripper.

5.2 Experiment 2: The pullback algorithm

Following on from our observation that the slow progress in Experiment 1 might be due to the stability of the clusters, we decided to modify the algorithm so that frisbees were not deposited hard up against one another, but some little way off. To do this, it was necessary to employ the pin-dropping mechanism to enable robots to pull the frisbees backwards for some distance before releasing them. We did not want to introduce too much noise into the situation, so we decided to use both red and yellow frisbees, but to apply the pull-back tactic only to the yellow frisbees. The rule set now became more complex:

Rule 1:
if (gripper pressed & *Object* ahead)
then make random turn away from object
Rule 2:
if (gripper pressed & no *Object* ahead)
if yellow
then lower pin
reverse for pull-back distance
raise pin
endif
reverse small distance
make random turn left or right
Rule 3:
go forward

The U-bot's behaviour with respect to reds is unchanged. However, if it is pushing a yellow and hits another frisbee, or if it collides with a yellow frisbee which is already on a cluster (perhaps at the boundary), the new version of rule 2 will cause it to drag the frisbee backwards and leave it the pullback distance away from the contact point.

22 red and 22 yellow frisbees were uniformly spaced throughout the environment; again, 6 robots were used, with the pull-back distance being set at 2.6 frisbee diameters. Figure 3 shows the distribution of frisbees after 7 hours 35 minutes, the earliest time at which 20 of the 22 red frisbees were in a single cluster. (We defined a cluster as a group of frisbees in which any member was within a

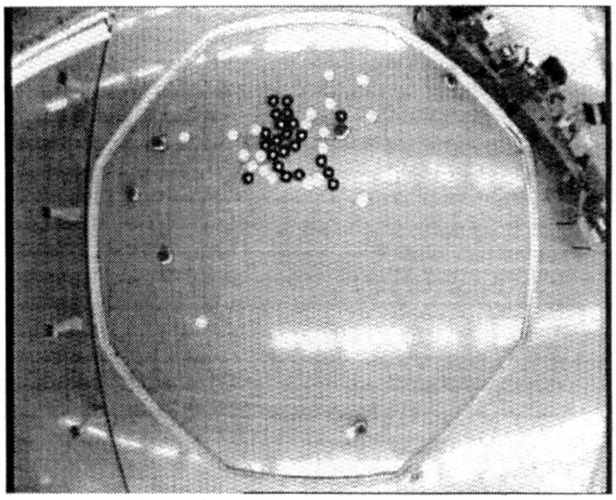

Figure 3: First appearance of sorting

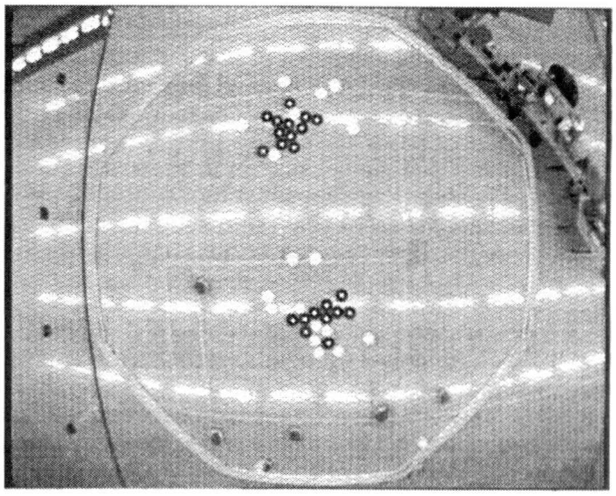

Figure 4: An intermediate stage

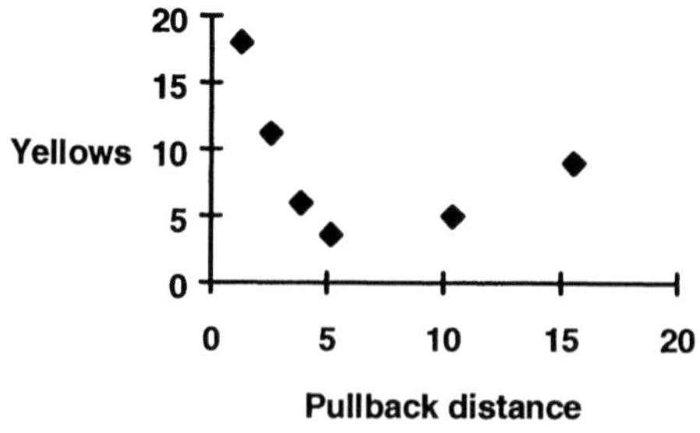

Figure 5: Number of yellows against pullback distance

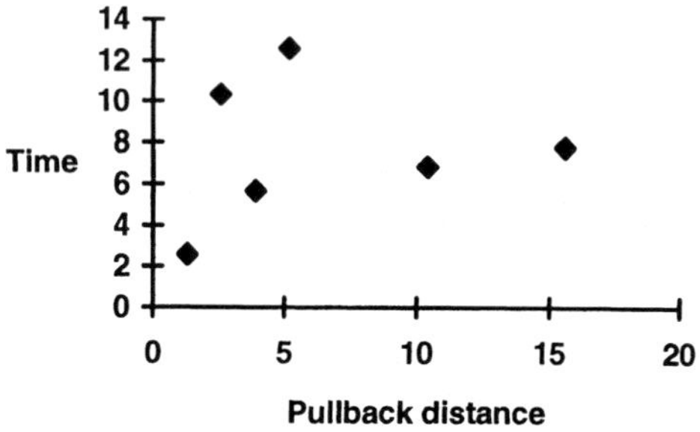

Figure 6: Time to completion against pullback distance

frisbee radius of at least one other member.) There is a central dense core of 16 reds, with 11 yellows and the other 4 reds being packed around this core, and the other yellows scattered more loosely nearby. This is clearly an outcome which can be regarded as sorting of some kind; within the classification proposed in the introduction, it is at least segregation, and inclines towards annular sorting. We had conceived the algorithm as a means of improving the rate of clustering, and had not thought that it might achieve sorting on its own. However, since the algorithm was probably simpler than anything we might have devised, we decided to explore it further.

The experiment was allowed to run on for half an hour; the cluster became more compact but the number of reds and yellows in the cluster at 8 hours 5 minutes was unchanged. The progress of the experiment over time reflected the final state in miniature, with a small number of tight red clusters each surrounded by yellows gradually giving way to the eventual pattern. Figure 4 shows the situation after 1 hour 45 minutes.

In order to check that this was not a fluke, we repeated the experiment 4 times. We decided to define the end of a trial as the time of occurrence of the first cluster containing at least 20 reds, and to measure both the elapsed time until that point, and the number of yellows included in the cluster. In every case, the outcome was similar, in that a single cluster containing 20 reds was formed. However, in the replications, the distribution of yellows within the cluster was generally less good, in that some yellows were trapped within the body of reds.

Trial	1	2	3	4	5
Time in hours	7.58	2.75	25.3	11.7	4.50
Number of yellows	11	12	11	10	12

Table 1

Table 1 shows the times to completion (when 20 reds were in the same cluster) of all 5 trials, and also the number of yellows included in the cluster. The number of yellows is almost constant; however the times are strikingly variable, with the slowest time (25 hours 20 minutes) being more than nine times the fastest (2 hours 45 minutes). Examination of the video records revealed that the very long times were associated with the formation of two clusters of the same size at opposite sides of the arena; these clusters could remain quite stable for hours. Other than this, we could see no reason for the differences in rates of progress, and can only conclude that, with these parameter settings, the variance of the time to completion is very high.

5.3 Experiment 3: Varying the pullback distance

The obvious next step was to vary the pull-back distance to establish any effects on the outcome, the time to completion, and the quality of sorting. Single trials were run using distances of 1.3, 3.9, 5.2, 10.4, 15.6, and 26.6 frisbee diameters. In every case a single cluster of reds was formed, except for the trial using 26.6 diameters, which appeared to be making no progress whatsoever after 6 hours, and was terminated. The number of yellows in the single cluster varied systematically with pull-back distance, decreasing from 18 (at 1.3 diameters) to a minimum of 3 (at 5.2 diameters) and increasing again to 10 (at 15.6 diameters). We added a further 4 trials at the 'best' value (5.2). Figure 5 shows the (average) number of yellows in the final cluster for each trial; Figure 6 shows the (average) times to completion.

The paucity of data points at some values makes interpretation difficult, but it may be reasonable to summarise the data as follows:

- for pullback distances of 15.6 or less, a single cluster containing at least 20 red frisbees is eventually formed

- the number of yellow frisbees forming part of the cluster decreases with increasing pullback distance to a minimum near 5.2, and then increases.
- for a given pullback distance, the time to completion can vary considerably

A key additional factor is the spatial distribution of the yellows which are not members of the central cluster. (Let us call them detached yellows). From observation three things are clear:

- detached yellows tend to be found near the final cluster
- as the pullback distance increases from 2.6, the detached yellows tend to become more widespread in the arena. (This is difficult to judge for values up to 10.4, simply because there are more detached yellows).
- detached yellows are often found near the two detached reds at termination

How should we judge the ability of the simple pullback algorithm to produce sorting and segregation? It is clear that the algorithm can produce good segregation (Figure 7) when the pullback distance is 5.2 frisbee diameters. It is also clear that a degree of annular sorting can occur, with reds being found in the centre of the cluster, and yellows being found at and beyond the edges of the cluster. However, the values giving the best segregation are not always those giving the best annular sorting.

At this stage several questions arose, which were answered by experiments which cannot be presented in this brief paper. If we ran the experiments for longer, would the yellow frisbees tend to form a better grouping? (Not noticeably). Would yellow frisbees show any degree of spatial association if there were no red clusters to act as foci? (Some, but not much). Could not segregation alone be achieved easily and economically simply by ignoring the yellow frisbees, and using the basic clustering algorithm on the reds? (No). And what would happen if, instead of determining the pullback distance by the colour of frisbee, it was determined randomly? (A single tight cluster is formed).

6. Discussion

The work reported here has shown that both segregation and also crude annular sorting can be achieved by systems of simple homogeneous agents with no ability to sense the local density of different types of objects. Previous work inspired by or concerned with ant behaviour had assumed that the sensing of local density was necessary. Could the results obtained here have any significance for the understanding of ant behaviour? The elements of the technique used here can certainly be found in the literature. For example, an account of *Leptothorax* building behaviour (Franks *et al*, 1992) mentions the possible use of an increased resistance to pushing a building block forwards against other building blocks as the cue to drop it: "...if the ants drop their granule only if they meet sufficient resistance...". He also reports a behaviour which is reminiscent of the pullback algorithm: "...workers individually carry granules into the nest. They walk head first towards the cluster of their nestmates, who are already installed in the nest, forming a fairly tight group. After

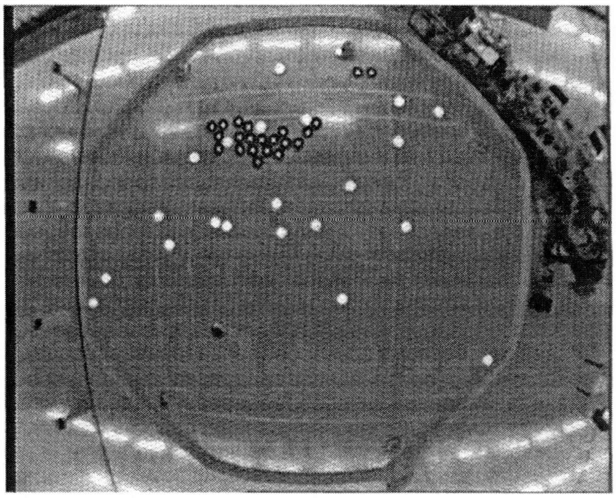

Figure 7: An example of good segregation

coming close to the group of ants, the builder then turns through 180° to face outwards from the nest. The worker then actively pushes the granule it is carrying into other granules already in the nest or, after a short time, if no other granules are encountered it simply drops its load."

The underlying phenomenon behind our sorting and segregation mechanism may be a variant of the self-sorting or self-sieving process, the involvement of which in brood sorting was suspected by Franks: "(In sand)...self-sorting occurs under the influence of shaking and gravity as the smaller particles are able to move down the crevices that lie between the larger particles (see Barker and Grimson 1990)...A similar process may occur when the ants sort their brood. Essentially the tendency of the ants to cluster their brood provides a centripetal force that serves instead of gravity." (Franks and Sendova-Franks, 1992). The shaking is clearly provide by the random movements of the ants (or robots). However, Franks sees the brood items as exerting some influence on the space around them (possibly via a pheromone) which affects the distance at which other items will be dropped; in our algorithm, the spacing is determined solely by the type of object being moved.

It is therefore fair to say that the pullback algorithm contains elements which are found in ant behaviours, that it implements a natural sorting process which has previously been thought to be used by ants for brood sorting, and that it requires less sensory information than any suggested alternative mechanism. We cannot ourselves establish whether it is in fact used by ants, but perhaps it should be added to the list of candidates for study. It must be admitted that it does not always produce very good segregation or sorting; on the other hand, the adaptive advantages of brood sorting as set out by Franks and Sendova-Franks are so strong that a variant of this simple method may well have been evolved at some stage, to be superseded later by a more complex but better solution.

Although the system studied here is a collective robot system, we have said little about the collective aspect. The main reason for this is that the system is so close to that of Beckers *et al* (1994) that their remarks will serve quite well for this investigation. The basis of the system is stigmergy: any agent may act on the environment to change it; any other agent may respond to this change; and all will respond in the same way in the same situation. Holland (1996a) distinguished between two fundamental types of stigmergy. Active stigmergy occurs when the effect of an environmental change due to one agent is to influence the *choice of behaviour* of the second agent through changed sensory input. For example, if in the present system a robot leaves a red frisbee against another red frisbee which is in the path of an approaching robot, the second robot will stop, reverse, and turn away when it hits the red frisbee, whereas if the first robot had not dropped its frisbee, the second robot would have scooped the original frisbee up and carried on in a straight line. The first robot affected the second by active stigmergy. Passive stigmergy occurs when the environmental change affects only the *outcome of the behaviour* of the second agent. In this system, as in Beckers', objects are usually collected by a robot scooping them up while moving forwards according to the default behaviour; if one robot dropped a yellow frisbee in the path of another, the second robot would not change its behaviour, but the outcome of the behaviour would have changed. The first robot would have affected the second by passive stigmergy.

If one is interested in minimalist collective systems, then the use of passive stigmergy where possible is attractive; it also appeals from the engineering point of view, since it is pure physics, and is therefore intrinsically reliable. The robot system described here uses some passive stigmergy, as mentioned above; the ant equivalent would have to actively grasp objects, rather than simply push them, and would therefore exhibit active rather than passive stigmergy in those instances. However, the two systems would be functionally equivalent.

As engineers, one of our main interests is in finding ways for systems of many homogeneous robots to work together to perform certain useful tasks, because the multiple robot paradigm offers many potentially useful features - for example, spatially distributed sensing and action, robustness, reliability through redundancy, and the ability to control the rate of work by controlling the number of agents. The robots in this study used a minimal algorithm, but clearly did not need to - the platform is capable of supporting rich sensors and significant computation. Why should we handicap ourselves by using minimally capable robots?

The answer is that technical progress is advancing towards the production of very small robots (micro- and nano-robots) which *will* have severely limited sensing and computation, and which will need to operate in very large groups or swarms to affect the macroworld. The only hope of managing such systems lies in using some variant of the methods used by social insects (Holland, 1996b; Holland and Melhuish, 1996). However, the limitations of the social insect paradigm for use in engineered systems are many, ranging from the lack of any principled design methods, through the impossibility of being sure that some combination of circumstances might not trigger an unwanted emergent behaviour, to the lack of adequate techniques for externally directing the operations of the agents. From the engineering point of view, these deficiencies are so serious that these methods will only be used where there is no choice; the animat community should be aware of this when discussing possible applications.

Acknowledgments

The authors are grateful to Nigel Franks and Ana Sendova-Franks for their advice and assistance, and to Ian Horsfield and Jason Welsby for hundreds of hours of robot-minding. This work was partially supported by Royal Mail.

References

Barker G. and Grimson M. (1990), 'The physics of muesli', New Scientist 126 (1718): 37-40

Beckers R. Holland O.E. and Deneubourg J.L. (1994), 'From local actions to global tasks: Stigmergy and collective robotics', in Brooks R. and Maes P. Artificial Life IV, MIT Press

Cao Y.U. Fukanaga A.S. Kahng A.B. and Meng F. (1995) 'Cooperative mobile robotics: Antecedents and directions', IEEE International Conference on Intelligent Robots and Systems

Deneubourg J.L. Goss S. Franks N. Sendova-Franks A. Detrain C. and Chrétien L. (1991), 'The dynamics of collective sorting: Robot-like ants and ant-like robots', in Meyer J.A. and Wilson S.W. (eds): Simulation of Adaptive Behaviour; From Animals to Animats, MIT Press

Franks N.R. and Sendova-Franks A.B. (1992), 'Brood sorting by ants: distributing the workload over the work-surface', *Behav. Ecol. Sociobiol.* 30: 109-123

Franks N.R. Wilby A. Silverman B.W. and Tofts C., (1992), 'Self-organising nest construction in ants: sophisticated building by blind bulldozing' *Animal Behaviour*, 44, 357-375

Holland O.E. (1996a), 'Multi-agent systems: lessons from social insects and collective robotics', AAAI Spring Symposium on Adaptation, Coevolution and Learning in Multiagent Systems, Stanford

Holland O.E. (1996b) 'The use of social insect control methods with multiple mobile robots' IEE Conference on Control '96, Exeter

Holland O.E. & Melhuish C.R. (1996) 'Getting the most from the least: lessons for the nanoscale from minimal mobile agents' Artificial Life 5, Nara, Japan

Maris M. and te Boekhurst R. (1996), 'Exploiting physical constraints: heap formation through behavioural error in a group of robots' (Manuscript)

C.R.Melhuish, O.Holland, and S.E.J.Hoddell, (1998) 'Using chorusing for the formation of travelling groups of minimal agents', 5th International Conference on Intelligent Autonomous Systems, Sapporo, Japan.

Tough guys don't dance: Intention movements and the evolution of signalling in animal contests

Jason Noble
School of Cognitive and Computing Sciences
University of Sussex, Brighton BN1 9QH, U.K.
jasonn@cogs.susx.ac.uk

Abstract

Two animals contesting possession of a resource sometimes fight, but more often engage in aggressive displays until one or the other retreats. Game theory predicts that signals of strength or aggressive intent will not be heeded unless necessity or high cost make them reliable, but some animal threat displays appear to contradict this prediction. An evolutionary simulation model is presented in which two animats of variable fighting ability compete for a resource and have access to their opponent's intention movements, i.e., their real-time behavioural choices. Reliable communication of fighting ability does not evolve. The results of the simulation are used to assess conflicting theories of signalling in competitive interactions.

1. Introduction

Animals contesting the possession of a resource are often observed to settle the dispute by exchanging signals or threat displays rather than engaging in an all-out fight. For example, mantis shrimps *Gonodactylus bredini* contest the ownership of small cavities in their coral reef habitat. These contests sometimes result in physical combat, but often an opponent is deterred by a claw-spreading threat display (Adams and Caldwell, 1990). Red deer stags *Cervus elaphus* compete for control of groups of females, but unless two stags are closely matched in strength, the weaker will usually retreat after a roaring contest and/or a parallel walk display (Clutton-Brock et al., 1979).

There is some controversy in the biological literature about how best to explain these observations. The early ethologists suggested that threat displays were honest signals of aggressive intent that served to prevent costly fights, but the group-selectionist overtones of this idea mean that it is no longer taken seriously. Subsequently, game-theoretic models of aggressive interaction have predicted that signals of aggressive intent or high fighting ability will be vulnerable to invasion by bluffers and thus not evolutionarily stable (Maynard Smith, 1982)—receivers of the signal will have no reason to take it seriously. There is widespread agreement that if a signal is for some reason unfakeable (such as the necessary connection between stamina and energetically costly roaring in red deer) then it will be evolutionarily stable for animals to use the signal to settle contests, escalating only when opponents are well-matched. However, the similar use of fakeable signals is not expected to be stable. This latter conclusion is apparently at odds with the ethological data (Hinde, 1981). Sometimes animals do seem to pay attention to threat displays that could be (and are) faked, as when a mantis shrimp, its exoskeleton soft and vulnerable after moulting, successfully drives off an intruder that could in fact defeat it in combat.

Several authors have tried to address this explanatory gap, constructing models of contests in which the strength and intentions of one animal cannot be directly perceived by the other, but might be more or less reliably communicated through a threat display. Adams and Mesterton-Gibbons (1995) argue that the use of a threat by a weak animal involves a "vulnerability cost". This is one version of the handicap principle (Zahavi, 1975), which asserts a necessary relationship between a signal's cost and its reliability. A vulnerability cost exists because, if the threat does not work, a weak animal is likely to be seriously injured by its probably-stronger opponent. On the other hand, weak animals stand to gain proportionately more if the threat is successful. Adams and Mesterton-Gibbons predict that bluffing will be stable: threats will be made and sometimes heeded, but should be expected from the very strong *and* the very weak. Wheeler and de Bourcier (1995) describe a simulation of aggressive territorial signalling. They also invoke the handicap principle, but to different effect: their simulation suggests that if the production costs associated with a threat display are sufficiently high, then the honest signalling of "aggressiveness", i.e., aggressive intent, will be evolutionarily stable.

In contrast, Enquist (1985) presents a game-theoretic model in which contestants are either strong or weak, and cost-free, binary signals are exchanged before the decision to fight or flee is taken. Enquist concludes that, under certain conditions[1], the honest signalling of fighting

[1] $0.5v - c > v - d$, where v is the value of the resource, c is the cost of an escalated fight between two equally matched opponents, and d is the cost to a weak animal of being

ability will be evolutionarily stable, even though the signals are cost free. Fights will occur only between evenly-matched opponents, and weak animals will defer to signals denoting strength. In a second model, Enquist goes on to claim that the cost-free signalling of "local strategy", i.e., aggressive intent, can also be evolutionarily stable. Hurd (1997) has recently extended Enquist's first model and also concludes that the cost-free signalling of fighting ability is possible. Furthermore, if only two signals are available, and if they vary in cost, it will be evolutionarily stable for weak animals to use the more costly of the two signals: not as a bluff, but as an honest advertisement of low fighting ability. These paradoxical results clearly run against the grain of most game-theoretic predictions about signalling in cases where the interests of the interacting parties conflict: such signals are either expected to be uninformative (Maynard Smith, 1982), and thus not signals at all, or stabilized by being too costly for poor-quality signallers to produce (Grafen, 1990). Caryl (1987) has criticized Enquist's methodology and argues that reconsideration of his first model leads to the opposite conclusion.

In an attempt to resolve the controversy, the current paper presents a simulation model of contests over an indivisible resource. The results will be compared with the conflicting predictions of the models outlined above. The aim of using a simulation is to avoid oversimplification. In particular, time will be modelled in an approximately continuous fashion: game-theoretic models of aggressive signalling rarely allow for more than two time-steps—an exchange of signals followed by a choice of actions—and thus may fail to capture critical aspects of real-time interactions.

There is a second motivation for the work presented here. Whilst the honest signalling of intentions looks unlikely from a game-theoretic perspective, it has been cogently argued (Tinbergen, 1952; Krebs and Dawkins, 1984) that *intention movements* (i.e., movements necessarily preceding an action, such as a dog baring its teeth to bite) probably function as "seeds" in signal evolution. Rather than incorporating an exchange of artificial, discrete signals, the current model seeks to explore the plausibility of the intention-movements idea by using such movements as the medium of potential information transmission.

2. The simulation model

The model involves two simulated animals (henceforth animats) contesting the possession of a resource. Due to a shortage of hard data on the metabolic costs of fighting, the simulation does not model contests in a particular species. However, the mantis shrimp will serve as a useful illustration: assume that two shrimps have discovered a desirable cavity. Each shrimp knows its own fighting ability, but is unable to perceive the ability of its opponent. Each shrimp can perceive movements towards aggression or retreat on the part of its opponent. Based on this information, shrimps can elect to attack or to flee or to do something in between. If the shrimps engage in an all-out fight, both shrimps will suffer costly injuries but the stronger one is likely to win and gain possession of the cavity. Over generational time, the shrimps may or may not evolve a signalling system that allows them to settle contests without fighting.

The model is based on a single behavioural continuum between attacking and fleeing. An animat is located at some point Θ on this continuum; contests involve two animats randomly selected from a population. Animats begin the contest at $\Theta = 0$ and, each time-step, can move a maximum of δ units in either direction. An animat with $\Theta > A$ has crossed the "attack line", and is deemed to be physically attacking its opponent. An animat with $\Theta < F$ has crossed the "flee line": the contest ends and the animat has abandoned the contested resource to its opponent. In the runs presented here, $F = -A$ and $\delta = A/4$. The simulation captures intention movements in that animats can neither immediately attack nor immediately flee: from the starting position, it takes at least four time-steps to do either. A plausible threat display might be to "hover" with Θ just less than A. Figure 1 shows a hypothetical contest.

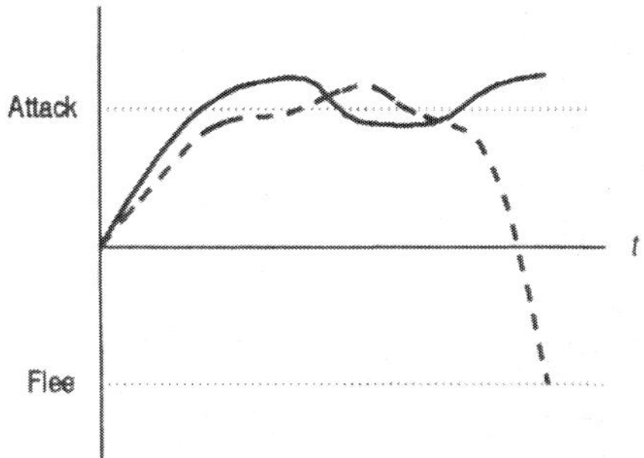

Figure 1 Course of a hypothetical contest: the horizontal axis is time and the vertical axis is the Θ-continuum. The weaker animal (dashed line) flees after the stronger animal shows a greater willingness to back up "threat displays" by attacking.

Animats have associated with them an energy level e that is set to 0 at birth, and a fighting ability f ($5 \leq f \leq 15$) that is randomly assigned at the beginning of each contest.[2] The two costs and one benefit of the contest affect the value of e. The first cost is due to being attacked: at each time-step that $\Theta_{opponent} > A$, the ani-

attacked by a strong one.

[2] Heritable fighting ability would quickly lead to uninteresting fixation. The model selects for responsiveness to varying values of f_{self}, as presumably exists in the mantis shrimp which is periodically vulnerable due to moulting.

mat suffers an injury cost of $-f_{opponent}$. The second cost is an energy cost for aggressive display or attack: at each time-step that $\Theta > 0$, the animat pays a cost of $-k\Theta/A$, with $k = 1$. So, for example, an animat attacking its opponent with $\Theta = 1.1 \times A$ would endure -1.1 units of energy cost per time-step; note that the cost of attacking is always much less than the cost of being attacked, for any f. Note also that any activity where $\Theta < 0$ involves no energy cost—this is justified on the basis that backing off, preparatory to running away, is much less energetically expensive than aggressive behaviour. In principle, therefore, cost-free signalling should be possible when $\Theta < 0$. The only benefit in the contest is to gain control of the resource, which is worth $V = 100$ units.

The contest can end in one of three ways. One animat may flee, as discussed above. Secondly, one animat may win the contest through brute force: if an animat loses more than $C = 200$ units of energy *during any one contest*, it has been physically overcome by its opponent. The contest ends immediately and the opponent gains the resource. Regrettably, the value of C puts an artificial cap on the costs an animat can sustain in the course of a single contest; however, the values of C and V have been chosen such that, on the face of it, the resource is worth having but not worth suffering serious injury for. In Maynard Smith's (1982) much simpler Hawk-Dove game, these relative cost and benefit values resulted in a mixed-strategy equilibrium where half of the population played Hawk and the other half played Dove. The contest described here can be thought of as a souped-up continuous-time version of the Hawk-Dove game, in which players of varying strength[3] can perceive and respond to their opponent's tendency towards a hawkish or dovish strategy. Finally, the contest can end because a time limit, $t_{max} = 50$ time-steps, has been reached, in which case neither animat gains the resource. The values of t_{max}, C, k, and δ, and the range of values of f, have been coordinated such that it is possible for even the weakest animat to overcome an opponent (assuming the opponent does not retaliate) within the time limit. Although a high degree of arbitrariness is inevitable in setting parameter values for an abstract simulation, it is hoped that this coordination of values will at least prevent such disasters as, for example, enforcing honesty amongst the weak by having C too high, or t_{max} too low, for a weak animat to ever win by fighting.

The animats have as inputs f_{self}, Θ_{self}, and $\Theta_{opponent}$. Informally, they know their own strength, they can see what they're doing, and they can see what their opponent is doing. The animats also have access to a random input, to allow for probabilistic strategies. The animats produce a continuous output in the range $\pm\delta$ which is applied to their Θ-position for the next time-step.

[3] In the Hawk-Dove game all players are of equal fighting ability.

The animats were implemented as five-neuron fully inter-connected continuous-time recurrent neural nets (CTRNNs), with the activity of neuron 0 taken as the output. CTRNNs are the among the most general of artificial neural network architectures; they have been treated here as a black-box control system. The recurrent aspect of the nets makes it possible for the animats to evolve some form of short-term memory rather than being purely reactive. All parameter values for the nets were taken from Yamauchi and Beer (1994); an animat's genotype specified the bias, connection weights, and tau-value for each of its five neurons.

The evolutionary engine was a real-valued genetic algorithm with a population size of 100, run for 5000 generations. In each generation, animats were randomly selected to play out 500 contests; each animat could thus expect to participate in 10 contests in its lifetime and was guaranteed at least 5. An animat's fitness score was equal to $e_{total}/n_{contests}$, i.e., its average energy return per contest. For breeding purposes, these fitness scores were normalized as deviations from the mean: animats with negative scores were discarded, and roulette-wheel selection was applied to those that remained.

Two control conditions were devised: in the "blind" condition, the animats are denied access to $\Theta_{opponent}$ and therefore any communication is impossible. In the "unfakeable" condition, the animats are given access to $f_{opponent}$, and can be thought of as exchanging unfakeable signals of strength. In order to facilitate comparisons with previous models, both continuous and discrete distributions of fighting ability were used. In the continuous fighting ability (CFA) case, f was uniformly randomly distributed between 5 and 15, while in the discrete fighting ability (DFA) case animats were either weak ($f = 5$) or strong ($f = 15$). The experimental and control conditions were crossed with the two f-distributions to make six conditions in all.

3. Results

Each of the six experimental conditions was run 10 times with a different random seed. The final 1000 generations of each 5000-generation run were used as a window period for statistical analysis; all of the results below refer to average behaviour within this period. Standard errors refer to error across the 10 trials.

In observing the progress of the contests on a computer monitor, the overwhelming impression was that most of the animats wasted no time in moving rapidly towards either the attack or the flee line. Contests were generally resolved quickly: the mean duration was 10.14 time-steps overall, and less than 0.4% of contests reached t_{max}. If threat displays were occurring, they did not involve sustained action.

Figure 2 shows the mean fitness values across conditions. Several points can be made based on this graph alone. Mean fitness is determined by the value V of the resource, which an animat can expect to obtain about half the time, less the mean costs of aggression and being injured. In an ideal signalling system, where cost-free signals of strength were exchanged and the weaker animat always retreated immediately, the mean fitness would be $V/2 = 50$ (assume that equally matched animats allocate the resource randomly). In the experimental condition, such ideal situations clearly did not evolve—the negative mean fitness implies that the $V/2$ expected payoff was balanced against greater energy and injury costs. The unfakeable condition provides an index of how efficiently the animats can allocate V when they have reliable, cost-free information about their opponent's strength, and in the CFA case their efficiency approaches 60% of the ideal. The fact that mean fitness in the experimental condition was significantly lower than in the unfakeable condition (CFA: $t_{18} = 8.03$, $p < 0.001$; DFA: $t_{18} = 19.30$, $p < 0.001$) is evidence that if fighting ability is being signalled, the signals are costly. Differences in mean fitness between the blind condition and the experimental condition were of marginal statistical significance. Communication is impossible in the blind condition; therefore the similarity in fitness scores suggests that if signals are being exchanged in the experimental condition, their costs cancel out their benefits.

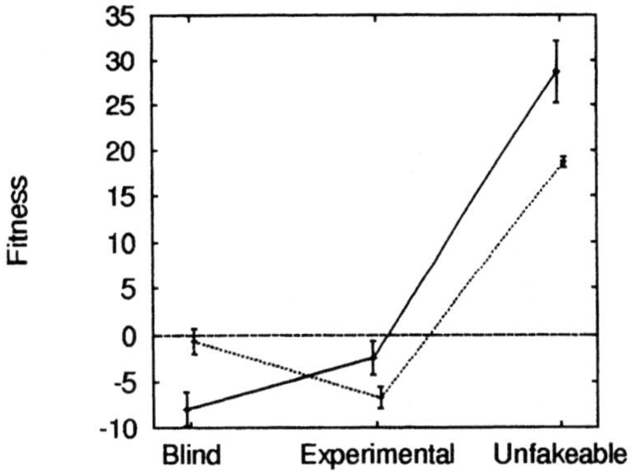

Figure 2 Mean fitness ±1 s.e., for CFA (solid line) and DFA (dashed line), across the three conditions.

Figure 3 shows the percentage of contests that were resolved by all-out fights, i.e., by one animat overcoming the other (the remainder were almost all resolved by one animat fleeing). In the unfakeable condition, as might be expected, fights were relatively rare. In the experimental condition (CFA) they occurred 24.5% of the time, and of those fights, 41.0% were between animats with closely matched fighting ability ($|f_a - f_b| < 2$). At first glance this might suggest that the animats had evolved a signalling system that allowed them to avoid costly fighting much of the time. However, there was no significant difference in the proportion of fights between the experimental condition and the blind condition, and fights in the blind condition were just as likely to be against closely matched opponents. If a reliable signalling system were in place, it would presumably lead to fewer fights overall and a greater proportion of well-matched opponents when fights did take place.

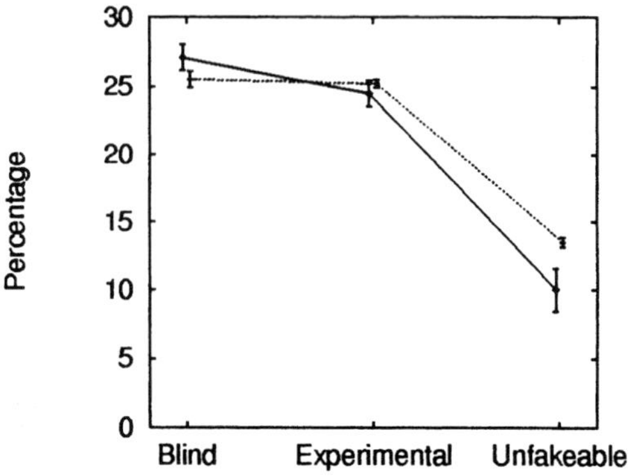

Figure 3 Mean percentage of contests resolved by all-out fights ±1 s.e., for CFA (solid line) and DFA (dashed line), across the three conditions.

Overall, stronger animats were more aggressive: they were likely to have a higher Θ-values than weak animats at any particular time-step. This suggests that weak animats should have been able to infer their opponent's strength from their Θ-value and respond accordingly. Figure 4 shows the mean correlation coefficient between Θ_{self} and $f_{opponent}$ at time-step 4, when all contests were still in progress (the correlation was calculated for each generation). In the unfakeable condition there is a negative relationship: unsurprisingly, animats that could reliably perceive their opponent's strength were likely to flee from stronger opponents. In the blind condition there is of course no relationship at all. In the experimental condition, there is the suggestion of a negative relationship, but it accounts for less than 1% of the variance in Θ_{self}.

A general link between f_{self} and Θ_{self} also suggests the possibility of bluffing, i.e., deception by moving to a Θ-value usually characteristic of higher f. Given the brevity of the contests and the likely importance of first impressions, bluffing was investigated by tabulating the animats' opening moves (their $\Delta\Theta$ for time-step 1). Figures 5 and 6 show the frequencies of opening moves by fighting ability for CFA and DFA respectively. Substantial variation existed in these relationships across trials, and presenting mean values would obscure the situation—figures 5 and 6 show data from typical runs. In both cases there is evidence of stereotypical, extreme responses. Stronger animats almost always make an aggressive first move. Weaker animats are more likely to play a mixed strategy of sometimes advancing and some-

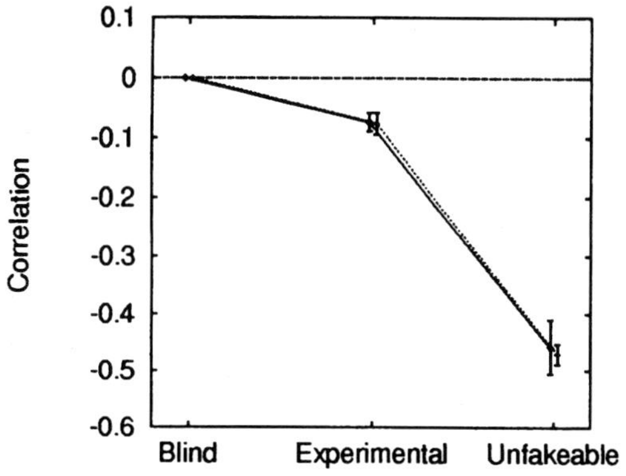

Figure 4 Mean correlation between Θ_{self} and $f_{opponent}$ ± 1 s.e., for CFA (solid line) and DFA (dashed line), across the three conditions.

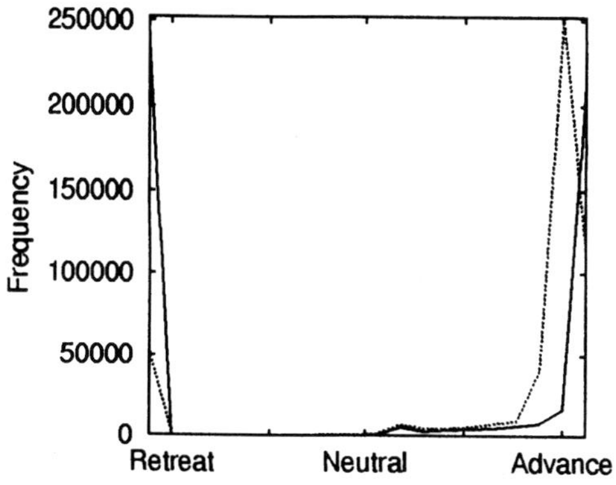

Figure 6 Total frequencies of opening moves by weak (solid line) and strong (dashed line) fighting ability, with the move data grouped into 20 bins for plotting. The plot shows data for the window period of run 5 in the experimental condition with DFA.

times retreating, and it seems reasonable to describe the advances as attempted bluffs.

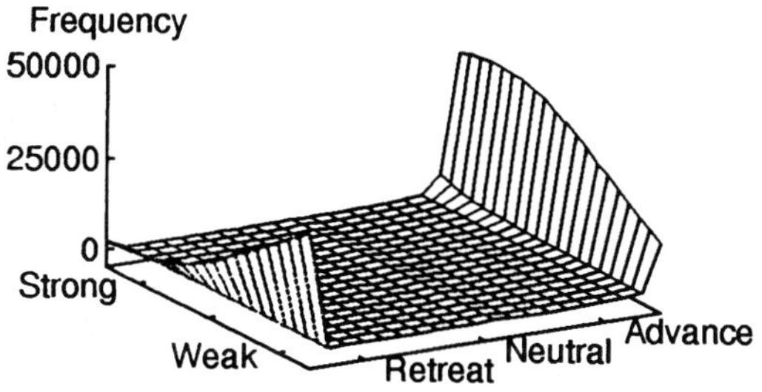

Figure 5 Total frequencies of opening moves by fighting ability, with each variable grouped into 20 bins for plotting. The plot shows data for the window period of run 6 in the experimental condition with CFA.

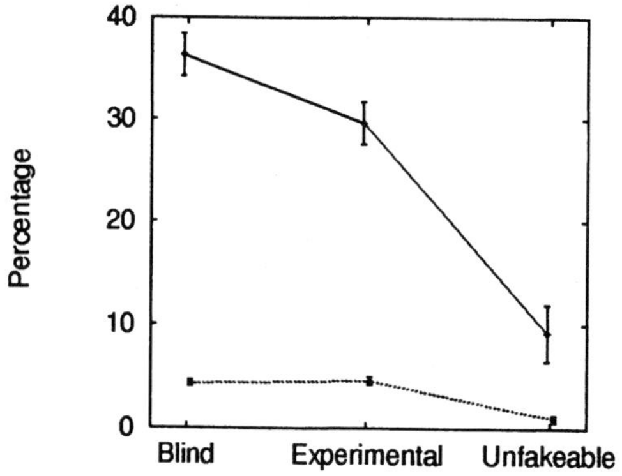

Figure 7 Mean percentage of contests won by the weaker animat ± 1 s.e., for CFA (solid line) and DFA (dashed line), across the three conditions.

If bluffing in the experimental condition was occurring successfully, then this would presumably result in a relatively high proportion of contests being won by weaker animats: bluffing is impossible in the blind control and pointless in the unfakeable. Figure 7 shows the percentage of contests in which the weaker animat gains the resource (note that the DFA results are not directly comparable as half of the time there *was* no weaker animat). While the experimental condition leads to more "upset wins" than the unfakeable condition (CFA case, $t_{18} = 5.93$, $p < 0.001$), the difference between the experimental and the blind condition is of marginal significance.

4. Discussion

Evolved behaviour in the experimental condition was more like the behaviour of "blind" animats than that of animats able to perceive the strength of their opponent. The data suggest that if any signalling systems evolved in the experimental condition they were both inherently costly and no more effective in preventing "unnecessary" fights than blind strategies. The most obvious interpretation is that no signalling systems evolved.[4] The mutual ability to perceive the intention movements of one's opponent in an aggressive interaction is, in the long run, of no fitness benefit.

The predictions of Enquist (1985) and Hurd (1997) , that cost-free signalling of fighting ability could be an evolutionarily stable strategy, were certainly not supported. This was true whether or not fighting ability

[4] Future work could incorporate more detailed time-series analysis of the contests as a way of verifying this claim.

was discretely distributed. Caryl's (1987) charge that Enquist's model simply fails to capture the details of animal conflict may be the best explanation.

Wheeler and de Bourcier's (1995) model suggests that evolution will favour costly, honest signals of aggressive intent. The current simulation allowed for the possibility that stronger animats would use a costly signal to display their greater willingness to engage in all-out fights: e.g., by moving to $\Theta \gg A$. If such a signalling system had evolved, however, and if receivers were paying attention to it, we would expect there to be a reduction in the number of contests resolved by fighting (compared to the blind control). This was not observed, and the conclusion must be that Wheeler and de Bourcier's rather specific simulation does not generalize to the current contest model.

Adams and Mesterton-Gibbons's (1995) prediction, that the strongest animals would signal their strength and the weakest would attempt to bluff, was not borne out either, although high levels of bluffing were observed. Indeed, the data on bluffing strongly support the conventional game-theoretic view (Maynard Smith, 1982; Krebs and Dawkins, 1984) that participants in aggressive interactions will eventually come to pay little attention to each other's manipulative "signals". For example, in the CFA case (figure 5), the weakest animats start with an aggressive move about a third of the time. These bluffs at least occasionally result in the animat gaining the resource (figure 7), but to explain this we do not need to propose that the animat's opponent is paying any attention, because in the blind condition similar results are observed. The blind condition forces animats to develop strategies based only on knowledge of their own fighting ability, and the fact that they manage to avoid all-out conflicts most of the time reminds us that communication is not always necessary for coordination.

The findings also serve as a reminder that accounting for the use of apparently fakeable signals in aggressive interactions between animals stands as an open problem for biological optimization modelling. Either our models or our interpretation of the ethological data have to change.

As no signalling systems evolved, it is difficult to assess the implications of the work for the idea that intention movements can serve as raw material for signal evolution. However, in nature animals are obviously not given dedicated, artificial communication channels *deus ex machina*. Future simulation models of signalling will, by incorporating phenomena like intention movements, investigate what must happen when evolution co-opts an existing behaviour for the new purpose of signalling.

Acknowledgements

I am grateful to the Association of Commonwealth Universities and the British Council for financial support. I would also like to thank Seth Bullock, Ezequiel di Paolo, and many others at Sussex for valuable discussions.

References

Adams, E. S. and Caldwell, R. L. (1990). Deceptive communication in asymmetric fights of the stomatopod crustacean *Gonodactylus bredini*. *Animal Behaviour*, 39:706–716.

Adams, E. S. and Mesterton-Gibbons, M. (1995). The cost of threat displays and the stability of deceptive communication. *Journal of Theoretical Biology*, 175:405–421.

Caryl, P. G. (1987). Acquisition of information in contests: The gulf between theory and biology. Paper presented at the ESS Workshop on Animal Conflicts, Sheffield, UK, July.

Clutton-Brock, T., Albon, S. D., Gibson, R. M., and Guinness, F. E. (1979). The logical stag: adaptive aspects of fighting in red deer *(Cervus elaphus L.)*. *Animal Behaviour*, 27:211–225.

Enquist, M. (1985). Communication during aggressive interactions with particular reference to variation in choice of behaviour. *Animal Behaviour*, 33:1152–1161.

Grafen, A. (1990). Biological signals as handicaps. *Journal of Theoretical Biology*, 144:517–546.

Hinde, R. A. (1981). Animal signals: Ethological and games-theory approaches are not incompatible. *Animal Behaviour*, 29:535–542.

Hurd, P. L. (1997). Is signalling of fighting ability costlier for weaker individuals? *Journal of Theoretical Biology*, 184:83–88.

Krebs, J. R. and Dawkins, R. (1984). Animal signals: Mind reading and manipulation. In Krebs, J. R. and Davies, N. B., editors, *Behavioural Ecology: An Evolutionary Approach*, pages 380–402. Blackwell, Oxford, second edition.

Maynard Smith, J. (1982). *Evolution and the Theory of Games*. Cambridge University Press, Cambridge.

Tinbergen, N. (1952). "Derived" activities; their causation, biological significance, origin, and emancipation during evolution. *Quarterly Review of Biology*, 27(1):1–32.

Wheeler, M. and de Bourcier, P. (1995). How not to murder your neighbor: Using synthetic behavioral ecology to study aggressive signaling. *Adaptive Behavior*, 3(3):273–309.

Yamauchi, B. M. and Beer, R. D. (1994). Sequential behavior and learning in evolved dynamical neural networks. *Adaptive Behavior*, 2(3):219–246.

Zahavi, A. (1975). Mate selection—a selection for a handicap. *Journal of Theoretical Biology*, 53:205–214.

Collective Learning for Spatial Collaboration

Cyril Panatier, Hervé Luga, Yves Duthen
IRIT, University Paul Sabatier,
118 rte Narbonne
31062 Toulouse CEDEX
France
Phone: (33) 5 61 55 83 29
Fax: (33) 5 61 55 62 58
E-mail: {panatier | luga | duthen}@irit.fr

Abstract

A learning model is presented to evolve the collective behavior of a group of autonomous entities in virtual reality applications. The entities are autonomous in that they have no means of direct communication with each other. The learning model uses a module called a "cognitive map". The state of advancement of a task being performed by the group is given by this map. The group receives a reward according to the evolution of the task performed. Through an egalitarian distribution of the reward within the group, an automatic evolution of cooperation is attempted in order to complete the task successfully. At each time interval, an entity must choose its behavior from a basis set of options. Rewards the entity has received both in the distant and in the recent past influence the choice made by the entity.

1. Introduction

The first notion of behavioral simulation in virtual reality appeared in the works of Reynolds (Reynolds, 89). By the implementation of repulsive force fields around the different characters, his model produced remarkable animations of a bird's flight. Using a different approach, Whilhems (Whilhems, 90) used connections between sensors and effectors to model animation. Interpretation of stimuli was allowed by control parameters in the sensors. The Terzopoulos fish (Terzopoulos, 94) is an example of realistic autonomous entities. The motivational level of each entity is taken into account by a module called a "generator of intention", in addition to a sensor/effector system. Blumberg (Blumberg, 90) showed how the autonomy of a virtual entity "dog" could be directed at different levels of its architecture.

In contrast to predefined behaviors, evolutionist methods allow new behaviors to emerge and develop. Genetic algorithms (Holland, 75)(Goldberg, 89) are one such evolutionist method. Ngo (Ngo, 93) used genetic algorithms to optimize the walking behavior of a skeleton between two locations. Sims (Sims, 94.1) used genetic methods to evolve the form and behavior of a population of entities in a virtual environment. He also applied this method to the field of coevolution (Sims, 94.2). Genetic programming (Koza, 92)(Koza, 94) is an evolutionist method whereby a program may evolve to solve dynamic problems. Gritz (Gritz, 95) used genetic programming to generate controllers that caused characters to execute the required movements.

Situated as they are in the biological domain, the works of Deneubourg could be shelved with those of behavioral simulation. He showed how entities with only simple individual behaviors could accomplish collective tasks without any mechanism of direct communication and without any hierarchy. The works deal more specifically with the sorting behavior of ants (Deneubourg, 90).

In the multi-agent domain, Arkin (Arkin, 92) demonstrated the efficiency of schema-based navigation for an objet retrieval where the agents interact in spite of absence of direct communication. Mataric (Mataric, 94) described a methodology based on reinforcement learning, whereby, from basis behavior, a group of mobile robots learned a foraging task.

Here, we present a learning model of collective behavior for an autonomous entity group. It is required to perform a specific task without any direct communication between the entities. Experiments with this model are implemented in virtual reality.

The domain of our simulation is presented in section 2, where our given hypotheses are described along with our desired objectives. In section 3 the calculation of the state of a group with respect to the task is presented. Section 4 deals with how entities use the state of the group to condition their own behavior. The movement of an articulated arm is described in section 5. Section 6 presents results of our experiments in virtual reality, and finally, in section 7, some conclusions are drawn along with possible improvements to be devised.

2. Simulation setting

Our simulation follows behavioral simulations developed within the IN VITRAM project (Mouli, 93). In the initial project the behavior of each individual entity was described by rules (Prolog) that an inference's motor followed. This cognitive approach to the description of the behavior requires quite difficult programming.

With our new approach, emergent behaviors arise in a group of autonomous entities facing a collective task. The individual entities in this study are articulated arms. The task for the articulated arms consists in pushing a puck towards a target. The task is set in two dimensions for two scenes. In the first, one group of articulated arms has the common goal to push the puck towards the target while avoiding some static obstacles (figure 1.a). In the second, two groups of articulated arms are in opposition so that each group must push the puck towards different targets (figure 1.b). In both scenes each articulated arm has a stationary point (attachment point) that limits its movements to within a circle centered at that point. Hence, in many cases the task cannot be performed by one articulated arm alone. The head of an articulated arm allows to push the puck as well as block an opposing articulated arm.

It is stated that there are no means of direct communication between articulated arms. Their perception is limited to the recognition of geometric shapes (number, position and color).

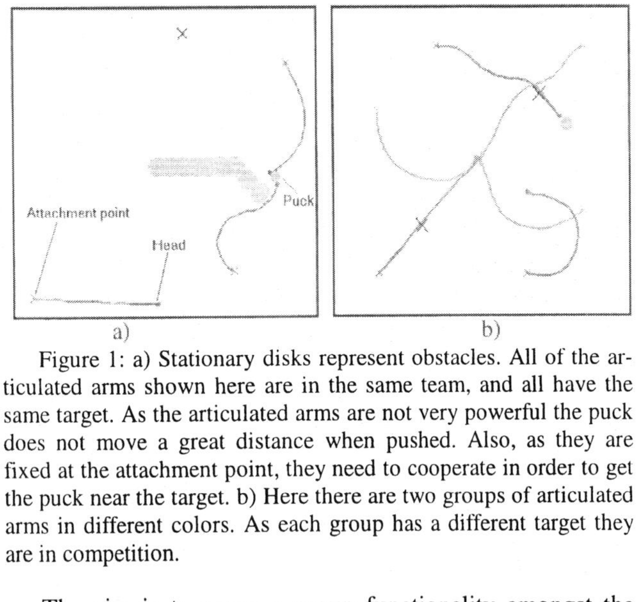

a) b)

Figure 1: a) Stationary disks represent obstacles. All of the articulated arms shown here are in the same team, and all have the same target. As the articulated arms are not very powerful the puck does not move a great distance when pushed. Also, as they are fixed at the attachment point, they need to cooperate in order to get the puck near the target. b) Here there are two groups of articulated arms in different colors. As each group has a different target they are in competition.

The aim is to cause a group functionality amongst the articulated arms to emerge. As these autonomous entities have no means of direct communication amongst themselves, they may be assimilated to susceptible components to cause a functionality to emerge (Steels, 90). However, the emerging functionality must be that required to perform the task. That is, the group of articulated arms knows how to push the puck, avoiding obstacles, towards the target (either on their own or facing an opposing group). For each scene of our simulation a "temporary structure" is established (Steels, 95). This structure is defined by the state of advancement of the group towards the accomplishment of the task. The state of advancement of the group is used by the entities in order to evolve a group behavior that will bring the puck to the target.

3. Computing the state of a group

The state of the group depends on its functionality. Thus, in order to make estimates of the state of the group at a given time, a measure in terms of cost of the work necessary to complete the task must be taken. This requires knowledge of the current situation of the group and its final situation. For a group of articulated arms the current situation is given by the location of the puck. The final situation is when the puck is on the target. Thus, the final situation is the location of the target. Between these two points a path is determined and the work along this path must be measured. A path between the two points is defined by a set of *np* intermediate points. As many different paths may exist between the two points, the path along which the work is minimal has to be found. This corresponds to solving the following optimization problem:

$$Inf \left\{ T((P_i)_{i \in \{1..np\}}) : (P_1, \ldots, P_{np}) \in \Re^{2*np} \right\}$$

A measure of work along a path is defined by:

$$T((P_i)_{i \in \{1..np\}}) = \sum_{i=0}^{np} C(P_i, P_{i+1})$$

$$p_0 = position of\ the\ puck$$

$$P_{np+1} = position of\ the\ target$$

C is the cost function between two points. The cost function is used to measure the work between successive pairs of points along a path.

Measuring work requires a group of entities to construct a "cognitive map". We recall terms used by Yeap and Handley (Yeap, 90) to make a representation of the world for any autonomous system. In our representation the cognitive map is described by a resistance function $R: \Re^2 \rightarrow \Re$ for all point of space. The cost function for two intermediate points, P_1 and P_2, is written as the curvilinear integral of the resistance function, R, along the segment $[P_1, P_2]$.

In our simulation high resistance is put into the circular obstacles and in zones out of range of the group of articulated arms. Thus in zones around the articulated arms the resistance is lowest. That is why the computation of the minimal path tends to place intermediate points close to these attraction zones. The presence of each articulated arm of a group is thus taken into account in the cognitive map to find the minimal path. In other words the presence of an articulated arm allows the work to be reduced along the minimal path (figure 2.a). This gives rise to the existence of a "team spirit". Conversely, resistance is added in zones near the heads of opposing articulated arms. The minimal path tends to avoid the opposing articulated arms (figure 2.b). The presence of each opposing articulated arm is taken into account in a negative fashion, thus reinforcing the notion of team spirit.

At each time interval, the minimal path representing the state of the group is computed using a genetic algorithm. The successive points of a path define the coding of chromosome.

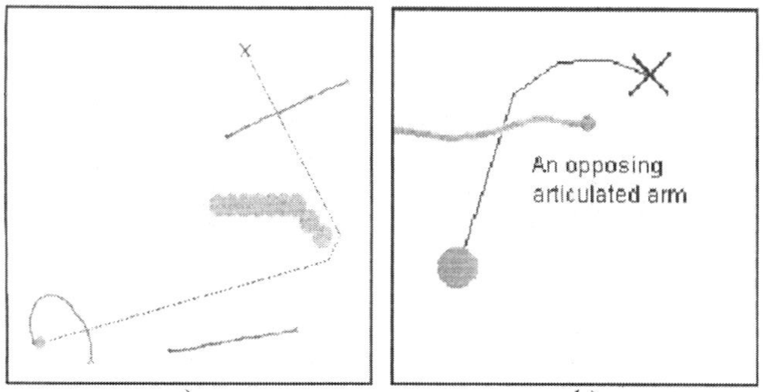

Figure 2: a) The broken line represents an approximation of the minimal path between the puck and the target. The influence of the articulated arms on the path defined by three intermediate points may be observed. b) Influence of the opposing articulated arms on the minimal path.

4. Use of the group state by member entities

The behavior of an articulated arm is defined by a set of basis behaviors. The behavior of the arm is given at any moment by the activation of one behavior only. To achieve this, the articulated arm must learn to select the best behavior from the basis set by a reinforcement learning method.

Such a method may be found in Maes and Brooks (Maes, 90) where techniques for immediate reinforcement are presented to learn behavior selection to walk to a six-legged robot.

Here, a *progress estimator* is used (Mataric, 94) to measure the progress accomplished with respect to the collective task. Rewards are then provided to the entities according to the evolution of the state of the group.

4.1 Global reward of a group

In the scene with static obstacles, only one basis behavior: "push the puck towards the target" is necessary for each entity to perform the collective task of getting the puck to the target. To avoid obstacles, each entity of the group uses the first point of the minimal path as an intermediate target. Hence if entity A, pushing the puck, is unable to reach the final target directly, whereas a neighboring entity B can, A will transmit the puck to B. This action is simply due to the fact that the intermediate target is located within an arm's reach of B.

In the scene evolving an opposing group, obstacles may have individual or collective behaviors. It is therefore necessary to provide a behavior that can deal with an attack from the opposing group. The problem is to know whether an entity should have a defensive behavior (to deal with an opposing entity) or an offensive behavior (to push the puck towards the target). The entities are designed to learn to select the best choice from their set of basis behaviors. Moreover, an entity in a group must learn consistently with its team-mates. For example, an entity must protect the puck from an opponent while a team-mate pushes the puck towards the target. The right behavior of an individual must be appreciated throughout the entire group and be reflected in the global behavior of the group. Hence the same reward is distributed at each moment to all entities in the group (see figure 3.a), so that when the group performs a right action all the entities share in the benefit.

The reward provided by the group to each entity is calculated by measuring its evolution at each step of the simulation. This evolution is defined as the difference between the length of the minimal path of the cognitive map at the present step and the length at the previous step.

$$Rew = \Delta_t\left(\sum_{i=0}^{np} \|P_{i+1} - P_i\|\right)$$

The reward can thus be positive, negative or zero. The reward is positive when the puck is brought closer to the target. Conversely, the reward is negative when, either by the voluntary action of an opposing entity or by the accidental action of a friend entity, the puck is moved further away from the target. The mere fact that an opposing entity approaches the puck causes the reward to be negative. As described earlier, the computation of the minimal path between the puck and the target is influenced by the presence of opposing entities.

Now, according to the received rewards, each entity has to learn how to choose the right basis behavior. To perform this choice, each entity is endowed with a decision center.

4.2 Designing the decision center of an entity

The decision center of an entity allows it, at every time step, to choose the basis behavior that must be activated. At each time step the decision center receives a reward from the state of the group. According to some criteria, this reward is shared among the various basis behaviors (figure 3.b).

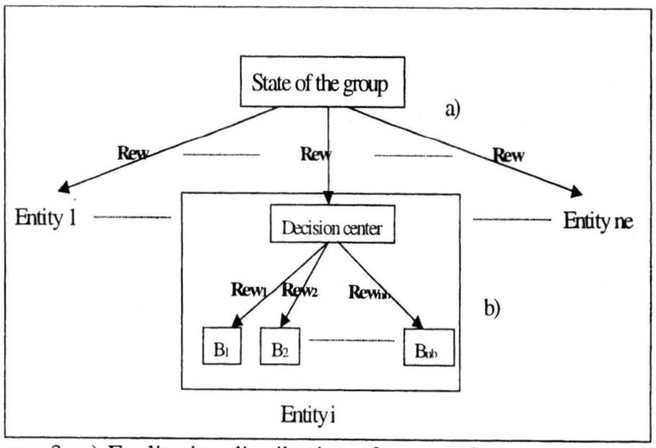

Figure 3: a) Egalitarian distribution of a reward among the entities of a group. b) Sharing a reward among the different basis behaviors in an entity.

From the allocated rewards, a judgement value is given to each basis behavior. The basis behavior designated as active is that which gets the best judgement value. Different types of decision centers may exist according to the criteria of reward sharing among the various basis behaviors, and according to the judgement value function.

For our first model, the reward received at the decision center is allocated entirely to the active basis behavior of the previous time step. This ensures that a behavior receives the fruits of its participation. The other passive basis behaviors from the previous time step receive nothing.

The judgement of a basis behavior must take into account the history of rewards received. This is done by taking the mean of the past rewards received by the basis behavior. There are, however, some unforeseen situations where the active basis behavior fails, such as a shock (successive arrival of negative rewards). The entity must quickly react by changing its active basis behavior. Rewards received relatively recently should have more importance on the judgement of basis behavior than those received further back in the past. A weight is therefore assigned to each reward. The size of the weight decreases as its time of reception recedes into the past. The weights are set so they are equal in the distant past and increase in the recent past.

The judgement value of a basis behavior j with respect to its distant past and recent past is expressed as:

$$Val^{j}_{distant\_past} = \sum_{i=0}^{n-p} \frac{1}{n-p+1} Rew^{i}_{j}$$

$$Val^{j}_{recent\_past} = \sum_{i=1}^{p} f(i) Rew^{i+n-p}_{j}$$

n is the period of learning
p is the duration of the recent past
f is an increasing function.

The decision center makes a final judgement by weighting distant past and recent past judgement values by a constant value $\alpha \in [0,1]$. The fixed value of α depends on how much the decision center is meant to take into account the recent past over the distant past. The judgement value of a basis behavior is written as:

$$Val = (1-\alpha).val_{distant\_past} + \alpha.val_{recent\_past}$$

The decision center reacts quite well when submitted to a shock (figure 4.1). When the active basis behavior is inefficient it becomes passive after a certain time. However, in the case of a constant flux of negative rewards, a new active basis behavior has only one time interval within which to remain active.

In a second model, the persistence of an active basis behavior is taken into account in the allocation of rewards to the basis behaviors. A genetic algorithm that uses a numbered population of basis behaviors is implemented now. The reward allocated to each basis behavior is:

$$Rew_j = \frac{N_j}{N}.Rew$$

N_j is the number of basis behavior j in the population, and N is the total number of behaviors in the genetic population.

Consequently the greater the presence of a basis behavior in the population, the larger the share of the rewards it will receive. This genetic population evolves through phases of selection, reproduction and mutation. The value of selection for a particular basis behavior in this genetic population is given by its judgement value from the previous round. This implies that the more a basis behavior is judged, the more its presence in the genetic population increases, and vice versa in the case of a negative judgement. In the case of a shock, if the new active basis behavior is intended to have time to act despite the incoming flux of negative rewards, it should receive less of them. That is, the basis behavior should not take over the genetic population. The propagation of this basis behavior can be slowed by taking a small value of *Pcross*. Varying the value of *Pcross* in the genetic algorithm enables to control the action time of the new active basis behavior (as shown in figure 4.2). This basis behavior was admittedly chosen to remedy the incompetent performance of the old basis behavior when the shock arrived.

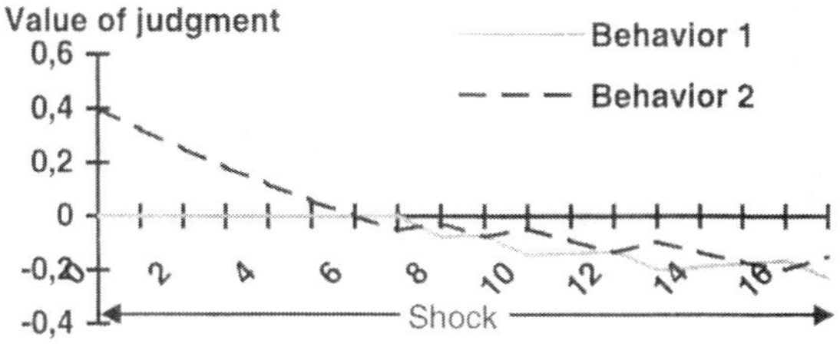

Figure 4.1: Temporal evolution of the judgement value with two behaviors. Numerical simulation with α=0,8; p=20; n=+∞; $val^{1}_{distant\_past}$=0; $val^{2}_{distant\_past}$=2. *f* is a linear function. The decision center receives a negative rewards flux Rew=-1. It can be observed that basis behavior 1 becomes active at the 7th period and becomes passive again at the 8th period.

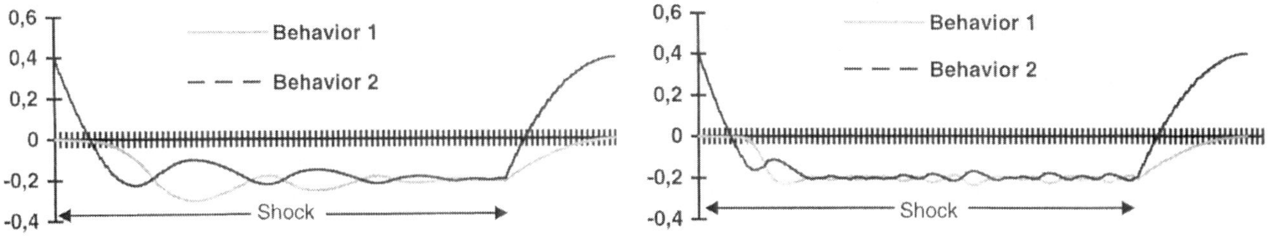

a)*Pcross*=0,1. b)*Pcross*=0,4 .

Figure 4.2: The value of mutation in the genetic algorithm is *Pmut* = 0.01. The total number of chromosomes is N=500. During 80 periods the decision center receives the reward Rew=-1 followed by 20 periods of Rew=0. Basis behavior 2 is succeeded by basis behavior 1 at the 7th period. Notice that the smaller the value of *Pcross*, the greater the number of active rounds of basis behavior 1. To *Pcross* values of a) 0.4 and b) 0.1 correspond action times of 6 and 11. When the received rewards return to zero (the shock is over), the judgement values of the two basis behaviors return to their initial values.

The value of *Pcross* however should not be too small. If the new active basis behavior acts very quickly and effectively, the rewards received by the entity becomes positive. The active basis behavior should thus receive some share of the profits. As long as the entity receives positive rewards, this basis behavior keeps on being active and showing its value (figure 4.3).

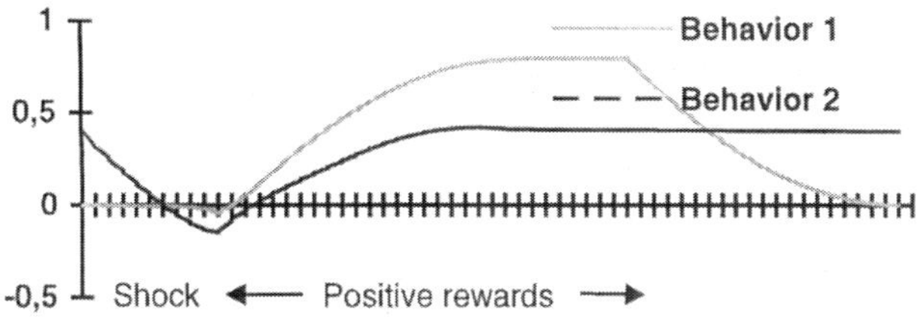

Figure 4.3: Over ten periods the decision center receives the negative reward Rew=-1. Basis behavior 1 becomes active at the 7th period. From the 10th period Rew=1. We wish to simulate the fact that basis behavior 1 has acted well. We thus assert that basis behavior keeps on being active as long as there is a flux of positive rewards. Its judgement value reaches a value of 0.8. The judgement value of basis behavior 2 goes back to its initial value. When the incoming rewards go to zero (after 40 periods), the judgement value of basis behavior 1 returns to its initial value.

5. Movement mechanism of an articulated arm

Morphologically, an articulated arm is a set of linked segments of varying lengths. The position of a segment i is defined by the relative angle θ_i between itself and the previous segment i-1. Hence, varying these angles enables the arm to move. The first segment is anchored to an attachment point and may rotate about this point.

The movement mechanism uses the filters as input, and the effectors as output (figure 5). Changes perceived in the environment are converted into stimuli by the filters. For example, a filter can calculate how far the head of the articulated arm is from the puck. These stimuli are then transmitted to the active basis behavior, which computes control values. These control values are used to weight the effectors according to some basic rules.

Effectors are functions of the movement parameter $\theta=\{\theta_1,..,\theta_n\}$. The sum of the effector functions, weighted by the control values, is called the "motor function". The new form of the articulated arm is realized by minimizing the motor function according to some physical constraints. The motor function is given by:

$$motor\ (\theta) = \sum_{i=1}^{p} \mu_i E_i(\theta)$$ E_i are the effector functions.

A genetic algorithm is employed to minimize the motor function. In the algorithm, The movement parameters define the coding of chromosome (Balet, 97).

A characteristic matrix Cij defines a basic behavior. According to the sign and value of the stimuli, various characteristics of the basic behavior are taken into account to compute the control values. Some interesting results are obtained when this basic model is applied to simple behaviors such as "push the puck towards the target" or "obstruct an opposing articulated arm". An articulated arm, furnished with the first behavior, pushes the puck successfully towards the target, showing small oscillations when it gets near the target. Nevertheless, this basis behavior doesn't allow the articulated arm to go round an obstacle situated between its head and the puck.

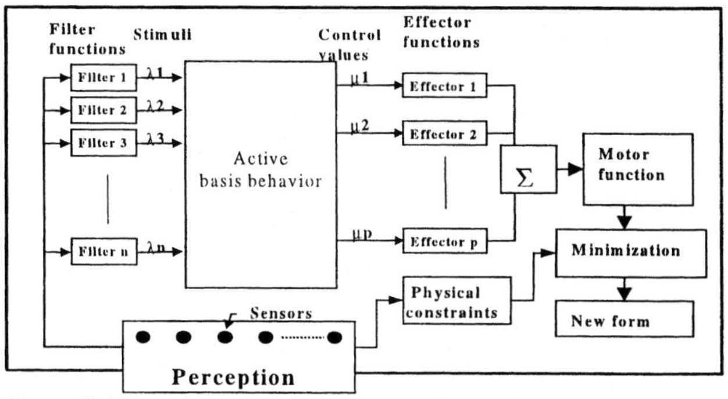

Figure 5: Flow chart of the movement of an articulated arm.

6. Analyze of simulations

In the scene with static obstacles, one conceives of an articulated arm pushing the puck towards the target, and passing the puck to a team-mate when blocked. If the target is sufficiently accessible, the articulated arm pushes the puck directly towards the target.

In the scene with two teams (one light and the other dark), each consisting of three articulated arms symmetrically distributed over the environment (figure 6), and each with their same color target, the simulation works as follows.

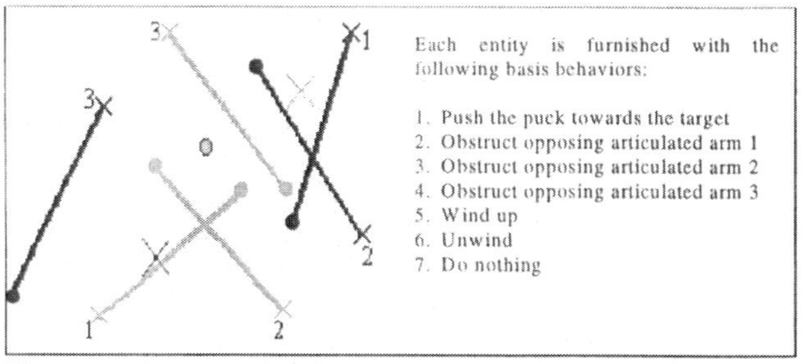

Figure 6: Scene of the simulation and set of basis behaviors for each entity. *Pcross*=0.4, *Pmut*=0.01, N=500, n=4000.

During the simulation, if the behavior "push the puck towards the target" is regarded as offensive ($val_{distant\_past}^{1} > val_{distant\_past}^{i}\ \forall i \in \{2..7\}$), then two offensive entities are observed to emerge in each team. The two offensive entities are those articulated arms that cover the entire environment between the two targets. They can thus push the puck towards their target or take control of the puck from near the opposing team's target. The other arm of each team has a relatively useless role in general. When performing a defensive behavior, it cannot protect the puck adequately from the most dangerous articulated arm of the opposing team.

A modification of the behaviors occurs when there is a shock to one of the teams, such as when the puck moves away from the target. This can happen when an articulated arm pushes the puck past the target. Some entities (the most sensible to a shock) change to a new behavior for several time steps, then resumes their old behavior. However, the new behavior of these entities is frequently not the best choice for the situation. When an opposing articulated arm takes control of the puck and pushes it towards its own target, the articulated arms respond by adopting new behaviors that do not react efficiently to the opposing team's attack. Moreover, when the selected behavior is the defensive "obstruct an opposing articulated arm", the chosen opposing articulated arm is often not the best choice.

7. Conclusion and perspectives

In this article, we have presented a model in which most of a group behavior is controlled by its cognitive map. This map enables the computation of the state of the group at each time step. Rewards are associated with the evolution of the state of the group, and these rewards are evenly allocated over the entities constituting the group. Thus what is a "good action" for the group is a "good action" for each entity. The decision center of an entity uses the incoming reward to change the behavior of the entity.

Our simulation has allowed to observe the emergence and evolution of a collective behavior in real time. This collective behavior is manifested by the accomplishment of the task "push the puck towards the target". However, the collective behavior is not optimal as some entities employ useless behavior throughout the simulation.

One way of addressing this problem is to change the system of reward allocation to encourage different levels of participation for each entity. However, the addition of such a conditioning would increase the complexity of our system, and effectively decrease the action of the cognitive map.

Another way would be to improve the cognitive map. In this simulation, the threat of a nearby head of an opposing articulated arm is taken into account at the level of the cognitive map. But the weakening of the threat by the nearby presence of an ally's head isn't taken into account. The cognitive map could thus be improved by taking into account not only these elements that create resistance, but also those that inhibit resistance. This would reduce the learning of inadequate or useless behaviors by some entities, and increase the learning of effective offensive and defensive behaviors.

We are currently working on a generic model that will help users efficiently to create cognitive map specific to the task. From this map a cooperative behavior will be automatically generated. This will be used with avatars in virtual reality. However, we are concerned only with tasks measurable in terms of progressive advancement.

References

Arkin R. (1992). *Cooperation without Communication: Multi-agent Based Robot Navigation.* Journal of Robotic Systems, vol. 9(3), pp 351-364.

Balet O., Luga H., Duthen Y. and Caubet R. (1997). *Provis: a platform for virtual prototyping and maintenance test.* Computer Animation, Genève.

Blumberg B. M. and Tinsley A. G. (1995). *Multi-Level Direction of Autonomous Creatures for Real-Time Virtual Environments.* Computer Graphics proceedings, Annual Conference Series, pp 47-54.

Deneubourg J. L., Goss S., Franks N., Sendova-Franks A., Detrain C., and Chretien L. (1990). *The dynamics of collective sorting robot-like ants and ant-like robots.* From Animals To Animats, pp 356-363, Paris.

Goldberg D.E. (1989). *Genetic algorithms in search, optimisation and machine learning.* Addison-Wesley, Reading.

Gritz L. and Hahn J. K. (1995). *Genetic Programming for Articulated Figure Motion.* Journal of visualisation and Computer Animation, Vol 6, pp 129-142.

Holland J. H. (1975). *Adaptation in natural and artificial systems.*Ann Arbor, University Of Michigan Press.

Koza J. R. (1992). *Genetic Programming.* MIT Press.

Koza J. R. (1994). *Genetic Programming II.* MIT Press.

Mataric M. J. (1994). *Interaction and Intelligent Behavior.* Technical Report AI-TR-1495, MIT Artificial Intelligence Lab.

Maes P. and Brooks R. A. (1990). *Learning to Coordinate Behaviors.* Proceedings AAA-90, pp 796-802, Boston.

Mouli R., Duthen Y., and Caubet R. (1993). *In VitrAm (in Vitro Animats), a behavioural simulation model.* 2nd IEEE International Workshop ROMAN'93 (robot human), pp 340-345, Tokyo.

Ngo J. T., Marks J. (1993). *Spacetime Constraints Revisited.* Computer Graphics proceedings, Annual Conference Series, pp 343-350.

Reynolds C. W. (1987). *Flocks, herbs and schools: a distributed beheviour model .* Computer Graphics, vol n°21.4, pp 25-34, Anaheim (California).

Sims K. (1994a). *Evolving Virtual Creatures.* Computer Graphics Proceedings, Annual Conference Series, SIGGRAPH'94, ACM-0-89791-667-0/94/007/0015.

Sims K. (1994b). *Evolving 3D Morphology and Behavior by Competition.* Artificial Life 4, pp 28-39.

Steels L. (1990). *Towards a theory of emergent functionality.* From Animals To Animats, pp 451-461, Paris.

Steels L. (1995). *The Artificial Life Roots of Artificial Intelligence.* In Artificial Life: An overview, edited by Langton C., MIT Press, pp 75-110.

Terzopoulos D., Tu X. and Grzeszczuk R. (1994). *Artificial Fishes with Autonomous Locomotion, Perception, Behavior, and Learning in a Simulated Physical World.* Artificial Life 4, pp 17-27.

Yeap W. K. Handley C. C. (1990). *Four Important Issues in Cognitive Mapping.* From Animals To Animats, pp 176-183, Paris.

Whilhems J., Skinner R. (1990). *A notion for interactive behavioral animation control.* IEEE Computer Graphics and Applications, 10(3), pp 14-22.

Adaptation to Multiple Robots Organization with Organizational Knowledge on Formation *

○ **Keiki Takadama**
ATR Human Information Processing Research Labs.
2-2 Hikaridai, Seika-cho, Soraku-gun, Kyoto 619-0288 Japan
Email: takadama@hip.atr.co.jp/
University of Tokyo
4-6-1 Komaba, Meguro-ku, Tokyo 153-8904 Japan
Email: keiki@ai.rcast.u-tokyo.ac.jp

Koichiro Hajiri
ATR Media Integration & Communications Research Labs.
2-2 Hikaridai, Seika-cho, Soraku-gun, Kyoto 619-0288 Japan
Email: hajiri@mic.atr.co.jp/
Ritsumeikan University
1916 Nojicho, Kusatsu, Shiga, 525-8577 Japan
Email: hajiri@airlab.cs.ritsumei.ac.jp

Tatsuya Nomura
ATR Human Information Processing Research Labs.
2-2 Hikaridai, Seika-cho, Soraku-gun
Kyoto 619-0288 Japan
Email: nomura@hip.atr.co.jp

Shinichi Nakasuka
University of Tokyo
4-6-1 Komaba, Meguro-ku, Tokyo
153-8904 Japan
Email: nakasuka@space.t.u-tokyo.ac.jp

Katsunori Shimohara
ATR Human Information Processing Research Labs.
2-2 Hikaridai, Seika-cho, Soraku-gun
Kyoto 619-0288 Japan
Email: katsu@hip.atr.co.jp

Abstract

This paper extends our organizational learning model by introducing the mechanism of utilizing organizational knowledge to improve the collective performance, and focuses on how this organizational knowledge works for supporting new robots in the process of their adaptation to an organization of multiple robots. Through intensive simulations of truss construction by multiple robots, the following experimental results have suggested: (1) Robots with organizational knowledge complete given tasks in fewer iterations than those without organizational knowledge; (2) The steps with organizational knowledge become fewer as the number of robots increases; and (3) Organizational knowledge enables robots to complete given tasks which cannot be completed without organizational knowledge.

Keywords: organizational learning, organizational knowledge, knowledge utilization, multiple robots, adaptive collective behavior

1 Introduction

In the context of adaptive behavior, a lot of researches focus on emergent processes of collective behaviors of robots, and discuss how to control these collective behaviors or how to cause the effective collective behaviors without a centralized control system. For example, Yuta et al. proposed an architecture which achieves the modest cooperation among robots [Yuta 92], and Asama et al. proposed a method of getting out of the deadlock situations with communication [Asama 91]. Furthermore, Mataric demonstrated that real robots learned social behaviors [Mataric 94], and Parker implemented a multirobot system that divided a given task [Parker 94].

However, few researches focus a significant of utilizing knowledge created in emergent processes of collective behaviors of robots, although the reuse of knowledge is important for practical and engineering use. Therefore, this paper focuses the knowledge on the division of work as one of the approaches, and investigates the method for utilizing this kind of knowledge with our novel organizational learning model [Takadama 98b, Takadama 98c]. This is because our model enables robots to divide the work for completing given tasks effectively. However, since our model does not have the mechanism of utilizing the knowledge on the division of work, this paper starts to extend our model by introducing this mechanism.

This paper is organized as follows. Next section starts to explain our organizational learning model, and section 3 proposes a new method for utilizing the knowledge on the division of work. An example of multiple robots for truss construction task is given in section 4, and section 5 discusses the experimental results. Finally, the conclusion is given in section 6.

2 Organizational-Learning Oriented Classifier System (OCS)

2.1 *Architecture of OCS*

Organizational-learning oriented Classifier System(OCS) [Takadama 98b, Takadama 98c] is a new multiagent-based learning model with Learning Classifier System (LCS) [Goldberg 89]. It is composed of a production system and a reinforcement learning mechanism together with introducing concepts of organizational learning. Although OCS is based on evolutionary approaches, OCS is different from the conventional evolutionary ones in terms of the followings: (1) OCS does not evaluate populations with a global evaluation function such a fitness function, but (2) robots self-evaluate their own behaviors with their own local evaluation functions. In OCS, robots acquire their own appropriate functions through local interactions among their neighbors, and form an organizational structure which completes given tasks without explicit control mechanisms.

* Paper submitted to the Fifth International Conference of the Society for Adaptive Behavior (SAB '98)

Although there are Michigan approach [Holland 78] and Pittsburgh approach [Smith 83] in LCS, we modify both approaches with introducing the concept of the organizational learning as follows.

- As shown in Fig. 1, robots do not employ the explicit global evaluation function, but employ their own local evaluation functions composed of the following four parts: (1) Rule base composed of a lot of rules (CFs: Classifiers), (2) Working Memory (WM) for storing information on the environmental state, (3) Memory for storing the sequence of the behavior rules, and (4) Evaluation Mechanism to evaluate the sequence of the behavior rules.

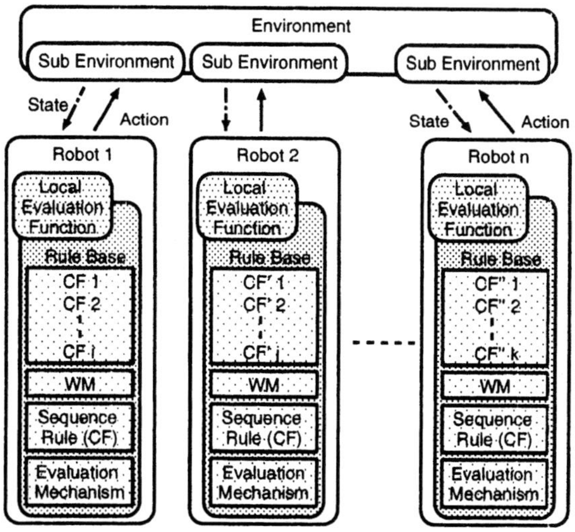

Figure 1: **Architecture of OCS**

- Robots can obtain local environmental states, but the total environmental states around robots cannot be obtained. This distributed recognition indicates that robots must determine their behaviors according to the local state. In OCS, a robot performs one behavior by selecting one behavior rule, and the rule is selected according to the local environmental state with a roulette selection used in a usual LCS.
- Robots do not have communication functions to reduce the communication costs which increase as the number of robots increases. Therefore, robots self-evaluate their own behaviors with their own local evaluation functions, and modify or acquire their own functions to cooperate among robots.
- At the beginning, each strength of the behavior rules is set to the same value.

2.2 Reinforcement learning for Multiple Robots

As an evaluation mechanism in OCS, we have proposed the reinforcement learning for multiple robots [Takadama 98a] as shown in Fig. 2. This method is an extended version of Profit Sharing [Grefenstette 88]. In this method, the positive rewards are distributed to all selected behavior rules according to Eq. (1) when robots complete given tasks, and the negative rewards are distributed to all selected behavior rules according to Eq. (2) when robots get into deadlock situations.

Unlike the conventional reinforcement learning, our reinforcement learning is designed for the multiple robots environment. For example, our method distributes the large positive/negative rewards to the first several selected behavior rules, because there are a lot of local minimum solutions in the multiple robot environment and the first several selected behaviors affect the behaviors of other robots strongly. As other features, the way of distributing the positive rewards in Fig. 2 (a) contributes to reducing the iteration counts with utilizing the previous sequence of behaviors, and the way of distributing the negative rewards in Fig. 2 (b) contributes to reducing the steps for completing tasks. In OCS, robots clear their memory for storing the sequence of behavior rules in Fig. 1, after they reinforce their behavior rules.

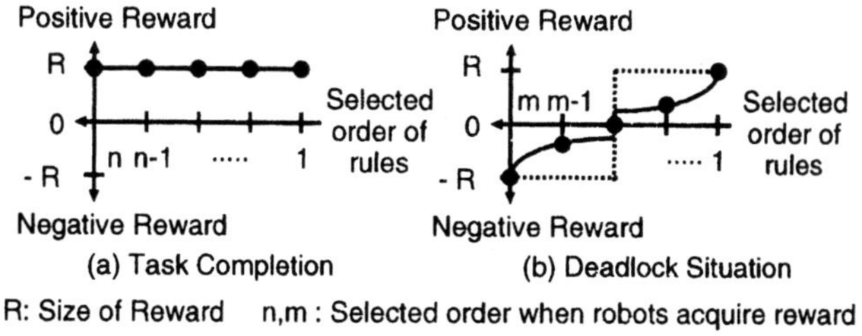

Figure 2: **Reinforcement learning**

- **Task Completion:**

$$ST(i) = ST(i) + R, where\ i = n, n-1, \cdots, 1 \quad (1)$$

- **Deadlock Situation:**

$$ST(i) = ST(i) - R \cdot G^{m-i},$$
$$where\ i = m, m-1, \cdots, m/2+1$$
$$ST(i) = ST(i) + R \cdot G^{i-1},$$
$$where\ i = m/2, m/2-1, \cdots, 1 \quad (2)$$

In the above figure or equations, ST represents the strength of CF, i represents the selected order of CF, n and m represent the max number of the selected CFs, R represents the size of the rewards, and G represents the geometric ratio with the range of $0 < G < 1$.

2.3 Definition in OCS

In OCS, the following technical terms are defined.

- **Function:** A sequence of behaviors determined by behavior rules.
- **Organizational Structure:** A structure in which robots acquire their own functions as shown in Fig.3.

From this definition, the aim of robots in OCS is defined to acquire their own function as a sequence of behaviors to complete given tasks. This is implemented by dividing the work with acquiring robot's own appropriate function. Actually, robots learn the order of using the behavior rules to complete given tasks.

Next, the organizational structure as shown in Fig. 3 is flat unlike the hierarchical structure, but this structure dynamically changes when robots acquire their new functions. This implies that robots address the same task and aim to form organizational structure which completes given tasks by acquiring their own appropriate functions. From this feature, an effective organizational structure is defined as the structure which completes given tasks more effectively than other structure.

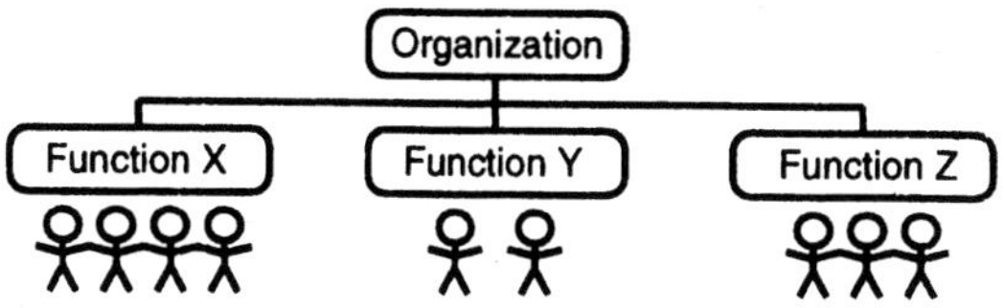

Figure 3: **Organizational Structure**

3 Organizational Knowledge

In OCS, organizational knowledge is defined as a method for the division of work, and is generated from an organizational structure which is converged in given tasks. This is because an organizational structure implies one method for the division of work. To introduce this kind of organizational knowledge into OCS, we extend OCS by introducing the mechanism that all robots store an effective organizational structure as an organizational knowledge when they complete given tasks. In this paper, to simplify the word of an effective organizational structure, we use "formation" in place of the above word.

Next, to introducing the reuse of knowledge into OCS, this section proposes six methods of utilizing organizational knowledge as shown in Fig. 4. Although the proposed methods do not cover all types of utilization of organizational knowledge, these methods can be applied to several tasks. Furthermore in multiple robots environment including multiagent environment, since there are few researches of the reusing knowledge unlike the single robot/agent environment, this paper starts to investigate the effect of the proposed methods to show one of design methods for multiple robots environment.

In Fig. 4, "A","B",···, and "E" indicate a robot which utilizes one function, "*" indicates a robot before acquiring its function, and the circle indicates one formation. In the following explanation of methods, we assume that 9 robots have already had organizational knowledge on formation of 3 robots, and starts to reform their structure for 9 robots with organizational knowledge on formation of 3 robots.

1. **Gathering Formation Method:**
   This method as shown in Fig. 4 (a) adds robots to the robot organization in a unit of formation. For example, when 6 new robots are added to 3 robots, Robot(n+3) and Robot(n+6) where n=1,2,3 utilize the same formation of 3 robots. In the above variables, Robot(n) indicates the nth robots. Especially

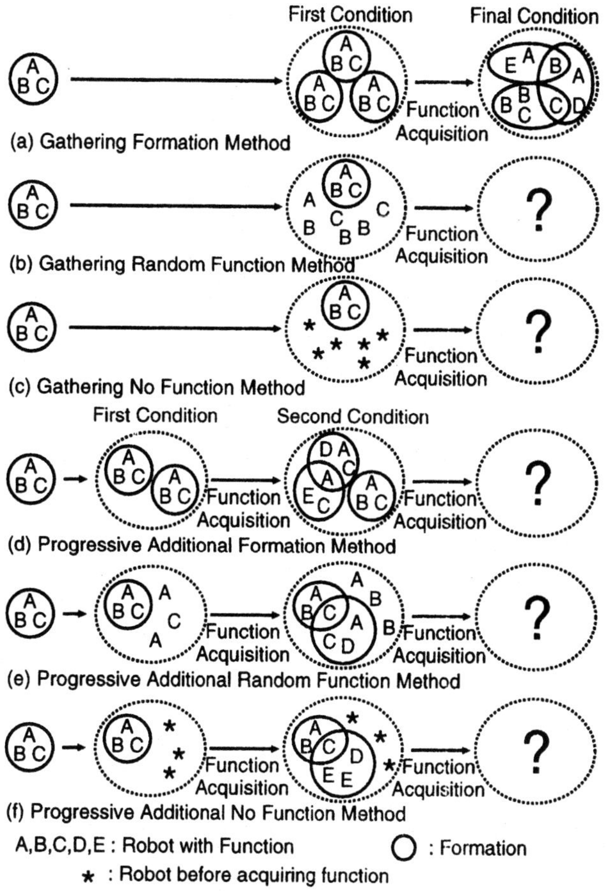

Figure 4: **Utilizing Formation Method**

   when the current total number of robots is not multiple of the number of robots in the stored organizational knowledge, some robots utilize functions which are components of formation.

2. **Gathering Selected Function Method:**
   This method as shown in Fig. 4 (b) adds robots to the robot organization with utilizing their own functions. Unlike the gathering formation method, this method does not utilize the formation directly, but utilizes the function in formation. In this figure, the functions A, B, B, B, C, C are selected at random, and robots with these functions are added to the robot organization.

3. **Gathering No Function Method:**
   This method as shown in Fig. 4 (c) adds robots without functions to the robot organization. Unlike the above methods, this method just adds the robots before acquiring their own functions.

4. **Progressive Additional Formation Method:**
   This method as shown in Fig. 4 (d) adds robots to the robot organization in a unit of formation when the organization structure is converged to be formed. Unlike the gathering methods, this method does not add robots at one time, but adds them one by one in a formation unit. In this figure, new func-

tion D and E are produced through the interactions between two formations of 3 robots.

5. **Progressive Additional Selected Function Method:**
This method as shown in Fig. 4 (e) adds robots to the robot organization with utilizing their own functions when the organization structure is converged to be formed. Unlike the gathering methods, this method adds robots one by one in formation unit.
6. **Progressive Additional No Function Method:**
This method as shown in Fig.4 (f) adds robots without functions to the robot organization when the organization structure is converged to be formed. Unlike the gathering methods, this method adds robots one by one in formation unit.

After adding robots to the robot organization, robots start to modify their own functions to improve the collective performance according to the reinforcement learning as shown in section 2.2.

4 Multiple Robots in Space

4.1 Example

To verify the effectiveness of the methods for utilizing organizational knowledge, we perform experiments of truss construction task completed by combining beams in space. Especially in this kind of space task, since a schedule not only dynamically changes but also is tight, it is important for robots to acquire their appropriate functions as quickly as possible. Therefore, this paper addresses the methods for utilizing organizational knowledge on division of work which minimizes the time for acquiring appropriate functions.

In this task, each robot has only one arm to reduce the weight of robots, and holds either a beam or a welding tool. From this condition, robots must perform the following two matters to complete this task.

- **Cooperation: Behavior order**
Robots must learn the order of selecting behaviors to cooperate with other robots. For example, two robots hold their own beams with setting a desired location in a desired angle between the two beams, and then the third robot welds the two beams.
- **Acquisition of Different Functions**
Robots must acquire different functions even if the situation is the same. For example, some robots hold their beams and others hold their welding tools to combine beams with cooperating each other.

4.2 Robots

In this truss construction task, robots at the station learn to decide whether they hold their own beams or go to the welding location. After beams are welded, robots with beams and robots with welding tools learn again to go to other welding locations to weld the next beams or return to the station to get other beams. Continuing this cycle of the learning process, robots acquire appropriate functions as a sequence of behaviors. Furthermore, there are the following assumptions in this task.

- All robots have their own welding tools, and they can put their tools back to the case attached on the side of their body in order to free their arm.
- Robots can move or rotate for setting a beam at the desired location and angle.
- Robots which wait at the beam constructing location with holding their own beams cannot release them until beams are welded by the welding robots.
- Robots receive their own beams at the space station.
- Robots cannot pass their own beams to other robots.
- Robots cannot weld beams at the welding location until the beams are set at the desired location with the desired angle.
- Robots do not know their appropriate functions beforehand.

From the above assumptions, it may be difficult to make a robot which satisfies all assumptions in the present manufacturing technology. However, our research does not aim to propose a technique with the present manufacturing technology, but aims to improve intelligent technology for future robots. Thus, we can now call a robot as an agent, but robots in our model satisfy the following two problems in physical environment: (1) communication traffic problem and (2) limited energy. To overcome the former problem, robots in our model do not employ a communication mechanism. Next, the latter problem is solved by employing our methods of utilizing organizational knowledge which minimizes the time for acquiring appropriate functions.

4.3 Deadlock Situation

In the truss construction task, robots get into deadlock situations when all robots hold their own beams or hold the welding tools at the same time, or when some robots cannot go to the beam constructing or welding location in the case that other robots wait in the course of the beam constructing or welding location.

4.4 Design of OCS and Task Environment

In this task, we design OCS and task environment as follows.

- Each robot has 74 classifiers (CFs), and the condition part of CF is composed of the following contents: (1) Previous action; (2) Place which distinguishes space station, beam constructing place, welding place and space; (3) Information on beam; (4) Information on welding; and (5) Flag which distinguishes whether robots stay near to be able to perform some action on a beam like welding beams or not. Furthermore, the action part of CF indicates some actions such as movements to the station.

- An action is indicated by a discrete number, but a movement distance is indicated by a real number. This implies that the task environment is not a grid world.
- The location of space station, the beam construction location and the welding location are determined beforehand.

5 Simulation

5.1 Experiments

We perform the following seven experiments: Method without organizational knowledge and Six methods with utilizing formation in Fig. 4 (a) ~ (f), and compare the results of the above experiments in term of both the steps and the iteration counts. In the experiments, we count one step when all robots perform one behavior, and count one iteration when robots complete a task or get into deadlock situations. In each case, 8, 12 and 16 robots construct the truss with combining 13 beams, and continue to complete the same task again and again with utilizing the organizational knowledge of 4 robots. Furthermore, R is defined as 1, and G is defined as 0.5.

5.2 Results

Fig. 5 shows the results of the final steps for completing tasks and the iteration counts. There results are the average of 10 experiments with changing the seed of random. As shown in Fig. 5 (1), robots with the method in Fig. 4 (a) ~ (f) cannot complete given tasks with fewer steps than those without organizational knowledge especially when the number of robots is small. However, the steps with these methods become fewer as the number of robots increases. Furthermore, robots with these methods except the gathering no function method can complete tasks which cannot be completed without organizational knowledge. From Fig. 5 (2), the iterations with organizational knowledge is fewer than those without organizational knowledge as the number of robots increases.

5.3 Findings and Discussions

From the experimental results in the above simulations of truss construction with multiple robots, the following findings are obtained.

- **Effectiveness of Organizational Knowledge :** Although 16 robots without organizational knowledge cannot complete given tasks, the methods with utilizing organizational knowledge on formation except the gathering no function method enable robots to complete tasks. Furthermore, the steps with these methods become fewer than those without organizational knowledge as the number of robots increases. This is because the search space without organizational knowledge is quite larger than the space with the organizational knowledge and this large search space prevents robots from finding the effective solutions which complete given tasks with fewer steps. Moreover, since the number of deadlock situations increases as the number of robots increases, robots without organizational knowledge tend to get into the deadlock situations before acquiring their own appropriate functions. From these results, this organizational knowledge contributes to completing tasks which cannot be completed without organizational knowledge, and also contributes to reducing the steps for completing given tasks especially when the number of robots is large.

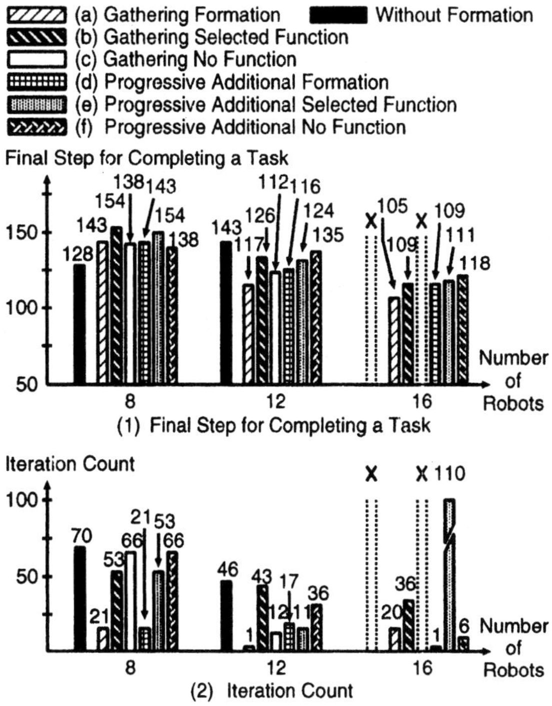

Figure 5: **Result**

- **Effectiveness of Gathering Formation Method :** Since the search space for the gathering formation method is smaller than the space with other methods, this method concentrates on finding the solution in the limited search space. This indicates that robots with this method do not have to acquire other appropriate functions newly, but just modify the assigned functions to improve their collective performance. Therefore, these robots do not get into the deadlock situations more than necessity and they form an organizational structure which completes given tasks with fewer steps and iteration counts as the number of robots increases.
- **Quality of collective behaviors :** Since the progressive additional methods extend the search space by adding new additional robots to the robot organization, this search space is bigger than the space with the gathering methods and is smaller

than the space without the organizational knowledge on formation. Therefore, robots with the progressive additional methods are expected to complete given tanks with fewer steps than those with the gathering methods. However, the results of the gathering methods and the progressive additional methods are not wildly different. This result indicates that the quality of solutions is almost decided by the formation which is used at the first condition. Furthermore, this result also indicates that the first several selected behaviors are quite important in multiple robot environments, comparing in the fact the last several selected behaviors are important in single robot/agent environments [Miyazaki 94].

- **Gathering Method vs. Progressive Additional Method :**
  Although the steps with the progressive additional methods are not quite different from those with the gathering methods, the iteration counts with the former methods expect that the selected function methods are smaller than those with the latter methods as the number of robots increases. Furthermore, in the case of adding robots before acquiring functions, the progressive additional methods enables 16 robots to complete given tasks which cannot be completed with gathering methods. From these factors, the progressive additional methods can be employed for the tasks which must be completed with few iterations. Furthermore, these methods enables robots to get out of deadlock situations even when the number of robots increases.

6 Conclusion

This paper has extended our organizational learning model by introducing the mechanism of utilizing organizational knowledge, and has shown its effectiveness in the truss construction task. From the experimental results, organizational knowledge contributes to completing given tasks which cannot be completed without organizational knowledge, and also contributes to reducing the steps as the number of robots increases. Furthermore, we have found that the gathering formation method enables robots to complete given tasks with fewer steps and iteration counts than those with other methods, and also found that the quality of collective behaviors is almost decided by the formation which is used at the first condition. Moreover, the progressive additional methods contribute to reducing the iterations and enables robots to get out of deadlock situation even when the number of robots increases. Future researches include:

- Proposition of the method for producing the formation which reduces both steps and iterations, and proposition of the method for selecting one effective formation from a lot of formations.
- Theoretical analysis of the effectiveness of the organizational knowledge.
- Investigation of the effectiveness of our method using physical robots with simple tasks.
- Verification of effectiveness of our model in applying to other task domains.

References

[Argyris 78] C. Argyris and D. A. Schon: "*Organizational Learning*", Addison-Wesley, 1978.

[Asama 91] K. Asama et al.: "Collision Avoidance among Multiple Mobile Robots Based on Rules and Communication", *IEEE Int'l Workshop on Intelligent Robots and Systems '91*, pp. 1215–1220, 1991.

[Duncan 79] R. Duncan and A. Weiss: "Organizational Learning: Implications for organizational design", in *Research in organizational behavior*, B. M. Staw (Ed.), Vol. 1, JAI Press, pp. 75–123, 1979.

[Espejo 96] R. Espejo et al: "*Organizational Transformation and Learning*", John Wiley & Sons, 1996.

[Grefenstette 88] J. J. Grefenstette: "Credit Assignment in Rule Discovery Systems Base on Genetic Algorithms", *Machine Learning*, Vol 3. pp. 225–245, 1988.

[Goldberg 89] D. E. Goldberg: "*Genetic Algorithms in Search, Optimization, and Machine Learning*", Addison-Wesley, 1989.

[Holland 78] J.H. Holland and J. Reitman: "Cognitive Systems Based on Adaptive Algorithms", in *Pattern Directed Inference System*, D.A. Waterman and F. Hayes-Roth (Eds.), Academic Press, 1978.

[March 91] J. G. March:, "Exploration and Exploitation in Organizational Learning", *Organizational Science*, Vol. 2, No. 1, pp. 71–87, 1991.

[Mataric 94] M. J. Mataric: "Learning to Behave Socially", *SAB '94*, pp. 453–462, 1994.

[Miyazaki 94] K. Miyazaki et al: "On the Rationality of Profit Sharing in Reinforcement Learning", *IIZUKA '94*, pp. 285–288, 1994.

[Parker 94] L. E. Parker: "ALLIANCE: An Architecture for Fault Tolerant, Cooperative Control of Heterogeneous Mobile Robots", *IROS '94*, pp. 776–783, 1994.

[Smith 83] S.F. Smith: "Flexible learning of problem solving heuristics through adaptive search", *IJCAI '83*, pp. 422–425, 1983.

[Takadama 98a] K. Takadama et al: "Reinforcement Learning for Multiple Robots with Organizational Learning", *AROB '98*, pp. 392–396, 1998.

[Takadama 98b] K. Takadama et al: "Printed Circuit Board Design via Organizational-Learning Agents", *Applied Intelligence*, 1998, to appear.

[Takadama 98c] K. Takadama et al: "Learning Model for Adaptive Behaviors as an Organized Group of Swarm Robots", *International Journal of Artificial Life and Robotics*, 1998, to appear.

[Yuta 92] S. Yuta and S. Premvuti: "Coordinating Autonomous and Centralized Decision making to Active Cooperative Behaviors Between Multiple Mobile Robots", *IEEE Int'l Workshop on Intelligent Robots and Systems '92*, pp. 1566–1575, 1992.

Robot Sheepdog Project achieves automatic flock control

Richard Vaughan*, **Neil Sumpter****, **Andy Frost*****, **Stephen Cameron***

* Oxford University Computing Laboratory,Parks Road, Oxford, UK
Richard.Vaughan@comlab.ox.ac.uk Stephen.Cameron@comlab.ox.ac.uk
** School of Computer Science, University of Leeds, Leeds, UK
neils@scs.leeds.ac.uk
*** Silsoe Research Institute, Wrest Park, Silsoe, Bedfordshire, UK
Andy.Frost@bbsrc.ac.uk

Abstract

This paper describes a mobile robot that can enter a circular arena, gather a flock of ducks and manouvre them safely to a specified goal position. A minimal simulation model of the ducks' flocking behaviour was developed and used as a tool to guide the design of a general flock-control algorithm. The algorithm was first tested in simulation then tranferred unchanged to a physical robot which succeeds in gathering a real flock of ducks. This is the first example of a robot system that exploits and controls an animal's behaviour to achieve a useful task.

Robots successfully manipulate many objects in factories and laboratories. Research continues on manipulating objects with complex or variable shapes and dynamics eg. food products (Juste et al., 1997). Recent work in mobile robotics has focused on 'adaptive behaviour' in animals, in order to extend the abilities of robots (Maes, 1990) (Hallam and Hayes, 1994), and to better understand the processes occurring in real creatures (Webb, 1994)
(McFarland and Bosser, 1993). The Robot Sheepdog Project examines the robotic manipulation of animals by exploiting their adaptive behaviour. In contrast, previous work combining robots and animals (Trevelyan's robot sheep-shearer (Trevelyan, 1992), Silsoe Research Institute's milking robot (Frost et al., 1993)) has deliberately minimised animal behaviour by physical restraint. We have demonstrated a mobile robot that can enter a circular arena, gather a flock of ducks and manouvre them safely to a specified goal position. This is the first example of a robot system that exploits and controls an animal's behaviour to achieve a useful task.

The sheepdog's gather-and-fetch task was chosen because of its familiarity and the strong interaction between the dog, shepherd and flock animals. Using ducks instead of sheep allows us to experiment on a conveniently small scale, in a controlled indoor environment. Duck flocking behaviour is recognised by shepherds as similar to sheep; ducks are often used to train sheepdogs because of their relatively slow movement.

Figure 1 Sheepdog with duck flock in Lancashire, 1996.

Flocking is considered an adaptive behaviour, as it affords various advantages in hazard-avoidance, mating and foraging. Models of flocking behaviour exist in the literature and are generally derived from Hamilton's observation that flocking may be produced by the mass action of individual animals, each seeking the proximity of its nearest neighbours (Hamilton, 1971). It was later suggested that this behaviour can be well modelled by an attractive 'force' acting between the animals, with the magnitude of the attraction varying with the inverse square of the animals' mutual distance (Partridge, 1982) (Warburton and Lazarus, 1991). It is argued that this relationship represents a linear response to sensory information which itself varies with the inverse square of distance. Similar models have produced realistic computer animations of bird flocks (Reynolds, 1987).

These ideas are familiar in robotics, where such *po-*

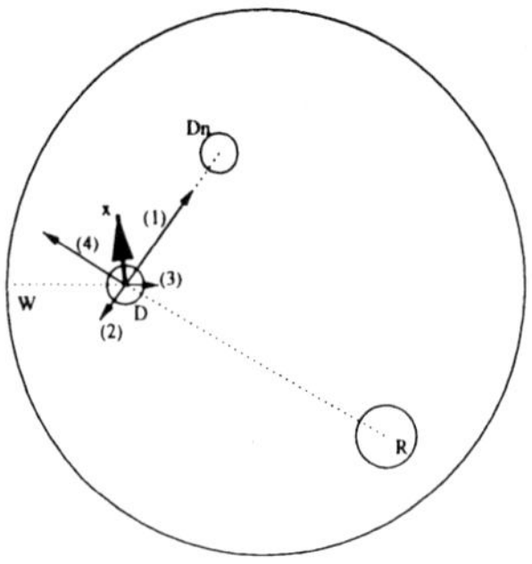

$$\mathbf{d} = \sum_{n=1}^{N}\left(\left(\frac{K_1}{|\mathbf{DD_n}|^2}\right)\widehat{\mathbf{DD_n}} - L - \left(\frac{K_2}{|\mathbf{DD_n}|^2}\right)\widehat{\mathbf{DD_n}}\right) - \left(\frac{K_3}{|\mathbf{DW}|^2}\right)\widehat{\mathbf{DW}} - \left(\frac{K_4}{|\mathbf{DR}|^2}\right)\widehat{\mathbf{DR}}$$

(1) (2) (3) (4)

Figure 2 Flock model (schematic not drawn to scale). Key: gain parameters $K_{1\to4}$; offset parameter L; number of ducklets N; ducklet position D, other ducklet D_n; Robot position R; Nearest point on wall W; algorithm terms $(1 \to 4)$ and resultant $\mathbf{d}$ (where $\widehat{\mathbf{a}}$ is the unit vector of $\mathbf{a}$).

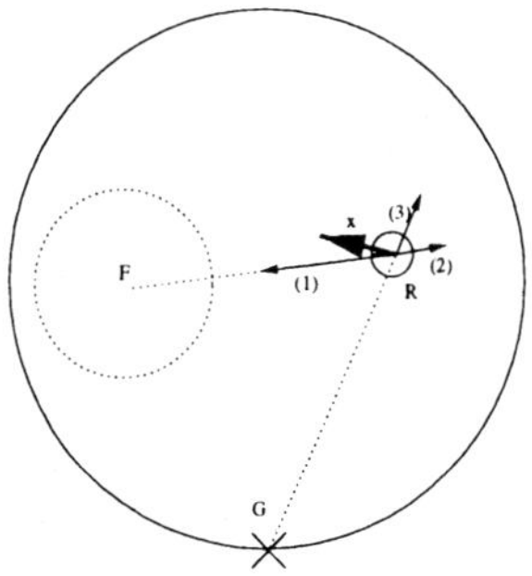

$$\mathbf{r} = K_1\mathbf{RF} - \left(\frac{K_2}{|\mathbf{RF}|^2}\right)\widehat{\mathbf{RF}} - K_3\widehat{\mathbf{RG}}$$

(1) (2) (3)

Figure 3 Flock control method (schematic not drawn to scale). Key: gain parameters $K_{1\to3}$; flock center F; Robot position R; algorithm terms $(1 \to 3)$ and resultant $\mathbf{r}$ (where $\widehat{\mathbf{a}}$ is the unit vector of $\mathbf{a}$)

tential field techniques are used for navigation (Cameron and Probert, 1994). This class of algorithm uses the analogy of forces acting on particles, such that the robot will move as if it were a particle attracted or repelled from features in its environment. A robot is typically attracted to a goal position and repelled from obstacles. The commonality of these animal and robot behaviour models forms the basis of an effective flock-gathering strategy.

A minimal simulation model of the duck-herding scenario was created, in which six model ducks (ducklets) move in a circular arena containing a model robot. A ducklet's movement vector $\mathbf{d} = (\frac{\delta x(t)}{\delta t}, \frac{\delta y(t)}{\delta t})$ is determined by the function shown in Figure 2.

The ducklets are (Figure 2, term 1) attracted to each other, aggregating the flock; (2) repelled from each other, preventing collisions and maintaining inter-ducklet spacing; (3) repelled from the arena wall, preventing collisions. A further term (4) which produces repulsion from the robot is proposed to model the aversive response of the ducklets to the robot. All these forces are scaled according to the inverse square of distance, and each ducklet moves according to the resultant of the forces acting upon it. The simulation produces a realistic-looking flock which can be manipulated by steering the model robot.

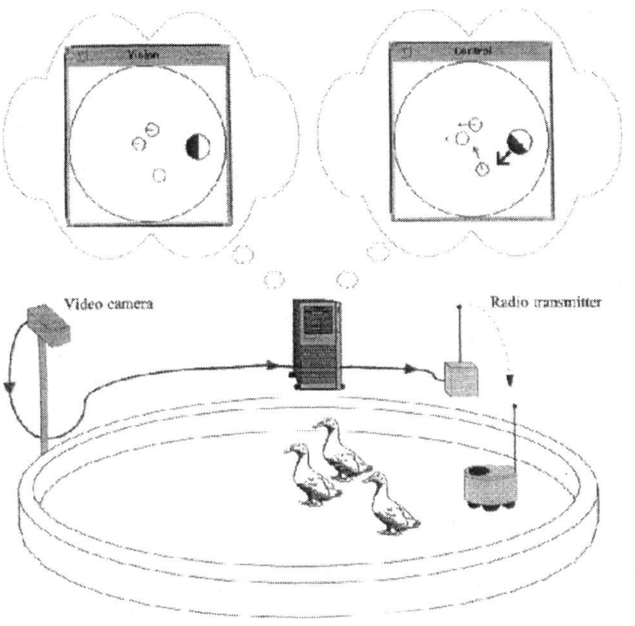

Figure 4 Robot Sheepdog system overview.

Figure 5 'Rover', the RSP vehicle.

By experimenting with the simulator, a closely related algorithm was developed which steers the robot to gather the flock and return it to a goal position on the edge of the arena. The robot's movement vector $\mathbf{r}$ is given by the function shown in Figure 3. The robot is (Figure 3, term 1) attracted to the center of the flock (defined as the average position of all the ducklets) with a magnitude proportional to their mutual distance. This force causes the robot to move towards the flock. A second force (2) repels the robot from the flock center with a magnitude proportional to the inverse square of their mutual distance. This prevents collisions. The resultant of these two forces creates a circular orbit of zero potential around the flock centre. A further force (3) repels the robot from the goal position with a constant magnitude. This has the effect of tilting the potential landscape such that the orbit around the flock now has a minimum behind the flock with respect to the goal. The robot will move towards this point; driving the flock away from it and towards the goal. A separate mechanism is used to prevent the robot hitting the walls (not described here).

This method was tested using the flock simulation. A representative experiment is shown in Figure 6: The robot starts at the goal position, approaches the ducks, moves around behind them with respect to the goal, driving them in the desired direction. The system stabilises with the ducklets near the goal, and the robot standing off some distance away. The paths of the robot and flock center are shown (Figure 6 top) and the distance of the flock from the goal (our success metric) is plotted over time (Figure 6 bottom).

The method was found to be successful and robust over a wide range of flock parameters (Vaughan et al., 1997). The algorithm was then implemented on a physical robot in order to test its performance with a real flock of ducks.

The physical experimental system comprises a robot vehicle, a workstation and a video camera (Figure 4). The vehicle is designed to work in a duck's environment: outdoors, on short grass, and in real time. Thus our robot has a top speed $\approx 4ms^{-1}$ (roughly twice as fast as our ducks) and acceleration $\approx 1ms^{-2}$. It is covered in a soft plastic bumper mounted on rubber springs, ensuring duck safety. In the tradition of mobile robotics, we call it 'Rover' (Figure 5).

The vehicle and the ducks are free to move in a visually uniform arena of 7m diameter, in view of the overhead camera. The position and orientation of the robot, and the position and size of the flock are determined by processing the video image stream . A standard background-differencing technique is used, to achieve an update rate of approximately 25Hz. The vision system's output is used to generate a desired vehicle trajectory using the algorithm described above.

The vehicle's current speed and heading are compared to the desired movement vector $\mathbf{r}$, and new wheel-speeds (R_{left}, R_{right}) are determined by the functions

$$\begin{pmatrix} R_{left} \\ R_{right} \end{pmatrix} = \begin{pmatrix} |\mathbf{r}| + \frac{D}{2}(\theta - \angle\mathbf{r}) \\ |\mathbf{r}| - \frac{D}{2}(\theta - \angle\mathbf{r}) \end{pmatrix}$$

where θ is the robot's current heading and D is the distance between the wheels. These wheel-speed demands are passed to the vehicle via a radio modem. The proportional controller

$$u(t) = K(R(t) - E(t))$$

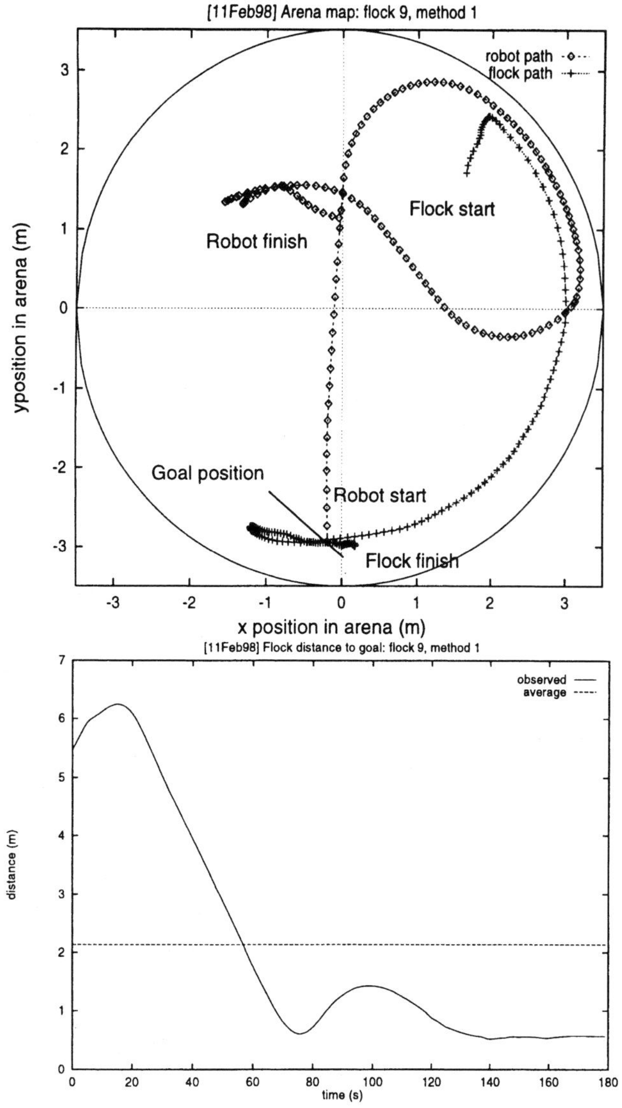

Figure 6 Simulation results. Top: paths of vehicle and flock (goal is at (0, -3.3). Bottom: distance from flock centre to goal position over time.

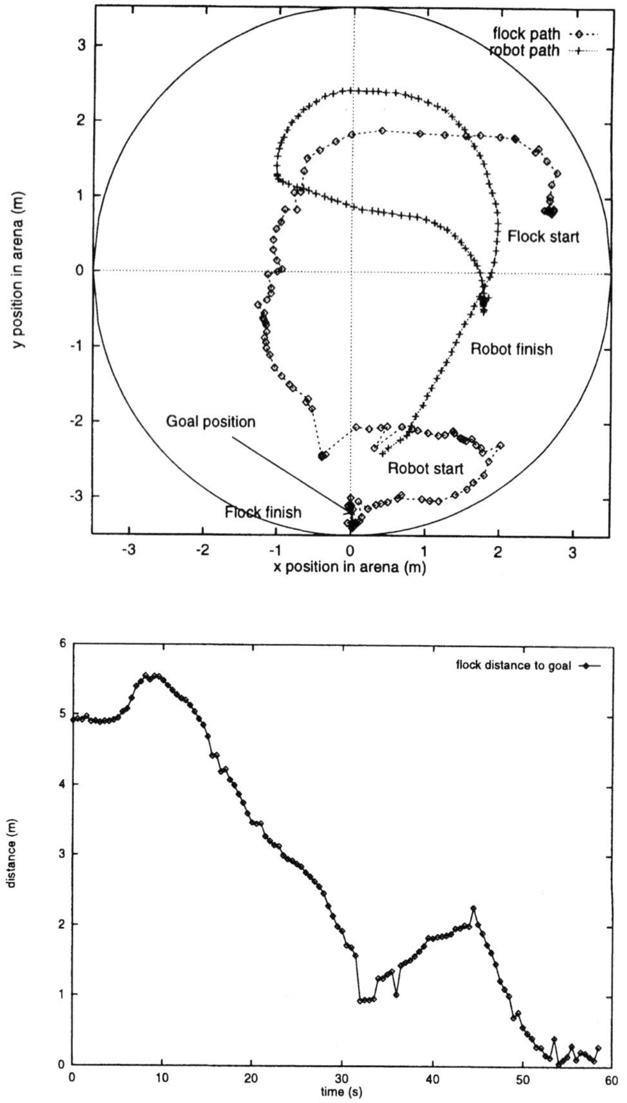

Figure 7 Real-world results. Top: paths of vehicle and flock (goal is at (0, -3.3). Bottom: distance from flock centre to goal position over time.

(where u is the output to the wheel, R is the desired speed, E is the speed error, and K is the controller gain) is implemented by the vehicle's on-board systems for each wheel at 100Hz. The controller gain was chosen by experiment and it is found that the vehicle's movement closely approximates the desired trajectory.

A random point along the arena boundary was chosen as the flock goal, corresponding to position (0, -3.3) on Figure 7 (top). With the robot inactive and positioned near the flock goal, the ducks were introduced into the arena and allowed to move freely. After 5 minutes accommodation time, the robot was activated. Figure 7 (top) shows the paths of the robot and the flock center over a 60 second trial, with samples taken at 0.5s intervals. Figure 7 (bottom) shows the distance of the flock center from the goal point over time; it can be seen that the flock is effectively moved to the goal position. The behaviour of the real-world system is subjectively similar to the simulation.

Since the first draft of this paper this pilot trial has been followed by extensive experimental trials, and a superior algorithm has been developed and tested. Papers describing this recent work will appear during 1998.

In this paper we have described a robot system that achieves a sheepdog-like task, gathering and fetching live animals to a pre-defined goal position. We believe this is the first automatic system to exploit an animals adaptive behaviour to achieve a useful task. The robot's controller was designed and tested using a minimal simulation model of the ducks' flocking behaviour, and successfully transferred to the real world. We conclude that behavioural simulations can be plausible engineering design tools, and that such a methodology is appropriate for such animal-interactive robotics experiments. These results also support the attraction-force model as a good description of flocking behaviour.

References

Cameron, S. and Probert, P. (1994). *Advanced Guided Vehicles: Aspects of the Oxford AGV project.* World Scientific.

Frost, A., Street, M., and Hall, R. (1993). The development of a pneumatic robot for attaching a milking machine to a cow. *Mechatronics*, 3(3):409–418.

Hallam, B. and Hayes, G. (1994). Comparing robot and animal behaviour. Research Paper 598, Department of Atrificial Intelligence, University of Edinburgh.

Hamilton, W. (1971). Geometry for the selfish herd. *Journal of Theoretical Biology*, 31:295–311.

Juste, F., Andreu, G., Valiente, J., and Benlloch, J., editors (1997). *Proceedings of the International Workshop on Robotics and automated machinery for bio-productions.*

Maes, P., editor (1990). *Designing Autonomous Agents.* The MIT Press: Cambridge, MA.

McFarland, D. and Bosser, T. (1993). *Intelligent behavior in animals and robots.* MIT Press.

Partridge, B. L. (1982). The structure and function of fish schools. *Scientific American*, pages 114–123.

Reynolds, C. W. (1987). Flocks, herds and schools: A distributed behavioural model. *Computer Graphics*, 21(4):25–34.

Trevelyan, J. P. (1992). *Robots for Shearing Sheep: Shear Magic.* Oxford University Press.

Vaughan, R., Henderson, J., and Sumpter, N. (1997). Introducing the robot sheepdog project. In *Proceedings of the International Workshop on Robotics and Automated Machinery for BioProductions.*

Warburton, K. and Lazarus, J. (1991). Tendency-distance models of social cohesion in animal groups. *Journal Of Theoretical Biology*, 150:473–488.

Webb, B. (1994). Robotic experiments in cricket phonotaxis. In Cliff, D., Husbands, P., Meyer, J., and Wilson, S., editors, *From Animals to Animats 3: Proceedings of the third international conference on the simulation of adaptive behaviour*, pages 45–54, Cambridge, Mass. MIT Press.

EVOLUTION

Emergence of Functional Modularity in Robots

Raffaele Calabretta[1,2], Stefano Nolfi[2], Domenico Parisi[2] and Günter P. Wagner[1]

[1] Department of Ecology and Evolutionary Biology, Yale University
New Haven, CT 06511, U.S.A.
e-mail: raffaele@peaplant.biology.yale.edu

[2] Department of Neural Systems and Artificial Life
Institute of Psychology, National Research Council
00137 Rome, Italy

Abstract

The origin and structural and functional significance of modular design in organisms represent an important issue debated in many different disciplines. To be eventually successful in clarifying the evolutionary mechanisms underpinning the emergence of modular design in complex organisms, one should be able to cover all different levels of biological hierarchy. Specifically, one should be able to investigate modularity at the behavioral level - the level on which natural selection operates - and understand how this level is related to the genetic level – the level at which natural selection works through mutation and recombination. We describe a simulation of the evolution of a population of robots that must execute a complex behavioral task to reproduce. During evolution modular neural networks, which control the robots' behavior, emerge as a result of genetic duplications. Simulation results show that the stepwise addition of structural units, in this case genetic and neural 'modules', can lead to a matching between specific behaviors and their structural representation, i.e., to functional modularity.

1. Introduction

The origin and structural and functional significance of modular design in organisms represent an important issue debated in many different disciplines such as evolutionary and developmental biology (e.g., Wagner, 1996; Raff, 1996; Wagner, 1995; Fontana and Buss, 1994; Needham, 1993; Bonner, 1988; Gould, 1977), the neurosciences, and cognitive science (e.g., Bates, 1994; Karmiloff-Smith, 1992; Fodor, 1983; Chomsky, 1957).

Many researchers tend to assume that human cognitive processes are accomplished by means of specialized modules (see e.g., Moscovitch and Umiltà, 1990) and, in fact, the modularity of mind is one of the most fundamental assumption of cognitivism (Fodor, 1983). The meaning of the word «module», however, is rather problematic since the word itself "is used in markedly different ways by neuroscientists and behavioral scientists, leading to considerable confusion and misunderstanding in interdisciplinary discussions" (Bates, 1994). This is just an example of a more general problem, that is, the existence of critical differences in the vocabularies and concepts used by the sciences which study organisms at different levels of organization: the biological sciences, the neurosciences, psychology, ecology, and the social sciences. The lack of a common conceptual apparatus represents a problem in that it makes more difficult to study the emergence of complex adaptations such as modularity, whose explanation may require the simultaneous analysis of phenomena and processes at many different levels (molecular, genetic, neural, behavioral, and population level).

In evolutionary biology, the concept of modularity is used to capture the fact that the bodies of higher organisms appear to be composed of semi-autonomous units (Raff, 1996; Wagner, 1996; Bonner, 1988). This observation raises several problems (Wagner, in preparation). First of all, the problem of identifying these 'natural kinds' (Quine, 1969; see also Griffiths, 1997; Wagner, 1996; Wagner, 1995), that is, the problem of 'decomposing' the organism into the appropriate units of phenotypic adaptation. (Notice that a similar problem exists in behavior-based robotics, where one may want to break down the desired complex behavior into an appropriate set of behavioral modules in order to design an integrated robot control system; see Nolfi, 1997a; Nolfi, 1997b; see also below). It has been argued that modularity is a prerequisite for the adaptation of complex organisms: modularity would allow the adaptation of different functions with little or no interference with other functions (Bonner, 1988). However, there is little understanding of how modularity originates, works, and remains incorporated in the genome.

In the field of Artificial Life, some researchers have tried to exploit modular design for improving the performance of various artificial systems such as artificial neural networks, evolutionary algorithms, and robots (see e.g., Nolfi, 1997a; Cho and Shimohara, 1997; Gruau, 1994). Conversely, evolutionary biologists are starting to appreciate the insights that could come from using these simulation models for the study of complex adaptations (see e.g., Levin *et al.*, 1997; Wagner and Altenberg, 1996; Frank, 1996; Toquenaga and

Wade, 1996; Wagner, 1995). The cross-fertilization between Artificial Life and Real Life could take place since Artificial Life partially shares the theoretical apparatus and vocabulary of evolutionary biology and, therefore, could supply it with additional methodological tools.

More specifically, models based on neural networks and genetic algorithms allow us to reproduce in a computer both the organisms and the environment in which they live, behave and reproduce. An organism can be simulated as having a body with a specific size, external shape, sensory and motor organs, etc., and an internal structure made up of a genotype, the nervous system, and other organs. Artificial neural networks that model organisms can be analyzed at the genetic level, at the level of the mapping from genotype to phenotype (development), at the neural and behavioral level, at the level of the effects of the network's output on the environment, at the level of the reproductive chances of each individual (fitness), at the level of populations of individuals and of entire ecosystems. Examining organisms at various levels could be crucial for understanding their behavior, because often an explanation of what happens at one level can be found at another level (see, for example, Miglino *et al.*, 1996; Calabretta *et al.*, 1996). As a consequence, one could hypothesize that the evolution of modularity results from the interaction among processes at different levels.

To evolve a neural controller for a mobile robot, Nolfi (1997a) used a modular neural network architecture that clearly outperformed other architectures in performing a task of garbage collecting (see below). It was denoted as an "emergent modular architecture" because it allows the required behavior to be broken down into sub-components corresponding to different neural modules as a result of an evolutionary adaptive process. Interestingly, Nolfi pointed out that evolved modules were correlated with a 'proximal' description of behavior, i.e., a description from the point of view of the robot's sensory-motor system that accounts for how the agent itself reacts in different sensory situations, rather than with a 'distal' description of behavior, i.e. a high level description in which terms such as 'approach' or 'avoid' are used to describe, from the observer's point of view, an entire sequence of sensory-motor steps (Sharkey and Heemskerk, in press). In other words, differently than in biology where distal functions are mapped to developmental modules, in evolved robots there was no correspondence between distal functional tasks and their structural representation.

It is important to stress that Nolfi was mainly concerned with exploiting modular design for synthesizing control systems for physical robots able to produce complex behavior. As a consequence, in his simulations the architecture of neural networks was pre-designed as modular right from the beginning and it remained fixed throughout the evolutionary process (*hardwired modularity*).

To investigate the issue of how modularity can emerge in nature, we present a modification of Nolfi's model (Nolfi, 1997a) in which gene duplication is also included as part of the evolutionary process and, therefore, modular neural networks can evolve starting from a population of non-modular ones as a result of gene duplication (*constructed* or, more specifically, *duplication-based modularity*). Our preliminary simulation results show that duplication-based modular architecture outperforms non-modular architecture, which represents the starting architecture in our simulations. Moreover, an interaction between mutation and duplication rate emerges from our results. Interestingly enough, duplicated modules, which are identical when duplication first takes place, begin to differentiate across generations as a result of genetic mutations and, eventually, a match between 'distal' modules and their genetic representation tends to emerge.

Figure 1 The Khepera robot.

2. The model

We ran a set of simulations in which neural networks (Rumelhart and McClelland, 1986) are evolutionarily trained to control a mobile robot (see Figure 1) designed to keep an arena clear by picking up trash objects and releasing them outside the arena. The robot has to look for 'garbage', somehow grasp it, and take it out of the arena.

This task can be broken down into several sub-tasks: (a) to explore the environment, avoiding the walls; (b) to recognize a target object and to place oneself in a position such that the object can be grasped; (c) to pick up the target object; (d) to move toward the walls while avoiding other target objects; (e) to recognize a wall and to place oneself in a position with respect to the wall that allows the object to be dropped out of the arena; (g) to release the object. Moreover, these sub-tasks can be broken down into smaller components. It should be noticed that the task chosen is particularly well suited for studying the role of modularity because it involves different basic behaviors that may be implemented in different neural modules. The task requires a controller able to produce very different motor responses for similar sensory states. As an example consider the robot in front of a target object. The robot should avoid or approach the object as a function of the presence or absence of a target on the gripper. If there is already a target on the gripper, the target object should be ignored. If there is no target on the gripper, the target object should be picked up. The only difference between the two situations is the state of one sensor out of 7. A modular neural network that can use different neural modules in different environmental situations should have an advantage with respect to an uniformly connected network in learning to produce very different motor responses for very similar sensory patterns.

The organism is a miniature mobile robot called Khepera, developed at E.P.F.L. in Lausanne (Mondada *et al.*, 1994; see Figure 1). The robot is supported by two wheels that allow it to move in various directions by regulating the speed of each wheel. In addition, the robot is provided with a gripper module with two degrees of freedom. The two arms of the gripper can move in parallel through any angle from vertical to horizontal while the gripper can assume only the open or closed position. The robot is also provided with eight infrared proximity sensors (six sensors are positioned on the front of the robot and two on the back) and an optical barrier sensor on the gripper capable of detecting the presence of an object between the two arms of the gripper. The infrared sensors allow the robot to detect obstacles to a distance of about 4 cm.

The environment is a rectangular arena 60x35 cm surrounded by walls containing 5 target objects. The walls are 3 cm in height and target objects are cylindrical boxes with a diameter of 2.3 cm and a height of 3 cm. The targets are positioned randomly inside the arena. To speed-up the evolutionary process a simulator was used (see Nolfi, 1997b).

In this work we compare the results obtained with three different neural network architectures (see Figure 2). In all cases the robot has 7 sensor neurons and 4 motor neurons. The first 6 sensory neurons are used to encode the activation level of the corresponding 6 frontal sensors of Khepera (the two back sensors are ignored) and the seventh sensory neuron is used to encode the light sensor on the gripper. On the motor side the 4 neurons respectively codify for the speed of the left and right wheels and for the triggering of the 'object pick up' and 'object release' procedures.

The first architecture (a) is a simple feedforward network with 7 input units encoding the state of the 7 sensors and four output units encoding the state of the four effectors. The input units are directly connected to the output units through 28 connection weights (plus 4 biases). This architecture is not divided into modules.

The second architecture (b) is a modular one and it has been called emergent modular architecture (Nolfi, 1997a) because it allows the required behavior to be broken down into sub-components controlled by different neural modules, although it does not require the designer to do such a partition in advance. (Notice that in this paper the emergent architecture is referred to as *hardwired modular*

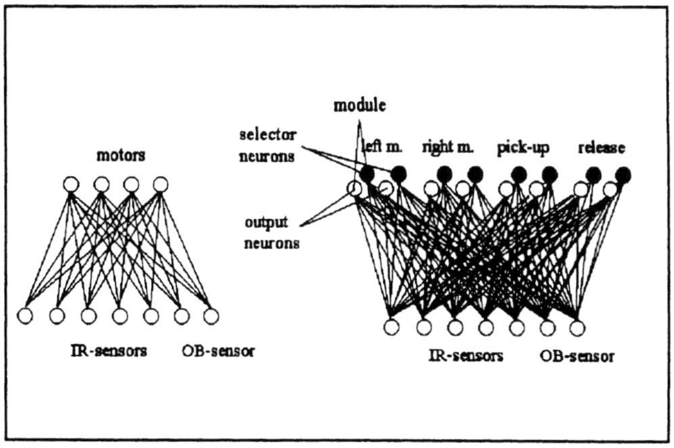

Figure 2 Architectures (a) and (b) are shown on the left and right side, respectively. Architecture (a) is used in the nonmodular population. Architecture (b) is used in the two modular populations (i.e., in both hardwired and duplication-based modular populations). However, in the hardwired modular population two modules compete to gain control of each of the four actuators. In the duplication-based modular population individuals of the initial generation have only one module for each motor, that is, they initially have architecture (a). A second competing module may be added in individuals of successive generations as a result of the duplication operator. Another difference is that in the hardwired modular population competing modules have different random weights from the beginning, while in the duplication-based modular population, when a second competing module is introduced, the two competing modules have identical weights.

architecture). There are two modules for each of the four outputs (two wheels, the object pick up procedure, and the object release procedure). In any particular input/output cycle only one of the two competing modules can control the corresponding output. Each module includes two output units: a motor output unit and a selector unit. The motor output unit determines the speed of the corresponding wheel or whether or not the two procedures are executed. The selector unit determines the probability that the module will control the corresponding output. In other words, which of the two competing modules determines the output depends on which of the two competing selector units is more activated. Both the motor output unit and the selector unit of each module receives 7 connections (plus one bias) from the 7 sensory neurons.

The third architecture (b) is also modular and is denoted as *duplication-based modular architecture* because, in this case, the modules are not hardwired in the architecture from the beginning of evolution but they can be added during the evolutionary process. Each module, as in the case of hardwired architecture, consists of two output units (one motor output unit and one selector unit) which receive connections from the 7 sensors. At the beginning of the evolutionary process there is only one module for each of the four outputs, i.e., always the same module controls the corresponding output. However, during reproduction, modules may be duplicated (see below). Duplicated modules, which are exactly the same when duplication takes place, can differentiate across generations because of genetic mutations.

A genetic algorithm (Holland, 1992) was used to evolve the connection weights of our neural networks. Each connection weight or bias is encoded as a sequence of 8 bits in the genotype. We begin with 100 randomly generated genotypes each representing a network with the same architecture and a different set of random connection weights. Each individual is allowed to 'live' for 15 epochs, each epoch consisting of 200 input-output cycles or actions. At the beginning of each epoch the robot and the target objects are randomly positioned in the arena. An epoch is terminated either after 200 actions or after the first object had been correctly released. At the end of life, the best individuals are selected for reproduction. The 20 individuals that have accumulated the highest 'score' during their lives generate 5 copies each of their neural networks. These 20x5=100 new robots constitute the next generation. The process is repeated for 1000 generations.

Reproduction consists in generating copies of an individual's genotype encoding the network's connection weights (we are assuming non-sexual reproduction in haploid populations) with the addition of random changes to some of the bits of the genotype sequence (genetic mutations) and, in the case of the duplication-based modular architecture, the duplication of a random selected neural module. Individuals were scored for their ability to perform the complete sequence of correct behaviors, i.e., for their ability to release objects correctly outside the arena. However, in order to facilitate the emergence of this ability individuals were also scored (even if with a much lower reward) for their ability to pick up targets (see Calabretta *et al.*, 1998).

In the present preliminary model the maximum number of duplicated modules allowed in the case of the duplication-based modular architecture is one for each motor output and no module-deletion operator was used. As a result, the hardwired modular architecture, already described in Nolfi (1997a), is the more complex architecture that can possibly evolve starting from architecture (a). However, the addition of competing modules during the course of evolution (instead than right from its beginning) that are initially identical to their competing module (instead of being completely unrelated) may produce qualitatively different results in the case of the hardwired and duplication-based modular architecture, respectively.

3. Results and discussion

We present the results of several simulations in which we compare a simple feedforward neural network, the hardwired modular architecture, and the duplication-based modular architecture (see Figure 2). In all simulations a mutation rate of 1% was used (i.e., 2% of the bits of the genotype randomly selected were replaced with a new randomly selected value). For the duplication-based modular architecture we investigated the performance obtained with a duplication rate of 0.02%, 0.03% and 0.04% (i.e., 0.02%, 0.03% and 0.04% of the modules were duplicated per replication). We ran 10 simulations for each of the 3 different architectures described above. Each simulation started with populations of 100 networks with randomly assigned connection weights and lasted 1000 generations.

Both populations with modules reach a higher fitness level than a population with only the basic architecture and no modules (Calabretta *et al.*, 1998). However, the two populations with modules do not differ in terms of overall fitness except that fitness growth is slightly slower in the population with duplication-based modules (Calabretta *et al.*, 1998). In order to demonstrate that modularity plays a critical role, we varied the duplication rate in the population with duplication-based modules, with the result that both average and peak performance decreased linearly with a decreased duplication rate until the advantage of modular design was lost (Calabretta *et al.*, 1998).

Let us now examine the behavior of a typical evolved individual with hardwired modularity and a typical evolved individual with duplication-based modularity.

Nolfi (1997a) pointed out that in hardwired modular architecture there was no correspondence between evolved modules and 'distal' description of behavior; rather, module switching and interaction was correlated with low-level sensory-motor mapping. He explained this by pointing out that the neural modules and the selection mechanism are represented homogeneously and evolve at the same time, and that, therefore, solutions in which both components are kept as simple as possible tend to be selected. Because of gene duplication (see Weiss, 1990; Ohno, 1970), a different evolutionary dynamic is involved in our model. In fact, if we look at the behavior of a typical evolved individual of Nolfi's simulations and a typical evolved individual of our simulations, an interesting difference emerges.

In the former case, the evolved individual described in Figure 5 always uses a single module to control the left wheel, the pick-up procedure, and the release procedure (LM, PU, and RL) while it uses both neural modules only for the right wheel (RM). Nolfi (1997a) stressed that "those two modules competing for the control of the right motor are both used in all the phases that can be described as distal sub-behaviors: when the gripper is empty and the robot has to look for a target (i.e., when sensor LB is off); when the gripper is carrying a target and the robot has to look for a wall (i.e., when sensor LB is on); when the robot perceives something and has to disambiguate between walls and targets (i.e., when the W/T graph shows the upper or bottom line); when the robot does not perceive anything (i.e., when the W/T graph does not show any line); when the robot is approaching a target (i.e., when sensor LB is off and the perceived object is a target); when the robot is approaching a wall (i.e., when sensor LB is on and the perceived object is a wall); when the robot is avoiding a target (i.e., when sensor LB is on and the perceived object is a target); when the robot is avoiding a wall (i.e., when the sensor LB is off and the perceived object is a wall)." In other words, he did not find one of the two neural modules competing for the

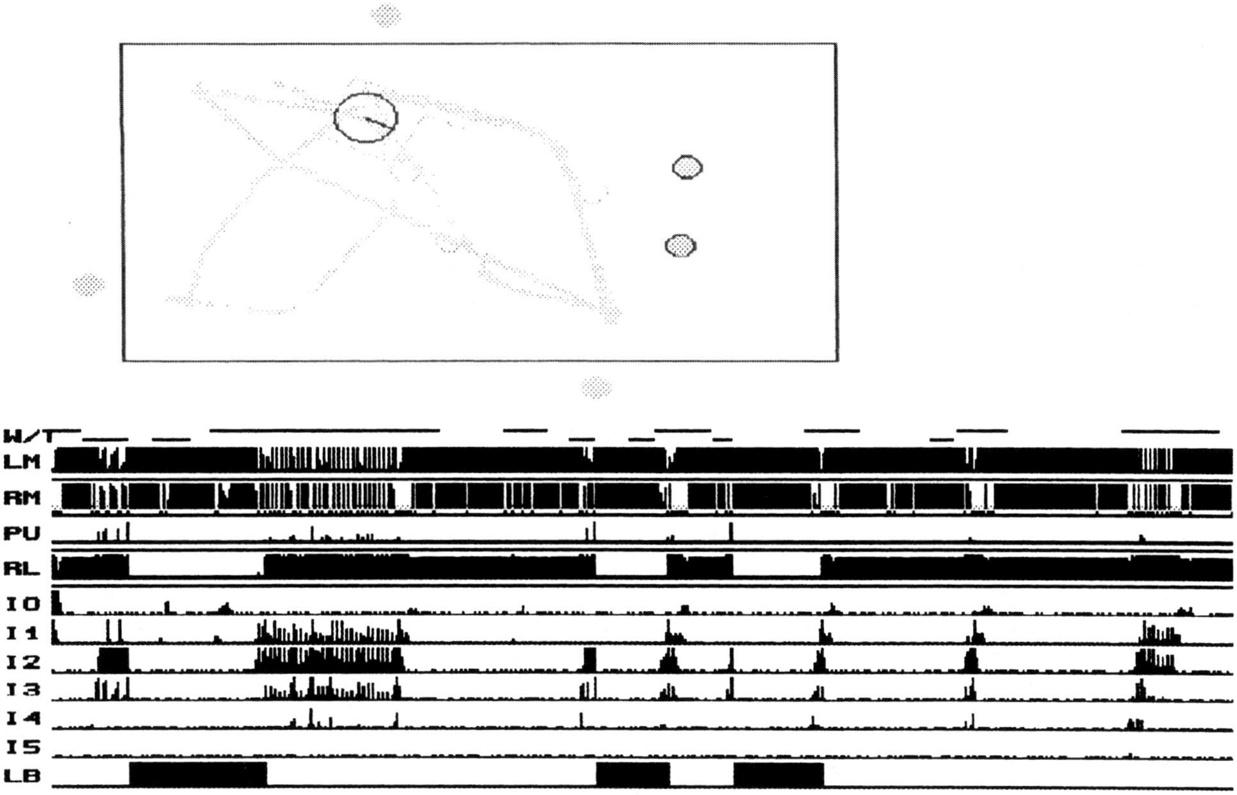

Figure 5 A typical evolved robot with a hardwired modular architecture. The top part of the figure represents the behavior of the robot in its environment. Lines represent walls, empty and full circles represent the original and the final position of the target objects, respectively, the trace on the terrain represents the trajectory of the robot. The bottom part of the figure represents the type of object currently perceived, the state of the motor of the two wheels, and the state of the sensors across 500 cycles. The 'W/T' graph shows whether the robot is currently perceiving a wall (top line), a target (bottom line), or nothing (no line). The 'LM', 'RM', 'PU', and 'RL' graphs show the state of the left and right motors and of the pick-up and release procedures. For each motor, the top portion of the graph shows the activation state (after the arbitration between component modules has been performed by the selector units) and the bottom part shows which of the two competing neural modules has control (the thickness of the black line at the bottom indicates whether the first or the second module has control: thin line = module 1; thick line = module 2). The graphs 'I0' to 'I5' show the state of the 6 infrared sensors. Finally, the 'LB' graph shows the state of the light-barrier sensor. The activation state of input and output units is represented by the height with respect to the baseline. (In the case of output units the activation state of the output units of the module that currently has control is shown). (Figure redrawn from Nolfi, 1997a).

control of the right motor to be responsible for specific distal sub-behaviors. On the contrary each sub-behavior was the result of the contribution of both neural modules (see Figure 5).

In the case of duplication-based modular architecture, on the other hand, the evolved individual described in Figure 6 uses both competing neural modules to control the left motor, the right motor, the pick-up procedure and the release procedure (LM, RM, PU, and RL). In this individual - a very successful one (it is able, in fact, to accomplish the correct sequence of behaviors for all the 5 objects present in the environment) - neural modules or combinations of neural modules appear to be responsible specific distal behaviors. In other words, this evolved individual tends to use different modules in different environmental situations. More precisely, both in the case of the left motor and in the case of the right motor, module switching is significantly correlated to the presence (i.e., when sensor LB is on) or absence (i.e., when sensor LB is off) of a target object on the gripper (see Figure 6). Furthermore, in the case of the pick-up procedure output unit (PU), module switching takes place every time the robot has to pick up an object, while in the case of the release procedure output unit (RL) it happens when the robot has to pick up and to release the object (see Figure 6).

4. Conclusions

Our results suggest that the stepwise addition of structural units, in this case genetic and neural 'modules', can lead to a matching between distal functional tasks and their structural representation, i.e., to functional modularity. A comparison with Nolfi's simulations (Nolfi, 1997a) indicates that such a match is functionally not necessary (performance level is similar in the two conditions; Calabretta *et al.*, 1998), but it results from the stepwise addition of structural units.

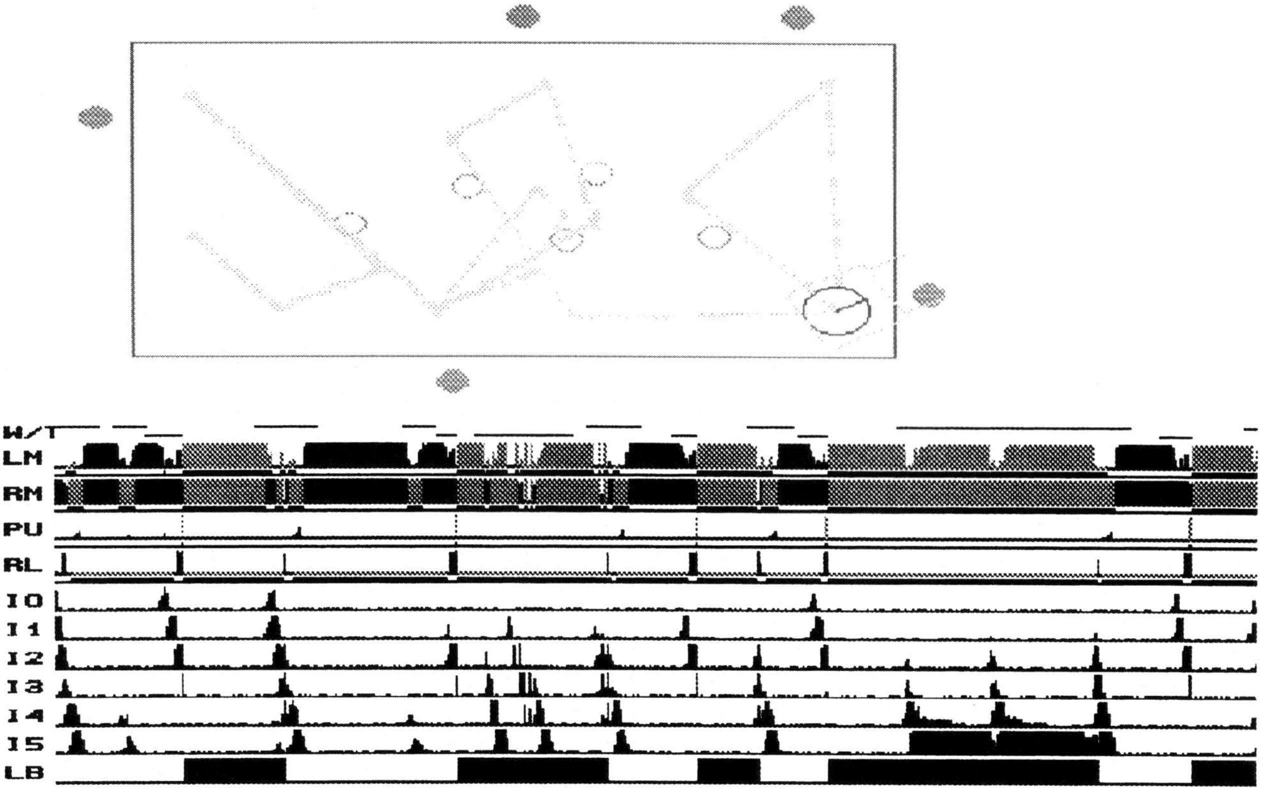

Figure 6 A typical evolved robot with a duplication-based modular architecture (see Figure 5).

A possible explanation of this effect is that the structural units have already adapted prior to the duplication event and most likely they have adapted to participate in multiple functions. After duplication the modules tend to specialize for one of these functions ultimately leading to a close match between functional tasks and structural modules. According to this interpretation modularity is a side-effect of this mode of evolution rather than dictated by the nature of the task to which the robots adapt. This is different from an earlier proposal (Wagner, 1996) in which the selection regime itself is defining the division of tasks. It is also different from Altenberg's constructional selection mechanism in which a structural unit increases in frequency if it has fewer pleiotropic effects (Altenberg, 1995). (Pleiotropy is «the influence of the same genes on different characters», Futuyma 1998, p. 429). In our model the duplicated units all have pleiotropic effects but they tend to reduce these effects by functional specialization after duplication. We conclude that our model suggests a third possible mode for the origin of modularity, besides constructional selection (Altenberg, 1995) and selection against pleiotropic effects (Wagner, 1996): modularity as a secondary effect of duplication. Functional modules originally emerge as duplicated modules which subsequently diverge and specialize for some specific function.

In the present work we have described some preliminary results in support of this model of evolution of modularity. We are currently engaged in further statistical analyses that should substantiate our hypothesis.

Acknowledgments

The financial support of the Italian National Research Council (National Advisory Committee for Biological and Medical Sciences: one-year sabbatical fellowship to RC and Bilateral Project 1998-1999 to RC and GPW) is gratefully acknowledged. RC would also like to acknowledge the support of Aurelio Romeo, Jeffrey R. Powell, Valerio Sbordoni, Riccardo Galbiati, Matthew Schlesinger and Cesare Giordano, the assistance of Ginger Booth and Jan Taschner, and the useful discussions with the members of the GPW's lab at Yale University during weekly meetings.

References

Altenberg, L. (1995). Genome growth and the evolution of the genotype-phenotype map. In Banzhaf, W. and Eeckman, F. H., editors, *Evolution and biocomputation. Computational models of evolution.* Springer Verlag, Berlino-Heidelberg.

Bates, E. (1994). Modularity, domain specificity and the development of language. *Discussions in Neuroscience,* 10:136-149.

Bonner, J. T. (1988). *The evolution of complexity*. Princeton University Press, Princeton.

Calabretta, R., Galbiati, R., Nolfi, S. and Parisi, D. (1996). Two is better than one: a diploid genotype for neural networks. *Neural Processing Letters*, 4:149-155.

Calabretta, R., Nolfi, S., Parisi, D. and Wagner, G. P. (1998). An artificial life model for investigating the evolution of modularity. To appear in Bar-Yam, Y., editor, *Proceedings of the First International Conference on Complex Systems*, Nashua, NH, September 21-26, 1997. Addison-Wesley, Reading.

Cho, S-B. and Shimohara, K. (1997). Emergence of Structure and Function in Evolutionary Modular Neural Networks. In Husbands, P. and Harvey, I., editors, *Proceedings of Fourth European Conference on Artificial Life*. MIT Press, Cambridge.

Chomsky, N. (1957). *Syntactic structures*. Mouton, The Hague.

Fodor, J. (1983). *Modularity of mind*. MIT Press, Cambridge.

Fontana, W. and Buss, L. (1994). The arrival of the fittest: toward a theory of biological organization. *Bull. Math. Biol.*, 56:1-64.

Frank, S. (1996). The design of natural and artificial adaptive systems. In Rose, R. and Lauder, G. V., editors, *Adaptation*. Academic Press , S. Diego.

Futuyma, D. J. (1998). *Evolutionary Biology*. Sinauer, Sunderland.

Gould, S. J. (1977). *Ontogeny and Phylogeny*. Harvard University Press, Cambridge.

Griffiths, P. E. (1997). What emotions really are. The problem of psychological categories. University of Chicago Press, Chicago.

Gruau, F. (1994). Automatic definition of modular neural networks. *Adaptive Behavior*, 2:151-183.

Holland, J. H. (1992). *Adaptation in natural and artificial systems: an introductory analysis with applications to biology, control, and artificial intelligence*. MIT Press, Cambridge.

Karmiloff-Smith, A. (1992). *Beyond modularity: A developmental perspective on cognitive science*. MIT Press, Cambridge.

Levin, S., Grenfell, B., Hastings, A. and Perelson, A. (1997). Mathematical and computational challenges in population biology and ecosystems science. *Science*, 275:334-343.

Miglino, O., Nolfi, S. and Parisi, D. (1996). Discontinuity in evolution: how different levels of organization imply pre-adaptation. In Belew, R. and Mitchell, M., editors, *Adaptive Individuals in Evolving Populations*. Addison-Wesley, Reading.

Mondada, F., Franzi, E. and Ienne, P. (1994). Mobile robot miniaturisation: a tool for investigation in control algorithms. In Yoshikawa, T. and Miyazaki, F. eds., *Experimental Robotics III*, Lecture Notes in Control and Information Sciences. Springer-Verlag, London.

Moscovitch, M. and Umiltà, C. (1990). Modularity and neuropsychology: implications for the organization of attention and memory in normal and brain-demaged people. In Schwartz, M. F., editors, *Modular Deficits in Alzheimer-type dementia*. MIT Press, Cambridge.

Needham, J. (1993). On the dissociability of the fundamental processes in ontogenesis. *Biol. Rev.*, 8:180-223.

Nolfi, S. (1997a). Using emergent modularity to develop control systems for mobile robots. *Adaptive Behavior*, 5:343-363.

Nolfi, S. (1997b). Evolving non-trivial behaviors on real robots: a garbage collecting robot. *Robotics and Autonomous Systems*, 22:187-198.

Ohno, S. (1970). *Evolution by gene duplication*. Springer Verlag, New York.

Quine, W. V. (1969). Natural kinds. In Quine, W. V. editor. *Ontological relativity and other essays*. Columbia University Press, New York.

Raff, R. (1996). The shape of life. Genes, development, and the evolution of animal form. University of Chicago Press, Chicago.

Rumelhart, D. and McClelland, J. (1986). *Parallel distributed processing: Explorations in the microstructure of cognition*. MIT Press, Cambridge.

Sharkey, N. and Heemskerk, J. (in press). The neural mind and the robot. In Browne, A. editor. *Current Perspective in Neural Computing*. IOP Press.

Toquenaga, Y. and Wade, M. (1996). Sewall Wright meets Artificial Life. *Trends in Ecology and Evolution* 11:478-482.

Wagner, G. P. (1995). The biological role of homologues: a

building block hypothesis. *Jb. Geol. Palont.* 195:279-288.

Wagner, G. P. (1996). Homologues, natural kinds and the evolution of modularity. *American Zoologist* 36:36-43.

Wagner, G. P. (in preparation). Appendages, fins, limbs and hands: homology and the hierarchical organization of the body.

Wagner, G. and Altenberg, L. (1996). Complex adaptations and the evolution of evolvability. *Evolution* 50:967-976.

Weiss, K. (1990). Duplication with variation: metameric logic in evolution from genes to morphology. *Yearbook of Physical Anthropology*, 33:1-23.

Coevolving Communicative Behavior in a Linear Pursuer-Evader Game

Sevan G. Ficici, Jordan B. Pollack
DEMO Lab
Computer Science Department
Volen National Center for Complex Systems
Brandeis University
Waltham, MA
http://www.demo.cs.brandeis.edu

Abstract

The pursuer-evader (PE) game is recognized as an important domain in which to study the coevolution of robust adaptive behavior and protean behavior (Miller and Cliff, 1994). Nevertheless, the potential of the game is largely unrealized due to methodological hurdles in coevolutionary simulation raised by PE; versions of the game that have optimal solutions (Isaacs, 1965) are closed-ended, while other formulations are opaque with respect to their solution space, for the lack of a rigorous metric of agent behavior. This inability to characterize behavior, in turn, obfuscates coevolutionary dynamics. We present a new formulation of PE that affords a rigorous measure of agent behavior and system dynamics. The game is moved from the two-dimensional plane to the one-dimensional bitstring; at each time step, the evader generates a bit that the pursuer must simultaneously predict. Because behavior is expressed as a time series, we can employ information theory to provide quantitative analysis of agent activity. Further, this version of PE opens vistas onto the communicative component of pursuit and evasion behavior, providing an open-ended serial communications channel and an open world (via coevolution). Results show that subtle changes to our game determine whether it is open-ended, and profoundly affect the viability of arms-race dynamics.

1. Introduction

The pursuer-evader (PE) game is argued to be an important domain for coevolutionary simulation (Miller and Cliff, 1994) due not only to its ubiquity in nature, but also because it provides a parsimonious, yet powerful, framework of investigation: robust adaptive behavior, open-ended coevolution, and adaptively unpredictable, or *protean*, behavior all fall within its purview. In particular, the PE game is believed to be rich enough to support the evolution of sophisticated perception, control, and predictive ability; insight gained by evolving agents for the game may impact disciplines in the biological and ethological communities. Nevertheless, the PE game also presents methodological hurdles that highlight shortcomings in the state of the art of coevolutionary simulation itself. While some of these hurdles have diminished, most notably the *red-queen effect* (Cliff and Miller, 1995), many others remain that demand a more lucid understanding of coevolutionary learning. Key issues involve the conditions that instigate *competitive arms races* or, to the contrary, *mediocre stable-states* (MSS) (Angeline and Pollack, 1994; Pollack et al., 1997), neither of which are well understood.

We posit that the spatially-oriented PE game brings at least three difficulties to coevolutionary research: 1) the lack of a canonical form, which leads to 2) the question of open-endedness, and 3) the lack of a rigorous metric of agent behavior. First, there exist many versions of the PE game. The earliest formulations (Isaacs, 1965) are games of perfect information where optimal pursuit and evasion strategies follow analytically from agent capabilities. More recent work casts PE into a purely discrete, non-kinematic reactive game (Koza, 1992; Reynolds, 1994), or into a continuous game of, essentially, imperfect information that incorporates a two-dimensional physical model (Cliff and Miller, 1996), for example. Second, these differences in game formulation impact the nature of the solution space; where PE formulations have optimal solutions, the game is closed-ended and, therefore, categorically less interesting than an open-ended domain. Further, these optimal solutions are not compelling examples of evolved complexity; a formulation where optimal pursuit is defined simply by reactively moving directly towards the evader does not even require the pursuer to hold state. Third, and finally, agent behavior in PE is currently described exclusively in subjective, qualitative terms; as (Cliff and Miller, 1996) point out, there does not yet exist a quantitative, operational measure by which one can characterize and judge agent behavior.

The hypothesized potential of the PE game is thus largely unrealized due to the opaqueness of the domain, of coevolutionary dynamics, and their interaction; we present a reformulation of the pursuer-evader game that helps clarify all three. Simply, we move the arena of action from the two-dimensional plane to the one-dimensional bitstring; at each time step, the evader generates a binary symbol that the pursuer must simultaneously predict. Games last one thousand time steps. A tally of correct guesses is used to derive players' scores. Because behavior is expressed as a binary time series we can take advantage of information theory to provide precise characterizations of agent behavior and evolutionary progress.

As importantly, recasting the PE game in this way opens vistas onto the *communicative* component of pursuit and evasion behavior. Recent work by (Di Paolo, 1997) raises a concern that much research on communication conflates functional utility with operational process: the utility of communicative behavior may very well be that it confers adaptiveness, yet communication so defined fails utterly to describe the behavior itself. Similarly, the denotative function of 'information' is often conflated with the process that generates it. In our PE game we can characterize pursuit and evasion behavior outside of the context of selective advantage. Further, we operationalize the behaviors and unify the notion of information with the structure of behavior. Effective pursuit requires behavioral coordination with the evader. Effective coordination requires the induction of the evader's behavioral dynamics by the pursuer. This process of induction represents our notion of communication.

Nevertheless, "communication" obviously encompasses a much broader range of phenomena than pursuit and evasion — presently too broad to be captured entirely within a single experimental framework. A coarse taxonomy might divide much research to-date into two classes: *convergent* systems and *competitive* systems. Examples of the former include work that focuses on establishing common conventions of behavior. In (Werner and Dyer, 1991), female and male agents converge onto a signaling convention that facilitates reproduction. Each utterance is fed from female to male via a three-bit parallel channel; in the most adaptive pairs, the utterance, produced by a "sighted" but stationary female, causes the "blind" male to move in one of four directions such that it draws nearer to the female.

Other work builds a self-organizing system that converges onto a convention of word/meaning associations (Steels, 1997). The universe of discourse is the set of agents themselves, each of which is uniquely defined by a finite set of meanings, or feature/value pairs. Experiments described have on the order of twenty agents, five meanings, and five words. The agents build lexicons that associate the finite set of words to the feature/value pairs and use word groups to signify particular agents. This domain is claimed to be open in the sense that it can assimilate new meanings (features or values) and agents at any time, even after a convention has emerged. Nevertheless, the semantic and denotative material used to establish a communicative convention is supplied externally from the system, not created by it; any convention that emerges is merely a reflection of a pre-existing, hand-built world.

In the second category of the taxonomy, we have systems that eschew convention; competitive games are designed to promote ever increasing "complexity," the precise meaning of which is inevitably domain-dependent. One game of imitation asks agents to observe another's behavior for some time and then reproduce that behavior (Kaneko and Suzuki, 1994); the observation and reproduction phases of the game last 255 and 32 times steps, respectively. Agents are coevolved to become better imitators and more difficult to imitate. Though behavior does become more complex, the primary cause of complexity likely derives from inductive bias in the substrate: agent behavior is synonymous with the logistic map. The evolutionary algorithm performs a one-dimensional search in the space of the logistic map's constant, which entirely determines agent behavior. While this system reveals what the adaptive advantages of various behaviors are according to the rules of the game, the substrate is so impoverished that complexity of behavior bears no meaningful relationship to agent complexity or evolutionary search.

A comparable situation is found in a system where grammars evolve to climb the Chomsky hierarchy through a competitive game of string generation and parsing (Hashimoto and Ikegami, 1996). The substrate in this case consists of explicit production rules, however, and the mutation operators allow the system to traverse the hierarchy easily. Indeed, the complexity of the evolved grammars is not determined by observing agents' behaviors (word length is limited to six symbols), but rather by looking directly "under the hood" at their production rules.

The above examples are emblematic of issues common to a great deal of work on communication, namely: powerful representations, small-scale communicative (behavioral) acts, fixed channels, and a fixed world. Viewed as an experiment in communication, our modified PE game provides 1) an open-ended serial communications channel (Gregory M. Saunders, 1996), which can lead to 2) substantial communicative acts, and 3) an open world (via coevolution). Too, information theory provides a rigorous and accepted vocabulary for describing signal properties. Further, the substrate used in our experiments is a recurrent artificial neural network; though powerful, the substrate is non-specific to our domain, and thus "weak." Finally, as described below, we seek to bridge the gap between convergent and competitive systems and explore the tension thus created.

Results show our domain to be an effective instrument for research in coevolution, pursuer-evader dynamics, and communication. Our methodology is built around the analytical tools made available by information theory. We are able to analyze agents that evolve and use hand-built agents of known ability to influence evolution. By characterizing generator behavior, we are able to identify variations of the game that lead to mediocre stable-states or define optimal solutions. These results lead us to a variation of our game that avoids both.

This paper is organized as follows: Section 2 details our version of the PE game and contrasts it with the more traditional spatial formulations; Section 3 describes the recurrent artificial neural network substrate used to represent our agents, and introduces the evolutionary algorithm; Section 4 defines concepts from information theory that provide our metric of agent behavior and coevolutionary progress; Section 5 places these concepts within the framework of a coevolutionary arms-race; Section 6 analyzes our results; Section 7 summarizes our work and points to future directions for our research.

2. Game Setup

2.1 Two-Player Pursuer/Evader

Illustrated in Figure 1, our basic PE game consists of two agents that play in discrete time and space. The *generator* (evader) is an agent that ballistically produces a binary time series, that is, its behavior is determined solely by its own internal dynamics; at each time step, the generator simply outputs a new bit. The *predictor* (pursuer) is an agent that, given its own internal state along with the generator's output from the previous time step as input, simultaneously tries to predict the generator's output for the current time step. Each match has a duration of one thousand time steps. Thus, our generator corresponds to a "blind" spatial evader, as its behavior is not modulated by that of its opponent. This modification "clamps" one side of the pursuer/evader dynamical system; indeed, a "sighted" evader would simply be a negated pursuer — an agent that predicts its opponent in order to perform the opposite action.

In addition, our game does not define a "capture" predicate; agent performance is measured strictly in terms of the number of correct and incorrect predictions made, which is analogous to measuring time-in-contact between pursuer and evader over the course of a spatial match. Too, we discover that, due to its discrete behavior, our generator corresponds to a particularly nimble spatial evader, as its current "location" in no way constrains where it may be in the next time step; we have at best a very rudimentary kinematic model.

Our variant game may thus appear quite foreign. Indeed, it resembles the *penny matching* game, a version of which is used in a non-predictive context in parasite/host gene-matching simulations (Hamilton et al., 1990). Nevertheless, we contend that our modifications distill the traditional PE game to its *informational* essence: given such an agile evader, the (equally agile) pursuer must induce, from observation of behavior, a model of the evader in order to be effective. The quality of the model is demonstrated through the pursuer's prediction abilities. This game, therefore, highlights the need to evolve perceptual, computational, and "motor" abilities in the pursuer. The search for effective strategies conducted by the simulated evolution traverses a wide range of such abilities; thus, while our game is objectively one of *perfect information* for the pursuer — it has the opportunity to know the evader's actions — pursuers are not necessarily endowed to make use of this knowledge. In effect, the game easily admits the possibility of information loss, albeit within the evolutionary substrate. In contrast, our game is invariably one of *imperfect information* for the evader; only through coevolutionary feedback do the evaders receive any information.

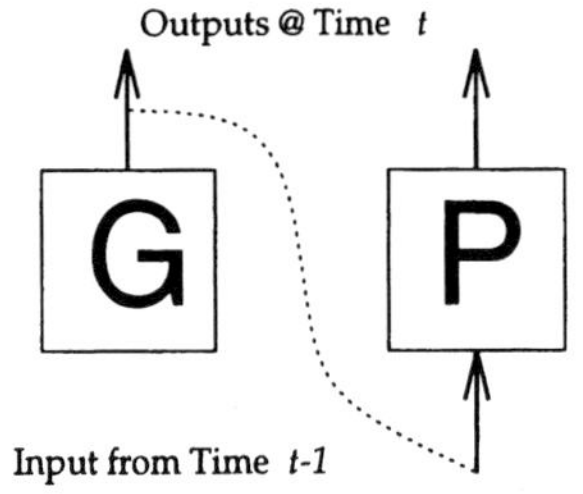

Figure 1 Basic Game Setup. At each time step, the predictor receives the output of the ballistic generator from the previous time step and outputs a prediction of the generator's behavior for the current time step.

2.2 Three-Player Pursuer/Evader

To synthesize convergent and competitive system dynamics, we introduce a third player into our game — a "friendly" predictor. We imagine the evaders to be individuals literally under constant predation who must, therefore, mate while "on the run." But, the mate must be able to keep up with the evader. The *three-player PE* game now consists of one ballistic generator that attempts to behave in a manner that is both predictable to the partner (friendly predictor), to allow mating, and simultaneously unpredictable to the pursuer (hostile predictor), to allow evasion.

This modification to the game is far from capricious. Prediction is intrinsically more difficult than generation; if we believe the Machine Learning aphorism that "you can only learn what you almost already know," then we must be mindful that generators might become too complex for predictors to learn against, breaking the coevolutionary arms race. Worse, generators might become inherently unpredictable. The friendly partner serves to dampen any such tendency by forcing generators to be predictable to someone. In terms of communication, this arrangement can be viewed as a competitive game between the two agent populations (generators and partners) trying to evolve a proprietary behavioral convention, and the third agent population (pursuers) trying to crack the convention.

For each generation of evolution, all generators are played against all partner and pursuer predictors. Scores across all games are averaged to derive fitness values. Game scores for all agents range between [0, 1]. The exact formulas used for scoring predictors are discussed below in the experiment descriptions. A generator's score is computed by subtracting the average scored by its pursuers from the average scored by its partners and normalizing the result to fall within the range [0, 1]; values above 0.5 thus indicate that a generator is able to make itself more predictable to partners than pursuers.

3. Substrate and Evolutionary Algorithm

Our agent substrate is an enhanced version of the discrete-time, deterministic recurrent artificial neural network used in the *GNARL* system (Angeline et al., 1994); the network enhancement consists of a set of nine new transfer functions, *min*, *max*, *sum*, *product*, *sig-prod* (sigmoid applied to product), *unit-time-delay*, *sign*, *uni-linear-truncate* (truncate outside range [-1, 1]), *dec-linear-truncate* (truncate outside range [-10, 10]), in addition to the traditional sigmoid function. These supplementary transfer functions increase the range of behavior significantly, though by no means do they guarantee agent success. Unfortunately, these new functions also make network analysis much more difficult.

GNARL coevolves the networks' internal architectures and weights. Generators may have up to 60 hidden nodes and 400 weights, while predictors are given more generous limits of 150 hidden nodes and 700 weights due to the relative difficulty of their task. These limits are not known to be optimal and are, in fact, never reached.

The input and output layers of the networks are fixed. All networks have a single, real-valued output that is thresholded to a binary value. Though the game formally defines friendly and hostile predictors to have a single input, we currently provide both predictor roles with a small buffer to enhance performance: the predictors have five binary-valued inputs, corresponding to the last five outputs of the generator at times $t-1, ..., t-5$. This enhancement in no way obviates the need for a recurrent network architecture: predictors must still induce generator behavior by observation over time.

The GNARL algorithm performs its search solely through mutation — crossover is not used. Five mutation operators are implemented: *change-weights*, *add-hidden-nodes*, *remove-hidden-nodes*, *add-weights* (connections), and *remove-weights* (connections). When a hidden node is removed, all efferent and afferent connections from and to that node are removed as well. New nodes are added with no connections. Only a single mutation operator is applied to a network when it is modified. Network weights are modified by adding a Gaussian to each weight. The overall severity of mutation performed to a network is determined by its *temperature*, which is an inverse function of its fitness. The higher the temperature of a network, the more severe the mutation will be.

Three distinct agent populations of size 75 are maintained, one each for the evaders, friendly partners, and hostile pursuers. In each generation of coevolution, all evaders are matched with all friendly and hostile predictors. After evaluation, approximately 75% of the worse half of each population is replaced by mutated versions of agents in the better half.

4. Metric of Behavior

Two key notions from the field of information theory provide our game with a rigorous and quantitative metric of agent and system behavior, namely *entropy* and *order*. Rather than give their formal mathematical definitions, we emphasize a more intuitive explanation of these concepts and their implications as they relate to our domain. Formal detail can be found in (Hamming, 1980).

4.1 Entropy

Information theory is concerned with characterizing signals and their transmission. A signal *source* produces some symbol, which is passed through a *channel* to a *receiver*. We assume, for our purposes, that the channel does not distort the signal. The *entropy*, h, of a source reflects the receiver's uncertainty as to what it will receive. The higher the entropy, the less certain the receiver is, and the more it learns once the symbol is actually received. Thus, entropy is a measure of the amount of *information* in a signal. More precisely, the entropy of a source is equal to the average number of bits of information produced (conveyed) per generated symbol.

By indicating the uncertainty of the receiver, entropy inversely indicates the degree to which the source can be *predicted* by the receiver, that is, the receiver's certainty. We must be careful to point out that the receiver's opinion of what the next symbol will be is based exclusively upon the *observed behavior* of the source — assumptions about the source's internal operation are not made.

4.2 Order

If the receiver's certainty is based upon observation of the source, we can ask "How much observation is required to maximize the receiver's certainty?" For example, let us consider some binary source, $\mathcal{S}$. If the receiver only tallies the number of occurrences of *0* and *1*, this source may be found to produce each 50% of the time. With this amount of behavioral context, the receiver's certainty of the next symbol is zero and entropy is measured at $h = 1.0$. Nevertheless, it may be that if the receiver keeps track of the previous symbol received, then the source will be found simply to be alternating between *0* and *1*; in this case, a behavioral context of one symbol makes the source completely predictable. Measured entropy is now $h = 0.0$. If the receiver keeps track of yet another symbol, now the previous two, no additional advantage is gained with respect to the source, $\mathcal{S}$.

The minimal amount of behavioral context needed to maximize the receiver's certainty of a source is the *order* of the source. The order is equal to the number of symbols that must be tracked, that is, the size of the *history window* needed to maximize receiver certainty. The entropy measured when using a window size equal to a source's order is the *true entropy* of the source; window sizes larger than a source's order will produce measurements equal to the source's true entropy, but not lower. Thus, a receiver cannot increase its certainty of a source by using a window size larger than the source's order.

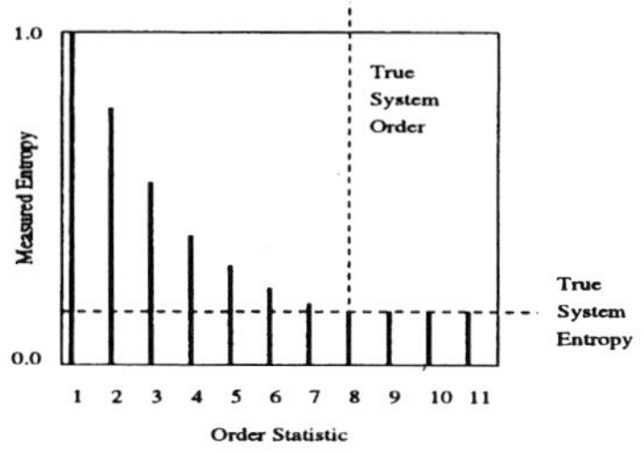

Figure 2 Measured vs. True System Entropy and Order.

4.3 Order Statistics and Measured Entropy

With our example source, $\mathcal{S}$, above, we first measured entropy without keeping track of the previously generated symbol; this is equivalent to measuring entropy with a window size of zero, or measuring with zero-order statistics. Our second measurement, then, used a window size of one, or first-order statistics. Our zero-order measurement gave us an entropy of $h = 1.0$, but the first-order measurement fell to the true entropy of $h = 0.0$. Indeed, measured entropy will always monotonically decrease as window size is increased, and eventually reach a source's true entropy, as illustrated in Figure 2.

A source with true entropy $h = 0.0$, such as $\mathcal{S}$, is completely predictable and regular. In contrast, a binary source with maximal true entropy of $h = 1.0$ entirely lacks structural regularity and cannot be predicted better than random, on average, without specific knowledge of its internal works. For a source with true entropy somewhere in between, $0.0 < h < 1.0$, there exists both a *regular component* and an *irregular component* to the source's signal. The regular component is that portion of the signal that can be reliably predicted, while the irregular component is that portion that cannot. By definition, the information content of a source must reside exclusively in the irregular component.

4.4 Complexity

System order is also equal to the logarithm of the maximal number of states required for a Markov model to reproduce behavior statistically identical to a source; entropy reflects the degree of certainty in the model's state transitions. Consider a randomly behaving binary source, with true entropy of $h = 1.0$. We find that the minimal window size needed to maximize a receiver's certainty of this source is *zero*. Since the order of such a source is zero, the equivalent Markov model requires $2^0 = 1$ state to reproduce statistically identical behavior. This result is understandable since there exists no signal structure to capture through state. Thus a random source is considered to be simpler than a completely predictable source of higher order; the size and structure of the Markov model is what counts, not the compressibility of the produced signal. This view is substantially similar to the notion of *statistical complexity* found in (Crutchfield, 1994).

5. Where's the Arms Race?

When considered together, order and entropy form the nexus between generator complexity and predictor power: if a signal has a regular component, then that component can be predicted assuming that the power of the predictor is sufficient; that is, the predictor must use an order statistic, i.e., history window, of size $m \geq n$, where n is the order of the signal being predicted. If the predictor's window size is m, such that $m < n$, then it will be able to predict only that portion of the signal's regular component that is detectable when measuring the signal's entropy with m^{th}-order statistics. Recall that as window size decreases, measured entropy increases; thus, predictors using smaller windows will necessarily be weaker than those using larger windows.

Because irregular signal components are inherently unpredictable, and our three-player game requires generators to be predictable to friendly predictors, generators must maintain substantial regular components in order to succeed. Nevertheless, generators need to be unpredictable to the hostile predictors. The only way both goals can be effectively met is for the generators and friendly predictors to evolve system order and predictive power that are closely matched, yet greater than the predictive power of the hostile predictors.

Regular signals allow for a general solution to the prediction task. A predictor of power n can predict any generator of order $m \leq n$; to escape prediction, therefore, a generator has no choice but to increase its order above n. The amount by which the generator increases its order and the unpredictability it exhibits at lower order-statistics determines how much the predictor's performance degrades. Of course, a generator may increase its true entropy instead; doing so, however, will also defeat any hopes of being predicted by the friendly predictor.

The assumption up to now has been that predictors will actually evolve such a general prediction algorithm. Of course, this represents an idealized solution; in reality, the issue of generalization vs. specialization is intimately tied to that of population diversity and domain "physics." Nevertheless, having some notion of what idealized predictors can and cannot do, and what an idealized arms race looks like, provides a useful framework in which to examine empirical results.

6. Results

6.1 Adaptive Pursuit Behavior

Our first experiments test the nominal capabilities of our recurrent artificial neural networks with respect to our problem domain. Due to the nature of our game, we can easily *hand-build* predictors of known power through a simple modification of the algorithm used to compute entropy. We thus craft an environment of two predictors of unequal power and evolve generators to become maximally unpredictable to the weaker predictor while remaining maximally predictable to the more powerful one. The generators that evolve successfully fill the niche between the two predictors. We use a variety of such hand-built predictors to evolve a set of 70 generators that range mostly over system-orders two through eight.

We then evolve predictors against this fixed set of generators. The operation of one such predictor is shown in Figure 3. The figure depicts the operation of our predictor over 800 time steps. During these 800 time steps, the predictor consecutively plays against eight different generators (against which it evolved), each for one hundred time steps. The generator's behavior is shown directly above the predictor's in each match. Each generator network has its activation levels reset to zero at the beginning of its match. In contrast, the activation levels of the predictor network is reset only once, at the beginning of the first match at time-step zero; thus, from the predictor's point of view, the generator changes its behavior every one hundred time steps.

During evolution, however, predictors *do* have their activation levels reset at the beginning of each match. Therefore, predictors are not explicitly evolved to adapt to such changes in generator behavior. Nevertheless, Figure 3 shows very compelling evidence of adaptive "pursuit" behavior against these generators; the sequence in which the generators are met does not affect the predictor's performance. Of special interest is the entrainment behavior exhibited by the predictor at the beginning of each match; in all matches (except the first), we see the predictor continue its pursuit behavior from the previous match for some period of time before it begins to entrain to the new generator's behavior.

From Figure 3, we see that the predictor is always "paying attention" to the generator, that is, the predictor does not settle into a single, immutable pattern of behavior. The predictor shows sensitivity to the frequency, phase, and duty-cycle of the first six generators. Further, generators 7 and 8 show that the predictor is capable of inducing more complex, "compound" signals, though generator 7 is not perfectly induced. In summary,

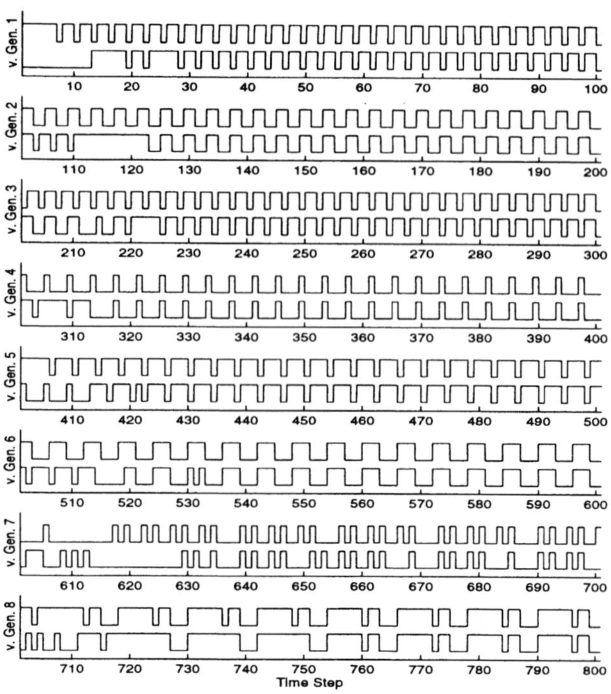

Figure 3 Adaptive Predictor Behavior. An evolved predictor consecutively plays against eight different generators, each for 100 time steps. Generator behavior is shown directly above the predictor's in each match.

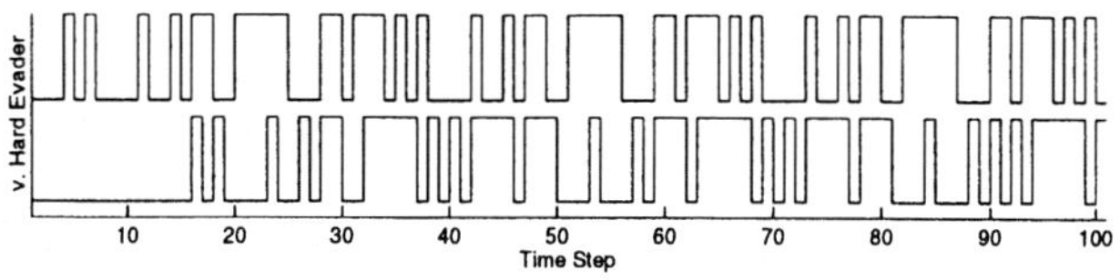

Figure 4 Predictor vs. Hard Generator. The evolved predictor is unable to entrain to this 5th-order generator.

these initial tests suggest a space of network behaviors rich enough to support coevolution.

6.2 Complexity and Computational Demand

Because the generators described above are ballistic and evolved to be predictable to someone, they settle quickly into regular (periodic) patterns of behavior. Though all regular generators have true entropies of zero, they can make very different demands of a predictor. For example, the generators in Figure 3 all have a true entropy of zero, but are of orders 2, 2, 2, 3, 3, 3, 13, and 6 (from top to bottom). The predictor in Figure 3 does very well against these generators. Yet, Figure 4 shows our predictor to perform no better than random guessing against a particular 5th-order generator. Since order indicates the minimal number of symbols that must be tracked to maximize prediction, this 5th-order generator should be easier to predict than either the 6th or 13th-order generator, against which the predictor does well.

When we measure the entropies of these three generators with window sizes zero through thirteen, we see why

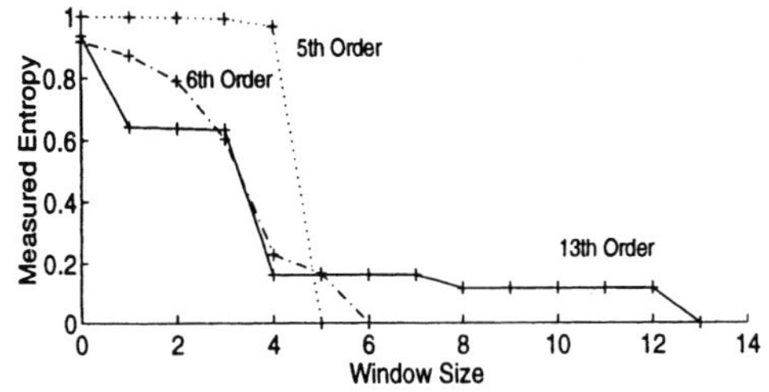

Figure 5 Entropy Curves for the 5th, 6th, and 13th-Order Generators. The X axis represents the window size used to measure entropy. The Y axis represents the measured entropy.

neither entropy nor order tells the complete story alone. Figure 5 graphs the *entropy contours* that are produced by measuring entropy with different window sizes. The solid line is the entropy contour for the 13th-order generator; we see that this generator is already substantially predictable with a much smaller window of four, but requires many more symbols to be tracked to be completely predictable. The dashed line is the contour for the 6th-order generator; this and the 13th-order generator are almost equally predictable with a window of size five. Yet, to be fully predictable, only one more symbol need be tracked for the 6th-order generator. The dotted line shows the entropy contour for the hard 5th-order generator; this generator's behavior appears random until the window size reaches five, whereupon it is completely predictable. That is, for *all* length-four subsequences of this generator's output, the next symbol could be *0* as easily as *1*; all 32 possible length-five subsequences thus appear in the output. For each length-five subsequence, however, the next symbol is uniquely determined.

Thus, while the 5th-order generator requires the fewest symbols to be tracked to be fully predictable, it also has the most contingencies to be resolved; representing and resolving these contingencies is computation that a generalized predictor must perform. All of the 32 possible five-bit patterns must be tracked to know unambiguously what the following bit will be. In contrast, the 13th-order generator requires only 17 of the 8192 possible 13-bit patterns (0.2%) to be tracked to determine subsequent bits.

The number of unique n-bit patterns found in an nth-order periodic sequence is equal to the period of the sequence. Thus, the 5th-order generator of Figure 4 has a period of 32, whereas the 13th-order generator of Figure 3 has a period of 17. Further, the period of a sequence indicates the number of functionally distinct states that a predictor must maintain in its steady-state behavior (i.e., after entrainment). Thus, we can say that the 5th-order generator is more difficult to predict simply because it requires more states. Nevertheless, the steady-state demands placed on a predictor tell us nothing about the computational "effort" the predictor spends to arrive at its steady-state; the measure of period does not allow us to distinguish between the $\Theta(2^n)$ distinct period-n sequences. A periodic sequence of 31 ones followed by a zero, a sequence of 16 ones followed by 16 zeros, and the sequence seen in Figure 4 do *not* place identical demands upon a predictor and are not equally difficult to evolve against. Information theory thus gives us a way to more effectively characterize "hardness." This ability is invaluable when investigating coevolutionary learning.

6.3 "Boilerplate" Strategies

The generality of a predictor strongly reflects the diversity of the generators against which it evolved. For example, when we evolve predictors exclusively against 4th-order and 5th-order generators, we find the predictors to be rather specialized; the ability to predict these generators does not necessarily confer an ability to predict lower-order behaviors. Indeed, despite its success, the predictor in Figure 3 performs poorly against a simple, period-eight square-wave behavior.

If we evolve predictors against generators of various long periods, but of near 50% duty cycle, we find that a simple *copier* prediction strategy becomes tenable — one that simply outputs what it saw in the previous time step; such a strategy only misses on the generator's transitions between zero and one and "predicts" the rest (majority) of the sequence correctly. Other copier variants seen appear tuned to particular periods, such that any pattern of the appropriate period will be matched after one cycle. Such mediocre solutions occur easily where (a lack of) generator diversity allows.

6.4 Mediocrity and Protean Behavior

We begin our analysis of *co*evolutionary results by questioning the earlier-stated need for a friendly partner predictor. We describe two versions of the two-player PE game in a coevolutionary setting; the friendly partners are omitted. The versions are differentiated solely by the scoring method used to reward the pursuer predictors. The first scoring method (A) gives predictors a point for each correct prediction. The tallies are then normalized to the percentage of correct predictions. A generator thus receives maximal reward when a predictor is unable to make any correct predictions. The second scoring method (B) tallies correct predictions, like method A, but predictors are now rewarded only to the extent that they perform better than random; predictors that get less than or equal to 50% correct prediction receive a score of zero. In this case, a generator receives maximal reward for causing a predictor to perform no better than random. In both cases, the stage appears set for an arms race. We find, however, that the scoring methods have important repercussions in coevolutionary dynamics.

We gauge generator behavior by again taking entropy measurements over a range of window sizes to get entropy contours. Figures 6 and 7 graph the contours for the best generators that evolve over evolutionary time in sample runs for scoring methods A and B, respectively. Both graphs show that generator behavior remains consistent over the course of the run; we have no indication of an arms race. The generators coevolved with method A are substantially predictable and regular, as evidenced by the rapidly dropping contours. The generators coevolved with method B, on the other hand, are considerably more irregular (unpredictable), as the contours decline gradually and consume a much greater volume of space in the graph. This contrast in outcome is found even in pairs of runs that start with identical initial populations.

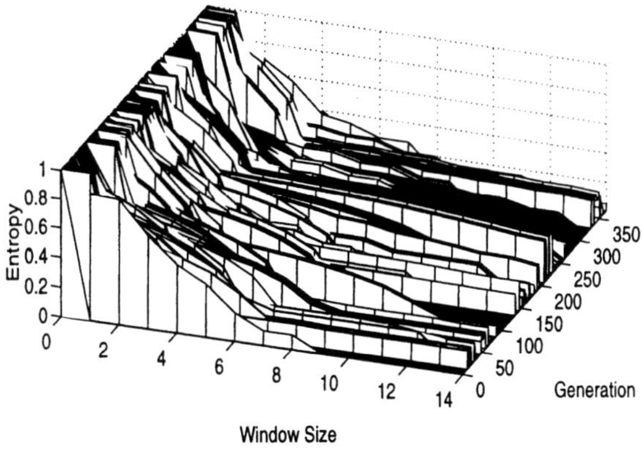

Figure 6 Scoring Method A — Reward for Each Correct Prediction. Rapidly declining entropy contours indicate substantially predictable behavior.

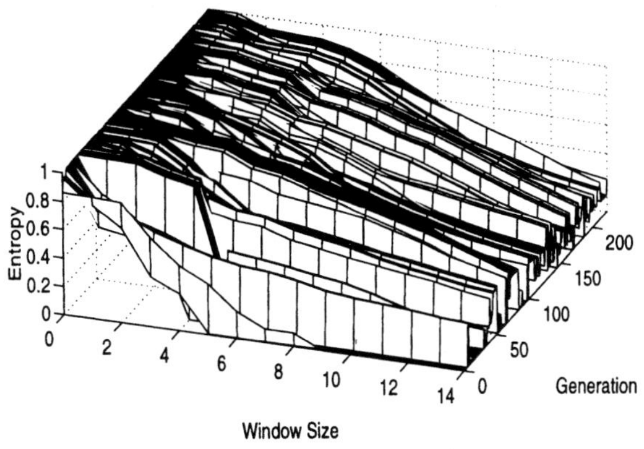

Figure 7 Scoring Method B — Reward for Predicting Better than Random. Gradually declining entropy contours indicate significantly more complex behavior than that seen by scoring method A.

The difference between scoring methods is manifested quickly, if not immediately. Since the initial conditions are similar (or, indeed, identical if we wish) we know that we have isolated a selection pressure that stems simply from how games are scored: regular behavior is more adaptive to scoring method A than is irregular behavior. That regular generators out-score irregular ones by method A indicates that simple prediction strategies can be elicited from the predictor population — prediction strategies against which the most adaptive generators act as potent ***anti-signals***, sequences that make predictors perform worse than random (Zhu and Kinzel, 1997). Indeed, the champion generators shown in Figure 6 receive scores that indicate exactly this. Nevertheless, if the ballistic generators are to be viable as anti-signals, then there must exist some sufficient amount of homogeneity within the predictor population. This homogeneity appears to reverberate through the system; rather than enter an arms race, the two populations fall into a circular pattern of mutual specialization, or *convention chasing* — a mediocre stable-state.

In contrast, scoring method B selects for an irregular behavior that reasonably captures the notion of proteanism in the PE game. Figure 8 provides a sample of behavior from a typical generator produced by this run. This behavior is adaptive because it guarantees a minimal predictor score, regardless of what the predictor does, short of specialized memorization. Indeed, while the generators in Figure 7 are not actually fully irregular, they are complex enough to cause the best predictors from this run to score only an average of 6% better than random guessing (53% correct prediction); that is, these generators are near optimal with respect to their predictor opponents. By defining an optimal strategy, namely random behavior, method B makes the game closed-ended.

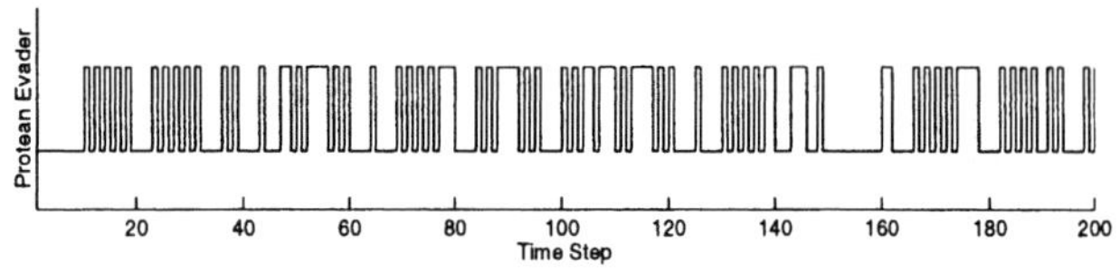

Figure 8 Typical Behavior of a Protean Generator.

6.5 Pursuing a Communicative Convention

The above results suggest that the two-player PE formulation may be problematic if we wish to develop an open-ended arms race. Particularly, the two-player game may oversimplify the ecology by inadequately placing constraints upon generator behavior. In the three-player PE game, we maintain scoring method B to promote complexity and simply add the friendly predictor partner. Now, the evader and friendly partner must establish a convention of behavior that allows the partner to entrain to (coordinate with) the evader, but not the hostile pursuer. We find that the tension created by the need to "mate" with (be predictable to) the partner profoundly affects evasion behavior. Far from extinguishing heated competition, the triangle of relationships gives rise to much more lively system dynamics than those found in the absence of friendly partners.

In contrast with Figures 6 and 7, Figure 9 shows that generator complexity in the three-player game is not at all consistent over evolutionary time; we see an overall

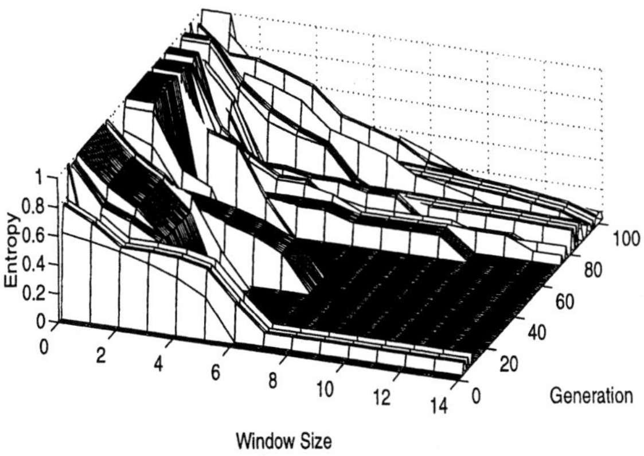

Figure 9 Entropy Contours from Full Game. Overall increase in complexity indicates features of an arms race.

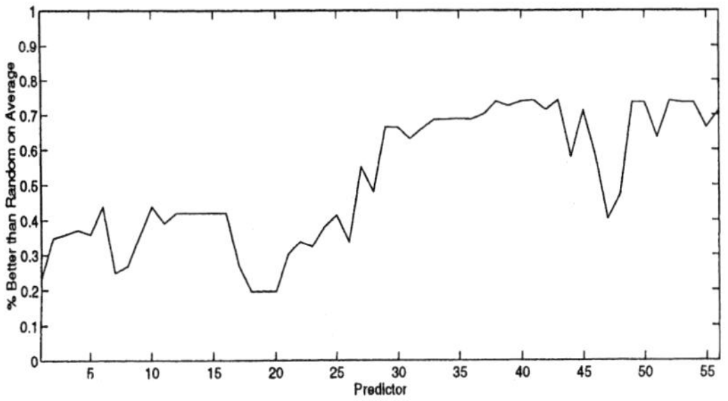

Figure 10 Skill Accumulation in Pursuer. Champion predictors that arise later in evolutionary time are more generalized and perform better against the set of champion generators.

increase in both order and entropy. Figure 10 graphs the performance of the champion pursuers from this run against the set of champion generators of Figure 9. We see clearly that the pursuers created later in evolutionary time are more skillful. The abilities of the friendly predictors likewise increase. Thus, we have reasonable indications of an arms race between the cooperating agents (evaders and friendly partners) and the hostile pursuers.

More typically, however, the three-player game produces punctuated shifts in generator behavior, from simple to complex and back again, rather than the monotonic increase in complexity indicative of an arms race. Indeed, Figure 9, used to argue for the presence of an arms race above, also shows abrupt retreats in generator complexity. What might the adaptive utility of these sudden changes be? Data from a different run suggests a hypothesis. The top three plots in Figure 11 graph the average population fitness of the friendly predictors, hostile predictors, and generators, respectively. Because average generator fitness values are almost always above 0.5, we know that the generators usually succeed in making themselves more predictable to the friendly partners than to the hostile pursuers. But, the degree of their success varies considerably during the run.

The bottom plot in Figure 11 depicts the order of the best generator in each generation. We see that generator

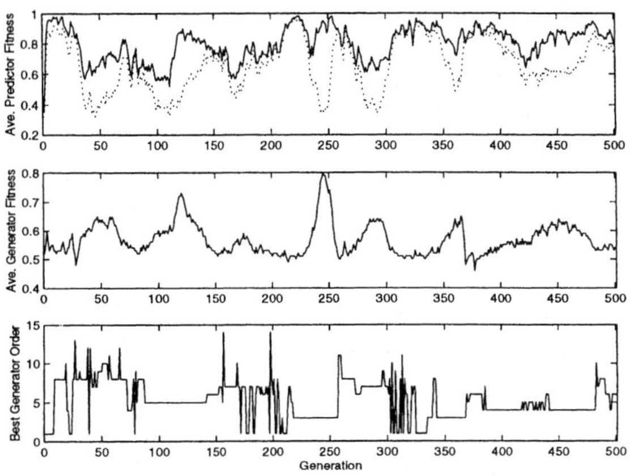

Figure 11 Punctuated Shifts in Generator Behavior in Full Game. From top to bottom, average fitness of friendly predictor partners, pursuer predictors, and generators, and order of best generator in generation, over evolutionary time.

order is most stably maintained over the onset and duration of periods of generator success — periods where the evaders and partners share a private convention for coordinated behavior. This very stability is what allows the hostile pursuers to eventually learn the convention, however. Once this happens, the propriety of the convention is lost and generator fitness reaches its nadir. This tends to be the moment at which the order of the best generators becomes unstable, often alternating between high-order and low-order behaviors. We suspect that this instability is adaptive because it forces the two, now similarly skilled, predictor populations to diverge in ability. Once sufficiently differentiated, some new behavioral convention can be found that only the friendly partner can predict. A new period of evader/partner coordination thus emerges.

The system dynamics of the three-player domain can resemble pursuit and evasion behavior; the *meta*-PE game exists between populations and is played over evolutionary time. The hostile pursuer population is "chasing" the behavioral convention of the evader and partner populations. To escape, the evader and partner populations must alter their convention. In this particular run, the specific escape strategy looks like a type of protean behavior on the part of the generator population; when the pursuer population "catches" the behavioral convention, the erratic changes in generator order serve to mislead the pursuer's evolutionary course. The earlier reference to cycling in host/parasite gene sequences is particularly apt here.

We must point out, however, that this run presents a particularly vivid example of the meta-PE phenomenon. Other runs show the predictor population to be, as a whole, unable to converge onto the behavioral convention of the evader and partner populations; in these cases, changes in generator behavior cannot so easily be tied to those of the pursuers. Nevertheless, we believe that our methodology will allow us to clarify our data by continuing work with hand-built environments.

7. Conclusions

We show that pursuer-evader can be reformulated as a one-dimensional, time-series prediction game. Information theoretic tools provide quantitative analyses and help operationalize the domain. Because the linear PE game emphasizes the informational (in the technical sense) aspects of pursuit and evasion behavior, it captures a fundamental aspect of communication.

Our behavioral metrics also yield informative views of coevolutionary dynamics. Though we can create good evaders and pursuers with simple evolution, we find that successful coevolution does not automatically follow. Subtleties in scoring method strongly influence the outcome of the two-player game; one method leads to mediocrity while the other defines an optimal strategy (protean behavior) and closes the world. The combination of competitive and convergent pressures of the three-player game is needed to avoid simple mediocre stable-states while keeping an open world.

Tools exist to characterize generator behavior and hand-build predictors of known power. This toolset has recently been augmented to include new methods that allow us to construct by hand generators with particular entropy contours and systematically analyze the generality and power of evolved predictors. Our future work will feature these techniques in a continued investigation of generator complexity, the computational demands generators place on predictors, and the ease with which predictors evolve against them. The final results we report are not fully understood and will require these tools to elucidate.

8. Acknowledgments

The authors gratefully acknowledge the many hours of conversation that have contributed to this work provided by Alan Blair, Marty Cohn, Paul Darwen, Pablo Funes, Greg Hornby, Ofer Melnik, Jason Noble, Elizabeth Sklar, and particularly Richard Watson.

References

Ackley, D. H. and Littman, M. L. (1994). Altruism in the evolution of communication. In (Brooks and Maes, 1994), pages 40–48.

Akiyama, E. and Kaneko, K. (1997). Evolution of communication and strategies in an iterated three-person game. In Langton, C. G. and Shimohara, K., editors, *Artificial Life V (1996)*, pages 150–158. MIT Press.

Angeline, P. J. and Pollack, J. B. (1994). Competitive environments evolve better solutions for complex tasks. In Forrest, S., editor, *Proceedings of the Fifth International Conference on Genetic Algorithms*. Morgan Kaufmann.

Angeline, P. J., Saunders, G. M., and Pollack, J. B. (1994). An evolutionary algorithm that constructs recurrent neural networks. *IEEE Transactions on Neural Networks*, 5:54–65.

Brooks, R. A. and Maes, P., editors (1994). *Artificial Life IV*. MIT Press.

Cliff, D. and Miller, G. F. (1995). Tracking the red queen: Measurments of adaptive progress in co-evolutionary simulations. In Moran, F. et al., editors, *Third European Conference on Artificial Life*, pages 200–218. Springer Verlag.

Cliff, D. and Miller, G. F. (1996). Co-evolution of pursuit and evasion 2: Simulation methods and results. In (Maes et al., 1996), pages 506–515.

Crutchfield, J. P. (1994). The calculi of emergence: Computation, dynamics, and induction. *Physica D*, 75:11–54.

Di Paolo, E. A. (1997). Social coordination and spatial organization: Steps towards the evolution of communication. In Husbands, P. and Harvey, I., editors, *Fourth European Conference on Artificial Life*, pages 464–473. MIT Press.

Gregory M. Saunders, J. B. P. (1996). The evolution of communication schemes over continuous channaels. In (Maes et al., 1996), pages 580–589.

Hamilton, W. D., Axelrod, R., and Tanese, R. (1990). Sexual reproduction as an adaptation to resist parasites (a review). *Proc. Natl. Acad. Sci. USA*, 87:3566–3573.

Hamming, R. W. (1980). *Coding and Information Theory*. Prentice-Hall, Inc., Englewood Cliffs, NJ.

Hashimoto, T. and Ikegami, T. (1996). Emergence of net-grammar in communicating agents. *BioSystems*, 38(1):1–14.

Isaacs, R. (1965). *Differential Games*. John Wiley and Sons, New York.

Kaneko, K. and Suzuki, J. (1994). Evolution to the edge of chaos in an imitation game. In Langton, C. G., editor, *Artificial Life III (1992)*, pages 43–54. Addison-Wesley.

Koza, J. (1992). *Genetic Programming*. MIT Press.

Langton, C. G. and Shimohara, K., editors (1997). *Artificial Life V (1996)*. MIT Press.

Maes, P. et al., editors (1996). *From Animals to Animats IV*. MIT Press.

Miller, G. F. and Cliff, D. (1994). Protean behavior in dynamic games: Arguments for the co-evolution of pursuit-evasion tactics. In Cliff, D. et al., editors, *From Animals to Animats III*, pages 411–420. MIT Press.

Oliphant, M. and Batali, J. (1996). Learning and the emergence of coordinated communication. *(Submitted)*.

Pollack, J. B., Blair, A., and Land, M. (1997). Coevolution of a backgammon player. In (Langton and Shimohara, 1997).

Reynolds, C. (1994). Competition, coevolution and the game of tag. In (Brooks and Maes, 1994), pages 59–69.

Steels, L. (1997). Self-organising vocabularies. In (Langton and Shimohara, 1997), pages 136–141.

Werner, G. M. and Dyer, M. G. (1991). Evolution of communication in artificial organisms. In Langton, C., Taylor, C., Farmer, J., and Rasmussen, S., editors, *Artificial Life II (1990)*, pages 659–687. Addison-Wesley.

Zhu, H. and Kinzel, W. (1997). Anti-predictable sequences: Harder to predict than a random sequence. *(Submitted)*.

Competitive Co-Evolutionary Robotics: From Theory to Practice

Dario Floreano*, **Stefano Nolfi****, **Francesco Mondada***
* Laboratory of Microcomputing
Swiss Federal Institute of Technology, CH-1015 Lausanne
{Name.Surname}@epfl.ch
** Department of Neural Systems and Artificial Life
National Research Council, Viale Marx 15, Roma, Italy
stefano@kant.irmkant.rm.cnr.it

Abstract

It is argued that competitive co-evolution is a viable methodology for developing truly autonomous and intelligent machines capable of setting their own goals in order to face new and continuously changing challenges. The paper starts giving an introduction to the dynamics of competitive co-evolutionary systems and reviews their relevance from a computational perspective. The method is then applied to two mobile robots, a predator and a prey, which quickly and autonomously develop efficient chase and evasion strategies. The results are then explained and put in a long-term framework resorting to a visualization of the Red Queen effect on the fitness landscape. Finally, comparative data on different selection criteria are used to indicate that co-evolution does not optimize "intuitive" objective criteria.

1. Competitive Co-Evolution

In a competitive co-evolutionary system the survival probability of a species is affected by the behavior of the other species. In the simplest scenario of only two competing species, such as a predator and a prey or a parasite and a host, the performance of each individual is tightly related to the performance of the competitor both on the ontogenetic and on the evolutionary time scale. Changes in one lineage might affect the selection pressure on the other lineage and, if the other lineage also responds with counter-adaptive features, the system might give rise to what biologists call a "a co-evolutionary arms race" (Dawkins and Krebs, 1979).

Within the framework of *artificial evolution*, the reciprocal effects of changes by one species on the selection pressure of the other species introduce novel complexities with respect to the case of a single species evolved in a static environment. In the latter case (figure 1, left), there is a unique relationship between the traits of organisms and its reproduction success, and evolution is often seen as a force driving the population towards combinations of traits that maximize reproduction success (Goldberg, 1989). Instead, in competitive co-evolution

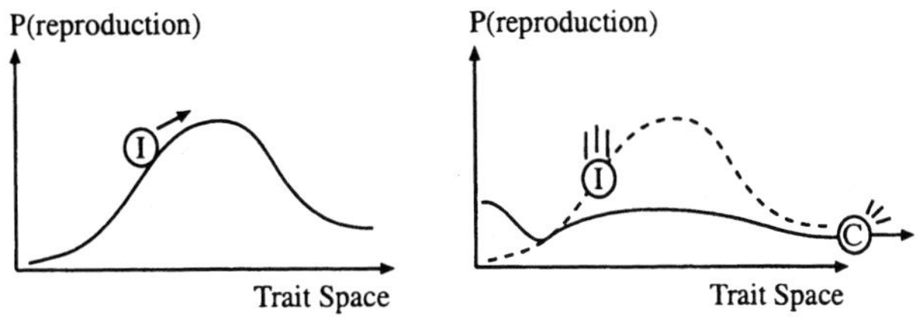

Figure 1 **Left**: Reproduction probability of a single species I under evolution in a static environment. Evolution drives the organisms towards zones (combinations of traits) corresponding to higher reproduction success. **Right**: Reproduction probability of species I under competitive co-evolution. The reproductive value (fitness) of certain trait combinations can be affected by adaptive changes in the competing species C, resulting in a continuous modification of the evolutionary surface. This phenomenon is often called the *Red Queen Effect* (van Valen, 1973).

the reproduction probability of an organism with certain traits can be modified by the competitors; that is, changes in one species affect the reproductive value of specific trait combinations in the other species (figure 1, right). It might thus happen that progress achieved by one lineage is reduced or eliminated by the competing species. This phenomenon is sometimes referred to as the "Red Queen Effect" (van Valen, 1973) (from the imaginary chess figure, invented by novelist Lewis Carroll, who was always running without making any advancement because the landscape was moving with her).

Under certain symmetry conditions, a competitive co-evolutionary system can display oscillatory dynamics whose outcome and stability are difficult to predict. Initial formal studies of predator-prey dynamics date back to the mid-20's when Lotka (1925) and Volterra (1926) independently developed a simple model composed of two differential equations based on the assumptions described above. The Lotka-Volterra model describes how the population densities N_1 and N_2 of two co-evolving competing species vary in evolutionary time t

$$\frac{dN_1}{dt} = N_1 (r_1 - b_1 N_2), \quad \frac{dN_2}{dt} = N_2 (-r_2 + b_2 N_1),$$

where r_1 is the increment rate of the prey population in the absence of predators, r_2 is the death rate of predators in the absence of prey, b_1 is the death rate of prey due to be eaten by predators, and b_2 is the ability of the predator in catching the prey. Probabilistic variations of this simple model have been used to explain several biological observations, such as the oscillatory dynamics of co-evolving host-parasite populations reported in figure 2 (Utida, 1957), and predict oscillation periods and stability conditions (Renshaw, 1991; Murray, 1993). However, it should be noticed that this model attempts to predict variations in population size assuming fixed performance for both predators and prey across generations.

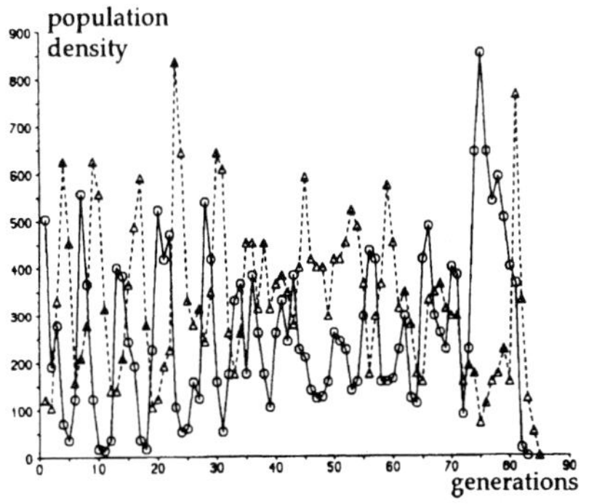

Figure 2 Population density of two biological co-evolving and competing species across generations. △=parasite; ○=host. Termination at generation 85 was caused by external factors. Adapted from (Utida, 1957).

1.1 Computational attractions

From a computational perspective, it would be interesting to know *whether competitive co-evolution might promote performance progress.* In other words, are predators of later generations better at catching prey of previous generations than predators belonging to the same generation of those prey (and viceversa)? Since data on population densities cannot be readily used as indicators of performance, theoretical models and biological observations do not tell us whether competitive co-evolution might be a powerful generator of increasingly better individuals. One possibility is that, since each species is selected only against competitors belonging to the same evolutionary period, there would be no selective pressure to evolve increasingly better performances across generations in either party which would rather display a cyclic alternation of the same strategies every n generations. Another possibility is that the ever-changing fitness landscape, caused by the struggle of each species to take profit of the competitors' weaknesses, might prevent stagnation of the two populations in local maxima.

Hillis (1990) reported a significative improvement in the evolution of sorting programs when parasites (programs deciding the test conditions for the sorting programs) were co-evolved , and similar results were found by Angeline and Pollack (1993) on co-evolution of players for the Tic Tac Toe game. More recently, Rosin and Belew (1997) compared various co-evolutionary strategies for discovering robust solutions to complex games.

In the context of adaptive autonomous agents, Koza (1991, 1992) applied Genetic Programming to the co-evolution of pursuer-evader behaviors, Reynolds (1994) observed in a similar scenario that co-evolving populations of pursuers and evaders display increasingly better strategies, and Sims used competitive co-evolution to develop his celebrated artificial creatures (Sims, 1994). Cliff and Miller realised the potentiality of co-evolution of pursuit-evasion tactics in evolutionary robotics. In a series of papers, they described a 2D simulation of simple robots with evolvable "vision morphology" (Miller and Cliff, 1994) and proposed a new set of performance and genetic measures in order to describe evolutionary progress which could not be otherwise tracked down due to the Red Queen Effect (Cliff and Miller, 1995). Recently, they described some results where simulated agents with evolved eye-morphologies could either evade or pursue their competitors from some hundred generations earlier and proposed some applications of this methodology in the entertainment industry (Cliff and Miller, 1996).

In recent work (Floreano and Nolfi, 1997b), we have re-evaluated competitive co-evolution in the framework of evolutionary robotics resorting to realistic (in the sense that the same setup could be used for physical robots) computer simulations of two mobile robots, a predator and a prey (figure 3). The results and analyses showed the emergence of a set of different behavioral strategies in very short time (shorter than single-agent evolution) without effort in fitness design, such as obstacle avoidance, straight navigation, visual tracking, object discrimination (robot vs. wall), object following, and others. In this paper we significantly extend those results describing what is, to the best of our knowledge, the first physical implementation of competitive co-evolutionary robotics. We then provide an explanation of the results based on the visualization of the Red Queen Effect on the fitness surface of the experimental data. Finally, we give comparative data on different selection criteria indicating that co-evolution does not "optimize" intuitive objectives.

2. Experimental method

Predators and prey often belong to different species with different sensory and motor characteristics. We thus em-

Figure 3 **Right**: The Predator is equipped with the vision module (1D-array of photoreceptors, visual angle of 36°). **Left**: The Prey has a black protuberance which can be detected by the predator everywhere in the environment, but its maximum speed is twice that of the predator. Both Predator and Prey are equipped with 8 infrared proximity sensors (max detection range was 3 cm in our environment).

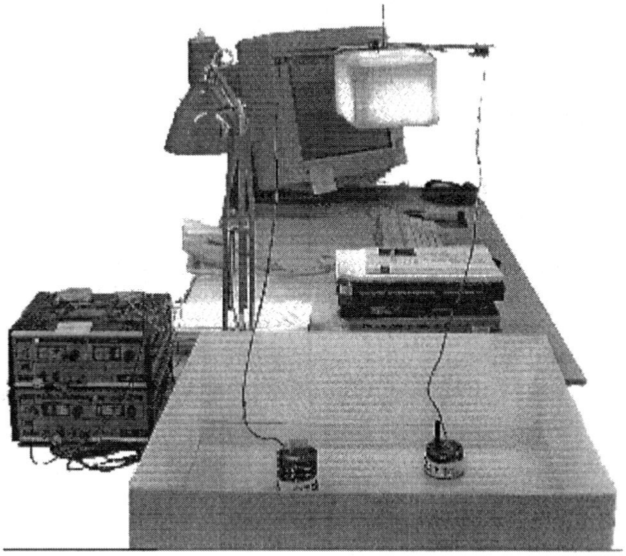

Figure 4 Setup to run co-evolutionary experiments on the physical robots. **Top**: the suspended bar with the three rotating contacts and a white box casting light over the arena. **Left**: two voltage transformers for powering the robots and the halogen lamp. **Background**: the workstation on which the two genetic algorithms run. The robots in the arena are equipped with contact detectors and wrapped in white paper to increase the reflection of infrared light.

ployed two Khepera robots, one of which (the *Predator*) was equipped with a vision module while the other (the *Prey*) had a maximum available speed set to twice that of the predator (figure 3). Both species were also provided with eight infrared proximity sensors (six on the front side and two on the back): a wall could be detected at a distance of 3 cm approx., but the other robot can be detected at only half that distance because it offers a smaller reflection surface to infrared rays. The two species evolved within a square arena of size 47 x 47 cm with high white walls so that the predator could always see the prey (if within the visual angle) as a black spot on a white background.

The two robots were connected to a desktop workstation equipped with two serial ports through a double aerial cable (figure 4). Aerial cables provided the robots with electric power and data communication to/from the workstation. The two cables ended up in two separate rotating contacts firmly attached to the far ends of a suspended thin bar. Both wires then converged into a single and thicker rotating contact at the center of the bar and ended up in the serial ports of the workstation and in two voltage transformers (on the left of figure 4). The thick rotating contact allowed the bar to freely rotate around its own center while the remaining two contacts allowed free rotations of the two robots. Attached under the bar was also a halogen lamp (20 W output) providing illumination over the arena.[1] Both robots were also fitted with a conductive metallic ring around their base to detect collisions. An additional general input/output module provided a digital signal any time the two robots hit each other (but not when they hit the walls). The motor bases of both robots were also wrapped by white paper in order to improve reflection of infrared light emitted by the other robot (approx. 1.5 cm against 0.5 cm). These two solutions are displayed in figure 4.

The vision module K213 of Khepera is an additional turret which can be plugged directly on top of the basic platform. It consists of a 1D-array of 64 photoreceptors which provide a linear image composed of 64 pixels of 256 gray-levels each, subtending a view-angle of 36°. The optics are designed to bring into focus objects at distances between 5cm and 50cm while an additional sensor of light intensity automatically adapts the scanning speed of the chip to keep the image stable and exploit at best the sensitivity of receptors under a large variety of illumination intensities. However, a reliable image at lower illumination comes at the cost of a slower scanning speed of the 64 receptor values. This means that the image would be updated less frequently, thus giving an advantage to the prey (which indeed exploited it during exploratory experiments). Although one might decide to vary ambient illumination as a further independent variable, we decided to keep the halogen lamp switched on over the arena. In the simple environment employed for these experiments, the prey is at the eye of the predator like a valley whose width is proportional to the distance from the predator (figure 5) and whose position indicates the relative position of the prey with respect to the predator.

[1] No special care was taken to protect the system against external light variations between day and night.

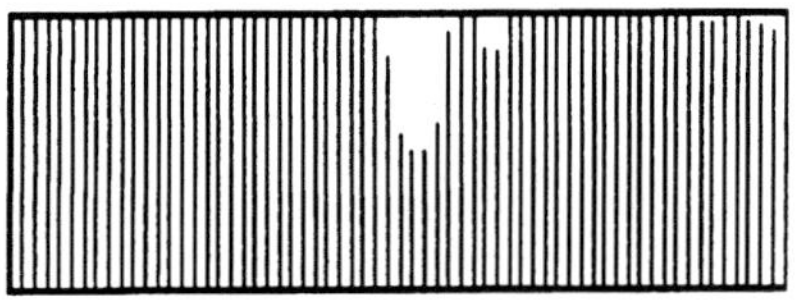

Figure 5 A snapshot of the visual field of the predator looking at the prey (taken at the position illustrated in figure 4). The heights of vertical bars represent the activations of corresponding photoreceptors. The black protuberance of the prey corresponds to a large valley. The small dip nearby corresponds to the cable. In standard illumination conditions, the image was refreshed at a rate of approximately 15-20 Hz.

2.1 Co-evolving neural controllers

In order to keep things simple, both the predator and the prey controllers were simple neural networks of sigmoid neurons. The input signals coming from the eight infrared proximity sensors (and from the vision module for the predator) were fed into two motor neurons with lateral and self-connections. The activation of each output unit was used to update the speed value of the corresponding wheel. Only the connection strengths were evolved. The maximum speed available for the prey was set to twice that of the predator.

In the case of the predator, the input layer was extended to include information coming from the vision module. The activation values of the 64 photoreceptors were fed into a layer of five *center off/surround on* neurons uniformely distributed over the retinal surface. The spatial sensitivity of each neuron was approximately 13° and the center/surround ratio filtered out low contrast features, such as those generated by weak shadows, the cable of the prey, and other imperfections of the walls. Each neuron generated a binary output of 1 when the prey was within its sensitivity field, and 0 otherwise. These five outputs were fed into the motor neurons along with the signals coming from the infrared sensors.

Given the small sizes of the neural controllers under co-evolution, we used *direct genetic encoding* of the connection strengths: each connection (including recurrent connections and threshold values of output units) was encoded on five bits, the first bit determining the sign of the synapse and the remaining four bits its strength. Therefore, the genotype of the predator was 5 x (30 synapses + 2 thresholds) bits long while that of the prey was 5 x (20 synapses + 2 thresholds) bits long. Two separate populations of N individuals each were co-evolved for g generations. Each individual was tested against the best competitors from k previous generations (a similar procedure was used in (Sims, 1994; Reynolds, 1994; Cliff and Miller, 1995)) in order to improve co-evolutionary stability. At generation 0, competitors were randomly chosen from the initial population, whereas in the remaining $k - 1$ initial generations they were randomly chosen from the pool of available best individuals (2 at generation 3, 3 at generation 4, etc.).

In our previous work (e.g., (Floreano and Mondada, 1996)) both the genetic operators and the robot controllers run on the workstation CPU and the serial cable was used to exchange sensory and motor information with the robot every 100 ms or longer. This method could not work in the current setup because transmission duration of visual signals and serial processing of the controller states for both robots on the same CPU significantly delayed and disturbed the interaction dynamics between the two robots. Here the adopted solution consisted in running the genetic operators on the workstation CPU and the neural controllers on the microcontrollers of the two Khepera robots. Each microcontroller, a Motorola MC68331 equipped with 128K ROM and 128K RAM was largely sufficient to store the set of instructions and variables necessary to handle all input/output routines and neural states. The speed of the sensorimotor cycles was set to approximately 15 Hz for both prey and predator. In the predator image acquisition and low-level visual preprocessing were handled by a private 68HC11 processor installed on the K213 vision turret.[2]

Each neurocontroller architecture, with the connection strengths initialized to zero, was downloaded into the corresponding robot. The two genetic algorithms were then started on the workstation CPU where each genetic string was decoded into a set of connection strengths and sent through the serial cable to the corresponding robot. Upon receipt of the connection strengths, each robot began to move and the internal clock (a cycle counter) of the prey was reset to zero. A tournament ended either when the predator hit the prey or when 500 sensorimotor cycles (corresponding to approximately 35 seconds) were performed by the prey without being hit by the predator. Upon termination, the prey sent back to the workstation CPU the value of the internal clock (ranging between 0 and 499) which was used as fitness value for both prey and predator. Upon receipt of the prey message, the workstation decoded the next pair of individuals and sent them back to both the predator and prey. In order to distantiate the two competitors at the beginning of each tournament, a simple random motion with obstacle avoidance was implemented by both robots for 5 seconds.

The fitness function Φ_i for species i was based only on the average *time to contact* over K tournaments,

$$\Phi_{py} = \frac{1}{K}\sum_{k=1}^{K}\frac{x_k}{500}, \quad \Phi_{pr} = \frac{1}{K}\sum_{k=1}^{K}\left(1 - \frac{x_k}{500}\right),$$

[2] More details on this architecture and its relevance for evolutionary robotics are given in (Floreano and Mondada, 1998).

that is the number x_k of sensorimotor cycles performed in tournament k normalized by the maximum number of sensorimotor cycles available (500) in the case of the prey py, and the complement in the case of the predator pr, further averaged over the number of tournaments K. This fitness function rewarded prey capable of resisting longer before being hit by predators, and predators capable of quickly hitting prey. The fitness values were always between 0 and 1, where 0 means worst. Individuals were ranked after fitness performance in descending order and the best 20% were allowed to reproduce by making an equal number of offspring in order to keep the population size constant. One-point crossover was applied on randomly paired strings with probability $pc = 0.6$, and random mutation (bit switching) was applied to each bit with constant probability $pm = 0.05$.

2.2 Analysis tools

The simple setup described above allowed us to develop a realistic simulation software (figure 6) that was used to perform initial explorations described in (Floreano and Nolfi, 1997b). This software has been used here only as a complementary analysis tool to investigate computationally expensive hypotheses on the dynamics of competitive co-evolution, and help to assess the results of the experiments on the physical robots. More details on the

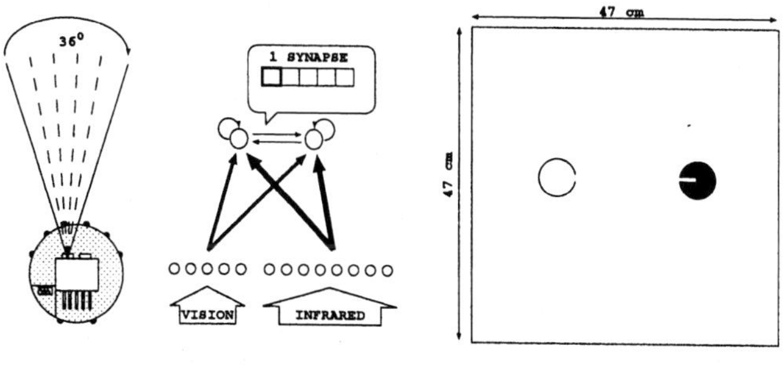

Figure 6 **Left and center**: Details of simulations. The prey differs from the predator in that it does not have 5 input units for vision. Each synapse in the network is coded by five bits, the first bit determining the sign of the synapse and the remaining four the connection strength. **Right**: Initial starting position for Prey (left, empty disk with small opening corresponding to frontal direction) and Predator (right, black disk with line corresponding to frontal direction) in the arena. For each competition, the initial orientation is random.

simulator and initial results are given in (Floreano and Nolfi, 1997b). Here it is sufficient to mention that the simulator was based on real sensory values sampled from the two robots (Miglino et al., 1996), not on a mathematical model of the environment.

3. Results

An exploratory set of experiments were performed in simulation to understand the influence of various parameters, such as the number of tournaments with opponents from previous generations, crossover and mutation probabilities, replicability of the experiments, etc. A detailed analysis of these data is provided in (Floreano and Nolfi, 1997b). Here we provide only a summary of the basic results and compare them to the results obtained with the real robots. Two populations of 100 individuals each were co-evolved for 100 generations; each individual was tested against the best opponents from the most recent 10 generations. Figure 7 shows the average population fitness (left graph) and the fitness of the best individual at each generation. For each generation, the fitness values of the two species do not sum to one because each individual is tested against the best opponents recorded from the previous 10 generations.

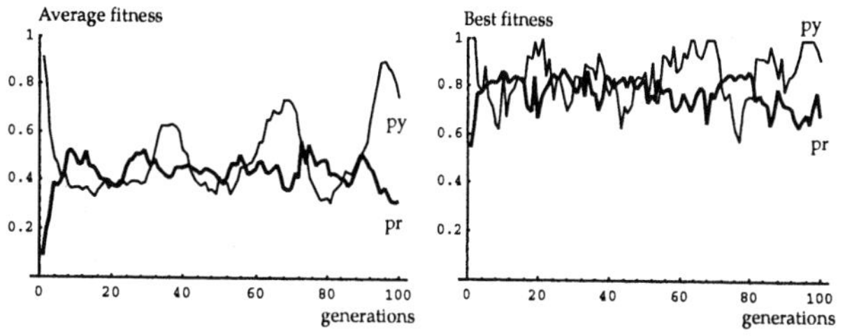

Figure 7 Co-evolutionary fitness measured in simulation. **Left**: Average population fitness. **Right**: Fitness of the best individuals at each generation. pr=predator; py=prey.

As expected, initially the prey score very high, whatever they might do, because the predators are not good at catching them; for the same reason, initially the predators score very low. Very quickly a set of counter-phase oscillations emerge in the two populations, as also reported by other authors (Sims, 1994, p. 36), but we never observed dominance of one population on the other in any of our evolutionary runs (even when continued for 500 generations). However, the fitness for the prey always tended to generate higher peaks due to position advantage (even in the case of the worst prey and best predator, the latter will always need some time to reach the prey). A similar pattern is observed for the fitness of the best individuals (right graph). However, these data cannot be taken as a measure of progress. The only information that they provide is the relative performance of the two species within a (moving) window of ten generations. They indicate that progress in one species is quickly counter-balanced by progress in the competing species, but do not tell us whether evolutionary time corresponds to true progress, or how to choose the best prey and the best predator from the point of view of optimization.

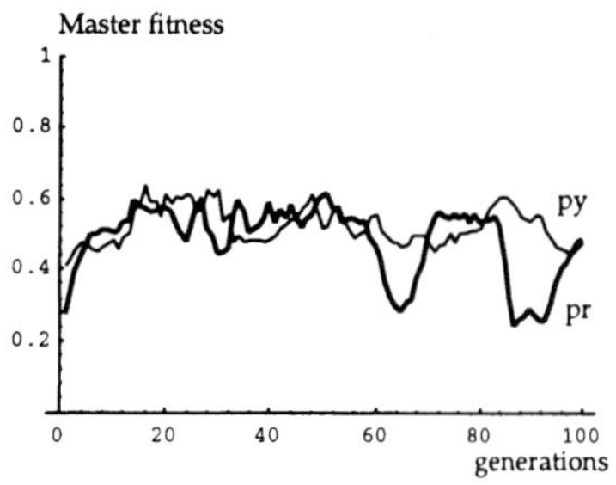

Figure 8 Master Fitness for species evolved in simulation. Each data point is the average fitness of all tournaments of the correponding individual against all the best 100 opponents recorded during co-evolution.

A simple way to learn more about absolute performance of the two species is to organize a Master Tournament where the best individuals for each generation are tested against the best competitors from all generations. For example, the best prey of generation 1 is tested against the 100 best predators and the scores of these tournaments are averaged. The Master fitness values reported in figure 8 indicate that *–in absolute terms–* individuals of the later generations are not necessarily better than those from previous ones. However, these measures can be used to tell *a)* at which generation we can find the best prey and the best predator; *b)* at which generation we are guaranteed to observe the most interesting tournaments. The first aspect is important for optimization purposes and applications, the latter for pure entertainment. The best individuals are those reporting the highest fitness when also the competitor reports the highest fitness (here the best prey and predators are to be found at generation 20, 50, and 82). Instead, the most entertaining tournaments are those that take place between individuals that report the same fitness level, because these are the situations where both species have the same level of ability to win over the competitor (here the most entertaining tournaments are guaranteed around generation 20 and around generation 50).

The results with the real robot displayed a trend similar to that observed in simulations. Two populations (one for the prey, the other for the predator) of 20 individuals each were co-evolved for 25 generations (P(crossover)=0.6; P(mutation)=0.05 per bit) in approximately 40 hours of continuous operation (time might vary in different replications, depending on the relative performances of the two species). Each individual was tested against the best competitors from the most recent 5 generations. Figure 9 shows the average fitness of the population (left graph) and the fitness of the best individual (right graph) along generations for both species. Very quickly the two scores become closer and closer until after generation 15 they diverge again. A similar trend is observed for the fitness of the best individuals at each generation.

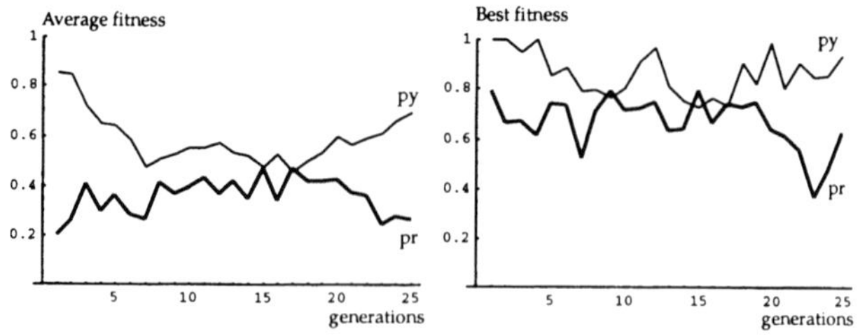

Figure 9 Co-evolutionary fitness measured on the real robots. **Left**: Average population fitness. **Right**: Fitness of the best individuals at each generation. pr=predator; py=prey.

25 generations are sufficient to display one oscillatory cycle. Once the relative fitness values of the two species reach the same value, one party improves over the other for some generations until the other counter-adapts (the best predators of the last three generations already show a fitness gain). Figure 10 shows the Master Fitness values for the two robot species. The best prey and predators can be found at generation 20 which also hosts the most entertaining tournaments. It can also be noticed that fitness oscillations of the best individuals between generation 9 and 16 (figure 9, right) do not show up in the Master Fitness, indicating that they are due to tight interactions between the two competing species which can amplify the effects of small behavioral differences.

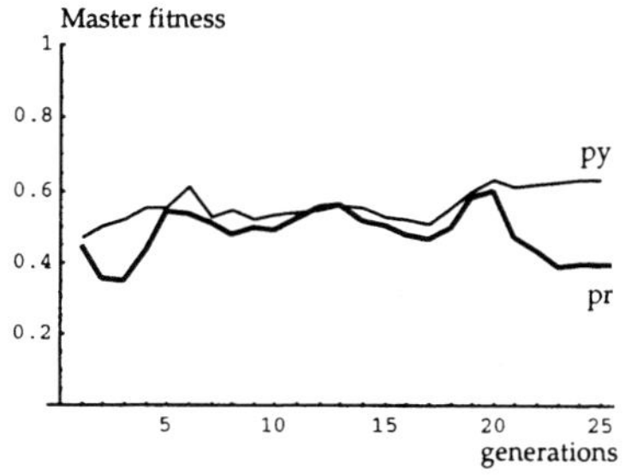

Figure 10 Master Fitness for species evolved on the real robots. Each data point is the average fitness of all tournaments of the correponding individual against all the best 25 opponents recorded during co-evolution.

The behaviors displayed by the two physical robots at significative points of co-evolution (for example, those corresponding to the overall best individuals and to the most entertaining tournaments) are only a subset of those recorded in simulation. The presence of much larger noise in the real environment filters out brittle solutions. Nevertheless, the best strategies displayed by the real robots can be found also in the experiments performed in simulation.[3] Figure 11 shows some typical tournaments recorded from individuals at generation 13, 20, and 22. At generation 13 the prey moves quickly around the environment and the predator attacks only when the prey is at a certain distance. Later on, at generation 20, the prey spins in place and, when the predator gets closer, it rapidly avoids it. Prey that move too fast around the environment sometimes cannot avoid an approaching predator because they detect it too late (infrared sensors have lower sensitivity for a small cylindrical object than for a large white flat wall). Therefore, it pays off for the prey to wait for the slower predator and accurately avoid it. However, the predator is smart enough to perform a small circle after having missed the target and re-attack until, by chance, the prey is caught on one of the two sides (where wheels and motors do not leave space for sensors). The drop in performance

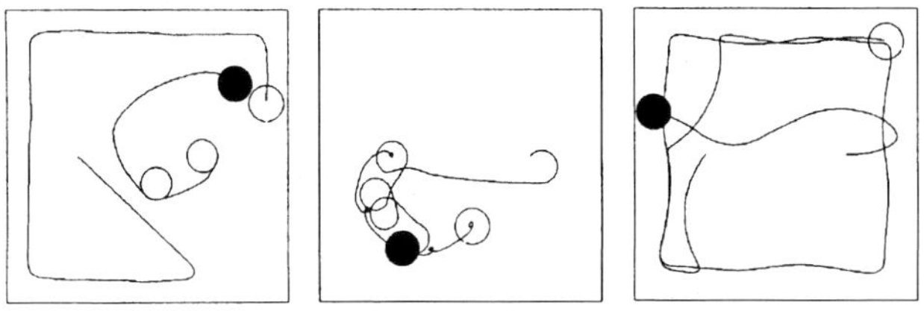

Figure 11 Typical strategies of the best predator and prey in the experiments with real robots. Black disk is the predator, white disk is the prey. Trajectories have been plotted running a tournament with simulated individuals who display the same behavioral strategies observed with the real robots.

of the predator in the following generations is due to a temporary loss of the ability to avoid walls (which was not needed in the few previous generations because the predator soon localized and approached the prey). At the same time the prey resumes a rapid wall following and obstacle avoidance which forces the predator to get closer to walls and collide if the prey is missed (right of figure 11).

4. Discussion

Co-evolution between competing species with a relatively short generational overlap does not seem to guarantee the type of monotonic progress over time expected from an optimization algorithm. This does not mean that competitive co-evolution is not an viable approach for the purpose of evolving efficient behavioral competencies. In fact, the rapid development, variety, and complexity of the behaviors observed, together with appropriate methods to pick them out (such as the Master Tournament), hint at the computational advantages of competitive co-evolution. In a companion paper, we report results showing that predators evolved against a fixed co-evolved prey (from another run) do not reach the same performance levels obtained when both prey and predators are co-evolved (Nolfi and Floreano, 1998). It has also been shown that, by including all the best opponents evolved so far as test cases for each individual (*Hall of Fame* method), co-evolution becomes very similar to a very robust optimization technique (Rosin and Belew, 1997).

However, the basic issue that the results presented above raise is to what extent natural evolution is an optimization process (see also (Gould and Lewontin, 1979)) and to what extent should we use artificial evolution as an optimization technique in a traditional fashion. In contrast to mainstream "evolutionary computation", there is not necessarily one or more fixed maxima to be found in competitive co-evolution. The landscape continuously changes. Furthermore, the way it changes cannot be predicted, and most of the time it will change towards the worst for each competing species. Nevertheless, after an initial period during which the two populations settle into a regime of tight interactions (corresponding to the generation when the two fitness measures become equal), the best individuals of the two populations are always optimal, or almost always, optimal with respect to the environment (competitor) that they are facing. In other words, the optimum is always now.

Other authors have criticized evolutionary computation for its interpretation of evolution as an optimization process (Atmar, 1993) and pointed out its limitations within the framework of incremental evolution (Harvey, 1997). We agree with Harvey that one should not look for the optimization properties of evolution, but rather for its "adaptive improving properties". To this extent, co-evolution of predator-prey scenarios, such as the one described in this paper, not only displays these properties, but also provides a framework for understanding some of the issues brought in by open-ended evolution and testing different methodologies. Indeed, elsewhere (Nolfi and Floreano, 1998) we have shown that under certain conditions "plain competitive co-evolution" can provide better adaptativity than single-agent evolution and methods aimed at improving the (traditional) optimization properties of co-evolution, such as the Hall of Fame mentioned above.

The results described above in simulation and on the real robots are very similar from the point of view of the dynamics of the co-evolutionary system. The main difference is that the larger amount of noise in the real experiments does not allow the persistence of behaviors which instead we observe in simulation. These behaviors are based on the synchronous and time-locked activity of the two individuals which does not hold when each physical robot moves around independently using its own

[3] Although individuals *evolved in simulation* do not behave in the same way when *downloaded into the real robots*.

controller and updates the wheel speed at a slightly different frequency. Noisier conditions may also explain the reason why when the system is run on the physical robots it takes slightly longer to reach the lock-phase where oscillations begin. In effect, noise may be seen as a way of increasing the variety of behaviors "observed" by each opponent. Therefore, individuals must search for more "general solutions". These results, which are in accordance with other data on simulations with high noise levels (Floreano and Nolfi, 1997a), also explain why fitness values recorded during evolution are a better indicator of "true" performance levels (as indicated by the Master Tournament).

As well-known in the Adaptive Behavior community, though, the advantages of running the experiments on the real robots show up as soon as we introduce more complexities, such as a less constrained visual preprocessing (which here was very easy to simulate) and, for example, a vision system for the prey too. At that point not only computer simulations become slower than the physical implementation, but also competitive co-evolution might become a powerful tool to discover solutions that would not be evolvable otherwise. For the time being, the rough similarity between our (careful) simulations and physical implementations allow us to exploit simulations for exploring other computationally expensive features of competitive co-evolution.

4.1 Exposing the Red Queen

The Red Queen effect illustrated in figure 1 is suspected to be the main actor behind the dynamics, complexities, and computational advantages of competitive co-evolution, but how exactly it operates is not known. Capitalizing on the fact that our simple experiment with the robots displayed dynamics similar to those measured in experiments carried out in simulation, we exploited our computer CPUs to study how the fitness surface of one species is affected by the co-evolving competitor.

Given its shorter genotype length, we analyzed how the fitness surface of the prey was changed when confronted with the best predators recorded at different generations. The genotype of the prey was composed of 5 bits x 22 synapses. Assuming that the most significative bits are those coding the sign of the synapses, we are left with 22 bits.[4]. Each of the corresponding 4,194,304 prey was separately tested against the best predators of the first eight generations and against the best predator of generation 20, yelding a total of almost 40 million tournaments. At the beginning of each tournament, both the prey and the predator were positioned at the same location (illustrated on the right of figure 6) facing north. The best predators were selected from the simulation run depicted in figure 7. Figure 12 plots the fitness surface of

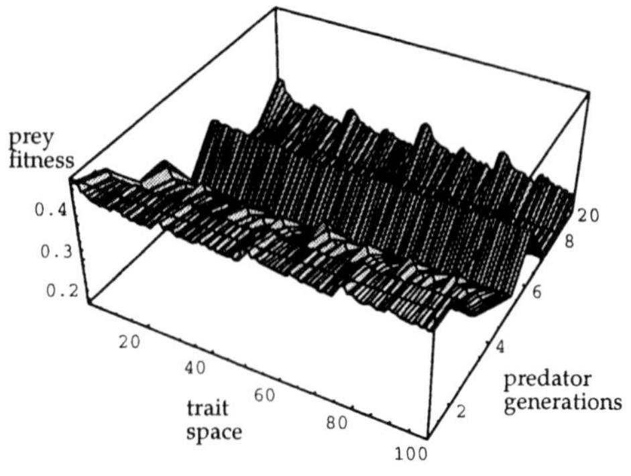

Figure 12 Fitness landscape for the prey when tested against the best predators from generation 1-8 and 20. Each data point is the average over the fitness values reported by 4,194 contiguous individuals.

the prey tested against best predators from generations 1-8 and generation 20. Since there are not enough many pixels on this page to show the results of all tournaments, the fitness values for each generation were grouped into 100 bins of 4,194 values each (discarding remainders) and the average value of each bin was plotted on the graph.

Despite these approximations, one can see that co-evolution of predators during initial generations cause a general decrement of the performance of the prey. However, it should be noticed that these are average values and that for every bin there are always several prey reporting maximum fitness 1.0. The Red Queen effect is clearly visible in the temporary and periodic smoothing of the fitness landscape, as highlighted in figure 13. For example, the best predator of generation 3 causes a redistribution of the fitness values, stretching out the relative gain of some trait combinations with respect to others. This smoothing effect is always temporary and roughly alternates with recovery of a rough landscape.

It should be noticed that some regions corresponding to better fitness remain relatively better also during periods of stretching, whereas others are canceled out. That implies that individuals sitting on these latter regions would disappear from the population. If we view these regions as minima or brittle solutions, our data show the potentials of the Red Queen for optimization problems. Furthermore, it can be noticed that the steepness of the surface around the maxima becomes more accentuated along generations. If we assume that steeper regions are harder to climb, competitive co-evolution might facilitate progressive development of abilities that would be difficult to achieve in the scenario of a single species evolved in a static environment. Although more analysis is to be carried out on these data, especially on the displacement of the prey population on this changing surface, these preliminary results support the hypotheses on the

[4] The remaining 4 bits for each synapse were set at 0101, a pattern that represents the expected number of on/off bits per synapse and also codes for the average synaptic strength.

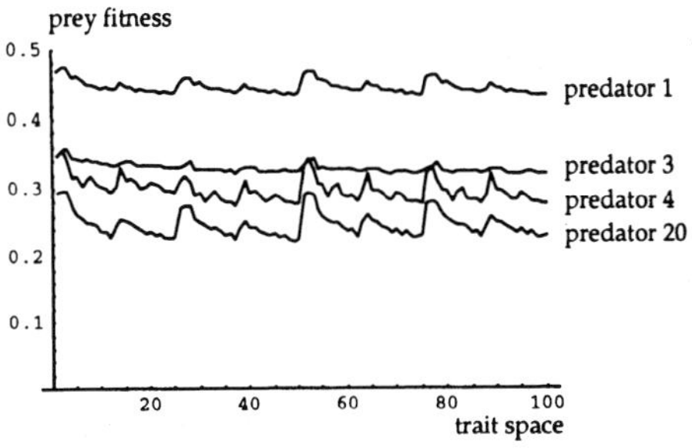

Figure 13 The Red Queen in action. See also figure 12.

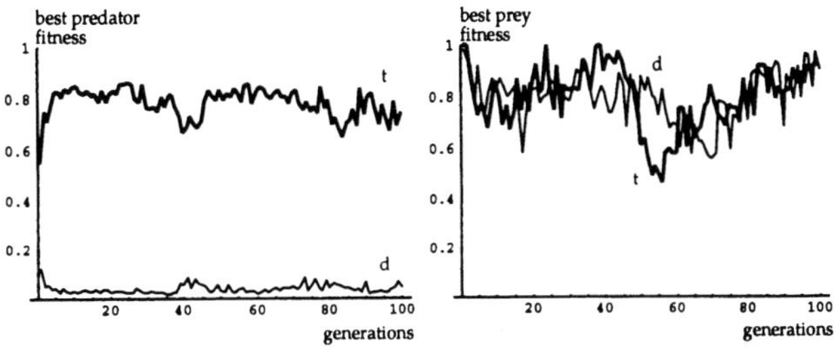

Figure 14 Comparisons between fitness of the best individuals measured as *time to contact* (t) and as *distance.* (d). Both species have been evolved using fitness t. **Left**: Best evolved predators do not attempt to minimize distance (predator fitness is 1 - d). **Right**: Best evolved prey attempt to maximize distance (prey fitness is d).

Red Queen effect outlined in section 1. and indicate its efficacy from an optimization perspective.

4.2 Selection criteria

In artificial evolution the choice of the selection criterion (fitness function) can make the difference between trivial parameter optimization and generation of creative and "life-like" solutions (Floreano, 1997). From an engineering point of view, one might think that it makes sense to pursue an optimization approach by including several constraints in the function in the attempt to describe the expected behavior. However, by doing so one might also include wrong assumptions that derive from an insufficient understanding of the environment and/or of the interactions that might arise. For example, from an engineering point of view, a successful predator should aim at the prey and approach it minimizing the distance, whereas the prey should attempt to maximize this distance, as in (Cliff and Miller, 1996). Did our robots (simulated and real) indirectly optimize this objective although the selection criterion employed was simply *time to contact*? We run a new set of simulations where each individual was selected and reproduced according to the fitness function described in section 2.1, but was also evaluated according to a fitness function based on the *distance* between the two competitors (namely, the *distance* for the prey, and *1 - distance* for the predator).

The fitness values computed according to the two methods (figure 14) did not overlap for the predators, but they did for the prey. In other words, predators selected to hit prey in the shortest possible time did not attempt to minimize the distance from the prey, as it might be expected. On the other hand, the prey did so. The strategy employed by the predators was more subtle. Rather than simply approaching the prey, they tended to "wait for the right moment" and, only then, attack. The behaviors shown in the first two insets of figure 11 are an example of this strategy. The best predator of generation 13 attacks only when the prey is within a certain range, and rotates in place in the other cases (neuroethological analyses showed that the predator infers the distance from the prey by observing how fast the prey moves on its own visual field). Similarly, the best predator of generation 20 does not follow the prey once the latter has moved away; rather, it performs one more turn on itself and re-attacks. By doing so, it has higher probability of hitting the prey on the side of the motors where there are no sensors. Competitive co-evolution is a methodology capable of discovering innovative and unexpected –yet efficient– solutions provided that one does not attempt to force it into specified directions and reduce it to a mere optimization technique.

5. Conclusion

Competitive co-evolution is not necessarily optimization in the sense of solution discovery for a pre-specified problem. The teleological interpretation of artificial evolution that underlies most research in genetic algorithms (Atmar, 1993) and some research in evolutionary theory (Gould and Lewontin, 1979) leaves space to a more complex and richer scenario where robust solutions and true innovation can endogenously arise from simple interactions between parts of the (co-evolutionary) system. In this paper we have shown that this methodology can be applied in evolutionary robotics in order to generate powerful controllers with minimal human effort. At the same time we have tried to give indications that competitive co-evolution can be a viable methodology for developing truly autonomous and intelligent machines capable of setting their own goals to face new and continuously changing challenges. Much work remains to be done to check and further extend the indications that arise from these results. Our current research is focused on the interactions between learning and co-evolution (Floreano and Nolfi, 1997a), and on the extension of formal models of competitive co-evolution that could establish a link between reproduction success and behavioral performance.

Acknowledgements

We would like to thank the two anonymous referees for rich and constructive comments.

References

Angeline, P. J. and Pollack, J. B. (1993). Competitive environments evolve better solutions for complex tasks. In Forrest, S., editor, *Proceedings of the Fifth International Conference on Genetic Algorithms*, pages 264–270, San Mateo, CA. Morgan Kaufmann.

Atmar, W. (1993). Notes on the Simulation of Evolution. *IEEE Transactions on Neural Networks*, 5:130–148.

Cliff, D. and Miller, G. F. (1995). Tracking the Red Queen: Measurements of adaptive progress in co-evolutionary simulations. In Morán, F., Moreno, A., Merelo, J. J., and Chacón, P., editors, *Advances in Artificial Life: Proceedings of the Third European Conference on Artificial Life*, pages 200–218. Springer Verlag, Berlin.

Cliff, D. and Miller, G. F. (1996). Co-evolution of Pursuit and Evasion II: Simulation Methods and Results. In Maes, P., Matarić, M., Meyer, J., Pollack, J., Roitblat, H., and Wilson, S., editors, *From Animals to Animats IV: Proceedings of the Fourth International Conference on Simulation of Adaptive Behavior*. MIT Press-Bradford Books, Cambridge, MA.

Dawkins, R. and Krebs, J. R. (1979). Arms races between and within species. *Proceedings of the Royal Society London B*, 205:489–511.

Floreano, D. (1997). Ago Ergo Sum. In Mulhauser, G., editor, *Evolving Consciousness*. Benjamins Press, New York.

Floreano, D. and Mondada, F. (1996). Evolution of homing navigation in a real mobile robot. *IEEE Transactions on Systems, Man, and Cybernetics-Part B*, 26:396–407.

Floreano, D. and Mondada, F. (1998). Hardware Solutions for Evolutionary Robotics. In Husbands, P. and Meyer, J., editors, *Proceedings of the first European Workshop on Evolutionary Robotics*. Springer Verlag, Berlin.

Floreano, D. and Nolfi, S. (1997a). Adaptive behavior in competing co-evolving species. In Husbands, P. and Harvey, I., editors, *Proceedings of the 4th European Conference on Artificial Life*, Cambridge, MA. MIT Press.

Floreano, D. and Nolfi, S. (1997b). God Save the Red Queen! Competition in Co-evolutionary Robotics. In Koza, J., Deb, K., Dorigo, M., Fogel, D., Garzon, M., Iba, H., and Riolo, R. L., editors, *Proceedings of the 2nd International Conference on Genetic Programming*, San Mateo, CA. Morgan Kaufmann.

Goldberg, D. E. (1989). *Genetic algorithms in search, optimization and machine learning*. Addison-Wesley, Redwood City, CA.

Gould, S. J. and Lewontin, R. C. (1979). The spandrels of San Marco and the Panglossian paradigm: a critique of the adaptationist program. *Proceedings of the Royal Society London, B*, 205:581–598.

Harvey, I. (1997). Cognition is not computation; evolution is not optimization. In Gerstner, W., Germond, A., Hasler, M., and Nicoud, J., editors, *Artificial Neural Networks - ICANN97*, pages 685–690, Berlin. Springer Verlag.

Hillis, W. (1990). Co-evolving parasites improve simulated evolution as an optimization procedure. *Physica D*, 42:228–234.

Koza, J. R. (1991). Evolution and co-evolution of computer programs to control independently-acting agents. In Meyer, J. and Wilson, S., editors, *From Animals to Animats. Proceedings of the First International Conference on Simulation of Adaptive Behavior*. MIT Press, Cambridge, MA.

Koza, J. R. (1992). *Genetic programming: On the programming of computers by means of natural selection*. MIT Press, Cambridge, MA.

Lotka, A. J. (1925). *Elements of Physical Biology*. Williams and Wilkins, Baltimore.

Miglino, O., Lund, H. H., and Nolfi, S. (1996). Evolving Mobile Robots in Simulated and Real Environments. *Artificial Life*, 2:417–434.

Miller, G. F. and Cliff, D. (1994). Protean behavior in dynamic games: Arguments for the co-evolution of pursuit-evasion tactics. In Cliff, D., Husbands, P., Meyer, J., and Wilson, S. W., editors, *From Animals to Animats III: Proceedings of the Third International Conference on Simulation of Adaptive Behavior*. MIT Press-Bradford Books, Cambridge, MA.

Murray, J. D. (1993). *Mathematical Biology*. Springer Verlag, Berlin. Second, Corrected Edition.

Nolfi, S. and Floreano, D. (1998). Co-evolving predator and prey robots: Do "arms races" arise in artificial evolution? *Artificial Life*. To be published.

Renshaw, E. (1991). *Modeling Biological Populations in Space and Time*. Cambridge University Press, Cambridge.

Reynolds, C. W. (1994). Competition, Coevolution and the Game of Tag. In Brooks, R. and Maes, P., editors, *Proceedings of the Fourth Workshop on Artificial Life*, pages 59–69, Boston, MA. MIT Press.

Rosin, C. and Belew, R. (1997). New methods for competitive co-evolution. *Evolutionary Computation*, 5(1):1–29.

Sims, K. (1994). Evolving 3D Morphology and Behavior by Competition. In Brooks, R. and Maes, P., editors, *Proceedings of the Fourth Workshop on Artificial Life*, pages 28–39, Boston, MA. MIT Press.

Utida, S. (1957). Population fluctuations. In *Proceedings of the XXII CSHL Symposium on Population Studies: Animal Ecology and Demography*, Cold Spring Harbor Laboratory, NY.

van Valen, L. (1973). A new evolutionary law. *Evolution Theory*, 1:1–30.

Volterra, V. (1926). Variazioni e fluttuazioni del numero di individui in specie animali conviventi. *Memorie dell'Accademia dei Lincei*, 2:31–113. Variations and fluctuations of the number of individuals in animal species living together. Translation in: R. N. Chapman, Animal Ecology. New York: McGraw Hill 1931, pp. 409–448.

Animal-Animat Coevolution:
Using the Animal Population as Fitness Function

Pablo Funes, Elizabeth Sklar, Hugues Juillé and Jordan Pollack
Volen Center for Complex Systems
Brandeis University
415 South St., Waltham MA 02254 USA
{pablo,sklar,hugues,pollack}@cs.brandeis.edu

Abstract

We show an artificial world where animals (humans) and animats (software agents) interact in a coevolutionary arms race. The two species each use adaptation schemes of their own.

Learning through interaction with humans has been out of reach for evolutionary learning techniques because too many iterations are necessary. Our work demonstrates that the Internet is a new environment where this may be possible through an appropriate setup that creates mutualism, a relationship where human and animat species benefit from their interactions with each other.

1. Introduction

1.1 Humans vs. Agents on the Internet.

In nature, an *ecosystem* is a community of organisms and their environment functioning together. The Internet can be seen as a virtual ecosystem — a community of human users and artificial environments where complex phenomena are taking place.

In the future, more and more such environments will contain software agents that interact with human users and adapt according to the behavior displayed in those interactions (Lieberman, 1997). Such a bi-adaptive relationship could be considered a form of mutualism (Gilbert and Raven, 1975), as both humans and agents participating have their own goals and adaptation strategies.

We have built a coevolutionary environment with a real-time computer game, implemented in Java, that matches artificial agents ("animats" or "robots") against human ("animal") Internet users. While humans participate for fun — and for outperforming other humans — our artificial agents try to win as many games as they can, learning via an evolutionary algorithm.

1.2 Coevolution in Natural and Artificial Systems

In nature, organisms and species coexist in an ecosystem; each species has its own place or *niche* in the system. The environment contains a limited number and amount of resources, and the various species must compete for access to those resources. Through these interactions, species grow and change, each influencing the others' evolutionary development. This process of reciprocal adaptation is known as *coevolution*.

In evolutionary computation, the term "coevolution" has been used to describe any iterated adaptation involving "arms races", either between learning species or between a learner and its learning environment. Examples of coevolutionary learning include the pioneering work by Hillis on sorting networks (Hillis, 1992), Backgammon learning (Tesauro, 1992, Pollack *et al., 1*996, Pollack and Blair, 1997), predator/prey games (Reynolds, 1994, Miller and Cliff, 1994, Miller and Cliff, 1996) and spacial distribution problems (Juillé and Pollack, 1996, Juillé and Pollack, 1996b).

We use coevolutionary programming techniques to maintain our robot populations. Two types of coevolution are involved: robot vs. robot in our background server and robot vs. human in our foreground server. The idea is that, within a species — the animat species — selection occurs and only the fittest are sent out to coevolve with the opposing "animal" population.

1.3 Evolutionary Learning

Evolutionary learning methods such as Genetic Algorithms (GA's) provide general-purpose approaches to the problem of machine learning. With them we can build engines that create a succession of partial results whose success at dealing with a problem (hopefully) increases over time.

Evolutionary learning calls for three basic ingredients: (a) a representation that is capable of encoding each candidate solution, (b) reproductive operators that can be applied to such a representation, and (c) a *fitness function* that is the standard measure against which to test each candidate solution during the iterative process (Holland, 1975, Goldberg, 1989).

In natural as in artificial evolution, a population moves toward fitness optimality while maintaining variation over all the dimensions of the genetic space, including those dimensions that are not being selected. This iterated generation/selection process is regulated in genetic algorithms by the fitness function, which must be applied to each single individual as it is produced.

1.4 Too Many Fitness Evaluations

The need to evaluate the fitness of a large number of individuals is a critical factor that restricts the range of application of GA's. In many domains, a computer can do these evaluations very fast; but in others, the time spent by this process may render the GA solution impractical. Examples of the latter case include a computer playing a game with people and trying to learn from experience or a robot attempting to complete a task in the physical world.

Robots that are reliable enough can run repeated trials of the same experiment over a long time in order to learn using evolutionary computation techniques. Floreano and Mondada (Floreano and Mondada, 1994, 1996) run their robots for several days in order to evolve controllers for basic tasks. Most evolutionary roboticists have preferred to rely on computer simulations to provide them with faster evaluations, but the crafting of appropriate simulations is also very difficult (Mataric and Cliff, 1996).

1.5 Using Humans for Fitness

Evolution of interactive adaptive software faces similar difficulties. On the one hand, it is nearly impossible to design a fitness function whose virtual environment will prepare it for meeting the enormous variation of human responses. On the other hand, if users themselves are asked to provide fitness evaluations, then hundreds of trials will be necessary and the process will take a long time. Humans — unlike robots — get tired of repetitive tasks. Humans act irregularly; they may react differently each time when faced with the same situation more than once. If users provide fitness evaluations, adaptive software would need to be able to filter out such sources of "noise" provided naturally by human users.

Our theory is that the Internet, with millions of human users, could be fertile ground for the evolution of interactive adaptive software. Instead of relying on a few selected testers, the whole community of users together constitutes a viable gauge of fitness for an evolutionary algorithm that is searching to optimize its behavior.

1.6 The problem of generalization

Given that it is an arduous task to evaluate fitness for multitudes of individuals, one might ask: is it possible to limit the search to just a few? The problem of lack of generalization or lack of transfer to a more general environment defeats this idea. The fact that an algorithm performs well in a certain group of test cases does not usually mean that it will generalize to a wider range of situations.

Neural networks are thought to have generalization capabilities (Wolpert, 1990, Darwen, 1996), successfully inducing, for example, a good backgammon player from a set of suggested moves (Tesauro, 1989). Supporters of the Genetic Programming (GP) paradigm (Koza, 1992) suggest that this may be the case for GP as well (Rosca, 1996). (Juillé and Pollack, 1996b) argues that the dynamics of coevolutionary fitness help to get "perspicacious" solutions to a problem of recognizing 194 points arranged in a spiral pattern. The GP function they obtain indeed defines two roughly spiral surfaces that continue outside the boundary of the original test points.

1.7 Learning game playing

Game playing is one of the traditional domains of AI research. Ever since Samuel's early experiments with checkers (Samuel, 1959), we have hoped that the computer would be able to make good use of experience, improving its skills by learning from its mistakes and successes.

In his work with the game of backgammon, Tesauro began collecting samples from human games to provide a fitness measure for training neural networks (Tesauro, 1989). Later, he abandoned this methodology and used introspective *self-play* (Tesauro, 1995). Of the earlier approach, he argued that "building human expertise into an evaluation function [...] has been found to be an extraordinarily difficult undertaking" (p. 59).

Learning to play a game by self-play involves a problem of transfer as well. The fitness landscape (even in the coevolutionary case, where the "landscape" is redefined in every generation) might be an insufficient sample of the larger problem defined by the whole game and the way humans approach it. While learning backgammon (Tesauro, 1995, Pollack *et al., 1996*) is a success for coevolution, the same approach has failed in most other cases.

Real-time, interactive games (e.g. video games) have distinctive features that differentiate them from the better known board games. (Koza, 1992) and others (Rosca, 1996) evolved players for the game of Pacman. There has been important research in pursuer-evader games (Reynolds, 1994, Miller and Cliff, 1994, 1996) as well as contests in simulated physics environments (Sims, 1994). But these games do not have human participants, as their environments are either provided by the game itself, or emerge from coevolutionary interactions inside a population of agents.

1.8 A space where agents can thrive and evolve

In this paper, we propose that learning complex behaviors can be achieved in a coevolutionary environment where one population consists of the human users of an interactive software tool and the "opposing" population is artificial, generated by a coevolutionary learning engine. A niche must be created in order for the arms race phenomenon to take place, requiring that:

1. A sufficiently large number of potential human users must participate.
2. The artificial population must provide a useful environment for the human users, even when — in the early stages — many instances perform poorly.

3. A (crude) estimation of the artificial population's performance must be measurable from its interaction with the human users.

As a prototype, we have created an experimental learning environment on the Internet for the game called **Tron**, meeting the requirements mentioned above. First, the game is played in a Java applet window on our web site. We know that there is considerable interest in Java-based games in the Internet community, so advertising our site in some Java games lists should attract visitors. Second, our earlier experiments with Tron have shown us that, by self-play, we can produce players that are not entirely uninteresting when faced by humans. And third, each round of Tron results in a performance measure: a win, a loss or (rarely) a tie.

This means that a form of mutualism is necessary in order for an experiment such as we are proposing to be successful. Both human users and learning agents must have their own reasons and rewards for interacting. Both being adaptive species, this relationship between software agents and human users implies coadaptation of both species.

From the point of view of the learning agent species, the human environment is very noisy and adapts very fast. The evolutionary process finds its way through this noise as the better agents are being selected, slowly but surely. The motivations in this case are very simple, as rewards come from the fitness function we programmed.

People have reasons for participating that are much more complex. Fun, curiosity, competition against an unfamiliar intelligence, and competition against each other to appear in the 'ranking' page are among them. The evolutionary properties of the animat population create a changing artificial environment that makes people come back searching for a renewed challenge.

Figure 1. Still from the movie *Tron*

2. Tron

Tron, a 1982 movie from Walt Disney Studios, shows a game in a virtual world where two futuristic motorcycles run at constant speeds, making only right angle turns and leaving solid wall trails behind them. As the game advances, the arena fills with walls and eventually one opponent dies by crashing into a wall. This popular game has been implemented on all kinds of computers with varying rules and configurations.

We have built a version of Tron using a Java applet and released it on the Internet. In our interpretation, the motorcycles are abstracted and represented only by their trails. Two players (one human, one animat) start in the middle region of the screen, moving in the same direction (fig. 2). The edges of the arena are not considered "walls"; players move past them and reappear on the opposite side, thus creating a "wraparound", or toroidal, game arena. The size of our arena is 256×256 pixels.

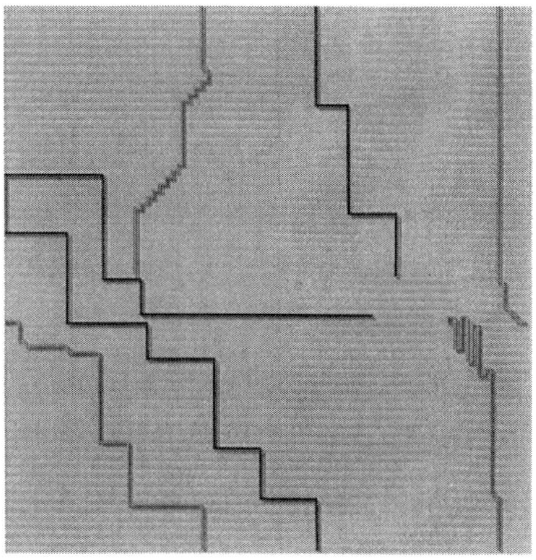

Figure 2. The Tron game arena. Both players need to avoid walls. The edges are "wrap around".

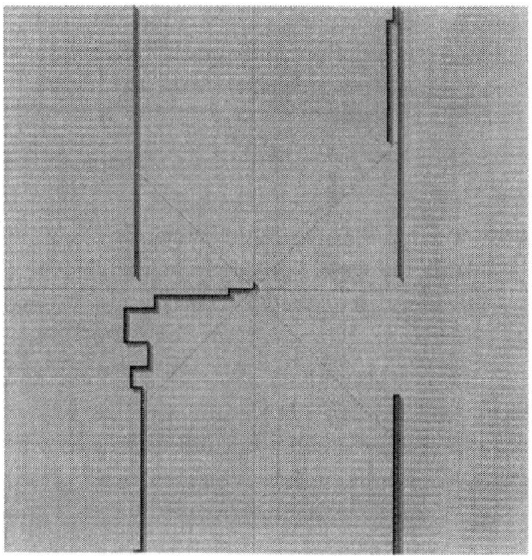

Figure 3. A Tron robot has eight sensory inputs.

Our robots, provided with simple "sensors", perceive the world in eight directions. Each sensor evaluates the distance in pixels from the current position to the nearest obstacle in these directions: Front, Back, Left, Right, FrontLeft, FrontRight, BackLeft and BackRight (fig. 3). Every sensor returns a maximum value of 1 for an immediate obstacle (i.e. a wall in an adjacent pixel), a lower number for an obstacle further away, and 0 when there are no walls in sight.

In earlier exploratory experiments (Funes, 1996), we used a Genetic Algorithm (GA) to learn the weights of a perceptron network that played Tron. It became evident that while this simple architecture is capable of coding players that could perform interestingly when facing human opponents, such "good" weights were difficult to find in evolutionary or coevolutionary scenarios. *Collusion* (Pollack and Blair, 1997) was likely to appear in most evolutionary runs in the form of "live and let live" strategies such as that shown in Figure 4.

Figure 4. "Live and let live": Two artificial Tron players make tight spirals in order to stay as far from the opponent as possible. This form of collusion is a frequent suboptimal equilibrium that prohibits artificial learning through self-play.

2.1 System Overview

In our present work, we are using Genetic Programming (GP) (Koza, 1992) as a means for coding artificial Tron players. The set of terminals is {**_A**,**_B**,..., **_H** (the eight sensors) and $\Re$ (random constants between 0 and 1)}. The functions are {+, -, * (arithmetic operations), % (safe division), **IFLTE** (if a ≤ b then-else), **RIGHT** (turn right) and **LEFT** (turn left)}. A maximum depth of 7 and a maximum length of 512 limit the valid s-expressions. A robot reads its sensors and evaluates its s-expression every three steps during a game. If a **RIGHT** or **LEFT** function is output, the robot makes the corresponding turn; otherwise, it will keep going straight.

We have written a Java applet and launched our game on the Internet. The architecture of the system takes advantage of Java's ability to run a "client" on the user's local machine and a "server" on our host (Web server) machine. As shown in Figure 5, the Java Applet runs on the user's local machine; the Foreground and Background servers execute on our machine.

The Tron applet receives a GP s-expression (from our server), representing a Tron-playing strategy. The applet runs, playing one game with the human user, until a crash occurs. When the game ends, the applet opens a connection to our server, reports the results of the game and receives a new s-expression for the next game. This cycle continues until the human decides to quit playing.

We use a two layer server architecture to maintain two separate Tron-playing artificial populations simultaneously, as illustrated in Figure 5. The Foreground Server plays games with humans, while the Background Server engages in self-play to create brand new artificial players that will be incorporated into the foreground population when the foreground process is ready for a new generation.

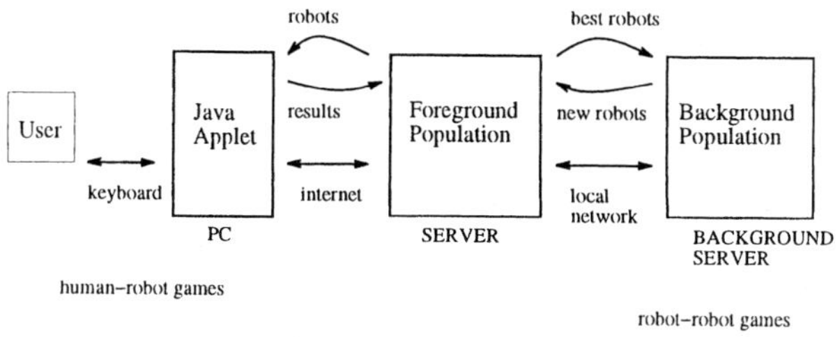

Figure 5. Scheme of information flow

The interaction between robots and humans occurs at a relatively slow pace: only a few hundred games may happen every day. An evolving population idles, waiting while these games are accumulated, until there is enough information to evaluate its individuals. The role of the background process is to replace the usual reproduction stage, exploiting all this idle time to produce the best robot players that self-play can give us. Instead of raw crossover and mutations, the new individuals of the population will have been trained and filtered through self-play.

2.2 Foreground Server

The foreground portion of the Tron system controls the interplay between robots and humans. It maintains a population of 100. Every time a player requests a new game, the server supplies one of these robots at random. When a user finishes a game, the foreground process saves the outcome in its database.

A generation in the foreground process lasts until all 100 agents have played a minimum number of games. The new agents that are playing for the first time in the current generation play a minimum of 10 games, while the "veterans" that have survived from previous generations play only 5 games. When all agents have completed their minimum number of games, the generation is finished and the next generation is started.

To start a new generation, the 100 current agents are sorted by fitness. The worst 10 are eliminated and replaced by 10 fresh ones, supplied by the background process. A new generation begins.

The fitness of robots is a shared fitness measure designed to promote speciation (Beasley *et al.*, 1993; Juillé and Pollack, 1996) by giving points for doing better than average against a human player, and negative points for doing worse than average. For each agent a, the fitness is calculated as

$$F(a) = \sum_{\{h:p(h,a)>0\}} \left(\frac{l(h,a)}{p(h,a)} - \frac{l(h)}{p(h)}\right)\left(1 - e^{-\frac{p(h)}{10}}\right) \quad (1)$$

where $l(h,a)$ is the number of games lost by each human opponent h against a; $p(h,a)$ is the total number of games between the two; $l(h)$ is the total games lost by h; and $p(h)$ is the number of games that h has played. The measure is summed across all games played (not just those that belong to the current generation). The exponential factor on the right is a confidence measure that devalues the average scores of those humans that have played only a few games.

An inherent exploitation/exploration bias occurs when we make 10 new robots play 10 times per generation and 90 veteran agents play 5 times each. This means that 18% of the games are played by the rookie agents, who have not been evaluated yet.

2.3 Background Server

The role of the background process is to supply the foreground process with good artificial players for each new generation, the best that can be produced given the results that have been accumulated. We proceed as follows: every time the foreground population begins a new generation, the background receives the 15 fittest agents from the foreground. This group of 15 agents comprises part of a training set against which a new population of 1000 random agents is generated and evolved. When the foreground process finishes one generation, it receives the 10 best agents that emerge from this procedure, adding them to its own population, and the cycle restarts (Figure 6).

The background process plays all the individuals in its population against the training set of 25 agents. Fitness is evaluated, and the bottom half of the population is replaced by random mating with crossover of the best half. The fitness function is defined as follows:

$$F_T(a) = \sum_{\{a' \in T:\, pt(a,a')>0\}} \frac{pt(a,a')}{l(a')} \quad (2)$$

where T is the training set, $pt(a, a') = \{0$ if a loses against a', 0.5 if they tie and 1 if a wins$\}$ and $l(a')$ is the number of games lost by a'. Thus we give more points for defeating good players than bad players.

As indicated above, the training set consists of two parts. The first 15 members are fetched from the foreground process. The remaining 10 members of the training set are replaced each generation with a fitness sharing criteria. The new training set T' is initialized to the empty set and then new members are added one at a time, choosing the highest according to the following shared fitness function:

$$F_{T,T'}(a) = \sum_{a' \in T} \frac{pt(a,a')}{(1 + \sum \{pt(a'',a') : a'' \in T'\})} \quad (3)$$

This selection function is adapted from (Rosin, 1997) and acts to decrease the relevance of a case that has already been "covered", that is, when there is already a player in the training set that beats it.

When the best players from the foreground population reenter the background as members of the training set, their genotype is isolated: they do not reproduce explicitly. They do so only implicitly as they disfavor players they can beat, and favor players that beat them.

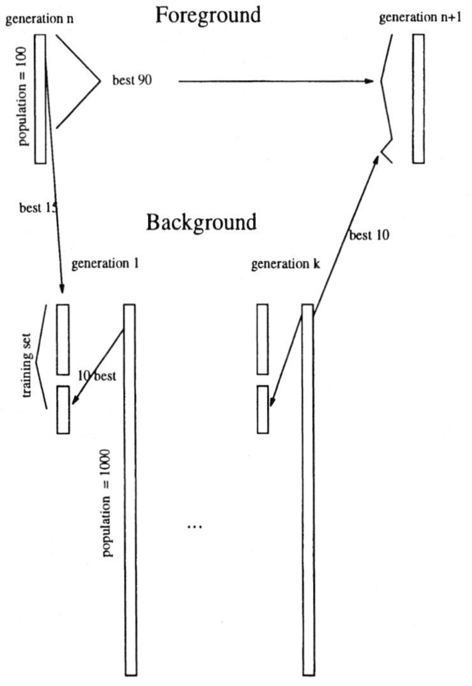

Figure 6. Scheme of foreground and background evolutionary populations.

3. Results

Our server has been operational since September 1997, and as of this writing we have collected the results of over 35,000 (animal vs. animat) games. This is a small amount compared with the number of games played by the background process, which plays 25,000 games per generation, approximately one generation per hour. We are keeping the system running, continuing to make adjustments and experiment with different ideas. The results presented here are based on the first 82 days of data, or 37,295 games.

Our basic performance measure is the *win rate*, that is, the fraction of games that the artificial players have won. The average win rate over the total number of games played is 0.42, meaning that 42% of all games completed have resulted in animat victories.

The graph in fig. 7 uses a sampling rate of 1000 to plot the evolution of the win rate over time. It illustrates two important factors. First, there are oscillations. This is a natural phenomenon in a coevolutionary environment, and occurs here more noticeably since one of the evolving populations consists of randomly selected human players. Each of the 416 individuals sampled here has a different level of expertise[1], has played a different number of games, and so on[2].

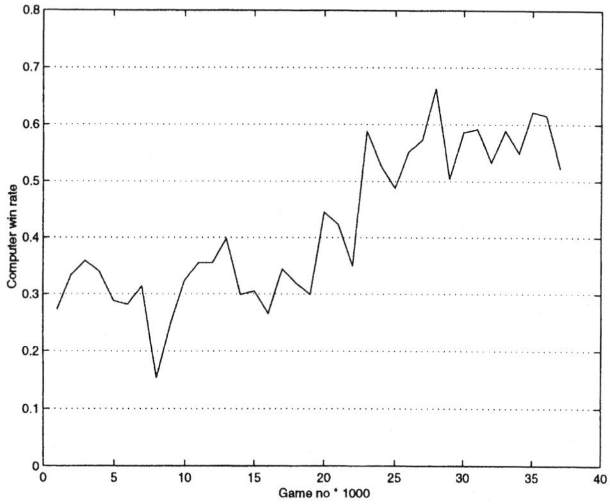

Figure 7. Evolution of the win rate.

The second important feature is that there is a visible trend toward improvement. The winning rate has gone up roughly from 28% to 55% over the time that the system has been operational.

3.1 Adaptation in humans

Humans learn very fast, so the Tron animats are chasing a moving target. The graph in Figure 8 illustrates the average human learning rate. A human playing his first game against our robots will have a 73% chance of losing; but by the time he has played 10 games, this expectation will be reduced to 53%. The animats win 67% of the first games, but then only 52% of the second games, etc. By the 10th game, the rate descends to 44%. (We smoothed with a moving average to show the trend over the right side of the graph.)

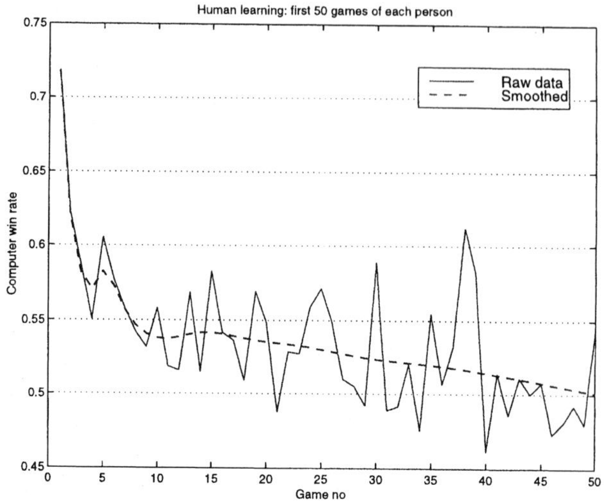

Figure 8. Averaged learning of human players[3].

3.2 Adaptation in animats

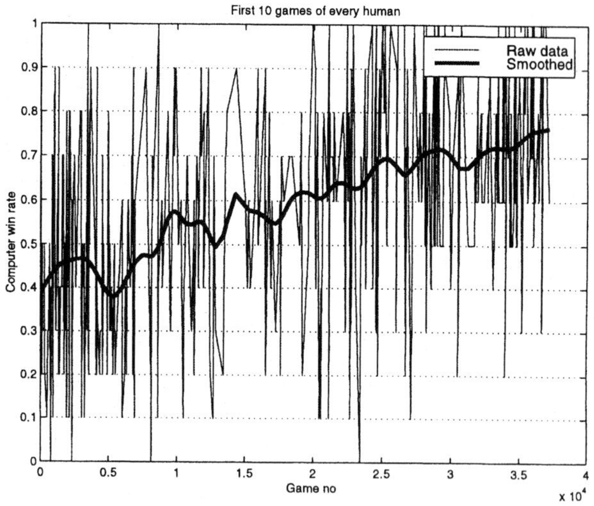

Figure 9. Performance of Tron during first 10 games with each opponent[4].

It would be desirable to factor out the influence of human

1. Another variable factor is the speed of the game in the client machine, which goes down due to the low speed of Java interpreters inside Internet browsers.
2. We have called the low peak at 8000 games the "khith anomaly" because it consists mostly of games by the same one person, with login name "khith".

3. Smoothing obtained by convolution with $e^{-\left(\frac{\ln x}{\alpha}\right)^2}$, normalized. $\alpha = 0.25$.

4. Smoothing obtained by convolution with $e^{-\left(\frac{x}{\alpha}\right)^2}$, normalized. $\alpha = 1024$.

learning from the data in fig. 7 in order to visualize the raw improvement of the animats. The technique of sampling suggested by (Cliff and Miller, 1995) cannot be used in this case; it is not possible to play for example 'player *H0001* from December 10th against robot *R230001*' because the human of that day is not available. Instead, we apply an averaging technique. We consider that a 'virgin' user — a human that plays for the first time — is at the same average level of expertise no matter when he plays his first game. Examining only the first 10 games of every person who has played 10 or more games, we obtain the graph shown in Figure 9.

The raw data is very noisy since each point is the score of a single human: some players log into the Tron system for the first time and win their first 10 consecutive games, showing a computer winning ratio of 0. Others lose their first 10 games and end up with a computer winning ratio of 1. Most players perform somewhere in between. But smoothing with a moving average shows a steady trend towards robot improvement. Starting at around 40%, the computer winning ratio has increased to nearly 74%.

Table 1: Best robots

Ranking	Robot ID No.	Wins/ Games	Games Played
1	510006	0.87	63
2	460003	0.82	94
3	460008	0.80	94
4	400010	0.78	132
5	480001	0.78	85
6	540004	0.77	52
7	460002	0.71	94
8	330003	0.74	167
9	540006	0.74	54
10	460007	0.73	91
...			
40	100009	0.60	288
...			
100	150001	0.45	269

Using our raw approximate measure (the win ratio), we obtain a score for each robot. But since luck determines the opponents of each artificial player, this ratio will only be an estimation that approaches a true value as the number of games played increases. In Table 1, we have listed the best artificial players, out of those who have played at least 50 games.

The two first digits in an robot's ID number represent the generation where it first appeared in the foreground population. The current 10 best robots belong to generations 51, 46, 40, 48, 54 and 33. Among the most veteran agents there are a few strategies born in generation 10, each with around 300 games played. The worst robot presently alive has a win rate of 0.45. Strategy 100009 is noteworthy as it maintains a steady win rate of 60% after nearly 300 games. The longest-lived agent from the original population (generation 1) lasted 334 games before being retired, with a final win rate of 39.5%.

3.3 Analysis of sample robots

Following is the GP s-expression for the current champion, R.510006:

```
(* _H (IFLTE _A 0.92063 _H (- (% _D (- (+ 0.92063 (IFLTE
0.92063 _F 0.92063 (LEFT_TURN))) (IFLTE (- (IFLTE _C
_G (RIGHT_TURN) (LEFT_TURN)) (IFLTE
(+(LEFT_TURN) (LEFT_TURN)) _G _F _G)) _H
(RIGHT_TURN) _G))) (RIGHT_TURN))))
```

This can be roughly reduced to pseudocode as:

```
if FRONT < 0.92063 go straight
else if 0.92063 >= REAR_RIGHT turn left
else if LEFT < RIGHT turn right
else turn left
```

This robot will always go straight, unless there is an obstacle in front of him closer than 8% of the size of the arena. At this point, he will turn right or left. The use of the rear_right sensor is confusing, and its is difficult to infer from the code the actual behavior of this expression, as complex variations arise from its interactions with the Tron environment.

When inserted in a Tron arena, this code shows an interesting behavior. It will avoid obstacles, get out of dead ends and do tight turns to maximize space when in a confined space.

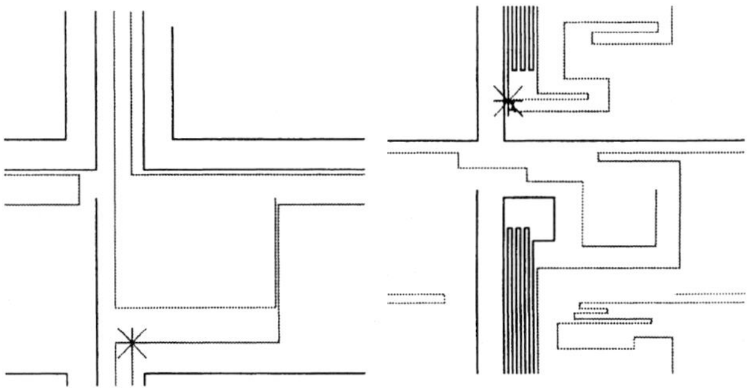

Figure 10. Sample games of robot 510006 (black) vs. a human opponent (grey). It quickly beats a novice player (left). It fights against an expert player, making tight turns when confined in a small space (right).

3.4 Analysis of Humans

We are also studying the characteristics of the human population that is playing our Tron. At present, we want to know if our Tron Web site has created an "egosystem", attracting

Tron fans to our site (as opposed to other Tron sites) and enticing them to keep coming back in order to improve their ranking in our "Hall of Fame".

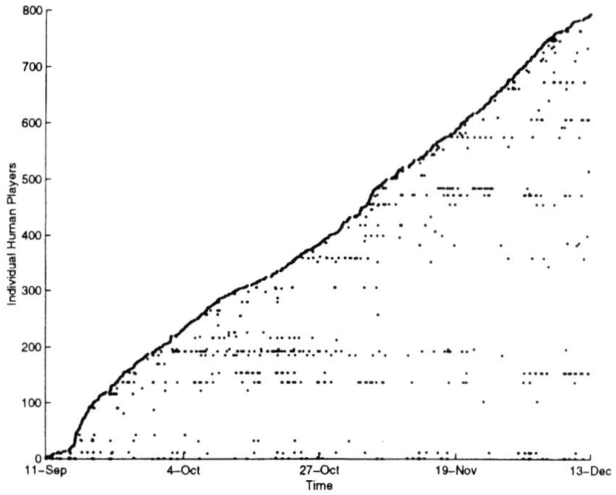

Figure 11. Longevity of human players (see text).

Figure 11 shows the longevity of human players, highlighting "return" players. Time is plotted along the horizontal axis, for the duration of the data analysis period. The vertical axis represents each individual human player that played any games during the time period shown. Each dot on the graph stands for one or a few consecutive games played (13 games on average).

This graph shows that players who came to the site at the beginning of the experiment are still coming back, since some horizontal lines start in the left edge of the graph and continue (though sporadically) to the right edge. The sporadicism is a positive feature because it means that players go away for a while and periodically come back to check the artificial players that have evolved since the last time and to monitor their ranking in the table of scores.

The continual rise of the left edge of the data points indicates that new players keep coming to the site at a steady pace. The implication here is that there is enough new interest in the site so that our coevolutionary space will continue to thrive.

4. Conclusions

Our experiment has succeeded in establishing an interactive space where animal and animat adaptation happen simultaneously.

Once an ecological niche is created, and by this we mean that a piece of adaptive software like our Tron game succeeds in attracting a collectivity of users, a coevolutionary arms race will occur where human and artificial learning chase each other. We can observe human learning rates by averaging all performances of players at the same levels of experience, and we can observe agent learning rates by looking at their performance over time against new incoming opponents.

The exploration vs. exploitation trade-off is linked to the problem of creating a niche. On one hand we want to offer users our best level of performance, but on the other we need to take risks with untested versions. We address this problem in two ways. First, we set our parameters so that only 18% of the games are played by rookies. Second, we exploit the slow pace of human interactions, evolving new robots in the background, so that our novice robots will be as good as possible.

Regarding the algorithmic setup, several details were chosen arbitrarily. The way the background population is reset every time the front end completes a generation is questionable, and other approaches may be better, such as not resetting it at all, or doing so when the dynamics of its own evolutionary process indicate. Genetic isolation between front and back end populations is an arguable factor, since it prevents new players from capitalizing on older genotypes.

The issue of diversity is a central point throughout our coevolutionary setup that needs to be emphasized. Shared fitness functions stimulate diversity by promoting original candidates in every new generation. Resetting the background population and genetic isolation are also promoting diversity, creating new solutions from scratch in each cycle. Diversity and speciation are key in addressing the problem of transfer, or lack of generalization, by widening the spectrum of solutions produced by an introspective learner that needs strategies to cope with a wider environment than the one defined by a limited fitness function that describes only a few points of a larger problem.

The shared fitness function of robots vs. humans (eq. 1) addresses the problem of diversity of levels of expertise among the user population, but also attempts to prevent cooperative phenomena at this level, as robots are given points not for winning or losing, only for doing better or worse than the average of the other robots. The factor of human adaptation has not been accounted for in this function, and future versions should try to adjust accordingly, perhaps using an exponential decay factor.

As perceived by each individual user, our virtual Tron is playing differently every time, since we randomly switch Tron programs for each individual game. Simultaneously, the overall level of play is going up over time. This heterogenous behavior is part of the success; when we say that a certain robot is winning 87% of its games, this is a valid indicator of its level of play *given the collective behavior of all other robots*. If we were to send the same identical robot over and over, humans would quickly learn strategies against it, and the system as a whole would be boring.

There are severe representational limitations to our Tron agents due to their limited sensory perception. Whereas humans observe the entire state of the game in every screenshot, robots only "see" the nearest object in eight fixed directions.

They have no perception of the position and heading of their opponents, and are incapable of analyzing the game board as a whole in order to make their decisions. Future research should contemplate a more powerful version of their sensors.

We have shown that evolutionary techniques can be successfully applied in the context of an environment of connected human users. The variable input of random players with diverse expertise and interests exhibits exploitable average trends. The trend toward collusion through cooperation in co-evolutionary agents can be broken in adaptive software environments by incorporating the raw fitness evaluation emerging from its end-users.

References

Axelrod, R. M. (1984). *The Evolution of Cooperation.* New York, Basic Books.

Beasley, D., Bull, D. R. and Martin, R. R. (1993). A sequential niche technique for multimodal function optimization. *Evolutionary Computation 1(2).* 101-125.

Gilbert, L. E. and Raven, P. H. (eds.) (1975) *Coevolution of animals and Plants.* University of Texas Press.

Cliff, D. and Miller, G. F. (1995). Tracking the Red Queen: Measurements of adaptive progress in co-evolutionary simulations. In F. Moran, A. Moreno, J. J. Merelo and P. Cachon (eds.) *Advances in Artificial Life: Proceedings of the Third European Conference on Artificial Life (ECAL95).* Lecture Notes in Artificial Intelligence 929, Springer-Verlag, pp.200-218.

Darwen, P. J. (1996). *Co-evolutionary Learning by Automatic Modularisation with Speciation.* University of New South Wales, 1996.

Floreano, D. and Mondada, F. (1994). Automatic Creation of an Autonomous Agent: Genetic Evolution of a Neural Network Driven Robot. In D. Cliff, P. Husbands, J.-A. Meyer, and S. Wilson (Eds.), *From Animals to Animats III,* Cambridge, MA. MIT Press.

Floreano, D. and Mondada, F. (1996). Evolution of Homing Navigation in a Real Mobile Robot. *IEEE Transactions on Systems, Man, and Cybernetics--Part B: Cybernetics,* 26(3), 396-407.

Funes, Pablo (1996). The Tron Game: An experiment in Artificial Life and Evolutionary Techniques. Unpublished.

Goldberg, David E. (1989). *Genetic Algorithms in Search, Optimization, and Machine Learning.* Addison-Wesley.

Hillis, W. D. (1992). Co-evolving parasites improve simulated evolution as an optimization procedure. In Langton, C. et al. (Eds.) *Artificial Life II,* Addison-Wesley. pp. 313-324.

Holland, John H. (1975). *Adaptation in Natural and Artificial Systems.* The University of Michigan Press.

Juillé, H. and Pollack, J. (1996). Dynamics of Co-evolutionary Learning. In *Proceedings of the Fourth International Conference on Simulation of Adaptive Behavior.* MIT Press. pp 526-534.

Juillé, H. and Pollack, J. (1996b). Co-evolving Intertwined Spirals. in *Proceedings of the Fifth Annual Conference on Evolutionary Programming,* MIT Press.

Koza, John R. (1992). *Genetic Programming: On the Programming of Computers by Means of Natural Selection.* Cambridge, MA: The MIT Press.

Lieberman, H. (1997). Autonomous Interface Agents, *ACM Conference on Human-Computer Interface [CHI-97],* Atlanta.

Mataric, M and Cliff, D. (1996). Challenges In Evolving Controllers for Physical Robots. In *Evolutionary Robotics,* special issue of *Robotics and Autonomous Systems,* Vol. 19, No. 1. 67-83.

Miller, G. F. and Cliff, D. (1994). Protean Behavior in Dynamic Games: Arguments for the Co-Evolution of Pursuit-Evasion Tactics. *From Animals to Animats 3: Proceedings of the Third International Conference on Simulation of Adaptive Behavior (SAB94).* D Cliff, P. Husbands, J.-A Meyer and S W Wilson, eds. MIT Press Bradford Books, pp.411--420.

Miller, G. F. and Cliff, D. (1996). Co-evolution of Pursuit and Evasion II: Simulation Methods and Results. In *Proceedings of the Fourth International Conference on Simulation of Adaptive Behavior.* MIT Press. pp 506-515.

Pollack, J. B., and Blair, A.D. (1997). Why did TD-Gammon work? *Advances in Neural Information Processing Systems 9.* 10-16.

Pollack, J. B., Blair, A. and Land, M.(1996). Coevolution of A Backgammon Player. *Proceedings Artificial Life V,* C. Langton, (Ed), MIT Press.

Reynolds, C.W. (1994). Competition, Coevolution and the Game of Tag", *Proceedings of Artificial Life IV.* R. Brooks and P. Maes, eds. MIT Press.

Rosca, J. P. (1996). Generality versus Size in Genetic Programming. *Proceedings of the Genetic Programming 1996 Conference (GP-96).* The MIT Press.

Rosin, C. D. (1997). *Coevolutionary Search Among Adversaries.* Ph.D. thesis, University of California, San Diego.

Samuel, A. L. (1959). Some Studies in Machine Learning Using the Game of Checkers. *IBM Journal of Research and Development,* vol. 3, 210-229.

Sims, K. (1994). Evolving 3D Morphology and Behavior by Competition. *Artificial Life IV Proceedings,* MIT Press.

Tesauro, G. (1989). Neurogammon Wins Computer Olympiad. *Neural Computation I,* 321-323.

Tesauro, G. (1992). Practical issues in temporal difference learning. In *Machine Learning.* 8:257-277.

Tesauro, G. (1995) Temporal difference learning and TD-Gammon. *Communications of the ACM,* 38(3): 58-68.

Wolpert, D. H. (1990). A Mathematical Theory of Generalization. *Complex Systems 4*: 151-249.

Perpetuating evolutionary emergence

A.D. Channon and R.I. Damper
Image, Speech & Intelligent Systems Research Group
University of Southampton, Southampton, SO17 1BJ, UK
http://www.soton.ac.uk/~adc96r
adc96r@soton.ac.uk rid@ecs.soton.ac.uk

Abstract

Perpetuating evolutionary emergence is the key to artificially evolving increasingly complex systems. In order to generate complex entities with adaptive behaviors beyond our manual design capability, long-term incremental evolution with continuing emergence is called for. Purely artificial selection models, such as traditional genetic algorithms, are argued to be fundamentally inadequate for this calling and existing natural selection systems are evaluated. Thus some requirements for perpetuating evolutionary emergence are revealed. A new environment containing simple virtual autonomous organisms has been created to satisfy these requirements. Resulting evolutionary emergent behaviors are reported alongside of their neural correlates. In one example, the collective behavior of one species clearly provides a selective force which is overcome by another species, demonstrating the perpetuation of evolutionary emergence via naturally arising coevolution.

1. Evolutionary emergence

Emergence is related to qualitatively novel structures and behaviors which are not reducible to those hierarchically below them. It poses an attractive methodology for tackling Descartes' Dictum: "how can a designer build a device which outperforms the designer's specifications?" (Cariani, 1991, page 776). Most importantly, it is necessary for the generation of complex entities with behaviors beyond our manual design capability.

Cariani identified the three current tracts of thought on emergence, calling them "computational", "thermodynamic" and "relative to a model" (Cariani, 1991). Computational emergence is related to the manifestation of new global forms, such as flocking behavior and chaos, from local interactions. Thermodynamic emergence is concerned with issues such as the origins of life, where order emerges from noise. The emergence relative to a model concept deals with situations where observers need to change their model in order to keep up with a system's behavior. This is close to Steels' concept of emergence, which refers to ongoing processes which produce results invoking vocabulary not previously involved in the description of the system's inner components – "new descriptive categories" (Steels, 1994, section 4.1).

Evolutionary emergence falls into the 'emergence relative to a model' category. Consider a virtual world of organisms that can move, reproduce and kill according to rules sensitive to the presence of other organisms, evolving under natural selection. Should flocking manifest itself in this system, we could classify it as emergent in two senses: firstly in the 'computational' sense from the interaction of local rules, flocking being a collective behavior, and secondly in the 'relative to a model' sense from the evolution, the behavior being novel to the system. While the first is also relevant to our goal, in that complex adaptive systems will involve such emergence, the second is the key to understanding *evolutionary* emergence.

Harvey's Species Adaptation Genetic Algorithm (SAGA) theory (Harvey, 1992) provides a framework for incremental evolution, necessary for evolutionary emergence. In this paradigm a population, with possibly just a few tens of members, evolves for many thousands of generations, with gradual changes in genotype information content. Increases in complexity must therefore result from evolution itself. This is in contrast to the common use of the Genetic Programming (GP) paradigm, where a population of millions may be evolved for less than a hundred generations (Harvey, 1997, section 5). In the GP case, recombination effectively mixes the random initial population, exhausting variation in few generations. Because (genetic codings of) computer programs result in rugged fitness landscapes, there can be little further evolution of this converged population. Here we see one of the requirements of SAGA: a smooth fitness landscape.

Having specified what is meant by evolutionary emergence, we will now explore the two types of selection which might be used to bring evolutionary emergence about. Packard referred to these as "*extrinsic* adaptation, where evolution is governed by a specified fitness function, and *intrinsic* adaptation, where evolution occurs "automatically" as a result of the dynamics of a system caused by the evolution of many interacting subsystems" (Packard, 1989, abstract). We will refer to them as artificial and natural selection respectively, because the first involves the imposition of an artifice crafted for some cause external to a system beneath it while the second relies solely on the innate dynamics of a system.

2. Artificial selection

Within the artificial evolution field, variants of the optimization paradigm have proven fruitful. Even where the concepts of SAGA theory are dominant, practice still holds to the use of fitness functions. But as the complexity of behaviors attempted increases, flaws in the artificial selection approach are appearing. Zaera, Cliff and Bruten's failed attempts at evolving schooling behavior in artificial 'fish' (Zaera et al., 1996) provide an account of the difficulties faced. An extract from the abstract of their paper still yields an excellent summary of the state of artificial selection work within the field:

> "The problem appears to be due to the difficulty of formulating an evaluation function which captures what schooling is. We argue that formulating an effective fitness evaluation function for use in evolving controllers can be at least as difficult as hand-crafting an effective controller design. Although our paper concentrates on schooling, we believe that this is likely to be a general issue, and is a serious problem which can be expected to be experienced over a variety of problem domains."

Zaera et al. considered possible reasons for their failure. The argument which most convinced them was that real schooling arises through complex interactions, and that their simulations lacked sufficient complexity (Zaera et al., 1996, section 5). They cited two promising works: Reynolds' evolution of coordinated group motion in 'prey' animats pursued by a hard-wired 'predator' (Reynolds, 1992), and Rucker's 'ecosystem' model (Rucker, 1993) in which Boid-like animat controllers (or rather their parameters) were evolved. Both of these are moves towards more intrinsic, automatic evolution.

The use of coevolutionary models is fast becoming a dominant approach in the adaptive behavior field. This is essentially a response to the problems encountered when trying to use artificial selection to evolve complex behaviors. However, artificial selection has kept its hold so far – most systems still use fitness functions. The reasoning given for imposing coevolution is often that it helps in overcoming problems arising from the use of static fitness landscapes.

From the discussion so far, one might assume our argument to be that evolutionary emergence is not possible in a system using artificial selection. This is not quite so, although we do argue that artificial selection is neither sufficient nor necessary. *In the context of evolutionary emergence, any artificial selection used constitutes just one of the parts of a system.* Artificial selection can only select for that which it is specified to. Therefore anything that emerges during evolution *must* be due to another aspect of selection, which must in turn be due to the innate dynamics of the system – natural selection.

3. Natural selection

As noted in section 1, genetic codings of computer programs result in rugged fitness landscapes, making them unsuitable for incremental evolution. However, most natural selection work has been program code evolution, following the initial success of 'Tierra' (Ray, 1991).

3.1 Natural selection of program code

Tierra is a system of self-replicating machine code programs, initialized as a single manually designed self-replicating program. To make evolution possible, random bit-flipping was imposed on the memory. A degree of artificial selection was imposed by *the system* deleting the oldest programs in order to free memory, with an added bias against programs that generated error conditions.

Tierra was implemented as a virtual computer, allowing Ray to design a machine language with some properties suiting it to evolution. One aspect of this language was that it contained no numeric constants (such as 13). Thus direct memory addressing was not possible. Instead, the manually designed program used consecutive NOP (No-OPeration) instructions which acted as templates that could be found by certain machine code instructions. This 'addressing by templates' is how the program determined the points at which to begin and end copying. Another aspect of the system was that computational errors were introduced at random. Such errors could lead to genetic changes by affecting replication.

When Tierra was run, various classes of programs evolved. 'Parasites' had shed almost half of their code; they replicated by executing the copy loop from neighboring organisms, which could easily be found by template matching instructions as before. Because the parasites depended on their 'hosts', they could not displace them and the host and parasite populations entered into Lotka-Volterra population cycles. Ray reported that coevolution occurred as the hosts became immune to the parasites, which overcame these defenses, and so on. 'Hyper-parasite' hosts emerged containing instructions that caused a parasite to copy the host rather than the parasite; this could lead to the rapid elimination of parasites. Ray also reported cooperation (symbiosis) in replication followed by 'cheaters' (social parasites) which took advantage of the cooperators.

The above are examples of ecological adaptations. Another class of adaptations found was "optimizations". For example, non-parasitic replicators almost a quarter the length of the initial replicator were found, as were programs with 'unrolled' copy loops which copied two or three bytes per loop, reducing the overhead of looping. By adding 'split' and 'join' instructions, which allowed a program to split into a multi-threaded process and join back into a single one, the evolution of efficient parallel-processing replicators was later achieved. While the results of Tierra are impressive, there have been no new reports of emergent phenomena during the last few years. It is generally accepted that not much more will occur unless further alterations are made to the system.

The evolvability of the code would seem to stem largely from the template matching system. This could account for all of the ecological adaptations reported but would be of little use for much other than replication. To see how this could be, consider the following pseudo-code of the initial, manually designed program:

```
T1111 a=address(T1110)+1 ; b=address(T1111) ; c=a-b
T1101 allocate memory for child, length c =>a=start
      call T1100 ; divide from child ;jump to T1101
T1100 push a,b,c onto stack              }
T1010 memory(a)=memory(b)                }   copy
      c-- ; if c=0 then jump to T1011    } procedure
      a++ ; b++ ; jump to T1010          }
T1011 pop c,b,a off stack ; return       }
T1110
```

This pseudo-code is based on the initial program as listed in (Ray, 1992, appendix C). The T????s on the left denote four-bit templates. Now, as reported in (Ray, 1992, section 3.1.1), a parasite can be obtained by simply mutating one bit of the T1100 template to produce T1110; this would then be the same as the end template, reducing the length to be copied (c), and the "call T1100" statement would search outwards in memory until it found a 'host' containing the copy procedure. Further, a 'host' that is immune to such a parasite can also be produced by a single-bit mutation in the template-comparison 1011 in the "if c=0 then jump to T1011" statement of the copy procedure; by mutating it to 1111, the program will re-evaluate its address and length after every reproduction. Thus, should a parasite try to use this program's copy routine, it will be copied just once but ever after that it will be copying the host; this host is also a 'hyper-parasite' as defined above. In just two template bit-flips, we have reproduced the most impressive ecological results of Tierra! While the actual evolution might have taken a slightly different route, it is clear that these phenomena are due to the flexibility of the template matching. It is also clear that the same argument applies to the other ecological adaptations reported.

While this demonstrates the flexibility of template matching at the four-bit level, it also shows the ecological results to be somewhat more trivial than we might have first hoped. For we must hold to our previous argument that programs are not suitable for incremental evolution; complex programs would necessarily contain many more templates, which would have to be significantly longer. Thus we would pass far beyond the trivial four-bit templates that random search can operate on, to a stage where evolution is impractical.

There is still the issue of the 'optimization' adaptations to address. Two of the results, 'unrolled' copy loops and efficient replication by parallel-processing, only needed evolution to insert (probably by the action of the random computational errors) repetitions of neighboring code; the local functionality was not changed and the efficiencies are from 'more of the same' solutions. These examples should not alter our perception of the brittleness in mutating code or inserting *different* instructions. The final results to be explained are the non-parasitic replicators almost a quarter the length of the initial replicator. Comparing the shortest self-replicating program (Ray, 1992, appendix D) with the initial program (Ray, 1992, appendix C) shows that the transition can be made by simply deleting instructions (mostly NOPs – cutting out the redundancy in the templates) and just six mutations, three of which are unnecessary. So apart from three mutations, this is a 'less of the unnecessary' solution – certainly not sufficient to challenge our argument.

3.2 Natural selection of more suitable entities

Although the approach of most natural selection work to date has been to evolve program code, there are two notable exceptions. The second (chronologically) is Channon's own work, which was conceived independently of and initially created in ignorance of both Tierra (including its derivatives) and the first exception: "PolyWorld" (Yaeger, 1993). PolyWorld simulates many aspects of the real world including energy conservation (an implicit fitness function), food, movement (on a 2D plane), vision, neural networks and learning.

PolyWorld organisms have seven pre-programmed behaviors: eating, mating, fighting, moving, turning, focusing and lighting. An organism's chromosome determines its physiology (size, strength, maximum speed, green coloration, mutation rate, number of crossover points and life span) and some basic characteristics of its neural network, including the number of neurons devoted to vision (in three groups: red, green and blue), the number of internal neuronal groups, a connectivity density (between groups) matrix and Hebbian learning rates. The neural networks are constructed stochastically. Yaeger reported a variety of emergent behaviors, including 'fleeing', 'fighting back', 'grazing', 'foraging' and 'following'.

One criticism of PolyWorld, in the context of perpetuating evolutionary emergence, is that (Hebbian) learning may well be overwhelmingly responsible for the results. There is little evidence of significant evolution at the genetic level; the genomes had very limited control over the small number of neural groups. It is conceivable that if comparison PolyWorld experiments were run with 'birth' networks specified by random parameters, then the same results might emerge; new organisms would either learn such that they become adapted to the evolving environment or die, and so evolution would occur without genetic change. While this is valid evolutionary emergence, it is not sufficient for perpetuating evolutionary emergence. Unless learning is evolvable or what is learnt can be passed on, a maximal level will be reached at which organisms are not capable of learning more.

4. A new environment – model definition

A system believed to be better suited to incremental artificial evolution by natural selection has been created. 'Geb' (named after the Egyptian god of the Earth) is a two-dimensional toroidal virtual world containing autonomous organisms each controlled by a neural network. The killing of organisms is under their own control.

Geb's world is divided into a grid of squares; 20 × 20 of them in most runs. No two individuals may occupy the same square at any one time. This gives the organisms a 'size' and puts a limit on their number. They are otherwise free to move within and between squares.

Initialization: Individuals have single-bit genotypes '0'.
Main Loop: For each individual:

1. Update network inputs.
2. Development – one iteration.
3. Update all neural activations, including network outputs.
4. Carry out actions associated with network outputs.

4.1 The neural networks

The neural networks used in Geb are recurrent networks of nodes as used successfully by Cliff, Harvey and Husbands in their evolutionary robotics work (figure 1).

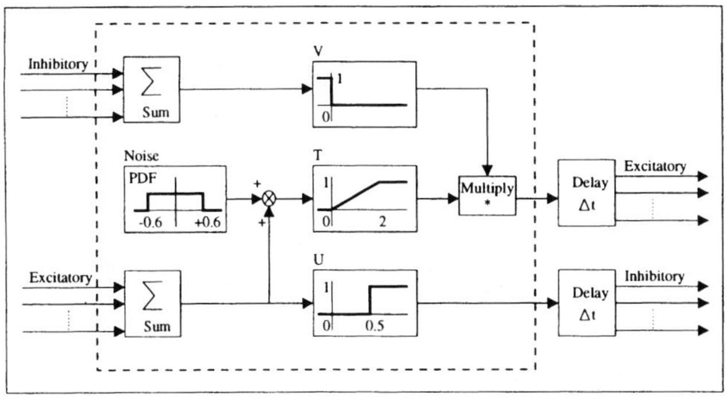

Figure 1 Schematic of a neuron, from (Cliff et al., 1992)

All links have unit weight; no lifetime learning is used. This is to avoid criticism that lifetime learning may be the main factor, as levelled at PolyWorld in section 3.2. Each node has a bit-string 'character' (label) attached to it. This is used to match organisms' inputs, outputs and actions, and to determine the node's development.

4.2 Organism ⟷ environment interactions

There are five built-in actions available to each organism. Each is associated with network output nodes whose characters start with a particular bit-string:

1. `01*` Try to *reproduce* with organism in front
2. `100*` *Fight*: Kill organism in front (if there is one)
3. `101*` *Turn anti-clockwise*
4. `110*` *Turn clockwise*
5. `111*` *Move forward* (if nothing in the way)

For example, if a network output node has the character 1101001, the organism will turn clockwise by an angle proportional to the node's excitatory output. If an action has more than one matching output node then the relevant output is the sum of these nodes' excitatory outputs, bounded by unity as within any node. If an action has no matching output node, then the relevant output is noise, at the same level as in the (other) nodes.

An organism's input nodes have their excitatory inputs set to the weighted sum of 'matching' output nodes' excitatory outputs from other individuals in the neighborhood. If the first bit of an input node's character is 1 then its input is from individuals to the right, otherwise from individuals to the left. An input node 'matches' an output node if the rest of its character is the same as the start of the output node's character. Weighting is inversely proportional to distance between individuals.

When an organism reproduces with a mate in front of it, the child is placed in the square beyond the mate if that square is empty. If it is not then the child replaces the mate unless the mate is fighting, in which case the child is killed. Reproduction involves crossover, with the cut point always offset by one gene (either way), and a single gene-flip (bit-flip) mutation.

4.3 The developmental system

A context-free L-system was designed in which children's networks resemble aspects of their parents'. The axiom network consists of three nodes with two excitatory links:

$$\text{network input } 001 \longmapsto 000 \longmapsto 01 \text{ network output}$$

The production rules have the following form:

$$\mathcal{P} \rightarrow \mathcal{S}_r, \mathcal{S}_n \; ; \; b_1, b_2, b_3, b_4, b_5, b_6 \qquad \text{where:}$$

$\mathcal{P}$	Predecessor (initial bits of node's character)
$\mathcal{S}_r$	Successor 1: *replacement* node's character
$\mathcal{S}_n$	Successor 2: *new* node's character
bits:	link details [0=no,1=yes]:
(b_1, b_2)	reverse types [inhibitory/excitatory] of (input, output) links on $\mathcal{S}_n$
(b_3, b_4)	(inhibitory, excitatory) link from $\mathcal{S}_r$ to $\mathcal{S}_n$
(b_5, b_6)	(inhibitory, excitatory) link from $\mathcal{S}_n$ to $\mathcal{S}_r$

For each node, the production rule with the longest matching predecessor is applied. Thus ever more specific rules can evolve. If a successor has no character then that node is not created. A 'replacement' successor is just the old (predecessor) node with its character changed. A 'new' successor inherits a copy of the old node's input links unless it has a link from the old node (b_3 or b_4). It inherits a copy of the old node's output links unless it has a link to the old node (b_5 or b_6).

New network input nodes are (only) produced from network input nodes and new output nodes (only) from output nodes. The character-based method of matching up network inputs and outputs ensures that the addition or removal of an input/ output node at a later stage of development or evolution will not damage the relationships of previously adapted inputs and outputs.

Details of the theory behind the choice of developmental system, and of the genetic decoding of production rules, can be found in (Channon and Damper, 1998).

5. Results

When two Geb organisms (with networks developed from more than just a couple of production rules each) reproduce, the child's network almost always resembles a combination of the two parents' networks. Examination of the networks from Geb's population, at any time, shows similarities between many of the networks. The population remains nearly-converged, in small numbers of species, throughout the evolution.

5.1 Emergent collective behavior

Once Geb has started, there is a short period while genotype lengths increase until capable of containing a production rule. For the next ten to twenty thousand time steps (in typical runs), networks resulting in very simple strategies such as 'do everything' and 'always go forwards and kill' dominate the population.

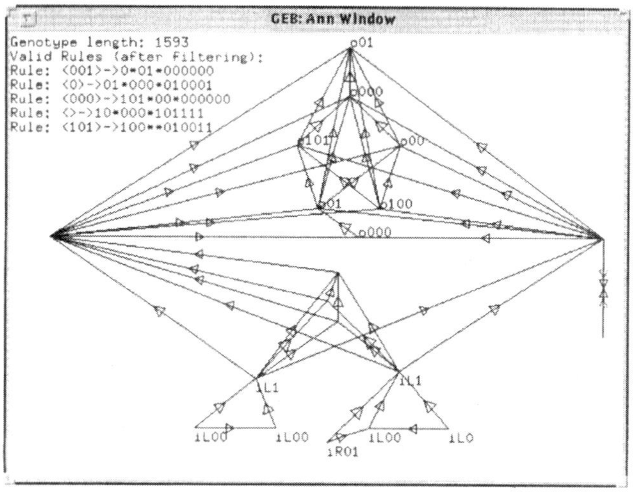

Figure 2 A Dominant Organism

In every run to date, the first dominant species that emerges has been one whose individuals turn in one direction while trying to fight and reproduce at the same time. Figure 2 shows an example of such an individual. Note the network outputs o101, o01 [x2] and o100 (turn anti-clockwise, reproduce and fight). Note also the large number of links necessary to pass from network inputs to outputs, and the network input characters which match non-action output characters of the same network (o000 [x2], o00). Individuals of this species use nearby members of the same species, who are also turning in circles, as sources of activation (so keeping each other going).

Although a very simple strategy, watching it in action makes its success understandable. The individuals keep each other moving quickly, in tight circles. Any attacking organism would have to either get its timing exactly right or approach in a faster spiral – both relatively advanced strategies. These individuals also mate just before killing. The offspring (normally) appear beyond the individual being killed, away from the killer's path.

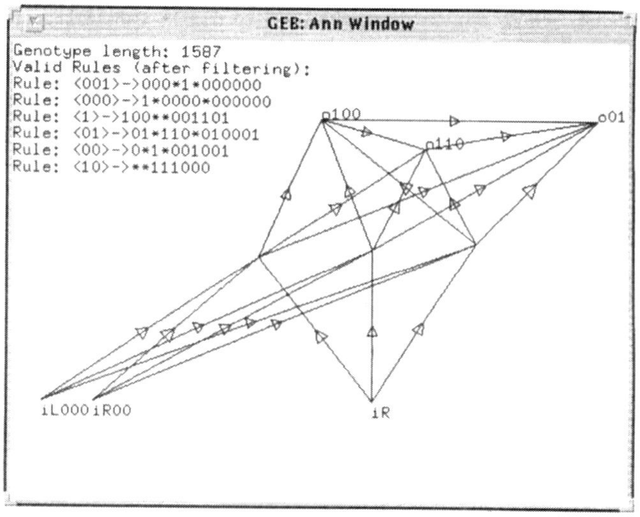

Figure 3 A rebel

5.2 Naturally arising coevolution

Because of the success of this first dominant species, the world always has enough space for other organisms to exist. Such organisms tend not to last long; just about any movement will bring them into contact with one of the dominant organisms, helping that species in its reproduction as much as themselves. Hence these organisms share some of the network morphology of the dominant species. However, they can make some progress: Individuals have emerged that are successful at turning to face the dominant species and holding their direction while trying to kill and reproduce. An example of such a rebel (from the same run as figure 2) is shown in figure 3. Note that most rebels have many more nodes and links; this one was picked for its clarity.

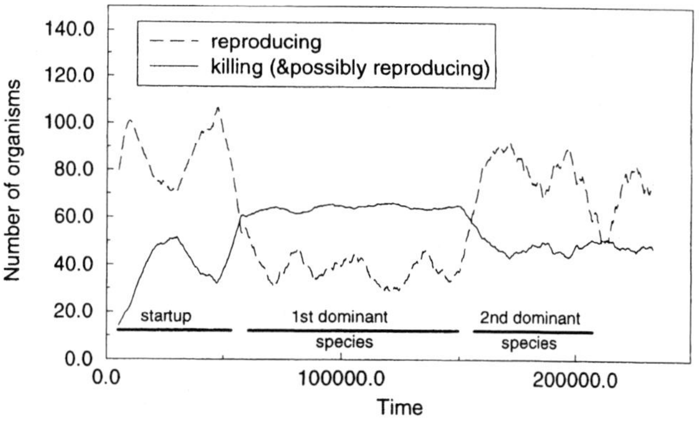

Figure 4 Typical run (running averages)

Further, running averages of the number of organisms reproducing and killing (figure 4) suggest that further species emerge, indicating perpetuating evolutionary emergence. However, organisms have proved difficult to analyze beyond the above, even at the behavioral level. All that can currently be said is that they share characteristics of the previous species but are different.

6. Conclusions

While computational emergence can arise via artificial selection, evolutionary emergence requires natural selection (by our definitions). The logical progression or aim is the perpetuation of evolutionary emergence via naturally arising coevolution. However, this requires long-term incremental evolution and so what we evolve and how we evolve it must be chosen accordingly. The initial groundwork on 'how' has already been covered by SAGA theory – by using low enough mutation rates that the population evolves as nearly-converged species. As for what to evolve, program code is too brittle. Even the use of template matching cannot overcome that fact. Neural networks are a clear choice because of their graceful degradation (high degree of neutrality).

The implication for animat research is that it should be leading the way, through the natural selection of neural controllers, towards the emergence of ever more complex and impressive behaviors. The work presented in this paper has started down that route, with some success. Geb organisms satisfy the convergence criterion, because offspring resemble their parents (but are not identical). So Geb is suited to long-term incremental artificial evolution. The behaviors identified are encouraging too, for the increases in complexity were in ways not specified by the design – evolutionary emergence. These are the two cardinal quality targets for any such work.

Whether on not emergence is continuing in Geb is hard to tell, for it soon becomes difficult to identify behaviors. This was a less significant problem in the evolution of program code but evolved neural networks are hard to understand and so offer little help. Constructing systems such that behaviors will be more transparent is likely to be the most productive way forward.

A further problem is that specifying the available 'actions' constrains organisms around these actions and so limits evolution. Despite showing the important new result of evolutionary emergent behaviors (not specified within the initial system) from a system suited to long-term incremental evolution, all basic (inter-)actions were as specified within the initial system and not evolvable. At a later stage, alternatives in which the evolvable embodiment of an organism gives rise to its actions will need to be considered. Only then will we have a truly open stage on which to watch the perpetuation of evolutionary emergence.

Acknowledgements

This work is supported by an award from the United Kingdom's Engineering and Physical Sciences Research Council to author ADC. It is a continuation of previous work (Channon, 1996) also supported by an award from the EPSRC (supervisor Inman Harvey, University of Sussex).

References

Cariani, P. (1991). Emergence and artificial life. In *Artificial Life II, Santa Fe Institute Studies in the Sciences of Complexity, Vol. X*, pages 775–797, Redwood City, CA.

Channon, A. (1996). The evolutionary emergence route to artificial intelligence. Master's thesis, School of Cognitive and Computing Sciences, University of Sussex. Revision October 1996. http://www.soton.ac.uk/~adc96r/.

Channon, A. and Damper, R. (1998). Evolving novel behaviors via natural selection. In Adami, C., Belew, R., Kitano, H., and Taylor, C., editors, *Proceedings of "Artificial Life VI", Los Angeles, June 26-29, 1998.* MIT Press. http://www.soton.ac.uk/~adc96r/.

Cliff, D., Harvey, I., and Husbands, P. (1992). Incremental evolution of neural network architectures for adaptive behaviour. Technical Report CSRP256, University of Sussex School of Cognitive and Computing Sciences.

Harvey, I. (1992). Species adaptation genetic algorithms: A basis for a continuing SAGA. In Varela, F. and Bourgine, P., editors, *Towards a Practice of Autonomous Systems: Proceedings of the First European Conference on Artificial Life*, pages 346–354, Cambridge, MA. MIT Press/ Bradford Books.

Harvey, I. (1997). Cognition is not computation: Evolution is not optimisation. In *forthcoming Proceedings of ICANN97. Special Session on Adaptive Autonomous Agents at ICANN97.*

Packard, N. H. (1989). Intrinsic adaptation in a simple model for evolution. In Langton, C., editor, *Artificial Life, Santa Fe Institute Studies in the Sciences of Complexity, Vol. VI*, pages 141–155. Addison-Wesley.

Ray, T. S. (1991). An approach to the synthesis of life. In Langton, C., Taylor, C., Farmer, J., and Rasmussen, S., editors, *Artificial Life II, Santa Fe Institute Studies in the Sciences of Complexity, Vol. X*, pages 371–408, Redwood City, CA. Addison-Wesley.

Ray, T. S. (1992). Evolution, ecology and optimization of digital organisms. Technical Report 92-08-042, Santa Fe Institute.

Reynolds, C. W. (1992). An evolved, vision-based behavioral model of coordinated group motion. In Meyer, Roitblat, and Wilson, editors, *From Animals to Animats 2: Proceedings of the Second International Conference on Simulation of Adaptive Behavior (SAB92)*, pages 384–392, Cambridge, MA. MIT Press.

Rucker, R. (1993). *Artificial Life Lab.* The Waite Group Press, Corte Madera, CA.

Steels, L. (1994). The artificial life roots of artificial intelligence. *Artificial Life Journal*, 1(1):89–125. MIT Press.

Yaeger, L. (1993). Computational genetics, physiology, metabolism, neural systems, learning, vision, and behavior or polyworld: Life in a new context. In Langton, C. G., editor, *Artificial Life III, Santa Fe Institute Studies in the Sciences of Complexity, Vol. XVII*, pages 263–298.

Zaera, N., Cliff, D., and Bruten, J. (1996). (Not) evolving collective behaviours in synthetic fish. In Maes, P., Mataric, M., Meyer, J.-A., Pollack, J., and Wilson, S., editors, *From Animals to Animats 4: Proceedings of the Fourth International Conference on Simulation of Adaptive Behavior (SAB96)*, pages 635–644. MIT Press Bradford Books.

Evolutionary Design of Morphology and Intelligence in Robotic System Using Genetic Programming

Kohki Kikuchi and Fumio Hara

Mechanical Engineering Department., Science University of Tokyo
1-3 Kagurazaka, Shinjuku-ku, Tokyo 162-8601, Japan
Email: kikut@hafu0103.me.kagu.sut.ac.jp, hara@hafu0103.me.kagu.sut.ac.jp

Abstract

This paper deals with the evolutionary design of morphology and intelligence in robotic systems and the characteristics of emerged robot behaviors. A robot performs a given task in a realistic virtual world including physical conditions such as gravity, collision and friction, and is assigned fitness according to its performance. Fitness is improved by genetic programming operations, and therein the robot evolves to a reasonably optimal morphology and control architecture. The behaviors of the robot undertaking three kinds of tasks differing in the number of objects to be picked up and task limit time are investigated. We find various morphologies and interesting intelligence emerged according to the differences in the tasks and environmental conditions. We examine the relation between robot morphology and behavior, and demonstrate the capability and flexibility of these evolutionary robotic systems.

1. Introduction

Lately, artificial systems simulating the biological principles have been highly interested in various fields of science and technology. Holland has proposed the genetic algorithm (GA) (Holland, 1975), which is the computational method of optimization developed by imitating the evolution of living organisms in nature. Koza has applied it to develop genetic programming (GP) (Koza, 1994), which is the computational technique to make computer programs to evolve. These algorithms enable evolutionary computation (EC), which is widely used in, for example, optimization problems (Hara and Kikuchi, 1994, 1997) and artificial life (AL) (Langton, 1989). As examples, Kikuchi and Hara studied the method of evolutionary design of biomimetic tree structures that change their morphology so as to adapt themselves to the environment conditions (Kikuchi and Hara, 1997). They used GA in their study to find reasonably optimal solutions. Sims has succeeded in the generation and evolution of virtual creatures in a computer by making use of competition for a prey and has shown the emergence process of their morphology and behavior (Sims, 1994). Ventrella also presented the evolutionary emergence of morphology and locomotion behavior in animated characters (Ventrella, 1994). Gruau and Quatramaran showed a cellular encoding method for interactive evolutionary robotics using artificial neural networks (Gruau and Quatramaran, 1997).

When we design a robotic system which must perform a given task under changing environment conditions, the robotic system must be fully adaptive. The motor function of a robotic system is generated by the dynamic interaction between the robotic body and its environment, and thus the morphology of the robotic system is considered one of the important aspects in designing robotic systems. The sensory function is also considered influential to the robotic intelligence since the intelligence is represented in terms of sensory-motor coordination. This means that a sensory morphology is also important design aspect. From these we may say that the balance between body morphology of a robotic system and its intelligence is thought a key issue in designing adaptive robotic systems.

From this viewpoint of design principles for adaptive robotic systems, in this paper we describe the evolutionary concurrent design of morphology and intelligence in robotic systems using GP. In the evolutionary concurrent design, many robotic individuals are evolving in their chromosome pool and they receive different fitness as a result of performing tasks in a realistic virtual arena including physical conditions such as gravity, collision and friction. An individual robot has morphology genes representing its body shape and intelligence genes determining its behavior through dynamic interaction with the given environment. The robot body is created by connecting round cells, each of which has a single motor for locomotion. Intelligence genes are represented by a decision tree that can coordinate sensor-motor relations and determine a single robot behavior for one given set of environmental conditions.

We focus on the simultaneous emergence of morphology and intelligence in the robotic system and also on the adaptive behavior characteristics of the evolved robots. We investigate three cases of tasks different in the number of objects to be picked up and the task limit time. We show the reasonably optimized robot morphology and intelligence evolved for each of the three tasks and discuss the characteristic features in their morphology and intelligence. Finally we discuss the capability and flexibility of their behaviors appeared in executing the task under various unknown environmental conditions.

2. Design Problem

Fig. 1 shows the 2-dimensional physical arena in which a single robot will collect N objects in the goal area. N objects are placed in the central region of the arena enclosed by a wall, and the robot starts to move from the center of the goal area located at the right hand bottom of the arena. The robot moves by using its motor and its intelligence constructed by GP. Then the robot is assigned fitness according to how well it performs the task. To seek for the optimal morphology and intelligence under the given task and environment conditions, GP is applied.

At first, the final distance of each object to the center of the goal area is summed up. If C_n is the final position of the nth object, G_p is the center of the goal area, N is the total number of objects and D_g is the diameter of the goal area, f_{dis} is defined by

$$f_{dis} = \sum_{n=0}^{N} \begin{cases} |C_n - G_p| & (|C_n - G_p| > \frac{D_g}{2}) \\ 0 & (|C_n - G_p| \le \frac{D_g}{2}) \end{cases}$$

Then the fitness of the individual robot, f, is given by

$$f = \begin{cases} 1 + f_{dis} & (if \;\; f_{dis} \neq 0) \\ \frac{t}{T} & (if \;\; f_{dis} = 0) \end{cases}$$

where t is the time at witch the task is achieved and T is the given task limit time. This formulation confines the fitness values within a limited range of positive numbers. Note that, if all objects are moved into the goal area within half the given task limit time, the individual robot is given a fitness of 0.5, which is about the smallest in the fitness values.

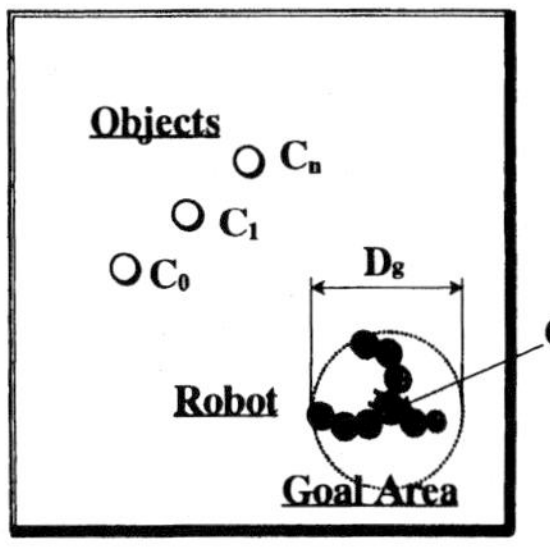

Fig. 1 Arena in which a single robot gathers objects into the goal

3. Representation of Morphology and Intelligence in GP

Each robot is represented by a phenotype which is encoded into a genotype consisting of morphology genes and intelligence genes in GP. An example of a genotype is shown in Table 1. Morphology genes determine the phenotypic robot body feature and intelligence genes specify the robot behavior under given environmental conditions. Each of two genes is written in S-expressions of LISP language (Steel Jr., 1990). This is convenient to evolve the morphology and intelligence genes by GP.

Table 1 Example of genotypes

Morphology genes	
Links	(0 (1 (2 (3 4)) (5 (6 7))))
Diameters(m)	(0.3 (0.3 (0.35 (0.4 0.35)) (0.3 (0.35 0.35))))
Intelligence genes	
ADF	*(defun adf00 (arg0) (+ PAI (sin arg0)))*
Program	*(if (START_DIS < 0.5) (parallel_move 1) (Stop))*

3.1 Morphology

The robot is modeled as a 2D body comprised of inter-connected cells. Each cell has a circular ring which can change its diameter (0.1-0.4m), a motor to drive the robot in all directions and a pin at the top for interconnection with other cells, as shown in Fig. 2 (a). All cells are of the same weight and modeled as a point mass when considered their dynamics equations.

Morphology genes represent the cell connection status and cell diameters. For instance, the morphology genes shown in Table 1 generates the phenotype of robotic body morphology shown in Fig. 2 (b). Each cell has its ID number and the 0th cell is named as the root cell. Internal state sensors such as position and velocity encoders are equipped with each cell, and a visual sensor such as CCD-camera is installed in the root cell.

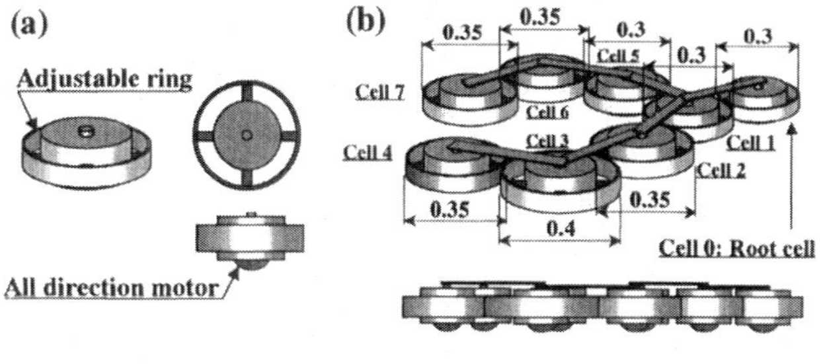

Fig. 2 Cell type robot component and robot morphology corresponding to the genotype shown in Table 1

3.2 Intelligence

Intelligence genes represent a program that determines the robot behavior and are expressed by a decision tree and written in S-expressions of LISP language.

The decision tree is generated using the variable nodes

and the function nodes shown in Table 2. In our GP, the variable nodes express sensing variables including environment data obtained via a CCD-camera as shown in Fig. 3, state variables such as motor force or velocity obtained by virtual encoders as shown in Fig. 4 and constants such as PAI and random value. The function nodes contain behavior functions such as "%", ADF (automatically defined function) suggested by Koza (Koza 1994) and the standard Common LISP commands (Steel Jr., 1990) such as "+, -, *sin*, *if*, *and*, ...". The following behavior functions are a priori set in our paper since our interest is not in generating these primitive behavior functions, but in the global morphology of a robot body and its whole intelligence feature. "*Parallel_move*" gives force to each motor in the direction of the argument, "*Open_limbs*" gives force to each terminal cell in the positive perpendicular direction to the line connecting the center of the *0*th cell and the center of the terminal cell, "*Close_limbs*" gives force to each terminal cell in the negative perpendicular direction to the line connecting them, "*Expand_limbs*" gives force to each terminal cell in the opposite direction to the *0*th cell, "*Point_move*" gives force to the motor of the cell indicated by the first argument in the direction specified by the second argument, which is similar to that employed in the study of Sims (Sims 1994), "*Direction_change*" changes the "Robot Direction" to the direction of the argument, "*Stop*" switches off all motors and "*Stop_cell*" turns off the motor of the cell indicated by the argument. These behavior functions are used for controlling cell motors and are illustrated in Fig. 5. Note that ADFs are subtrees or strategies taken out partially from the decision tree or intelligence genes, and the number of ADFs is allowed to increase up to 32.

Table 2 GP nodes used in this paper

Variable nodes	
Sensing Variables	*GOAL_DIS, GOAL_ANGLE, G_OBJ_D[n], G_OBJ_A[n], OBJ_D[n], OBJ_A[n],*
State Variables	*C_SIZE, STAT_TOTAL_LIMB, 0_CELL_D[n], 0_CELL_A[n], NCELL_VEL_A[n], NCELL_VEL_D[n]*
Constants	*REGION, PAI, PAI/2, PAI/4, -PAI, -2PAI, -4PAI, RANDOM*
Function nodes	
Behavior Functions	*Parallel_move, Expand_limbs, Open_limbs, Close_limbs, Direction_change, Stop, Stop_cell Point_move*
Lisp Commands	*+, -, *, %, pow, sin, cos, <, <=, >, >=, max, min, if, and, or,*
ADFs	*Adf00, Adf01, ... Adfxx*

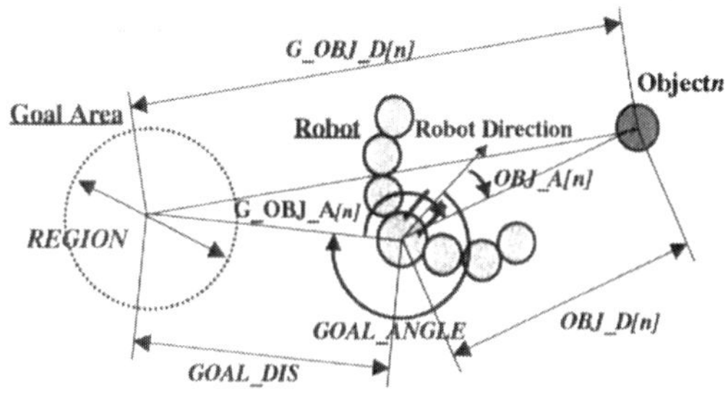

Fig. 3 Sensing variables: "*REGION*" = diameter of goal, "*GOAL_DIS*" = distance from center of goal area to *0*th cell, "*GOAL_ANGLE*" = angle from "*Robot direction*" to line joining centers of goal area and *0*th cell, "*G_OBJ_D[n]*" = distance from *n*th object to the center of goal area, "*G_OBJ_A[n]*" = angle between lines joining the center of *0*th cell, and those of goal area and *n*th object, "*OBJ_D[n]*" = distance from the *0*th cell to the *n*th object and "*OBJ_A[n]*" = angle from the robot direction to line joining centers of *0*th cell and *n*th object.

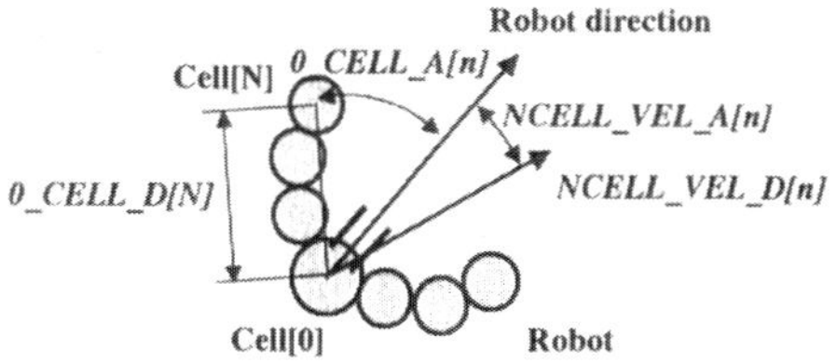

Fig. 4 State variables: "*0_CELL_D[n]*" = distance from the *n*th cell to the *0*th cell, "*0_CELL_A[n]*" = angle of the vector from the center of the *0*th cell to the center of the *n*th cell, "*NCELL_VEL_A[n]*" = angle of *n*th cell's velocity and "*Robot direction*", "*NCELL_VEL_D[n]*" = *n*th cell's velocity and the robot direction.

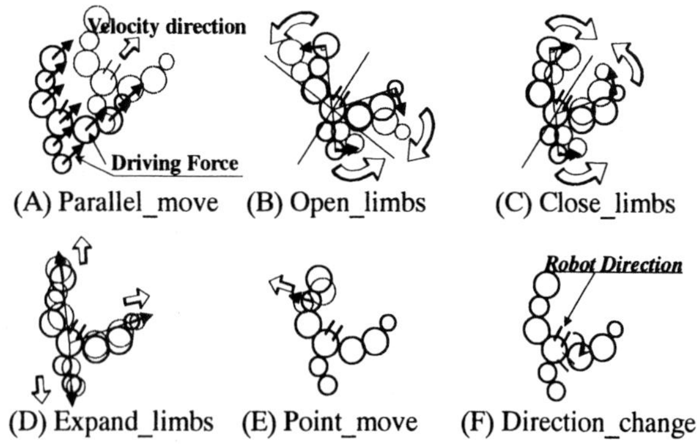

Fig. 5 Primitive behavior functions used in this paper

4. Genetic Operators

The GP enables individual robots to evolve its morphology and intelligence according to the fitness function. Under this artificial evolution, low-fitness individuals are eliminated by selection, and new individuals are produced using mutation and crossover. Then they develop their morphology and intelligence and converge to a reasonably optimal solution from generation to generation.

Each individual robot is generated according to the following two processes: The robot cell is divided N times at random generating a robot body comprised of the interconnected cells having various diameters. Intelligence is induced by splitting off the basic decision tree at random using a mutation. This basic decision tree represents the basic strategy shown in Fig. 6, which determines whether it goes to the goal (Action A) or to the object (Action B) depending on the geometrical relation between the positions of the robot, goal and object (Condition I). This basic strategy was found by our intuition.

Crossover is the operation to exchange the subtrees of two parent genes randomly selected. For mutation are used 3 kinds of operations in our GP. Mutation A is usual mutation to exchange a chosen node such as "sin", "if" and "1.5" to another node. Mutation B is almost same as Angeline's compression operation (Angeline, 1994), which is used to make a new ADF. Mutation C is the one for modifying the parent behavior function to increase its variety.

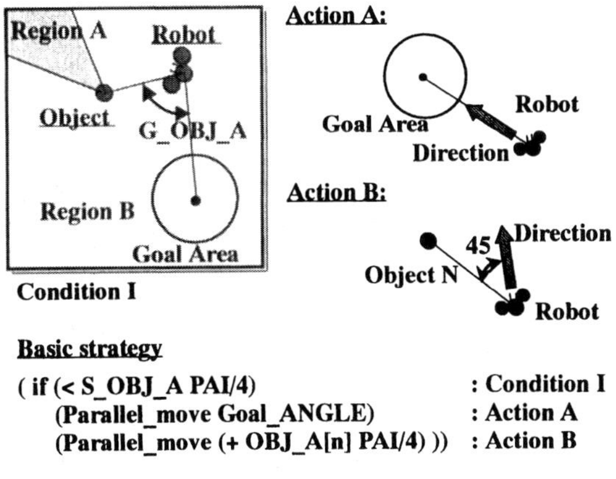

Fig. 6 Basic strategy

5. Robot Dynamics

Robot dynamics is used to calculate the movements of the robot and objects under the dynamic interaction between the robot and the environment including the objects. We took into account the effects of collision and friction in the virtual arena. The equations of motion for the connected cells were automatically generated from the morphology genes, where the connection conditions of cells, the impact collision force, the tensile force, the dynamic friction force and the driving force of the motor acting on the cell center are taken into account. Equations of motion for each of the cells and objects are numerically solved by a fourth order Runge-Kutta-Gill method (Abramowitz, and Stegun, 1972). In this work, the mass of a cell is 1.0 kg, the driving force of the motor is 3.0 N, the mass of an object is 1.0 kg, the diameter of an object is 0.2 m, the coefficient of friction is 0.2 and the number of cells is in the range of 2 to 32. The time step in Runge-Kutta-Gill calculations was carefully selected and was 33msec.

6. Simulation

We have simulated the three tasks shown in Table 3. In Pattern A, the robot is given the task to collect 3 objects to the goal area within 60 seconds. In Pattern B, the number of objects is 5, and in Pattern C, the task limit time is 120 seconds. Fig. 7 shows the object allocation in the initial stage. The diameter of the goal area is 1.2 m and the arena size is 7.0m x 7.0m. For taking into account the uncertainty in the object allocation, the initial position of objects is randomly perturbed with 5% of the object diameter. Sensing and state variables are sampled at 100 msec interval.

In each of three simulation cases, 10 trials were performed using the GP parameters such as population size 240, generation 120, selection ratio 0.3, crossover ratio 0.1 and mutation ratio 0.2. The roulette rule was used for selection. Note that individuals having over 1,000 nodes in their genotypes were eliminated by selection.

Table 3 Simulation cases

	Object number	*Task time[s]*
Pattern A	3	60
Pattern B	5	60
Pattern C	5	120

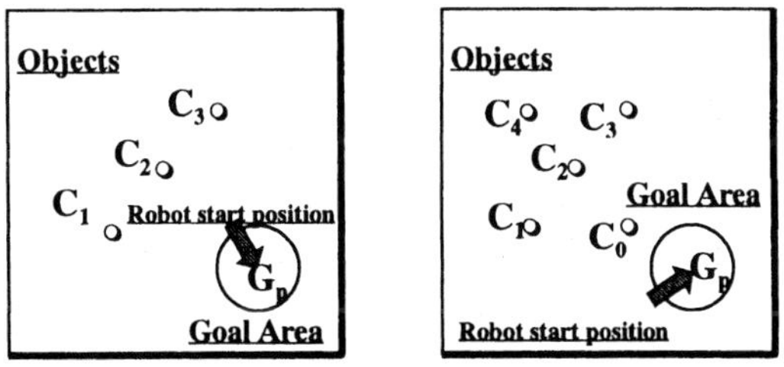

Fig. 7 Object arrangement: C_0 *(0.6, 0.6)*, C_1 *(-1.0, 0.6)*, C_2 *(-0.2, -0.2)*, C_3 *(0.6, -1.0)*, C_4 *(-1.0, -1.0) and* G_p *(1.4, 1.4).*

7. Results and Discussion

Fig. 8 shows the initial object allocation and the robot body morphology (see 0 sec), and the stroboscopic pictures (time step is 7 sec) of the robot trajectory. The robot morphology and behavior are the optimal one obtained by GP for Pattern A. The robot achieved the task using two basic strategies. In the first, the robot moved behind object C_1 (see Fig. 7), and pushed it to the goal area. In the next step, it collected the remaining two objects C_2, C_3 simultaneously using its whole body. This result suggests the following: Since it is physically difficult to push either one object or multiple objects using a single cell in a robot body, the intelligence that enables a robot to simultaneously push several objects by using just one cell is rarely gained. The robot tends to gain the intelligence to enclose and push objects by using several cells, resulting in the stable transport of objects. Fig. 9 shows the examples of development of robot body morphology through generations. The evolution in which the robot has gained a long limb explains the above behavior very well.

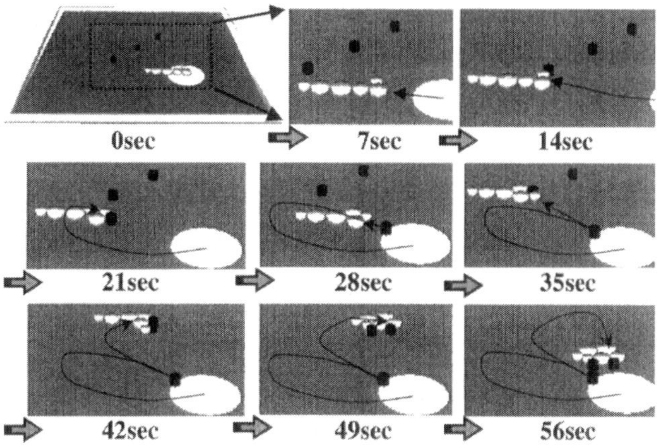

Fig. 8 Morphology and trajectory obtained for GP in Pattern A: Fitness = 0.961.

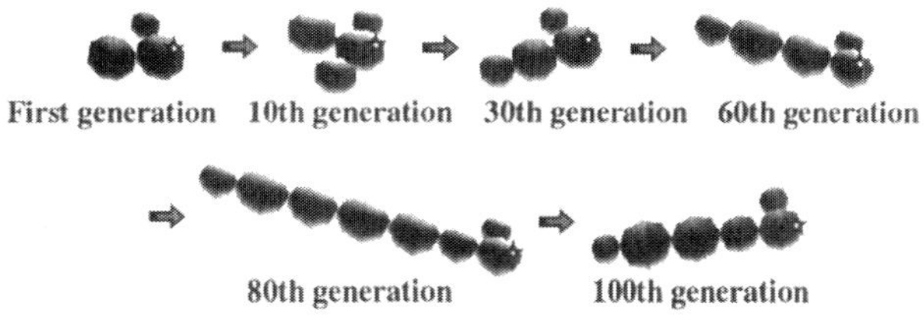

Fig. 9 Robot body morphology obtained at the highest fitness. Root cell is marked by star.

In Pattern B, the time simply to reach the 5 objects and return to the goal is estimated as about 78sec. Thus the robot can not accomplish the given task within the task limit time, 60sec, by collecting the objects one by one. The robot must generate different strategies to collect multiple objects simultaneously. When the best fitness was 4.18 in Pattern B, the robot gained the behavior to collect four objects simultaneously in such a way as shown in Fig. 10 (b), where the trajectory of the robot is shown by the arrow and the objects in the dotted rectangle were collected simultaneously. Fig. 11 shows how stable the gathering and pushing behavior resulted from enclosing the objects is.

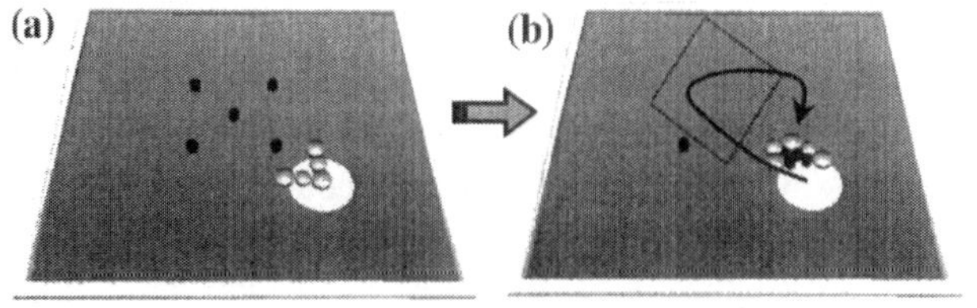

Fig. 10 Robot morphology and trajectory obtained by GP for Pattern B: (a) initial stage, (b) evolved stage, Fitness = 4.18.

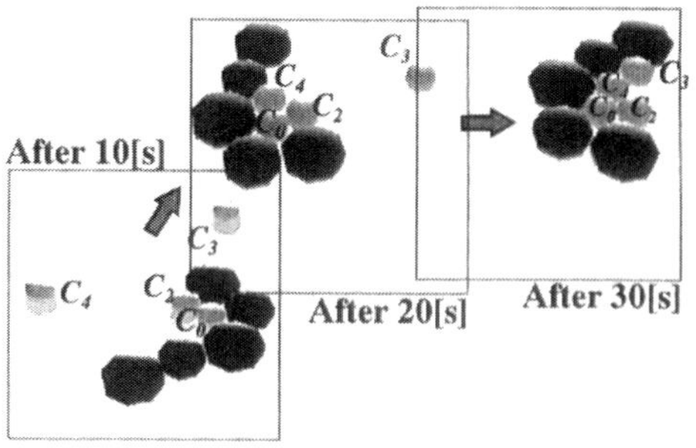

Fig. 11 Motion of gathering objects obtained for pattern B. Robot is black, and objects are gray. Fitness = 4.18.

Fig. 12 shows the result for Pattern C, where the robot accomplished the given task through the behavior of gathering objects one by one. In this case, the time constraint is not severe to the robot when considered the above discussion since the task limit time was 120sec. Thus in Pattern C, we found the various morphologies and intelligence due to the effect of a small uncertainty imposed on the initial object allocation. This suggests that the difference in the task conditions or environment conditions influences the dynamic interaction between the robot and environments and leads to various morphologies and intelligence in a robotic system.

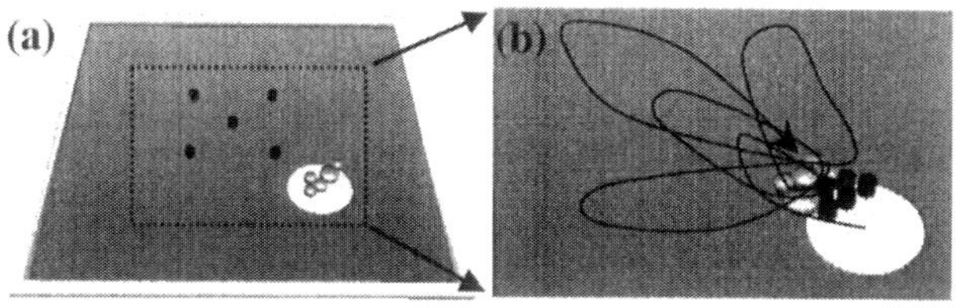

Fig. 12 Morphology and trajectory gained for Pattern C: Fitness = 0.984.

To examine the capability and flexibility of the robotic system obtained in the case of Pattern C, we set the unknown tasks to the robotic system in which the number of objects and their initial allocation were different from those in Pattern C. Fig. 13 shows the initial arrangements of the objects (see (a)) and the trajectories of the robot to collect the objects (see (b)) for 3 cases A, B and C. Case A is the task to collect three objects, which can be thought as a subtask of Pattern C. The robot accomplished the task in shorter time about 81sec. In Case B, the arrangement of the 5 objects is randomly shifted 200% of the object diameter. The robot accomplished the task for 74 trials among 90, which gives a high success percentage comparing with the success ratio of 76 trials over 90 in Pattern C. This indicates that the robot possesses a high adaptability against changes in environment conditions. Case C is the task to collect the 6 objects arranged differently. This was the most different task for this robot, but it succeeded in collecting the whole objects for 60 trials among 90.

Finally it should be pointed out the following: It is

essentially important for the robotic system to learn the interaction dynamics between itself and the environment for evolving the robot morphology and intelligence. But this is extremely difficult to achieve it when the evolutionary design starts from very primitive behavior functions or intelligence. Thus we provide the robot with the 6 primitive behavior functions and the basic decision tree shown in Fig. 5 and 6 which were built by hand from considering the interaction dynamics between the objects and the robot. Although the robot, composed of two cells and having only this basic intelligence, could not initially gather even two objects, the evolved robot did succeed eventually in gathering all objects.

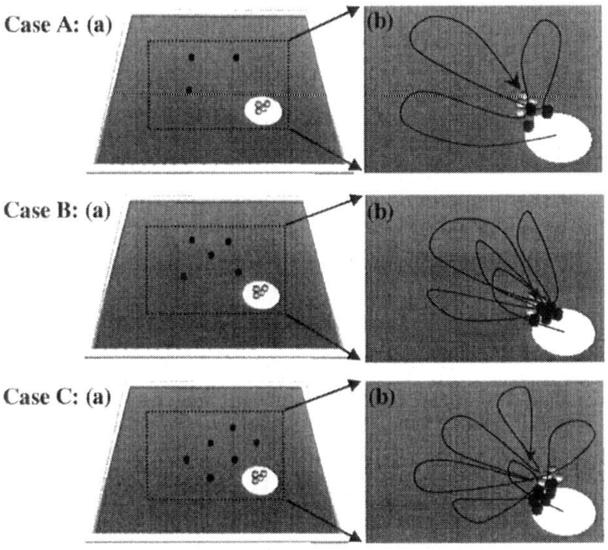

Fig. 13 Performance of 3 different tasks using the robot obtained for Pattern C: Fitness = (A) 0.680, (B) 0.937, (C) 0.910.

8. Conclusion

This work has demonstrated the simultaneous emergence of morphology and intelligence of robotic systems for a given task in a simulated physical world and the characteristics of the robot behaviors. The evolved robot has clearly exhibited various and interesting morphologies and intelligence generated through the dynamic interaction between itself and the environment. Then based on the robot performance in carrying out unknown tasks, we have shown the capability and flexibility of this evolved robot.

As some of future works, we will expand the evolutionary design to the case including sensory morphology in a robotic system. And then we will investigate the characteristics of emergence of the morphology and intelligence through analyzing the evolution history, those of the robotic system having behavior memories, and the balance of the complexity in the morphology and intelligence to establish the design principles of robotic systems.

Acknowledgments

This research work was carried out as part of the Research for the Future Project sponsored by the JSPS in 1996 – 1997 and the authors express their thanks to those involved in the program.

References

J.H.Holland, (1975), *Adaptation in Natural and Artificial Systems*, Univ. of Michigan Press.

J.Koza, (1994), *Genetic Programming II: Automatic Discovery of Reusable Problems*, ISBN, MIT Press, Cambridge, Massachusetts.

F.Hara and K.Kikuchi, (1994), Topology Optimization by Genetic Algorithm for Smart Structure Element, *Fifth International Conference on Adaptive Structures*, pages 439-442

K.Kikuchi and F.Hara, (1997), Simultaneous Optimization of Configuration and Assembly Process of Bridge Structure using a Genetic Algorithm. *JSME Trans.* 63-608C:1159-1166

K.Kikuchi and F.Hara, (1997), A Study of the Evolutionary Design of Structures Adapting to the Environment. *JSME Trans.* 63-613C: 3030-3037

C.Langton, editor, (1989), *Artificial Life*, Addison-Wesley,

K.Sims, (1994), Evolving 3D Morphology and Behavior by Competition, *Artificial Life 4*:28-39

J.Ventrella, (1994), Exploration in The Emergence of Morphology and Locomotion Behavior in Animated Characters, *Artificial Life 4*:436-441

J.Ventrella, (1996), Sexual Swimmers – Emergent Morphology and Locomotion Without a Fitness Function -, *From animals to animates 4*: 484-493

F.Gruau and K.Quatramaran, (1997), Cellular Encoding for Interactive Evolutionary Robots. *ECAL97*, MIT Press.

G.L.Steel Jr. (1990), *Common Lisp Language, second edition*, ISBN 1-55558-042-4, Digital Press,

P.J.Angeline, (1994), Genetic Programming and Emergent Intelligence, *In Advances in Genetic Programming*, MIT Press, :75-95

M.Abramowitz, and C.A.Stegun, (1972), *Handbook of Mathematical Functions with Formulas, Graphs, and Mathematical Tables*, 9th printing. New York

Mate Choice and NK-Landscapes

Patricio Lerena, Michèle Courant
Institut d'Informatique, Université de Fribourg
Chemin du Musée 3, CH-1700 Fribourg, Suisse
lerena@iname.com

Abstract

This paper presents preliminary results of a new attempt to understand the interaction between sexual and natural selection. Based on a simple model, computer simulations (in the form of "adaptive walks") are used to study that interaction, depending on the adaptive landscape characteristics of both a global viability-function and an individual-dependent preference-function, which are described using Kauffman's *NK* model. A first set of results is shown and discussed. Based on the special case presented the suggestion made is that, depending essentially on the relative weights given to viability constraints and to sexual-selection-imposed constraints, there is an optimal number of ("epistatic") interactions affecting the contribution of single preferred traits to the overall preference value of a potential mate.

1. Introduction

Sexual selection, as defined by Darwin, is the struggle between the individuals of one sex for the possession of the other sex (Darwin, 1859; Darwin, 1871). Two forms of sexual selection have been identified: (1) intrasexual: competition within a sex, such as males fighting for access to females, and (2) intersexual: selection between sexes, such as females choosing males for mating.

As Darwin already pointed out: "Amongst many animals, sexual selection will have given its aid to ordinary selection, by assuring to the most vigorous and best adapted males the greatest number of offspring." But this is not the only possibility. Maladaptive traits can evolve due to the effects of sexual selection. A well known argument was proposed by the population geneticist R.A. Fisher (Fisher, 1930), as an explanation for the stability of preferences for maladaptive traits (as it seems to be the case with the peacock's tail): suppose there is a polygamous population in which the majority of females have evolved a preference for long tails. A new mutant female that selects males with short tails will have sons with short tails, which are probably more viable. However they will have low fitness, because typical females will not choose them. This means that the presence of males with long tails, and of females preferring them, should be stable.

Numerous models have been proposed, that consider different factors relevant to the interaction between sexual and natural selection. A good review is given in (Markpatrick and Ryan, 1991). In that article, the authors present a classification of the proposed mechanisms for the evolution of mating preferences in polygynous animals, where three main classes of mechanisms are defined:

- Direct selection of preferences
  - Males provide resources to females or offspring
  - Costs of searching for mates
  - Selection against hybridization
  - Males differ in sperm fertility
  - Pleiotropic effects of preference genes
  - Disease or parasite transmission
- Indirect selection of preferences
  - Runaway process
  - Good genes: (i) Host-parasite coevolutionary cycles; (ii) Unconditionally advantageous mutations; (iii)Unconditionally deleterious mutations;
  - Genetic epistasis and dominance
  - Social system
  - Mutation pressure on trait
- Other mechanisms:
  - Random genetic drift
  - Group selection
  - Mutation pressure on preference

In the a-life community, individual based models have been proposed helping us to understand important characteristics of the coevolution of preferences and traits (Todd and Miller, 1996; Werner, 1996; Werner and Todd, 1997).

Many questions related to sexual selection are relevant to nearly all natural or artificial populations of genetically determined autonomous systems that use sex.

In this paper we present a model, that differs from the models mentioned above, together with some preliminary results. The main novelty introduced here is the modeling of the way single preferences for each of the different kinds of traits relevant to mate-choice interact to yield an overall mate-preference value. This work constitutes a preparatory phase of a project that attempts to capture self-organizing aspects in the interaction between natural and sexual selection and follows the line initiated by S. Kauffman who has shown the importance of self-organization in evolution. Our approach is based on the *NK* model (Kauffman, 1993) and it is also strongly inspired by the works of Todd, Miller and Werner.

After the model has been introduced, the simulation setup is described and a first set of experimental results is presented and discussed. The last section draws some conclusions and gives an outlook for future work.

2. The Model

Our aim is to address the problem of how the degree of epistasis[1] in both survival traits and, especially, in sexual preferences influences the course of evolution. The *NK* model constitutes an excellent framework for this question.

The *NK* model assigns to each of the *N* traits *t* of an organism, *K* other traits (among these *N*) which epistatically influence the fitness contribution of *t*, and assigns to every combination of values of *t* and its *K* "epistatic inputs" a random fitness contribution. The fitness of a whole genotype is then defined as the mean of the contributions of all traits.

This model is simple and powerful. Among the many properties of fitness landscapes it is able to consistently describe are: the number of fitness peaks in genotype space, the lengths of walks via fitter neighbors to fitness optima, the total number of mutants tried before an optimum is reached, the ratio of accepted to tried mutations on a walk, the number of alternative optima to which one genotype can climb, the number of genotypes which can climb to the same optimum, the rate at which the fraction of fitter neighbors falls to zero during walks to local maxima, the similarity of local optima.

For the purposes of this paper, the following facts about *NK*-landscapes are the most relevant. With increasing values of *K*, the ruggedness of the landscape increases, correlation among neighbors decreases and in particular, for neighbors of optima (its one-mutant variants), fitness drops become more and more precipitous. Values of *K* between *2* and *8* yield landscapes with the highest peaks and as *K* increases as a constant proportion of *N*, values may transiently increase but finally fall toward *0.5*. For example, in the case *N=24*, the mean values of local optima varying *K* are as follows (with standard deviations in parentheses): *.67(.03)* for *K=0*, *.71(.03)* for *K=2*, *.71(.04)* for *K=4*, *.69(.04)* for *K=8* and *.63(.03)* for *K=23*.

In the model introduced here, individuals are selected for both their viability and their mate-choice related features. An individual is represented as having two independent parts, taking the form of lists: (1) the traits and (2) the preferences. Both parts evolve. Traits are "evaluated" by a viability function and by the *preference* function of individuals in the population (see below). This yields, for each trait, a viability value and a *sexiness* value. As we will see, both functions are modeled using the *NK* model.

We can now present in detail the parameters of the model (see also fig. 1). N_v specifies the number of traits

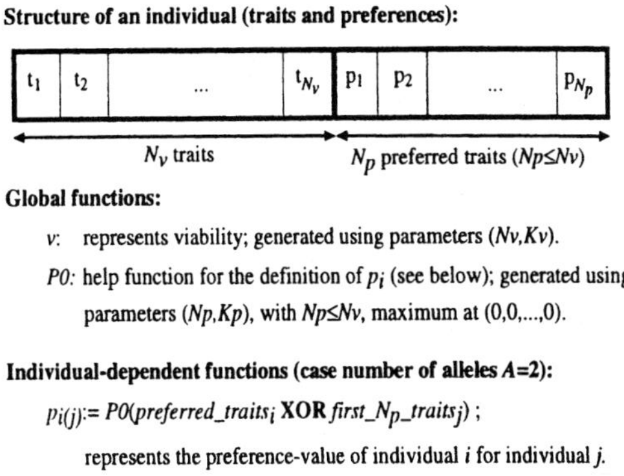

Figure 1 Elements of the model

every individual has. Each trait has K_v epistatic inputs (from the other traits). N_v and K_v are used to determine the viability function v, in the same way N and K are used to determine the fitness function in the original *NK*-model. This function v is static (does not change in time) and global (uniform for all individuals). In addition to that, individuals have N_p (with $N_p \leq N_v$) preferred trait-values (the preferences list mentioned above) each having K_p epistatic inputs. Parameters (N_p, K_p) are used to determine the intermediate function *P0*, which is used to define, for each individual, its *preference-function p* (this is explained in detail below). *P0* is also static and uniform (in the preliminary version presented here) and is chosen so that its global maximum is at the "zero list". Function *p* evaluates the sexual attractiveness by one individual to another by applying *P0* to the list of differences between the first N_p traits of the former and the respective preferred trait-values of the latter, where *0* represents equality and *1* represents a difference. As an example in the case of two-allele systems with $N_p=3$ and $K_p=0$, if we take *P0* as the zeros-counting function divided by N_p (*P0(000)=1*, *P(001)=P(010)=P(100)=2/3*, *P(011)=P(101)=P(110)=1/3*, *P(111)=0*), given a couple of individuals *j* with traits *000*, and *i* with preferences *010* (the list of differences is *000 xor 010 = 010*), this function would assign the value *2/3* to the preference $p_i(j)$ (because *P0(000 xor 010)=2/3*). The *sexiness* value *s* of an individual *i* is computed as the mean value over the population of preference-function applied to the traits of *i*.

Note that this way, even if the *landscape characteris-*

[1] In this paper, the term *epistasis* is used to refer, in any given system and with respect to any given function defined on it, to interactions of non-alternative elementary parts (in analogy to its sense of nonallelic gene interactions).

tics of the preference-functions p_i are static and uniform (the same for all individuals), preference-functions may differ by a "translation" in genotype space (individual preferences determine the position of the global maximum).

This first version of the model assumes that preferences are not under direct individual-level natural selection (changes in preferences of an individual do not affect its survival); there is no epistatic input from traits to preferences nor viceversa. Future extensions will be more realistic in this respect.

For simplicity reasons, in the rest of this paper, we concentrate on the case where

- the number of alleles is *2*,
- $N_p = N_v$ (and we can use N for short, omitting subscripts).

This last assumption is unrealistic, the number of preferred traits is too high with respect to the number of total traits (all traits can be "seen" by any potential mate), putting all traits under the direct action of sexual selection. (One could think of this case as modeling the evolution of the traits that are under the influence of both natural and sexual selection, under the assumption that the other traits are stable.) We think that the results obtained this way (for the experiments presented here) should qualitatively hold for lower values of N_p (relative to N_v). This will be tested in the immediate future.

We will consider that fitness contribution of the traits f_t, is given by the prouuct $v \times s$ (or its square root). The contribution of preferences will also be considered in the experiment presented and in our discussions below.

The model is probably too simple but it should capture at least some flavor of the idea that constraints in the form of the landscapes of the considered evolutionary forces can have a big influence in the final outcome.

Some remarks are opportune. If preferences present heritable variation with respect to K_p, then selection will probably favor values of K_p between *2* and *8*. Individuals having choosing-systems with such values of K_p, can discriminate better among their mates, because these values (1) offer good "contrast": peaks are high and (due to symmetry) valleys are deep; (2) allow for good discrimination in a neighborhood of a reasonable size and (3) yield a high enough correlation among neighbors, so that discrimination with respect to small differences is moderate. We will not discuss here about the range of values of K_v for which this argument holds, but what should be allways true, and for any value of K_v, is that the favored values of K_p are not greater than *8*.

Considering now the evolution of traits under no (or extremely weak) viability-constraints and assuming all values of K_p are equally represented in the population, they will tend to climb preferences with K_p values yielding the best discrimination (this way they obtain a good compromise between climbing-time and sexiness stability). This force (when natural selection is not present) would give traits a tendency to adapt to choosing-systems with K_p between *2* and *8*.

Finally, interference by sexual selection is expected to lead to a sub-optimal viability. There will be selective forces that favor values of K_p minimizing this interference.

3. The Experiments

Based on the experimental setup used in (Kauffman, 1993) which models the evolution of uniform populations, simulations (adaptive walks) have been made for different random examples of couples of NK landscapes $((N_v, K_v), (N_p, K_p))$. Considering the kind of phenomena we are modeling, this is an extreme simplification, but one can at least consider this setup as a basis for further reflexion.

At this point we should mention that in the experiments reported below, neither the effects of sex (sexual reproduction) nor of the potential presence of distinct sexes are taken into account. More detailed setups will be used to consider them in a near future.

To be able to integrate the concept of mate discrimination by preferences, trying to keep things simple at the same time, we will think of populations, during their adaptive walks (discrete steps), as being composed of two homogeneous classes respectively represented by the "previous" traits (the traits in the previous step) and the "new" traits (the traits in the current step). Previous traits will be still present in the population during the preference-step (described below), so that discrimination can take place.

Adaptive walks simulating the evolution of a given population, start with an individual having random initial traits and preferences. Every step consists of the following two successive sub-steps: (1) trait-step: keeping preferences unchanged, a randomly fitter one-mutant variant of the traits is chosen to become the new traits (should this not be possible, the sub-step is said to have *failed*) and then (2) preference-step (takes place only if the trait-step succeeds): a randomly zero- or one-mutant variant of the preferences is chosen among those (if any) that prefer the current traits (the ones just obtained) to the previous traits, and is taken as the new preferences (traits are kept unchanged during this sub-step); should that be impossible, the preferences minimizing the difference between the sexiness of the last step traits and the current ones are chosen as the new preferences. The walk continues until the trait-step fails. It is not excluded preferences being kept unchanged in a successful preference-step. Notice that in sub-step (2), we take into consideration the traits of the previous step.

The first sub-step seems natural and needs no further justification. The second sub-step, for simplicity, assigns a fixed and simple "evolutionary behavior" to changes in preferences, that represents a local optimization of the list of preferred traits with respect to the discrimination between low and high fitness individuals (with previous and present generations as unique basis of comparison).

Even if this simulation models a limit case (K_p cannot evolve, preference-evolutionary behavior is fixed, etc.), we consider it an interesting starting point. Let us remark that the fact that only small local optimizations (in cases where an optimization is being performed) are possible, together with the importance given to the history of traits in the co-adaptation of preferences and traits, should nevertheless belong to a general pattern. Understanding this kind of walks will possibly help us to build a more pertinent model.

For each set of model parameters experimented (N was set to *24* for all runs; K_v and K_p vary from *0* to *23*), 100 walks were carried out for 100 couples of NK landscapes, where tables representing viability are all different and randomly chosen and tables representing *P0* are randomly chosen from a pool of 10 tables.

Mean values over all runs have been taken for: viability, sexiness, fitness contribution of traits, and preferred-traits viability. They are shown in tables 1 to 4.

4. Results

The mean viability at the end of the walk decreases monotonically and quickly as K_p increases (see table 1); the maximum, varying only K_p, can be found for $K_p=0$; the maximum, varying only K_v keeping K_p fixed, is at $K_v=4$ or higher values of K_v; the highest values of K_v are the most "robust" with respect to the characteristics of preference-landscapes (viability is less sensible to it).

K_p	$K_v=0$	$K_v=1$	$K_v=2$	$K_v=4$	$K_v=8$	$K_v=16$	$K_v=23$
-	.67(.04)	.69(.05)	.70(.04)	.71(.04)	.69(.03)	.66(.03)	.63(.03)
0	.62(.05)	.65(.05)	.68(.04)	.70(.03)	.69(.04)	.66(.04)	.63(.03)
1	.61(.05)	.65(.04)	.67(.04)	.69(.04)	.69(.04)	.65(.03)	.63(.03)
2	.58(.05)	.64(.04)	.66(.04)	.68(.04)	.67(.04)	.65(.04)	.63(.03)
4	.59(.06)	.63(.04)	.65(.04)	.67(.04)	.67(.03)	.64(.03)	.62(.03)
8	.57(.05)	.61(.05)	.63(.05)	.65(.04)	.65(.04)	.63(.04)	.62(.03)
16	.54(.05)	.58(.06)	.59(.05)	.61(.06)	.63(.05)	.62(.04)	.60(.04)
23	.54(.06)	.56(.06)	.58(.06)	.58(.05)	.60(.05)	.61(.05)	.60(.04)

Table 1 Mean final viability values. The first row shows the results for walks without sexual selection.

Mean sexiness final values are shown in table 2. The maxima, varying only K_p, can be found for K_p between *2* and *8* for low and moderate values of K_v and between *4* and *16* for higher values of K_v. This is not very surprising, given the characteristics of local maxima of NK landscapes for these values of K and given that the level of sexiness is reflecting how good current traits and preferences match to each other. Increasing the values of K_v affects the sexiness value. This is because of the interference with viability constraints (see above).

Mean final f_t values are shown in table 3. Maxima varying only K_p are found between *1* and *4* for low and moderate values of K_v and *2* and *8* for high values of K_v.

K_p	$K_v=0$	$K_v=1$	$K_v=2$	$K_v=4$	$K_v=8$	$K_v=16$	$K_v=23$
0	.63(.06)	.61(.05)	.59(.05)	.58(.05)	.56(.06)	.53(.06)	.52(.07)
1	.65(.04)	.64(.04)	.63(.04)	.61(.04)	.58(.05)	.56(.05)	.55(.05)
2	.68(.05)	.67(.04)	.65(.04)	.64(.04)	.60(.05)	.57(.06)	.56(.06)
4	.69(.03)	.67(.04)	.67(.04)	.64(.04)	.62(.04)	.59(.05)	.58(.05)
8	.67(.04)	.66(.03)	.65(.04)	.64(.03)	.62(.05)	.60(.05)	.58(.05)
16	.64(.03)	.63(.03)	.63(.03)	.63(.03)	.61(.04)	.60(.04)	.58(.05)
23	.62(.03)	.62(.03)	.61(.03)	.62(.03)	.61(.04)	.59(.04)	.57(.04)

Table 2 Mean final sexiness values

K_p	$K_v=0$	$K_v=1$	$K_v=2$	$K_v=4$	$K_v=8$	$K_v=16$	$K_v=23$
0	.62(.03)	.63(.03)	.63(.03)	.63(.03)	.62(.04)	.59(.04)	.57(.04)
1	.63(.03)	.65(.03)	.65(.02)	.65(.03)	.63(.03)	.60(.03)	.59(.03)
2	.63(.03)	.65(.03)	.66(.03)	.66(.03)	.64(.03)	.60(.03)	.59(.03)
4	.63(.03)	.65(.03)	.66(.02)	.65(.03)	.64(.03)	.61(.03)	.60(.03)
8	.62(.03)	.63(.03)	.64(.03)	.64(.03)	.63(.03)	.61(.03)	.60(.03)
16	.59(.03)	.60(.03)	.61(.03)	.62(.03)	.62(.03)	.61(.02)	.59(.02)
23	.58(.04)	.59(.03)	.59(.03)	.60(.03)	.60(.03)	.60(.03)	.58(.02)

Table 3 Mean final values of (the square root of) the fitness contribution of traits.

Viability of preferred individuals (see table 4) falls quickly to *0.5* when K_p is increased. The degree of "alignment" preferences can have with respect to natural selection decreases when the complexity of the landscape imposed by sexual preferences increases. In the limit case, the "ideal" list of traits, the sexiest possible individual, is of extremely low viability.

K_p	$K_v=0$	$K_v=1$	$K_v=2$	$K_v=4$	$K_v=8$	$K_v=16$	$K_v=23$
0	.53(.06)	.54(.06)	.54(.06)	.53(.06)	.51(.06)	.51(.05)	.50(.06)
1	.52(.06)	.52(.06)	.51(.06)	.50(.06)	.49(.06)	.51(.06)	.50(.05)
2	.49(.06)	.50(.06)	.51(.07)	.50(.06)	.51(.06)	.50(.06)	.51(.07)
4	.49(.06)	.49(.07)	.50(.05)	.50(.06)	.50(.05)	.50(.06)	.49(.06)
8	.49(.06)	.51(.06)	.51(.06)	.50(.06)	.50(.05)	.49(.06)	.50(.06)
16	.48(.06)	.50(.05)	.49(.07)	.50(.05)	.51(.07)	.50(.06)	.51(.06)
23	.50(.06)	.49(.06)	.50(.06)	.51(.06)	.51(.06)	.49(.06)	.50(.06)

Table 4 Mean final preferred-traits-viability values

Notice that even if viability decreases with increasing K_p, this is not the case of f_t (the contribution of traits to fitness). Due to the contribution of the sexiness, f_t increases first reaching its maximum for K_p between *1* and *8*.

Some other forms adaptive walks are currently being tested, that vary the way preferences evolve (the preference-step) and the halting condition of the walk. These variants will allow us to better situate and interpretate these results. It is too early at the moment of writing this paper to correctly present them, so that we will only make a few comments on the results obtained so far.

For these other walks, the general patterns seen in the tables presented above are qualitatively the same. This includes the fact that with increasing K_p viability decreases (optimum for $K_p=0$); sexiness gets its maximum for moderate values of K_p and of course, f_t is maximized somewhere in between.

Furthermore, the results we are obtaining this way indicate that mean viability values reached at the end of the walks are reasonably insensitive to precise way preferences evolve; the values obtained in the experiment we have seen, are essentially valid for the adaptive walks we have tried. This property is important because it means we can obtain information about the value of K_p being used on the basis of the reached viability (and of the other parameters of the model) without detailed information about the evolution of preferences.

One should be careful with the interpretation of this results and with the significance given to them. The statements we make are restricted to the experiments presented and should not be extrapolated. In this paper, we do not intend to present general rules.

Returning to the tables, the simplest and most obvious observation is that high values of K_p are bad. Because both, viability and sexiness are implied, evolution will certainly avoid this values.

With respect to viability, we see that a very simple mate-choosing system, i.e. $K_p = 0$ (or even no mate-choice at all) is best. This is also a simple result.

With respect to f_t, it is important to see that, if all K_p were equally represented in the population, traits should tend to "follow" the preferences having the value of K_p that allows them to increase f_t most (K_p near 4).

Preferences, as already mentioned, will be selected at the individual level for having good discrimination capabilities ($K_p = 2,..,8$). But, in most situations, selection at the population level will tend to favor individual viability (when mate discrimination is not important for population survival), by selecting low values of K_p. In extreme cases, it could eliminate mate choice altogether supporting random mate, leaving a minimal capability to identify potential mates. This leads us to the conclusion that values of K_p between 0 and 8 can be selected for, depending on the relative weight of viability constraints and sexually imposed constraints (K_p between 2 and 4 being the optimal for reasonably "balanced" weights).

5. Conclusion

In this paper a first model was described, that attempts to simultaneously consider the effects of both the landscape characteristics of viability and of constraints due to sexual preferences. Preliminary arguments and results were presented suggesting that there can be an important influence not only of the landscape characteristics of viability constraints, but also of the characteristics of the landscape imposed by sexual preferences.

In particular, we have argued that even if viability decreases with increasing K_p, there should be selection for values of K_p usually ranging from about 2 to about 4. Lower values (between 0 and 2) are possible if viability constraints are extremely strong and higher (but not greater than about 8), if viability constraints are very weak.

The model presented and especially the simulation based on it are probably too simple and it is difficult, for the moment, to evaluate the relevance of the results obtained. A deeper study of the significant variants of the model is being carried out. Guided by this preparatory experiments, we are setting up an individual based model in which we will be able to consider, among other things, the detailed effects of sexual reproduction and will allow for traits and (especially) preferences to evolve freely. Extensions that consider the coevolution of two distinct sexes will be part of this study. The level of detail introduced will help us to determine in which situations the results obtained with models of the kind presented in this paper are qualitatively relevant and robust.

We think the landscape characteristics of viability constraints and of reproduction constraints should play an important role in the process of "speciation through sexual selection" (Todd and Miller, 1996). Using models based on the one introduced in this paper we could complement the experimentation carried out in those works. Finally, we are interested in adding parasites to our model to study their effects on sexual selection as an extension to the ideas presented in (Hamilton and Zuk, 1982).

References

Darwin, C. (1859). *On The Origin of Species*. London: John Murray.

Darwin, C. (1871). *The descent of man, and selection in relation to sex*. London: John Murray, 1871.

Fisher, R. (1930). *The Genetical Theory of Natural Selection*. Clarendon Press, Oxford.

Hamilton, W. and Zuk, M. (1982). Heritable true fitness and bright birds: a role for parasites? *Science*, 218:384–387.

Kauffman, S. (1993). *Origins of Order: Self-organization and selection in evolution*. New York: Oxford University Press.

Markpatrick, M. and Ryan, M. (1991). The evolution of mating preferences and the paradox of the lek. *Nature*, 350.

Todd, P. and Miller, G. (1996). Biodiversity through sexual selection. In *Proceedings of Artificial Life V*, pages 23–32.

Werner, G. (1996). Why the peacock's tail is so short: Limits to sexual selection. In *Proceedings of Artificial Life V*, pages 23–32.

Werner, G. and Todd, P. (1997). Too many love songs: Sexual selection and the evolution of communication. In *Proceedings of ECAL 97*.

Component-Based Adaptive Architecture with Classifier Systems

Sotaro Shimada, Yuichiro Anzai
Department of Computer Science, Keio University,
3-14-1 Hiyoshi, Kohoku-ku, Yokohama, 223, JAPAN
{shimada, anzai}@aa.cs.keio.ac.jp

Abstract

This paper proposes an architecture, called *genetic architecture*, for adaptive organisms to evolve its structures dynamically with environmental changes. This study is inspired by characteristics of animals' genetic systems that can be regarded as a set of switches to change their body structures through interactions with the environment. The genetic architecture consists of two parts; a task execution module and a structure decision module that correspond to *phenotype* and *genotype* respectively. The task execution module is responsible for tasks that the organism is faced with, whereas the structure decision module selects and produces components of the task execution module appropriately for the current situation. The task execution module employs software objects for structural components. The structure decision module consists of a classifier system to learn associations between situations and organisms' structures. We conducted some experiments, and results show validity of the genetic architecture.

1. Introduction

Structure represents functionality. Adaptation usually occurs as a result of structural changes of organisms. For example, a car's performance is improved by exchanging its engine. Computer software also will be better by upgrading its components. For animals, learning in central nerve systems is led by reinforcement of connections among synapses, or improvement of a physical action is caused by muscle enhancement. The difference between cars and animals is whether structural changes occur automatically or not. Our purpose of this study is to investigate an architecture that organisms enable to evolve its structure dynamically and automatically with environmental changes.

As for animals, sources of adaptation are embedded in genetic systems. Genome is not just a static blueprint of an animal, but a complicated set of switches to change its structure through interactions with the environment. The genetic system detects environmental states and produces appropriate proteins (or components). A cascade of producing proteins, finally, changes the animal's body structure to fit to the environment.

Classifier systems (Holland, 1992) have similarity with the genetic systems since basic ideas of the classifier systems are derived from "operon-operator model" (Jacob and Monod, 1961), and classifiers are evolved by using genetic algorithms. Classifier systems are parallel, message-passing, rule-based systems designed to avoid *brittleness* of traditional symbolic production systems. Classifier systems employ genetic-algorithm-based learning, and have been applied to various reinforcement learning problems, including gas pipeline control (Goldberg, 1983), the animat problem (Willson, 1987), sequential decision problems (Grefenstette et al., 1990), and adaptive control systems for real robots (Dorigo, 1995). However, these studies are mainly aimed at learning stimulus-response rules, not at representing organisms' structure.

On the other hand, Component-Based Software Engineering (CBSE) (Brown, 1996), where components are designed to replace with other components independently of remains of a system, is useful for creating structural components of artificial adaptive organisms. Object-oriented programming is the most popular approach for CBSE, and provides both selectivity and flexibility for connections among components. Class definition defines not only object functionality but also its connectivity to other classes. This connectivity has some flexibility since *polymorphism* and *dynamic binding* allow an object to connect with an instance of a subclass of the designated class at run-time. This works effectively when a certain method of a superclass is overridden in some subclasses: system behavior can be changed by replacing an object with one of other subclasses. This technique is called *object composition*, and realized by the following steps: (1) preparing components that implement a particular interface and (2) plugging these components into application frameworks (Fayad and Schmidt, 1997). Because of its modularity, combinational flexibility and selectivity, and message-passing property, objects (components) can be regarded as computational cells. Most of these studies, however, do not explicitly address the autonomous and adaptive composition of an organism.

We propose an architecture, called *genetic architecture (GEAR)*, for artificial adaptive organisms to evolve its structures dynamically with environmental changes. Genetic architecture employs a classifier system and software objects as the genetic system and components of organisms respectively. The remainder of this paper is organized as follows: Section 2 describes genetic architecture in detail. Section 3 demonstrates some experiments with the architecture, and also describes Intelligent Sensor-Actuator Networks that is an application extensively studied in our laboratory. Section 4 discusses some issues and possible future directions.

2. Genetic Architecture

2.1 An Overview

Genetic architecture (GEAR) borrows basic concepts from genetic systems of animals. Genetic architecture consists of task execution module (TE) and structure decision module (SD) that correspond to *phenotype* and *genotype* respectively (Figure 1). TE is composed of a set of objects that are used to solve problems that the organism is faced with. SD selects and produces objects of TE appropriately for the current situation. A classifier system is used as a reasoning and learning mechanism embedded in SD.

Our basic scenario is as follows:

1. State changes of objects in TE are detected by SD and encoded into messages. The state changes are due to both external stimuli (environmental changes) and internal interaction among objects.
2. Messages are manipulated by the classifier system. SD selects appropriate objects according to outputs of the classifier system and produces them in TE.
3. TE adds produced objects at relevant places in it or replaces older ones with the objects.
4. Productions of objects are also detected by SD and encoded into messages. A cycle continues (return to 2) until no object is produced. As a result, TE structure changes for the current situation.
5. The classifier system learns by a genetic algorithm with payoffs from the environment. Payoffs are given at each structural change. Finally, SD obtains appropriate rules of relations between structures and situations.

To accomplish this, TE must provide a framework to exchange objects dynamically and flexibly, and the classifier system of SD must represent structures of an organism. The following subsections describe each in more detail.

2.2 Structure Decision Module

This subsection describes some strategies used in the classifier system.

Representation of classifiers Classifiers are designed to represent object-oriented structures of TE that changes as a result of object productions. To accomplish cascades of object productions, we prepare five kinds of messages for the classifier system (Figure 2):

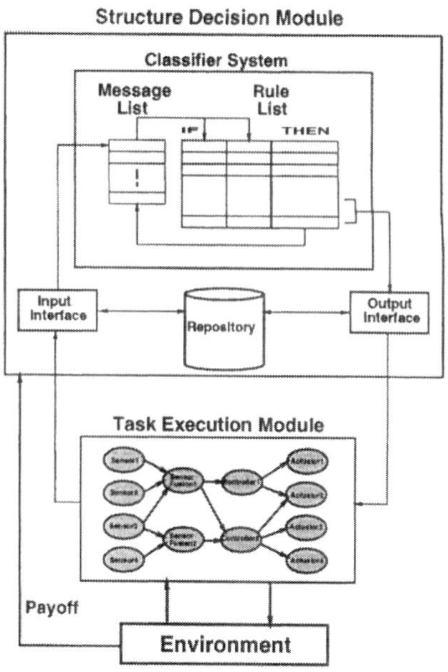

Figure 1 Genetic Architecture

- *Input messages* represent state changes of TE objects detected by the input interface.
- *Internal messages* mediate communication among classifiers, and are the same as of the standard classifier system.
- *Output messages* represent a pair of a certain object and a certain domain; the object will be produced in the domain[1] (See Section 2.3). Mappings between an object ID in the output message and an object reference (or pointer) in TE are stored in *repository* (Figure 1) as well as a domain and its corresponding ID.
- *Production messages* represent a production of an object and are made by the output interface. A production message has the same string as an output message that causes the production, with a tag changed to indicate the production message (Figure 2).
- *Stop messages* stops production cycles. That is, if a stop message is posted to the message list, then the cycle finishes and the message list is reset.

Each Condition of a classifier is one of {*input, internal, production*}, while an action, {*output, internal, stop*}.

Production cycle Basic execution cycle is the same as of the standard classifier systems (Holland, 1986)(Willson, 1987). The difference is that the input interface must ob-

[1] It is correct to say that an output message means that a binding between an object and a domain is established, rather than an object is created in a domain.

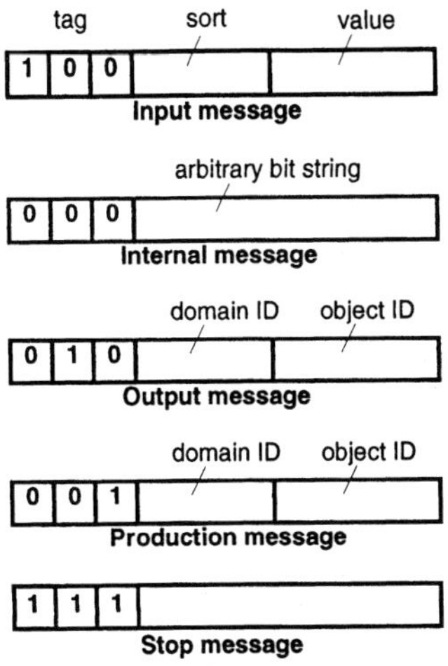

Figure 2 Message formats

C1:	input1	input2	→	internal1
C2:	input3	internal1	→	internal2
C3:	internal2	internal2	→	output1
C4:	internal2	production1	→	output2
C5:	internal2	production2	→	stop

Figure 3 A schematic example (a classifier consists of two conditions and an action). Suppose that an environmental state is {input1, input2, input3}. As a result of the cascade ($C1 \to C2 \to C3 \to C4 \to C5$), two objects are created for this situation. Thus, cascades of producing objects, like cascades of gene activation in animals' genetic systems, can be realized.

serve TE objects instead of the environment[2]. The input interface posts messages only when one or more objects change their state. Another difference is that the output interface creates an object in a certain domain instead of executing some action directly to the environment, and posts a production message to the message list. Object production cycle continues as far as objects are created. Figure 3 shows a schematic example.

2.3 Task Execution Module

Task execution module (TE) provides an object-oriented framework based on object composition to exchange its components flexibly. However, we must solve some problems when object composition is applied to genetic architecture:

- When an object is added to TE, connections between the object and an existing object around it must be established. But how does the object know which objects should be connected to, especially when their class relation is many-to-many?
- Or what should the object do if the class relation has cardinality constraints, meaning that the number of objects that can be connected at the same time is restricted?

We introduce *domain classes* to solve these problems. Domain classes are defined for each class relation, or *pattern*, and manage local connections among objects taking account of cardinality constraints. An object must be produced in a certain domain, for establishing appropriate connections among objects. Objects can belong to several domains at the same time as shown in Figure 4. Note that domains have no effect for task execution: domains are responsible only for the assignment of object references. Thus, domains make no restriction on organisms' structures.

Consider, for example, the situation illustrated in Figure 4. There are Sensor objects and SensorFusion objects that can be connected with. Since their class relation is many-to-many, domains must manage connections among them. In this case, domains are assigned to each SensorFusion object to divide the many-to-many class relation into one-to-many. Suppose that a new Sensor object (*new_sensor*) is produced in *domain1*. Then, *domain1* takes care of the Sensor object and makes connection between the Sensor object and the SensorFusion object, while *domain2* does nothing.

3. Experiments

The experiments described here concern a particular adaptive system called *Intelligent Sensor-Actuator Networks (ISANs)*. This section provides background knowledge of ISANs, and demonstrates some experiments with the genetic architecture.

3.1 Intelligent Sensor-Actuator Networks

There is a growing interest in distributed sensor-actuator networks, including autonomous building management

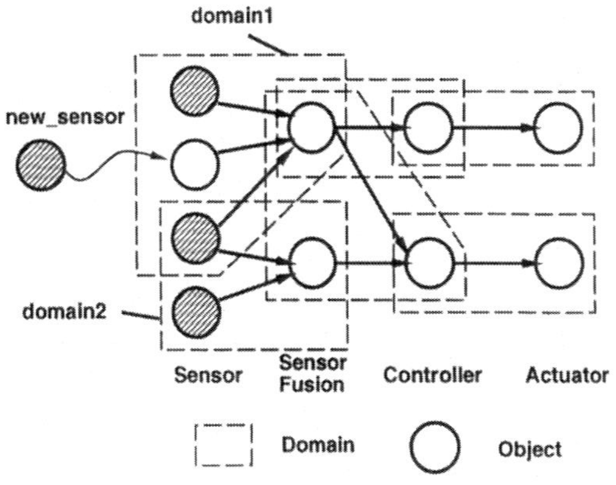

Figure 4 An example of domains

[2] It is too costly to check every object at each cycle, or it is not necessary to do that. So the genetic architecture allows to register some objects to be checked by the input interface.

systems, chemical plant control systems, and reconfigurable traffic systems[3] (Williams and Nayak, 1996). Intelligent Sensor-Actuator Networks (ISANs), that is an application extensively studied in our laboratory, are widely distributed adaptive sensor-actuator networks (Kohno and Anzai, 1997) (Shimada et al., 1997).

An ISAN is a network of many kinds of sensors and actuators equipped in the real-world environment, such as a room, an office, a building, or a city. An ISAN is an object-oriented system, and must be adaptive and reactive since it is faced with the real-world uncertainty. Behavior of the ISAN is defined by combinations of objects, i.e., Sensors, SensorFusions, Actuators, and Controllers. From the animats viewpoint, an ISAN can be regarded as a body consisting of robotic nerves, muscles, cells, and so on, that is too large and complicated to design relevant combinations a priori. To adapt to environmental changes, ISANs' structure (object combination) must be modified dynamically. Thus, learning appropriate combinations of objects is an essential problem for ISANs.

3.2 Example: Elevator Group Control

Here we consider a relatively simple example of ISANs: an elevator group control system (EGC). Complex needs for the transportation in large-scale buildings require an adaptive control of a limited number of elevator cars (Imasaki et al., 1995). The objectives of EGC are to control appropriate elevator cars to reduce waiting time, to save energy, and to satisfy users' needs, such as preferential floors (Fujino et al., 1995). We examined validity of the genetic architecture with simulations of EGC.

3.3 Simulation Environment

We suppose a lunch-time traffic at a typical office building. User arrival rates are assumed beforehand, and an arrival is generated by using random numbers. The simulation duration is 60 minutes in which the elevator utilization situation changes each 20 minutes as shown in Table 1. The simulation step is 1 second.

Details of the simulation settings are as follows:

- number of elevator cars: 4 cars
- number of floors: 15 floors
- speed of elevator cars: 1 second to move one floor
- capacity of an elevator car: 20 persons

3.4 TE Components

TE Components are Controllers, Elevators, and a Sensor. There are three kinds of Controllers:

- *1st floor preferential.* This Controller controls elevators to give priority to 1F.
- *Top floor preferential.* This Controller controls elevators to give priority to the top floor.

[3] Also robots can be considered as a special case of sensor-actuator networks.

Table 1 Elevator utilization situation changed with time zones.

Time (min)	Src (F)	Dest (F)	Rate (persons/min)
0 - 20	2 - 15	1	100
20 - 40	1 - 15	1 - 15	10
40 - 60	1	2 - 15	100

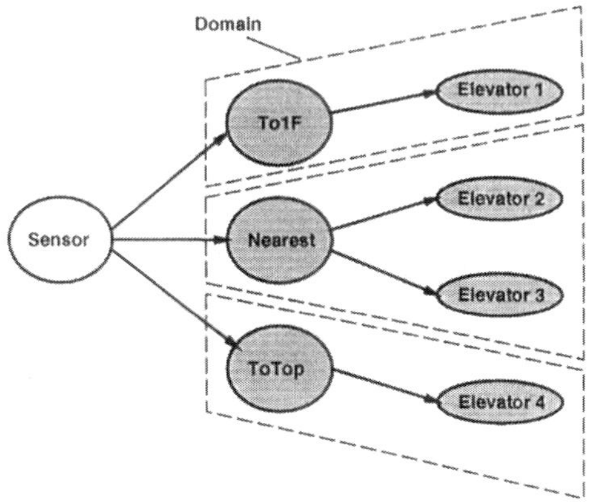

Figure 5 Components of EGC

- *Nearest hall-call floor preferential.* This Controller controls elevators to give priority to the nearest hall-call floor.

Each controller controls elevator cars that are connected to it. The objective of learning is to acquire representation of structural patterns of Controllers and Elevators appropriate to elevator utilization situations.

Domains are designed to manage connections among Controllers and Elevators, and assigned to each Controller (Figure 5). Elevators are assigned exclusively to domains; they *are not* shared by two or more domains. A Sensor is connected to every Controller in advance and checked by the input interface to generate messages for the classifier system.

3.5 Implementation Issues in SD

This subsection discusses some implementation issues of the classifier system.

Message coding To simplify, the Sensor object provides information only about time zones, ranging from 0 to 60 in increments of 10. The message length is 7 bit contained tags for 3bit. Output and production messages contains a domain ID for 2bit, and an object ID for 2 bit. Internal messages are not used in the experiments. Each classifier has two conditions and an action.

Learning algorithms GA is called at the end of the day. Elitist and binary tournament selection (Back, 1996) are used for GA strategies.

Every graph shows the mean performance of 10 runs.

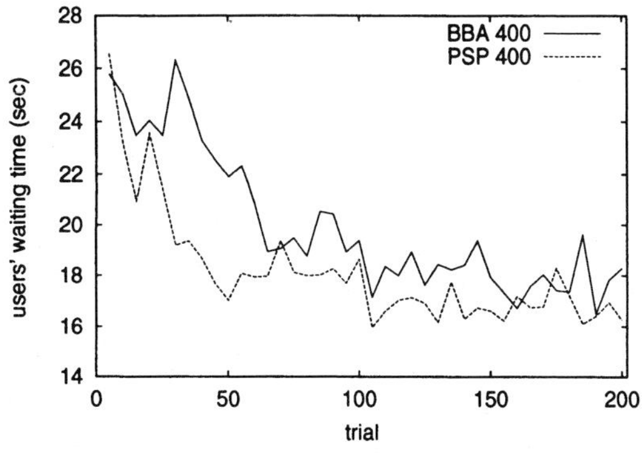

Figure 6 Effects of credit assignment algorithms

3.6 Results

Effects of credit assignment We conducted an experiment to compare performance of a bucket brigade algorithm (BBA) and a profit sharing plan (PSP) as credit assignment[4]. Since the performance of the genetic architecture is heavily upon how fast classifier chains are obtained, it is essential to examine which credit assignment algorithm is suitable.

The evaluation functions are defined as follows:

$$BBA : P(t+1) = C - wait(t)/user(t)$$
$$PSP : P(t+1) = C - wait(t)/user(t) + 50$$

where wait(t) is users' waiting time, user(t) is a number of arrival users, and C is a constant value (in the experiments, 40). The population of classifiers is 400, and initial strength of a classifier is 50.

Figure 6 shows average users' waiting time of both BBA and PSP during 200 days. PSP demonstrates better performance and converges faster than BBA. This means that PSP obtains relevant chains through the execution. The reason is that PSP reinforces a chain of classifiers as a line while BBA do as a sum of local connections (between two classifiers). Also this result is consistent of previous studies, such as (Grefenstette, 1988).

Chain of classifiers Figure 7 is an extracted set of classifiers after 200 days using PSP as credit assignment. First 3bit of each message represents a tag and next 4 bit contains information. An input message (begin with "100") represents a time zone, for example, "1000011" means "the current time zone is 30-40". An output message (begin with "010") consists of a domain ID (next 2bit) and an object ID (last 2bit), for example, "0101000" means "elevator car 0 is assigned for domain 2". A production message is created by the output interface with a tag "001", for example, an output message "0101000" is transferred to an production message "0011000", and posted to the message list. We see from Figure 7 that a cascade of object productions is triggered by input messages, such as "1000000". Note that a cascade may not be finished with a stop message when there is no classifier that matches with production messages.

[4] Detailed explanation can be found in (Grefenstette, 1988).

Table 2 The structure of EGC changes with time zones. 1st means 1st-floor preferential controller, Top as top-floor preferential, and Near as nearest hall-call floor preferential. # means one of them.

Time (minutes)	EL0	EL1	El2	EL3
0 - 20	Near	Top	#	Near
20 - 40	Near	Near	Near	Top
40 - 60	Near	Near	1F	1F

```
100#00# 100#### → 0101000 dm2 el0
100#00# 0011000 → 0100101 dm1 el1
100#00# 0010101 → 0101011 dm2 el3
100#00# 0011111 → 1110000 stop

100#01# 100#### → 0100111 dm1 el3
100#01# 0010111 → 0101010 dm2 el2
100#01# 0011010 → 0101000 dm2 el0
100#01# 0011000 → 0101001 dm2 el1

100#10# 100#### → 0100011 dm0 el3
100#10# 0010011 → 0100010 dm0 el2
100#10# 0011111 → 1110100 stop
```

Figure 7 Chain of classifiers obtained after 200 days (using PSP for credit assignment): (dm2 el0), for example, means that elevator0 is assigned to domain2.

Table 2 demonstrates that the structure of EGC changes with time zones, and corresponds to Figure 7. In the time zone 0-20, it is suitable to disperse elevator cars because users arrive at 2nd-15th floors. Top floor preferential is effective because nearest hall-call floor preferential tends to serve over lower floors. (Note that users are going to 1F). On the other hand, 1st-floor preferential is suitable in the time zone 40-60, since users arrive only at 1F. The results show that genetic architecture learns rules to change the structure of EGC appropriate to elevator utilization situations.

Under dynamic environment Genetic architecture's scalability for the dynamic environment is, then, examined. Basic settings are the same as of the previous experiments, but after 200th day, the utilization situation of time zone 0-20 is replaced with 40-60, and vice versa (Table 3). Then, it changes back as before after 300th day (Table 1). Note that the classifier system is not aware of the change of the environment itself. PSP is used as credit assignment, and classifier populations are 200 and 400.

Figure 8 shows results of the experiment. The genetic architecture with classifier population 400 can follow the

Table 3 Elevator utilization situation after 200th day

Time (min)	Src (F)	Dest (F)	Rate (persons/min)
0 - 20	1	2 - 15	100
20 - 40	1 - 15	1 - 15	10
40 - 60	2 - 15	1	100

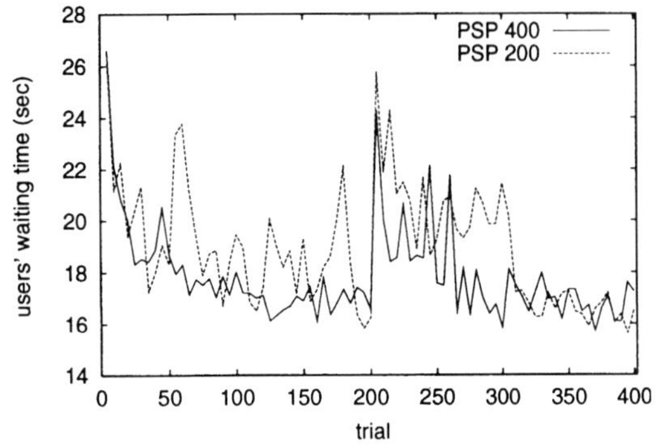

Figure 8 Performance in dynamic environment

change of the environment itself, while the population 200 cannot. The reason is that the classifier system requires enough population size (more than 200) to adapt to another environment. Note that even the population 200 can adapt to the environment when it changes back after 300th day, since it maintains rules for the previous environment.

4. Discussion

To see the basic performance of the presented genetic architecture, we conducted experiments with a relatively simplified EGC. However, adaptive systems are usually faced with more complex environments. Thus, further study is required to investigate the scalability of the genetic architecture for the message length, numbers of components and domains, the number of sorts of input messages, and so on. As we see from Figure 8, the ability of the genetic architecture depends heavily upon the population size of classifiers. Thus, we need more exploration of effective use of classifiers or classifier representation that enhances reusability for other environments. We also should develop a systematic methodology to analyze classifiers.

Genetic architecture employs the classifier system to obtain representation of adaptive object-oriented systems, while other classifier system studies generally aim at obtaining stimulus-response rules. The major difference is that a classifier system of other studies is a rule base that is a part of the organism, whereas the classifier system of genetic architecture is an entire design of the organism: the structure of the organism is completely represented in the classifier system. This indicates possibility to classify organisms' types by comparing structures of the classifiers.

Acknowledgment

This work is supported by the grants from "Research for the Future" Program of Keio University, under the auspices of Japanese Society for the Promotion of Science.

References

Back, T. (1996). *Evolutionary Algorithms in Theory and Practice.* Oxford University Press.

Brown, A. (1996). *Component-Based Software Engineering.* IEEE Computer Society Press.

Dorigo, M. (1995). Alecsys and the autonomouse: Learning to control a real robot by distributed classifier systems. *Machine Learning*, 19:209–240.

Fayad, M. and Schmidt, D. (1997). Object-oriented application frameworks. *Communications of the ACM*, 40:32–38.

Fujino, A., Tobita, T., Segawa, K., Yoneda, K., and Togawa, A. (1995). An elevator group control system with floor attribute control method and system optimization using genetic algorithms. In *IECON*, pages 1502–1507. IEEE.

Goldberg, D. (1983). Computer-aided gas pipeline operation using genetic algorithms and machine learning. Doctoral dessertation, Department of Computer and Communication Sciences, University of michigan.

Grefenstette, J. (1988). Credit assignment in rule discovery systems based on genetic algorithms. *Machine Learning*, 3:225–245.

Grefenstette, J., Ramsey, C., and Schultz, A. (1990). Learning sequential decision rules using simulation models and competition. *Machine Learning*, 5:355–381.

Holland, J. (1986). Escaping brittleness: The possibilities of a general purpose machine learning algorithm applied to parallel rule-based systems. In *Machine Learning II*, pages 593–624.

Holland, J. (1992). *Adaptation in Natural and Artificial Systems.* The MIT Press.

Imasaki, N., Kubo, S., Nakai, S., and Yoshitsugu, T. (1995). Elevator group control system tuned by a fuzzy neural network applied method. In *Proceedings of IEEE International Conference on Fuzzy Systems*, pages 1735–40.

Jacob, F. and Monod, J. (1961). Genetic regulatory mechanisms in the sysnthesis of proteins. *Mol. Biol.*, 3:318–356.

Kohno, M. and Anzai, Y. (1997). Intelligent sensor network as nervous system of artifacts. In *Proceedings of International Symposium on System Life*, pages 109–115.

Shimada, S., Yamamoto, Y., and Anzai, Y. (1997). A fundamental mechanism of intelligent sensor-actuator networks. In *Proceedings of the Seventh International Conference on Human-Computer Interaction*, pages 201–204.

Williams, B. and Nayak, P. (1996). Immobile robots: Ai in the new millennium. *AI Magazine*, 17(3):16–35.

Willson, S. (1987). Classifier systems and the animat problem. *Machine Learning*, 2:199–228.

Evolution of protean behavior in pursuit-evasion contests

Mattias Wahde*, **Mats G. Nordahl****
* Nordita, Blegdamsvej 17, DK-2100 Copenhagen, Denmark
wahde@nordita.dk
** Institute of Theoretical Physics,
Chalmers University of Technology,
S-412 96 Gothenburg, Sweden
and
Santa Fe Institute, 1399 Hyde Park Road,
Santa Fe, NM 87501, USA
tfemn@fy.chalmers.se

Abstract

We have studied the evolution of evasion behavior in games of pursuit and evasion, with asymmetries between the pursuer and evader. The introduction of such asymmetries favors the emergence of more complex evasion strategies. With a pursuer capable of slightly higher top speed but smaller maximal acceleration than the evader, examples of complex and unpredictable (protean) behavior were found for successful evaders. In particular, some of the evaders made repeated sudden direction changes which confused the pursuer and allowed the evader to escape. Our simulation methods are described in some detail, and we also analyze and discuss the evasion strategies discovered.

1. Introduction

Protean (i.e., unpredictable) behavior occurs frequently in nature, and especially when animals are chasing each other. The ability to carry out seemingly random motions is crucial for the survival of animals which function as prey to other animals since, in many cases, predators are capable of moving faster than their prey.

Protean behavior is also desirable in several technical applications. For instance, unpredictable behavior (to the extent that it is possible) can increase the ability of a fighter jet to outmaneuver a pursuing missile. A missile using a fixed strategy will clearly have great trouble tracking an unpredictably moving target.

Thus, the study of protean behavior is interesting from several points of view.

Clearly, the neural architectures of even the most primitive animals are much more complex than those one can, at present, study by means of computer simulations. Nevertheless, the different types of dynamics and neural architectures obtained in evolutionary simulations can hopefully shed some light on the basic features of proteanism.

Pursuit-evasion behavior has previously been studied in several different contexts: differential game theory (Isaacs, 1965), behavioral ecology and neuroethology, and robotics. A nice review which argues convincingly for the importance of pursuit-evasion problems can be found in Miller and Cliff (1994). Coevolutionary simulations of pursuit-evasion have been performed, e.g., by Cliff and Miller (1996), and the problem has also been studied with real robots (Floreano and Nolfi, 1997). However, the possibility of emergent unpredictable behavior clearly needs to be studied further, in particular its dependence on asymmetries between the pursuer and evader. This contribution is intended as a first step in this direction.

Unpredictable behavior has been found to evolve in simpler settings such as discrete games — one example is the rock-paper-scissors game with deterministic strategies of finite memory, where simulations similar to those in Lindgren (1991) show an evolutionary trend towards higher memory strategies, which attempt to deterministically emulate a simple random choice.

In this contribution we concentrate on evasion behavior, by using fixed pursuer strategies and letting the evaders evolve. The pursuers are either extracted from coevolutionary simulations, or designed by hand.

2. Methods

In order to study protean behavior, we have written a simulation code in which evading creatures (hereafter *evaders*) with neural network brains are evolved against pursuing creatures (hereafter *pursuers*).

2.1 Dynamics and genetic representation

The artificial creatures used in the simulations are controlled by recurrent continuous-time neural networks, with neurons described by the equation

$$\tau \frac{\mathrm{d}x_i}{\mathrm{d}t} = -x_i + \sigma\left(\sum w_{i,j} x_j\right) + I_i, \qquad (1)$$

where x_i denotes the activation level of neuron i, and I_i is the visual input (if any). σ is the neuron activation function, which takes values between 0 and 1. τ is a time constant, which takes the same value for all neurons. Information about the surroundings of a creature is fed to it by means of visual input neurons. The input signals propagate through the network and produce motor signals, which are fed to muscles that control the movements of the creature. The equations of motion are

$$m\frac{\mathrm{d}v}{\mathrm{d}t} + c_l v = k_l \frac{(M_\mathrm{l} + M_\mathrm{r})}{2}, \quad (2)$$

$$\frac{\mathrm{d}s}{\mathrm{d}t} = v, \quad (3)$$

$$I\frac{\mathrm{d}\dot{\theta}}{\mathrm{d}t} + c_a\dot{\theta} = k_a \frac{(M_\mathrm{l} - M_\mathrm{r})}{2}, \quad (4)$$

and

$$\frac{\mathrm{d}\theta}{\mathrm{d}t} = \dot{\theta}, \quad (5)$$

where m, c_l, k_l, I, c_a, and k_a are constants specified in the input data, and M_l and M_r are the left and right motor signals, respectively. s denotes the curvilinear coordinate along the direction of motion of the creature, which is denoted by θ. The neurons are arranged in layers. However, not only forward connections are allowed, but also connections within a layer and even backward connections. This allows the network to have nontrivial internal dynamical behavior.

The neural network brains of the creatures are generated by means of a growth process reminiscent of that in Cliff and Miller (1996). The parameters of this process are obtained from chromosomes consisting of 13 genes (which take integer values between 0 and 9) per neuron. The first bit codes for the layer in which the neuron is situated, and bits 2-4 determine its horizontal position. For input neurons, bits 5-6 code for the opening angle, i.e. the circle sector in which the neuron receives visual input. The visual input equals the fraction of the circle sector occupied by the body of the other creature. In Fig. 6, the visual neurons (bottom row) are situated along a line, which is mapped to the circular body of the creature. Thus, visual input neurons can point in any direction. For non-input neurons, bits 5-6 code for the direction of neural growth (which is vertical for input neurons). Bits 7-8 code for the opening angle of the neural growth cone, i.e. the circle sector in which the neuron makes connections to other neurons. This circle sector is centered on the growth direction defined by bits 5-6. The connection weight at distance r from the neuron body is given by the parameters w_0 (determined by bits 9-11) and l_{scale} (determined by bits 12-13) according to

$$w = w_0 \mathrm{e}^{-r/l_{\mathrm{scale}}}. \quad (6)$$

Thus, the weights of the network are *not* coded in the chromosome. Instead, they result from the growth process described above. The parameter w_0 can be either positive or negative, corresponding to an excitatory neuron or an inhibitory neuron, respectively.

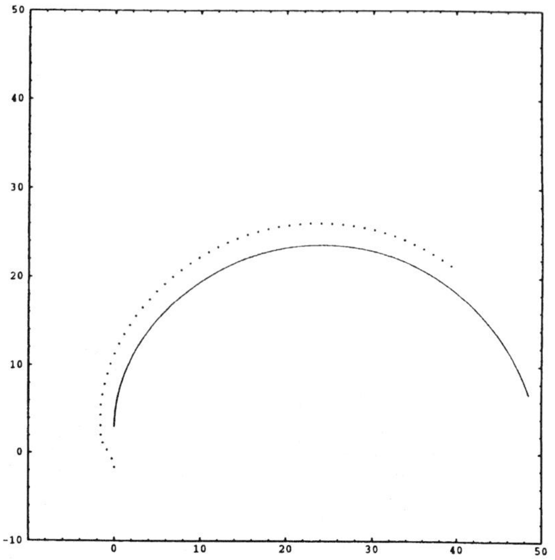

Figure 1 The orbit of an evader (solid line) escaping from a neural network controlled pursuer (dotted line).

2.2 Pursuer strategies

We tried two different methods of generating pursuers for our pursuit-evasion contests.

In the first method, pursuers and evaders were coevolved, and the pursuers with the highest fitness values were tested until a pursuer (the *neural network pursuer*) capable of tracking evaders on several different test trajectories was obtained.

In the second method (the *fixed strategy pursuer*), the pursuers were not controlled by a neural network. Instead, they were given high-level information (such as positions and velocities) about their prey. More specifically, given the velocity components of the evader, the pursuer predicted its position N_p time steps later according to

$$x_p = x + v_x N_p \mathrm{d}t, \quad (7)$$

$$y_p = y + v_y N_p \mathrm{d}t, \quad (8)$$

where (x, y) and (x_p, y_p) denote the actual and predicted positions of the evader, (v_x, v_y) denotes the velocity of the evader, and $\mathrm{d}t$ is the time step. The pursuer then accelerated towards the evader such that the acceleration was larger the further away it was from its target. The top speed of the pursuer could be specified as an input parameter, and was usually taken to be a few per cent higher than the speed of the evader. Clearly, this is not the only possible pursuer strategy based on high-level information, but it is one that works well and can (by varying parameters) be adapted to track any type of evader.

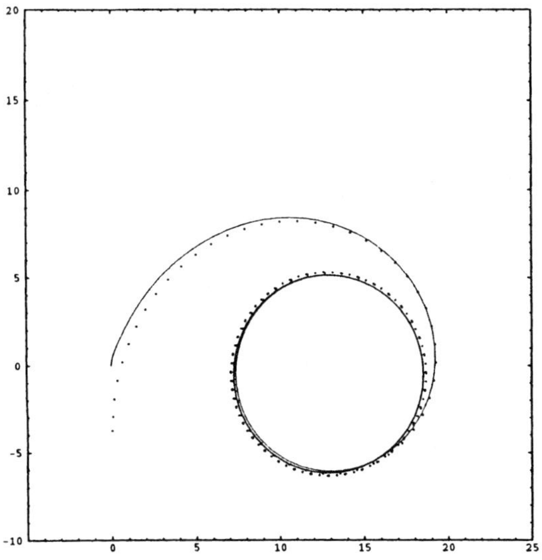

Figure 2 An evader escaping a pursuer by running in a tight circle. The pursuer followed the same circular trajectory, but was not able to reach the evader.

2.3 The simulations

At the start of each simulation, an evader population of typically 20-100 creatures was generated by assigning random numbers to the bits in their chromosomes. The maximum number of neurons of a creature was equal to 30. The individuals were then evaluated by releasing them some distance away from their pursuer, and allowing the pursuer to chase its prey during a maximum of N time steps, where N was taken to be between 5,000 and 30,000. However, if the pursuers managed to catch the evader, the chase was terminated. An evader was considered to be caught if the distance between the two creatures was smaller than the *encounter radius* d_{min}, which was given as an input parameter. The creatures had unit radius, and d_{min} was taken to be either 1.0 or 2.0.

When all evaders had been evaluated and fitness values had been assigned to them (see below), the next generation was formed using a genetic algorithm with fitness-proportional selection and mutation. The parameters of the pursuer were unchanged throughout the simulation.

2.4 Fitness assignment

The task of the evaders was to avoid getting caught by staying as far away from the pursuer as possible. As a fitness measure (type I) we therefore used the sum (over time) of the distances between the pursuer and the evader during the chase. However, as we shall see below, it turned out that due to the limited acceleration capability of the pursuers, an often (i.e. for many parameter values) successful evader strategy (*the elliptic strategy*)

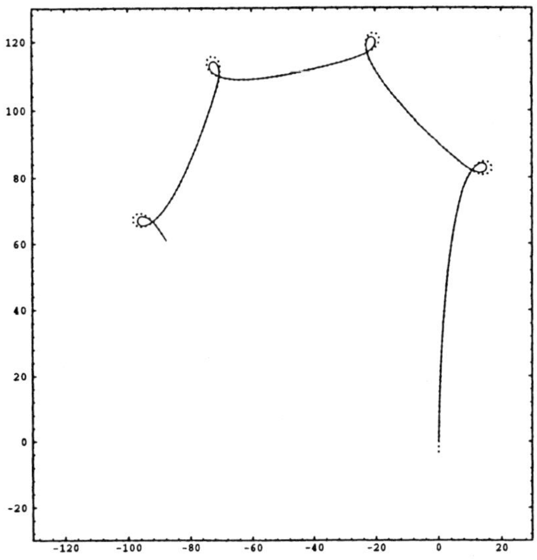

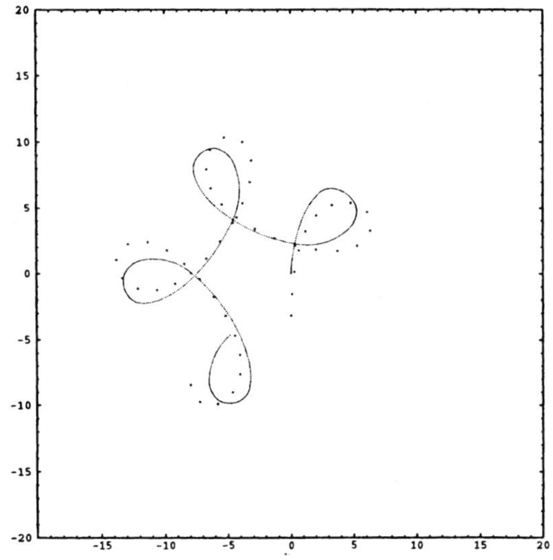

Figure 3 Two examples of evasion behavior obtained with the type I fitness measure. The top speed of the pursuer in the upper panel was 3% higher than that of the evader, and the top speed of the pursuer in the lower panel was 1.5% higher than that of its prey.

was to run perpetually in a very tight circle or ellipse.

In order to decrease the frequency of such solutions we also used another fitness measure (type II): We covered the two-dimensional world with a grid of 'food pellets'. A food pellet was considered to be eaten (and was removed from the grid) if the evader passed sufficiently close to it, and we defined the fitness of the evaders as the number of food pellets eaten during the chase (which, again, was terminated if the evader was caught). Thus, with this fitness measure, it was favorable for the evaders to visit any given position only once.

3.2 Fixed strategy pursuers

Type I fitness measure With the fixed strategy pursuers, and the type I fitness measure, several types of complex evasion behavior evolved. In order to obtain interesting dynamics, the top speed of the pursuer needed to exceed that of the evader — if not, the latter could escape by running along a straight line trajectory. If the top pursuer speed was much higher than the top evader speed, the limited acceleration capabilities of the former caused it too 'overshoot', and the evader could often escape by running in a closed, elliptic orbit. Normally, the most interesting dynamics was obtained when the pursuer speed was taken to be a few per cent higher that the evader speed, but even in this case the evaders often attempted the elliptic strategy. An example of such an orbit is shown in Fig. 2. However, some features of protean behavior were also discovered in several runs. Two examples are shown in Fig. 3.

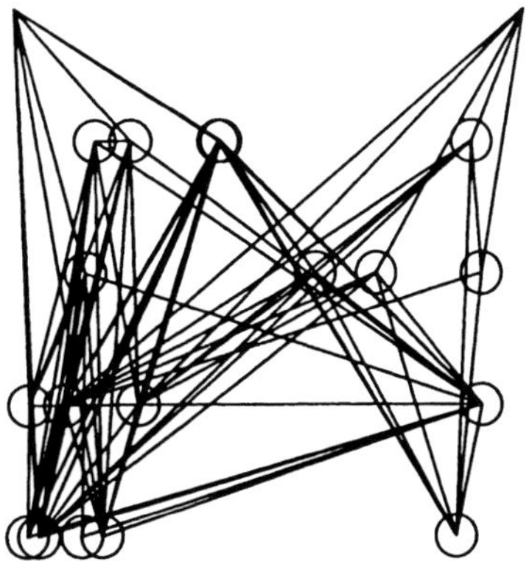

Figure 6 The neural network structure of the evader shown in the upper panel of Fig. 4.

These evaders start with the sensible tactics of running as fast as possible away from the pursuer. The pursuer is faster, though, and at the moment when it seems that the evader will be caught, it makes a rapid turn and (temporarily) shakes off the pursuer. The turns are of similar size and direction each time, which makes the behavior quasiperiodic rather than chaotic. However, this behavior is unpredictable from the point of view of the pursuer, in that it deceives its predictive algorithm. From the point of the evader the evolutionary process has found a sufficiently unpredictable solution. The sudden changes of speed and direction are also reminiscent of protean behavior occurring in nature.

Type II fitness measure With this fitness measure, the elliptic strategy was not favored, and almost all runs provided examples of rapidly turning evaders. Two examples are shown in Fig. 4. The first of these (upper panel) shows quasiperiodic behavior similar to that found for the type I fitness measure, with repeated sudden turns. Fig. 5 shows the distance between the pursuer and the evader as a function of time in this case.

The second example in Fig. 4 illustrates more irregular behavior that also was found in several runs with this choice of fitness function. This example captures the non-repeating nature of protean behavior, which depends on a more complex internal dynamics for the evader. In future work, we intend to characterize the degree of unpredictability in the simulations in a quantitative way.

One should note that in all cases, the chase takes place in the plane without any boundaries. Since the pursuer is faster than the evader, the strategy of running away in a straight line towards infinity is not favorable, and we in fact find solutions where the pursuer and evader remain in a bounded region.

4. Conclusions and discussion

We have demonstrated how complex evasion strategies can arise in pursuit-evasion games when the game is made slightly asymmetric between pursuer and evader. The strategies illustrate several aspects of protean behavior, such as repeated sudden direction changes that confuse the pursuer. Irregular unpredictable trajectories were also observed for some choices of fitness function.

An important topic for further study is the precise quantitative definition of protean behavior. Definitions based on methods of quantifying deterministic chaotic behavior, such as entropies or Lyapunov exponents (e.g., Ruelle and Eckmann, 1985) would be a natural choice. These are however properties of the dynamical system consisting of both pursuer and evader, and must be averaged over some set of pursuers to characterize an evader. The measures should of course depend only on the trajectories, and not on the internal dynamics of the agents.

An example of the neural network architecture of a successful evader is shown in Fig. 6. One property of the networks which could be investigated and related to the behavior would be the size and connectivity of the network. The probability of finding unpredictable, or chaotic, internal dynamics in the network is likely to be higher for a more connected network, in analogy with the transition to chaotic behavior in random Boolean networks when connectivity is increased (Kauffman, 1969).

Coevolution of pursuit-evasion strategies is discussed in Wahde and Nordahl (1998), but needs to be explored further, in particular to study the effects of the parameters in the game that determine the asymmetry between pursuer and evader. Pursuit-evasion games with more than two players also provide a test bed for exploring phenomena such as the evolution of flocking behavior (Hamilton, 1971).

References

Cliff, D. and Miller, G.F. (1996). Co-Evolution of Pursuit and Evasion II: Simulation Methods and Results, in Maes, P. et al, editors *From Animals to Animats 4*, pages 506–515, MIT Press.

Floreano, D. and Nolfi, S. (1997). God Save the Red Queen! Competition in Co-Evolutionary Robotics, in Koza, J.R. et al, editors, *Genetic Programming 1997: Proceedings of the Second Annual Conference*, pages 398–406, Morgan Kaufmann.

Hamilton, W.D. (1971). Geometry for the Selfish Herd, *Journal of Theoretical Biology*, **31**:295–311.

Isaacs, R. (1965). *Differential Games: A Mathematical Theory with Applications to Warfare and Pursuit, Control and Optimization*, John Wiley.

Kauffman, S.A. (1969). Metabolic Stability and Epigenesis in Randomly Connected Nets, *Journal of Theoretical Biology*, **22**:437.

Lindgren, K. (1991). Evolutionary Phenomena in Simple Dynamics, in Langton, C. et al, editors, *Artificial Life II*, pages 295–312, Addison-Wesley, Redwood City,.

Miller, G.F. and Cliff, D. (1994). Co-Evolution of Pursuit and Evasion I: Biological and Game-Theoretic Foundation, Technical Report CSRP311, School of Cognitive and Computing Sciences, University of Sussex.

Ruelle, D. and Eckmann, J.-P. (1985). *Reviews of Modern Physics*, **54**:617.

Wahde, M. and Nordahl, M.G. (1998). Coevolving Pursuit-Evasion Strategies in Open and Confined Regions, to appear in C. Adami, R. Belew, H. Kitano, and C. Taylor, eds., *Proc. of "Artificial Life VI"*, Los Angeles, MIT Press.

AUTHOR INDEX

Made in United States
North Haven, CT
04 November 2026

21730437R10317